P9-AER-362

For All Practical Purposes

Mathematical Literacy in Today's World

SIXTH EDITION

Project Director

Solomon Garfunkel, *Consortium for Mathematics and Its Applications*

Contributing Authors

Part I Management Science

Joseph Malkevitch, *York College, CUNY*

Part II Statistics: The Science of Data

David S. Moore, *Purdue University*

Part III The Digital Revolution

Joseph Gallian, *University of Minnesota–Duluth*
Ran Libeskind-Hadas, *Harvey Mudd College*

Part IV Social Choice and Decision Making

Alan D. Taylor, *Union College*
Bruce P. Conrad, *Temple University*
Steven J. Brams, *New York University*

Part V On Size and Shape

Paul J. Campbell, *Beloit College*

Part VI Modeling in Mathematics

Paul J. Campbell, *Beloit College*

For All Practical Purposes

Mathematical Literacy in Today's World

SIXTH EDITION

W. H. Freeman and Company • New York

Executive Editor: Craig Bleyer
Senior Development Editor: Randi Rossignol
Marketing Manager: Jeffrey Rucker
Project Editor: Vivien Weiss
Cover and Text Designer: Diana Blume
Illustrations: Academy Artworks, Inc., Burmar Technical Corporation, Progressive Information Technologies
Photo Editors: Patricia Marx, Julie Tesser
Production Manager: Julia DeRosa
Media and Supplements Editors: Mark Santee, Brian Donnellan
Composition: Progressive Information Technologies, Circa 86
Manufacturing: RR Donnelly & Sons Company

Cover image: Getty

Library of Congress Cataloging-in-Publication Data
For all practical purposes : mathematical literacy in today's world. - 6th ed.
p. cm.
By the Consortium for Mathematics and Its Applications.
Includes index.
ISBN 07167-4782-0 (hardcover)
ISBN 07167-4783-9 (paperback)
1. Mathematics I. Consortium for Mathematics and Its Applications (U.S.)
QA7 .F68 2003
510-dc21

2002021479

Printed in the United States of America

First printing 2002

Brief Contents

Contents

B. Daemmrich/ The Image Works

PART II Statistics: The Science of Data

Scott Camazine/Sue Trainor

Rob Crandall/ The Image Works

American Museum of Natural History

PART V On Size and Shape

Fidelity Investments

PART VI Modeling in Mathematics

PREFACE

Our goal for this new edition of *For All Practical Purposes* remains what it was for the five previous editions: to bring the excitement of contemporary mathematical thinking to every student and to help students think logically and critically about the mathematical information that abounds in our society. We are reminded always of Thomas Jefferson's notion of an "enlightened citizenry" in which people, having acquired a broad knowledge of topics, can use sound judgment in making personal and political decisions.

What's New

Because *For All Practical Purposes* stresses the connections between contemporary mathematics and modern society, our text must be flexible enough to accommodate new ideas in mathematics and their new applications to our daily lives.

Four New Chapters

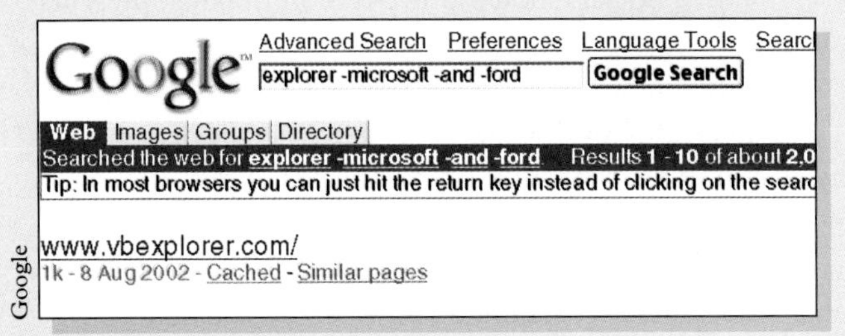

Google

- **Chapter 11, The Internet, The Web, and Logic,** shows students how the basic tenets of Boolean logic underpin the workings of search engines, such as Google, that have become so essential to modern life, both at home and at school. The chapter also examines the ways in which computers represent and send data and the application of circuits to computer/router networks and web search queries.

- **Chapter 17, Electing the President,** is the new capstone chapter for Part IV, Social Choice and Decision Making. Presidential elections, most notably the 2000 election, are presented as games of strategy. Spatial models, which use the basic statistics concepts of distribution, mean, and median, are used to visualize the positions candidates take and how a change in a position might affect the voters' response.

Mark Wilson/Newsmakers

EXAMPLE **Earnings on Those Bonds**

Suppose that you bought one of those 30-year U.S. Treasury bonds in April 2002 for \$10,000. How much interest would it earn by April 2032, and what would be the total amount accumulated? We have $P = \$10{,}000$, $r = 3.375\% = 0.03375$, and $t = 30$ years:

$$I = Prt = \$10{,}000 \times 0.0375 \times 30 = \$10{,}125$$

- **Chapter 21, Consumer Finance Models I: Saving,** will appeal to students' need and desire to learn more about managing and understanding their personal finances. This chapter covers the various ways to save money, including bonds, sinking funds, and retirement funds. Other topics include real versus relative growth, evaluating stock prices, and the Consumer Price Index.
- **Chapter 22, Consumer Finance Models II: Borrowing,** examines the various ways that consumers borrow money, including bonds, add-on loans, discounted loans, credit card interest, home equity loans and mortgages, and annuities.

PhotoEdit

Updated Coverage Throughout

EXAMPLE **Making a Histogram**

One of the most striking findings of the 2000 census was the growth of the Hispanic population in the United States. Table 6.1 presents the percent of adult residents (age 18 and over) in each of the 50 states who identified themselves in the 2000 census as "Spanish/Hispanic/Latino." The *individuals* in this data set are the 50 states. The *variable* is the percent of Hispanic adults in a state. To make a histogram of the distribution of this variable, proceed as follows:

1. Divide the range of the data into classes of equal width. The data in Table 6.1 range from 0.6 to 38.7, so we decide to choose these classes:

$$0.0 < \text{percent Hispanic} \le 5.0$$
$$5.0 < \text{percent Hispanic} \le 10.0$$
$$\vdots$$

10 states have populations that are between 5.1% and 10% Hispanic.
New Mexico is an outlier, 38.7% Hispanic.
Number of states
Percent adults who call themselves Hispanic

- Dozens of new Examples and topics, reflecting recent events such as the 2000 presidential election and the 2000 census, are found throughout the text. For instance, Chapter 6, Exploring Data, includes Examples of making and interpreting histograms using the 2000 census. Part III, Social Choice and Decision Making, presents many Examples of how different systems of voting might have affected the outcome of the 2000 presidential election.
- **Part III, The Digital Revolution,** presents new systems of cryptography as well as a new section on digital signatures.
- Newsworthy events, such as the merger between the pharmaceutical company giants SmithKline Beecham and Glaxo Wellcome and a recent national AIDS survey (Chapter 8), have replaced dated Examples. Chapter 3 now discusses why the September 11, 2001, tragedy disrupted but did not destroy New York City's subway and communication systems.

Simplified and Streamlined Presentation

We have worked hard in this edition to address our users' concerns, particularly the oft-stated desire that the book speak in a single voice. **We are very fortunate to welcome Francis Su of Harvey Mudd College, to the *For All Practical Purposes* author team.** Professor Su has worked closely with all the authors to bring consistency of style, level, and tone to the text. As an invested teacher and enthusiastic user of ***FAPP***, he is very familiar with the areas of the text that students find most challenging. His goal was to make those sections more

Although this apportionment may seem fair enough, President Washington vetoed that bill.[1] Washington came from Virginia, a state that lost something in the apportionment bill. Was he biased in favor of his home state, as the field hockey coach was in favor of her team? We will return to this question.

In the following example, we will see how to set up an apportionment problem.

EXAMPLE The High School Mathematics Teacher

A high school has one mathematics teacher who teaches all geometry, pre-calculus, and calculus classes. She has time to teach a total of five sections. One hundred students are enrolled as follows: 52 for geometry, 33 for pre-calculus, and 15 for calculus. How many sections of each course should be scheduled?

The number of students enrolled in each course is called the population. Thus, the population for geometry is 52, the population for pre-calculus is 33, and the population for calculus is 15. There are five sections for the 100 students, so the average section will have $100 \div 5 = 20$ students. We will call this average section size the *standard divisor,* because each quota can be determined by dividing the corresponding population by this number. Table 15.3 displays these calculations.

accessible to students and to make the presentation lean and lively throughout.

- Many topics are now more fully explained and illustrated with simpler Examples. These include counting Borda scores (Chapter 12), equivalent voting systems (Chapter 13), and rounding fractions and apportionment (Chapter 15).
- Those sections identified by our users as being beyond the scope of the course have been taken out of the text and moved to the text's Web site at www.whfreeman.com/fapp. These sections include Statistical Process Control and Perils of Data Analysis, Systems with Large Numbers of Voters, Manipulability, Truels, Problems of Envy in Fair Division Schemes, and Radioactive Decay.

Analysis of voting systems with voters numbering in the thousands or millions is only possible if all the voters, with only a few exceptions, are equally powerful. For further discussion of systems with large numbers of voters, go to www.whfreeman.com/fapp.

- Annotations in many of the text's illustrations, particularly graphs, lead students through the art.

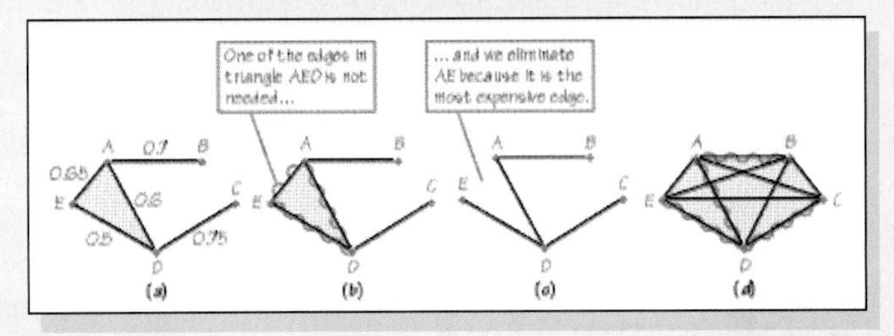

Many More Exercises

SKILLS CHECK

1. What are all the values that a standard deviation s can possibly take?
(a) $0 \le s \le 1$
(b) $-1 \le s \le 1$
(c) $0 \le s$

2. What are all the values that a correlation r can possibly take?
(a) $0 \le r \le 1$
(b) $-1 \le r \le 1$
(c) $0 \le r$

- **Twice as many Skills Check exercises** are now included to give students more practice solving basic problems.

EXERCISES ▲ Optional. ■ Advanced. ◆ Discussion.

Sampling

1. Different types of writing can sometimes be distinguished by the lengths of the words used. A student interested in this fact wants to study the lengths of words used by Danielle Steele in her ...els. She opens a Steele novel at random and ...rds the lengths of the first 250 words on the .... What is the population in this study? What ...e sample?

2. The American Community Survey (ACS) will contact 3 million households, including some in every county in the United States. This new Census Bureau survey will ask each household questions about their housing, economic, and social status. The new survey will replace the census "long form." What is the population for the ACS?

◆ 3. A member of Congress is interested in whether her constituents favor a proposed gun-control bill. Her staff reports that letters on the bill

- **More than 125 new Exercises** give students practice solving problems with a wide range of difficulty and variety.

APPLET EXERCISES

To do these exercises, go to www.whfreeman.com/fapp

Simple Random Sample

The purpose of this exercise is to see the results of sampling variability. Suppose that a researcher wants to test the effectiveness of a new breakfast food. He takes a group of 30 rats and assigns them randomly into a treatment group of 15 and a control group of 15.

(a) Use the Simple Random Sample applet to choose the 15 rats for the treatment group, and record the labels of the rats chosen.

(b) Suppose that the 15 even-numbered rats among the 30 rats available are (unknown to the experimenters) a fast-growing variety. We hope

- **New Applet Exercises** engage students interactively, to solve problems using Web-based applications.

Media and Supplements

The media and supplements package for the new edition has been updated to reflect the changes in the book and the growing importance of electronic media. Both students and instructors will benefit from the innovative materials available to them. All supplements described can be found at **www.whfreeman.com/fapp.**

- A printed **Study Guide** has been developed to help students better understand the materials in the book and to enhance their performance in the course. Each chapter offers section-by-section summaries of the text coverage, including explanations of challenging concepts and numerous questions (with answers) designed to ensure that students fully grasp a concept before moving on. The *Study Guide* also contains chapter objectives and a practice quiz for each chapter. The practice quiz questions are similar to the questions students are likely to encounter on a test.

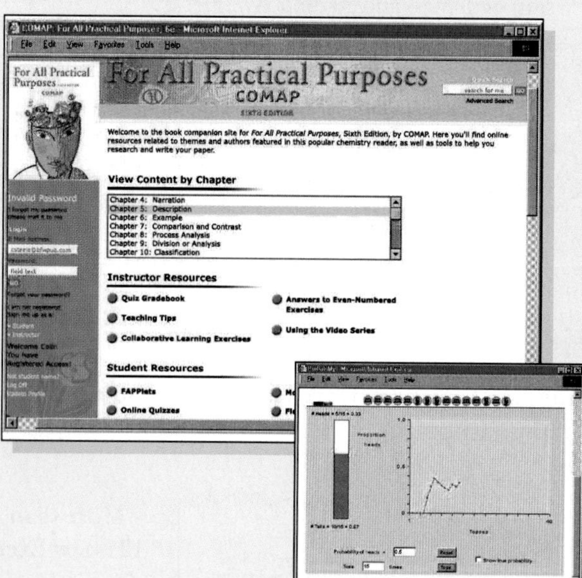

- Our ***For All Practical Purposes*** **Web site** has been carefully developed to serve both students and instructors. New to this edition's site are interactive **Java Applets** and **Spreadsheet Projects.** The applets allow students to manipulate data or graphics in order to better understand the mathematical concepts involved. The Spreadsheet Projects require that the students use Excel spreadsheets, which can be downloaded by users, to solve problems. On the Web site students will also find **chapter objectives** to help organize study, **Flash Card exercises** to aid in the mastery of the text's Review Vocabulary, chapter **self-quizzes** for test preparation, and **Web links** to additional real-world applications of text topics.

  For instructors, the site features an **Online Instructor's Guide,** which includes extensive class-tested teaching hints, detailed summaries, and suggested skill objectives. Also featured are quizzes for each chapter of the text. With these quizzes, instructors can easily and securely quiz students online using prewritten, multiple-choice questions for each text chapter, separate from those appearing in the *Test Bank.* Because the quiz questions can be randomized, no two students receive the same series of questions. Students receive instant feedback and can take the quizzes multiple times. Instructors can view the results by quiz, student, question, or can get weekly results via email. Also available on the Web site are videos and clips from the Annenberg Series, "For All Practical Purposes."

- Our **Instructor's CD-ROM** offers all the images from the textbook in two formats–as part of our **Presentation Manager Pro** software and in **JPEG files**. Presentation Manager Pro allows instructors to prepare playlists of images for display during lectures. The JPEG files are provided for users who prefer to use commercially available presentation software. Our *Online Instructor's Guide* can also be found on the Instructor's CD-ROM.

- Our **Test Bank** comes in Windows and Macintosh formats. It offers 2000 questions–50 multiple-choice and 25 short-answer questions per chapter.

- **Online Testing,** powered by Diploma from the Brownstone Research Group, is also available. With Diploma, instructors can easily create and administer secure exams over a network and over the Internet, with questions that incorporate multimedia and interactive exercises. The program lets you restrict tests to specific computers or time blocks, and includes an impressive suite of gradebook and result-analysis features.

- **Video clips** illustrating key topics from the textbook, from the Annenberg series "For All Practical Purposes," are available on the Web site. There is also an Annenberg video series of 26 half-hour programs called **For All Practical Purposes Telecourse.** For more information on the telecourse preview, purchase, or rental, please call 1-800-Learner, or write The Annenberg CPB Project, P.O. Box 2345, South Burlington, VT 05407-2345.

We thank the many people who have contributed to the supplements package, including the authors of supplements to previous editions. Following is a list of all the instructors who have been involved, some of whom worked on more than one title in this package:

Annette Burden, Youngstown State University
John Emert, Ball State University
Chris Leary, St. Bonaventure College
Jeannette Martin, Washington State University
Eli Passow, Temple University
Dan Reich, Temple University
Kay Meeks Roebuck, Ball State University
Sandra H. Savage, Orange Coast College

Acknowledgements

For All Practical Purposes continues to improve and evolve due in great part to our many friends and colleagues who have offered suggestions, comments, and corrections. We are grateful to them all.

Gisela Ahlbrandt, Eastern Michigan University
Janet Heine Barnett, University of Southern Colorado
Terence Blows, Northern Arizona University
Annette M. Burden, Youngstown State University
Jim Cliber, Nebraska Wesleyan University
Dan Dreibelbis, University of North Florida
Sandra Fillebrown, St. Joseph's University
John Fink, Kalamazoo College
Marilyn Hasty, Southern Illinois University at Edwardsville
John Henry, Lincoln Land Community College
Anton Kaul, Tufts University
James Kays, Auburn University
Michael Keller, St. Johns River Community College
W. Thomas Kiley, George Mason University
Don Krug, Northern Kentucky University
Jan LaTurno, Rio Hondo College
Jeannette Martin, Washington State University
Christopher McCord, University of Cincinnati
Donna Lee McCracken, University of Florida
Margaret A. Michener, University of Nebraska at Kearney
Patrick Morandi, New Mexico State University
Michael G. Neubauer, California State University, Northridge
Soula O'Bannon, Louisiana State University
John Oprea, Cleveland State University
Richard Patterson, University of North Florida
Stephen Schecter, North Carolina State University
Gene Schlereth, University of Tennessee
Connie S. Schrock, Emporia State University
Marilyn Seman, Norwalk Community College
Alu Srinivasan, Temple University
Pat Stanley, Ball State University
David Urion, Winona State University
W. D. Wallis, Southern Illinois University at Carbondale
Fredric Zerla, University of South Florida

We owe our appreciation to the people at W. H. Freeman and Company who participated in the development and production of this edition. We wish especially to thank the editorial staff for their tireless efforts and support. Among them are Craig Bleyer, Acquisitions Editor; Randi Rossignol, Development Editor; Vivien Weiss, Project Editor; Julia DeRosa, production manager; Diana Blume, designer; Patricia Marx and Julie Tesser, photo editors; Mark Santee and Brian Donnellan, media and supplements editors; Karen Osborne, copy editor; and Walter Hadler, proofreader.

The efforts of the COMAP staff must be recognized. We thank our production and administrative staff, George Ward and Roland Cheney, and Laurie Aragon, the *manager.*

To everyone who helped make our purposes practical, we offer our appreciation for an exciting and exhilirating time.

Solomon Garfunkel, COMAP

CHAPTER

1

Street Networks

In this chapter we cover...

...and at www. whfreeman.com/fapp:

Flashcards

Quizzes

Spreadsheet Projects

Representing and analyzing graphs

Applet Exercises

Eulerizing a graph

Euler Circuits

The underlying theme of **management science,** also called **operations research,** is finding the best method for solving some problem—what mathematicians call the **optimal solution.** In some cases, the goal may be to finish a job as quickly as possible. In other situations, the objective might be to maximize profit or minimize cost. In this chapter, our goal is to save time in traversing a street network while checking parking meters, delivering mail, or removing snow.

Let's begin by assisting the parking department of a city government. Most cities and many small towns have parking meters that must be regularly checked for parking violations or emptied of coins. We will use an imaginary town to show how management science techniques can help to make parking control more efficient.

1.1 Euler Circuits

The street map in Figure 1.1 is typical of many towns across the United States, with streets, residential blocks, and a village green. Our job, or that of the commissioner of parking, is to find the most efficient route for the parking-control officer, who travels on foot, to check the meters in an area. Efficient routes save money. Our map shows only a small area, allowing us to start with an easy problem. But the problem occurs on a larger scale in all cities and towns and, for larger areas, there are almost unlimited possibilities for parking-control routes.

The commissioner has two goals in mind: (1) The parking-control officer must cover all the sidewalks that have parking meters without retracing any more steps than are necessary; and (2) the route should end at the same point from which it began, perhaps where the officer's patrol car is parked. To be specific, suppose there are only two blocks that have parking meters, the two blue-shaded blocks that are side by side toward the top of Figure 1.1. Suppose further that the parking-control officer must start and end at the upper left corner of the left-hand block. You might enjoy working out some routes by trial and error and evaluating their good and bad points. We are going to leave this problem for the

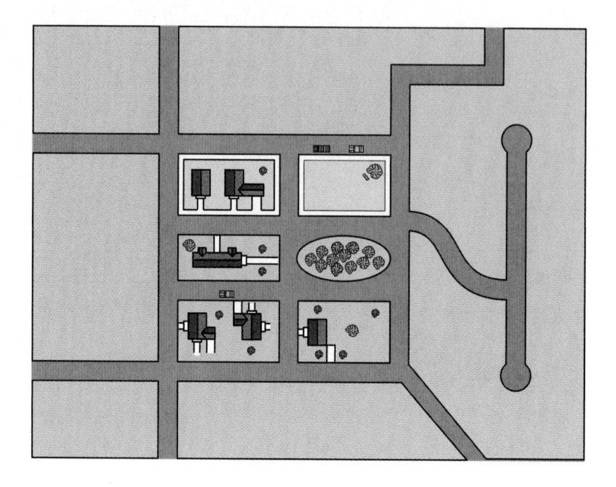

Figure 1.1 A street map for part of a town.

moment and establish some concepts that will give us a better method than trial and error to deal with this problem.

A **graph** is a finite set of dots and connecting links. The dots are called **vertices** (a single dot is called a ***vertex***), and the links are called **edges.** Each edge must connect two different vertices. A **path** is a connected sequence of edges showing a route on the graph that starts at a vertex and ends at a vertex; a path is usually described by naming in turn the vertices visited in traversing it. A path that starts and ends at the same vertex is called a **circuit.** A graph can represent our city map, a communications network, or even a system of air routes.

EXAMPLE

Parts of a Graph

We can see examples of these technical terms in Figure 1.2. The vertices represent cities, and the edges represent nonstop airline routes between them. We see that there is a nonstop flight between Berlin and Rome, but no such flight between New York and Berlin. There are several paths that describe how a person might travel with this airline from New York to Berlin. The path that seems most direct

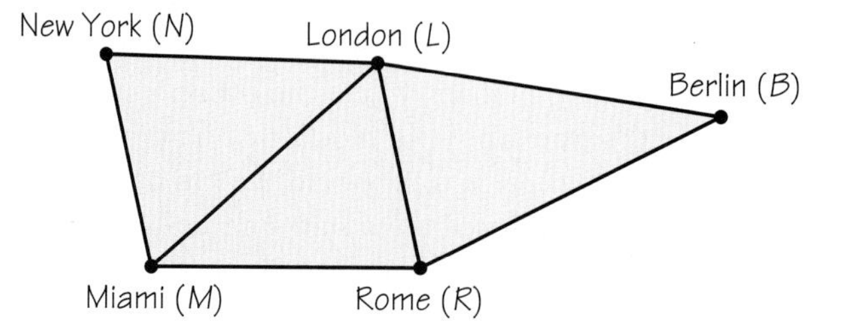

Figure 1.2 The edges of the graph show nonstop routes that an airline might offer.

is New York, London, Berlin, but New York, Miami, Rome, Berlin is also such a path. We can describe these two paths as: *NLB* and *NMRB*. An example of a circuit is Miami, Rome, London, Miami. It is a circuit because the path starts and ends at the same vertex. This circuit can best be described in symbols by *MRLM*. In this chapter, we are especially interested in circuits, just as we are in real life; most of us end our day in the same location where we start it – at home! ■

Returning to the case of parking control in Figure 1.1, we can use a graph to represent the whole territory to be patrolled: think of each street intersection as a vertex and each sidewalk that contains a meter as an edge, as in Figure 1.3. Notice in Figure 1.3b that the width of the street separating the blocks is not explicitly represented; it has been shrunk to nothing. In effect, we are simplifying our problem by ignoring any distance traveled in crossing streets.

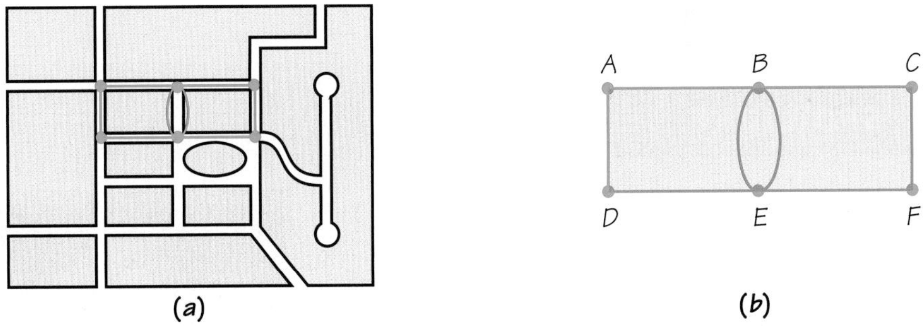

Figure 1.3 (a) A graph superimposed upon a street map. The edges show which sidewalks have parking meters. (b) The same graph enlarged.

The numbered sequence of edges in Figure 1.4a shows one circuit that covers all the meters (note that it is a circuit because its path returns to its starting point). However, one edge is traversed three times. Figure 1.4b shows another solution that is better because its circuit covers every edge (sidewalk) exactly once. In Figure 1.4b, no edge is covered more than once, or *deadheaded* (a term borrowed from shipping, which means making a return trip without a load).

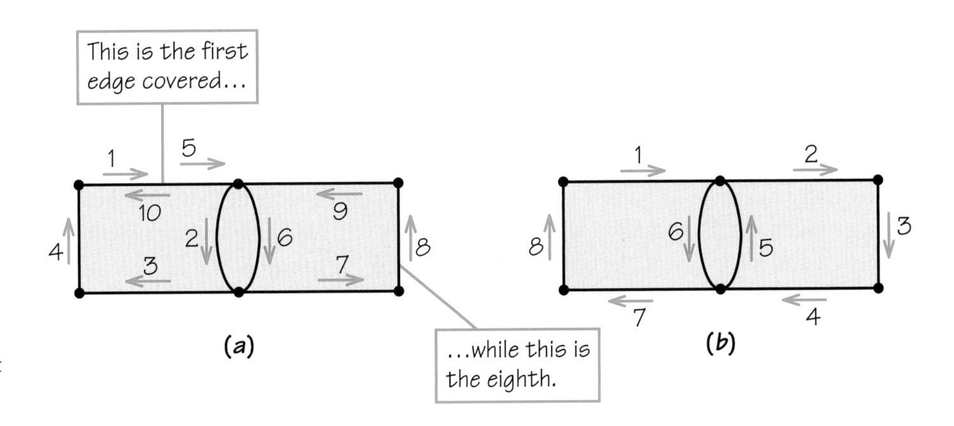

Figure 1.4 (a) A circuit and (b) an Euler circuit.

Circuits that cover every edge only once are called **Euler circuits.**

Figure 1.4b shows an Euler circuit. These circuits get their name from the great eighteenth-century mathematician Leonhard Euler (pronounced oy′ lur), who first studied them (see Spotlight 1.1). Euler was the founder of the theory of graphs, or graph theory. One of his first discoveries was that some graphs have no Euler circuits at all. For example, in the graph in Figure 1.5b, it would be impossible to start at one point and cover all the edges without retracing some steps: If we try to start a circuit at the leftmost vertex, we discover that once we have left the vertex, we have "used up" the only edge meeting it. We have no way to return to our starting point except to reuse that edge. But this is not allowed in an Euler circuit. If we try to start a circuit at one of the other two vertices, we likewise can't complete it to form an Euler circuit.

SPOTLIGHT 1.1 Leonhard Euler

Leonhard Euler (1707 – 1783) was one of those rare individuals who was remarkable in many ways. He was extremely prolific, publishing over 500 works in his lifetime. But he wasn't devoted just to mathematics; he was a people person, too. He was extremely fond of children and had thirteen of his own, of whom only five survived childhood. It is said that he often wrote difficult mathematical works with a child or two in his lap.

Human interest stories about Euler have been handed down through two centuries. He was a prodigy at doing complex mathematical calculations under less than ideal conditions, and he continued to do them even after he became totally blind later in life. His blindness diminished neither the quantity nor the quality of his output. Throughout his life, he was able to mentally calculate in a short time what would have taken ordinary mathematicians hours of pencil-and-paper work. A contemporary claimed that Euler could calculate effortlessly, "just as men breathe, as eagles sustain themselves in the air."

Leonhard Euler

Euler invented the idea of a graph in 1736 when he solved a problem in "recreational mathematics." He showed that it was impossible to stroll a route visiting the seven bridges of the German town of Königsberg exactly once. Ironically, in 1752 he discovered that three-dimensional polyhedra obey the remarkable formula $V - E + F = 2$ (i.e., number of vertices − number of edges + number of faces = 2) but failed to give a proof because he did not analyze the situation using graph theory methods.

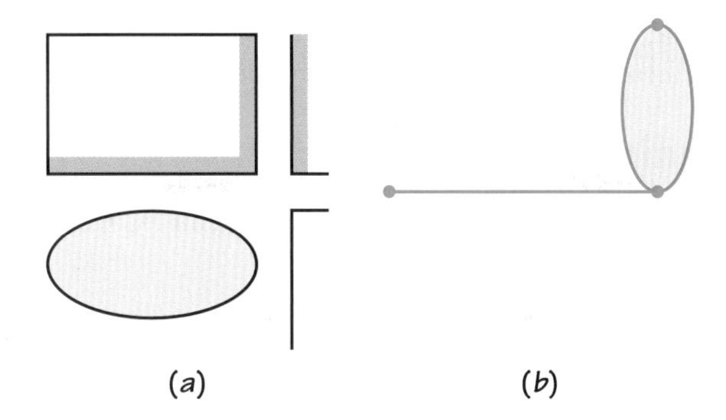

Figure 1.5 (a) The three shaded sidewalks cannot be covered by an Euler circuit. (b) The graph of the shaded sidewalks in part (a).

As mentioned in Spotlight 1.2, realistic problems of this type will involve larger neighborhoods that might require the use of a computer. In addition, there may be other complications that might take us beyond the simple mathematics we want to stick to.

Because we are interested in finding circuits, and Euler circuits are the most efficient ones, we will want to know how to find them. If a graph has no Euler

SPOTLIGHT 1.2 The Human Aspect of Problem Solving

Thomas Magnanti, professor of operations research and management, heads the Department of Management Science at MIT's Sloan School of Management. Here are some of his observations:

> Typically, a management science approach has several different ingredients. One is just structuring the problem – understanding that the problem is an Euler circuit problem or a related management science problem. After that, one has to develop the solution methods.
>
> But one should also recognize that you don't just push a button and get the answer. In using these underlying mathematical tools, we never want to lose sight of our common sense, of understanding, intuition, and judgment. The computer provides certain kinds of insights. It deals with some of the combinatorial complexities of these problems very nicely. But a model such as an Euler circuit can never capture the full essence of a decision-making problem.

Thomas Magnanti

> Typically, when we solve the mathematical problem, we see that it doesn't quite correspond to the real problem we want to solve. So we make modifications in the underlying model. It is an interactive approach, using the best of what computers and mathematics have to offer and the best of what we, as human beings, with our own decision-making capabilities, have to offer.

circuit, we will want to develop the next best circuits, those having minimum deadheading. These topics make up the rest of this chapter.

1.2 Finding Euler Circuits

Now that we know what an Euler circuit is, we are faced with two obvious questions:

1. Is there a way to tell by calculation, not by trial and error, if a graph has an Euler circuit?
2. Is there a method, other than trial and error, for finding an Euler circuit when one exists?

Euler investigated these questions in 1735 by using the concepts of valence and connectedness.

The **valence** of a vertex in a graph is the number of edges meeting at the vertex.

Figure 1.6 illustrates the concept of valence, with vertices A and D having valence 3, vertex B having valence 2, and vertex C having valence 0. Isolated vertices such as vertex C are an annoyance in Euler circuit theory. Because they don't occur in typical applications, we henceforth assume that our graphs have no vertices of valence 0.

Figure 1.3b has four vertices of valence 2, namely, A, C, F, and D. This graph also has two vertices, B and E, of valence 4. Notice that each vertex has a valence that is an even number. We'll soon see that this is very significant.

A graph is said to be **connected** if for every pair of its vertices there is at least one path connecting the two vertices.

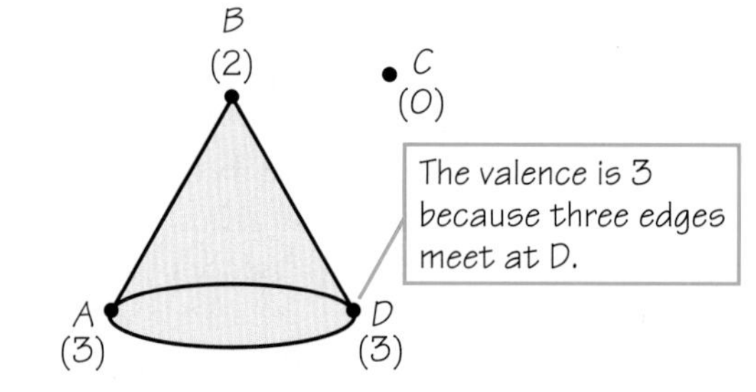
B
(2)
C
(0)
The valence is 3 because three edges meet at D.
A
(3)
D
(3)

Figure 1.6 Valences of vertices.

Figure 1.7 A nonconnected graph.

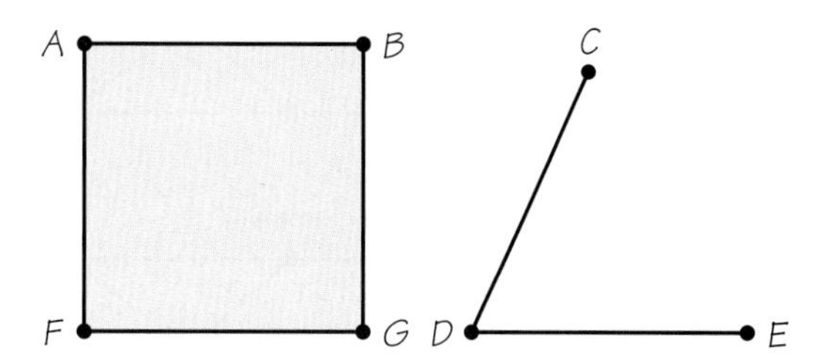

Given a graph, if we can find even one pair of vertices not connected by a path, then we say that the graph is not connected. For example, the graph in Figure 1.7 is not connected because we are unable to join A to D with a path of edges. However, the graph does consist of two "pieces" or connected components, one containing the vertices A, B, F, and G, the other containing C, D, and E. A connected graph will contain a single connected component. Notice that the parking-control graph of Figure 1.3b is connected.

We can now state Euler's theorem, his simple answer to the problem of detecting when a graph G has an Euler circuit:

1. If G is connected and has all valences even, then G has an Euler circuit.
2. Conversely, if G has an Euler circuit, then G must be connected and all its valences must be even numbers.

In the optional section "Proving Euler's Theorem," you will find an outline of a proof of this theorem.

Because the parking-control graph of Figure 1.3b conforms to the connectedness and even-valence conditions, Euler's theorem tells us that it has an Euler circuit. We already have found an Euler circuit for Figure 1.4b by trial and error. For a very large graph, however, trial and error may take a long time. It is usually quicker to check whether the graph is connected and even-valent than to find out if it has an Euler circuit.

Once we know there is an Euler circuit in a certain graph, how do we find it? Many people find that, after a little practice, they can find Euler circuits by trial and error, and they don't need detailed instructions on how to proceed. At this point you should see if you can develop this skill by trying to find Euler circuits in Figure 1.8a, Figure 1.9a, and Figure 1.10. In doing your experiments, draw your graph in ink and the circuit in pencil so you can erase if necessary. Make your graph big and clear so you won't get confused.

If you would like more guidance on how to find an Euler circuit without trial and error, here is a method that works: Never use an edge that is the only link between two parts of the graph that still need to be covered. Figure 1.9b illustrates this. Here we have started the circuit at A and gotten to D via B and C, and we want to know what to do next. Going to E would be a bad idea because the uncovered part of the graph would then be disconnected into left and right portions. You will never be able to get from the left part back to the right part

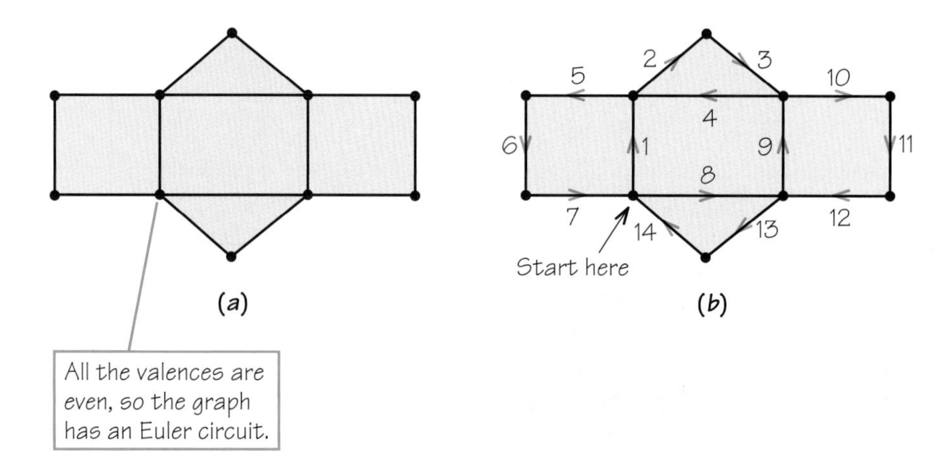

Figure 1.8 (a) A graph having (b) an Euler circuit.

because you have just used the last remaining link between these parts. Therefore, you should stay on the right side and finish that before using the edge from D to E. This kind of thinking needs to be applied every time you need to choose a new edge.

Let's see how this works, starting at the beginning at A. From vertex A there are two possible edges, and neither of them disconnects the unused portion of

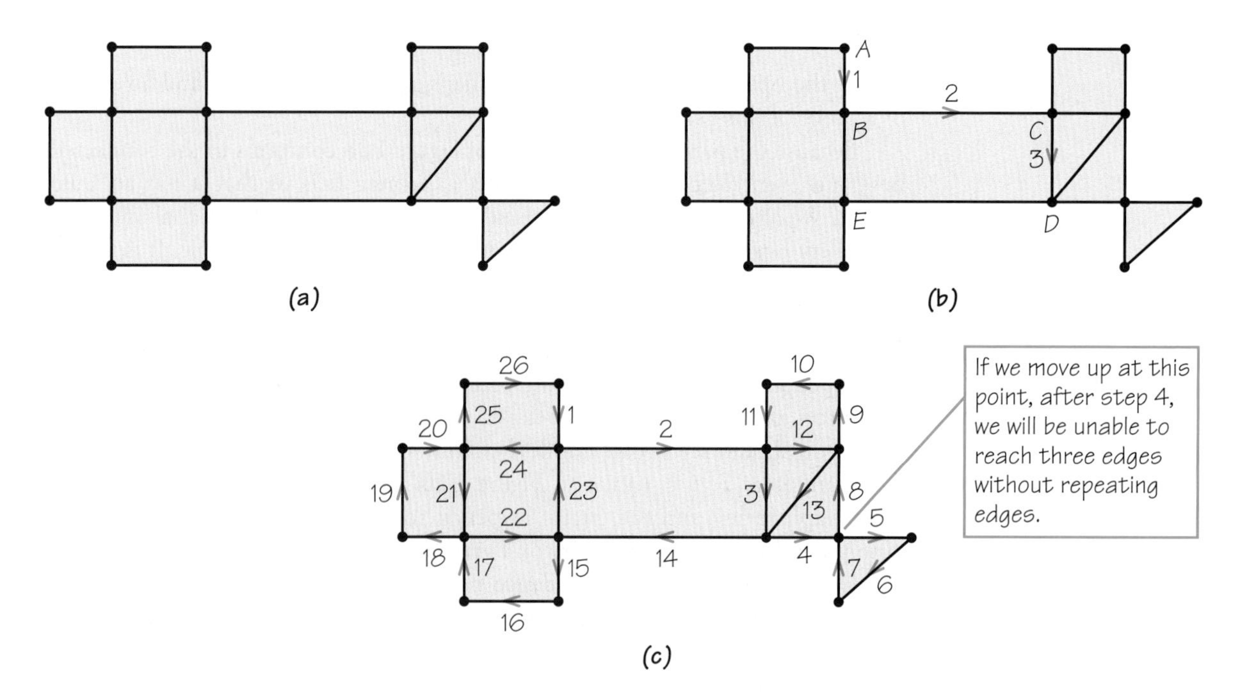

Figure 1.9 (a) A graph that has an Euler circuit. (b) A critical junction in finding an Euler circuit in this graph, starting from vertex A. (c) A description of a full Euler circuit for this graph.

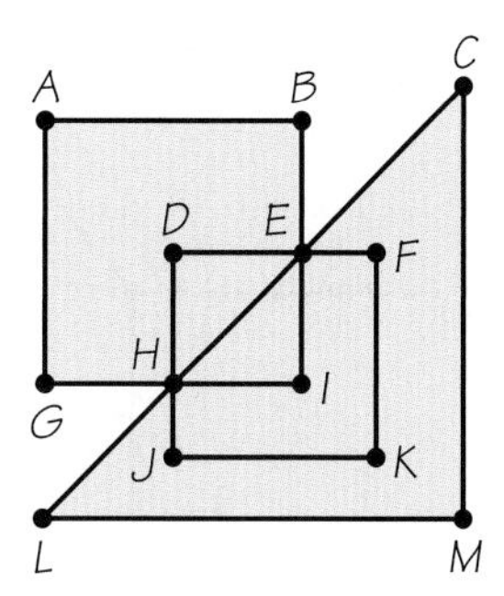

Figure 1.10 A graph with an Euler circuit.

the graph. Thus, we could have gone either to the left or down. Having gone down to B, we now have three choices, none of which disconnects the unused part of the graph. After choosing to go from B to C, we find that any of the three choices at C is acceptable. Can you complete the Euler circuit? Figure 1.9c shows one of many ways to do this.

The method just described leaves many edge choices up to you. When there are many acceptable edges for your next step, you can pick one at random. You might even flip a coin. When computers carry out algorithms of this sort, they use random-number generators, which mimic the flipping of a coin.

EXAMPLE

Finding an Euler Circuit

Check the valences of the vertices and the connectivity of the graph in Figure 1.8a to verify that the graph does have an Euler circuit. Now try to find an Euler circuit for that graph. You can start at any vertex. When you are done, compare your solution with the Euler circuit given in Figure 1.8b. If your path covers each edge exactly once and returns to its original vertex (is a circuit), then it is an Euler circuit, even if it is not the same as the one we give. ■

OPTIONAL

Proving Euler's Theorem

We'll start by proving that if a graph has an Euler circuit, then it must have only even-valent vertices and it must be connected. Let X be any vertex of the graph. We will show that the edges at X can be paired up, and this will prove that the valence is even. Every edge at X is used by the Euler circuit as an outgoing edge (leaving from X) or an incoming edge (arriving at X). If the Euler circuit starts at X, then pair up the first edge used by the circuit with the last one (when the circuit returns to X for the last time). In addition, each other edge at X that is used by the circuit as an incoming edge will be paired with the outgoing edge that is used next. Because all edges at X are used by the Euler circuit, none more than once, this pairs up the edges. But what if X is not the start of the Euler circuit? Then do the pairing like this: The first incoming edge at X is paired with the outgoing one used next, the second incoming edge at X is paired with the outgoing one used next, and so on. For example, in Figure 1.11 at vertex B we would pair up edges 2 and 3 and edges 9 and 10. At vertex C we would pair up edges 4 and 5

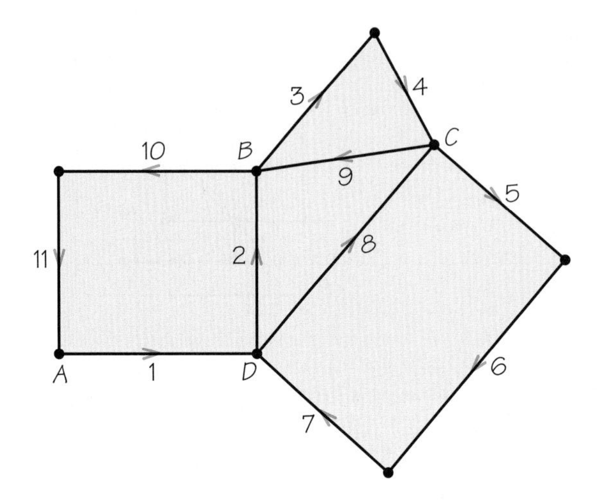

Figure 1.11 An Euler circuit starting and ending at A.

and edges 8 and 9. Can you see how the pairings would work at D? How about vertex A? (In studying this particular example, you might think it would be simpler to count the edges at a vertex to see that the valence is even. True, but our pairing method works for a graph about which we know nothing except that it has an Euler circuit.)

To see that a graph with an Euler circuit is connected, note that by following the Euler circuit around we can get from one edge to any other edge (it covers them all) using a portion of the Euler circuit. Because every vertex is on an edge (there are no vertices of 0 valence), we can get from any vertex to any other using a portion of the Euler circuit.

So far, this is not a complete proof of Euler's theorem. We also need to prove that if a graph has all vertices even-valent and is connected, then an Euler circuit can be found for it. Euler's original paper did not have such a proof, though it is not terribly complicated. (See the Suggested Readings for where a proof can be found.)

1.3 Circuits with Reused Edges

Now let's see what Euler's theorem tells us about the three-block neighborhood with parking meters, represented by dots in Figure 1.12a. Figure 1.12b shows the corresponding graph. (Because we only use edges to represent sidewalks along which the officer must walk, the sidewalk with no meters is not represented by any edge in the graph.) This graph has odd valences (at vertices C and G), so Euler's theorem tells us that there is no Euler circuit for this graph.

Because we must reuse some edges in this graph in order to cover all edges in a circuit, for efficiency we need to keep the total length of reused edges to a minimum. This type of problem, in which we want to minimize the length of a circuit by carefully choosing which edges to retrace, is often called the **Chinese postman problem** (like parking-control routes, mail routes need to be efficient).

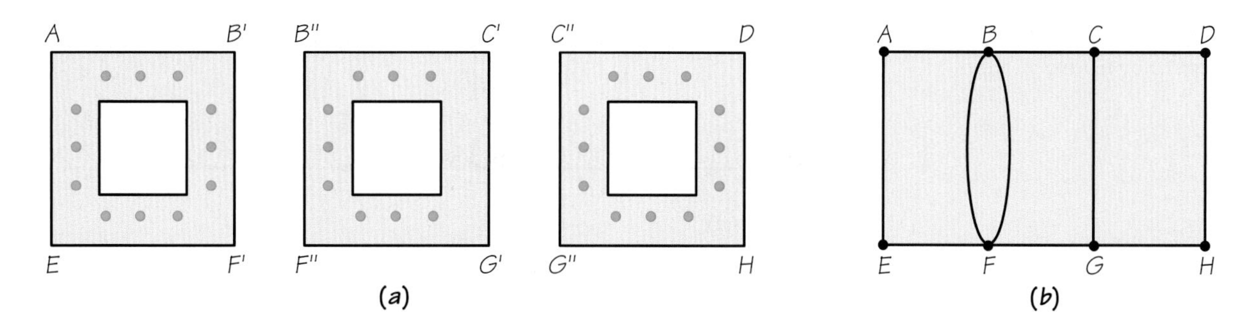

Figure 1.12 (a) A street network and (b) its graphic representation. Locations such as B′ and B″, C′ and C″, F′ and F″, and G′ and G″ are merged to form the vertices *B, C, F,* and *G*.

The problem was first studied by the Chinese mathematician Meigu Guan in 1962—hence the name. Although the Euler circuit theory doesn't deal directly with reused edges or edges of different lengths, we can extend the theory to help solve the Chinese postman problem. The remainder of this chapter is dedicated to solving the Chinese postman problem and discussing applications besides parking control.

In a realistic Chinese postman problem, we need to consider the lengths of the sidewalks, streets, or whatever the edges represent, because we want to minimize the total length of the reused edges. However, to simplify things at the start, we can suppose that all edges represent the same length. (This is often called the *simplified* Chinese postman problem.) In this case, we need only count reused edges and need not add up their lengths. To solve the problem, we want to find a circuit that covers each edge and that has the minimal number of reuses of edges already covered.

To follow the procedure we are going to develop, look at the graph in Figure 1.13a, which is the same graph as in Figure 1.12b. This graph has no Euler circuit, but there is a circuit that has only one reuse of an edge (*CG*), namely, *ABCDHGCGFBFEA*. Let's draw this circuit so that when edge *CG* is about to be reused, we install a new, extra, blue edge in the graph for the circuit to use. By duplicating edge *CG*, we can avoid reusing the edge. To duplicate an edge, we must add an edge that joins the two vertices that are already joined by the edge we want to duplicate. (It makes no sense to join vertices that are not already connected by an edge, because such edges would not represent sidewalk sections with meters; see Figure 1.15.) We have now created the graph of Figure 1.13b. In the graphs we draw, the edges that are added will be shown in color to distinguish them from the original edges, which are shown in black. (You may want to use a similar scheme to help you remember which edges are the originals and which are duplicated in the graphs you draw.) In the graph of Figure 1.13b, the original circuit can be traced as an Euler circuit, using the new edge when

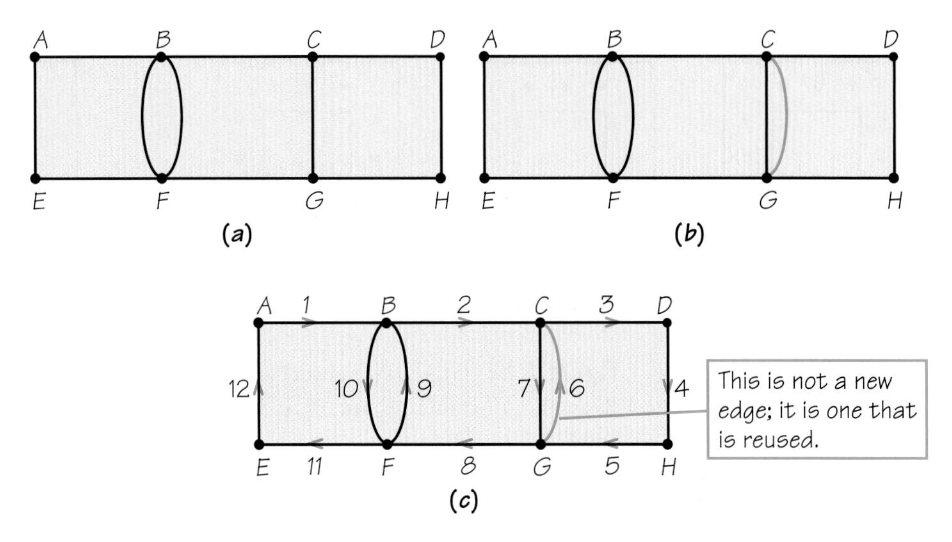

Figure 1.13 Making a circuit by reusing an edge.

needed. The circuit is shown in Figure 1.13c. Our theory will be based on using this idea in reverse, as follows:

1. Take the given graph and add edges by duplicating existing ones, until you arrive at a graph that is connected and even-valent.

Adding edges to a graph to make all valences even is called **eulerizing** the graph.

Note that after a graph is eulerized, the new graph produced will have an Euler circuit.

2. Find an Euler circuit on the eulerized graph.
3. "Squeeze" this Euler circuit from the eulerized graph onto the original graph by reusing an edge of the original graph each time the circuit on the eulerized graph uses an added edge.

EXAMPLE

Eulerizing a Graph

Suppose we want to eulerize the graph of Figure 1.14a. When we eulerize a graph, we first locate the vertices with odd valence. The graph in Figure 1.14a has two, B and C. Next, we add one end of an edge at each such vertex, matching the new edge up with an existing edge in the original graph. Figure 1.14b shows one way to eulerize the graph. Note that B and C have even valence in the second graph. After eulerization, each vertex has even valence. To see an Euler circuit on the eulerized graph in Figure 1.14c, simply follow the edges in numerical order and in the direction of the arrows, beginning and ending at vertex A. The final

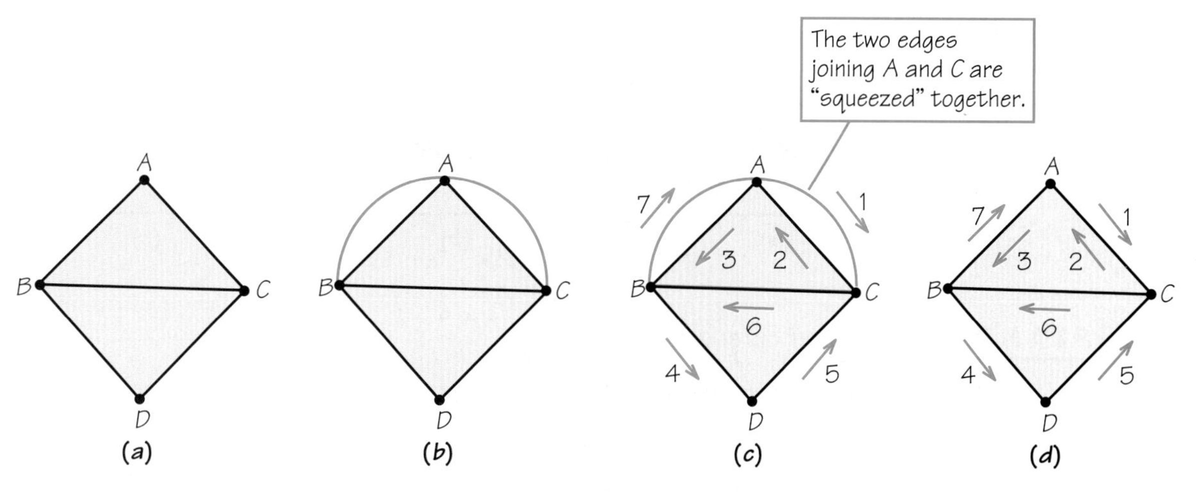

Figure 1.14 Eulerizing a graph.

step, shown in Figure 1.14d, is to "squeeze" our Euler circuit into the original graph. There are two reuses of previously covered edges. Notice that each reuse of an edge corresponds to an added edge. ■

In the previous example, we noticed that we could count how many reuses we needed by counting added edges. This is generally true in this type of problem: *If you add the new edges correctly, the number of reuses of edges equals the number of edges added during eulerization.*

Adding new edges correctly means adding only edges that are duplicates of existing edges. Doing this makes the rule, just stated in italics, always true, and so it is easy to count the needed reuses.

To see why we add only duplicate edges, examine Figure 1.15a. We need to alter the valences of vertices X and Y by adding edges so that they become even-valent. Adding one long edge from X to Y (Figure 1.15b) might seem like an attractive idea, but adding this edge is equivalent to asking a snowplow, say, to get from X to Y without moving along existing streets. At times it is necessary to traverse sections of the graph that have been previously traversed. This is the significance of the duplicated edges. Here the structure of the graph forces us to repeat some edges. We can get away with no fewer than three repeats—the three edges XU, UV, and VY (Figure 1.15c). The duplicated edges are shown in color.

Now that we have learned to eulerize, the next step is to try to get a best eulerization we can—one with the fewest added edges. It turns out that there are many ways to eulerize a graph. It is even possible that the smallest number of added edges can be achieved with two different eulerizations. This is the reason we use the phrase "a best eulerization" rather than "the best eulerization." Remember, we want a best eulerization because this enables us to find the circuit for the original graph that has the minimum number of reuses of edges.

Figure 1.15 Eulerizing when the vertices are more than one edge apart.

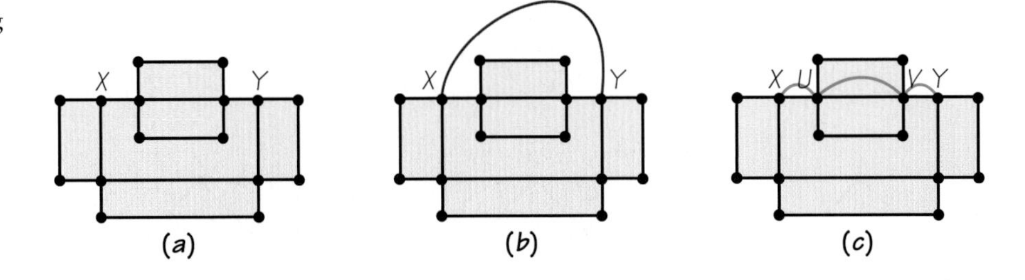

EXAMPLE

A Better Eulerization

In Figure 1.16a, we begin with the same graph as in Figure 1.14, but we eulerize it in a different way—by adding only one edge (see Figure 1.16b). Figure 1.16c shows an Euler circuit on the eulerized graph, and in Figure 1.14d we see how it is squeezed onto the original graph. There is only one reuse of an edge, because we added one edge during eulerization. ■

The solution in Figure 1.16 is better than the solution in Figure 1.14 because one reuse is better than two. These examples suggest the following addition to our solution procedure: Try to find the eulerization with the smallest number of added edges. This extra requirement makes the problem both more interesting and more difficult. For large graphs, a best eulerization may not be obvious. We can try out a few and pick the best among the ones we find, but there may be an even better one that our haphazard search does not turn up.

A systematic procedure for finding a best eulerization does exist, but the process is complicated. There is an especially easy technique for eulerizing the following special category of networks often found in our neighborhoods.

If a street network is composed of a series of rectangular blocks that form a large rectangle a certain number of blocks high by a certain number of blocks wide, the network is called *rectangular*.

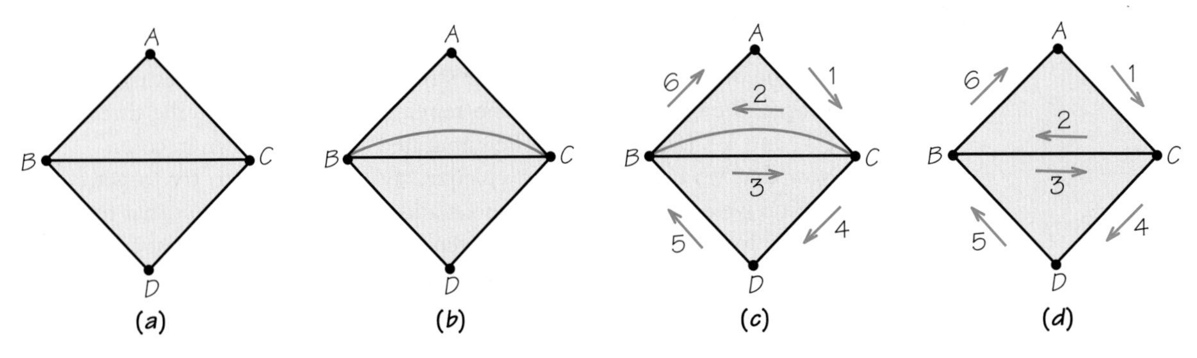

Figure 1.16 A better eulerization of Figure 1.14.

Examples of rectangular street networks (a 3-by-3, a 3-by-4, and a 4-by-4) are shown in Figure 1.17. The graph on the right in each pair shows a best eulerization for the rectangular street network on the left. There appear to be three different eulerization patterns, depending upon whether the rectangle height and width in the original graph are odd or even numbers. In Figure 1.17a, both lengths are 3, both odd; in Figure 1.17b, one length is odd (3) and one is even (4); in Figure 1.17c, both lengths are 4, an even number.

Although the patterns appear different, one technique can be used to create all of them. This technique can be thought of as involving an "edge walker" who walks around the outer boundary of the large rectangle in some direction, say, clockwise. He starts at any corner, say, the upper left corner. As he goes around, he adds edges by the following rules. When he comes to an odd-valent vertex, he links it to the next vertex with an added edge. This next vertex now becomes either even or odd. If it became even, he skips it and continues around, looking for an odd vertex. If it became odd (this could only happen at a corner of the big rectangle), the edge walker links it to the next vertex and then checks this vertex to see whether it is even or odd. Each of the three parts of Figure 1.17 has been eulerized by this method.

In a street network that is not rectangular, the eulerization process is started by locating all the vertices with odd valence and then pairing these vertices with each other and finding the length of the shortest path between each pair. We look for the shortest paths, because each edge on the connecting paths will be duplicated. The idea is to choose the pairings cleverly so that the sum of the lengths of those paths is the smallest it can be. With a little practice, most people can find a best or nearly best eulerization using only this idea together with trial

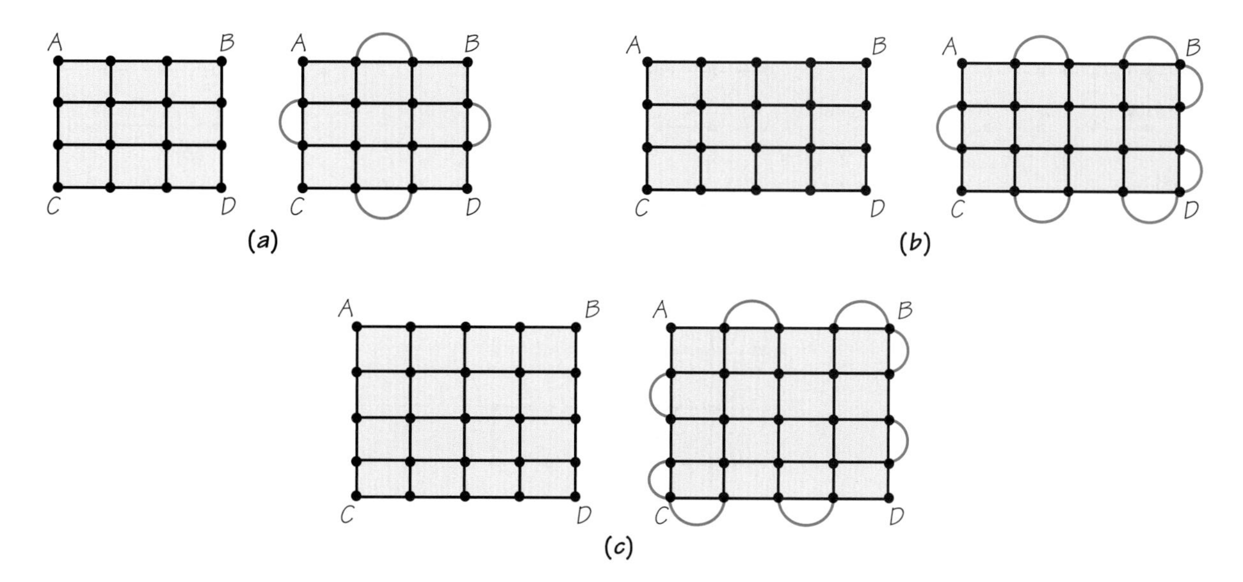

Figure 1.17 Eulerizations of three rectangular networks.

and error and some ingenuity. Those interested in a further discussion can read the following optional section, "Finding Good Eulerizations."

OPTIONAL

Finding Good Eulerizations

Suppose we want a perfect procedure for eulerizing a graph. What theoretical ideas and methods could we use to build such a tool?

One building block we could use is a method for finding the shortest path between two given vertices of a graph. For example, let us focus on vertices X and Y in Figure 1.18a; both have odd valence. We can connect them with a pattern of duplicate edges, as in Figure 1.18b. The cost of this is the length of the path we duplicated from X to Y. A shorter path from X to Y, such as the one shown in Figure 1.18c, would be better. Fortunately, the *shortest-path problem* has been well studied, and we have many good procedures for solving it exactly, even in large, complex graphs. These procedures are discussed in some of the Suggested Readings given at the end of this chapter but are beyond the scope of this text.

But there is more to eulerizing the graph in Figure 1.18a than dealing with X and Y. Notice that we have odd valences at Z and W. Should we connect X and

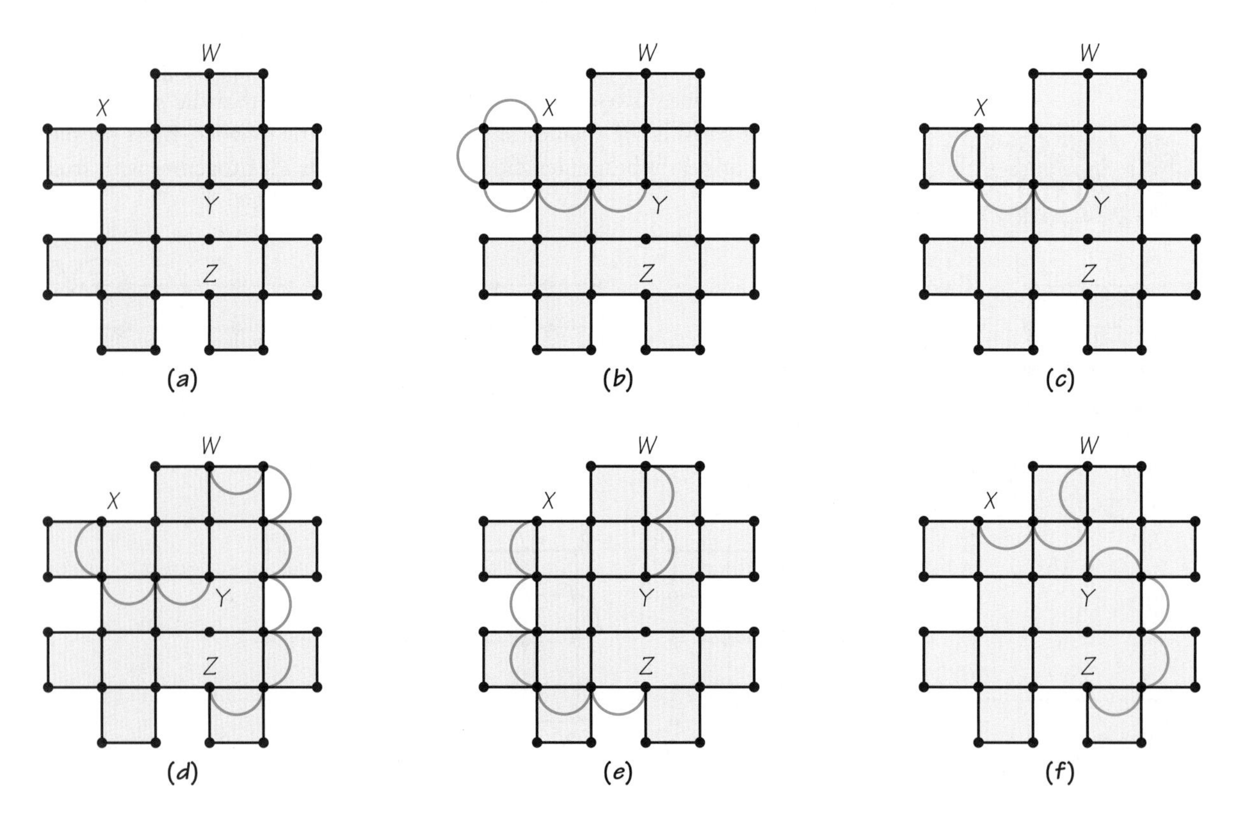

Figure 1.18 Choosing among eulerizations.

Y with a path, and then connect Z and W, as in Figure 1.18d? Or should we connect X to Z and Y to W, as in Figure 1.18e? Another alternative is to use connections X to W and Y to Z, as in Figure 1.18f. It turns out that the alternatives in both Figures 1.18e and 1.18f are preferable to the one in Figure 1.18d, because they involve seven added edges, whereas Figure 1.18d uses nine.

At the start, it is often not clear which alternatives are best. The problem is how to pair up the vertices that are odd-valent for connection to get a set of paths whose total length is minimal. This problem, called the *matching problem*, has also been studied and solved. (For details, see the Suggested Readings.)

1.4 Circuits with More Complications

Euler circuits and eulerizing have many more practical applications than just checking parking meters. Almost any time services must be delivered along streets or roads, our theory can make the job more efficient. Examples include collecting garbage, salting icy roads, plowing snow, inspecting railroad tracks, and reading electric meters (see Spotlight 1.3).

Each of these problems has its own special requirements that may call for modifications in the theory. For example, in the case of garbage collection, the edges of our graph will represent streets, not sidewalks. If some of the streets are one-way, we need to put arrows on the corresponding edges, resulting in a

SPOTLIGHT 1.3 Israel Electric Company Reduces Meter-Reading Task

The Beersheba branch of Israel's major electric company wanted to make the job of meter reading more efficient. When the branch managers decided to minimize the number of people required to read the electric meters in the houses of one particular neighborhood, they set a precedent by applying management science. Formerly, each person's route had been worked out by trial and error and intuition, with no help from mathematics. The whole job required 24 people, each doing a part of the neighborhood in a five-hour shift.

At first, it looks as though one would find a more efficient way of doing the work the same way as in the Chinese postman problem, but there are two important differences. First, the neighborhood was big enough to negate any possibility of having only one route assigned to one person. Instead, it was necessary to find a number of routes that, taken together, covered all the edges (sidewalks). Second, a meter reader who was done with a route was allowed to return home directly. Thus, there was no reason for the individual routes to return to their starting points; therefore, routes could be paths instead of circuits.

The Beersheba researchers found solutions to these problems by modifying the basic ideas we have described in this chapter. They managed to cover the neighborhood with 15 five-hour routes, a 40% reduction of the original 24 five-hour routes. Altogether, these routes involve a total of 4338 minutes of walking time, of which 41 minutes (less than 1%) is deadheading.

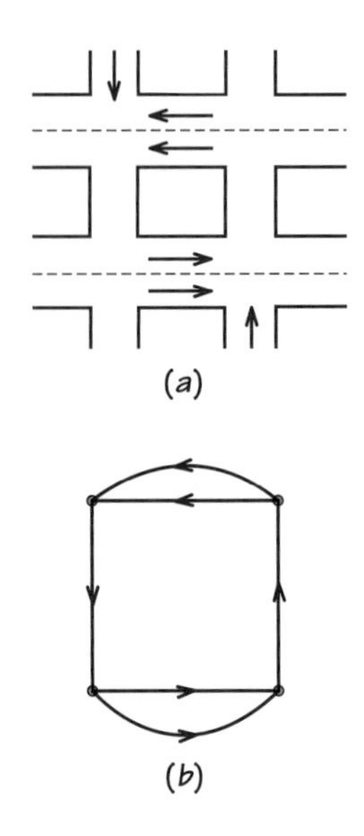

Figure 1.19 (a) Salt-spreading route, where each west–west street has two traffic lanes in the same direction, and (b) an appropriate digraph model.

directed graph, or **digraph.** The circuits we seek will have to obey these arrows. In the case of salt spreaders and snowplows, each lane of a street needs to be modeled as a directed edge, as shown in Figure 1.19. Note that the arrows on the map and digraph are not in color because these arrows denote restrictions in traversal possibilities, not parts of circuits.

Like salt spreaders, street-sweeping trucks can travel in only one lane at a time and must obey the direction of traffic. Street sweepers, however, have an additional complication: parked cars. It is very difficult to clean the street if cars are parked along the curb. Yet for overall efficiency, those who are responsible for routing street sweepers want to interfere with parking as little as possible. The common solution is to post signs specifying times when parking is prohibited. Because the parking-time factor is a constraint on street sweeping, it is important not only to find an Euler circuit, or a circuit with very few duplications, but a circuit that visits streets when they are free of cars. Once again, mathematicians have developed techniques to handle this constraint.

Finally, because towns and cities of any size will have more than one street sweeper, parking officer, or garbage truck, a single best route may not suffice. Instead, it becomes necessary to divide the territory into multiple routes. The general goal is to find optimal solutions while taking into account traffic direction, number of lanes, parking-time restrictions, and divided routes (see Figure 1.20).

Management science makes all this possible. For example, a pilot study done in the 1970s in New York City showed that applying these techniques to street sweepers in just one district could save about $30,000 per year. With 57 sanita-

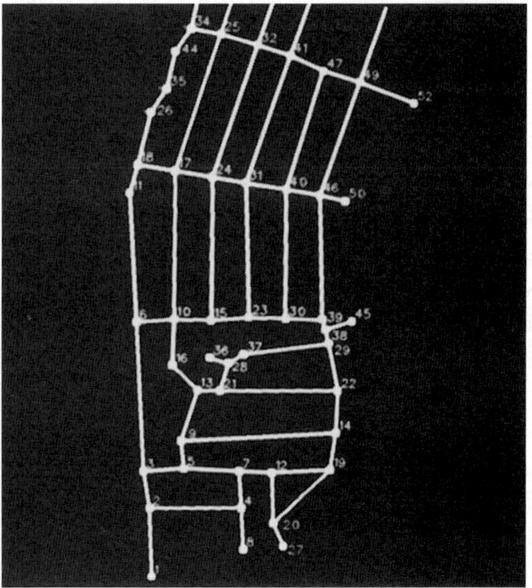

Figure 1.20 (a) Fairfield, California, USA. Today, finding optimal routes within complex street networks is often done with sophisticated computer-based color graphic systems. (b) A computer-generated street network. (a: Superstock.)

tion districts in New York, this would amount to a savings of more than $1.5 million in a single year. In addition, the same principles could be extended to garbage collection, parking control, and other services carried out on street networks.

This plan was not adopted when first proposed. Because city services take place in a political context, several other factors come into play. For example, union leaders try to protect the jobs of city workers, bureaucrats might try to keep their departmental budgets high, and elected politicians rarely want to be accused of cutting the jobs of their constituents. Thus, political obstacles can overrule management science. As mentioned in Spotlight 1.2, such human factors often arise when applying management science. Perhaps a more acceptable street-sweeping plan would have been devised for New York City if more attention had been paid to the human factors earlier.

Despite the complications of real-world problems, management science principles provide ways to understand these problems by using graphs as models. We can reason about the graphs and then return to the real-world problem with a workable solution. The results we get can have a lasting effect on the efficiency and economic well-being of any organization or community.

REVIEW VOCABULARY

Chinese postman problem The problem of finding a circuit on a graph that covers every edge of the graph at least once and that has the shortest possible length.

Circuit A path that starts and ends at the same vertex.

Connected graph A graph is connected if it is possible to reach any vertex from any specified starting vertex by traversing edges.

Digraph A graph in which each edge has an arrow indicating the direction of the edge. Such directed edges are appropriate when the relationship is "one-sided" rather than symmetric (e.g., one-way streets as opposed to regular streets).

Edge A link joining two vertices in a graph.

Euler circuit A circuit that traverses each edge of a graph exactly once.

Eulerizing Adding new edges to a graph so as to make a graph that possesses an Euler circuit.

Graph A mathematical structure in which points (called vertices) are used to represent things of interest and in which links (called edges) are used to connect vertices, denoting that the connected vertices have a certain relationship.

Management science A discipline in which mathematical methods are applied to management problems in pursuit of optimal solutions that cannot readily be obtained by common sense.

Operations research Another name for management science.

Optimal solution When a problem has various solutions that can be ranked in preference order (perhaps according to some numerical measure of "goodness"), the optimal solution is the best-ranking solution.

Path A connected sequence of edges in a graph.

Valence (of a vertex) The number of edges touching that vertex.

Vertex A point in a graph where one or more edges end.

SUGGESTED READINGS

BELTRAMI, EDWARD J. *Models for Public Systems Analysis,* Academic Press, New York, 1977. Section 5.4 deals with material similar to that in this chapter. The rest of the book gives a nice selection of applications of mathematics to plant location, manpower scheduling, providing emergency services, and other public service areas. The mathematics is somewhat more advanced than in this chapter.

COZZENS, MARGARET B., and RICHARD P. PORTER. *Mathematics and Its Applications,* Heath, Lexington, Mass., 1987. Includes a nice discussion of Euler circuit ideas applied to DNA fragments.

MALKEVITCH, JOSEPH, and WALTER MEYER. *Graphs, Models, and Finite Mathematics,* Prentice Hall, Englewood Cliffs, N.J., 1974. An introductory text that includes much the same material as in this chapter, but with a little more detail. A different algorithm for finding Euler circuits is given.

The following two references discuss the shortest-path and matching problems in depth; they are suitable for advanced students and for faculty.

ROBERTS, FRED S. *Applied Combinatorics,* Prentice Hall, Englewood Cliffs, N.J., 1984.

TUCKER, ALAN. *Applied Combinatorics,* 2nd ed., Wiley, New York, 1984.

SUGGESTED WEB SITE

www.math.harvard.edu/~hmb/isue2.2/euler/euler.htmlwww.informs.org This Web site is maintained by the Institute for Operations Research and the Management Sciences, the main professional organization in these fields in the United States. It contains information on (and/or links to) news items about operations research and management science as well as employment opportunities and summer internships; it also has a student newsletter. Much of the material is written in a nontechnical style.

SKILLS CHECK

1. What is the valence of vertex *A* in the graph below?

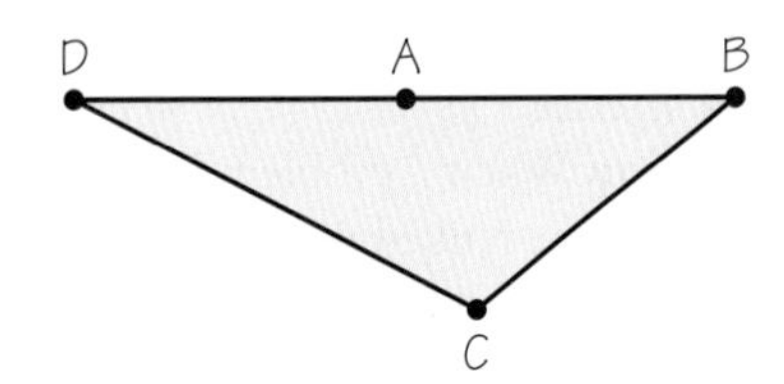

(a) 2
(b) 1
(c) 4

2. For the graph below, which statement is correct?

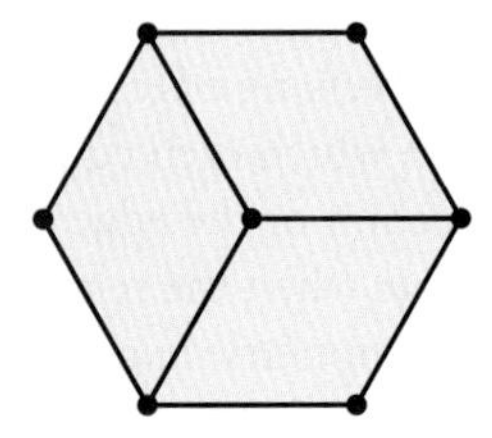

(a) The graph has an Euler circuit.
(b) One new edge is required to eulerize this graph.
(c) Three new edges are required to eulerize this graph.

3. For which of the situations below is it most desirable to find an Euler circuit or an efficient eulerization of the graph?

(a) Sweeping the sidewalks of a small town
(b) Planning a new highway
(c) Planning a parade route in Muncie, Indiana

4. Consider the path represented by the sequence of numbered edges on the graph below. Which statement is correct?

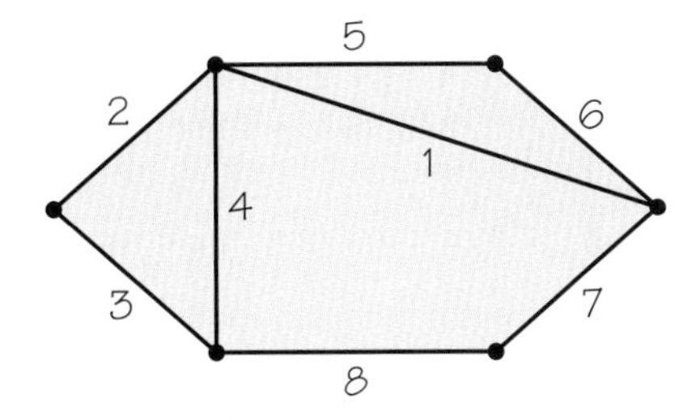

(a) The sequence of numbered edges forms an Euler circuit.
(b) The sequence of numbered edges traverses each edge exactly once but is not an Euler circuit.
(c) The sequence of numbered edges forms a circuit but not an Euler circuit.

5. What is the minimum number of duplicated edges needed to create a good eulerization for the graph here?

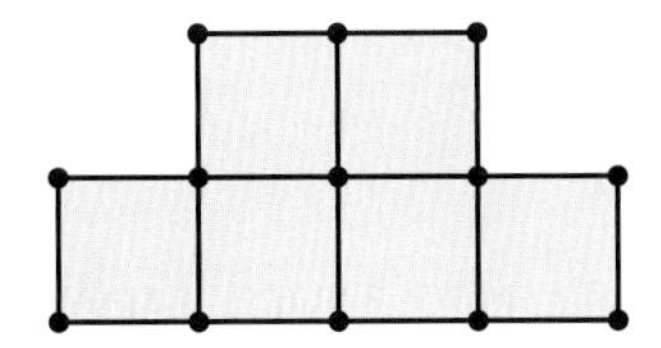

(a) Four edges
(b) Three edges
(c) Two edges

6. Suppose the edges of a graph represent streets that must be plowed after a snowstorm. In order to eulerize the graph, four edges must be added. The real-world interpretation of this is that

(a) four streets will not be plowed.
(b) four streets will be traversed twice.
(c) four new streets would be built.

7. If a graph has six vertices of odd valence, what is the absolute minimum number of edges that must be added (duplicated) to eulerize the graph?

(a) Six
(b) Zero
(c) Three

8. Which of the following statements is true about a *path?*

(a) A path always forms a circuit.
(b) A path is always connected.
(c) A path can visit any vertex only once.

9. Which of the following situations is best modeled by a *digraph?*

(a) A system of walking trails
(b) An electrical wiring plan for a home
(c) A bus route map

10. If a graph consists of four vertices and every pair of vertices is connected by a single edge, how many edges are in the graph?

(a) Four
(b) Five
(c) Six

11. If a graph is connected and has six vertices, what can be said about the number of edges in the graph?

(a) There are at least five edges in the graph.
(b) There are exactly five edges in the graph.
(c) There are at least six edges in the graph.

12. If the vertices of a graph represent cities and the edges of a graph represent flight routes for a particular airline, then which of the objects below best models a pilot's daily schedule?

(a) A path
(b) An Euler circuit
(c) A graph

13. Suppose each vertex of a graph represents a baseball team and each edge represents a game played by two baseball teams. If the resulting graph is not connected, which of the following statements must be true?

(a) At least one team never played a game.
(b) At least one team played every other team.
(c) The teams play in distinct leagues.

14. Suppose a civic club offers several craft courses, and each club member can choose to participate in up to two different courses. Let each vertex of a graph represent one of these courses and each edge represent a club member who wants to take the two courses represented by the vertices at its endpoints. What can be said about the vertices in the resulting graph, whose valence is zero?

(a) These are no vertices whose valence is zero.
(b) These vertices represent courses that can occur at the same time without displeasing any club member.
(c) These vertices represent the least popular courses.

15. If the valences of the vertices of a graph are added together, what can be said about their sum?

(a) The sum is always even.
(b) The sum is always odd.
(c) The sum is sometimes even and sometimes odd.

EXERCISES

▲ Optional. ■ Advanced. ◆ Discussion.

Basic Concepts

1. In the graph below, the vertices represent cities and the edges represent roads connecting them. What are the valences of the vertices in this graph? (Keep in mind that *E* is part of the graph.) What might the valence of city *E* be showing about the geography?

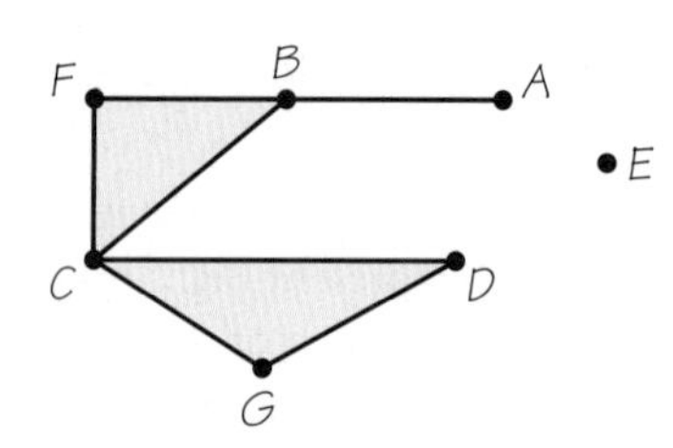

2. In the two graphs above right, the vertices represent cities and the edges represent roads connecting them. In which graphs could a person located in city *A* choose any other city and then find a sequence of roads to get from *A* to that other city?

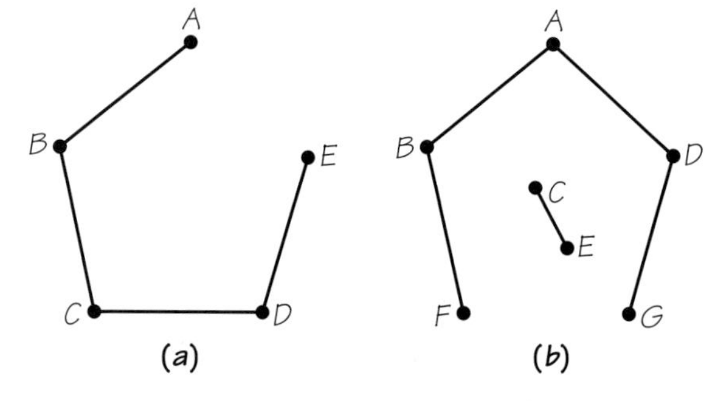

3. Is the figure below a graph? Explain your answer.

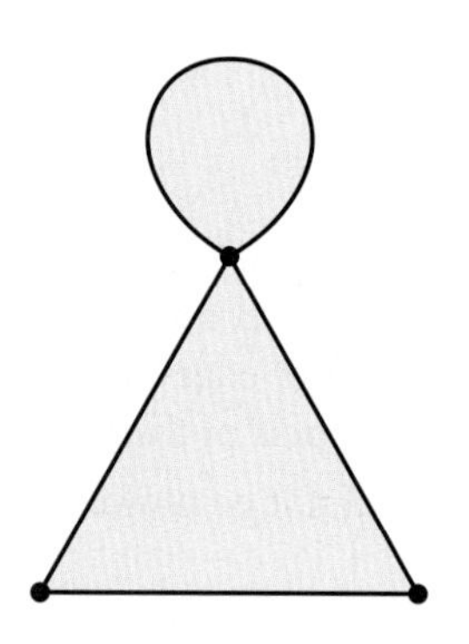

4. Jack and Jill are located in Miami and want to fly to Berlin (see Figure 1.2). Jill says she can think of three paths to get them there. Can you? Jack says that a path can repeat edges so there are more than three. Is Jack right?

5. Draw a graph with six vertices that is connected where

(a) each vertex has valence 3.
(b) each vertex has valence 4.

6. (a) Add up the numbers you get for the valences of the vertices in Figure 1.6.

(b) Add up the numbers you get for the valences of the vertices in Figure 1.8a.
(c) Add up the numbers you get for the valences of the vertices in Figure 1.9a.
(d) Do you see a pattern in the answers you got for parts (a) through (c)?
(e) Can you show that the pattern describes a fact that is true for any graph? (*Hint:* How many endpoints does an edge have?)

7. In the graph in Figure 1.8, find the smallest possible number of edges you could remove that would disconnect the graph.

8. In the graphs in Figure 1.17, find the smallest possible number of edges you could remove that would disconnect the graph.

9. Is it possible that a street network gives rise to a disconnected graph? If so, draw such a network of blocks and streets and parking meters (in the style of Figure 1.12a). Then draw the disconnected graph it gives rise to.

10. (a) Can you draw a graph where every vertex has valence of at least 3 but where removing a single edge disconnects the graph?

(b) In what urban settings might a road network be represented by a graph that has an edge whose removal would disconnect the graph?

11. (a) Can you find a graph where the valences of the seven vertices of the graph are 1, 2, 2, 3, 3, 3, 4?

(b) Can you find another graph with the same valences as above that is "different" from the one you found for part (a)?

12. (a) Give examples where it might be reasonable to disregard the issue of one-way versus two-way streets in constructing a graph model for the service to be provided along that stretch of street.

(b) Give examples where there might be a difference in whether or not one could disregard the issue of one-way versus two-way streets in constructing a graph model if one is dealing with an urban area as contrasted with a suburban area.

◆ **13.** A postal worker is supposed to deliver mail on all streets represented by edges in the graph below by traversing each edge exactly once. The first day the worker traverses the numbered edges in the order shown in (a), but the supervisor is not satisfied—why? The second day the worker follows the path indicated in (b), and the worker is unhappy—why? Is the original job description realistic? Why?

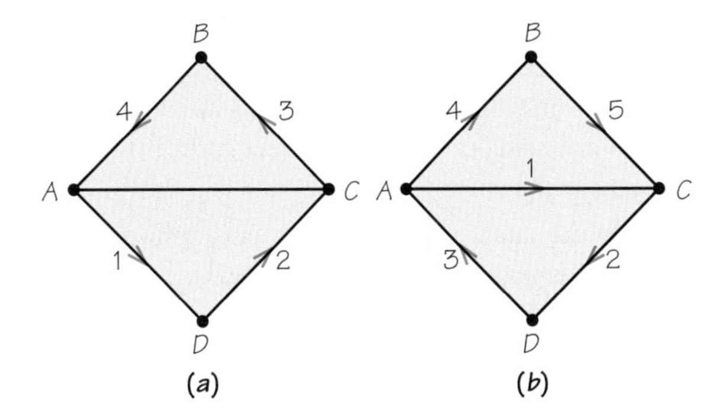

▲ **14.** For the street network in Exercise 13, draw the graph that would be useful for routing a snowplow. Assume that all streets are two-way, one lane in each direction, and that you need to pass down each lane separately.

15. Can you find an efficient route for the snowplow to follow in the graph you drew in Exercise 14?

16. (a) Give examples of services that could be performed by a vehicle that moved in the direction of traffic down either lane of a two-way street.

(b) Give examples of services that would probably require a vehicle to travel down each of the lanes of a two-way street (in the direction of traffic for that lane) in order to perform the service.

▲ **17.** For the street network shown below, draw the graph that would be useful for finding an efficient route for checking parking meters. (*Hint:* Notice that not every sidewalk has a meter; see Figure 1.12.)

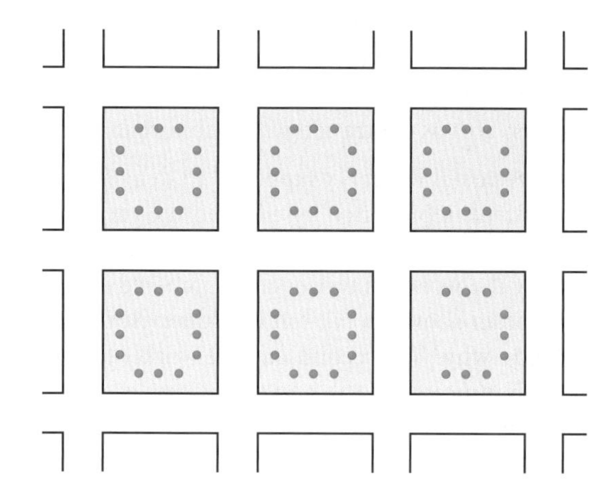

▲ **18. (a)** For the street network in Exercise 17, draw a graph that would be useful for routing a garbage truck. Assume that all streets are two-way and that passing once down a street suffices to collect from both sides.

(b) Do the same problem on the assumption that one pass down the street only suffices to collect from one side.

19. (a) In the graph below, find the largest number of paths from A to H that do not have any edges in common.

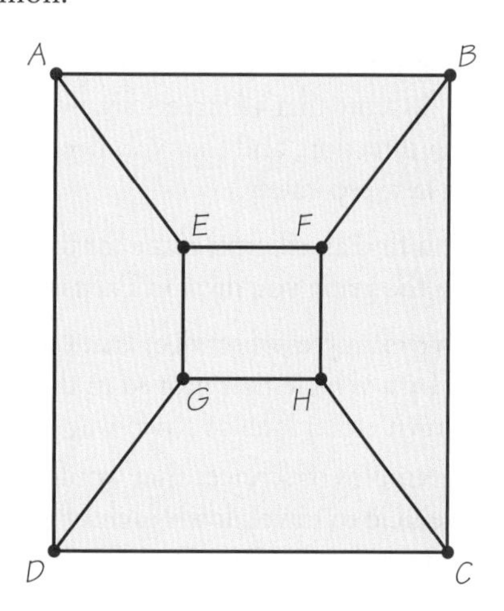

(b) Verify that the largest number of paths with no edges in common between any pair of vertices in this graph is the same.

(c) Why might one want to be able to design graphs such that one can move between two vertices of the graph using paths that have no edges in common?

Euler Circuits

20. Examine the paths represented by the numbered sequences of edges in both parts of the figure below. Determine whether each path is a circuit. If it is a circuit, determine if it is an Euler circuit.

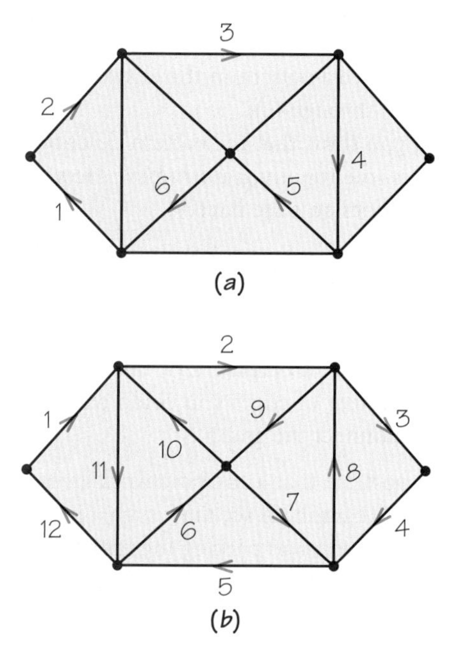

21. In Figure 1.13c, suppose we started an Euler circuit using this sequence of edges: 6, 7, 8, 9 (ignore existing arrows on the edges). What does our guideline for finding Euler circuits tell you *not* to do next?

22. In Figure 1.8b, suppose we started an Euler circuit using this sequence of edges: 14, 13, 8, 1, 4 (ignore existing arrows on the edges). What does our guideline for finding Euler circuits tell you *not* to do next?

23. Find an Euler circuit on the graph of Figure 1.15c (including the blue edges).

24. Find Euler circuits in the right-hand graphs in Figures 1.17a and 1.17b.

25. In the graph below, we see a territory for a parking-control officer that has no Euler circuit. How many sidewalks (edges) need to be dropped in order to enable us to find an Euler circuit? What effect would this have in the associated real-world situation?

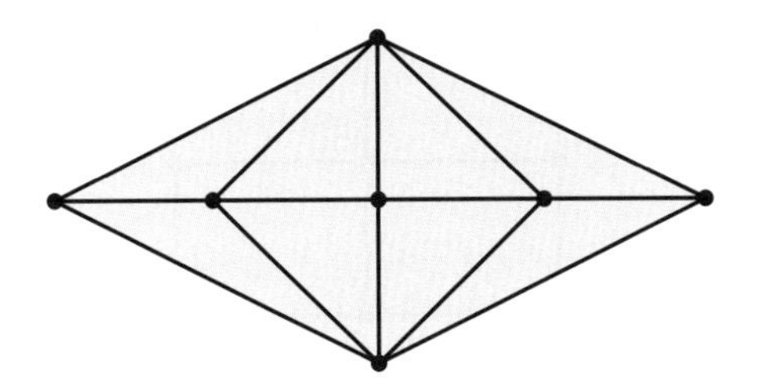

26. An Euler circuit visits a four-valent vertex *X*, such as the one in the graph below, by using the edges *AX* and *XB* consecutively, and then using *CX* and *XD* consecutively. When this happens, we say that the Euler circuit cuts through at *X*.

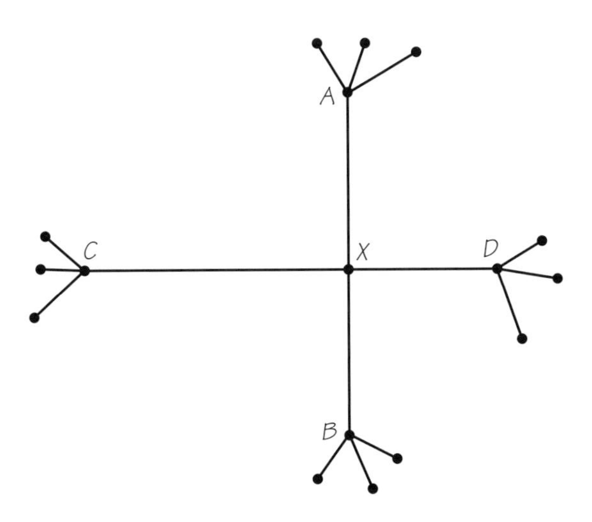

Suppose *G* is a four-valent graph such as that in the accompanying diagram. Is it possible to find an

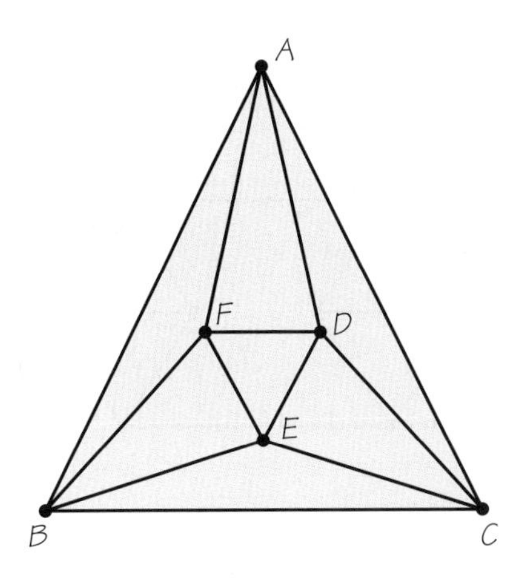

Euler circuit of this graph that never cuts through any vertex? Can you see why it might be desirable to find an Euler circuit of this special kind in an applied situation?

Eulerization and Squeezing

27. Find an Euler circuit on the eulerized graph (b) of the following figure. Use it to find a circuit on the original graph (a) that covers all edges and only reuses edges five times.

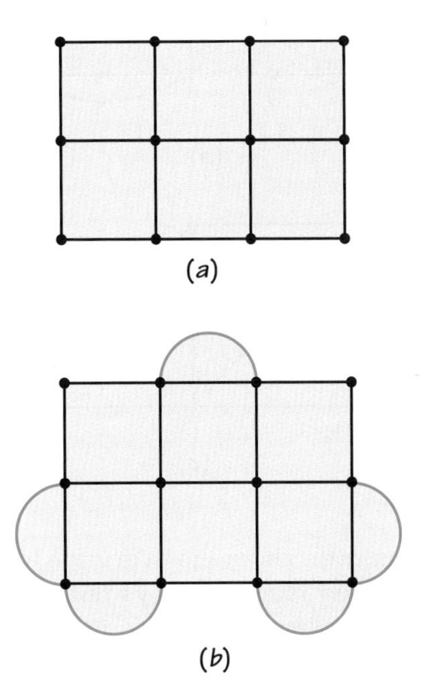

28. In the graph below, add one or more edges to produce a graph that has an Euler circuit.

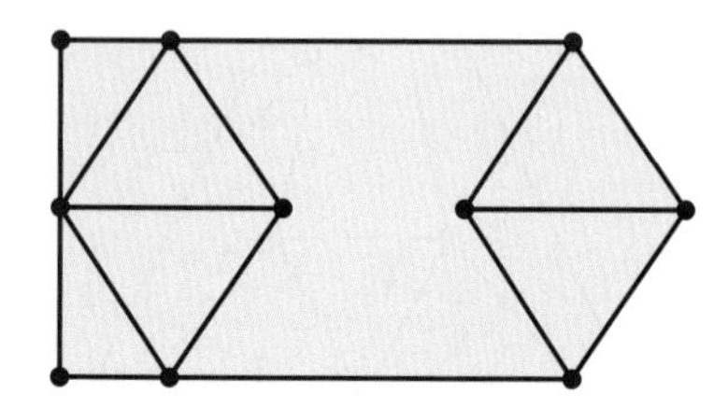

29. Can you find an eulerization with nine added edges for a 2-by-7-block rectangular street network? Can you do better than nine added edges?

30. Squeeze the circuit shown in graph (a) below onto graph (b). Show your answers by writing numbered arrows on the edges and by listing a sequence of vertices (e.g., *ABEB . . . A*).

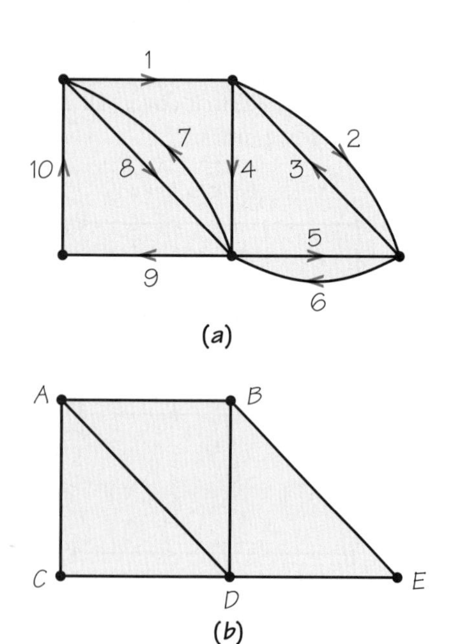

(a)

(b)

Then squeeze the circuit shown in graph (c) onto graph (d). Show your answers by writing numbered arrows on the edges and by listing a sequence of vertices.

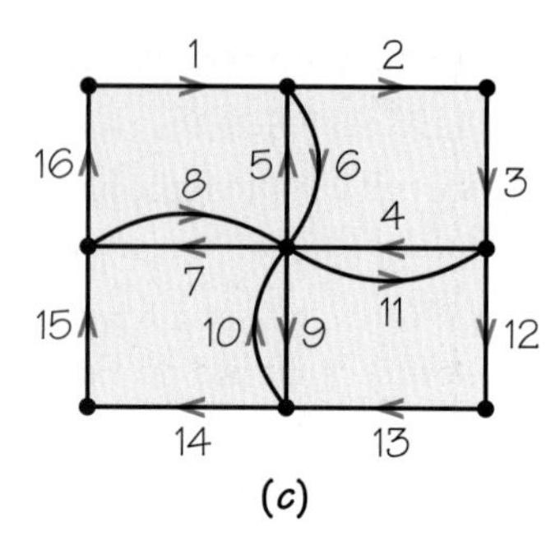

(c)

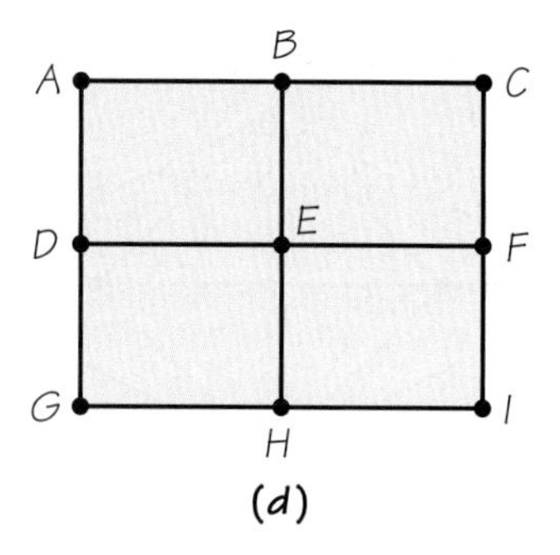

(d)

▲ **31.** Eulerize these rectangular street networks using the same patterns that would be used by the "edge walker" described in the text.

(a) A 5×5 rectangle

(b) A 5×4 rectangle

(c) A 6×6 rectangle

▲ **32.** Find good eulerizations for these graphs, using as few duplicated edges as you can. See the optional section "Finding Good Eulerizations" for hints.

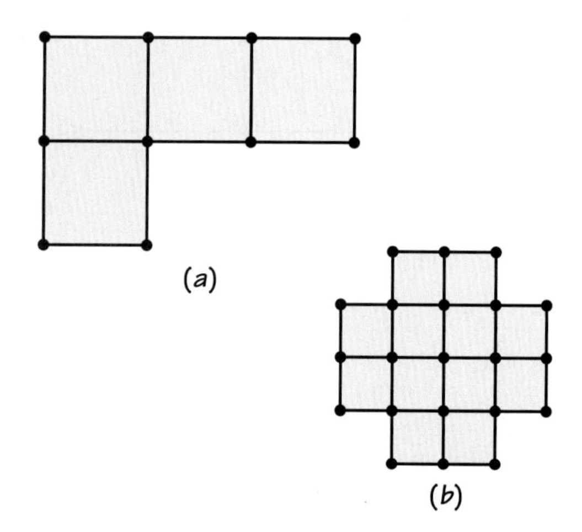

(a)

(b)

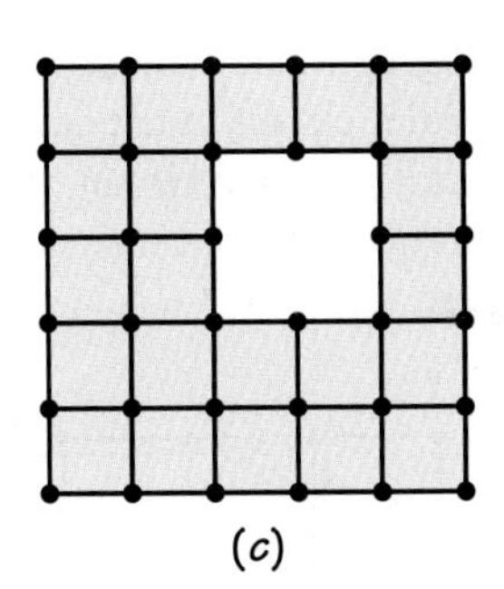

33. A college campus has a central square with sides arranged as shown by the edges in the graph below. Show how all these sidewalks can be traversed at least once in a tour that starts and ends at the same vertex.

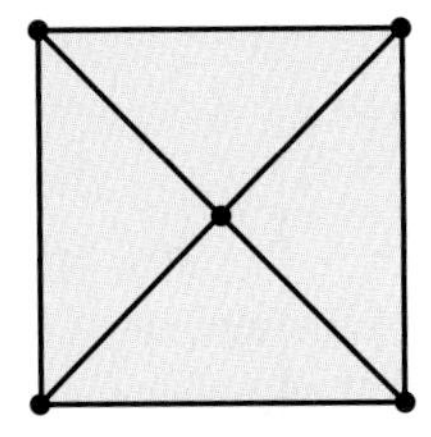

34. The figure below shows a river, some islands, and bridges connecting the islands and riverbanks. A charity is sponsoring a race in which entrants have to start at A, go over each bridge at least once, and end at A. Draw a graph that would be useful for finding a route that requires the least recrossing of bridges. Show what that route would be. (*Historical note:* This situation resembles the one that inspired Leonhard Euler's 1736 "recreational mathematics" problem that resulted in the first work in graph theory.)

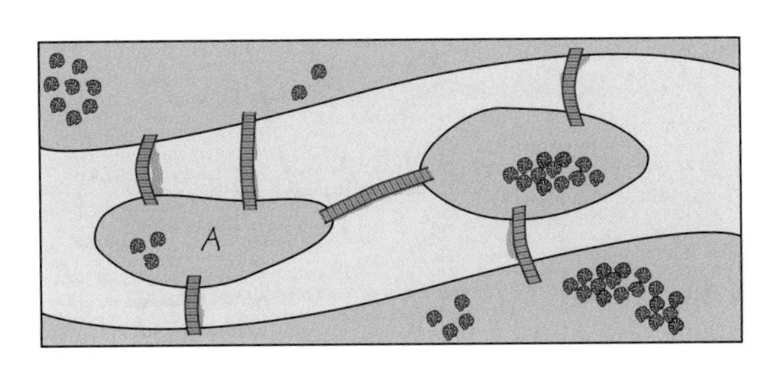

▲ **35.** Find a circuit in the graph below that covers every edge and has as few reuses as possible. See the optional section "Finding Good Eulerizations" for hints.

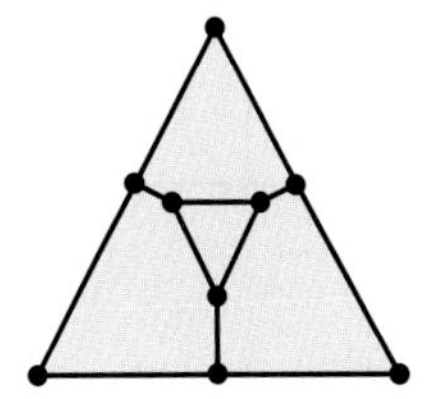

▲ **36.** In the figure below, all blocks are 1000-by-1000 feet, except for the middle column of blocks, which are 1000-by-4000 feet. Find a circuit of minimum total length that covers all edges.

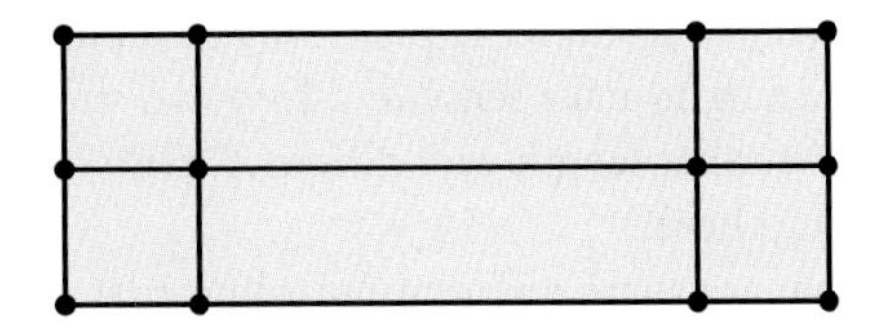

▲ **37.** In the figure below, all blocks are 1000-by-1000 feet, except for the middle column of blocks, which are 1000-by-4000 feet. Find a circuit of minimum total length that covers all edges.

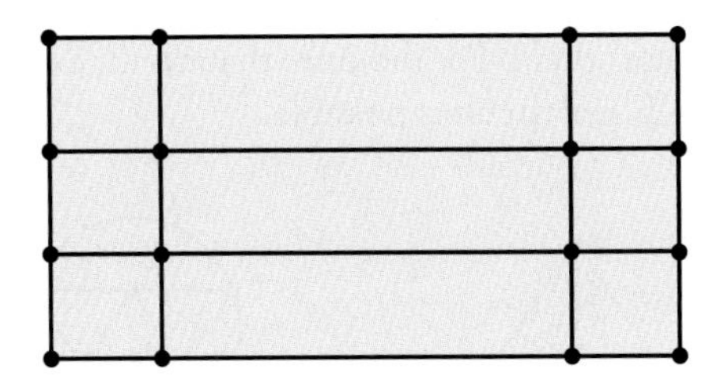

38. (a) Find the cheapest route in the accompanying graph, where one starts at vertex A, finishes at vertex A, and traverses each edge at least once. The cost of a route is computed by summing the numbers along the edges that one uses.

(b) How many edges are repeated in the minimal-cost route?

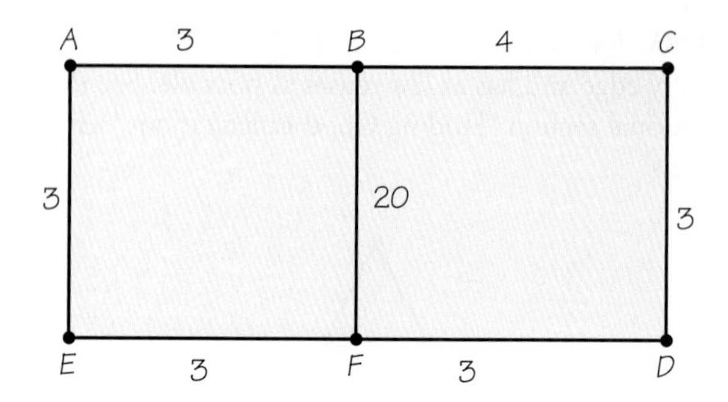

(c) Discuss the implications of this example for the relation between finding good eulerizations of graphs and the problem of finding cheap routes that start and end at the same vertex and traverse each edge at least once.
(d) The physical edge with cost 20 in the diagram is not physically longer than other edges with lower costs attached to them. Explain why in an urban setting it might make sense to assign two stretches of street of similar length very different "costs" for traversing them.
(e) What are some different meanings that "weights" (e.g., traffic volume) potentially assigned to edges in a graph might have in an urban setting?

Additional Exercises

39. Which graphs (see figures below) have Euler circuits? In the ones that do, find the Euler circuits by numbering the edges in the order the Euler circuit uses them. For the ones that don't, explain why no Euler circuit is possible.

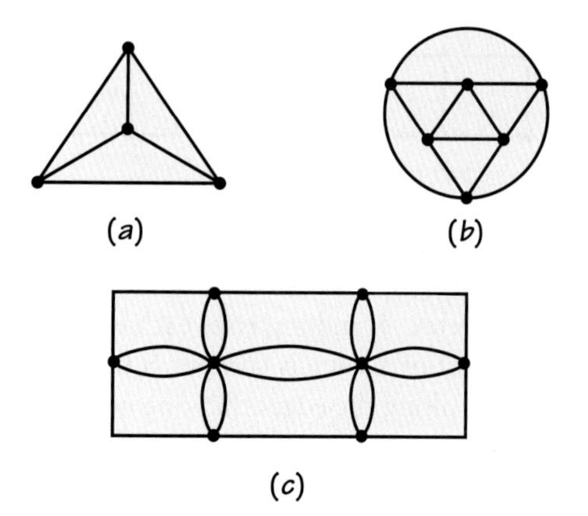

40. Eulerize the graph below by using four new edges. Find an Euler circuit in the eulerized graph and use that circuit to find a circuit of the original graph that covers all edges but only reuses edges four times.

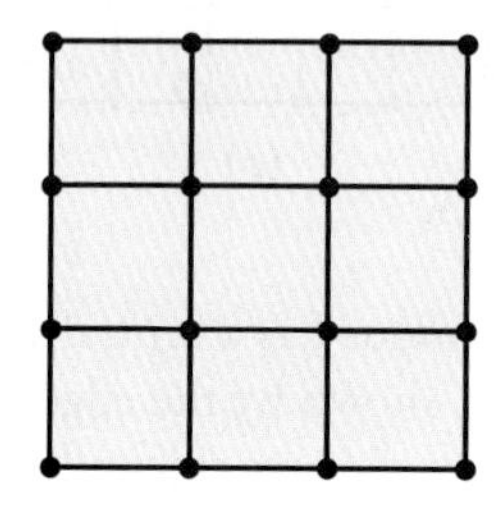

41. (a) What are the valences of the vertices in the graphs below?

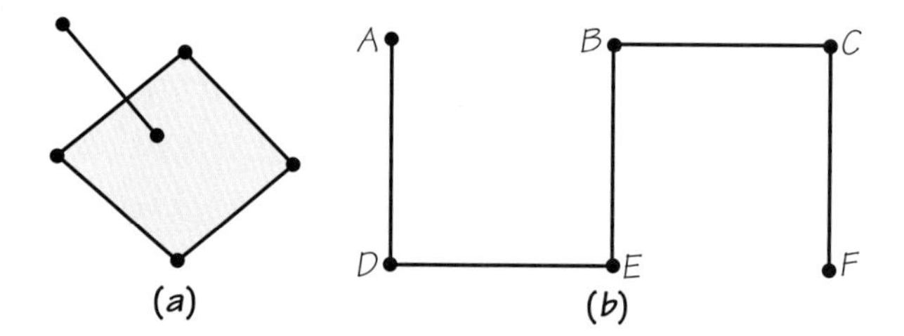

(b) Are these graphs both connected?

▲ **42.** In the graph below, find a circuit that covers every edge and has as few reuses as possible. See the optional section "Finding Good Eulerizations" for hints.

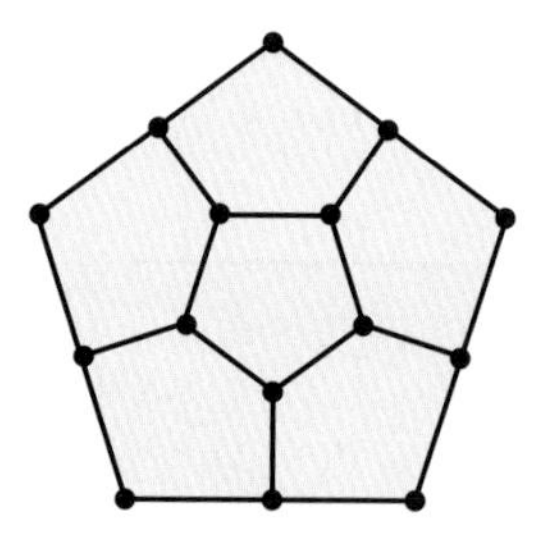

■ **43.** A graph G represents a street network to be traveled by a postal worker who must traverse every street twice, once for each side of the street. In graph G, the edges represent sidewalks. Does such a graph always have an Euler circuit? Explain your answer.

■ **44.** Suppose that for a certain graph it is possible to disconnect it by removing one edge. Explain why such a graph (before the edge is removed) must have at least one vertex of odd valence. (*Hint:* Show that it cannot have an Euler circuit.)

45. (a) Find the best eulerizations you can for the two graphs below.

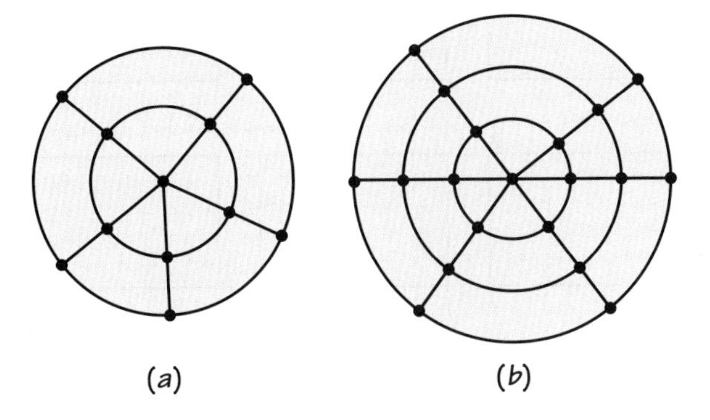

(a) (b)

(b) Graph (a) can be thought of as having five rays and two circles, and graph (b) as having six rays and three circles. Draw a graph with four rays and four circles and find the best eulerization you can for this graph.
(c) Can you find a "formula" involving r and s for the smallest number of edges needed to eulerize a graph of this type having r rays and s circles?

46. Each graph below represents the sidewalks to be cleaned in a fancy garden (one pass over a

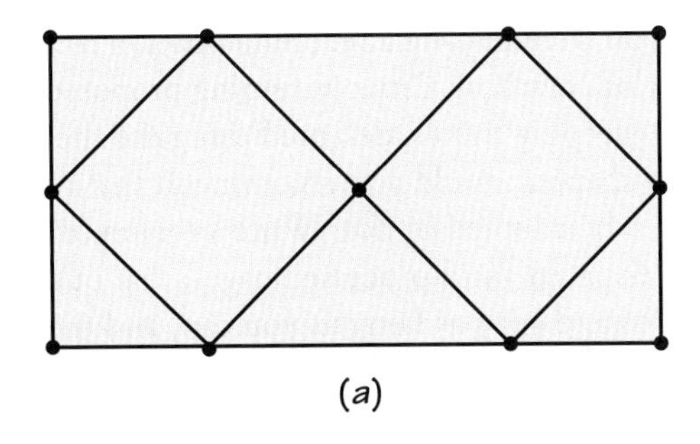

(a)

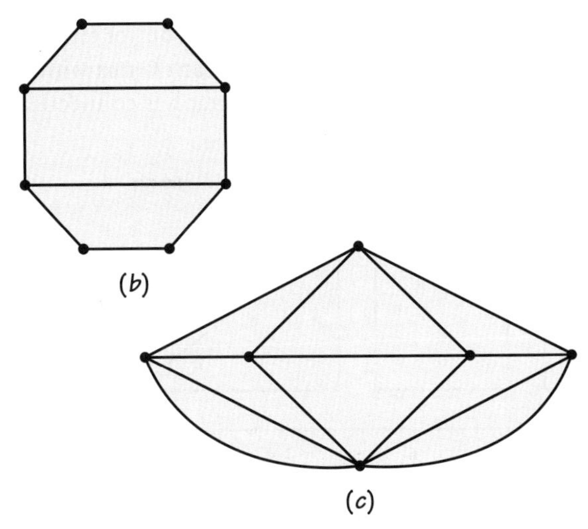

(b) (c)

sidewalk will clean it). Can the cleaning be done using an Euler circuit? If so, show the circuit in each graph by numbering the edges in the order the Euler circuit uses them. If not, explain why no Euler circuit is possible.

47. Can you draw a graph with six vertices where the valence of each vertex is 5?

48. Find a minimum duplication circuit in a 3-by-5-block rectangular street network.

■ **49.** If an edge is added to an already existing graph, connecting two vertices already in the graph, explain why the number of vertices with odd valence has the same parity before and after. (This means if it was even before, it is even after, while if it was odd before, it remains odd.)

■ **50.** Any graph can be built in the following fashion: Put down dots for the vertices, then add edges connecting the dots as needed. When you have put down the dots, and before any edges have been added, is the number of vertices with odd valence an even number or an odd number? What can you say about the number of vertices with odd valence when all the edges have been added (see Exercise 49)?

51. Draw the graph for the parking-control territory shown in the figure below. Label each vertex with its valence and determine if the graph is connected.

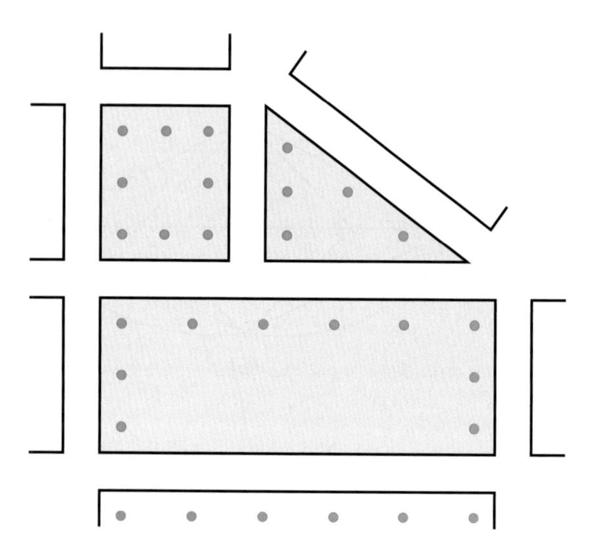

52. If a rectangular street network is r blocks by s blocks, can you find a formula for the minimum number of edges that must be added to eulerize a graph representing the network in terms of r and s? (*Hint:* Treat the case $r = 1$ separately. Test your formula with the cases: 6 blocks by 5 blocks; 6 blocks by 6 blocks; 5 blocks by 3 blocks.)

◆ **53.** The word *valence* is also used in chemistry. Find out what it means in chemistry and explain how this usage is similar to the use we make of it here.

■ **54.** For the street network below, draw a graph that represents the sidewalks with meters. Then find the minimum-length circuit that covers all sidewalks with meters. If you drew the graph as we recommended in this section, you would find that the shortest circuit has length 18 (it reuses every edge).

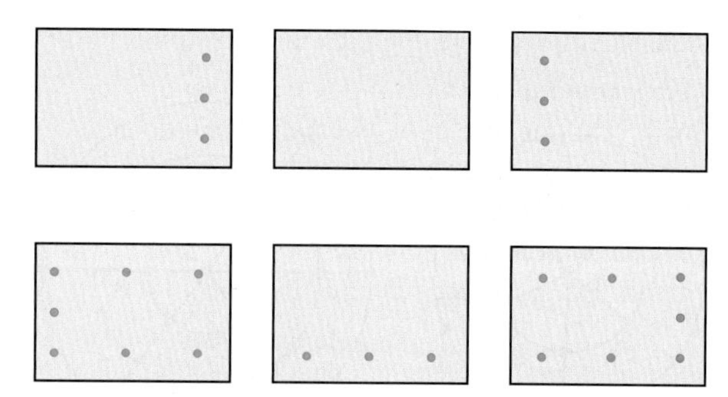

But the meter checker comes to you and says: "I don't know anything about your theories, but I have found a way to cover the sidewalks with meters using a circuit of length 10. My trick is that I don't rule out walking on sidewalks with no meters." Explain what he means and discuss whether his strategy can be used in other problems.

WRITING PROJECTS

1. Write a memo of three double-spaced word-processed pages to your local department of parking control (or police department) in which you suggest that management science techniques like the ones in this chapter be used to plan routes. Assume that the person to whom you are writing is not extensively trained in mathematics but is willing to read through some technical material, provided you make it seem worth the trouble.

2. Do the same as in Writing Project 1, but to the department in charge of spreading salt on roads after snowstorms.

3. If you were making a recommendation to the mayor of New York City concerning proposed new street-sweeping routes, designed using the theory of this chapter, would you recommend that the changes be adopted or not? Write a memo (three double-spaced word-processed pages) that outlines the pros and cons as fairly as you can, and then conclude with your recommendation.

SPREADSHEET PROJECTS

To do these projects, go to www.whfreeman.com/fapp.

Graphs can be represented in ways other than drawings. For example, computers analyze graphs by tabulating the connections formed by the edges. This chapter's projects introduce methods to represent and analyze graphs using spreadsheets.

APPLET EXERCISES

To do these exercises, go to www.whfreeman.com/fapp.

Eulerizing a Graph

We learned that if a graph has exactly two vertices with odd valences, then an Euler circuit does not exist–but an Euler path does. It is also possible to produce an Euler circuit through the process of eulerization, by duplicating certain edges of the graph. But how many duplications are necessary to obtain an Euler circuit? Investigate this problem and more general related topics using the Eulerizing a Graph applet.

Euler Circuits

We know that if all the vertices have even valence, then an Euler circuit exists. Try your hand at finding such circuits in the Euler Circuit applet.

CHAPTER 2

Visiting Vertices

In this chapter we cover...

2.1 Hamiltonian Circuits

2.2 Traveling Salesman Problem

2.3 Strategies for Solving the Traveling Salesman Problem

2.4 Minimum-Cost Spanning Trees

2.5 Critical-Path Analysis

...and at www.whfreeman.com/fapp:

Flashcards

Quizzes

Spreadsheet Projects

- Low-cost spanning trees
- Low-cost Hamiltonian circuits

Applet Exercises

- Nearest-neighbor algorithm
- Sorted-edges algorithm
- Kruskal's algorithm

In the last chapter, we saw that it is relatively easy to determine whether there is a circuit traversing the edges of a graph exactly once—for example, a route for street sweepers that covers the streets in a section of a city exactly once. However, the situation changes radically if we make a seemingly small change in the problem: When is it possible to find a route along distinct edges of a graph that visits each *vertex* once and only once in a simple circuit?

This problem is called the Hamiltonian circuit problem, and, like the Euler circuit problem, it is a graph theory problem. The Hamiltonian circuit problem has many applications. Suppose inspections or deliveries need to be made at each vertex (rather than along each edge) of a graph. An "efficient" tour of the graph would be a route that started and ended at the same vertex and passed through all the vertices without reuse, or repetition; that is, the route would be a Hamiltonian circuit. Such routes would be useful for inspecting traffic signals or for delivering mail to drop-off boxes, which hold heavy loads of mail so that urban postal carriers do not have to carry them long distances.

For example, the wiggly line in Figure 2.1a shows a circuit we can take to tour that graph, visiting each vertex once and only once. This tour can be written *ABDGIHFECA*. Note that another way of writing the same circuit would be

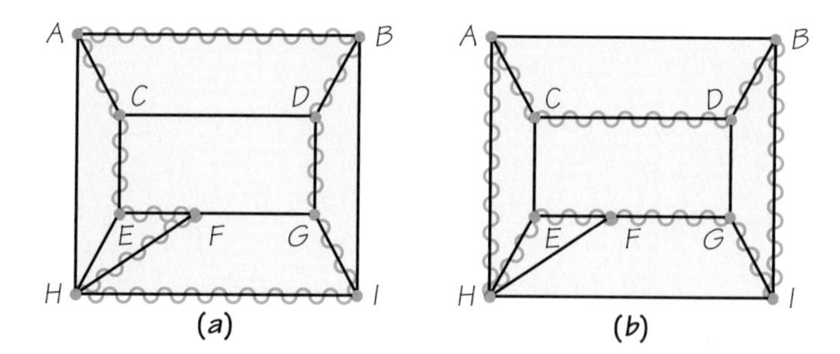

Figure 2.1 Wiggly edges illustrate Hamiltonian circuits.

This photograph suggests some of the complexities of efficient mail delivery in a suburban environment. Sometimes services must be provided along edges (streets) and sometimes at vertices (corners). (Debra Hershkowitz/Bruce Coleman.)

EFHIGDBACE. A different circuit visiting each vertex once and only once would be *CDBIGFEHAC* (Figure 2.1b). Do not be confused because *C* is written twice when we write down this list of vertices. We can think of the circuit as starting at any of its vertices, but we do start and end at the same vertex.

A tour, like the ones marked by wiggly edges in Figure 2.1, that starts at a vertex of a graph and visits each vertex once and only once, returning to where it started, is called a **Hamiltonian circuit.**

2.1 Hamiltonian Circuits

The concept is named for the Irish mathematician William Rowan Hamilton (1805–1865), who was one of the first to study it. (We now know that the concept was discovered somewhat earlier by Thomas Kirkman [1806–1895], a British minister with a penchant for mathematics.)

The concepts of Euler and Hamiltonian circuits are similar in that both forbid reuse: the Euler circuit of edges, the Hamiltonian circuit of vertices. However, it is far more difficult to determine which connected graphs admit a Hamiltonian circuit than to determine which connected graphs have Euler circuits. As we saw in Chapter 1, looking at the valences of vertices tells us whether a

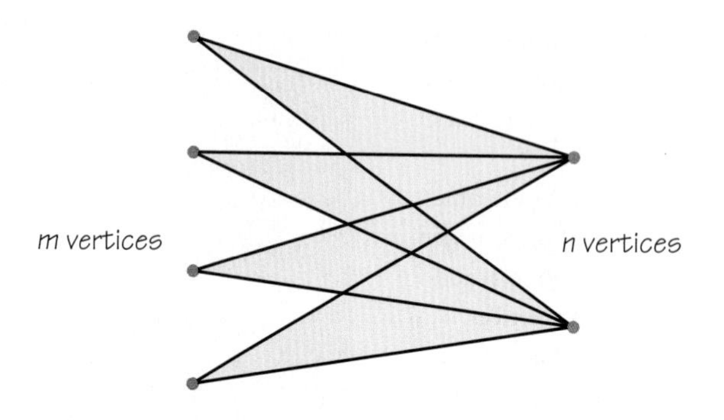

Figure 2.2 An example of one graph from a family of graphs that has no Hamiltonian circuit. The number of vertices m on the left is chosen to be greater than the number of vertices n on the right. The case $m = 4$ and $n = 2$ is shown.

connected graph has an Euler circuit, but we have no such simple method for telling whether or not a graph has a Hamiltonian circuit.

Some special classes of graphs are known to have Hamiltonian circuits, and some special classes of graphs are known to lack them. For example, here is a method for constructing an infinite family of graphs where each graph in the family cannot have a Hamiltonian circuit. Construct a vertical column of m vertices and a parallel column of n vertices, where m is bigger than n, as shown in Figure 2.2. The figure illustrates the typical case where $m = 4$ and $n = 2$. Now join each vertex on the left in the figure to every vertex on the right. As m and n vary, we get a family of different graphs.

Any graph obtained in this manner cannot have a Hamiltonian circuit. If a Hamiltonian circuit existed, it would have to alternately include vertices on the left and right of the figure. This is not possible, because the number of vertices on the left and right are not the same. It is unlikely that a method will ever be found to easily determine whether or not a graph has a Hamiltonian circuit. If Hamiltonian circuits were easy to find in any graph at all, many applied problems could be solved in a less costly way.

The Hamiltonian circuit problem and the Euler circuit problem are both examples of graph theory problems. Although we posed the Hamiltonian circuit problem merely as a variant of another graph theory problem with many applications (i.e., the Euler circuit problem), the Hamiltonian circuit problem itself has many applications. This is not unusual in mathematics. Often mathematics used to solve a particular real-world problem leads to new mathematics that suggests applications to other real-world situations. One class of problems to which we can apply Hamiltonian circuits is vacation planning.

EXAMPLE Vacation Planning

Let's imagine that you are a college student studying in Chicago. During spring break you and a group of friends have decided to take a car trip to visit other friends in Minneapolis, Cleveland, and St. Louis. There are many choices as to the order of visiting the cities and returning to Chicago, but you want to design a route that minimizes the distance you have to travel. Presumably, you also want

Regional routes for one airline. (Young-Wolff/PhotoEdit, Inc.)

a route that cuts costs, and you know that minimizing distance will minimize the cost of gasoline for the trip. (Similar problems with different complications would arise for bus, railroad, or airplane trips.)

Imagine now that the local automobile club has provided you with the intercity driving distances between Chicago, Minneapolis, Cleveland, and St. Louis. We can construct a graph model with this information, representing each city by a vertex and the legs of the journey between the cities by edges joining the vertices. To complete the model, we add a number called a **weight** to each graph edge, as in Figure 2.3. In this example, the weights represent the distances between the cities, each of which corresponds to one of the endpoints of the edges in the graph. (In other examples the weight might represent a cost, time, or profit.) We want to find a minimum-cost tour that starts and ends in Chicago and visits each other city only once. Using our earlier terminology, what we wish to find is a **minimum-cost Hamiltonian circuit**—a Hamiltonian circuit with the lowest possible sum of the weights of its edges.

How can we determine which Hamiltonian circuit has minimum cost? There is a conceptually easy **algorithm,** or mechanical step-by-step process, for solving this problem:

1. Generate all possible Hamiltonian tours (starting from Chicago).
2. Add up the distances on the edges of each tour.
3. Choose the tour of minimum distance.

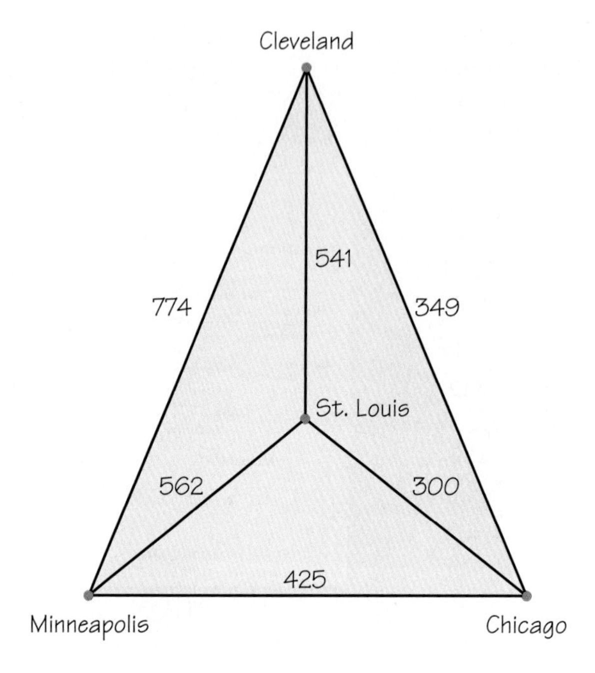

Figure 2.3 Road mileages between four cities.

Steps 2 and 3 of the algorithm are straightforward. Thus, we need worry only about step 1, generating all the possible Hamiltonian circuits in a systematic way. To find the Hamiltonian tours, we will use the **method of trees,** as follows. Starting from Chicago, we can choose any of the three cities to visit after leaving Chicago. The first stage of the enumeration tree is shown in Figure 2.4. If Minneapolis is chosen as the first city to visit, then there are two possible cities to visit next, Cleveland and St. Louis. The possible branchings of the **tree** at this stage are shown in Figure 2.5. In this second stage, however, for each choice of first city to visit, there are two choices from this city to the second city to visit. This would lead to the diagram in Figure 2.6.

Having chosen the order of the first two cities to visit, and knowing that no revisits (reuses) can occur in a Hamiltonian circuit, there is only one choice left for the next city. From this city we return to Chicago. The complete tree diagram showing the third and fourth stages for these routes is given in Figure 2.7. Notice, however, that because we can traverse a circular tour in either of two directions, the paths shown in the tree diagram of Figure 2.7 do not correspond to different

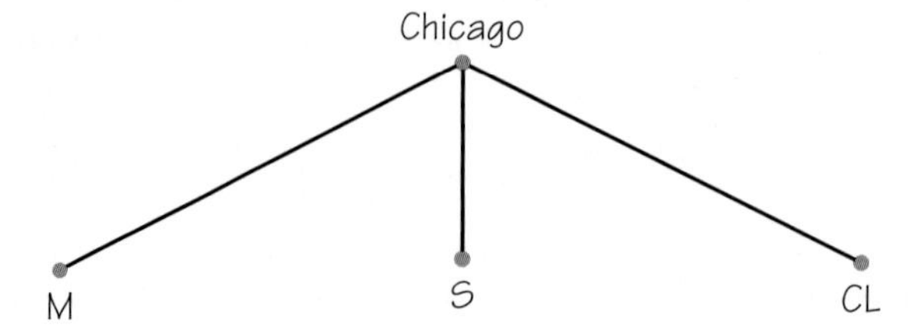

Figure 2.4 First stage in finding vacation-planning routes.

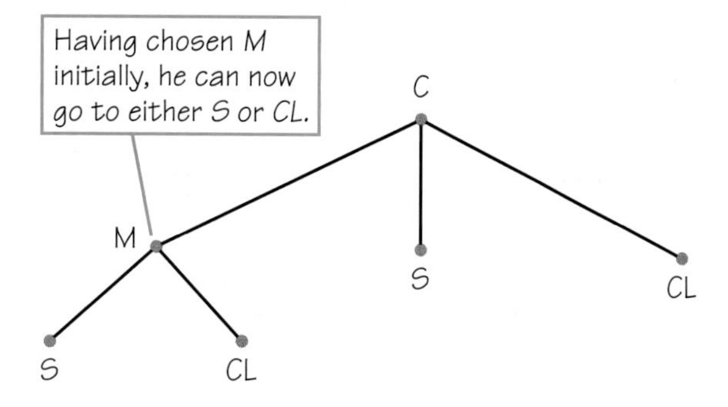

Figure 2.5 Part of the second stage in finding vacation-planning routes.

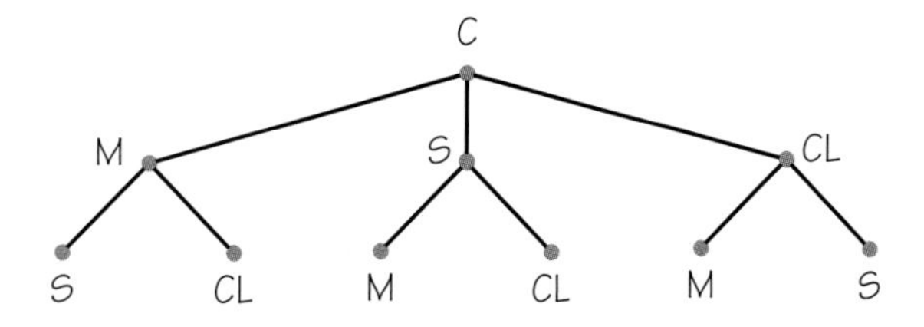

Figure 2.6 Complete second stage in finding vacation-planning routes.

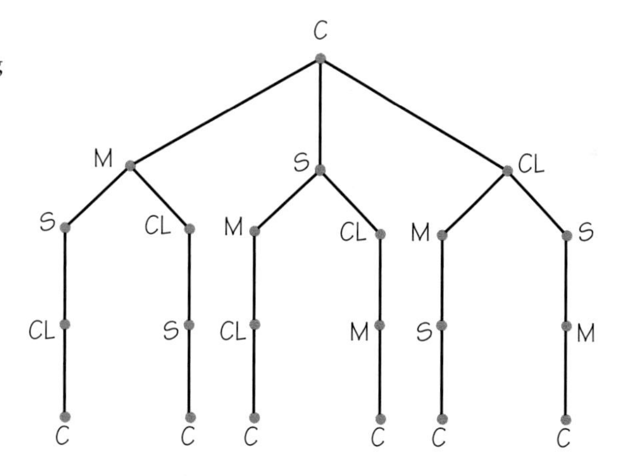

Figure 2.7 Completed enumeration of routes using the method of trees for the vacation-planning problem.

Hamiltonian circuits. For example, the leftmost path (C–M–S–CL–C) and the rightmost path (C–CL–S–M–C) represent the same Hamiltonian circuit. Thus, among what appear to be six different paths in the tree diagram, only three in fact correspond to different Hamiltonian circuits. These three distinct Hamiltonian circuits are shown in Figure 2.8.

Note that in generating the Hamiltonian circuits we disregard the distances involved. We are concerned only with the different patterns of carrying out the visits. To find the optimal route, however, we must add up the distances on the edges to get each tour's length. Figure 2.8 shows that the optimal tour is Chicago, Minneapolis, St. Louis, Cleveland, Chicago. The length of this tour is 1877 miles, which saves 163 miles over the longest choice of tour. ■

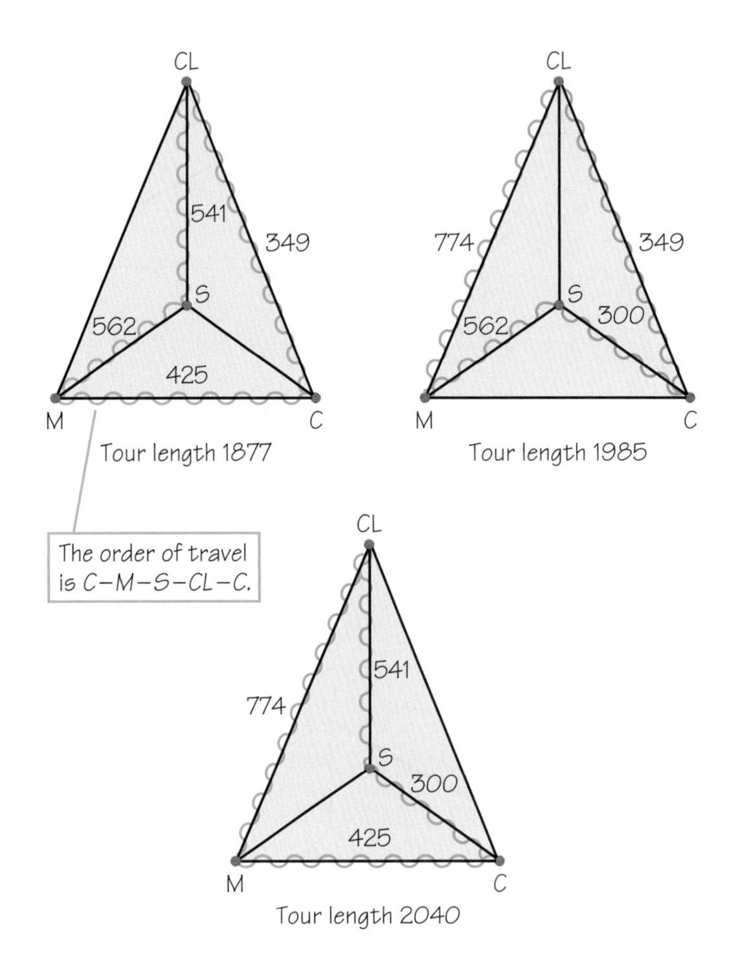

Figure 2.8 The three Hamiltonian circuits for the vacation-planning problem of Figure 2.3.

The method of trees is not always as easy to use as our example suggests. Instead of doing our analysis for four cities, consider the general case of n cities. The graph model similar to that in Figure 2.3 would consist of a weighted graph with n vertices, with every pair of vertices joined by an edge. Such a graph is called **complete** because the edge between any pair of vertices is present in the graph. A complete graph with five vertices is illustrated in Figure 2.9.

Fundamental Principle of Counting

How many Hamiltonian circuits are in a complete graph of n vertices? We can solve this problem by using the same type of analysis that we used in the method of trees. The method of trees is a visual application of the fundamental principle of counting, a procedure for counting outcomes in multistage processes. Using this procedure, we can count how many patterns occur in a situation by looking at the number of ways in which the component parts can occur. For example, if Jack has 9 shirts and 4 pairs of trousers, he can wear $9 \times 4 = 36$ shirt–pants

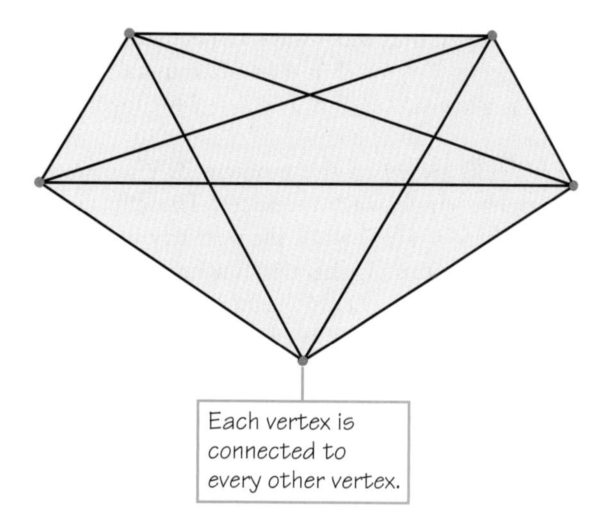

Each vertex is
connected to
every other vertex.

Figure 2.9 A complete graph with five vertices. Every pair of vertices is joined by an edge.

outfits. Each shirt can be worn with any of the pants. (This can be verified by drawing a tree diagram, but such a diagram is cumbersome for big numbers.)

EXAMPLE Counting

In general, the **fundamental principle of counting** can be stated this way: If there are a ways of choosing one thing, b ways of choosing a second after the first is chosen, . . . , and z ways of choosing the last item after the earlier choices, then the total number of choice patterns is $a \times b \times c \times \ldots \times z$.

Here are some other examples of how to use the fundamental principle of counting:

1. In a restaurant there are 4 kinds of soup, 12 entrees, 6 desserts, and 3 drinks. How many different four-course meals can a patron choose from? The four choices can be made in 4, 12, 6, and 3 ways, respectively. Hence, applying the fundamental principle of counting, there are $4 \times 12 \times 6 \times 3 = 864$ possible meals.
2. In a state lottery a contestant gets to pick a four-digit number that contains no zero followed by an uppercase or lowercase letter. How many such sequences of digits and a letter are there? Each of the four digits can be chosen in 9 ways (that is, 1, 2, . . . , 9), and the letter can be chosen in 52 ways (that is, A, B, . . . , Z plus a, b, . . . , z). Hence, there are $9 \times 9 \times 9 \times 9 \times 52 = 341,172$ possible patterns.
3. A corporation is planning a musical logo consisting of four different ordered notes from the scale C, D, E, F, G, A, and B. How many logos are there to chose from? The first note can be chosen in 7 ways, but because reuse is not allowed, the next note can be chosen in only 6 ways. The remaining two notes

can be chosen in 5 and 4 ways, respectively. Using the fundamental principle of counting, $7 \times 6 \times 5 \times 4 = 840$ musical logos are possible. If reuse of notes is allowed, $7 \times 7 \times 7 \times 7 = 2401$ logos are possible. ■

Let's now return to the problem of enumerating Hamiltonian circuits for the complete graph with n vertices. The city visited first after the home city can be chosen in $n - 1$ ways, the next city in $n - 2$ ways, and so on, until only one choice remains. Using the fundamental principle of counting, there are $(n - 1)! = (n - 1)(n - 2) \times \ldots \times 3 \times 2 \times 1$ routes. The exclamation mark in $(n - 1)!$ is read "factorial" and is shorthand notation for the product $(n - 1)(n - 2) \times \ldots \times 3 \times 2 \times 1$. For example, $5! = 5 \times 4 \times 3 \times 2 \times 1 = 120$.

As we saw in Figure 2.7, pairs of routes correspond to the same Hamiltonian circuit because one route can be obtained from the other by traversing the cities in reverse order. Thus, although there are $(n - 1)!$ possible routes, there are only half as many, or $(n - 1)!/2$, different Hamiltonian circuits. Now, if we have only a few cities to visit, $(n - 1)!/2$ Hamiltonian circuits can be listed and examined in a reasonable amount of time. Analysis of a six-city problem would require generation of $(6 - 1)!/2 = 5!/2 = 120/2 = 60$ tours. But for, say, 25 cities, $24!/2$ is approximately 3×10^{23}. Even if these tours could be generated at the rate of 1 million a second, it would take 10 billion years to generate them all. Because it would take so long to solve large vacation-planning problems using this method, it is sometimes referred to as the **brute force method** (that is, trying all the possibilities).

2.2 Traveling Salesman Problem

If the only benefit were saving money and time in vacation planning, the difficulty of finding a minimum-cost Hamiltonian circuit in a complete graph with n vertices for large values of n would not be of great concern. However, the problem we are discussing is one of the most common in *operations research,* the branch of mathematics concerned with getting governments and businesses to operate more efficiently. This problem is usually called the **traveling salesman problem (TSP)** because of its early formulation: Determine the trip of minimum cost that a salesperson can make to visit the cities in a sales territory, starting and ending the trip in the same city.

Many situations require solving a TSP:

1. A lobster fisherman has set out traps at various locations and wishes to pick up his catch.
2. The telephone company wishes to pick up the coins from its pay telephone booths. (To avoid the high cost of picking up these coins, phone companies in many countries have adopted a system that uses prepurchased phone cards to operate phones. This means that there are no coins to collect!)
3. The electric (or gas) company needs to design a route for its meter readers.

4. A minibus must pick up six day campers and deliver them to camp, and later in the day return them home.
5. In drilling holes in a series of plates, the drill press operator (perhaps a robot) must drill the holes in a predetermined order.
6. Physical records generated at automated teller machine (ATM) locations – as backup in case of failure of the electronic systems – must be picked up periodically.
7. A limousine service with a van located at an airport must pick up five customers and deliver them to the airport in time to catch their planes.

Perhaps surprisingly, TSP problems are also solved regularly in the design of computer chips. The components must be located so that the machines involved in the assembly can insert them on the chips in as efficient a manner as possible. Because many chips are manufactured, even a small improvement in the time needed to make a chip can save a lot of money.

The meaning of *cost* can vary from one formulation of a TSP to another. We can measure cost as distance, airplane ticket prices, time, or any other factor that is to be optimized. In many situations, the TSP arises as a subproblem of a more complicated problem. For example, a supermarket chain may have a very large number of stores to be served from a single large warehouse. If there are fewer trucks than stores, the stores must be grouped into clusters so that one truck can serve each cluster. If we then solve the TSP for every truck, we can minimize total costs for the supermarket chain. Similar vehicle-routing problems– for dial-a-ride services that transport senior citizens to activity centers, for example, or that deliver children to their schools or camps – often involve solving the TSP as a subproblem.

2.3 Strategies for Solving the Traveling Salesman Problem

Because the traveling salesman problem arises so often in situations where the associated complete graphs would be very large, we must find a faster method than the brute force method we have described. We need to look at our original problem in Figure 2.3 and try to find an alternative algorithm for solving it. Recall that our goal is to find the minimum-cost Hamiltonian circuit.

Nearest-Neighbor Algorithm

Let's try a new approach: Starting from Chicago, first visit the nearest city, then visit the nearest city that has not already been visited. We return to the start city when no other choice is available. This approach is called the **nearest-neighbor algorithm.**

Applying this algorithm to the TSP in Figure 2.3 quickly leads to the tour of Chicago, St. Louis, Cleveland, Minneapolis, and Chicago, with a length of 2040 miles. Here is how this tour is determined. Because we are starting in Chicago, there is a choice of going to a city that is 425, 300, or 349 miles away. Because the smallest of these numbers is 300, we next visit St. Louis, which is the nearest neighbor of Chicago not already visited. At St. Louis, we have a choice of visiting next cities that are 541 or 562 miles away. Hence, Cleveland, which is nearer (541 miles), is visited. To complete the tour, we visit Minneapolis and return to Chicago, thereby adding 774 and 425 miles to the length of the tour.

The nearest-neighbor algorithm is an example of a **greedy algorithm,** because at each stage a best (greedy) choice, based on an appropriate criterion, is made. Unfortunately, this is not the optimal tour, which we saw was C–M–S–CL–C, for a total length of 1877 miles. Making the best choice at each stage may not yield the best "global" solution. However, even for a large TSP, one can always find a nearest-neighbor route quickly.

Figure 2.10 again illustrates the ease of applying the nearest-neighbor algorithm, this time to a weighted complete graph with five vertices. Starting at vertex *A*, we get the tour *ADECBA* (cost 2800) (Figure 2.10a). Note that the nearest-neighbor algorithm starting at vertex *B* yields the tour *BCADEB* (cost 3050) (Figure 2.10b).

This example illustrates that a nearest-neighbor tour can be computed for each vertex of the complete graph being considered and that different nearest-neighbor tours can be obtained starting at different vertices. Thus, even though we may seek a tour starting at a particular vertex—say, *A* in Figure 2.10—because a Hamiltonian circuit can be thought of as starting at any of its vertices, we can just as easily apply the nearest-neighbor procedure, starting at vertex *B* (rather than at *A*). The Hamiltonian circuit we get can still, be thought of as beginning at vertex *A* rather than *B*. Even for complete graphs with a large number of

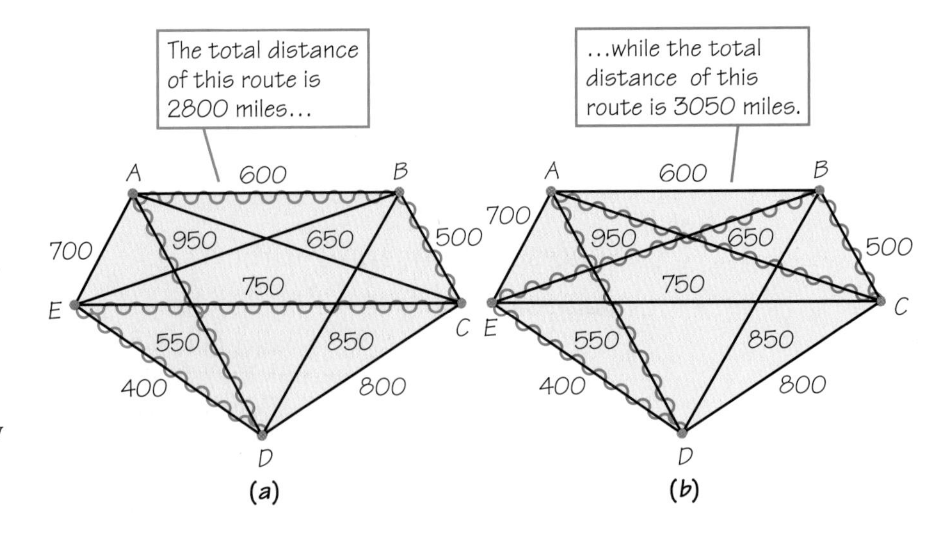

Figure 2.10 (a) A weighted complete graph with five vertices that illustrates the use of the nearest-neighbor algorithm (starting at *A*). (b) TSP tour generated by the nearest-neighbor algorithm (starting at *B*).

vertices, it would still be faster to apply the nearest-neighbor algorithm for each vertex and pick the cheapest of the tours generated (though such a tour might not be optimal) than to apply the brute force method.

Sorted-Edges Algorithm

We might start by sorting or arranging the edges of the complete graph in order of increasing cost (or, equivalently, arranging the intercity distances in order of increasing distance). Then at each stage we can select that edge of least cost that (1) never requires that three used edges meet at a vertex (because a Hamiltonian circuit uses up exactly two edges at each vertex) and that (2) never closes up a circular tour that doesn't include all the vertices. This algorithm is called the **sorted-edges algorithm.**

Applying the sorted-edges algorithm to the TSP in Figure 2.3 works as follows: First, the six weights on the edges listed in increasing order would be 300, 349, 425, 541, 562, and 774. Because the cheapest edge in this sorted list is 300, this is the first edge we select for the tour we are building. Next we add the edge with weight 349 to the tour. The next-cheapest edge would be 425, but using this edge together with those already selected would result in having three edges at a vertex (Figure 2.11a), which is not consistent with having a Hamiltonian circuit. Hence, we do not use this edge. The next-cheapest edge, 541, used together with the edges already selected, would create a circuit (see Figure 2.11b) that does not

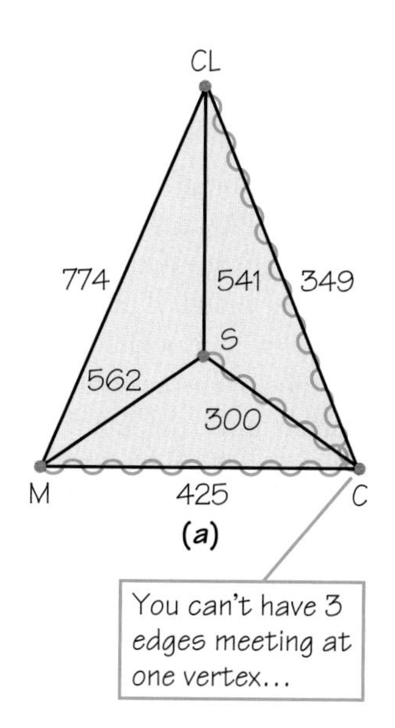

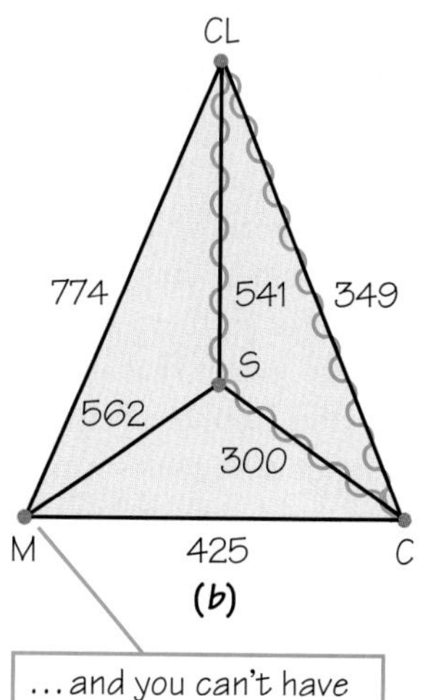

Figure 2.11 (a) When three shortest edges are added in order of increasing distance, three edges at a vertex are selected, which is not allowed as part of a Hamiltonian circuit. (b) When the edges of distances 300, 349, and 541 are selected, a circuit that does not include all vertices results.

include all the vertices. Thus, this edge, too, would be skipped over. However, we are able to add the edges 562 and 774 without either creating a circuit shorter than one including all the vertices or having three edges at a vertex. Hence, the tour we arrive at is Chicago, St. Louis, Minneapolis, Cleveland, and Chicago. Again, this solution is not optimal because its length is 1985. Note that this algorithm, like the nearest-neighbor, is greedy.

Although the edges selected by applying the sorted-edges method to the example in Figure 2.3 are connected to each other at every stage, this does not always happen. For example, if we apply the sorted-edges algorithm to the graph in Figure 2.10a, we build up the tour first with edge *ED* (400) and then edge *BC* (500), which do not touch. The edges that are then selected are *AD*, *AB*, and *EC*, giving the circuit *EDABCE*, which is the same as the nearest-neighbor circuit starting at vertex *A*.

Many "quick-and-dirty" methods for solving the TSP have been suggested; while some methods give an optimal solution in some cases, none of these methods guarantees an optimal solution. Surprisingly, most experts believe that no efficient method that guarantees an optimal solution for the TSP will ever be found (see Spotlight 2.1).

Recently, mathematical researchers have adopted a somewhat different strategy for dealing with TSP problems. If finding a fast algorithm to generate optimal solutions for large problems is unlikely, perhaps we can show that the quick-

SPOTLIGHT 2.1 NP-Complete Problems

Steven Cook, a computer scientist at the University of Toronto, showed in 1971 that certain computational problems are equivalently difficult. This class of problems, now referred to as **NP-complete problems,** has the following characteristic: If a "fast" algorithm for solving one of these problems could be found, then a fast method would exist for all these problems.

In this context, "fast" means that as the size n of the problem grows (the number of cities gives the problem size in the traveling salesman problem), the amount of time needed to solve the problem grows no more rapidly than a polynomial function in n. (A polynomial function has the form $a_k n^k + a_{k-1} n^{k-1} + \dots + a_1 n + a_0$.) On the other hand, if it could be shown that any problem in the class of NP-complete problems required an amount of time that grows faster than any polynomial (an exponential function, such as 3^n, is an example of a function that grows faster than any polynomial) as the problem size increased, then all problems in the NP-complete class would share this characteristic. The TSP, along with a wide variety of other practical problems, is known to be NP-complete. It is widely believed that large versions of none of these problems can be solved quickly. Furthermore, the security of some recent cryptographical systems relies on the hope that large NP-complete problems are actually as time consuming to solve as they appear to be. The Clay Foundation is offering a $1 million prize for determining whether NP-complete problems are truly computationally hard.

and-dirty methods, usually called **heuristic algorithms,** come close enough to giving optimal solutions to be important for practical use. For example, suppose we could prove that the nearest-neighbor heuristic was never off by more than 25% in the worst case or by more than 15% in the average case. For a medium-sized TSP, we would then have to choose whether to spend a lot of time (or money) to find an optimal solution or instead to use a heuristic algorithm to obtain a fairly good solution. Investigators at AT&T Research have developed many remarkably good heuristic algorithms. The best-known guarantee for a heuristic algorithm for a TSP is that it yields a cost no worse than one and a half times the optimal cost. Interestingly, this heuristic algorithm involves solving a Chinese postman problem (see Chapter 1), for which a "fast" algorithm is known to exist.

Throughout our discussion of the TSP, we have concentrated on the goal of minimizing the cost (or time) of a tour that visited each of a variety of sites once and only once. One of the things that makes mathematical modeling exciting, however, is the subtle issues that arise in specific real-world situations (or that provide a contrast between seemingly similar situations). For example, suppose the TSP situation is that of picking up day campers and taking them to and from the camp. The camp wants to minimize the total length of time that the bus needs to pick up the campers. The parents of the campers, however, may want to minimize the time their children spend on the bus. For some problems, the tour that minimizes the mean (average) time that a child spends on the bus may not be the same tour that minimizes the total time of the tour. (Specifically, if the bus first picks up the child who lives the farthest from the camp, and then picks up the other children, this may yield a relatively short time on the bus for the kids but a relatively long time for the tour itself.) It is these subtleties between problems that mathematicians return to examine at a later time, after the basic structure of the main problem itself is well understood. It is in this way that mathematics continues to grow, explore new ideas, and find new applications.

2.4 Minimum-Cost Spanning Trees

The traveling salesman problem is but one of many graph theory optimization problems that have grown out of real-world problems in both government and industry. Here is another.

EXAMPLE Pictaphone Service

Imagine that Pictaphone service (i.e., telephone service that provides a video image of the callers) is to be set up on an experimental basis between five cities. The graph in Figure 2.12 shows the possible links that might be included in the Pictaphone network, with each edge showing the cost in millions of dollars to create that particular link. To send a Pictaphone message between two cities, a direct communication link is not necessary because it is possible to send a message indirectly via another city. Thus, in Figure 2.12, sending a message from *A* to *C* could be achieved by sending the message from *A* to *B*, from *B* to *E*, and from *E* to *C*, provided the links *AB*, *BE*, and *EC* are part of the network. We assume

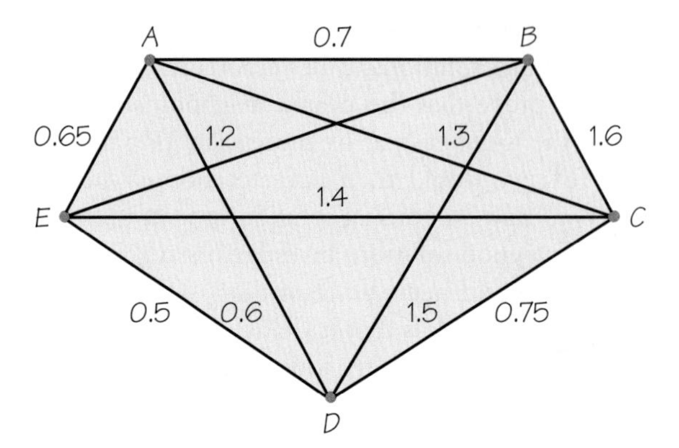

Figure 2.12 Costs (in millions of dollars) of installing Pictaphone service between five cities.

that the cost of relaying a message, compared with the cost of the direct communication link, is so small that we can neglect this amount. The problem that concerns us, therefore, is to provide service between any pair of cities in a way that minimizes the total cost of the links.

Our first guess at a solution is to put in the cheapest possible links between cities first, until all cities could send messages to any other city. Such an approach would be analogous to the sorted-edges method that was used to study the traveling salesman problem. In our example, if the cheapest links are added until all cities are joined, we obtain the connections shown in Figure 2.13a.

The links were added in the order *ED*, *AD*, *AE*, *AB*, *DC*. However, because this graph contains the circuit *ADEA* (wiggly edges in Figure 2.13b), it has redundant edges: We can still send messages between any pair of cities using relays after omitting the most expensive edge in the circuit—*AE*. After deleting an edge of a circuit, a message can still be relayed among the cities of the circuit by send-

Businesspeople holding a videoconference.
(Steven Peters/Tony Stone Images.)

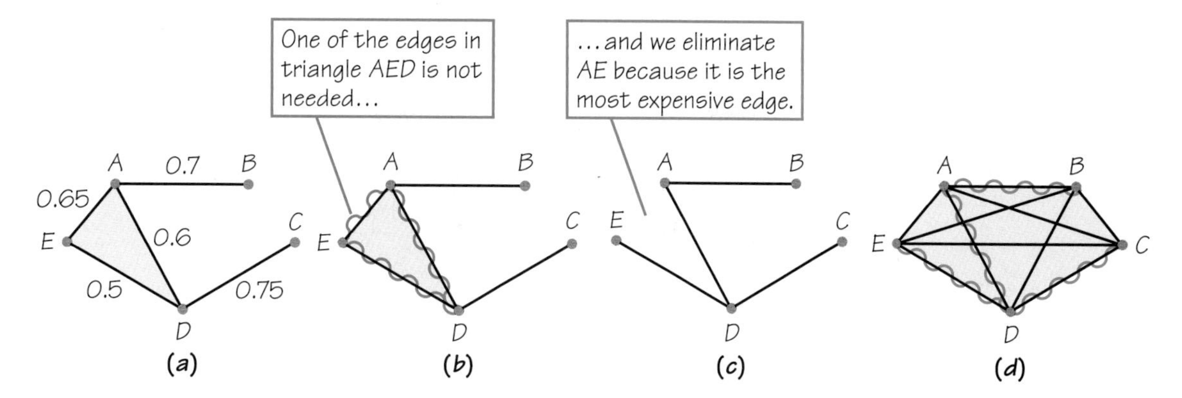

Figure 2.13 (a) Cities are linked in order of increasing cost until all cities are connected. (b) Circuit in part (a) highlighted. (c) Most expensive link in circuit in part (a) deleted. (d) Highlighted edges show, as a subgraph of the original graph, those links connecting the cities with minimum cost, obtained using Kruskal's algorithm.

ing signals the long way around. After *AE* is deleted, messages from *A* to *E* can be sent via *D* (Figure 2.13c).

This procedure suggests a modified algorithm for our problem, **Kruskal's algorithm:** Add the links in order of cheapest cost so that no circuits form and so that every vertex belongs to some link added (Figure 2.13d).

As in the sorted-edges method for the TSP, the edges that are added need not be connected to each other until the end.

A subgraph formed in this way will be a tree; that is, it will consist of one piece and contain no circuits. It will also include all the vertices of the original graph. A subgraph that is a tree and that contains all the vertices of the original graph is called a **spanning tree** of the original graph.

To understand these concepts better, consider the graph *G* in Figure 2.14a. The wiggly edges in Figure 2.14b would constitute a subgraph of *G* that is a tree (because it is connected and has no circuit), but this tree would not be a spanning tree of *G* because the vertices *D* and *E* would not be included. On the other hand, the wiggly edges in Figure 2.14c and 2.14d show subgraphs of *G* that include all the vertices of *G* but are not trees because the first is not connected and the second contains a circuit. Figure 2.14e shows a spanning tree of *G*; the wiggly edges are connected and contain no circuit, and every vertex of the original graph is an endpoint of some wiggly edge. ■

Finding a **minimum-cost spanning tree**—that is, a spanning tree whose edge weights sum to a minimum value—solves the Pictaphone problem. Note that having a different goal in the Pictaphone problem led to a different mathematical

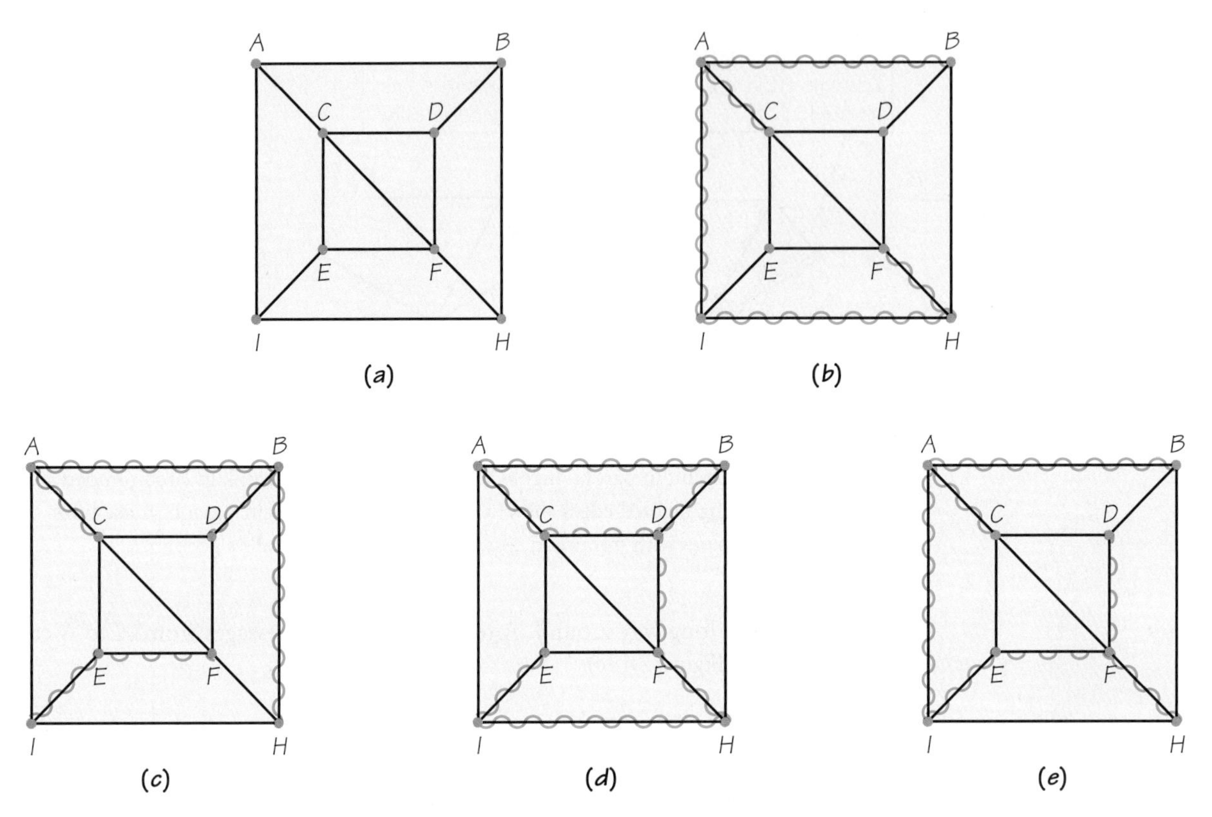

Figure 2.14 (a) A graph to help illustrate the concept of a spanning tree. (b) The wiggly edges are a tree, but not a spanning tree, because vertices *D* and *E* are not part of the tree. (c) The vertices of the graph are, however, endpoints of wiggly edges. (d) The wiggly edges are not a tree, because they contain the edges of the circuit *BDCAB*. All the vertices of the graph are, however, endpoints of wiggly edges. (e) The wiggly edges form a tree and include all of the vertices of the graph as endpoints of wiggly edges.

question from that of finding a Chinese postman tour of TSP tour. This application required that we find a minimum-cost spanning tree. In Figure 2.15a we have a graph model showing the costs of putting in roads to connect new houses in a suburban land-development project. Applying Kruskal's algorithm—adding the edges in the order of increasing cost, but avoiding the creation of a circuit—yields a minimum-cost spanning tree, indicated by the wiggly edges in Figure 2.15b. This tree is the cheapest one that makes it possible to drive between any pair of homes, though the driving distance between some of the homes will be relatively large, because only roads corresponding to wiggly edges will be built.

Remember that the weights on the edges of the graph in Figure 2.15a represent the costs of building roads, not the driving distance between the houses. Note that Figure 2.15a is not a complete graph, one in which all possible edges are included. Edges that correspond to roads that would be economically prohibitive to build have not been shown in the graph model. Also, in Figure 2.15b,

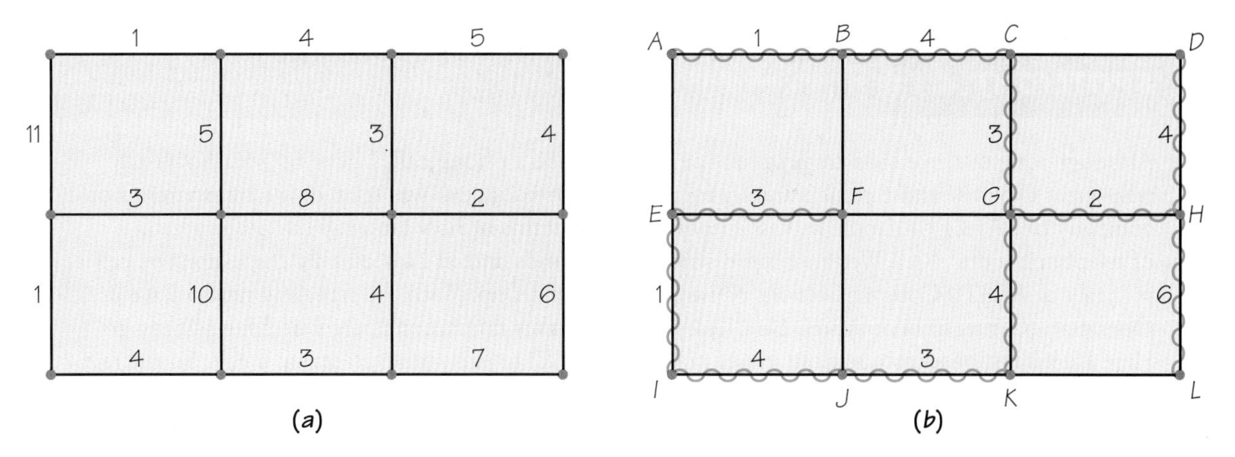

Figure 2.15 (a) A graph showing costs for construction of roads between houses. (b) Wiggly edges show a minimum-cost spanning tree for the graph in part (a).

the two edges of weight 5 (shown in Figure 2.15a) do not become part of the minimum-cost spanning tree, because they would create circuits with edges already chosen.

Although Kruskal's algorithm worked in our example, how do we know that the spanning tree found by this algorithm will always achieve the minimum possible cost? While this sounds very plausible, our experience with the TSP should suggest caution. Remember that for the TSP, the sorted-edges algorithm did not necessarily give an optimal solution even though it is a greedy algorithm like Kruskal's. On what basis should we have more faith in Kruskal's algorithm?

Kruskal's Algorithm

This algorithm was first suggested by Joseph Kruskal (AT&T Research) in the mid-1950s to solve a problem in pure mathematics proposed by a Czechoslovakian mathematician. In mathematics it is surprising but not uncommon to find that ideas used to solve problems with no apparent application often turn out to have many real-world uses. Kruskal's solution to the problem of finding a minimum-cost spanning tree in a graph with weights is a good example of this phenomenon. Kruskal showed that the greedy algorithm described does yield the minimum answer, and his work led to applications of these and related ideas in designing minimum-cost computer networks, phone connections, and road and railway systems. For additional discussion of operations research in the communications industry, see Spotlight 2.2. In order to explore how one can reconstruct full information from partial information using the tree concept, see Spotlight 2.3.

In our discussion of routing problems in graphs, we have not touched on one of the most obvious ones: finding the path between two specified, distinct vertices while keeping the sum of the weights of the edges in the path as small as possible. (Here there is no need to cover all vertices or to cover all edges.)

SPOTLIGHT 2.2 AT&T Manager Explains How Long-Distance Calls Run Smoothly

Although long-distance calls are now routine, it takes great expertise and careful planning for a company like AT&T to handle its vast amounts of telephone traffic. Rich Wetmore was district manager of AT&T's Communications Network Operations Center in Bedminster, New Jersey. Here are his responses to questions about how AT&T handles its huge volumes of long-distance traffic and how it tracks its operations to keep things running smoothly.

How do you make sure that a customer doesn't run into a delayed signal when attempting a long-distance call?

We monitor the performance of our AT&T network by displaying data collected from all over the country on a special wallboard. The wallboard is configured to tell us if a customer's call is not going through because the network doesn't have enough capacity to handle it.

That's when we step in and take control to correct the problem. The typical control we use is to reroute the call. Instead of sending the customer's call directly to its destination, we'll route it via a third city–to someplace else in the country that has the capacity to complete the call.

It would seem that routing via another city would take longer. Is the customer aware of this process?

Routing a call via a third city is entirely transparent [imperceptible] to the customer. I'm an expert about the network, and even when I make a phone call, I have no idea how that individual call was routed. It's transparent both in terms of how far away the other person sounds and in how quickly the telephone call gets set up. With the signaling network we use, it takes milliseconds for switching systems to "talk" to each other to set up a call. So the fact that you are involved in a third switch in some distant city is something you would never know.

You want to be sure to keep costs down while supplying enough sevice to customers. So how do you balance company benefits with customer benefit?

BXN

In terms of making the network efficient, we want to do two things. First, we want our customers to be happy with our service and for all their calls to go through, which means we must build enough capacity in the network to allow that to happen. Second, we want to be efficient for stockholders and not spend more money than we need to for the network to be at the optimum size.

There are basically two costs in terms of building the network. There is the cost of switching systems and the cost of the circuits that connect the switching systems. Basically, you can use operations-research techniques and mathematics to determine cost trade-offs. It may make sense to build direct routing between two switching systems and use a lot of circuits, or maybe to involve three switching systems, with fewer circuits between the main two, and so on.

SPOTLIGHT 2.3 Common Ancestors?

In the study of ancient manuscripts, different manuscripts of the same book are available, even though the original manuscript upon which they are based has been lost. Examples of this include Euclid's *Elements* and Chaucer's *Canterbury Tales.* What interests scholars is reconstructing the relationships between the manuscripts and the common ancestors of the manuscripts, even when some of the ancestors are now missing.

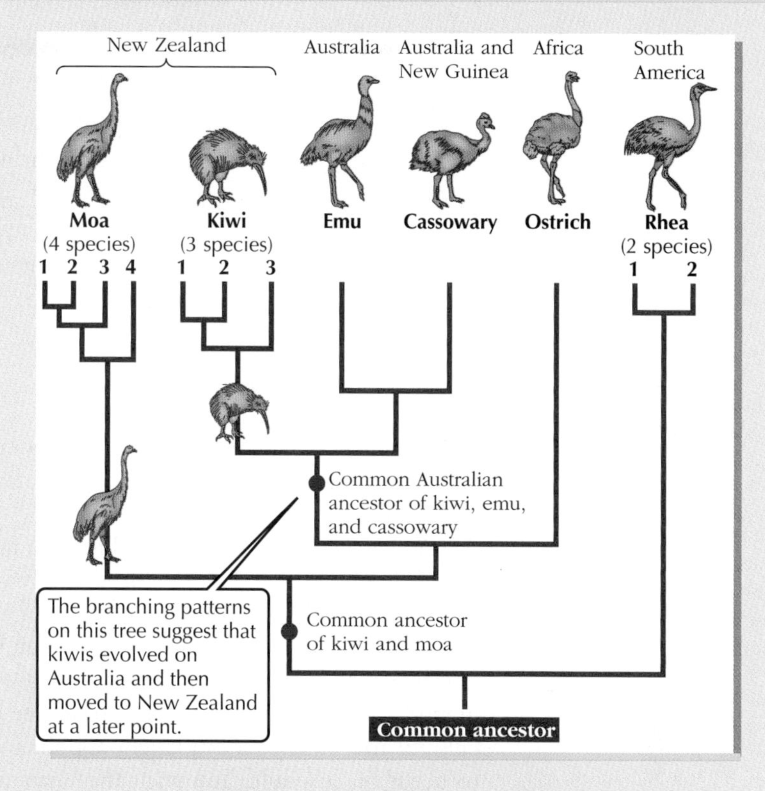

Similarly, perceptual psychologists may be interested in which colors people perceive as being closely related and comparing these perceptions with those of people who are color-blind. Finally, in studying different species, biologists are interested in determining which species are more closely related to each other, including species known only in fossil form, and constructing a "tree" of life that shows which species were ancestors of others.

Reconstructions of this kind are made possible by using graph theory, specifically using the graph theory concept of a tree. The value of the graph theory in these and many other situations lies in using the distance between pairs of vertices in the tree as a way of reflecting the closeness of the relationships that pairs of manuscripts, pairs of colors, or pairs of species have. The distance between two vertices in a tree is the sum of the weights along the one path that joins the two vertices. If there are no weights on the edges, the distance is the number of edges in the path. In some reconstruction problems, a special vertex of the tree called the *root* is singled out. This root plays the role of the original common ancestor, and distances to the root are of critical interest.

In the case of species, trees of family relatedness were traditionally constructed based on similarities of bones and physical appearance. With the discovery of molecular biology, many new avenues have been opened up. We can now draw trees of relatedness based on an organism's genetic material, DNA, or the proteins that the DNA codes for. The traditional trees based on physical traits often show different species as being more closely related than trees based on newer molecular biological approaches. These differences focus scholars on how to resolve the discrepancies and thereby reach a deeper understanding of the unity of life.

We have seen that the weights on the edges have many possible interpretations, including time, distance, and cost. Following are some of the many possible applications:

1. Design routes to be used by an ambulance, police car, or fire engine to get to an emergency as quickly as possible.
2. Design delivery routes that minimize gasoline use.
3. Design routes to bring soldiers to the front as quickly as possible.
4. Design a route for a truck carrying nuclear waste.

The need to find shortest paths seems natural. Next we investigate a situation in which finding a *longest* path is the right tool.

2.5 Critical-Path Analysis

Mathematics can confirm the obvious in certain situations while showing that our intuition is wrong in other circumstances. Our next group of applications will illustrate this point.

A characteristic of American life is its fast pace. People are interested in getting things done quickly and efficiently. This means that when you take your car in to be repaired before going to work, you want to know for sure that the repairs will be done when you pick the car up. You want the trains and the bus that take you to your doctor's appointment to run on time; and when you arrive at the doctor's office, you want a nurse to be free to take a blood sample and a throat culture. You want your outpatient appointment for an X-ray at the local hospital to occur on schedule. You want the X-ray to be interpreted quickly and the results reported back to your internist.

Scheduling machines and people is a big part of modern life. It is involved in running a school, a hospital, an airline, or in landing a person on Mars, and modern mathematics plays a big part in solving scheduling problems.

Part of what makes scheduling complicated is that the tasks that make up a job usually cannot be done in a random order. For example, to make Thanksgiving dinner you must buy and prepare the turkey before putting it in the oven, and you must set the table before serving the food.

If the tasks cannot be performed in a random order, we can specify the order in an **order-requirement digraph.** The term *digraph* is short for "directed graph." A digraph is a geometrical tool similar to a graph except that each edge has an arrow on it to indicate a direction for that edge. Digraphs can be used to illustrate that traffic on a street must go in one direction or that certain tasks in a job must be completed prior to other tasks. A typical example of an order-requirement digraph is shown in Figure 2.16. There is a vertex in this digraph for each task. If one task must be done immediately before another, we draw a directed edge, or arrow, from the prerequisite task to the subsequent task. The numbers within the circles representing vertices are the times it takes to complete the tasks. In

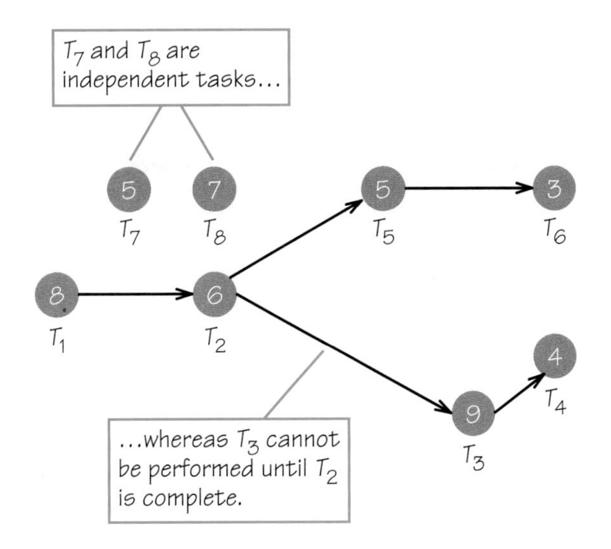

Figure 2.16 A typical order-requirement digraph.

Figure 2.16 there is no arrow from T_1 to T_5 because task T_2 intervenes. Also, T_1, T_7, and T_8 have no tasks that must precede them. Hence, if there are at least three processors (i.e., people or machines) available, tasks T_1, T_7, and T_8 can be worked on simultaneously at the start of the job.

Let's investigate a typical scheduling problem faced by a business.

EXAMPLE

Turning a Plane Around

Consider an airplane that carries both freight and passengers. The plane must have its passengers and freight unloaded and new passengers and cargo loaded

A phase of turning an airplane around. (Young-Wolff/PhotoEdit, Inc.)

before it can take off again. Also, the cabin must be cleaned before departure can occur. Thus, the job of "turning the plane around" requires completion of five tasks:

TASK *A*	Unload passengers	13 minutes
TASK *B*	Unload cargo	25 minutes
TASK *C*	Clean cabin	15 minutes
TASK *D*	Load new cargo	22 minutes
TASK *E*	Load new passengers	27 minutes

The order-requirement digraph for the problem of turning an airplane around is shown in Figure 2.17. The presence or absence of an edge in the order-requirement digraph depends on the analysis made as part of the modeling process for the problem. It seems natural that we need an arrow between task *A* and task *C*, because the passengers have to be unloaded before the cabin is cleaned. Other arrows may not seem natural—say, perhaps the arrow from task *B* (unload the cargo) to task *E* (load new passengers). This arrow may be due to government rules or requirements. What matters is that the mathematics of solving the problem does not depend on the reason that the order-requirement digraph looks the way that it does. The person solving the problem constructs the order-requirement digraph and then the mathematical techniques we will develop can be applied, regardless of whether or not some business faced with a similar problem might model the problem in a different way. Because we want to find the earliest completion time, it might seem that finding the shortest path in the digraph (i.e., path *BD* with time length $25 + 22 = 47$) would solve the problem. But this approach shows the danger of ignoring the relationship between the mathematical model (the digraph) and the original problem.

The time required to complete all the tasks, *A* through *E*, must be at least as long as the time necessary to do the tasks on any particular path. Consider the path *BD*, which has length $25 + 22 = 47$. Recall that here *length* of a path refers to the sum of the times of the tasks that lie along the path. Because task *B* must be done before task *D* can begin, the two tasks *B* and *D* cannot be completed before time 47. Hence, even if work on other tasks (such as *A*, *C*, and *E*) proceeds during this period, all the tasks cannot be finished before the tasks on path *BD* are finished. The same statement is true for every other path in the order-requirement digraph. Thus, the earliest completion time actually corresponds to the length of the longest path. In the airplane example, this earliest completion

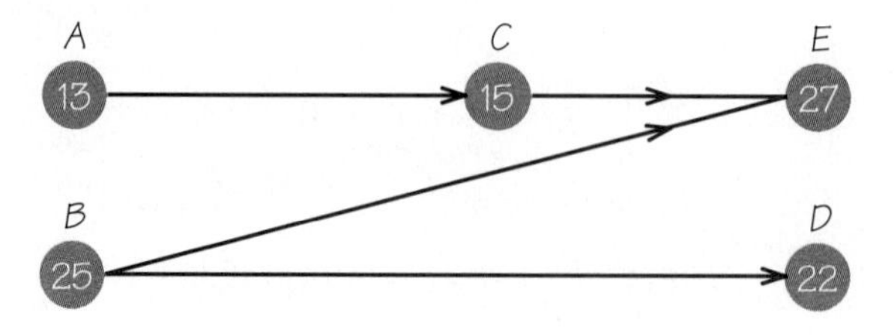

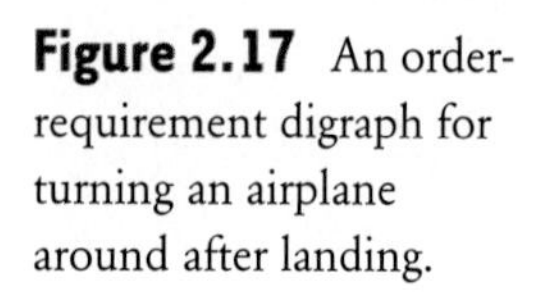

Figure 2.17 An order-requirement digraph for turning an airplane around after landing.

time is 55 (= 13 + 15 + 27) minutes, corresponding to the path ACE. We call ACE the *critical path* because the times of the tasks on this path determine the earliest completion time.

A **critical path** in an order-requirement digraph is a longest path. The length is measured in terms of summing the task times of the tasks making up the path.

Note that if none of these tasks could go on simultaneously, the time to complete all the tasks would be 13 + 25 + 15 + 22 + 27 = 102 minutes. However, even though some tasks may be performed simultaneously, the fact that the length of the critical path is 55 means that completion of the tasks in less than 55 minutes is not possible. Only by speeding up the times to complete the critical-path tasks themselves can a completion time less than 55 time units be achieved.

Suppose it were desirable to speed the turnaround of the plane to less than 55 minutes. One way to do this might be to build a second jetway to help unload passengers more quickly. For example, we could unload passengers (task A) in 7 minutes instead of 13. However, reducing task A to 7 minutes does not reduce the completion time by 6 minutes, because in the new digraph (Figure 2.18) ACE is no longer the critical (i.e., longest) path. The longest path is now BE, which has a length of 52 minutes. Thus, shortening task A by 6 minutes results in only a 3-minute saving in completion time. This may mean that building a new jetway is uneconomical. Note also that shortening the time to complete tasks that are not on the original critical path ACE will not shorten the completion time at all. Speeding tasks on the critical path will shorten completion time of the job only up to the point where a new critical path is created. Also note that a digraph may have more than one longest path. ■

Not all order-requirement digraphs are as simple as the one shown in Figure 2.17. The order-requirement digraph in Figure 2.19 has 12 paths, which can be found by exhaustive search. Examples of such paths are $T_1T_2T_3$, $T_1T_5T_9$, $T_4T_5T_9$, and $T_7T_5T_3$. (Although we have not discussed them here, fast algorithms for finding longest and shortest paths in graphs are known.) The critical path is $T_7T_8T_6$ (length 21), and the earliest completion time for all nine tasks is time 21.

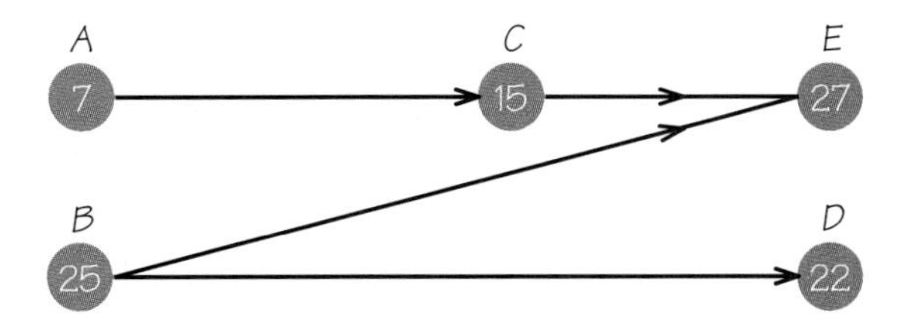

Figure 2.18 An order-requirement digraph for turning an airplane around in reduced time due to construction of a new jetway.

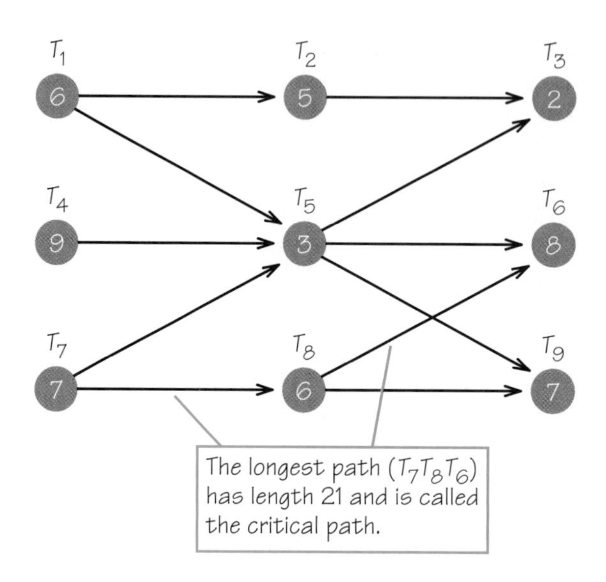

Figure 2.19 An order-requirement digraph with 12 paths, to examine how to find the length of the longest path.

These examples are typical of many scheduling problems that occur in practice (see Spotlight 2.4). Perhaps the most dramatic use of critical-path analysis is in the construction trades. No major new building project is now carried out without a critical-path analysis first being performed to ensure that the proper personnel and materials are available at the right times in order to have the project finished as quickly as possible. Many such problems are too large and complicated to be solved without the aid of computers.

The critical-path method was popularized and came into wider use as a consequence of the *Apollo* project. This project, which aimed at landing a man on the moon within 10 years of 1960, was one of the most sophisticated projects in planning and scheduling ever attempted. The dramatic success of the project can be attributed partly to the use of critical-path ideas and the related program evaluation and review technique (PERT), which helped keep the project on schedule.

In Chapter 3 we will see how mathematical ideas drawn from outside of graph theory can be used to gain insight into scheduling problems.

REVIEW VOCABULARY

Algorithm A step-by-step description of how to solve a problem.

Brute force method The method that solves the traveling salesman problem (TSP) by enumerating all the Hamiltonian circuits and then selecting the one with minimum cost.

Complete graph A graph in which every pair of vertices is joined by an edge.

Critical path The longest path in an order-requirement digraph. The length of this path gives the earliest completion time for all the tasks making up the job consisting of the tasks in the digraph.

SPOTLIGHT 2.4 Every Moment Counts in Rigorous Airline Scheduling

When people think of airline scheduling, the first thing that comes to mind is how quickly a particular plane can safely reach its destination. But using ground time efficiently is just as important to an airline's timetable as the time spent in flight. Bill Rodenhizer was the manager of control operations for an airline that provided shuttle service between Boston and New York. He is considered to be an expert on airplane turnaround time, the process by which an airplane is prepared for almost immediate takeoff once it has landed. He tells us how this well-orchestrated effort works:

> Scheduling, to the airline, is just about the whole ball game. Everything is scheduled right to the minute. The whole fleet operates on a strict schedule. Each of the departments responsible for turning around an aircraft has an allotted period of time in which to perform its function. Manpower is geared to the amount of ground time scheduled for that aircraft. This would be adjusted during off-weather or bad-weather days or during heavy air-traffic delays.
>
> Most of our aircraft in Boston are scheduled for a 42- to 65-minute ground time. Boston is the end of the line, so it is a "terminating and originating station." In plain talk, that means almost every aircraft that comes in must be fully unloaded, refueled, serviced, and dispatched within roughly an hour's time.
>
> This is how the process works: In the larger aircraft, it takes passengers roughly 20 minutes to load and 20 minutes to unload. During this period, we will have completely cleaned the aircraft and unloaded the cargo, and the caterers will have taken care of the food. The ramp service may take 20 to 30 minutes to unload the baggage, mail, and cargo from underneath the plane, and it will take the same amount of time to load it up again. We double-crew those aircraft with heavier weights so that the workload will fit the time it takes passengers to load and unload upstairs.
>
> While this has been going on, the fueler has fueled the aircraft. As to repairs, most major maintenance is done during the midnight shift, when [most of our] several hundred aircraft are inactive. We all work under a very strict time frame.

New security requirements in the wake of the World Trade Center attack (9/11/2001) have increased the difficulty of adhering to timetables in operating shuttle services between East Coast cities such as New York and Boston. This makes it even more important to use analytical tools in keeping operations on schedule.

Fundamental principle of counting A method for counting outcomes of multistage processes.

Greedy algorithm An approach for solving an optimization problem, where at each stage of the algorithm the best (or cheapest) action is taken. Unfortunately, greedy algorithms do not always lead to optimal solutions.

Hamiltonian circuit A circuit using distinct edges of a graph that starts and ends at a particular vertex of the graph and visits each vertex once and only once. A Hamiltonian circuit can start at any one of its vertices.

Heuristic algorithm A method of solving an optimization problem that is "fast" but does not guarantee an optimal answer to the problem.

Kruskal's algorithm An algorithm developed by Joseph Kruskal (AT&T Research) that solves the minimum-cost spanning-tree problem by selecting edges in order of increasing cost, but in such a way that no edge forms a circuit with edges chosen earlier. It can be proved that this algorithm always produces an optimal solution.

Method of trees A visual method of carrying out the fundamental principle of counting.

Minimum-cost Hamiltonian circuit A Hamiltonian circuit in a graph with weights on the edges, for which the sum of the weights of the edges of the Hamiltonian circuit is as small as possible.

Minimum-cost spanning tree A spanning tree of a weighted connected graph having minimum cost. The cost of a tree is the sum of the weights on the edges of the tree.

Nearest-neighbor algorithm An algorithm for attempting to solve the TSP that begins at a "home" vertex and visits next that vertex not already visited that can be reached most cheaply. When all other vertices have been visited, the tour returns to home. This method may not give an optimal answer.

NP-complete problems A collection of problems, which includes the TSP, that appear to be very hard to solve quickly for an optimal solution.

Order-requirement digraph A directed graph that shows which tasks precede other tasks among the collection of tasks making up a job.

Sorted-edges algorithm An algorithm for attempting to solve the TSP where the edges added to the circuit being built up are selected in order of increasing cost, but no edge is added that would prevent a Hamiltonian circuit's being formed. These edges must all be connected at the end, but not necessarily at earlier stages. The tour obtained may not have the lowest possible cost.

Spanning tree A subgraph of a connected graph that is a tree and includes all the vertices of the original graph.

Traveling salesman problem (TSP) The problem of finding a minimum-cost Hamiltonian circuit in a complete graph where each edge has been assigned a cost (or weight).

Tree A connected graph with no circuits.

Weight A number assigned to an edge of a graph that can be thought of as a cost, distance, or time associated with that edge.

SUGGESTED READINGS

BODIN, LAWRENCE. Twenty years of routing and scheduling, *Operations Research,* 38 (1990): 571–579. A survey of real-world situations where routing and scheduling were used, written by a pioneer in this area.

DOLAN, ALAN, and JOAN ALDUS. *Networks and Algorithms: An Introductory Approach,* Wiley, Chichester, England, 1993. An excellent introduction to graph theory algorithms.

GUSFIELD, DAN. *Algorithms on Strings, Trees, and Sequences,* Cambridge University Press, New York, 1997. Details applications of graph theory in pattern recognition and reconstruction problems.

LAWLER, EUGENE, J. LENSTRA, RINNOY KAN, and D. SHMOYS, eds. *The Traveling Salesman Problem,* Prentice Hall, Englewood Cliffs, N.J., 1985. Includes survey and technical articles on all aspects of the TSP.

LUCAS, WILLIAM, FRED ROBERTS, and ROBERT THRALL, eds. *Discrete and Systems Models,* vol. 3: *Modules in Applied Mathematics,* Springer-Verlag, New York, 1983. Chapter 6, "A Model for Municipal Street Sweeping Operations," by A. Tucker and L. Bodin, describes street-sweeping and related models in detail. Other chapters detail many recent applications of mathematics.

ROBERTS, FRED. *Applied Combinatorics,* Prentice Hall, Englewood Cliffs, N.J., 1984. The chapters on graphs and related network-optimization problems are excellent.

ROBERTS, FRED. *Graph Theory and Its Applications to Problems of Society,* Society for Industrial and Applied Mathematics, Philadelphia, 1978. A very readable account of how graph theory is finding a wide variety of applications.

SUGGESTED WEB SITES

www.informs.org The Web page of the major professional society in operations research.

mat.gsia.cmu.edu A "hub" for resources about operations research.

SKILLS CHECK

1. Which of the following describes a Hamiltonian circuit for the graph below?

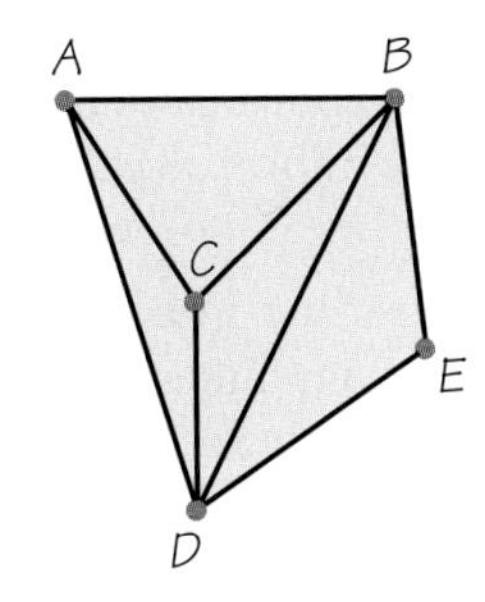

(a) *ABCDEA*
(b) *ABEDCBDAC*
(c) *ADEBCA*

2. Using the nearest-neighbor algorithm and starting at vertex *A*, find the cost of the Hamiltonian circuit for the graph below.

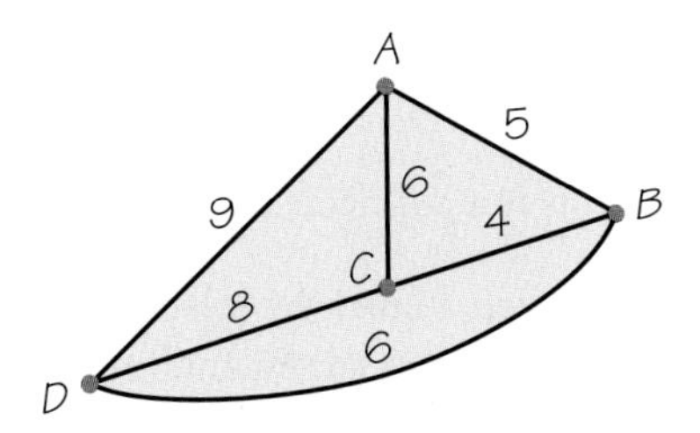

(a) 25
(b) 26
(c) Another answer

3. Using the sorted-edges algorithm, find the cost of the Hamiltonian circuit for the graph below.

(a) 25
(b) 26
(c) Another answer

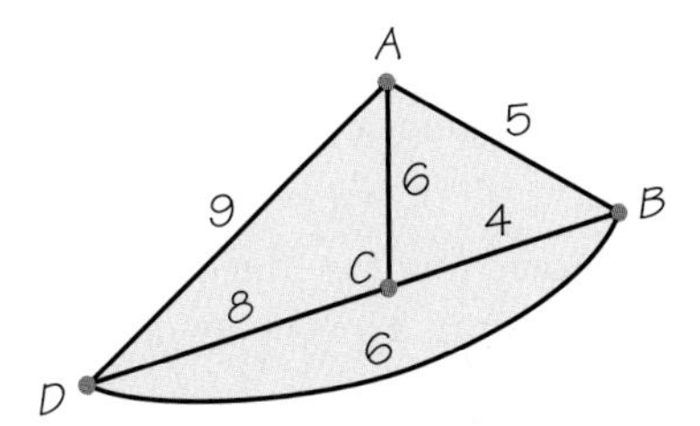

4. Using Kruskal's algorithm, find the minimum-cost spanning tree for the graph below. Which statement is true?

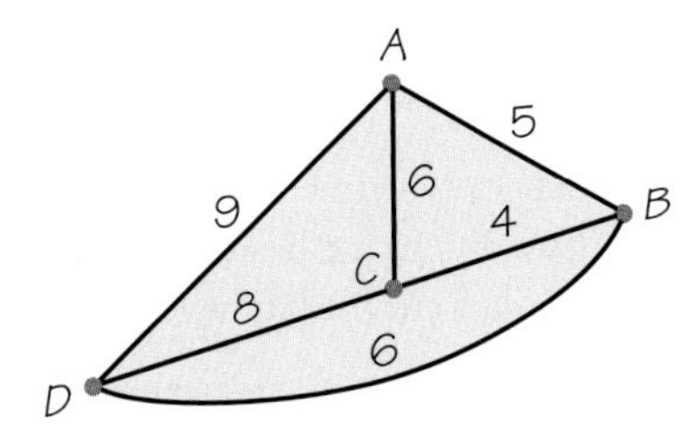

(a) Edges *AC* and *BD* are included in the minimum-cost spanning tree.
(b) Edges *AB* and *BD* are included in the minimum-cost spanning tree.
(c) Edges *CD* and *BD* are included in the minimum-cost spanning tree.

5. What is the earliest completion time (in minutes) for a job with the following order-requirement digraph?

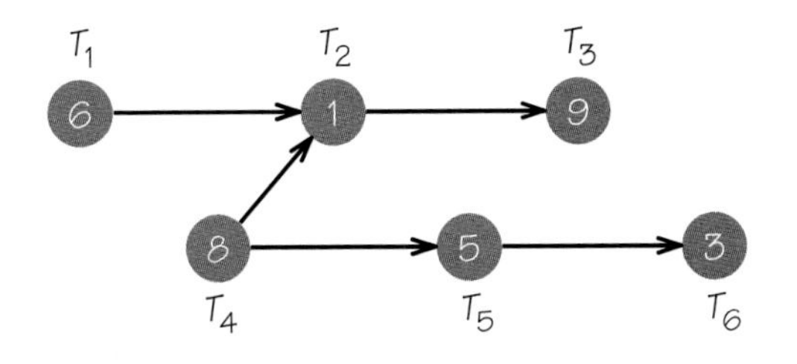

(a) 16 minutes

(b) 17 minutes
(c) 18 minutes

6. Suppose that after a hurricane, a van is dispatched to pick up five nurses at their homes and bring them to work at the local hospital. Which of these techniques is most likely to be useful in solving this problem?

(a) Finding an Euler circuit in a graph
(b) Solving a TSP (traveling salesman problem)
(c) Finding a minimum-cost spanning tree in a graph

7. Paul has packed four ties, three shirts, and two pairs of pants for a trip. How many different outfits can he create?

(a) Fewer than 10
(b) Between 10 and 25
(c) More than 25

8. Assuming a graph with E edges and V vertices has a Hamiltonian circuit, which of the following statements *must* be true?

(a) The graph has only one Hamiltonian circuit.
(b) The Hamiltonian circuit has exactly V edges.
(c) The Hamiltonian circuit has at least $E/2$ edges.

9. Assuming a graph with E edges and V vertices has a minimum-cost spanning tree T, which of the following statements *must* be true?

(a) The tree T has exactly V edges.
(b) The tree T includes every minimum-cost edge.
(c) The graph is connected.

10. Assume that every edge of a graph G has a different cost. If Kruskal's algorithm is used to find the minimum-cost spanning tree T for a graph G, which of the following statements *must* be true?

(a) Any other spanning tree for the graph G will have more edges than T.
(b) Any other spanning tree for the graph G will have a greater cost than T.
(c) The edge of G having greatest weight is included in T.

11. Assume a job has an order-requirement digraph consisting of six tasks ranging in time from 5 to 12 minutes. The earliest possible completion time is at least

(a) 12 minutes.
(b) 30 minutes.
(c) 51 minutes.

12. Assume a job has an order-requirement digraph with five tasks whose critical path is 25 minutes in length. Based on this information, what can be said about the tasks?

(a) Each task takes exactly 5 minutes.
(b) Some task takes 25 minutes.
(c) The five tasks in total take at least 25 minutes.

13. Which of the following statements about algorithms is true?

(a) An algorithm *always* finds the *best* solution.
(b) An algorithm *sometimes* finds the *best* solution.
(c) An algorithm *always* finds a *best or nearly best* solution.

14. If a graph contains a circuit, which of the following statements is true?

(a) The graph cannot be a tree.
(b) The graph must have the same number of vertices as edges.
(c) The graph is not connected.

15. Janet can design a lunch by selecting one of three meats, one of three salads, and one of six vegetables. How many different lunches can be designed?

(a) Fewer than 16
(b) Between 17 and 50
(c) More than 50

EXERCISES

▲ **Optional.** ■ **Advanced.** ◆ **Discussion.**

Hamiltonian Circuits

1. For the accompanying graphs (a) through (e) write a Hamiltonian circuit starting at X_4.

2. If the edge $X_3 X_4$ is erased from each of the graphs in Exercise 1, does the resulting graph still have a Hamiltonian circuit?

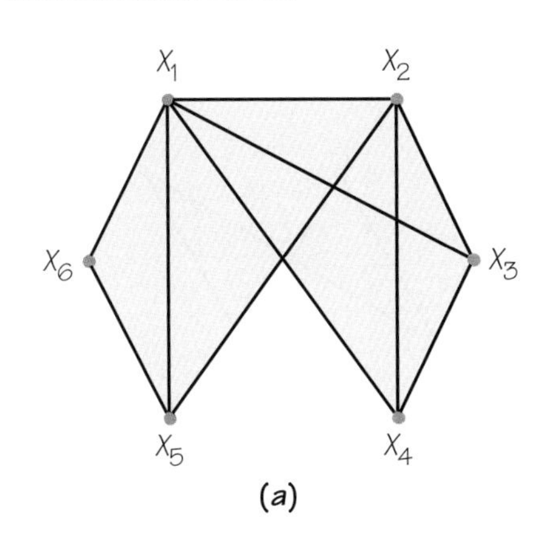

(a)

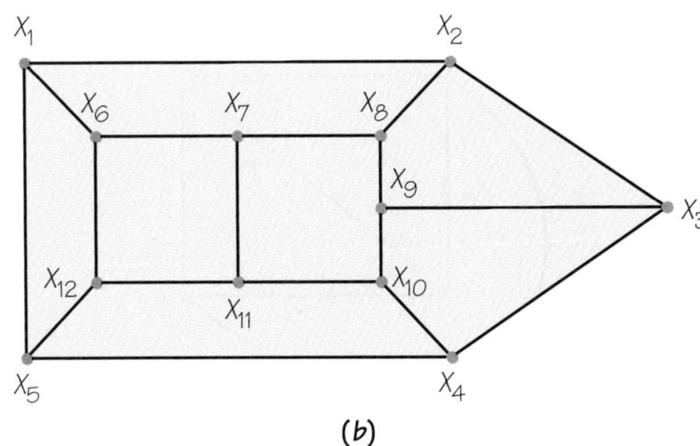

(b)

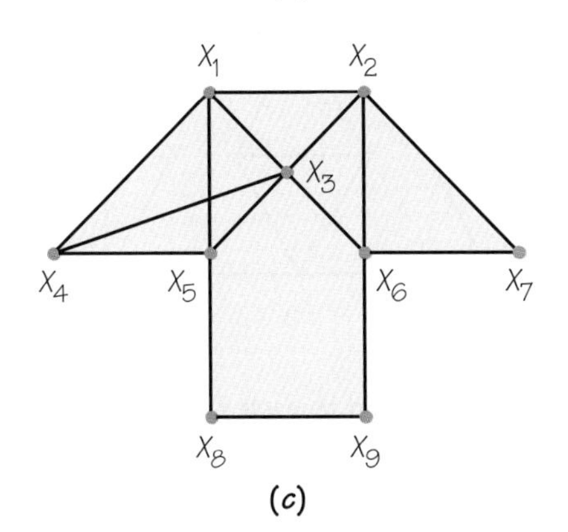

(c)

3. **(a)** If the vertex X_1 and the edges attached to X_1 are removed from the graphs in Exercise 1, do the new graphs that result still have Hamiltonian circuits? **(b)** If you think of the graphs in Exercise 1 as communications networks, what interpretation might be given to the "removal" of a vertex and the edges attached as described in part (a)?

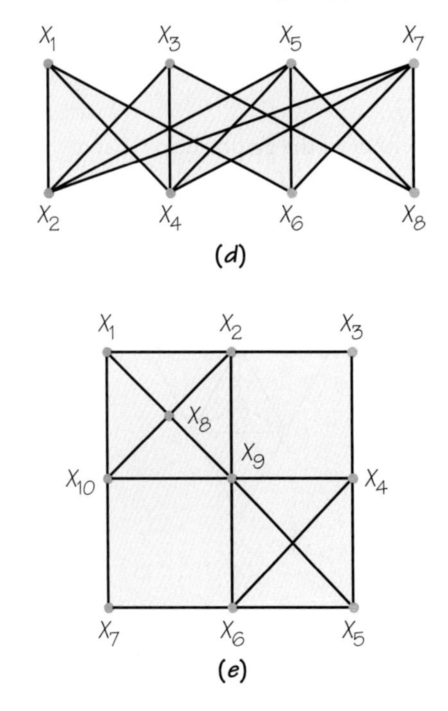

(d)

(e)

4. For each of the following graphs add wiggly edges to indicate a Hamiltonian circuit.

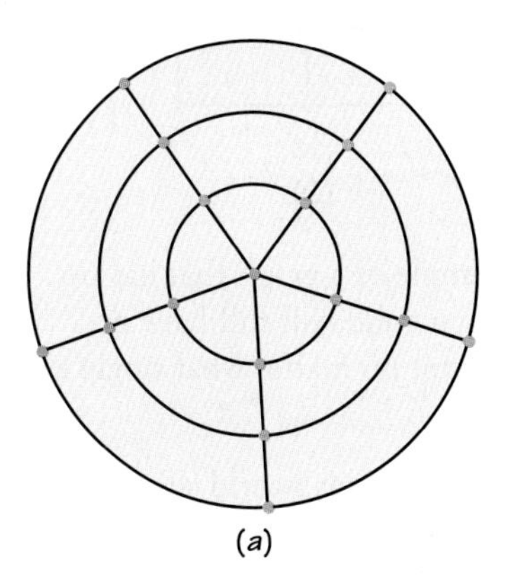

(a)

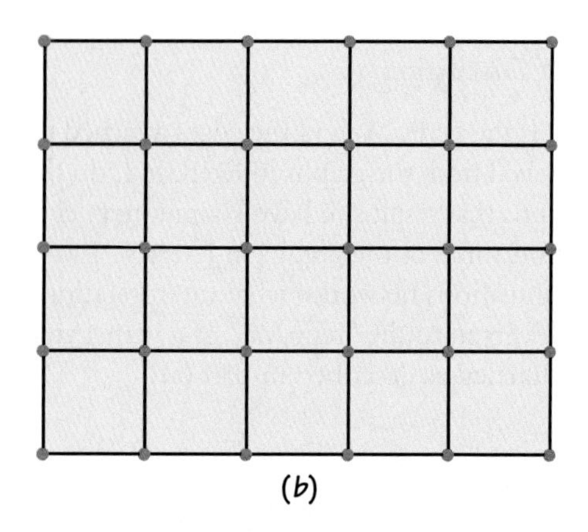

(b)

5. **(a)** Neither of the graphs below has a Hamiltonian circuit. Is it possible to add a single new edge to these graphs to obtain a new graph that has a Hamiltonian circuit?

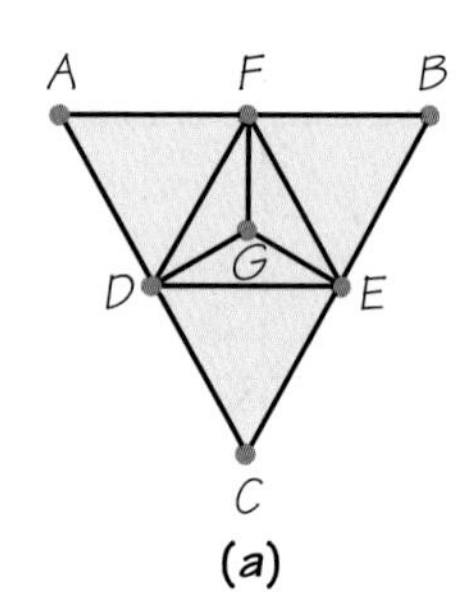

(a)

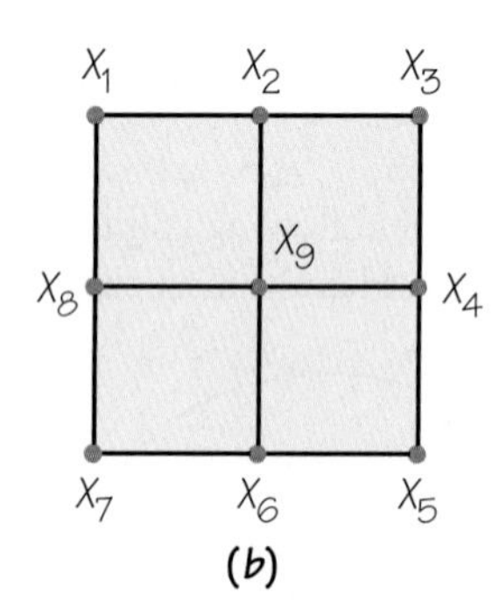

(b)

(b) Find an example of a graph that has no Hamiltonian circuit and will still have no Hamiltonian circuit no matter what single edge is added to it.

◆ **6.** Give examples of real-world situations that can be modeled using a graph and for which finding a Hamiltonian circuit in the graph would be of interest.

7. Suppose two Hamiltonian circuits are considered different if the collections of edges that they use are different. How many other Hamiltonian circuits can you find in the graph in Figure 2.1 that are different from the two discussed?

■ **8.** Explain why the tour *ACEDCBA* is not a Hamiltonian circuit for the graph below. Does this graph have a Hamiltonian circuit?

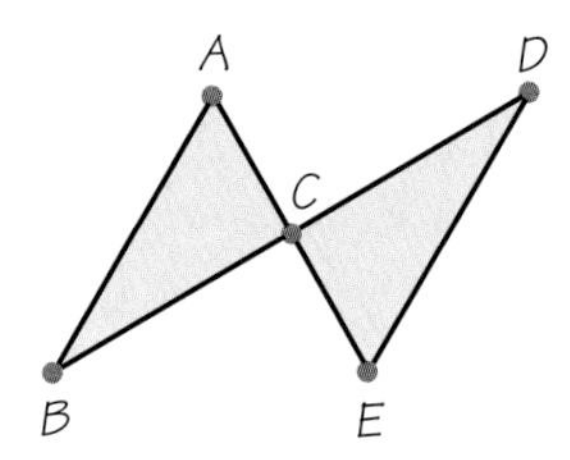

■ **9.** Do the following graphs have Hamiltonian circuits? If not, can you demonstrate why not?

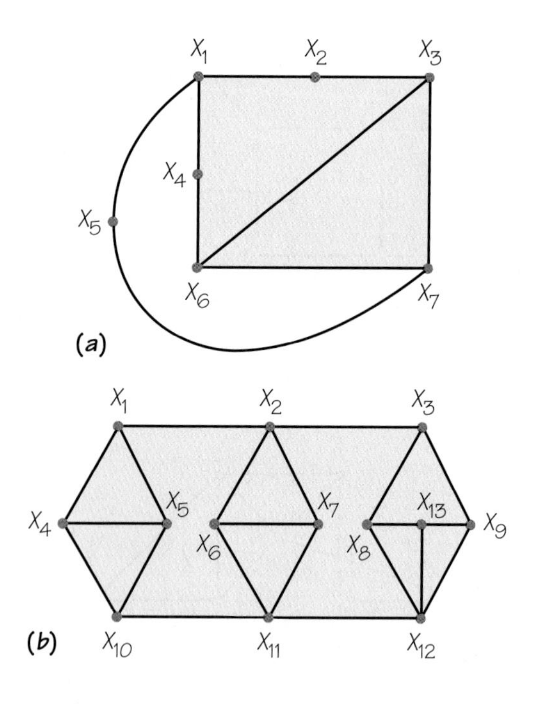

10. If an edge is added to each graph in Exercise 9 from X_2 to X_4, do the new graphs that result have a Hamiltonian circuit?

11. For each of the graphs below, determine whether there is a Hamiltonian circuit.

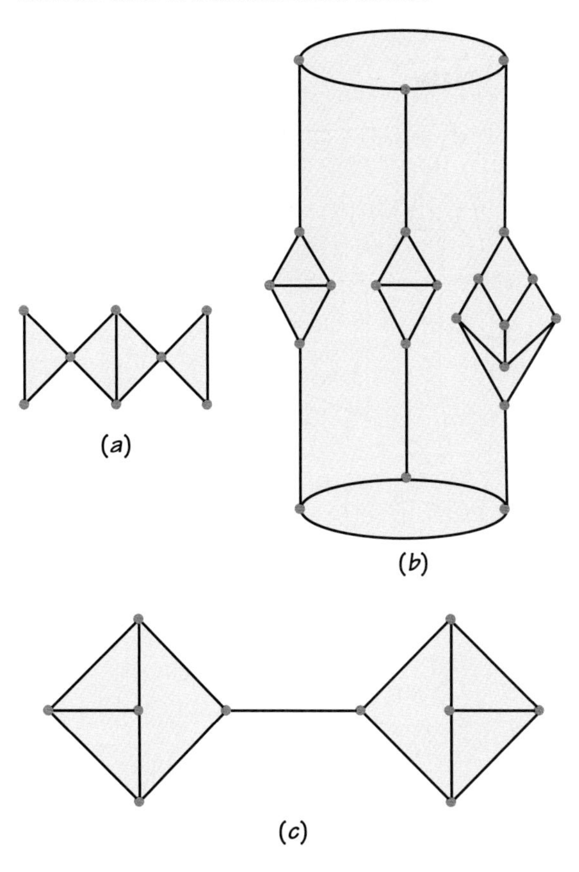

12. (a) The graph below is known as a four spokes and three concentric circles graph. What conditions on m and n guarantee that an m spokes and n concentric circles graph has a Hamiltonian circuit? (Assume $m \geq 2$, $n \geq 1$.)

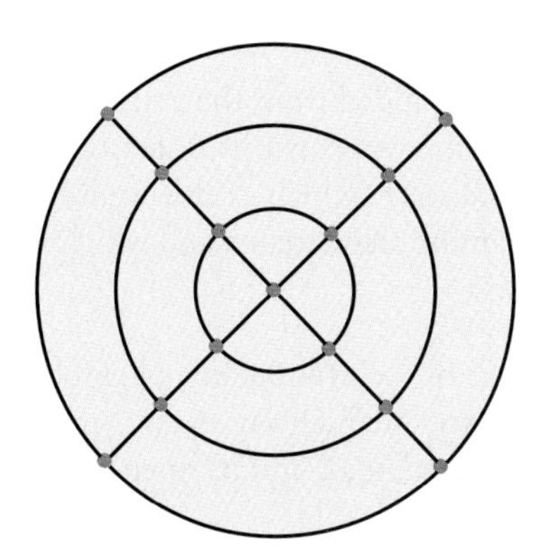

(b) The graph below is known as a 3×4 grid graph. What conditions on m and n guarantee that an $m \times n$ grid graph has a Hamiltonian circuit?

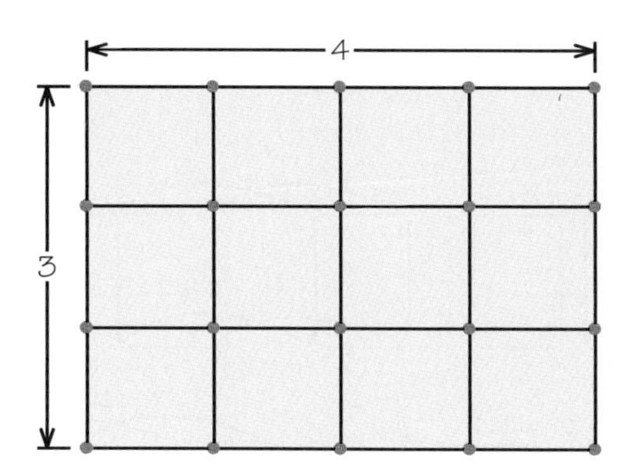

Can you think of a real-world situation in which finding a Hamiltonian circuit in an $m \times n$ grid graph would represent a solution to the problem? If an $m \times n$ grid graph has no Hamiltonian circuit, can you find a tour that repeats a minimum number of vertices and starts and ends at the same vertex?

13. A Hamiltonian path in a graph is a tour of the vertices of the graph that visits each vertex once and only once and starts and ends at different vertices.

(a) For each of the graphs shown in Exercise 9, does the graph have a Hamiltonian path?
(b) Does each of these graphs have a Hamiltonian path that starts at X_1 and ends at X_2?
(c) Can you think of real-world situations where finding a Hamiltonian path in a graph would be required?

14. Using the terminology of Exercise 13, can you draw a graph that has

(a) a Hamiltonian path but no Hamiltonian circuit?
(b) an Euler circuit but no Hamiltonian path?
(c) a Hamiltonian path but no Euler circuit?

■ **15.** The n-dimensional cube is obtained from two copies of an $(n - 1)$-dimensional cube by joining corresponding vertices. (The process is illustrated for the 3-cube and the 4-cube in the figure below.) Find formulas for the number of vertices and the number of edges of an n-cube. Can you show that every n-cube has a Hamiltonian circuit? [*Hint:* Show that if you know how to find a Hamiltonian circuit on an $(n - 1)$-cube, then you can use two copies of this to build a Hamiltonian circuit on an n-cube.]

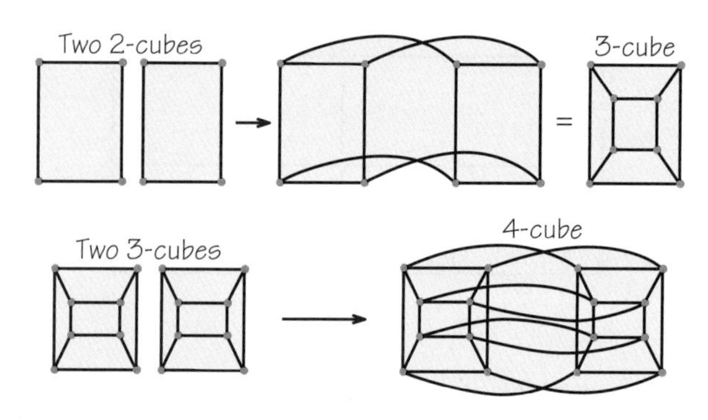

16. To practice your understanding of the concepts of Euler circuits and Hamiltonian circuits, determine for graphs (a) through (d) below whether there is an Euler circuit and/or a Hamiltonian circuit. If so, write it down.

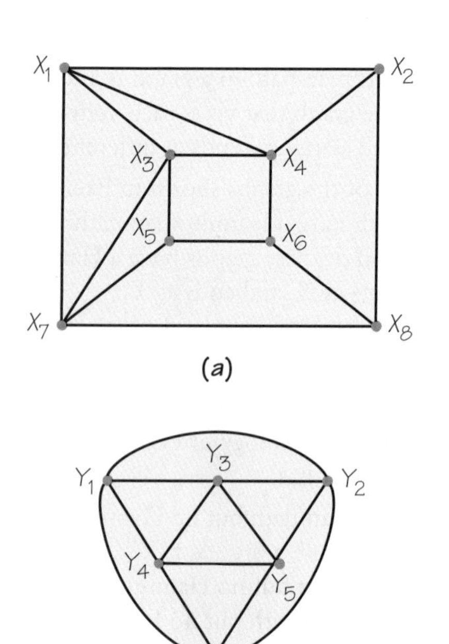

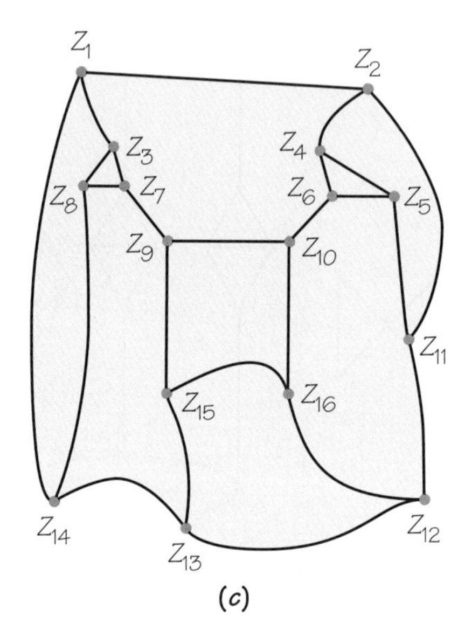

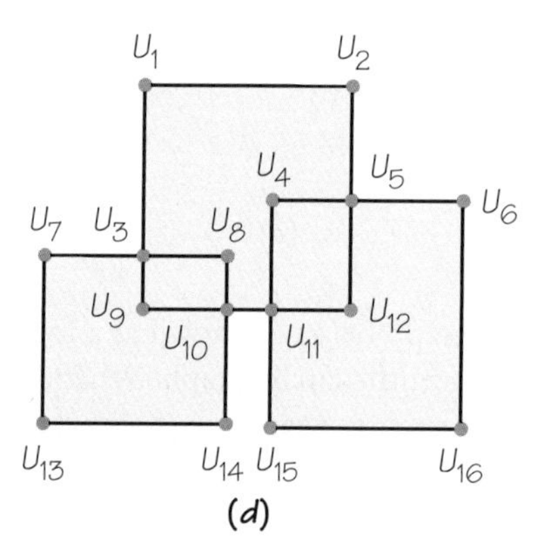

17. If an edge is added from the vertex with subscript 1 to the vertex with subscript 5 in each graph in Exercise 16, which of the resulting graphs will have Hamiltonian circuits and which will have Euler circuits?

18. Each edge of the accompanying graph represents a two-lane highway. A grass-mowing machine is located at A, and its operator has the job of cutting the grass along each of the edges of road shown. Can you find a tour for the mowing machine that begins and ends at A? Can you find such a tour that begins and ends at A and, as the mowing is done, moves along the edge of the road in the same direction as the traffic is going?

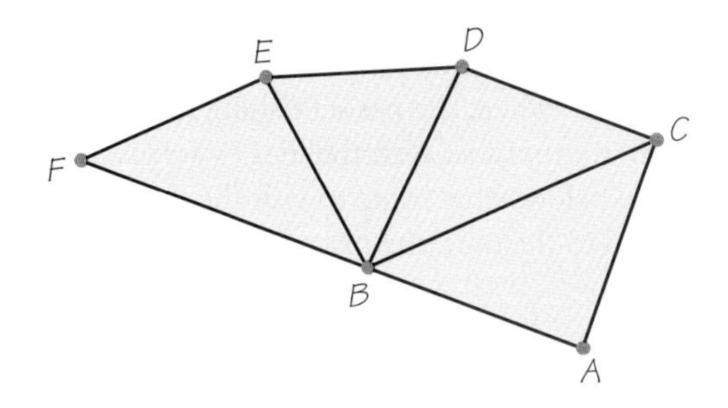

Counting Problems

19. A large corporation has found that it has "outgrown" its current code system for routing interoffice mail. The current system places a code of three ordered, distinct nonzero digits on the mail. The new proposal calls for the use of two ordered capital letters. Is the new system an improvement? If so, how many more locations will the new system enable the company to encode over the current system?

20. A lottery game requires that a person select an upper- or lowercase letter followed by five different two-digit numbers (where the digits cannot both be zero). How many different ways are there to fill out a lottery ticket?

21. (a) In designing a security system for its accounts, a bank asks each customer to choose a five-digit number, all the digits to be distinct and nonzero. How many choices can a customer make? **(b)** A suitcase with a liquid-crystal display allows one to unlock it with a specific combination of three capital letters that are not necessarily different. How many choices would a thief have to go through to be sure that all the possibilities had been tried? How does this compare to a "standard" combination lock?

22. (a) When going outside on a cold winter day, Jill can choose from three winter coats, four wool scarves, four pairs of boots, and three ski hats. How many outfits might her friends see her in?
(b) If Jill always insists on wearing her green wool scarf, how many outfits might her friends see her in?

23. The notes C, D, E, F, G, A, and B are to be used to form an ordered five-note musical logo. In how many ways can this be done if (a) no note can be repeated; (b) notes can be repeated; (c) notes can be repeated but all the notes cannot be the same?

24. Repeat Exercise 23, except that exactly one of the notes in the musical logo must be a sharp and the note chosen to be sharped cannot appear elsewhere (e.g., BCD#AG, where D# denotes D sharp).

◆ **25. (a)** In New York State, one type of license plate has three letters followed by three numbers. Suppose the digits from 0, 1, . . . , 9 can be used, except that all three digits cannot be zero, and that any letter from A to Z (repeats allowed) can be used. How many plates are possible?
(b) Investigate what schemes for license plates are used in your state and determine how many different plates are possible.

26. A restaurant offers 3 soups, 10 entrees, and 8 desserts. How many different choices for a meal can a customer make if one selection is made from each category? If 3 of the desserts are pies and the customer will never order pie, how many meals can the customer choose?

Traveling Salesman Problems

27. Draw complete graphs with four, five, and six vertices. How many edges do these graphs have? Can you generalize to n vertices? How many TSP tours would these graphs have? (Tours yielding the same Hamiltonian circuit are considered the same.)

28. Calculate the values of 5!, 6!, 7!, 8!, 9!, and 10!. Then find the number of TSP tours in the complete graph with 10 vertices.

29. The table below shows the mileage between four cities: Springfield, Ill. (S); Urbana, Ill. (U); Effingham, Ill. (E); and Indianapolis, Ind. (I).

	E	*I*	*S*	*U*
E	–	147	92	79
I	147	–	190	119
S	92	190	–	88
U	79	119	88	–

(a) Represent this information by drawing a weighted complete graph on four vertices.
(b) Use the weighted graph in part (a) to find the cost of the three distinct Hamiltonian circuits in the graph. (List them starting at U.)
(c) Which circuit gives the minimum cost?
(d) Would there be any difference in parts (b) and (c) if the start vertex were at I?
(e) If one applies the nearest-neighbor method starting at U, what circuit would be obtained? Does the answer change if one applies the nearest-neighbor algorithm starting at S? At E? At I?
(f) If one applies the sorted-edges method, what circuit would be obtained? Does one get the optimal answer?

30. After a party at her house, Francine (F) has agreed to drive Mary (M), Rachel (R), and Constance (C) home. If the times (in minutes) to drive between her friends' homes are shown below, what route gets Francine back home the quickest?

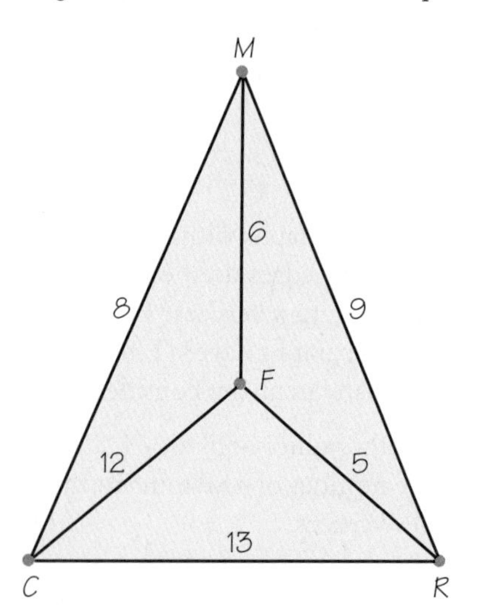

31. In Exercise 30, what route would Francine have to follow to get home as quickly as possible, assuming she promised to drive Mary home first?

32. In Exercise 30, Francine is planning to deliver her friends home and then spend the night at Rachel's house. What would her fastest route be?

33. Starting from the location where she moors her boat (M), a fisherwoman wishes to visit three areas – A, B, and C – where she has set fishing nets. If the times (in minutes) between the locales are given in the figure below, what route to visit the three sites and return to the mooring place would be optimal?

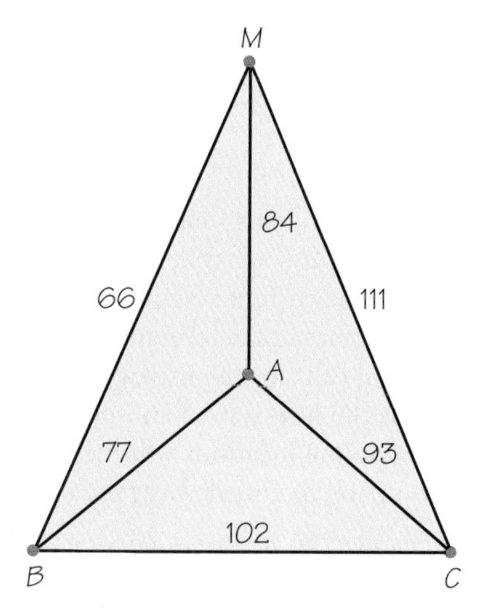

34. (a) For the two complete graphs that follow, find the costs of the nearest-neighbor tour starting at A and of the tour generated by the sorted-edges algorithm.

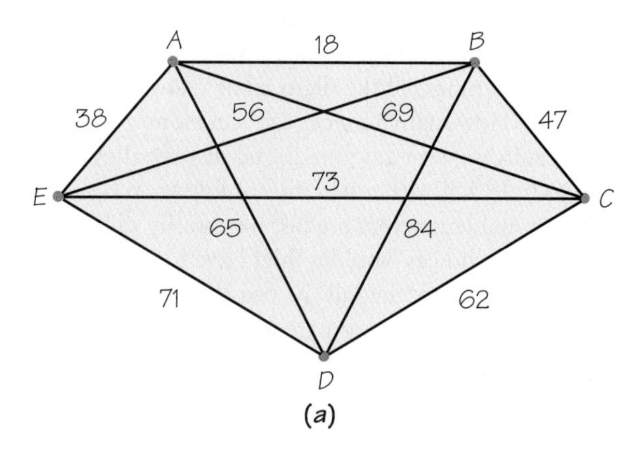

(*a*)

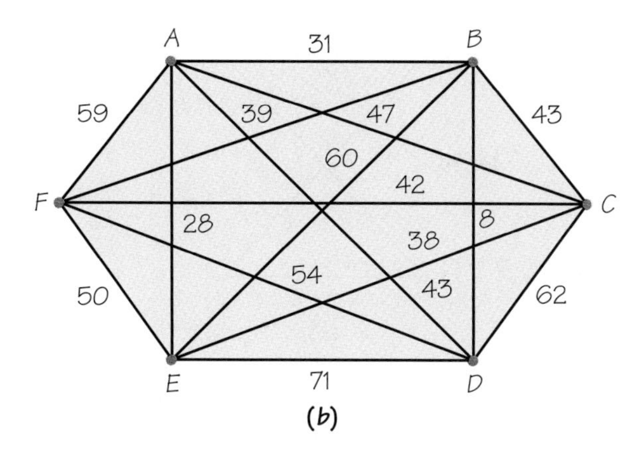

(b)

(b) How many Hamiltonian circuits would have to be examined to find a shortest route for part (a) by the brute force method?
(c) Can you invent an algorithm different from the sorted-edges and nearest-neighbor algorithms that is easy to apply for finding TSP solutions? (See Lawler et al. in the Suggested Readings.)

35. An airport limo must take its six passengers from the airport to different downtown hotels. Is this a traveling salesman problem, a Chinese postman problem, or an Euler circuit problem?

36. (a) Solve the six-city TSP shown in the diagram using the nearest-neighbor algorithm starting at vertex A and starting at vertex B.
(b) Apply the sorted-edges method.

37. Construct an example of a complete graph of five vertices, with distinct weights on the edges for which the nearest-neighbor algorithm starting at a particular vertex and the sorted-edges algorithm yield different solutions for the traveling salesman problem. Can you find a five-vertex complete graph with weights on the edges in which the optimal solution, the nearest-neighbor solution, and the sorted-edges algorithm solution are all different?

38. If the brute force method of solving a 20-city TSP is employed, use a calculator to determine how many Hamiltonian circuits must be examined. How long would it take to determine the minimum-cost tour if the cost of tours could be computed at the rate of 1 billion per second? (Convert your answer to years by seeing how many years are equivalent to a billion seconds!)

39. Suppose one has found an optimal tour for a given 10-city TSP problem to have weight 4200. Now suppose the weights on the edges of the complete graph are increased by 30. What can you say about the optimal tour and its weight?

Trees and Spanning Trees

40. Which of the graphs below are trees?

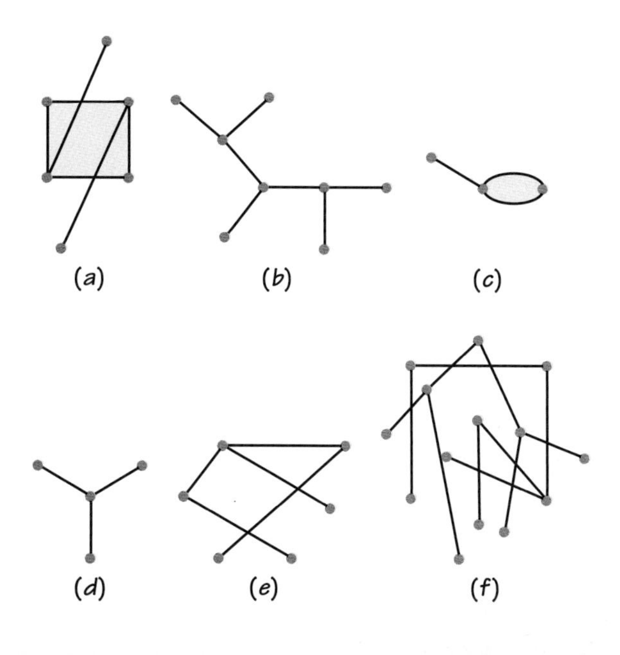
(a) (b) (c)
(d) (e) (f)

41. For each of the diagrams below, explain why the wiggly edges are not

(a) a spanning tree.
(b) a Hamiltonian circuit.

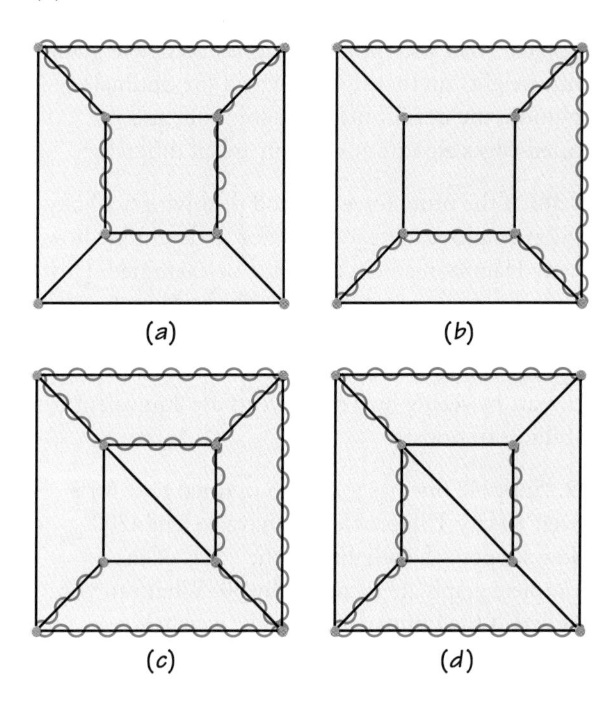

42. Find all the spanning trees in the graphs below.

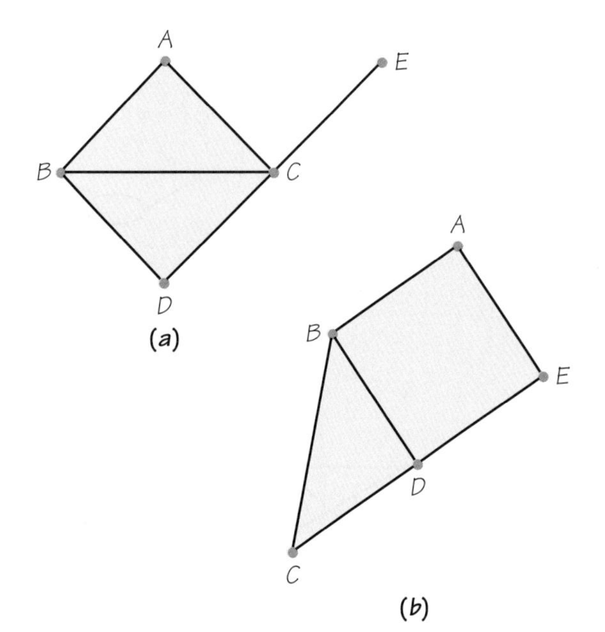

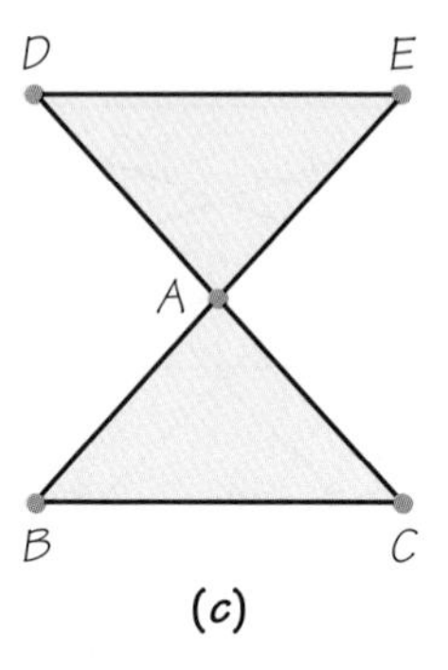

43. Use Kruskal's algorithm to find a minimum-cost spanning tree for graphs (a), (b), (c), and (d). In each case, what is the cost associated with the tree?

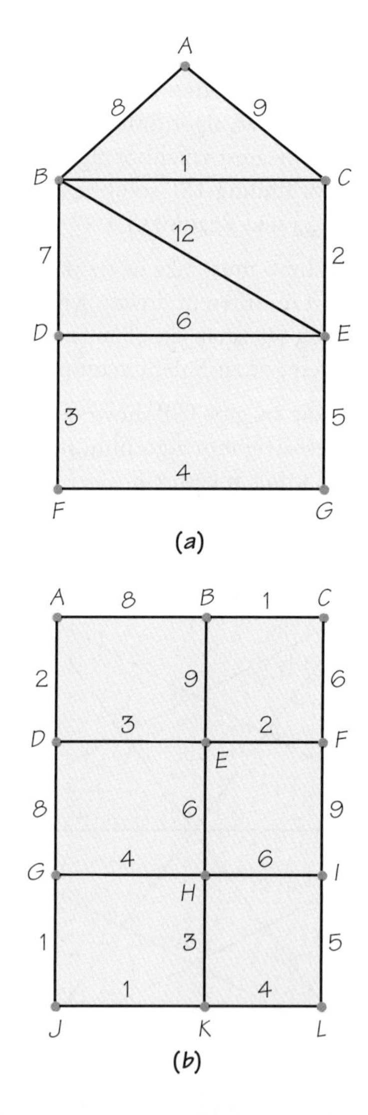

(c)

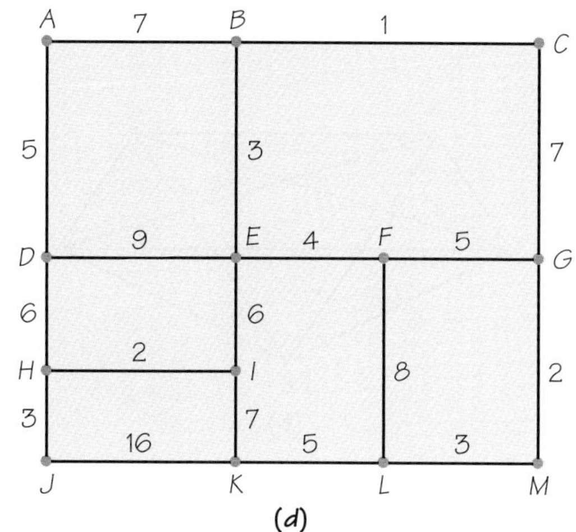

(d)

44. A connected graph G has 20 vertices. How many edges does a spanning tree of G have? How many vertices does a spanning tree of G have? What can one say about the number of edges G has?

45. A connected graph H has a spanning tree with 25 edges. How many vertices does the spanning tree have? How many vertices does H have? What can one say about the number of edges H has?

46. A large company wishes to install a pneumatic tube system that would enable small items to be sent between any of 10 locales, possible by relay. If the nonprohibitive costs (in $100) are shown in the graph model below left, between which sites should the tube be installed to minimize the total cost?

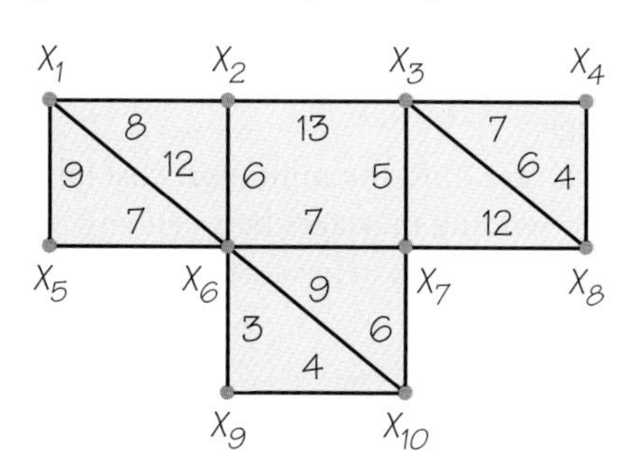

47. If the weight of each edge in Exercise 46 is increased by 3, will the tree that achieves minimum cost for the new collection of weights be the same as the one that achieves minimum cost for the original set of weights?

48. (a) The table shown gives the "closeness" or distance values between four objects. Can you construct a four-vertex tree with weights on its edges such that the distances between pairs of vertices of the tree (as measured by the sum of the weights on the path in the tree between these vertices) give rise to this table?

0	3	10	14
3	0	7	11
10	7	0	4
14	11	4	0

(b) Can you produce several real-world contexts that might give rise to the situation described here?

49. The figure below represents four objects using a tree with weights on the edges. Construct a table with 4 rows and 4 columns recording how "close" pairs of vertices in the tree are to each other. To find how close a pair of objects is, add together the weights along the path that joins these two objects.

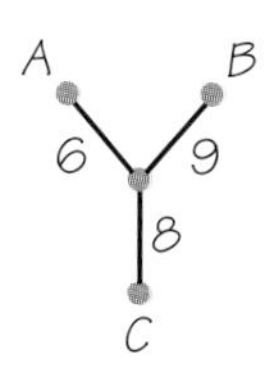

◆ **50.** Give examples of real-world situations that can be modeled using a weighted graph and for which finding a minimum-cost spanning tree for the graph would be of interest.

■ **51.** Can Kruskal's algorithm be modified to find a maximum-weight spanning tree? Can you think of an application for finding a maximum-weight spanning tree?

◆ **52.** Find the cost of providing a relay network between the six cities with the largest populations in your home state, using the road distances between the cities as costs. Does it follow that the same solution would be obtained if air distances were used instead?

■ **53.** Would there ever be a reason to find a minimum-cost spanning tree for a weighted graph in which the weights on some of the edges were negative? Would Kruskal's algorithm still apply?

54. Let G be a graph with weights assigned to each edge. Consider the following algorithm:

(a) Pick any vertex V of G.
(b) Select an edge E with a vertex at V that has a minimum weight. Let the other endpoints of E be W.
(c) Contract the edge VW so that edge VW disappears and vertices V and W coincide (see the following figures).

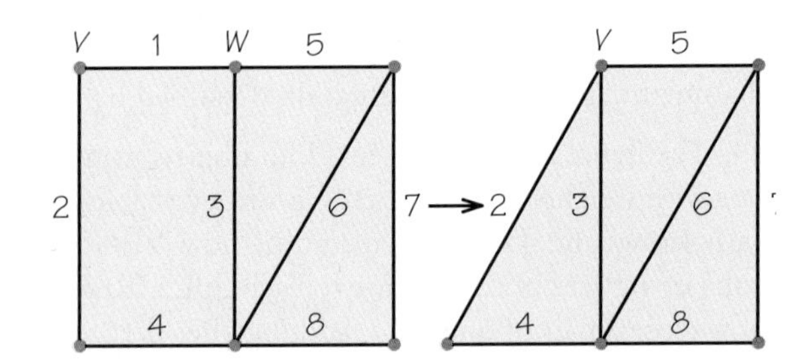

If in the new graph two or more edges join a pair of vertices, delete all but the cheapest. Continue to call the new vertex V.

(d) Repeat steps (b) and (c) until a single point is obtained. The edges selected in the course of this algorithm (called Prim's algorithm) form a minimum-cost spanning tree. Apply this algorithm to the following graphs.

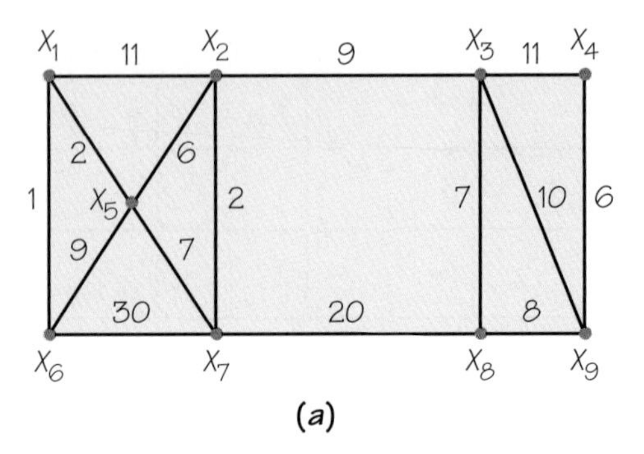

(a)

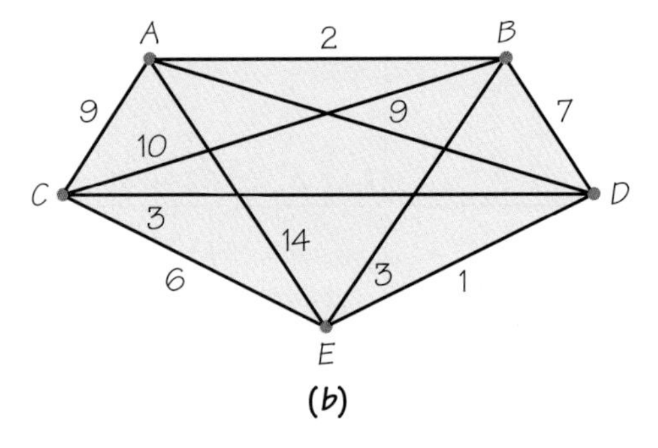

(b)

55. Determine whether each of the following statements is true or false for a minimum-cost spanning tree T for a weighted connected graph G:

(a) T contains a cheapest edge in the graph.
(b) T cannot contain a most expensive edge in the graph.
(c) T contains one fewer edge than there are vertices in G.
(d) There is some vertex in T to which all others are joined by edges.
(e) There is some vertex in T that has valence 3.

■ **56.** In the following graphs, the number in the circle for each vertex is the cost of installing equipment at the vertex if relaying must be done at the vertex, while the number on an edge indicates the cost of providing service between the endpoints of the edge.

In each case, find the minimum cost (allowing relays) for sending messages between any pair of vertices, taking vertex relay costs into account.

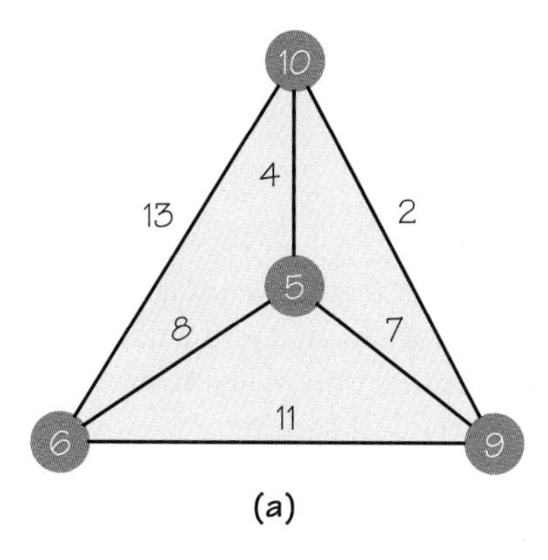

(a)

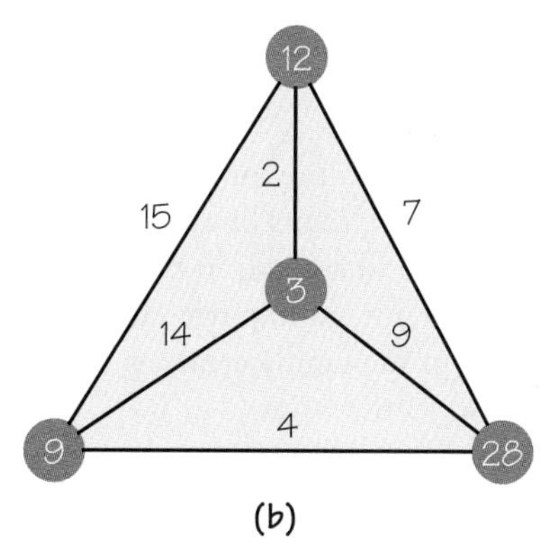

(b)

Would your answer be different if vertex relay costs were neglected? (*Warning:* Kruskal's algorithm cannot be used to answer the first question. This problem illustrates the value of having an algorithm over-relying on "brute force.")

57. Two spanning trees of a (weighted) graph are considered different if they use different edges. Show that the graph below has different minimum-cost spanning trees, though all these different trees have the same cost.

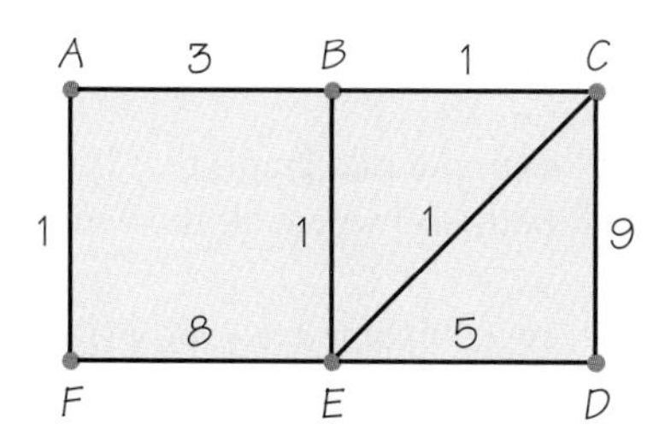

■ **58.** Suppose G is a graph such that all the weights on its edges are different numbers. Show that there is a unique minimum-cost spanning tree.

59. Find a minimum-cost spanning tree for the complete graphs in Exercise 34.

Scheduling

60. Find the earliest completion time and critical paths for the three order-requirement digraphs below.

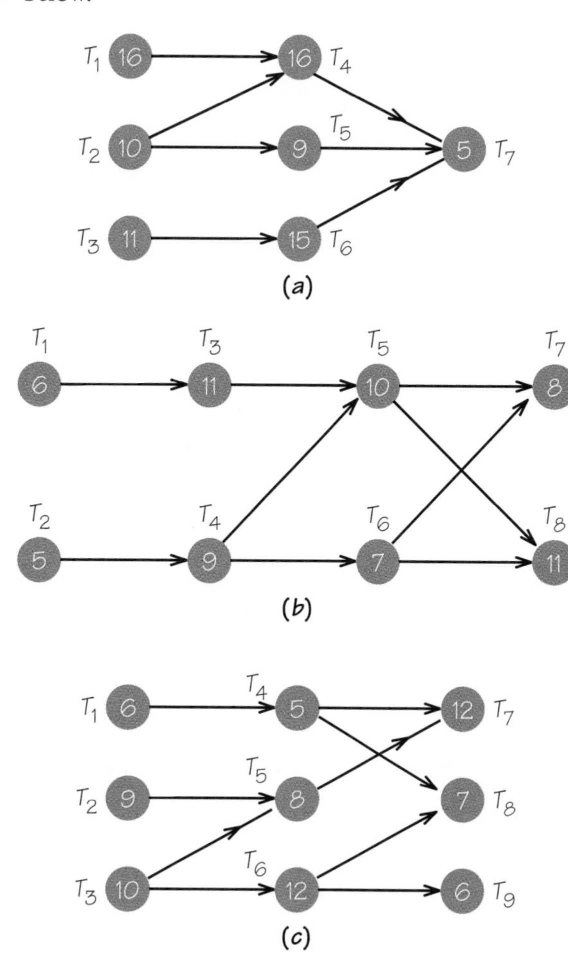

61. Construct an example of an order-requirement digraph with two different critical paths.

■ **62.** In the order-requirement digraph below, determine which tasks, if shortened, would reduce the earliest completion time and which would not. Then find the earliest completion time if task T_5 is reduced to time length 7. What is the new critical path?

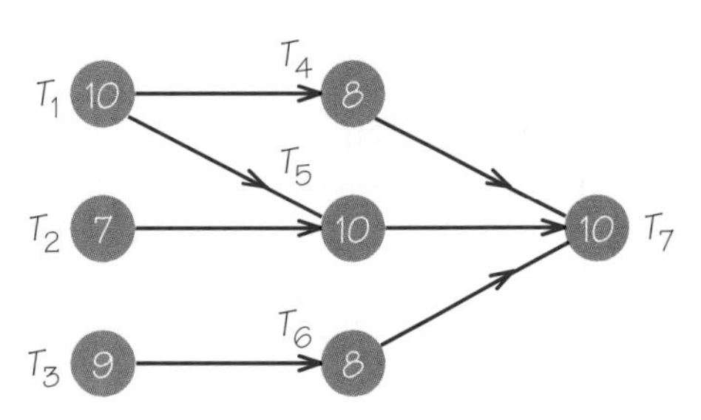

63. To build a new addition on a house, the following tasks must be completed:

(a) Lay foundation.
(b) Erect sidewalls.
(c) Erect roof.
(d) Install plumbing.
(e) Install electric wiring.
(f) Lay tile flooring.
(g) Obtain building permits.
(h) Put in door that adjoins new room to existing house.
(i) Install track lighting on ceiling.
(j) Install wall air-conditioner.

Construct reasonable time estimates for these tasks and a reasonable order-requirement digraph. What is the fastest time in which these tasks can be completed?

64. At a large toy store, scooters arrive unassembled in boxes. To assemble a scooter, the following tasks must be performed:

TASK 1. Remove parts from the box.
TASK 2. Attach wheels to the footboard.
TASK 3. Attach vertical housing.
TASK 4. Attach handlebars to vertical housing.
TASK 5. Put on reflector tape.
TASK 6. Attach bell to handlebars.
TASK 7. Attach decals.
TASK 8. Attach kickstand.
TASK 9. Attach safety instructions to handlebars.

Give reasonable time estimates for these tasks and construct a reasonable order-requirement digraph. What is the earliest time by which these tasks can be completed?

65. For the order-requirement digraph in Exercise 62, find the critical path and the task(s) in the critical path whose time, when reduced the least, creates a new critical path.

66. Construct an order-requirement digraph with six tasks that has three critical paths of length 24.

Additional Exercises

67. Can you find a family of graphs of which none have Hamiltonian circuits but for which adding a single edge to the first graph in the family creates a Hamiltonian circuit, adding two edges to the second graph in the family creates a Hamiltonian circuit, and so forth?

68. In order to encourage her son to try new things, a mother offers to take him for a dish of ice cream with a topping once a week, for as many weeks as he does not get the same choice as on a previous occasion. If the store offers 12 flavors and 6 toppings, for how many weeks will she have to do this if her son never picks either of the 2 types of chocolate ice cream or the 3 types of nut topping that the store carries?

69. A Hamiltonian path in a graph is a tour of the vertices that visits each vertex once and only once and that starts and ends at different vertices.

(a) Do the graphs in Exercise 11 have Hamiltonian paths?
(b) Draw an example of a graph that has no Hamiltonian path and where all the vertices are 3-valent.
(c) Can you suggest some real-world situations in which it would be natural to look for a Hamiltonian path in a graph?
(d) Can you draw a graph that has no Hamiltonian path but that does have an Euler circuit?

70. The following figure represents a town where there is a sewer located at each corner (where two or more streets meet). After every thunderstorm, the department of public works wishes to have a truck start at its headquarters (at vertex H) and make an inspection of sewer drains to be sure that leaves are not clogging them. Can a route start and end at H that visits each corner exactly once? (Assume that all the streets are two-way streets.) Does this problem involve finding an Euler circuit or a Hamiltonian circuit?

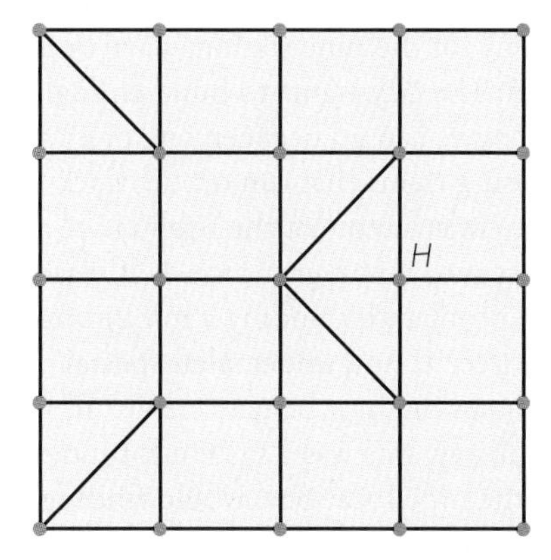

Assume that at equally spaced intervals along the blocks in this graph there are storm sewers that must be inspected after each thunderstorm to see whether they are clogged. Is this a Hamiltonian circuit problem, an Euler circuit problem, or a Chinese postman problem? Can you find an optimal tour to do this inspection?

71. In the last several years, heavily populated regions that had had only one area code have had to be divided into service areas with more than one area code. What is the largest number of different phone numbers that can be served using one area code? If an area code cannot begin with a zero, how many different area codes are possible?

72. For each of the graphs with weights (at right), apply the nearest-neighbor method (starting at vertex A) and the sorted-edges method to find (it is hoped) a cheap tour.

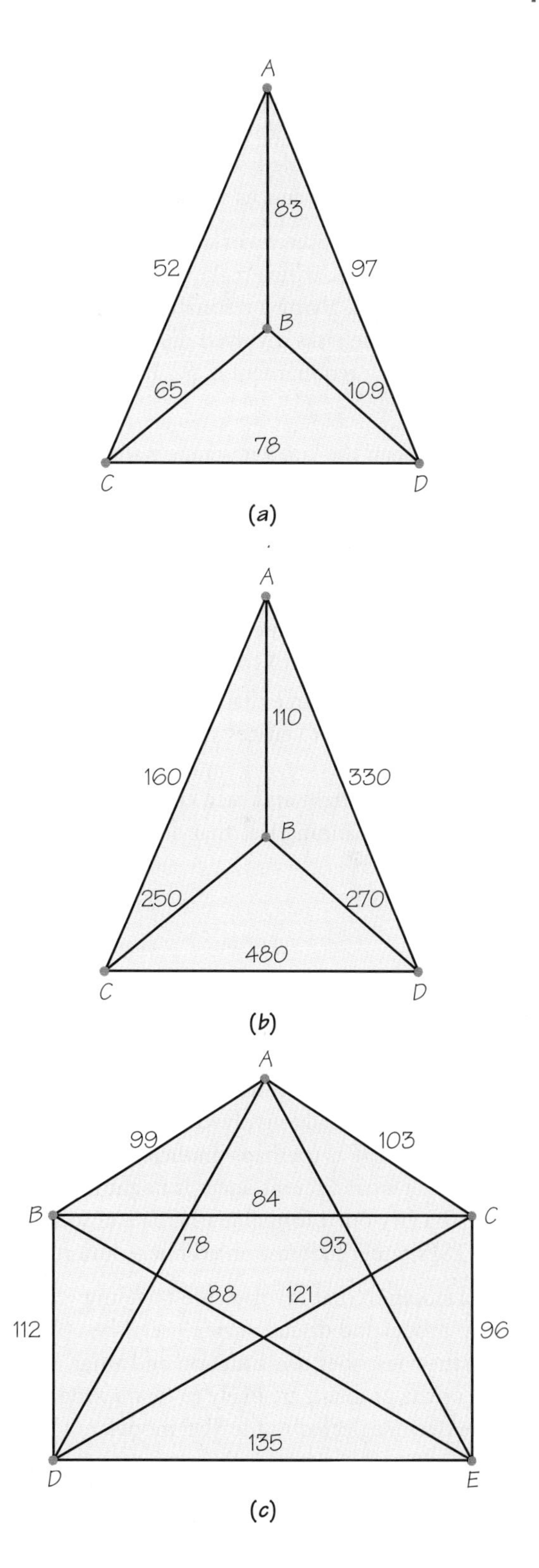

73. Suppose one has found an optimal tour for a given 10-city TSP problem to have weight 4520. Now suppose that the weights on the edges of the complete graph are doubled. What can you say about the optimal tour and its weight?

74. Draw an order-requirement digraph for the following set of tasks, which make up a kitchen remodeling project, giving a resonable estimate for the times to do the tasks involved. Find a critical path for the order-requirement digraph that you obtain.

TASKS: Clean the kitchen, scrape walls to remove old paint, prime walls, install wallpaper on walls, scrape paint on ceiling, paint ceiling, replace old floor with new floor tiles, install new stove, install new sink, install new refrigerator.

75. **(a)** Show that for each edge of graph J at right there is a spanning tree of J that avoids that edge.

(b) For each spanning tree that you found in graph J, count the number of vertices and edges. Do you notice any pattern?

(c) For graph H at right and each edge e in the graph, is there a spanning tree that does not include edge e of H?

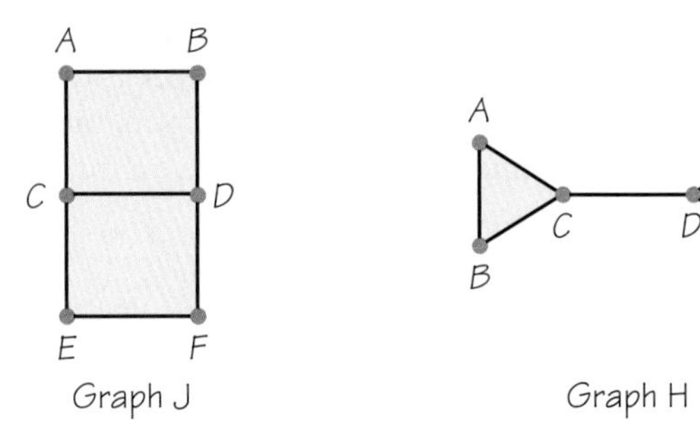

Graph J

Graph H

76. There are plans to construct a new subway system in city X. The distances between every pair of locations for the nine stations have been computed. The city wants to build enough track so that passengers can go between any pair of the nine stations using trains that run on the track that is built. The constructors of the subway are anxious to keep the cost to taxpayers down. Is it reasonable to solve this problem by finding a minimum-cost spanning tree? If not, under what special circumstances might it be reasonable? If, instead of being a subway, this was a system for moving freight between the nine locations, would finding a minimum-cost spanning tree seem reasonable? Explain your reasons.

✎ WRITING PROJECTS

1. Write an essay about a variety of situations in which you are personally involved for which a solution of the TSP is (perhaps implicity) required. Explain under what circumstances it might be valuable to carry out a formal mathematical solution to such TSPs rather than use an ad hoc solution.

2. Pick a situation that involves the traveling salesman model and discuss how closely the mathematics describes this situation and what features of the problem are likely to be important but have been neglected in the TSP model.

3. Construct an example, of the kind suggested on page 45, that shows that in a situation where three day campers must be picked up and brought to camp, it may make a difference if the optimization criterion is minimizing distance traveled by the camp bus versus minimizing average time that the children spend on the bus.

4. Determine the six largest cities in the state in which you live. By consulting a road atlas (or by some other means) construct the graph that represents the road distances between your hometown and these six other cities. Now apply (a) the nearest-neighbor method, (b) the sorted-edges method, and (c) the nearest neighbor from each city, and pick the minimum tour method to solve the associated TSP. Do you have reason to believe that the answers you get might include an optimum solution among them?

5. Give the pros and cons of converting between a phone system in which calls are paid for by placing coins into the phone versus a phone system that operates using prepaid cards.

6. If a housing developer must create a new system of roads to drive between any pair of houses (currently there are no roads in the area) and the paved access road, do you think the people in the new housing tract will be pleased if a minimal-cost spanning-tree approach is used by the developer in solving the problem?

SPREADSHEET PROJECTS

To do these projects, go to www.whfreeman.com/fapp.

Graphs with weighted edges can easily be represented by spreadsheets. Such technology can help to locate low-cost spanning trees and Hamiltonian circuits. These spreadsheet projects provide such examples.

APPLET EXERCISES

To do these exercises, go to www.whfreeman.com/fapp.

1. TSP: Nearest-Neighbor Algorithm. There is an extended version of the nearest-neighbor algorithm, in which you compare the total distances of the Hamiltonian circuits produced by applying the ordinary nearest-neighbor algorithm starting at each of the vertices of the graph (rather than just a specific one). Explore the effectiveness of this algorithm using the applet TSP: Nearest-Neighbor.

2. TSP: Sorted-Edges Algorithm. Go to the applet TSP: Sorted-Edges, where you can apply the sorted-edges algorithm to see if it solves the traveling salesman problem for the following graphs (and others):

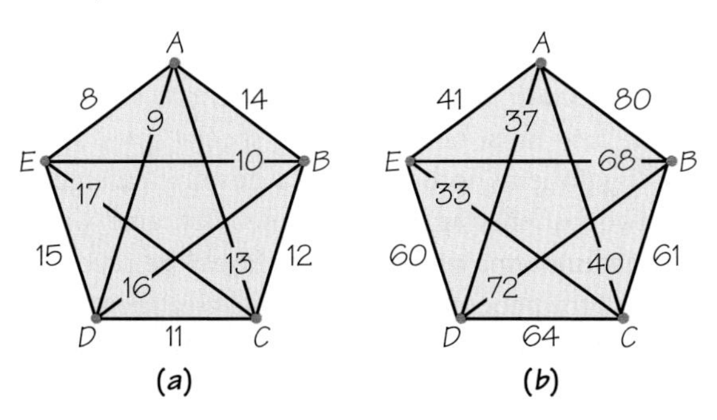

(a) (b)

3. Kruskal's Algorithm. Go to the Kruskal's Algorithm applet, where you can apply Kruskal's algorithm to find the minimum-cost spanning trees in the following graphs (and others):

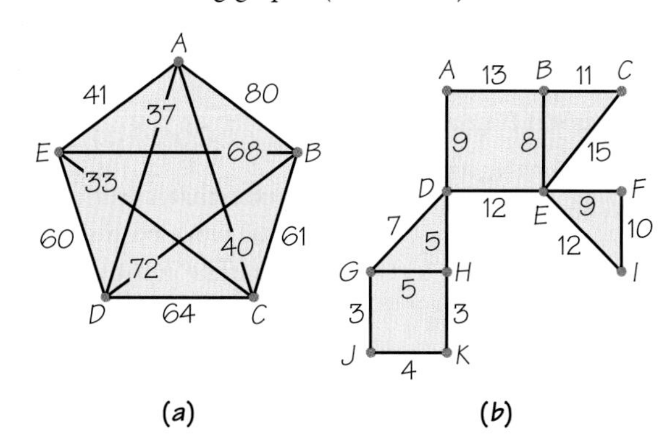

(a) (b)

Planning and Scheduling

...and at www.whfreeman.com/fapp:

Flashcards
Quizzes
Spreadsheet Projects
Bin-packing algorithms
Applet Exercises
Graph coloring
Scheduling

In a society as complex as ours, everyday problems such as providing services efficiently and on time require accurate planning of both people and machines. Take the example of a hospital in a major city. Around-the-clock scheduling of nurses and doctors must be provided to guarantee that people with particular expertise are available during each shift. The operating rooms must be scheduled in a manner flexible enough to deal with emergencies. Equipment used for X-ray, CAT, or MRI scans must be scheduled for maximum efficiency.

Although many scheduling problems are often solved on an ad hoc basis, we can also use mathematical ideas to gain insight into the complications that arise in scheduling. The ideas we develop in this chapter have practical value in a relatively narrow range of applications, but they throw light on many characteristics of more realistic, and hence more complex, scheduling problems.

3.1 Scheduling Tasks

Assume that a certain number of identical **processors** (machines, humans, or robots) work on a series of tasks that make up a job. Associated with each task is a specified amount of time required to complete the task. For simplicity, we assume that any of the processors can work on any of the tasks. Our problem, known as the **machine-scheduling problem,** is to decide how the tasks should be scheduled so that the completion time for the tasks collectively is as early as possible.

Even with these simplifying assumptions, complications in scheduling will arise. Some tasks may be more important than others and perhaps should be scheduled first. When "ties" occur, they must be resolved by special rules. As an example, suppose we are scheduling patients to be seen in a hospital emergency room staffed by one doctor. If two patients arrive simultaneously, one with a bleeding foot, the other with a bleeding arm, which patient should be processed first? Suppose the doctor treats the arm patient first, and, while treatment is going on, a person in cardiac arrest arrives. Scheduling rules must establish appropriate priorities for cases such as these.

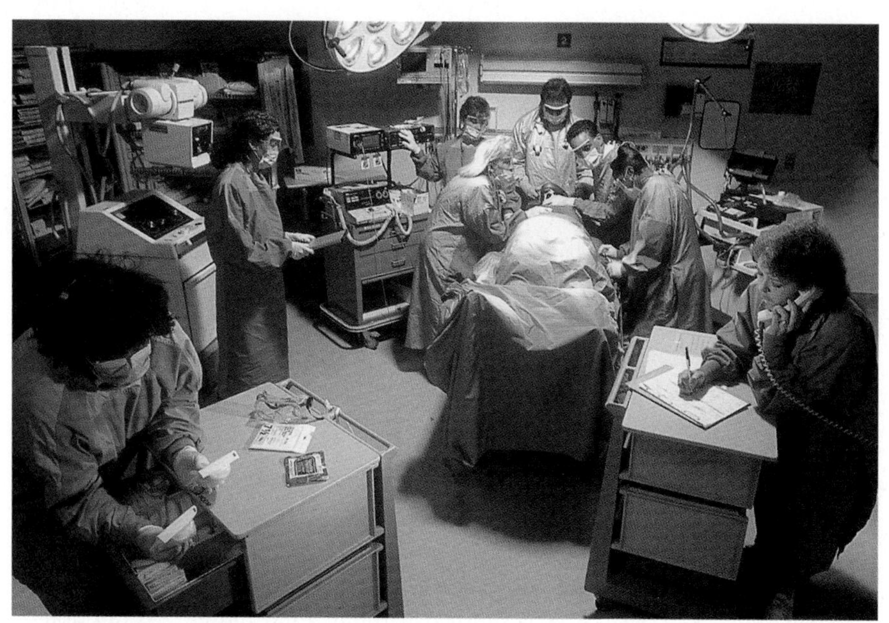

(Charles Gupton/Stock Boston.)

Another common complication arises with jobs consisting of several tasks that cannot be done in an arbitrary order. For example, if the job of putting up a new house is treated as a scheduling problem, the task of laying the foundation must precede the task of putting up the walls, which in turn must be completed before work on the roof can begin.

Assumptions and Goals

To simplify our analysis, we need to make clear and explicit assumptions:

1. If a processor starts work on a task, the work on that task will continue without interruption until the task is completed.
2. No processor stays voluntarily idle. In other words, if there is a processor free and a task available to be worked on, then that processor will immediately begin work on that task.
3. The requirements for ordering the tasks are given by an order-requirement digraph. (A typical example is shown in Figure 3.1, with task times highlighted within each vertex. The ordering of the tasks imposed by the order-requirement digraph represents constraints of physical reality. For example, you cannot fly a plane until it has taken fuel on board.)
4. The tasks are arranged in a priority list that is independent of the order requirements. (The priority list is a ranking of the tasks according to some criterion of "importance.")

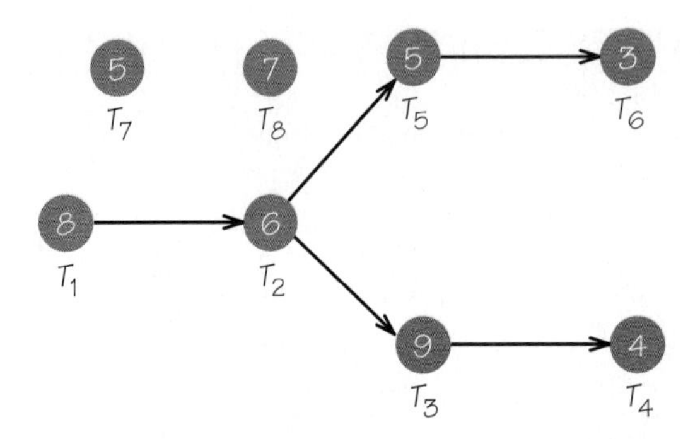

Figure 3.1 A typical order-requirement digraph.

EXAMPLE

Home Construction

Let's see how these assumptions might work for an example involving a home construction project. In this case, the processors are human workers with identical skills. Assumption 1 means that once a worker begins a task, the work on this task is finished without interruption. Assumption 2 means that no worker will stay idle if there is some task for which the predecessors are finished. Assumption 3 requires that the ordering of the tasks is summarized in an order-requirement digraph. This digraph would code facts such as the site must be cleared before the foundation is begun. Assumption 4 requires that the tasks be ranked in a list from some perspective, perhaps a subjective view. The task with highest priority rank is listed first in the list, followed left to right by the other tasks in priority rank. The priority list might be based on the size of the payments made to the construction company when a task is completed, even though these payments have no relation to the way the tasks must be done, as indicated in the order-requirement digraph. Alternatively, the priority list might reflect an attempt to help an algorithm to schedule the tasks needed to complete the whole job more quickly. ■

When considering a scheduling problem, there are various goals we might want to achieve. Among these are:

Goal 1. Minimizing the completion time of the job

Goal 2. Minimizing the total time that processors are idle

Goal 3. Finding the minimum number of processors necessary to finish the job by a specified time

In the context of the construction example, goal 1 would complete the home as quickly as possible. Goal 2 would ensure that workers, who are perhaps paid by the hour, were not paid for doing nothing. One way of accomplishing this would be to hire one fewer worker even if it means the house takes longer to finish. Goal 3 might be reasonable if the family wants the house done before the first day of school, even if they have to pay a lot more workers to get the house done by this time.

For now we will concentrate on goal 1, finishing all the tasks at the earliest possible time. Note, however, that optimizing with respect to one goal may not optimize with respect to another. Our discussion here goes beyond what was discussed in Chapter 2 (see Critical-Path Analysis) by dealing with how to assign tasks in a job to the processors that do the work.

List-Processing Algorithm

The scheduling problem we have described sounds more complicated than the traveling salesman problem (TSP). Indeed, like the TSP, it is known to be NP-complete. This means that it is unlikely that anyone will ever find a computationally fast algorithm that can find an optimal solution. Thus, we will be content to seek a solution method that is computationally fast and gives only approximately optimal answers.

The algorithm we use to schedule tasks is the **list-processing algorithm.** In describing it, we will call a task **ready** at a particular time if all its predecessors as indicated in the order-requirement digraph have been completed at that time. In Figure 3.1 at time 0, the ready tasks are T_1, T_7, and T_8, while T_2 cannot be ready until 8 time units after T_1 is started. The algorithm works as follows: At a given time, assign to the lowest-numbered free processor the first task on the priority list that is *ready* at that time and that hasn't already been assigned to a processor.

In applying this algorithm, we will need to develop skill at coordinating the use of the information in the order-requirement digraph and the priority list. It will be helpful to cross out the tasks in the priority list as they are assigned to a processor to keep track of which tasks remain to be scheduled. Let's apply this algorithm to one possible priority list—T_8, T_7, T_6, . . . , T_1—using two processors and the order-requirement digraph in Figure 3.1. The result is the schedule shown in Figure 3.2, where idle processor time is indicated by white. How does the list-processing algorithm generate this schedule?

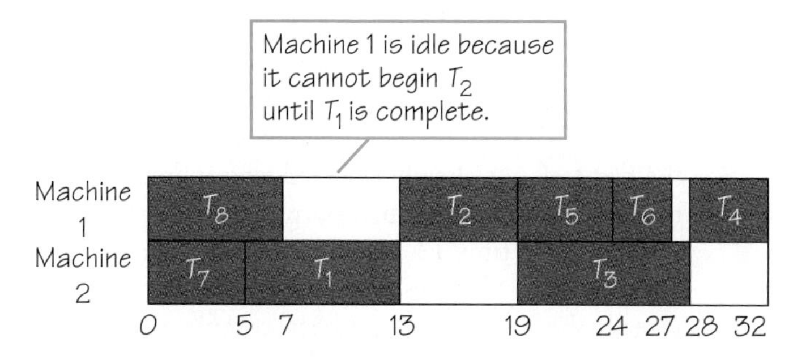

Figure 3.2 The schedule produced by applying the list-processing algorithm to the order-requirement digraph in Figure 3.1 using the list T_8, T_7, . . . , T_1.

Because T_8 (task 8) is first on the priority list and ready at time 0, it is assigned to the lowest-numbered free processor, processor 1. Task 7, next on the priority list, is also ready at time 0 and thus is assigned to processor 2. The first processor to become free is processor 2 at time 5. Recall that by assumption 1, once a processor starts work on a task, its work cannot be interrupted until the task is complete. Task 6, the next unassigned task on the list, is not ready at time 5, as can be seen by consulting Figure 3.1. The reason task 6 is not ready at time 5 is that task 5 has not been completed by time 5. In fact, at time 5, the only ready task on the list is T_1, so that task is assigned to processor 2. At time 7, processor 1 becomes free, but no task becomes ready until time 13. Thus, processor 1 stays idle from time 7 to time 13. At this time, because T_2 is the first ready task on the list not already scheduled, it is assigned to processor 1. Processor 2, however, stays idle because no other ready task is available at this time. The remainder of the scheduling shown in Figure 3.2 is completed in this manner.

As the priority list is scanned from left to right to assign a processor at a particular time, we pass over tasks that are not ready to find ones that are ready. If no task can be assigned in this manner, we keep one or more processors idle until such time that, reading the priority list from the left, there is a ready task not already assigned. After a task is assigned to a processor, we resume scanning the priority list, starting over at the far left, for unassigned tasks.

When Is a Schedule Optimal?

The schedule in Figure 3.2 has a lot of idle time, so it may not be optimal. Indeed, if we apply the list-processing algorithm for two processors to another possible priority list $T_1, \ldots, T_8$, using the digraph in Figure 3.1, the resulting schedule is that shown in Figure 3.3.

Here are the details of how this schedule was arrived at. Remember that we must coordinate the list $T_1, T_2, \ldots, T_8$ with the information in the order-requirement digraph shown in Figure 3.3a. At time 0, task T_1 is ready, so this task is assigned to processor 1. However, at time 0, tasks $T_2, T_3, \ldots, T_6$ are not ready because their predecessors are not done. For example, T_2 is not ready at time 0 because T_1, which precedes it, is not done at time 0. The first ready task on the list, reading from left to right, that is not already assigned is T_7, so task T_7 gets assigned to processor 2. Both processors are now busy until time 5, at which point processor 2 becomes idle (Figure 3.3b).

Tasks T_1 amd T_7 have been assigned; reading from left to right along the list, the first task not already assigned whose predecessors are done by time 5 is T_8, so this task is started at time 5 on processor 2; processor 2 will continue to work on this task until time 12, because the task time for this task is 7 time units. At time 8, processor 1 becomes free, and reading the list from left to right we find that T_2 is ready (because T_1 has just been completed). Thus, T_2 is assigned to processor 1,

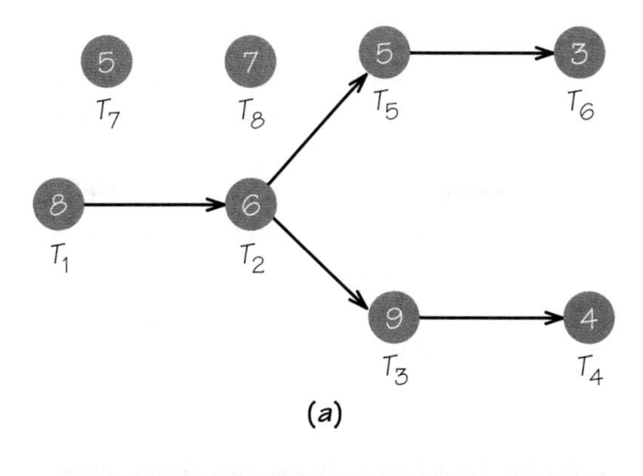

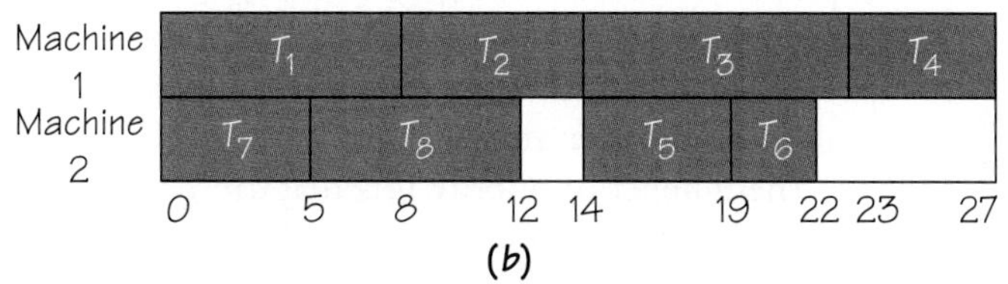

Figure 3.3 (a) A typical order-requirement digraph (repeat of Figure 3.1). (b) The schedule produced by applying the list-processing algorithm to the order-requirement digraph in Figure 3.3a using the list $T_1, T_2, \ldots, T_8$.

which will stay busy on this task until time 14. At time 12, processor 2 becomes free, but the tasks that have not already been assigned from the list, T_3, T_4, T_5, T_6 are not ready, because they depend on T_2 being completed before these tasks can start. Thus, processor 2 must stay idle until time 14. At this time, T_3 and T_5 become ready. Since both processors 1 and 2 are idle at time 14, the lower numbered of the two, processor 1, gets to start on T_3 because it is the first ready task left to be assigned on the list scanned from left to right. Task T_5 gets assigned to processor 2 at time 14. The remaining tasks are assigned in a similar manner.

The schedule shown in Figure 3.3b is optimal because the path T_1, T_2, T_3, T_4, with length 27, is the critical path in the order-requirement digraph. As we saw in Chapter 2, the earliest completion time for the job made up of all the tasks is the length of the longest path in the order-requirement digraph.

There is another way of relating optimal completion time to the completion time that is yielded by the list-processing algorithm. Suppose that we add all the task times given in the order-requirement digraph and divide by the number of processors. The completion time using the list-processing algorithm must be at least as large as this number. For example, the task times for the order-requirement digraph in Figure 3.3a sum to 47. Thus, if these tasks are scheduled on two processors, the completion time is at least $47/2 = 23.5$ (in fact, 24, because the list-processing algorithm applied to integer task times must yield an integer solution), while for three processors the completion time is at least 47/3 (in fact, 16).

Why is it helpful to take the total time to do all the tasks in a job and divide this number by the number of processors? Think of each task that must be scheduled as a rectangle that is 1 unit high and t units wide, where t is the time allotted for the task. Think of the scheduling diagram with m processors as a rectangle that is m units high and whose width, W, is the completion time for the tasks. The scheduling diagram is to be filled up by the rectangles that represent the tasks. How small can W be? The area of the rectangle that represents the scheduling diagram must be at least as large as the sum of all the rectangles representing tasks that are "packed" into it. The area of the scheduling diagram rectangle is mW. The combined areas of all the tasks, plus the area of rectangles corresponding to idle time, will equal mW. Width W is smallest when the idle time is zero. Thus, W must be at least as big as the sum of all the task times divided by m.

Sometimes the estimate for completion time given by the list-processing algorithm from the length of the critical path gives a more useful value than the approach based on adding task times; sometimes the opposite is true. For the order-requirement digraph in Figure 3.1, except for a schedule involving one processor, the critical-path estimate is superior. For some scheduling problems, both these estimates may be poor.

The number of priority lists that can be constructed if there are n tasks is $n!$ and can be computed using the fundamental principle of counting. For example, for eight tasks, $T_1, \ldots, T_8$, there are $8 \times 7 \times 6 \ldots \times 1 = 40{,}320$ possible priority lists. For different choices of the priority list, the list-processing algorithm may schedule the tasks, subject to the constraints of the order-requirement digraph, in different ways. More specifically, two different lists may yield different completion times or the same completion time, but the order in which the tasks are carried out will be different. It is also possible that two different lists produce identical ordering of the assignments of the tasks to processors and completion time. A little later we will see a method that can be used to select a list that, if we are lucky, will give a schedule with a relatively good completion time. In fact, no method is known, except for very specialized cases, of how to choose a list that can be guaranteed to produce an optimal schedule when the list algorithm is applied to it.

Strange Happenings

The list-processing algorithm involves four factors that affect the final schedule. The answer we get depends on the

1. Times of the tasks
2. Number of processors
3. Order-requirement digraph
4. Ordering of the tasks on the list

To see the interplay of these four factors, consider another scheduling problem, this time associated with the order-requirement digraph shown in Figure 3.4 (the highlighted numbers are task time lengths). The schedule generated by the list-processing algorithm applied to the list $T_1, T_2, \ldots, T_9$, using three processors, is given in Figure 3.5.

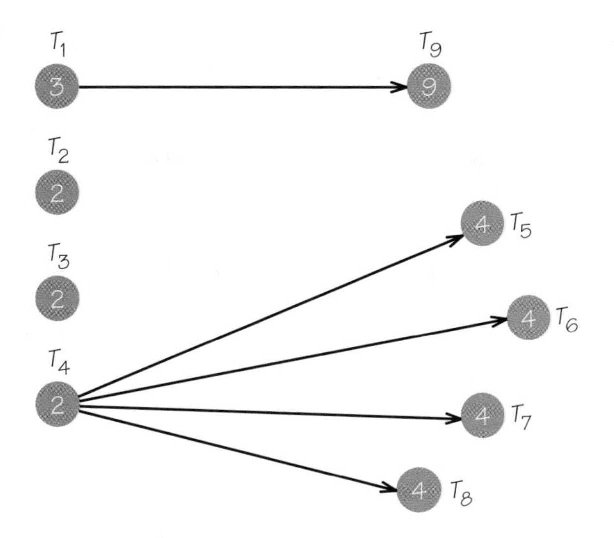

Figure 3.4 An order-requirement digraph designed to help illustrate some paradoxical behavior produced by the list-processing algorithm.

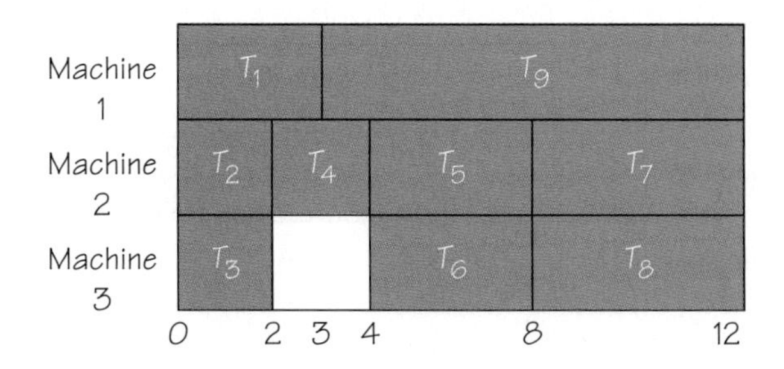

Figure 3.5 The schedule produced by applying the list-processing algorithm to the order-requirement digraph in Figure 3.4 using the list $T_1, T_2, \ldots, T_9$ with three processors.

Treating the list $T_1, \ldots, T_9$ *as fixed,* how might we make the completion time earlier? Our alternatives are to pursue one or more of these strategies:

1. Reduce task times.
2. Use more processors.
3. "Loosen" the constraints of the order-requirement digraph.

Let's consider each alternative in turn, changing one feature of the original problem at a time, and see what happens to the resulting schedule. If we use strategy 1 and reduce the time of each task by one unit, we would expect the completion time to go down. Figure 3.6 shows the new order-requirement digraph, and Figure 3.7 shows the schedule produced for this problem, using the list-processing algorithm with three processors applied to the list $T_1, \ldots, T_9$. The completion time is now 13–longer than the completion time of 12 for the case (Figure 3.5) with longer task times. Here is something unexpected! Let's explore further and see what happens.

Next we consider strategy 2, increasing the number of machines. Surely this should speed matters up. When we apply the list-processing algorithm to the

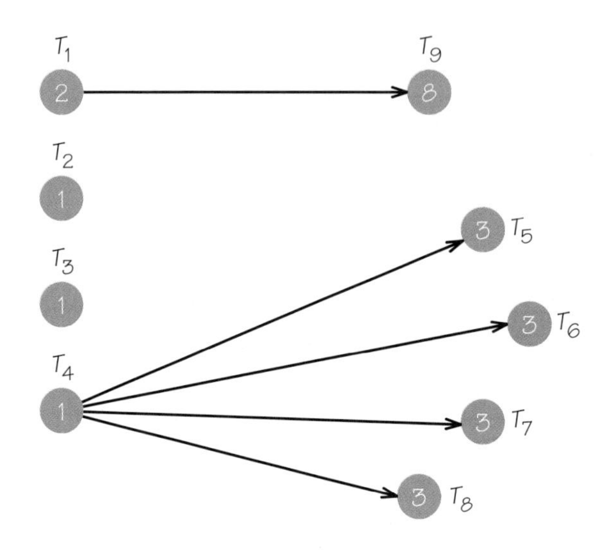

Figure 3.6 The order-requirement digraph obtained from the one in Figure 3.4 by reducing by one unit each of the task times shown there.

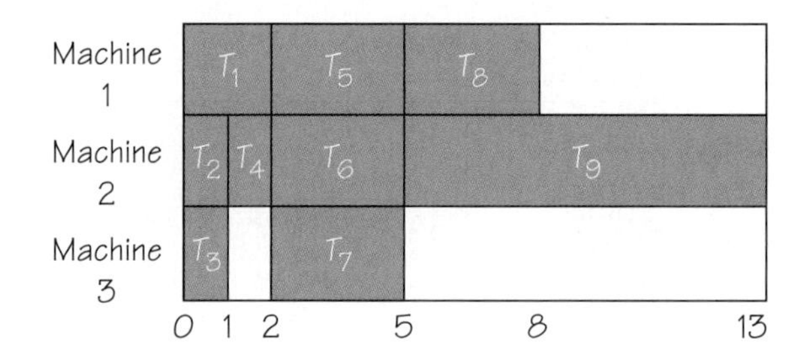

Figure 3.7 The schedule produced by applying the list-processing algorithm to the order-requirement digraph in Figure 3.6 using the list $T_1, T_2, \ldots, T_9$ with three processors.

original graph in Figure 3.4, using the list $T_1, \ldots, T_9$ and four machines, we get the schedule shown in Figure 3.8. The completion time is now 15 – an even later completion time than for the previous alteration!

Finally, we consider strategy 3, trying to shorten completion time by erasing all constraints (edges with arrows) in the order-requirement digraph shown in Figure 3.4. By increasing flexibility of the ordering of the tasks, we might guess we could finish our tasks more quickly. Figure 3.9 shows the schedule using the list

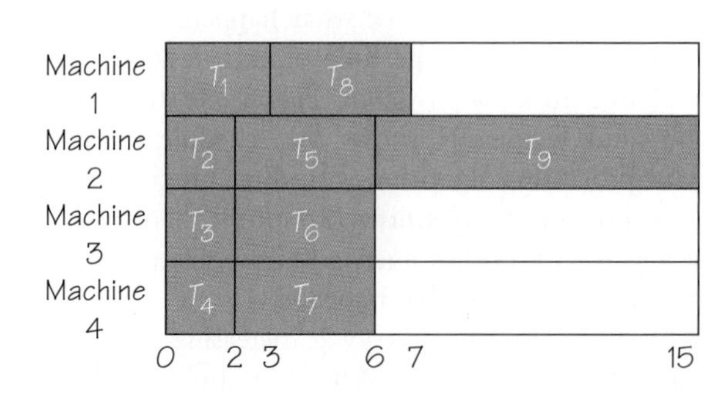

Figure 3.8 The schedule produced by applying the list-processing algorithm to the order-requirement digraph in Figure 3.4 using the list $T_1, T_2, \ldots, T_9$ with four processors.

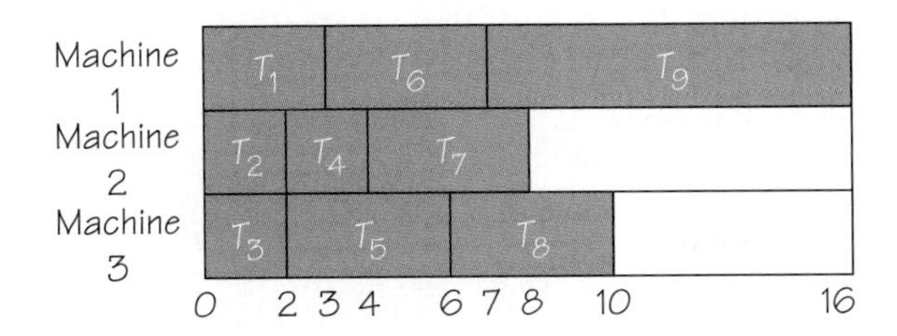

Figure 3.9 The schedule produced by applying the list-processing algorithm to the order-requirement digraph in Figure 3.4, modified by erasing all its directed edges, using the list $T_1, T_2, \ldots, T_9$ with three processors.

$T_1, \ldots T_9$; now it takes 16 units! This is the worst of our three strategies to reduce completion time.

The failures we have seen here are surprising at first glance, but they are typical of what can happen when a situation is too complex to analyze with naïve intuition. The value of using mathematics rather than intuition or trial and error to study scheduling and other problems is that it points up flaws that can occur in unguarded intuitive reasoning.

The unexpected results we see here are due to the rules we set up for generating schedules. Such paradoxical behavior for the list-processing algorithm will not occur for every example you try. In fact, one has to be quite clever to design such examples. The list-processing algorithm has many nice features, including the fact that it is easy to understand and fast to implement. However, the results of the model in some cases can appear strange. Because we have been explicit about our assumptions, we could go back and make changes in these assumptions in hopes of eliminating the strange behavior. But the price we might pay is more time spent constructing schedules and perhaps even discovering new types of strange behavior. Unfortunately for modern society, with its increasing concern with economical and efficient scheduling (see Spotlight 3.1), recent mathematical research suggests that scheduling is an intrinsically hard problem.

3.2 Critical-Path Schedules

In our discussion so far, we have acted as though the priority list used in applying the list-processing algorithm was given to us in advance based on external considerations. Let's now consider the question of whether there is a systematic method of *choosing* a priority list that yields optimal or nearly optimal schedules. We will show how to construct a specific priority list based on this principle, to which the list-processing algorithm can then be applied.

Recall from our discussion of critical-path analysis in Chapter 2 that no matter how a schedule is constructed, the finish time cannot be earlier than the length of the longest path in the order-requirement digraph. This suggests that we should try to schedule first those tasks that occur early in long paths, because they might be a bottleneck for the other tasks.

SPOTLIGHT 3.1 Management Science and Disaster Recovery

The city of New York depends on the public transportation system of subways and roads to bring hundreds of thousands of people who live in the four outer boroughs (i.e., Queens, Brooklyn, the Bronx, and Staten Island) into Manhattan to work and "play." New York City also has a communication system of telephones, radio and television stations, and computer networks. These systems speed information between New York's citizens and people outside the city and around the world. The area in southern Manhattan, in the vicinity of the World Trade Center (WTC), was a center for banking, insurance, financial markets, and domestic and international commerce. The attack on the World Trade Center on September 11, 2001, disrupted these networks and markets but did not destroy them, partly because the principles of operations research and management science were used in the design and development of these systems over a long period of time.

The diagram below shows a very simple subway (train) system between an eastern and western terminus.

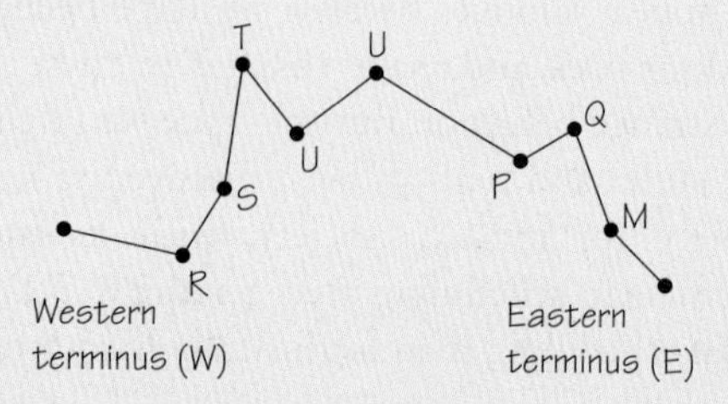

There are two tracks, each dedicated for use by westbound or eastbound trains to run between the two termini. The only place where trains can be turned around is at these termini. Simple graph theory tells us that in such a system, if a vertex is "destroyed" or out of service, or an edge is "destroyed" or out of service, the system totally breaks down. However, the simple provision that trains can be turned around at *U*, even though this is usually only one stop on the way from *W* to *E*, gives much greater flexibility to the system if there is a water main break, or a gas leak, etc. Thanks to simple

(AP/World Wide Photo.)

principles of this kind and creating routes that use independent lines with many transfer points, New Yorkers were able to use the subway system in a flexible way after the World Trade Center disaster. In the days right after the WTC collapsed, trains were not allowed past the geographic area near the WTC for fear that the tunnels' structural foundation had been weakened and that subway vibrations could collapse damaged buildings. After it was ascertained that running the subways was safe both for partially damaged buildings and for the subways themselves, routes were altered several times to give rescue workers and people returning to their daily routines maximum support. One line's tunnels did collapse, and several stations had to be closed for extended periods, but due to the redundancy and flexibility of the design of the system, a remarkable amount of service was quickly restored.

Unfortunately, with one major exception, the antennae that supported New York's broadcast television system were located atop the WTC. The exception is WCBS, Channel 2, which has its antenna atop the Empire State Building. This meant that weeks after the WTC was destroyed, New Yorkers who did not have cable television were able to watch only one major channel and a few minor ones. The cable television system was largely unaffected.

Good planning and wise application of the principles of management science make it possible to minimize the effects of natural and manmade disasters.

EXAMPLE

Scheduling Two Processors

To illustrate this method, consider the order-requirement digraph in Figure 3.10a. Suppose we wish to schedule these tasks on two processors. Initially, there are two critical paths of length 64: T_1, T_2, T_3 and T_1, T_4, T_3. Thus, we place T_1 first on the priority list. With T_1 "gone," there is a new critical path of length 60 (i.e., T_5, T_6, T_4, T_3) that starts with T_5, so T_5 is placed second on the priority list. At this stage, with T_1 and T_5 removed, we have the residual order-requirement digraph shown in Figure 3.10b. In this diagram are paths of length 50 (T_2, T_3), 56 (T_6, T_4, T_3), 36 (T_6, T_4, T_7), and 24 (T_8, T_4, T_{10}). Because T_6 heads the path that is currently longest in length, it gets placed third in the priority list. Once T_6 is removed from Figure 3.10b, there is a tie for which is the longest path remaining, because both T_2, T_3 and T_4, T_3 are paths of length 50.

When there is a tie between two longest paths, we place next on the priority list in the lowest-numbered task heading a longest path. In the example shown here, this means that T_2 is placed next into the priority list, to be followed by T_4. Continuing in this fashion, we obtain the priority list T_1, T_5, T_6, T_2, T_4, T_3, T_8, T_9, T_7, T_{10}. Note that the order of T_7 and T_{10} was decided using the rule for breaking ties. The list-processing algorithm is now applied using this priority list and the order-requirement digraph in Figure 3.10a. We obtain the schedule in Figure 3.11. ■

This example illustrates what is called **critical-path scheduling.** The algorithm for critical-path scheduling can be described for use with the order-requirement digraph defining any particular scheduling problem. The algorithm applies the list-processing algorithm using the priority list L obtained as follows:

1. Find a task that heads a critical (longest) path in the order-requirement digraph. If there is a tie, choose the task with the lower number.
2. Place the task found in step 1 next on the list L. (The first time through the process this task will head the list.)
3. Remove the task found in step 1 and the edges attached to it from the current order-requirement digraph, obtaining a new (modified) order-requirement digraph.
4. If there are no vertices left in the new order-requirement digraph, the procedure is complete; if there are vertices left, go to step 1.

This procedure will terminate when all the tasks in the original order-requirement digraph have been placed on the list L.

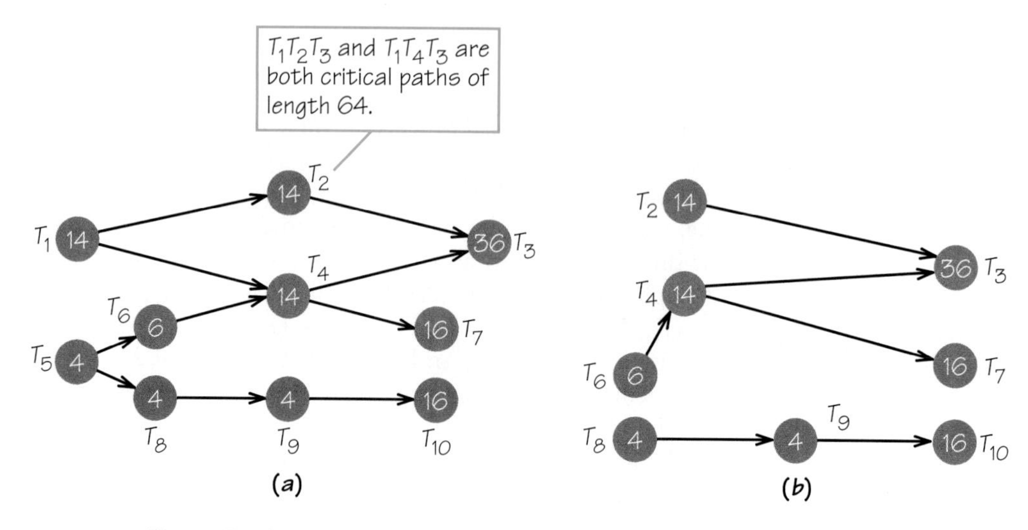

Figure 3.10 (a) An order-requirement digraph used to illustrate the critical-path scheduling method. (b) Residual order-requirement digraph after tasks T_1 and T_5 have been removed.

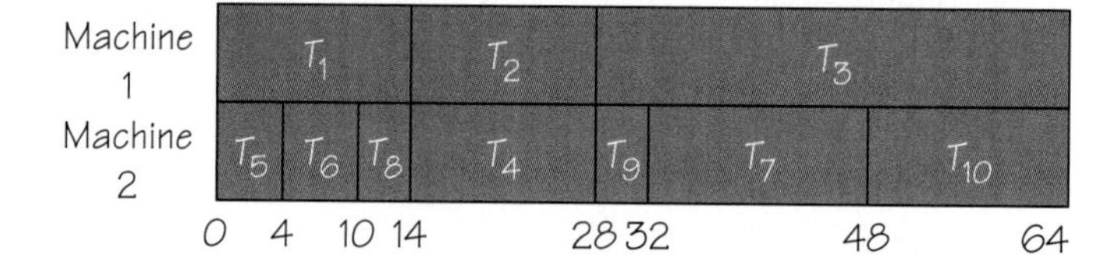

Figure 3.11 The optimal schedule produced by applying the critical-path scheduling method to the order-requirement digraph in Figure 3.10. The list used was T_1, T_5, T_6, T_2, T_4, T_3, T_8, T_9, T_7, T_{10}.

The preceding example shows that critical-path scheduling can sometimes yield optimal solutions. Unfortunately, this algorithm does not always perform well. For example, the critical-path method employing four processors applied to the order-requirement digraph shown in Figure 3.12 yields the list T_1, T_8, T_9, T_{10}, T_{11}, T_5, T_6, T_7, T_{12}, T_2, T_3, T_4 and then the schedule in Figure 3.13. (Note that T_5, T_6, T_7 are thought of as heading paths of length 10.) In fact, there can be no worse schedule than this one. An optimal schedule is shown in Figure 3.14.

Many of the results we have examined so far are negative because we are dealing with a general class of problems that defy our using computationally efficient algorithms to find an optimal schedule. But we can close on a more positive note. Consider an arbitrary order-requirement digraph, but assume all the tasks take equal time. It turns out that we can always construct an optimal schedule using two processors in this situation. Ironically, we can choose among many algorithms to produce these optimal schedules. The algorithms are easy to understand (though not easy to prove optimal) and have all been discovered since 1969! Many people think that mathematics is a subject that is no longer

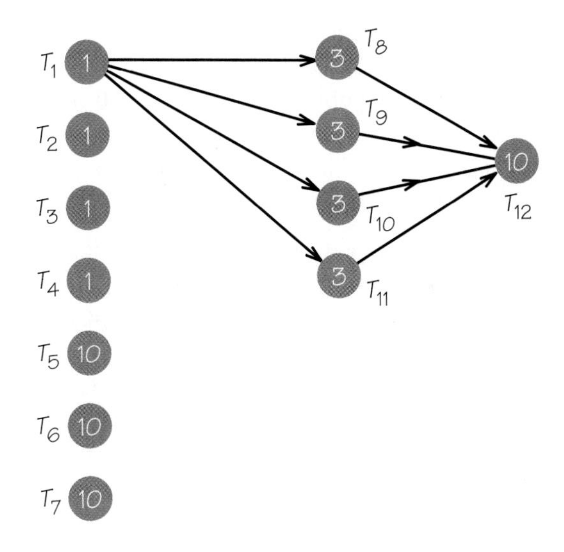

Figure 3.12 An order-requirement digraph used to illustrate how poorly the critical-path scheduling method can sometimes behave.

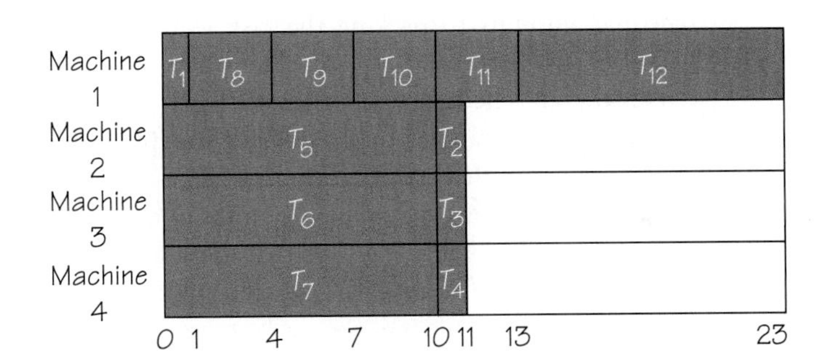

Figure 3.13 The schedule produced by applying the critical-path scheduling method to the order-requirement digraph in Figure 3.12 using four processors. The list used was T_1, T_8, T_9, T_{10}, T_{11}, T_5, T_6, T_7, T_{12}, T_2, T_3, T_4.

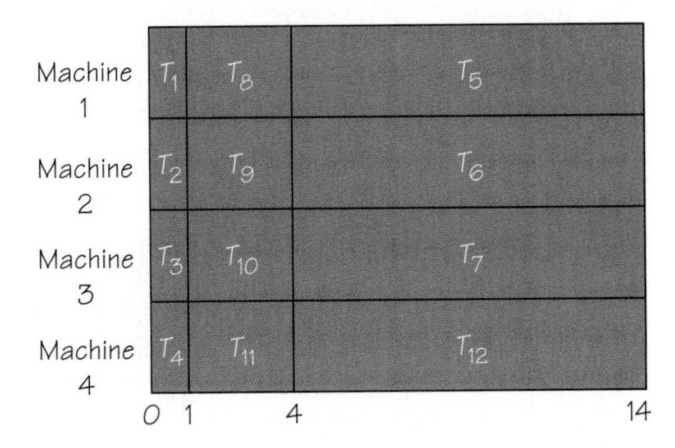

Figure 3.14 An optimal schedule for the order-requirement digraph in Figure 3.12 using four processors.

alive, and that all its ideas and methods were discovered hundreds of years ago—but as we have just seen, this is not true. In fact, more new mathematics has been discovered and published in the last 30 years than during any previous 30-year period.

3.3 Independent Tasks

Mathematicians suspect that no computationally efficient algorithm for solving general scheduling problems optimally will ever be found. Owing to our limited success in designing algorithms for finding optimal schedules for general order-requirement digraphs, we will consider a special class of scheduling problems for which the order-requirement digraph has no edges. In this case, we say that the tasks are *independent* of one another, because they can be performed in any order. (No edges in the order-requirement digraph indicates that no tasks need to precede others; that is, the tasks can be done in any order.) In this section we consider the problem of scheduling **independent tasks.**

Geometrically, we can think of the independent tasks as rectangles of height 1 whose lengths are equal to the time length of the task. Finding an optimal schedule amounts to packing the task rectangles into a longer rectangle whose height equals the number of machines. For example, Figure 3.15 shows two different ways to schedule tasks of length 10, 4, 5, 9, 7, 7 on two machines. (For convenience, the rectangles in the case of independent tasks are labeled with their task times rather than their task numbers.) Scheduling basically means efficiently packing the task rectangles into the machine rectangle. Finding the optimal answer among all possible ways to pack these rectangles is like looking for a needle in a haystack. The list-processing algorithm produces a packing, but it may not be a good one.

There are two approaches we can consider. To study **average-case analysis,** we might ask if the average (mean) of the completion times arrived at by using the list-processing algorithm with all the possible different lists is close to the

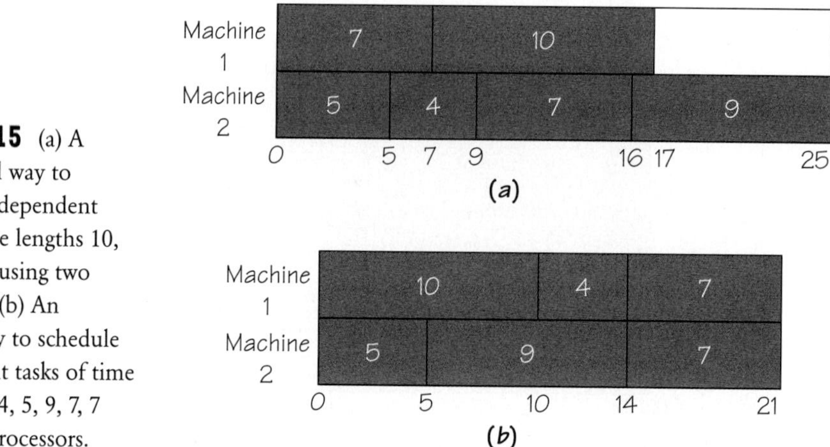

Figure 3.15 (a) A nonoptimal way to schedule independent tasks of time lengths 10, 4, 5, 9, 7, 7 using two processors. (b) An optimal way to schedule independent tasks of time lengths 10, 4, 5, 9, 7, 7 using two processors.

optimal possible completion time. To study **worst-case analysis,** we might ask *How far from optimal is a schedule obtained using the list-processing algorithm with one particular priority list?* What is being contrasted with these two points of view is that an algorithm may work well most of the time (i.e., give an answer close to optimal) even though there may be a few cases in which it performs very badly. Average-case analysis is amenable to mathematical solution but requires methods of great sophistication. For independent tasks, the worst-case analysis can be answered using a surprisingly simple argument developed by Ronald Graham, formerly of AT&T Bell Research. The idea is that if the tasks are independent, no processor can be idle at a given time and then busy on a task at a later time.

What Graham's worst-case analysis showed for independent tasks is that no matter which list L we use, if the optimal schedule requires time T, then the completion time for the schedule produced by the list-processing algorithm applied to list L with m processors is less than or equal to $(2 - 1/m)T$. For example, for two machines ($M = 2$), if an optimal schedule yields completion at time 30, then no list would ever yield a completion later than

$$\left(2 - \frac{1}{2}\right)\left(30\right) = 45$$

Although it is of great theoretical interest, Graham's result does not provide much comfort to those who are trying to find good schedules for independent tasks.

Decreasing-Time Lists

Is there some way of choosing a priority list that consistently yields relatively good schedules? The surprising answer is yes! The idea is that when long tasks appear toward the end of the list, they often seem to "stick out" on the right end, as in Figure 3.15a.

This suggests that before one tries to schedule a collection of tasks, the tasks should be placed in a list where the that longest tasks are listed first. The list-processing algorithm applied to a list arranged in this fashion is called the **decreasing-time-list algorithm.**

If we apply it to the set of tasks listed previously (10, 4, 5, 9, 7, 7), we obtain the times 10, 9, 7, 7, 5, 4 and the schedule (packing) shown in Figure 3.16. This packing is again optimal, but it is different from the optimal scheduling in Figure 3.15b. It is worth noting that the decreasing-time list and the list obtained by the critical-path method discussed earlier will coincide in the case of independent tasks. The decreasing-time list can also be constructed for the case in which the tasks are not independent; but for general order-requirement digraphs, the decreasing-time list does not produce particularly good schedules.

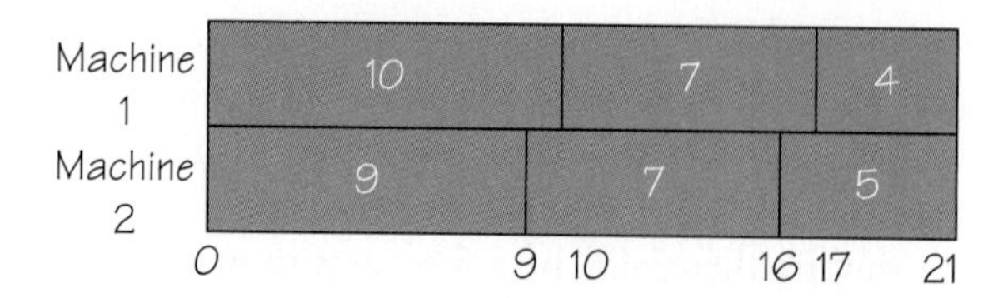

Figure 3.16 The optimal schedule resulting from applying the decreasing-time-list algorithm to a collection of independent tasks. The list used, written in terms of task times only, is 10, 9, 7, 7, 5, 4.

It is important to remember that the decreasing-time-list algorithm does not *guarantee* optimal solutions. This can be seen by scheduling the tasks with times 11, 10, 9, 6, 4 (Figure 3.17). The schedule has a completion time of 21. However, the rearranged list 9, 4, 6, 11, 10 yields the schedule in Figure 3.18, which finishes at time 20. This solution is obviously optimal because the machines finish at the same time and there is no idle time. Note that when tasks are independent, if there are m machines available, the completion time cannot be less than the sum of the task times divided by m.

The problems we have encountered in scheduling independent tasks seem to have taken us a bit far from our goal of applying the mathematics we have developed. Sometimes mathematicians will pursue their mathematical ideas even though they have reached a point where there appear to be no applications. Fortunately, it is very common to be able to find applications for the "abstract" extensions—as is the case here.

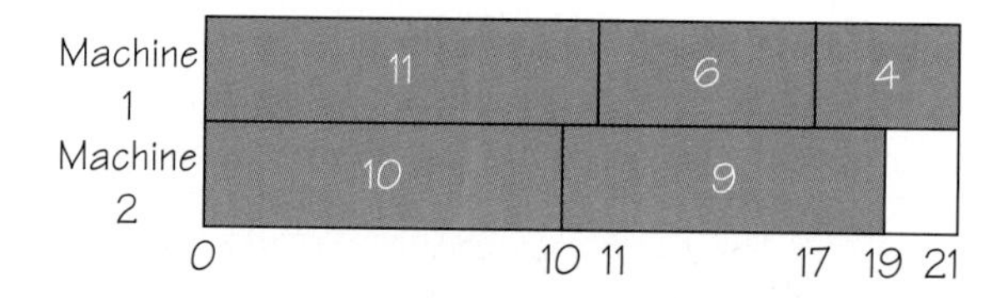

Figure 3.17 The nonoptimal schedule resulting from applying the decreasing-time-list algorithm to a collection of independent tasks. The list used, written in terms of task times only, is 11, 10, 9, 6, 4.

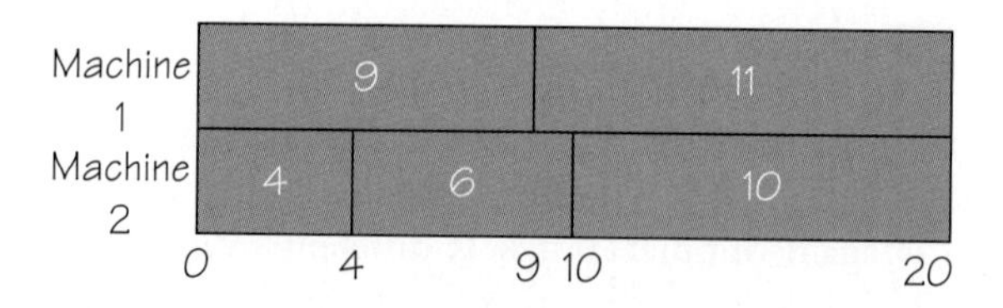

Figure 3.18 The optimal schedule resulting from applying the list-processing algorithm to a collection of independent tasks. The list used, written in terms of task times only, is 9, 4, 6, 11, 10.

EXAMPLE Photocopy Shop and Data Entry Problems

Imagine a photocopy shop with three photocopiers. Photocopying tasks that must be completed overnight are accepted until 5 P.M. The tasks are to be done in any manner that minimizes the finish time for all the work. Because this problem involves scheduling machines for independent tasks, the decreasing-time-list algorithm would be a good heuristic to apply.

For another example, consider a data entry pool at a large corporation or college, where individual entry tasks can be assigned to any data entry specialist. In this setting, however, the assumption that the data entry workers are identical in skill is less likely to be true. Hence, the tasks might have different times with different processors. This phenomenon, which occurs in real-world scheduling problems, violates one of the assumptions of our mathematical model.

Graham's result for the list-processing algorithm – that the finishing time is never more than

$$\left(2 - \frac{1}{m}\right)T$$

(where T represents optimal completion time and m the number of processors) – offers us the small comfort of knowing that even the worst choice of priority list will not yield a completion time worse than twice the optimal time. Compared with the list-processing algorithm, the decreasing-time-list algorithm seems to improve completion times. Thus, it is not surprising that an improved bound or

(Alexander Farnsworth/The Image Works.)

time estimate can be given for this case: The decreasing-list algorithm gives a completion time of no more than

$$\left[\frac{4}{3} - \frac{1}{(3m)}\right]T$$

where m is the number of processors and T is the optimal time in which the tasks can be completed. In particular, when the number of processors is 2, the schedule produced by the decreasing-time-list algorithm is never off by more than 17%! Usually, the error is much less. This result is a remarkable instance of the value of mathematical research into applied problems. Note that the optimal completion time T depends on m and that Graham's theoretical analysis is necessary precisely because there is no known algorithm to compute T easily. ■

3.4 Bin Packing

Suppose you plan to build a wall system for your books, records, and stereo set. It requires 24 wooden shelves of various lengths: 6, 6, 5, 5, 5, 4, 4, 4, 4, 2, 2, 2, 2, 3, 3, 7, 7, 5, 5, 8, 8, 4, 4, and 5 feet. The lumberyard, however, sells wood only in boards of length 9 feet. If each board costs $8, what is the minimum cost to buy sufficient wood for this wall system?

Because all shelves required for the wall system are shorter than the boards sold at the lumberyard, the largest number of boards needed is 24, the precise number of shelves needed for the wall system. Buying 24 boards would, of course, be a waste of wood and money because several of the shelves you need could be cut from one board. For example, pieces of length 2, 2, 2, and 3 feet can be cut from one 9-foot board.

To be more efficient, we think of the boards as bins of capacity W (9 feet in this case) into which we will pack (without overlap) n weights (in this case, lengths) whose values are $w_1, \ldots, w_n$, where each $w_i \leq W$. We wish to find the minimum number of bins into which the weights can be packed. In this formulation, the problem is known as the **bin-packing problem.** Thus, *bin packing* refers to finding the minimum number of bins of weight capacity W into which weights $w_1, \ldots, w_n$ (each less than or equal to W) can be packed.

At first glance, bin-packing problems may appear unrelated to the machine-scheduling problems we have been studying; however, there is a connection.

Let's suppose we want to schedule independent tasks so that each machine working on the tasks finishes its work by time W. Instead of fixing the number of machines and trying to find the earliest completion time, we must find the minimum number of machines that will guarantee completion by the fixed completion time (W). Despite this similarity between the machine-scheduling problem and the bin-packing problem, the discussion that follows will use the traditional terminology of bin packing.

By now, it should come as no surprise to learn that no one knows a fast algorithm that always picks the optimal (smallest) number of bins (boards). In fact, the bin-packing problem belongs to the class of NP-complete problems (see Spotlight 2.1), which means that most experts think it unlikely that any fast optimal algorithm will ever be found.

Bin-Packing Heuristics

We will think of the items to be packed, in any particular order, as constituting a list. In what follows we will use the list of 24 shelf lengths given for the wall system. We will consider various **heuristic** algorithms, namely, methods that can be carried out quickly but cannot be guaranteed to produce optimal results. Probably the easiest approach is simply to put the weights into the first bin until the next weight won't fit, and then start a new bin. (Once you open a new bin, don't use leftover space in an earlier, partially filled bin.) Continue in the same way until as many bins as necessary are used. The resulting solution is shown in Figure 3.19. This algorithm, called **next fit (NF),** has the advantage of not requiring knowledge of all the weights in advance; only the remaining space in the bin currently being packed must be remembered. The disadvantage of this heuristic is that a bin packed early on may have had room for small items that come later in the list.

Our wish to avoid permanently closing a bin too early suggests a different heuristic – **first fit (FF):** Put the next weight into the first bin already opened that has room for this weight; if no such bin exists, start a new bin. Note that a computer program to carry out first fit would have to keep track of how much room was left in all the previously opened bins. For the 24 wall-system shelves, the first-fit algorithm would generate a solution that uses only 14 bins (see Figure 3.20) instead of the 17 bins generated by the next-fit algorithm.

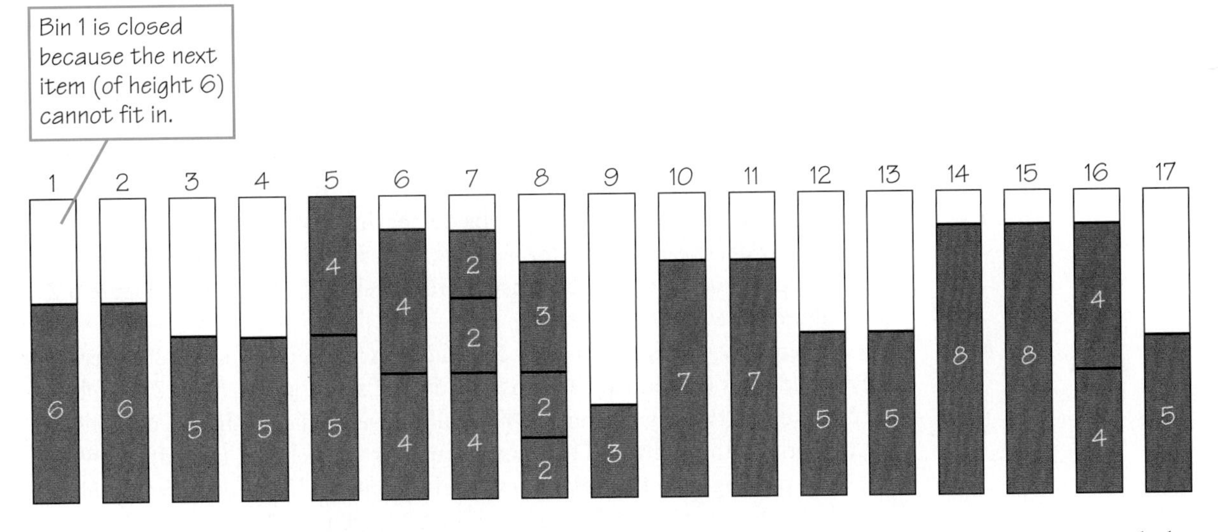

Figure 3.19 The list 6, 6, 5, 5, 5, 4, 4, 4, 4, 2, 2, 2, 2, 3, 3, 7, 7, 5, 5, 8, 8, 4, 4, 5 packed in bins using next fit.

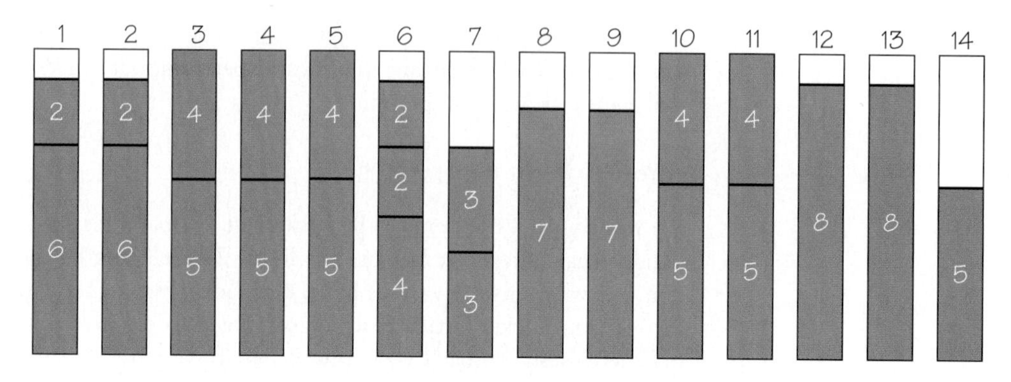

Figure 3.20 The list 6, 6, 5, 5, 5, 4, 4, 4, 4, 2, 2, 2, 2, 3, 3, 7, 7, 5, 5, 8, 8, 4, 4, 5 packed in bins using first fit. Worst fit would yield a packing that would look identical.

If we are keeping track of how much room remains in each unfilled bin, we can put the next item to be packed into the bin that currently has the most room available. This heuristic will be called **worst fit (WF).** The name *worst fit* refers to the fact that an item is packed into a bin with the most room available, that is, into which it fits "worst," rather than into a bin that will leave little room left over after it is placed in that bin (i.e., "best fit"). The solution generated by this approach looks the same as that shown in Figure 3.20. Although this heuristic also leads to 14 bins, the items are packed in a different order. For example, the first item of size 2, the tenth item in the list, is put into bin 6 in worst fit, but into bin 1 in first fit.

Decreasing-Time Heuristics

One difficulty with all three of these heuristics is that large weights that appear late in the list can't be packed efficiently. Therefore, we should first sort the items to be packed in order of decreasing size, assuming that all items are known in advance. We can then pack large items first and then the smaller items into leftover spaces. This approach yields three new heuristics: **next-fit decreasing (NFD), first-fit decreasing (FFD),** and **worst-fit decreasing (WFD).** Here is the original list sorted by decreasing size: 8, 8, 7, 7, 6, 6, 5, 5, 5, 5, 5, 5, 4, 4, 4, 4, 4, 4, 3, 3, 2, 2, 2, 2. Packing using first-fit-decreasing order yields the solution in Figure 3.21. This solution uses only 13 bins.

Is there any packing that uses only 12 bins? No. In Figure 3.21, there are only 2 free units (1 unit each in bins 1 and 2) of space in the first 12 bins, but 4 occupied units (two 2's) in bin 13. We could have predicted this by dividing the total length of the shelves (110) by the capacity of each bin (board): $\frac{110}{9} = 12\frac{2}{9}$. Thus, no packing could squeeze these shelves into 12 bins—there would always be at least 2 units left over for the 13th bin. (In Figure 3.21, there are 4 units in bin 13 because of the 2 wasted empty spaces in bins 1 and 2.) Even if this

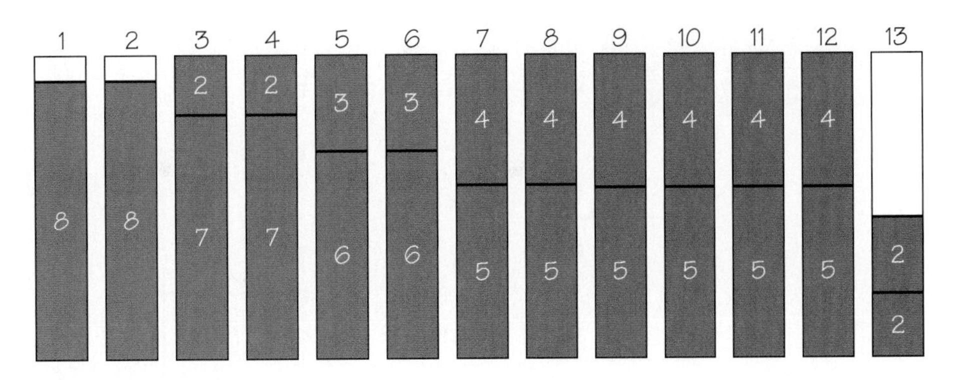

Figure 3.21 The bin packing resulting from applying first-fit decreasing to the wall-system numbers. The list involved, which uses the original list sorted in decreasing order, is 8, 8, 7, 7, 6, 6, 5, 5, 5, 5, 5, 5, 4, 4, 4, 4, 4, 4, 3, 3, 2, 2, 2, 2.

division created a zero remainder, there would still be no guarantee that the items could be packed to fill each bin without wasted space. For example, if the bin capacity is 10 and there are weights of 6, 6, 6, 6, and 6, the total weight is 30; dividing by 10, we get 3 bins as the minimum requirement. Clearly, however, 5 bins are needed to pack the five 6's.

None of the six heuristic methods shown will necessarily find the optimal number of bins for an arbitrary problem. How can we decide which heuristic to use? One approach is to see how far from the optimal solution each method might stray. Various formulas have been discovered to calculate the maximum discrepancy between what a bin-packing algorithm actually produces and the best possible result. For example, in situations where a large number of bins are to be packed, FF can be off by as much as 70%, but FFD is never off by more than 22%. Of course, FFD doesn't give an answer as quickly as FF, because extra time for sorting a large collection of weights may be considerable. Also, FFD requires knowing the whole list of weights in advance, whereas FF does not. It is important to emphasize that 22% margin of error is a worst-case figure. In many cases, FFD will perform much better. Results obtained by computer simulation indicate excellent average-case performance for this algorithm.

When solving real-world problems, we always have to look at the relationship between mathematics and the real world. Thus, first-fit decreasing usually results in fewer bins than next fit, but next fit can be used even when all the weights are not known in advance. Next fit also requires much less computer storage than first fit, because once a bin is packed, it need never be looked at again. Fine-tuning of the conditions of the actual problem often results in better practical solutions and in interesting new mathematics as well. (See Spotlight 3.2 for a discussion of some of the tools mathematicians use to verify and even extend mathematical truths.)

SPOTLIGHT 3.2 Using Mathematical Tools

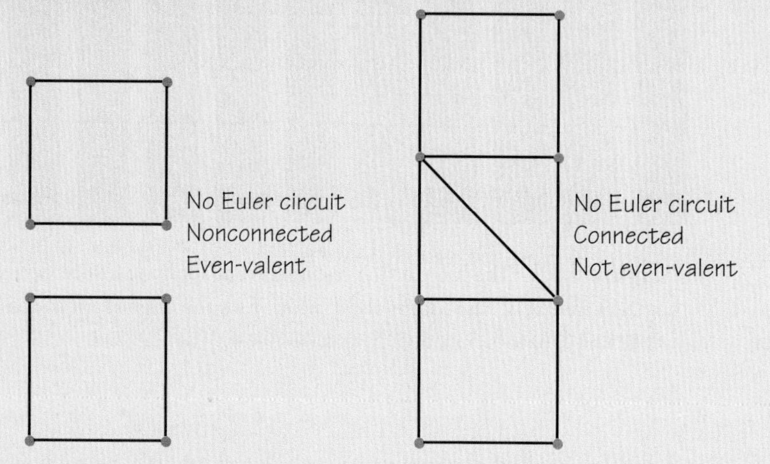

The tools of a carpenter include the saw, T square, level, and hammer. A mathematician also requires tools of the trade. Some of these tools are the proof techniques that enable verification of mathematical truths. Another set of tools consists of strategies to sharpen or extend the mathematical truths already known. For example, suppose that if A and B hold, then C is true. What happens if only A holds? Will C still be true? Similarly, if only B holds, will C still be true?

This type of thinking is of value because such questions will result either in more general cases where C holds or in examples showing that B alone and/or A alone can't imply C. For example, we saw that if a graph G is connected (hypothesis A) and even-valent (hypothesis B), then G has a tour of its edges using each edge only once (conclusion C). If either hypothesis is omitted, the conclusion fails to hold. The figures illustrate this point. On the left is an even-valent but nonconnected graph; on the right, a connected graph with two odd-valent vertices. Neither graph has an Euler circuit.

Here is another way that a mathematician might approach extending mathematical knowledge. If A and B imply C, will A and B imply both C and D, where D extends the conclusion of C? For example, not only can we prove that a connected, even-valent (hypotheses A and B) graph has an Euler circuit, but we can also show that the first edge of the Euler circuit can be chosen arbitrarily (conclusions C and D). It turns out that being able to specify the first two edges of the Euler circuit may not always be possible. Mathematicians are trained to vary the hypotheses and conclusions of results they prove, in an attempt to clarify and sharpen the range of applicability of the results.

We have seen that machine scheduling and bin packing are probably computationally difficult to solve because they are NP-complete. A mathematician could then try to find the simplest version of a bin-packing problem that would still be NP-complete: What if the items to be packed can have only eight weights? What if the weights are only 1 and 2? Asking questions like these is part of the mathematician's craft. Such questions help to extend the domain of mathematics and hence the applications of mathematics.

3.5 Resolving Conflict

In attempting to understand situations that involve scheduling, one might desire to achieve a wide variety of goals. For example, in certain types of scheduling problems, as we have seen here, one is interested in optimization issues. What is the earliest completion time for getting a collection of tasks done on two identical processors? However, in other situations a different goal may arise. For example, in sports, consider a league of baseball teams. Each team has to play some games during the day, some at night, some at home, and some away from home. In the interests of *equity*, it may be desirable for each team to play the same number of day games and night games both at home and away against each of the other teams in the league. If, for example, team A plays 8 games away against team B and 2 games at home against B, then if A wins both home games but loses 7 out of 8 away games, it may appear that B had an advantage due to the way its games against A were scheduled.

Another goal of scheduling, other than optimization and equity, may be to prevent conflicts from occurring. We can use our knowledge of graph theory to solve some interesting scheduling problems where the goal is "conflict resolution." For example, at most colleges, every semester and summer session final examinations must be scheduled. From the point of view of students and faculty both, it would be desirable to schedule these examinations so that (1) no two examinations are scheduled at the same time when a student is enrolled in both of the courses; (2) the examinations are scheduled in as "compact" a way as possible, that is, in as few time slots or days as possible. The administration of the college may share the desire for these two features and want still another property for the scheduling: (3) no more than five examinations are scheduled for any time slot. The reason for a condition such as the last might be that during the summer only five rooms with reliable enough air conditioning are available

Graph theory is used to resolve scheduling conflicts, such as those that occur in a newsroom. (Roger Ressmeyer/Corbis.)

(or there might be only five rooms large enough to hold all the students taking the common final for multiple-section courses).

EXAMPLE

Scheduling Examinations

Small State is offering eight courses during its summer session. The table shows with an X which pairs of courses have two or more students in the same course. Only two air-conditioned lecture halls are available for use at any one time. To design an efficient way to schedule the final examinations, we can represent the information in this table by using a graph, as shown in Figure 3.22a. In the graph, courses are represented by vertices and two courses are joined by an edge if there is any student enrolled in both courses.

	F	M	H	P	E	I	S	C
French (F)		X		X	X	X		X
Mathematics (M)	X				X	X		
History (H)						X	X	X
Philosophy (P)	X							X
English (E)	X	X				X		
Italian (I)	X	X	X		X		X	
Spanish (S)			X			X		
Chemistry (C)	X		X	X				

The graph theory problem we are faced with is the following: Can we assign labels to the vertices of the graph in such a way that vertices that are joined by an edge get different labels? We think of the labels as the time slots the courses are assigned for final examinations. Traditionally, in graph theory such labels are referred to as *colors*. In this language we seek to color the vertices of the graph so that vertices that are joined by an edge get different colors. Such a coloring is called a **vertex coloring.**

Figure 3.22b shows one way to color the vertices of the graph so that each vertex gets a different color. Note that numbers are being used to represent the different colors. This solution is not very valuable, however, because it means that each course be given its own time slot.

In order to minimize the number of time slots used, we assign colors so that no two vertices that are joined by an edge get the same color. Thus, vertices F, M, I, and E must get four different colors. These four colors can then be used to color the remaining vertices, ensuring that no two connected vertices have the same color.

The coloring in Figure 3.22c is a major improvement over the one in Figure 3.22b. It uses only four colors. In fact, this is the smallest number of colors that can be used. To see this, notice that the vertices F, M, I, E in Figure 3.22a are all joined by edges to each other. Thus, in any coloring of this graph they would

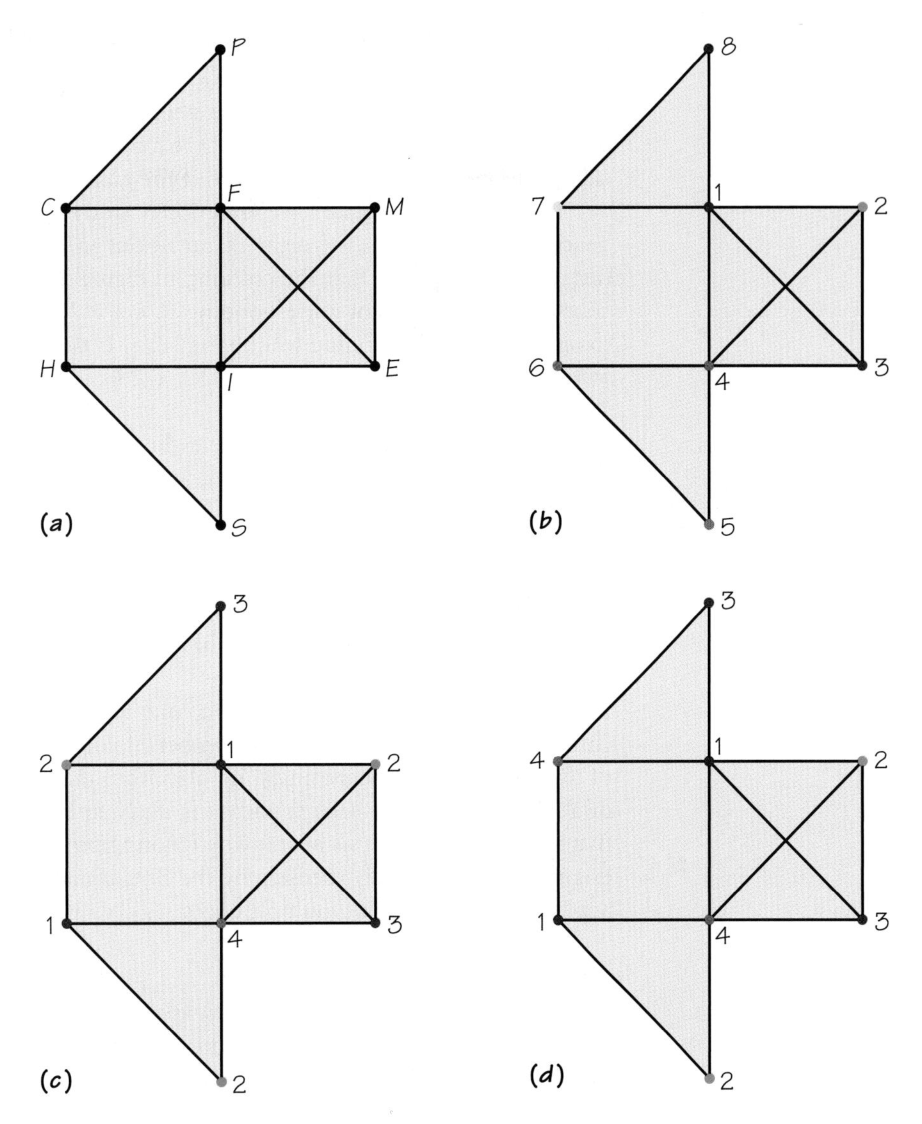

Figure 3.22 (a) A graph used to represent conflict information about courses. When two courses have a common student, an edge is drawn between the vertices that represent these courses. (b) A coloring of the scheduling graph with 8 colors, representing 8 time slots. Using this coloring would lead to a schedule where 8 time slots are used to schedule the examinations. This number is far from optimal. (c) A coloring of the scheduling graph with 4 colors. This translates into a way of scheduling the examinations during 4 time slots, and it is not possible to design a schedule with fewer time slots. However, this schedule calls for the use of three different rooms, because three examinations are scheduled during time slot 2. (d) A coloring of the scheduling graph with 4 colors. This means that the examinations can be scheduled in 4 time slots. However, because each color appears only twice, all the examinations can be scheduled in two air-conditioned rooms.

require four different colors. The improved coloring in Figure 3.22c was found by trial and error.

The minimum number of colors needed to label the vertices of a graph so that no two vertices of the graph that are joined by an edge get the same color is called the (vertex) **chromatic number** of the graph.

The examination graph we have been studying has chromatic number 4; hence, we can schedule the eight examinations in four time slots without a conflict. Notice, however, that the coloring in Figure 3.22c schedules three different courses for the time slot corresponding to color 2. This means that not enough rooms with air conditioning will be available. Is there a way to recolor the graph with four colors so that each of the four colors is only used twice? Figure 3.22d shows that the answer is yes.

Thus, we are able to schedule the eight final examinations in four time slots, using only two air-conditioned rooms, and no student will have a conflict under this schedule! ■

Realistic problems to schedule government committees, high school and university final examinations, and job interviews (see Spotlight 3.3) are usually so large that graph coloring algorithms have to be incorporated into elaborate software packages to solve them.

Given any particular small graph, one can use trial and error to find the chromatic number, the minimum number of colors needed to color the vertices of that graph. Rather surprisingly, no algorithm that finds the chromatic number of a graph quickly has been found or is likely to be found. This does not mean that mathematical analysis in studying coloring problems has not been made. For example, here is a lovely theorem by the British mathematician R. L. Brooks that applies to graphs without multiple edges:

If G is a graph (other than a graph where each vertex is joined to every other or a circuit with odd length) with the property that its maximal valence is Δ, then the chromatic number of the graph is at most Δ.

Brooks's theorem means that if a graph has a million vertices and no vertex of valence more than 3, it can be colored with 3 or fewer colors.

Mathematicians have examined many kinds of coloring problems. One can study problems that involve the coloring of the edges of a graph rather than its vertices. Using techniques that have emerged from the study of coloring problems, problems involving such diverse contexts as scheduling government committees, using runways at airports efficiently, assigning frequencies for use by mobile pagers and cell phones, and designing timetables for public transportation have been solved—all these benefits from a problem that at first glance looks as if it belongs to recreational mathematics!

SPOTLIGHT 3.3 Scheduling Job Interviews

A group of companies is coming to campus for job interviews. The companies have been assigned the number of hours they need to cover the number of students they hope to interview during a block of consecutive hours wherein a representative from each company can hold interviews. Due to the fact that classes are going on at the same time, five departmental conference rooms have been made available to the companies to conduct their interviews.

The interviews will follow the school's regular hourly periods, which start at 9 A.M. and end at 4 P.M. (Companies will be scheduled for continuous interviews during lunch-hour times. Interviews cannot be scheduled beyond the end of the period that starts at 4 P.M. and ends at 5 P.M.)

Company	Time Slot Requested
A (Apricot Computers)	7
B (Big Green)	1
C (Challenge Insurance)	4, 5
D (Daisy Printers)	7, 8
E (Earnest Engine)	4, 5, 6
F (Flexible Systems)	2, 3
G (Gutter Leaders)	1, 2
H (Halley's Combs)	6, 7
I (Indelible Ink Corporation)	7, 8
J (Jay's Produce)	4, 5
K (Kelly's Detective Agency)	2, 3
L (Large Clothes)	4, 5, 6
M (Metropolitan TV)	1, 2
N (Nationwide Bank)	4, 5, 6, 7

Look at the list of time blocks that the companies requested (where 1 = 9–10 A.M., . . . , 8 = 4–5 P.M.). Is it possible to accommodate all the companies that wish to do interviewing in the five rooms available while meeting their desired schedule times?

Problems of this kind seem simple enough, and you should try your hand at solving this particular one, for which a schedule does exist! However, this situation is not simple at all. The following facts are known about problems of this kind.

FACT 1. Suppose there are i interviewers, p time periods, and r rooms where interviews can be scheduled. Each interviewer has specified periods during which he or she wishes to conduct interviews. Is it possible to design a schedule that meets the desired specifications? It turns out that this problem is NP-complete (see Spotlight 2.1), that is, it belongs to a large group of problems for which, among other things, the fastest known algorithms run very slowly on large-problem versions.

FACT 2. The problem just described remains NP-complete even for the case where only three rooms have to be scheduled (i.e., $p = 3$).

The moral is *surprisingly simple:* Scheduling problems are very hard to solve.

However, the situation is not as hopeless as it might seem. If you look at the list of time requests for the corporations, you will note that, not surprisingly, each company has requested a contiguous block of times. It turns out that when this condition holds, it is possible to determine whether there is a feasible schedule using an algorithm that works relatively quickly.

REVIEW VOCABULARY

Average-case analysis The study of the list-processing algorithm (more generally, any algorithm) from the point of view of how well it performs in all the types of problems it may be used for and seeing on average how well it does. *See also* worst-case analysis.

Bin-packing problem The problem of determining the minimum number of containers of capacity W into which objects of size $w_1, \ldots, w_n$ ($w_i \leq W$) can be packed.

Chromatic number The chromatic number of a graph G is the minimum number of colors (labels) needed in any vertex coloring of G.

Critical-path scheduling A heuristic algorithm for solving scheduling problems where the list-processing algorithm is applied to the priority list obtained by listing next in the priority list a task that heads a longest path in the order-requirement digraph. This task is then deleted from the order-requirement digraph, and the next task placed in the priority list is obtained by repeating the process.

Decreasing-time-list algorithm The heuristic algorithm that applies the list-processing algorithm to the priority list obtained by listing the tasks in decreasing order of their time length.

First fit (FF) A heuristic algorithm for bin packing in which the next weight to be packed is placed in the lowest-numbered bin already opened into which it will fit. If it fits in no open bin, a new bin is opened.

First-fit decreasing (FFD) A heuristic algorithm for bin packing where the first-fit algorithm is applied to the list of weights sorted so that they appear in decreasing order.

Heuristic algorithm An algorithm that is fast to carry out but that doesn't necessarily give an optimal solution to an optimization problem.

Independent tasks Tasks are independent when there are no edges in the order-requirement digraph.

List-processing algorithm A heuristic algorithm for assigning tasks to processors: Assign the first ready task on the priority list that has not already been assigned to the lowest-numbered processor that is not working on a task.

Machine scheduling The problem of assigning tasks to processors so as to complete the tasks by the earliest time possible.

Next fit (NF) A heuristic algorithm for bin packing in which a new bin is opened if the weight to be packed next will not fit in the bin that is currently being filled; the current bin is then closed.

Next-fit decreasing (NFD) A heuristic algorithm for bin packing where the next-fit algorithm is applied to the list of weights sorted so that they appear in decreasing order.

Priority list An ordering of the collection of tasks to be scheduled for the purpose of attaining a particular scheduling goal. One such goal is minimizing completion time when the list algorithm is applied.

Processor A person, machine, robot, operating room, or runway whose time must be scheduled.

Ready task A task is called ready at a particular time if its predecessors, as given by the order-requirement digraph, have been completed by that time.

Vertex coloring A vertex coloring of a graph G is an assignment of labels, which can be thought of as "colors," to the vertices of G so that vertices joined by an edge get different labels (colors).

Worst-case analysis The study of the list-processing algorithm (more generally, any algorithm) from the point of view of how well it performs on the hardest problems it may be used on. *See also* average-case analysis.

Worst fit (WF) A heuristic algorithm for bin packing in which the next weight to be packed is placed into the open bin with the largest amount of room remaining. If the weight fits in no open bin, a new bin is opened.

Worst-fit decreasing (WFD) A heuristic algorithm for bin packing where the worst-fit algorithm is applied to the list of weights sorted so that they appear in decreasing order.

SUGGESTED READINGS

BRUCKER, P. *Scheduling Algorithms,* Springer-Verlag, Heidelberg, Germany, 1995. A detailed mathematical look at scheduling.

FRENCH, SIMON. *Sequencing and Scheduling,* Wiley, New York, 1982. A detailed account of a wide variety of scheduling models, most of them different from the ones treated in this chapter.

GRAHAM, RONALD. Combinatorial scheduling theory, in Lynn Steen (ed.), *Mathematics Today,* Springer-Verlag, New York, 1978, pp. 183–211. This essay on scheduling is one of many excellent accounts of recent developments in mathematics in this book.

GRAHAM, RONALD. The combinatorial mathematics of scheduling. *Scientific American,* March 1978, pp. 124–132. A very readable introduction to scheduling and bin packing.

JENSEN, T. R., and BJARNE TOFT. *Graph Coloring Problems,* Wiley, New York, 1995. A detailed summary of what is known about coloring problems and many questions that await answering.

LAWLER, E., et al. Sequencing and scheduling algorithms and complexity, in S. C. Graves, et al. (eds.), *Handbooks in OR and MS,* vol. 4, Elsevier, New York, 1993, pp. 445–522. A recent survey of results about scheduling.

PARKER, R. GARY. *Deterministic Scheduling Theory,* Chapman & Hall, London, 1995. A wide-ranging look at scheduling methods and their applications.

SUGGESTED WEB SITES

www.informs.org The Web page of the major professional society in operations research.

www.mat.gsia.cmu.edu A "hub" for resources about operations research.

SKILLS CHECK

1. Given the order-requirement digraph below (time in minutes) and the priority list T_1, T_2, T_3, T_4, T_5, T_6, apply the list-processing algorithm to construct a schedule using two processors. How much time does the resulting schedule require?

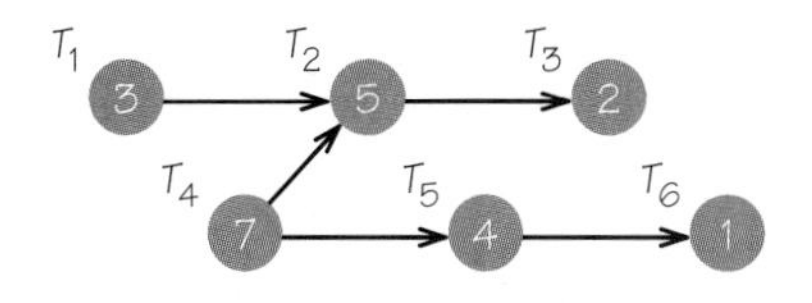

(a) 11 minutes
(b) 13 minutes
(c) 14 minutes

2. A radio announcer has 10 songs of various lengths to schedule into several segments. The announcer must identify the station at least once every 15 minutes, so the segments cannot be longer than 15 minutes. This job can be solved using the

(a) list-processing algorithm for independent tasks.
(b) critical-path scheduling algorithm.
(c) worst-fit algorithm for bin packing.

3. What is the minimum time required to complete 8 independent tasks with a total task time of 64 minutes on 4 machines?

(a) Less than 5 minutes
(b) Between 5 and 10 minutes
(c) More than 10 minutes

4. Use the decreasing-time-list algorithm to schedule these independent tasks on two machines: 6 minutes, 7 minutes, 4 minutes, 3 minutes,

6 minutes. How much time does the resulting schedule require?

(a) 13 minutes
(b) 14 minutes
(c) More than 14 minutes

5. Use the first-fit (FF) bin-packing algorithm to pack the following weights into bins that can hold no more than 10 lb: 6 lb, 7 lb, 4 lb, 3 lb, 6 lb. How many bins are required?

(a) 3 bins
(b) 4 bins
(c) 5 bins

6. Use the worst-fit-decreasing (WFD) bin-packing algorithm to pack the following weights into bins that can hold no more than 10 lb: 6 lb, 7 lb, 4 lb, 3 lb, 6 lb. How many bins are holding a full 10 lb?

(a) 0 bins
(b) 1 bin
(c) 2 bins

7. Suppose that a crew can complete in a minimum amount of time the job whose order-requirement digraph is shown below. If task T_2 is shortened from 5 minutes to 2 minutes, then what is the maximum amount by which the completion time for the entire job can be shortened?

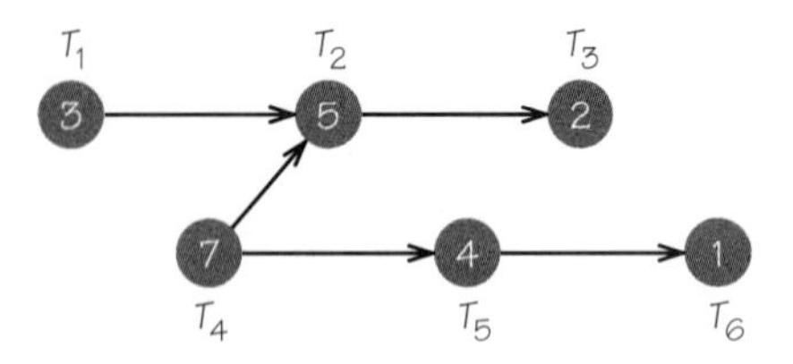

(a) 3 minutes
(b) 2 minutes
(c) 1 minute

8. Assume an order-requirement digraph requires 15 minutes when scheduled on two machines. Based on this information, when the digraph is instead scheduled on three machines, how much time will be required?

(a) Exactly 10 minutes
(b) At least 10 minutes
(c) Exactly 15 minutes

9. Assume an order-requirement digraph has a critical path with length 20 minutes. Based on this information, when the digraph is scheduled on two machines, how much time will be required?

(a) Exactly 10 minutes
(b) Exactly 20 minutes
(c) At least 20 minutes

10. Assume a job consists of six independent tasks ranging in time from 2 to 10 minutes and totaling 27 minutes. Efficiently scheduled on three machines, how much time will the job require?

(a) Exactly 9 minutes
(b) Exactly 10 minutes
(c) More than 10 minutes

11. Which of the following statements about the decreasing-time-list algorithm is true?

(a) The decreasing-time-length algorithm *always* yields an *optimal* schedule.
(b) The decreasing-time-length algorithm *often* yields an *optimal* schedule.
(c) The decreasing-time-length algorithm *always* yields an *optimal* or *near-optimal* schedule.

12. A vertex coloring seeks to color the vertices of a graph in order to ensure which of the following traits?

(a) Every color is used.
(b) Every edge connects vertices of the same color.
(c) Vertices of the same color are never connected by an edge.

13. Assume the 8 corners of a cube represent vertices of a graph and the 12 edges of a cube represent the cube's edges. What is the chromatic number of this graph?

(a) 2
(b) 3
(c) More than 3

EXERCISES

▲ Optional. ■ Advanced. ◆ Discussion.

Scheduling

1. List as many scheduling situations as you can for these environments:

(a) Hospital
(b) Railroad station
(c) Airport
(d) Automobile repair garage
(e) Restaurant
(f) Your home
(g) Your school
(h) Police station
(i) Firehouse

Compare and contrast the scheduling issues involved in these situations.

2. For the situation where a family with three children is preparing a Thanksgiving meal for 10 guests, list tasks that must be completed and the types of processors that are involved. Can any of these tasks be done simultaneously?

◆ **3.** In order to get to a ski resort for a weekend vacation, Jocelyn must accomplish a variety of things. She will leave work early at 1 P.M. and must get to the airport to be on a 5 P.M. shuttle to Boston. She then hopes to take a bus to get to the resort. Discuss some of the tasks that must be accomplished to get Jocelyn to the resort by 10 P.M. What are the different types of processors that are involved in getting these tasks done? Can any of these tasks be done simultaneously?

4. **(a)** Use the order-requirement digraph at top right to schedule the 6 tasks T_1, T_2, T_3, T_4, T_5, T_6 on two processors with the priority lists:

(i) $T_1, T_2, T_3, T_4, T_5, T_6$
(ii) $T_1, T_6, T_3, T_5, T_4, T_2$

(b) Are either of the schedules produced from these lists optimal? If not, can you find a priority list which will result in an optimal schedule?
(c) Find the critical path and its length. Explain why no schedule has earliest completion time equal to the length of the critical path.

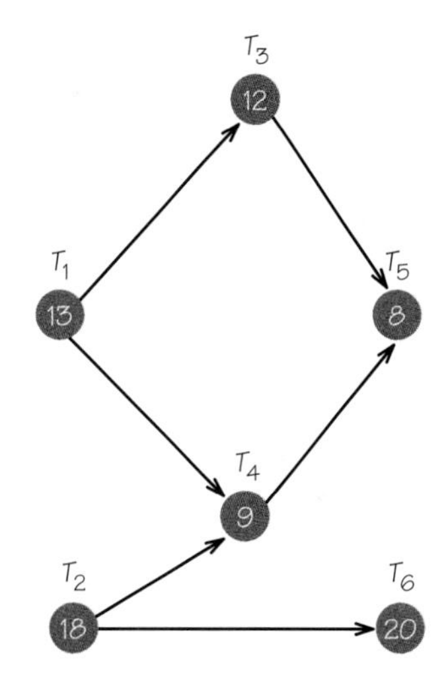

5. **(a)** Repeat Exercise 4, but interchange the task times of tasks T_2 and T_6.
(b) How does the completion time for an optimum schedule for this situation compare with the optimum schedule for Exercise 4?

◆ **6.** Discuss scheduling problems for which it is not reasonable to assume that once a processor starts a task, it would always complete that task before it works on any other task. Give examples for which this approach would be reasonable.

7. Use the list-processing algorithm to schedule the tasks in the following order-requirement digraph on

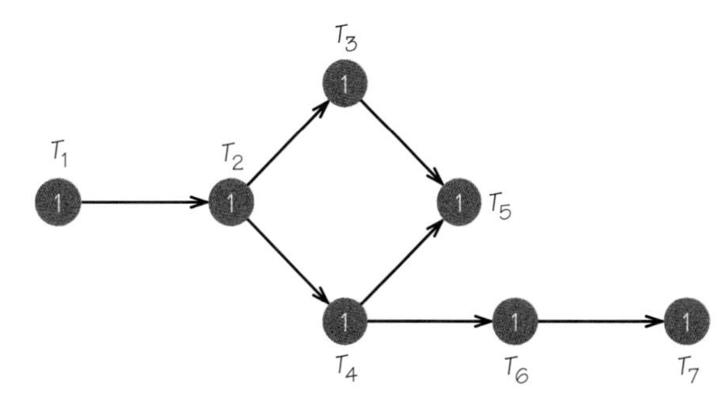

(a) two processors using the list $T_1, \ldots, T_7$.
(b) two processors using the list T_1, T_2, T_3, T_4, T_6, T_5, T_7.

(c) Is either of the schedules that you obtain optimal?
(d) Will adding a third processor enable the tasks to be finished earlier?

8. Use the list-processing algorithm to schedule the tasks in the following order-requirement digraph on

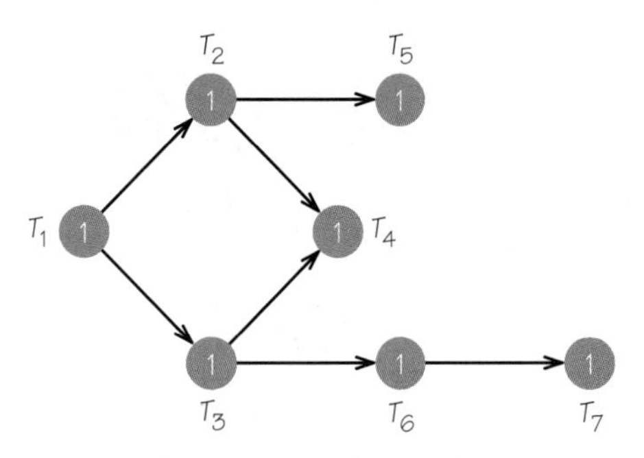

(a) two processors using the list $T_1, \ldots, T_7$.
(b) two processors using the list $T_1, T_2, T_3, T_4, T_6, T_5, T_7$.
(c) Is either of the schedules that you obtain optimal?

Using the List-Processing Algorithm

9. **(a)** Making use of the order-requirement digraph below, determine at time 0 which tasks are ready.

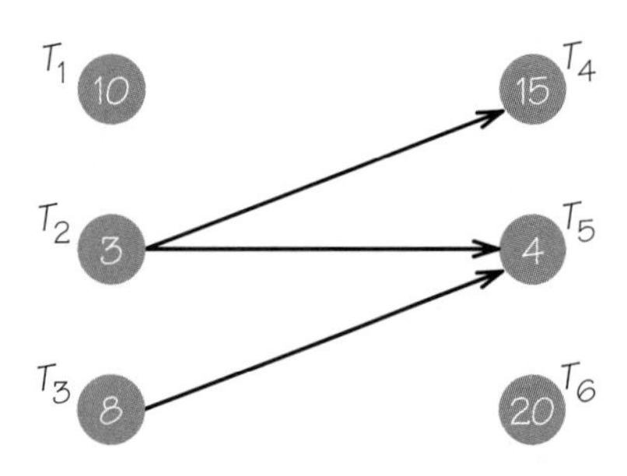

(b) What is special about tasks T_1 and T_6?
(c) What is the critical path and what is its length?
(d) Schedule the tasks on three processors with the priority list $T_1, \ldots, T_6$.
(e) Is the schedule found in part (d) optimal?
(f) Schedule the tasks on three processors using the priority list $T_6, \ldots, T_1$.
(g) Is the schedule found in part (f) optimal?
(h) Can you find a priority list which yields an optimal schedule?

10. **(a)** In Exercise 9, what priority list would be used if you apply the critical-path scheduling method?
(b) Use this priority list to schedule the tasks on three processors. Is this schedule optimal?
(c) How does this schedule compare with the schedules that you found using the lists in Exercise 9?

11. For the accompanying order-requirement digraph, apply the list-processing algorithm, using three processors for lists (a) through (c). How do the completion times obtained compare with the length of the critical path?

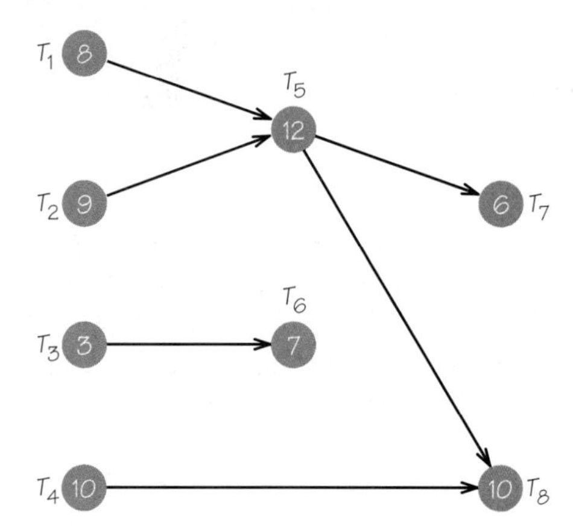

(a) $T_1, T_2, T_3, T_4, T_5, T_6, T_7, T_8$
(b) $T_1, T_3, T_5, T_7, T_2, T_4, T_6, T_8$
(c) $T_8, T_6, T_4, T_2, T_1, T_3, T_5, T_7$

12. Consider the following order-requirement digraph:

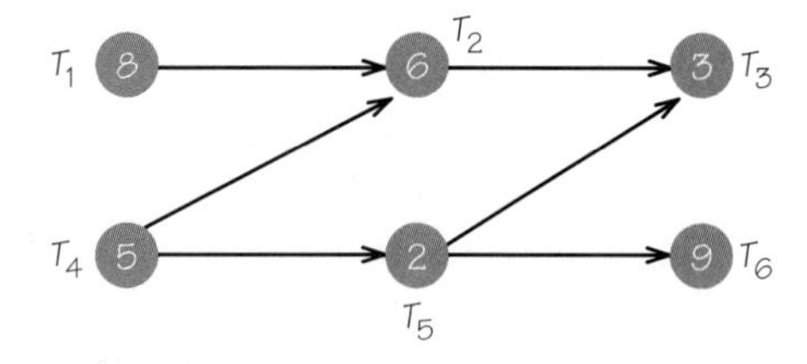

(a) Find the critical path(s).

(b) Schedule these tasks on one processor using the critical-path scheduling method.

(c) Schedule these tasks on one processor using the priority list obtained by listing the tasks in order of decreasing time.

(d) Does either of these schedules have idle time? How do their completion times compare?

(e) If two different schedules have the same completion time, what criteria can be used to say one schedule is superior to the other?

(f) Schedule these tasks on two processors using the order-requirement digraph shown and the priority list from part (b).

(g) Does the schedule produced in part (f) finish in half the time that the schedule in part (b) did, which might be expected, since the number of processors has doubled?

(h) Schedule the tasks on (i) one processor and (ii) two processors (using the decreasing-time list), assuming that each task time has been reduced by one. Do the changes in completion time agree with your expectations?

13. (a) Can all the processors being used to schedule tasks be simultaneously idle at a time prior to the completion time of a collection of tasks scheduled using the list-processing algorithm?

(b) Explain why the list-processing algorithm cannot give rise to the schedule below, regardless of what priority list was used to schedule the tasks on the two processors.

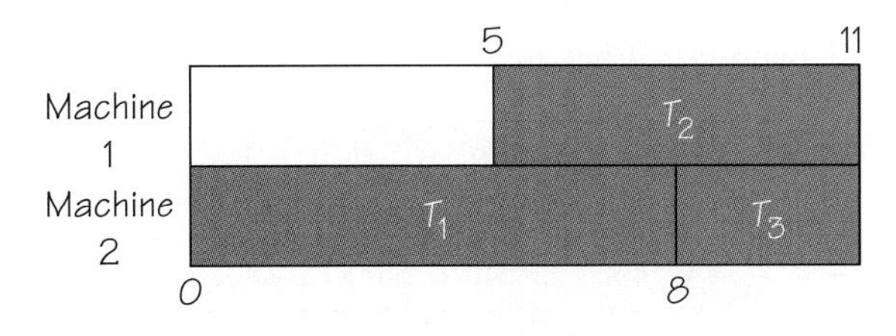

(c) Construct an order-requirement digraph and a priority list which will give rise to the following schedule on two processors.

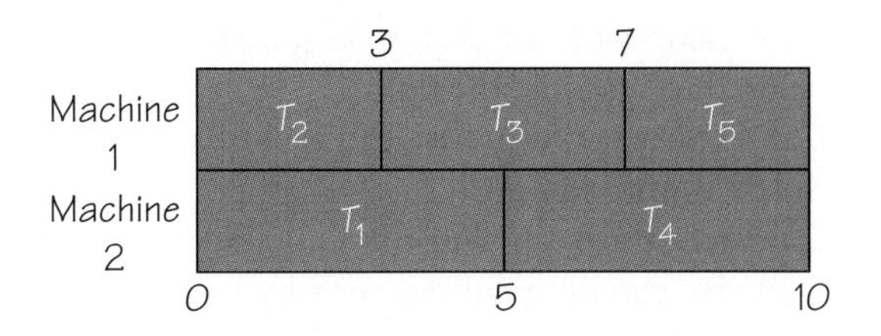

14. To prepare a meal quickly involves carrying out the tasks shown (time lengths in minutes) in the following order-requirement digraph:

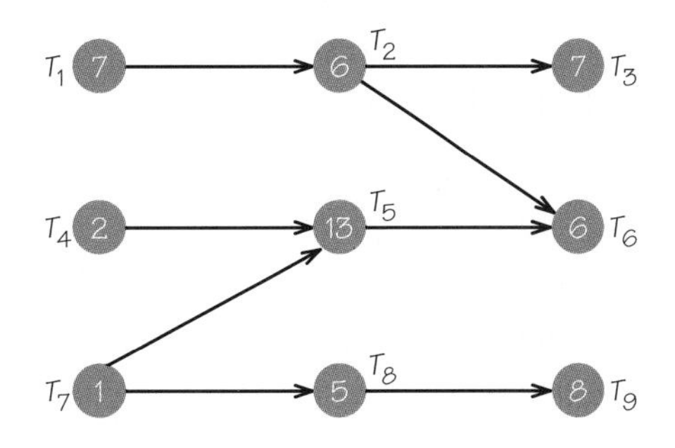

(a) If Mike prepares the meal alone, how long will it take?

(b) If Mike can talk Mary into helping him prepare the meal, how long will it take them if the tasks are scheduled using the list T_5, T_9, T_1, T_3, T_2, T_6, T_8, T_4, T_7 and the list-processing algorithm?

(c) If Mike can talk Mary and Jack into helping him prepare the meal, how long will it take if the tasks are scheduled using the same list as in part (b)?

(d) What would be a reasonable set of criteria for choosing a priority list in this situation?

15. Consider the order-requirement digraph below. Suppose one plans to schedule these tasks on two identical processors.

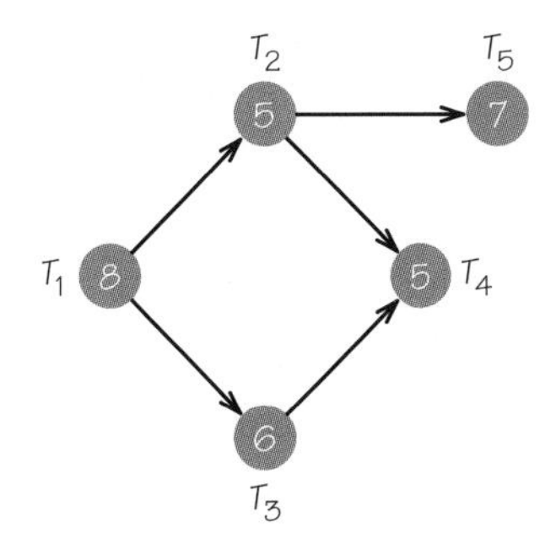

(a) How many different priority lists are there that can be used to schedule the tasks?
(b) Can all these priority lists lead to different schedules? If not, why not?
(c) Can an optimal schedule have no idle time? Can you give two different reasons why an optimal schedule must have some idle time?
(d) Is there any list that produces a schedule where the second processor has no idle time?

16. **(a)** In Exercise 15, how many different lists are there that do not list T_1 first?
(b) Would it make any sense not to list T_1 first in a list?
(c) Construct a list and schedule the tasks on two processors.
(d) Can you find another list that leads to a different completion time than the schedule you found for part (c)?
(e) Find a list that leads to an optimal schedule.

17. Can you find an order-requirement digraph with five tasks for which every possible list yields exactly the same schedule?

18. Can you find an order-requirement digraph such that the schedule corresponding to every list is different?

19. At a large toy store, scooters arrive unassembled in boxes. To assemble a scooter, the following tasks must be performed:

TASK 1. Remove parts from the box.
TASK 2. Attach wheels to the footboard.
TASK 3. Attach vertical housing.
TASK 4. Attach handlebars to vertical housing.
TASK 5. Put on reflector tape.
TASK 6. Attach bell to handlebars.
TASK 7. Attach decals.
TASK 8. Attach kickstand.
TASK 9. Attach safety instructions to handlebars.

(a) Give reasonable time estimates for these tasks and construct a reasonable order-requirement digraph. What is the earliest time by which these tasks can be completed?
(b) Schedule this job on two processors (humans) using the decreasing-time-list algorithm.

20. If two schedules for the same number of processors have the same completion time, can one schedule have more idle time than the other?

21. Could the schedule below be obtained by applying the list-scheduling algorithm to some order-requirement digraph?

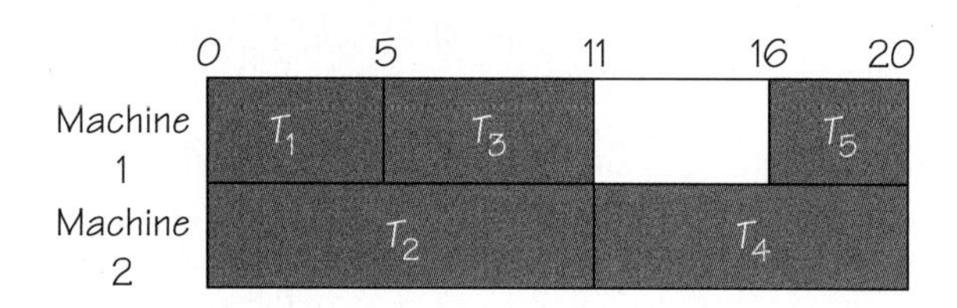

22. Could the schedule below be obtained by applying the list-scheduling algorithm to some order-requirement digraph?

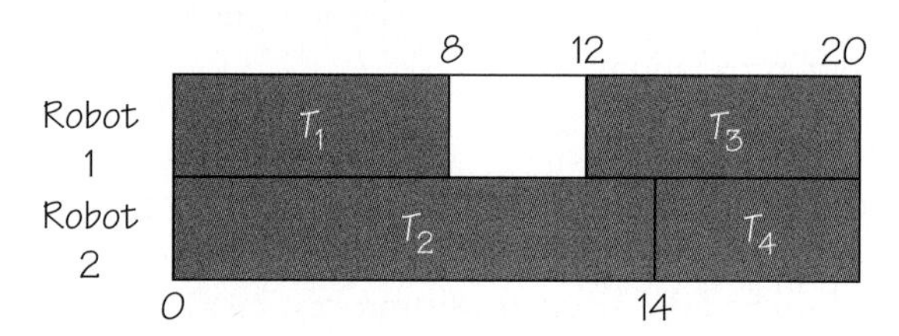

Independent Tasks and Other Issues

23. For the following schedules, can you produce a list so that the list-processing algorithm produces the schedule shown when the tasks are independent? What are the task times for each task?

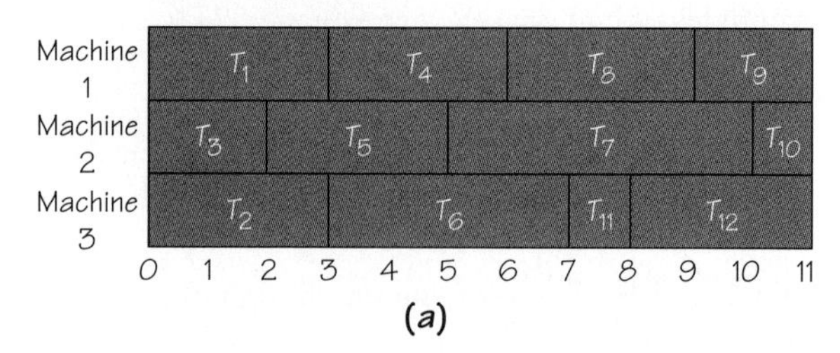

(a)

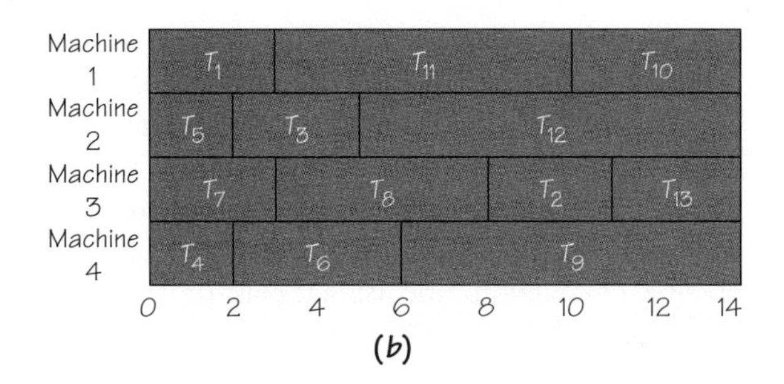

(b)

◆ **24.** Once an optimal schedule has been found for independent tasks (e.g., see diagrams in Exercise 23), usually the scheduling of the tasks can be rearranged and the same optimal time achieved (i.e., one can, among other things, reorder the tasks done by a particular processor). Discuss criteria that might be used to implement the rearrangement process.

25. The task times of eight independent tasks T_1 to T_8 are 1, 2, 3, 4, 5, 6, 7, 8.

(a) Schedule the tasks on two processors using the lists: (i) $T_1, T_2, \ldots, T_8$ and (ii) $T_8, T_7, \ldots, T_1$.
(b) Is either of the schedules you get in part (a) optimal? If not, find a list that gives an optimal schedule.

26. Repeat Exercise 25, but schedule the tasks (with the same lists) on three processors. If the schedules you get are not optimal, find a list that gives an optimal schedule.

◆ **27.** Discuss different criteria that might be used to construct a priority list for a scheduling problem.

◆ **28.** Some scheduling projects have due dates for tasks (i.e., times by which a given task should be completed) and release dates (i.e., times before which a task cannot have work begun on it). Give examples of circumstances where these situations might arise.

29. Using the lists you found in Exercise 23 and the task times you computed for those independent tasks, schedule the tasks for (a) on four processors and the tasks for (b) on five processors. Can you see why for any schedule you may produce for (a) on four processors and (b) on five processors there must be some idle time for one or more processors?

30. Given the accompanying order-requirement digraph:

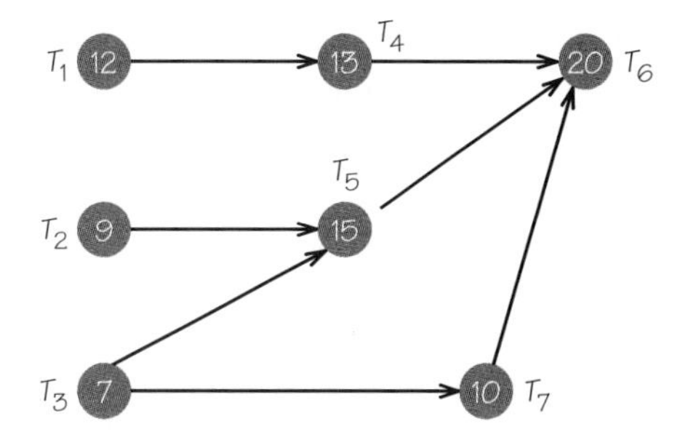

(a) Use the list-processing algorithm to schedule these seven tasks on two processors using these lists:
(i) $T_1, T_3, T_7, T_2, T_4, T_5, T_6$
(ii) $T_1, T_3, T_2, T_4, T_5, T_6, T_7$
(iii) The list obtained by listing the tasks in order of decreasing time
(b) Try to determine if any of the resulting schedules are optimal.
(c) Schedule the tasks using the critical-path scheduling method. Try to determine if this schedule is optimal.

31. Repeat the questions in Exercise 30 using the order-requirement digraph obtained by erasing all the (directed) edges shown there. How do the schedules you get compare with the ones you originally got?

32. (a) Find the completion time for independent tasks of length 8, 11, 17, 14, 16, 9, 2, 1, 18, 5, 3, 7, 6, 2, 1 on three processors, using the list-processing algorithm.
(b) Find the completion time for the tasks in part (a) on three processors, using the decreasing-time-list algorithm.
(c) Does either algorithm give rise to an optimal schedule?
(d) Repeat for tasks of lengths 19, 19, 20, 20, 1, 1, 2, 2, 3, 3, 5, 5, 11, 11, 17, 18, 18, 17, 2, 16, 16, 2.

33. A photocopy shop must schedule independent batches of documents to be copied. The times for the different sets of documents are (in minutes): 12, 23, 32, 13, 24, 45, 23, 23, 14, 21, 34, 53, 18, 63, 47, 25, 74, 23, 43, 43, 16, 16, 76.

(a) Construct a schedule using the list-processing algorithm on three machines.
(b) Construct a schedule using the list-processing algorithm on four machines.
(c) Repeat parts (a) and (b), but use the decreasing-time-list algorithm.
(d) Suppose union regulations require that an 8-minute rest period be allowed for any photocopy task over 45 minutes. Use the decreasing-time-list algorithm, with the preceding times modified to take into account the union requirement, to schedule the tasks on three human-operated machines.

34. Find a list that produces the following optimal schedule when the list-processing algorithm is applied to this list. (Assume the tasks are independent.)

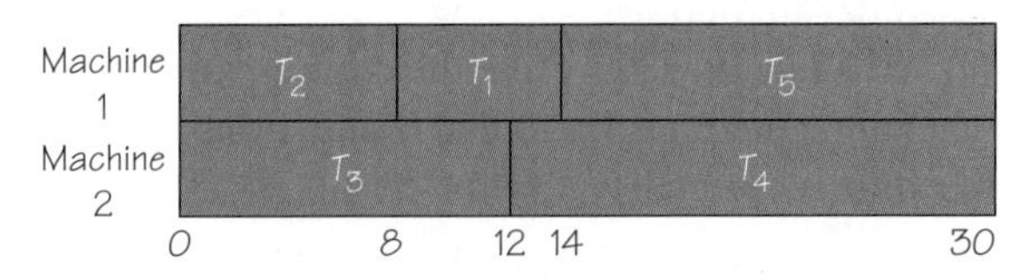

What completion time and schedule are obtained when the decreasing-time-list algorithm is applied to this list?

35. Can you think of situations other than those mentioned in the text where scheduling independent tasks on processors occurs?

36. Can you think of real-world scheduling situations in which all the tasks have the same time and are independent? Can you find an algorithm for solving this problem optimally? (If there are n independent tasks of time length k, when will all the tasks be finished?)

37. **(a)** Show that when tasks to be scheduled are independent, the critical-path method and the decreasing-time-list method are identical.

(b) The (usually unknown) optimal time to complete a specific collection of independent tasks on three machines turns out to be 450 minutes. Estimate the worst possible completion time when the list-processing algorithm is used with the worst choice of priority list. Now estimate the longest possible completion time using the list-processing algorithm and the decreasing-time list.

Bin Packing

38. Two wooden wall systems are to be made of pieces of wood with lengths shown in the accompanying diagram. If wood is sold in 10-foot planks and can be cut with no waste, what number of boards would be purchased if one uses the first-fit-decreasing, next-fit-decreasing, and worst-fit-decreasing heuristics, respectively?

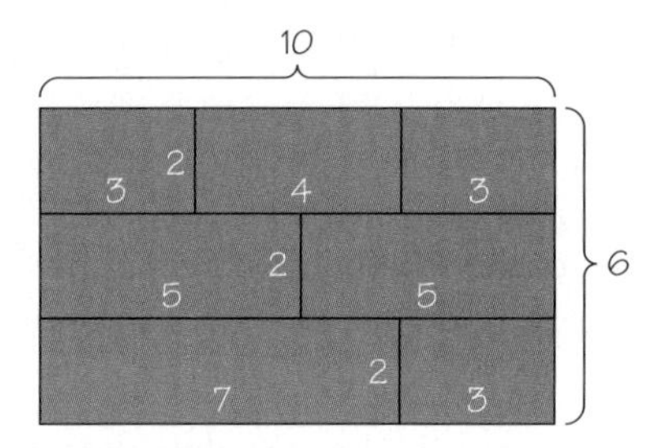

In solving this problem, does it make a difference if the 10-foot horizontal shelves and 6-foot vertical boards employ single-length pieces as compared with using pieces of boards that add up to 10- and 6-foot lengths?

39. It takes 4 seconds to photocopy one page. Manuscripts of 10, 8, 15, 24, 22, 24, 20, 14, 19, 12, 16, 30, 15, and 16 pages are to be photocopied. How many photocopy machines would be required, using the first-fit-decreasing algorithm, to guarantee that all manuscripts are photocopied in 2 minutes or less? Would the solution differ if worst-fit decreasing were used?

40. A radio station's policy allows advertising breaks of no longer than 2 minutes, 15 seconds. Using first-fit and first-fit-decreasing algorithms, determine the minimum number of breaks into which the following ads will fit (lengths given in seconds): 80, 90, 130, 50, 60, 20, 90, 30, 30, 40. Can you find the optimum solution? Do the same for these ads: 60, 50, 40, 40, 60, 90, 90, 50, 20, 30, 30, 50.

41. Fiberglass insulation comes in 36-inch precut sections. A plumber must install insulation in a basement on piping that is interrupted often by joints. The distances between the joints on the stretches of pipe that must be insulated are 12, 15, 16, 12, 9, 11, 15, 17, 12, 14, 17, 18, 19, 21, 31, 7, 21, 9, 23, 24, 15, 16, 12, 9, 8, 27, 22, 18 inches. How many precut sections would he have to use to provide the insulation if he bases his decision on

(a) next fit?
(b) next-fit decreasing?
(c) worst fit?
(d) worst-fit decreasing?

42. The files that a company has for its employees dealing with utilities occupy 100, 120, 60, 90, 110, 45, 30, 70, 60, 50, 40, 25, 65, 25, 55, 35, 45, 60, 75, 30, 120, 100, 60, 90, 85 sectors. If, after operating systems are installed, a disk can store up to 480 sectors, determine the number of disks to store the utilities if each of these heuristics is used to pack the disk with files:

(a) NF
(b) NFD
(c) FF
(d) FFD

◆ **43.** We have described two algorithms for bin packing called "worst fit" and "best fit" (see page 96). The words *best* and *worst* have connotations in English. However, the performance of algorithms depends on their merits as algorithms, not on the names we give them.

(a) On the basis of experiments you perform with the best-fit and worst-fit algorithms, which one do you think is the "better" of the two?
(b) Can you construct an example where worst fit uses fewer bins than best fit?

44. The best-fit heuristic (see page 96) also has a "decreasing" version, where the list is first sorted in decreasing order. Using bins of capacity 10, apply the best-fit heuristic and its decreasing version to the following list: 6, 9, 5, 8, 3, 2, 1, 9, 2, 7, 2, 5, 4, 3, 7, 6, 2, 8, 3, 7, 1, 6, 4, 2, 5, 3, 7, 2, 5, 2, 3, 6, 2, 7, 1, 3, 5, 4, 2, 6.

■ **45.** One pianist's recording of the complete Mozart piano sonatas takes the following times (given in minutes and seconds): 13:46, 6:15, 3:29, 5:37, 7:52, 2:55, 5:00, 4:28, 4:21, 7:39, 7:55, 6:42, 4:23, 3:52, 4:21, 4:20, 5:46, 6:29, 5:34, 6:23, 6:39, 7:19, 5:54, 6:54, 2:58, 5:22, 1:42, 5:00, 1:29, 5:47, 7:30, 8:19, 4:44, 4:57, 4:09, 14:31, 3:55, 4:04, 4:01, 6:06, 6:50, 5:27, 4:28, 5:40, 2:52, 5:16, 5:34, 3:10, 7:22, 4:40, 3:08, 6:32, 4:47, 6:59, 5:38, 7:57, 3:38. If the maximum time that can be recorded on a compact disk is 70:30, can all the music be performed on four compact disks? Can all the music be performed on five compact disks?

■ **46.** In the wall-system example in the text, first fit and worst fit required equal numbers of bins (see Figure 3.20). Can you find an example where first fit and worst fit yield different numbers of bins? Can you find an example where first fit, worst fit, and next fit yield answers with different numbers of bins?

◆ **47.** A common suggestion for heuristics for the bin-packing problem with bins of capacity W involves finding weights that sum to exactly W. Discuss the pros and cons of a heuristic of this type.

■ **48.** A record company wishes to record all the Beethoven string quartets (16 quartets, each consisting of several consecutive parts called movements) on LPs. It wishes to complete the project on as few records as possible. Recording can be done on two sides as long as the movements are consecutive. Is this an example of a bin-packing problem? (Defend your answer.) If the project were to record the quartets on (standard) tape cassettes or compact disks, would your answer be different?

■ **49.** Can you find an example of weights that, when packed into bins using first fit, use fewer bins than the number of bins used when the first-fit algorithm is applied with the first weight on the list removed?

◆ **50.** Can you formulate "paradoxical" situations for bin packing that are analogous to those we found for scheduling processors?

Coloring Problems

51. For each of the graphs below:

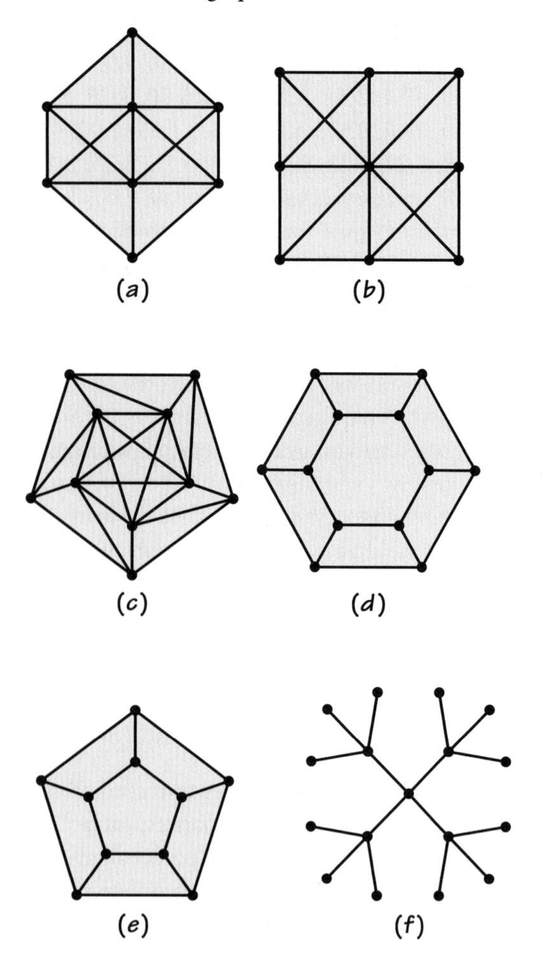

(a) Color the vertices (if possible) with three different colors.
(b) Color the vertices (if possible) with four different colors.
(c) Find the chromatic number of the graph.

52. For each of the following graphs:

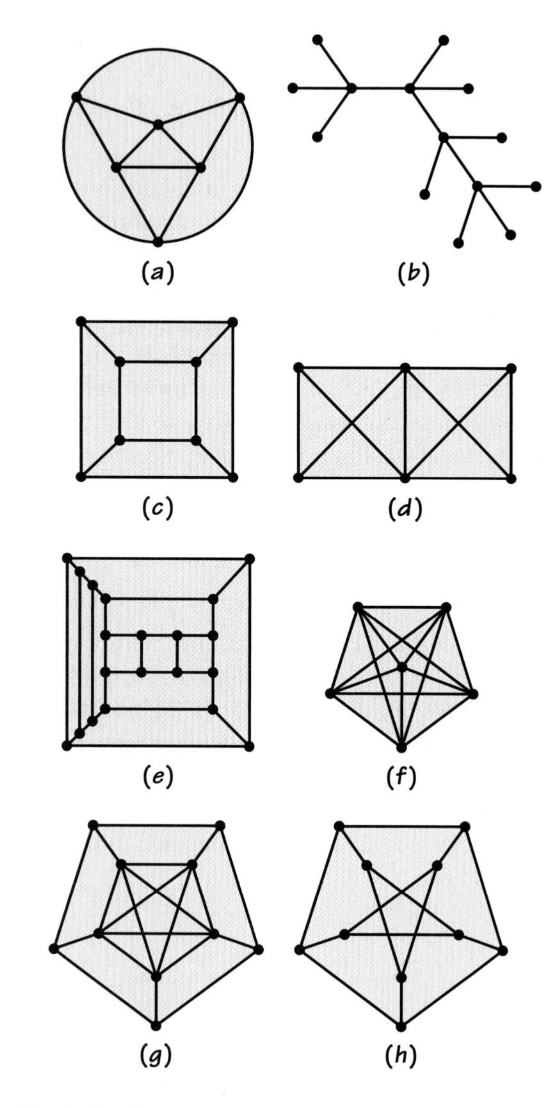

(a) Color the vertices (if possible) with two different colors.
(b) Color the vertices (if possible) with three different colors.
(c) Find the chromatic number of the graph.

53. The owner of a new pet store wishes to display tropical fish in display tanks. The following table shows the incompatibilities between the species, in the sense that an X indicates that it is unwise to allow those species in the row and column that meet at the X to be in the same tank.

	A	B	C	D	E	F	G	H	I
A						X	X		X
B			X					X	
C		X			X			X	
D					X	X		X	
E			X	X			X		
F	X			X			X		X
G	X				X	X		X	X
H		X	X	X			X		
I	X					X	X		

(a) What is the minimum number of tanks needed to display all the fish she wishes to sell?
(b) Is it possible to display the species so that the number of species in each tank is as nearly equal as possible?

54. The managers of a zoo are planning to open a small satellite branch. The animals are to be in enclosures in which compatible animals are displayed together. The accompanying table indicates those pairs of animals that are compatible. (Thus, an X in a particular row and column means that the animals that label this row and column can share an enclosure.)

	A	B	C	D	E	F	G	H	I	J
A	X	X		X	X	X	X			
B	X	X			X	X	X		X	X
C			X		X	X	X			
D	X			X	X	X	X		X	X
E	X	X	X	X	X			X	X	
F	X	X	X	X		X	X	X	X	
G	X	X	X	X		X	X	X		
H					X	X	X	X		
I		X		X	X	X			X	
J		X		X						X

(a) What is the minimum number of enclosures needed to avoid housing incompatible animals in the same enclosure?
(b) Is it possible to enclose the animals in such a way that each enclosure contains the same number of animals?
(c) Why might that be desirable? Why might this approach to grouping the animals not be ideal?

55. The nine standing committees of a state legislature are designing a schedule for when the committees can meet. The matrix shown in the table below has an X in a position where the committees corresponding to the row and column have a common member and, hence, should not be scheduled to meet at the same hour. The committees involved are Agriculture (A), Commerce (C), Consumer Affairs (CA), Education (E), Forests (F), Health (H), Justice (J), Labor (L), and Rules (R).

	A	C	CA	E	F	H	J	L	R
A		X	X			X			
C	X		X	X	X				
CA	X	X					X		X
E		X			X	X			
F		X		X		X	X		
H	X			X	X			X	
J			X		X			X	X
L						X	X		X
R			X				X	X	

(a) Draw a graph that will be of value in determining the minimum number of time slots the committees can meet in without any legislator having to be in two places at one time.
(b) What is the minimum number of time slots in which the committees can be scheduled without a conflict?
(c) How many different rooms are needed at any time that a committee is scheduled to meet? (Why might this issue matter?)

56. Determine the minimum number of colors, and how often each color is used, in a vertex coloring of the graphs below.

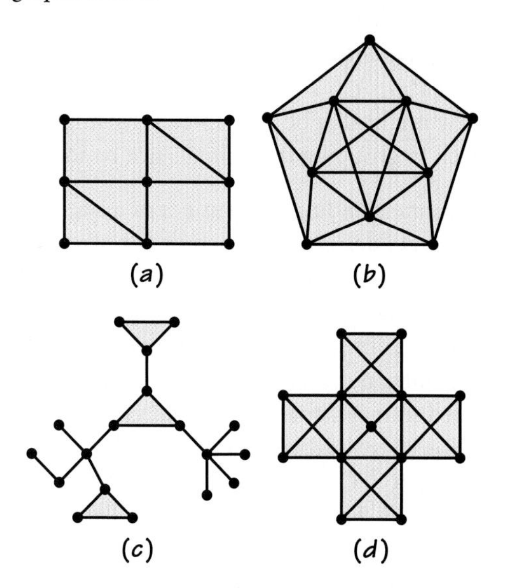

57. The faculty-student governing council at All State College has nine standing committees (e.g., Curriculum, Academic Standards, Campus Life) that are designated A, B, C, D, . . . , I for convenience. The following table shows which committees have no member in common.

	A	B	C	D	E	F	G	H	I
A		X		X		X	X		X
B	X				X	X		X	X
C				X		X	X	X	X
D	X		X			X		X	
E		X					X	X	X
F	X	X	X	X					
G	X		X		X			X	
H		X	X	X	X		X		X
I	X	X	X		X			X	

(a) What is the minimum number of time slots in which all the committee meetings can be scheduled?

(b) How many rooms are needed during each time slot to accommodate the committees that are scheduled to meet in that time slot?

58. When two towns are within 145 miles of each other, the frequency used by a certain type of emergency response system for the towns would require that they be on different frequencies to avoid possible interference with each other. The table below shows the mileage distances between six towns.

	E	F	G	I	S	T
Evansville (E)		290	277	168	303	113
Ft. Wayne (F)	290		132	83	79	201
Gary (G)	277	132		153	58	164
Indianapolis (I)	168	83	153		140	71
South Bend (S)	303	79	50	140		196
Terre Haute (T)	113	201	164	71	196	

(a) What would be the minimum number of frequencies that are needed for each town to have its emergency broadcasts not conflict with those of any other town using this system?

(b) How many different towns would be assigned to each frequency used?

59. Show that the vertices of any tree can be colored with two colors.

60. Can you find a family of graphs H_n that requires n colors to color its vertices?

61. The edge-coloring number of a graph G is the minimum number of colors needed to color the edges of G so that edges that share a common vertex get different colors. Determine the edge-coloring number for each of the graphs in Exercise 52. Can you make a conjecture about the value of the minimum number of colors needed to color the edges of any graph?

62. Can you think of any applications that require determining the minimum number of colors to color the edges of a graph?

63. When a graph has been drawn on a piece of paper so that edges only meet at vertices, the graph divides the paper up into regions called *faces*. The faces include one called the "infinite" face, which surrounds the whole graph. The face-coloring number of a graph G (which can be drawn in this special way) is the minimum number of colors needed to color the faces of G so that two faces that share an edge receive different colors. (Note that if two faces meet only at a vertex, they can be colored the same color.)

(a) Determine the minimum number of colors to color the faces of the graphs below. In each case, remember to color the infinite face, which is labeled I (for "infinite").

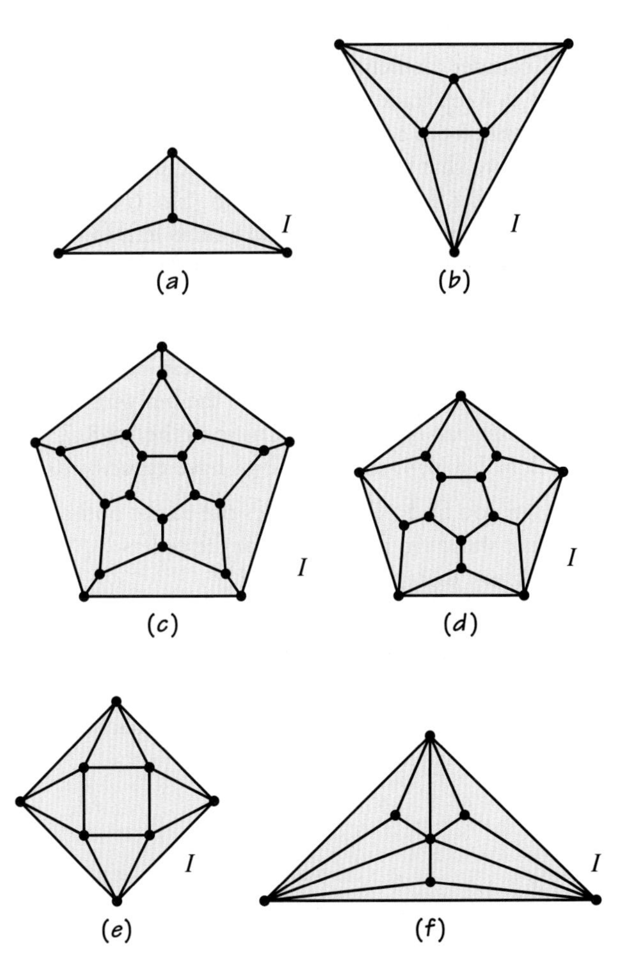

(b) Can you think of an application of the problem of coloring the faces of a graph with a minimum number of colors?

Additional Exercises

64. Consider the accompanying order-requirement diagraph:

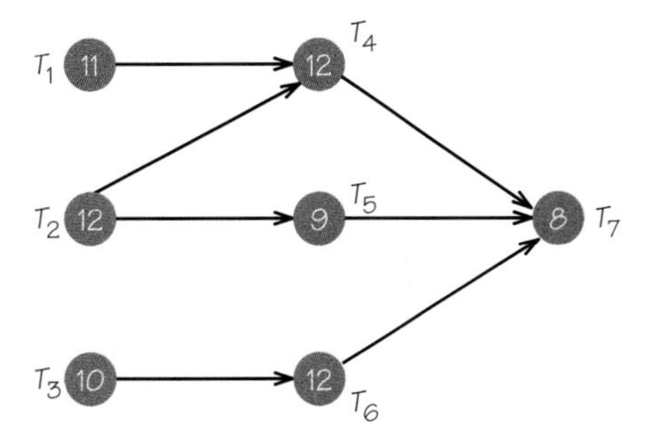

(a) Find the length of the critical path.
(b) Schedule these seven tasks on two processors using the list algorithm and the lists.
(i) $T_1, T_2, T_3, T_4, T_5, T_6, T_7$
(ii) $T_2, T_1, T_3, T_6, T_5, T_4, T_7$
(c) Does either list lead to a completion time that equals the length of the critical path?
(d) Show that no list can ever lead to a completion time equal to the length of the critical path (providing the schedule uses two processors).

■ **65.** Could the following schedule have arisen from the list-processing algorithm? Could it have arisen from the application of the list-processing algorithm to a collection of independent tasks?

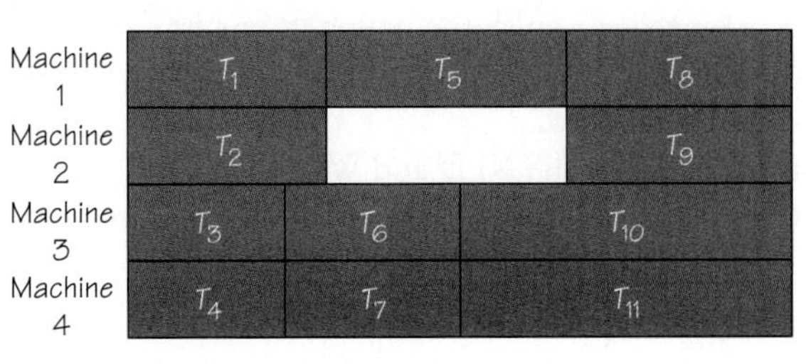

66. Can you give examples of scheduling problems for which it seems reasonable to assume that all the task times are the same?

■ **67.** Two-dimensional bin packing refers to the problem of packing rectangles of various sizes into a minimum number of $m \times n$ rectangles, with the sides of the packed rectangles parallel to those of the containing rectangle.

(a) Suggest some possible real-world applications of this problem.
(b) Devise a heuristic algorithm for this problem.
(c) Give an argument to show that the problem is at least as hard to solve as the usual bin-packing problem.
(d) If you have $1 \times m$ rectangles with total area W to be packed into a single rectangle of area $p \times q = W$, can the packing always be accomplished?

◆ **68.** In what situations would packing bins of different capacities be the appropriate model for real-world situations? Suggest some possible algorithms for this type of problem.

69. A data entry group must handle 30 (independent) tasks that will take the following amounts of time (in minutes) to type: 25, 18, 13, 19, 30, 32, 12, 36, 25, 17, 18, 26, 12, 15, 31, 18, 15, 18, 16, 19, 30, 12, 16, 15, 24, 16, 27, 18, 9, 14. Using these times as a priority list:

(a) Use the list-processing algorithm to find the completion time for scheduling these tasks with four secretaries; with five secretaries.
(b) Repeat the scheduling using the decreasing-time-list algorithm.
(c) Can you show that any of the schedules that you get are optimal?

If one needs to finish the typing in one hour:
(d) Use the FFD heuristic to find how many typists would be needed.
(e) Repeat for the NFD and WFD heuristics.
(f) Can you show that any of the solutions you get are optimal?

70. Find the minimum number of bins necessary to pack items of size 8, 5, 3, 4, 3, 7, 8, 8, 6, 5, 3, 2, 1, 2, 1, 2, 1, 3, 5, 2, 4, 2, 6, 5, 3, 4, 2, 6, 7, 7, 8, 6, 5, 4, 6, 1, 4, 7, 5, 1, 2, 4 in bins of capacity (a) through (d) using the first-fit and first-fit-decreasing algorithms. Can you determine if any of the packings you get are optimal?

(a) 9
(b) 10
(c) 11
(d) 12

71. Advertisements for the TV show *Q* are permitted to last up to a total of 8 minutes, and each group of ads can last up to 2 minutes. If the ads slated for *Q* last 63, 32, 11, 19, 24, 87, 64, 36, 27, 42, 63 seconds, determine if FF and FFD yield acceptable configurations for the ads.

72. Consider the heuristic for packing bins known as *best fit.* Keep track of how much room remains in each unfilled bin and put the next item to be packed into that bin that would leave the least room left over after the item is put into the bin. (For example, suppose that bin 4 had 6 units left, bin 7 had 5 units, and bin 9 had 8 units left. If the next item in the list had size 5, then first fit would place this item in bin 4, worst fit would place the item in bin 9, while best fit would place the item in bin 7.) If there is a tie, place the item into the bin with the lowest number. Apply this heuristic to the list 8, 7, 1, 9, 2, 5, 7, 3, 6, 4, where the bins have capacity 10.

73. Can you find a list that gives rise to the optimal schedule shown in Figure 3.14 for the order-requirement digraph in Figure 3.12?

74. Give an example to show that scheduling to minimize completion time may not minimize total idle time. (*Hint:* Assume independent tasks and use two machines for one schedule and three machines for the other schedule.)

WRITING PROJECTS

1. Scheduling is important for hospitals, schools, transportation systems, police services, and fire services. Pick one of these areas and write an essay about the different scheduling situations that come up, types of processors, and extent to which the assumptions of the list-processing model hold for the area you pick.

2. Write an essay that compares and contrasts the basic scheduling problem we investigated with the scheduling version of the bin-packing problem.

3. One of the oversimplifications made in our discussion of scheduling was that there were no "due dates" involved for the tasks making up a job. Develop an algorithm for solving a scheduling problem under the assumption that each task has a due date as well as a time length. You will probably want to decide on a penalty amount that will occur when a due date is exceeded.

4. Consider the problem of scheduling tasks on a single machine. Design different algorithms for achieving different goals. You will probably wish to assume that each task has a due date such that if the task is not finished by this date, some penalty payment must be made.

5. Suppose that one has found that the optimal solution to a bin-packing problem with bins of size W requires p bins. What can one say about the number of bins needed when the bin size is $2W$? (*Hint:* Be careful!)

6. Discuss the role of graph colorings for scheduling committee meetings so as to avoid conflicts. Research whether or not these ideas are used in the legislature of your home state.

SPREADSHEET PROJECTS

To do these projects, go to www.whfreeman.com/fapp.

Spreadsheets provide a handy way to monitor the assignment of jobs to people or processors. This chapter's projects use spreadsheets to analyze bin-packing algorithms.

APPLET EXERCISES

To do these exercises, go to www.whfreeman.com/fapp.

Graph coloring

Solving a scheduling problem such as the one below can be accomplished by constructing a related graph and then coloring it in a way that adjacent vertices have different colors. Explore the problem of graph coloring in the applet Graph Coloring.

Scheduling

A mathematics department has seven faculty committees – A, B, C, D, E, F and G. Because there is overlap in the composition of the committees, the chairman of the department is attempting to work out a schedule that will avoid conflicts among the committees. The following chart indicates the overlapping committee structure:

	A	B	C	D	E	F	G
A		X		X		X	
B	X		X			X	
C		X			X		X
D	X						X
E			X			X	X
F	X	X			X		
G			X	X	X		

Help the chairman arrange a schedule without conflicts in the applet Scheduling.

Linear Programming

In this chapter we cover...

...and at www.whfreeman.com/fapp:

Flashcards

Quizzes

Spreadsheet Projects

Mixture problems

A manager's job often calls for making very complicated decisions. One set of decisions involves planning what products the business is to make and determining what resources are needed. In the modern business world, diversification of products provides a company with stability in a climate of changing tastes and needs. So it is not surprising that companies would produce many products, some of which share resource needs. For example, any bakery uses many resources—like butter, sugar, eggs, and flour—to make its products: cookies, cakes, pies, and breads.

Resources can include more than just raw materials. A labor force with appropriate skills, farmland, time, and machinery are also resources. Typically, resources are limited: A farmer owns only so much land; there are only so many hours in a day; in a year of drought the wheat crop is very small. Resource availability is also limited by location and competition.

Because resources are limited, management faces important questions: How should the available resources be shared among the possible products? One goal of management is to maximize profit. How can that determine how much of each product should be produced? There are usually so many alternative product mixes that it is impossible to evaluate them all individually. Despite this complexity, millions of dollars may ride on management's decision.

In this chapter, we learn about **linear programming,** a management science technique that helps a business allocate the resources it has on hand to make a particular mix of products that will maximize profit. Linear programming is a tool for maximizing or minimizing a quantity, typically a profit or a cost, subject to constraints. The technique is so powerful that linear programming is said to account for over 50% and perhaps as much as 90% of all computing time used for management decisions in business.

Linear programming is an example of "new" mathematics. It came into being, along with many other management science techniques, during and shortly after World War II, in the 1940s; it is quite young as intellectual ideas go. Yet, during its short history, linear programming has changed the way businesses and governments make decisions, from "seat-of-the-pants" methods based on guess-

(Mike Greenlar/The Image Works.)

work and intuition to using an algorithm based on available data and guaranteed to produce an optimal decision.

Linear programming is but one operations research tool belonging to a family of tools known as mathematical programming. Another such tool is integer programming. The difference between linear programming and integer programming is that for linear programming, the quantities being studied can take on values such as pi = 3.14159 . . . or $7\frac{1}{8}$; in integer programming, the values are confined to whole numbers such as 8, 50, or 1,102,362. Whole numbers are conceptually easier than the broader group consisting of all numbers that can be represented by decimals (e.g., 1.32, 1.455555 . . .).

Yet integer programming problems have proved much harder to solve. In the discussions that follow, we often describe "relaxed" versions of integer programming problems as linear programming problems. For example, it would make no sense to produce 3.24 dolls to sell. So, strictly speaking, we must find an optimum whole number of dolls to produce. If we are "lucky," the linear programming problem associated with an integer programming problem has an integer solution. In this case, we have also found the correct answer to the integer

SPOTLIGHT 4.1 Case Studies in Linear Programming

Linear programming is not limited to mixture problems. Here are two case studies that do not involve mixture problems, yet where applying linear programming techniques produced impressive savings:

- The Exxon Corporation spends several million dollars per day running refineries in the United States. Because running a refinery takes a lot of energy, energy-saving measures can have a large effect. Managers at Exxon's Baton Rouge plant had over 600 energy-saving projects under consideration. They couldn't implement them all because some conflicted with others, and there were so many ways of making a selection from the 600 that it was impossible to evaluate all selections individually.

  Exxon used linear programming to select an optimal configuration of about 200 projects. The savings are expected to be about $100 million over a period of years.

- American Edwards Laboratories uses heart valves from pigs to produce artificial heart valves for human beings. Pig heart valves come in different sizes. Shipments of pig heart valves often contain too many of some sizes and too few of others; however, each supplier tends to ship roughly the same imbalance of valve sizes on every order, so the company can expect consistently different imbalances from the different suppliers. Thus, if they order shipments from all the suppliers, the imbalances could cancel each other out in a fairly predictable way. The amount of cancellation will depend on the sizes of the individual shipments. Unfortunately, there are too many combinations of shipment sizes to consider all combinations individually.

  American Edwards used linear programming to figure out which combination of shipment sizes would give the best cancellation effect. This reduced the company's annual cost by $1.5 million.

programming problem. Some other examples that fall into this category are discussed below.

Linear programming has saved businesses and governments billions of dollars. Of all the management science techniques presented in this book, linear programming is far and away the most frequently used. It can be applied in a variety of situations, in addition to the one we study in this chapter. Some of the problems studied in Chapters 1, 2, and 3 can be viewed as linear programming problems, and examples of other uses are in Spotlight 4.1. Linear programming is an excellent example of a mathematical technique useful for solving many different kinds of problems that at first do not seem to be similar problems at all. It has been suggested that without linear programming, management science would not exist.

4.1 Mixture Problems

In this chapter, we study how to use linear programming to solve a special kind of problem—a mixture problem. Realistic versions of such problems would be much more involved. Our discussion is designed to give you the flavor of what is actually done.

In a **mixture problem,** limited resources are combined into products so that the profit from selling those products is a maximum.

Mixture problems are widespread because nearly every product in our economy is created by combining resources. A typical example would be how different kinds of aviation fuel are manufactured using different kinds of crude oil.

Let's analyze small versions of the kinds of problems that might confront a toy or a beverage manufacturer. Both manufacturers can sell many different products on which each company can make a profit. There could be dozens of possible products and many resources. A manufacturer must periodically look at the quantities and prices of resources and then determine which products should be produced in which quantities in order to gain the greatest, or optimum, profit. This is an enormous task that usually requires a computer to solve.

What does it mean to find a solution to a linear programming mixture problem? A solution to a mixture problem is a production policy that tells us how many units of each product to make.

An **optimal production policy** has two properties. First it is possible; that is, it does not violate any of the limitations under which the manufacturer operates, such as availability of resources. Second, the optimal production policy gives the maximum profit.

Having studied the previous chapters on management science, you may sense that there must be some algorithm that will give us the optimal production policy. Indeed there are such algorithms. At the heart of every algorithm for linear programming are geometric ideas. This is somewhat surprising, as there seem to be no geometric ideas used to describe a mixture problem. We can see these geometric ideas clearly, however, if we solve small linear programming problems involving just one or two products and one or two resources, and draw some appropriate pictures, or graphs.

We start our discussion with some very simple examples and work up to more involved ones, all of which we can solve by drawing graphs and making some relatively simple calculations. We end this chapter with a discussion of what larger problems look like and how they are typically solved.

4.2 Mixture Problems Having One Resource

One Product and One Resource: Making Skateboards

Suppose a toy manufacturer has 60 containers of plastic and wants to make and sell skateboards. The "recipe" for one skateboard requires five containers of plastic, plus paint and decals, which for simplicity we assume are available in essentially unlimited quantities. The profit on one skateboard is $1.00, and in order to keep things simple, we will assume that there will be customers for every skateboard produced. So the manufacturer must decide how many skateboards to make.

We see that the manufacturer can make $\frac{60}{5} = 12$ skateboards. And there seems to be no particular reason not to do exactly that, earning a profit of $1.00(12) = $12.00. We use the variable x to stand for the number of skateboards made; we see that x could be any value between 0 and 12, or, algebraically, $0 \leq x \leq 12$. (Strictly speaking, skateboards can be sold only in positive integer numbers.) Those values are the *feasible set,* the particular values of our variable x that are feasible, or possible, given the available resources. In the number line, or x-axis, shown below, the feasible set is indicated by a thick blue line. The problems discussed in this chapter are simple enough that we will always be able to draw a picture of the feasible set. Sometimes the feasible set is called the *feasible region,* so we will use the terms *feasible set* and *feasible region* interchangeably.

Feasible region for the skateboard problem

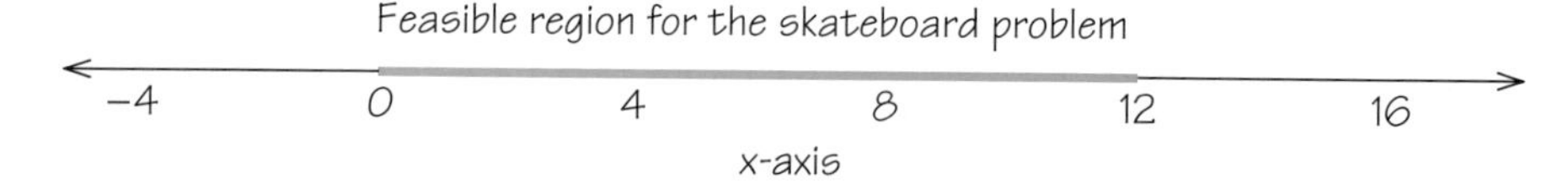

> The **feasible set,** also called the **feasible region,** is the set of all possible solutions to a linear programming problem.

There are several features to note about our feasible region and the point within it that gives the maximum profit:

1. There are no negative values of x in the feasible region. That makes physical sense: How could one make a negative number of skateboards?
2. Any point within the feasible region represents a possible **production policy**—that is, it gives the number of skateboards (product) that it is possible to produce with the limited supply of containers of plastic (resource). The manufacturer could close shop early and spend less time making skateboards, making only 7, for a profit of $1.00(7) = $7.00.
3. The point $x = 0$ of the feasible region represents the manufacturer making no skateboards at all, having no product to sell. Eliminating the only product does not seem sensible; when we consider problems with more than one product, we will again consider whether eliminating a product is a desirable action.
4. The point where the profit is greatest, $x = 12$, happens to be an endpoint, or "corner," of the feasible region. Although this observation may not seem earthshaking in the context of this example, the realization that maximum profit always occurs at a corner point of the feasible region, in both simple and complicated problems, was a crucial insight in the development of linear programming.

The profit made can be described in terms of x: Profit is the number of skateboards, x, multiplied by the profit per skateboard, \$1.00, or \$1.00x. We call the formula \$1.00x a **profit formula,** because it describes how to calculate the profit when we know the number of units (skateboards) to be made.

Common Features of Mixture Problems

Although our first mixture problem has only one product and one resource, it does contain the essential features that are common to *all* mixture problems:

Resources. Definite resources are available in limited, known quantities for the time period in question. The resource here is containers of plastic.
Products. Definite products can be made by combining, or mixing, the resources. In this example, the product is skateboards.
Recipes. A recipe for each product specifies how many units of each resource are needed to make one unit of that product. Here, each skateboard uses five containers of plastic.
Profits. Each product earns a known profit per unit. (We assume that every unit produced can be sold. More complicated mathematical models, beyond the scope of this book, are needed if we want to consider the possibility of items being produced but not sold.)
Objective. The objective in a mixture problem is to find how much of each product to make so as to maximize the profit without exceeding any of the resource limitations.

The method we used to solve the skateboards problem is the same method we will use for more involved situations. That is, we analyze the problem, determining the resources and the products. We draw a picture, a graph, of the feasible region, and then we find a point in the feasible region that gives the maximum profit.

EXAMPLE

Making Lemonade

One batch of lemonade powder requires 5 lemons, and we have 40 lemons on hand. Draw the feasible region for making lemonade powder. What is the maximum profit we can make if we clear a profit of \$1.50 per batch?

Solution: The feasible region corresponds to $0 \leq x \leq \frac{40}{5} = 8$, shown as the thick blue portion of the number line below. Using all the lemons and making 8 batches gives a maximum profit of \$1.50(8) = \$12.00. ■

Feasible region for the lemonade problem

−4 0 4 8 12

x-axis

As our manufacturing situations become more realistic, with more than one product and more than one resource, we will need a bit more algebra to draw the graph and a bit more calculation to find the point we want. But as we become engrossed in the details, it is important to remember the purpose of mixture problems: Given limited resources and fixed recipes for making products from the resources, we want to find a mix of products that will result in a maximum profit. Our next problem has two products and one resource.

Two Products and One Resource: Skateboards and Dolls, Part 1

Any good businessperson looks for new ways to make money. The toy manufacturer wants to expand and produce two products: skateboards and dolls. In order to keep our story simple, we will assume that most of the resources needed, such as labor, fasteners, and paint, are available in essentially unlimited quantities. The only limited resource is containers of plastic, which are needed in both products. We continue to assume that everything produced will sell. The recipe for one doll calls for 2 containers of plastic. Again, it is not difficult to see that if the original supply is again 60 containers of plastic, then the manufacturer could make $\frac{60}{2} = 30$ dolls. But then there would be no containers of plastic for skateboards. And if they make 12 skateboards, using 5 containers of plastic per board, there would be no containers of plastic for dolls. But what if they made some of each? For example, they could make 2 skateboards, using up $5(2) = 10$ containers of plastic and leaving $60 - 10 = 50$ containers of plastic. That would allow for $\frac{50}{2} = 25$ dolls.

But wait, the manufacturer is in this for profit. The company needs to know how much profit it will get from one doll. Suppose it is \$0.55. Now, because we used x for the number of skateboards made, we will use y for the number of dolls. So the profit from the dolls will be $\$0.55y$. The total profit will come from the sale of x skateboards plus y dolls, so the profit formula is $\$1.00x + \$0.55y$. We want to find a pair of numbers (x, y) that makes that profit formula as high as possible. But we don't know what (x, y) pairs are even possible! We need to locate those **feasible points,** the points that make up the feasible set, before we even think about maximizing profit. In order to construct that region, it is helpful to summarize our problem in a *mixture chart.*

Mixture Charts

The most important skill required when solving a mixture problem is the ability to understand and model its underlying structure. This skill is as important as being able to do the subsequent algebra and arithmetic. In fact, because linear programming problems can be solved using readily available computer software, extracting the important data from the underlying structure may be the only part of the problem-solving process that must be done by a human being. Understanding the underlying structure means being able to answer these questions:

1. What are the resources?
2. What quantity of each resource is available?
3. What are the products?
4. What are the recipes for creating the products from the resources?
5. What are the unknown quantities?
6. What is the profit formula?

We will display the answers to these questions in a diagram called a **mixture chart.** Then we will translate information in our mixture chart into mathematical statements that we can use to solve the mixture problem. Figure 4.1 shows the mixture chart for Skateboards and Dolls, Part 1.

Some features are present in every mixture chart. There is a row of the chart for every product, for every type of item on which the business can make a profit. Here the products are skateboards and dolls. All the entries on a row give information about the one product to which the row belongs. There is a column for every resource, every input to the business that comes only in limited quantities. Each resource column has information about just one of the limited resources. This problem has only one resource, containers of plastic, so it has only one resource column. (As we progress to problems with more than one resource, our mixture charts will have more columns.) There is also a column for the profit data. We will formulate a mathematical statement corresponding to the profit column and to each of the resource columns.

Each problem has specifics that are put into the mixture chart. In filling in the mixture chart, we have labeled the number of skateboards as x and the number of dolls as y, just as we did before. The recipes for the two products in terms of the number of containers of plastic they use have been entered in the column for the containers of plastic. Because each skateboard uses 5 containers of plastic, there is a 5 where the row for product skateboards meets the column for the container of plastic resource. Similarly, there is a 2 where the row for the product dolls meets the column for the container of plastic resource.

PRODUCTS	RESOURCE(S) Containers of plastic 60	PROFIT
Skateboards (x units)	5	$1.00
Dolls (y units)	2	$0.55

Figure 4.1 Mixture chart for Skateboards and Dolls, Part 1.

We have also entered the profit numbers in the chart. For example, in the row for dolls and the column for profits, we put \$0.55 to indicate that each doll brings in a \$0.55 profit. We can use the mixture chart to determine a profit formula of $\$1.00x + \$0.55y$. The \$1.00 and the x are both on the row for skateboards, and the \$0.55 and the y are both on the row for dolls.

EXAMPLE

Making a Mixture Chart

Make a mixture chart to display this situation. A clothing manufacturer has 60 yards of cloth available to make shirts and decorated vests. Each shirt requires 3 yards of material and provides a profit of \$5. Each vest requires 2 yards of material and provides a profit of \$2.

Solution: See the mixture chart in Figure 4.2. ■

PRODUCTS	RESOURCE(S) Yards of cloth 60	PROFIT
Shirts (x units)	3	\$5
Vests (y units)	2	\$3

Figure 4.2 Mixture chart for the clothing manufacturer.

Resource Constraints

Every resource in our mixture problems gives us a *resource constraint*, an algebraic statement that says what is obvious in the physical world: "You can't use more of a resource than the amount you have available." Each resource column in the mixture chart gives us one resource constraint.

We put together the information we have about the resource, containers of plastic. There are just 60 containers of plastic, so "the number of containers of plastic used must be less than or equal to 60." That statement can be rewritten as "the number of containers of plastic used" ≤ 60. To translate the words in quotation marks, we reason in much the same way we did for finding our profit formula. If the manufacturer makes x skateboards, and each skateboard requires five containers of plastic, $5x$ containers of plastic are used. Similarly, in making y dolls, each requiring two containers of plastic, $2y$ containers of plastic are used up. So making x skateboards *plus* y dolls uses up $5x$ *plus* $2y$ containers of plastic, or $5x + 2y$, which cannot exceed 60. Thus, we get the resource constraint $5x + 2y \leq 60$. Note that the numbers 5, 2, and 60 are all in the "container of plastic resource column" of the mixture chart in Figure 4.1.

A **resource constraint** is an inequality in a mixture problem that reflects the fact that no more of a resource can be used than what is available.

The resource constraint $5x + 2y \leq 60$ is really a combination of two mathematical statements: $5x + 2y < 60$ and $5x + 2y = 60$. The first statement, $5x + 2y < 60$, is an *inequality* that tells us that the number of containers of plastic used to make the two products is *less than* the total number of containers of plastic available. The second statement, $5x + 2y = 60$, is an *equality,* or *equation,* that tells us that the number of containers of plastic used to make the two products is *equal to* the total number of containers of plastic available. So $5x + 2y \leq 60$ tells us that the number of containers of plastic used to make the two products must be *less than or equal to* the total number of containers of plastic available.

EXAMPLE

Writing a Resource Constraint and a Profit Formula

Using the numbers in the mixture chart in Figure 4.2, write a resource constraint for the cloth resource. Also write the profit formula.

Solution: The resource constraint is $3x + 2y \leq 60$. The profit formula is $\$5x + \$3y$. ■

Graphing the Constraints to Form the Feasible Region

When we have two products in a mixture problem, we use two variables, x and y. So the feasible region for a problem having two products will be a portion of the xy, or *Cartesian,* plane.

Every point in the feasible region is a possible solution to the linear programming problem because it satisfies every constraint of that problem. For a two-product problem, the feasible region is a part of the plane bounded by pieces of lines. If the problem has n products, the feasible region is a portion of n-dimensional space, bounded by flat surfaces, which are the analog of lines.

How can we use a resource constraint to help us find the feasible region? In particular, how do we graph an inequality such as $5x + 2y \leq 60$? It is not difficult to draw a graph of $5x + 2y = 60$. That equation, and all the other ones we will get from resource constraints, is the equation of a straight line. Any equation having either an x term, a y term, or both, and some numerical constant, like the 60, but no other kinds of terms, always represents a line. (The equation cannot have any squared, square root, or other kind of algebraic combination of x or y.)

Constraint inequalities are always associated with equations for lines; hence the term *linear programming.* The programming does not refer to a computer but to a well-defined sequence of steps, or program of action, that solves the kinds of problems we are exploring. Here we are using *program* as a synonym for *algorithm.* With the introduction of computers into business settings, linear programming is usually carried out by running a computer program.

In order to draw a straight line, we need to know two of the points on the line. In fact, we already know two useful points, but we may not have thought of them as points. One point says that if the manufacturer makes zero skateboards, then there are enough containers of plastic to make 30 dolls. That point, expressed in terms of the x and y coordinates of the plane, would take the general form of (x, y). Note that the order of the numbers matters. (3, 10) is not the same as (10, 3). The x value is always written first. Because x is the number of skateboards and y is the number of dolls, the point we want is (0, 30). Another point we have already considered is the point representing 12 skateboards and 0 dolls, which we write as (12, 0).

In general, when we want to draw the graph of the line portion of a resource constraint, we can substitute $x = 0$ into the equation part of the resource constraint, and find the corresponding value for y. Here's how that algebra looks:

$5x + 2y = 60$	Substitute $x = 0$ into the equation.
$5(0) + 2y = 60$	Multiply 5 by 0 and simplify.
$2y = 60$	Divide both sides of the equation by 2.
$y = 30$	

So one (x, y) point on the line is (0, 30). This is the point representing our making $x = 0$ skateboards and $y = 30$ dolls.

We now follow a similar procedure for making zero dolls; we substitute $y = 0$ into the equation $5x + 2y = 60$ and find the corresponding value of x. Starting with $5x + 2(0) = 60$ and following the same steps as in the algebra above, we get $x = 12$. So another (x, y) point on the line is (12, 0). This point represents our making $x = 12$ skateboards and $y = 0$ dolls.

In Figure 4.3a we have a graph of the xy-plane showing the points (0, 30) and (12, 0) and a segment of the line $5x + 2y = 60$ connecting them. Every point on the line segment represents a production policy for the two products that uses up all the containers of plastic available. Some of the points, like (2, 25), give us whole products, and some of the points represent fractional products. We verify that (2, 25) is on the line by substituting 2 for x and 25 for y into the equation $5x + 2y = 60$, getting $5(2) + 2(25) = 60$, which simplifies to $10 + 50 = 60$, which is a true statement. Remember, points on the xy-plane are always expressed in the form (x, y), with the x value, or coordinate, written before the y value, or coordinate.

The xy-plane has four portions, called *quadrants*. In the graph in Figure 4.3a, we show only the quadrant in which both x and y are nonnegative because, in reality, we can never make negative quantities of our products. Reflecting that reality, we have **minimum constraints** of $x \geq 0$ and $y \geq 0$. The line segment in the

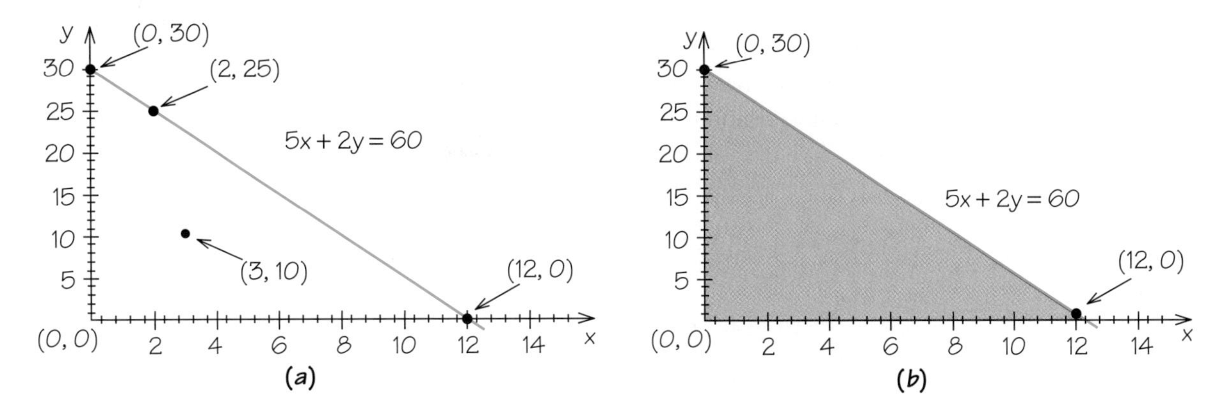

Figure 4.3 The feasible region for Skateboards and Dolls, Part 1. (a) Graph of $5x + 2y = 60$. (b) Shading of the half plane $5x + 2y < 60$.

graph represents all the points, or production policies, for which these properties are true: All the containers of plastic are used up, $5x + 2y = 60$, and both x and y are nonnegative. These points are part of our feasible region. We also need to identify those points corresponding to the inequality $5x + 2y < 60$; these are points corresponding to production policies that do not use up all the containers of plastic.

Any line, such as $5x + 2y = 60$, divides the xy-plane into two parts, called *half planes*. Each of those half planes corresponds to one of two inequalities, in this case $5x + 2y < 60$ and $5x + 2y > 60$. We can determine which half plane goes with which inequality by testing one point not on the line and seeing which inequality it makes true. For example, (3, 10) is on the "down" side of the line segment. When we substitute that point into the inequality $5x + 2y < 60$, which we do by replacing the x by 3 and the y by 10, we get $5(3) + 2(10) < 60$. Simplifying gives us $15 + 20 < 60$, or $35 < 60$, which is true. Substituting the same point into the other inequality, $5x + 2y > 60$, would give $35 > 60$, which is false. So we know that the down side of the line segment corresponds to the inequality $5x + 2y < 60$, and the down side plus the line segment itself corresponds to the combination inequality $5x + 2y \leq 60$. The "up" side of the line corresponds to the inequality $5x + 2y > 60$, and thus is *not* part of the feasible region. In practice, the point (0, 0) is often used as a test point. (Can you see why?) In Figure 4.3b we show the feasible region for the skateboards and dolls problem as a shaded region in the quadrant where both x and y are nonnegative.

EXAMPLE

Drawing a Feasible Region

In the earlier clothing manufacturer example, we developed a resource constraint of $3x + 2y \leq 60$. Draw the feasible region corresponding to that resource constraint, using the reality minimums of $x \geq 0$ and $y \geq 0$.

Solution: First we find the two points where the line, $3x + 2y = 60$, crosses the axes. When $x = 0$, we get $3(0) + 2y = 60$, giving $y = \frac{60}{2} = 30$, yielding the point

(0, 30). For $y = 0$, we get $3x + 2(0) = 60$, or $x = \frac{60}{3} = 20$, so we have the point (20, 0). We draw the line connecting those points. Testing the point (0, 0), we find that the down side of the line we have drawn corresponds to $3x + 2y < 60$. The feasible region is shown in Figure 4.4. ■

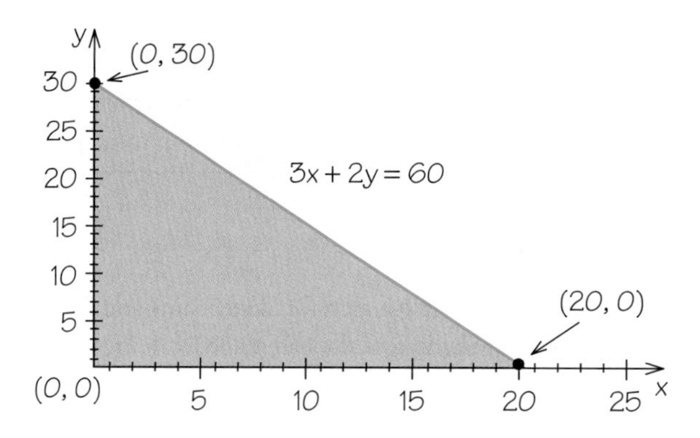

Figure 4.4 Feasible region for the clothing manufacturer.

Finding the Optimal Production Policy

After all this work we may think we are done, but in fact we have learned to draw just one type of feasible region. We still must find the *optimal production policy,* a point within that region that gives a maximum profit. There are a lot of points in that region. If you consider points with only whole numbers as values for x or y, there are many points, but in fact either x or y or both of them could be some fractional number. There are so many points in this feasible region that to consider the profit at each one of them would require us to calculate profits from now until we grow very old, and still the calculations would not be done. Here is where the genius of the linear programming technique comes in, with the *corner point principle,* which we define in terms of our mixture problems.

The **corner point principle** states that in a linear programming problem, the maximum value for the profit formula always corresponds to a **corner point** of the feasible region. (Later in this chapter we discuss why the principle works; for now we will accept its validity.)

The corner point principle is probably the most important insight into the theory of linear programming. The geometric nature of this principle explains the value of creating a geometric model from the data in a mixture chart.

The corner point principle gives us the following method to solve a mixture problem:

1. Determine the corner points of the feasible region.
2. Evaluate the profit at each corner point of the feasible region.
3. Choose the corner point with the highest profit as the production policy.

Let's look at the feasible region we drew in Figure 4.3. It is a triangle having three corners, namely, (0, 0), (0, 30), and (12, 0). Now all we need to do is find out which of these three points gives us the highest value for the profit formula, which in this problem is 1.00x$ + 0.55y$. We display our calculations in Table 4.1. The maximum profit for the toy manufacturer is $16.50, and that happens if the manufacturer makes 0 skateboards and 30 dolls. The point (0, 30) is called the *optimal production policy*.

An **optimal production policy** corresponds to a corner point of the feasible region where the profit formula has a maximum value.

EXAMPLE

Finding the Optimal Production Policy

Our analysis of the clothing manufacturer problem resulted in a feasible region with three corner points, (0, 0), (0, 30), and (20, 0). Which of these maximizes the profit formula, 5x$ + 3y$, and what does that corner represent in terms of how many shirts and vests to manufacture?

Solution: The evaluation of the profit formula at the corner points is shown in Table 4.2. The maximum profit of $100 occurs at the corner point (20, 0), which represents making 20 shirts and no vests. ■

TABLE 4.1 Calculation of the Profit Formula for Skateboards and Dolls, Part 1

Corner Point	Value of the Profit Formula: 1.00x$ + 0.55y$
(0, 0)	$1.00(0) + $0.55(0) = $0.00 + $0.00 = $0.00
(0, 30)	$1.00(0) + $0.55(30) = $0.00 + $16.50 = $16.50
(12, 0)	$1.00(12) + $0.55(0) = $12.00 + $0.00 = $12.00

TABLE 4.2 Evaluating the Profit Formula in the Clothing Example

Corner Point	Value of the Profit Formula: 5x$ + 3y$
(0, 0)	$5(0) + $3(0) = $0 + $0 = $0
(0, 30)	$5(0) + $3(30) = $0 + $90 = $90
(20, 0)	$5(20) + $3(0) = $100 + $0 = $100

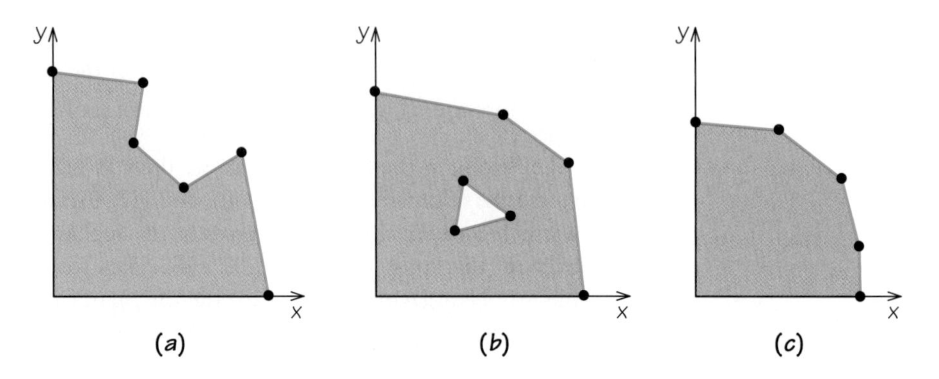

Figure 4.5 A feasible region may not have (a) dents or (b) holes. Graph (c) shows a typical feasible region.

General Shape of Feasible Regions

The shape of a feasible region for a linear programming mixture problem has some important characteristics, without which the corner point principle would not work:

1. The feasible region is a polygon in the first quadrant, where both $x \geq 0$ and $y \geq 0$. This is because the minimum constraints require that both x and y be nonnegative.
2. The region is a polygon that has neither dents (as in Figure 4.5a) nor holes (as in Figure 4.5b). Figure 4.5c is a typical example. Such polygons are called *convex*.

The Role of the Profit Formula: Skateboards and Dolls, Part 2

In practice, there are often different amounts of resources available in different time periods. The selling price for the products can also change. For example, if competition forces us to cut our selling price, the profit per unit can decrease. In order to maximize profit, it is usually necessary for a manufacturer to redo the mixture problem calculations whenever any of the numbers change.

Suppose that business conditions change and now the profits per skateboard and doll are, respectively, \$1.05 and \$0.40. Let us keep everything else about the skateboards and dolls problem the same. The change in profits would give us a new profit formula of $\$1.05x + \$0.40y$. When we evaluate the new profit formula at the corner points, we get the results shown in Table 4.3. This time the optimal production policy, the point that gives the maximum value for the profit formula, is the point (12, 0). To get the maximum profit of \$12.60, the toy manufacturer should now make 12 skateboards and 0 dolls.

TABLE 4.3 A Different Profit Formula: Skateboards and Dolls, Part 2

Corner Point	Value of the Profit Formula: $1.05x + $0.40y
(0, 0)	$1.05(0) + $0.40(0) = $0.00 + $0.00 = $0.00
(0, 30)	$1.05(0) + $0.40(30) = $0.00 + $12.00 = $12.00
(12, 0)	$1.05(12) + $0.40(0) = $12.60 + $0.00 = $12.60

We see from this example that the shape of the feasible region, and thus the corner points we test, are determined by the constraint inequalities. The profit formula is used to choose an optimal point from among the corner points, so it is not surprising that different profit formulas might give us different optimal production policies.

We started the exploration of skateboard and doll production with the idea that the toy manufacturer wanted to expand the product line from one to two products. But both linear programming solutions we have found tell the manufacturer that to maximize profit, just make one product. This is probably not an acceptable result for the manufacturer, who might want to produce both products for business reasons other than profit, such as establishing brand loyalty. And it certainly would be very difficult for the manufacturer to be ready to switch back and forth between producing either skateboards or dolls every time the profit formula changed. Linear programming is a flexible enough technique that it can accommodate the desire for there to be both products in the optimal production policy. The way this is done is by specifying that there be nonzero minimum quantities for each period.

Setting Minimum Quantities for Products: Skateboards and Dolls, Part 3

Suppose the toy manufacturer has kept track of the sales of the two products, and has discovered that no matter what, every day there has been demand for at least 4 skateboards and at least 10 dolls. It seems reasonable to set the minimum number of skateboards as 4 and the minimum number of dolls as 10. We keep the same recipes and the 60 containers of plastic. We will redo the mixture chart to include these minimums, draw a new feasible region, and find its corner points. Then we will use each of the earlier profit formulas to see which corner point is the optimal production policy in each case.

Figure 4.6 gives the mixture chart for our expanded problem. A column for minimums has been added to our mixture chart. Note that there are two sets of profits. This time when we draw the feasible region, we have the same resource constraint as we did before, namely, $5x + 2y \leq 60$, so we get the same line as we

PRODUCTS	RESOURCE(S) Containers of plastic 60	MINIMUMS	PROFIT
Skateboards (x units)	5	4	(1) \$1.00; (2) \$1.05
Dolls (y units)	2	10	(1) \$0.55; (2) \$0.40

Figure 4.6 Mixture chart for Skateboards and Dolls, Part 3 (with nonzero minimums).

did before, and the desired inequality still is on the down side of that line, as in Figure 4.3b. But now we have minimum constraints that are nonzero. Let us first see what that means in terms of the skateboards. The skateboards row has a 4 in the column for minimums. That says to us "Make a minimum of four skateboards." Another way to say this is, "The number of skateboards must be equal to or greater than 4." Because x represents the number of skateboards, we get "x must be equal to or greater than 4," which becomes the mathematical statement $x \geq 4$. Instead of the "reality minimum constraint" of $x \geq 0$, which we used earlier, now we have the nonzero minimum $x \geq 4$. Similarly, the minimum for dolls translates into $y \geq 10$.

Drawing a Feasible Region When There Are Nonzero Minimum Constraints

Figure 4.7a shows the feasible region we constructed with minimum constraints reflecting the reality that $x \geq 0$ and $y \geq 0$. We now need to incorporate the nonzero minimum constraints into that feasible region. As we did with the resource constraint, we can split a minimum constraint into two parts—an equation and an inequality. We can draw the line that corresponds to the equality and then determine which side of that line matches the inequality. First we follow these steps for the skateboards minimum. The constraint $x \geq 4$ has two parts, $x = 4$ and $x > 4$. What sort of line do we draw in the *xy*-plane for $x = 4$? When is a point (x, y) on the line $x = 4$? Well, clearly, x must be equal to 4, so the points look like $(4, y)$.

But what about the y value? We note that any y value will "work," because there is no y in the equation $x = 4$, so no matter what y value we choose we can never substitute it into the equation and get a false statement. So not only will any y value work, every possible y value works. If we pick two y values, we get two points, and then we can draw a line. Suppose we choose $y = 0$ and $y = 30$. We can draw the two points (4, 0) and (4, 30) and then draw the line they determine. We do that in Figure 4.7b. In general, any line whose equation is of the form

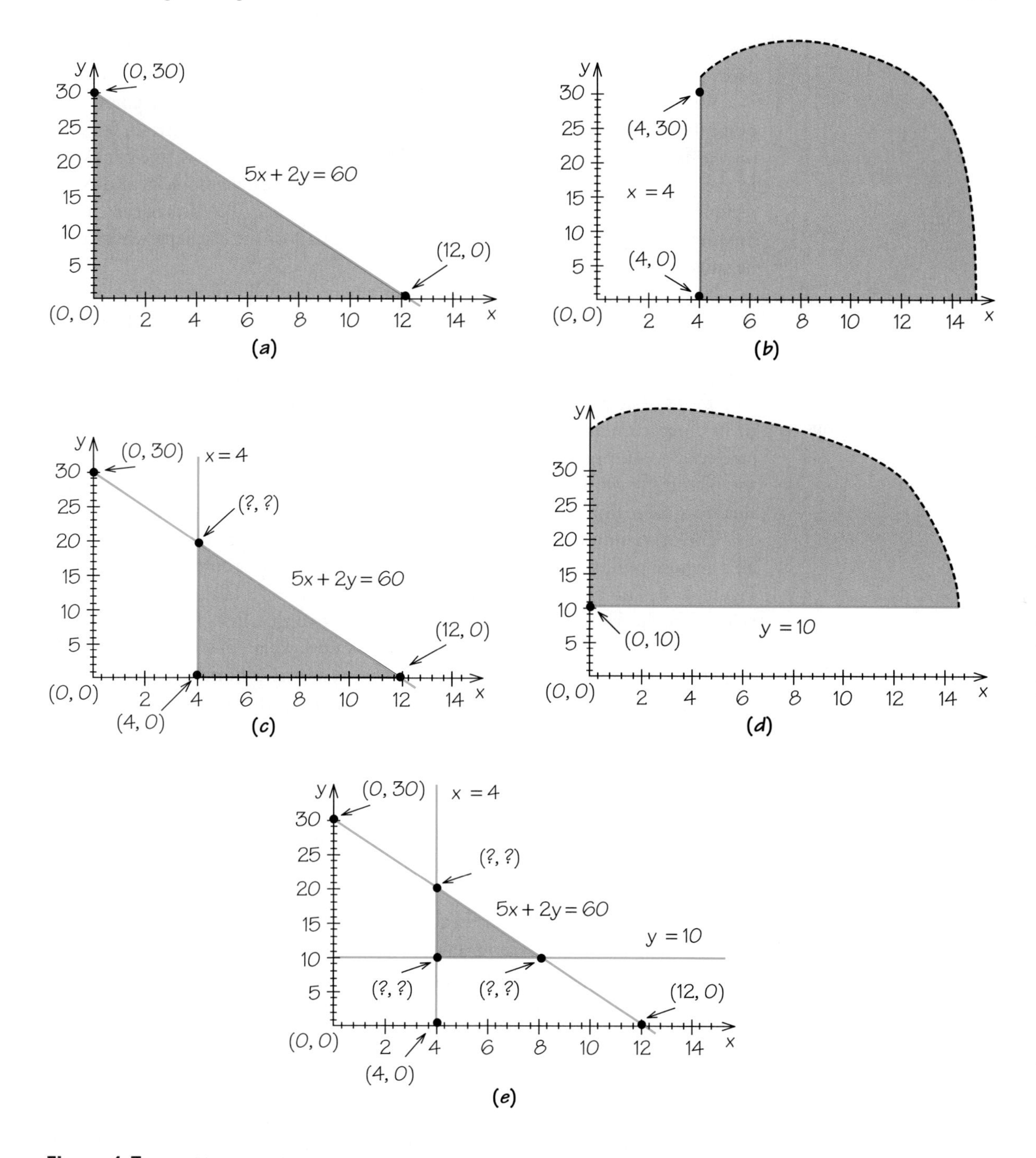

Figure 4.7 Feasible region for Skateboards and Dolls, Part 3 (with nonzero minimums). (a) Region where $5x + 2y \leq 60$. (b) Region where $x \geq 4$ is true (and also $y \geq 0$). (c) Region where both inequalities $x \geq 4$ and $5x + 2y \leq 60$ are true. (d) Region where $y \geq 10$ is true (and also $x \geq 0$). (e) Region where $x \geq 4$, $y \geq 10$, and $5x + 2y \leq 60$.

x = *some number* is a vertical line passing through the point (*that number,* 0). The y-axis is a special case, with equation $x = 0$.

When we substitute a point into the inequality $x > 4$—for example, the point (0, 0)—we get $0 > 4$, which is false. So we know that the point (0, 0) is not on the side of the line for which $x > 4$ is true. That inequality is true for points to the right of the line $x = 4$. Thus, the shaded area of Figure 4.7b shows the region of the plane for which $x \geq 4$ is true. Because $y \geq 0$ is always the case in mixture problems, the figure does not show the part of the xy-plane where y has negative values.

Now we need to draw the region where both inequalities $x \geq 4$ and $5x + 2y \leq 60$ are true. We need to find the shape that is shaded in both Figures 4.7a and 4.7b, to find the shape that is shaded twice. That shape is the intersection, or overlapping, of the two shaded regions and is shown in Figure 4.7c.

We see that the new region is a triangle, and we know the coordinates of two of its three corners. However, before we calculate the coordinates of the point labeled (?, ?) in Figure 4.7c, we should finish incorporating the minimums into the picture, so we know that our calculations are really of corner points in the final feasible region.

The other minimum constraint is $y \geq 10$. We need to graph the line $y = 10$, all of whose points are of the form (x, 10). This line is horizontal; it is graphed in Figure 4.7d. The point (0, 0) makes the inequality $y > 10$ false, so the region we want is the horizontal line and all the points above that line. Figure 4.7d shows the region where $y \geq 10$ is true for just those values where $x \geq 0$. In general, any line whose equation is of the form y = *some number* is a horizontal line passing through the point (0, *that number*). A special case is the x-axis, which has the equation $y = 0$.

Now we find the points that lie in both of the shaded regions in Figures 4.7c and 4.7d, giving us, in Figure 4.7e, the feasible region for our problem.

Finding Corner Points of a Feasible Region Having Nonzero Minimums

In Figure 4.7e, we see that the feasible region for the problem is a triangle with three corners, but as yet we do not know the coordinates of any of those corners. Not to worry. We can get lots of help from the vertical and horizontal lines from the minimum constraints. First, let us work out the coordinates of the lower left corner. Because that point is on the line $x = 4$, we know its x-coordinate must be 4. Similarly, its y-coordinate must be 10, because the point is on the line $y = 10$. So the point at the lower left is (4, 10).

We proceed clockwise around the boundary of the feasible region. The next corner point of the feasible region is directly above (4, 10). Its x-coordinate is also 4, but we need to do some calculation to find its y-coordinate. The point is on the intersection of the lines $x = 4$ and $5x + 2y = 60$. We need to find the y-coordinate of the point that has $x = 4$ and lies on that second line. We do this by substituting 4 for x in the equation of the second line. Here is the algebra:

$5(4) + 2y = 60$	Substitute $x = 4$ into the equation.
$20 + 2y = 60$	Multiply 5 by 4.
$2y = 40$	Substract the 20 from both sides of the equation.
$y = 20$	Divide both sides of the equation by 2.

So the coordinates of the point are (4, 20).

Finally, we find the coordinates of the third corner point of the triangle. This point has $y = 10$, so we make that substitution into the equation of the line $5x + 2y = 60$, because the point lies on that line. We get $5x + 2(10) = 60$, and then follow the same steps as we did earlier in a similar calculation. From $5x + 20 = 60$, or $5x = 40$, we get $x = 8$. So the coordinates of the point are (8, 10).

EXAMPLE

Incorporating Nonzero Minimums

Suppose the clothing manufacturer needs to make at least four shirts and six vests. Incorporate these minimums into the feasible region shown in Figure 4.4.

Solution: The minimum constraint $x \geq 4$ gives us a vertical line at $x = 4$. The part of the old feasible region to the left of that line is no longer feasible. Similarly, the minimum constraint $y \geq 6$ gives us a horizontal line. The part of the old feasible region below that line is no longer feasible. The new feasible region is shown in Figure 4.8. To find the coordinates of the new corner points, we use the equations of the two lines that meet at that corner. The point (4, 6) is on the lines $x = 4$ and $y = 6$. Substituting $x = 4$ (one line) into $3x + 2y = 60$ (other line), we get $3(4) + 2y = 60$, or $2y = 60 - 12 = 48$, so $y = 24$, giving us (4, 24). We get the coordinates (16, 6) by substituting $y = 6$ (one line) into $3x + 2y = 60$ (other line). ■

Evaluating the Profit Formula at the Corners of a Feasible Region with Nonzero Minimums

Now that we have the corner points of the feasible region for making skateboards and dolls, we can find out which point gives the maximum profit. The original

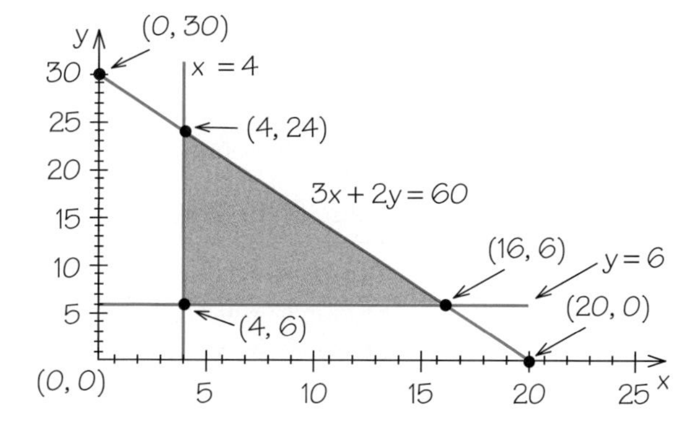

Figure 4.8 Feasible region for clothing manufacturer, with nonzero minimums.

TABLE 4.4 Evaluating One Profit Formula When There Are Nonzero Minimums

Corner Point	Value of the Profit Formula: $1.00x + $0.55y
(4, 10)	$1.00(4) + $0.55(10) = $4.00 + $5.50 = $9.50
(4, 20)	$1.00(4) + $0.55(20) = $4.00 + $11.00 = $15.00
(8, 10)	$1.00(8) + $0.55(10) = $8.00 + $5.50 = $13.50

profit formula, 1.00x$ + 0.55y$, gave a production policy to make no skateboards. This time all the feasible production policies have nonzero minimums, so that kind of result cannot occur. As we see from Table 4.4, the optimal production policy in this case is to make 4 skateboards and 20 dolls, for a maximum profit of $15.00. The policy calls for the absolute minimum number of skateboards, but not zero. And the "price paid for having the minimums" in this case is $16.50 (the old profit with minimums that are zeroes) − $15.00 (the profit now) = $1.50.

The second profit formula, 1.05x$ + 0.40y$, resulted in a production policy to make no dolls. With the nonzero minimums, that profit formula gives a production policy that says to make 8 skateboards and 10 dolls (check it by using the second formula in a table like Table 4.4), the minimum number, for a profit of $12.40. And the "price paid for having the minimums" in this case is $12.60 (the old profit with the minimums that are zeroes) − $12.40 (the profit now) = $0.20, a very small amount.

EXAMPLE

Evaluating a Profit Formula

Finish the clothing manufacturer problem by finding the corner point that maximizes the profit formula 5x$ + 3y$.

Solution: From Table 4.5 we see that the maximum profit of $98 is obtained by making 16 shirts and 6 vests. Note that the optimal production policy is still slanted toward shirts, with the manufacturer making just the minimum number of vests. ■

TABLE 4.5 Evaluating the Clothing Profit Formula When There Are Nonzero Minimums

Corner Point	Value of the Profit Formula: $5x + $3y
(4, 6)	$5(4) + $3(6) = $20 + $18 = $38
(4, 24)	$5(4) + $3(24) = $20 + $72 = $92
(16, 6)	$5(16) + $3(6) = $80 + $18 = $98

Summary of the Pictorial Method

Before we proceed to more involved linear programming problems, let's stop and summarize the steps we are following to find the optimal production policy in a mixture problem:

1. Read the problem carefully. Identify the resources and the products.
2. Make a mixture chart showing the resources (associated with limited quantities), the products (associated with profits), the recipes for creating the products from the resources, the profit from each product, and the amount of each resource on hand. If the problem has nonzero minimums, include a column for those as well.
3. Assign an unknown quantity, x or y, to each product. Use the mixture chart to write down the resource constraints, the minimum constraints, and the profit formula.
4. Graph the line corresponding to each resource constraint and determine which side of the line is in the feasible region. If there are nonzero minimum constraints, graph lines for them also, and determine which side of each is in the feasible region. Sketch the feasible region by finding the common points in the half planes from all the resource constraints plus the minimum constraints. (This process is called finding the "intersection" of the half planes.)
5. Find the coordinates of all the corner points of the feasible region. Some of these may have been calculated in order to graph the individual lines. Proceed in order around the boundary of the feasible region. Be sure that every point you consider is part of the feasible region.
6. Evaluate the profit formula for each of the corner points. The production policy that maximizes profit is the one that gives the biggest value to the profit formula.

4.3 Mixture Problems Having Two Resources

Two Products and Two Resources: Skateboards and Dolls, Part 4

We return for one last time to the toy manufacturer, now to consider two limited resources instead of one. The second limited resource will be time, the number of person-minutes available to prepare the products. Suppose that there are 360 person-minutes of labor available and that making one skateboard requires 15 person-minutes and making one doll requires 18 person-minutes. We will continue to use the original figures regarding containers of plastic, the first of our two profit formulas, and to keep the problem relatively simple, we will return to the zero minimum constraints: $x \geq 0$ and $y \geq 0$. We need a new mixture chart. In general, we will only include a column for minimums in a mixture chart if

PRODUCTS	Containers of plastic 60	Person-minutes 360	PROFIT
Skateboards (x units)	5	15	$1.00
Dolls (y units)	2	18	$0.55

RESOURCE(S)

Figure 4.9 Mixture chart for Skateboards and Dolls, Part 4 (two resources).

there are any nonzero minimum constraints. In Figure 4.9 we have the mixture chart for this problem. Using the mixture chart, we can write the two resource constraints:

$$5x + 2y \leq 60 \qquad \text{for containers of plastic}$$

and

$$15x + 18y \leq 360 \qquad \text{for person-minutes}$$

We can also write the profit formula: $\$1.00x + \$0.55y$.

The half plane corresponding to the plastic resource is shown in Figure 4.10a. We now need to graph the half plane corresponding to the time constraint. We find where the line $15x + 18y = 360$ intersects the two axes by substi-

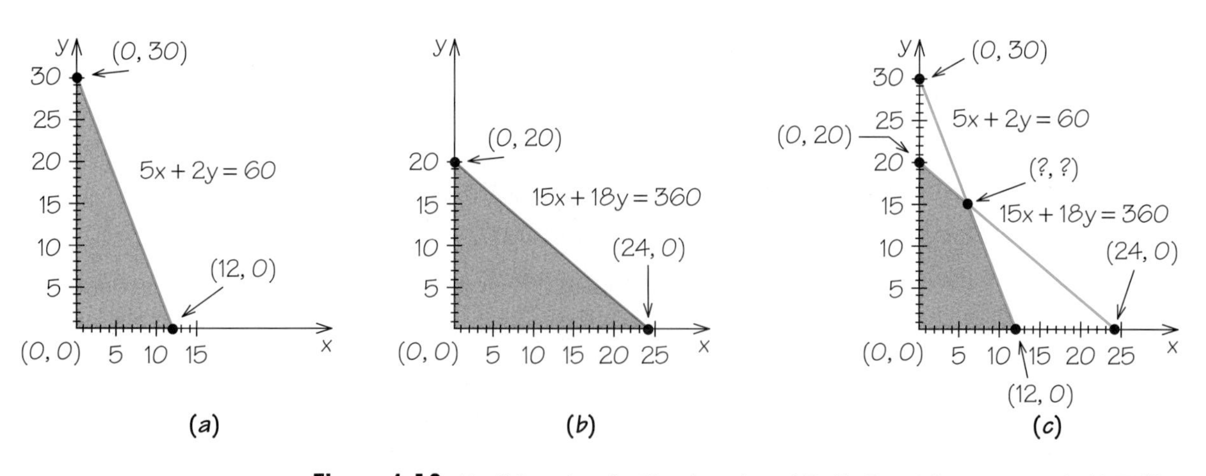

Figure 4.10 Feasible region for Skateboards and Dolls, Part 4 (two resources). (a) Half plane for the plastic resource constraint. (b) Half plane for the time resource constraint. (c) Intersection for the two half planes.

tuting first $x = 0$ and then $y = 0$ into that equation. Here's how that algebra looks:

$5(0) + 18y = 360$	Substitute $x = 0$ into the equation.
$18y = 360$	Simplify.
$y = 20$	Divide both sides of the equation by 18.

So one (x, y) point on the line is (0, 20). This is the point representing our making $x = 0$ skateboards and $y = 20$ dolls.

We now follow a similar procedure for making zero dolls. We substitute $y = 0$ into the equation $15x + 18y = 360$, getting $15x + 18(0) = 360$, and carry out similar calculations to the earlier ones, getting $x = 24$. So another (x, y) point on the line is (24, 0). This point represents our making $x = 24$ skateboards and $y = 0$ dolls.

The line corresponding to the time constraint contains the two points (0, 20) and (24, 0). When we substitute the point (0, 0) into the inequality $15x + 18y < 360$, we get $15(0) + 18(0) < 360$, or $0 < 360$, which is true, so (0, 0) is on the side of the line that we shade. Putting all this together, we get the half plane in Figure 4.10b as the correct half plane for the time resource constraint.

We are not permitted to exceed the supply of even a single resource; therefore, the feasible region must be made up of points that are shaded twice—both in the half plane for the plastic resource constraint, shown in Figure 4.10a, and in the half plane for the time resource constraint in Figure 4.10b. As we did before with the half planes from nonzero minimum constraints, we build our feasible region by finding the intersection, or overlap, of the individual half planes in the problem. In Figure 4.10c we show the result of intersecting the half planes from the two resource constraints. Because this problem has minimums that are zeroes, the shaded region in Figure 4.10c is in fact the feasible region for the problem.

We now need to find the coordinates of the corner points of the feasible region. In general, there are points that assist us in drawing the graph of the feasible region that are not corner points of the feasible region. One easy way to keep track is to shade in the feasible region and then systematically move around its boundary, working from one corner point to the next. We start at the origin, (0, 0), and proceed clockwise. The next corner point is one whose coordinates we calculated in order to draw a line, so we know that it is (0, 20). For the point labeled (?, ?) we need to use a bit of algebra. And the last of the four corner points is again one whose coordinates are already known to us, namely, (12, 0).

To find the coordinates of the point that lies on the intersection of the lines $5x + 2y = 60$ and $15x + 18y = 360$, we take these two equations with two variables, or unknowns, and eliminate one of the unknowns, leaving us with an equation having just one unknown. We have solved such an equation in earlier problems. Once we know the value of one of the unknowns, we simply substitute it into the equation of the line to get the other coordinate. Here are the worked-out details:

Write the two equations with "like terms" in columns:

$$5x + 2y = 60$$
$$15x + 18y = 360$$

Pick the variable to eliminate. Here we will eliminate y. Multiply the top equation by the coefficient of that variable in the bottom equation (the top equation gets multiplied by the 18 from the $18y$). Multiply the bottom equation by the coefficient of that variable in the top equation (the bottom equation gets multiplied by the 2 from the $2y$). Change one of the two multipliers to a negative number (the 2 became a -2):

$$18(5x + 2y = 60)$$
$$-2(15x + 18y = 360)$$

Do the actual multiplication. Note that the coefficients of the variable to be eliminated are now the same except for sign; now one is + and one is − :

$$90x + 36y = 1080$$
$$-30x - 36y = -720$$

Add the two equations by adding "like terms to like terms."

$$(90 - 30)x + (36 - 36)y = (1080 - 720)$$

$60x + 0y = 360, \quad \text{so} \quad 60x = 360$ — Simplify; the y variable "drops out."

$x = 6$ — Divide both sides by 60, getting the coordinate for x.

Now that we know the value of one of the coordinates of the point, we are in the same place algebraically as we were when we needed to find the coordinates of a point where a resource line meets a minimum-constraint line. We just substitute the value we have, in this case $x = 6$, into either of the two *original* equations and find the corresponding value of the other coordinate. For example, if we use the equation $5x + 2y = 60$, the substitution gives us $5(6) + 2y = 60$, which yields $y = 15$. So the point at which the two lines from resource constraints cross is (6, 15). You can verify that if you picked the other original equation to find the y-coordinate, you would still get $y = 15$. In fact, substituting (6, 15) into the equation $15x + 18y = 360$ and seeing that you get a true statement is a good way to check your calculations.

Some readers may notice that other pairs of numbers will work as multipliers that will cause elimination of the y variable. That is true, and if you know ways to find pairs that work, you of course may use them. Many "tricks" involving getting the smallest values for such numbers were very useful when all calculations were done by hand. Today, because calculators are available to most of us, these tricks are not as important.

TABLE 4.6 The Profit at the Four Corner Points

Corner Point	Value of the Profit Formula: $1.00x + $0.55y
(0, 0)	$1.00(0) + $0.55(0) = $0.00 + $0.00 = $0.00
(0, 20)	$1.00(0) + $0.55(20) = $0.00 + $11.00 = $11.00
(6, 15)	$1.00(6) + $0.55(15) = $6.00 + $8.25 = $14.25
(12, 0)	$1.00(12) + $0.55(0) = $12.00 + $0.00 = $12.00

Now we are ready to finish the problem. In Table 4.6 we have evaluated the profit formula at the four corner points of the feasible regions. The optimal production policy for the toy manufacturer would be to make 6 skateboards and 15 dolls, for a maximum profit of $14.25.

EXAMPLE

Mixtures of Two Fruit Juices: Beverages, Part 1

A juice manufacturer produces and sells two fruit beverages: 1 gallon of cranapple is made from 3 quarts of cranberry juice and 1 quart of apple juice; and 1 gallon of appleberry is made from 2 quarts of apple juice and 2 quarts of cranberry juice. The manufacturer makes a profit of 3 cents on a gallon of cranapple and 4 cents on a gallon of appleberry. Today, there are 200 quarts of cranberry juice and 100 quarts of apple juice available. How many gallons of cranapple and how many gallons of appleberry should be produced to obtain the highest profit without exceeding available supplies? We use zeroes as "reality minimums." The mixture chart for this problem is shown in Figure 4.11.

For each resource, we develop a resource constraint reflecting the fact that the manufacturer cannot use more of that resource than is available. The number of quarts of cranberry juice needed for x gallons of cranapple is $3x$. Similarly, $2y$ quarts of cranberry are needed for making y gallons of appleberry. So if the manufacturer makes x gallons of cranapple and y gallons of appleberry, then $3x + 2y$ quarts of cranberry juice will be used. Because there are only 200 quarts of cranberry available, we get the cranberry resource constraint $3x + 2y \leq 200$. Note that the numbers 3, 2, and 200 are all in the "cranberry" column. We get

PRODUCTS	RESOURCE(S): Cranberry 200 quarts	Apple 100 quarts	PROFIT
Cranapple (x gallons)	3 quarts	1 quart	3 cents/gallon
Appleberry (y gallons)	2 quarts	2 quarts	4 cents/gallon

Figure 4.11 A mixture chart for Beverages, Part 1.

another resource constraint from the column for the apple juice resource: $1x + 2y \leq 100$. We also have these minimum constraints: $x \geq 0$ and $y \geq 0$.

Finally, we have the profit formula. Because $3x$ is the profit from making x units of cranapple and $4y$ is the profit from making y units of appleberry, we get the profit formula: $3x + 4y$.

We summarize our analysis of the juice mixture problem. Maximize the profit formula, $3x + 4y$, given these constraints:

cranberry:	$3x + 2y \leq 200$		
apple:	$1x + 2y \leq 100$		
minimums:	$x \geq 0$	and	$y \geq 0$

Remember, in a mixture problem, our job is to find a production policy, (x, y), that makes all the constraints true and maximizes the profit.

Focus first on the cranberry juice constraint, $3x + 2y \leq 200$, and the associated equation $3x + 2y = 200$. We remember that when the line crosses, or intersects, the x-axis, the value of y is 0. Substituting $y = 0$ into the equation $3x + 2y = 200$ gives $3x = 200$, or $x = 200/3$, which we approximate by 66.7. So one point on the line $3x + 2y = 200$ is (66.7, 0). On the y-axis, $x = 0$. Substituting $x = 0$ into our equation $3x + 2y = 200$, we get $2y = 200$, or $y = 200/2$, giving us $y = 100$. So we have a second point on our line, namely, (0, 100). In Figure 4.12a these two points are shown and the line segment labeled $3x + 2y = 200$ was drawn by connecting them. To find out which half plane corresponds to the inequality $3x + 2y < 200$, we need only test the point (0, 0), which is not on the line. We see that $3(0) + 2(0) < 200$ is a true statement, so the inequality corresponds to the down side of the line, and that is the portion shaded in Figure 4.12a.

In Figure 4.12b we show the graph of the apple constraint inequality $1x + 2y \leq 100$. Note that the two points used to draw that line are (0, 50) and (100, 0),

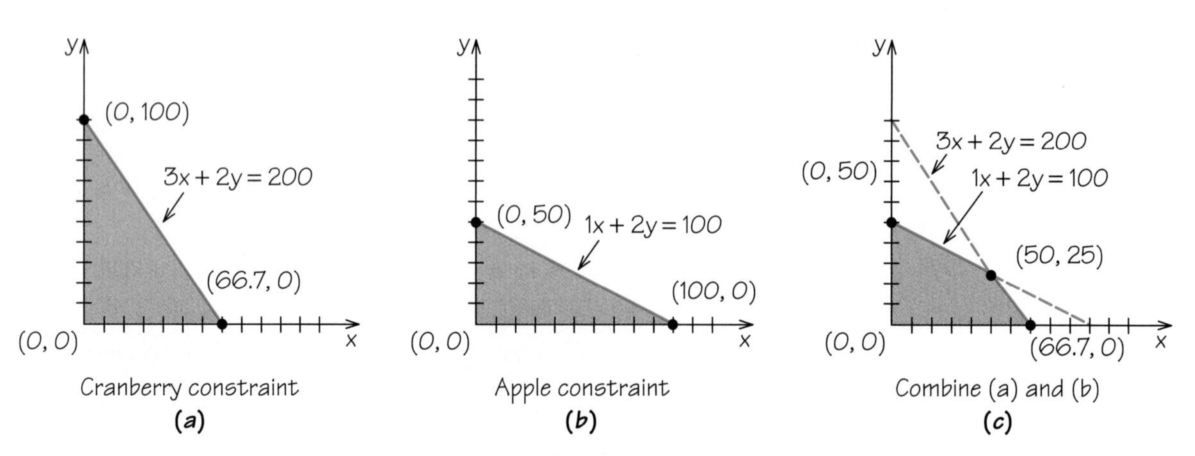

Figure 4.12 Feasible region for Beverages, Part 1.

the two intercepts. (Can you find those two intercepts by substituting first $x = 0$ and then $y = 0$ into the inequality $1x + 2y \leq 100$?) You can check that the correct half plane, the one for which $1x + 2y < 100$, has been shaded by substituting the point (0, 0) into that inequality and seeing that the result you get is a true statement.

We know that a point in the feasible region must satisfy, or make true, every constraint inequality in the problem. Which points in the plane make both the cranberry and the apple constraint inequalities true? They are the points that are in the overlap of the shaded regions in Figures 4.12a and 4.12b. The feasible region satisfying both resource constraints is shown in Figure 4.12c. In this problem we also have two minimum-constraint inequalities, $x \geq 0$ and $y \geq 0$, so our graph is drawn just in the quadrant in which both x and y are nonnegative.

We now find the corner points of the feasible region in Figure 4.12c. We start at the origin, (0, 0), and work our way clockwise around the boundary of the feasible region. Although we know the coordinates of the origin, it is useful to note that it is the intersection of two lines having equations $x = 0$, the y-axis, and $y = 0$, the x-axis, and corresponding to minimum constraints. These equations "solve" the problem of finding the coordinates. In general, we are trying to solve for values of x and y that satisfy both linear equations; then we will have the coordinates of the intersection.

The next corner is the intersection of the lines $x = 0$ and $1x + 2y = 100$. We found this point to be (0, 50) when we sketched the line $1x + 2y = 100$.

Continuing clockwise, we come to the intersection of lines $1x + 2y = 100$ and $3x + 2y = 200$. This more general type of intersection can be solved by using multiplication and addition to eliminate one unknown, solving for the remaining unknown, and then substituting that value into an original equation to get the value of the eliminated unknown.

First, we multiply one equation by a positive value and the other by a negative value so that when the two equations are added together, one unknown gets a coefficient of zero:

$$(-3)(1x + 2y = 100) = -3x - 6y = -300$$
$$(1)(3x + 2y = 200) = 3x + 2y = 200$$

Now we add the new equations together:

$$0x - 4y = -100 \qquad \text{or} \qquad -4y = -100$$

Dividing both sides of the equation by -4, we get $y = 25$. Substituting $y = 25$ into $1x + 2y = 100$, we get $1x + 2(25) = 100$, which gives $1x + 50 = 100$, and that simplifies to $x = 50$. The point of intersection therefore seems to be (50, 25). We can check our work in the other original equation: $3(50) + 2(25) = 200$ is a true statement.

The last corner point of this feasible region comes from the intersection of $3x + 2y = 200$ and $y = 0$. As with the other intersection of a resource constraint

TABLE 4.7 Finding the Optimal Production Policy for Beverages, Part 1

Corner Point	Value of the Profit Formula: $3x = 4y$ cents
(0, 0)	$3(0) + 4(0) = 0$ cents
(0, 50)	$3(0) + 4(50) = 200$ cents
(50, 25)	$3(50) + 4(25) = 250$ cents
(66.7, 0)	$3(66.7) + 4(0) = 200$ cents (rounded)

line and an axis, we already know the coordinates: (66.7, 0). Note that some points we used to draw the resource constraint lines are *not* corner points of the feasible region.

When we evaluate the profit formula at these four corner points (see Table 4.7), we see that the optimal production policy is to make 50 gallons of cranapple and 25 gallons of appleberry for a profit of 250 cents. ■

4.4 The Corner Point Principle

In finding solutions to our mixture problems, we have been using the corner point principle, which says that the highest profit value on a polygonal feasible region is always at a corner point. A feasible region has infinitely many points, making it impossible to compute the profit for each point. The corner point principle gives us a finite set of points, making the calculation possible.

You can visualize a mathematical proof of the corner point principle by imagining that each point of the plane is a tiny light bulb that is capable of lighting up. For the juice mixture example, whose feasible region is shown in Figure 4.12c, imagine what would happen if we ask this question: Will all points with profit = 360 please light up? What geometric figure do these lit-up points form?

In algebraic terms, we can restate the profit question in this way: Will all points (x, y) with $3x + 4y = 360$ please light up? As it happens, this version of the profit question is one mathematicians learned to answer hundreds of years before linear programming was born. The points that light up make a straight line because $3x + 4y = 360$ is the equation of a straight line. Furthermore, it is a routine matter to determine the exact position of the line. We call this line the **profit line** for 360; it is shown in Figure 4.13. For numbers other than 360, we would get different profit lines. Unfortunately, there are no points on the profit line for 360 that are feasible, that is, which lie in the feasible region. Therefore, the profit of 360 is impossible. *If the profit line corresponding to a certain profit doesn't touch the feasible region, then that profit isn't possible.*

Because 360 is too big, perhaps we should ask the profit line for a more modest amount, say, 160, to light up. You can see that the new profit line of 160 in Figure 4.13 is parallel to the first profit line and closer to the origin. This is no accident: All profit lines for the profit formula $3x + 4y$ have the same coeffi-

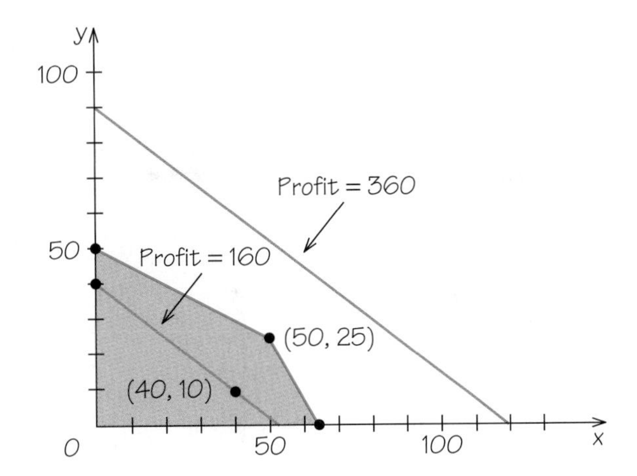

Figure 4.13 The profit line for 360 lies outside the feasbile region, whereas the profit line for 160 passes through the region.

cients for x and y—namely, 3 for x and 4 for y. Because the slope of the line is determined by those coefficients, they all have the same slope. Changing the profit value from 360 to 160 has the effect of changing where the line intersects the y-axis, but it does not affect the slope.

The most important feature of the profit line for 160 is that it has points in common with the interior of the feasible region. For example, (40, 10) is on that profit line because $3(40) + 4(10) = 160$, and in addition (40, 10) is a feasible point. This means that it is possible to make 40 gallons of cranapple and 10 gallons of appleberry and that if we do so, we will have a profit of 160.

Can we do better than a 160 profit? As we slowly increase our desired profit from 160 toward 360, the location of the profit line that lights up shifts smoothly upward away from the origin. As long as the line continues to cross the feasible region, we are happy to see it move away from the origin, because the more it moves, the higher the profit represented by the line. We would like to stop the movement of the line at the last possible instant, while the line still has one or more points in common with the feasible region. It should be obvious that this will occur when the line is just touching the feasible region either at a corner point (Figure 4.14a) or along a line segment joining two corners (Figure 4.14b). That point or line segment corresponds to the production policy or policies with the maximum achievable profit. This is just what the corner point principle says: The maximum profit always occurs at a corner or along an edge of the feasible region.

EXAMPLE

Adding Nonzero Minimums: Beverages, Part 2

Suppose that in the beverage example the profit for cranapple changes from 3 cents per gallon to 2 cents and the profit for appleberry changes from 4 cents per gallon to 5 cents. You can verify that this change moves the optimal production policy to the point (0, 50)—no cranapple is produced. This result is not surprising: Appleberry is giving a higher profit and the policy is to produce as much

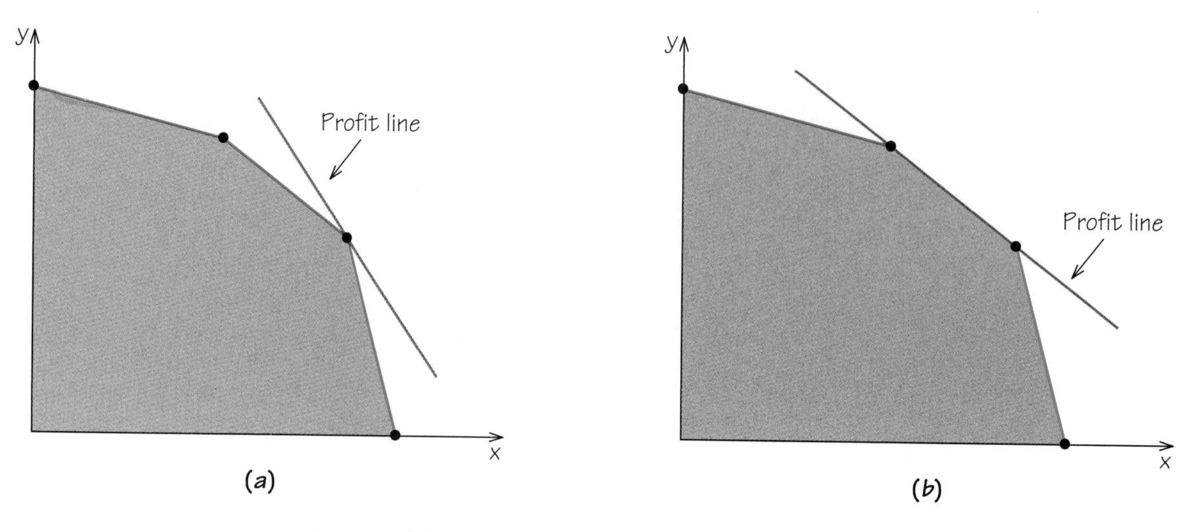

Figure 4.14 The highest profit will occur when the profit is just touching the feasible region, either (a) at the corner point or (b) along a line segment.

of it as possible. But suppose the manufacturer wants to incorporate nonzero minimums into the linear programming specifications so that there will always be both cranapple, x, and appleberry, y, produced. Specifically, they decide that $x \geq 20$ and $y \geq 10$ are desirable minimums. Figure 4.15 is the mixture chart showing the new profit formula and the nonzero minimums along with the unchanged rest of the beverage problem.

The feasible region for Beverages, Part 1, is shown in Figure 4.16a. The feasible region for Beverages, Part 2, is shown in Figure 4.16b. You can verify that, starting at the lower left corner of the new feasible region and moving clockwise around its boundary, we have corner points (20, 10), (20, 40), (50, 25), and (60, 10). (One of those points was also a corner point of the old feasible region. Can you explain why?) Table 4.8 shows the evaluation of the profit formula at these corner points. For this modified problem the optimal production policy is

	RESOURCE(S)			
PRODUCTS	Cranberry juice 200 quarts	Apple juice 100 quarts	MINIMUMS	PROFIT
Cranapple (x gallons)	3	1	20	2 cents
Appleberry (y gallons)	2	2	10	5 cents

Figure 4.15 Mixture chart for Beverages, Part 2.

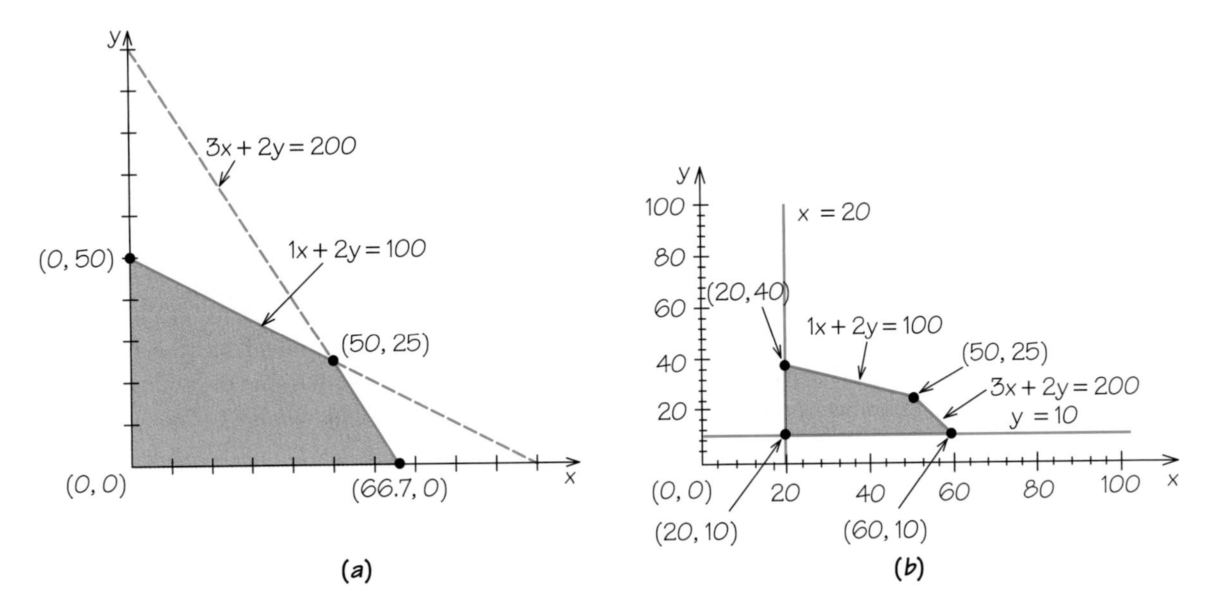

Figure 4.16 Feasible region for Beverages, Part 2. (a) Zero minimums. (b) Nonzero minimums.

TABLE 4.8 Profit Evaluation for Beverages, Part 2

Corner Point	Value of the Profit Formula: $2x + 5y$
(20, 10)	2(20) + 5(10) = 40 + 50 = 90 cents
(20, 40)	2(20) + 5(40) = 40 + 200 = 240 cents
(50, 25)	2(50) + 5(25) = 100 + 125 = 225 cents
(60, 10)	2(60) + 5(10) = 120 + 50 = 170 cents

to produce 20 gallons of cranapple and 40 of appleberry for a maximum profit of 240 cents.

One final note about this solution concerns the resources. The point (20, 40) is on the resource constraint line for the apple juice resource, so it represents using up all the available apple juice. We can see this by substituting into the apple juice resource constraint: $1(20) + 2(40) = 100$ is true. However, (20, 40) is *below* the line for the cranberry juice resource, indicating that there will be *slack,* or leftover, amounts of cranberry juice. Specifically, substituting (20, 40) into the cranberry juice constraint gives $3x + 2y = 3(20) + 2(40) = 60 + 80 = 140$, which is 60 quarts less than the 200 quarts available. The slack is 60 quarts of cranberry juice. Dealing with slack can be an important consideration for manufacturers. Can you see why? ■

Using a Computer to Calculate Profit at the Corner Points

Computers can do all of the calculations in a linear programming problem by following an algorithm. We discuss linear programming algorithms in the next section.

4.5 Linear Programming: The Wider Picture

Characteristics of Linear Programming Algorithms

Every algorithm for solving a linear programming problem has the following three characteristics, which hold true regardless of the number of products or the number of resources in the problem:

1. The algorithm can distinguish between "good" production policies – those in the feasible set that satisfy all the constraints – and those that violate some constraint(s) and are thus not feasible. There are usually many good points, each of which corresponds to some production policy; for example, "Make x units of product 1 and y units of product 2."
2. The algorithm makes use of some geometric principles – one such principle is the corner point principle – to select a special subset of the feasible set.
3. The algorithm evaluates the profit formula at points in the special subset to find which corner point actually gives the maximum profit.

The various algorithms for linear programming differ in how they find the feasible set and in how quickly the algorithm finds the production policy – corner point – that gives the optimal profit. The better algorithms are faster at finding the optimal production policy.

In practical linear programming problems, the feasible region will not be as simple as the ones we have examined here. There are two ways the feasible region can be more complex:

1. Sometimes, as in Figure 4.17, we have a great many corners. The more corners there are, the more calculations we need to determine the coordinates of all of them and the profit at each one. The number of corners

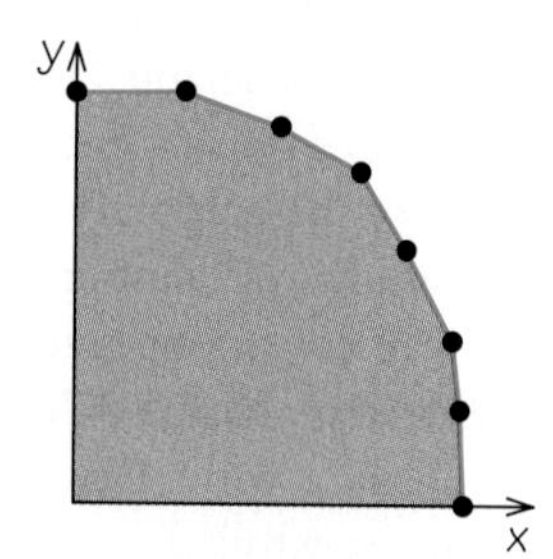

Figure 4.17 A feasible region with many corners.

literally can exceed the number of grains of sand on the earth. Even with the fastest computer, computing the profit of every corner is impossible.

2. It is not possible to visualize the feasible region as a part of two-dimensional space when there are more than two products. Each product is represented by an unknown, and each unknown is represented by a dimension of space. If we have 50 products, we would need 50 dimensions and couldn't visualize the feasible region.

Another type of complication can occur even in simple two-dimensional regions: Corner points can have fractional coordinates, not the integer ones we see in the specially constructed problems in this text. Making 3.75 skateboards and 5.45 dolls is not possible. Integer programming, a special type of linear programming, is used when it is not possible to use fractional answers.

The Simplex Method

Several methods are used to solve the typically large linear programming problems solved in practice. The oldest method is the **simplex method,** which is still the most commonly used. Devised by the American mathematician George Dantzig (see Spotlight 4.2), this ingenious mathematical invention makes it possible to find the best corner point by evaluating only a tiny fraction of all the corners. With the use of the simplex method, a problem that might be impossible to solve if each corner point had to be checked can be solved in a few minutes or even a few seconds on a typical business computer.

The operation of the simplex method may be likened to the behavior of an ant crawling on the edges of a polyhedron (a solid with flat sides) looking for an optimal corner point–one that gives the highest profit (Figure 4.18). The ant cannot see where the optimal corner is. As a result, if it were to wander along the edges randomly, it might take a long time to reach that corner. The ant will do much better if it has a temperature clue to let it know it is getting warmer (closer to the optimal corner) or colder (farther from the optimal corner).

Think of the simplex method as a way of calculating these temperature hints. We begin at any corner. All neighboring corners are evaluated to see which ones

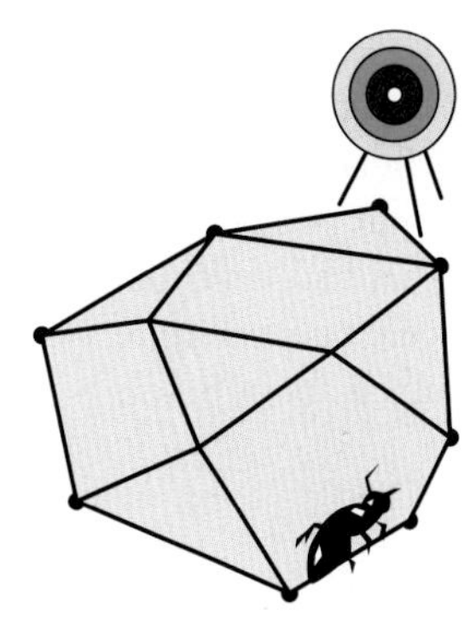

Figure 4.18 The simplex method can be compared to an ant crawling along the edges of a polyhedron, looking for the "target"–the optimal corner point.

SPOTLIGHT 4.2 Father of Linear Programming Recalls Its Origins

George Dantzig is professor of operations research and computer science at Stanford University. He is credited with inventing the linear programming technique called the simplex method. Since its invention in the 1940s, the simplex method has provided solutions to linear programming problems that have saved both industry and the military time and money. Dantzig talks about the background of his famous technique:

Initially, all the work we did had to do with military planning. During World War II, we were planning on a very extensive scale. The civilian population and the military were all performing scheduling and planning tasks, perhaps on a larger scale than at any time in history. And this was the case up until about 1950. From 1950 on, the whole emphasis shifted from military planning to practical planning for the civilian population, and industry picked it up.

The first areas of industry to use linear programming were the petroleum refineries. They used it for blending gasoline. Nowadays, all of the refineries in the world (except for one) use linear programming methods. They are one of the biggest users of it, and it's been picked up by every other industry you can think of – the forestry industry, the steel industry – you could fill up a book with all the different places it's used.

The question of why linear programming wasn't invented before World War II is an interesting one. In the postwar period, various technologies just evolved that had never been there before. Computers were one example. These technologies were talked about before. You can go back in history and you'll find papers on them, but these were isolated cases that never went anywhere.

In the immediate postwar period, everything just fermented and began to happen. One of the things that began to happen was linear programming. Mathematicians as well as economists and others who do practical planning and scheduling began to ask questions: How could you formulate the process as a sort of mathematical system? How could computers be used to make this happen?

The problems we solve nowadays have thousands of equations, sometimes a million variables. One of the things that still amazes me is to see a program run on the computer – and to see the answer come out. If we think of the number of combinations of different solutions that we're trying to choose the best of, it's akin to the stars in the heavens. Yet we solve them in a matter of moments. This, to me, is staggering. Not that we can solve them – but that we can solve them so rapidly and efficiently.

The simplex method has been used now for roughly 50 years. There has been steady work going on trying to use different versions of the simplex method, nonlinear methods, and interior methods. It has been recognized that certain classes of problems can be solved much more rapidly by special algorithms than by using the simplex method. If I were to say what my field of specialty is, it is in looking at these different methods and seeing which are more promising than others. There's a lot of promise in this – there's always something new to be looked at.

are warmer and which are colder. A new corner is chosen from among the warmer ones, and the evaluation of neighbors is repeated—this time checking neighbors of the new corner. The process ends when we arrive at a corner all of whose neighbors are colder than it is.

Part of what the simplex method has going for it is that it works faster in practice than its worst-case behavior would lead us to believe. Although mathematicians have devised artificial cases for which the simplex method bogs down in unacceptable amounts of arithmetic, the examples arising from real applications are never like that. This may be the world's most impressive counterexample to Murphy's law, which says that if something can go wrong, it will.

Although the simplex method usually avoids visiting every corner, it may require visiting many intermediate ones as it moves from the starting corner to the optimal one. The simplex method has to search along edges on the boundary of the polyhedron. If it happens that there are a great many small edges lying between the starting corner and the optimal one, the simplex method must operate like a slow-moving bus that stops on every block.

In the introduction to this chapter, we noted that linear programming accounts for over 50% and possibly as much as 90% of nonroutine computer time used for management decisions. Although there are alternatives to the simplex method, much of that computer time is spent using the simplex method.

Many computer programs are available that will use the simplex method to produce an optimal production policy if we just supply the computer with the constraint inequalities and profit formula. Simplex method programs can be found in a variety of places, among which are spreadsheets, packages of mathematics programs designed for business applications or finite mathematics courses, and large "all-purpose" mathematics packages. (See Note at the end of the Suggested Readings.) We hope you have access to such a program and try to use it. For this purpose, we have included some exercises (51 through 54) in this chapter that are larger and more realistic than the ones we have been solving graphically. A graphical solution is only possible for problems limited to two products; these special exercises involve more than two products.

An Alternative to the Simplex Method

In 1984, Narendra Karmarkar (see Figure 4.19), a mathematician working at Bell Laboratories, devised an alternative method for linear programming that finds the optimal corner point in fewer steps than the simplex algorithm by making use of search routes through the interior of the feasible region. The potential applications of Karmarkar's algorithm are important to a lot of industries, including telephone communications and the airlines (see Spotlight 4.3). Routing millions of long-distance calls, for example, means deciding how to use the resources of long-distance landlines, repeater amplifiers, and satellite terminals to best advantage. The problem is similar to the juice company's need to find the best use of its stocks of juice to create the most profitable mix of products.

Figure 4.19 Narendra Karmarkar, a researcher at AT&T Bell Laboratories, invented a powerful new linear programming algorithm that solves many complex linear programming problems faster and more efficiently than any previous method. (Courtesy of AT&T Labs.)

American Airlines worked with Karmarkar to see if his algorithm could cut fuel costs. According to Thomas Cook, director of operations research for American Airlines, "It's big dollars. We're hoping we can solve harder problems faster, and we think there's definite potential."

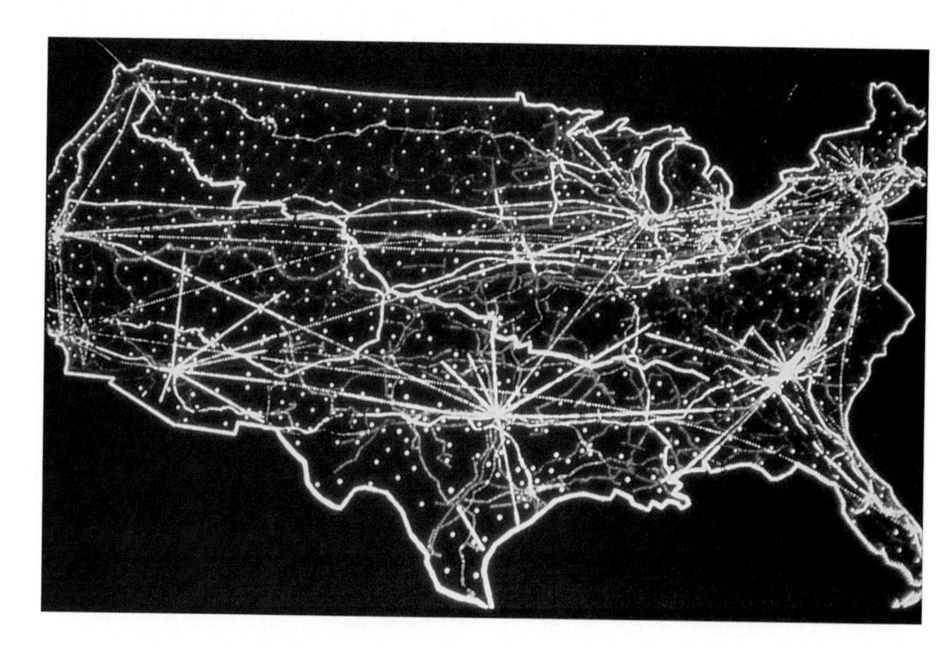

Figure 4.20 A map of the United States showing one conceivable network of major communication lines connecting major cities. Routing millions of calls over this immense network requires sophisticated linear programming techniques and high-speed computers. (Courtesy of AT&T Labs.)

SPOTLIGHT 4.3 Finding Fast Algorithms Means Better Airline Service

Linear programming techniques have a direct impact on the efficiency and profitability of major airlines. Thomas Cook, director of operations research at American Airlines, was interviewed in 1985 concerning his ideas on why optimal solutions are essential to his business:

> Finding an optimal solution means finding the best solution. Let's say you are trying to minimize a cost function of some kind. For example, we may want to minimize the excess costs related to scheduling crews, hotels, and other costs that are not associated with flight time. So we try to minimize that excess cost, subject to a lot of constraints, such as the amount of time a pilot can fly, how much rest time is needed, and so forth.
>
> An optimal solution, then, is either a minimum-cost solution or a maximizing solution. For example, we might want to maximize the profit associated with assigning aircrafts to the schedule; so we assign large aircraft to high-need segments and small aircraft to low-load segments.
>
> The simplex method, which was developed some 50 years ago by George Dantzig, has been very useful at American Airlines and, indeed, at a lot of large businesses. The difference between his method and Narendra Karmarkar's is speed. Finding fast solutions to linear programming problems is also essential. With an algorithm like Karmarkar's, which is 50 to 100 times faster than the simplex method, we could do a lot of things that we couldn't do otherwise. For example, some applications could be real-time applications, as opposed to batch applications. So instead of running a job overnight and getting an answer the next morning, we could actually key in the data or access the database, generate the matrix, and come up with a solution that could be implemented a few minutes after keying in the data.
>
> A good example of this kind of application is what we call a major weather disruption. If we get a major weather disruption at one of the hubs, such as Dallas or Chicago, then a lot of flights may get canceled, which means we have a lot of crews and airplanes in the wrong places. What we need is a way to put that whole operation back together again so that the crews and airplanes are in the right places. That way, we minimize the cost of the disruption as well as passenger inconvenience.

In the 1980s, scientists at Bell Labs applied Karmarkar's algorithm to a problem of unprecedented complexity: deciding how to economically build telephone links between cities so that calls can get from any city to any other, possibly being relayed through intermediate cities. Figure 4.20 shows one such linking. The number of possible linkings is unimaginably large, so picking the most economical one is difficult. For any given linking, there is also the problem of deciding how to economically route calls through the network to reach their destinations.

Although difficult, these problems are definitely worth solving. Nat Levine, director of the transmission facilities planning center at Bell Labs, speculated that if one found the best solution, "The savings could be in the hundreds of millions of dollars." Work on these problems at Bell Labs involved a linear programming problem with about 800,000 variables, which Karmarkar's algorithm solved in 10 hours of computer time. Scientists involved believe that the problem might have taken weeks to solve if the simplex method had been used. It appears that for some kinds of linear programming problems, Karmarkar's algorithm is a big improvement over the simplex method.

REVIEW VOCABULARY

Corner point principle The principle states that there is a corner point of the feasible region that yields the optimal solution.

Feasible point A possible solution (but not necessarily the best) to a linear programming problem. With just two products, we can think of a feasible point as a point on the plane.

Feasible region The set of all feasible points, that is, possible solutions to a linear programming problem. For problems with just two products, the feasible region is a part of the plane.

Feasible set Another term for **feasible region.**

Linear programming A set of organized methods of management science used to solve problems of finding optimal solutions, while at the same time respecting certain important constraints. The mathematical formulations of the constraints in linear programming problems are linear equations and inequalities. Mixture problems are usually solved by some type of linear programming.

Minimum constraint An inequality in a mixture problem that gives a minimum quantity of a product. Negative quantities can never be produced.

Mixture chart A table displaying the relevant data in a linear programming mixture problem. The table has a row for each product and a column for each resource, for any nonzero minimums, and for the profit.

Mixture problem A problem in which a variety of resources available in limited quantities can be combined in different ways to make different products. It is usually desired to find the way of combining the resources that produces the most profit.

Optimal production policy A corner point of the feasible region where the profit formula has a maximum value.

Production policy A point in the feasible set, interpreted as specifying how many units of each product are to be made.

Profit formula The expression, involving the unknown quantities such as x and y, that tells how much profit results from a particular production policy.

Profit line In a two-dimensional, two-product, linear programming problem, the set of all feasible points that yield the same profit.

Resource constraint An inequality in a mixture problem that reflects the fact that no more of a resource can be used than what is available.

Simplex method One of a number of algorithms for solving linear programming problems.

SUGGESTED READINGS

ANDERSON, DAVID R., DENNIS J. SWEENEY, and THOMAS A. WILLIAMS. *An Introduction to Management Science: Quantitative Approaches to Decision Making,* West, St. Paul, Minn., 1985. A business management text with seven chapters on linear programming.

BARNETT, RAYMOND A., and MICHAEL R. ZIEGLER. *Finite Mathematics for Business, Economics,*

Life Sciences and Social Sciences, 7th ed., Prentice Hall, Upper Saddle River, N.J., 1996. Chapter 5 presents both the geometric approach done here and the simplex method, with attention given to the geometric aspects of the simplex method.

GASS, SAUL I. *Decision Making, Models, and Algorithms,* Krieger, Melbourne, Fla., 1991. This unique book combines serious applied mathematics, including 12 chapters on linear programming, with a wonderful chatty style.

GASS, SAUL I. *An Illustrated Guide to Linear Programming,* McGraw-Hill, New York, 1970. An engagingly written beginner's approach that emphasizes the formulation of problems more than algebraic technique.

MEYER, WALTER. *Concepts of Mathematical Modeling,* McGraw-Hill, New York, 1984. Chapter 4 discusses several other types of linear programming problems, including minimization problems and the "transportation problem," for which there is a special algorithm, and linear programming problems in which the corner points must have coordinates that are whole numbers, not fractions.

Note: Simplex software can be found in *Maple* (keyword is *simplex*), *Mathematica* (keyword is *Linear Programming*), in both *Lotus 1-2-3* and *MSExcel* via *Solver,* and in other software packages, especially those intended for quantitative mathematics courses focusing on business applications.

SUGGESTED WEB SITE

www.informs.org This Web site is maintained by the Institute for Operations Research and the Management Sciences, the main professional organization in these fields in the United States. It contains information on (and/or links to) news items about operations research and management science and employment opportunities and summer internships; it also has a student newsletter. Much of the material is written in a nontechnical style.

SKILLS CHECK

1. Where do the lines $2x + 6y = 36$ and $y = 3$ intersect?

(a) At the point (3, 5)
(b) At the point (9, 3)
(c) At the point (0, 6)

2. Where do the lines $3x + y = 13$ and $2x + 3y = 18$ intersect?

(a) At the point (3, 4)
(b) At the point (6, 2)
(c) At the point (3, 2)

3. Which of these points lie in the region $4x + 3y \leq 24$?

(a) Points (5, 2) and (3, 4)
(b) Points (2, 5) and (3, 4)
(c) Points (5, 2) and (2, 5)

4. Producing a bench (x) requires 2 boards, and producing a table (y) requires 5 boards. There are 25 boards available. What is the resource constraint for this situation?

(a) $x + y \leq 25$
(b) $2x + 5y \leq 25$
(c) $5x + 2y \leq 25$

5. A tart requires 3 oz. of fruit and 2 oz. of dough; a pie requires 13 oz. of fruit and 7 oz. of dough. There are 140 oz. of fruit and 90 oz. of dough available. Each tart earns 6 cents profit; each pie earns 25 cents profit. What are the resource inequalities of this situation?

(a) $3x + 2y \leq 140$
$13x + 7y \leq 90$
$x \geq 0, y \geq 0$

(b) $3x + 13y \leq 140$
$2x + 7y \leq 90$
$x \geq 0, y \geq 0$

(c) $3x + 2y \leq 6$
$13x + 7y \leq 25$
$x \geq 0, y \geq 0$

6. A tart requires 3 oz. of fruit and 2 oz. of dough; a pie requires 13 oz. of fruit and 7 oz. of dough. There are 140 oz. of fruit and 90 oz. of dough available. Each tart earns 6 cents profit; each pie earns 25 cents profit. What is the profit formula for this situation?

(a) $P = 140x + 90y$
(b) $P = 70/3x + 18/5y$
(c) $P = 6x + 25y$

7. Graph the feasible region identified by the following inequalities:

$2x + 4y \leq 20$
$4x + 2y \leq 16$
$x \geq 0, y \geq 0$

Which of these points is *not* in the feasible region of the graph drawn?

(a) (2, 4)
(b) (1, 1)
(c) (10, 0)

8. Suppose the feasible region has four corners, at points (0, 0), (4, 0), (0, 3), and (3, 2). If the profit formula is $\$3x - \$2y$, what is the maximum profit possible?

(a) \$4
(b) \$5
(c) \$12

9. Suppose the feasible region has four corners, at points (0, 0), (4, 0), (0, 3), and (3, 2). For which of these profit formulae is the profit maximized by producing a mix of products?

(a) $\$x - \$y$
(b) $\$x + \$2y$
(c) $\$2x - \$y$

10. Suppose the feasible region has five corners, at points (1, 1), (2, 1), (3, 2), (2, 4), (1, 5). If the profit formula is $\$5x - \$3y$, which point maximizes the profit?

(a) (2, 1)
(b) (3, 2)
(c) (1, 5)

11. Suppose the feasible region has five corners, at points (1, 1), (2, 1), (3, 2), (2, 4), (1, 5). If the profit formula is $\$5x - \$3y$, what is the maximum profit possible?

(a) \$13
(b) \$10
(c) \$9

12. Suppose the feasible region has five corners, at points (1, 1), (2, 1), (3, 2), (2, 4), (1, 5). Which of these points is *not* in the feasible region?

(a) (1, 3)
(b) (2, 2)
(c) (0, 0)

13. Consider the feasible region identified by the inequalities below.
$x \geq 0$; $y \geq 0$; $3x + y \leq 10$; $x + 2y \leq 5$
Which point is *not* a corner of the region?

(a) (0, 2.5)
(b) (3, 1)
(c) (5, 0)

14. How does the line representing the maximum feasible profit intersect the feasible region?

(a) No points of intersection
(b) Only one point of intersection
(c) At least one point of intersection, and sometimes more than one point of intersection

EXERCISES

▲ Optional. ■ Advanced. ◆ Discussion.

Graphing

1. Using intercepts, the points where the lines cross the axes, graph each line.

(a) $2x + 3y = 12$
(b) $3x + 5y = 30$
(c) $4x + 3y = 24$
(d) $7x + 4y = 42$
(e) $x = 15$
(f) $y = 4$

2. Using intercepts, the points where the lines cross axes, graph each line.

(a) $4x + 5y = 20$
(b) $7x + 6y = 84$
(c) $5x + 4y = 60$
(d) $6x + 5y = 15$
(e) $y = 3$
(f) $x = 10$

3. Graph both lines on the same axes. Put a dot where the lines intersect. Use algebra to find the x- and y-coordinates of the point of intersection.

(a) $4x + 3y = 18$ and $x = 3$
(b) $5x + 3y = 45$ and $y = 5$

4. Graph both lines on the same axes. Put a dot where the lines intersect. Use algebra to find the x- and y-coordinates of the point of intersection.

(a) $3y + 5x = 60$ and $x = 9$
(b) $5x + 2y = 30$ and $y = 10$

5. Graph both lines on the same axes. Put a dot where the lines intersect. Use algebra to find the x- and y-coordinates of the point of intersection.

(a) $x + y = 10$ and $x + 2y = 14$
(b) $y - 2x = 0$ and $x = 4$

6. Graph both lines on the same axes. Put a dot where the lines intersect. Use algebra to find the x- and y-coordinates of the point of intersection.

(a) $x + y = 10$ and $x + 2y = 14$
(b) $y - 4x = 0$ and $y = 8$

Resource-Constraint Inequalities

7. Graph the line and half plane corresponding to the inequality, a typical constraint from a mixture problem.

(a) $x \geq 8$
(b) $y \geq 5$
(c) $5x + 3y \leq 15$
(d) $4x + 5y \leq 30$

8. Graph the line and half plane corresponding to the inequality, a typical constraint from a mixture problem.

(a) $x \geq 4$
(b) $y \geq 9$
(c) $3x + 2y \leq 18$
(d) $7x + 2y \leq 42$

In Exercises 9–13, for each description, write one or more appropriate resource-constraint inequalities. The unknown to use for each product is given in parentheses.

9. One bridesmaid's bouquet (x) requires 2 roses, and one corsage (y) requires 4 roses. There are 28 roses available.

10. Maintaining a large tree (x) takes 2 hours of pruning time and 30 minutes of shredder time; maintaining a small tree (y) takes 30 minutes of pruning time and 15 minutes of shredder time. There are 40 hours of pruning time and 2 hours of shredder time available.

11. Manufacturing one package of hot dogs (x) requires 6 ounces of beef, and manufacturing one package of bologna (y) requires 4 ounces of beef. There are 300 ounces of beef available.

12. It takes 30 feet of 12-inch board to make one bookcase (x); it takes 72 feet of 12-inch board to make one table (y). There are 420 feet of 12-inch board available.

13. Manufacturing one salami (x) requires 12 ounces of beef and 4 ounces of pork. Manufacturing one bologna (y) requires 10 ounces of beef and 3 ounces of pork. There are 40 pounds of beef and 480 ounces of pork available.

Graphing the Feasible Region

Graph the feasible region, label each line segment bounding it with the appropriate equation, and give the coordinates of every corner point.

14. $x \geq 0$; $y \geq 0$; $x + 2y \leq 12$

15. $x \geq 0$; $y \geq 0$; $2x + y \leq 8$

16. $x \geq 0; \quad y \geq 0; \quad 2x + 5y \leq 60$

17. $x \geq 10; \quad y \geq 0; \quad 3x + 5y \leq 120$

18. $x \geq 0; \quad y \geq 4; \quad x + y \leq 20$

19. $x \geq 2; \quad y \geq 6; \quad 3x + 2y \leq 30$

20. $x \geq 8; \quad y \geq 5; \quad 5x + 4y \leq 60$

Finding Points in the Feasible Region

Determine whether the points (a) (2, 4) and/or (b) (10, 6) are points of the given feasible regions of:

21. Exercises 15, 17, and 19.

22. Exercises 16, 18, and 20.

23. In the toy problem, x represents the number of skateboards and y the number of dolls. Using the version of that problem whose feasible region is presented in Figure 4.3b, with the profit formula $\$2.30x + \$3.70y$, write a sentence giving the maximum profit and describing the production policy that gives that profit.

24. In the toy problem, x represents the number of skateboards and y the number of dolls. Using the version of that problem whose feasible region is presented in Figure 4.3b, with the profit formula $\$5.50x + \$1.80y$, write a sentence giving the maximum profit and describing the production policy that gives that profit.

Finding the Point of Intersection of Two Resource Constraints

25. Graph both lines on the same axes. Put a dot where the lines intersect. Use algebra to find the x- and y-coordinates of the point of intersection.

(a) $5x + 4y = 22$ and $5x + 10y = 40$
(b) $x + y = 7$ and $3x + 4y = 24$

26. Graph both lines on the same axes. Put a dot where the lines intersect. Use algebra to find the x- and y-coordinates of the point of intersection.

(a) $x + 2y = 10$ and $5x + y = 14$
(b) $5x + 10y = 130$ and $7x + 4y = 112$

Graphing the Feasible Region

Graph the feasible region, label each line segment bounding it with the appropriate equation, and give the coordinates of every corner point.

27. $x \geq 0; \quad y \geq 0; \quad 3x + y \leq 9; \quad x + 2y \leq 8$

28. $x \geq 0; \quad y \geq 0; \quad 2x + y \leq 4; \quad 4x + 4y \leq 12$

29. $x \geq 0; \quad y \geq 2; \quad 5x + y \leq 14; \quad x + 2y \leq 10$

30. $x \geq 4; \quad y \geq 0; \quad 5x + 4y \leq 60; \quad x + y \leq 13$

Finding Points in the Feasible Region

Determine whether the points (a) (4, 2) and/or (b) (1, 3) are points of the given feasible regions of:

31. Exercises 27 and 29.

32. Exercises 28 and 30.

Linear Programming and its Geometry

33. Find the maximum value of P where $P = 3x + 2y$ subject to the constraints:
$x \geq 3; \quad y \geq 2; \quad x + y \leq 10; \quad 2x + 3y \leq 24$

34. Find the maximum value of P where $P = 3x - 2y$ subject to the constraints:
$x \geq 2; \quad y \geq 3; \quad 3x + y \leq 18; \quad 6x + 4y \leq 48$

35. Find the maximum value of P where $P = 5x + 2y$ subject to the constraints:
$x \geq 2; \quad y \geq 4; \quad x + y \leq 10$

36. Find the maximum value of P where $P = 5x + 10y$ subject to the constraints:
$x \geq 0; \quad y \geq 0; \quad 5x + 3y \leq 10$

(*Warning:* The corner point where the optimal solution occurs will not have integer values for both x and y.)

37. Given profit $P = 21x + 11y$ subject to the constraints:
$x \geq 0; \quad y \geq 0; \quad 7x + 4y \leq 13$

(a) Graph the feasible region.
(b) Determine a corner point where there is an optimal solution.

(*Warning:* The corner point where the optimal solution occurs will not have integer values for both x and y.)

38. (a) Referring to Exercise 37, use the usual rounding rule to round the x-coordinate and the y-coordinate of the point where the optimal linear programming solution occurs. Call the point with these coordinates Q.
(b) Determine if Q's coordinates define a feasible point by checking them against the constraints.
(c) Evaluate the profit value P at point Q. How does the profit value compare with the point where the optimal value occurred in Exercise 37?
(d) Let R be the point with coordinates (0, 3). Is R in the feasible region? Evaluate P at point R and compare the result with the answer at Q and where the optimum linear programming value occurred.
(e) Explain the significance of the situation here for solving maximization problems where $P = ax + by$ (a and b are known in advance) is subject to linear constraints but where the variables must be nonnegative integers rather than arbitrary nonnegative decimal numbers.

Mixture Problems

Exercises 39–50 each have several steps leading to a complete solution to a mixture problem. Practice a specific step of the solution algorithm by working out just that step for several problems. The steps are:

(a) Make a mixture chart for the problem.
(b) Using the mixture chart, write the profit formula and the resource- and minimum-constraint inequalities.
(c) Draw the feasible region for those constraints and find the coordinates of the corner points.
(d) Evaluate the profit information at the corner points to determine the production policy that best answers the question.
(e) (Requires technology) Compare your answer with the one you get from running the same problem on a simplex algorithm computer program.

39. A clothing manufacturer has 600 yards of cloth available to make shirts and decorated vests. Each shirt requires 3 yards of material and provides a profit of \$5. Each vest requires 2 yards of material and provides a profit of \$2. The manufacturer wants to guarantee that under all circumstances there are minimums of 100 shirts and 30 vests produced. How many of each garment should be made in order to maximize profit? If there are no minimum quantities, how, if at all, does the optimal production policy change?

40. A car maintenance shop must decide how many oil changes and how many tune-ups can be scheduled in a typical week. The oil change takes 20 minutes. The tune-up requires 100 minutes. The maintenance shop makes a profit of \$15 on an oil change and \$65 on a tune-up. What mix of services should the shop schedule if the typical week has available 8000 minutes for these two types of services? How, if at all, do the maximum profit and optimal production policy change if the shop is required to schedule at least 50 oil changes and 20 tune-ups?

41. A clerk in a bookstore has 90 minutes at the end of each workday to process orders received by mail or on voice mail. The store has found that a typical mail order brings in a profit of \$30 and a typical voice-mail order brings in a profit of \$40. Each mail order takes 10 minutes to process and each voice-mail order takes 15 minutes. How many of each type of order should the clerk process? How, if at all, do the maximum profit and optimal processing policy change if the clerk must process at least 3 mail orders and 2 voice-mail orders?

42. In a certain medical office, a routine office visit requires 5 minutes of doctors' time and a comprehensive office visit requires 25 minutes of doctors' time. In a typical week, there are 1800 minutes of doctors' time available. If the medical office clears \$30 from a routine visit and \$50 from a comprehensive visit, how many of each should be scheduled per week? How, if at all, do the maximum profit and optimal production policy change if the office is required to schedule at least 20 routine visits and 30 comprehensive ones?

43. A bakery makes 600 specialty breads–multigrain or herb–each week. Standing orders from restaurants are for 100 multigrain breads and 200 herb breads. The profit on each multigrain bread is $8 and on herb bread, $10. How many breads of each type should the bakery make in order to maximize profit? How, if at all, do the maximum profit and optimal production policy change if the bakery has no standing orders?

44. A student has decided that passing a mathematics course will, in the long run, be twice as valuable as passing any other kind of course. The student estimates that to pass a typical math course will require 12 hours a week to study and do homework. The student estimates that any other course will require only 8 hours a week. The student has available 48 hours for study per week. How many of each kind of course should the student take? (*Hint:* The profit could be viewed as 2 "value points" for passing a math course and 1 "value point" for passing any other course.) How, if at all, do the maximum value and optimal course mix change if the student decides to take at least 2 math courses and 2 other courses?

Problems 45–50 require finding the point of intersection of two lines, each corresponding to a resource constraint.

45. Webs-R-Us creates and maintains Web sites for client companies. There are two types of Web sites: "Hot" sites change their layout frequently but keep their content for long times; "cool" sites keep their layout for a while but frequently change their content. To maintain a hot site requires 1.5 hours of layout time and 1 hour for content changes. To maintain a cool site requires 1 hour of layout time and 2 hours for content changes. Every day, Webs-R-Us has available 12 hours for layout changes and 16 hours for content changes. Net profit is $50 for a set of changes on a hot site and $250 for a set of changes on a cool site. In order to maximize profit, how many of each type of site should Webs-R-Us maintain daily? How, if at all, do the maximum profit and optimal policy change if the company must maintain at least 2 hot and 3 cool sites daily?

46. A paper recycling company uses scrap cloth and scrap paper to make two different grades of recycled paper. A single batch of grade A recycled paper is made from 25 pounds of scrap cloth and 10 pounds of scrap paper, whereas one batch of grade B recycled paper is made from 10 pounds of scrap cloth and 20 pounds of scrap paper. The company has 100 pounds of scrap cloth and 120 pounds of scrap paper on hand. A batch of grade A paper brings a profit of $500, whereas a batch of grade B paper brings a profit of $250. What amounts of each grade should be made? How, if at all, do the maximum profit and optimal production policy change if the company is required to produce at least 1 batch of each type?

47. Jerry Wolfe has a 100-acre farm that he is dividing into one-acre plots, on each of which he builds a house. He then sells the house and land. It costs him $20,000 to build a modest house and $40,000 to build a deluxe house. He has $2,600,000 to cover these costs. The profits are $25,000 for a modest house and $60,000 for a deluxe house. How many of each type of house should he build in order to maximize profit? How, if at all, do the maximum profit and optimal production policy change if Wolfe is required to build at least 20 of each type of house?

48. The maximum production of a soft-drink bottling company is 5000 cartons per day. The company produces regular and diet drinks, and must make at least 600 cartons of regular and 1000 cartons of diet per day. Production costs are $1.00 per carton of regular and $1.20 per carton of diet. The daily operating budget is $5400. How many cartons of each type of drink should be produced if the profit is $0.10 per regular and $0.11 per diet? How, if at all, do the maximum profit and optimal bottling policy change if the company has no minimum required production?

49. Wild Things raises pheasants and partridges to restock the woodlands and has room to raise 100 birds during the season. The cost of raising one bird is \$20 per pheasant and \$30 per partridge. The Wildlife Foundation pays Wild Things for the birds; the latter clears a profit of \$14 per pheasant and \$16 per partridge. Wild Things has \$2400 available to cover costs. How many of each type of bird should they raise? How, if at all, do the maximum profit and optimal restocking policy change if Wild Things is required to raise at least 20 pheasants and 10 partridges?

50. Lights Afire makes desk lamps and floor lamps, on which the profits are \$2.65 and \$4.67, respectively. The company has 1200 hours of labor and \$4200 for materials each week. A desk lamp takes 0.8 hour of labor and \$4 for materials; a floor lamp takes 1.0 hour of labor and \$3 for materials. What production policy maximizes profit? How, if at all, do the maximum profit and optimal production policy change if Lights Afire wants to produce at least 150 desk lamps and 200 floor lamps per week?

In Exercises 51–54, there are more than two products in the problem. Although you cannot solve these problems using the two-dimensional graphical method, you can follow these steps:

(a) Make a mixture chart for each problem.
(b) Using the mixture chart, write the resource- and minimum-constraint inequalities. Also write the profit formula.
(c) (Requires technology) If you have a simplex method program available, run the program to obtain the optimal production policy.

51. A toy company makes three types of toys, each of which must be processed by three machines: a shaper, a smoother, and a painter. Each Toy A requires 1 hour in the shaper, 2 hours in the smoother, and 1 hour in the painter, and brings in a \$4 profit. Each Toy B requires 2 hours in the shaper, 1 hour in the smoother, and 3 hours in the painter, and brings in a \$5 profit. Each Toy C requires 3 hours in the shaper, 2 hours in the smoother, and 1 hour in the painter, and brings in a \$9 profit. The shaper can work at most 50 hours per week, the smoother 40 hours, and the painter 60 hours. What production policy would maximize the toy company's profit?

52. A rustic furniture company handcrafts chairs, tables, and beds. It has three workers, Chris, Sue, and Juan. Chris can only work 80 hours per month, but Sue and Juan can each put in 200 hours. Each of these artisans has special skills. To make a chair takes 1 hour of Chris's time, 3 from Sue, and 2 from Juan. A table needs 3 hours from Chris, 5 from Sue, and 4 from Juan. A bed requires 5 hours from Chris, 4 from Sue, and 8 from Juan. Even artisans are concerned about maximizing their profit, so what product mix should they stick with if they get \$100 profit per chair, \$250 per table, and \$350 per bed?

53. A candy manufacturer has 1000 pounds of chocolate, 200 pounds of nuts, and 100 pounds of fruit in stock. The Special Mix requires 3 pounds of chocolate, 1 pound each of nuts and fruit, and brings in \$10. The Regular Mix requires 4 pounds of chocolate, 0.5 pound of nuts, and no fruit, and brings in \$6. The Purist Mix requires 5 pounds of chocolate, no nuts or fruit, and brings in \$4. How many boxes of each type should be produced to maximize profit?

54. A gourmet coffee distributor has on hand 17,600 ounces of African coffee, 21,120 ounces of Brazilian coffee, and 12,320 ounces of Colombian coffee. It sells four blends – Excellent, Southern, World, and Special – on which it makes these per-pound profits, respectively: \$1.80, \$1.40, \$1.20, and \$1.00. One pound of Excellent is 16 ounces of Colombian; it is not a blend at all. One pound of Southern consists of 12 ounces of Brazilian and 4 ounces of Colombian. One pound of World requires 6 ounces of African, 8 of Brazilian, and 2 of Colombian. One pound of Special is made up of 10 ounces of African and 6 ounces of Brazilian. What product mix should the gourmet coffee distributor prepare in order to maximize profit?

55. Explain why finding a point of the feasible region that gives the maximum profit would be time-consuming and nearly impossible if we did not have the corner point principle.

56. Which steps of the pictorial method are required of a person who is solving a linear programming problem by using a simplex method computer program?

Additional Exercises

In Exercises 57 and 58, use the fact that the corner point approach can also solve minimization problems to minimize the given expression for cost *C*.

57. Minimize C given by $C = 4x + 7y$ over the feasible region for Exercise 33.

58. Minimize C given by $C = 3x + 11y$ over the feasible region for Exercise 34.

59. Show by example that a feasible region which has the nonnegativity constraints $x \geq 0$ and $y \geq 0$ can have no feasible points with integer coordinates other than (0, 0).

60. Find the maximum value of P given by $P = 0.6x + 3.8y$ where $x \geq 0$ and $y \geq 0$ and $0.9x + 1.2y \leq 3$.

61. Courtesy Calls makes telephone calls for businesses and charities. A profit of $0.50 is made for each business call and $0.40 for each charity call. It takes 4 minutes (on average) to make a business call and 6 minutes (on average) to make a charity call. If there are 240 minutes of calling time to be distributed each day, how should that time be spent so that Courtesy Calls makes a maximum profit? What changes, if any, occur in the maximum profit and optimal production policy if every day they must make at least 12 business and 10 charity calls?

62. A refinery mixes high-octane and low-octane fuels to produce regular and premium gasolines. The profits per gallon on the two gasolines are $0.30 and $0.40, respectively. One gallon of premium gasoline is produced by mixing 0.5 gallon of each of the fuels. One gallon of regular gasoline is produced by mixing 0.25 gallon of high octane with 0.75 gallon of low octane. If there are 500 gallons of high octane and 600 gallons of low octane available, how many gallons of each gasoline should the refinery make? How, if at all, do the maximum profit and optimal production policy change if the refinery is required to produce at least 100 gallons of each gasoline?

63. A toy manufacturer makes bikes, for a profit of $12, and wagons, for a profit of $10. To produce a bike requires 2 hours of machine time and 4 hours painting time. To produce a wagon requires 3 hours machine time and 2 hours painting time. There are 12 hours of machine time and 16 hours of painting time available per day. How many of each toy should be produced to maximize profit? How, if at all, do the maximum profit and optimal production policy change if the manufacturer must daily produce at least 2 bikes and 2 wagons?

64. The planner for the office holiday party needs to get the maximum "happiness points" out of the choices made for the foods. There are two kinds of foods that could be purchased: fancy foods and junk foods. Each order of fancy food costs $50 per order and adds 60 happiness points to the party. Each order of junk food costs $30 and adds 40 happiness points to the party. To satisfy both the formal and the casual types at the office, it is necessary to order at least 6 fancy foods and 5 junk foods. If there is $900 to spend, how many of each kind of food should the planner order? If there are no minimum quantities, how, if at all, does the food order change?

Exercises 65–70 are designed to highlight areas where the interface between the linear programming model and the reality of specific situations presents us with interesting discussion possibilities. These exercises do not have "right answers." Exercise 65 gives a concrete situation that could be used in the other questions.

◆ **65.** You are in charge of a business that produces sandwiches for snack bars. Make a list of your products (kinds of sandwiches), the resources you would need (sandwich ingredients), and the profit you might expect to get for each product. You need not specify the recipes numerically or the amounts of the resources available.

◆ **66.** Discuss the validity of the corner point principle if the solution (x, y) to the mixture problem is required not only to lie in the feasible region but also to have integer coordinates. In the sandwich problem (Exercise 65), would there be a useful meaning to a fractional number of some kind of sandwich?

◆ **67.** A linear programming analysis tells the sandwich company to stop making 7 of the 30 sandwiches in its product line. What are some business considerations that might suggest that this would be a bad business decision? How could linear programming still be used to help the sandwich company if it decided that it did not want to drop any of the 30 sandwiches?

◆ **68.** In a mixture problem we view recipes as fixed, but in practice this is not always true. For example, bolognas having only slightly different percentages of beef and pork all taste the same to the customers. What might prompt the manufacturer to vary a recipe? What effect might the varying have on the profit function, the feasible region, and the optimal product mix?

◆ **69.** Firms often give discounts for large-volume purchases. Does this necessarily contradict the assumption of a fixed (constant) profit on each unit sold? In examining this situation, you should note that discounts for large-volume purchases might apply not only to the products sold by a manufacturer but also to the prices paid by the manufacturer for resources.

◆ **70.** We learn in economics that prices are determined by the interplay of supply and demand. For example, the price of a product may fall if a large quantity of it is available. In mixture problems, however, we assume a fixed (constant) profit regardless of how much is produced. Is there a contradiction here? Could the model be adjusted to incorporate this economic fact of life?

71. **(a)** Determine the feasible region for the following collection of constraints: $x \geq 0$, $y \geq 0$, $x \leq 4$, $y \leq 7$.
(b) Determine the feasible region if the constraint $2x + y \leq 15$ is added to those in part (a).
(c) Determine the feasible region if the constraint $x + y \leq 8$ is added to those in part (b).
(d) How many corner points are there for the feasible regions in parts (a), (b), and (c) above?
(e) The feasible region R has arisen from a collection of two variable constraints. What is the largest number of corner points that can be added to the feasible region R if a single new (two-variable) constraint is added to those constraints that gave rise to R?

(*Note:* Part (c) of this Exercise illustrates the problem that one or more constraints may not change the corner points of a feasible region.)

✎ WRITING PROJECTS

1. Interview a local businessperson who is in charge of deciding the product mix for that business. Must this business take into consideration situations other than minimum and resource constraints? If so, what are these considerations? Find out what methods the person uses to make production policy decisions. Is linear programming used? Are other methods used? If so, what are they? Write a report of your findings, and add some of your own conclusions about the usefulness of linear programming for this business.

2. In economics, it is often useful to distinguish between a firm that has a monopoly (for example, is the only supplier of a product) and firms that supply only a small share of the market. How would the presence of a monopoly affect the relation between production and price? Would the presence of a monopoly tend to ensure the fixed-profit assumption of linear programming, or would it make it more likely that the interplay of supply and demand would have to be considered in order to have a truly realistic model?

SPREADSHEET PROJECTS

To do these projects, go to www.whfreeman.com/fapp.

Mixture problems are easily simulated by spreadsheets. Estimating optimal solutions by trying various proportions makes good use of the dynamic nature of spreadsheet calculations.

Producing Data

In this chapter we cover...

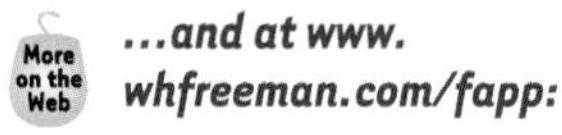

...and at www.whfreeman.com/fapp:

You can't read the newspaper or watch TV news without encountering data and statistical studies. You hear that last month's unemployment rate was 4.5%. A newspaper article says that only 21% of people aged 18 to 29 say that they always vote, as opposed to 59% of people 65 or older. A longer article says that low-income children who received high-quality day care as infants are more likely to go to college and hold good jobs than other similar children.

Another day, another headline. This one says, "Study Shows Aspirin Prevents Heart Attacks." We read on and learn that the study looked at 22,000 middle-aged doctors. Half took an aspirin every other day, and the other half took a dummy pill. In the aspirin group, 139 doctors suffered a heart attack. The dummy-pill group had 239 heart attacks in the same period. Is that difference large enough to show that aspirin really does prevent heart attacks?

To escape the unpleasantness of unemployment and heart attacks, we turn to our favorite advice columnist. Ann Landers has asked her female readers whether they would be content with affectionate treatment by men with no sex ever. More than 90,000 women wrote in, with 72% answering "yes." Can it really be true that 72% of women feel that way?

How trustworthy numbers are depends first of all on where they came from. You can trust the unemployment rate, but you ought to disbelieve Ann Landers's 72%. This chapter explains why. You will learn to recognize good and bad methods of producing numerical facts, which we call *data*. Understanding how to produce trustworthy data is the first—and the most important—step toward the ability to judge whether conclusions based on data are reliable or not. The design of trustworthy methods for producing data is our entry into the subject of statistics, the science of data.

5.1 Sampling

The National Center for Health Statistics wants to know what percent of Americans are covered by health insurance. The Gallup poll wants to know what percent of the public bought a state lottery ticket in the past year. A quality engineer must estimate what percent of the bearings rolling off an assembly line are defective. In all these situations we want to gather information about a large group of people or things. It is too expensive and time-consuming to contact every worker or inspect every bearing. So we gather information about only part of the group in order to draw conclusions about the whole.

The **population** in a statistical study is the entire group of individuals about which we want information.

A **sample** is a part of the population from which we actually collect information, used to draw conclusions about the whole.

We often draw conclusions about a whole on the basis of a sample. Everyone has sipped a spoonful of soup and judged the entire bowl on the basis of that taste. But a bowl of soup is homogeneous, so that the taste of a single spoonful represents the whole. Choosing a representative sample from a large and varied population is not so easy. The first step is to say carefully just what population we want to describe. The second step is to say exactly what we want to measure. These preliminary steps can be complicated, as this example illustrates.

EXAMPLE The Current Population Survey

Much government data are collected by sample surveys (see Figure 5.1). For example, the unemployment rate comes from the Current Population Survey (CPS), a sample of about 55,000 households each month. To measure unemployment, we must first specify the population we want to describe. Which age groups will we include? Will we include illegal aliens or people in prisons? What about full-time students? The CPS defines its population as all U.S. residents (whether citizens or not) 16 years of age and over who are civilians and are not in an institution such as a prison. The civilian unemployment rate announced in the news refers to this specific population.

Figure 5.1 The Web site for the National Health and Nutrition Examination Survey at the National Center for Health Statistics. This survey asks 5000 people each year about all aspects of their health and nutrition.

The second question is harder: What does it mean to be "unemployed"? Someone who is not looking for work–for example, a full-time student–should not be called unemployed just because she is not working for pay. If you are chosen for the CPS sample, the interviewer first asks whether you are available to work and whether you actually looked for work in the past four weeks. If not, you are neither employed nor unemployed–you are not in the labor force.

If you are in the labor force, the interviewer goes on to ask about employment. Any work for pay or in your own business the week of the survey counts you as employed. So does at least 15 hours of unpaid work in a family business. You are also employed if you have a job but didn't work because of vacation, being on strike, or other good reason. An unemployment rate of 4.7% means that 4.7% of the sample was unemployed, using the exact CPS definitions of both "labor force" and "unemployed." ■

5.2 Bad Sampling Methods

How can we choose a sample that is truly representative of the population? The easiest–but not the best–way to select a sample is to choose individuals close at hand. If we are interested in finding out how many people have jobs, for example, we might go to a shopping mall and ask people passing by if they are employed. A sample selected by taking the members of the population that are easiest to reach is called a **convenience sample.** Convenience samples often produce unrepresentative data.

EXAMPLE Convenience Samples

A sample of mall shoppers is fast and cheap. But people at shopping malls tend to be more prosperous than typical Americans. They are also more likely to be teenagers or retired. What is more, when we decide which people to question, we

(B. Daemmrich/The Image Works.)

will tend to choose well-dressed, respectable people and we will tend to avoid poorly dressed, unfriendly, or tough-looking individuals. In short, our shopping mall interviews will not contact a sample that is representative of the entire population and so will not accurately reflect the nation's rate of unemployment.

Our shopping mall sample will almost surely overrepresent middle-class and retired people and underrepresent the poor. This will happen every time we take such a sample. That is, it is a systematic error caused by a bad sampling method, not just bad luck on one sample. This is *bias:* The outcomes of mall surveys will repeatedly miss the truth about the population in the same ways. ■

The design of a statistical study is **biased** if it systematically favors certain outcomes.

EXAMPLE

Call-In Polls

Television news programs like to conduct call-in polls of public opinion. The program announces a question and asks viewers to call one telephone number to respond "yes" and another for "no." Telephone companies charge for these calls. The ABC program *Nightline* once asked whether the United Nations should continue to have its headquarters in the United States. More than 186,000 callers responded, and 67% said "no."

People who spend the time and money to respond to call-in polls are not representative of the entire adult population. In fact, they tend to be the same people who call radio talk shows. People who feel strongly, especially those with strong negative opinions, are more likely to call. It is not surprising that a properly designed sample showed that 72% of adults wanted the UN to stay in the United States. ■

Call-in opinion polls are an example of *voluntary response sampling.* A voluntary response sample can easily produce 67% "no" when the truth about the population is close to 72% "yes."

A **voluntary response sample** consists of people who choose themselves by responding to a general appeal. Voluntary response samples are biased because people with strong opinions, especially negative opinions, are most likely to respond.

5.3 Simple Random Samples

In a voluntary response sample, people choose whether to respond. In a convenience sample, the interviewer makes the choice. In both cases, personal choice produces bias. The statistician's remedy is to allow impersonal chance to choose the sample. A sample chosen by chance allows neither favoritism by the sampler nor self-selection by respondents. Choosing a sample by chance avoids bias by giving all individuals an equal chance to be chosen. Rich and poor, young and old, black and white – all have the same chance to be in the sample.

The simplest way to use chance to select a sample is to place names (the population) in a hat and draw out a handful (the sample). This is the idea of *simple random sampling*.

A **simple random sample (SRS)** of size n consists of n individuals from the population chosen in such a way that every set of n individuals has an equal chance to be the sample actually selected.

Picturing drawing names from a hat helps us understand what an SRS is. The same picture helps us see that an SRS is a better method of choosing samples than convenience or voluntary response sampling because it doesn't favor any part of the population. But writing names on slips of paper and drawing them from a hat is slow and inconvenient. That's especially true if, like the Current Population Survey, we must draw a sample of size 55,000. We can speed up the process by using a *table of random digits*. In practice, samplers use computers to do the work, but we can do it by hand for small samples.

A **table of random digits** is a long string of the digits 0, 1, 2, 3, 4, 5, 6, 7, 8, 9 with these two properties:

1. Each entry in the table is equally likely to be any of the 10 digits 0 through 9.
2. The entries are independent of one another. That is, knowledge of one part of the table gives no information about any other part.

Table 5.1 is a table of random digits. The digits in the table appear in groups of five to make the table easier to read, and the rows are numbered so we can refer to them, but the groups and row numbers are just for convenience. The entire table is one long string of randomly chosen digits. There are two steps in using the random-digit table to choose a simple random sample.

Step 1. **Label** Give each member of the population a numerical label of the *same length*. Up to 100 items can be labeled with two digits: 01, 02, . . . , 99, 00. Up to 1000 items can be labeled with three digits, and so on.

Step 2. **Table** To choose a simple random sample, read from Table 5.1 successive groups of digits of the length you used as labels. Your sample contains the individuals whose labels you find in the table. This gives all individuals the same chance because all labels of the same length have the same chance to be found in the table. For example, any pair of digits in the table is equally likely to be any of the 100 possible labels 01, 02, . . . , 99, 00. Ignore any group of digits that was not used as a label or that duplicates a label already in the sample.

TABLE 5.1 Random Digits

101	19223	95034	05756	28713	96409	12531	42544	82853
102	73676	47150	99400	01927	27754	42648	82425	36290
103	45467	71709	77558	00095	32863	29485	82226	90056
104	52711	38889	93074	60227	40011	85848	48767	52573
105	95592	94007	69971	91481	60779	53791	17297	59335
106	68417	35013	15529	72765	85089	57067	50211	47487
107	82739	57890	20807	47511	81676	55300	94383	14893
108	60940	72024	17868	24943	61790	90656	87964	18883
109	36009	19365	15412	39638	85453	46816	83485	41979
110	38448	48789	18338	24697	39364	42006	76688	08708
111	81486	69487	60513	09297	00412	71238	27649	39950
112	59636	88804	04634	71197	19352	73089	84898	45785
113	62568	70206	40325	03699	71080	22553	11486	11776
114	45149	32992	75730	66280	03819	56202	02938	70915
115	61041	77684	94322	24709	73698	14526	31893	32592
116	14459	26056	31424	80371	65103	62253	50490	61181
117	38167	98532	62183	70632	23417	26185	41448	75532
118	73190	32533	04470	29669	84407	90785	65956	86382
119	95857	07118	87664	92099	58806	66979	98624	84826
120	35476	55972	39421	65850	04266	35435	43742	11937
121	71487	09984	29077	14863	61683	47052	62224	51025
122	13873	81598	95052	90908	73592	75186	87136	95761
123	54580	81507	27102	56027	55892	33063	41842	81868
124	71035	09001	43367	49497	72719	96758	27611	91596
125	96746	12149	37823	71868	18442	35119	62103	39244
126	96927	19931	36809	74192	77567	88741	48409	41903
127	43909	99477	25330	64359	40085	16925	85117	36071
128	15689	14227	06565	14374	13352	49367	81982	87209
129	36759	58984	68288	22913	18638	54303	00795	08727
130	69051	64817	87174	09517	84534	06489	87201	97245
131	05007	16632	81194	14873	04197	85576	45195	96565
132	68732	55259	84292	08796	43165	93739	31685	97150
133	45740	41807	65561	33302	07051	93623	18132	09547
134	27816	78416	18329	21337	35213	37741	04312	68508
135	66925	55658	39100	78458	11206	19876	87151	31260
136	08421	44753	77377	28744	75592	08563	79140	92454
137	53645	66812	61421	47836	12609	15373	98481	14592
138	66831	68908	40772	21558	47781	33586	79177	06928
139	55588	99404	70708	41098	43563	56934	48394	51719

(*Continued*)

TABLE 5.1 Random Digits *(Continued)*

140	12975	13258	13048	45144	72321	81940	00360	02428
141	96767	35964	23822	96012	94591	65194	50842	53372
142	72829	50232	97892	63408	77919	44575	24870	04178
143	88565	42628	17797	49376	61762	16953	88604	12724
144	62964	88145	83083	69453	46109	59505	69680	00900
145	19687	12633	57857	95806	09931	02150	43163	58636
146	37609	59057	66967	83401	60705	02384	90597	93600
147	54973	86278	88737	74351	47500	84552	19909	67181
148	00694	05977	19664	65441	20903	62371	22725	53340
149	71546	05233	53946	68743	72460	27601	45403	88692
150	07511	88915	41267	16853	84569	79367	32337	03316

SPOTLIGHT 5.1 Is It Really Random?

Are the random digits in Table 5.1 really random? Not a chance. They were produced by a computer program. A computer program implements an algorithm that does exactly what you tell it to do. Give the program the same input and it will produce exactly the same "random" digits. You can get quite respectable random digits by calculating $\pi = 3.14159265358979\ldots$ to more and more decimal places. Go to www.cecm.sfu.ca/pi/pi.html and you will see these digits stream by. You get the same digits on every visit, of course. Clever people have devised algorithms that produce output that *looks* like random digits. These are called "pseudo-random numbers," and that's what Table 5.1 contains. Pseudo-random numbers work fine for statistical randomizing, but they have hidden nonrandom patterns that can mess up more refined uses.

For purists, the RAND Corporation long ago published a book titled *One Million Random Digits*. The book lists 1 million digits that were produced by a very elaborate physical randomization and really are random. An employee of RAND once told me that this is not the most boring book that RAND has ever published.

Cryptologists and computer scientists would like an endless supply of really random digits. Really random digits must come from nature, not from a computer program. Radioactive decay is really random, and so is the "thermal noise" in an amplifier, which you can hear as a soft whoosh if you turn up the volume with no music playing. Alas, extracting random digits from these really random sources requires various human devices, and these often impose subtle patterns. As of now, in fact, pseudo-random numbers from the best algorithms actually look more random than numbers refined from the randomness in nature by some human apparatus. "Easy to say, hard to do" applies to making random digits as well as to many other human aspirations.

EXAMPLE

Sampling Autos

An auto manufacturer wants to select 5 of the last 50 cars produced on an assembly line for a very detailed quality inspection. Can you see why allowing the workers to choose 5 cars is likely to cause bias? To avoid bias, we will choose a simple random sample.

Step 1. **Label** Give each car a numerical label. Because two digits are needed to label 50 cars, all labels will have two digits. The labels are 01 to 50, as shown in Figure 5.2. Always say how you labeled the members of the population. To sample from the day's production of 1240 cars, you would label the cars 0001, 0002, . . . , 1239, 1240.

Step 2. **Table** Now go to Table 5.1. Starting at line 130 (any line will do), we find

69051 64817 87174 09517 84534 06489 87201 97245

Because our labels are two digits long, we read successive two-digit groups from the table. Ignore groups not used as labels, like the initial 69. Also ignore any repeated labels, like the second 17 in this row, because we can't choose the same car twice. Our sample contains the cars labeled 05, 16, 48, 17, and 40. ■

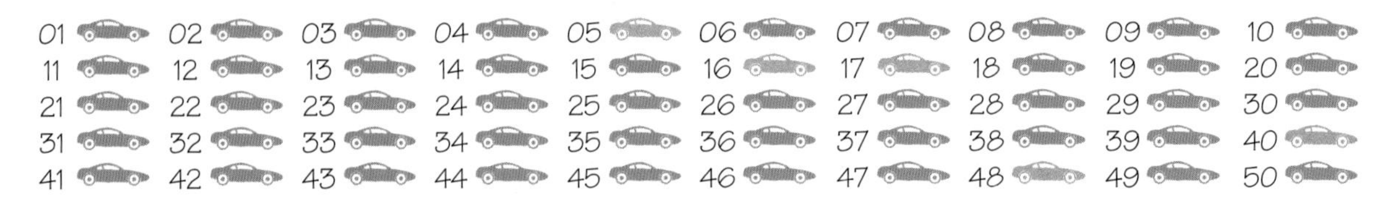

Figure 5.2 Random sampling: Assign labels to 50 cars, then use Table 5.1 to choose the 5 orange cars as a sample.

Call-in polls and mall interviews produce samples. We can't trust results from these samples, because they are chosen in ways that invite bias. We have more confidence in results from an SRS, because it uses impersonal chance to avoid bias. The first question to ask about any sample is whether it was chosen at random. Opinion polls and other sample surveys carried out by people who know what they are doing use random sampling. Many national samples use schemes more complex than an SRS. They may, for example, dial the last four digits of a telephone number at random separately within each exchange (the area code and first three digits). The national sample is pieced together from many smaller samples. The big idea remains the deliberate use of chance to choose the sample. Because simple random sampling is the essential principle behind all random sampling and because it is also the main building block for more complex samples, we will focus our study on simple random sampling.

EXAMPLE A Gallup Poll

A Gallup poll on smoking began with the question, "Have you, yourself, smoked any cigarettes in the past week?" The press release reported that "just 23% of Americans smoke." Can we trust this fact? Ask first how Gallup selected its sample. Later in the press release we read this: "The results are based on telephone interviews with a randomly selected national sample of 1039 adults, 18 years and older, conducted September 23–26, 1999."

This is a good start toward gaining our confidence. Gallup tells us what population it has in mind (people at least 18 years old living anywhere in the United States). We know that the sample from this population was of size 1039 and, most important, that it was chosen at random. There is more to say, and we will soon say it, but we have at least heard the comforting words "randomly selected." ■

5.4 Statistical Estimation

You know that lotteries are popular. How popular? Here's what the Gallup poll says: "They can offer massive jackpots–and a ticket just costs a dollar at your neighborhood store. For many Americans, picking up a lottery ticket has become routine, despite the massive odds against striking it rich. A new Gallup Poll Social Audit on gambling shows that 57% of Americans have bought a lottery ticket in the last 12 months, making lotteries by far the favorite choice of gamblers." Reading further (Figure 5.3), we find that Gallup talked with 1523 randomly selected adults to reach these conclusions.

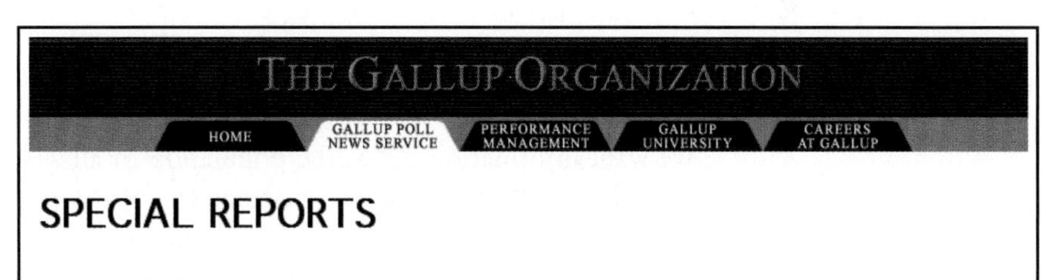

THE GALLUP ORGANIZATION

HOME | GALLUP POLL NEWS SERVICE | PERFORMANCE MANAGEMENT | GALLUP UNIVERSITY | CAREERS AT GALLUP

SPECIAL REPORTS

Gambling in America

Topline and Trends

Methodology

The study is based on telephone interviews with a randomly selected national sample of 1,523 adults (18+) and 501 teens (ages 13-17), conducted April 30 - May 23, 1999. For results based on a sample of this size, one can say with 95 percent confidence that the error attributable to sampling and other random effects could be plus or minus 3 percentage points for adults (18+), and plus or minus 5 percentage for teens (13-17 year olds).

Figure 5.3 The Gallup Organization's Web page for its poll on gambling in America.

(Ray Stubblebine/Reuters.)

EXAMPLE Do You Lotto?

Gallup asked a sample of 1523 adults, "Please tell me whether or not you bought a state lottery ticket in the past 12 months." Of these people, 868 said "yes." So the percent of the sample who said "yes" is

$$\frac{868}{1523} = 0.57 = 57\%$$

We want information about the population of all 210 million U.S. residents age 18 and over. We don't know what percent of the population would say "yes" if we asked them about lottery tickets. Because everyone had the same chance to be in the sample, we expect the sample to represent the population. So we estimate that about 57% of all adults bought lottery tickets in the past 12 months. This is a basic move in statistics: Turn the *fact* that 57% of the *sample* bought lottery tickets into an *estimate* that about 57% of *all adults* bought tickets. ■

If Gallup took a second random sample of 1523 adults, the new sample would have different people in it. It is almost certain that there would not be exactly 868 "yes" responses. That is, the percent who bought lottery tickets will *vary* from sample to sample. Could it happen that one random sample finds that 57% of adults recently bought a lottery ticket and a second random sample finds that only 37% had done so? Random samples eliminate *bias* from the act of choosing a sample, but they can still be wrong because of the *variability* that results when we choose at random. If the variation when we take repeat samples from the same population is too great, we can't trust the results of any one sample.

There are different kinds of variability. The answers obtained by sending an interviewer to a shopping mall vary in a haphazard and unpredictable way. Repeated random samples, however, vary in a regular manner because a specific chance mechanism is used to choose the sample. The long-run results are not haphazard. Think of tossing a coin 1523 times. The results will vary if we repeat the 1523 tosses, but we can say how much they will vary because they will show a regular pattern in the long run. Let's see what would happen if Gallup repeated its sample many times. This is another basic move in statistics: *To see how trustworthy one sample is likely to be, ask what would happen if we took many samples from the same population.*

EXAMPLE

Lots and Lots of Samples

Suppose that in fact (unknown to Gallup), exactly 60% of all adults have bought a lottery ticket in the past 12 months. Can we trust a sample of 1523 to come close to this truth? To find out, we took 1000 simple random samples from a population with exactly 60% "yes" and recorded the percent of "yes" responses in each sample.

Figure 5.4 shows the process of repeated sampling. The first sample gave 58.9%, the second 58.2%, the third 60.1%, and so on. The sample percents do vary. Collect all the sample percents and draw a **histogram.** The result appears at the right of Figure 5.4. The height of each bar in the histogram shows how often the outcomes covered by the base of that bar occurred. For example, the height of the bar covering 59.5% to 60.0% is 177 because 177 of our 1000 samples had between 59.5% and 60.0% "yes" responses. ■

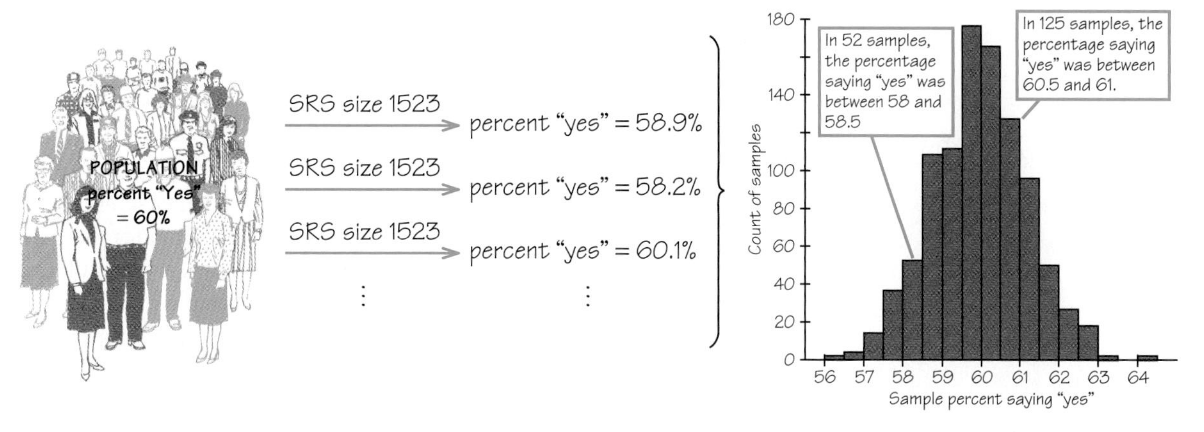

Figure 5.4 The results of many SRSs have a regular pattern. Here, we draw 1000 SRSs of size 1523 from the same population. The population has 60% "yes" responses. The sample percents vary from sample to sample, but their values center at the truth about the population.

Look carefully at Figure 5.4. We flow from the population, to many samples from the population, to the many values of the percent of "yes" responses from these many samples. Gather these values together and study the histogram that displays them.

- The values of the sample percent do vary from sample to sample, but their values are centered at 60%, the truth about the population. Some samples give an estimate that is too high and others give an estimate that is too low, but there is *no bias,* no tendency to be systematically high or systematically low.
- The responses follow a regular pattern. Results that are near the truth about the population are most common, and the bar heights fall off on either side. Most samples give estimates quite close to the population truth. In fact, 95% of our 1000 samples of size 1523 have between 57.6% and and 62.4% "yes" responses. That is, these samples give an estimate within $\pm 2.4\%$ of the truth about the population.

The upshot is that we can rely on a sample of size 1523 to almost always give an estimate that is close to the truth about the population. Figure 5.4 illustrates this fact for just one value of the population proportion, but it is true for any population. This is wonderful–there are 210 million adults in the country, and choosing just 1523 of them at random allows us to describe their opinions quite accurately.

The regular pattern of the histogram in Figure 5.4 isn't an accident. Using chance to select samples forces the pattern of the results of a large number of samples to have the regular shape that the figure displays. We don't actually have to take thousands of samples to learn what the shape is. The mathematics of chance allows us to calculate it in advance. We'll learn more about that in Chapter 8. Here are the basic facts that explain why we can trust sample estimates:

- If we take many random samples, *the pattern of results is centered about the population truth*. The center of the histogram reflects the lack of bias in random sampling.
- *The spread of the pattern is controlled by the size of the sample.* Larger samples give results that cluster closer to the population truth than do smaller samples. So the larger the sample, the more confident we can be that it estimates the population truth accurately. Opinion polls usually interview between 1000 and 2000 people. The Current Population Survey uses a sample of 55,000 households because the government wants to know the unemployment rate very accurately.

Samplers usually tell us how accurate their results are by giving a *margin of error*. We can't be *certain* that the sample results are as close to the population truth as the margin of error says. After all, chance chooses the sample, so it's possible to have terribly bad luck. An opinion poll about gambling *might* have the bad luck to choose 1523 people who think gambling is immoral and never buy lottery tickets. Figure 5.4 shows that this will almost never happen if the truth about the whole population is 60%. The usual margin of error comes from looking at the central 95% of the outcomes in histograms such as Figure 5.4.

The **margin of error** announced by most national samples says how close to the truth about the population the sample result would fall in 95% of all samples drawn by the method used to draw this one sample. We say that we have **95% confidence** that the truth about the population lies within the margin of error.

A news report on a Gallup poll says, "More than one in three women, 37%, say they are dissatisfied with society's treatment of women. The poll's margin of error was plus or minus 3%." That means, "We got this result using a method that comes within plus or minus three percent of the truth 95% of the time." This particular sample might be one of the 5% of all samples that miss by more, but knowing that we will land within the margin of error 95% of the time gives us a good idea of the poll's accuracy.

Finding the margin of error exactly is a job for statisticians. You can, however, use a "quick method" to get a rough idea of the size of a sample survey's margin of error.

The **quick method for margin of error:** Use the sample percent from a simple random sample of size n to estimate an unknown population percent. The margin of error for 95% confidence is roughly equal to $100/\sqrt{n}$.

EXAMPLE

What Is the Margin of Error?

The Gallup poll in our example interviewed 1523 people. The margin of error for 95% confidence will be about

$$\frac{100}{\sqrt{1523}} = \frac{100}{39.03} = 2.6\%$$

Gallup actually announced a margin of error of 3%. Our result differs a bit from Gallup's for two reasons. First, polls usually round their announced margin of error to the nearest whole percent to keep their press releases simple. Second, our rough formula works for an SRS. The Gallup poll, like the CPS, uses a more complex random sampling design. Nonetheless, in practice, our quick method comes pretty close. ■

The quick method also reveals an important fact about how margins of error behave. Because the sample size n appears in the denominator of the fraction, larger samples have smaller margins of error. We knew that. Because the formula uses the square root of the sample size, however, *to cut the margin of error in half, we must use a sample four times as large.*

5.5 Experiments

Sample surveys gather information on part of the population in order to draw conclusions about the whole. When the goal is to describe a population, statistical sampling is the right tool to use.

Suppose, however, that we want to study the response to a stimulus, to see how one variable affects another when we change existing conditions. Will a new mathematics curriculum improve the scores of sixth-graders on a standard test of mathematics achievement? Will taking small amounts of aspirin daily reduce the risk of suffering a heart attack? Does a mother's smoking during pregnancy reduce the IQ of her child? Studies that simply *observe and describe* are ineffective tools for answering these questions. *Experiments* give us clearer answers.

An **observational study,** such as a sample survey, observes individuals and measures variables of interest but does not attempt to influence the responses. The purpose of an observational study is to describe some group or situation.

An **experiment,** on the other hand, deliberately imposes some *treatment* on individuals in order to observe their responses. The purpose of an experiment is to study whether the treatment causes a change in the response.

Experiments are the preferred method for examining the effect of one variable on another. By imposing the specific treatment of interest and controlling other influences, we can pin down cause and effect. A sample survey may show that two variables are related, but it cannot demonstrate that one causes the other. Statistics has something to say about how to arrange experiments, just as it suggests methods for sampling.

EXAMPLE An Uncontrolled Experiment

A college regularly offers a review course to prepare candidates for the Graduate Management Admission Test (GMAT) required by most graduate business schools. This year, it offers only an online version of the course. The average GMAT score of students in the online course is 10% higher than the long-time average for those who took the classroom review course. Is the online course more effective?

This experiment has a very simple design. A group of subjects (the students) were exposed to a treatment (the online course), and the outcome (GMAT scores) was observed. Here is the design:

Online course → Observe GMAT scores

or, in general form

Treatment → Observe response ■

Most laboratory experiments use a design like that in the example: Apply a treatment and measure the response. In the controlled environment of the laboratory, simple designs often work well. But field experiments and experiments with human subjects are exposed to more variable conditions and deal with more variable subjects. It isn't possible to control outside factors that can influence the outcome. With greater variability comes a greater need for statistical design.

A closer look at the GMAT review course showed that the students in the online review course were quite different from the students who in past years took the classroom course. In particular, they were older and more likely to be employed. An online course appeals to these mature people, but we can't compare their performance with that of the undergraduates who previously dominated the course. The online course might even be less effective than the classroom version. The effect of online versus in-class instruction is hopelessly mixed up with influences lurking in the background. Figure 5.5 shows the mixed-up influences in picture form. We say that student age and background is *confounded* with whatever effect the change to online instruction may have.

Variables, whether part of a study or not, are said to be **confounded** when their effects on the outcome cannot be distinguished from each other.

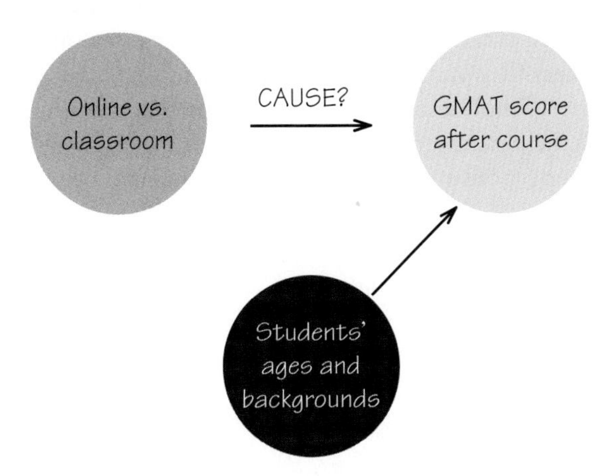

Figure 5.5
Confounding. We can't distinguish the effects of the treatment from those of other influences.

5.6 Randomized Comparative Experiments

The remedy for confounding is to do a *comparative experiment* in which some students are taught in the classroom and other similar students take the course online. The first group is called a **control group.** Most well-designed experiments compare two or more treatments. Of course, comparison alone isn't enough to produce results we can trust. If the treatments are given to groups that differ markedly when the experiment begins, bias will result. For example, if we allow students to elect online or classroom instruction, older employed students are likely to sign up for the online course. Personal choice will bias our results in the same way that volunteers bias the results of call-in opinion polls. The solution to the problem of bias is the same for experiments and for samples: Use impersonal chance to select the groups.

EXAMPLE A Randomized Comparative Experiment

The college decides to compare the progress of 25 on-campus students taught in the classroom with that of 25 students taught the same material online. Select the students who will be taught online by taking a simple random sample of size 25 from the 50 available subjects. The remaining 25 students form the control group. They will receive classroom instruction. The result is a **randomized comparative experiment** with two groups. Figure 5.6 outlines the design in graphical form.

The selection procedure is exactly the same as it is for sampling: Label and table. First, tag all 50 students with numerical labels, say, 01 to 50. Then go to the table of random digits and read successive three-digit groups. The first 25 labels encountered select the online group. As usual, ignore repeated labels and groups of digits not used as labels. For example, if you begin at line 125 in Table 5.1, the first five students chosen are those labeled 21, 49, 37, 18, and 44. ■

Randomized comparative experiments are a relatively new idea. They were introduced in the 1920s by Sir Ronald Fisher (see Spotlight 5.2). The GMAT experiment is *comparative* because it compares two treatments (the two instructional settings). It is *randomized* because the subjects are assigned to the treatments by chance. Randomization creates groups that are similar to each other before we start the experiment. Comparison means that possible confounding variables act on both groups at once. The only difference between the groups is the online

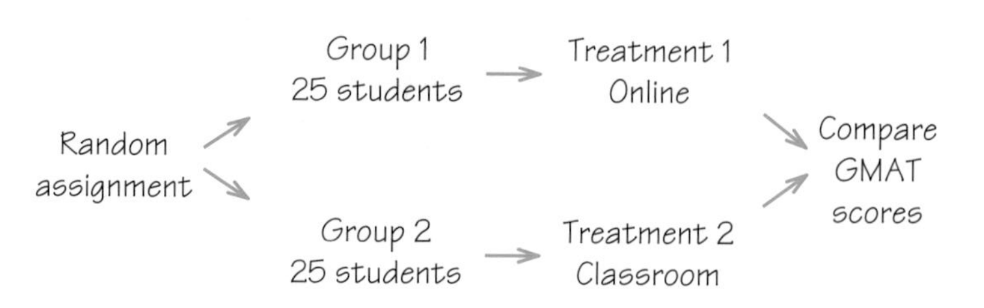

Figure 5.6 Outline of the design of a randomized comparative experiment to evaluate an online course.

SPOTLIGHT 5.2 Sir Ronald A. Fisher, 1890–1962

The ideas and methods that we study as "statistics" were invented in the nineteenth and twentieth centuries by people working on problems that required analysis of data. Astronomy, biology, the social sciences, and even surveying can claim a role in the birth of statistics. But if anyone can claim to be "the father of statistics," that honor belongs to Sir Ronald A. Fisher.

Fisher's writings organized statistics as a distinct field of study whose methods apply across many disciplines. He systematized the mathematical theory of statistics and invented many new techniques. The randomized comparative experiment is perhaps Fisher's greatest contribution.

Like other statistical pioneers, Fisher was driven by the demands of practical problems. Beginning in 1919, he worked on agricultural field experiments at Rothamsted in England. How should we arrange the planting of different crop varieties or the application of different fertilizers to get a fair comparison among them? Because fertility and other variables change as we move across a field, experimenters used elaborate checkerboard planting arrangements to obtain fair comparisons. Fisher had a better idea: "Arrange the plots deliberately at random."

(University of London.)

versus in-class setting. So if we see a difference in performance, it must be due to the different setting. That is the basic logic of randomized comparative experiments. We will see later that there are some fine points to worry about, but this basic logic shows why experiments can give good evidence that the different treatments really *caused* different outcomes. Here is another example, this time comparing three treatments.

EXAMPLE Conserving Energy

Many utility companies have introduced programs to encourage energy conservation among their customers. An electric company considers placing electronic indicators in households to show what the cost would be if the electricity use at that moment continued for a month. Will indicators reduce electricity use? Would cheaper methods work almost as well? The company decides to design an experiment.

One cheaper approach is to give customers a chart and information about monitoring their electricity use. The experiment compares these two approaches (indicator, chart) and also a control. The control group of customers receives

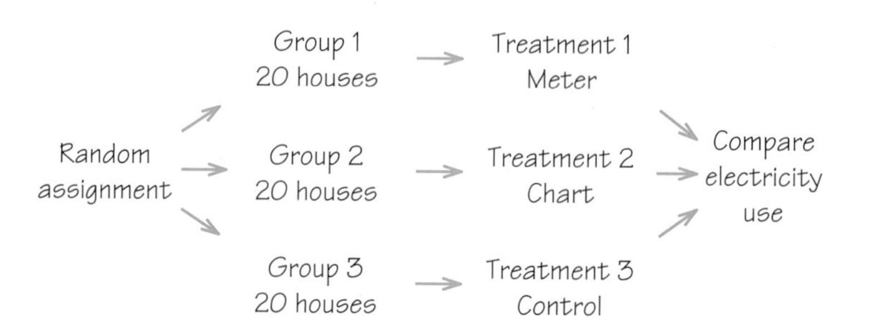

Figure 5.7 The design of a randomized comparative experiment to compare three ways of encouraging households to conserve electricity.

information about energy conservation but no help in monitoring electricity use. The outcome is measured by total electricity used in a year. The company finds 60 single-family residences in the same city willing to participate, so it assigns 20 residences at random to each of the three treatments. Figure 5.7 outlines the design.

To carry out the random assignment, label the 60 households 01 to 60. Enter Table 5.1 to select an SRS of 20 to receive the indicators. Continue in Table 5.1, selecting 20 more to receive charts. The remaining 20 form the control group. ■

Randomized comparative experiments are common tools of industrial and academic research. They are also widely used in medical research. For example, federal regulations require that the safety and effectiveness of new drugs be demonstrated by randomized comparative experiments. Let's look at an important medical experiment.

EXAMPLE

The Physicians' Health Study

There is some evidence that taking aspirin regularly will reduce the risk of heart attacks. Some people also suspect that regular doses of beta-carotene (which the body converts into vitamin A) will help prevent some types of cancer. The Physicians' Health Study was a large experiment designed to test these claims. The subjects of this study were 22,000 male physicians at least 40 years old. Each subject took a pill every day over a period of several years. There were four treatments: aspirin alone, beta-carotene alone, both, and neither. The subjects were randomly assigned to one of these treatments at the beginning of the experiment. ■

The Physicians' Health Study example introduces several new ideas about the design of experiments. The first is the importance of the **placebo effect,** a special kind of confounding. A placebo is a fake treatment, a dummy pill that contains no active ingredient but looks and tastes like the real thing. The placebo effect is the tendency of subjects to respond favorably to any treatment, even a placebo. If subjects given aspirin, for example, are compared with subjects who receive no treatment, the first group gets the benefit of both aspirin and the placebo effect. Any beneficial effect that aspirin may have is confounded with the placebo effect. To prevent confounding, it is important that some treatment be given to all subjects in any medical experiment. In the Physicians' Health

Study, all subjects took pills that looked alike. Some pills contained aspirin or beta-carotene and some contained a placebo. Figure 5.8 shows the design of the experiment.

The Physicians' Health Study was a **double-blind experiment:** Neither the subjects nor the experimenters who worked with them knew which treatment any subject received. Subjects might react differently if they knew they were getting "only a placebo." Knowing that a particular subject was getting "only a placebo" could also influence the researchers who interviewed and examined the subjects. So both subjects and workers were kept "blind." Only the study's statistician knew which treatment each subject received.

Finally, the Physicians' Health Study is a more elaborate experiment than our earlier examples. In the GMAT and energy conservation experiments, we compared values of a single variable (which teaching setting? which conservation program?). The Physicians' Health Study looks at two distinct experimental variables: aspirin or not and beta-carotene or not. A two-variable experiment, usually called a *two-factor experiment,* allows us to study the *interaction,* or joint effect, of the two drugs as well as the separate effects of each. For example, beta-carotene may reinforce the effect of aspirin on future heart attacks. By comparing these four groups, we can study all these possible interactions. Nonetheless, the outline of the design in Figure 5.8 is similar to our earlier examples because the basic ideas of randomization and comparison of several treatments remain.

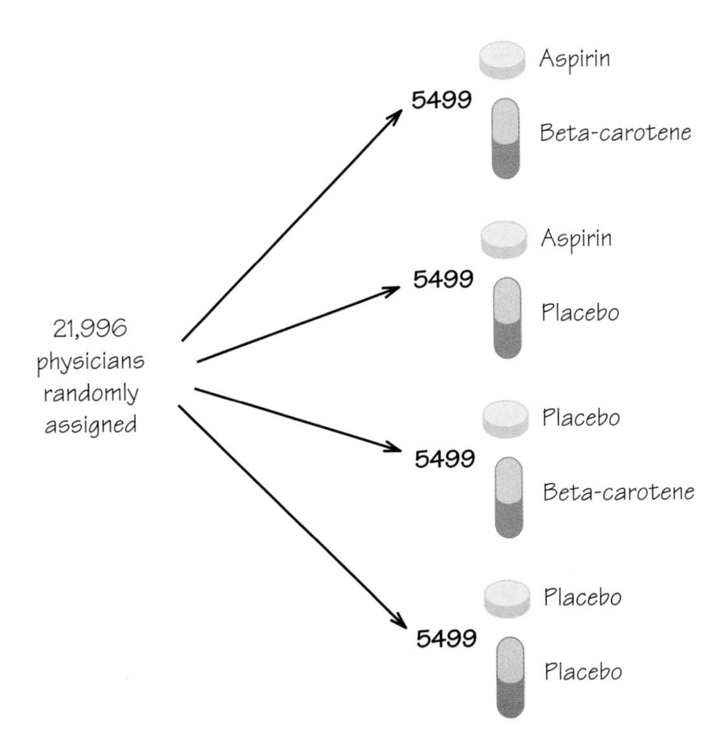

Figure 5.8 The design of the Physicians' Health Study, an experiment with two factors.

5.7 Statistical Evidence

A properly designed experiment, in the eyes of a statistician, is an experiment employing the principles of *comparison* and *randomization:* comparison of several treatments and randomization in assigning subjects to the treatments.

The future health of the subjects in the Physicians' Health Study, for example, may depend on age, past medical history, emotional status, smoking habits, and many other variables known and unknown. Randomization will, on the average, balance the groups simultaneously in all such variables. Because the groups are exposed to exactly the same environment, except for the actual content of the pills, we can say that any differences in heart attacks or cancer among the groups are caused by the medication. That is the logic of randomized comparative experiments.

Let's be a bit more careful: Any difference among the groups is due *either* to the medication *or* to the accident of chance in the random assignment of subjects. It could happen, for example, that men about to have a heart attack were, just by chance, overrepresented in one of the groups. The problem is exactly the same as in random sampling, where it could happen just by chance that an SRS chooses only Republicans. Just as in sampling, we are saved by the regular pattern of chance behavior.

If we repeat the random assignment of subjects to groups many times, differences among the groups follow a regular pattern if we don't apply different treatments. This regular pattern tells us how large the differences among the four groups are likely to be if nothing but chance is operating. If we observe differences so large that they would almost never occur just by chance, we are confident that we are seeing the effects of the treatments. So it is not *any* differences that show the results of the treatments, just differences so large that chance cannot easily account for them. Differences among the treatment groups that are so large that they would rarely occur just by chance are called *statistically significant.*

An observed effect so large that it would rarely occur by chance is called **statistically significant.**

Again as in sampling, larger numbers of subjects increase our confidence in the results. The Physicians' Health Study followed 22,000 subjects in order to be quite certain that any medically important differences among the groups would be detected and that these differences could be attributed to aspirin or to beta-carotene. In fact, there were significantly fewer heart attacks among the men who took aspirin than among men who took the placebo. As a result of the

Physicians' Health Study, doctors often recommend that men over age 50 take small amounts of aspirin regularly. Beta-carotene, on the other hand, did not significantly reduce cancer.

The logic of experimentation, the statistical design of experiments, and the laws that govern chance behavior combine to give compelling evidence of cause and effect. Only experimentation can produce fully convincing evidence of causation.

EXAMPLE

Smoking and Health

By way of contrast, consider the statistical evidence linking cigarette smoking to lung cancer. We can't assign groups of people to smoke or not, so a direct experiment isn't possible. The most careful studies have selected samples of smokers and nonsmokers, then followed them for many years, eventually recording the cause of death. These are called *prospective studies* because they follow the subjects forward in time. Prospective studies are comparative, but they are not experiments because the subjects themselves choose whether or not to smoke. A large prospective study of British doctors found that the death rate from lung cancer among cigarette smokers was 20 times that among nonsmokers. Another study of American men aged 40 to 79 found that the lung cancer death rate was 11 times higher among smokers than among nonsmokers. These and many other observational studies show a strong connection between smoking and lung cancer. ■

The connection between smoking and lung cancer is statistically significant. That is, it is far stronger than would occur by chance. We can be confident that something other than chance links smoking to cancer. But observation of samples cannot tell us *what* factors other than chance are at work. Perhaps there is something in the genetic makeup of some people that predisposes them both to nicotine addiction and to lung cancer. In that case, we would observe a strong link even if smoking itself had no effect on the lungs.

The statistical evidence that points to cigarette smoking as a cause of lung cancer is about as strong as nonexperimental evidence can be. First, the connection has been observed in many studies in many countries. This eliminates factors peculiar to one group of people or to one specific study design. Second, there is a *dose-response relationship:* People who smoke more are more likely to get lung cancer than those who smoke less, and quitting cigarettes reduces the cancer risk. Third, specific ways in which smoking could cause cancer have been identified—cigarette smoke contains tars that have been shown by experiment to cause tumors in animals. Finally, no plausible alternative explanation is available. For example, the genetic hypothesis cannot explain the increase in lung cancer among women that occurred as more and more women became smokers. Lung cancer, which has long been the leading cause of cancer deaths in men, has now passed breast cancer as the most fatal cancer for women.

This evidence is convincing, but it is not quite as strong as the conclusive statistical evidence we get from randomized comparative experiments.

5.8 Statistics in Practice

There is more to the wise use of statistics than a knowledge of such statistical techniques as simple random samples and randomized comparative experiments. These designs for data production avoid the pitfalls of voluntary response samples or uncontrolled experiments. But there are other pitfalls that can reduce the usefulness of data even when we use a sound statistical design.

EXAMPLE Nonresponse in Sampling

Choosing a sample at random is only the first step in carrying out a sample survey of a large human population. You must then contact the people in the sample and persuade them to cooperate. This isn't easy. Some people are rarely at home. Others don't want to talk with an interviewer. *Nonresponse* occurs when an individual chosen for the sample can't be contacted or refuses to cooperate. Ask the rate of nonresponse before putting too much trust in a sample result.

How bad is nonresponse? The Current Population Survey has the lowest nonresponse rate of any poll we know: Only about 6% or 7% of the households in the CPS sample don't respond. People are more likely to respond to a government survey such as the CPS, and the CPS contacts its sample in person before doing later interviews by phone.

What about polls done by the media and by market research and opinion-polling firms? We don't know their rates of nonresponse, because they don't say. That itself is a bad sign. The Pew Research Center imitated a careful telephone survey and published the results: Out of 2879 households called, 1658 were never at home, refused to participate, or would not finish the interview. That's a nonresponse rate of 58%. The Pew researchers were more thorough than are many polls. Insiders say that nonresponse often reaches 75% or 80% of an opinion poll's original sample. ■

Nonresponse is the most serious problem facing sample surveys. People are increasingly reluctant to answer questions, particularly over the phone. The rise of telemarketing, answering machines, and caller ID drives down response to telephone surveys. Gated communities and buildings guarded by doormen prevent face-to-face interviews. Nonresponse can bias sample survey results, because different groups have different rates of nonresponse. Refusals are higher in large cities and among the elderly, for example. Bias due to nonresponse can easily overwhelm the error due to random sampling described by a survey's margin of error.

> The announced margin of error for a sample survey covers only random sampling error. Nonresponse and other practical difficulties can cause large bias that is not covered by the margin of error.

EXAMPLE Is the Experiment Realistic?

The Physicians' Health Study gave pills to middle-aged men going about their everyday lives. Many experiments, however, take place in artificial environments. A psychologist studying the effects of stress on teamwork observes teams of students carrying out tasks in a psychology laboratory under different conditions. The students know it's "just an experiment" and that the stress will only last an hour. Do the conclusions of such experiments apply to real-life stress? An engineer uses a small pilot production process in a laboratory to find the choices of pressure and temperature that maximize yield from a complex chemical reaction. Do the results apply to a full-scale manufacturing plant?

These are not statistical questions. The psychologist and the engineer must use their understanding of psychology and engineering to judge how far their results apply. The statistical design enables us to trust the results for the students and the pilot process but not to generalize the conclusions to other settings. ■

SPOTLIGHT 5.3 Experiments and Ethics

Dr. Charles Hennekens, director of the Physicians' Health Study, had to concern himself with the goals, design, and implementation of his large-scale study. But other questions also arise in the course of such an experiment. Dr. Hennekens was asked about the ethics of experimenting on human health:

> Much has been made of the ethical concerns about randomized trials. There are instances where it would not be ethical to do a randomized trial. When penicillin was introduced for the treatment of pneumococcal pneumonia, which was virtually 100% fatal, the mortality rate plummeted significantly. Certainly it would have been unethical to do a randomized trial, to withhold effective treatment from people who need it.
>
> There's a delicate balance between when to do or not to do a randomized trial. On the one hand, there must be a sufficient belief in the agent's potential to justify exposing half the subjects to it. On the other hand, there must be sufficient doubt about its efficacy to justify withholding it from the other half of subjects who might be assigned to the placebos, the pills with inert ingredients. It was just these circumstances that we felt existed with regard to the aspirin and the beta-carotene hypotheses.

Charles Hennekens

(AP/World Wide Photo.)

When we are planning a statistical study, we must also face some *ethical questions*. Does the knowledge gained from an experiment or study justify the possible risk to the subjects? In the Physicians' Health Study, doctors gave their informed consent to take either aspirin, beta-carotene, or a placebo, in any combination, as prescribed by the study designers. When it became clear that men taking aspirin had fewer heart attacks, the experiment was stopped so that all the subjects could take advantage of this new knowledge. In Spotlight 5.3 the director of the Physicians' Health Study explains why randomized comparative experiments are a mainstay of medical research and when such clinical trials are justified. Even the GMAT experiment faces ethical issues: Some students may object to being told they must take the review course online, and others may object to being denied access to the online version. Practical and ethical problems are never far from the surface when statistics is applied to real problems.

REVIEW VOCABULARY

Bias A systematic error that tends to cause the observations to deviate in the same direction from the truth about the population whenever a sample or experiment is repeated.

Confounding Two variables are confounded when their effects on the outcome of a study cannot be distinguished from each other.

Control group A group of experimental subjects who are given a standard treatment or no treatment (such as a placebo).

Convenience sample A sample that consists of the individuals who are most easily available, such as people passing by in the street. A convenience sample is usually biased.

Double-blind experiment An experiment in which neither the experimental subjects nor the persons who interact with them know which treatment each subject received.

Experiment A study in which treatments are applied to people, animals, or things in order to observe the effect of the treatments.

Histogram A graph that displays how often various outcomes occur by means of bars. The height of each bar is the number of times an outcome or group of outcomes occurred in the data.

Margin of error As announced by most national polls, the margin of error says how close to the truth about the population the sample result would fall in 95% of all samples drawn by the method used to draw this one sample. A **quick method** to estimate this margin of error for a simple random sample of size n is $100/\sqrt{n}$.

Observational study A study (such as a sample survey) that observes individuals and measures variables of interest but does not attempt to influence the responses.

Placebo effect The effect of a dummy treatment (such as an inert pill in a medical experiment) on the response of subjects.

Population The entire group of people or things that we want information about.

Randomized comparative experiment An experiment to compare two or more treatments in which people, animals, or things are assigned to treatments by chance.

Sample A part of the population that is actually observed and used to draw conclusions, or inferences, about the entire population.

Simple random sample (SRS) A sample chosen by chance, so that every possible sample of the same size has an equal chance to be the one selected.

Statistical significance An observed effect is statistically significant if it is so large that it is unlikely to occur just by chance in the absence of a real effect in the population from which the data were drawn.

Table of random digits A table whose entries are the digits 0, 1, 2, 3, 4, 5, 6, 7, 8, 9 in a completely random order. That is, each entry is equally likely to be any of the 10 digits and no entry gives information about any other entry.

Voluntary response sample A sample of people who choose themselves by responding to a general invitation to write or call with their opinions. Such a sample is usually strongly biased.

SUGGESTED READINGS

COBB, GEORGE W. *Design and Analysis of Experiments,* Springer, New York, 1998. Chapter 1 of this more advanced text is a nice essay on designing experiments.

KALTON, GRAHAM. *Introduction to Survey Sampling,* Sage, Newbury Park, Calif., 1983. A detailed but relatively nontechnical introduction to the statistics of sample surveys.

MOORE, DAVID S. *Statistics: Concepts and Controversies,* 5th ed., Freeman, New York, 2001, part I. Written for liberal arts students, this book provides more extensive discussion at about the same level as *For All Practical Purposes.*

MOORE, DAVID S. *The Basic Practice of Statistics,* 2nd ed., Freeman, New York, 1999, chap. 3. Clear treatment of data production in a text on practical statistics at about the same level as *For All Practical Purposes.*

TANUR, JUDITH M. Samples and surveys. In David C. Hoaglin and David S. Moore (eds.), *Perspectives on Contemporary Statistics,* Mathematical Association of America, Washington, D.C., 1992, pp. 55–70. This essay describes the practice of sample surveys at a relatively nontechnical level.

SUGGESTED WEB SITES

Several of the most important sample surveys in the United States are conducted by the Bureau of Labor Statistics **(www.bls.gov)** and the Bureau of the Census **(www.census.gov).** The Gallup Organization **(www.gallup.com)** conducts the Gallup poll; the section on "How Gallup Polls Are Conducted" is excellent. Important medical studies – many based on randomized comparative experiments – appear in the *Journal of the American Medical Association* **(jama.ama-assn.org)** and the *New England Journal of Medicine* **(content.nejm.org).**

SKILLS CHECK

1. An opinion poll contacts 1161 adults and asks them, "Which political party do you think has better ideas for leading the country in the twenty-first century?" In all, 696 of the 1161 say, "The Democrats." The sample in this setting is

(a) all 210 million adults in the United States.
(b) the 1161 people interviewed.
(c) the 696 people who chose the Democrats.

2. You must choose an SRS of 10 of the 440 retail outlets in New York that sell your company's products. How would you label this population in order to use Table 5.1?

(a) 001, 002, 003, . . . , 439, 440
(b) 000, 001, 002, . . . , 439, 440
(c) 1, 2, . . . , 439, 440

3. The Gallup poll asked 501 teenagers whether they approved of legal gambling; 52% said they did. Use the quick method to estimate the margin of error for conclusions about all teenagers.

(a) 22%
(b) 13.9%
(c) 4.5%

4. Suppose the population of a small county includes 10,404 registered voters. The local newspaper polled a random sample of 225 registered voters on a referendum issue and found 144 (64%) to be in favor of the proposal. What is the approximate margin of error for this poll?

(a) 6.7%
(b) 8.3%
(c) 12.5%

5. A local café invites its clientele to complete comment cards. Which of these statements is most likely true?

(a) The comment cards constitute a random sample.
(b) The comment cards constitute an observational study.
(c) The comment cards constitute a voluntary response sample.

6. Using a random digit table to choose a sample is useful to minimize which of the following?

(a) Bias
(b) The placebo effect
(c) The control group

7. Suppose a newspaper survey reports that "35% of college students currently smoke, with a margin of error of 3%." Which of the following statements follows from this?

(a) One is 97% sure that the actual percentage of college students currently smoking is exactly 35%.
(b) One is 95% sure that the actual percentage of college students currently smoking is between 32% and 38%.
(c) One is absolutely sure that the actual percentage of college students currently smoking is between 32% and 38%.

8. Increasing the sample size has which of the following effects?

(a) The margin of error is reduced.
(b) Confounding is reduced.
(c) Bias is reduced.

9. Suppose a survey client requires a margin of error of approximately 2%. Using the quick method, approximately how many people must be polled?

(a) 50
(b) 700
(c) 2500

10. A control group is often used to reduce

(a) confounding.
(b) bias.
(c) the placebo effect.

EXERCISES

▲ **Optional.** ■ **Advanced.** ◆ **Discussion.**

Sampling

1. Different types of writing can sometimes be distinguished by the lengths of the words used. A student interested in this fact wants to study the lengths of words used by Danielle Steele in her novels. She opens a Steele novel at random and records the lengths of the first 250 words on the page. What is the population in this study? What is the sample?

2. The American Community Survey (ACS) will contact 3 million households, including some in every county in the United States. This new Census Bureau survey will ask each household questions about their housing, economic, and social status.

The new survey will replace the census "long form." What is the population for the ACS?

◆ **3.** A member of Congress is interested in whether her constituents favor a proposed gun-control bill. Her staff reports that letters on the bill have been received from 361 constituents and that 323 of these oppose the bill. What is the population of interest? What is the sample? Is this sample likely to represent the population well? Explain your answer.

Bad Sampling Methods

◆ **4.** You see a woman student standing in front of the student center, now and then stopping other students to ask them questions. She says that she is collecting student opinions for a class assignment. Explain why this sampling method is almost certainly biased.

◆ **5.** The Excite instant poll was found online at news.excite.com. The question appeared on the screen, and you simply clicked buttons to vote "yes," "no," or "don't know." One recent question was, "Should female athletes be paid the same as men for the work they do?" In all, 13,147 (44%) said "yes," another 15,182 (50%) said "no," and the remaining 1448 said "don't know."

(a) What was the sample size for this poll?
(b) That's a much larger sample than standard sample surveys. In spite of this, we can't trust the result to give good information about any clearly defined population. Why?
(c) It is still true that more men than women use the Web. How might this fact have affected the poll results?

◆ **6.** Highway planners made a main street in a college town a one-way street. Local businesses were against the change. The local newspaper invited readers to call a telephone number to record their comments. The next day, the paper reported:

> Readers overwhelmingly prefer two-way traffic flow to one-way streets. By nearly a 7–1 margin, callers to the newspaper's Express Yourself opinion line on Wednesday complained about the one-way streets that have been in place since May. Of the 98 comments received, all but 14 said no to one-way.

(a) What population do you think the newspaper wants information about?
(b) Is the proportion of this population who favor one-way streets almost certainly larger or smaller than the proportion 14/98 in the sample? Why?

◆ **7.** Ann Landers once asked her female readers whether they would be content with affectionate treatment by men with no sex ever. More than 90,000 women wrote in, with 72% answering "yes." Explain carefully why this sample is almost certainly biased. What is the *direction* of the bias? That is, is the percentage of all adult women who would be content with no sex ever lower or higher than the 72% in the sample?

Simple Random Sampling

8. You are planning a report on apartment living in a college town. You decide to select three apartment complexes at random for in-depth interviews with residents. Use Table 5.1, starting at line 117, to select a simple random sample of three of the following apartment complexes.

Ashley Oaks	Fowler
Bay Pointe	Franklin Park
Beau Jardin	Georgetown
Bluffs	Greenacres
Brandon Place	Lahr House
Briarwood	Mayfair Village
Brownstone	Nobb Hill
Burberry	Pemberly Courts
Cambridge	Peppermill
Chauncey Village	Pheasant Run
Country Squire	Richfield
Country View	Sagamore Ridge
Country Villa	Salem Courthouse
Crestview	Village Manor
Del-Lynn	Waterford Court
Fairington	Williamsburg
Fairway Knolls	

9. A firm wants to understand the attitudes of its minority managers toward its system for assessing management performance. Below is a list of all the firm's managers who are members of minority groups. Use Table 5.1 at line 110 to choose six to be interviewed in detail about the performance appraisal system.

Agarwal	Dewald	Huang	Puri
Anderson	Fernandez	Kim	Richards
Baxter	Fleming	Liao	Rodriguez
Bowman	Gates	Mourning	Santiago
Brown	Goel	Naber	Shen
Castillo	Gomez	Peters	Vega
Cross	Hernandez	Pliego	Wang

10. A student wishes to study the opinions of faculty at his college on the advisability of setting up a state board of higher education to oversee all colleges in the state. The college has 380 faculty members.

(a) What is the population in this situation?
(b) Explain carefully how you would choose a simple random sample of 50 faculty members.
(c) Use Table 5.1 starting at line 135 to choose the first 5 members of this sample.

11. The number of students majoring in political science at Ivy University has increased substantially without a corresponding increase in the number of faculty. The campus newspaper plans to interview 25 of the 450 political science majors to learn student views on class size and other issues. You suggest a simple random sample. Explain carefully how you would choose this sample. Then use Table 5.1 starting at line 120 to select the first 5 members of your sample.

■ **12.** Which of the following statements are true of a table of random digits and which are false? Explain your answers.

(a) There are exactly four 0's in each row of 40 digits.
(b) Each pair of digits has chance 1/100 of being 00.
(c) The digits 0000 can never appear as a group, because this pattern is not random.

Statistical Estimation

◆ **13.** The Ministry of Health in the Canadian province of Ontario wants to know whether the national health care system is achieving its goals in the province. The Ministry conducted the Ontario Health Survey, which interviewed a random sample of 61,239 people who live in Ontario.

(a) What is the population for this sample survey? What is the sample?
(b) The survey found that 76% of males and 86% of females in the sample had visited a general practitioner at least once in the past year. Do you think these estimates are close to the truth about the entire population? Why?

◆ **14.** Here is the language used by the Harris poll to explain the accuracy of its results: "In theory, with a sample of this size, one can say with 95 percent certainty that the results have a statistical precision of plus or minus 3 percentage points of what they would be if the entire adult population had been polled with complete accuracy." What does Harris mean by "95 percent certainty"?

◆ **15.** The Harris poll asked a sample of 1009 adults which causes of death they thought would become more common in the future. Topping the list was gun violence: 70% of the sample thought deaths from guns would increase.

(a) How many of the 1009 people interviewed thought deaths from gun violence would increase?
(b) Harris says that the margin of error for this poll is plus or minus 3 percentage points. Explain to someone who knows no statistics what "margin of error plus or minus 3 percentage points" means.

◆ **16.** In January of 2000, the Gallup poll asked a random sample of 1633 adults, "In general, are you satisfied or dissatisfied with the way things are going in the United States at this time?" It found that 1127 said that they were satisfied. Write a short report of this finding, as if you were writing for a newspaper. Be sure to include a margin of error.

17. Use the quick method to estimate the margin of error for the Harris poll survey of Exercise 15.

◆ **18.** Exercise 5 describes an online opinion poll with sample size 29,777. The poll found that 44% of the sample thought female athletes should be paid the same as male athletes. What margin of error does the quick method give for this poll? Explain clearly why it's wrong to apply the quick method here.

19. Exercise 15 describes a Harris poll that interviewed 1009 people. Suppose you want a margin of error half as large as the one you found in that exercise. How many people must you plan to interview?

20. The Ontario Health Survey interviewed a random sample of 61,239 residents of the Canadian province of Ontario; 49,164 of the people in the sample were over 12 years old. Estimate the percent of the province's population that is over the age of 12. Give a margin of error for your estimate. Explain in simple language what this margin of error tells us.

21. You must allocate 5 tickets to a rock concert among 25 clamoring members of your club. We will use this example to illustrate sampling variability.

(a) Choose 5 at random to receive the tickets, using line 135 of Table 5.1 (ignore the asterisks).

Agassiz	Gutierrez	Spencer*
Binet*	Herrnstein	Thomson
Blumenbach	Jimenez*	Toulmin
Chase*	Liang	Vogt*
Chen*	McKim*	Went
Darwin	Moll*	Wilson
Epstein	Montoya	Yerkes
Ferri	Perez*	Zimmer
Gupta*		

(b) In fact, 10 of the 25 club members are female. Their names are marked with asterisks in the list. Draw 5 at random 20 times, using a different row in Table 5.1 each time [include your sample from part (a)]. Record the number of females in each of your samples. Make a histogram to display your results. What is the average number of females in your 20 samples?

(c) Do you think the club members should suspect discrimination if none of the 5 tickets go to women?

22. An opinion poll asks a sample of 1324 adults whether they believe that life exists on other planets; 609 say "yes." What percent of the sample believes in extraterrestrial life? The polling organization announces a margin of error of ±3%. What conclusion can you draw about the percent of all adults who believe that life exists on other planets?

◆ **23.** A news article reports that in a recent Gallup poll, 78% of the sample of 1108 adults said they believe there is a heaven. Only 60% said they believe there is a hell. The news article ends, "The poll's margin of sampling error was plus or minus four percentage points." Can we be certain that between 56% and 64% of all adults believe there is a hell? Explain your answer.

Experiments

◆ **24.** A state starts a job-training program for manufacturing workers who lose their jobs. Critics claim that the program doesn't work because the state's unemployment rate was 4% when the program began and 7% five years later. Explain why confounding makes the effect of the training program on unemployment hard to see.

◆ **25.** The Example on page 185 describes an experiment to learn whether providing households with electronic indicators or charts will reduce their electricity consumption. An executive of the electric company objects to including a control group. He says, "It would be simpler to just compare electricity use last year [before the indicator or chart was provided] with consumption in the same period this year. If households use less electricity this year, the indicator or chart must be working." Explain clearly why this design is inferior to that in the Example.

◆ **26.** How should women diagnosed with breast cancer be treated? In some cases, surgeons remove only the tumor. In others, they remove the entire breast. In yet others, they remove the breast, underlying muscle, and nearby lymph nodes.

A study looks at the records of all breast cancer patients at 25 large medical centers and finds that a higher percent of "only the tumor" cases survived five years or more. Explain why we can't conclude that removing only the tumor is the best treatment for all patients.

◆ **27.** Could the magnetic fields from power lines cause leukemia in children? Investigators spent five years and $5 million comparing 638 children who had leukemia and 620 who did not. They went into the homes and actually measured the magnetic fields in the children's bedrooms, in other rooms, and at the front door. They recorded facts about nearby power lines for the family home and also for the mother's residence when she was pregnant. Result: no evidence of more than a chance connection between magnetic fields and childhood leukemia. Explain carefully why this study is *not* an experiment.

◆ **28.** An educational software company wants to compare the effectiveness of its computer animation for teaching about supply, demand, and market clearing with that of a textbook presentation. The company tests the economic knowledge of each of a group of first-year college students, then divides them into two groups. One group uses the animation, and the other studies the text. The company retests all the students and compares the increase in economic understanding in the two groups. Is this an experiment? Why or why not?

◆ **29.** Men and women differ in their choices for many product categories. Are there gender differences in preferences for health insurance plans as well? A market researcher interviews a large sample of consumers, both men and women. She asks each consumer which of two health plans he or she prefers. Is this an observational study or an experiment? Explain your answer.

Randomized Comparative Experiments

30. A university's Department of Statistics wants to attract more majors. It prepares two advertising brochures. Brochure A stresses the intellectual excitement of statistics. Brochure B stresses how much money statisticians make. Which will be more attractive to first-year students? You have a questionnaire to measure interest in majoring in statistics, and you have 50 first-year students to work with. Outline the design of an experiment to decide which brochure works better.

31. You can use your computer to make telephone calls over the Internet. How would cost affect the behavior of users of this service? You will offer the service to all 200 rooms in a college dormitory. Some rooms will pay a low flat rate. Others will pay higher rates at peak periods and very low rates off-peak. You are interested in the amount and time of use and in the effect on the congestion of the network. Outline the design of an experiment to study the effect of rate structure.

◆ **32.** A college allows students to choose either classroom or self-paced instruction in a basic mathematics course. The college wants to compare the effectiveness of self-paced and regular instruction. Someone proposes giving the same final exam to all students in both versions of the course and comparing the average score of those who took the self-paced option with the average score of students in regular sections.

(a) Explain why confounding makes the results of that study worthless.

(b) Given 30 students who are willing to use either regular or self-paced instruction, outline an experimental design to compare the two methods of instruction. Then use Table 5.1 starting at line 108 to carry out the randomization.

◆ **33.** Here is part of the summary of an article in the *Journal of the American Medical Association* (Volume 282, 1999, page 137) that asks whether flu vaccine works.

> **Design** Randomized, double-blind, placebo-controlled trial conducted from September 1997 through March 1998.

Doctors are supposed to understand this. Explain in one sentence each what *randomized, double-blind,* and *placebo-controlled* mean.

34. Below are the names of 20 patients who have consented to participate in a trial of surgical treatments for angina. Outline an experiment to compare surgical treatment with a placebo (sham surgery) and use Table 5.1, beginning at line 101, to do the required randomization. (Ignore the asterisks.)

Ashley	Cravens*	Lippmann	Strong*
Bean*	Dorfman	Mark*	Tobias
Block	Garcia	Morton*	Valenzuela*
Chavez*	Huang*	Popkin	Washington
Chen	Kidder	Sosa	Williams

35. Unknown to the researchers in Exercise 34, the eight subjects whose names are marked by asterisks will have a fatal heart attack during the study period. We can observe how sampling variability operates in a randomized experiment by keeping track of how many of these eight subjects are assigned to the group that will receive the new surgical treatment. Carry out the random assignment of 10 subjects to the treatment group 20 times, keeping track of how many asterisks are on the names you choose each time. Then make a histogram of the count of heart attack victims assigned to the treatment. What is the average number in your 20 tries?

◆ **36.** Explain clearly the advantage of using several thousand subjects, rather than just 20, in the experiment of Exercise 34.

■ **37.** Explain carefully how you would randomly assign the 20 subjects named in Exercise 34 to the four treatments in the Physicians' Health Study. Figure 5.8 describes the treatments. Assign 5 of the 20 to each group. Use Table 5.1 at line 120 to carry out the randomization.

Statistical Evidence

◆ **38.** The Nurses Health Study has interviewed a sample of more than 100,000 female registered nurses every two years since 1976. Beginning in 1980, the study asked questions about diet, including alcohol consumption. The researchers concluded that "light-to-moderate drinkers had a significantly lower risk of death" than either nondrinkers or heavy drinkers.

(a) Is the Nurses Health Study an observational study or an experiment? Why?
(b) What does "significant" mean in a statistical report?
(c) Suggest some confounding variables that might explain why moderate drinkers have lower death rates than nondrinkers. (The study adjusted for these variables.)

◆ **39.** The financial aid office of a university asks a sample of students about their employment and earnings. The report says that "for academic year earnings, a statistically significant difference was found between the sexes, with men earning more on the average. No significant difference was found between the earnings of black and white students." Explain both of these conclusions, for the effects of sex and of race on average earnings, in language understandable to someone who knows no statistics.

Statistics in Practice

◆ **40.** A survey of users of the Internet found that males outnumbered females by nearly 2 to 1. This was a surprise, because earlier surveys had put the ratio of men to women closer to 9 to 1. Later in the article we find that surveys were sent to 13,000 organizations and that 1468 of these responded. The survey report claims that "the margin of error is 2.8 percent, with 95 percent confidence."

(a) What was the *response rate* for this survey? (The response rate is the percent of the planned sample that responded.)
(b) Use the quick method (page 181) to estimate the margin of error of this survey. Is your result close to the 2.8% claimed?
(c) Do you think that the small margin of error is a good measure of the accuracy of the survey's results? Explain your answer.

◆ **41.** A common form of nonresponse in telephone surveys is "ring–no-answer." That is, a call is made to an active number but no one answers.

The Italian National Statistical Institute looked at nonresponse to a government survey of households in Italy during the periods January 1 to Easter and July 1 to August 31. All calls were made between 7 and 10 P.M., but 21.4% gave "ring–no-answer" in one period versus 41.5% "ring–no-answer" in the other period. Which period do you think had the higher rate of no answers? Why? Explain why a high rate of nonresponse makes sample results less reliable.

◆ **42.** The *wording of questions* can strongly influence the results of a sample survey. You are writing an opinion poll question about a proposed amendment to the Constitution. You can ask if people are in favor of "changing the Constitution" or "adding to the Constitution" by approving the amendment. One of these choices of wording will produce a much higher percent in favor. Which one? Why?

◆ **43.** Do those high center brake lights, required on all cars sold in the United States since 1986, really reduce rear-end collisions? Randomized comparative experiments with fleets of rental and business cars, done before the lights were required, showed that the third brake light reduced rear-end collisions by as much as 50%. Alas, requiring the third light in all cars led to only a 5% drop. Explain why the experiment did not realistically imitate conditions after the lights were required.

Additional Exercises

◆ **44.** A poll of 586 adults who had used the Internet in the past week asked whether "the Internet has made your life much better, somewhat better, somewhat worse, much worse, or has not affected your life either way." In all, 152 of the 586 subjects said "much better."

(a) What is the population for this sample survey?
(b) Use the quick method to find a margin of error. Then write a short summary of the poll results, as if you were writing the script for a TV newscast.

◆ **45.** Should the government spend money to help low-income families who want to send their children to private or religious schools? A poll of 1006 adults found that 362 said "yes."

(a) What is the population for this sample survey?
(b) Use the quick method to find a margin of error. Then write a short summary of the poll results, as if you were writing the script for a TV newscast.

46. Will people spend less on health care if their health insurance requires them to pay some part of the cost themselves? An experiment on this issue asked if the percent of medical costs that are paid by health insurance has an effect either on the amount of medical care that people use or on their health. The treatments were four insurance plans. Each plan paid all medical costs above a ceiling. Below the ceiling, the plans paid 100%, 75%, 50%, or 0% of costs incurred. Outline the design of a randomized comparative experiment suitable for this study.

◆ **47.** People who eat lots of fruits and vegetables have lower rates of colon cancer than those who eat little of these foods. Fruits and vegetables are rich in antioxidants such as vitamins A, C, and E. Will taking antioxidants help prevent colon cancer? A clinical trial studied this question with 864 people who were at risk of colon cancer. The subjects were divided into four groups: daily beta-carotene, daily vitamins C and E, all three vitamins every day, and daily placebo. After four years, the researchers were surprised to find no significant difference in colon cancer among the groups.

(a) Outline the design of the experiment. Use your judgment in choosing the group sizes.
(b) Assign labels to the 864 subjects and use Table 5.1 starting at line 118 to choose the first 5 subjects for the beta-carotene group.
(c) The study was double-blind. What does this mean?
(d) What does "no significant difference" mean in describing the outcome of the study?
(e) Suggest some lurking variables that could explain why people who eat lots of fruits and vegetables have lower rates of colon cancer. The experiment suggests that these variables, rather than

the antioxidants, may be responsible for the observed benefits of fruits and vegetables.

◆ **48.** The advice columnist Ann Landers regularly invites her readers to respond to questions asked in her newspaper column. On one occasion, she asked, "If you had it to do over again, would you have children?" Almost 10,000 parents wrote in, of whom 70% said "no." Shortly afterward, a national poll asked a random sample of 1400 parents the same question; 90% of this sample said "yes." Which of these polls is more trustworthy, and why?

49. Joan's small accounting firm serves 30 business clients. Joan wants to interview a sample of 5 clients in detail to find ways to improve client satisfaction. To avoid bias, she chooses an SRS of size 5. Use Table 5.1 at line 123 to choose the SRS from Joan's client list:

A-1 Plumbing
Accent Printing
Action Sport Shop
Anderson Construction
Bailey Trucking
Balloons Inc.
Bennett Hardware
Best's Camera Shop
Blue Print Specialties
Central Tree Service
Classic Flowers
Computer Answers
Darlene's Dolls
Fleisch Realty
Hernandez Electronics
JL Records
Johnson Commodities
Keiser Construction
Liu's Chinese Restaurant
MagicTan
Peerless Machine
Photo Arts
River City Books
Riverside Tavern
Rustic Boutique
Satellite Services
Scotch Wash
Sewer's Center
Tire Specialties
Von's Video Store

50. The Census Bureau divides the entire country into "census tracts" that contain about 4000 people. Each tract is in turn divided into small "blocks," which in urban areas are bounded by local streets. An SRS of blocks from a census tract is often the next-to-last stage in a national sample. Figure 5.9 shows part of census tract 8051.12, in Cook County, Illinois, west of Chicago. The 44 blocks in this tract are divided into three "block groups."

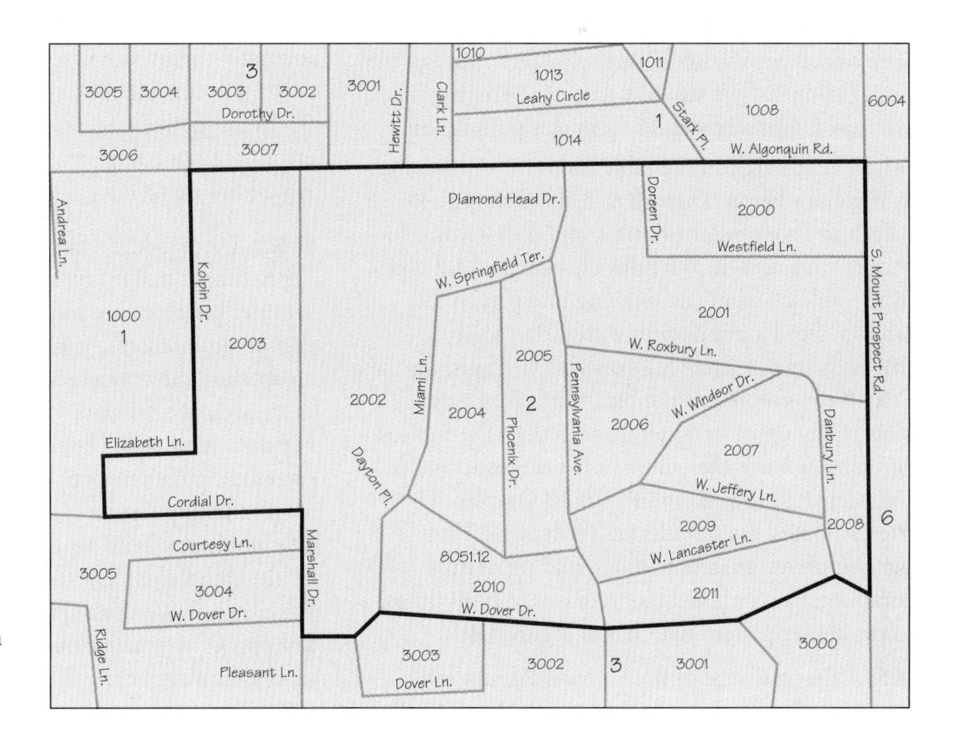

Figure 5.9 Census blocks in Cook County, Illinois. The outlined area is a block group. (From factfinder.census.gov.)

Group 1 contains 6 blocks numbered 1000 to 1005; group 2 (outlined in Figure 5.9) contains 12 blocks numbered 2000 to 2011; group 3 contains 26 blocks numbered 3000 to 3025. Use Table 5.1 beginning at line 125 to choose an SRS of 5 of the 41 blocks in this census tract. Explain carefully how you labeled the blocks.

■ **51.** The Internal Revenue Service plans to examine an SRS of individual federal income tax returns from each state. One variable of interest is the proportion of returns claiming itemized deductions. The total number of individual tax returns in a state varies from 14 million in California to 227,000 in Wyoming.

(a) Will the margin of error for estimating the proportion change from state to state if an SRS of 2000 tax returns is selected in each state? Explain your answer.
(b) Will the margin of error change from state to state if an SRS of 1% of all tax returns is selected in each state? Explain your answer.

■ **52.** You can use a table of random digits to *simulate* sampling from a population. Suppose that 60% of the population bought a lottery ticket in the last 12 months. We will simulate the behavior of random samples of size 40 from this population.

(a) Let each digit in the table stand for one person in this population. Digits 0 to 5 stand for people who find shopping frustrating, and 6 to 9 stand for people who do not. Why does looking at one digit from Table 5.1 simulate drawing one person at random from a population with 60% "yes"?
(b) Each row in Table 5.1 contains 40 digits. So the first 10 rows represent the results of 10 samples. How many digits between 0 and 5 does the top row contain? What is the percent of "yes" responses in this sample? How many of your 10 samples overestimated the population truth 60%? How many underestimated it? You could program a computer to continue this process, say 1000 times, to produce a pattern like that in Figure 5.4.

■ **53.** The last stage of the Current Population Survey uses a *systematic random sample*. An example will illustrate the idea of a systematic sample. Suppose that we must choose 4 rooms out of the 100 rooms in a dormitory. Because 100/4 = 25, we can think of the list of 100 rooms as 4 lists of 25 rooms each. Choose 1 of the first 25 rooms at random, using Table 5.1. The sample will contain this room and the rooms 25, 50, and 75 places down the list from it. If 13 is chosen, for example, then the systematic random sample consists of the rooms numbered 13, 38, 63, and 88.

(a) Use Table 5.1 to choose a systematic random sample of 5 rooms from a list of 200. Enter the table at line 120.
(b) Your sample gives every room the same chance to be chosen. Explain why. Yet this systematic sample is not a simple random sample. Explain why.

■ **54.** Though opinion polls usually make 95% confidence statements, some sample surveys use other confidence levels. The monthly unemployment rate, for example, is based on the Current Population Survey of about 55,000 households. The margin of error in the unemployment rate is announced as about ±0.15% with 90% confidence. Is the margin of error for 90% confidence larger or smaller than the margin of error for 95% confidence? Why? (*Hint:* Look at Figure 5.4 again.)

■ **55.** The quick method gives only the approximate margin of error for a simple random sample. Here's a fine point: The approximation is conservative, that is, it gives a margin of error a bit larger than the correct value. The quick method is quite accurate for sample outcomes roughly between 30% and 70% and less accurate for more extreme outcomes. Suppose that in fact the truth about the population is 0%. That is, absolutely no one would say "yes" to your question. What would be the sample percent for any sample from this population? What would be the margin of error for samples of any size? You see that in this extreme (and unrealistic) case, the quick method overestimates the correct margin of error by a lot.

56. Give an example of a question about college students, their behavior, or their opinions that would best be answered by

(a) a sample survey.
(b) an experiment.

57. Your college wants to gather student opinion about parking for students on campus. It isn't practical to contact all students.

(a) Give an example of a way to choose a sample of students that is poor practice because it depends on voluntary response.
(b) Give another example of a bad way to choose a sample that doesn't use voluntary response.

◆ **58.** A large study used records from Canada's national health care system to compare the effectiveness of two ways to treat prostate disease. The two treatments are traditional surgery and a new method that does not require surgery. The records described many patients whose doctors had chosen each method. The study found that patients treated by the new method were more likely to die within eight years.

(a) Explain why this study is not an experiment. Then explain what influences on doctors' decisions may be confounded with the effects of the two treatments.
(b) Outline an experiment to compare the two ways to treat prostate disease.

◆ **59.** In a test of the effects of persistent pesticides, researchers will feed a diet contaminated with DDT to rats for 60 days after weaning. Then they will measure the rats' nerve responses to assess the effects of the DDT.

(a) Explain why the experimenters should also study a control group of rats that are fed the same diet uncontaminated with DDT.
(b) If 20 newly weaned male rats are available, outline the design of the experiment and use Table 5.1 starting at line 123 to carry out the randomization.

◆ **60.** The National Institute of Mental Health wants to know whether intense education about the risks of AIDS will help change the behavior of people who now report sexual activities that put them at risk of infection. Investigators screened 38,893 people to identify 3706 suitable subjects. The subjects were assigned to a control group (1855 people) or an intervention group (1851 people). The control group attended a one-hour AIDS education session; the intervention group attended seven single-sex discussion sessions, each lasting 90 to 120 minutes. After 12 months, 64% of the intervention group and 52% of the control group said they used condoms. (None of the subjects used condoms regularly before the study began.)

(a) Because none of the subjects used condoms when the study started, we might just offer the intervention sessions and find that 64% used condoms 12 months after the sessions. Explain why this greatly overstates the effectiveness of the intervention.
(b) Outline the design of this experiment.
(c) You must randomly assign 3706 subjects. How would you label them? Use line 119 of Table 5.1 to choose the first 5 subjects for the intervention group.

■ **61.** Is the number of days a letter takes to reach another city affected by the day of the week it is mailed and whether or not the ZIP code is used? Briefly describe the design of a two-factor experiment to investigate this question. Be sure to specify the treatments exactly and to tell how you will handle outside variables such as the time of day the letter is mailed.

■ **62.** What are the effects of repeated exposure to an advertising message? The answer may depend both on the length of the ad and on how often it is repeated. An experiment investigates this question using undergraduate students as subjects. All subjects view a 40-minute television program that includes ads for a digital camera. Some subjects see a 30-second commercial; others, a 90-second version. The same commercial is repeated either one, three, or five times during the program. After viewing, all of the subjects answer questions about their recall of the ad, their attitude toward the camera, and their intention to purchase it. Here are the names of the student subjects:

Alomar	Farouk	Liang	Solomon
Asihiro	Fleming	Maldonado	Trujillo
Bennett	George	Marsden	Tullock
Bikalis	Han	Montoya	Valasco
Chen	Howard	O'Brian	Vaughn
Clemente	Hruska	Ogle	Wei
Denman	Imrani	Padilla	Wilder
Durr	James	Plochman	Willis
Edwards	Kaplan	Rosen	Zhang

(a) This experiment has two factors. How many treatments are being compared? What are these treatments?

(b) Outline the design of the experiment.

(c) Carry out the random assignment called for by your design. Enter Table 5.1 at line 107.

◆ **63.** A psychologist studies how much people disclose about themselves to other people met at a party. He arranges for student subjects to be introduced to new people. The subjects are both female and male and both black and white. The results show that "there were no significant race effects, but self-disclosure was significantly higher among females than among males." Explain what this means in language understandable to someone who knows no statistics. Do not use the word *significance* in your answer.

WRITING PROJECTS

1. Go to the Web site of the Gallup Organization (www.gallup.com) and click on "Gallup Poll News Service." Under "Poll Analyses" you will find archives of recent press releases put out by the Gallup poll. Choose a poll topic of interest to you and summarize the poll results. Now examine the press release in detail and report your findings: Does Gallup give the exact questions asked? The margin of error? Warnings about nonresponse and other sources of additional errors? You can find more detailed information about the polls under "Poll Surveys" for comparison with the press release.

2. Go to the Web site of a medical journal such as the *Journal of the American Medical Association*, the *New England Journal of Medicine*, or the *British Medical Journal*. *JAMA* and *NEJM* make the full text of some particularly important articles available free of charge. *BMJ* is unusual in making all its articles available. Select an article from the current issue or from a past issue that describes a study whose topic interests you. Write a newspaper article that summarizes the design and the findings of the study. (Be sure to include statistical aspects, such as observational study versus experiment and any randomization used. News accounts often neglect these facts.)

3. Choose an issue of current interest to students at your school. Prepare a short (no more than five questions) questionnaire to determine opinions on this issue. Choose a sample of about 25 students, administer your questionnaire, and write a brief description of your findings. Also write a short discussion of your experiences in designing and carrying out the survey.

(Although 25 students are too few for you to be statistically confident of your results, this project centers on the practical work of a survey. You must first identify a population; if it is not possible to reach a wider student population, use students enrolled in this course. Did the subjects find your questions clear? Did you write the questions so that it was easy to tabulate the responses? At the end, did you wish you had asked different questions?)

4. Locate a news discussion of an ethical issue that concerns statistical studies. Write your own brief summary of the debate and any conclusions you feel you can reach.

Here is an example of one way to approach this project. A debate continues as to whether it is ethical to study AIDS in African countries without offering all subjects the expensive treatments available in rich nations. Searching the archives at the Web site of the *New York Times* (www.nytimes.com) for "AIDS and Ethics" finds many articles, including a promising one in the issue of March 30, 2000. To read the article, you must either pay $2.50 or go to the library.

5. Any institution that receives federal funds must have an *Institutional Review Board* (IRB) that reviews in advance all studies that use human subjects. The IRB is charged with protecting the welfare of the subjects. What is the name of your college's IRB? Who are the members? Are there representatives from outside the college, and, if so, how are they chosen? What guidelines does your IRB follow? Do you have any suggestions for strengthening the protection of subjects offered by the IRB's review process?

SPREADSHEET PROJECTS

To do these projects, go to www.whfreeman.com/fapp.

There are several ways to generate numbers randomly. Spreadsheets provide a convenient alternative to flipping coins, rolling dice, or using a random number table. This spreadsheet project will introduce the spreadsheet's random number generator. Spreadsheet-created histograms are then created to compare these random numbers to data generated in other ways.

APPLET EXERCISES

To do these exercises, go to www.whfreeman.com/fapp.

Simple Random Sample

The purpose of this exercise is to see the results of sampling variability. Suppose that a researcher wants to test the effectiveness of a new breakfast food. He takes a group of 30 rats and assigns them randomly into a treatment group of 15 and a control group of 15.

(a) Use the Simple Random Sample applet to choose the 15 rats for the treatment group, and record the labels of the rats chosen.
(b) Suppose that the 15 even-numbered rats among the 30 rats available are (unknown to the experimenters) a fast-growing variety. We hope that these rats will be roughly equally distributed between the two groups. How many of the rats chosen in part (a) are fast-growing?
(c) Take 10 additional samples of size 15 from the 30 rats. (Be sure to click "Reset" after each sample.) Record the counts of even-numbered rats in each of your 10 samples. You see that there is considerable chance variation, but no systematic bias in favor of one or the other group in assigning the fast-growing rats. Larger samples from a larger population on the average do a better job of making the two groups equivalent.

Exploring Data

In this chapter we cover...

... and at www.whfreeman.com/fapp:

Flashcards
Quizzes
Spreadsheet Projects
  Analyzing data
Applet Exercises
  Correlation and regression

A flood of data is a prominent feature of modern society. Data, or numerical facts, are essential for making decisions in almost every area of life and work. Like other great floods, the flood of numbers threatens to overwhelm us. We must control the flood by careful organization and interpretation. A corporate database, for example, contains an immense volume of data—on employees, sales, inventories, customer accounts, equipment, taxes, and other topics. These data are useful only if we can organize them and present them so that their meaning is clear. The penalties for ignoring data can be severe—several banks have suffered billion-dollar losses from unauthorized trades in financial markets by their employees, trades that were hidden in a mass of data that the banks' management did not examine carefully.

Any set of data contains information about some group of *individuals*. The information is organized in *variables*.

> **Individuals** are the objects described by a set of data. Individuals may be people, but they may also be animals or things.
>
> A **variable** is any characteristic of an individual. A variable can take different values for different individuals.

EXAMPLE A Corporate Data Set

Figure 6.1 displays a small part of the data set in which CyberStat Corporation records information about its employees. The *individuals* described are the employees. Each row records data on one individual. Each column contains the values of one *variable* for all the individuals. In addition to the person's name, there are five variables. Gender, race, and job type are variables that classify the

Figure 6.1 Part of a data set displayed by the Excel spreadsheet program.

	A	B	C	D	E	F
1	Name	Job Type	Age	Gender	Race	Salary
2	Cedillo, Jose	Technical	27	Male	White	52,300
3	Chambers, Tonia	Management	42	Female	Black	112,800
4	Childers, Amanda	Clerical	39	Female	White	27,500
5	Chen, Huabang	Technical	51	Male	Asian	83,600
6						

Ready NUM

employees but do not take numerical values. Age and salary do take numerical values. You can see that age is measured in years and salary in dollars.

Most data tables follow this format–each row is an individual, and each column is a variable. This data set appears in a *spreadsheet* program that has rows and columns ready for your use. Spreadsheets are commonly used to enter and transmit data, and spreadsheet programs also have functions for basic statistics. ■

Statistical tools and ideas help us examine data in order to describe their main features. This examination is called **exploratory data analysis.** Like an explorer crossing unknown lands, we want first to simply describe what we see. In this chapter, we use both numbers and graphs to explore data. Here are two principles that provide the tactics for exploratory analysis of data.

1. Begin by examining each variable by itself. Then move on to study the relationships among the variables.
2. Begin with a graph or graphs. Then add numerical summaries of specific aspects of the data.

These principles also organize the material in this chapter. We start with data on a single variable, then move to relations among several variables. In each setting, we first display the data in graphs, then add numerical summaries.

6.1 Displaying Distributions: Histograms

Data analysis begins with graphical displays of the values of a single variable. For example, CyberStat Corporation may wish to compare the salaries of its female and male employees. Because individual salaries vary so much, we are interested in the *distribution* of salaries of all female and all male employees.

The **distribution** of a variable tells us what values it takes and how often it takes these values.

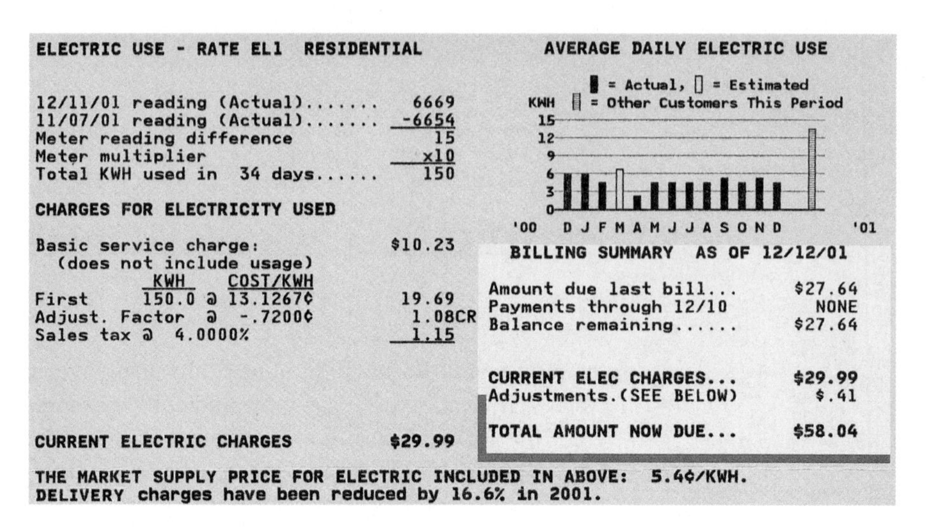

ELECTRIC USE - RATE EL1 RESIDENTIAL

12/11/01 reading (Actual)....... 6669
11/07/01 reading (Actual)....... -6654
Meter reading difference 15
Meter multiplier x10
Total KWH used in 34 days...... 150

CHARGES FOR ELECTRICITY USED

Basic service charge: $10.23
(does not include usage)

	KWH	COST/KWH	
First	150.0 @	13.1267¢	19.69
Adjust. Factor	@	-.7200¢	1.08CR
Sales tax @		4.0000%	1.15

CURRENT ELECTRIC CHARGES $29.99

AVERAGE DAILY ELECTRIC USE

BILLING SUMMARY AS OF 12/12/01

Amount due last bill...	$27.64
Payments through 12/10	NONE
Balance remaining......	$27.64
CURRENT ELEC CHARGES...	$29.99
Adjustments.(SEE BELOW)	$.41
TOTAL AMOUNT NOW DUE...	$58.04

THE MARKET SUPPLY PRICE FOR ELECTRIC INCLUDED IN ABOVE: 5.4¢/KWH.
DELIVERY charges have been reduced by 16.6% in 2001.

This electricity bill displays electricity usage in a histogram. (Courtesy of Trish Marx.)

Numerical variables often take many values. A graph of the distribution is clearer if nearby values are grouped together. The most common graph of the distribution of one numerical variable is a **histogram.**

EXAMPLE Making a Histogram

One of the most striking findings of the 2000 census was the growth of the Hispanic population in the United States. Table 6.1 presents the percent of adult residents (age 18 and over) in each of the 50 states who identified themselves in the 2000 census as "Spanish/Hispanic/Latino." The *individuals* in this data set are the 50 states. The *variable* is the percent of Hispanic adults in a state. To make a histogram of the distribution of this variable, proceed as follows:

1. Divide the range of the data into classes of equal width. The data in Table 6.1 range from 0.6 to 38.7, so we decide to choose these classes:

$$0.0 < \text{percent Hispanic} \leq 5.0$$
$$5.0 < \text{percent Hispanic} \leq 10.0$$
$$\vdots$$
$$35.0 < \text{percent Hispanic} \leq 40.0$$

Be sure to specify the classes precisely so that each individual falls into exactly one class. A state with 5.0% Hispanic residents would fall into the first class, but a state with 5.1% falls into the second.

TABLE 6.1 Percent of Hispanics in Adult Population, by State (2000)

State	Percent	State	Percent	State	Percent
Alabama	1.5	Louisiana	2.4	Ohio	1.6
Alaska	3.6	Maine	0.6	Oklahoma	4.3
Arizona	21.3	Maryland	4.0	Oregon	6.5
Arkansas	2.8	Massachusetts	5.6	Pennsylvania	2.6
California	28.1	Michigan	2.7	Rhode Island	7.0
Colorado	14.9	Minnesota	2.4	South Carolina	2.2
Connecticut	8.0	Mississippi	1.3	South Dakota	1.2
Delaware	4.0	Missouri	1.8	Tennessee	2.0
Florida	16.1	Montana	1.6	Texas	28.6
Georgia	5.0	Nebraska	4.5	Utah	8.1
Hawaii	5.7	Nevada	16.7	Vermont	0.8
Idaho	6.4	New Hampshire	1.4	Virginia	4.2
Illinois	10.7	New Jersey	12.3	Washington	6.0
Indiana	3.1	New Mexico	38.7	West Virginia	0.6
Iowa	2.3	New York	13.8	Wisconsin	2.9
Kansas	5.8	North Carolina	4.3	Wyoming	5.5
Kentucky	1.3	North Dakota	1.0		

SOURCE: Census Bureau, www.census.gov.

2. Count the number of individuals in each class.

Class	Count	Class	Count
0.1 to 5.0	30	20.1 to 25.0	1
5.1 to 10.0	10	25.1 to 30.0	2
10.1 to 15.0	4	30.1 to 35.0	0
15.1 to 20.0	2	35.1 to 40.0	1

3. Draw the histogram. First mark the scale for the variable whose distribution you are displaying on the horizontal axis. That's the percent of adults who are Hispanic. The scale runs from 0 to 40 because that is the span of the classes we chose. The vertical axis contains the scale of counts. Each bar represents a class. The base of the bar covers the class, and the bar height is the class count. There is no horizontal space between the bars unless a class is empty, so that its bar has height zero. Figure 6.2 is our histogram. ■

The bars of a histogram should cover the entire range of values of a variable. When the possible values of a variable have gaps between them, extend the bases of the bars to meet halfway between two adjacent possible values. For example,

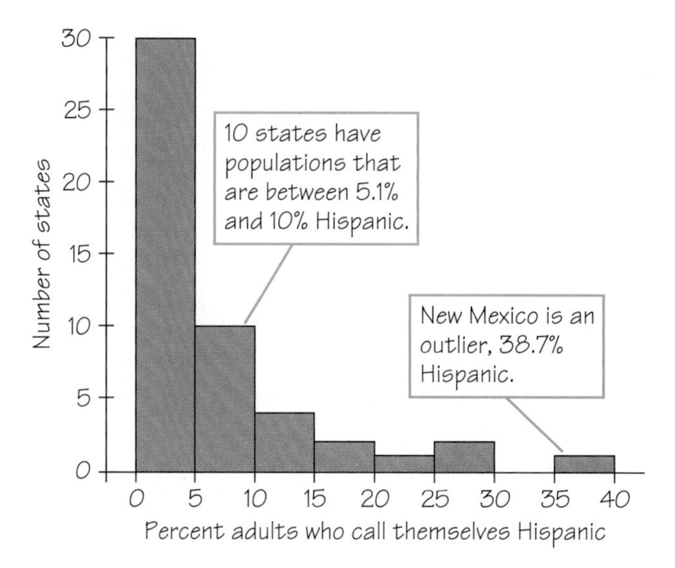

Figure 6.2 Histogram of the percent of Hispanics among the adult residents of the states.

in a histogram of the ages in years of university faculty, the bars representing 25 to 29 years and 30 to 34 years would meet at 29.5.

Our eyes respond to the *area* of the bars in a histogram. Because the classes are all the same width, area is determined by height and all classes are fairly represented. There is no one right choice of the classes in a histogram. Too few classes will give a "skyscraper" graph, with all values in a few classes with tall bars. Too many will produce a "pancake" graph, with most classes having one or no observations. Neither choice will give a good picture of the shape of the distribution. You must use your judgment in choosing classes to display the shape. Statistics software will choose the classes for you. The computer's choice is usually a good one, but you can change it if you want.

6.2 Interpreting Histograms

Making a statistical graph is not an end in itself. The purpose of the graph is to help us understand the data. After you make a graph, always ask, "What do I see?" Once you have displayed a distribution, you can see its important features as follows.

> In any graph of data, look for the **overall pattern** and for striking **deviations** from that pattern.
>
> You can describe the overall pattern of a histogram by its **shape, center,** and **spread.** We will soon learn how to describe center and spread numerically.
>
> An important kind of deviation is an **outlier,** an individual value that falls outside the overall pattern.

EXAMPLE

Describing a Distribution

Look again at the histogram in Figure 6.2. *Shape:* The distribution is *skewed to the right.* Most states have no more than 10% Hispanics, but some states have much higher percentages, so that the graph trails off to the right. *Center:* For a rough measure of center, note that about half the states have less than 4% Hispanics among their adult residents and half have more. So the midpoint of the distribution is close to 4%. *Spread:* The spread is from about 0% to 40%, but only one state falls above 30%.

Outliers: The one state that stands out is New Mexico, with 38.7% Hispanics. Other states (Arizona, California, Texas) with high percents are part of the long right tail of the distribution, but New Mexico is 10% higher than any other state. Once you have spotted outliers, look for an explanation. Some outliers are due to mistakes, such as typing 3.8 as 38. Other outliers point to the special nature of some observations. New Mexico is heavily Hispanic by history and location. ■

When you describe a distribution, concentrate on the main features. Look for major peaks, not for minor ups and downs in the bars of the histogram. Look for clear outliers, not just for the smallest and largest observations. Look for rough *symmetry* or clear *skewness.*

A distribution is **symmetric** if the right and left sides of the histogram are approximately mirror images of each other.

A distribution is **skewed to the right** if the right side of the histogram (containing the half of the observations with larger values) extends much farther out than the left side. It is **skewed to the left** if the left side of the histogram extends much farther out than the right side.

Here are more examples of describing the overall pattern of a histogram.

EXAMPLE

Iowa Test Scores

Figure 6.3 displays the scores of all 947 seventh-grade students in the public schools of Gary, Indiana, on the reading part of the Iowa Test of Basic Skills. The distribution is *single-peaked* and *symmetric.* In mathematics, the two sides of symmetric patterns are exact mirror images. Real data are almost never exactly symmetric. We are content to describe Figure 6.3 as symmetric. The center (half above, half below) is close to 7. This is seventh-grade reading level. The scores range from 2.0 (second-grade level) to 12.1 (twelfth-grade level). ■

EXAMPLE

College Tuition

Jeanna plans to attend college in her home state of Massachusetts. In the College Board's *Annual Survey of Colleges,* she finds data on college tuition and fees for

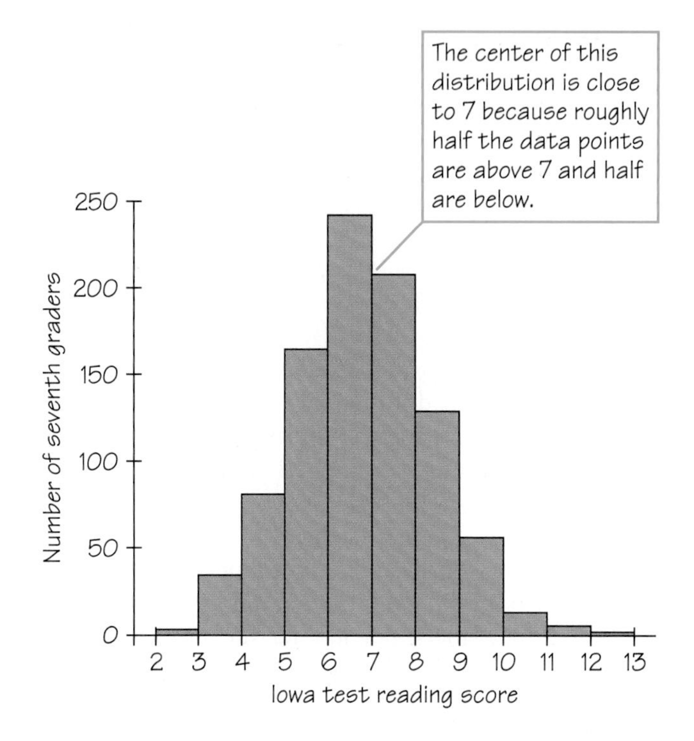

Figure 6.3 Histogram of the Iowa Test of Basic Skills reading scores of 947 seventh-grade students.

the 2000–2001 academic year. Figure 6.4 displays the costs for all 56 four-year colleges in Massachusetts (omitting art schools and other special colleges). As is often the case, we can't call this irregular distribution either symmetric or skewed. It does show two separate *clusters* of colleges, 11 with tuition less than

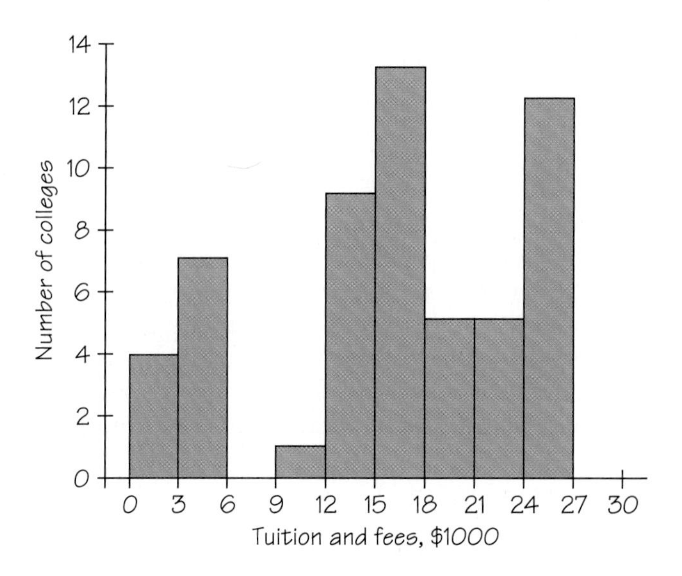

Figure 6.4 Histogram of the tuition and fees charged by four-year colleges in Massachusetts.

$6000 and the remaining 45 costing more than $9000. Clusters suggest that two types of individuals are mixed in the data set. In fact, the histogram distinguishes the 11 state colleges in Massachusetts from the 45 private colleges, which charge much more. ■

6.3 Displaying Distributions: Stemplots

Histograms are not the only way to graphically display distributions. For small data sets, a *stemplot* is quicker to make and presents more detailed information.

To make a **stemplot:**

1. Separate each observation into a **stem** consisting of all but the final (rightmost) digit and a **leaf,** the final digit. Stems may have as many digits as needed, but each leaf contains only a single digit.
2. Write the stems in a vertical column with the smallest at the top, and draw a vertical line at the right of this column.
3. Write each leaf in the row to the right of its stem, in increasing order out from the stem.

EXAMPLE

Making a Stemplot

For the "percent Hispanic" percents in Table 6.1, take the whole number part of the percent as the stem and the final digit (tenths) as the leaf. The Alaska entry, 3.6%, has stem 3 and leaf 6. Indiana, at 3.1%, places leaf 1 on the same stem. These are the only observations on this stem. We then arrange the leaves in order, as 16, so that 3|16 is one row in the stemplot. Figure 6.5 is the complete stemplot for the data in Table 6.1. We left out New Mexico, the high outlier with stem 38 and leaf 7. ■

A stemplot looks like a histogram turned on end. Comparing the stemplot in Figure 6.5 with the histogram in Figure 6.2 reveals the strengths and weaknesses of stemplots. The stemplot, unlike the histogram, preserves the actual value of each observation. But you can choose the classes in a histogram, whereas the classes (the stems) of a stemplot are forced on you. Whether the large number of classes in Figure 6.5 is an improvement over Figure 6.2 is a matter of taste. Histograms are preferable for large data sets like the 947 Iowa Test scores in Figure 6.3.

When the observed values have many digits, it is often best to round the numbers to just a few digits before making a stemplot. For example, a stemplot of data like

3.468 2.567 2.981 1.095 . . .

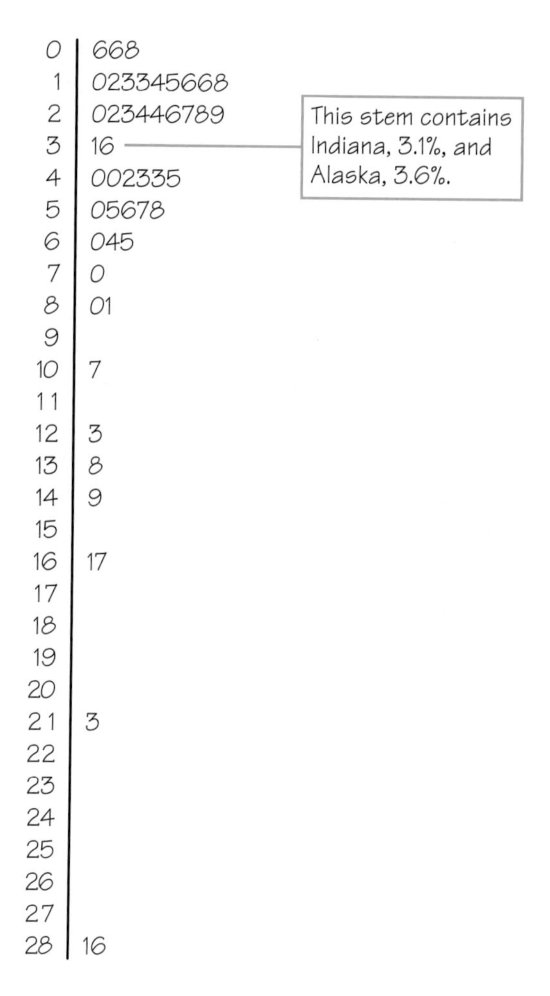

Figure 6.5 Stemplot of the percent of Hispanics among the adult residents of the states.

would have very many stems and no leaves or just one leaf on most stems. You can round these data to

3.5 2.6 3.0 1.1 . . .

before making a stemplot.

6.4 Describing Center: Mean and Median

A description of a distribution almost always includes a measure of its center or average. The most common measure of center is the ordinary arithmetic average, or *mean*.

To find the **mean** of a set of observations, add their values and divide by the number of observations. If the n observations are $x_1, x_2, \ldots, x_n$, their mean is

$$\bar{x} = \frac{x_1 + x_2 + \cdots + x_n}{n}$$
$$= \frac{1}{n} \sum x_i$$

The summation sign Σ means "add up all the x-values." The bar over the x indicates the mean of all the x-values. Pronounce the mean $\bar{x}$ as "x-bar." This notation is very common. When writers who are discussing data use $\bar{x}$ or $\bar{y}$, they are talking about a mean.

EXAMPLE

Calculating the Mean

Interested in an exotic car? The Environmental Protection Agency lists most such vehicles in its "minicompact" or "two-seater" categories. Table 6.2 gives the city and highway gas mileage for all cars in these groups. (The mileages are for the basic engine and transmission combination for each car.)

A stemplot shows that the distribution of highway mileage for minicompact cars is irregular in shape, as is often the case when we have few observations:

Stem	Leaves
22	00
23	0
24	00
25	0
26	
27	0
28	00
29	
30	0
31	0

The mean gas highway mileage for the 11 minicompacts is

$$\bar{x} = \frac{x_1 + x_2 + \cdots + x_n}{n}$$
$$= \frac{31 + 27 + 28 + 23 + 24 + 22 + 28 + 24 + 30 + 25 + 22}{11}$$
$$= \frac{284}{11} = 25.8 \text{ miles per gallon}$$

TABLE 6.2 Fuel Economy (Miles per Gallon) for Model Year 2001 Cars

Minicompact Cars			Two-Seater Cars		
Model	City	Highway	Model	City	Highway
Audi TT Coupe	22	31	Acura NSX	17	24
BMW 325CI Convertible	19	27	Audi TT Roadster	22	30
BMW 330CI Convertible	20	28	BMW Z3 Coupe	21	28
BMW M3 Convertible	16	23	BMW Z3 Roadster	20	27
Jaguar XK8 Convertible	17	24	BMW Z8	13	21
Jaguar XKR Convertible	16	22	Chevrolet Corvette	18	26
Mercedes-Benz CLK320	20	28	Dodge Viper	11	21
Mercedes-Benz CLK430	18	24	Ferrari Modena	11	16
Mitsubishi Eclipse	22	30	Ferrari Maranello	8	13
Porsche 911 Carrera	17	25	Honda S2000	20	26
Porsche 911 Turbo	15	22	Lamborghini Diablo	10	13
			Mazda Miata	22	28
			Mercedes-Benz SL500	16	23
			Mercedes-Benz SL600	13	19
			Mercedes-Benz SLK320	21	27
			Plymouth Prowler	17	23
			Porsche Boxster	19	27
			Toyota MR2	25	30

SOURCE: Environmental Protection Agency, *2001 Fuel Economy Guide*, www.fueleconomy.gov.

In practice, you can key the data into your calculator and hit the $\bar{x}$ key. You don't have to actually add and divide. But you should know that this is what the calculator is doing. ■

The mean is the average value. Another way to measure center is to give the midpoint, the value with half the observations below it and half above. This is the idea of the *median*. Here is the full rule for finding the median.

The **median** M is the midpoint of a distribution, the number such that half the observations are smaller and the other half are larger. To find the median of a distribution:

1. Arrange all observations in order of size, from smallest to largest.
2. If the number n of observations is odd, the median M is the center observation in the ordered list. The location of the median is found by counting $(n + 1)/2$ observations up from the bottom of the list.
3. If the number n of observations is even, the median M is the average of the two center observations in the ordered list. The location of the median is again $(n + 1)/2$ from the bottom of the list.

Be sure to write down each individual observation in the data set, even if several observations repeat the same value. And be sure to arrange the observations in order of size before locating the median. Note that the recipe $(n + 1)/2$ gives the *position* of the median in the ordered list of observations, *not* the median itself.

EXAMPLE

Calculating the Median

To find the median highway mileage for the 11 minicompacts, first arrange the observations in order:

22 22 23 24 24 **25** 27 28 28 30 31

The boldface 25 is the midpoint, with 5 observations to its left and 5 to its right. When the number of observations n is odd, there is always one observation in the center of the ordered list. This is the median, $M = 25$ miles per gallon (mpg). The rule for the location of the median agrees that the median is the 6th observation in the ordered list:

$$\frac{n+1}{2} = \frac{11+1}{2} = \frac{12}{2} = 6$$

What about the 18 two-seaters? Arrange their highway mileages in order:

13 13 16 19 21 21 23 23 **24** **26** 26 27 27 27 28 28 30 30

When n is even, there is no one middle observation. But there is a middle pair—the boldface 24 and 26 have 8 observations on either side. We take the median to be halfway between this middle pair. That is,

$$M = \frac{24+26}{2} = \frac{50}{2} = 25 \text{ mpg}$$

Our rule says that the location of the median is

$$\frac{n+1}{2} = \frac{18+1}{2} = \frac{19}{2} = 9.5$$

Location 9.5 means "halfway between the ninth and tenth observations in the ordered list." The median is the average of these two observations.

The two groups of cars have the same median highway mileage, confirming that they are quite similar. ■

Comparing these groups of cars illustrates an important difference between the mean and the median. *The mean is pulled more strongly toward a few extreme observations.* The two-seaters include a Ferrari and a Lamborghini that get only 13 miles per gallon. You can check that the mean highway mileage for two-seaters is 23.4 mpg, well below the 25.8 mpg of the minicompacts. The two low observations pull down the mean. They do not affect the median, which would remain at 25 even if the two leftmost observations in the list were only 1 mpg.

Many distributions of monetary values are strongly skewed to the right. In this case, the mean is pulled away from the median toward the long right tail. For example, the distribution of house prices is skewed to the right. There are many moderately priced houses and a few very expensive mansions. The mansions pull the mean up but do not affect the median. The mean price of existing single-family houses sold in 2000 was $177,000, but the median price for these same houses was only $139,100.

6.5 Describing Spread: The Quartiles

The mean and median provide two different measures of the center of a distribution. But a measure of center alone can be misleading. Two neighborhoods with median house price $139,000 are very different if one has both mansions and modest homes and the other has little variation among houses. We are interested in the spread or variability of house prices as well as their centers. *The simplest useful numerical description of a distribution consists of both a measure of center and a measure of spread.*

One way to measure spread is to give the smallest and largest observations. For example, the percents of Hispanic adults in the states range from 0.6% in Maine and West Virginia to 38.7% in New Mexico. These single observations show the full spread of the data, but they may be outliers. We can improve our description of spread by also looking at the spread of the middle half of the data. The *quartiles* mark out the middle half. Count up the ordered list of observations, starting from the smallest. The *first quartile* lies one-quarter of the way up the list. The *third quartile* lies three-quarters of the way up the list. In other words, the first quartile is larger than 25% of the observations, and the third quartile is larger than 75% of the observations. The second quartile is the median, which is larger than 50% of the observations. That is the idea of quartiles. We need a rule to make the idea exact. The rule for calculating the quartiles uses the rule for the median.

To calculate the **quartiles:**

1. Arrange the observations in increasing order and locate the median M in the ordered list of observations.
2. The **first quartile Q_1** is the median of the observations whose position in the ordered list is to the left of the location of the overall median.
3. The **third quartile Q_3** is the median of the observations whose position in the ordered list is to the right of the location of the overall median.

EXAMPLE

Calculating Quartiles

The highway gas mileages of the 11 minicompact cars in Table 6.2 (arranged in increasing order) are:

22 22 23 24 24 **25** 27 28 28 30 31

There is an odd number of observations, so the median is the middle one, the boldface 25 in the list. The first quartile is the median of the 5 observations to the left of the median. This is the 3rd of these 5 observations, $Q_1 = 23$. If you want, you can use the recipe for the location of the median with $n = 5$:

$$\frac{n+1}{2} = \frac{5+1}{2} = 3$$

The third quartile is the median of the 5 observations to the right of the median, $Q_3 = 28$. The overall median is left out of the calculation of the quartiles when there is an odd number of observations.

For the 18 two-seaters, the data (again arranged in order) are:

13 13 16 19 21 21 23 23 24 | 26 26 27 27 27 28 28 30 30

There is an even number of observations, so the median lies midway between the middle pair. Its location is between the 9th and 10th values, marked by | in the list. The first quartile is the median of the first 9 observations, because these are the observations to the left of the location of the median. Check that $Q_1 = 21$ and $Q_3 = 27$. When the number of observations is even, all the observations enter into the calculation of the quartiles. ■

Some software packages use a slightly different rule to find the quartiles, so computer results may be a bit different from your own work. Don't worry about this. The differences will always be too small to be important.

6.6 The Five-Number Summary and Boxplots

We can combine the median and quartiles with the smallest and largest observations to obtain a more complete description.

The **five-number summary** of a distribution consists of the smallest observation, the first quartile, the median, the third quartile, and the largest observation, written in order from smallest to largest. In symbols, the five-number summary is

Minimum Q_1 M Q_3 Maximum

The smallest and largest observations give information about the tails of the distribution that is missing if we know only Q_1, M, and Q_3. These five numbers offer a reasonably complete description of center and spread. The five-number summary for highway gas mileages for minicompacts is

22 23 25 28 31

For two-seaters, it is

13 21 25 27 30

The five-number summary of a distribution leads to a new graph, the *boxplot*. Figure 6.6 shows boxplots for both city and highway gas mileages for our two groups of cars.

A **boxplot** is a graph of the five-number summary.

- A central box spans the quartiles Q_1 and Q_3.
- A line in the box marks the median M.
- Lines extend from the box out to the smallest and largest observations.

Because boxplots show less detail than histograms or stemplots, they are best used for side-by-side comparison of more than one distribution, as in Figure 6.6. When you look at a boxplot, first locate the median, which marks the center of the distribution. Then look at the spread. The quartiles show the spread of the middle half of the data, and the extremes (the smallest and largest observations)

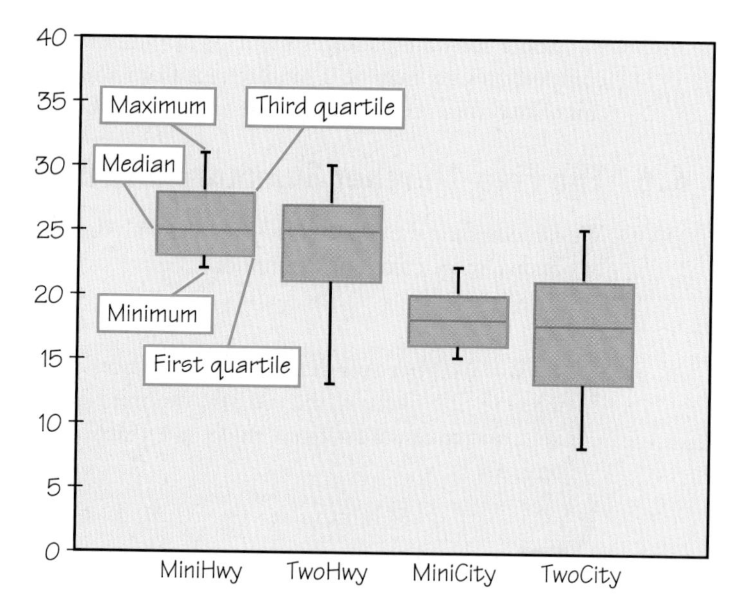

Figure 6.6 Side-by-side boxplots comparing the highway gas mileages of minicompact and two-seater cars.

show the spread of the entire data set. We see at once that city mileages are lower than highway mileages. The median gas mileages for the two groups of cars are very close together, but the two-seaters are more spread out.

6.7 Describing Spread: The Standard Deviation

Although the five-number summary is the most generally useful numerical description of a distribution, it is not the most common. That distinction belongs to the combination of the mean with the *standard deviation.* The mean, like the median, is a measure of center. The standard deviation, like the quartiles and extremes in the five-number summary, measures spread. The standard deviation and its close relative, the *variance,* measure spread by looking at how far the observations are from their mean.

EXAMPLE Understanding the Standard Deviation

A person's metabolic rate is the rate at which the body consumes energy. Metabolic rate is important in studies of weight gain, dieting, and exercise. Here are the metabolic rates of 7 men who took part in a study of dieting. (The units are calories per 24 hours. These are the same calories used to describe the energy content of foods.)

1792 1666 1362 1614 1460 1867 1439

Figure 6.7 displays the data as points above the number line, with their mean marked by an asterisk (*). The arrows mark two of the deviations from the mean. These deviations show how spread out the data are about their mean. Some of the deviations are positive and some are negative. Squaring the deviations makes them all positive. Observations far from the mean in either direction will have large positive squared deviations. So a reasonable measure of spread is the average of the squared deviations. This average is called the *variance.* The variance is large if the observations are widely spread about their mean; it is small if the observations are all close to the mean.

But the variance has the wrong units: If we measure metabolic rate in calories, the variance of the metabolic rates is in squared calories. Taking the square

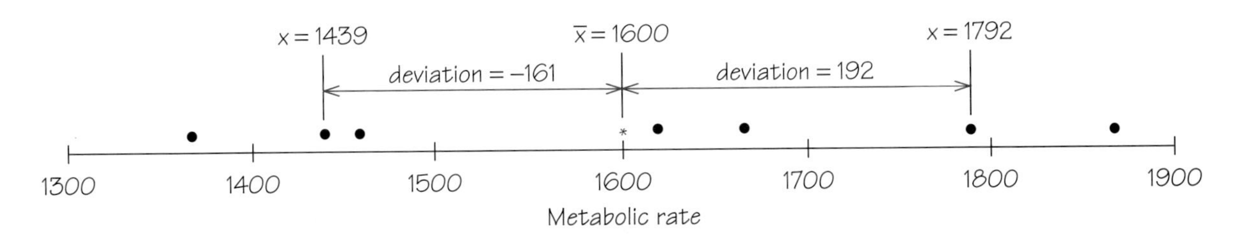

Figure 6.7 The variance and standard deviation measure spread by looking at the deviations of observations from their mean.

root of the variance gets us back to calories. The square root of the variance is the *standard deviation.* ■

The **variance** s^2 of a set of observations is an average of the squares of the deviations of the observations from their mean. In symbols, the variance of n observations $x_1, x_2, \ldots, x_n$ is

$$s^2 = \frac{(x_1 - \bar{x})^2 + (x_2 - \bar{x})^2 + \cdots + (x_n - \bar{x})^2}{n - 1}$$

$$= \frac{1}{n-1} \sum (x_i - \bar{x})^2$$

The **standard deviation** s is the square root of the variance s^2.

Notice that it is usual to divide by $n - 1$ rather than by n when we average the squared deviations to get the variance. The reason is a bit subtle: The deviations $x_i - \bar{x}$ always add to exactly 0, so if you know the value of all but one, you know them all. In practice, use software or your calculator to obtain The standard deviation from keyed-in data. Doing an example step by step, however, will help you understand how the variance and standard deviation work.

EXAMPLE

Calculating the Standard Deviation

To find the standard deviation of the 7 metabolic rates, first find the mean:

$$\bar{x} = \frac{1792 + 1666 + 1362 + 1614 + 1460 + 1867 + 1439}{7}$$

$$= \frac{11{,}200}{7} = 1600 \text{ calories}$$

The deviations shown in Figure 6.7 are the starting point for calculating the variance and the standard deviation.

Observations x_i	Deviations $x_i - \bar{x}$	Squared deviations $(x_i - \bar{x})^2$
1792	$1792 - 1600 = 192$	$192^2 = 36{,}864$
1666	$1666 - 1600 = 66$	$66^2 = 4{,}356$
1362	$1362 - 1600 = -238$	$(-238)^2 = 56{,}644$
1614	$1614 - 1600 = 14$	$14^2 = 196$
1460	$1460 - 1600 = -140$	$(-140)^2 = 19{,}600$
1867	$1867 - 1600 = 267$	$267^2 = 71{,}289$
1439	$1439 - 1600 = -161$	$(-161)^2 = 25{,}921$
	sum = 0	sum = 214,870

The variance is the sum of the squared deviations divided by one less than the number of observations:

$$s^2 = \frac{214{,}870}{6} = 35{,}811.67$$

The standard deviation is the square root of the variance:

$$s = \sqrt{35{,}811.67} = 189.24 \text{ calories}$$

■

More important than the details of hand calculation are the properties that determine the usefulness of the standard deviation:

- s measures spread about the mean $\bar{x}$. Use s to describe the spread of a distribution only when you use $\bar{x}$ to describe the center.
- $s = 0$ only when there is *no spread*. This happens only when all observations have the same value. Otherwise $s > 0$. As the observations become more spread out about their mean, s gets larger.
- s has the same units of measurement as the original observations. For example, if you measure metabolic rates in calories, s is also in calories. This is one reason to prefer s to the variance s^2, which is in squared calories.
- Like the mean $\bar{x}$, s is strongly influenced by a few extreme observations. For example, the standard deviation of the highway gas mileages for two-seaters is 5.34 mpg. (Use your calculator to verify this.) If we omit the two cars with 13 mpg, the standard deviation drops to 3.99 mpg.

We now have a choice between two descriptions of the center and spread of a distribution: the five-number summary, or $\bar{x}$ and s. Because $\bar{x}$ and s are sensitive to extreme observations, they can be misleading when a distribution is strongly skewed or has outliers. In fact, because the two sides of a skewed distribution have different spreads, no single number such as s describes the spread well. The five-number summary, with its two quartiles and two extremes, does a better job.

The five-number summary is usually better than the mean and standard deviation for describing a skewed distribution or a distribution with outliers. Use $\bar{x}$ and s for only reasonably symmetric distributions that are free of outliers.

Although the standard deviation is widely used, it is not a natural or convenient measure of the spread of a distribution. The real reason for the popularity of the standard deviation is that it is the natural measure of spread for *normal distributions,* an important class of distributions that we will discuss in the next chapter.

Do remember that a graph gives the best overall picture of a distribution. Numerical measures of center and spread report specific facts about a distribution, but they do not describe its entire shape. Numerical summaries do not disclose the presence of clusters, for example. *Always start with a graph of your data.*

6.8 Displaying Relations: Scatterplots

The examples we have looked at so far considered only a single variable, such as the highway gas mileage of a car. Now we will examine data for two variables, emphasizing the nature and strength of the relationship between the variables. To study a relationship, we measure both variables on the same individuals. Often, we think that one of the variables explains or influences the other.

A **response variable** is a variable that measures an outcome or result of a study. An **explanatory variable** is a variable that we think explains or causes changes in the response variables.

EXAMPLE Natural Gas Consumption

Sue is about to install solar panels to reduce the cost of heating her house in the Midwest. In order to know how much the solar panels help, she records her consumption of natural gas before the panels are installed. Gas consumption is higher in cold weather, so the relationship between outside temperature and gas consumption is important.

Table 6.3 gives data for 9 months. The response variable is the average amount of natural gas consumed each day during the month, in hundreds of cubic feet. The explanatory variable is the average number of heating degree-days each day during the month. (Heating degree-days are the usual measure of demand for heating. One degree-day is accumulated for each degree a day's average temperature falls below 65°F. An average temperature of 20°F, for example, corresponds to 45 degree-days.)

TABLE 6.3 Natural Gas Consumption of a Household

Month	Oct.	Nov.	Dec.	Jan.	Feb.	Mar.	Apr.	May	June
Degree-days per day	15.6	26.8	37.8	36.4	35.5	18.6	15.3	7.9	0.0
Gas consumed per day (100 cubic feet)	5.2	6.1	8.7	8.5	8.8	4.9	4.5	2.5	1.1

SOURCE: Robert Dale, Purdue University.

(Corbis.)

Looking at the numbers in the table, we can see that more degree-days (colder temperatures) go with higher gas consumption. But the shape and strength of the relationship are not fully clear. To display and interpret these data, we need a graph. Figure 6.8a is a *scatterplot* of Sue's data. ■

A **scatterplot** shows the relationship between two numerical variables measured on the same individuals. The values of one variable appear on the horizontal axis, and the values of the other variable appear on the vertical axis. Each individual in the data appears as the point in the plot fixed by the values of both variables for that individual.

Always plot the explanatory variable, if there is one, on the horizontal axis (the x-axis) of a scatterplot. As a reminder, we usually call the explanatory variable x and the response variable y. If there is no explanatory–response distinction, either variable can go on the horizontal axis. Degree-days appear on the horizontal scale and gas consumption on the vertical scale in Figure 6.8a because degree-days is the explanatory variable. The weather affects gas consumption; gas consumption does not explain the weather.

To interpret a scatterplot, apply the usual strategies of data analysis.

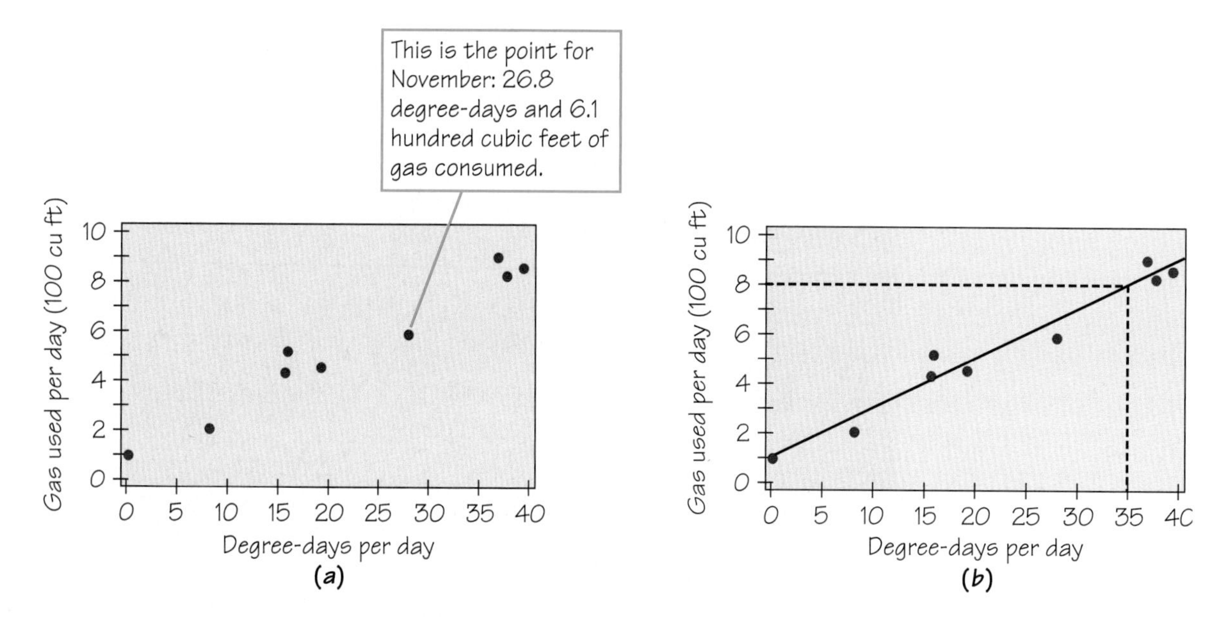

Figure 6.8 Natural gas consumption versus degree-days. (a) A scatterplot. (b) A regression line and its use for prediction.

In any graph of data, look for the **overall pattern** and for striking **deviations** from that pattern.

You can describe the overall pattern of a scatterplot by the **form, direction,** and **strength** of the relationship.

An important kind of deviation is an **outlier,** an individual value that falls outside the overall pattern of the relationship.

The *form* of the relationship between degree-days and gas consumption is clear: The points have a straight-line pattern. We can represent the overall pattern of the relationship by drawing a straight line through the points of the scatterplot. Figure 6.8b shows such a line. As degree-days increase, gas consumption also increases; that is the *direction* of the relationship. The points in the plot lie quite close to the line, so the relationship is quite *strong*. The number of degree-days explains most of the variation in gas consumption. A weaker straight-line relationship would show more scatter of the points about a generally straight-line pattern. The scatter reflects the effects of other factors, such as use of gas for cooking or turning down the thermostat when the family is away from home. These effects are relatively small. There are no *outliers* (points far outside the overall straight-line pattern) or other important deviations.

Of course, not all relationships have a straight-line form. Some relationships do not even have a clear direction. Exercise 22 gives an example.

6.9 Regression Lines

Sue wants to use her data to predict how much gas she will use for any outside temperature (in degree-days). She can do this by drawing a line through the straight-line pattern on the scatterplot, as in Figure 6.8b.

A **regression line** is a straight line that describes how a response variable y changes as an explanatory variable x changes. We often use a regression line to predict the value of y for a given value of x.

The points in Figure 6.8 lie so close to a line that it is easy to draw a regression line on the graph using a transparent straightedge. This gives us a line on the graph but not an equation for the line. There is also no guarantee that the line we fit by eye is the best line for predicting gas consumption. There are statistical techniques for finding from the data the equation of the best line (with various meanings of "best"). We will soon discuss the most common of these techniques, called *least-squares regression*. The line in Figure 6.8b is the least-squares regression line for Sue's data. All statistical software packages and many calculators will calculate the least-squares line for you, so that a line is often available with little work. You should therefore know how to use a fitted line even if you do not learn the details of how to get the equation from the data.

In writing the equation of a line, we use x for the explanatory variable, because this is plotted on the horizontal x-axis, and y for the response variable. Any line has an equation of the form

$$y = a + bx$$

The number b is the *slope* of the line, the amount by which y changes when x increases by one unit. The slope is usually important in a statistical setting, because it is the rate of change of the response y as x increases. The number a is the *intercept,* the value of y when $x = 0$.

EXAMPLE

Interpreting Slope and Intercept

A computer program tells us that the least-squares regression line computed from Sue's data is

$$y = 1.23 + 0.202x$$

The slope of this line is $b = 0.202$. This means that gas consumption increases by 0.202 hundred cubic feet per day when there is one more degree-day per day. The intercept is $a = 1.23$. When there are no degree-days (that is, when the average temperature is 65°F or above), gas consumption will be 1.23 hundred cubic feet per day. The slope and intercept are estimates based on fitting a line to the data in Table 6.3. We do not expect every month with no degree-days to average exactly 1.23 hundred cubic feet of gas per day. The line represents only the overall pattern of the data. ■

The purpose of a regression line is to predict the value of the response variable for a given value of the explanatory variable. A line drawn on a scatterplot can be used for making predictions with a straightedge and pencil. If the equation of the line is available, we can simply substitute the given value of the explanatory variable into the equation.

After installing solar panels, Sue wants to know how much she has saved in heating costs. She cannot simply compare before-and-after gas usage, because the winters before and after will not be equally severe. Instead, she can use the regression line to predict how much gas she would have used without the solar panels. Comparing this prediction with the actual amount used will show her savings.

EXAMPLE

Predicting Gas Consumption

The next February averages 35 degree-days per day. How much gas would Sue have used without the solar panels? Figure 6.8b illustrates using the regression line for prediction. First locate 35 on the horizontal axis. Go up to the regression line and then over to the gas consumption scale. We predict that slightly more than 800 cubic feet per day will be consumed.

We can give a more exact prediction using the equation of the regression line. This equation is

$$y = 1.23 + 0.202x$$

In this equation, x is the number of degree-days per day during a month and y is the predicted gas consumption per day, in hundreds of cubic feet. Our predicted gas consumption for a month with $x = 35$ degree-days per day is

$$\begin{aligned} y &= 1.23 + (0.202)(35) \\ &= 8.3 \text{ hundred cubic feet per day} \end{aligned}$$

This prediction will almost certainly not be exactly correct for the next month that has 35 degree-days per day. But the past data points lie so close to the line that we can be confident that gas consumption in such a month will be quite close to 830 cubic feet per day. ■

6.10 Correlation

A scatterplot displays the form, direction, and strength of the relationship between two numerical variables. *Linear* (straight-line) relationships are important because a line is a simple pattern that is quite common. We say a linear relation is strong if the points lie close to a line and weak if they are widely scattered about a line. Our eyes are not good judges of how strong a relationship is. We need to follow our strategy for data analysis by using a numerical measure to supplement the graph. *Correlation* is the measure we use.

The **correlation** measures the direction and strength of the linear relationship between two numerical variables. Correlation is usually written as r.

Suppose that we have data on variables x and y for n individuals. The values for the first individual are x_1 and y_1, the values for the second individual are x_2 and y_2, and so on. The means and standard deviations of the two variables are $\bar{x}$ and s_x for the x-values and $\bar{y}$ and s_y for the y-values. The correlation r between x and y is

$$r = \frac{1}{n-1} \sum \left(\frac{x_i - \bar{x}}{s_x} \right) \left(\frac{y_i - \bar{y}}{s_y} \right)$$

The summation sign Σ means "add these terms for all the individuals." The formula for the correlation r helps us see what correlation is, but in practice you should use software or a calculator that finds r from keyed-in values of two variables x and y. Exercise 31 asks you to calculate a correlation step by step from the definition to solidify its meaning.

The correlation uses the deviations of the x and y observations from their means. So the sign of r shows the direction of the relationship between x and y. Height and weight, for example, tend to move together. People who are above average in height tend also to be above average in weight. People who are below average in height tend also to have below-average weight. So the deviations from the mean that are multiplied to form the sum for r are mostly both positive or both negative. Their products are therefore mostly positive, so that r is positive.

More detailed study of the formula gives more detailed properties of r. Here is what you need to know in order to interpret correlation.

1. Correlation makes no distinction between explanatory and response variables. It makes no difference which variable you call x and which you call y in calculating the correlation.
2. Correlation requires that both variables have meaningful numerical values, so that it makes sense to do the arithmetic called for by the formula for r. We cannot calculate a correlation between the incomes of a group of people and what city they live in, because cities do not have numerical values.
3. Correlation measures the strength only of linear relationships. Correlation does not describe curved relationships between variables, no matter how strong those relationships are.
4. The sign of r describes the direction of the relationship. Positive r indicates a *positive association:* The variables tend to move together. Negative r indicates *negative association:* The variables tend to move in opposite directions.

5. The correlation r is always a number between -1 and 1. Values of r near 0 indicate a very weak linear relationship. The strength of the linear relationship increases as r moves away from 0 toward either -1 or 1. Values of r close to -1 or 1 indicate that the points lie close to a straight line. The extreme values $r = -1$ and $r = 1$ occur only in the case of a perfect linear relationship, when the points in a scatterplot lie exactly along a straight line.
6. The correlation r does not change when we change the units of measurement of x, y, or both. Measuring height in inches rather than centimeters and weight in pounds rather than kilograms does not change the correlation between height and weight. The correlation r itself has no unit of measurement; it is just a number.
7. Like the mean and standard deviation, the correlation is strongly affected by a few outlying observations.

The scatterplots in Figure 6.9 illustrate how values of r closer to 1 or -1 correspond to stronger linear relationships. To make the meaning of r clearer, the standard deviations of both variables in these plots are equal and the horizontal

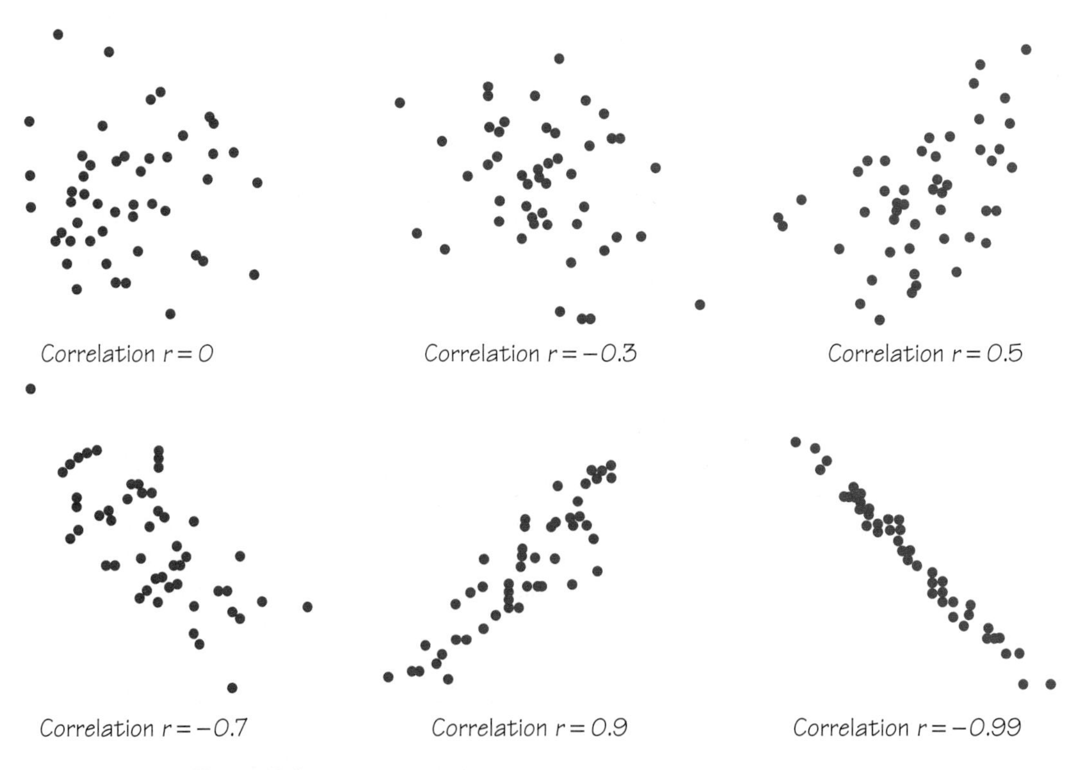

Figure 6.9 How correlation measures the strength of a straight-line relationship. Patterns closer to a straight line have correlations closer to 1 or -1.

and vertical scales are the same. In general, it is not so easy to guess the value of r from the appearance of a scatterplot. Changing the plotting scales in a scatterplot can alter the appearance of the graph, but it does not change the correlation. Exercise 50 illustrates the effect of changing scales.

Figure 6.8 shows a very strong positive linear relationship between degree-days and natural gas consumption. The correlation is $r = 0.989$, close to the $r = 1$ of a perfect straight line. Check this on your calculator using the data in Table 6.3.

Correlation and regression *describe* relationships. *Interpreting* relationships is more subtle. In particular, a strong relationship between two variables does not always mean that one of the variables *causes* changes in the other. The basic meaning of causation is that by changing x we can bring about a change in y. Lower the outside temperature and natural gas consumption goes up when gas is used for heating. Many strong relationships, however, involve no direct causation. Measure the number of television sets per person x and the life expectancy y in the world's nations. There is a high positive correlation: Nations with many TV sets have higher life expectancies. Could we lengthen the lives of people in Botswana by shipping them TV sets? No. Rich nations have more TV sets than poor nations. Rich nations also have longer life expectancies because they offer better nutrition, clean water, and better health care. Both TV sets and long lives are consequences of prosperity, and this joint consequence creates a relationship that has nothing to do with "x causes y."

6.11 Least-Squares Regression

When a scatterplot shows a linear relationship between an explanatory variable x and a response variable y, we want to draw a line to describe the relationship. The points will rarely lie exactly on a line, so our problem is to find the line that best fits the points. To do this, we must first say what we mean by the "best-fitting" line.

Suppose that we want to use our line to predict y for given values of x, as Sue used degree-days to predict gas consumption. The error in our prediction is measured in the y, or vertical, direction. So we want to make the vertical distances of the points from the line as small as possible. A line that fits the data well does not pass entirely above or below the plotted points, so some of the errors will be positive and some negative. Their squares, however, will all be positive. The *least-squares regression line* makes the sum of the squares of the errors as small as possible.

The **least-squares regression line** of y on x is the line that makes the sum of the squares of the vertical distances of the data points from the line as small as possible.

The least-squares idea says what we mean by the best-fitting line. We must still learn how to find this line from the data. Given n observations on variables x and y, what is the equation of the least-squares line? Here is the solution to this mathematical problem.

> We have data on an explanatory variable x and a response variable y for n individuals. From the data, calculate the means $\bar{x}$ and $\bar{y}$ and the standard deviations s_x and s_y of the two variables, as well as their correlation r. The **least-squares regression line** is the line
>
> $$y = a + bx$$
>
> with **slope**
>
> $$b = r\frac{s_y}{s_x}$$
>
> and **intercept**
>
> $$a = \bar{y} - b\bar{x}$$

This equation gives insight into the behavior of least-squares regression by showing that it is related to the means and standard deviations of the x and y observations and to the correlation between x and y. In practice, you don't need to calculate the means, standard deviations, and correlation first. Statistical software or your calculator will give the slope b and intercept a of the least-squares line from keyed-in values of the variables x and y. Use your calculator to verify that the equation for the least-squares line from Sue's gas consumption data is indeed $y = 1.23 + 0.202x$, as we claimed earlier. Your calculator will report more decimal places for the intercept and slope.

EXAMPLE Do Heavy People Burn More Energy?

Metabolic rate, the rate at which the body consumes energy, is important in studies of weight gain, dieting, and exercise. Table 6.4 gives data on the lean body mass and resting metabolic rate for 12 women and 7 men who are subjects in a study of dieting. Lean body mass, given in kilograms, is a person's weight leaving out all fat. Metabolic rate is measured in calories burned per 24 hours, the same calories used to describe the energy content of foods. The researchers believe that lean body mass is an important influence on metabolic rate.

Figure 6.10 is a scatterplot of the data. Because we think that body mass helps explain metabolic rate, we plot body mass on the horizontal x-axis. We have also added a feature to the scatterplot: Two different colors distinguish the female and male subjects. This allows us to see that, while the women as a group have lower

TABLE 6.4 Lean Body Mass and Metabolic Rate

Subject	Sex	Mass (kg)	Rate (cal)	Subject	Sex	Mass (kg)	Rate (cal)
1	M	62.0	1792	11	F	40.3	1189
2	M	62.9	1666	12	F	33.1	913
3	F	36.1	995	13	M	51.9	1460
4	F	54.6	1425	14	F	42.4	1124
5	F	48.5	1396	15	F	34.5	1052
6	F	42.0	1418	16	F	51.1	1347
7	M	47.4	1362	17	F	41.2	1204
8	F	50.6	1502	18	M	51.9	1867
9	F	42.0	1256	19	M	46.9	1439
10	M	48.7	1614				

body mass than the men, the nature of the relationship is similar for both genders. We will therefore do calculations for all 19 subjects together.

The scatterplot shows a moderately strong positive linear relationship. The correlation $r = 0.865$ describes the strength of the relationship more exactly. The line on the plot is the least-squares regression line for predicting metabolic rate from lean body mass. The equation of this line is

$$y = 113.165 + 26.879x$$

The slope of the line tells us that on the average the subjects burn about 27 more calories per day for every additional kilogram of body mass. The intercept $a = 113.165$ is needed to draw the line but has no statistical interpretation. Body mass $x = 0$ is impossible, so we can't speak of the value of metabolic weight when $x = 0$. ■

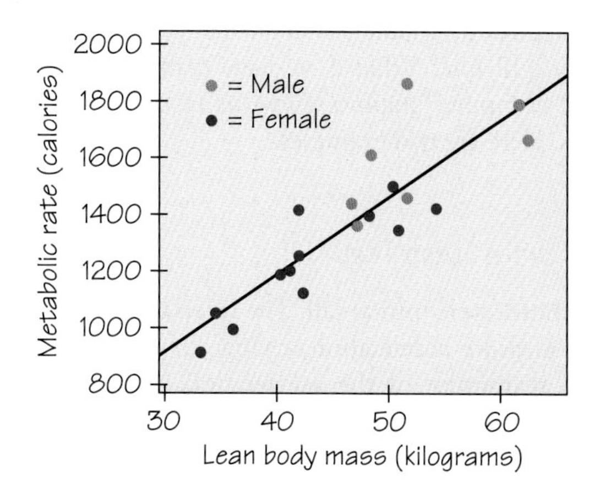

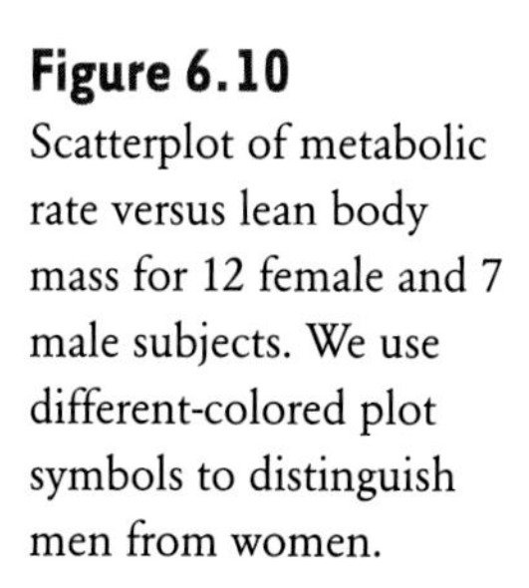

Figure 6.10
Scatterplot of metabolic rate versus lean body mass for 12 female and 7 male subjects. We use different-colored plot symbols to distinguish men from women.

SPOTLIGHT 6.1 Regression in Action

No statistical method is used as much as regression. Here are some applications of regression that are a bit unusual.

Did the vote counters cheat? Republican Bruce Marks was ahead of Democrat William Stinson when the voting machines were tallied in their Pennsylvania election. But Stinson was ahead after absentee ballots were counted by the Democrats who controlled the election board. A court fight followed. The court called in a statistician, who used regression with data from past elections to predict the counts of absentee ballots based on the results from the voting machines. Marks's lead of 564 votes from the machines predicted that he would get 133 more absentee votes than Stinson. In fact, Stinson got 1025 more absentee votes than Marks. This looks suspicious.

Is regression garbage? No – but garbage can be the setting for regression. The Census Bureau once asked if weighing a neighborhood's garbage would help count its people. So 63 households had their garbage sorted and weighed. It turned out that pounds of plastic in the trash gave the best garbage prediction of the number of people in a neighborhood. Alas, the prediction wasn't good enough to help the Census Bureau.

How much is this injury worth? Jury Verdict Research makes money from regression. The company collects data on more than 20,000 jury verdicts in personal-injury lawsuits each year, recording more than 30 variables describing each case. Multiple regression extends regression to allow more than one explanatory variable to contribute to explaining a response variable. Jury Verdict Research uses multiple regression to predict how much a jury will award in a new case. Lawyers pay for these predictions and use them to negotiate settlements with insurance companies.

6.12 Modern Data Analysis

Scatterplots, correlation, and regression are basic tools for describing the relationship between two variables. What if the form of the relationship is more complex than a straight line? What if we have many variables rather than just two? Software and computer graphics allow us to display and describe complicated relationships. Here are two examples.

EXAMPLE A Motorcycle Crash Test

Crash a motorcycle into a wall. The rider, fortunately, is a dummy with an instrument to measure acceleration (change of velocity) mounted in its head. Figure 6.11 is a scatterplot of the acceleration of the dummy's head against time in milliseconds. Acceleration is measured in g's, or multiples of the acceleration due

Figure 6.11 Plot of the acceleration of the head of a crash-test dummy as a motorcycle hits a wall, with the overall pattern calculated by a scatterplot smoother.

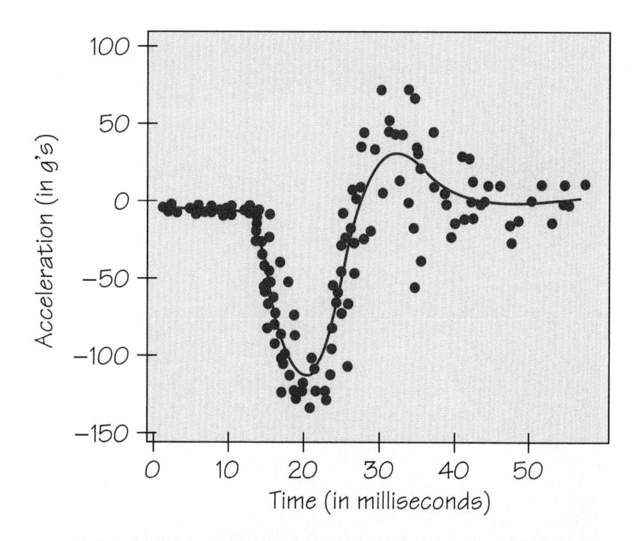

(H. Osendorfer and S. Rascher, BMW AG.)

to gravity at the earth's surface. The motorcycle approaches the wall at a constant speed (acceleration near 0). As it hits, the dummy's head snaps forward and decelerates violently (negative acceleration reaching more than 100 g's), then snaps back again (up to 75 g's) and wobbles a bit before coming to rest.

The scatterplot has a clear overall pattern, but it does not obey a simple form such as linear. Moreover, the strength of the pattern varies, from quite strong at the left of the plot to weaker (much more scatter) at the right. Statistical software includes a *scatterplot smoother* that deals with this complexity and draws a line on the plot to represent the overall pattern. ■

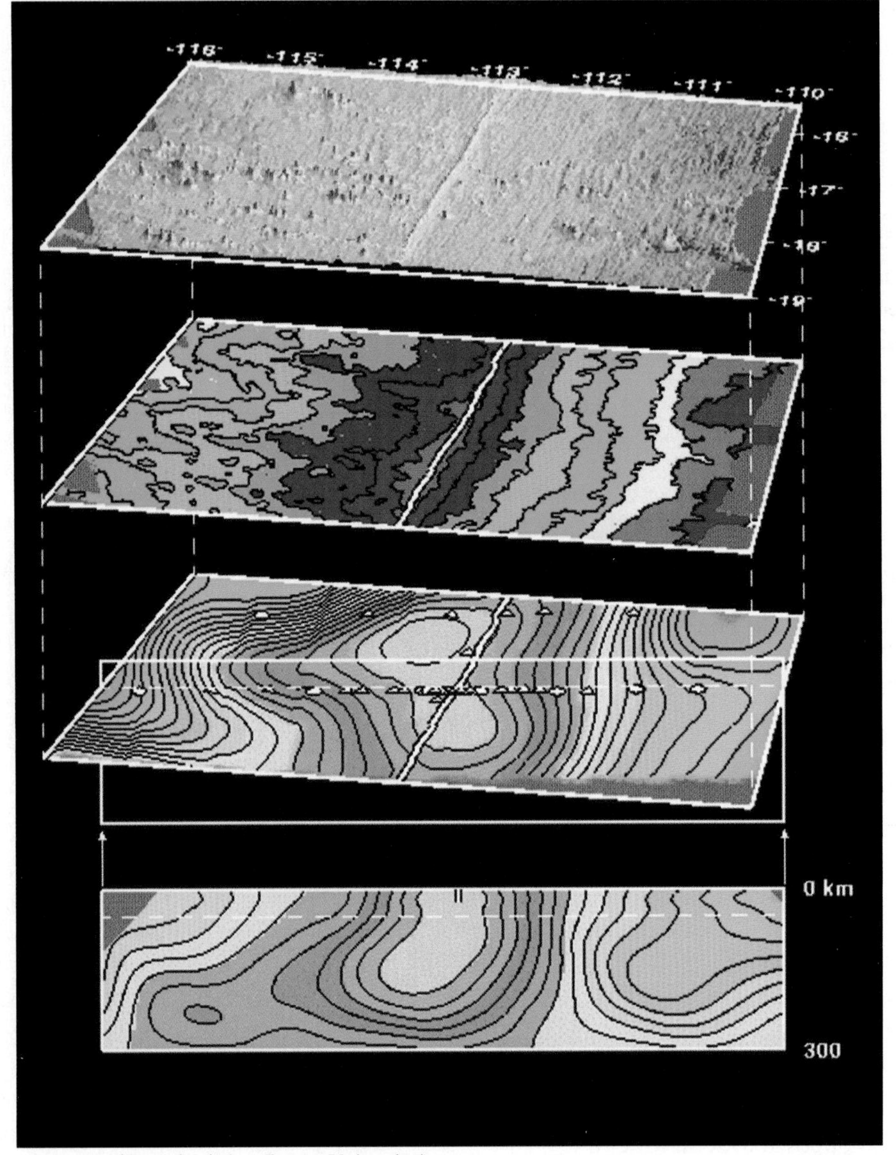

Figure 6.12 The seafloor in the South Pacific near a mid-ocean ridge. This computer graphic shows the topography and measurements on gravity and earthquake-wave velocity as well as location on the earth's surface. (This image was provided by D. S. Scheirer of Brown University. The study is reported in several papers in the May 22, 1998, issue of *Science*.)

(Courtesy of Daniel Scheirer, Brown University.)

So far we have looked only at plots for two variables. What if we want to display a third variable in the same plot? Because we have already used the horizontal and vertical directions of the graph, there is only one direction left – moving out of and into the page. Three-dimensional plots are hard to see clearly unless we use color or motion (or both) to help us gain perspective. Computer graphics can supply color and motion, allowing us to see data on several variables at once.

EXAMPLE Images of the Earth

The large plates that make up the earth's crust are pushed apart at ridges in mid-ocean where hot magma (melted rock) wells up from below. Scientists study this seafloor spreading by combining data from several sources, including instruments planted two miles down at the bottom of the ocean. Figure 6.12 is a computer graphic image of the data from a study in the South Pacific.

The top panel shows the topography of the ocean floor. Latitude and longitude coordinates locate the area. The ridge separating two plates runs down the center, and small underwater volcanoes can be seen on either side. The second panel displays small variations in gravity that help distinguish different kinds of rock. The third panel adds data from tracking the velocity of earthquake waves, showing the structure of the magma underneath the earth's crust. The panels are lined up so that scientists can visually compare several variables at many locations to understand a geological process that shapes our planet. ■

REVIEW VOCABULARY

Boxplot A graph of the five-number summary. A box spans the quartiles, with an interior line marking the median. Lines extend out from this box to the extreme high and low observations.

Correlation A measure of the direction and strength of the linear relationship between two variables. Correlations take values between 0 (no linear relationship) and ± 1 (perfect straight-line relationship).

Distribution The pattern of outcomes of a variable. The distribution describes what values the variable takes and how often each value occurs.

Exploratory data analysis The practice of examining data for unanticipated patterns or effects, as opposed to seeking answers to specific questions.

Five-number summary A summary of a distribution of values consisting of the median, the first and third quartiles, and the largest and smallest observations.

Histogram A graph of the distribution of outcomes (often divided into classes) for a single variable. The height of each bar is the number of observations in the class of outcomes covered by the base of the bar. All classes should have the same width.

Individuals The people, animals, or things described by a data set.

Least-squares regression line A line drawn on a scatterplot that makes the sum of the squares of the vertical distances of the data points from the line as small as possible. The regression line can be used to predict the response variable y for a given value of the explanatory variable x.

Mean The ordinary arithmetic average of a set of observations. To find the mean, add all the observations and divide the sum by the number of observations summed.

Median The midpoint of a set of observations. Half the observations fall below the median and half fall above.

Outlier A data point that falls clearly outside the overall pattern of a set of data.

Quartiles The first quartile of a distribution is the point with 25% of the observations falling below it; the third quartile is the point with 75% below it.

Regression line Any line that describes how a response variable y changes as we change an explanatory variable x. The most common such line is the least-squares regression line.

Response variable, explanatory variable A response variable measures an outcome of a study. An explanatory variable attempts to explain the observed outcomes.

Scatterplot A graph of the values of two variables as points in the plane. Each value of the explanatory variable is plotted on the horizontal axis, and the value of the response variable for the same individual is plotted on the vertical axis.

Skewed distribution A distribution in which observations on one side of the median extend notably farther from the median than do observations on the other side. In a right-skewed distribution, the larger observations extend farther to the right of the median than the smaller observations extend to the left.

Standard deviation A measure of the spread of a distribution about its mean as center. It is the square root of the average squared deviation of the observations from their mean.

Stemplot A display of the distribution of a variable that attaches the final digits of the observations as leaves on stems made up of all but the final digit.

Symmetric distribution A distribution with a histogram or stemplot in which the part to the left of the median is roughly a mirror image of the part to the right of the median.

Variable Any measured characteristic of an individual.

Variance A measure of the spread of a distribution about its mean. It is the average squared deviation of the observations from their mean. The square root of the variance is the standard deviation.

SUGGESTED READINGS

CLEVELAND, WILLIAM S. *The Elements of Graphing Data,* rev. ed., Hobart Press, Summit, N.J., 1994, books@hobart.com. A careful study of the most effective elementary ways to present data graphically, with much sound advice on improving simple graphs.

MOORE, DAVID S. *The Basic Practice of Statistics,* 2nd ed., Freeman, New York, 1999. The first two chapters of this text provide a more extensive treatment of displaying and describing data for one and two variables. They cover the material of this chapter in more detail and present much new material on both technique and interpretation.

ROSSMAN, ALLAN J., and BETH L. CHANCE. *Workshop Statistics: Discovery with Data,* 2nd ed., Springer, New York, 2000. An excellent source for hands-on activities.

TUFTE, EDWARD R. *Envisioning Information,* Graphics Press, Cheshire, Conn., 1990. Tufte's three beautifully printed books on data graphics blend art and science.

VELLEMAN, PAUL F., and DAVID C. HOAGLIN. Data analysis. In David C. Hoaglin and David S. Moore (eds.), *Perspectives on Contemporary Statistics,* Mathematical Association of America, Washington, D.C., 1992, pp. 19–39. A conceptual essay that presupposes knowledge of the basic techniques described in this chapter and in the text by Moore.

SUGGESTED WEB SITES

Want data? The Census Bureau (**www.census.gov**) is a prime source. You can click quickly to information on your home state and county or explore the details of money income in the United States by clicking on "Income." The Chance Web site (**www.dartmouth.edu/~chance/**) features current news items that involve statistics and also an archive of data. The electronic *Journal of Statistics Education* (**www.amstat.org/publications/jse/**) has articles about teaching statistics and has a data archive. Now is the time to visit the interactive applets that accompany this book (**www.whfreeman.com/fapp**). The Mean and Median and Correlation and Regression applets demonstrate the behavior of these measures better than any number of words.

SKILLS CHECK

1. What are all the values that a standard deviation s can possibly take?

(a) $0 \leq s \leq 1$
(b) $-1 \leq s \leq 1$
(c) $0 \leq s$

2. What are all the values that a correlation r can possibly take?

(a) $0 \leq r \leq 1$
(b) $-1 \leq r \leq 1$
(c) $0 \leq r$

3. Fred keeps his savings in his mattress. He began with $500 from his mother and adds $100 each year. His total savings y after x years are given by which equation?

(a) $y = 500 + 100x$
(b) $y = 100 + 500x$
(c) $y = 500 + x$

4. Here are six measured lengths (in feet): 7, 8, 5, 6, 9, 9. Find their median.

(a) 5.5
(b) 6
(c) 7.5

5. Here are six measured lengths (in feet): 7, 8, 5, 6, 9, 9. Find their quartiles.

(a) 6 and 9
(b) 5.5 and 9
(c) 8 and 9

6. Suppose the points of a scatterplot lie very close to the line $y = 4 - 3x$. The correlation for these points is

(a) almost 1.
(b) almost -1.
(c) approximately -3.

7. Suppose data from a random-number table are plotted as points on a scatterplot. The correlation for this data is

(a) approximately 0.
(b) approximately 1.
(c) completely random.

8. A stemplot is most similar to

(a) a histogram.
(b) a boxplot.
(c) a scatterplot.

9. If the quartiles both have the value 15, then which of the following statements must be true?

(a) The mean has the value 15.
(b) The median has the value 15.
(c) The mean and median are equal.

10. If the standard deviation of a distribution is 4, what is the variance of the distribution?

(a) 2
(b) 4
(c) 16

11. Which of the following is least affected by the addition of an outlying observation?

(a) Standard deviation
(b) Correlation
(c) The quartiles

EXERCISES

▲ Optional. ■ Advanced. ◆ Discussion.

Many exercises require use of a calculator (or software) that will find mean, standard deviation, correlation, and the slope and intercept of the least-squares regression line from keyed-in data.

Displaying Distributions

1. Here are the first lines of a professor's data set at the end of a mathematics course:

Name	Major	Points	Grade
ADVANI, SURA	COMM	397	B
BARTON, DAVID	HIST	323	C
BROWN, ANNETTE	LIT	446	A
CHIU, SUN	PSYC	405	B
CORTEZ, MARIA	PSYC	461	A

What are the individuals and the variables in these data?

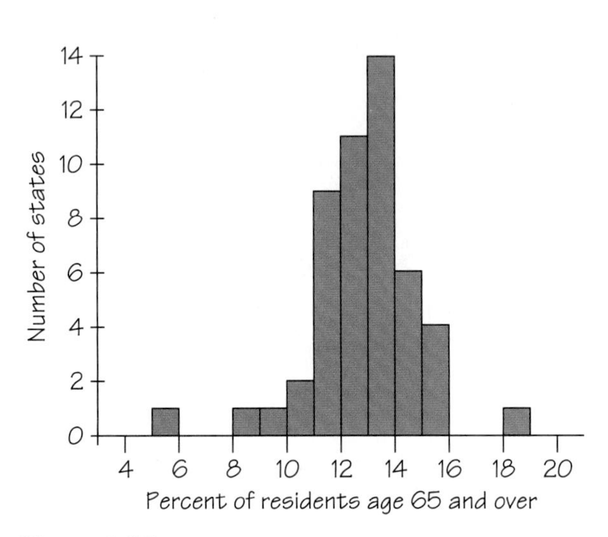

Figure 6.13 Histogram of the percent of state residents aged 65 or older, for Exercise 3.

2. Here is a small part of a data set that describes mutual funds available to the public:

Fund	Category	Net assets ($ million)	Year-to-date return	Largest holding
⋮				
Fidelity Low-Priced Stock	Small value	6,189	4.56%	Dallas Semiconductor
Price International Stock	International stock	9,745	−0.45%	Vodafone
Vanguard 500 Index	Large blend	89,394	3.45%	General Electric
⋮				

What individuals does this data set describe? In addition to the fund's name, how many variables does the data set contain?

◆ **3.** The histogram in Figure 6.13 displays the percent of residents aged 65 or over in the states. Describe the approximate shape, center, and spread of this distribution, ignoring the two outliers. The outliers are Alaska and Florida. Which state is the low outlier and which is the high outlier? Explain your reasoning.

◆ **4.** Figure 6.14 is a histogram of the lengths of words used in Shakespeare's plays. Because there are so many words in the plays, the vertical axis of the graph is the percent that are of each length, rather than the count. What is the overall shape of this distribution? What does this shape say about word lengths in Shakespeare? Do you expect other authors to have word-length distributions of the same general shape? Why?

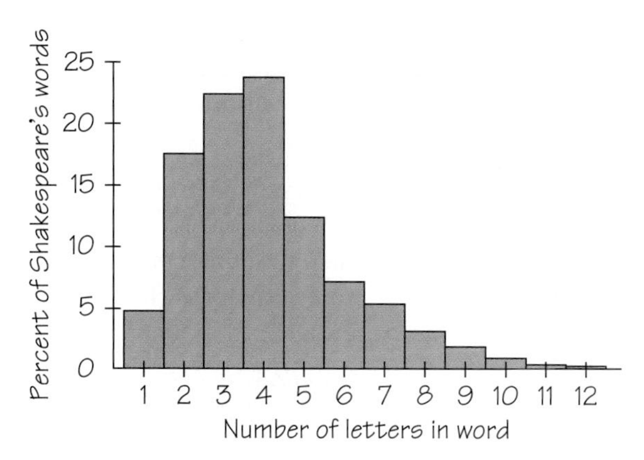

Figure 6.14 Histogram of the lengths of words used in Shakespeare's plays, for Exercise 4.

5. At top right is a stemplot of the percent of residents aged 25 to 34 in each of the 50 states. The stems are whole percents and the leaves are tenths of a percent.

```
10 | 9
11 | 0
12 | 1344677889
13 | 00124555667899999
14 | 11222344445789
15 | 24478999
```

(a) The mountain states Montana and Wyoming have the smallest percent of young adults, perhaps because they lack job opportunities. What are the percents for these two states?

(b) Ignoring Montana and Wyoming, describe the shape, center, and spread of this distribution.

◆ **6.** Table 6.5 gives the number of medical doctors per 100,000 people in each state.

(a) Why is the number of doctors per 100,000 people a better measure of the availability of health care than is a simple count of the number of doctors in a state?

TABLE 6.5 Medical Doctors per 100,000 Population, by State (1998)

State	Doctors	State	Doctors	State	Doctors
Alabama	198	Louisiana	246	Ohio	235
Alaska	167	Maine	223	Oklahoma	169
Arizona	202	Maryland	374	Oregon	225
Arkansas	190	Massachusetts	412	Pennsylvania	291
California	247	Michigan	224	Rhode Island	338
Colorado	238	Minnesota	249	South Carolina	207
Connecticut	354	Mississippi	163	South Dakota	184
Delaware	234	Missouri	230	Tennessee	246
Florida	238	Montana	190	Texas	203
Georgia	211	Nebraska	218	Utah	200
Hawaii	265	Nevada	173	Vermont	305
Idaho	154	New Hampshire	237	Virginia	241
Illinois	260	New Jersey	295	Washington	235
Indiana	195	New Mexico	212	West Virginia	215
Iowa	173	New York	387	Wisconsin	227
Kansas	203	North Carolina	232	Wyoming	171
Kentucky	209	North Dakota	222	D.C.	737

SOURCE: *Statistical Abstract of the United States*, 2000.

(b) Make a histogram of the distribution of doctors per 100,000 people. Write a brief description of the distribution. Are there any outliers? If so, can you explain them?

7. In 1798 the English scientist Henry Cavendish measured the density of the earth in a careful experiment with a torsion balance. Here are his 29 repeated measurements of the same quantity (the density of the earth relative to that of water) made with the same instrument. [S. M. Stigler, Do robust estimators work with real data? *Annals of Statistics,* 5 (1977): 1055–1078.]

5.50	5.61	4.88	5.07	5.26	5.55
5.36	5.29	5.58	5.65	5.57	5.53
5.62	5.29	5.44	5.34	5.79	5.10
5.27	5.39	5.42	5.47	5.63	5.34
5.46	5.30	5.75	5.68	5.85	

Make a stemplot of the data. Describe the distribution: Is it approximately symmetric or distinctly skewed? Are there gaps or outliers?

Describing Distributions

8. Table 6.5 gives the number of medical doctors per 100,000 people in each state. You made a graph of these data in Exercise 6. The District of Columbia (D.C.) is a high outlier. Because D.C. is a city rather than a state, we will omit it here.

(a) Calculate both the five-number summary and $\bar{x}$ and s for the number of doctors per 100,000 people in the 50 states. Based on your graph, which description do you prefer?

(b) What facts about the distribution can you see in the graph that the numerical summaries don't reveal? Remember that measures of center and spread are not complete descriptions of a distribution.

9. Find the five-number summary of Cavendish's measurements of the density of the earth in Exercise 7. How is the symmetry of the distribution reflected in the five-number summary?

10. The mean of the 29 measurements in Exercise 7 was Cavendish's best estimate of the density of the earth. Find this mean. Then find the standard deviation. (Because of the symmetry of the distribution, it can be summarized by $\bar{x}$ and s.)

11. Table 6.4 gives the lean body masses and metabolic rates for 7 men and 12 women. Compare the distributions of body mass for men and women, using the five-number summary and side-by-side boxplots. What do the data show?

12. Below are the percents of the popular vote won by the successful candidate in each of the presidential elections from 1948 to 2000.

(a) Make a stemplot of the winners' percents.

(b) What is the median percent of the vote won by the successful candidate in presidential elections?

(c) Call an election a landslide if the winner's percent falls at or above the third quartile. Find the third quartile. Which elections were landslides?

■ **13.** The level of various substances in the blood influences our health. Here are measurements of the level of phosphate in the blood of a patient, in milligrams of phosphate per deciliter of blood, made on six consecutive visits to a clinic.

5.6	5.2	4.6	4.9	5.7	6.4

A graph of only six observations gives little information, so we proceed to compute the mean and standard deviation.

(a) Find the mean from its definition. That is, find the sum of the six observations and divide by 6.

(b) Find the standard deviation from its definition. That is, find the deviations of each observation

Year	1948	1952	1956	1960	1964	1968	1972	1976	1980	1984	1988	1992	1996	2000
Percent	49.6	55.1	57.4	49.7	61.1	43.4	60.7	50.1	50.7	58.8	53.9	43.2	49.2	47.9

from the mean, square the deviations, then obtain the variance and the standard deviation.
(c) Now enter the data into your calculator and use the mean and standard deviation buttons to obtain $\bar{x}$ and s. Do the results agree with your hand calculations?

◆ **14.** Figure 6.15 displays side-by-side boxplots of the 2000–2001 tuition and fee charges for all four-year colleges in Massachusetts and in Michigan. For state schools, we used the in-state tuition.

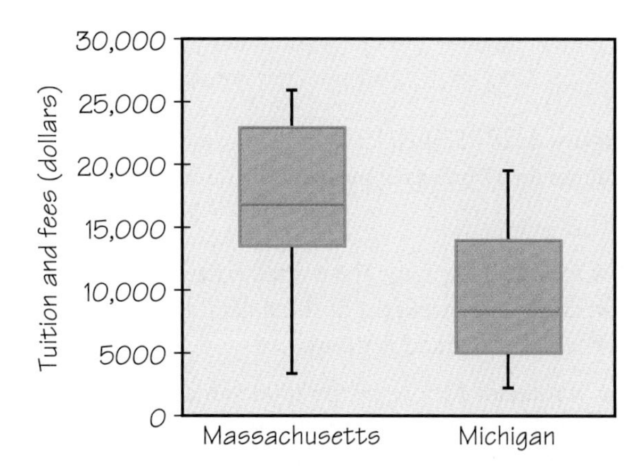

Figure 6.15 Side-by-side boxplots comparing the tuition and fee charges of four-year colleges in Massachusetts and Michigan, for Exercise 14.

(a) The histogram in Figure 6.4 and the left-hand boxplot in Figure 6.15 display the same 56 observations. What does the histogram show about the distribution that the boxplot hides?
(b) Use the boxplots to write a brief comparison of the two distributions.
(c) Eastern states had many private colleges before state-supported higher education became common. Public higher education is more prevalent in the Midwest, which was settled later. How does this historical pattern help explain the differences between the Massachusetts and Michigan distributions of college charges?

◆ **15.** Some people worry about how many calories they consume. *Consumer Reports* magazine (June 1986) measured the calories in 20 brands of beef hot dogs, 17 brands of meat hot dogs, and 17 brands of poultry hot dogs. Here is computer output describing the beef hot dogs:

```
Mean = 156.8      Standard deviation = 22.64
Min = 111      Max = 190      N = 20
Median = 152.5      Quartiles = 140, 178.5
```

For meat hot dogs:

```
Mean = 158.7      Standard deviation = 25.24
Min = 107      Max = 195      N = 17
Median = 153      Quartiles = 139, 179
```

And for poultry hot dogs:

```
Mean = 122.5      Standard deviation = 25.48
Min = 87      Max = 170      N = 17
Median = 129      Quartiles = 102, 143
```

Use this information to make side-by-side boxplots of the calorie counts for the three types of hot dogs. Write a brief comparison of the distributions. Will eating poultry hot dogs usually lower your calorie consumption compared with eating beef or meat hot dogs?

◆ **16.** The distribution of household incomes in the United States is skewed to the right. In 1999, the mean and median incomes of American households were \$40,816 and \$54,842. Which of these numbers is the mean and which is the median? Explain your reasoning.

◆ **17.** The business magazine *Forbes* (March 5, 2001) reports that 4567 companies sold their first stock to the public between 1990 and 2000. The *mean* change in the stock price of these companies since the first stock was issued was +111%. The *median* change was −31%. Explain how this could happen. (*Hint:* Start with the fact that Cisco Systems stock went up 60,600%.)

18. Scores of adults on the Stanford-Binet IQ test have mean 100 and standard deviation 15. What is the variance of scores on this test?

Displaying Relations

◆ **19.** Figure 6.16 is a scatterplot of data from the World Bank. The individuals are all the world's nations for which data are available. The explanatory variable is a measure of how rich a country is, the gross domestic product (GDP) per person. GDP is the total value of the goods and services produced in a country, converted into dollars. The response variable is life expectancy at birth. Three African nations are outliers, with lower life expectancy than usual for their GDP. A full study would ask what special circumstances explain these outliers.

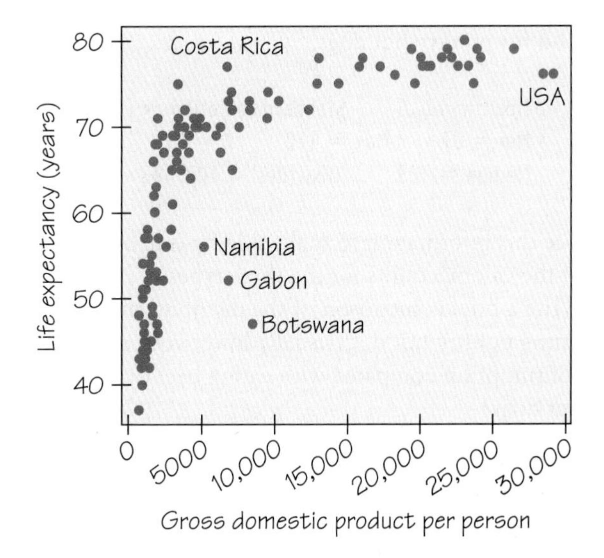

Figure 6.16 Scatterplot of the life expectancy of people in many nations against each nation's gross domestic product per person, for Exercise 19. (Data from the World Bank's *1999 World Development Indicators.*)

(a) Describe the direction and form of the relationship. Aside from the outliers, it is moderately strong.
(b) Explain why the direction and form of this relationship make sense.

20. Figure 6.17 shows the calories and salt content (milligrams of sodium) in 17 brands of meat hot dogs. (These are the same brands described in Exercise 15.) Describe the overall pattern (form, direction, and strength) of these data. In what way is the point marked A unusual?

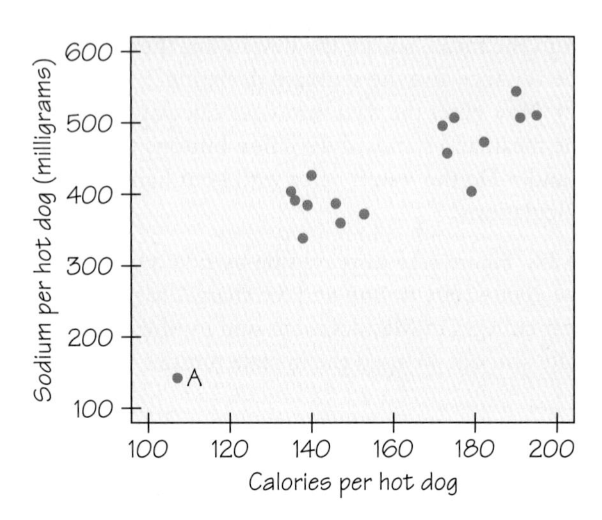

Figure 6.17 Scatterplot of sodium content versus calories for 17 brands of meat hot dogs, for Exercise 20.

21. Professor Moore swims 2000 yards regularly in a vain attempt to undo middle age. At the top of the next page are his times (in minutes) and his pulse rate (in beats per minute) after swimming for 23 sessions in the pool.

(a) Make a scatterplot. (Which is the explanatory variable?)
(b) Is the association between these variables positive or negative? Explain why you expect the relationship to have this direction.
(c) Describe the form and strength of the relationship.

◆ **22.** How does the fuel consumption of a car change as its speed increases? At the bottom of the next page are given data for a British Ford Escort. Speed is measured in kilometers per hour, and fuel consumption is measured in liters of gasoline used per 100 kilometers traveled. [T. N. Lam, Estimating fuel consumption from engine size, *Journal of Transportation Engineering,* 111 (1985): 339–357.]

(a) Make a scatterplot. (Which is the explanatory variable?)

Time	34.12	35.72	34.72	34.05	34.13	35.72	36.17	35.57
Pulse	152	124	140	152	146	128	136	144

Time	35.37	35.57	35.43	36.05	34.85	34.70	34.75	33.93
Pulse	148	144	136	124	148	144	140	156

Time	34.60	34.00	34.35	35.62	35.68	35.28	35.97
Pulse	136	148	148	132	124	132	139

(b) Describe the form of the relationship. Explain why the form of the relationship makes sense.
(c) How would you describe the direction of this relationship?
(d) Is the relationship reasonably strong or quite weak? Explain your answer.

Regression Lines

23. Exercise 21 gives Professor Moore's times (in minutes) to swim 2000 yards and his pulse rate (in beats per minute) after swimming for 23 sessions in the pool. The least-squares regression line for predicting pulse from time is

$$\text{pulse} = 479.9 - (9.695 \times \text{time})$$

(a) What is the slope of this line? What does the slope say about how pulse rate changes with swim time?
(b) The next day's time is 34.30 minutes. Predict Professor Moore's pulse rate. In fact, his pulse rate was 152. How accurate is your prediction?

24. Researchers studying acid rain measured the acidity of precipitation in a Colorado wilderness area for 150 consecutive weeks. Acidity is measured by pH. Lower pH values show higher acidity. The acid rain researchers observed a linear pattern over time. They reported that the least-squares regression line

$$\text{pH} = 5.43 - (0.0053 \times \text{weeks})$$

fit the data well. [W. M. Lewis and M. C. Grant, Acid precipitation in the western United States, *Science*, 207 (1980): 176–177.]

(a) Draw a graph of this line. Explain in plain language what the line says about how pH was changing over time.
(b) According to the regression line, what was the pH at the beginning of the study (weeks = 1)? At the end (weeks = 150)?
(c) What is the slope of the regression line? Explain clearly what this slope says about the rate of change in pH.

Speed (km/h)	10	20	30	40	50	60	70	80
Fuel (liters/100 km)	21.00	13.00	10.00	8.00	7.00	5.90	6.30	6.95

Speed (km/h)	90	100	110	120	130	140	150
Fuel (liters/100 km)	7.57	8.27	9.03	9.87	10.79	11.77	12.88

Correlation

25. Exercise 21 gives data on a middle-aged professor's time to swim 2000 yards and his pulse rate after swimming.

(a) Use a calculator to find the correlation r. Explain from looking at the scatterplot why this value of r is reasonable.
(b) Suppose that the times had been recorded in seconds. For example, the time 34.12 minutes would be 2047 seconds. How would the value of r change?

26. Exercise 22 gives data on gas used versus speed for a small car. Make a scatterplot if you did not do so in Exercise 22. Calculate the correlation (use your calculator or software). Explain why r is close to 0 despite a strong relationship between speed and gas used.

27. If women always married men who were two years older than they are, what would be the correlation between the ages of husband and wife? (*Hint:* Draw a scatterplot for several ages.)

28. For each of the following pairs of variables, would you expect a substantial negative correlation, a substantial positive correlation, or a small correlation?

(a) The age of secondhand cars and their prices
(b) The weight of new cars and their gas mileages in miles per gallon
(c) The heights and the weights of adult men
(d) The heights and the IQ scores of adult men

29. Make a scatterplot of the following data:

x	1	2	3	4	10	10
y	1	3	3	5	1	11

Use your calculator to confirm that the correlation is about 0.5. What feature of the data is responsible for reducing the correlation to this value despite a strong straight-line association between x and y in most of the observations?

30. A study shows that there is a positive correlation between the size of a hospital (measured by its number of beds x) and the median number of days y that patients remain in the hospital. Does this mean that you can shorten a hospital stay by choosing a small hospital? Explain your answer.

■ **31.** *Archaeopteryx* is an extinct beast having feathers like a bird but teeth and a long bony tail like a reptile. Only six fossil specimens are known. Because these specimens differ greatly in size, some scientists think they are different species rather than individuals from the same species. If the specimens belong to the same species and differ in size because some are younger than others, there should be a linear relationship between the lengths of a pair of bones from all individuals. An outlier from this relationship would suggest a different species. Here are data on the lengths in centimeters of the femur (a leg bone) and the humerus (a bone in the upper arm) for the five specimens that preserve both bones. [M. A. Houck et al., Allometric scaling in the earliest fossil bird, *Archaeopteryx lithographica*, *Science*, 247 (1990): 195–198.]

Femur length x	38	56	59	64	74
Humerus length y	41	63	70	72	84

(a) Make a scatterplot. Do you think that all five specimens come from the same species?
(b) Find the correlation r step by step. *Step 1:* Find the mean $\bar{x}$ and standard deviation s_x of the five femur lengths. Find the mean $\bar{y}$ and the standard deviation s_y of the five humerus lengths. (Use your calculator.) *Step 2:* Find the deviations $x - \bar{x}$ of each of the five femur lengths from their mean, and do the same for the five humerus lengths. *Step 3:* Substitute your numbers from steps 1 and 2 into the formula for r.
(c) Now enter the x and y data into your calculator and use the calculator's correlation function to find r. Check that you get the same result as in part (b).

Least-Squares Regression

32. Table 6.2 gives the city and highway gas mileage for 29 rather exotic cars. We expect that fuel economy in the city and on the highway will be closely related.

(a) Make a scatterplot, taking city mileage as the explanatory variable. Describe the direction, form, and strength of the relationship. Are there any conspicuous outliers?
(b) Find the equation of the least-squares regression line for predicting highway mileage from city mileage. (Use your calculator.) Add this line to your scatterplot.
(c) Another similar car gets 14 miles per gallon in the city. What do you predict will be its highway mileage?

33. Find the equation of least-squares regression line for the gas versus speed data in Exercise 22. Make a scatterplot and draw your line on the plot. This is the line that best fits these data (in the sense of least squares), but you would not use it for prediction.

34. We gave the equation of the regression line of gas consumption y on degree-days x for the data in Table 6.3 as

$$y = 1.23 + 0.202x$$

Enter the data from Table 6.3 into your calculator.
(a) Use your calculator's regression function to find the equation of the least-squares regression line.
(b) Use your calculator to find the mean and standard deviation of both x and y and their correlation r. Find the slope b and intercept a of the regression line from these, using the equation of the least-squares regression line.
(c) Verify that in both part (a) and part (b) you get the equation in the example. (Results may differ slightly because of rounding off.)

Additional Exercises

35. How much oil wells in a given field will ultimately produce is key information in deciding whether to drill more wells. At top right are the estimated total amounts of oil recovered from 64 wells in the Devonian Richmond Dolomite area of the Michigan basin, in thousands of barrels. [J. Marcus Jobe and Hutch Jobe, A statistical approach for additional infill development, *Energy Exploration and Exploitation,* 18 (2000): 89–103.]

21.7	53.2	46.4	42.7	50.4	97.7	103.1	51.9
43.4	69.5	156.5	34.6	37.9	12.9	2.5	31.4
79.5	26.9	18.5	14.7	32.9	196.0	24.9	118.2
82.2	35.1	47.6	54.2	63.1	69.8	57.4	65.6
56.4	49.4	44.9	34.6	92.2	37.0	58.8	21.3
36.6	64.9	14.8	17.6	29.1	61.4	38.6	32.5
12.0	28.3	204.9	44.5	10.3	37.7	33.7	81.1
12.1	20.1	30.5	7.1	10.1	18.0	3.0	2.0

(a) Graph the distribution and describe its main features.
(b) Find the mean and median of the amounts recovered. Explain how the relationship between the mean and the median reflects the shape of the distribution.
(c) Give the five-number summary and explain briefly how it reflects the shape of the distribution.

◆ **36.** Each March, the Bureau of Labor Statistics adds a Demographic Supplement to the monthly Current Population Survey. This supplement asks many additional questions, including highest level of education and personal income. Figure 6.18 is based on the March 1999 Demographic Supplement, which included the incomes of 71,512

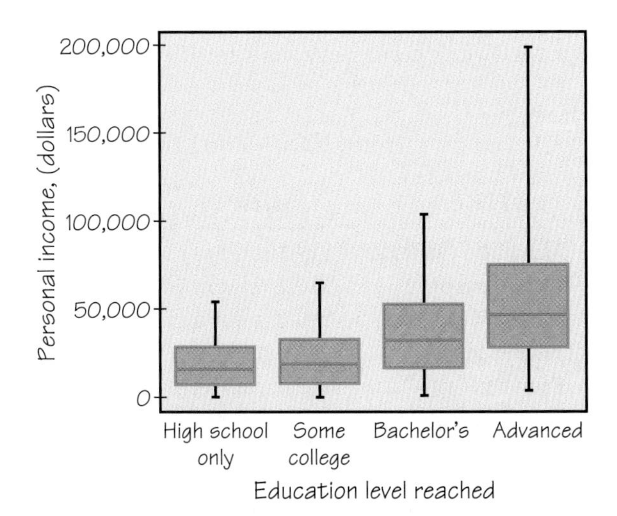

Figure 6.18 Side-by-side boxplots comparing the distributions of income for people with four levels of education, for Exercise 36.

people with the four education levels shown. This figure is a variation on the boxplot: The lines extend not to the smallest and largest individual incomes but to the points that mark off the highest and lowest 5% of all the incomes in the group.

(a) Why do you think the 5% and 95% points are more informative than the smallest and largest values in a graph based on 72,512 incomes?
(b) Write a brief description of how income varies with education. Be sure to mention spread as well as center.

37. Table 6.6 gives the approximate salaries (in millions of dollars) of the members of the NBA champion Los Angeles Lakers basketball team for the year 2000–2001. Describe the salary distribution with a graph, a numerical summary, and a brief written summary.

■ **38.** Here are the numbers of home runs that Babe Ruth hit in his 15 years with the New York Yankees, 1920 to 1934:

54 59 35 41 46 25
47 60 54 46 49 46
41 34 22

The leading contemporary home run hitter is Mark McGwire, who retired after the 2001 season. Here are McGwire's home run counts for 1987 to 2001:

49 32 33 39 22 42
9 9 39 52 58 70
65 32 29

A *back-to-back stemplot* helps us compare two distributions. Write the stems as usual, but with a vertical line both to their left and to their right. On the right, put leaves for Ruth. On the left, put leaves for McGwire. Arrange the leaves on each stem in increasing order out from the stem. Now write a brief comparison of Ruth and McGwire as home run hitters. Include numerical summaries. McGwire was injured in 1993 and there was a baseball strike in 1994. How do these events appear in the data?

39. What are the mean salary $\bar{x}$ and the median salary M for the Los Angeles Lakers (Table 6.6)? How much do $\bar{x}$ and M change when you omit the salaries of the stars, Shaquille O'Neal and Kobe Bryant? What fact about the mean and median do your results illustrate?

40. The single-season home run record is now held by Barry Bonds of the San Francisco Giants, who hit 73 in 2001. Here are Bonds's home run totals from 1986 (his first year) to 2001:

16 25 24 19 33 25
34 46 37 33 42 40
37 34 49 73

TABLE 6.6 Year 2000–2001 Salaries for the Los Angeles Lakers

Player	Salary	Player	Salary
Shaquille O'Neal	$19.3 million	Greg Foster	$1.8 million
Kobe Bryant	$10.1 million	Tyronn Lue	$0.9 million
Horace Grant	$ 7.0 million	Devean George	$0.9 million
Robert Horry	$ 4.8 million	Sam Jacobsen	$0.8 million
Rick Fox	$ 3.5 million	Mark Madsen	$0.7 million
Derek Fisher	$ 3.4 million	Isiah Rider	$0.6 million
Brian Shaw	$ 2.3 million	Stan Medvedenko	$0.3 million
Ron Harper	$ 2.2 million	Corey Hightower	$0.3 million

SOURCE: Approximate salaries from www.insidehoops.com. The NBA does not make player salaries public.

Bonds's record year is a high outlier. How do his career mean and median number of home runs change when we drop the record 73? What general fact about the mean and median does your result illustrate?

■ **41.** A common criterion for detecting suspected outliers in a set of data is as follows:

1. Find the quartiles Q_1 and Q_3 and the *interquartile range* $IQR = Q_3 - Q_1$. The interquartile range is the spread of the central half of the data.
2. Call an observation an outlier if it falls more than $1.5 \times IQR$ above the third quartile or below the first quartile.

Which states are suspected outliers in the distribution of percent of Hispanics among adult residents, Table 6.1?

42. Table 6.7 gives the survival times (in days) of 72 guinea pigs after they were infected by tubercle bacilli in a medical study. Make a histogram of these data. Is the survival-time distribution approximately symmetric or strongly skewed? Based on the shape of the distribution, would you prefer the five-number summary or $\bar{x}$ and s as a numerical description? Compute the numerical description you chose.

■ **43.** Are any of the survival times in Table 6.7 suspected outliers by the $1.5 \times IQR$ criterion of Exercise 41? Because the criterion is based on the central half of the distribution, it sometimes calls observations in the long tail of a skewed distribution outliers. Use your judgment: Do any suspected outliers in this distribution really stand apart from the overall pattern?

44. Give an example of a small set of data for which the mean is larger than the third quartile.

■ **45.** Create a set of five positive numbers (repeats allowed) that have median 10 and mean 7. What thought process did you use to create your numbers?

■ **46.** Look at the histogram of lengths of words in Shakespeare's plays, Figure 6.14. The heights of the bars tell us what percent of words have each length. The median length is the midpoint, the length with half of all words shorter and half longer. What is the median length of words used by Shakespeare? Similarly, what are the quartiles? Give the five-number summary for Shakespeare's word lengths.

■ **47.** This is a standard deviation contest. You must choose four numbers from the whole numbers 0 to 10, with repeats allowed.

(a) Choose four numbers that have the smallest possible standard deviation.
(b) Choose four numbers that have the largest possible standard deviation.
(c) Is more than one choice possible in either part (a) or part (b)? Explain.

TABLE 6.7 Guinea Pig Survival Times (Days)

43	45	53	56	56	57	58	66	67	73
74	79	80	80	81	81	81	82	83	83
84	88	89	91	91	92	92	97	99	99
100	100	101	102	102	102	103	104	107	108
109	113	114	118	121	123	126	128	137	138
139	144	145	147	156	162	174	178	179	184
191	198	211	214	243	249	329	380	403	511
522	598								

SOURCE: T. Bjerkedal, Acquisition of resistance in guinea pigs infected with different doses of virulent tubercle bacilli, *American Journal of Hygiene,* 72 (1960): 130–148.

48. Your data consist of observations on the age of several subjects (measured in years) and the reaction times of these subjects (measured in seconds). In what units are each of the following descriptive statistics measured?

(a) The mean age of the subjects
(b) The standard deviation of the subjects' reaction times
(c) The correlation between age and reaction time
(d) The median age of the subjects

◆ **49.** For a biology project, you measure the length (centimeters) and weight (grams) of 12 crickets.

(a) Explain why you expect the correlation between length and weight to be positive.
(b) If you measured length in inches, how would the correlation change? (There are 2.54 centimeters in an inch.)

■ **50.** Changing the units of measurement can greatly alter the appearance of a scatterplot. Consider the following data:

x	-4	-4	-3	3	4	4
y	0.5	-0.6	-0.5	0.5	0.5	-0.6

(a) Draw x- and y-axes, each extending from -6 to 6. Plot the data on these axes.
(b) Calculate the values of new variables $x^* = x/10$ and $y^* = 10y$, starting from the values of x and y. Plot y^* against x^* on the same axes using a different plotting symbol. The two plots are very different in appearance.
(c) Use your calculator to find the correlation between x and y. Then find the correlation between x^* and y^*. How are the two correlations related? Explain why this isn't surprising.

◆ **51.** A mutual fund company's newsletter says, "A well-diversified portfolio includes assets with low correlations." The newsletter includes a table of correlations between the returns on various classes of investments. For example, the correlation between municipal bonds and large-cap stocks is 0.50 and the correlation between municipal bonds and small-cap stocks is 0.21.

(a) Rachel invests heavily in municipal bonds. She wants to diversify by adding an investment whose returns do not closely follow the returns on her bonds. Should she choose large-cap stocks or small-cap stocks for this purpose? Explain your answer.
(b) If Rachel wants an investment that tends to increase when the return on her bonds drops, what kind of correlation should she look for?

■ **52.** Use the definition of the mean $\bar{x}$ to show that the sum of the deviations $x - \bar{x}$ of the observations from their mean is always zero. This is one reason why the variance and standard deviation use squared deviations.

■ **53.** Use the equation for the least-squares regression line to show that this line always passes through the point $(\bar{x}, \bar{y})$. That is, set $x = \bar{x}$ and show that the line predicts that $y = \bar{y}$.

54. How is the flow of investors' money into stock mutual funds related to the flow of money into bond mutual funds? On the next page are data on the net new money flowing into stock and bond mutual funds in the years 1985 to 2000, in billions of dollars. "Net" means that funds flowing out are subtracted from those flowing in. If more money leaves than arrives, the net flow will be negative. To eliminate the effect of inflation, all dollar amounts are in "real dollars" with constant buying power equal to that of a dollar in the year 2000.

(a) Make a scatterplot using cash flow into stock funds as the explanatory variable.
(b) Find the least-squares line for predicting net bond investments from net stock investments. Add this line to your plot.
(c) What conclusion do the data suggest?

◆ **55.** Metatarsus adductus (call it MA) is a turning in of the front part of the foot that is common in adolescents and usually corrects itself. Hallux abducto valgus (call it HAV) is a deformation of the

Year	1985	1986	1987	1988	1989	1990	1991	1992
Stocks	12.8	34.6	28.8	−23.3	8.3	17.1	50.6	97.0
Bonds	100.8	161.8	10.6	−5.8	−1.4	9.2	74.6	87.1

Year	1993	1994	1995	1996	1997	1998	1999	2000
Stocks	151.3	133.6	140.1	238.2	243.5	165.9	194.3	309.0
Bonds	84.6	−72.0	−6.8	3.3	30.0	79.2	−6.2	−48.0

big toe that is not common in youth and often requires surgery. Perhaps the severity of MA can help predict the severity of HAV. Table 6.8 gives data on 38 consecutive patients who came to a medical center for HAV surgery. Using X rays, doctors measured the angle of deformity for both MA and HAV. They speculated that there is a positive correlation—more serious MA is associated with more serious HAV.

(a) Make a scatterplot of the data in Table 6.8. (Which is the explanatory variable?)

(b) Describe the form, direction, and strength of the relationship between MA angle and HAV angle. Are there any clear outliers in your graph?

(c) Do you think the data confirm the doctors' speculation?

◆ **56.** A news report says that minority children who take algebra and geometry in high school succeed in college about as well as whites. We should hesitate to conclude that requiring more math in high school would cause more minority students to succeed in college. Why? What factors

TABLE 6.8 Angle of Deformity (Degrees) for Two Types of Foot Deformity

HAV Angle	MA Angle	HAV Angle	MA Angle	HAV Angle	MA Angle
28	18	21	15	16	10
32	16	17	16	30	12
25	22	16	10	30	10
34	17	21	7	20	10
38	33	23	11	50	12
26	10	14	15	25	25
25	18	32	12	26	30
18	13	25	16	28	22
30	19	21	16	31	24
26	10	22	18	38	20
28	17	20	10	32	37
13	14	18	15	21	23
20	20	26	16		

SOURCE: Alan S. Banks et al., Juvenile hallux abducto valgus association with metatarsus adductus, *Journal of the American Podiatric Medical Association,* 84 (1994): 219–224.

might influence both the choice of high school courses and college ambitions, creating a relationship between these variables?

◆ **57.** People who use artificial sweeteners in place of sugar tend to be heavier than people who use sugar. Does this mean that artificial sweeteners cause weight gain? Give a more plausible explanation for this association.

◆ **58.** Each of the following statements contains a blunder. Explain what is wrong in each case.

(a) "There is a high correlation between the gender of American workers and their income."
(b) "We found a high correlation ($r = 1.09$) between students' ratings of faculty teaching and ratings made by other faculty members."
(c) "The correlation between age and income was found to be $r = 0.53$ years."

■ **59.** Figure 6.17 displays the relationship between the number of calories x and the amount of sodium y in each of 17 brands of meat hot dogs. Here are some facts about these data:

Mean calories = 158.7

Mean sodium = 418.5 mg

Standard deviation of calories = 25.24

Standard deviation of sodium = 93.87 mg

Correlation between calories and sodium = 0.8634

(a) What is the equation of the least-squares regression line for predicting sodium from calories? If another brand of meat hot dogs has 160 calories, what do you predict to be its sodium content?
(b) What is the equation of the least-squares regression line for predicting calories from sodium? If another brand of meat hot dogs has 423 mg of sodium, how many calories do you predict it to have?
(c) Explain carefully why we can get two different regression lines from the same data.

60. Exercise 55 describes a medical problem. The equation of the least-squares regression line for predicting HAV angle from MA angle is

$$y = 19.72 + 0.338x$$

(a) Add this line to the scatterplot you made in Exercise 55.
(b) A new patient has MA angle 25 degrees. What do you predict this patient's HAV angle to be?
(c) Does knowing MA angle allow doctors to predict HAV angle accurately? Explain your answer from the scatterplot.

61. Table 6.9 presents four sets of data prepared by the statistician Frank Anscombe to illustrate the dangers of calculating without first plotting the data.

(a) Without making scatterplots, find the correlation and the least-squares regression line for all four data sets. What do you notice? Use the regression line to predict y for $x = 10$.
(b) Make a scatterplot for each of the data sets and add the regression line to each plot.
(c) In which of the four cases would you be willing to use the regression line to describe the dependence of y on x? Explain your answer in each case.

WRITING PROJECTS

1. Coins are stamped with the year in which they were minted. Collect data from at least 50 coins of each denomination: pennies, nickels, dimes, and quarters. Write a description of the distribution of dates on coins now in circulation, including graphs and numerical descriptions. Are there differences among the denominations? Did you find any outliers?

2. Choose two variables that you think have a roughly straight-line relationship. Gather data on these variables and do a statistical analysis: Make a scatterplot, find the correlation, find the regression

TABLE 6.9 Four Data Sets for Exploring Correlation and Regression

Data Set A

x	10	8	13	9	11	14	6	4	12	7	5
y	8.04	6.95	7.58	8.81	8.33	9.96	7.24	4.26	10.84	4.82	5.68

Data Set B

x	10	8	13	9	11	14	6	4	12	7	5
y	9.14	8.14	8.74	8.77	9.26	8.10	6.13	3.10	9.13	7.26	4.74

Data Set C

x	10	8	13	9	11	14	6	4	12	7	5
y	7.46	6.77	12.74	7.11	7.81	8.84	6.08	5.39	8.15	6.42	5.73

Data Set D

x	8	8	8	8	8	8	8	8	8	8	19
y	6.58	5.76	7.71	8.84	8.47	7.04	5.25	5.56	7.91	6.89	12.50

SOURCE: Frank J. Anscombe, Graphs in statistical analysis, *The American Statistician,* 27 (1973): 17–21.

line (use a statistical calculator or software), and draw the line on your plot. Then write a report on your work. Some examples of suitable pairs of variables are:

(a) The height and arm span of a group of people
(b) The height and walking stride length of a group of people
(c) The price per ounce and bottle size in ounces for several brands of shampoo and several bottle sizes for each brand

3. Write a factual report on high school dropouts in the United States. Some examples of questions you might address are: What states have the highest percentages of adults who did not finish high school? How do the earnings and employment rates of dropouts compare with those of other adults? Is the percentage who fail to finish high school higher among blacks and Hispanics than among whites?

The *Statistical Abstract,* available in the library or online at www.census.gov/statab/www/, will supply you with data. Look in the index under "education" for an entry on "high school dropouts." You may want to look at other parts of the *Statistical Abstract* for data on earnings and other variables broken down by education.

4. Write a snappy, attention-getting article on the theme that "association is not causation." Use pointed but not-too-serious examples like TV sets and life expectancy (page 231) or this one: There is an association between long hair and being female, but cutting a woman's hair will not turn her into a man. Be clear, but don't be technical. Imagine that you are writing for high school students. (For more information on association versus causation, see pages 292–297 of David S. Moore, *Statistics: Concepts and Controversies,* 5th ed., Freeman, New York, 2001.)

SPREADSHEET PROJECTS

To do these projects, go to www.whfreeman.com/fapp.

Spreadsheets provide an easy means to analyze data. This project uses spreadsheet commands to replicate formulae for the mean, standard deviation, and least-squares line for actual data. Using the graphing capabilities of spreadsheets, the scatterplot graph and least-squares line for this data can be shown together.

APPLET EXERCISES

To do these exercises, go to www.whfreeman.com/fapp.

Correlation and Regression

Use the Correlation and Regression applet to see how much information about the distribution of data you get from a specific value of the correlation.

You are going to make scatterplots with 10 points that have correlation close to 0.7. The lesson is that many patterns can have the same correlation. Always plot your data before you trust a correlation.

(a) Stop after adding the first two points. What is the correlation? Why does it have this value?

(b) Make a lower left to upper right pattern of 10 points with correlation about $r = 0.7$. (You can drag points up or down to adjust r after you've marked 10 points.) Make a rough sketch of your scatterplot.

(c) Make another scatterplot with 9 points in a vertical stack at the left of the plot. Add one point far to the right and move it until the correlation is close to 0.7. Make a rough sketch of your scatterplot.

(d) Make yet another scatterplot with 10 points in a curved pattern that starts at the lower left, rises to the right, then falls again at the far right. Adjust the points up or down until you have a smooth curve with correlation close to 0.7. Make a rough sketch of this scatterplot also.

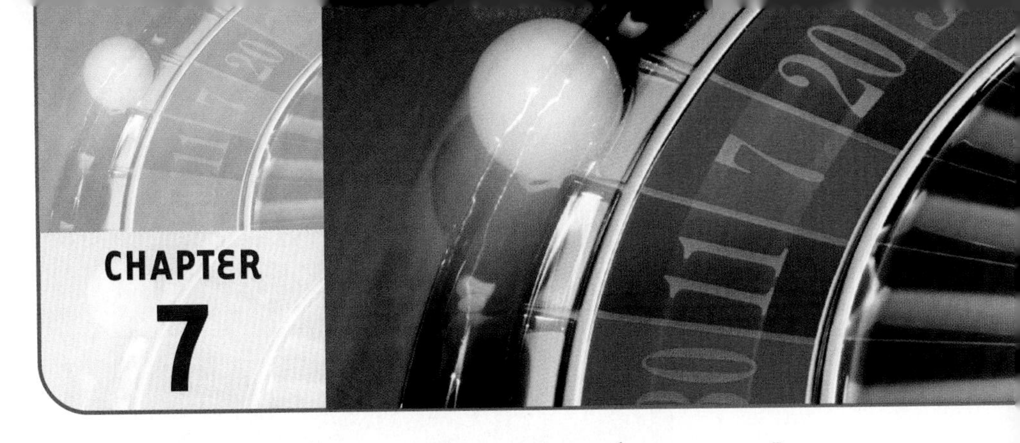

CHAPTER 7

Probability: The Mathematics of Chance

In this chapter we cover...

...and at www.whfreeman.com/fapp:

Flashcards

Quizzes

Spreadsheet Projects

A simple game of chance

Applet Exercises

Normal curve density calculator

Have you ever wondered how gambling, which is a recreation or an addiction for individuals, can be a business for the casino? A business requires predictable revenue from the service it offers, even when the service is a game of chance. Individual gamblers may win or lose. They can never say whether a day at the casino will turn a profit or a loss. But the casino isn't gambling. Casinos are consistently profitable, and state governments make money both from running lotteries and from selling licenses for other forms of gambling.

It is a remarkable fact that the aggregate result of many thousands of chance outcomes can be known with near certainty. The casino need not load the dice, mark the cards, or alter the roulette wheel. It knows that in the long run each dollar bet will yield its five cents or so of revenue. It is therefore good business to concentrate on free floor shows or inexpensive bus fares to increase the flow of dollars bet. The flow of profit will follow.

Gambling houses are not alone in profiting from the fact that a chance outcome many times repeated is firmly predictable. For example, although a life insurance company does not know *which* of its policyholders will die next year, it can predict quite accurately *how many* will die. It sets its premiums according to this knowledge, just as the casino sets its jackpots.

> A phenomenon is said to be **random** if individual outcomes are uncertain but the long-term pattern of many individual outcomes is predictable.

To a statistician, "random" does not mean "haphazard." Randomness is a kind of order, an order that emerges only in the long run, over many repetitions.

Casino dice. (Superstock.)

Many phenomena, both natural and of human design, are random. The life spans of insurance buyers and the hair colors of children are examples of natural randomness. Indeed, quantum mechanics asserts that at the subatomic level the natural world is inherently random. Probability theory, the mathematical description of randomness, is essential to much of modern science.

Games of chance are examples of randomness deliberately produced by human effort. Casino dice are carefully machined, and their drilled holes are filled with material equal in density to the plastic body. This guarantees that the side with six spots has the same weight as the opposite side, which has only one spot. Thus, each side is equally likely to land upward. All the odds and payoffs of dice games rest on this carefully planned randomness.

Statisticians also rely on planned randomness, although they use tables of random digits rather than dice and cards. The reasoning of statistical inference rests on random samples and randomized experiments–and on the mathematics of probability, the same mathematics that guarantees the profits of casinos and insurance companies. The mathematics of chance is the topic of this chapter.

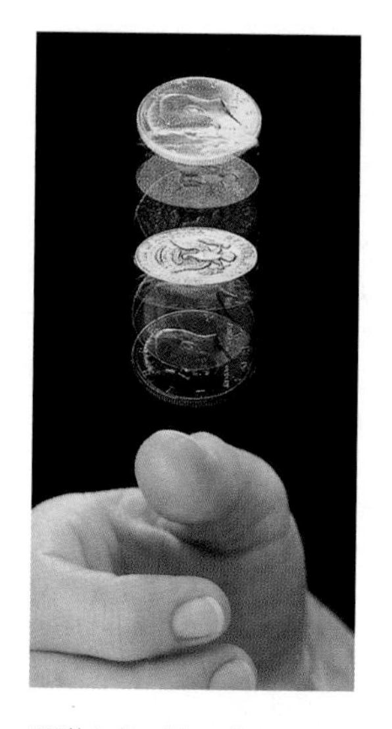

Will it land heads or tails? (Superstock.)

7.1 What Is Probability?

The mathematics of chance, the mathematical description of randomness, is called *probability theory*. Probability describes the predictable long-run patterns of random outcomes.

EXAMPLE Coin Tossing

When you toss a coin, there are only two possible outcomes, heads or tails. Figure 7.1 shows the results of tossing a coin 1000 times. For each number of tosses from 1 to 1000, we have plotted the proportion of those tosses that produced a head. The first toss was a head, so the proportion of heads starts at 1. The second toss was a tail, reducing the proportion of heads to 0.5 after two tosses. The next three tosses produced a tail followed by two heads, so the proportion of heads after five tosses is 3/5, or 0.6.

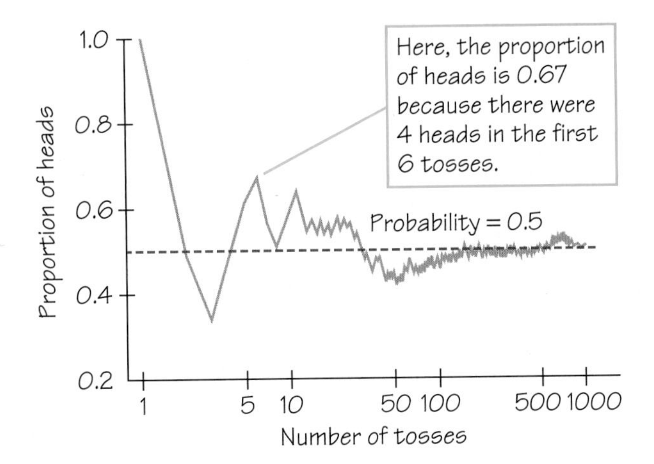

Figure 7.1 Proportion of heads versus number of tosses in tossing a coin. The proportion of heads eventually settles down to the probability of a head.

The proportion of tosses that produce heads is quite variable at first, but it settles down as we make more and more tosses. Eventually this proportion gets close to 0.5 and stays there. We say that 0.5 is the *probability* of a head. The probability 0.5 appears as a horizontal line on the graph. ■

> The **probability** of any outcome of a random phenomenon is the proportion of times the outcome would occur in a very long series of repetitions.

We might suspect that a coin has probability 0.5 of coming up heads just because the coin has two sides. As Exercises 1 and 2 illustrate, such suspicions are not always correct. The idea of probability is *empirical.* That is, it is based on observation rather than theorizing. Probability describes what happens in very many trials, and we must actually observe many trials to pin down a probability.

7.2 Probability Models

Gamblers have known for centuries that the fall of coins, cards, and dice displays clear patterns in the long run. The idea of probability rests on the observed fact that the average result of many thousands of chance outcomes can be known with near certainty. But a definition of probability as "long-run proportion" is vague. Who can say what "the long run" is? We can always toss the coin another 1000 times. Instead, we give a mathematical description of *how probabilities behave,* based on our understanding of long-run proportions. To see how to proceed, think first about a very simple random phenomenon, tossing a coin once. When we toss a coin, we cannot know the outcome in advance. What do we know? We are willing to say that the outcome will be either heads or tails. We believe that each of these outcomes has probability 1/2. This description of coin tossing has two parts:

- A list of possible outcomes
- A probability for each outcome

This description is the basis for all probability models. Here is the vocabulary we use.

The **sample space** S of a random phenomenon is the set of all possible outcomes.

An **event** is any outcome or any set of outcomes of a random phenomenon. That is, an event is a subset of the sample space.

A **probability model** is a mathematical description of a random phenomenon consisting of two parts: a sample space S and a way of assigning probabilities to events.

The sample space S can be very simple or very complex. When we toss a coin once, there are only two outcomes, heads and tails. So the sample space is $S = \{H, T\}$. If we draw a random sample of 1500 U.S. residents aged 18 and over, as Gallup polls do, the sample space contains all possible choices of 1500 of the more than 210 million adults in the country. This S is extremely large. Each member of S is a possible Gallup poll sample, which explains the term *sample space.*

EXAMPLE

Rolling Dice

Rolling two dice is a common way to lose money in casinos. There are 36 possible outcomes when we roll two dice and record the up faces in order (first die, second die). Figure 7.2 displays these outcomes. They make up the sample space S. "Roll a 5" is an event, call it A, that contains four of these 36 outcomes:

$A = \{$ (1,4) (2,3) (3,2) (4,1) $\}$

If the dice are carefully made, experience shows that each of the 36 outcomes in Figure 7.2 comes up equally often. So a reasonable probability model assigns probability 1/36 to each outcome.

In craps and other games, all that matters is the *sum* of the spots on the up faces. Let's change the random outcomes we are interested in: Roll two dice and count the spots on the up faces. Now there are only 11 possible outcomes, from a sum of 2 for rolling a double 1 through 3, 4, 5, and on up to 12 for rolling a double 6. The sample space is now

$$S = \{2, 3, 4, 5, 6, 7, 8, 9, 10, 11, 12\}$$

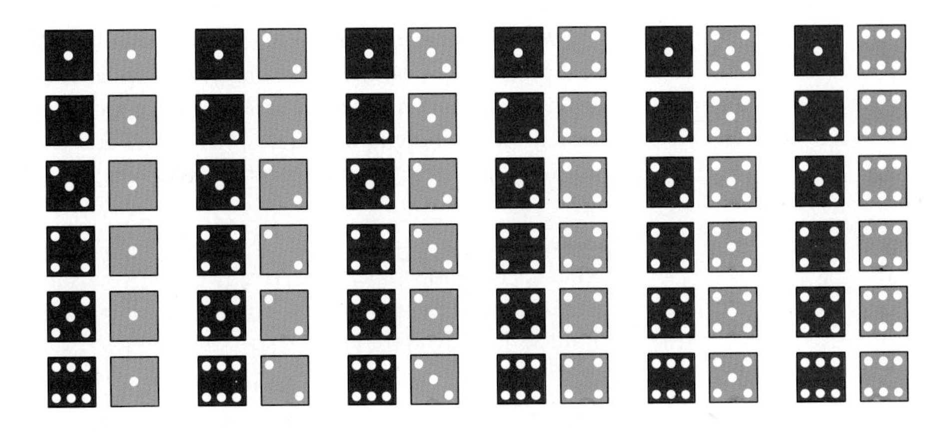

Figure 7.2 The 36 possible outcomes for rolling two dice.

Comparing this S with Figure 7.2 reminds us that we can change S by changing the detailed description of the random phenomenon we are describing. The outcomes in this new sample space are *not* equally likely, because there are six ways to roll a 7 and only one way to roll a 2 or a 12. ■

7.3 Probability Rules

There are many ways to assign probabilities, so it is convenient to start with some general rules that any assignment of probabilities to outcomes must obey. These facts follow from the idea of probability as "the long-run proportion of repetitions on which an event occurs."

1. **Any probability is a number between 0 and 1.** Any proportion is a number between 0 and 1, so any probability is also a number between 0 and 1. An event with probability 0 never occurs, and an event with probability 1 occurs on every trial. An event with probability 0.5 occurs in half the trials in the long run.
2. **All possible outcomes together must have probability 1.** Because some outcome must occur on every trial, the sum of the probabilities for all possible outcomes must be exactly 1.
3. **The probability that an event does not occur is 1 minus the probability that the event does occur.** If an event occurs in (say) 70% of all trials, it fails to occur in the other 30%. The probability that an event occurs and the probability that it does not occur always add to 100%, or 1.
4. **If two events have no outcomes in common, the probability that one or the other occurs is the sum of their individual probabilities.** If one event occurs in 40% of all trials, a different event occurs in 25% of all trials, and the two can never occur together, then one or the other occurs on 65% of all trials because 40% + 25% = 65%.

We can use mathematical notation to state facts 1 to 4 more concisely. We use capital letters near the beginning of the alphabet to denote events. If A is any event, we write its probability as $P(A)$. Here are our probability facts in formal language. As you apply these rules, remember that they are just another form of intuitively true facts about long-run proportions.

Rule 1. The probability $P(A)$ of any event A satisfies $0 \leq P(A) \leq 1$.
Rule 2. If S is the sample space in a probability model, then $P(S) = 1$.
Rule 3. For any event A,

$$P(A \text{ does not occur}) = 1 - P(A)$$

Rule 4. Two events A and B are **disjoint** if they have no outcomes in common and so can never occur simultaneously. If A and B are disjoint,

$$P(A \text{ or } B) = P(A) + P(B)$$

This is the **addition rule for disjoint events.**

EXAMPLE

Probabilities for Rolling Dice

Figure 7.2 displays the 36 possible outcomes of rolling two dice. For casino dice it is reasonable to assign the same probability to each of the 36 outcomes in Figure 7.2. Because all 36 outcomes together must have probability 1 (rule 2), each outcome must have probability 1/36.

What is the probability of rolling a 5? Because the event "roll a 5" contains the four outcomes displayed in the previous Example, the addition rule (rule 4) says that its probability is

$$\begin{aligned} P(\text{roll a 5}) &= P(⚀⚃) + P(⚁⚂) + P(⚂⚁) + P(⚃⚀) \\ &= \frac{1}{36} + \frac{1}{36} + \frac{1}{36} + \frac{1}{36} \\ &= \frac{4}{36} = 0.111 \end{aligned}$$

What is the probability of rolling anything other than a 5? By rule 3,

$$\begin{aligned} P(\text{roll does not give a 5}) &= 1 - P(\text{roll a 5}) \\ &= 1 - 0.111 = 0.889 \end{aligned}$$

What about the probability of rolling a 7? Figure 7.2 shows that there are six outcomes for which the sum of the spots is 7. The probability is 6/36, or about 0.167. Continue in this way to get the full probability model (sample space and assignment of probabilities) for rolling two dice and summing the spots on the up faces. Here it is:

Outcome	2	3	4	5	6	7	8	9	10	11	12
Probability	1/36	2/36	3/36	4/36	5/36	6/36	5/36	4/36	3/36	2/36	1/36

Figure 7.3 is a **probability histogram** of this probability model. The height of each bar shows the probability of the outcome at its base. Because the heights are probabilities, they add to 1. Think of Figure 7.3 as an idealized picture of the results of very many rolls of a die. As an idealized picture, it is perfectly symmetric.

This model assigns probabilities to individual outcomes. To find the probability of an event, just add the probabilities of the outcomes that make up the event. For example:

$$\begin{aligned} P(\text{outcome is odd}) &= P(3) + P(5) + P(7) + P(9) + P(11) \\ &= \frac{2}{36} + \frac{4}{36} + \frac{6}{36} + \frac{4}{36} + \frac{2}{36} \\ &= \frac{18}{36} = \frac{1}{2} \end{aligned}$$

This example illustrates one way to assign probabilities to events: Assign a probability to every individual outcome, then add these probabilities to find the probability of any event. If such an assignment is to satisfy the rules of probability, the probabilities of all the individual outcomes must sum to exactly 1.

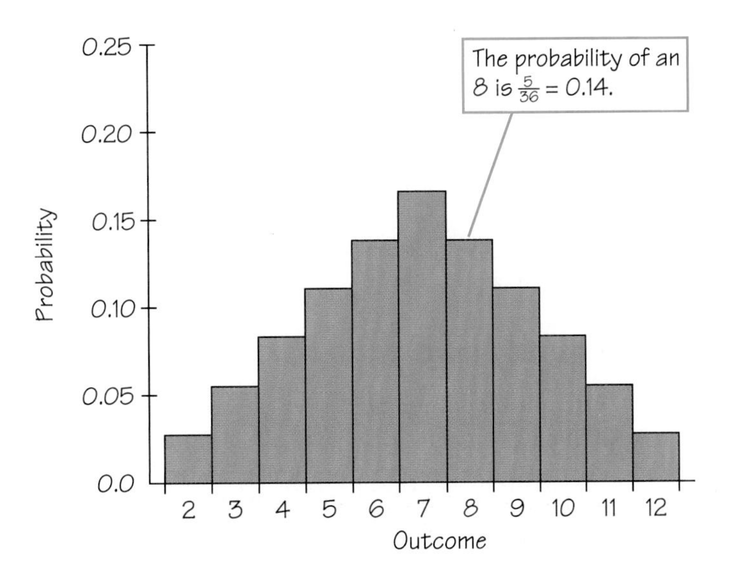

Figure 7.3 Probability histogram showing the probability model for rolling two balanced dice and counting the spots on the up faces.

To give a **probability model for a finite sample space,** assign a probability to each individual outcome. These probabilities must be numbers between 0 and 1 and must have sum 1.

The probability of any event is the sum of the probabilities of the outcomes making up the event.

EXAMPLE

Marital Status of Young Women

Draw a woman aged 25 to 29 years old at random and record her marital status. The probability of any marital status is just the proportion of all women aged 25 to 29 who have that status—if we drew many women, this is the proportion we would get. Thanks to the Census Bureau, we know these proportions. Here is the probability model:

Marital status	Never married	Married	Widowed	Divorced
Probability	0.389	0.553	0.003	0.055

Each probability is between 0 and 1. Check that these probabilities add to 1. These two facts tell us that this is a legitimate assignment of probabilities to the four outcomes in this sample space.

The probability that the woman we draw is not married is, by rule 3,

$$\begin{aligned} P(\text{not married}) &= 1 - P(\text{married}) \\ &= 1 - 0.553 = 0.447 \end{aligned}$$

That is, if 55.3% are married, then the remaining 44.7% are not married. "Never married" and "divorced" are disjoint events, because no woman can be both never married and divorced. So the addition rule says that

$$\begin{aligned} P(\text{never married or divorced}) &= P(\text{never married}) + P(\text{divorced}) \\ &= 0.389 + 0.055 = 0.444 \end{aligned}$$

That is, 44.4% of women in this age group either have never been married or are divorced. ■

7.4 Equally Likely Outcomes

A simple random sample gives all possible samples an equal chance to be chosen. Rolling two casino dice gives all 36 outcomes the same probability. When randomness is the product of human design, it is often the case that the outcomes in the sample space are all equally likely. Rules 1 and 2 force the assignment of probabilities in this case.

If a random phenomenon has k possible outcomes, all **equally likely,** then each individual outcome has probability $1/k$. The probability of any event A is

$$P(A) = \frac{\text{count of outcomes in } A}{\text{count of outcomes in } S} = \frac{\text{count of outcomes in } A}{k}$$

SPOTLIGHT 7.1 Probability and Risk

High exposures to asbestos are dangerous. Low exposures, such as that experienced by teachers and students in schools where asbestos is present in the insulation around pipes, are not very risky. The probability that a teacher who works for 30 years in a school with typical asbestos levels will get cancer from the asbestos is around 15/1,000,000. The probability of dying in a car accident during a lifetime of driving is about 15,000/1,000,000. That is, driving regularly is 1000 times more risky than teaching in a school where asbestos is present.

Risk does not stop us from driving. Yet the much smaller risk from asbestos launched massive cleanup campaigns and a federal requirement that every school inspect for asbestos and make the findings public. Why do we ignore the probabilities when we think about risk? Why do we take asbestos so much more seriously than driving? Why do we worry about improbable threats from tornadoes and mad cow disease more than we worry about heart attacks? It appears that psychology rather than probability governs our judgment of risk. We feel safer when a risk seems under our control than when we cannot control it. We are in control (or so we imagine) when we are driving, but we can't control the risk from asbestos or tornadoes.

Our reaction to risk also reflects the weakness of our intuition about how probability operates. It is hard to grasp the meaning of very small probabilities. Probabilities of 15 per million and 15,000 per million are both so small that our intuition cannot distinguish between them. Psychologists have shown that we generally overestimate very small risks and underestimate higher risks. Despite our bias toward things we think we can control, we should at least look at the probabilities when we choose what risks to take.

(Bob Daemmrich/The Image Works.)

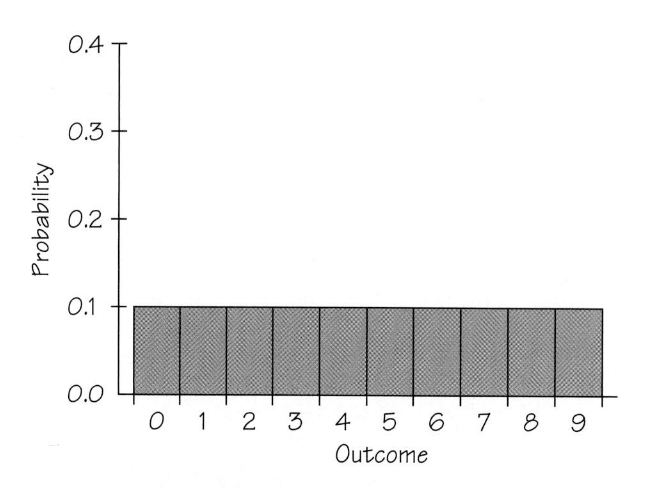

Figure 7.4 Probability histogram showing the probability model for generating a random digit between 0 and 9.

EXAMPLE

Random Digits

The successive digits in Table 5.1 (page 174) were produced by a careful randomization that makes each entry equally likely to be any of the 10 candidates. Because the total probability must be 1, the probability of each of the 10 outcomes must be 1/10. That is, the probability model is:

Outcome	0	1	2	3	4	5	6	7	8	9
Probability	0.1	0.1	0.1	0.1	0.1	0.1	0.1	0.1	0.1	0.1

Figure 7.4 displays this model in a probability histogram.

We can find the probability of any event by counting its outcomes. Here are two events, with the outcomes that they contain:

$$A = \{\text{odd outcome}\} = \{1, 3, 5, 7, 9\}$$
$$B = \{\text{outcome less than or equal to } 3\} = \{0, 1, 2, 3\}$$

We see that $P(A) = 0.5$ and $P(B) = 0.4$. The event $\{A \text{ or } B\}$ contains 7 outcomes,

$$\{A \text{ or } B\} = \{0, 1, 2, 3, 5, 7, 9\}$$

so it has probability 0.7. This is *not* the sum of $P(A)$ and $P(B)$, because A and B are *not* disjoint events. Outcomes 1 and 3 belong to both A and B. ■

When outcomes are equally likely, we find probabilities by counting outcomes. The study of counting methods is called **combinatorics.** We will start with the multiplication method that we called the *fundamental principle of counting* in Chapter 2 (page 39).

EXAMPLE Code Words

A computer system assigns log-in identification codes to users by choosing three letters at random. All three-letter codes are therefore equally likely. What is the probability that the code assigned to you has no x in it?

First count the total number of code words. There are 26 letters that can occur in each position in the word. Any of the 26 letters in the first position can be combined with any of the 26 letters in the second position to give 26×26 choices. (This is true because the order of the letters matters, so that *ab* and *ba* are different choices.) Any of the 26 letters can then follow in the third position. The number of different codes is

$$26 \times 26 \times 26 = 17{,}576$$

Now count the number of code words that have no x. These codes are made up of the other 25 letters. So there are

$$25 \times 25 \times 25 = 15{,}625$$

such codes. The probability that your code has no x is therefore

$$\begin{aligned} P(\text{no } x) &= \frac{\text{number of codes with no } x}{\text{number of codes}} \\ &= \frac{15{,}625}{17{,}576} = 0.889 \end{aligned}$$

Suppose that the computer is programmed to avoid repeated letters in the identification codes. Any of the 26 letters can still appear in the first position. But only the 25 remaining letters are allowed in the second position, so that there are 26×25 choices for the first two letters in the code. Any of these choices leaves 24 letters for the third position. The number of different codes without repeated letters is

$$26 \times 25 \times 24 = 15{,}600$$

Codes with no x are allowed one fewer choice in each position. There are

$$25 \times 24 \times 23 = 13{,}800$$

such codes. The probability that your code has no x is then

$$\begin{aligned} P(\text{no } x) &= \frac{\text{number of codes with no } x}{\text{number of codes}} \\ &= \frac{13{,}800}{15{,}600} = 0.885 \end{aligned}$$

Eliminating repeats slightly lowers your chance of avoiding an x. ■

The example makes use of two rules about counting that we often apply in finding probabilities:

Counting Rule A. Suppose we have a collection of n distinct items. We want to arrange k of these items in order, and the same item can appear several times in the arrangement. The number of possible arrangements is

$$n \times n \times \cdots \times n = n^k$$

Counting Rule B. Suppose we have a collection of n distinct items. We want to arrange k of these items in order, and any item can appear no more than once in the arrangement. The number of possible arrangements is

$$n \times (n - 1) \times \cdots \times (n - k + 1)$$

In the example, n (the number of letters available) is first 26, then 25, and k (the number of letters to be arranged to make a code) is 3. It is easier to think your way through the counting than to memorize the recipes.

EXAMPLE

How Many Orderings?

A jury of 7 students is seated in a row of 7 chairs to judge a speaking competition. In how many orders can the students sit?

Because each student sits in just 1 chair, no repeats are allowed. This is the case described by Counting Rule B, with n and k both equal to 7. To think through the problem, proceed like this: Any of the 7 students can sit in the first chair; then any of the 6 who remain can sit in the second chair; and so on. The number of arrangements is therefore

$$7 \times 6 \times 5 \times 4 \times 3 \times 2 \times 1 = 5040$$

(This number is often called 7!, read "seven factorial.") ■

7.5 The Mean of a Probability Model

Suppose you are offered this choice of bets, each costing the same: Bet A pays \$10 if you win and you have probability 1/2 of winning, while bet B pays \$10,000 and offers probability 1/10 of winning. You would very likely choose B even though A offers a better chance to win, because B pays much more if you win. It would be foolish to decide which bet to make just on the basis of the probability of winning. How much you can win is also important. When a random phenomenon has numerical outcomes, we are concerned with their amounts as well as with their probabilities.

What will be the average payoff of our two bets in many plays? Recall that the probabilities are the long-run proportions of plays in which each outcome occurs. Bet A produces \$10 half the time in the long run and nothing half the time. So the average payoff should be

$$\left(\$10 \times \frac{1}{2}\right) + \left(\$0 \times \frac{1}{2}\right) = \$5$$

Bet B, on the other hand, pays out $10,000 on 1/10 of all bets in the long run. Bet B's average payoff is

$$\left(\$10{,}000 \times \frac{1}{10}\right) + \left(\$0 \times \frac{9}{10}\right) = \$1000$$

If you can place many bets, you should certainly choose B. Here is a general definition of the kind of "average outcome" we used to compare the two bets.

Suppose that the possible outcomes $x_1, x_2, \ldots, x_k$ in a sample space S are numbers and that p_j is the probability of outcome x_j. The **mean** μ of this probability model is

$$\mu = x_1p_1 + x_2p_2 + \cdots + x_kp_k$$

Earlier, we met the mean $\bar{x}$, the average of n observations that we actually have in hand. The mean μ, on the other hand, describes the probability model rather than any one collection of observations. You can think of μ as a theoretical mean that gives the average outcome we expect in the long run.

EXAMPLE

Mean Household Size

What is the mean size of an American household? Here is the distribution of the size of households according to the Census Bureau:

Inhabitants	1	2	3	4	5	6	7
Proportion of households	0.26	0.33	0.17	0.14	0.07	0.02	0.01

If we imagine selecting a single household at random, the size of the household chosen has the probability model given by the table. The mean μ is the mean household size in the population. This mean is

$$\begin{aligned}\mu &= (1)(0.26) + (2)(0.33) + (3)(0.17) + (4)(0.14) \\ &\quad + (5)(0.07) + (6)(0.02) + (7)(0.01) \\ &= 2.53\end{aligned}$$

Figure 7.5 is a probability histogram of the distribution of household size, with the mean $\mu = 2.53$ marked.

In this case, the mean μ is the average size of all American households. If we took a random sample of, say, 100 households and recorded their sizes, we would call the average size for this sample $\bar{x}$. A second random sample would no doubt give a somewhat different value of $\bar{x}$. So $\bar{x}$ varies from sample to sample, but μ, which describes the distribution of probabilities, is a fixed number. ■

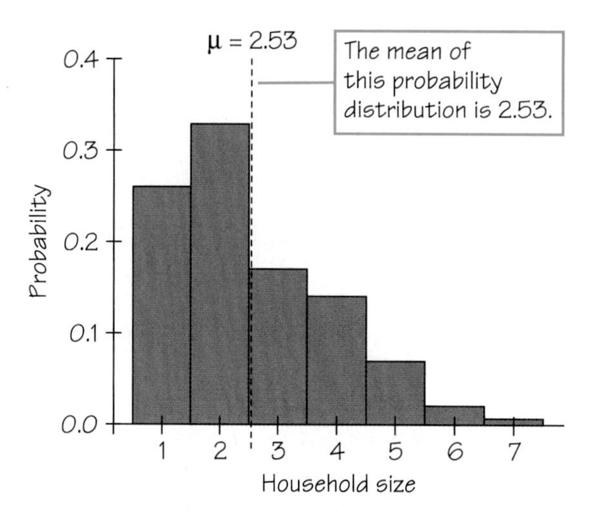

Figure 7.5 Probability histogram showing the probability model for the number of people in a randomly chosen household. The mean household size is $\mu = 2.53$ people.

The mean μ is an average outcome in two senses. The definition says that it is the average of the possible outcomes, not weighted equally but weighted by their probabilities. More likely outcomes get more weight in the average. An important fact of probability, the *law of large numbers,* says that μ is the average outcome in another sense as well.

Observe any random phenomenon having numerical outcomes with finite mean μ. According to the **law of large numbers,** as the random phenomenon is repeated a large number of times

- The proportion of trials on which each outcome occurs gets closer and closer to the probability of that outcome, and
- The mean $\bar{x}$ of the observed values gets closer and closer to μ.

These facts can be stated more precisely and then proved mathematically. The law of large numbers brings the idea of probability to a natural completion. We first observed that some phenomena are random in the sense of showing long-run regularity. Then we used the idea of long-run proportions to motivate the basic laws of probability. Those laws are mathematical idealizations that can be used without interpreting probability as proportion in many trials. Now the law of large numbers tells us that in many trials the proportion of trials on which an outcome occurs will always approach its probability.

The law of large numbers also explains why gambling can be a business. The winnings (or losses) of a gambler on a few plays are uncertain—that's why gambling is exciting. It is only *in the long run* that the mean outcome is predictable. The house plays tens of thousands of times. So the house, unlike individual gamblers, can count on the long-run regularity described by the law of large numbers. The average winnings of the house on tens of thousands of plays will be very close to the mean of the distribution of winnings. Needless to say, gambling games have mean outcomes that guarantee the house a profit.

SPOTLIGHT 7.2 Cancer Clusters

Distressing events bring out the desire to pinpoint a cause. Yet sometimes misfortune is just the play of chance. Consider the recurring worry about cancer clusters. A cancer cluster will quite likely appear in a neighborhood near you. You may find, for example, your local newspaper full of worry about a "larger than expected" number of cancer cases among people who have been students or teachers at a large elementary school.

Exactly what is "larger than expected"? Cancer is common. At any moment, more than 8 million people in the United States have cancer. The disease accounts for more than 23% of all deaths. So it is not surprising that cancer cases sometimes occur in clusters in the same neighborhood. There are bound to be clusters *somewhere* simply by chance. Think of tossing a large number of marbles at random onto a bed. Some of them will cluster together just by chance, and some spots on the bed will be bare, also just by chance. How many marbles do we expect to see in a square foot of the bed? We could construct a probability model for "tossing at random" and find a mean number from that model. That's the number we would expect to see. Now someone worried about "marble clusters" looks at the bed, singles out a spot where the marbles lie close together, and says, "Aha! There are more marbles than expected here." But the count in a spot singled out exactly because marbles were clustered there can't be compared with the mean count in a random area.

When a cancer cluster occurs in *our* neighborhood, we naturally tend to suspect the worst and look for someone to blame. State authorities get several thousand calls a year from people worried about "too much cancer" in their area. Some clusters do have specific causes, such as a toxic substance in the local water supply. This is most likely to be true when the cluster involves cases of a specific and relatively rare form of cancer. But most cancer clusters are simply the result of chance.

7.6 Sampling Distributions

Sampling, in a way, is a lot like gambling. Both rely on the deliberate use of chance. We want to apply probability to describe the results of sampling. At first glance, this is a formidable task. Suppose that we choose a simple random sample of size 1523 from the more than 210 million adults in the United States. All possible samples are equally likely – that's the definition of simple random sampling. There is an immense number of possible samples, so that finding probabilities by counting is not appealing. Here's another way: Rather than counting, we can actually choose a large number of samples and observe the outcomes. In practice, we program a computer to imitate (the formal word is *simulate*) drawing many samples.

EXAMPLE A Sampling Experiment

That's what we did in our discussion of a Gallup opinion poll in Chapter 5. Gallup asked a sample of 1523 people, "Please tell me whether or not you bought a state lottery ticket in the past 12 months." We simulated drawing 1000 random samples of size 1523 from a population in which (unknown to Gallup) 60% would say "yes" if we asked them about buying a lottery ticket. Now we have a new interest: What's the difference between small samples (say, of size 100) and the larger sample (size 1523) that Gallup actually used?

Figure 7.6 displays the results of two simulations. Figure 7.6a is a histogram of the percents saying "yes" in 1000 samples of size 100. Figure 7.6b displays the distribution of the percents saying "yes" in 1000 samples of size 1523 from the same population. Both histograms use the same scales so that we can easily compare them. (In fact, Figure 7.6b presents the same 1000 outcomes as Figure 5.4 on page 179. We changed the scale to allow comparison with the results of the smaller samples.) Both histograms show the regular pattern of outcomes that is characteristic of random sampling. Now we can use the language of probability to describe this pattern.

The shaded bar in Figure 7.6a covers the range

$$60\% < \text{percent saying "yes"} \leq 62\%$$

Exactly 150 of the 1000 samples of size 100 had outcomes in this range. Because 1000 samples is a large number of repetitions of the random sampling, the proportion of outcomes that fall in this class is close to the probability of the class. So we estimate that the probability of getting an outcome greater than 60% but no larger than 62% is 150/1000, or 0.15. The height of this bar in the histogram is 0.15. ■

Statisticians call a number that is computed from a sample a **statistic.** The percent of a sample who say "yes" to a poll question is a statistic. The probability histograms in Figure 7.6 show the assignment of probabilities to sample outcomes for samples of two sizes, 100 and 1523. These probabilities make up the *sampling distributions* of the statistic in these two settings.

The **sampling distribution** of a statistic is the distribution of values taken by the statistic in all possible samples of the same size from the same population.

Strictly speaking, the sampling distribution is the ideal pattern that would emerge if we looked at all possible samples of the same size from our population. A distribution obtained from a fixed number of trials, like the 1000 trials in Figure 7.6, is only an approximation to the sampling distribution. One of the uses of probability theory in statistics is to obtain sampling distributions without actually drawing many samples. The interpretation of a sampling distribution is the same, however, whether we obtain it by actual sampling or by the mathematics of probability.

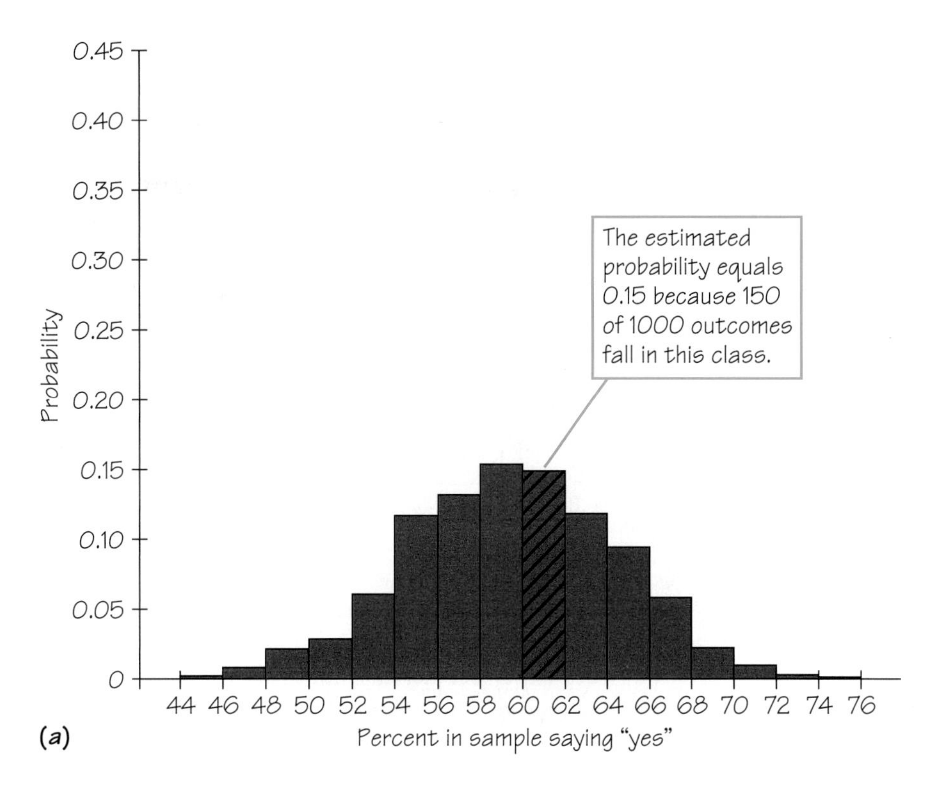

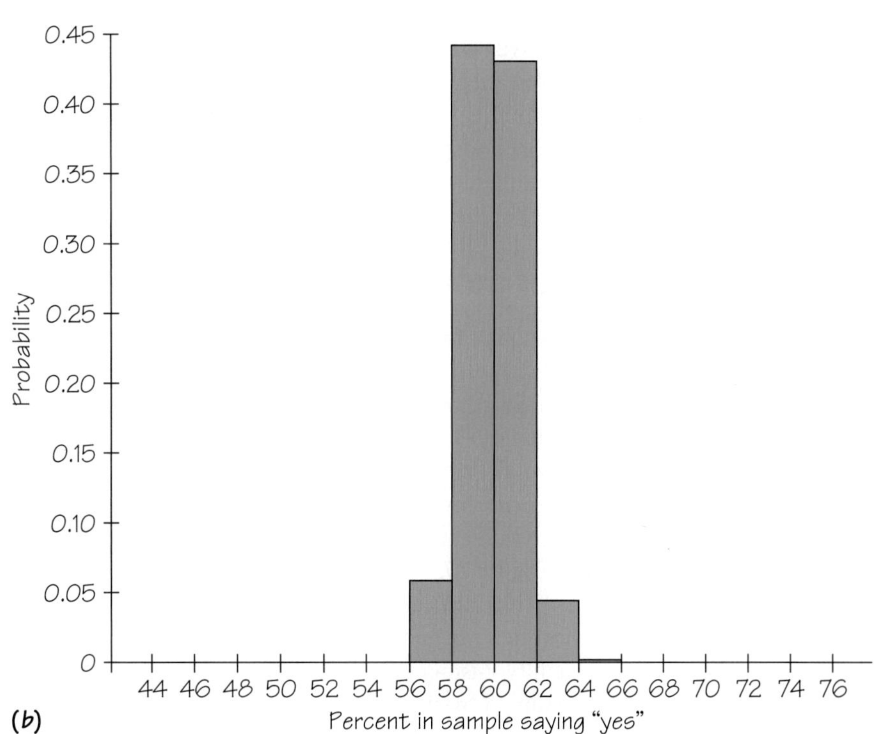

Figure 7.6 Sampling distributions that show the behavior of the percent of a sample who say "yes" to an opinion poll question in many simple random samples from the same population. The truth about the entire population is 60% "yes." (a) Sample size 100. (b) Sample size 1523.

EXAMPLE Examining Sampling Distributions

We can apply our tools for describing distributions to the two sampling distributions in Figure 7.6. We will examine the shape, center, and spread of these distributions.

The two distributions share a distinctive *shape*. They are quite symmetric, with a single peak in the center. There are no outliers to disturb the pattern. The mean outcomes are 60.016 for samples of size 100 and 60.007 for samples of size 1523. The median is exactly 60 for both distributions. That is, both sampling distributions are *centered* at the truth about the population, which is 60% "yes." As we noted in Chapter 5, this reflects the lack of bias in simple random sampling.

What is new here is the comparison of the *spread* for smaller and larger samples. The distribution of results for samples of size 1523 is much less spread out than the distribution for samples of 100 people – the histogram in Figure 7.6b is taller and narrower than that in Figure 7.6a. The standard deviations of the 1000 sample results are 5.036 for the smaller samples and 1.229 for the larger samples. *The spread of the outcomes of random samples from the same population goes down as we take larger samples.* A statistic from any random sample estimates without bias the corresponding truth about the population. Statistics from large samples take values close together in almost all samples. So large samples usually give results close to the truth about the population. ■

Our sampling experiment has both produced an approximate assignment of probabilities (without counting) and showed us how the sampling distribution behaves when we increase the size of the sample. Our goal is to learn enough about the mathematics of probability to get more exact results than sampling experiments can provide. In the next chapter, we'll learn specific recipes for the mean and standard deviation of the sampling distributions that Figure 7.6 approximates. The first step is to study the distinctive shape of these distributions. They are *normal distributions*.

7.7 Normal Distributions

Although they differ in spread, the histograms in Figures 7.6a and 7.6b have similar shapes in other respects. Both are symmetric, with centers at 60%. The tails fall off smoothly on either side, with no outliers. Suppose that we represent the shape of each histogram by drawing a smooth curve through the tops of the bars. If we do this carefully – using the actual probabilities of the outcomes rather than estimates from only 1000 samples – the two curves we obtain will be quite close to two members of the family of *normal curves*. These two normal curves appear in Figure 7.7.

Normal curves introduce a new way of assigning probabilities to events. We can describe an assignment of probabilities to the values of a statistic by a probability histogram. The height of any bar is the probability of the outcomes spanned by the base of that bar. Because all bars have the same width, their area (height times width) is proportional to the probability. You can think of a normal

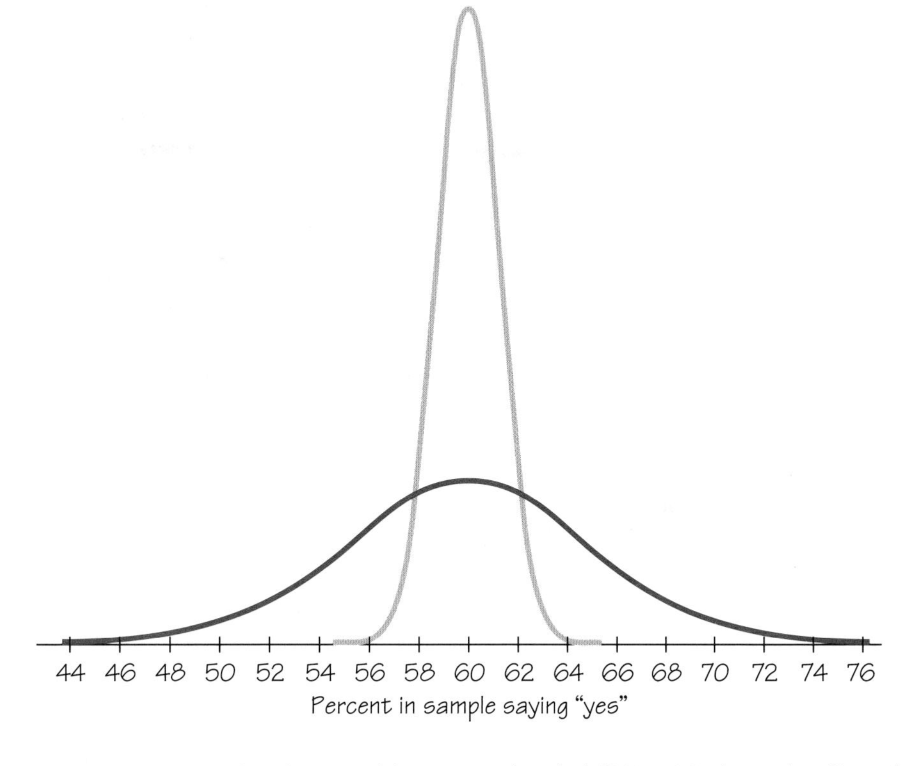

Figure 7.7 The normal curves that approximate the sampling distributions in Figure 7.6. The taller curve is for sample size 1523, the flatter curve for smaller samples of size 100. Each curve has area exactly 1 beneath it.

curve as an approximation to a histogram of probabilities with the scale adjusted so that area is exactly equal to probability. Normal curves are easier to work with than histograms because many bars are replaced by a single smooth curve. Normal curves have the property that the total area under the curve is exactly 1, corresponding to the fact that all outcomes together have probability 1.

> A normal curve assigns probabilities to outcomes as follows: The probability of any interval of outcomes is the area under the normal curve above that interval. The total area under any normal curve is exactly 1.

EXAMPLE

Probability as Area Under a Curve

Figure 7.8 repeats part of Figure 7.7. This is the normal curve that approximates the probabilities for the percent of a random sample of size 100 who say "yes" when asked if they bought a lottery ticket in the past year. The shaded area is the area under this normal curve between 60% and 62%. This area is 0.158. So the probability that between 60% and 62% of the people in a randomly chosen sample from this population will say "yes" is 0.158. ■

You should compare Figure 7.8 with Figure 7.6a, where the shaded bar also spans the range from 60% to 62%. We now have two estimates of the probability

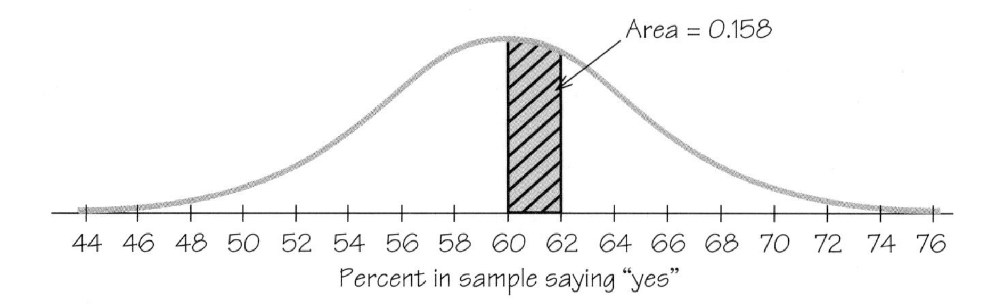

Figure 7.8 Probability as area under a normal curve. The area 0.158 is the probability of an outcome between 60 and 62.

of this set of outcomes. Simulation of 1000 samples estimates the probability to be 0.150 because 150 of the 1000 outcomes were greater than 60% and less than or equal to 62%. The normal curve estimates the probability to be 0.158. There is a subtle difference. The histogram class includes outcomes exactly at 62% and leaves out those exactly at 60%. If we had included 60% and left out 62%, we would get a different count and a different estimated probability. The smooth normal curve has no area exactly at 60% or 62%, so we need not worry about whether or not the event "between 60% and 62%" includes the endpoints. Nonetheless, the two estimates are quite close.

Our first method of assigning probabilities was to give a probability to each individual outcome, then add these probabilities to get the probability of any event. *Probability as area under a curve is the second important method of assigning probabilities.* It is easier when there are many individual outcomes falling close together. Curves of different shapes describe different assignments of probability. We will emphasize the normal curves, because they describe probability in several important situations. An assignment of probabilities to outcomes by a normal curve is a **normal probability distribution.**

Figures 7.6 and 7.7 demonstrate that the sampling distribution of a sample proportion from a simple random sample is close to a normal distribution. This is not just a matter of artistic judgment. It is a mathematical fact, first proved by Abraham DeMoivre in 1718. Some other common statistics, such as the mean $\bar{x}$ of a large sample, also have sampling distributions that are approximately normal. A normal curve will not exactly describe a specific set of outcomes, such as our 1000 sample percentages. It is an idealized distribution that is convenient to use and gives a good approximation to the actual distribution of outcomes.

There is a close connection between describing an assignment of probabilities to numerical outcomes and describing a set of data. We have used histograms for both tasks. Similarly, smooth curves such as the normal curves can replace histograms for describing large sets of data as well as for assigning probabilities. Many sets of data are approximately described by normal distributions. The normal distributions therefore deserve more detailed study.

7.8 The Shape of Normal Curves

Normal curves can be specified exactly by an equation, but we will be content with pictures like Figures 7.7 and 7.8. All normal curves are symmetric and bell-shaped, with tails that fall off rapidly. The center of the symmetric normal curve is the center of the distribution in several senses. It is the mean μ for the assignment of probabilities. It is also the median in the sense that half the probability (half the area under the curve) lies on each side of the center. When probabilities are assigned as areas under a symmetric curve, the mean μ is also the median of the distribution.

The mean and median of a skewed distribution are not equal. Figure 7.9, for example, shows a *right-skewed distribution.* The right tail of the curve is much longer than the left. The selling prices of houses are an example of a skewed distribution—there are many moderately priced houses and a few mansions out in the right tail. The very high prices of the mansions pull the mean (the average price) up, so that it is greater than the median (the midpoint). The mean price of existing single-family houses sold in 2000 was $177,000, but the median price for these same houses was only $139,100.

As we saw in Chapter 6, even the most cursory description of data on a single variable should include a measure of spread in addition to a measure of center or location. What about the spread of a normal curve? *Normal curves have the special property that their spread is completely measured by a single number, the standard deviation.* We learned in the last chapter how to calculate the standard deviation from a set of observations. For normal distributions, the standard deviation, like the mean, can be found directly from the curve.

The **mean** of a normal distribution lies at the center of symmetry of the normal curve.

To find the **standard deviation** of a normal distribution, run a pencil along the normal curve from the center (the mean) outward. At first, the curve falls ever more steeply as you go out; farther from the mean it falls ever less steeply. The two points where the curvature changes are located 1 standard deviation on either side of the mean.

With a little practice you can locate the change-of-curvature points quite accurately. For example, Figure 7.10 shows the distribution of heights of American women ages 18 to 24. The shape of the curve is normal, with mean (and median) height $\mu = 64.5$ inches. The two change-of-curvature points are at

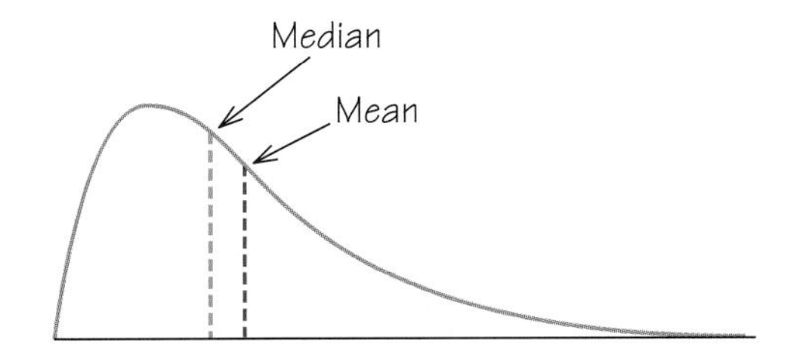

Figure 7.9 The mean of a skewed distribution is located farther toward the long tail than the median.

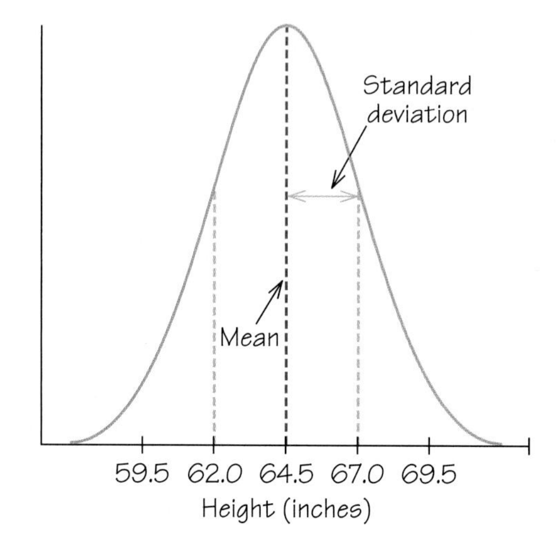

Figure 7.10 Locating the mean and standard deviation on a normal curve. For this normal curve, $\mu = 64.5$ and $\sigma = 2.5$.

62 inches and 67 inches. The standard deviation of the distribution is the distance of either of these points from the mean, 2.5 inches.

The usual notation for the standard deviation of a probability distribution is σ, the Greek letter sigma. Just as for the mean μ, it is possible to find σ for any distribution directly from the assignment of probabilities. Again, just as for the mean, we distinguish between s, the standard deviation of a given set of observations, and σ, the standard deviation of a probability distribution.

In Chapter 6, we often used the quartiles to indicate the spread of a distribution. Because the standard deviation completely describes the spread of any normal distribution, it tells us where the quartiles are. Here are the facts.

> The first quartile of any normal distribution is located 0.67σ below the mean; the third quartile is 0.67σ above the mean.

EXAMPLE

Heights of Young Women

The distribution of heights of young women, shown in Figure 7.10, is approximately normal with mean $\mu = 64.5$ inches and standard deviation $\sigma = 2.5$ inches. The quartiles lie 0.67σ, or

$$(0.67)(2.5) = 1.7 \text{ inches}$$

on either side of the mean. The first quartile is 64.5 − 1.7, or 62.8 inches. The third quartile is 64.5 + 1.7, or 66.2 inches. Figure 7.11 marks the quartiles on the normal curve. They contain between them the middle 50% of women's heights. ■

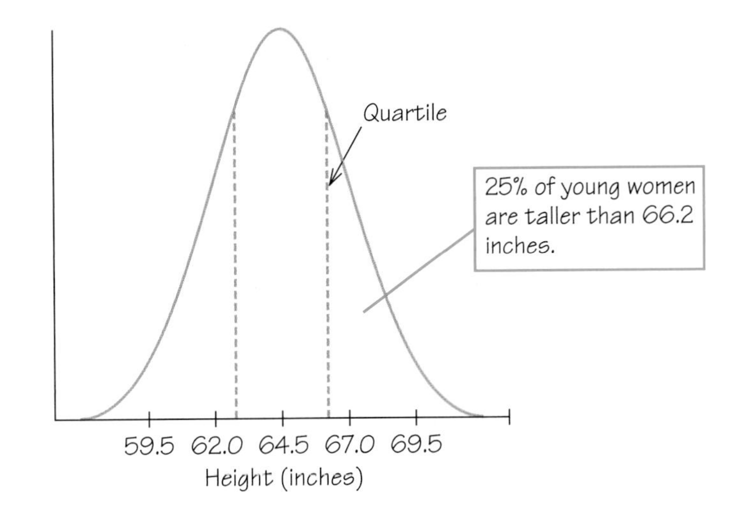
Quartile
25% of young women are taller than 66.2 inches.
59.5 62.0 64.5 67.0 69.5
Height (inches)

Figure 7.11 The quartiles of a normal distribution are located 0.67 standard deviation on either side of the mean. For this normal curve, $\mu = 64.5$ and $\sigma = 2.5$.

The mean and standard deviation of normal curves have a special property: *The shape of a normal distribution is completely specified by giving μ and σ.* A measure of center and a measure of spread are not enough to determine the exact shape of most distributions of data, but the mean and standard deviation are enough when the distribution is normal. Changing the mean of a normal curve does not change its shape; it only moves the curve to a new location. Changing the standard deviation does change the shape. A normal curve with a smaller standard deviation is taller and narrower (has less spread) than one with a larger standard deviation. You can see this by comparing the two normal curves for our random sampling experiments in Figure 7.7. Both normal curves have the same mean, and the area under each curve is exactly 1, but the curve for samples of size 1523 has the smaller standard deviation.

7.9 The 68–95–99.7 Rule

One consequence of the fact that the mean and standard deviation completely specify a normal distribution is that all normal distributions are the same when we record observations in terms of how many standard deviations they lie from the mean. In particular, the probability that an observation falls within 1, 2, or 3 standard deviations of the mean is the same for all normal distributions. The probability of an outcome falling within 1 standard deviation on either side of the mean is 0.68. If we go out 2 standard deviations from the mean, the probability is 0.95. Finally, the probability of falling within 3 standard deviations of the mean is almost 1, or 0.997 to be more exact. These facts can be derived mathematically from the equation of a normal curve. They need not be true for distributions with other shapes.

Figure 7.12 illustrates these facts expressed in terms of percents. Together, we call them the *68–95–99.7 rule* for normal distributions.

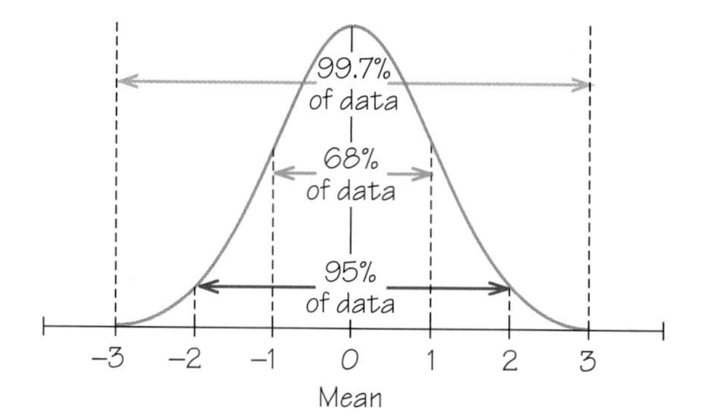

Figure 7.12 The 68–95–99.7 rule for normal distributions.

Using the three numbers in the 68–95–99.7 rule, we can quickly derive helpful information about any normal distribution. More detailed information can be gleaned from tables of areas under the normal curves, but the 68–95–99.7 rule is adequate for our purposes.

According to **the 68–95–99.7 rule,** in any normal distribution:

- 68% of the observations fall within 1 standard deviation of the mean.
- 95% of the observations fall within 2 standard deviations of the mean.
- 99.7% of the observations fall within 3 standard deviations of the mean.

EXAMPLE

Heights of Young Women

The heights of women between the ages of 18 and 24 are roughly normally distributed, with mean $\mu = 64.5$ inches and standard deviation $\sigma = 2.5$ inches. One standard deviation below the mean is $64.5 - 2.5$, or 62 inches. Similarly, 1 standard deviation above the mean is $64.5 + 2.5$, or 67 inches. The "68" part of the 68–95–99.7 rule says that about 68% of women are between 62 and 67 inches tall. Because 2 standard deviations are 5 inches, we know that 95% of young women are between $64.5 - 5$ and $64.5 + 5$, that is, between 59.5 and 69.5 inches tall. Almost all women have heights within 3 standard deviations of the mean, or between 57 and 72 inches. Few women are 6 feet (72 inches) tall or over. ■

EXAMPLE

SAT Scores

The distribution of scores on tests such as the SAT college entrance examination is close to normal. SAT scores are adjusted so that the mean score is about $\mu = 500$ and the standard deviation is about $\sigma = 100$. This information allows us to answer many questions about SAT scores.

- *How high must a student score to fall in the top 25%?*
  The third quartile is $(0.67)(100) = 67$ points above the mean. So scores above 567 are in the top 25%.

- *What percent of scores fall between 200 and 800?*
  Scores of 200 and 800 are 3 standard deviations on either side of the mean. The "99.7" part of the 68–95–99.7 rule says that 99.7% of all scores lie in this range. (In fact, 200 and 800 are the lowest and highest scores that are reported on the SAT. The few scores higher than 800 are reported as 800.)
- *What percent of scores are above 700?*
  A score of 700 is 2 standard deviations above the mean. By the "95" part of the 68–95–99.7 rule, 95% of all scores fall between 300 and 700 and 5% fall below 300 or above 700. Because normal curves are symmetric, half of this 5% are above 700. So a score above 700 places a student in the top 2.5% of test-takers.

Sketching a normal curve with the points 1, 2, and 3 standard deviations from the mean marked can help you use the 68–95–99.7 rule. Figure 7.13 shows the distribution of SAT scores with the areas needed to find the percent of scores above 700. ■

7.10 The Central Limit Theorem

The significance of normal distributions is explained in part by a key fact in probability theory known as the *central limit theorem*. This theorem says that the distribution of any random phenomenon tends to be normal if we average it over a large number of independent repetitions. The central limit theorem allows us to analyze and predict the results of chance phenomena if we average over many observations.

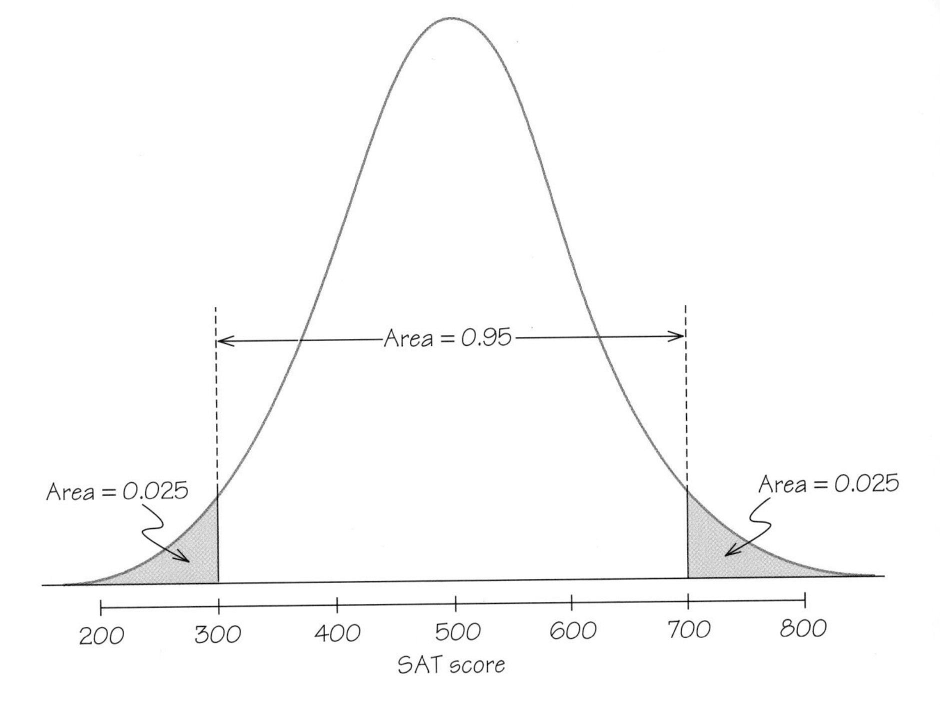

Figure 7.13 Using the 68–95–99.7 rule to find the percent of SAT scores that are above 700. For this normal curve, $\mu = 500$ and $\sigma = 100$.

We have already seen the central limit theorem at work in our random sampling experiment. A single person drawn at random says either "yes" or "no" to the opinion poll question. Only two outcomes are possible, and there is no normal curve in sight. However, the percent of "yes" answers when 100 people are drawn at random roughly follows a normal distribution. You can think of the percent of "yes" answers as an average of "yes" or "no" answers over the 100 people. When we sample 1523 people, the percent of "yes" responses represents an average over a larger number of people and is even closer to a normal curve.

Our sampling experiment showed that samples of size 1523 have much less spread than samples of size 100. We describe spread by the standard deviation of the normal distribution of outcomes. The central limit theorem makes the relationship of standard deviation to sample size explicit. Here is a more exact statement.

The **central limit theorem** states:

- A sample mean or sample proportion from n trials on the same random phenomenon has a distribution that is approximately normal when n is large.
- The mean of this normal distribution is the same as the mean for a single trial.
- The standard deviation of this normal distribution is the standard deviation for a single trial divided by $\sqrt{n}$.

Pay attention to the fact that the standard deviation of a mean or proportion decreases with the square root of the number of observations, $\sqrt{n}$. This is true for all values of n, not just when n is large enough that the central limit theorem says that the distribution is close to normal.

EXAMPLE

Averages Are Less Spread Out Than Individuals

Choose a single young woman at random. The resulting height varies in repeated random selections, with standard deviation $\sigma = 2.5$ inches.

Now choose 5 young women at random and take the mean $\bar{x}$ of their heights. The mean $\bar{x}$ varies in repeated samples, with standard deviation

$$\frac{\sigma}{\sqrt{n}} = \frac{2.5}{\sqrt{5}}$$

$$= \frac{2.5}{2.236} = 1.118 \text{ inches}$$

The standard deviation σ describes the variation when we measure many individual women. The standard deviation $\sigma/\sqrt{n}$ of the distribution of $\bar{x}$ describes the variation in the average heights of samples of women when we take many samples. Pay close attention to this distinction. ■

That averages $\bar{x}$ of several observations are less spread out than individual observations is an important statistical fact. For example, the average of 25 observations ($\sqrt{25} = 5$) from the same population has standard deviation 1/5 as large as the standard deviation for an individual. The average of 100 observations ($\sqrt{100} = 10$) has a standard deviation 1/10 as large as the standard deviation of a single observation. Because of the square root, to cut the standard deviation of a sample mean or proportion in half, we must multiply the sample size by 4, not just by 2.

7.11 Applying the Central Limit Theorem

We can use the central limit theorem to see just how good a business gambling can be for a casino. Let's look at just one of the many bets that a casino offers.

EXAMPLE

Red or Black in Roulette

An American roulette wheel has 38 slots, of which 18 are black, 18 are red, and 2 are green (see Figure 7.14). When the wheel is spun, the ball is equally likely to come to rest in any of the slots. Gamblers can place a number of different bets in roulette. One of the simplest wagers chooses red or black. A bet of $1 on red will

Figure 7.14 A gambler may win or lose at roulette, but in the long run the casino always wins. (Ken Whitmore/ Tony Stone Images.)

pay off an additional \$1 if the ball lands in a red slot. Otherwise, the player loses his \$1. When gamblers bet on red or black, the two green slots belong to the house.

If we decide to bet on red, there are only two possible outcomes: win or lose. We win if the ball stops in one of the 18 red slots. We lose if it lands in one of the 20 slots that are black or green. Because casino roulette wheels are carefully balanced so that all slots are equally likely, the probabilities are

$$P(\text{win } \$1) = 18/38$$
$$P(\text{lose } \$1) = 20/38$$

The mean outcome of a single bet on red is found in the usual way:

$$\mu = (1)\left(\frac{18}{38}\right) + (-1)\left(\frac{20}{38}\right)$$
$$= -\frac{2}{38} = -0.053$$

The law of large numbers says that the mean μ is the average outcome of a very large number of individual bets. In the long run, gamblers will lose (and the casino will win) an average of 5.3 cents per bet. ■

Just as when we ask only one person's opinion, there is no normal curve in sight when a gambler makes only one bet on red in roulette. But the central limit theorem ensures that the average outcome of many bets follows a distribution that is close to normal. Suppose that Lou places 50 bets in an evening's play. The mean outcome $\bar{x}$ is the overall gain (or loss) divided by 50. If Lou wins 30 and loses 20 times, his overall gain is \$10, an average winnings of $\bar{x} = 10/50$, or \$0.20 per bet. If he wins 10 and loses 40 of 50 bets, his average is negative, $\bar{x} = -\$0.60$. If Lou continues to gamble night after night, placing 50 bets each

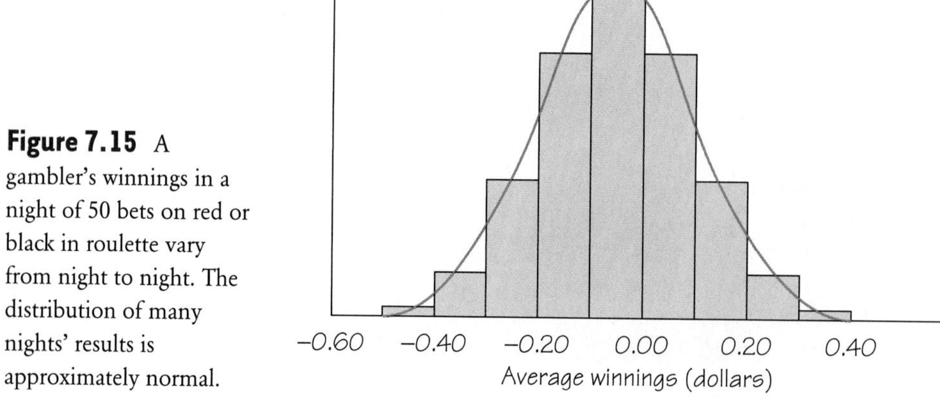

Figure 7.15 A gambler's winnings in a night of 50 bets on red or black in roulette vary from night to night. The distribution of many nights' results is approximately normal.

night, his average winnings per bet will vary from night to night. A histogram of these values will follow a normal distribution. Figure 7.15 shows the results of many trials of 50 bets each. The normal curve superimposed on the histogram is the distribution given by the central limit theorem in this case.

We know that the mean of the normal distribution in Figure 7.15 is the same as the mean of a single bet, -0.053. What is the standard deviation? We know that it is $\sigma/\sqrt{50}$, where σ is the standard deviation of the distribution of individual bets. Individual bets do not have a normal distribution, but their distribution is so simple that we can find the standard deviation starting from the idea of the variance as the average squared deviation from the mean. Just as in the case of the mean, "average" here is found by using the probabilities of the outcomes.

Suppose that the possible outcomes $x_1, x_2, \ldots, x_k$ in a sample space S are numbers, and that p_j is the probability of outcome x_j. The **variance** σ^2 of this probability model is

$$\sigma^2 = (x_1 - \mu)^2 p_1 + (x^2 - \mu)^2 p_2 + \cdots + (x_k - \mu)^2 p_k$$

The **standard deviation** σ is the square root of the variance.

EXAMPLE

More About Red or Black in Roulette

We saw that the mean for betting red or black is $\mu = -0.053$. The variance and standard deviation of the outcome of a single bet are

$$\begin{aligned} \sigma^2 &= (1 - (-0.053))^2 \frac{18}{38} + (-1 - (-0.053))^2 \frac{20}{38} \\ &= (1.053)^2 \frac{18}{38} + (-0.947)^2 \frac{20}{38} \\ &= 0.9972 \\ \sigma &= \sqrt{0.9972} = 0.9986 \end{aligned}$$

The standard deviation of the mean outcome $\bar{x}$ for 50 bets is therefore

$$\frac{\sigma}{\sqrt{n}} = \frac{0.9986}{\sqrt{50}} = 0.14$$

Check this by locating the change-of-curvature points of the normal curve in Figure 7.15. ■

What will be the experience of a habitual gambler like Lou who places 50 bets per night? Almost all average nightly winnings will fall within three standard deviations of the mean, that is, between

$$-0.053 + (3)(0.14) = 0.367$$

and

$$-0.053 - (3)(0.14) = -0.473$$

The total winnings after 50 bets will then fall between

$$(0.367)(50) = 18.35$$

and

$$(-0.473)(50) = -23.65$$

Lou may win as much as $18.35 or lose as much as $23.65. Gambling is exciting because the outcome, even after an evening of bets, is uncertain. It is possible to walk away a winner. It's all a matter of luck.

The casino, however, is in a different position. It doesn't want excitement, just a steady income. The house bets with all its customers—perhaps 100,000 individual bets on black or red in a week. The distribution of average customer winnings on 100,000 bets is very close to normal, and the mean is still the mean outcome for one bet, −0.053, a loss of 5.3 cents per dollar bet. The standard deviation is much smaller when we average over 100,000 bets. It is

$$\frac{\sigma}{\sqrt{n}} = \frac{0.9986}{\sqrt{100{,}000}} = 0.003$$

Here is what the spread in the casino's average result looks like after 100,000 bets:

$$\begin{aligned} \text{Spread} &= \text{mean} \pm 3 \text{ standard deviations} \\ &= -0.053 \pm (3)(0.003) \\ &= -0.053 \pm 0.009 \\ &= -0.044 \text{ to } -0.062 \end{aligned}$$

Because the casino covers so many bets, the standard deviation of the average winnings per bet becomes very small. And because the mean is negative, almost all outcomes will be negative. The gamblers' losses and the casino's winnings are almost certain to average between 4.4 and 6.2 cents for every dollar bet.

The gamblers who collectively place those 100,000 bets will lose money. The probable range of their losses is:

$$(-0.044)(100{,}000) = -4400$$
$$(-0.062)(100{,}000) = -6200$$

The gamblers are almost certain to lose–and the casino is almost certain to take in–between \$4400 and \$6200 on those 100,000 bets. What's more, we have seen from the central limit theorem that the more bets that are made, the narrower is the range of possible outcomes. That is how a casino can make a business out of gambling. The more money that is bet, the more accurately the casino can predict its profits.

REVIEW VOCABULARY

Addition rule for disjoint events If two events are disjoint, the probability that one or the other occurs is the sum of their individual probabilities.

Central limit theorem The average of many independent random outcomes is approximately normally distributed. When we average n independent repetitions of the same random phenomenon, the resulting distribution of outcomes has mean equal to the mean outcome of a single trial and standard deviation proportional to $1/\sqrt{n}$.

Combinatorics The branch of mathematics that counts arrangements of objects.

Disjoint events Events that have no outcomes in common.

Equally likely outcomes occur when every possible outcome of a random phenomenon has the same probability. If there are k outcomes, each has probability $1/k$.

Event Any collection of possible outcomes of a random phenomenon. An event is a subset of the sample space.

Law of large numbers As a random phenomenon is repeated many times, the mean $\bar{x}$ of the observed outcomes approaches the mean μ of the probability model.

Mean of a probability model The average outcome of a random phenomenon with numerical values. When possible values $x_1, x_2, \ldots, x_k$ have probabilities $p_1, p_2, \ldots, p_k$, the mean is the average of the outcomes weighted by their probabilities, $\mu = x_1p_1 + x_2p_2 + \cdots + x_kp_k$.

Normal distributions A family of probability models that assign probabilities to events as areas under a curve. The normal curves are symmetric and bell-shaped. A specific normal curve is completely described by giving its mean μ and its standard deviation σ.

Probability A number between 0 and 1 that gives the long-run proportion of repetitions of a random phenomenon on which an event will occur.

Probability histogram A histogram that displays a probability model when the outcomes are numerical. The height of each bar is the probability of the outcome or group of outcomes at the base of the bar.

Probability model A sample space S together with an assignment of probabilities to events. If probabilities $P(s)$ are assigned to individual outcomes s in S, they must be numbers between 0 and 1 that add to exactly 1. A probability model can also assign probabilities to events as areas under a curve. In this case, the total area under the curve must be exactly 1.

Random phenomenon A phenomenon is random if it is uncertain what the next outcome will be but each outcome nonetheless tends to occur in a fixed proportion of a very long sequence of repetitions. These long-run proportions are the probabilities of the outcomes.

Sample space A list of all possible outcomes of a random phenomenon.

Sampling distribution The distribution of values taken by a statistic when many random samples are drawn under the same circumstances. A sampling distribution consists of an assignment of probabilities to the possible values of a statistic.

68–95–99.7 rule In any normal distribution, 68% of the observations lie within 1 standard deviation on either side of the mean; 95% lie within 2 standard deviations of the mean; and 99.7% lie within 3 standard deviations of the mean.

Standard deviation of a normal curve The standard deviation σ of a normal curve is the distance from the mean to the change-of-curvature points on either side.

Standard deviation of a probability model A measure of the variability of a probability model. When the possible values $x_1, x_2, \ldots, x_k$ have probabilities $p_1, p_2, \ldots, p_k$, the variance is the average (weighted by probabilities) of the squared deviations from the mean, $\sigma^2 = (x_1 - \mu)^2 p_1 + (x_2 - \mu)^2 p_2 + \cdots + (x_k - \mu)^2 p_k$. The **standard deviation** σ is the square root of the variance.

Statistic A number computed from a sample, such as a sample mean or sample proportion. In random sampling, the value of a statistic will vary in repeated sampling.

SUGGESTED READINGS

COMAP. *Principles and Practices of Mathematics,* Springer, New York, 1997. Chapter 4 of this team-authored text presents combinatorics, and Chapter 8 discusses probability. This is a good choice for going beyond *For All Practical Purposes* in these areas.

MOSTELLER, FREDERICK, ROBERT E. K. ROURKE, and GEORGE B. THOMAS. *Probability with Statistical Applications,* Addison-Wesley, Reading, Mass., 1970. A rich treatment of basic probability that requires only high school algebra but is somewhat sophisticated. Although out of print, this book is a classic that nonetheless deserves mention.

PITMAN, JIM. *Probability,* Springer, New York, 1997. Although this text for math majors requires calculus, it is a model of clear exposition and offers excellent problems.

SUGGESTED WEB SITES

Both probabilities and means describe what happens in many repetitions of a random phenomenon. The best way to understand these ideas is to actually watch many repetitions. The What Is Probability? and Mean of a Random Phenomenon applets **(www.whfreeman.com/fapp)** allow you to watch as more repetitions are done and the long-term pattern emerges.

You may be interested in the debate over legalized gambling. For the case against, visit the National Coalition Against Legalized Gambling **(www.ncalg.org).** For the defense by the casino industry, visit the American Gaming Association **(www.americangaming.org).** State lotteries make their case via the National Association of State and Provincial Lotteries **(www.naspl.org).** The report of a commission established by Congress to study the impact of gambling is also available online **(www.ngisc.gov).** You'll find lots of facts and figures at all of these sites.

SKILLS CHECK

1. You read in a book on poker that the probability of being dealt three of a kind in a five-card poker hand is 1/50. What does this mean?

(a) If you deal thousands of poker hands, the fraction of them that contain three of a kind will be very close to 1/50.

(b) If you deal 50 poker hands, exactly one of them will contain three of a kind.
(c) If you deal 10,000 poker hands, exactly 200 of them will contain three of a kind.

2. Government data on causes of death show that the probability is 0.45 that a randomly chosen death was due to heart disease and 0.23 that it was due to cancer. What is the probability that the death was due to some other cause?

(a) 0.77
(b) 0.55
(c) 0.32

3. "Pick 3" games in many state lotteries work like this: You choose a three-digit number, like 456 or 999. The lottery chooses a winning number at random, so that each of the 1000 three-digit numbers has probability 0.001 of winning. You win $500 if your number is chosen; otherwise you lose the $1 cost of a ticket. What is the mean outcome of one bet?

(a) $5
(b) $0.50
(c) −$0.499

4. The law of large numbers implies which conclusion?

(a) When the sample size is large, the mean of the observed outcomes is equal to the mean of the probability model.
(b) The difference between the mean of the observed outcomes and the mean of the probability model is very large.
(c) As an experiment is repeated, the proportion of trials in which an outcome occurs approaches the probability of that outcome.

5. The central limit theorem implies which of these conclusions?
I. When the sample size is large, the mean and median are approximately equal.
II. When the sample size is large, the distribution is approximately symmetric.

(a) I only
(b) II only
(c) Both I and II

6. If events A and B are disjoint, then the probability that A or B occurs, $P(A \text{ or } B)$, must be

(a) equal to $P(A) + P(B)$.
(b) less than or equal to $P(A) + P(B)$.
(c) the maximum of $P(A)$ and $P(B)$.

7. If a distribution is skewed to the right, then

(a) the mean is less than the median.
(b) the mean is greater than the median.
(c) the mean and median are equal.

8. A fair coin is equally likely to land heads or tails. Represent heads by +1 and tails by −1. What is the mean of this probability model?

(a) 0
(b) 0.5
(c) 1

9. A fair coin is equally likely to land heads or tails. Represent heads by +1 and tails by −1. What is the standard deviation of this probability model?

(a) 0
(b) 0.5
(c) 1

10. For a casino, how does the distribution of a single day's red/black roulette bets compare to the distribution of a month's red/black roulette bets?

(a) The standard deviations are approximately equal.
(b) The means are approximately equal.
(c) Neither the means nor standard deviations are approximately equal.

11. The working life of a lightbulb is normally distributed with mean 1.3 years and standard deviation 0.3 year. What is the probability that it will last at least a year?

(a) 66%
(b) 68%
(c) 84%

EXERCISES

▲ Optional. ■ Advanced. ◆ Discussion.

What Is Probability?

1. Hold a penny upright on its edge under your forefinger on a hard surface, then snap it with your other forefinger so that it spins for some time before falling. Based on 50 spins, estimate the probability of heads.

2. You may feel that it is obvious that the probability of a head in tossing a coin is about 1/2 because the coin has two faces. Such opinions are not always correct. The previous exercise asked you to spin a coin rather than toss it—that changes the probability of a head. Now try another variation. Stand a nickel on edge on a hard, flat surface. Pound the surface with your hand so that the nickel falls over. What is the probability that it falls with heads upward? Make at least 50 trials to estimate the probability of a head.

3. Open your local telephone directory to any page and note whether the last digit of each of the first 100 telephone numbers on the page is odd or even. How many of the digits were odd? What is the approximate probability that the last digit of a telephone number is odd?

4. The table of random digits (Table 5.1 on page 174) was produced by a random mechanism that gives each digit probability 0.1 of being a 0. What proportion of the first 200 digits in the table are 0's? This proportion is an estimate of the true probability, which in this case is known to be 0.1.

◆ **5.** Suppose that the first six tosses of a coin give six tails and that tosses after that are exactly half heads and half tails. (Exact balance is unlikely, but it illustrates what happens in the long run.) What is the proportion of heads after the first six tosses? What is the proportion of heads after 100 tosses if the last 94 produce 47 heads? What is the proportion of heads after 1000 tosses if the last 994 produce half heads? What is the proportion of heads after 10,000 tosses if the last 9994 produce half heads? Someone says, "This illustrates the fact that probability works not by compensating for imbalances but by overwhelming them in the long run." Explain this statement.

6. Probability is a measure of how likely an event is to occur. Match one of the probabilities that follow with each statement about an event. (The probability is usually a much more exact measure of likelihood than is the verbal statement.)

0, 0.01, 0.3, 0.6, 0.99, 1

(a) This event is impossible. It can never occur.
(b) This event is certain. It will occur on every trial of the random phenomenon.
(c) This event is very unlikely, but it will occur once in a while in a long sequence of trials.
(d) This event will occur more often than not.

Probability Models and Rules

In each of Exercises 7 to 9, describe a reasonable sample space for the random phenomena mentioned. In some cases, more than one choice is possible.

7. Toss a coin 10 times.

(a) Count the number of heads observed.
(b) Calculate the percent of heads among the outcomes.
(c) Record whether or not at least five heads occurred.

8. Choose a student in your class at random. Ask how much time that student spent studying during the past 24 hours.

9. A basketball player shoots four free throws.

(a) You record the sequence of hits and misses.
(b) You record the number of baskets she makes.

10. Choose a new car or light truck at random and note its color. Here are the probabilities of the most popular colors:

Color	Silver	White	Black	Dark green	Dark blue	Medium red
Probability	0.176	0.172	0.113	0.089	0.088	0.067

What is the probability that the car you choose has any color other than the six listed? What is the probability that a randomly chosen car is either silver or white?

11. Ask a randomly chosen married woman whether her husband does his fair share of household chores. Here are the probabilities for her response:

Outcome	Probability
Does more than his fair share	0.12
Does his fair share	0.61
Does less than his fair share	?

(a) What must be the probability that the woman chosen says that her husband does less than his fair share? Why?
(b) The event "I think my husband does at least his fair share" contains the first two outcomes. What is its probability?

12. If you draw an M&M at random from a bag of the candies, the candy you draw will have one of six colors. The probability of drawing each color depends on the proportion of each color among all candies made. Table (a) shows the probabilities of each color for a randomly chosen milk chocolate M&M.

(a) What must be the probability of drawing a blue candy?
The probabilities for peanut M&M's are a bit different, shown in Table (b).
(b) What is the probability that a peanut M&M chosen at random is blue?
(c) What is the probability that a plain M&M is any of red, yellow, or orange? What is the probability that a peanut M&M has one of these colors?

Color	Brown	Red	Yellow	Green	Orange	Blue
Probability	0.3	0.2	0.2	0.1	0.1	?

(a)

Color	Brown	Red	Yellow	Green	Orange	Blue
Probability	0.2	0.2	0.2	0.1	0.1	?

(b)

13. In each of the following situations, state whether or not the given assignment of probabilities to individual outcomes is legitimate, that is, satisfies the rules of probability. If not, give specific reasons for your answer.

(a) When a coin is spun, $P(\text{H}) = 0.55$ and $P(\text{T}) = 0.45$.
(b) When two coins are tossed, $P(\text{HH}) = 0.4$, $P(\text{HT}) = 0.4$, $P(\text{TH}) = 0.4$, and $P(\text{TT}) = 0.4$.
(c) Milk chocolate M&M's have not always had the mixture of colors given in Exercise 12. In the past there were no red candies and no blue candies. Tan had probability 0.10, and the other four colors had the same probabilities that are given in Exercise 12.

14. A company that offers courses to prepare would-be MBA students for the GMAT examination has the following information about its customers: 20% are currently undergraduate students in business; 15% are undergraduate students in other fields of study; 60% are college graduates who are currently employed; and 5% are college graduates who are not employed.

(a) This is a legitimate assignment of probabilities to customer backgrounds. Why?
(b) What percent of customers are currently undergraduates?

■ **15.** You have two balanced, six-sided dice. One is a standard die, with faces having 1, 2, 3, 4, 5, and 6 spots. The other die has three faces with 0 spots and three faces with 6 spots. Find the probability model for the total number of spots on the up faces when you roll these two dice. (The method is illustrated for standard dice in the Example on page 258.)

Equally Likely Outcomes

16. Exercise 12 gives the probabilities that an M&M candy is each of brown, red, yellow, green, orange, and blue. "Crispy Chocolate" M&M's are equally likely to be any of these colors. What is the probability of any one color?

17. Psychologists sometimes use tetrahedral dice to study our intuition about chance behavior. A tetrahedron (Figure 7.16) is a pyramid with four faces, each a triangle with all sides equal in length. Label the four faces of a tetrahedral die with 1, 2, 3, and 4 spots. Give a probability model for rolling such a die and recording the number of spots on the down face.

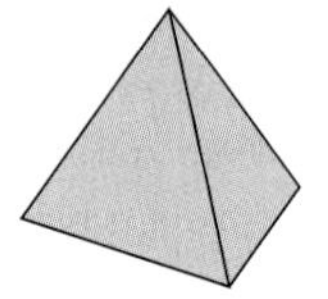

Figure 7.16 A tetrahedron. Exercises 17 and 25 concern dice with this shape.

18. Abby, Deborah, Julie, Sam, and Roberto work in a firm's public relations office. Their employer must choose two of them to attend a conference in Paris. To avoid unfairness, the choice will be made by drawing two names from a hat. [This is a simple random sample (SRS) of size 2.]

(a) Write down all possible choices of two of the five names. This is the sample space.
(b) The random drawing makes all choices equally likely. What is the probability of each choice?
(c) What is the probability that Julie is chosen?
(d) What is the probability that neither of the two men (Sam and Roberto) is chosen?

19. A couple plans to have three children. There are eight possible arrangements of girls and boys. For example, GGB means the first two children are girls and the third child is a boy. All eight arrangements are (approximately) equally likely.

(a) Write down all eight arrangements of the sexes of three children. What is the probability of any one of these arrangements occurring?
(b) Starting from this probability model, find the probability model for the number of girls the couple has. Make a probability histogram of this model.
(c) Use your model from part (b) to find the probability of at least two girls.
(d) Return to your model in part (a). Use this model to find the probability of at least two girls. You should get the same result as in part (c).

20. The Korean language, unlike Chinese and Japanese, uses an alphabet. The Korean alphabet has 10 vowels and 14 consonants. A computer assigns three-letter log-in codes in Korean at random, with repeated letters allowed.

(a) How many different codes are there?
(b) What is the probability that a code contains only consonants?

■ **21.** What is the probability that a code in Korean assigned by the computer in the previous exercise has the form consonant–vowel–consonant?

■ **22.** Suppose that a computer assigns three-character log-in codes that may contain the digits 0 to 9 as well as the letters *a* to *z*, with repeats allowed. What is the probability that your code contains no *x*? What is the probability that your code contains no digits?

■ **23.** The personal identification numbers (PINs) for automatic teller machines and telephone calling cards usually consist of four digits. You notice that most of your PINs have at least one 0, and you wonder if the issuers use lots of 0's to make the numbers easy to remember. What is the probability that a PIN chosen at random has at least one 0?

The Mean of a Probability Model

◆ **24.** You roll dice balanced so that all six sides are equally likely to land facing upward.

(a) What is the mean number of spots observed in rolling a single die?

(b) The Example on page 261 gives a probability model for rolling two balanced dice. What is the mean number of spots observed?

(c) Explain why your result in part (a) leads you to expect the result you found in part (b).

25. Exercise 17 introduces tetrahedral dice. Give a probability model for rolling two such dice and adding the spots on their down faces. What is the mean number of spots?

26. In Exercise 19 you found a probability model for the genders of three children. Compute the mean number of girls such couples will have.

27. The distribution of grades (A = 4, B = 3, and so on) in Professor Lopez's economics course is:

Grade	0	1	2	3	4
Probability	0.10	0.15	0.30	0.30	0.15

Find the average (that is, the mean) grade in this course. Make a probability histogram for the distribution of grades and mark the mean on your histogram.

28. Gain Communications sells aircraft communications units. Next year's sales depend on market conditions that cannot be predicted exactly. Gain follows the modern practice of using probability estimates of sales. The sales manager estimates next year's sales as follows:

Units sold	1000	3000	5000	10,000
Probability	0.1	0.3	0.4	0.2

These are "personal probabilities" that express the informed opinion of the sales manager. What is the mean value of next year's sales?

■ **29.** A state lottery "Pick 3" game offers a choice of several bets. You choose a three-digit number and bet $1. The lottery commission announces the winning three-digit number, chosen at random, at the end of each day. The "box" pays $82.33 if the number you chose has the same digits as the winning number, in any order. Otherwise, you lose your dollar. Find the mean winnings for a bet on the box, taking into account that you paid $1 to play. (Assume that you chose a number having three distinct digits.)

30. An American roulette wheel has 38 slots numbered 0, 00, and 1 to 36. The ball is equally likely to come to rest in any of these slots when the wheel is spun. The slot numbers are laid out on a board on which gamblers place their bets. One column of numbers on the board contains multiples of 3, that is, 3, 6, 9, . . . , 36. Joe places a $1 column bet that pays out $3 if any of these numbers comes up.

(a) What is Joe's probability of winning?

(b) What are Joe's mean winnings for one play, taking into account the $1 cost of each play?

(c) Joe plays roulette every day for years. What does the law of large numbers tell us about his results?

31. A life insurance company sells a term insurance policy to a 21-year-old male that pays $100,000 if the insured dies within the next five years. The probability that a randomly chosen male will die each year can be found in mortality tables. The company collects a premium of $250 each year as payment for the insurance. The amount that the company earns on this policy is $250 per year, less the $100,000 that it must pay if the insured dies. Here is the distribution of the company's earnings.

Age at death	21	22	23	24	25	≥26
Payout	−$99,750	−$99,500	−$99,250	−$99,000	−$98,750	$1250
Probability	0.00183	0.00186	0.00189	0.00191	0.00193	?

Fill in the missing probability in the table and calculate the mean earnings.

Sampling Distributions

32. Let us illustrate the idea of a sampling distribution in the case of a very small sample from a very small population. The population is the scores of 10 students on an exam:

Student	0	1	2	3	4	5	6	7	8	9
Score	82	62	80	58	72	73	65	66	74	62

We want to estimate the mean score μ in this population. The sample is an SRS of size $n = 4$ drawn from the population. Because the students are labeled 0 to 9, a single random digit from Table 5.1 chooses one student for the sample.

(a) Find the mean of the 10 scores in the population. This is the population mean μ.
(b) Use Table 5.1 to draw an SRS of size 4 from this population. Write the four scores in your sample and calculate the mean $\bar{x}$ of the sample scores. This statistic is an estimate of μ.
(c) Repeat this process 10 times using different parts of Table 5.1. Make a histogram of the 10 values of $\bar{x}$. You are constructing the sampling distribution of $\bar{x}$. Is the center of your histogram close to μ?

33. Table 6.7 (page 249) gives the survival times of 72 guinea pigs in a medical experiment. Consider these 72 animals to be the population of interest.

(a) Make a histogram of the 72 survival times. This population is strongly skewed to the right.
(b) Find the mean of the 72 survival times. This is the population mean μ. Mark μ on the x-axis of your histogram.
(c) Label the members of the population 01 to 72 and use Table 5.1 to choose an SRS of size $n = 12$. What is the mean survival time $\bar{x}$ for your sample? Mark the value of $\bar{x}$ with a point on the axis of your histogram from part (a).
(d) Choose four more SRSs of size 12, using different parts of Table 5.1. Find $\bar{x}$ for each sample and mark the values on the axis of your histogram from part (a). Would you be surprised if all five $\bar{x}$'s fell on the same side of μ? Why?
(e) If you chose a large number of SRSs of size 12 from this population and made a histogram of the $\bar{x}$ values, where would you expect the center of this sampling distribution to lie?

Normal Distributions

34. High levels of blood cholesterol are a problem for many middle-aged people. The distribution of cholesterol for adult American men aged 45 to 54 is approximately normal, with mean 215 milligrams per deciliter (mg/dL) and standard deviation 40 mg/dL. Draw a normal curve on which this mean and standard deviation are correctly located. (*Hint:* Draw the curve first, then mark the horizontal axis.)

35. Using the normal distribution described in Exercise 34 and the 68–95–99.7 rule, answer the following questions about the cholesterol levels of middle-aged men.

(a) Between what levels do the middle 95% of middle-aged men fall?
(b) What percent of men have cholesterol levels above 295 mg/dL?
(c) What percent of men have cholesterol levels below 175 mg/dL?

36. What are the quartiles of the distribution of blood cholesterol levels for middle-aged men in Exercise 34?

37. Newly manufactured automobile radiators may have small leaks. Most have no leaks, but some have one, two, or more. The number of leaks in radiators made by one supplier has mean 0.15 and standard deviation 0.4. The distribution of the number of leaks cannot be normal because only whole-number counts are possible. The supplier ships 400 radiators per day to an auto assembly plant. Take $\bar{x}$ to be the mean number of leaks in these 400 radiators. Over several years of daily shipments, what range of values will contain the middle 95% of the many $\bar{x}$'s?

◆ **38.** Figure 7.17 shows a probability distribution that is not symmetric. The mean and median do not coincide. Which of the points marked is the mean of the distribution, and which is the median? Explain your answer.

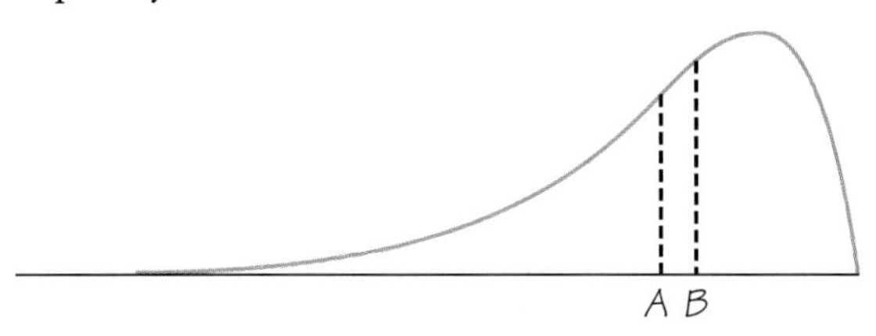

Figure 7.17 A skewed distribution, for Exercise 38.

39. Scores on the National Assessment of Educational Progress (NAEP) twelfth-grade mathematics test for the year 2000 were approximately normal with mean 300 points (out of 500 possible) and standard deviation 35 points.

(a) About what percent of twelfth-grade students had scores above 300?
(b) About what percent had scores above 370?
(c) What are the quartiles of the distribution of NAEP test scores? Explain in simple language what these two numbers tell us.

40. The Army reports that the distribution of head circumference among soldiers is approximately normal with mean 22.8 inches and standard deviation 1.1 inches.

(a) What percent of soldiers have head circumference greater than 23.9 inches?
(b) The Army plans to make helmets in advance to fit the middle 95% of head circumferences. Soldiers who fall outside this range will get custom-fitted helmets. What head circumferences are small enough or big enough to require custom-fitting?

41. Suppose that 47% of all adult women think they do not get enough time for themselves. An opinion poll interviews 1025 randomly chosen women and records the sample proportion who don't feel they get enough time for themselves. This statistic will vary from sample to sample if the poll is repeated. The sampling distribution is approximately normal with mean 47% and standard deviation about 1.6%. Sketch this normal curve and use it to answer the following questions.

(a) The truth about the population is 47%. In what range will the middle 95% of all sample results fall?
(b) What is the probability that the poll gets a sample in which fewer than 45.4% say they do not get enough time for themselves?

The Central Limit Theorem

42. The scores of twelfth-grade students on the NAEP year 2000 mathematics test have a distribution that is approximately normal with mean $\mu = 300$ and standard deviation $\sigma = 35$.

(a) Choose one twelfth-grader at random. What is the probability that his or her score is higher than 300? Higher than 335?
(b) Now choose an SRS of four twelfth-graders. What is the probability that their mean score is higher than 300? Higher than 335?

◆ **43.** Juan makes a measurement in a chemistry laboratory and records the result in his lab report. The standard deviation of students' lab measurements is $\sigma = 10$ milligrams. Juan repeats the measurement three times and records the mean $\bar{x}$ of his three measurements.

(a) What is the standard deviation of Juan's mean result? (That is, if Juan kept on making three measurements and averaging them, what would be the standard deviation of all his $\bar{x}$'s?)
(b) How many times must Juan repeat the measurement to reduce the standard deviation of $\bar{x}$ to 5? Explain to someone who knows no statistics the advantage of reporting the average of several measurements rather than the result of a single measurement.

44. A student organization is planning to ask a sample of 50 students if they have noticed AIDS education brochures on campus. The sample percentage who say "yes" will be reported. Their statistical adviser says that the standard deviation of this percentage will be about 7%. What would the standard deviation be if the sample contained 100 students rather than 50?

◆ **45.** How large a sample is required in the setting of Exercise 44 to reduce the standard deviation of the percentage who say "yes" from 7% to 3.5%? Explain to someone who knows no statistics the advantage of taking a larger sample in a survey of opinion.

■ **46.** In Exercise 30 you found the mean winnings for a $1 column bet in roulette. They are of course negative – in the long run, players lose and the house wins. Now find the standard deviation for this type of bet. Compare your results with those in the Examples on pages 281 and 283 for a bet on red or black. Is there any reason to prefer one bet over the other?

■ **47.** In the previous exercise you found the mean and standard deviation of the winnings for a column bet in roulette. What is the spread (mean ± 3 standard deviations) for a gambler's average winnings after 100 bets? After 1000 bets?

Additional Exercises

48. Choose a student at random and record the number of dollars in bills (ignore change) that he or she is carrying. Give a reasonable sample space S for this random phenomenon. (We don't know the largest amount that a student could reasonably carry, so you will have to make a choice in stating the sample space.)

◆ **49.** The 2000 census allowed each person to choose from a long list of races. That is, in the eyes of the Census Bureau, you belong to whatever race you say you belong to. "Hispanic/Latino" is a separate category; Hispanics may be of any race. If we choose a resident of the United States at random, the 2000 census gives these probabilities:

	Hispanic	Not Hispanic
Asian	0.000	0.036
Black	0.003	0.121
White	0.060	0.691
Other	0.062	0.027

(a) Verify that this is a legitimate assignment of probabilities.
(b) What is the probability that a randomly chosen American is Hispanic?
(c) Non-Hispanic whites are the historical majority in the United States. What is the probability that a randomly chosen American is not a member of this group?
(d) It does not make sense to speak of the mean of this distribution. Why not?

◆ **50.** A bridge deck contains 52 cards, four of each of the 13 face values ace, king, queen, jack, ten, nine, . . . , two. You deal a single card from such a deck and record the face value of the card dealt. Give an assignment of probabilities to these outcomes that should be correct if the deck is thoroughly shuffled. Give a second assignment of probabilities that is legitimate (that is, obeys the rules of probability) but differs from your first choice. Then give a third assignment of probabilities that is *not* legitimate and explain what is wrong with this choice.

■ **51.** Automobile license plates in Hawaii consist of three letters followed by three digits.

(a) How many different license plates are possible in Hawaii?
(b) A visitor to Honolulu observes that all license plates seem to begin with one of E, F, G, or H. How many license plates are possible if all plates begin with one of these letters?
(c) Suppose that a state allowed license plates consisting of any six letters or digits in any order. How many different license plates would then be possible?

■ **52.** A computer assigns three-letter log-in identification codes at random as in the example on page 264. If we take the vowels to be *a*, *e*, *i*, *o*, *u*, and *y*, what is the probability that your code contains no vowels if repeated letters are allowed? If no repeats are allowed?

53. A monkey at a keyboard presses three keys and hits the letters *a*, *g*, and *s* in random order. How many possible three-letter "words" can the monkey type using only these letters? Which of these are meaningful English words? What is the probability that the word the monkey typed is meaningful?

54. You are about to visit a new neighbor. You know that the family has four children, but you do not know their ages or sex. Write down all possible arrangements of girls and boys in order from youngest to oldest, such as BBGG (the two youngest are boys, the two oldest girls). The laws of genetics say that all of these arrangements are equally likely.

(a) What is the probability that the oldest child is a girl?
(b) What is the probability that the family has at least three boys?
(c) What is the probability that the family has at least three children of the same sex?

55. Choose an American household at random and ask how many cars they own. Here is the probability model if we ignore the few households who own more than five cars:

Cars	0	1	2	3	4	5
Probability	0.09	0.36	0.35	0.13	0.05	0.02

(a) Verify that this is a legitimate probability model.
(b) What outcomes make up the event "The household owns more than two cars"? What is the probability of this event?
(c) What is the mean number of cars owned by American households?

56. You have two balanced, six-sided dice. The first has 1, 3, 4, 5, 6, and 8 spots on its six faces. The second die has 1, 2, 2, 3, 3, and 4 spots on its faces.

(a) What is the mean number of spots on the up face when you roll the first die?
(b) What is the mean number of spots on the up face when you roll the second die?
(c) The sum of these means is the mean number of spots when you roll both dice. What is this mean? How does it compare with the mean when you roll two standard dice, found in Exercise 24?

■ **57.** Here is a simple way to create a probability model that has specified mean μ and standard deviation σ: There are only two outcomes, $\mu - \sigma$ and $\mu + \sigma$, each with probability 0.5. Use the definition of the mean and variance for probability models to show that this model does have mean μ and standard deviation σ.

58. The concentration of the active ingredient in capsules of a prescription painkiller varies according to a normal distribution with $\mu = 10\%$ and $\sigma = 0.2\%$.

(a) What is the median concentration? Explain your answer.
(b) What range of concentrations covers the middle 95% of all the capsules?
(c) What range covers the middle half of all capsules?

59. Answer the following questions for the painkiller in Exercise 58.

(a) What percent of all capsules have a concentration of active ingredient higher than 10.4%?
(b) What percent have a concentration higher than 10.6%?

60. The length of human pregnancies from conception to birth varies according to a distribution that is approximately normal with mean 266 days and standard deviation 16 days.

(a) Between what values do the lengths of the middle 95% of all pregnancies fall?
(b) How short are the shortest 2.5% of all pregnancies?

61. Bigger animals tend to carry their young longer before birth. The length of horse pregnancies from conception to birth varies according to a roughly normal distribution with mean 336 days and standard deviation 3 days. Use the 68–95–99.7 rule to answer the following questions.

(a) Almost all (99.7%) of horse pregnancies fall in what range of lengths?

(b) What percent of horse pregnancies are longer than 339 days?

62. The *deciles* of a distribution are the points having 10% (lower decile) and 90% (upper decile) of the observations falling below them. The lower and upper deciles contain between them the central 80% of the data. The lower and upper deciles of any normal distribution are located 1.28 standard deviations on either side of the mean. What score is needed to place you in the top 10% of the distribution of SAT scores (normal with mean 500 and standard deviation 100)?

■ **63.** Many companies "grade on a bell curve" to compare the performance of their managers and professional workers. This forces the use of some low performance ratings, so that not all workers are listed as "above average." Ford Motor Company's "performance management process" for a time assigned 10% A grades, 80% B grades, and 10% C grades to the company's 18,000 managers. Suppose that Ford's performance scores really are normally distributed. This year, managers with scores less than 25 received C's and those with scores above 475 received A's. What are the mean and standard deviation of the scores? (Use the information about deciles in the previous exercise.)

■ **64.** An ancient Korean drinking game involves a 14-sided die. The players roll the die in turn and must submit to whatever humiliation is written on the up face: something like "Keep still when tickled on face." Six of the 14 faces are squares. Let's call them A, B, C, D, E, and F for short. The other eight faces are triangles, which we will call 1, 2, 3, 4, 5, 6, 7, and 8. Each of the squares is equally likely. Each of the triangles is also equally likely, but the triangle probability differs from the square probability. The probability of getting a square is 0.72. Give the probability model for the 14 possible outcomes.

◆ **65.** You can use random digits to *simulate* repeated random sampling. Suppose that you are drawing simple random samples of size 25 from a large number of college students and that 20% of the students are unemployed during the summer. To simulate this SRS, let 25 consecutive digits in Table 5.1 stand for the 25 students in your sample. The digits 0 and 1 stand for unemployed students, and other digits stand for employed students. This is an accurate imitation of the SRS because 0 and 1 make up 20% of the 10 equally likely digits.

Simulate the results of 50 samples by counting the number of 0's and 1's in the first 25 entries in each of the 50 rows of Table 5.1. Make a histogram that displays the percents unemployed in your 50 samples. Is the truth about the population (20% unemployed) near the center of your histogram? What are the smallest and largest percents of unemployed students you obtained in your 50 samples? What percent of your samples had between 16% and 24% unemployed?

WRITING PROJECTS

1. The history of probability as the mathematical study of chance behavior began in seventeenth-century France, when gamblers turned to mathematicians for advice. Two of the mathematicians in question were Pierre de Fermat and Blaise Pascal, both interesting characters. Do some reading to learn more about the origins of probability and write a brief essay describing the role of Fermat and Pascal. (One good source is Carl B. Boyer, *A History of Mathematics,* Wiley, New York, 1991.)

2. Most American states and Canadian provinces operate lotteries. How much does your state or province take in from its lottery? Where does the money go? Write an essay on these issues. Find out what percent of the money bet is returned to the bettors in the form of prizes. What percent of the money bet is used by the state to pay lottery expenses? What percent is net revenue to the state? For what purposes does the state use lottery revenue? Your state lottery's Web site is the place to look for some facts.

3. Psychologists have shown that our intuitive understanding of chance behavior is rather poor. Amos Tversky (1937–1996) was a leader in the study of how we make decisions in the face of uncertainty. In its obituary of Tversky, the *New York Times* cited the following example:

Tversky asked subjects to choose between two public health programs that affect 600 people. One has probability 1/2 of saving all 600 and probability 1/2 that all 600 will die. The other is guaranteed to save exactly 400 of the 600 people. Most people chose the second program. He then offered a different choice. One program has probability 1/2 of saving all 600 and probability 1/2 of losing all 600, while the other will definitely lose exactly 200 lives. Most people chose the first program.

Discuss this example carefully. What is the difference between the two choices offered? What is the mean number of people saved by the two options in each choice? What do the reactions of most subjects to these choices show about how people make decisions?

SPREADSHEET PROJECTS

To do these projects, go to www.whfreeman.com/fapp.

This spreadsheet project models a simple game of chance. Because spreadsheets actively update their values, one can change prize and probabilities and immediately see their impact on the game. Using a random number generator, this project then simulates many plays of the game to model one's actual experience with the game.

APPLET EXERCISES

To do these exercises, go to www.whfreeman.com/fapp.

Normal Curve Density Calculator

In 1999, the scores of the students who took the mathematics part of the SAT exam followed a normal distribution with mean 511 and standard deviation 114.

(a) Use the Normal Curve Density Calculator applet to find out what scores are in the top 10%. (SAT scores are reported in whole numbers, so round your results to the nearest whole number.)
(b) In what range do the middle 90% of all scores fall?

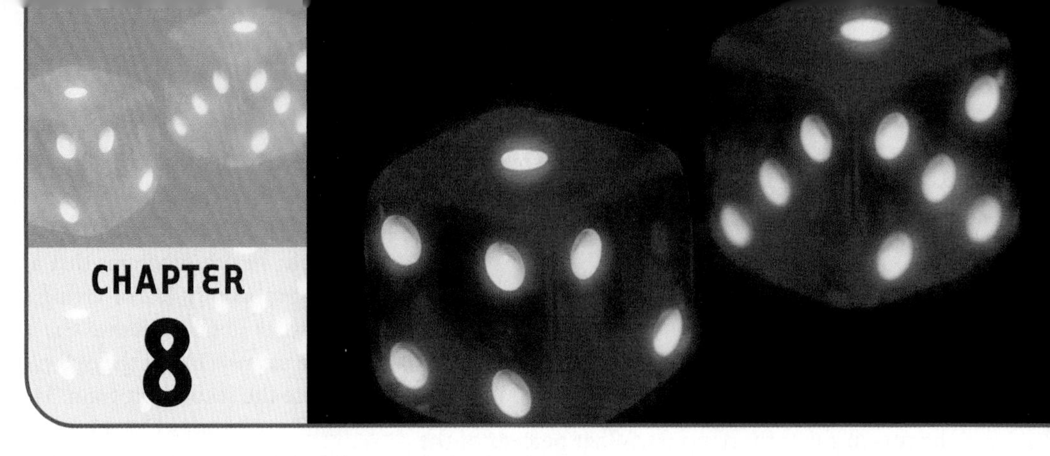

Statistical Inference

In this chapter we cover...

... and at www.whfreeman.com/fapp:

Flashcards

Quizzes

Spreadsheet Projects

The random number generator

Applet Exercises

Confidence intervals

Inference is the process of reaching conclusions based on evidence. Evidence can come in many forms. In a murder trial, evidence might be presented by the testimony of a witness, by a record of telephone conversations, or by DNA analysis of blood samples. In statistical inference, the evidence is provided by data. When we select a sample, we know the responses of the individuals in the sample. Often we are not content with information about the sample. We want to infer from the sample data some conclusion about a wider population that the sample represents.

> **Statistical inference** provides methods for drawing conclusions about a population from sample data.

We cannot be certain that our conclusions are correct—a different sample might lead to different conclusions. Statistical inference uses the language of probability to say how trustworthy its conclusions are.

EXAMPLE Do You Lotto?

Lotteries are popular. How popular? We would like to know the proportion of American adults that bought a state lottery ticket in the past year. Our *population* is all residents of the United States who are at least 18 years old. The percent who bought a lottery ticket—call it p, for percent—is a fact about the population. We can't afford to ask every adult, but the Gallup poll did ask a *random sample* of 1523 adults. Of these people, 868 said "yes." So the percent of the sample who said "yes" is

$$\hat{p} = \frac{868}{1523} = 0.57 = 57\%$$

Read the **sample proportion** $\hat{p}$ as "p-hat."

Because Gallup chose a random sample, we expect the sample result to reflect the truth about the population. That is, we infer that the unknown population proportion p is somewhere around 57%. ■

Drawing conclusions in mathematics is a matter of starting from a hypothesis and using logical argument to prove without doubt that the conclusion follows. This is a *deductive argument* from hypothesis to consequences. Statistics argues in almost the reverse order. Gallup's sample result reflects the truth about the population. We expect $\hat{p}$ to be small if the population truth p is 5% and we expect $\hat{p}$ to be larger if p is 60%. So we argue back from the sample to the population: Because the sample result was 57%, we think the population truth is close to that value. This is an *inductive argument* from consequences back to a hypothesis.

Inductive arguments do not produce proof. Gallup might have had the bad luck to get a sample with 57% "yes" from a population with only 20% "yes." Of course, a sample result that far from the population truth is unlikely. How unlikely? How close to the sample result can we expect the population truth to be? A really useful estimate of the unknown population percent p isn't just a number like 57%. Statistical estimates start with the sample outcome (57% "yes" in this example) and add a *margin of error* that tells us how accurately we have pinned down the truth about the population. We discussed statistical estimation and margins of error briefly in Chapter 5. Now we will fill in some details.

The lottery example illustrates a basic move in statistics: Use a fact about a sample to estimate the truth about the whole population. To think about such moves, we must keep straight whether a number describes a sample or a population. Here is the vocabulary we use.

A **parameter** is a number that describes the **population.** A parameter is a fixed number, but in practice we don't know its value.

A **statistic** is a number that describes a **sample.** The value of a statistic is known when we have taken a sample, but it can change from sample to sample. We often use a statistic to estimate an unknown parameter.

So parameter is to population as statistic is to sample. Want to estimate an unknown parameter? Choose a sample from the population and use a sample statistic as your estimate. That's what Gallup did.

8.1 Estimating with Confidence

We want to estimate the proportion p of the individuals in a population who have some characteristic – they are employed, or they have bought a lottery ticket in the past year, for example. Let's call the characteristic we are looking for a "success." We use the proportion $\hat{p}$ of successes in a simple random sample (SRS) to estimate the proportion p of successes in the population. How good is the statistic $\hat{p}$ as an estimate of the parameter p? To find out, we ask, "What would happen if we took many samples?" We know that $\hat{p}$ would vary from sample to sample. We also know that this sampling variability isn't haphazard. It has a clear pattern in the long run, a pattern that is pretty well described by a normal curve. Here are the facts.

The **sampling distribution** of a statistic is the distribution of values taken by the statistic in all possible samples of the same size from the same population.

Choose a simple random sample of size n from a large population of which the percent p are successes. Let $\hat{p}$ be the percent of the sample that are successes. Then:

- The sampling distribution of $\hat{p}$ is *approximately normal* and is closer to a normal distribution when the sample size n is large.
- The *mean* of the sampling distribution is exactly p.
- The *standard deviation* of the sampling distribution is

$$\sqrt{\frac{p(100 - p)}{n}}$$

These facts can be proved by mathematics, so they are a solid starting point. Figure 8.1 summarizes them in a form that also reminds us that a sampling distribution describes the results of lots of samples from the same population.

EXAMPLE

Risky Behavior in the Age of AIDS

How common is behavior that puts people at risk of AIDS? The National AIDS Behavioral Surveys interviewed a random sample of 2673 adult heterosexuals. *Suppose that the truth is that 6% of adult heterosexuals have multiple partners.* Of course,

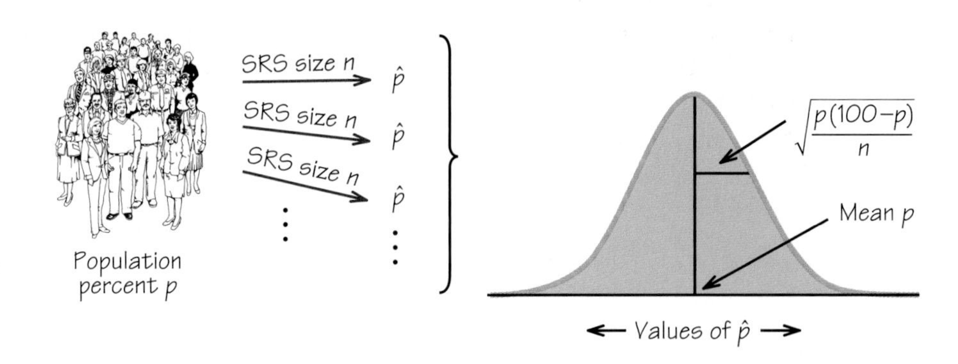

Figure 8.1 Repeat many times the process of selecting an SRS of size n from a population of which the proportion p are successes. The values of the sample proportion of successes $\hat{p}$ have this normal sampling distribution.

in practice we don't know the value of p, but we want to see how the sample reflects the truth about the population. So we will temporarily suppose that $p = 6\%$. The National AIDS Behavioral Surveys' sample of size $n = 2673$ would, if repeated many times, produce sample proportions $\hat{p}$ that closely follow the normal distribution with

$$\text{Mean} = p = 6\%$$

and

$$\begin{aligned}\text{Standard deviation} &= \sqrt{\frac{p(100-p)}{n}} \\ &= \sqrt{\frac{(6)(94)}{2673}} \\ &= \sqrt{0.211} = 0.46\%\end{aligned}$$

The center of this normal distribution is at the truth about the population. That's the absence of bias in random sampling once again. The standard deviation is small because the sample is quite large. So almost all samples will produce a statistic $\hat{p}$ that is close to the true p. In fact, the "95" part of the 68–95–99.7 rule says that 95% of all sample outcomes will fall between

$$\text{Mean} - 2 \text{ standard deviations} = 6.0 - 0.92 = 5.08$$

and

$$\text{Mean} + 2 \text{ standard deviations} = 6.0 + 0.92 = 6.92$$

Figure 8.2 displays these facts. ■

So far, we have just put numbers on what we already knew: We can trust the results of large random samples because almost all such samples give results that are close to the truth about the population. The numbers say that in 95% of all samples of size 2673, the statistic $\hat{p}$ and the parameter p are within 0.92 of each other. We can put this another way: 95% of all samples give an outcome $\hat{p}$ such that the population truth p is captured by the interval from $\hat{p} - 0.92$ to $\hat{p} + 0.92$. This is a *confidence interval.*

A 95% **confidence interval** is an interval obtained from the sample data by a method that in 95% of all samples will produce an interval containing the true population parameter.

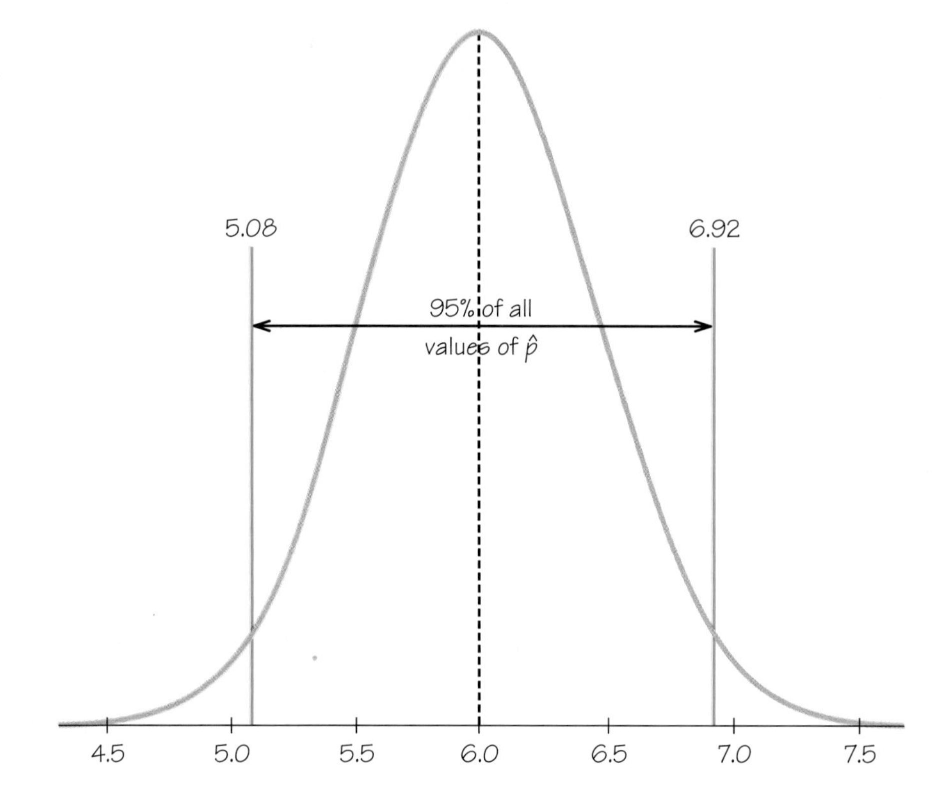

Figure 8.2 Repeat many times the process of selecting an SRS of size 2673 from a population of which $p = 6\%$ are successes. The middle 95% of the values of the sample proportion $\hat{p}$ will lie between 5.08% and 6.92%.

8.2 Confidence Intervals for a Population Proportion

We saw that if the true population parameter is $p = 6\%$ and we choose an SRS of size $n = 2673$, then the sample percent $\hat{p}$ will fall within $\pm 0.92\%$ of the truth in 95% of all samples. The margin or error 0.92% came from substituting $p = 6\%$ into the formula for the standard deviation of $\hat{p}$ and using the 95 part of the 68–95–99.7 rule.

We can repeat this reasoning for any value of the parameter p and the sample size n. It is always true that 95% of all samples catch p in the interval extending 2 standard deviations on either side of $\hat{p}$. That's the interval

$$\hat{p} \pm 2\sqrt{\frac{p(100 - p)}{n}}$$

Is this the 95% confidence interval we want? Not quite. We can't calculate this interval from the data because the standard deviation involves the population percent p, and in practice we don't know p. In the example, we applied the formula for $p = 6\%$, but this may not be the true p.

What to do? The standard deviation of the statistic $\hat{p}$ does depend on the parameter p, but it doesn't change a lot when p changes. Go back to the example and redo the calculation for other values of p. Here's the result:

Value of p	4	5	6	7	8
Standard deviation	0.38	0.42	0.46	0.49	0.52

You see that if we guess a value of p reasonably close to the true value, the standard deviation found from the guessed value will be about right. We know that when we take a large random sample, the statistic $\hat{p}$ is almost always close to the parameter p. So we will use $\hat{p}$ as the guessed value of the unknown p. Now we have an interval that we can calculate from the sample data.

Choose an SRS of size n from a large population that contains an unknown percent p of successes. A **95% confidence interval for p** is

$$\hat{p} \pm 2\sqrt{\frac{\hat{p}(100 - \hat{p})}{n}}$$

Here $\hat{p}$ is the proportion of successes in the sample. Both p and $\hat{p}$ are measured in percent. This recipe is only approximately correct, but is quite accurate when the sample size n is large.

This interval is only approximately a 95% confidence interval. It isn't exact for two reasons. The sampling distribution of the sample proportion $\hat{p}$ isn't exactly normal. And we don't get the standard deviation of $\hat{p}$ exactly right because we used $\hat{p}$ in place of the unknown p. Both of these difficulties go away as the sample size n gets larger. So our recipe is good only for large samples. What is more, the recipe assumes that the population is really big—at least 10 times the size of the sample. Professional statisticians use more elaborate methods that take the size of the population into account and work even for small samples. But our method works well enough for many practical uses. More important, it shows how we get a confidence interval from the sampling distribution of a statistic. That's the reasoning behind any confidence interval.

EXAMPLE

A Confidence Interval for Risky Behavior

The National AIDS Behavioral Surveys' sample of 2673 adult heterosexuals actually found that 170 people in the sample had more than one sexual partner in the past year. The sample proportion who admit to multiple partners is

$$\hat{p} = \frac{170}{2673} = 6.36\%$$

A 95% confidence interval for the proportion p of all adult heterosexuals with multiple partners is therefore

$$\hat{p} \pm 2\sqrt{\frac{\hat{p}(100 - \hat{p})}{n}} = 6.36 \pm 2\sqrt{\frac{(6.36)(93.64)}{2673}}$$
$$= 6.36 \pm 0.94$$
$$= 5.42\% \text{ to } 7.30\%$$

Interpret this result as follows: We got this interval by using a recipe that catches the true unknown population proportion in 95% of all samples. The shorthand is: We are **95% confident** that the true proportion of heterosexuals with multiple partners lies between 5.42% and 7.30%. ■

The length of a confidence interval depends on the size n of the sample; larger samples give shorter intervals. But the interval does *not* depend on the size of the population. This is true as long as the population is much larger than the sample. The confidence interval in the Example works for a sample of 2673 from a city with 100,000 adults as well as for a sample of 2673 from a nation of 210 million. What matters is how many people we interview, not what percent of the population we contact.

Any confidence interval has two essential pieces: the interval itself and the confidence level. The interval usually has the form

estimate ± margin of error

The estimate is a sample statistic such as $\hat{p}$ that estimates the unknown parameter. The margin of error indicates how accurate this estimate is. In the AIDS survey example, the estimate is 6.36% and the margin of error is ±0.94%.

The *confidence level* states how confident we are that our interval captures the true parameter. Although 95% confidence is common, you can hold out for higher confidence, such as 99%, or be satisfied with lower confidence, such as 90%. Our 95% confidence interval was based on the middle 95% of a normal distribution. A 99% confidence interval requires the middle 99% of the distribution and so is wider (has a larger margin of error). Similarly, a 90% confidence interval is shorter than a 95% interval obtained from the same data. There is a trade-off between how closely we can pin down the parameter (the margin of error) and how confident we can be in the result.

EXAMPLE

Understanding the News

Here's what the TV news announcer says: "A new Gallup poll finds that 57% of American adults bought a lottery ticket in the last 12 months. The margin of error for the poll was 3 percentage points." Plus or minus 3 percent starting at 57% is 54% to 60%. Most people think Gallup claims that the truth about the entire population lies in that range.

This is what Gallup actually said: "For results based on a sample of this size, one can say with 95% confidence that the error attributable to sampling and other random effects could be plus or minus 3 percentage points for adults." That is, Gallup tells us that the margin of error only works for 95% of all its samples. "95% confidence" is shorthand for that. The news report left out the "95% confidence." In fact, *almost all margins of error in the news are for 95% confidence.* If you don't see the confidence level in a report, it's usually safe to assume 95%. ■

8.3 Understanding Confidence Intervals

Confidence intervals use the central idea of probability: Ask what would happen if we repeated the sampling many times. The 95% in a 95% confidence interval is a probability, the probability that the method produces an interval that does capture the true parameter.

EXAMPLE How Confidence Intervals Behave

The National AIDS Behavioral Surveys' sample of 2673 heterosexuals found 170 with multiple partners, so the sample proportion was

$$\hat{p} = \frac{170}{2673} = 0.0636 = 6.36\%$$

and the 95% confidence interval was

$$\hat{p} \pm 2\sqrt{\frac{\hat{p}(100 - \hat{p})}{n}} = 6.36 \pm 0.94$$

Draw a second sample from the same population. It finds 148 of its 2673 respondents who have multiple partners. For this sample,

$$\hat{p} = \frac{148}{2673} = 0.0554 = 5.54\%$$

$$\hat{p} \pm 2\sqrt{\frac{\hat{p}(100 - \hat{p})}{n}} = 5.54 \pm 0.88$$

Draw another sample. Now the count is 152 and the sample proportion and confidence interval are

$$\hat{p} = \frac{152}{2673} = 0.0569 = 5.69\%$$

$$\hat{p} \pm 2\sqrt{\frac{\hat{p}(100 - \hat{p})}{n}} = 5.69 \pm 0.90$$

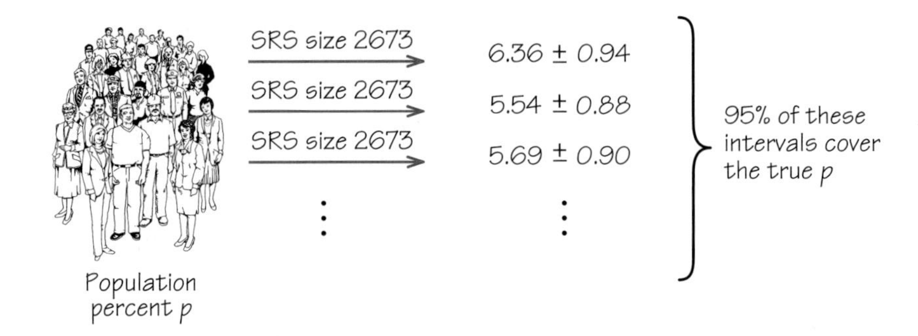

Figure 8.3 Repeated samples from the same population give different 95% confidence intervals, but 95% of these intervals capture the true population proportion p.

Keep sampling. Each sample yields a new estimate $\hat{p}$ and a new confidence interval. *If we sample forever, 95% of these intervals capture the true parameter.* This is true no matter what the true value is. Figure 8.3 summarizes the behavior of the confidence interval in graphical form. ■

On the ground that two pictures are better than one, Figure 8.4 gives a different view of how confidence intervals behave. Figure 8.3 reminds us that repeated samples give different results and that we are only guaranteed that 95% of the samples give a correct result. Figure 8.4 goes behind the scenes. The vertical line is the true value of the population proportion p. The normal curve at the top of the figure is the sampling distribution of the sample statistic $\hat{p}$, which is centered at the true p. We are behind the scenes because in real-world statistics we don't know p.

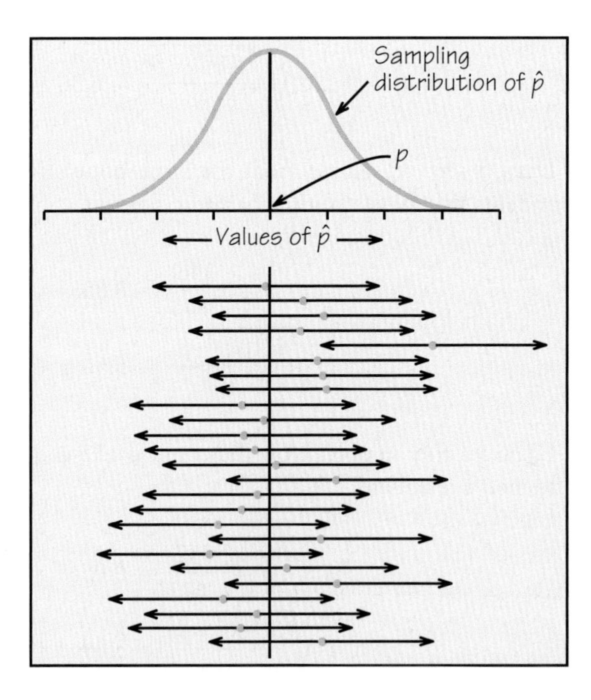

Figure 8.4 Twenty-five samples from the same population give these 95% confidence intervals. In the long run, 95% of all such intervals cover the true population proportion, marked by the vertical line.

The 95% confidence intervals from 25 SRSs appear below, one after the other. The central dots are the values of $\hat{p}$, the centers of the intervals. The arrows on either side span the confidence interval. In the long run, 95% of the intervals will cover the true p and 5% will miss. Of the 25 intervals in Figure 8.4, 24 hit and 1 misses. (Remember that probability describes only what happens in the long run—we don't expect exactly 95% of 25 intervals to capture the true parameter.)

8.4 The Limitations of Inference

Inference is the part of statistics that rests on mathematics. It is possible to prove mathematically that the sampling distribution of a sample proportion $\hat{p}$ is close to normal when the sample is large. Our confidence interval for the unknown population proportion follows from this. You should, however, remember the wise saying, "Mathematical theorems are true; statistical methods are sometimes effective when used with judgment." Any formula for inference is correct only in specific circumstances. If statistical procedures carried warning labels like those on drugs, most inference methods would have long labels indeed. Our handy formula $\hat{p} \pm 2\sqrt{\hat{p}\,(100 - \hat{p})/n}$ for estimating a population proportion comes with two important warnings for the user.

The data must be an SRS from the population. There are different recipes for data from more complex random sampling designs. But there is no correct method for inference from data haphazardly collected with bias of unknown size. You cannot base statistical inference on voluntary response samples or convenience samples. Whenever you do inference, you are acting as though you have a random sample. If you did not actually select subjects at random, your conclusions are open to criticism. In many cases, actual random selection isn't possible. Educational studies, for example, can't choose schoolchildren at random because they must work with children already grouped in schools and classrooms. Can we think of the children in one third-grade classroom as a random sample of all third-graders in the school? In the city? In the country? That's a matter of judgment.

The margin of error ignores nonresponse and other practical difficulties. The margin of error in our confidence interval comes from the sampling distribution of the statistic. The sampling distribution describes the variation of the statistic due to chance in repeated random samples. This random variation is the *only* source of error covered by the margin of error. Real-life samples also suffer from not starting with a list of the entire population (think of homeless people), nonresponse by some people chosen for the sample, and false responses by some who do respond. Errors from these practical difficulties are usually more serious than random sampling error. The actual error in sample surveys is often much larger than the announced margin of error. Worse, we can't say how much larger (see Spotlight 8.1).

SPOTLIGHT 8.1 How Accurate Is Our Poll?

Responsible polling organizations tell the public something about the accuracy of their poll results. The margin of error is basic, and so is a warning that there are other sources of error not covered by the announced margin of error. Here are the statements that accompanied poll results from three sources, found on their Web sites.

The Gallup Poll. "For results based on this sample, one can say with 95 percent confidence that the maximum error attributable to sampling and other random effects is plus or minus 3 percentage points. In addition to sampling error, question wording and practical difficulties in conducting surveys can introduce error or bias into the findings of public opinion polls."

The Harris Poll. "In theory, with a probability sample of this size, one can say with 95 percent certainty that the results have a statistical precision of plus or minus 3 percentage points of what they would be if the entire adult population had been polled with complete accuracy. Unfortunately, there are several other possible sources of error in all polls or surveys that are probably more serious than theoretical calculations of sampling error. They include refusals to be interviewed (non-response), question wording and question order, interviewer bias, weighting by demographic control data and screening (e.g., for likely voters). It is difficult or impossible to quantify the errors that may result from these factors."

The New York Times. "In theory, in 19 cases out of 20 the results based on such samples will differ by no more than three percentage points in either direction from what would have been obtained by seeking out all American adults. For smaller subgroups, the margin of sampling error is larger. In addition to sampling error, the practical difficulties of conducting any survey of public opinion may introduce other sources of error into the poll. Variation in the wording and order of questions, for example, may lead to somewhat different results."

Statisticians work hard to overcome the limitations of inference. Nonetheless, statistical conclusions are approximations to a complicated truth, not mathematical results that are simply true.

EXAMPLE Measuring Risky Behavior

What about the National AIDS Behavioral Surveys? The interviews were carried out by telephone. This is acceptable for surveys of the general population, because about 94% of American households have telephones. However, some groups at high risk for AIDS, such as intravenous drug users, often don't live in settled households and are underrepresented in the sample. About 30% of the people reached refused to cooperate. A nonresponse rate of 30% is not unusual in large sample surveys, but it may cause some bias if those who refuse differ systematically from those who cooperate. The survey used statistical methods that adjust for unequal response rates in different groups. Finally, some respondents may not have told the truth when asked about their sexual behavior. The survey team tried hard to make respondents feel comfortable. For example, Hispanic women were interviewed by Hispanic women, and Spanish speakers were interviewed by Spanish speakers with the same regional accent (Cuban, Mexican, or Puerto Rican). Nonetheless, the survey report says that some bias is probably present:

> It is more likely that the present figures are underestimates; some respondents may underreport their numbers of sexual partners and intravenous drug use because of embarrassment and fear of reprisal, or they may forget or not know details of their own or of their partner's HIV risk and their antibody testing history. [Joseph H. Catania et al., Prevalence of AIDS-related risk factors and condom use in the United States, *Science,* 258 (1992), p. 1104.]

Reading the report of a large study like the National AIDS Behavioral Surveys reminds us that statistics in practice involves much more than recipes for inference. ■

8.5 Estimating a Population Mean

The statistician's toolkit contains many different confidence intervals, matching the many different population parameters we may wish to estimate. We have met the confidence interval for estimating a population proportion p. Now we want to estimate a population mean. We have used the **sample mean $\bar{x}$** of a sample of observations to describe the center of a set of data. Now we will use the sample mean $\bar{x}$ to estimate the unknown mean μ of the entire population from which the sample is drawn. We use μ, the symbol for the mean of a probability distribution, for the population mean because it is the mean of the distribution of the results of drawing one individual at random from the population. The sample mean $\bar{x}$ is a statistic that will vary in repeated samples, while the population mean μ is a parameter, a fixed number. Fortunately, the new confidence interval for estimating μ is quite similar to the familiar confidence interval for estimating p, because both intervals are based on a normal sampling distribution.

EXAMPLE School Math in Texas

The National Assessment of Educational Progress (NAEP) regularly tests random samples of school children. The NAEP year 2000 Mathematics Assessment sample included 2171 fourth-graders in Texas. Their mean score was $\bar{x} = 233$ out of 500. This is better than the national average. On the basis of this sample, what can we say about the mean score μ in the population of all Texas fourth-graders? ■

The law of large numbers tells us that the sample mean $\bar{x}$ from a large random sample will be close to the unknown population mean μ. Because $\bar{x} = 233$, we guess that μ is somewhere around 233. To make "somewhere around 233" more precise, we ask: "How would the sample mean $\bar{x}$ vary if we took many samples of 2171 fourth-graders from this same population?" The answer is given by the *central limit theorem* (page 279).

Choose a simple random sample of size n from a population in which individuals have mean μ and standard deviation σ. Let $\bar{x}$ be the mean of the sample. Then:

- The sampling distribution of $\bar{x}$ is *approximately normal* when the sample size n is large.
- The *mean* of the sampling distribution is equal to μ.
- The *standard deviation* of the sampling distribution is $\sigma/\sqrt{n}$.

We can add a new fact to this familiar statement. *If the distribution of individuals in the population is normal, then the sampling distribution of $\bar{x}$ is exactly normal.* This is true for samples of any size. Figure 8.5 shows the relation between the distribution of a single observation drawn from a normally distributed population and the distribution of the mean of several (in this case 10) observations. The mean of several observations is less variable than individual observations.

To make use of the sampling distribution of $\bar{x}$, we must know the standard deviation σ for the population. In practice, we don't. Sometimes we have a good estimate of σ based on past experience with similar data. If we have a large sample, the standard deviation s of the observed data will be close to the standard deviation of the entire population from which the data were drawn. Just as we replaced the population proportion p by the sample proportion $\hat{p}$ in the margin of error for a proportion, we can replace the population standard deviation σ by the sample statistic s in estimating a population mean.

EXAMPLE

Estimating the Mean Texas Math Score

The standard deviation of the math scores of the NAEP sample of 2171 Texas fourth-graders was $s = 25$. Because the sample is large, the standard deviation σ of the scores all Texas fourth-graders would get if they all took the test is close to 25.

The normal sampling distribution of $\bar{x}$ has mean equal to the unknown population mean μ. The standard deviation of the sampling distribution is

$$\frac{\sigma}{\sqrt{n}} = \frac{\sigma}{\sqrt{2171}}$$

We don't know σ. We do know the standard deviation of the scores in the sample, $s = 25$. So the approximate standard deviation of $\bar{x}$ in many samples is

$$\frac{s}{\sqrt{n}} = \frac{25}{\sqrt{2171}} = 0.54$$

By the 95 part of the 68–95–99.7 rule, $\bar{x}$ will fall within 2 standard deviations of μ in 95% of all samples. Two standard deviations is 2×0.54, or about 1.1

Figure 8.5 The sampling distribution of the sample mean $\bar{x}$ from an SRS of 10 observations compared with the distribution of a single observation.

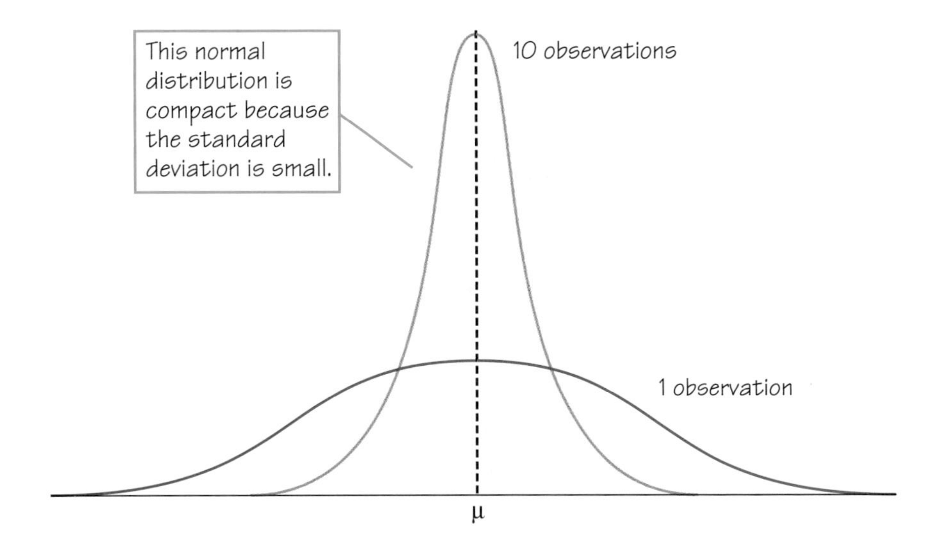
This normal distribution is compact because the standard deviation is small.
10 observations
1 observation
μ

points. We observed $\bar{x} = 233$ in our sample. So we are 95% confident that the population mean μ lies in the interval

$$233 \pm 1.1$$

or between 231.9 and 234.1. ■

Here is a recipe that summarizes our development. The confidence interval again has the form

$$\text{estimate} \pm \text{margin of error}$$

The estimate is now the sample mean $\bar{x}$.

Choose an SRS of size n from a large population of individuals having mean μ. A **95% confidence interval for μ** is

$$\bar{x} \pm 2\frac{s}{\sqrt{n}}$$

Here $\bar{x}$ and s are the mean and standard deviation of the n observations in the sample. This recipe is only approximately correct, but is quite accurate when the sample size n is large.

Just as for our earlier confidence interval, there are two reasons why this interval does not give exactly confidence 95%. First, the sampling distribution of $\bar{x}$ is not exactly normal unless the population is exactly described by a normal distribution. Second, the sample standard deviation s is not exactly equal to the unknown population parameter σ. The interval becomes more accurate for populations that look roughly normal and for large samples. Here is another example of estimating a population mean.

EXAMPLE

Estimating Dust in Coal Mines

Because the mean of several observations is less variable than a single observation, it is good practice to take the average of several observations when accuracy is important. The amount of dust in the atmosphere of coal mines is measured by exposing a filter in the mine and then weighing the dust collected by the filter. The weighing is not perfectly precise. Repeated weighings of the same filter will vary according to a normal distribution. The values that would be obtained in many weighings form the population we are interested in. The mean μ of this population is the true weight (that is, there is no bias in the weighing). The population standard deviation describes the precision of the weighing. It is known to be $\sigma = 0.08$ milligram (mg). Each filter is weighed three times, and the mean weight is reported.

For one filter the three weights are

123.1 mg   122.5 mg   123.7 mg

What is the 95% confidence interval for the true weight μ? First compute the sample mean

$$\bar{x} = \frac{123.1 + 122.5 + 123.7}{3}$$
$$= \frac{369.3}{3} = 123.1 \text{ mg}$$

The 95% confidence interval is

$$\bar{x} \pm 2\frac{\sigma}{\sqrt{n}} = 123.1 \pm 2\frac{0.08}{\sqrt{3}}$$
$$= 123.1 \pm (2)(0.046) = 123.1 \pm 0.09$$

We are 95% confident that the true weight is between 123.01 mg and 123.19 mg. In this example we knew σ, so we of course use the known value rather than a value estimated from the data. ■

Statistics does not produce proof—that's why confidence intervals have both margins of error and confidence levels. Statistics reflects reality: Conclusions are never quite certain, variation is always present, proof is always lacking. In such a world, statistical evidence is often the best evidence available.

For discussions of statistical process control and the perils of data analysis, go to **www.whfreeman.com/fapp.**

REVIEW VOCABULARY

Confidence interval An interval computed from a sample by a method that has a known probability of producing an interval containing the unknown parameter. This probability is called the *confidence level.* Confidence intervals usually have the form

$$\text{estimate} \pm \text{margin of error}$$

Parameter A number that describes the population. In statistical inference, the goal is often to estimate an unknown parameter or make a decision about its value.

Sample mean The mean (arithmetic average) $\bar{x}$ of the observations in a sample. The sample mean from a simple random sample is used to estimate the unknown mean μ of the population from which the sample was drawn.

Sample proportion The proportion $\hat{p}$ of the members of a sample having some characteristic (such as agreeing with an opinion poll question). The sample proportion from a simple random sample is used to estimate the corresponding proportion p in the population from which the sample was drawn. In this chapter, we express all proportions as percents.

Sampling distribution The distribution of values taken by a statistic when all possible random samples of the same size are drawn from the same population. The sampling distributions of sample proportions and sample means are approximately normal.

Statistic A number that describes a sample. A statistic can be calculated from the sample data alone; it does not involve any unknown parameters of the population.

SUGGESTED READINGS

MOORE, DAVID S. *The Basic Practice of Statistics,* 2nd ed., Freeman, New York, 1999. Chapter 6 of this text presents the reasoning of inference in detail. Chapters 7 and 8 discuss practical inference about means and proportions.

MOSES, LINCOLN E. The reasoning of statistical inference, in David C. Hoaglin and David S. Moore (eds.), *Perspectives on Contemporary Statistics,* Mathematical Association of America, Washington, D. C. , 1992, pp. 107–122. This broad essay on the nature of inference requires some knowledge of probability and should be read after an introductory text on statistics.

SUGGESTED WEB SITES

The Confidence Intervals applet on the Web site for this book **(www.whfreeman.com/fapp)** is an animated and interactive version of Figure 8.4. It is an excellent aid to understanding confidence intervals. The National Council on Public Polls **(www.ncpp.org)** has a statement on "20 Questions a Journalist Should Ask About a Poll" that makes interesting reading. The American Association for Public Opinion Research **(www.aapor.org/main1.html)** describes "Best Practices for Survey and Public Opinion Research" in its ethics section.

SKILLS CHECK

1. The Bureau of Labor Statistics announces that last month it interviewed all members of the labor force in a sample of 55,000 households; **4.9%** of the people interviewed were unemployed. The boldface number is a

(a) margin of error.
(b) parameter.
(c) statistic.

2. A poll of 1190 adults finds that 702 prefer balancing the federal budget over cutting taxes. The sample proportion $\hat{p}$ who prefer balancing the budget is

(a) 169.5%.
(b) 59%.
(c) 41%.

3. A simple random sample (SRS) can base statistical inference on

(a) voluntary response samples.
(b) convenience samples.
(c) small samples.

4. How do the margins of error computed using p and $\hat{p}$ compare?

(a) The margins of error are approximately the same.
(b) The margins of error are exactly the same.
(c) The margins of error are unrelated.

5. A random sample of 200 residents indicates that 45% are in favor of building a new school. Of the sample, how many are in favor?

(a) 45
(b) 90
(c) 200

6. A random sample of 200 residents indicates that 45% are in favor of building a new school. What is the margin of error for this survey?

(a) 3.5%
(b) 4.7%
(c) 7%

7. Suppose you suspect that approximately 5% of the county population own a goldfish. If 100 people are polled, what approximate margin of error would you expect?

(a) 2%
(b) 5%
(c) 10%

8. Suppose you suspect that approximately 15% of the county population own a dog. Approximately what size sample would be required to ensure a margin of error of approximately 5%?

(a) 50
(b) 250
(c) 2500

9. When a survey reports a 95% confidence interval of 7.5% ± 1.4%, what is the sample proportion?

(a) 8.9%
(b) 7.5%
(c) 1.4%

10. When a survey reports a 95% confidence interval of 7.5% ± 1.4%, what is the approximate standard deviation?

(a) 1.4%
(b) 2.8%
(c) 0.7%

11. Suppose a state exam has a mean score of 24 and a standard deviation of 4. If a random sample of 50 scores are selected from each school in the state and school averages are computed, what is the mean score of the school averages?

(a) Approximately 24
(b) Approximately 50
(c) More than 100

12. Suppose a state exam has a mean score of 24 and a standard deviation of 4. If a random sample of 50 scores are selected from each school in the state and school averages are computed, what is the standard deviation of the school averages?

(a) Approximately 4
(b) Approximately 0.6
(c) Approximately 0.1

EXERCISES

▲ Optional. ■ Advanced. ◆ Discussion.

Estimating with Confidence

Identify each of the boldface numbers in Exercises 1 to 3 as either a *parameter* or a *statistic.*

1. Voter registration records show that **68%** of all voters in Indianapolis are registered as Republicans. To test a random-digit dialing device, you use the device to call 150 randomly chosen residential telephones in Indianapolis. Of the registered voters contacted, **73%** are registered Republicans.

2. A researcher carries out a randomized comparative experiment with young rats to investigate the effects of a toxic compound in food. She feeds the control group a normal diet. The experimental group receives a diet with 2500 parts per million of the toxic material. After eight weeks, the mean weight gain is **335** grams for the control group and **289** grams for the experimental group.

3. A telemarketer in Los Angeles uses a random-digit dialing device to dial residential phone numbers in that city at random. Of the first 100 numbers dialed, **43** are unlisted numbers. This is not surprising, because **52%** of all Los Angeles residential phones are unlisted.

4. A college president says, "99% of the alumni support my firing of Coach Boggs." You contact an SRS of 200 of the college's 15,000 living alumni and find that 76 of them support firing the coach. What is the population here? Explain in words what the parameter p is. What is the value of the statistic $\hat{p}$?

5. Tonya wants to estimate what proportion of the students in her dormitory like the dorm food. She interviews an SRS of 50 of the 680 students living in the dormitory. She finds that 14 think the dorm food is good.

(a) Describe the population and explain in words what the parameter p is.
(b) Give the value (in percent) of the statistic $\hat{p}$ that estimates p.
(c) If in fact 25% of all students like the food, what are the mean and standard deviation of the sampling distribution of $\hat{p}$?

6. PTC is a substance that has a strong bitter taste for some people and is tasteless for others. The ability to taste PTC is inherited. About 75% of Italians can taste PTC, for example. You want to estimate the proportion of Americans with at least one Italian grandparent who can taste PTC. Suppose that the 75% estimate for Italians holds true in this population and that you test 500 people. Sketch the normal curve that shows how the proportion $\hat{p}$ in your sample that can taste PTC will vary if you take many samples.

7. Harley-Davidson motorcycles make up 14% of all the motorcycles registered in the United States. You plan to interview an SRS of 500 motorcycle owners.

(a) What is the approximate distribution of the proportion of your sample who own Harleys?
(b) If you took many samples, in what range would the central 95% of the proportion of Harley owners fall?

8. The standard deviation $\sqrt{p(100 - p)/n}$ of a sample proportion $\hat{p}$ varies with the true value of the population proportion p. Fortunately, it does not vary greatly unless p is near 0% or 100%. Suppose that the size of the sample is $n = 1500$. Evaluate this standard deviation for $p = 30\%, 40\%, 50\%, 60\%$, and 70%. Then evaluate it for $p = 0\%, 10\%$, and 20%. In which range does the standard deviation of $\hat{p}$ change most rapidly as p changes? Make a graph of this standard deviation against p.

Confidence Intervals for a Population Proportion

National sample surveys do not use SRSs. Their sample designs are, however, such that in the exercises below you can assume the sample is an SRS without serious error.

9. A news article reports that in a recent opinion poll, 78% of a sample of 1108 adults said they believe there is a heaven. Only 60% said they believe there is a hell.

(a) How many of the 1108 people in the sample believe there is a hell?
(b) Give a 95% confidence interval for the proportion of all adults who believe there is a hell.

10. In a recent year, 73% of first-year college students responding to a national survey identified "being very well-off financially" as an important personal goal. A state university finds that 132 of an SRS of 200 of its first-year students say that this goal is important. Give a 95% confidence interval for the percent of all first-year students at the university who would identify being well-off as an important personal goal.

11. The National Collegiate Athletic Association (NCAA) requires colleges to report the academic progress of their athletes. At one Big Ten university, 95 of the 147 athletes admitted one year graduated within six years. We are willing to consider one year's athletes as a random sample of all athletes who would be admitted under current policies. Give a 95% confidence interval for the graduation rate of athletes at this university.

◆ **12.** In January 2000, the Gallup poll asked a random sample of 1633 adults, "In general, are you satisfied or dissatisfied with the way things are going in the United States at this time?" It found that 1127 said that they were satisfied. Write a short report of this finding, as if you were writing for a newspaper. Be sure to include a margin of error.

◆ **13.** Should the government spend money to help low-income families who want to send their children to private or religious schools? A poll of 1006 adults found that 362 said "yes." Write a short summary of the poll results, as if you were writing the script for a TV newscast. Be sure to include a margin of error.

◆ **14.** If your memory is sharp, you will recall that the two previous exercises repeat exercises from Chapter 5. In Chapter 5, you used a "quick method" (page 183) for the margin of error. Now you know a more detailed recipe. The margin of error from the quick method is a bit larger than needed. It differs most from the more accurate method of this chapter when $\hat{p}$ is close to 0% or 100%. Let's illustrate this by comparing two results for two samples.

(a) How do the two margins of error for the data in Exercise 13 compare?
(b) An SRS of 500 motorcycle registrations finds that 68 of the motorcycles are Harley-Davidsons. Give a 95% confidence interval for the proportion of all motorcycles that are Harleys by the quick method and then by the method of this chapter. How much larger is the quick method margin of error?

■ **15.** The "quick method" margin of error for 95% confidence (page 183) is $100/\sqrt{n}$. The full recipe is $2\sqrt{\hat{p}(100 - \hat{p})/n}$. We can derive the quick method from the full recipe.

(a) Graph the function $\hat{p}(100 - \hat{p})$ for values of $\hat{p}$ between 0% and 100%. For what value of $\hat{p}$ is this function largest? What is the largest value that $\hat{p}(100 - \hat{p})$ takes?
(b) Show that if we replace $\hat{p}(100 - \hat{p})$ in the full recipe by the largest value it can take, we get the quick method. You now see why the quick method margin of error is larger than is needed.

Understanding Confidence Intervals

◆ **16.** A Gallup poll in December 2000 asked, "Do you think this country would be governed better or governed worse if more women were in political office?" Of the 1026 adults in the sample, 57% said "better." Gallup added, "For results based on the total sample of National Adults, one can say with 95% confidence that the margin of error is ±3 percentage points." Explain to someone who knows no statistics what the phrase "95% confidence" means here.

◆ **17.** Just before a presidential election, a national opinion poll increases the size of its weekly sample from the usual 1500 people to 4000 people. Does the larger random sample reduce the bias of the poll result? Does it reduce the variability of the result?

◆ **18.** The Gallup poll asked a random sample of 1005 adults whether they favored bilingual education or immersion training in English for non-English-speaking students in public schools. Sixty-three percent of the sample favored immersion. The press release stated that this poll has a 3% margin of error. Explain carefully to someone who knows no statistics what is meant by a "3% margin of error."

◆ **19.** A student reads that a 95% confidence interval for the mean NAEP Mathematics Assessment score for Texas fourth-graders is 231.9 to 234.1. Asked to explain the meaning of this interval, the student says, "95% of all Texas fourth-graders have scores between 231.9 and 234.1." Is the student right? Justify your answer.

◆ **20.** Here is an explanation from the Associated Press concerning one of its opinion polls. Explain briefly but clearly in what way this explanation is incorrect.

> For a poll of 1,600 adults, the variation due to sampling error is no more than three percentage points either way. The error margin is said to be valid at the 95 percent confidence level. This means that, if the same questions were repeated in 20 polls, the results of at least 19 surveys would be within three percentage points of the results of this survey.

◆ **21.** The *New York Times* and CBS News conducted a nationwide poll of 1048 randomly selected 13- to 17-year-olds. Of these teenagers, 692 had a television in their room and 189 named Fox as their favorite television network. We will act as if the sample were a simple random sample.

(a) Give 95% confidence intervals for the proportion of all people in this age group who have a TV in their room and the proportion who would choose Fox as their favorite network.
(b) The news article says, "In theory, in 19 cases out of 20, the poll results will differ by no more than three percentage points in either direction from what would have been obtained by seeking out all American teenagers." Explain how your results agree with this statement.

Exercises 22 to 25 are based on the following situation. A Gallup poll conducted by telephone on March 5 to 7, 2001, asked 1060 randomly selected adults, "How would you rate the overall quality of the environment in this country today—as excellent, good, only fair, or poor?" In all, 46% of the sample rated the environment as good or excellent. Gallup said that "one can say with 95% confidence that the margin of sampling error is ± 3 percentage points."

◆ **22.** Would a 90% confidence interval based on the poll results have a margin of error less than ± 3 percentage points, equal to ± 3 percentage points, or greater than ± 3 percentage points? Explain your answer.

◆ **23.** If the poll had interviewed 1500 persons rather than 1060 (and still found 46% rating the environment as good or excellent), would the margin of error for 95% confidence be less than ± 3 percentage points, equal to ± 3 percentage points, or greater than ± 3 percentage points? Explain your answer.

◆ **24.** Suppose that the poll had obtained the outcome 46% by a similar random sampling method from all adults in New York State (population 19 million) instead of from all adults in the United States (population 281 million). Would the margin of error for 95% confidence be less than ± 3 percentage points, equal to ± 3 percentage points, or greater than ± 3 percentage points? Explain your answer.

The Limitations of Inference

◆ **25.** Which of the following sources of error are included in the Gallup environment poll's margin of error?

(a) The poll dialed telephone numbers at random and so missed all people without phones.
(b) Nonresponse—some people whose numbers were chosen never answered the phone in several calls or answered but refused to participate in the poll.
(c) There is chance variation in the random selection of telephone numbers.

◆ **26.** A television news program conducts a call-in poll about a proposed city ban on handgun ownership. Of the 2372 calls, 1921 oppose the ban. The station, following recommended practice, makes a confidence statement: "81% of the Channel 13 Pulse Poll sample opposed the ban. We can be 95% confident that the true proportion of citizens opposing a handgun ban is within 1.6% of the sample result." Show that the calculation of the confidence interval is correct. Nonetheless, the conclusion is not justified. Why?

◆ **27.** Joe is writing a report on the backgrounds of American presidents. He looks up the ages of all 43 presidents when they entered office. Because Joe took a statistics course, he uses these 43 numbers to get a 95% confidence interval for the mean age of all men who have been president. This makes no sense. Why not?

◆ **28.** A recent Gallup poll found that 68% of adult Americans favor teaching creationism along with evolution in public schools. The Gallup press release says:

> For results based on samples of this size, one can say with 95 percent confidence that the maximum error attributable to sampling and other random effects is plus or minus 3 percentage points.

Give one example of a source of error in the poll result that is *not* included in this margin of error.

◆ **29.** A survey of users of the Internet found that males outnumbered females by nearly 2 to 1. This was a surprise, because earlier surveys had put the ratio of men to women closer to 9 to 1. Later in the article we find this information:

> Detailed surveys were sent to more than 13,000 organizations on the Internet; 1,468 usable responses were received. According to Mr. Quarterman, the margin of error is 2.8 percent, with a confidence level of 95 percent.

(a) What was the response rate for this survey? (The response rate is the percent of the planned sample that responded.)
(b) Do you think that the small margin of error is a good measure of the accuracy of the survey's results? Explain your answer.

◆ **30.** Many subjects don't give honest answers to questions about activities that are illegal or sensitive in some other way. One study divided a large group of white adults into thirds at random. All were asked whether they had ever used cocaine. The first group was interviewed by telephone: 21% said "yes." In the group visited at home by an interviewer, 25% said "yes." The final group was interviewed at home but answered the question on an anonymous form that they sealed in an envelope. Of this group, 28% said they had used cocaine.

(a) Which result do you think is closest to the truth? Why?
(b) Give two other examples of behavior you think would be underreported in a telephone survey.

Estimating a Population Mean

31. Figure 8.1 shows the idea of the sampling distribution of a sample proportion $\hat{p}$ in picture form. Draw a similar picture that shows the idea of the sampling distribution of a sample mean $\bar{x}$.

32. The scores of more than 1 million students on the ACT college entrance examination in 2001 had approximately the normal distribution with mean $\mu = 26$ and standard deviation $\sigma = 4.7$.

(a) What range of scores contains the middle 95% of all scores?
(b) If the ACT scores of 25 randomly selected students are averaged, what range contains the middle 95% of the averages $\bar{x}$?

33. A study of the career paths of hotel general managers sent questionnaires to an SRS of 160 hotels belonging to major U.S. hotel chains. There were 114 responses. The average time these 114 general managers had spent with their current company was $\bar{x} = 11.78$ years. We do not know the population standard deviation σ, but the sample standard deviation was $s = 3.2$ years. Because the sample is large, s will be close to σ. Give a 95% confidence interval for the mean number of years general managers of major-chain hotels have spent with their current company.

34. A laboratory scale is known to have a standard deviation of $\sigma = 0.001$ gram in repeated weighings. Suppose that scale readings in repeated weighings are normally distributed, with mean equal to the true weight of the specimen. Three weighings of a specimen give (in grams)

3.412 3.414 3.415

Give a 95% confidence interval for the true weight of the specimen. What are the estimate and the margin of error in this interval?

◆ **35.** Here are the IQ test scores of 31 seventh-grade girls in a midwest school district:

114	100	104	89	102	91	114	114
103	105	108	130	120	132	111	128
118	119	86	72	111	103	74	112
107	103	98	96	112	112	93	

(a) We expect the distribution of IQ scores to be close to normal. Make a stemplot of the distribution of these 31 scores. Does your plot show outliers, clear skewness, or other nonnormal features? Using a calculator, find the mean and standard deviation of these scores.
(b) Treat the 31 girls as a simple random sample of all seventh-grade girls in the school district. Give a 95% confidence interval for the mean score in the population.
(c) In fact, the scores are those of all seventh-grade girls in one of the several schools in the district. Explain carefully why we cannot trust the confidence interval from part (b).

◆ **36.** Find the margin of error for 95% confidence in Exercise 34 if we weigh each specimen 12 times rather than 3 times. Check that your result is half as large as the margin of error you found in Exercise 34. Explain why you knew without calculating that the new margin of error would be half as large.

◆ **37.** The NAEP mathematics test was also given to a sample of 1077 women of ages 21 to 25 years. The mean of these 1077 scores was 275 and the standard deviation was 60.

(a) Give a 95% confidence interval for the mean score μ in the population of all young women.
(b) Suppose that a sample of 250 women had produced the same mean and standard deviation. Give the 95% confidence interval for the population mean μ in this case.
(c) Suppose that a sample of 4000 women had produced the sample mean and standard deviation. Again, give the 95% confidence interval for μ.
(d) What are the margins of error for samples of size 250, 1077, and 4000? How does increasing the sample size affect the margin of error of a confidence interval?

38. Table 6.8 (page 251) gives data on 38 consecutive patients who came to a medical center for treatment of Hallux abducto valgus (HAV), a deformation of the big toe. It is reasonable to consider these patients as an SRS of people suffering from HAV. The seriousness of the deformity is measured by the angle (in degrees) of deformity. Give a 95% confidence interval for the mean angle of deformity in the population.

◆ **39.** A radio talk show invites listeners to enter a dispute about a proposed pay increase for city council members. "What yearly pay do you think council members should get? Call us with your number." In all, 958 people call. The mean pay they suggest is $\bar{x} = \$8740$ per year, and the standard deviation of the responses is $s = \$1125$. For a large sample such as this, s is very close to the unknown population σ. The station calculates the 95% confidence interval for the mean pay μ that all citizens would propose for council members to be \$8667 to \$8813.

(a) Show that the station's calculation is correct.
(b) Nonetheless, its conclusion does not describe the population of all the city's citizens. Explain why.

40. The HAV data in Exercise 38 follow a normal distribution quite closely except for one patient with HAV angle 50 degrees, a high outlier.

(a) Find the 95% confidence interval for the population mean based on the 37 patients who remain after you drop the outlier.

(b) Compare your interval in part (a) with your interval from Exercise 38. What is the most important effect of removing the outlier?

Additional Exercises

◆ **41.** Explain in your own words the advantages of bigger random samples in a sample survey.

42. The eighteenth-century French naturalist Count Buffon tossed a coin 4040 times. He got 2048 heads. Give a 95% confidence interval for the probability that Buffon's coin lands heads up. Are you confident that this probability is not 1/2? Why?

43. We know that the proportion of 0's among a large set of random digits is $p = 10\%$ because all 10 possible digits are equally probable. The entries in a table of random digits are a sample from the population of all random digits. To get an SRS of 200 random digits, look at the first digit in each of the 200 five-digit groups in lines 101 to 125 of Table 5.1 (page 174). How many of these 200 digits are 0's? Give a 95% confidence interval for the proportion of 0's in the population from which these digits are a sample. Does your interval cover the true parameter value, $p = 0.1$?

◆ **44.** A *New York Times* poll on women's issues interviewed 1025 women randomly selected from the United States, excluding Alaska and Hawaii. The poll found that 47% of the women said they do not get enough time for themselves.

(a) The poll announced a margin of error of ±3 percentage points for 95% confidence in its conclusions. What is the 95% confidence interval for the percent of all adult women who think they do not get enough time for themselves?
(b) Explain to someone who knows no statistics why we can't just say that 47% of all adult women do not get enough time for themselves.
(c) Then explain clearly what "95% confidence" means.

■ **45.** When the statistic that estimates an unknown parameter has a normal distribution, a 95% confidence interval for the parameter has the form

$$\text{estimate} \pm 2\sigma_{\text{estimate}}$$

In a complex sample survey design, estimates and their standard deviations require elaborate computations. But when we are given the estimate and its standard deviation, we can calculate a confidence interval for μ without knowing the formulas that led to the numbers given.

A report based on the Current Population Survey estimates the unemployment rate in the United States in August 2001 as 4.9%. The Bureau of Labor Statistics says that the standard deviation of this estimate is 0.09%. The Current Population Survey uses an elaborate multistage sampling design to select a sample of about 55,000 households. The sampling distribution of the estimated unemployment rate is approximately normal. Give a 95% confidence interval for the unemployment rate in the entire labor force.

46. In December 2000, following a presidential election that was decided by a disputed count of votes in Florida, the Gallup poll asked random samples of 297 black adults and 867 white adults, "Do you think black voters were—or were not—less likely to have their votes counted fairly in Florida than whites?" Here are the counts of responses:

	Were less likely	Not less likely	No opinion
Blacks	202	71	24
Whites	260	503	104

(a) Give a 95% confidence interval for the percent of all black adults who thought black votes were less likely to be counted fairly.
(b) Give a 95% confidence interval for the percent of all white adults who thought black votes were less likely to be counted fairly.

◆ **47.** Sulfur compounds cause "off-odors" in wine, so winemakers want to know the odor threshold, the lowest concentration of a compound that the human nose can detect. The odor threshold for dimethyl sulfide (DMS) in trained wine tasters is about 25 micrograms per liter of wine (μg/L). The untrained noses of consumers may be less sensitive, however. Here are the DMS odor thresholds for 10 untrained students:

31 31 43 36 23 34 32 30 20 24

Assume that the standard deviation of the odor threshold for untrained noses is known to be $\sigma = 7$μg/L.

(a) Make a stemplot to verify that the distribution is roughly symmetric with no outliers. (More data confirm that there are no systematic departures from normality.)
(b) Give a 95% confidence interval for the mean DMS odor threshold among all students.
(c) Are you confident that the mean odor threshold for students is higher than the published threshold, 25 μg/L? Why?

◆ **48.** U.S. Treasury bills are safe investments, but how much do they pay investors? Below are data on the total return (in percent) on Treasury bills for the years 1970 to 2000.

(a) Make a histogram of these data, using bars 2 percentage points wide. What kind of deviation from normality do you see? Thanks to the central limit theorem, we can nonetheless treat $\bar{x}$ as approximately normal.
(b) Suppose that we can regard these 31 years' results as a random sample of returns on Treasury bills. Give a 95% confidence interval for the long-term mean return.
(c) The rate of inflation during these years averaged about 5.1%. Are you convinced that Treasury bills have a mean return higher than 5.1%? Why?

◆ **49.** When the Current Population Survey (CPS) asked the adults in its sample of 55,000 households if they voted in the 1996 presidential election, 54% said they had. In fact, only 49% of the adult population voted in that election. Why do you think the CPS result missed by much more than the margin of error?

■ **50.** We used the sampling distribution of $\hat{p}$ and the 68–95–99.7 rule to give a 95% confidence interval for a population proportion p.

(a) Explain carefully why

$$\hat{p} \pm \sqrt{\frac{\hat{p}(100 - \hat{p})}{n}}$$

is a 68% confidence interval for p.
(b) Give the recipe for a 99.7% confidence interval for p.

■ **51.** Use the result of Exercise 50, part (a), and the data in Exercise 12 to give a 68% confidence interval for the percent of adults who were satisfied with

Year	Rate	Year	Rate	Year	Rate	Year	Rate
1970	6.52	1978	7.19	1986	6.16	1994	3.91
1971	4.39	1979	10.38	1987	5.47	1995	5.60
1972	3.84	1980	11.26	1988	6.36	1996	5.20
1973	6.93	1981	14.72	1989	8.38	1997	5.25
1974	8.01	1982	10.53	1990	7.84	1998	4.85
1975	5.80	1983	8.80	1991	5.60	1999	4.69
1976	5.08	1984	9.84	1992	3.50	2000	5.69
1977	5.13	1985	7.72	1993	2.90		

how things were going in January 2000. Compare the width of the 68% interval with that of the 95% interval from Exercise 12 and explain in plain language the reason for the difference.

■ **52.** Use the result of Exercise 50, part (b), and the data in Exercise 13 to give a 99.7% confidence interval for the percent of all adults who favor government subsidies for private school tuition. Compare the width of the 99.7% interval with that of the 95% confidence interval from Exercise 13. What is the reason for the difference in widths?

■ **53.** The upper and lower deciles of any normal distribution are located 1.28 standard deviations above and below the mean. (The lower decile is the point with probability 10% below it; the upper decile has probability 90% below it.)

(a) Use this information to give a recipe for an 80% confidence interval for a population proportion p based on the sample proportion $\hat{p}$ that is accurate for large sample sizes n.
(b) Give an 80% confidence interval for the proportion of adults who support government subsidies for private school tuition, using the poll result in Exercise 13.

■ **54.** The upper and lower deciles of any normal distribution are located 1.28 standard deviations above and below the mean.

(a) Use this information to give a recipe for an 80% confidence interval for the mean μ of a normal population based on the sample mean $\bar{x}$ of a simple random sample of size n.
(b) Give an 80% confidence interval for the mean IQ score in Exercise 35.

WRITING PROJECTS

1. Traditional opinion polls choose a random sample of adults, beginning with telephone numbers dialed at random. Harris Interactive thinks it has a better idea: Sign up lots of Web users, then email surveys to a random sample of this list. Each member receives one or two survey questionnaires per month. The Harris Web site claims that "unlike old-fashioned surveys, the Harris Poll Online produces accurate, reliable information at Internet speed."

Visit the site vr.harrispollonline.com. Write a brief description of how the poll works. Then compare the reliability of data from the Harris Poll Online with data from traditional polls. What new sources of bias does the online poll introduce? How do you think the size of the list of members affects the reliability of the data? A statement on Web-based surveys issued by the American Association for Public Opinion Research may still be available at www.aapor.org/press/.

2. The margin of error announced for a sample survey takes into account the chance variation due to random sampling. In practice, survey results can be in error for other reasons. Some subjects can't be contacted, others lie or don't remember information, and the wording of the questions will influence the responses.

Write a brief discussion of the most important practical difficulties encountered in opinion polls and other surveys of human populations.
You will want to read more on the subject. Some sources are Chapter 4 of *Statistics: Concepts and Controversies* (see the Suggested Readings in Chapter 5) and the article by P. E. Converse and M. W. Traugott, Assessing the accuracy of polls and surveys, *Science*, 234, 1986: 1094–1098.

3. If you toss a thumbtack on a hard surface, what is the probability that it will land point up? Estimate this probability p by tossing a thumbtack 500 times. Your 500 tosses are an SRS of size 500 from the population of all tosses. The proportion of these 500 tosses that land point up is the sample proportion $\hat{p}$. Use the result of your tosses to give a 95% confidence interval for p. Write an explanation of your findings for someone who knows no statistics but wonders how often a thumbtack will land point up.

SPREADSHEET PROJECTS

To do these projects, go to www.whfreeman.com/fapp.

We have relied on the spreadsheet's random number generator throughout this unit. This spreadsheet project will help you judge if this generator is fair or biased. Sample standard deviations and confidence intervals using multiple measurements were computed in this chapter. As shown in this spreadsheet project, spreadsheets also support these computations.

APPLET EXERCISES

To do these exercises, go to www.whfreeman.com/fapp.

Confidence Intervals

Confidence tells us how often our method will produce an interval that captures the true population parameter if we use the method a very large number of times. The Confidence Interval applet allows us to actually use the method many times.

(a) Set the confidence level in the applet to 90%. Click the "Sample 50" to choose 50 SRSs and calculate the confidence intervals. How many captured the true population mean μ? Keep clicking "Sample 50" until you have 1000 samples. What percent of the 1000 confidence intervals hit?
(b) Now choose 95% confidence. Look carefully when you first click "Sample 50." Are these intervals longer or shorter than the 90% confidence intervals? Again, take 1000 samples. What percent of the intervals captured the true μ?
(c) Do the same for 99% confidence. What percent of 1000 samples gave confidence intervals that caught the true mean? Did the behavior of many intervals for the three confidence levels closely reflect the choice of confidence level?

Identification Numbers

In this chapter we cover...

... and at www.whfreeman.com/fapp:

Flashcards
Quizzes
Spreadsheet Projects
Check digits

Modern identification numbers serve at least two functions. Obviously, an identification number should unambiguously identify the person or thing with which it is associated. Not obvious is a "self-checking" aspect of the number.

9.1 Check Digits

Look at the ISBN printed on the back of this book. The number 0-7167-4782-0 (0-7167-4783-9 for paperback version) distinguishes this book from all others. The last digit 0 is there solely to detect errors that may occur when the ISBN is entered into a computer. Look at the bottom of the airline ticket shown in Figure 9.1. Notice the letters CK (for "check") above the last digit of the stock control number and above the last digit of the document number. These digits are also there for the purpose of error detection. Grocery items, credit cards, overnight mail, magazines, personal checks, traveler's checks, soft-drink cans, automobiles, and many other items you encounter daily have identification numbers that code data and include an extra digit called a **check digit** for error detection. In this chapter, we examine some of the methods that are used to assign identification numbers and check digits.

Let us begin by considering the U.S. Postal Service money order shown in Figure 9.2. The first 10 digits of the 11-digit number 63024383845 simply identify the money order. The last digit, 5, serves as an **error-detecting** mechanism. Let us see how this mechanism works. The eleventh (last) digit of a Postal Service money order number is the remainder obtained when the sum of the first 10 digits of the number is divided by 9. In our example, the last digit is 5 because $6 + 3 + 0 + 2 + 4 + 3 + 8 + 3 + 8 + 4 = 41$ and the remainder when 41 is divided by 9 is 5.

Now suppose instead of the correct number, the number 63054383845 (an error in the fourth position) were entered into a computer programmed for error detection of money orders. The machine would divide the sum of the first 10 digits of the entered number, 44, by 9 and obtain a remainder of 8. Because the last digit of the entered number is 5 rather than 8, the entered number cannot be correct. This crude method of error detection will not detect the mistake of replacing

Figure 9.1 Airline ticket with identification number 127881879532 and check digit 1; stock control number 3026164775 and check digit 4.

Figure 9.2 Money order with identification number 6302438384 and appended check digit 5. The check digit is the remainder upon dividing the sum of the digits by 9.

a 0 with a 9, or vice versa. Nor will it detect the transposition of digits, such as 63204383845 instead of 63024383845 (the digits in positions three and four have been transposed).

American Express traveler's checks, VISA traveler's checks, and Euro banknotes (see Figure 9.3) also utilize a check digit determined by division by 9. In these cases, the check digit is chosen so that the sum of the digits, including the check digit, is evenly divisible by 9.

EXAMPLE

The American Express Travelers Cheque

The American Express Travelers Cheque with the identification number 387505055 has check digit 7 because $3 + 8 + 7 + 5 + 0 + 5 + 0 + 5 + 5 = 38$ and $38 + 7$ is evenly divisible by 9. ■

The scheme used on airline tickets, Federal Express and UPS packages, and Avis and National rental cars assigns the remainder upon division by 7 of the number itself as the check digit (see Figure 9.1) rather than dividing the sum of

Figure 9.3 A Euro with identification number X1183806538 and check digit 4. To calculate the check digit, the letter (which indicates in which country the note was issued) is converted to a number. The check digit is then chosen so that the sum of all the digits, including the check digit, is evenly divisible by 9. For the note shown here, the letter X is converted to 7.

the digits by 7. For example, the check digit for the number 540047 is 4 because $540047 = 7 \times 7149 + 4$. This method will not detect the substitution of 0 for a 7, 1 for an 8, 2 for a 9, or vice versa. However, unlike the Postal Service method, it will detect transpositions of adjacent digits with the exceptions of the pairs 0, 7; 1, 8; or 2, 9. For example, if 5400474 were entered into a computer as 4500474 (the first two digits are transposed), the machine would determine that the check digit should be 3 since $450047 = 7 \times 64292 + 3$. Because the last digit of the entered number is not 3, the error has been detected.

The scheme used on grocery products, the so-called **Universal Product Code (UPC),** is more sophisticated. Consider the number 0 38000 00127 7 found on the bottom of a box of corn flakes. The first digit identifies a broad category of goods, the next five digits identify the manufacturer, the next five the product, and the last is a check digit. Suppose this number were entered into a computer as 0 58000 00127 7 (a mistake in the second position). How would the computer recognize the mistake?

For any UPC number $a_1a_2a_3a_4a_5a_6a_7a_8a_9a_{10}a_{11}a_{12}$, the computer is programmed to carry out the following computation: $3a_1 + a_2 + 3a_3 + a_4 + 3a_5 + a_6 + 3a_7 + a_8 + 3a_9 + a_{10} + 3a_{11} + a_{12}$. If the result doesn't end with a 0, the computer knows the entered number is incorrect.

For the incorrect corn flakes number, we have $3 \cdot 0 + 5 + 3 \cdot 8 + 0 + 3 \cdot 0 + 0 + 3 \cdot 0 + 0 + 3 \cdot 1 + 2 + 3 \cdot 7 + 7 = 62$. Because 62 doesn't end with 0, the error is detected. Notice that had we used the correct digit 3 in the second position instead of 5, the sum would have ended in a 0 as it should. This simple scheme detects *all* single-position errors and about 89% of all other kinds of errors.

The U.S. banking system uses a variation of the UPC scheme that appends check digits to the numbers assigned to banks. Each bank has an eight-digit identification number $a_1a_2 \ldots a_8$ together with a check digit a_9 so that a_9 is the last digit of $7a_1 + 3a_2 + 9a_3 + 7a_4 + 3a_5 + 9a_6 + 7a_7 + 3a_8$. The numbers 7, 3, and 9 used in this formula are called the **weights.** (The weights for the UPC scheme are 3 and 1.) The weights were carefully chosen so that all single-digit errors and most transposition errors are detected. (The use of different weights in adjacent positions permits the detection of most transposition errors.)

EXAMPLE

Bank Identification Number

The First Chicago Bank has the number 071000013 on the bottom of all its checks (see Spotlight 9.1). The check digit 3 is the last digit of $7 \cdot 0 + 3 \cdot 7 + 9 \cdot 1 + 7 \cdot 0 + 3 \cdot 0 + 9 \cdot 0 + 7 \cdot 0 + 3 \cdot 1 = 33$. ■

One of the most efficient error-detection methods is one used by all major credit card companies as well as by many libraries, blood banks, photofinishing companies, German banks, and the South Dakota driver's license department. It is called **Codabar.** Say a bank intends to issue a credit card with the identification number 312560019643001. It must then add an extra digit for error detection. This is done as follows. Add the digits in positions 1, 3, 5, 7, 9, 11, 13, and 15 and double the result: $(3 + 2 + 6 + 0 + 9 + 4 + 0 + 1) \times 2 = 50$. Next, count the number of digits in positions 1, 3, 5, 7, 9, 11, 13, and 15 that exceed 4 and add this to the total. For our example, only 6 and 9 exceed 4, so the count is 2 and our running total is 52. Now add in the remaining digits: $52 + (1 + 5 + 0 + 1 + 6 + 3 + 0) = 68$.

The check digit is whatever is needed to bring the final tally to a number that ends with 0. Because $68 + 2 = 70$, the check digit for our example is 2. This digit is appended to the end of the number the bank issues for identification purposes. Errors in input data are detected by applying the same algorithm to the input, including the check digit. If the correct number is entered into a computer, the

SPOTLIGHT 9.1 Bank Checks

What do the string of numbers at the bottom of a check represent? Here is the answer.

0710	the bank's Federal Reserve District, office, and state or special collection arrangement
0001	the bank's identification number
3	the check digit
22 633 78	the checking account number
0134	the check number

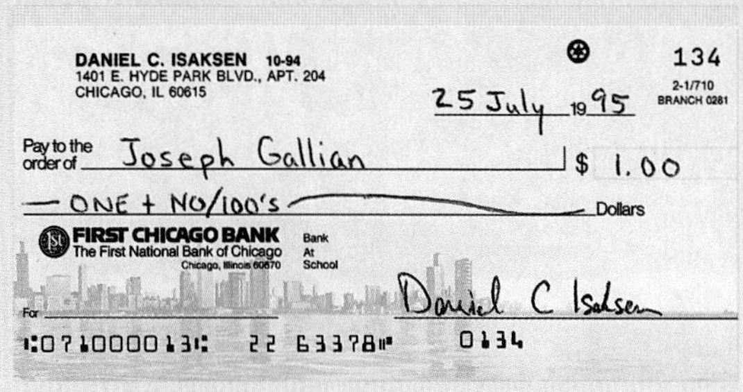

result will end in 0. If the result doesn't end with 0, a mistake has been made. The credit card shown in Figure 9.4 is reproduced from an ad promoting the Citibank VISA card. Notice that the check digit on the card is not valid because the Codabar algorithm yields

$$(4 + 2 + 0 + 1 + 3 + 5 + 7 + 9) \times 2 + 3 + (1 + 8 + 0 + 2 + 4 + 6 + 8) + 0 = 94$$

which does not end in 0. This method allows computers to detect 100% of single-position errors and about 98% of other common errors.

Besides detecting errors, the check digit offers partial protection against fraudulent numbers. A person who wanted to create a phony credit card, bank account number, or driver's license number would have to know the appropriate check-digit scheme for the number to go unchallenged by the computer.

Thus far we have not discussed any schemes that detect 100% of single errors and 100% of transposition errors. The **International Standard Book Number (ISBN)** method used throughout the world is one that detects all such errors. (See the copyright page or the back of this book.)

A correctly coded 10-digit ISBN $a_1a_2 \ldots a_{10}$ has the property that $10a_1 + 9a_2 + 8a_3 + 7a_4 + 6a_5 + 5a_6 + 4a_7 + 3a_8 + 2a_9 + a_{10}$ is evenly divisible by 11. Consider the ISBN of the book you are now reading: 0-7167-4782-0 (0-7167-4783-9 for paperback version). The initial digit 0 indicates that the book is published in an English-speaking country (*not* that the book is written in English). The next block of digits—7167—identifies the publisher, W. H. Freeman and Company. The third block—4782 for the hardback version—is assigned by the publisher and identifies the particular book. The last digit 0 (for the hardback version) is the check digit. Let us verify that this number is a legitimate possibility. We must compute $10 \cdot 0 + 9 \cdot 7 + 8 \cdot 1 + 7 \cdot 6 + 6 \cdot 7 + 5 \cdot 4 + 4 \cdot 7 + 3 \cdot 8 + 2 \cdot 2 + 0 = 231$. Because $231 = 11 \cdot 21$, it is evenly divisible by 11, and no error has been detected.

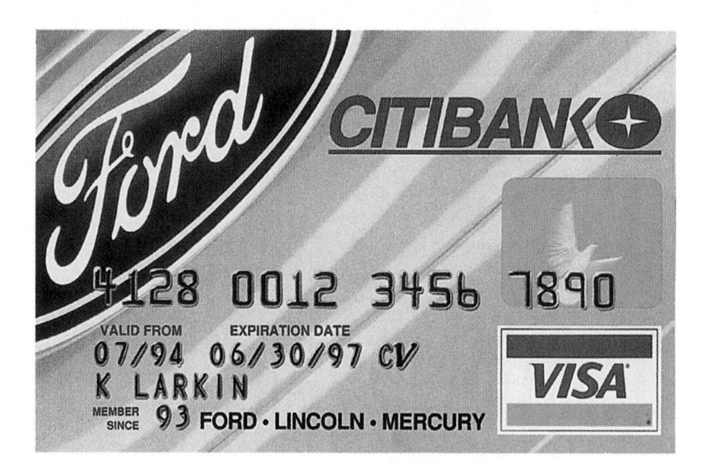

Figure 9.4 VISA card with an invalid Codabar number.

How can we be sure that this method detects 100% of the single-position errors? Well, let us say that a correct number is $a_1a_2a_3a_4a_5a_6a_7a_8a_9a_{10}$ and that a mistake is made in the second position. (The same argument applies equally well in every position.) We may write this incorrect number as $a_1a'_2a_3a_4a_5a_6a_7a_8a_9a_{10}$, where $a'_2 \neq a_2$. Now in order for this error to go undetected, it must be the case that $10 \cdot a_1 + 9 \cdot a'_2 + 8 \cdot a_3 + 7 \cdot a_4 + 6 \cdot a_5 + 5 \cdot a_6 + 4 \cdot a_7 + 3 \cdot a_8 + 2 \cdot a_9 + a_{10}$ is evenly divisible by 11. Then, because both $10a_1 + 9a_2 + 8a_3 + 7a_4 + 6a_5 + 5a_6 + 4a_7 + 3a_8 + 2a_9 + a_{10}$ and $10a_1 + 9a'_2 + 8a_3 + 7a_4 + 6a_5 + 5a_6 + 4a_7 + 3a_8 + 2a_9 + a_{10}$ are divisible by 11, so is their difference:

$$(10 \cdot a_1 + 9 \cdot a_2 + 8 \cdot a_3 + \ldots + a_{10}) - (10 \cdot a_1 + 9 \cdot a'_2 + 8 \cdot a_3 + \ldots + a_{10}) = 9 \cdot (a_2 - a'_2)$$

Because a_2 and a'_2 are distinct digits between 0 and 9, their difference must be one of $\pm 1, \ldots, \pm 9$. Thus, the only possibilities for the number $9 \cdot (a_2 - a'_2)$ are ± 9, ± 18, ± 27, ± 36, ± 45, ± 54, ± 63, ± 72, ± 81, and none of these is divisible by 11. So, a single-position error cannot go undetected.

If this method, in contrast to the others we have described, detects all single-position errors and all transposition errors, why is it not used more? Well, it does have a drawback. Say the next title published by Freeman is to have 1910 for the third block. (All Freeman books begin with 0-7167-.) What check digit should be assigned? Call it a. Then $10 \cdot 0 + 9 \cdot 7 + 8 \cdot 1 + 7 \cdot 6 + 6 \cdot 7 + 5 \cdot 1 + 4 \cdot 9 + 3 \cdot 1 + 2 \cdot 0 + a = 199 + a$. Because the next integer after 199 that is divisible by 11 is 209, we see that $a = 10$. But appending 10 to the existing 9-digit number would result in an 11-digit number instead of a 10-digit one. This is the only flaw in the ISBN scheme. To avoid this flaw, publishers use an X to represent the check digit 10. As a result, not all ISBNs consist solely of digits (some end with X). Publishers could avoid this inconsistency by simply refraining from using numbers that require an X.

It is possible to detect 100% of all single-digit errors and 100% of transposition errors involving adjacent digits without the drawback of introducing an extra character such as X, but to do so requires a noncommutative multiplication system—that is, a system in which a times b is not always the same as b times a. One situation in which such a system is used is described in Spotlight 9.2.

At this point the reader might naturally ask, "Why are there so many different methods for achieving the same purpose?" Like many practices in the "real world," historical accident and lack of knowledge about existing methods seem to be the explanation.

Many identification numbers utilize both alphabetic and numerical characters. The vehicle identification number (VIN) used to identify cars and trucks is one such example (see Spotlight 9.3). One of the most prevalent of these, developed in 1975, is called **Code 39.** Code 39 permits the 26 uppercase letters A through Z and the digits 0 through 9. Because Code 39 has been chosen by the Department of Defense, the automotive companies, and the health industry for use by their suppliers, it has become the workhorse of nonretail business.

SPOTLIGHT 9.2 German Banknotes

Before adopting the Euro in 2002, Germany was one of few countries in the world that included a check digit on their currency. Each German banknote had a 10-character serial number comprising letters and numbers. To compute the check digit, the letters were converted to numbers, and complicated schemes for weighing the numbers in each position and for calculating the check digit were employed. Interestingly, the method used to calculate the check digit was not commutative (that is, *a* times *b* need not be the same as *b* times *a*). For example, in this unusual mathematical system, 4 times 5 is 9 but 5 times 4 is 6. Using a noncommutative system of calculation results in greater error-detection capability than is possible with a commutative system. The banknote shown here features the mathematician Carl Gauss. The last digit is the check digit.

German banknote with serial number DZ6768309Y and check digit 9.

A typical example of a Code 39 number is 210SA0162322ZAY. The last character is the "check." The check character is determined by assigning the letters A through Z the numerical values 10 through 35, respectively. The original number, composed of the digits 0 through 9 and letters A through Z, is now converted to a string $a_1, a_2, \ldots, a_{14}, a_{15}$, where the a_i are integers between 0 and 35. The check character a_{15} is chosen so that $15a_1 + 14a_2 + 13a_3 + \ldots + 2a_{14} + a_{15}$ is divisible by 36. Finally, a_{15} is converted to its alphabetic counterpart if it is greater than 9 (for example, 13 is converted to D).

EXAMPLE Code 39 Number 210SA0162322ZA

Let us examine the Code 39 method for the number 210SA0162322ZA. Here is how we determine the check character. First we convert the alphabetic characters to their numerical counterparts: 210*SA*0162322*ZA* → 2, 1, 0, 28, 10, 0, 1, 6, 2, 3, 2, 2, 35, 10. Then we compute

$$15 \cdot 2 + 14 \cdot 1 + 13 \cdot 0 + 12 \cdot 28 + 11 \cdot 10 + 10 \cdot 0 + 9 \cdot 1 + 8 \cdot 6 + 7 \cdot 2 + 6 \cdot 3 + 5 \cdot 2 + 4 \cdot 2 + 3 \cdot 35 + 2 \cdot 10 = 30 + 14 + 0 + 336 + 110 + 0 + 9 + 48 + 14 + 18 + 10 + 8 + 105 + 20 = 722$$

SPOTLIGHT 9.3 VIN System

Automobiles and trucks are given a vehicle identification number (VIN) by the manufacturer. A typical VIN has 17 alphanumeric characters that code information, such as country where the vehicle was built, manufacturer, make, body style, engine type, plant where the vehicle was built, model year, model, type of restraint, a check digit, and a production sequence number. The check digit is calculated by converting the 26 consecutive letters of the alphabet, respectively, to the numbers 1, 2, 3, 4, 5, 6, 7, 8, 9, 1, 2, 3, 4, 5, 6, 7, 8, 9, 2, 3, 4, 5, 6, 7, 8, 9 (note the anomaly after the second 9) so as to obtain a 16-digit number $a_1a_2 \ldots a_{15}a_{16}$ that is weighted with 8, 7, 6, 5, 4, 3, 2, 10, 9, 8, 7, 6, 5, 4, 3, 2. The check digit is the remainder when the weighted sum $8 \cdot a_1 + 7 \cdot a_2 + \ldots + 3 \cdot a_{15} + 2 \cdot a_{16}$ is divided by 11 unless the remainder is 10, in which case an X is used instead. The check digit is inserted in position 9.

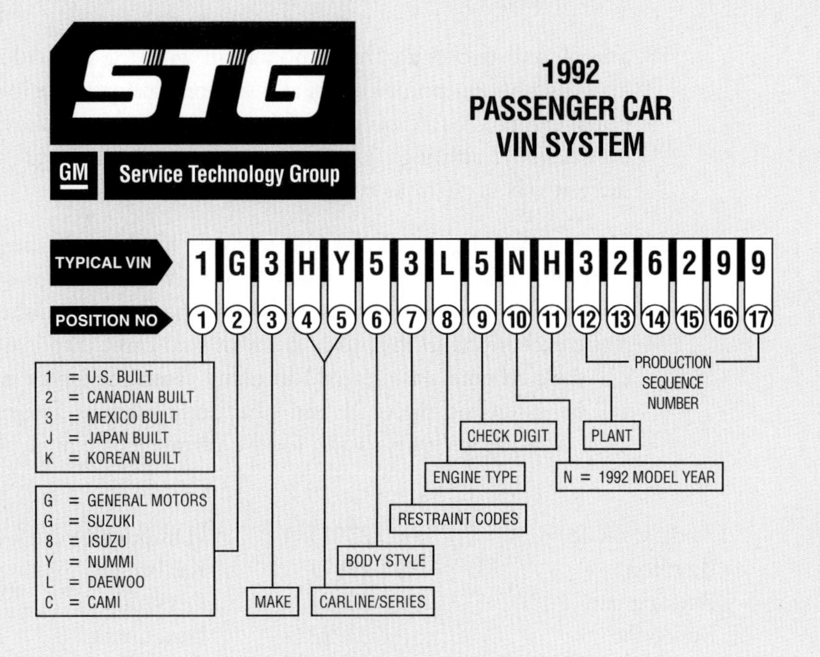

Now we select a_{15} so that $722 + a_{15}$ is divisible by 36. Because 722 divided by 36 has a remainder of 2 ($722 = 36 \cdot 20 + 2$), we choose a_{15} as 34. Finally, we convert 34 to Y. Thus the number becomes 210SA0162322ZAY. ■

In many applications of Code 39 the seven special characters -, ., space, \$, /, +, and % are permitted. These characters are assigned the numerical values 36 through 42, respectively. In these applications the check character is determined by the remainder upon division by 43 instead of 36.

9.2 The ZIP Code

Identification numbers occasionally **encode** geographic data. The ZIP code, Social Security numbers, and telephone numbers are prime examples. In 1963 the U.S. Postal Service numbered every American post office with a five-digit **ZIP code.** The numbers begin with 0's at the point farthest east—00601 for Adjuntas, Puerto Rico—and work up to 9's at the point farthest west—99950 for Ketchikan, Alaska (see Figure 9.5).

Let's use one of the ZIP codes for Duluth, Minnesota, as an example:

55812

5 The first digit represents one of 10 geographic areas, usually a group of states. The numbers begin at points farthest east (0) and end at the points farthest west (9).

58 The second two digits, in combination with the first, identify a central mail-distribution point known as a sectional center. The location of a sectional center is based on geography, transportation facilities, and population density; although just four centers serve the entire state of Utah, there are six of them to take care of New York City.

12 The last two digits indicate the town or local post office. The order is often alphabetical for towns within a delivery area—for example, towns with names beginning with A usually have low numbers. (There are many exceptions to this, such as towns that came into existence after the ZIP code scheme was created.) In many cases, the largest city in a region will be given the digits 01 and surrounding towns assigned succeeding digits alphabetically, as shown for the Farmville, Virginia, area.

Andersonville	23911	Hampden-Sydney	23943
Boydton	23917	Kenbridge	23944
Buckingham	23921	Keysville	23947
Burkeville	23922	Lunenburg	23952
Charlotte Court House	23923	Meherrin	23954
Chase City	23924	Nottoway	23955
Clarksville	23927	Pamplin	23958
Crewe	23930	Phenix	23959
Cullen	23934	Prospect	23960
Darlington Heights	23935	Red Oak	23964
Dillwyn	23936	Rice	23966
Drakes Branch	23937	Skipwith	23968
Dundas	23938	Victoria	23974
Farmville	23901	Wylliesburg	23976
Green Bay	23942		

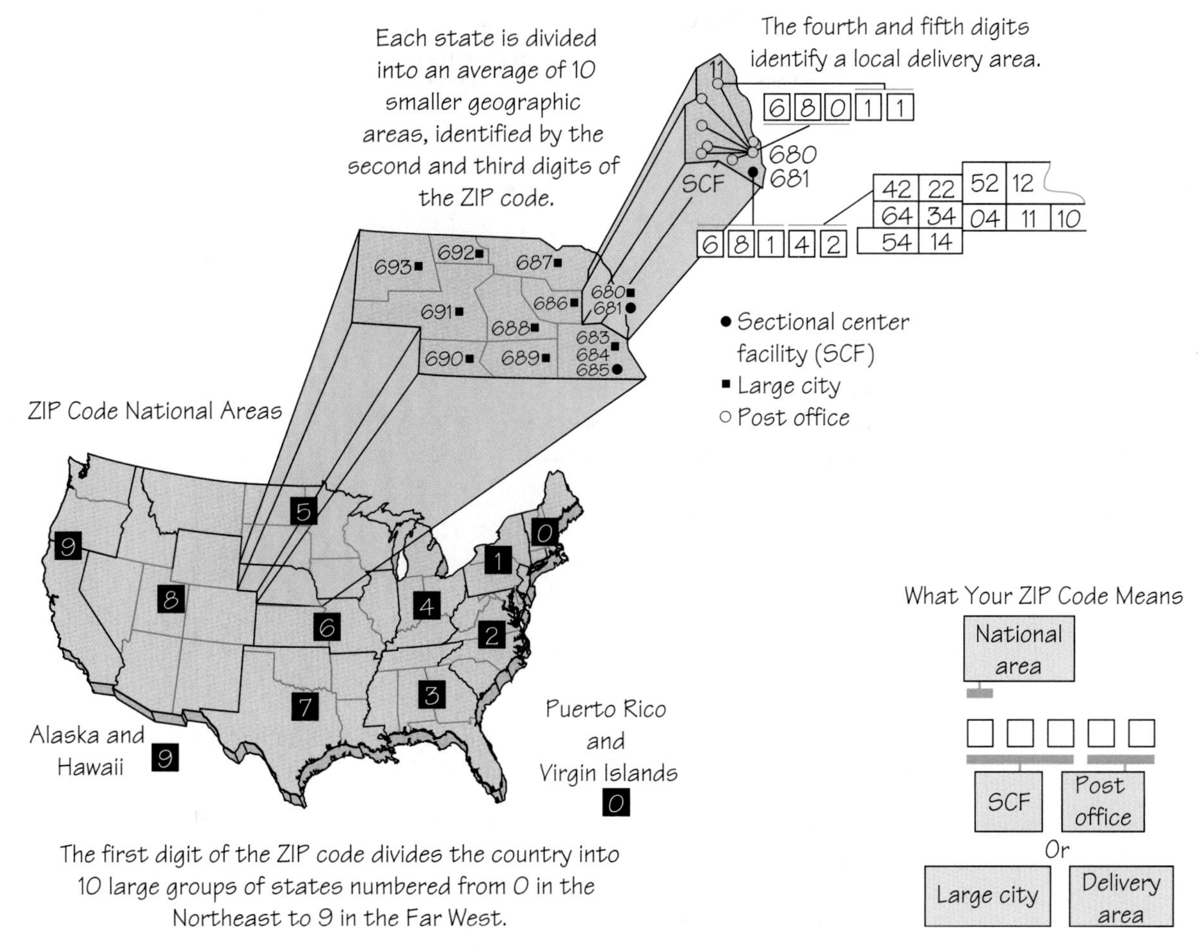
Each state is divided into an average of 10 smaller geographic areas, identified by the second and third digits of the ZIP code.
The fourth and fifth digits identify a local delivery area.
6 8 0 1 1
680
681
SCF
42 22 52 12
64 34 04 11 10
54 14
6 8 1 4 2
693
692
687
691
686
680
681
688
683
684
685
690
689
• Sectional center facility (SCF)
▪ Large city
○ Post office
ZIP Code National Areas
Alaska and Hawaii 9
Puerto Rico and Virgin Islands 0
The first digit of the ZIP code divides the country into 10 large groups of states numbered from 0 in the Northeast to 9 in the Far West.
What Your ZIP Code Means
National area
SCF
Post office
Or
Large city
Delivery area

Figure 9.5 ZIP code scheme.

In 1983 the U.S. Postal Service added four digits to the ZIP code. When four digits are added after a dash – for example, 68588-1234 – the number is called the **ZIP + 4 code.** Mail with ZIP + 4 coding is eligible for cheaper bulk rates, being easier to sort with automated equipment. It's also helpful for businesses that wish to sort the recipients of their mailings by geographic location. The first two numbers of the four-digit suffix represent a delivery sector, which may be several blocks, a group of streets, several office buildings, or a small geographic area. The last two numbers narrow the area further; they might denote one floor of a large office building, a department in a large firm, or a group of post office boxes.

For businesses that receive an enormous volume of mail, the ZIP + 4 code permits automation of in-house mailroom sorting. For example, the first seven digits of all mail sent to the University of Minnesota Duluth, are 55812-24. The school has designated nine pairs of digits for the last two positions to direct the mail to the appropriate dormitory or apartment complex.

Hand-held scanner reading the shipping bar code on a crate. (Stewart Cohen/Tony Stone Images.)

9.3 Bar Codes

In modern applications, bar codes and identification numbers go hand in hand. Bar coding is a method for automated data collection. It is a way to transmit information rapidly, accurately, and efficiently to a computer.

A **bar code** is a series of dark and light spaces that represent characters.

To **decode** the information in a bar code, a beam of light is passed over the bars and spaces via a scanning device, such as a handheld wand or a fixed-beam device. The dark bars reflect very little light back to the scanner, whereas the light spaces reflect much light. The differences in reflection intensities are detected by the scanner and converted to strings of 0's and 1's that represent specific numbers and letters. Such strings are called a *binary coding* of the numbers and letters.

Any system for representing data with only two symbols is a **binary code.**

ZIP Code Bar Code

The simplest bar code is the **Postnet code** used by the U.S. Postal Service and commonly found on business reply forms (see Figure 9.6). For a ZIP + 4 code there are 52 vertical bars of two possible lengths (long and short). The long bars

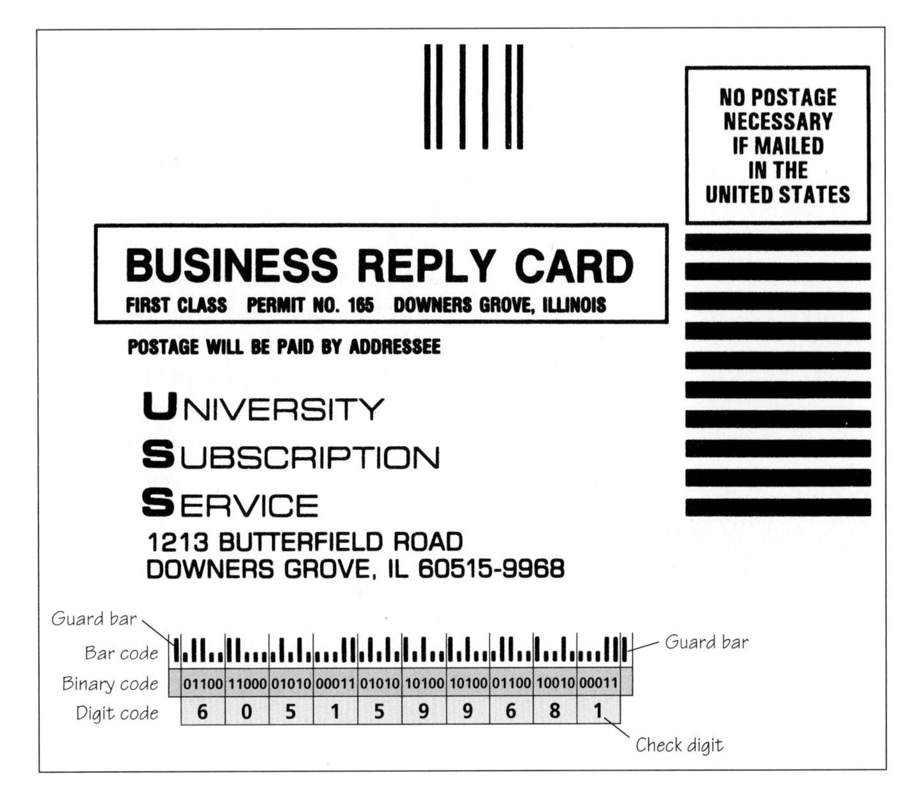

Figure 9.6 ZIP + 4 bar code.

at the beginning and end are called *guard bars* and together provide a frame for the remaining 50 bars. In blocks of five, the 50 bars within the guard bars represent the ZIP + 4 code and a tenth digit for error correction. Each block of five is composed of exactly two long bars and three short bars, according to the pattern shown below:

Decimal digit	Bar code
1	
2	
3	
4	
5	
6	
7	
8	
9	
0	

The tenth digit of a Postnet code number is a check digit chosen so that the sum of the nine digits of the ZIP + 4 code and the tenth one is evenly divisible by 10. That is, the check digit C for the ZIP + 4 code $a_1a_2 \ldots a_9$ is the digit with the property that the sum $a_1 + a_2 + \ldots + a_9 + C$ ends with 0. For example, the ZIP + 4 code 80321-0421 has the check digit 9, because 8 + 0 + 3 + 2 + 1 + 0 + 4 + 2 + 1 = 21 and 21 + 9 = 30 ends with 0.

Because each digit is represented by exactly two long bars and three short ones, any error in reading or printing a single bar would result in a block of five with only one long bar or three long bars. In either case, the error is detected. (This is the reason behind the choice of five bars to code each digit rather than four bars. With five bars per digit, there are exactly 10 arrangements composed of two long bars and three short bars. Any misreading of a single bar in such a block is therefore recognizable, because it does not match any other of the blocks for the 10 digits.) And because the block location of the error is known, the check digit permits the correction of the error. Let's look at an example of an incorrectly printed bar code and see how the error is correctable.

The scanner ignores the guard bars at the beginning and the end and reads the remaining bars in blocks of five as shown below. (We have inserted dashed dividing lines for readability.)

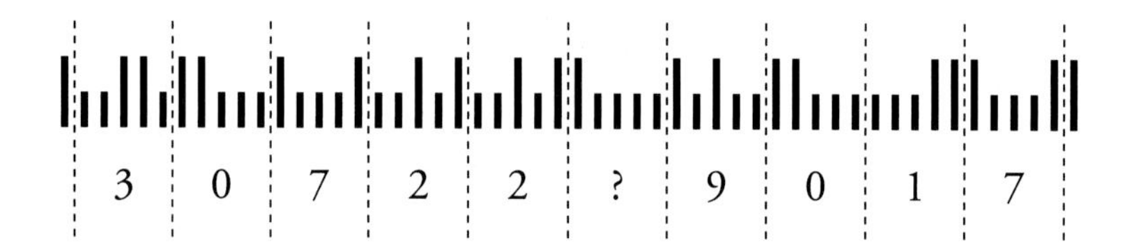

Figure 9.7
Entomologist Stephen Buchmann developed a reliable, inexpensive way to track bees using the same technology that supermarkets use to speed up the checkout lines and keep track of inventory. He glued bar-code labels onto the backs of 100 bees and placed a laser scanner above the hive. In the past, researchers marked bees with paint or tags, but the monitoring of activity required the presence of a human observer. (Scott Camazine/Sue Trainor.)

Because the sixth block has only one long bar, it is an incorrect one. To correct the error, the computer linked with the bar-code scanner sums the remaining 9 digits to obtain 31. Because the sum of all 10 digits ends with 0, the correct value for the sixth digit must be 9.

Beginning in 1993, large organizations and businesses that wanted to receive reduced rates for ZIP + 4 bar-coded mail were required to use a 12-digit bar code called the *delivery-point bar code.* This code permits machines to sort a letter into the order in which it will be delivered by the carrier. (Mail for the first location on a mail route occurs first, mail for the second location on a route occurs second, and so on.)

The 12-digit bar code uses the Postnet bar scheme to code the 12-digit string composed of the 9-digit ZIP + 4 number followed by the last two digits of the street address or box number and a check digit chosen so that the sum of all 12 digits is evenly divisible by 10. For example, a letter addressed to 1738 Maple Street with ZIP + 4 code 55811-2742 would have the Postnet bar code for the digits 558112742384 (38 is from the street address and 4 is the check digit).

The UPC Bar Code

The bar code that we encounter most often is the *Universal Product Code (UPC).* The UPC was first used on grocery items in 1973 and has since spread to most retail products. As Figure 9.7 shows, it has other applications as well. The UPC bar code translates a 12-digit number into bars that can be quickly and accurately read by a laser scanner. The number has four components – two five-digit numbers sandwiched between two single digits – as shown in Figure 9.8.

Here is what the four components represent:

5 The first digit identifies the kind of product. For example, a 2 signals random-weight items, such as cheese or meat; a 3 means drug and certain other health-related products; a 4 means products marked for price reduction by the retailer; a 5 signals cents-off coupons (see Figure 9.8).

13000 The next five digits identify the manufacturer. This number is assigned by the Uniform Code Council in Dayton, Ohio.

22020 The next five digits, assigned by the manufacturer to identify the product, can include size, color, or other important information (but not price).

5 The final digit is the check digit. This digit is often not printed, but it is always included in the bar code.

Figure 9.8 UPC identification number 5 13000 22020 5. The initial 5 indicates the number is a manufacturer's coupon. The block 13000 identifies the manufacturer as Heinz. The block 22020 identifies the product. The last digit, 5, is a check digit.

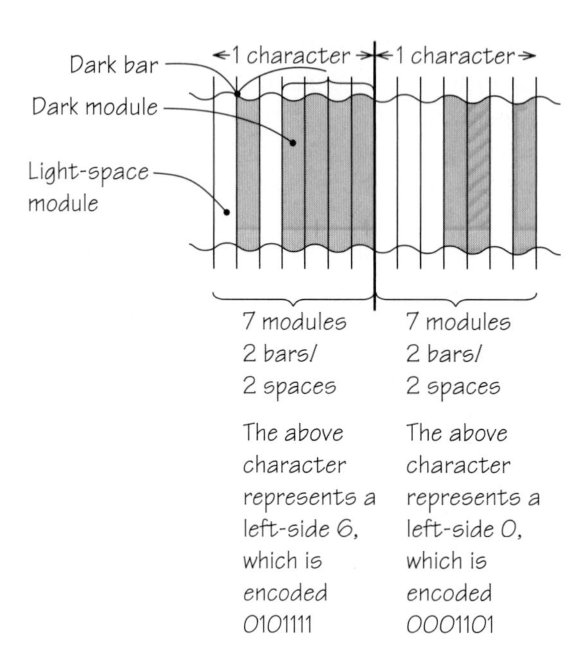
Dark bar
Dark module
Light-space module
←1 character→
←1 character→
7 modules
2 bars/
2 spaces
7 modules
2 bars/
2 spaces
The above character represents a left-side 6, which is encoded 0101111
The above character represents a left-side 0, which is encoded 0001101

Figure 9.9 UPC bar coding for a left-side 6 and left-side 0.

Each digit of the UPC code is represented by a space divided into seven modules of equal width, as illustrated in Figure 9.9. How these seven modules are filled depends on the digit being represented and whether the digit being represented is part of the manufacturer's number or the product number. In every case there are two light "spaces" and two dark bars of various thicknesses that alternate. A UPC code has on each end two long bars of one-module thickness separated by a light space of one-module thickness. These three modules are called the *guard bar patterns* (Figure 9.10). The guard bar patterns define the thickness of a single module of each type. They are not part of the identification number. The manufacturer's number and the product number are separated by a center bar pattern consisting of the following five modules: a light space, a (long) dark bar, a light space, a (long) dark bar, and a light space (see Figure 9.10). The center bar pattern is not part of the identification number but merely serves to separate the manufacturer's number and product number. Figure 9.9 shows how the digits 6 and 0 in a manufacturer's number are coded.

Observe the following pattern in Figure 9.9: a light space of one-module thickness, a dark bar of one-module thickness, a light space of one-module thickness, a dark bar of four-module thickness. Symbolically, such a pattern of light spaces and dark bars is represented as 0101111. Here each 0 means a one-module-thickness light space and each 1 means a one-module-thickness dark bar. Figure 9.11 illustrates how the thicknesses of the spaces and bars are translated into a binary code.

Table 9.1 shows the binary code for all digits. Notice that the code for the digits in the product number (the block of five digits on the right side) can be obtained from the code for the digits in the manufacturer's number (the block of

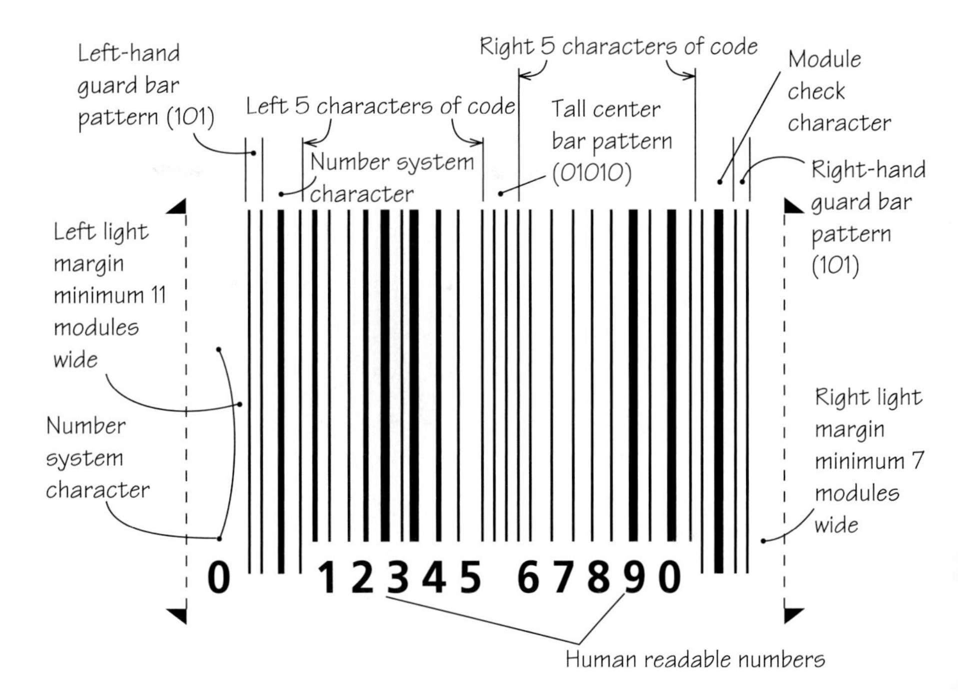

Figure 9.10 UPC bar-code format.

digits on the left side), and vice versa, by replacing each 0 by a 1 and each 1 by a 0. Thus, the code 0111011 for 7 in a manufacturer's number becomes 1000100 in the product number. Also notice that each manufacturer's number has an odd number of 1's, whereas each product number has an even number of 1's. This permits a computer linked with an optical scanner to determine whether the bar code was scanned left to right or right to left. (If the first block of digits has an even number of 1's for each digit, the scanning is being done right to left.) Thus, scanning can be done in either direction without ambiguity.

TABLE 9.1 Binary UPC Coding

Digit	Manufacturer's Number	Product Number
0	0001101	1110010
1	0011001	1100110
2	0010011	1101100
3	0111101	1000010
4	0100011	1011100
5	0110001	1001110
6	0101111	1010000
7	0111011	1000100
8	0110111	1001000
9	0001011	1110100

Figure 9.11 Translation of bars and space modules into binary code (see top of bars). The guard pattern defines a single-module thickness for a bar and a space.

OPTIONAL

Encoding Personal Data

Consider this Social Security number: 189-31-9431. What information about the holder can be deduced from the number? Only that the holder obtained it in Pennsylvania (see Spotlight 9.4). Figure 9.12 shows an Illinois driver's license number: M200-7858-1644. What information about the holder can be deduced from this number? This time we can determine the date of birth, sex, and much about the person's name.

These two examples illustrate the extremes in coding personal data. The Social Security number has no personal data encoded in the number. It is entirely determined by the place and time it is issued, not the individual to whom it is assigned. In contrast, in some states the driver's license numbers are entirely determined by personal information about the holders. It is no coincidence that the unsophisticated Social Security numbering scheme predates computers. Agencies that have large databases that include personal information such as

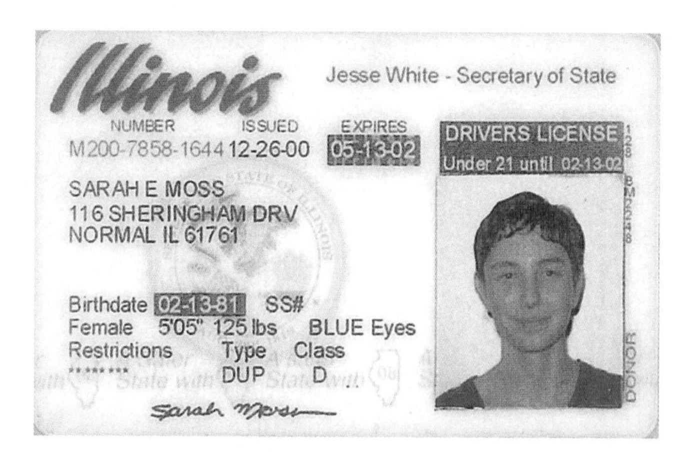

Figure 9.12 Illinois driver's license.

SPOTLIGHT 9.4 Social Security Numbers

The first three digits of Social Security numbers show where the number was applied for. Changes in population have forced some numbers to be moved or assigned out of sequence over the years.

Numbers	Location
001–003	New Hampshire
004–007	Maine
008–009	Vermont
010–034	Massachusetts
035–039	Rhode Island
040–049	Connecticut
050–134	New York
135–158	New Jersey
159–211	Pennsylvania
212–220	Maryland
221–222	Delaware
223–231	Virginia
232–236	West Virginia
232, 237–246	North Carolina
247–251	South Carolina
252–260	Georgia
261–267 & 589–595	Florida
268–302	Ohio
303–317	Indiana
318–361	Illinois
362–386	Michigan
387–399	Wisconsin
400–407	Kentucky
408–415	Tennessee
416–424	Alabama
425–428 & 587–588	Mississippi
429–432	Arkansas
433–439	Louisiana
440–448	Oklahoma
449–467	Texas
468–477	Minnesota
478–485	Iowa
486–500	Missouri
501–502	North Dakota
503–504	South Dakota
505–508	Nebraska
509–515	Kansas
516–517	Montana
518–519	Idaho
520	Wyoming
521–524	Colorado
525 & 585	New Mexico
526–527 & 600–601	Arizona
528–529	Utah
530	Nevada
531–539	Washington
540–544	Oregon
545–573 & 602–626	California
574	Alaska
575–576	Hawaii
577–579	District of Columbia
580	Virgin Islands
580–584 & 596–599	Puerto Rico
586	Guam
586	American Samoa
586	Philippines
700–728	*through July 1, 1963, reserved for railroad employees*

Source: Social Security Administration.

names, sex, and dates of birth find it convenient to encode these data into identification numbers. Examples of such agencies are the National Archives (where census records are kept), genealogical research centers, the Library of Congress, and state motor vehicle departments.

There are many methods in use to encode personal data such as name, sex, and date of birth. These methods are perhaps most widely used in assigning driver's license numbers in some states. Coding license numbers solely from personal data enables automobile insurers, government entities, and law enforcement agencies to determine the number from the personal data.

Many states encode the surname, first name, middle initial, date of birth, and sex by quite sophisticated schemes (see Figure 9.12).

In one scheme the first four characters of the license number are obtained by applying the **Soundex Coding System** to the surname as follows:

1. Delete all occurrences of *h* and *w*. (For example, *Schworer* becomes *Scorer* and *Hughgill* becomes *uggill.*)
2. Assign numbers to the remaining letters as follows:

$a, e, i, o, u, y \rightarrow 0$

$b, f, p, v \rightarrow 1$ $\qquad l \rightarrow 4$

$c, g, j, k, q, s, x, z \rightarrow 2$ $\qquad m, n \rightarrow 5$

$d, t \rightarrow 3$ $\qquad r \rightarrow 6$

3. If two or more letters with the same numeric value are adjacent, omit all but the first. (For example, *Scorer* becomes *Sorer* and *uggill* becomes *ugil*).
4. Delete the first character of the original name if still present. (*Sorer* becomes *orer*).
5. Delete all occurences of *a, e, i, o, u,* and *y*.
6. Retain only the first three digits corresponding to the remaining letters; append trailing 0's if fewer than three letters remain; precede the three digits obtained in step 6 with the first letter of the surname.

Figure 9.13 shows three examples.

What is the advantage of this method? It is an error-correcting scheme. Indeed, it is designed so that likely misspellings of a name nevertheless result in the correct coding of the name. For example, frequent misspellings of the name *Erickson* are *Ericksen, Eriksen, Ericson,* and *Ericsen*. Observe that all of these yield the same coding as *Erickson.* If a law enforcement official, a genealogical researcher, a librarian, or an airline reservation agent wanted to pull up the file from a data bank for someone whose name was pronounced "Erickson," the correct spelling isn't essential because the computer searches for records that are coded as E-625 for all spelling variations. (The Soundex Coding System was designed for the U.S. Census Bureau when much census information was obtained orally. Airlines use a somewhat different system called the *Davidson Consonant Code.*)

Figure 9.13 The Soundex Coding System.

Schworer →(Step 1) Scorer →(Step 2) Scorer 220606 →(Step 3) Sorer 20606 →(Step 4) orer 0606 →(Step 5) rr 66 →(Step 6) S-660

Hughgill →(Step 1) uggill →(Step 2) uggill 022044 →(Step 3) ugil 0204 →(Step 4) ugil 0204 →(Step 5) gl 24 →(Step 6) H-240

Schmidlapper →(Step 1) Scmidlapper →(Step 2) Scmidlapper 22503401106 →(Step 3) Smidlaper 250340106 →(Step 4) midlaper 50340106 →(Step 5) mdlpr 534116 →(Step 6) S-534

There are many schemes for encoding the date of birth and the sex in driver's license numbers. For example, the last five digits of Illinois and Florida driver's license numbers capture the year and date of birth as well as the sex. In Illinois, each day of the year is assigned a three-digit number in sequence beginning with 001 for January 1. However, each month is assumed to have 31 days. Thus, March 1 is given the number 063 because both January and February are assumed to have 31 days. These numbers are then used to identify the month and day of birth of male drivers. For females, the scheme is identical except 600 is added to the number. The last two digits of the year of birth, separated by a dash (probably to obscure that fact that they represent the year of birth), are listed in the fifth and fourth positions from the end of the driver's license number. Thus, a male born on October 13, 1940, would have the last five digits 4-0292 ($292 = 9 \cdot 31 + 13$), whereas a female born on the same day would have 4-0892. The scheme to identify birth date and sex in Florida is the same as in Illinois except each month is assumed to have 40 days and 500 is added for women. For example, the five digits 4-9585 belong to a woman born on March 5, 1949.

In this chapter, we have illustrated how mathematics is used to append a check digit to an identification number for the purpose of error detection. In the next chapter, we explain how codes consisting of 0's and 1's can be devised so that errors can be corrected.

REVIEW VOCABULARY

Bar code A code that employs bars and spaces to represent information.
Binary code A coding scheme that uses two symbols, usually 0 and 1.
Check digit A digit included in an identification number for the purpose of error detection.
Codabar An error-detection method used by all major credit card companies, many libraries, blood banks, and others.
Code A group of symbols that represent information together with a set of rules for interpreting the symbols.
Code 39 An alphanumeric code that is widely used on nonretail items.
Decoding Translating code into data.
Encoding Translating data into code.
Error-detecting code A code in which certain types of errors can be detected.
International Standard Book Number (ISBN) A 10-digit identification number used on books throughout the world that contains a check digit for error detection.
Postnet code The bar code used by the U.S. Postal Service for ZIP codes.
Soundex Coding System An encoding scheme for surnames based on sound.
Universal Product Code (UPC) A bar code and identification number that are used on most retail items. The UPC code detects 100% of all single-digit errors and most other types of errors.
Weights Numbers used in the calculation of check digits.
ZIP code A five-digit code used by the U.S. Postal Service to divide the country into geographic units to speed sorting of the mail.
ZIP + 4 code The nine-digit code used by the U.S. Postal Service to refine ZIP codes into smaller units.

SUGGESTED READINGS

GALLIAN, J. Error detection methods, *ACM Computing Surveys,* 28 (1996): 504–517. A detailed description of numerous error-detection methods.

GALLIAN, J. Assigning driver's license numbers, *Mathematics Magazine,* 64 (1992): 13–22. Discusses various methods used by the states to assign driver's license numbers. Several of these methods include check digits for error detection.

GALLIAN, J. The mathematics of identification numbers, *College Mathematics Journal,* 22 (1991): 194–202. A comprehensive survey of check-digit schemes that are associated with identification numbers.

GALLIAN, J., and S. WINTERS. Modular arithmetic in the marketplace, *American Mathematical Monthly,* 95 (1988): 548–551. A detailed analysis of the check-digit schemes presented in this chapter. In particular, the error-detection rates for the various schemes are given.

KIRTLAND, J. *Identification Numbers and Check Digit Schemes,* Mathematical Association of America, Washington, D.C., 2001. Provides more examples and exercises for the check-digit schemes discussed in the chapter.

PHILIPS, L. Hanging on the Metaphone, *Computer Language,* 7 (December 1990): 39–43. Describes a sound-based retrieval algorithm that in some respects is superior to Soundex.

ROUGHTON, K., and D. TYCKOSEN. Browsing with sound: Sound-based codes and automated authority control, *Information Technology and Libraries,* 4 (June 1985): 130–136. Explains the Soundex system and the Davidson Consonant Code and their uses. The Davidson Consonant Code is a sound-based retrieval algorithm used by airline passenger retrieval systems.

SUGGESTED WEB SITES

www.d.umn.edu/~jgallian/fapp6 This Web site enables users to calculate check digits using the various methods discussed in this chapter. Also included are the methods used by several states to assign driver's license numbers.

www.upl.cs.wisc.edu/cgi-bin/wilic This site enables users to determine a Wisconsin driver's license number from the person's name and date of birth. Moreover, after the user enters a valid Wisconsin driver's license number, the site will list many possible names that will produce the number, as well as the date of birth corresponding to the entered number.

SKILLS CHECK

1. Suppose a U.S. Postal Service money order is numbered 1012065994X, where X indicates that the last digit is obliterated. What is the missing digit?

(a) 1
(b) 8
(c) The missing digit cannot be determined.

2. Suppose an American Express Travelers Cheque is numbered X425036790, where X indicates that the first digit is obliterated. What is the missing digit?

(a) 0
(b) 9
(c) The missing digit cannot be uniquely determined.

3. Is the number 105408970012 a legitimate airline ticket number?

(a) Yes.
(b) No, but if the final digit is changed to a 5, the resulting number 105408970015 is legitimate.
(c) No, but if the final digit is changed to a 6, the resulting number 105408970016 is legitimate.

4. Determine the check digit that should be appended to the UPC code 0-14300-25433.

(a) 1
(b) 9
(c) Another number

5. Determine the check digit that should be appended to the bank identification number 01500085.

(a) 1
(b) 9
(c) Another number

6. Suppose the ISBN 0-1750-3549-0 is incorrectly reported as 0-1750-3540-0. Which of the following statements is true?

(a) This error will not be detected by the check digit.
(b) While this particular error will be detected, the check digit does not detect all single-digit errors in ISBNs.
(c) Any single-digit error in an ISBN is detectable by the check digit.

7. Suppose that the Postnet code is incorrectly reported as 20001-5800-7. You know that only the sixth digit is incorrectly reported. Which of these statements is true?

(a) It is impossible to determine the correct digit for the sixth position.
(b) It is possible to determine the correct digit in this case. However, it is necessary to know which digit was incorrectly reported.
(c) It is possible to determine the correct digit in this case. Moreover, it is not necessary to know which digit was incorrectly reported.

8. When a single incorrect digit is entered, an error-detecting code

(a) will sometimes detect the error.
(b) will always detect the error but may not be able to correct the error.
(c) will always detect and correct the error.

9. The conversion of data to code is called

(a) encoding.
(b) decoding.
(c) Codabar.

10. If the first five digits of a valid U.S. Postal Service money order number are rearranged, will the resulting number also be valid?

(a) Yes, always
(b) Sometimes
(c) Never

11. If the first two digits of a valid airline ticket identification number are exchanged, will the resulting number also be valid?

(a) Yes, always
(b) Sometimes
(c) Never

12. The U.S. banking system uses a system that detects

(a) all transpositions and some single-digit errors.
(b) all single-digit errors and some transpositions.
(c) all single-digit errors and all transpositions.

13. A correctly coded 10-digit ISBN has a weighted sum that is evenly divisible by what integer?

(a) 7
(b) 10
(c) 11

14. A correctly coded 12-digit UPC has a weighted sum that is evenly divisible by what integer?

(a) 7
(b) 10
(c) 11

15. The U.S. Postal Service Postnet code converts the ZIP + 4 code into

(a) a coding of 9 digits using long and short bars.
(b) a coding of 10 digits whose sum is a multiple of 10.
(c) a coding of 10 digits for which the sum of the first 9 digits minus the tenth digit is a multiple of 10.

EXERCISES

▲ Optional. ■ Advanced. ◆ Discussion.

Postnet Codes

1. Determine the ZIP + 4 code and check digit for each of the following Postnet bar codes:

(a)
(b)
(c)

2. Determine the ZIP + 4 code and check digit for each of the following Postnet bar codes:

(a)
(b)
(c)

3. In each Postnet bar code below, exactly one mistake occurs (that is, a long bar appears instead of a short one, or vice versa). Determine the correct ZIP code.

(a)
(b)
(c)

4. Below is a 12-digit delivery-point bar code. Determine the ZIP + 4 number, the last two digits of the street address, and the check digit.

5. Explain why any two errors in a particular block of five bars in a Postnet are always detectable. Explain why not all such errors can be corrected.

Identification Numbers

6. Determine the check digit for a money order with identification number 7234541780.

7. Determine the check digit for a money order with identification number 395398164.

8. Suppose a money order with the identification number and check digit 21720421168 is erroneously copied as 27750421168. Will the check digit detect the error? Explain your reasoning.

9. Determine the check digit for the United Parcel Service (UPS) identification number 873345672.

10. Determine the check digit for the Avis rental car with identification number 540047.

11. Determine the check digit for the airline ticket number 30860422052.

12. Determine the check digit for the UPC number 05074311502.

13. Determine the check digit for the UPC number 38137009213.

14. Determine the check digit for the ISBN 0-669-33907.

15. Determine the check digit for the ISBN 0-669-19493.

16. Determine the check digit for the bank number 09100001.

17. Determine the check digit for the bank number 09190204.

18. Determine whether the Master Card number 3541 0232 0033 2270 is valid.

19. Create a check digit for the UPC number 38137009213 using the weights 7, 1, 7, 1, 7, 1, . . . , 7, 1 instead of 3, 1, 3, 1, 3, 1, . . . , 3, 1. Test to see whether this check digit will detect single-digit errors by trying several examples.

20. Create a check digit for the UPC number 38137009213 using the weights 2, 1, 2, 1, 2, 1, . . . , 2, 1 instead of 3, 1, 3, 1, 3, 1, . . . , 3, 1. Is the error caused by replacing the 3 in the first position with an 8 detected? What about the error caused by replacing the 1 in the third position with a 6? Explain why or why not.

21. If the weights 5, 1, 5, 1, 5, 1, . . . , 5, 1 were used for the UPC code, which single-digit errors would go undetected?

22. Determine the check digit for the American Express Travelers Cheque with identification number 461212023.

23. Use the Codabar scheme to determine the check digit for the number 300125600196431.

24. Determine the check character for the Code 39 number 210SA0162305ZA. (Assume the code uses only 36 characters.)

25. Determine the check character for the number 3050-0000 HEAD using the 43-character Code 39 scheme. (Be sure to include the hyphen and the space as characters.)

26. Change 173 into Postnet code.

27. Is there any mathematical reason for a check digit to be at the end of an identification number? Explain your reasoning.

28. Suppose the first block of a UPC bar code following the guard bar pattern that a scanner reads is 1000100. Is the scanner reading left to right or right to left?

29. For some products, such as soft-drink cans and magazines, an 8-digit UPC number called Version E is used instead of the 12-digit number. The method of calculating the eighth digit, which is the check digit, depends on the value of the seventh digit. The check digit a_8 for a UPC Version E identification number $a_1a_2a_3a_4a_5a_6a_7$—where a_7 is 0, 1, or 2—is chosen so that $a_1 + a_2 + 3a_3 + 3a_4 + a_5 + 3a_6 + a_7 + a_8$ is divisible by 10. Use this fact to determine the check digit for the following Version E numbers:

(a) 0121690
(b) 0274551

(c) 0760022
(d) 0496580

30. The check digit a_8 for a UPC Version E identification number $a_1a_2a_3a_4a_5a_6a_7$—where a_7 is 4—is chosen so that $a_1 + a_2 + 3a_3 + a_4 + 3a_5 + 3a_6 + a_8$ is divisible by 10. Use this fact to determine the check digit for the following Version E numbers:

(a) 0754704
(b) 0774714
(c) 0724444

■ **31.** The ISBN 0-669-03925-4 is the result of a transposition of two adjacent digits not involving the first or last digit. Determine the correct ISBN.

32. Explain why the bank scheme will detect the error 751 · · · → 157 · · · but the UPC scheme will not.

33. Suppose the check digit a_9 for bank checks were chosen to be the last digit of $3a_1 + 7a_2 + a_3 + 3a_4 + 7a_5 + a_6 + 3a_7 + 7a_8$ instead of the way described in the chapter. How would this compare with the actual check digit?

34. Explain why an error caused by transposing the first two digits of a Postal Service money order is not detected by the check-digit scheme. Explain why the same is true for the second and third digits. What about the last two digits?

35. In computing the check digit for German banknotes, the letter Z is counted as having the numerical value 9 (see Spotlight 9.2). Criticize this choice of values. (You do not have to know how the scheme works to answer this question.)

Encoding Personal Data (Optional)

36. Determine the Soundex code for Smith, Schmid, Smyth, and Schmidt.

37. Determine the Soundex code for Skow, Sachs, Lennon, Lloyd, Ehrheart, and Ollenburger.

38. Determine the last five digits of an Illinois driver's license number for a male born on June 18, 1942.

39. In Florida, the last three digits of the driver's license number of a female with birth month m and birth date b are $40(m - 1) + b + 500$. For both males and females, the fourth and fifth digits from the end give the year of birth. Determine the last five digits of a Florida driver's license number for a female born on July 18, 1942.

40. Determine the birth date of a person whose Illinois driver's license number ends with 58818.

41. In Illinois one obtains the last three digits of the driver's license for a female by adding 600 to the number for a male with the same birthday. In Florida 500 is added to the number for a male. Why can't Florida use 600?

42. Explain why an Illinois driver's license number that ends with 77061 cannot be valid.

43. Explain why an Illinois driver's license number that ends with the last five digits 99817 cannot be valid.

Additional Exercises

44. Suppose two distinct adjacent digits of a Postal Service money order are transposed and the error is detected by the check-digit scheme. At which positions were the two digits located?

45. Which digits can occur as a check digit for an airline identification number?

46. In Florida the last three digits of the driver's license number of a male with birth month m and birth date b are $40(m - 1) + b$. For both males and females, the fourth and fifth digits from the end give the year of birth. Determine the birth dates of people with numbers whose last five digits are 42218 and 53953.

47. For driver's license numbers issued in New York prior to September 1992, the last two digits were the year of birth. The three digits preceding the year encoded the sex and the month and day of birth. For a woman with birth month m and birth date b, the three digits were $63m + 2b + 1$ (insert a 0 in front for numbers less than 100). For a man with birth month m and birth date b, the three digits were $63m + 2b$. Determine the birth months, birth dates, and sexes of drivers with the three digits 248 and 601 preceding the year.

■ **48.** The state of Utah appends a ninth digit a_9 to an eight-digit driver's license number $a_1a_2 \ldots a_8$ so that $9a_1 + 8a_2 + 7a_3 + 6a_4 + 5a_5 + 4a_6 + 3a_7 + 2a_8 + a_9$ is divisible by 10.

(a) If the first eight digits of a Utah driver's license number are 14910573, what is the ninth digit?
(b) Suppose a legitimate Utah driver's license number 149105767 were miscopied as 149105267. How would you know a mistake was made? Is there any way you could determine the correct number? If you know the error was in the seventh position, could you correct the mistake?
(c) If a legitimate Utah driver's license number 149105767 were miscopied as 199105767, would you be able to tell a mistake was made? Explain.
(d) Explain why any transposition error involving adjacent digits of a Utah driver's license number would be detected.

49. Form all possible strings consisting of exactly three *a*'s and two *b*'s and arrange the strings in alphabetical order (for example, the first two possibilities are *aaabb* and *aabab*). Do you see any relationship between your list and the Postnet code?

■ **50.** Suppose the check digit a_{10} of ISBN numbers were chosen so that $a_1 + 2a_2 + 3a_3 + 4a_4 + 5a_5 + 6a_6 + 7a_7 + 8a_8 + 9a_9 + 10a_{10}$ was divisible by 11 instead of the way described in the chapter. How would this compare with the actual check digit?

■ **51.** The Canadian province of Quebec assigns a check digit a_{12} to an 11-digit driver's license number $a_1a_2 \ldots a_{11}$ so that $12a_1 + 11a_2 + 10a_3 + 9a_4 + 8a_5 + 7a_6 + 6a_7 + 5a_8 + 4a_9 + 3a_{10} + 2a_{11} + a_{12}$ is divisible by 10. Criticize this method. Describe all single-digit errors that are undetected by this scheme.

52. Most recently published books include a bar code on the back cover that has the ISBN above the bars and a 13-digit identification number below the bars. Examine several books with a bar code on the back cover. How does the number below the bar code differ from the UPC code? How is the number below the bar code related to the ISBN? Given the fact that the last digit in the number below the bar code is a check digit, determine how it is calculated.

53. Determine the check digit for the VIN JM1GD222J1581570 (see Spotlight 9.3 for a description of the method to be used).

54. Below is an actual identification number and bar code from a roll of wallpaper. What appears to be wrong with them? Speculate on the reason for the apparent violation of the UPC format.

Building Regulations: 1985 Class 0
FINE ART WALLCOVERINGS LTD.
HOLMES CHAPEL, CHESHIRE
MADE IN ENGLAND
FABRIQUE EN ANGLETERRE

55. The state of Washington encodes the last two digits of the year of birth into driver's license numbers (in positions 8 and 9) by subtracting the two-digit number from 100. For example, a person born in 1942 has 58 in positions 8 and 9, whereas a person born in 1971 has 29 in positions 8 and 9. Speculate on the reason for subtracting the birth year from 100.

56. Driver's license number assignment schemes that utilize personal data occasionally produce the same number for different people. Speculate about circumstances under which this is more likely to occur.

■ **57.** Consider a UPC number in which the digits 7 and 2 appear consecutively (that is, the number has the form $\cdots 72 \cdots$). Will the error caused by transposing these digits (that is, the number is taken as $\cdots 27 \cdots$) be detected? What if the digits 6 and 2 are transposed instead? State the general criterion for the detection of an error of the form

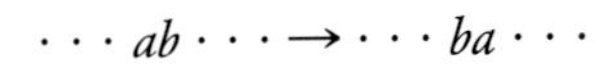

$$\cdots ab \cdots \rightarrow \cdots ba \cdots$$

by the UPC scheme.

▲ **58.** Apply the Soundex code to common ways to misspell your name. Do they give the same code as your name does?

◆ **59.** The Canadian postal system has assigned each geographic region a six-character code composed of alternating letters and digits, such as

P7B5E1 and K7L3N6. Discuss the advantages this scheme has over the five-digit ZIP code used in the United States.

60. Speculate on the reason why telephone numbers, Social Security numbers, and serial numbers on most currency do not have check digits.

61. Suppose a company uses a check-digit scheme similar to the UPC scheme but with different weights. If two of the ID numbers used by the company are 73215674 and 73215661, what can you conclude about the weight used for the seventh position?

62. If a publishing company has headquarters in both the United States and Germany and publishes the same book in both countries, it is likely that the ISBN for the book will be identical except for the first and last digits (because the first digit for U.S. publications is 0 and first digit for German publications is 3). If the last digit of the U.S. edition is 1, what is the last digit for the German publication?

WRITING PROJECTS

1. Prepare a report on coded information in your location. Possibilities for investigation include driver's license numbers in your state, student ID numbers and bar codes at your school, and bar codes used by your school library and city library. Identify the coding schemes and, when possible, determine whether a check digit is employed. Include samples. The Suggested Readings for this chapter contain information that will assist you.

2. Prepare a report on the driver's license coding schemes used by Minnesota, Michigan, Maryland, and Washington (the first three states use the same method). J. Gallian's "Assigning Driver's License Numbers" (see the Suggested Readings) has the information you will need.

3. Imagine that you are employed by a small company that doesn't use identification numbers and bar codes for its employees or products. As requested by your boss, prepare a report discussing the various methods and make a recommendation.

4. The Davidson Consonant Code and Metaphone are two text-retrieval algorithms based on sound that are alternatives to Soundex. Prepare a report on either of these encoding schemes. The article by K. Roughton and D. Tyckosen and the one by L. Philips (see the Suggested Readings) contain the information you will need.

5. Prepare a report on the check-digit method used on German currency prior to 2002. J. Gallian's article "Error Detection Methods" (see the Suggested Readings) has the information you will need.

More on the Web SPREADSHEET PROJECTS

To do these projects, go to www.whfreeman.com/fapp.

This project uses spreadsheets to compute check digits and verify the validity of identification numbers. Spreadsheets are particularly useful when this calculation uses weights, such as the 3131 . . . weighting of the UPC. Because the calculations are immediately updated, the impact of single-digit errors or transpositions can be easily analyzed using spreadsheets.

Transmitting Information

In this chapter we cover...

... and at www. whfreeman.com/fapp:

Flashcards

Quizzes

Spreadsheet Projects

Hamming codes

With the enormous volume of email, faxes, Internet traffic, and cellular phone calls, the Information Age has brought about many mathematical challenges. One is how to correct errors in data transmission. Another is how to electronically send and store information efficently. A third is how to ensure security of transmitted data. In this chapter, we illustrate some of the ways in which mathematicians and engineers have responded to these challenges.

10.1 Binary Codes

A system for coding data made up of two states (or symbols) is called a *binary code.* Binary codes are the hidden language of computers. The Postnet code and the UPC bar code are two examples of binary codes. Morse code and Braille are two more. Automatic 35mm cameras send a current across five silver or black squares on the film canister that code the film speed. The Ebert and Roeper "thumbs up/thumbs down" rating of films is a binary code with four messages. Compact disc players, fax machines, digital televisions, modems, cell phones, and space probes represent data as strings of 0's and 1's rather than the usual digits 0 through 9 and letters A through Z. In this section, we will illustrate one way binary codes can be devised so that errors in the transmission of the code can be corrected.

The idea behind error-correction schemes is simple and one you often use. To illustrate, suppose you are reading the employment section of a newspaper and you see the phrase "must have a minimum of bive years experience." Instantly you detect an error because *bive* is not a word in the English language. Moreover, you are fairly confident that the intended word is *five.* Why so? Because *five* is a word and it makes the phrase sensible. In other phrases, words such as *bike* or *give* might be sensible alternatives to *bive.* Using the extra information provided by the context, we are often able to infer the intended meaning when errors occur.

To demonstrate the way error-correcting schemes work, suppose that NASA sends a spacecraft to land at one of 16 possible landing sites on Mars. The spacecraft orbits Mars while surveying the sites for the most favorable landing conditions. NASA officials have coded the 16 landing sites with four-digit strings of 0's and 1's such as 0000, 0001, 0010, 0100 (see Table 10.1 for the complete list). Once the best site has been selected, NASA will inform the spacecraft where to land by sending the code for the site. However, signals sent through space are subject to interference called *noise.* The noise might cause the spacecraft to interpret the signal as 0001 when the signal actually sent was 1001. Fortunately, over the past 50 years mathematicians and engineers have devised highly sophisticated schemes to build extra information into messages composed of 0's and 1's that often permits one to infer the correct message even though the message may have been received incorrectly (see Spotlights 10.1 and 10.2).

SPOTLIGHT 10.1 The Ubiquitous Reed–Solomon Codes

One of the mathematical ideas underlying current error-correcting techniques for everything from computer hard disk drives to CD players was first introduced in 1960 by Irving Reed and Gustave Solomon. Reed–Solomon codes made possible the stunning pictures of the outer planets sent back by the space probes *Voyager 1* and *2*. They make it possible to scratch a compact disc and still enjoy the music.

"When you talk about CD players and digital audio tape and now digital television, and various other digital imaging systems that are coming–all of those Reed–Solomon [codes] as an integral part of the system," says Robert McEliece, a coding theorist at Caltech.

Why? Because digital information consists of 0's and 1's, and a physical device may occasionally confuse the two. *Voyager 2*, for example, was transmitting data at incredibly low power–barely a whisper–over billions of miles. Error-correcting codes are a kind of safety net, mathematical insurance against the vagaries of an imperfect material world.

In 1960, the theory of error-correcting codes was only about a decade old. Through the 1950s, a number of researchers began experimenting with a variety of error-correcting codes. But the Reed–Solomon paper, McEliece says, "hit the jackpot." "In hindsight it seems obvious," Reed recently said. However, he added, "Coding

Irving Reed and Gustave Solomon
At the Jet Propulsion Laboratory in 1989 to monitor the encounter of *Voyager 2* with Neptune. (Rex Ridenhouse.)

theory was not a subject when we published the paper." The two authors knew they had a nice result; they didn't know what impact the paper would have.

Four decades later, the impact is clear. The vast array of applications, both current and pending, has settled the questions of the practicality and significance of Reed–Solomon codes. Billions of dollars in modern technology depend on ideas that stem from Reed and Solomon's original work.

Source: Adapted from an article by Barry Cipra, with permission from *SIAM News,* January 1993, p. 1.

TABLE 10.1

Message	→	Code Word	Message	→	Code Word
0000	→	0000000	0110	→	0110010
0001	→	0001011	0101	→	0101110
0010	→	0010111	0011	→	0011100
0100	→	0100101	1110	→	1110100
1000	→	1000110	1101	→	1101000
1100	→	1100011	1011	→	1011010
1010	→	1010001	0111	→	0111001
1001	→	1001101	1111	→	1111111

SPOTLIGHT 10.2 Vera Pless

Vera Pless was born on March 5, 1931, to Russian immigrants on the West Side of Chicago. The neighborhood was intellectually stimulating, and there was a tradition of teaching each other things. At age 12, Pless was taught some calculus by a mathematics graduate student. She accepted a scholarship to attend the University of Chicago at age 15. The program at Chicago emphasized great literature but paid little attention to physics and mathematics. At age 18, with no more than one precalculus course in mathematics, she entered the prestigious graduate program in mathematics at Chicago, where, at that time, there were no women on the mathematics faculty nor even women colloquium speakers. After receiving her master's degree, Pless took a job as a research associate at Northwestern University while pursuing a Ph.D. there. In the midst of writing her thesis, she moved to Boston with her husband and continued to work on her thesis at home. She defended her thesis two weeks before her daughter was born.

Over the next several years, Pless stayed at home to raise her children and taught part time at Boston University. When she decided to work full time, she found that women were not welcome at most colleges and universities.

Vera Pless
A leader in the field of coding theory. (Courtesy of Vera Pless.)

Some people told her outright, "I would never hire a woman." Fortunately, there was an Air Force lab in the area that had a group working on error-correcting codes. Although she had never even heard of coding theory, she was hired because of her background in algebra. When the lab discontinued basic research, she took a position as a research associate at MIT. In 1975, she went to the University of Illinois – Chicago, where she remains. Having written more than 100 research papers and a widely used book on coding theory, Pless is a leader in the field.

As a simple example, let's assume our message is 1001. We will build extra information into this message with the aid of the diagram in Figure 10.1. Begin by placing the four message digits in the four overlapping regions I, II, III, IV, with the digit in the first position (starting at the left of the sequence) in region I, the digit in the second position in region II, and so on. For regions V, VI, and VII, assign 0 or 1 so that the total number of 1's in each circle is even. See Figure 10.2.

We have now encoded our message 1001 using the diagram as 1001101. Now suppose that this encoded message is received as 0001101 (an error in the first position). How would we know an error was made? We place each digit from the received message in its appropriate region, as in Figure 10.3.

Noting that in both circles A and B there is an odd number of 1's, we instantly realize that something is wrong, because the intended message had an even number of 1's in each circle. How do we correct the error? Because circles A and B have the wrong parity (parity refers to the oddness or evenness of a number—even integers have **even parity;** odd integers have **odd parity**) and C does not, the error is located in the portion of the diagram in circles A and B, but not in circle C; that is, region I (see Figure 10.4). Here we also see the advantage of using only 0's and 1's to encode data. If you have only two possibilities and one of them is incorrect, then the other one must be correct. Because the 0 in region I is incorrect, we know 1 is correct. This technique can be used to encode all 16 possible binary messages of length 4, as shown in the right-hand column of Table 10.1. The encoded messages are called *code words.* The extra three digits appended to each string of length 4 provides the "extra information" that is sufficient to infer the intended four-digit message as long as the received seven-digit message has at most one error. If a received message has two or more errors, this method will not always yield the correct message.

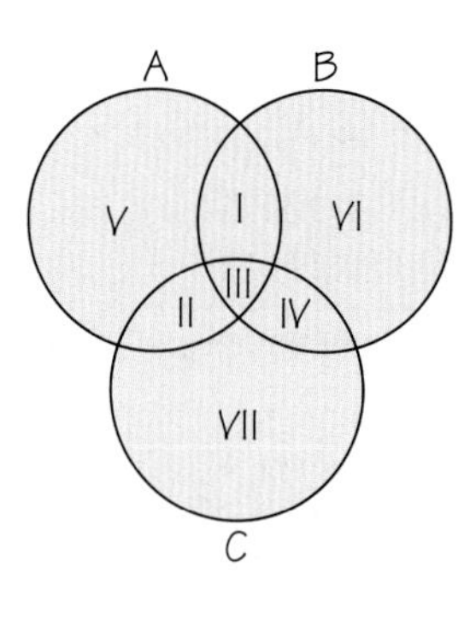

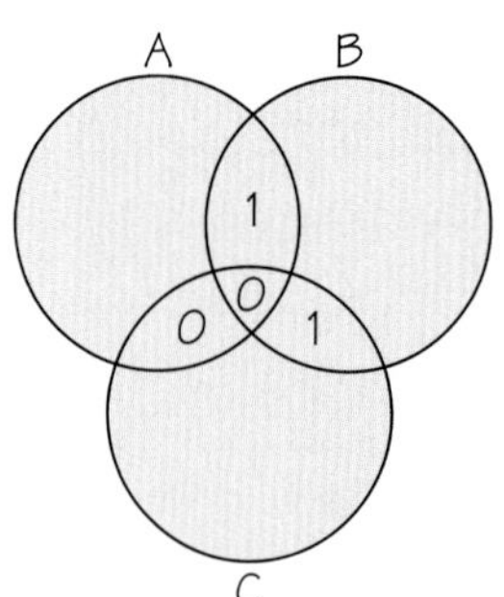

Figure 10.1 Diagram for message 1001.

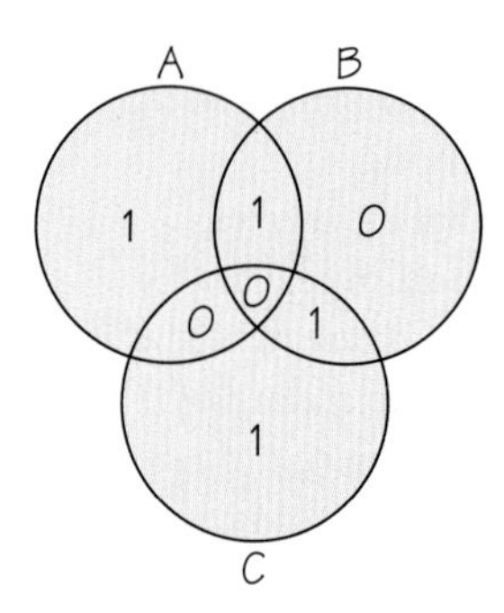

Figure 10.2 Diagram for encoded message 1001101.

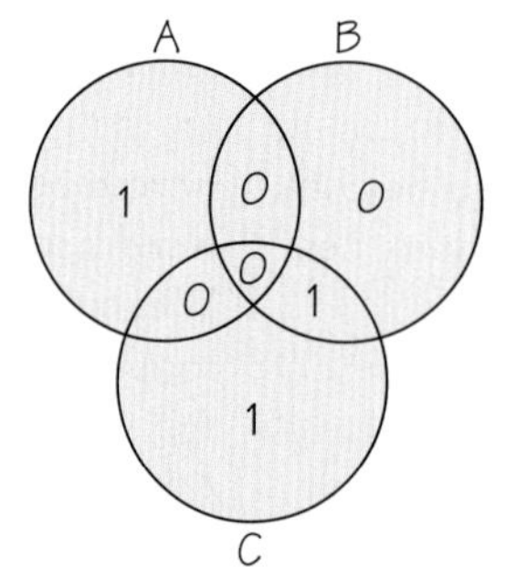

Figure 10.3 Diagram for received message 0001101.

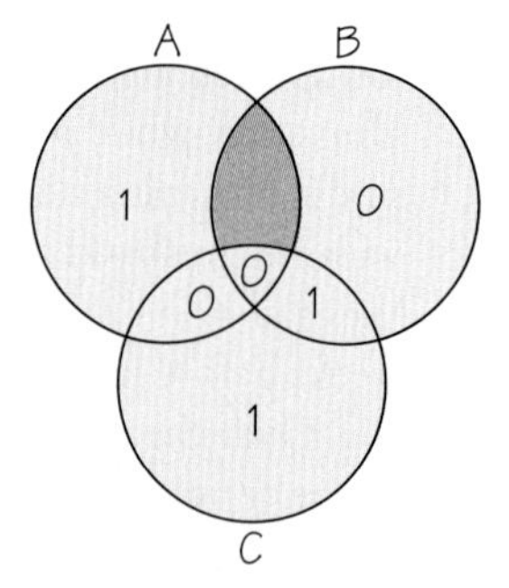

Figure 10.4 Circles A and B but not C have wrong parity.

10.2 Encoding with Parity-Check Sums

Strings of 0's and 1's with extra digits for error correction can be used to send full-text messages. A simple way to do this is to assign a space the string 00000, the letter a the string 00001, b the string 00010, c the string 00100, and so on (see Exercise 30). Because there are 32 possible binary strings of length 5, the five unassigned strings can be used for special purposes, such as indicating uppercase letters or numerals. For example, we might use the string 11111 to indicate a "shift" from lowercase to uppercase when it precedes the code for a letter (1111100010 represents B). This is analogous to the shift key on a keyboard. Similarly, we could use 11110 to indicate we are "shifting" from letters to numerals. Here 11110 followed by the code for a represents the numeral 0, 11110 followed by the code for b represents the numeral 1, and so on up to 9. Punctuation marks could be handled in the same fashion. However, our diagram method for assigning extra digits does not work for strings with five or more digits. Rather, the messages are encoded by appending extra digits determined by the parity of various sums of certain portions of the messages. We illustrate this method for the 16 messages shown in the left-hand column of Table 10.1. (See also Spotlight 10.3.)

Our goal is to take any binary string $a_1a_2a_3a_4$ and append three check digits $c_1c_2c_3$ so that any single error in any of the seven positions can be corrected. This is done as follows: Choose

$$c_1 = 0 \text{ if } a_1 + a_2 + a_3 \text{ is even}$$
$$c_1 = 1 \text{ if } a_1 + a_2 + a_3 \text{ is odd}$$
$$c_2 = 0 \text{ if } a_1 + a_3 + a_4 \text{ is even}$$
$$c_2 = 1 \text{ if } a_1 + a_3 + a_4 \text{ is odd}$$
$$c_3 = 0 \text{ if } a_2 + a_3 + a_4 \text{ is even}$$
$$c_3 = 1 \text{ if } a_2 + a_3 + a_4 \text{ is odd}$$

The sums $a_1 + a_2 + a_3$, $a_1 + a_3 + a_4$, and $a_2 + a_3 + a_4$ are called **parity-check sums.** They are so named because their function is to guarantee that the sum of various components of the encoded message is even. Indeed, c_1 is defined so that $a_1 + a_2 + a_3 + c_1$ is even. (Recall that this is precisely how the value in region V was defined.) Similarly, c_2 is defined so that $a_1 + a_2 + a_4 + c_2$ is even, and c_3 is defined so that $a_2 + a_3 + a_4 + c_3$ is even.

Let us revisit the message 1001 we considered in Figure 10.1. Then $a_1a_2a_3a_4 = 1001$ and

$$c_1 = 1 \text{ because } 1 + 0 + 0 \text{ is odd}$$
$$c_2 = 0 \text{ because } 1 + 0 + 1 \text{ is even}$$

and

$$c_3 = 1 \text{ because } 0 + 0 + 1 \text{ is odd}$$

So, because $c_1c_2c_3 = 101$, we have $1001 \rightarrow 1001101$.

SPOTLIGHT 10.3 Neil Sloane

In the middle of Neil Sloane's office, which is in the center of AT&T Bell Laboratories, which in turn is at the heart of the Information Age, there sits a tidy little pyramid of shiny steel balls stacked up like oranges at a neighborhood grocery. Sloane has been pondering different ways to pile up balls of one kind or another for most of his professional life. Along the way he has become one of the world's leading researchers in the field of sphere packing, a field that has become indispensable to modern communications. Without it, we might not have modems or compact discs or satellite photos of Neptune. "Computers would still exist," says Sloane. "But they wouldn't be able to talk to one another."

To exchange information rapidly and correctly, machines must code it. As it turns out, designing a code is a lot like packing spheres: Both involve cramming things together into the tightest possible arrangement. Sloane, fittingly, is also one of the world's leading coding theorists, not least because he has studied the shiny steel balls on his desk so intently.

Here's how a code might work. Imagine, for example, that you want to transmit a child's drawing that used every one of the 64 colors found in a jumbo box of Crayola crayons. For transmission, you could code each of those colors as a number–say, the integers from 1 to 64. Then you could divide the image into many small units, or pixels, and assign a code to each one based on the color it contains. The transmission would then be a steady stream of those numbers, one for each pixel.

In digital systems, however, all those numbers would have to be represented as strings of 0's and 1's. Because there are 64 possible combinations of 0's and 1's in a six-digit string, you could handle the entire Crayola palette with 64 different six-digit "code words." For example,

Neil Sloane
At work, wearing his famous "Codemart" T-shirt (952 points in a sphere). (Courtesy of Neil Sloane/AT&T Labs.)

000000 could represent the first color, 000001 the next color, 000010 the next, and so on.

But in a noisy signal, two different code words might look practically the same. A bit of noise, for example, might shift a spike of current to the wrong place, so that 001000 looks like 000100. The receiver might then wrongly color someone's eyes. An efficient way to keep the colors straight in spite of noise is to add four extra digits to the six-digit code words. The receiver, programmed to know the 64 permissible combinations, could now spot any other combination as an error introduced by noise and it would automatically correct the error to the "nearest" permissible color.

In fact, says Sloane, "If any of those ten digits were wrong, you could still figure out what the right crayon was."

Source: Adapted from an article by David Berreby, *Discover*, October 1990.

Now how is the intended message determined from a received encoded message? This process is called **decoding.** Say, for instance, that the message 1000, which has been encoded using parity-check sums as u = 1000110, is received as v = 1010110 (an error in the third position). We simply compare v with each of the 16 code words (that is, the possible correct messages) in Table 10.1 and decode it as the one that differs from v in the fewest positions. (Put another way, we decode v as the code word that agrees with v in the most positions.) This method works even if the error in the message is one of the check digits rather than one of the digits of the original message string. In a situation when there is more than one code word that differs from v in the fewest positions, we do not decode. To carry out this comparison, it is convenient to define the distance between two strings of equal length.

The **distance between two strings** of equal length is the number of positions in which the strings differ.

For example, the distance between v = 1010110 and u = 1000110 is 1, because they differ in only one position (the third). In contrast, the distance between 1000110 and 0111001 is 7, because they differ in all seven positions. Thus, our decoding procedure is simply to decode any received message v as the code word v' that is "nearest" to v in the sense that among all distances between v and code words, the distance between v and v' is a minimum. (If there is more than one possibility for v', we do not decode.) Table 10.2 shows the distance between v = 1010110 and all 16 code words. From this table we see that v will be decoded as u, because it differs from u in only one position whereas it differs from all others in the table in at least two positions. This method is called *nearest-neighbor decoding.*

The **nearest-neighbor decoding** method decodes a received message as the code word that agrees with the message in the most positions.

TABLE 10.2

v	1010110	1010110	1010110	1010110	1010110	1010110	1010110	1010110
code word	0000000	0001011	0010111	0100101	1000110	1100011	1010001	1001101
distance	4	5	2	5	1	4	3	4
v	1010110	1010110	1010110	1010110	1010110	1010110	1010110	1010110
code word	0110010	0101110	0011100	1110100	1101000	1011010	0111001	1111111
distance	3	4	3	2	5	2	6	3

Assuming that errors occur independently, the nearest-neighbor method decodes each received message as the one it most likely represents.

The scheme we have just described was first proposed in 1948 by Richard Hamming, a mathematician at Bell Laboratories. It is one of a family of codes that are called the Hamming codes.

Strings of 0's and 1's obtained from all possible k-tuples of 0's and 1's by appending extra 0's and 1's using parity-check sums, as illustrated earlier, are called *binary linear codes*. The strings with the appended digits are called *code words*.

A **binary linear code** consists of words composed of 0's and 1's obtained from all possible k-tuple messages by using parity-check sums to append check digits to the messages. The resulting strings are called **code words.**

You should think of a binary linear code as a set of n-tuples in which each n-tuple is composed of two parts: the message part, consisting of the original k-digit messages, and the remaining check-digit part.

The longer the messages are, the more check digits are required to correct errors. For example, binary messages consisting of six digits require four check digits to ensure that all messages with one error can be decoded correctly. Where there is no possibility of confusion, it is customary to denote an n-tuple $(a_1, a_2, \ldots, a_n)$ more concisely as $a_1a_2 \ldots a_n$, as we did in Table 10.1.

Given a binary linear code, how can we tell whether it will correct errors and how many errors it will detect? It is remarkably easy. We examine all the code words to find one that has the fewest number of 1's, excluding the code word consisting entirely of 0's. Call this minimum number of 1's in any nonzero code word the *weight* of the code and denote it by t.

The **weight of a binary code** is the minimum number of 1's that occur among all nonzero code words of that code.

If t is odd, the code will correct any $(t - 1)/2$ or fewer errors; if t is even, the code will correct any $(t - 2)/2$ or fewer errors. If we prefer simply to detect errors rather than to correct them (as is often the case in applications), the code will detect any $t - 1$ or fewer errors.

Applying this test to the code in Table 10.1, we see that the weight is 3, so it will correct any $(3 - 1)/2 = 1$ error or it will detect any $3 - 1 = 2$ errors. Be careful here. We must decide *in advance* whether we want our code to correct single errors or detect double errors. It can do whichever we choose, but not both. If we decide to detect errors, then we will not decode any message that was not among our original list of encoded messages (just as *bive* is not a word in the English language). Instead, we simply note that an error was made and, in most applications, request a retransmission. An example of this occurs when a bar-code reader at the supermarket detects an error and therefore does not emit a sound (in effect, requesting a rescanning). On the other hand, if we decide to correct errors, we will decode any received message as its nearest neighbor.

Here is an example of another binary linear code. Let the set of messages be {000, 001, 010, 100, 110, 101, 011, 111} and append three check digits c_1, c_2, and c_3 using

$$\begin{aligned}
c_1 &= 0 \text{ if } a_1 + a_2 + a_3 \text{ is even} \\
c_1 &= 1 \text{ if } a_1 + a_2 + a_3 \text{ is odd} \\
c_2 &= 0 \text{ if } a_1 + a_3 \text{ is even} \\
c_2 &= 1 \text{ if } a_1 + a_3 \text{ is odd} \\
c_3 &= 0 \text{ if } a_2 + a_3 \text{ is even} \\
c_3 &= 1 \text{ if } a_2 + a_3 \text{ is odd}
\end{aligned}$$

For example, if we take $a_1a_2a_3$ as 101, we have

$$\begin{aligned}
c_1 &= 0 \text{ because } 1 + 0 + 1 \text{ is even} \\
c_2 &= 0 \text{ because } 1 + 1 \text{ is even} \\
c_3 &= 1 \text{ because } 0 + 1 \text{ is odd}
\end{aligned}$$

So we encode 101 by appending 001, that is, 101 → 101001. The entire code is shown in Table 10.3.

Because the minimum number of 1's of any nonzero code word is three, this code will either correct any single error or detect any double error, whichever we choose.

It is natural to ask how the method of appending extra digits with parity-check sums enables us to detect or even correct errors. Error detection is obvious. Think of how a computer spell checker works. If you type *bive* instead of *five,* the spell checker detects the error because the string *bive* is not on its list of valid words. On the other hand, if you type *give* instead of *five,* the spell checker will not detect the error because *give* is on its list of valid words.

TABLE 10.3

Message	→	Code Word	Message	→	Code Word
000	→	000000	110	→	110011
001	→	001111	101	→	101001
010	→	010101	011	→	011010
100	→	100110	111	→	111100

Our error-detection scheme works the same way, except that if we add extra digits to ensure that our code words differ in many positions—say, t positions—then even as many as $t - 1$ mistakes will not convert one code word into another code word. And if every pair of code words differs from each other in at least three positions, we can correct any single error because the incorrect received word will differ from the correct code word in one position, but it will differ from all others in two or more positions. Thus, in this case, the correct word is the "nearest neighbor." So the role of the parity-check sums is to ensure that code words differ in many positions. For example, consider the code in Table 10.1. The messages 1000 and 1100 differ in only the second position. But the two parity-check sums $a_1 + a_2 + a_3$ and $a_2 + a_3 + a_4$ will guarantee that encoded words for these messages will have different values in positions 5 and 7 as well as in position 2. It is the job of mathematicians to discover the appropriate parity-check sums to correct several errors in long, complicated codes.

Data Compression

Binary linear codes are fixed-length codes. In a fixed-length code, each code word is represented by the same number of digits (or symbols). In contrast, the Morse code (see Figure 10.5), designed for the telegraph, is a **variable-length** code; that is, a code in which the number of symbols for each code word may vary.

Notice that in the Morse code the letters that occur most frequently have the shortest coding, whereas the letters that occur the least frequently have the longest coding. By assigning the code in this manner, telegrams could convey more information per line than would be the case for fixed-length codes or a randomly assigned variable-length coding of the letters. The Morse code is an example of data compression.

Data compression is the process of encoding data so that the most frequently occurring data are represented by the fewest symbols.

Figure 10.6 shows a typical frequency distribution for letters in English-language text material.

Data compression provides a means to reduce the costs of data storage and transmission. Let us illustrate the principles of data compression with a simple example. Biologists are able to describe genes by specifying sequences composed of the four letters A, T, G, and C, which represent the four nucleotides adenine, thymine, guanine, and cytosine, respectively. One way to encode a sequence such as AAACAGTAAC in fixed-length binary form would be to encode the letters as

$$A \rightarrow 00 \quad C \rightarrow 01 \quad T \rightarrow 10 \quad G \rightarrow 11$$

Figure 10.5 Morse code.

A	·—	N	—·
B	—···	O	———
C	—·—·	P	·——·
D	—··	Q	——·—
E	·	R	·—·
F	··—·	S	···
G	——·	T	—
H	····	U	··—
I	··	V	···—
J	·———	W	·——
K	—·—	X	—··—
L	·—··	Y	—·——
M	——	Z	——··

Figure 10.6 A widely used frequency table for letters in normal English usage.

	A	B	C	D	E	F	G	H	I	J	K	L	M
Percentage:	8	1.5	3	4	13	2	1.5	6	6.5	0.5	0.5	3.5	3

	N	O	P	Q	R	S	T	U	V	W	X	Y	Z
Percentage:	7	8	2	0.25	6.5	6	9	3	1	1.5	0.5	2	0.25

The corresponding binary code for the sequence AAACAGTAAC is then

$$00000001001110000001$$

On the other hand, if we knew from experience that the hierarchy of occurrence of the letters is A, C, T, and G (that is, A occurs most frequently, C second most frequently, and so on) and that A occurs much more frequently than T and G together, the most efficient binary encoding would be

$$A \rightarrow 0 \quad C \rightarrow 10 \quad T \rightarrow 110 \quad G \rightarrow 111$$

For this encoding scheme the sequence AAACAGTAAC is encoded as

$$0001001111100010$$

Notice that this binary sequence has 20% fewer digits than our previous sequence, in which each letter was assigned a fixed length of 2 (16 digits versus 20 digits). However, to realize this savings, we have made decoding more difficult. For the binary sequence using the fixed length of two symbols per character, we decode the sequence by taking the digits two at a time in succession and converting them to the corresponding letters. For the compressed coding, we can decode by examining the digits in groups of three.

EXAMPLE

Decode 0001001111100010

Consider the compressed binary sequence 0001001111100010. Look at the first three digits: 000. Since our code words have one, two, or three digits and neither 00 nor 000 is a code word, the sequence 000 can only represent the *three* code words 0, 0, and 0. Now look at the next three digits: 100. Again, because neither 1 nor 100 is a code word, the sequence 100 represents the *two* code words 10 and 0. The next three digits, 111, can only represent the code word 111 because the other three code words all contain at least one 0. Next consider the sequence 110. Because neither 1 nor 11 is a code word, the sequence 110 can only represent 110 itself. Continuing in this fashion, we can decode the entire sequence to obtain AAACAGTAAC.

The following observation can simplify the decoding process for compressed sequences. Note that 0 occurs only at the end of a code word. Thus, each time you see a 0, it is the end of the code word. Also, because the code words 0, 10, and 110 end in a 0, the only circumstances under which there are three consecutive 1's is when the code word is 111. So, to quickly decode a compressed binary sequence using our coding scheme, insert a comma after every 0 and after every three consecutive 1's. The digits between the commas are code words. ■

EXAMPLE

Code AGAACTAATTGACA and Decode the Result

Recall: A → 0, C → 10, T → 110, and G → 111. So

AGAACTAATTGACA → 01110010110001101101110100

To decode the encoded sequence, we insert commas after every 0 and after every occurrence of 111 and convert to letters:

0, 111, 0, 0, 10, 110, 0, 0, 110, 110, 111, 0, 10, 0
A, G, A, A, C, T, A, A, T, T, G, A, C, A ■

Modern data compression schemes were first invented in the 1950s (see Spotlight 10.4). They are now routinely used by modems, fax machines, and the Internet for data transmissions and by computers for data storage. In many cases, data compression results in a saving of up to 50% on transmission time or storage space.

10.3 Cryptography

Thus far, we have discussed ways in which data can be encoded to detect errors or correct errors in transmission. In many situations there is also a desire for

SPOTLIGHT 10.4 David Huffman

Large networks of IBM computers use it. So do high-definition televisions, modems, and a popular electronic device that takes the brainwork out of programming a videocassette recorder. All these digital wonders rely on the results of a 50-year-old term paper by an MIT graduate student–a data compression scheme known as Huffman encoding.

In 1951 David Huffman and his classmates in an electrical engineering graduate course on information theory were given the choice of a term paper or a final exam. For the term paper, Huffman's professor had assigned what at first appeared to be a simple problem. Students were asked to find the most efficient method of representing numbers, letters, or other symbols using binary code. Huffman worked on the problem for months, developing a number of approaches, but none that he could prove to be the most efficient. Finally, he despaired of ever reaching a solution and decided to start studying for the final. Just as he was throwing his notes in the garbage, the solution came to him. "It was the most singular moment of my life," Huffman says. "There was the absolute lightning of sudden realization. It was my luck to be there at the right time and also not have my professor discourage me by telling me that other good people had struggled with the problem," he says. When presented with his student's discovery, Huffman recalls, his professor exclaimed: "Is that all there is to it!"

"The Huffman code is one of the fundamental ideas that people in computer science and data communications are using all the time," says Donald Knuth of Stanford University. Although others have used Huffman's code to help make millions of dollars, Huffman's main compensation was dispensation from the final exam. He never tried to patent an invention from his work and experiences only a twinge of regret at not having used his creation to make himself rich. "If I had the best of both worlds, I would have had recognition as a scientist, and I would have gotten monetary rewards," he says. "I guess I got one and not the other."

David Huffman
(Matthew Mulbry.)

But Huffman has received other compensation. A few years ago an acquaintance told him that he had noticed that a reference to the code was spelled with a lowercase *h*. Remarked his friend to Huffman, "David, I guess your name has finally entered the language."

Source: Adapted from an article by Gary Stix, *Scientific American,* September 1991, pp. 54, 58.

security against unauthorized interpretation of coded data (that is, a desire for secrecy). The process of disguising data is called **encryption. Cryptology** is the study of methods to make and break secret codes.

Historically, encryption was used primarily for military and diplomatic transmissions. Today, encryption is essential for securing electronic transactions of all kinds. Cryptography is what allows you to have a Web site safely receive your credit card number. Cryptographic schemes prevent hackers from charging calls to your cellular phone. Crytography is also used for authenticating electronic transactions. In September 1998 history was made when former President Bill Clinton and Ireland's Prime Minister Bertie Ahren used digital signatures to sign an intergovernmental document. Each leader had a unique signing code and a digital certificate that served as a "digital ID," thereby ensuring that the document was approved by them. Although modern encryption schemes are extremely complex, we will illustrate the fundamental concepts involved with a few simple examples.

Among the first known cryptosystems is the so-called *Caesar cipher* used by Julius Caesar to send messages to his troops. To encrypt a message, Caesar used the following table to replace each letter in the top row with the letter below it.

A B C D E F G H I J K L M N O P Q R S T U V W X Y Z
D E F G H I J K L M N O P Q R S T U V W X Y Z A B C

For example, the message ATTACK AT DAWN is encrypted as DVVDFN DW GDZQ.

To decrypt the message, his soldiers replaced each letter with the letter above it in the table. Obviously, it would not require much effort for someone to "crack" this code.

To describe a more sophisticated scheme for transmitting messages secretly, it is convenient to introduce a special kind of arithmetic used in cryptography. For any positive integers a and n, we define a mod n (read: "a modulo n" or just "a mod n") to be the remainder when a is divided by n. Thus,

$$3 \bmod 2 = 1 \text{ because } 3 = 1 \cdot 2 + 1$$
$$6 \bmod 2 = 0 \text{ because } 6 = 3 \cdot 2 + 0$$
$$5 \bmod 3 = 2 \text{ because } 5 = 1 \cdot 3 + 2$$
$$37 \bmod 10 = 7 \text{ because } 37 = 3 \cdot 10 + 7$$
$$38 \bmod 26 = 12 \text{ because } 38 = 1 \cdot 26 + 12$$
$$342 \bmod 85 = 2 \text{ because } 342 = 4 \cdot 85 + 2$$
$$62 \bmod 85 = 62 \text{ because } 62 = 0 \cdot 85 + 62$$

Arithmetic involving mod n is called **modular arithmetic.** Although this arithmetic may appear unfamiliar, you often unconsciously use it. For example, if it is now September, what month will it be 25 months henceforth? Of course, you answer "October," but the interesting fact is that you didn't arrive at the

SPOTLIGHT 10.5 Modeling the Genetic Code

The way that genetic material is comprised can be conveniently modeled using modulo 4 arithmetic. A DNA molecule is comprised of two long strands in the form of a double helix. Each strand is made up of strings of the four nitrogen bases adenine (A), thymine (T), guanine (G), and cytosine (C). Each base on one strand binds to a complementary base on the other strand. Adenine is always bound to thymine and guanine is always bound to cytosine. To model this situation, we identify A with 0, T with 2, G with 1, and C with 3. Thus, the DNA segment ACGTAACAGGA and its complement segment TGCATTGTCCT are identified by 03120030110 and 21302212332.

Using modulo 4 arithmetic, $0 + 2 = 2$, $2 + 2 = 0$, $1 + 2 = 3$, and $3 + 2 = 1$, and we see that adding 2 to any of the integers 0, 1, 2, or 3 interchanges 0 and 2 and 1 and 3. So, for any DNA segment $a_1a_2 \ldots a_n$ represented by strings of 0's, 1's, 2's, and 3's, we see that its complementary segment is represented by $a_1a_2 \ldots a_n + 22 \ldots 2$, where we add the integers in each component using modulo 4. In particular, $03120030110 + 22222222222 = 21302212332$.

Source: Adapted from *Discrete Mathematics* by S. Washburn, T. Marlow, and C. Ryan, Addison-Wesley, 1999.

answer by starting with September and counting off 25 months. Instead, without even thinking about it, you simply observed that $25 = 2 \cdot 12 + 1$ so that 25 mod 12 = 1 and you added one month to September. Similarly, if it is now Wednesday, you know that in 23 days it will be Friday. This time, you arrived at your answer by noting that $23 = 3 \cdot 7 + 2$ (that is, 23 mod 7 = 2), so you added 2 days to Wednesday instead of counting off 23 days. Likewise, if your electricity is off for 26 hours, you know you must advance your clock 2 hours, because 26 mod 12 = 2. An application of modular arithmetic to genetics is described in Spotlight 10.5.

With modular arithmetic we can easily describe the Caesar cipher as follows. Begin by saying that the letter A is in position 0, B is in position 1, C is in position 2, and so on. Then the Caesar cipher replaces the letter in position i with the letter in position $(i + 3)$ mod 26. This formula expresses the fact that the Caesar cipher shifts each letter from A through W three positions to the right while X, Y, and Z are replaced with A, B, and C, respectively.

Modular arithmetic also provides the basis for a more sophisticated cryptosystem called the *Vigenère cipher.* For this method we first select a **key word,** which can be any word. The letters of the key word are then used to determine the amount of shifting for each letter of our message.

EXAMPLE Vigenère Cipher

We will use the Vigenère system to encrypt the message ATTACK AT DAWN. Choosing the key word MATH, we shift the first letter of the message by 12

TABLE 10.4

ATTACK AT DAWN	0	19	19	0	2	10	0	19	3	0	22	13
MATHMA TH MATH	12	0	19	7	12	0	19	7	12	0	19	7
MTMHOK TA PAPU	12	19	12	7	14	10	19	0	15	0	15	20

because M is in position 12; the second letter of the message is shifted by 0 (unchanged) because A is in position 0; the third letter of the message is shifted by 19 because T is in position 19, and so on. A shift of j means that the letter in position i is replaced by the letter in position $(i + j)$ mod 26. When we have used all the letters of the key word, we start over at the beginning. To encrypt ATTACK AT DAWN using the key word MATH, we first note that the letters in the key word MATH are in positions 12, 0, 19, and 7, respectively. So, the A in ATTACK is converted to M $(0 + 12 = 12)$, the first T in ATTACK is converted to T $(19 + 0 = 19)$, the second T in ATTACK is converted to M $((19 + 19) \bmod 26 = 12)$, and so on. The first two lines of Table 10.4 show the position numbers for the letters of the message and the key word. The third line of the table is obtained from the first two by adding the values in the columns mod 26 and converting the results back to letters. ■

Premium television channels such as HBO and Showtime also have a need for encryption to prevent their television signals to cable subscribers and satellite dish subscribers from being received free of charge.

Beginning in 1984, HBO scrambled its signal. To permit a cable system operator or dish owner who paid a monthly fee to unscramble the signal, HBO employed a **password** that was changed monthly. The password was transmitted along with the scrambled signal and a subscriber sequence called a *key*. To describe this process, we need to perform addition of binary sequences. We add two binary sequences $a_1a_2 \ldots a_n$ and $b_1b_2 \ldots b_n$ as follows:

$$\begin{array}{r} a_1a_2 \ldots a_n \\ +\ b_1b_2 \ldots b_n \\ \hline c_1c_2 \ldots c_n \end{array}$$

where $c_i = 0$ if $a_i = b_i$ and $c_i = 1$ if $a_i \neq b_i$. Equivalently, $c_i = (a_i + b_i) \bmod 2$. (Add a_i and b_i in the ordinary way but replace 2 by 0.)

EXAMPLE Sum of Binary Sequences

$$\begin{array}{r} 11000111 \\ +\ 01110110 \\ \hline 10110001 \end{array} \qquad \begin{array}{r} 00111011 \\ +\ 01100101 \\ \hline 01011110 \end{array} \qquad \begin{array}{r} 10011100 \\ +\ 10011100 \\ \hline 00000000 \end{array}$$

■

The data security method we describe hinges on the fact that the sum of two binary sequences $a_1a_2 \ldots a_n + b_1b_2 \ldots b_n = 00 \ldots 0$ if and only if the sequences are identical.

Although HBO used binary sequences of length 56, we will illustrate the method with sequences of length 8. Let us say that the password for the month was p. Each subscriber of the service was assigned a sequence, called a **key,** uniquely associated with him or her. Let us say that the list of keys issued by HBO to its customers was $k_1, k_2, \ldots$. HBO transmitted the password p, as well as the encrypted sequences $k_1 + p, k_2 + p, \ldots$ (that is, one sequence for each authorized user). A microprocessor in each subscriber's decoding box added its key, say k_i, to each of the encrypted sequences. That is, it calculated $k_i + (k_1 + p)$, $k_i + (k_2 + p), \ldots$. As it did so, the microprocessor compared each of these calculated sequences with the correct password p. When one of the sequences matched p, the microprocessor would unscramble the signal. Notice that the correct password p was produced precisely when k_i was added to $k_i + p$, because $k_i + (k_i + p) = (k_i + k_i) + p = 00 \ldots 0 + p = p$ and $k_i + (k_j + p) \neq p$ when $k_j \neq k_i$. (That is, key k_i "unlocked" the encrypted sequence $k_i + p$ and no other.) If a subscriber with key k_i failed to pay the monthly bill, HBO could terminate the service by not transmitting the sequence $k_i + p$ the next month.

EXAMPLE Encryption and Decoding

The password for some month was $p = 10101100$ and a subscriber key was $k = 00111101$. One of the sequences transmitted by HBO was $k + p$:

$$\begin{array}{r} 00111101 \\ +\ 10101100 \\ \hline 10010001 \end{array}$$

The subscriber's decoder box added his key $k = 00111101$ to each of the sequences received and compared the result with the password p. Eventually, it found the sequence obtained by adding the password to the subscriber's key (namely, $p + k = 10010001$) and calculated

$$\begin{array}{r} 00111101 \\ +\ 10010001 \\ \hline 10101100 \end{array}$$

to obtain the password p. Once the password was found, the decoder descrambled the signal. ■

OPTIONAL Public Key Cryptography

In the mid-1970s Ron Rivest, Adi Shamir, and Len Adleman devised an ingenious method that permits each person who is to receive a secret message to publicly tell how to scramble messages sent to him or her. And even though the method used to scramble the message is known publicly, only the person for whom it is intended will be able to unscramble the message.

To illustrate their method for transmitting messages secretly, we need the following property of modular arithmetic:

$$(ab) \bmod n = ((a \bmod n)(b \bmod n)) \bmod n$$

This property allows you to replace integers greater than or equal to n with integers less than n to simplify calculations. You should think of it as saying, "mod before you multiply."

EXAMPLE

Multiplication Property for Modular Arithmetic

$$\begin{aligned}
(17 \cdot 23) \bmod 10 &= ((17 \bmod 10)(23 \bmod 10)) \bmod 10 \\
&= (7 \cdot 3) \bmod 10 = 21 \bmod 10 = 1 \\
(22 \cdot 19) \bmod 8 &= ((22 \bmod 8)(19 \bmod 8)) \bmod 8 \\
&= (6 \cdot 3) \bmod 8 = 18 \bmod 8 = 2 \\
(100 \cdot 8) \bmod 85 &= ((100 \bmod 85)(8 \bmod 85)) \bmod 85 \\
&= (15 \cdot 8) \bmod 85 = 120 \bmod 85 = 35
\end{aligned}$$

■

We now describe the Rivest, Shamir, and Adleman method by way of a simple example that nevertheless illustrates the essential features of the method. Say we wish to send the message "IBM." We convert the message to digits by replacing A by 1, B by 2, . . . , and Z by 26. So the message IBM becomes 9213. The person to whom the message is to be sent has picked two primes p and q, say, $p = 5$ and $q = 17$. (Recall that a *prime* is an integer greater than 1 whose only divisors are 1 and itself.) The receiver has also picked a number r, such as 3, that has no divisors in common with the least common multiple m of $(p - 1) = 4$ and $(q - 1) = 16$ other than 1, and published $n = pq = 85$ and $r = 3$ in a public directory. To decode our message, the receiver must find a number s so that $r \cdot s = 1 \bmod m$ (this is where knowledge of p and q is necessary). That is, $3 \cdot s = 1 \bmod 16$. This number is 11. (The number s can be found by calculating successive powers of r mod m; when 1 is reached, the previous power of r is s. In our example we have $3 \bmod 16 = 3$, $3^2 \bmod 16 = 9$, $3^3 \bmod 16 = 11$, $3^4 \bmod 16 = 1$, so $s = 3^3 \bmod 16 = 11$.)

To send our message to this person, we consult the public directory to find $n = 85$ and $r = 3$, then send the "scrambled" numbers $9^3 \bmod 85$, $2^3 \bmod 85$, and $13^3 \bmod 85$ rather than 9, 2, and 13, and the receiver will unscramble them. Thus, we send

$$\begin{aligned}
9^3 \bmod 85 &= 49 \\
2^3 \bmod 85 &= 8 \\
13^3 \bmod 85 &= 72
\end{aligned}$$

Now the receiver must take the numbers he or she receives—49, 8, and 72—and convert them back to 9, 2, and 13 by calculating 49^{11} mod 85, 8^{11} mod 85, and 72^{11} mod 85.

The calculation of 49^{11} mod 85 can be simplified as follows:[1]

$$\begin{aligned}
49 \bmod 85 &= 49\\
49^2 \bmod 85 &= 2401 \bmod 85 = 21\\
49^4 \bmod 85 &= 49^2 \cdot 49^2 \bmod 85 = 21 \cdot 21 \bmod 85 = 441 \bmod 85 = 16 \bmod 85\\
49^8 \bmod 85 &= 49^4 \cdot 49^4 \bmod 85 = 16 \cdot 16 \bmod 85 = 1
\end{aligned}$$

So, $49^{11} \bmod 85 = (49^8 \bmod 85)(49^2 \bmod 85)(49 \bmod 85)$.

$$\begin{aligned}
&= (1 \cdot 21 \cdot 49) \bmod 85\\
&= 1029 \bmod 85\\
&= 9 \bmod 85
\end{aligned}$$

Thus, the receiver has correctly determined the code for *I*. The calculations for 8^{11} mod 85 and 72^{11} mod 85 are left as exercises for the reader. Notice that without knowing how $n = pq$ factors, one cannot find the least common multiple of $p - 1$ and $q - 1$ (in our case, 16), and therefore the s that is needed to determine the intended message.

The procedure just described is called the **RSA public key encryption scheme** in honor of Rivest, Shamir, and Adleman, who discovered it. The method is practical and secure because there exist efficient methods for finding very large prime numbers (say, about 100 digits long) and for multiplying large numbers, but no one knows an efficient algorithm for factoring large integers (say, about 200 digits long).

The algorithm is summarized below. (In practice, the messages are not sent one letter at a time. Rather, the entire message is converted to decimal form, with *A* represented by 01, *B* by 02, . . . , and a space by 00. The message is then broken up into blocks of uniform size and the blocks are sent. See step 2 under Sender below.)

Receiver

1. Pick very large primes p and q and compute $n = pq$.
2. Compute the least common multiple of $p - 1$ and $q - 1$; let us call it m.
3. Pick r so that it has no divisors in common with m other than 1 (any such r will do).

[1]To determine 49^2 mod 85 with a calculator, enter 49×49 to obtain 2401, then divide 2401 by 85 to obtain 28.247058. Finally, enter $2401 - (28 \times 85)$ to obtain 21.

4. Find s so that $rs = 1 \bmod m$. (To find s, simply compute $r^2 \bmod m$, $r^3 \bmod m$, $r^4 \bmod m$, . . . until you reach $r^t \bmod m = 1$. Then $s = r^{t-1} \bmod m$.)
5. Publicly announce n and r, but keep p, q, and s secret.

Sender

1. Convert the message to a string of digits.
2. Break up the message into uniformly sized blocks of digits, appending 0's in the last block if necessary; call them $M_1, M_2, \ldots, M_k$. For example, for a string such as 2105092315, we would use $M_1 = 2105$, $M_2 = 0923$, and $M_3 = 1500$.
3. Check to see that the greatest common divisor of each M_i and n is 1. If not, n can be factored and the code is broken. (In practice, the primes p and q are so large that they exceed all M_i, so this step may be omitted.)
4. Calculate and send $R_i = M_i^r \bmod n$.

Receiver

1. For each received message R_i, calculate $R_i^s \bmod n$.
2. Convert the string of digits back to a string of characters.

Why does this method work? It works because of a basic property of modular arithmetic and the choice of r. It so happens that the number m has the property that for each positive integer x having no common divisors with n except 1, we have $x^m = 1 \bmod n$. So, because each message M_i has no common divisors with n except 1, and r was chosen so that $rs = 1 + mt$ for some t, we have modulo n,

$$R_i^s = (M_i^r)^s = M_i^{rs} = M_i^{1+mt} = M_i M_i^{mt} = M_i (M_i^m)^t = M_i 1^t = M_i$$

OPTIONAL Digital Signatures

With so many electronic financial transactions now taking place over the Internet, the need for authenticity is paramount. How is a stockbroker to know that an electronic message he receives to sell one stock and buy another actually came from his client and not someone posing as the client? The technique used in public key cryptography allows for secure digital signatures as well.

Let us say that Alice wants to send a secret message to Bob in such a way that only Bob can decode the message and that Bob will know that only Alice could have sent it. Abstractly, let E_A and D_A denote the algorithms that Alice uses for encryption and decryption respectively, and let E_B and D_B denote the algorithms that Bob uses for encryption and decryption, respectively. (Just as with the RSA encryption scheme, we assume that the encryption methods E_A for Alice and E_B for Bob are available to the public, whereas the decryption algorithm D_A is known only to Alice and the decryption algorithm D_B is known only to Bob.)

For Alice to send a message M that only Bob can read in such a way that Bob will know that the message could only have been sent by Alice, Alice sends

M to Bob by first applying her decryption algorithm D_A to M, then applying his encryption algorithm E_B to the result. In symbols, she sends $E_B(D_A(M))$ to Bob. Bob, in turn, decodes the message $E_B(D_A(M))$ sent to him by first applying his decryption algorithm D_B (this undoes the E_B part of the encoded message) and then applying her encryption algorithm E_A to $D_A(M)$ to undo the decryption algorithm Alice applied to M. In symbols, Bob is calculating

$$(E_A D_B)(E_B(D_A(M))) = E_A(D_B E_B)(D_A(M)) = E_A(D_A(M)) = M$$

This works because whenever D_B and E_B appear back to back, they cancel each other so that $D_A(M)$ is unchanged. Then applying E_A to $D_A(M)$ simply undoes what D_A had done to M, because E_A and D_A applied in either order to M results in M. [For example, if E_A means add 3 to a number, then D_A subtracts 3 from a number. So, if we apply D_A to M and then apply E_A to the result, we obtain $(M - 3) + 3 = M$.]

Notice that only Alice can execute the first step [i.e., create $D_A(M)$] and only Bob can implement the last step (i.e., apply $E_A D_B$ to the received message). Moreover, Bob knows that Alice initiated the process, because no one but she could have begun it by applying D_A to M.

Transactions using digital signatures became legally binding in the United States in October 2000.

REVIEW VOCABULARY

Binary linear code A code consisting of words composed of 0's and 1's obtained by using parity-check sums to append check digits to messages.
Code words Words from a binary linear code.
Cryptography The study of how to make and break secret codes.
Data compression The process of encoding data so that the most frequently occurring data are represented by the fewest symbols.
Decoding The process of translating received data into code words.
Distance between two strings The distance between two strings of equal length is the number of positions in which they differ.
Encryption The process of encoding data to protect against unauthorized interpretation.
Even parity Even integers are said to have even parity.
Key A string used to encode and decode data.
Modular arithmetic Addition and multiplication involving modulo n.
Nearest-neighbor decoding A method that decodes a received message as the code word that agrees with the message in the most positions.
Odd parity Odd integers are said to have odd parity.
Parity-check sums Sums of digits whose parities determine the check digits.
Password A word used to encode data.
RSA public key encryption scheme A method of encoding that permits each person to announce publicly the means by which secret messages are to be sent to him or her.
Variable-length code A code in which the number of symbols for each code word may vary.
Weight of a binary code The minimum number of 1's that occur among all nonzero code words of a code.

SUGGESTED READINGS

DENEEN, L. Secret encryption with public keys. *UMAP Journal,* 8 (1987): 9–29. Describes several ways in which modular arithmetic can be used to code secret messages.

PETZOLD, C. *Code,* Microsoft Press, Redmond, Wash., 1999. The first three chapters of this book provide an excellent explanation of the Morse code and the Braille system of coding.

RICHARDS, I. The invisible prime factor, *American Scientist,* 70 (1982): 176–179. Explains how elementary number theory and modular arithmetic can be used to test whether an integer is prime and how prime numbers can be used to create secret codes that are extremely difficult to break.

THOMPSON, T. *From Error-Correcting Codes Through Sphere Packing to Simple Groups,* Mathematical Association of America, Washington, D.C., 1983. Chapter 1 of this award-winning book gives a fascinating historical account of the origins of error-correcting codes.

SUGGESTED WEB SITES

www.d.umn.edu/~jgallian/fapp6 This Web site implements the nearest-neighbor decoding method for seven-digit binary strings using the code given in Table 10.1.

hotwired.com This site features the latest news about encryption schemes and applications.

www.nyise.org/blind/britbrl2.htm This site describes the Braille writing system.

SKILLS CHECK

1. If you use the circular diagram method to encode the message 1011, what is the encoded message?

(a) 1011001
(b) 1011010
(c) 1010001

2. Suppose the message 1010010 is received and decoded using the nearest-neighbor method. What message is recovered?

(a) 1010
(b) 1011
(c) 1110

3. What is the distance between received words 1011001 and 1000101?

(a) 0
(b) 1
(c) 3

4. Use the encoding scheme A → 0, B → 10, C → 11 to decode the sequence 010110.

(a) ABCB
(b) ABCA
(c) ABACA

5. What is the sum of the binary sequences 1011001 and 1001101?

(a) 0100110
(b) 0010100
(c) 1011101

6. Using modular arithmetic, 3^5 mod 20 is equal to

(a) 3.
(b) 12.
(c) 15.

7. Use the RSA scheme with $n = 91$ and $s = 5$ to decode the message 4.

(a) 11
(b) 20
(c) 23

8. Use the Caesar cipher to encrypt GO HOME NOW.

(a) JR KRPH QRZ
(b) DL ELJB KLT
(c) Neither of these

9. Use the Caesar cipher to decrypt EIEIO.

(a) HLHLR
(b) BFBFL
(c) Neither of these

10. Use the Vigenère cipher to encrypt RUN NOW, using the key word ADAM.

(a) RXN ZOZ
(b) RXN NAW
(c) SXO ZPZ

11. Use the Vigenère cipher to decrypt EIEIO, using the key word ADAM.

(a) ELELR
(b) EFEFL
(c) EFEWO

12. A simple encoding scheme for binary numbers replaces each 0 with 0000 and each 1 with 1111. What is the weight of this code?

(a) 2
(b) 4
(c) 8

13. If every pair of code words differs in at least five positions, then nearest-neighbor decoding can accurately decode words that have

(a) two mistakes.
(b) three mistakes.
(c) four mistakes.

14. Which of the following codes uses a data compression process?

(a) Hamming codes
(b) RSA codes
(c) Morse code

15. Digital signatures are a way to verify

(a) the identity of a computer hacker.
(b) the identify of a client.
(c) the type of code used.

EXERCISES

▲ **Optional.** ■ **Advanced.** ◆ **Discussion.**

Binary Codes

1. Use the diagram method shown in Figures 10.1 and 10.2 to verify the code words in Table 10.1 for the messages 0101, 1011, and 1111.

2. Use the diagram method to decode the received messages 0111011 and 100010.

3. Find the distance between each of the following pairs of words:

(a) 11011011 and 10100110
(b) 01110100 and 11101100

4. Referring to Table 10.1, use the nearest-neighbor method to decode the received words 0000110 and 1110100.

5. If the code word 0110010 is received as 1001101, how is it decoded using the diagram method?

6. Suppose a received word has the Venn diagram arrangement shown at right:

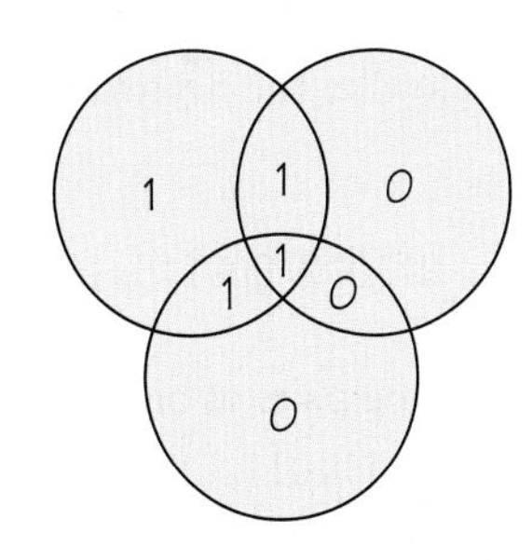

What can we conclude about the received word?

Parity-Check Sums

7. Determine the binary linear code that consists of all possible three-digit messages with three check digits appended using the parity-check sums $a_2 + a_3$, $a_1 + a_3$, and $a_1 + a_2$. (That is, $c_1 = 0$ if $a_2 + a_3$ is even, $c_1 = 1$ if $a_2 + a_3$ is odd, and similarily for c_2 and c_3.)

8. Let C be the code

$$\{0000000, 1110100, 0111010, 0011101, 1001110, 0100111, 1010011, 1101001\}$$

What is the error-correcting capability of C? What is the error-detecting capability of C?

9. Find all code words for binary messages of length 4 by adding three check digits using the parity-check sums $a_2 + a_3 + a_4$, $a_2 + a_4$, and $a_1 + a_2 + a_3$. Will this code correct any single error?

10. Consider the binary linear code

$$C = \{00000, 10011, 01010, 11001, 00101, 10110, 01111, 11100\}$$

Use nearest-neighbor decoding to decode 11101 and 01100. If the received word 11101 has exactly one error, can you determine the intended code word? Explain your reasoning.

11. Construct a binary linear code using all eight possible binary messages of length 3 and appending three check digits using the parity-check sums $a_1 + a_2$, $a_2 + a_3$, and $a_1 + a_3$. Decode each of the received words below by the nearest-neighbor method.

001001, 011000, 000110, 100001

12. Add the following pairs of binary sequences:

(a) 10111011 and 01111011
(b) 11101000 and 01110001

13. All binary linear codes have the property that the sum of two code words is another code word. Use this fact to determine which of the following sets cannot be a binary linear code.

(a) {0000, 0011, 0111, 0110, 1001, 1010, 1100, 1111}
(b) {0000, 0010, 0111, 0001, 1000, 1010, 1101, 1111}
(c) {0000, 0110, 1011, 1101}

Cryptography

14. Use the Caesar cipher to encrypt the message RETREAT. Determine the intended message corresponding to the message DWWDFN that was encrypted using the Caesar cipher.

15. Use the Vigenère cipher with the key word HELP to encypt the message PHONE HOME.

16. Given that BEATLES was used as the key word to encrypt SSLETRY TXOGPW, decrypt the message.

17. Use the Vigenère cipher with the key word CLUE to encypt the message THE WALRUS WAS PAUL.

▲ **18.** Use the RSA scheme with $p = 5$, $q = 17$, and $r = 3$ to determine the numbers sent for the message VIP.

▲ **19.** Use the RSA scheme with $p = 5$, $q = 17$, and $r = 3$ to decode the received numbers 52 and 72.

▲ **20.** In the RSA scheme with $p = 5$, $q = 17$, and $r = 5$, determine the value of s.

▲ **21.** Why can't we use the RSA scheme with $p = 7$, $q = 11$, and $r = 3$?

Data Compression

22. Suppose we code the four-symbol genetic set {A, C, T, G} into binary form as follows:

$$\text{A} \rightarrow 0, \text{C} \rightarrow 10, \text{T} \rightarrow 110, \text{G} \rightarrow 111$$

Convert the sequence ACAAGTAAC into binary code.

23. Use the code in the previous exercise to determine the sequence of symbols represented by the binary code 001100001111000.

■ **24.** Suppose we code a five-symbol set $\{A, B, C, D, E\}$ into binary form as follows:

$$A \rightarrow 0, B \rightarrow 10, C \rightarrow 110, D \rightarrow 1110, \text{and } E \rightarrow 1111$$

Convert the sequence to *AEAADBAABCB* into binary code. Determine the sequence of symbols represented by the binary code 0100011010001111110.

25. Use the code in the previous exercise to convert the sequence *EABAADABB* into binary code. Determine the sequence of letters represented by the binary code 00100011001111011010.

■ **26.** Devise a variable-length binary coding scheme for a six-symbol set {*A*, *B*, *C*, *D*, *E*, *F*}. Assume the *A* is the most frequently occuring symbol, *B* is the second most frequently occuring symbol, and so on.

27. Judging from the Morse code, what are the three most frequently occuring consonants in English text material? What is the most frequently occuring vowel?

28. In English, the letter *H* occurs more often than *D*, *G*, *K*, and *W*, but in Morse code *H* has a longer code than *D*, *G*, *K*, and *W*. Speculate on the reason for this apparent violation of data compression principles.

29. Explain why the Morse code must include a space after each letter but fixed-length codes do not.

Additional Exercises

30. For a space and for the 26 letters of the alphabet, assign the binary string listed beneath it in the table below.

(Think of a two-character alphabet comprised of 0 and 1, where 0 comes before 1. Then the 27 binary strings of length 5 below are listed in alphabetical order.) Use 11111 in front of a letter to indicate that the letter is uppercase. Use 11110 in front of the code for *a* to indicate the numeral 0, use 11110 front of the code for *b* to indicate the numeral 1, and so on up to 11110 in front of the code for *j* to indicate the numeral 9. Determine the code for Ma 4.

31. Use the code given in the previous exercise to decode the message 11110000110000011111000010001 0.

32. Let $v = a_1a_2 \ldots a_n$ and $u = b_1b_2 \ldots b_n$ be binary sequences. Explain why the number of 1's in $v + u$ is the same as the distance between u and v.

33. Extend the code words listed in Table 10.1 to eight digits by appending a 0 to words of even weight and a 1 to words of odd weight. What is the error-detecting and error-correcting capability of the new code?

34. Suppose the weight of a binary linear code is 6. How many errors can the code correct? How many errors can the code detect?

35. How many code words are there in a binary linear code that has all possible messages of length 5 with three check digits appended?

■ **36.** Explain why no binary linear code with all three possible message digits together with three check digits can correct all possible double errors.

■ **37.** A *ternary* code is formed by starting with all possible strings of a fixed length composed of 0's, 1's, and 2's and appending extra digits that are also

space	*a*	*b*	*c*	*d*	*e*	*f*	*g*	*h*
00000	00001	00010	00011	00100	00101	00110	00111	01000
i	*j*	*k*	*l*	*m*	*n*	*o*	*p*	*q*
01001	01010	01011	01100	01101	01110	01111	10000	10001
r	*s*	*t*	*u*	*v*	*w*	*x*	*y*	*z*
10010	10011	10100	10101	10110	10111	11000	11001	11010

0's, 1's, or 2's. Form a ternary code by appending to each message a_1a_2 the check digits c_1c_2 using:

$$c_1 = (a_1 + a_2) \bmod 3$$
$$c_2 = (2a_1 + a_2) \bmod 3$$

■ **38.** Use the ternary code in the previous exercise and the nearest-neighbor method to decode the received word 1211.

■ **39.** Suppose a ternary code is formed by starting with all possible 4-tuples of 0's, 1's, and 2's and appending two extra digits that are also 0's, 1's, and 2's. How many code words are there in this code? How many possible received words are there in this code?

▲ **40.** For each part above right, explain how modular arithmetic can be used to answer the question.

(a) If today is Wednesday, what day of the week will it be in 16 days?
(b) If a clock (with hands) indicates that it is now four o'clock, what will it indicate in 37 hours?
(c) If a military person says it is now 0400, what time would it be in 37 hours? (Instead of A.M. and P.M., military people use 1300 for 1:00 P.M., 1400 for 2:00 P.M., and so on.)
(d) If it is now July 20, what day will it be in 65 days?
(e) If the odometer of an automobile reads 97,000 now, what will it read in 12,000 miles?

41. Where can you often find the Braille code in public buildings?

42. Guess the precentage of the occurences of spaces in typical English text material.

WRITING PROJECTS

1. Prepare a report on cryptography. Discuss at least three methods of encryption. Discuss the interface between computers and cryptography.

2. Prepare a report on applications of modular arithmetic. Explain the calculation of the check digits described in Exercises 7, 9, and 11 with modular arithmetic. Use modular arithmetic to describe the error-detection schemes used in Chapter 9.

3. Prepare a report on the Braille system of coding. The Suggested Reading by Petzold has the information you will need.

4. Prepare a report on the Morse code. The Suggested Reading by Petzold has the information you will need.

5. Prepare a report on the early history of error-correcting codes. The Suggested Reading by Thompson has the information you will need.

6. A Smart Card is a card the size of a credit card that has a built-in microprocessor and memory. In the near future, consumers will be able to use a Smart Card in combination with a network like the Internet to make secure monetary transactions and to authenticate remote users accessing private intranets. Use the Internet to find information about Smart Card technology and prepare a report on your findings.

More on the Web

SPREADSHEET PROJECTS

To do these projects, go to www.whfreeman.com/fapp.

Hamming codes can be computed automatically, using simple spreadsheet calculations. This project also uses spreadsheets to find valid choices for the value of the exponent for RSA encryption, using modular calculations. Using the RSA spreadsheet, one can also track one's success in "breaking" an RSA code by guessing the values of the hidden numbers.

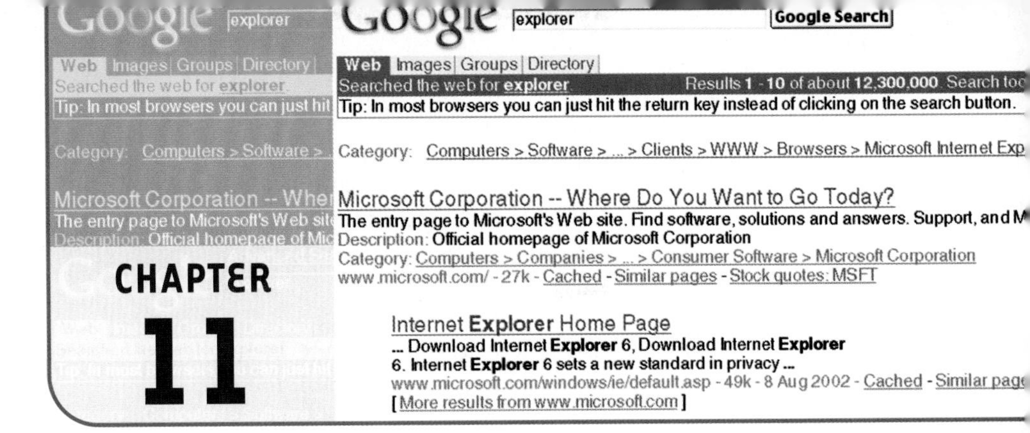

CHAPTER 11

The Internet, the Web, and Logic

In this chapter we cover...

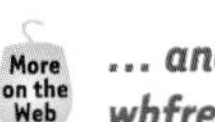

... and at www.whfreeman.com/fapp:

Flashcards
Quizzes
Spreadsheet Projects
Decimal and binary representations

The Internet is a global network comprising hundreds of millions of computers and growing daily. All these computers can communicate with one another, allowing a pair of pen pals in Lawrence, Kansas, and Ulan Bator, Mongolia, to exchange emails or a shopper in Hong Kong to buy a sweatshirt from a Web site in Wisconsin.

The vast size of the Internet offers a powerful medium for communication with many important and exciting applications. However, the complexity of the Internet also poses some challenges. For example, how is an email message routed from a computer in Lawrence, Kansas, to a computer in Ulan Bator, Mongolia? How is the vast amount of information on the World Wide Web organized so that a shopper can easily search for Web sites that sell a particular type of sweatshirt? When the shopper is ready to purchase the sweatshirt from the vendor, how is the credit card information kept secure? In this chapter we will examine several areas of mathematics, such as binary coding, search algorithms, and particularly Boolean logic, and show how they are used to solve some of the problems of the Internet.

11.1 The Internet

A computer on the Internet is typically connected to a device called a *router*. A router has connections not only to computers but also to one or more other routers. In this way, a large number of computers at a university or company, for example, can be connected into what is known as a *local area network (LAN)*. One or more routers in a local area network are, in turn, connected to larger routers belonging to an *Internet service provider (ISP)*. Many ISPs are then connected to routers belonging to other, larger ISPs. Eventually an entire country is interconnected, and finally different countries themselves are interconnected to form the Internet.

The Internet can be represented by a *graph* consisting of *vertices* and *edges*. The vertices represent computers and routers, and the *edges* represent connections between these computers and routers.

An example is shown in Figure 11.1, where the red vertices represent computers and the blue vertices represent routers.

11.2 Representing and Sending Data

Computers represent information as a sequence 0's and 1's because of the way their internal electronics work. A sequence of 0's and 1's is called a **binary sequence.** Fortunately, every number, letter, and symbol can be represented as a binary sequence.

The standard way of writing numbers is called the **base 10 representation** because we have 10 different digits available. Each number can also be expressed in its **base 2 representation** with just the two digits 0 and 1. Because computers use only 0's and 1's to represent and transmit their data, the base 2 representation is used.

For example, notice that 793 is the base 10 representation of the number $\mathbf{7} \times 100 + \mathbf{9} \times 10 + \mathbf{3} = \mathbf{7} \times 10^2 + \mathbf{9} \times 10^1 + \mathbf{3} \times 10^0$. The powers of 10 occur here because we are using base 10. By analogy, 101 is the base 2 representation of the number $\mathbf{1} \times 2^2 + \mathbf{0} \times 2^1 + \mathbf{1} \times 2^0$, which is 5 in base 10. The powers of 2 occur here because we are using base 2.

The sequence $a_k a_{k-1} \ldots a_1 a_0$ is the base 10 representation of the number $a_k \times 10^k + a_{k-1} \times 10^{k-1} + \ldots a_1 \times 10^1 + a_0 \times 10^0$, where each of a_k, a_{k-1}, $\ldots$, a_1, a_0 is a digit in the range from 0 to 9.

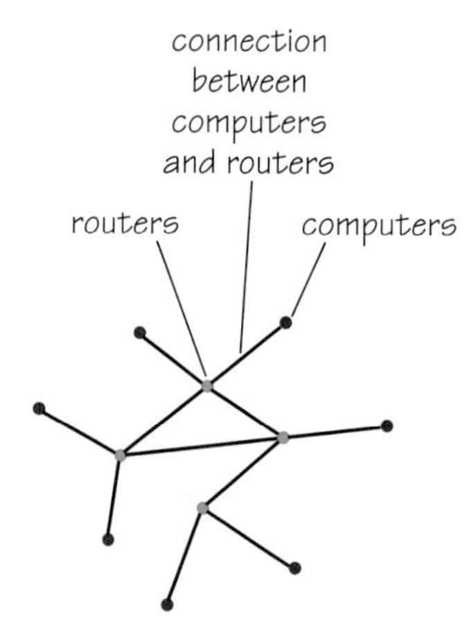

Figure 11.1 A graph representing a part of the Internet.

The binary sequence $b_k b_{k-1} \ldots b_1 b_0$ is the base 2 representation of the number $b_k \times 2^k + b_{k-1} \times 2^{k-1} + \ldots + b_1 \times 2^1 + b_0 \times 2^0$, where each of $b_k, b_{k-1}, \ldots, b_1, b_0$ is a 0 or a 1.

EXAMPLE

Converting a Number from One Base to Another

The number 1101 in base 2 represents the number $\mathbf{1} \times 2^3 + \mathbf{1} \times 2^2 + \mathbf{0} \times 2^1 + \mathbf{1} \times 2^0$, which is 13 in base 10. Notice that in applying the definition above, the leftmost 1 in 1101 is the term b_3, the next 1 is the term b_2, the 0 is the term b_1, and the rightmost 1 is the term b_0. This tells us that b_3, the leftmost 1, represents 1×2^3. Similarly, b_2, the second 1 from the left, represents 1×2^2, and so forth.

Converting from base 10 to base 2 requires only slightly more effort. The number 27 in base 10 can be represented in base 2 as follows: Notice first that the powers of 2 are $2^0 = 1$, $2^1 = 2$, $2^2 = 4$, $2^3 = 8$, $2^4 = 16$, $2^5 = 32$, and so forth. Because $32 > 27$, we know that 2^5 and all higher powers of 2 will not be needed to represent 27 in base 2. Because 16 is the first power of 2 less than or equal to 27, we begin with $27 = \mathbf{1} \times 2^4 + 11$. The number 11, in turn, is $\mathbf{1} \times 2^3 + 3$. Repeating this process, we have $3 = \mathbf{1} \times 2^1 + 1 = \mathbf{1} \times 2^1 + \mathbf{1} \times 2^0$. Therefore, $27 = \mathbf{1} \times 2^4 + \mathbf{1} \times 2^3 + \mathbf{0} \times 2^2 + \mathbf{1} \times 2^1 + \mathbf{1} \times 2^0$, which is 11011 in base 2. ■

A 0 or 1 is called a **bit,** which is an abbreviation of **b**inary dig**it**. A sequence of 8 bits is called a **byte.** There are 256 different patterns that can be represented using a byte. To see this, notice that there are 2 possibilities for the bit in the first position of the byte, 2 possibilities for the bit in the second position, and so forth up to the eighth position. Therefore, there are $2 \times 2 \times 2 \times 2 \times 2 \times 2 \times 2 \times 2 = 2^8 = 256$ different patterns possible.

Although base 2 provides us with a way to encode numbers as binary sequences, a different encoding system is required to represent letters, punctuation, and other symbols.

Typically, computers use a special code such as **Extended ASCII** (ASCII stands for the American Standard Code for Information Interchange), in which each symbol is represented by a binary sequence of 8 symbols, that is, a byte.

All the upper- and lowercase letters in the English alphabet, the digits 0 through 9, symbols from several other alphabets, and various punctuation marks and other symbols have unique Extended ASCII codes. For example, in Extended ASCII, the lowercase letter "a" is represented by 01100001, the uppercase letter "A" is represented by 01000001, the symbol "{" is represented by 01111011, and the Greek symbol "δ" (delta) is represented by 11101011.

Notice that a number, such as 14, can be represented in two different ways using 0's and 1's. One way to represent it is in base 2. The base 2 representation of 14 is 1110. The other way to represent 14 is to use the Extended ASCII representation for the symbol "1" followed by the Extended ASCII representation for the symbol "4." While the base 2 representation uses only 4 bits to represent the number 14, the Extended ASCII representation uses 16 bits to represent it, since each symbol is represented using 8 bits.

Both representations are used, depending on the application. When data are being transmitted over the Internet or stored in the computer, they are generally represented using Extended ASCII. The reason for this is that data may contain a mixture of letters, numbers, and other symbols, and it is easiest to use a single, consistent representation for all data. On the other hand, when a computer does computation using numbers, it is more efficient to have the numbers represented in base 2, as we will see shortly.

When an email message or a Web page is transmitted over the Internet, the data being sent are first partitioned into blocks called *packets*. Each packet contains a specially encoded address of the computer to which the data should be sent, followed by approximately 1500 bytes of data. The address is encoded in base 2 and the data are encoded in Extended ASCII.

Let's examine the way that addresses are encoded in base 2. As computer users, we typically employ easily remembered email addresses such as jane_smith@northstate.edu or Web addresses such as www.whitehouse.gov. A computer in the local area network translates these addresses into base 2 binary addresses that computers and routers on the Internet can understand.

The Internet uses a special address convention known as *IP addresses* (IP stands for "Internet protocol"). An IP address is comprised of 4 bytes. We have seen that each byte can have one of 256 different patterns. Each byte can be interpreted as the base 2 representation of a number. Interpreted as base 2 numbers, we see that 00000000 represents the number 0, 00000001 represents the number 1, and so forth up to 11111111, which represents the number $1 \times 2^7 + 1 \times 2^6 + 1 \times 2^5 + 1 \times 2^4 + 1 \times 2^3 + 1 \times 2^2 + 1 \times 2^1 + 1 \times 2^0 = 255$. So, the numbers 0 through 255 are the 256 different numbers that can be represented in base 2 using 1 byte. Because each byte can represent 256 different values, 4 bytes result in $256^4 =$ 4,294,967,296 different IP addresses. Each computer and router on the Internet has a unique IP address. You can see the IP address of a computer connected to the Internet by looking in the TCP/IP control panel on the computer.

> Each computer and router on the Internet has a unique IP address composed of 4 bytes of information.

EXAMPLE

Decoding an IP Address

An IP address typically looks something like 134.173.200.20. Because $134 = \mathbf{1} \times 2^7 + \mathbf{0} \times 2^6 + \mathbf{0} \times 2^5 + \mathbf{0} \times 2^4 + \mathbf{0} \times 2^3 + \mathbf{1} \times 2^2 + \mathbf{1} \times 2^1 + \mathbf{0} \times 2^0$, the first byte of this IP address is 10000110. The second byte is $173 = \mathbf{1} \times 2^7 + \mathbf{0} \times 2^6 + \mathbf{1} \times 2^5 + \mathbf{0} \times 2^4 + \mathbf{1} \times 2^3 + \mathbf{1} \times 2^2 + \mathbf{0} \times 2^1 + \mathbf{1} \times 2^0$. What are the last two bytes of this IP address? ■

So, an email message or Web page is sent from one computer to another by partitioning the data into packets. Each packet contains the IP address of the destination computer followed by some data. The sending computer sends each packet to the router to which it is connected. The router examines the IP address and determines how the packet should be forwarded en route to the destination. Eventually, the packet reaches the destination. When all packets have arrived at the destination computer, the packets are reassembled into the full message.

11.3 Web Searches

We now turn our attention to a part of the Internet called the World Wide Web, henceforth simply called "the Web." The Web comprises literally billions of documents, called *Web pages,* residing on computers throughout the Internet. Web pages are connected via *hyperlinks.* Typically hyperlinks, which are represented as blue text on a Web page, refer to another Web page. When you click on this hyperlink text with the mouse, the Web browser (such as Netscape or Microsoft Explorer) looks up the address of the referenced Web page. Your Web browser then sends a message to that address requesting the new Web page. Note that a Web page may refer to another Web page on the same computer, and thus at the same address, or to a Web page anywhere else in the world.

Information on the Web is not neatly organized in the way that books are in a library. Information on parasailing, for example, may exist on thousands of different Web pages throughout the Web and thus throughout the world. How does one find the information one is looking for in this morass of data? *Web search engines* provide the solution. Most search engines allow the Web user to type in a *query* comprising one or more words. A query is simply one or more words (such as *parasailing*) that indicate the type of information one is interested in finding. The search engine then displays a list of Web pages (along with hyperlinks to these pages) such that each Web page contains all the words in the query. But how does a search engine do this?

Search engines use systematic techniques, or *algorithms,* to visit millions of Web pages in order to gather information about them. These so-called *Web crawler* algorithms build up a database of all the Web pages on the Internet that they can find. This is a process that takes considerable time and is only performed periodically. This giant database is then used by the Web search engine to find Web pages that match each user's queries. The idea, therefore, is to use the precomputed database to avoid physically searching through the entire Web each time a user submits a query.

Let's examine how a Web crawler works to build up its database. In the next section we will investigate how the search engine uses its database to respond to queries. A general searching technique used by many Web crawlers is called **breadth-first search.** To see how breadth-first search works, let's represent the

Web by a directed graph in which each vertex represents a Web page and an edge from vertex u to vertex v means that the Web page represented by vertex u has a hyperlink to the Web page represented by vertex v. Note that two vertices in the Web may be connected by just one edge (one page has a hyperlink to the other), connected by two oppositely directed edges (both pages have hyperlinks to each other), or not connected at all (there are no hyperlinks from either page to the other).

To facilitate the search through this directed graph, the breadth-first search algorithm organizes its data using something called a **queue.**

A queue is like a line at a movie ticket window. On arriving at the theater, a person goes to the end of the line. On getting to the front of the line, the person buys a ticket and leaves the line. Breadth-first search employs a queue of vertices (Web pages) to keep track of the vertices from which it needs to continue expanding its search.

The search engine begins at an arbitrary vertex (Web page) in the graph. The search engine stores this Web page (or just its address) in its database and places this vertex on the queue. The algorithm now repeats the following two steps until the queue becomes empty:

Step 1: Remove the vertex at the front of the queue from the queue. This element is a vertex corresponding to a Web page. Let u denote this vertex.

Step 2: Next find each vertex (corresponding to a Web page) v such that there is an edge from u to v. This corresponds to a hyperlink from Web page u to Web page v. If v is not yet in the database, store it there and also place v at the end of the queue.

EXAMPLE Doing a Breadth-First Web Search

Consider a scaled-down version of the Web represented by the directed graph in Figure 11.2. Assume, for example, that the search begins at vertex 3. Figure 11.2a shows a representation of the queue and the database, both of which currently contain only vertex 3.

In our example, vertex 3 is the only element on the queue. In step 1 of the algorithm, vertex 3 is removed from the queue. In step 2 of the algorithm, we discover that vertices 1 and 2 are reachable from vertex 3. Because neither of these vertices is yet in the database, they are now added to the database and inserted at the end of the queue. The order in which they are placed on the queue does not matter, and we will assume that vertex 1 was placed on the queue before vertex 2. Figure 11.2b shows the queue and the database at this point. Notice that the queue has a front

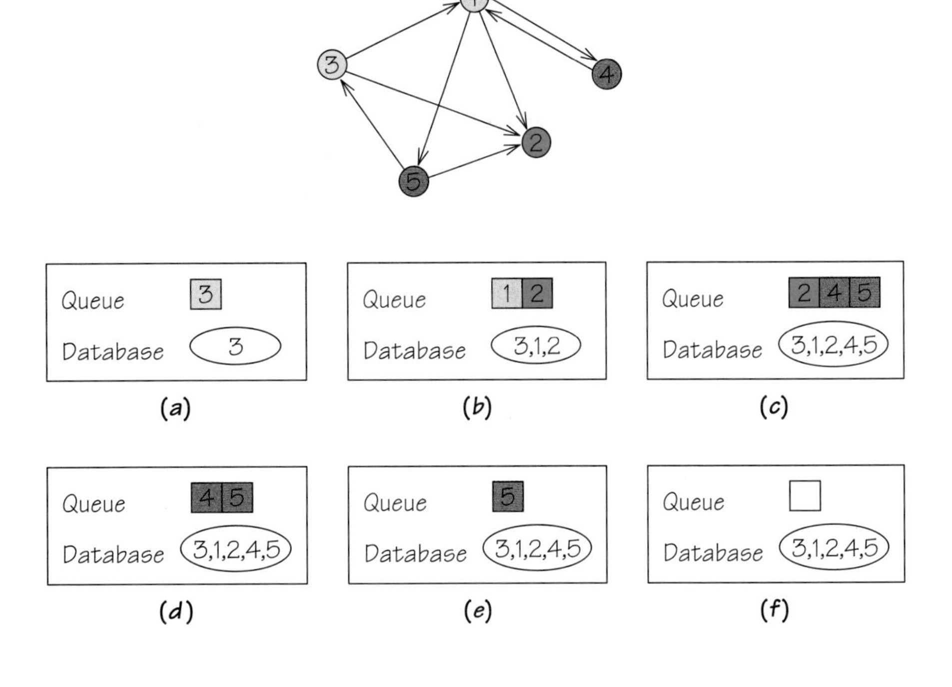

Figure 11.2 The states of the queue and database during a breadth-first search of a small part of the Web, as represented by the directed graph.

and back. Items are removed from the front and added to the back. In the figure, the left side of the queue depicts the front and the right side depicts the back.

The two steps of the algorithm are now repeated once again. Vertex 1 is removed from the queue in step 1. In step 2 we discover that vertex 1 has edges to vertices 4, 2, and 5. Vertex 2 is already in the database, but vertices 4 and 5 are not. Vertices 4 and 5 are therefore added to the database and inserted at the end of the queue. The state of the queue and database at this point are shown in Figure 11.2c. In the next iteration, vertex 2 is removed from the queue. It has no outgoing edges, and thus nothing happens at step 2 of the algorithm. The queue and database at this point are shown in Figure 11.2d. In the next iteration, vertex 4 is removed from the queue in step 1. Vertex 4 has just one outgoing edge, to vertex 1, which is already in the database. The state of the queue and database at this point are shown in Figure 11.2e. Finally, vertex 5 is removed from the queue. It has edges to vertices 2 and 3, both of which are already in the database. Therefore, nothing happens in step 2 of the algorithm. At this point, the queue is empty and the database contains all vertices reachable from the start vertex, as shown in Figure 11.2f. The fact that the queue is empty indicates that the algorithm has completed its search. ■

> At the end of the breadth-first search, the database will contain all of the vertices (Web pages) that are reachable from the start vertex. Therefore, the choice of start vertex can impact the outcome of this algorithm.

In the previous Example, we started at vertex 3. Because every vertex in the graph can be reached from vertex 3 via some path, at the end of the algorithm, all vertices were in the database. If we had started at vertex 2, on the other hand, the algorithm would end with only vertex 2 in its database, because no vertices are reachable from vertex 2. To help mitigate against this kind of problem, a Web crawler could perform a breadth-first search multiple times, using different start vertices each time. Nevertheless, because many Web pages have no hyperlinks to them from other Web pages, these pages may not be discovered by search engines.

Once the breadth-first search is completed, the Web pages are automatically analyzed and catalogued. The search engine maintains an enormous dictionary of words, often even more than appear in a Webster's dictionary. Each word has a list of the addresses of all Web pages that contain that word. When a Web page is discovered during the breadth-first search, every word on that Web page is examined. If the word is in the dictionary, then that dictionary entry is updated to include this Web page.

11.4 Web Search Queries

You have seen that a user can search for Web pages on a particular topic by providing the search engine with one or more keywords. If the user provides just one keyword, the search engine simply looks that word up in its dictionary and finds all the Web pages that contain that word. If the query contains two or more words, the search engine finds every Web page that contains all of those words. One way to do this is to find the list of Web pages that contain each of the words in the query and then find those pages that appear on all of the lists. The process of finding Web pages that appear on each of the lists is potentially time-consuming, but search engines typically employ efficient algorithms for doing this reasonably quickly.

For example, if the Google search engine (www.google.com) is given the query word *explorer,* it finds well over 11 million Web pages that contain this word and reports that it does so in about 0.1 second. An example of this search using Google is shown in Figure 11.3a. If the query comprises the two words *ancient explorer,* then Google finds about 200,000 pages that contain both words – one-fiftieth of the previous number of pages – but it now takes more time due to the need to find pages that contain both words. An example of this search using Google is shown in Figure 11.3b. If you try these searches on Google, you may obtain somewhat different results because Google's Web crawler searches the Web occasionally, building up a larger and larger database of Web pages.

A user certainly doesn't want to see the names of 11 million Web pages (and 200,000 isn't acceptable either!). The search engine must therefore present the user with only the "best" pages. Typically, a search engine assigns "ranks" to each Web page and presents the user with 50 to 100 pages that match the query, beginning with those of highest rank. How does a search engine rank a Web page? Each search engine does this differently, and the exact details of each method are well-kept secrets. If Google, for example, divulged the details of how its ranking

Figure 11.3 Example Google search on (a) query "explorer" and (b) query "ancient explorer."

algorithm works, an unscrupulous party could use that information to make sure that its Web site was ranked highly and thus appeared at the top of the list for a search. This would amount to free advertising for that party and could render the ranking useless.

Although the details are proprietary, there are certain features that many search engines use in computing the rank of a page. One feature is word frequency. For example, if the query is on the word *explorer,* a Web page that contains that word 500 times may be assigned a higher rank than one that contains the word only twice. Alternatively, the rank might be based not on the absolute number of times that the word *explorer* appears, but rather on the relative number of occurrences of that word—that is, the number of occurrences of that word divided by the total number of words in the document. Thus, short documents with a high frequency of occurrences of the keyword would be given a high rank.

Another way to rank a Web page is to determine the number of other Web pages with hyperlinks to that page. The idea here is that if a Web page has a hyperlink to page p, then this is a recommendation for page p. Thus, the total number of recommendations for page p can be used as a method of ranking the page. For example, using this method of ranking for the pages represented in Figure 11.2, vertices 3, 4, and 5 have rank 1, vertex 1 has rank 2, and vertex 2 has rank 3. In graph theoretic language, the number of edges entering a vertex is called the *in-degree* of that vertex.

> Search engines use a number of different methods for ranking a Web page. Among these are the absolute or relative number of occurrences of the keywords in the Web page and the number of Web pages with hyperlinks to this Web page.

How can the ranks of the page—that is, the in-degrees of the corresponding vertices—be computed? Fortunately, this can be done very easily at the time that the breadth-first search algorithm is crawling through the Web in search of all the pages. Notice that the in-degree of a vertex v is simply the number of vertices that have outgoing edges directed toward v. Thus, we can maintain a counter for each vertex v that keeps track of times that v was visited during the breadth-first search.

EXAMPLE

Computing the Ranks in a Breadth-First Web Search

Consider again the breadth-first search on the graph in Figure 11.2. We now keep a "counter" associated with each vertex in the database. The vertex at which we began our search, vertex 3 in our earlier Example, would initially have its counter set to 0. When any other vertex is first placed in the database, its counter is set to 1. Each time a vertex is visited again, its counter is incremented by 1. For example, when vertex 3 is removed from the queue and vertices 1 and 2 are inserted into the queue and the database, those two vertices have their counters set to 1. When vertex 1 is then removed from the queue, it has edges to vertices 4, 2, and 5. Vertices 4 and 5 are added to the database and their counters are set to 1. Vertex 2 is already in the database, so we simply increment its counter. Notice that later in the process, vertex 5 will be removed from the queue. Vertex 5 has edges to vertices 2 and 3, both of which are in the database at that point. Thus, vertex 3's counter will be incremented from 0 to 1 and vertex 2's counter will be incremented from 2 to 3. At the end of the algorithm, the counter for each vertex v will be the in-degree of v, because the counter will get a contribution of 1 for each edge entering v. Note that this assumes that every vertex in the graph eventually gets entered into the database. This is true if there is a path from the start vertex to every vertex in the graph. ■

Most search engines use some combination of features, such as word counts, in-degrees of nodes, and others, to rank the pages.

11.5 Web Search Queries Using Boolean Logic

Imagine that you would like to use a Web search engine to learn more about some of the famous explorers in history. If we simply use the word *explorer* as a query, the search engine is likely to report many high-ranking Web pages on Microsoft's Internet Explorer, the Ford Explorer, the Explorer Channel, the *Explorer 1* satellite, and other entities that are not relevant to this search. Many search engines therefore allow the use of **Boolean logic** in order to formulate more precise queries.

An *expression* in Boolean logic is simply a statement that is either true or false. For example, consider the expression "*For All Practical Purposes* weighs a ton." This expression is either true or false. In this case, the expression is false (you may wish to use a scale to verify this). Another expression is "This Web page contains the word *explorer*." For each Web page, this expression is either true or false. When we enter a word such as *explorer* as a query to a search engine, the search engine automatically interprets it as the expression "This Web page contains the word *explorer*." The search engine then returns the list of Web pages for which this expression is true. More complex expressions can be constructed by connecting expressions with the *connectives* AND, OR, and NOT.

For example, to obtain Web pages containing the word *explorer* but not pages that also contain *Microsoft* or *Ford*, we could formulate the query using the expression "explorer AND (NOT Microsoft) AND (NOT Ford)" (Figure 11.4a). The search engine interprets this as "This Web page contains the word *explorer* AND it is NOT the case that this Web page contains the word *Microsoft* AND it is NOT the case that this Web page contains the word *Ford*" (Figure 11.4b). The parentheses are not necessary but are sometimes useful, as we will see shortly. Most search engines are also not *case sensitive*—no distinction is made between uppercase and lowercase letters. It should be noted that each search engine has slightly different conventions for formulating queries. Virtually every search engine has a hyperlink on its Web page that explains how to formulate queries. The Boolean logic queries used here, for example, are formulated in the way the search engine www.northernlight.com expects. You are encouraged to try these queries out for yourself!

Is the expression "explorer AND (NOT Microsoft) AND (NOT Ford)" equivalent to the "explorer AND NOT (Microsoft AND Ford)"? Is it equivalent

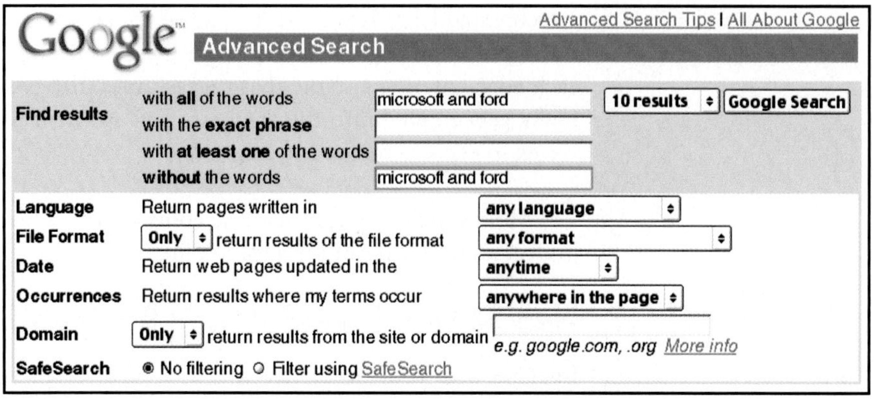

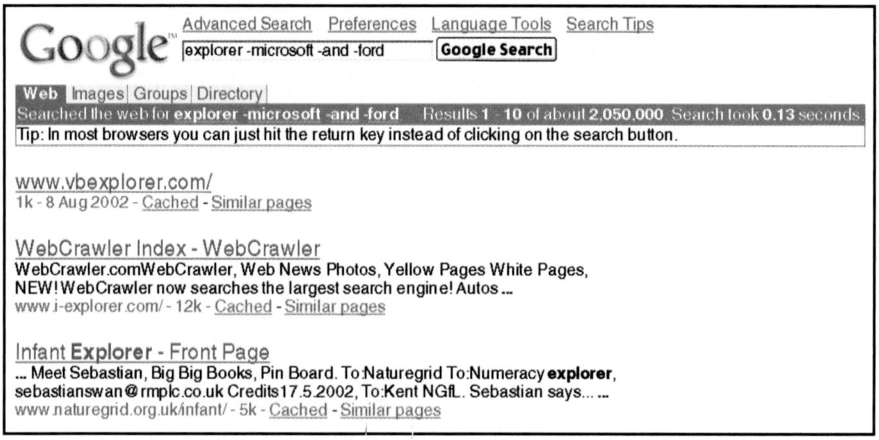

Figure 11.4 Example Google advanced search on (a) query "explorer" and (b) query "explorer" – "microsoft" and "ford."

to the expression "explorer AND NOT (Microsoft OR Ford)"? To answer these questions, we will now take a closer look at the connectives AND, OR, and NOT.

The NOT connective allows us to take an expression P and create a new expression NOT P, called the *negation* of P. If P is true, then NOT P is false; if P is false, then NOT P is true. Rather than writing NOT P, we will use the more standard mathematical notation $\neg P$. The negation relationship can be summarized in the following format, known as a **truth table:**

P	$\neg P$
T	F
F	T

Notice that T and F are used here as shorthand for true and false, respectively. The left column of the truth table shows the two possible values of P: T and F. The right column shows the values of $\neg P$ for each of the corresponding values of P.

The AND connective allows us to combine two expressions, P and Q, into a new expression P AND Q, called the *conjunction* of P and Q. The new expression is true when both statements P and Q are true and is otherwise false. The mathematical notation for P AND Q is $P \wedge Q$. This relationship can also be summarized in a truth table as follows:

P	Q	$P \wedge Q$
T	T	T
T	F	F
F	T	F
F	F	F

Here, the first two columns are used to show all possible values of P and Q, and the right column shows the value of $P \wedge Q$.

Finally, the OR connective allows us to combine two expressions, P and Q, into a new expression, P OR Q, called the *disjunction* of P and Q, which is true if either P or Q or both are true and is otherwise false. The mathematical notation for this P OR Q is $P \vee Q$. This relationship is summarized by the following truth table:

P	Q	$P \vee Q$
T	T	T
T	F	T
F	T	T
F	F	F

Symbols such as P, Q, and R are sometimes called *Boolean variables.* (See Spotlight 11.1 for some background on these logic applications.)

It is interesting to note that given three expressions P, Q, and R, the expression $P \wedge Q \wedge R$ is ambiguous. Does this expression mean that P and Q are first combined into a new expression $P \wedge Q$ and this new expression is then combined with R? In other words, should we interpret this expression as $(P \wedge Q) \wedge R$? Alternatively, perhaps the intention was to first combine Q and R into a single expression $Q \wedge R$ and then combine P with this new expression. In other words, the expression might be interpreted as $P \wedge (Q \wedge R)$. It turns out that both interpretations are the same! For example, let P be the statement "He is tall," let Q be the statement "He has short hair," and let R be the statement "He has pink hair." The expression $P \wedge (Q \wedge R)$ can be interpreted as "He is tall and also his hair is short and pink," and the expression $(P \wedge Q) \wedge R$ can be intrepreted as "He is tall with short hair and also his hair is pink."

SPOTLIGHT 11.1 A Brief History of Logic

Web searching is just one of many applications of logic. Perhaps the first systematic study of logic appears in the writings of the Greek philosopher Aristotle (384–322 B.C.). Aristotle proposed rules and principles to provide a foundation for logical reasoning. Aristotle's formulations were taught through the Middle Ages.

The German mathematician Gottfried Wilhelm Leibniz (1646–1716) is considered by many to be the founder of logic as a serious mathematical discipline. Leibniz was a brilliant mathematician and philosopher who made numerous significant contributions to mathematics. In particular, Leibniz developed calculus in Germany at the same time as, but independently from, Sir Isaac Newton in England. At age 20, Leibniz developed the foundations of mathematical logic. Unfortunately, Leibniz's work on logic went largely unnoticed for well over 100 years.

In the mid-1800s the English mathematician George Boole (1815–1864) reinvigorated logic as a discipline for mathematical research. Boole was born into a very low social and economic class in a very class-conscious society. The opportunities available to him were limited, at best. He was

Notice that both of these descriptions are effectively the same. Both expressions are only true in the event that all of P, Q, and R are true. One way to verify this is to construct the truth table for expression $(P \wedge Q) \wedge R$ and the truth table for $P \wedge (Q \wedge R)$ and show that they give the same values for every possible value of P, Q, and R. Since the *order of operations* does not matter in this case, we may simply write $P \wedge Q \wedge R$ without worrying about any possible ambiguity. The same thing is true for $P \vee Q \vee R$. We indicate this fact by saying that the connectives $\wedge$ and $\vee$ are **associative.**

In some cases, the ambiguity is not quite so easy to resolve. For example, consider the expression $P \wedge Q \vee R$. This can be interpreted as either $(P \wedge Q) \vee R$ or as $P \wedge (Q \vee R)$. Using the statements P, Q, and R above, $(P \wedge Q) \vee R$ can be interpreted as "He is tall with short hair or his hair is pink," whereas $P \wedge (Q \vee R)$ can be interpreted as "He is tall and also his hair is short or pink." These two descriptions are certainly not the same! One way to see this is largely self-educated, and he primarily studied Latin and Greek. At age 16 he stopped his studies in order to help support his family. Boole got a job as a teacher's aide, and in the evenings after work he continued his self-education. At age 20, he opened his own school. In order to train his students properly, he had to teach them some mathematics, a subject he had previously known little about.

Boole read and mastered several difficult mathematics books. His interest in mathematics was instantly aroused. While reading one particularly difficult text by the French mathematician Joseph-Louis Lagrange, Boole made an important mathematical discovery that resulted in his first publication and, undoubtedly, some confidence that he could make further contributions to the field.

Over time, Boole got to know a number of the leading English mathematicians, both personally and by correspondence. In particular, he befriended the famous mathematician Augustus De Morgan and became aware of De Morgan's work on logic. Boole published his own thin volume, entitled *Mathematical Analysis of Logic,* which gave him instant recognition. Although he continued on as a schoolteacher, he also enrolled at Cambridge University for his first formal training in mathematics. Shortly thereafter, he received a professorship at Queen's College in Ireland. At Queen's College, Boole made several other major contributions to logic and became increasingly famous. He died at age 50 from pneumonia.

Boole's contributions greatly influenced a number of later mathematicians, including the German mathematicians Ernst Schroder (1841 – 1902) and David Hilbert (1862 – 1943) and the English mathematicians Alfred North Whitehead (1861 – 1947) and Bertrand Russell (1872 – 1970). Today, logic is an integral part of mathematics and philosophy and is applied to many disciplines such as electrical engineering and computer science, among others.

to compare the truth table for $(P \wedge Q) \vee R$ to the truth table for $P \wedge (Q \vee R)$. The truth table for $(P \wedge Q) \vee R$ is as follows:

P	Q	R	$(P \wedge Q)$	$(P \wedge Q) \vee R$
T	T	T	T	T
T	T	F	T	T
T	F	T	F	T
T	F	F	F	F
F	T	T	F	T
F	T	F	F	F
F	F	T	F	T
F	F	F	F	F

The truth table for $P \wedge (Q \vee R)$ is as follows:

P	Q	R	$(Q \vee R)$	$P \wedge (Q \vee R)$
T	T	T	T	T
T	T	F	T	T
T	F	T	T	T
T	F	F	F	F
F	T	T	T	F
F	T	F	T	F
F	F	T	T	F
F	F	F	F	F

These two truth tables differ for some values of P, Q, and R. For example, notice that if P is false and Q and R are both true, then $(P \wedge Q)$ is false; but, because R is true, $(P \wedge Q) \vee R$ is true. On the other hand, $P \wedge (Q \vee R)$ is false in this case, because P is false. Thus, regardless of whether $(P \wedge Q)$ is true or false (it turns out to be true), the expression will certainly be false. For this reason it is often desirable to use parentheses to avoid ambiguity.

Another way to avoid ambiguity, without using parentheses, is to adopt a convention on the order of operations. For example, in arithmetic the convention is that multiplication takes precedence over addition. Therefore, $3 + 4 \times 5$ is evaluated by first evaluating 4×5 and then adding 3. Of course, we could have just written $3 + (4 \times 5)$ to avoid the ambiguity altogether. Similarly, in Boolean logic we adopt the convention that $\wedge$ (AND) takes precedence over $\vee$ (OR). Therefore, the expression $P \wedge Q \vee R$, by convention, is to be interpreted as $(P \wedge Q) \vee R$. Furthermore, the convention states that $\neg$ (NOT) takes the highest precedence of all. Thus, $\neg P \wedge \neg Q \vee R$ is interpreted as $((\neg P) \wedge (\neg Q)) \vee R$.

EXAMPLE

Applying Boolean Logic to a Web Search

Let us revisit the Web queries. Let P represent the query "explorer," which corresponds to the expression "This Web page contains the word *explorer*"; let Q represent the expression "This Web page contains the word *Microsoft*"; and let R represent the expression "This Web page contains the word *Ford*." We now translate the query "*explorer* AND (NOT *Microsoft*) AND (NOT *Ford*)" as $P \wedge (\neg Q) \wedge (\neg R)$ and write the truth table as follows:

P	Q	R	$\neg P$	$\neg Q$	$P \wedge (\neg Q) \wedge (\neg R)$
T	T	T	F	F	F
T	T	F	F	T	F
T	F	T	T	F	F
T	F	F	T	T	T
F	T	T	F	F	F
F	T	F	F	T	F
F	F	T	T	F	F
F	F	F	T	T	F

For every possible value of P, Q, and R, the truth table gives us the value of our expression. The fourth and fifth columns of the table are not strictly necessary, but they are helpful to us in determining the values in the last column. As expected, this table tells us that the expression $P \wedge (\neg Q) \wedge (\neg R)$ is true precisely when P is true, Q is false, and R is false. ■

Two expressions are said to be **logically equivalent** if they have the same value, true or false, for each possible assignment of the Boolean variables. One way of testing if two expressions are logically equivalent is by using truth tables. Given two expressions, we construct the truth tables for each one and then check if they have the same values for each of the possible assignments of the Boolean variables. If so, the expressions are logically equivalent. If they differ for even one case, however, then the expressions are not equivalent.

EXAMPLE

Applying Boolean Logic to a Web Search

Compare the query from the previous Example to the query "explorer AND NOT (Microsoft AND Ford)", which is represented mathematically as $P \wedge \neg(Q \wedge R)$. The truth table for this expression is as follows:

P	Q	R	$Q \wedge R$	$\neg(Q \wedge R)$	$P \wedge \neg(Q \wedge R)$
T	T	T	T	F	F
T	T	F	F	T	T
T	F	T	F	T	T
T	F	F	F	T	T
F	T	T	T	F	F
F	T	F	F	T	F
F	F	T	F	T	F
F	F	F	F	T	F

This truth table differs in the last column from the truth table in the previous Example. For instance, when P is true, Q is true, and R is false, we see that $P \wedge (\neg Q) \wedge (\neg R)$ is false but $P \wedge \neg(Q \wedge R)$ is true. Therefore, we must conclude that $P \wedge (\neg Q) \wedge (\neg R)$ is not logically equivalent to $P \wedge \neg(Q \wedge R)$. Therefore, we would expect that these two queries would result in different Web pages being found by the search engine. ■

11.6 From Boolean Logic to Digital Logic

Boolean logic is used in a variety of applications for the Internet and for computing in general. As we've seen, computers use 0's and 1's to represent data. Boolean logic also has just two values, false and true. By using 0 to represent false and 1 to represent true, we can simply re-express all the rules of logic by replacing every occurrence of F with a 0 and every occurrence of T with a 1. This is sometimes called **digital logic.**

EXAMPLE

The AND Connective in Digital Logic

Rewriting the AND connective in terms of 0's and 1's gives us the following truth table:

P	Q	$P \wedge Q$
1	1	1
1	0	0
0	1	0
0	0	0

■

We now examine two applications of digital logic.

Digital Logic in Message Routing

Earlier in this chapter, we mentioned that when a router receives a packet, it looks at the IP address of the destination and determines how the packet should be forwarded so that it will eventually reach its destination. At first glance, it might appear that the router would need to know about all the billions of possible IP addresses of all computers on the Internet. Fortunately, this is not the case!

IP addresses use a hierarchy convention similar to standard mailing addresses. Mailing addresses typically contain country, city, street, and house number. Imagine that a postal worker in a small town sees a letter addressed to Singapore. Clearly, she doesn't need to look at any other part of the address–she probably just deposits the letter in the international mail bag. The international mail bag is then forwarded to a larger post office that separates the mail according to country, and so forth. On the other hand, if the destination address is in the same town, she may simply put the letter in the bin for that address.

Analogously, all computers in a local area network, such as those in a university or company, will have IP addresses that have many bits in common. Recall that an IP address is a sequence of 4 numbers, each in the range from 0 to 255. Each of these four numbers can be represented by 8 bits, that is, 1 byte.

For example, 129.128.3.15 is a valid IP address. Imagine that a small company has 200 computers on its local area network. For simplicity, assume that these 200 computers are connected to a single router. The IP addresses of all these computers might begin with 129.128.3, which in base 2 is simply 10000001 10000000 00000011.

The router for this local area network stores two binary numbers–the *base address,* which is the part of the IP address common to all computers in the local area network, and the *subnet mask,* which is used to help the router determine whether an IP address belongs to the local area network. In our example, the base address might be the base 2 representation of 129.128.3.0, which is 10000001 10000000 00000011 00000000. The subnet mask might be the base 2 representation of 255.255.255.0, which is 11111111 11111111 11111111 00000000.

Assume that a computer in the local area network sends a message to another computer at IP address 129.128.3.15, which in base 2 is 10000001 10000000 00000011 00001111. The router takes the IP address of the computer to which the message is destined and the subnet mask and lines them up, one above the other. It then performs the AND of each pair of 0's and 1's in the same column. This is called the *bitwise AND* of the two numbers. In this case we get

Destination IP address	10000001	10000000	00000011	00001111
Subnet mask	11111111	11111111	11111111	00000000
Bitwise AND	10000001	10000000	00000011	00000000

Notice that the bitwise AND is exactly the base address stored by the router. This indicates to the router that the destination IP address is in the local area network or subnet. Thus, the router simply forwards the packet to the appropriate local computer.

On the other hand, if the IP address of the destination was something that did not begin with 129.128.3, then the result of performing the bitwise AND of this IP address and the subnet mask would result in a pattern different from the base address stored at the router. This would indicate to the router that the packet must be forwarded to the "higher-up" router belonging to the Internet service provider that will handle the delivery of this packet.

EXAMPLE

An IP Address Outside the Local Area Network

Using the base address and subnet mask above, what is the result of performing the bitwise AND when a computer in the local area network sends a message to another computer at IP address 130.65.5.1? The base 2 representation of 130.65.5.1 is 10000010 01000001 000000101 00000001. Thus, the bitwise AND is

Destination IP address	10000010	01000001	00000101	00000001
Subnet mask	11111111	11111111	11111111	00000000
Bitwise AND	10000010	01000001	00000101	00000000

This differs from the base address. Consequently, the router knows that the message is destined for a computer outside of the local area network. It therefore forwards the packet to the router at the next-higher level of the Internet (the router belonging to the ISP). ■

In this Example, the subnet mask was 255.255.255.0, which effectively means that the router should compare the first 3 numbers in the IP address of the destination to the base address. However, by using a different subnet mask, we could have specified a different relationship between the IP address of the destination and the base address of the router.

EXAMPLE

An IP Address Outside the Local Area Network

Consider a router using the base address 129.128.3.128 and subnet mask 255.255.255.192. What is the result of performing the bitwise AND when a computer in the local area network sends a message to another computer at IP address 129.128.3.155? The base 2 representation of the subnet mask 255.255.255.192 is 11111111 11111111 11111111 1100000. The base 2 representation of the destination address 129.128.3.155 is 10000001 10000000 00000011 10011011. The base 2 representation of the base address 129.128.3.128 is 10000001 10000000 00000011 1000000. Thus, the bitwise AND of the destination address and the subnet mask is

Destination IP address	10000001	10000000	00000011	10011011
Subnet mask	11111111	11111111	11111111	11000000
Bitwise AND	10000001	10000000	00000011	10000000

This is the same as the base address. Consequently, the router knows that the message is destined for a computer within the same local area network. ■

Encrypting Credit Card Data on the Web

Millions of purchases are made every day over the Web. Frequently, these purchases require that the buyer provide the seller with confidential information such as a credit card number and home address. These data are sent over the Internet from the client's computer to the seller's computer. Recall that these data are sent in packets that are likely to travel through many routers before reaching their final destination. Each router may be owned by a different organization, and it is virtually impossible to guarantee the security of all the routers in the Internet. In theory, a hacker with access to one of the intermediate routers could make copies of the packets as they pass through the router. In this way, a malicious party could electronically eavesdrop on the transaction and obtain the credit card number and other confidential information.

Mathematicians and computer scientists have devised a number of solutions to this problem. One solution is based on *public key encryption schemes* such as the *RSA algorithm* described in Chapter 10. While we will omit the details of the RSA algorithm in our discussion, you may want to refer to Chapter 10 to understand precisely how this algorithm works.

The main idea of a public key encryption scheme is that each computer maintains two numbers: a *private key* and a *public key*. Each computer keeps its private key secret but willingly sends its public key to anyone who requests it.

When the buyer's Web browser wants to send a secret message to the seller, such as a credit card number, the buyer requests the seller's public key. It then uses the seller's public key to encode the message and sends this encoded message to the seller. The seller uses its private key to decode the message. These schemes have the property that a message encoded with the public key can only be decoded with the matching private key and that knowledge of the public key does not allow one to determine the private key. Thus, because the seller holds its private key secret, only the seller is able to decode the message. Analogously, the seller can send the buyer a secret message by encoding the message using the buyer's public key. The buyer then decodes the message using its private key.

Although public key encryption schemes are secure, one disadvantage is that each interaction between the buyer and seller is relatively slow due to the computation required to perform the encoding and decoding using the public and private keys. An alternative is to use a simpler and more efficient *symmetric scheme.* In these schemes, the buyer and seller agree on a single key that is used to both encode and decode the message. Before describing one such scheme, we note that there seems to be a serious problem with this idea. How will the buyer and seller agree on the single key? One of the two parties could choose the key, but it would then need to be sent to the other party. If a malicious hacker intercepts the packets containing the key, then we have achieved no security at all! This problem is solved by using a public key encryption scheme just once to transmit the key for the symmetric scheme. This transmission is secure. Although the encryption and decryption of the key may be somewhat slow, now the buyer and seller have the symmetric scheme key and can henceforth communicate using the faster symmetric scheme!

A **symmetric encryption scheme** is one in which the sender and receiver have the same key. This key is used to both encode and decode the message.

One symmetric scheme, called the *one-time pad scheme,* is based on logic. This scheme uses a new connective called XOR (pronounced ex-or), which stands for "exclusive or." The mathematical symbol for this connective is $\oplus$. This connective is similar to $\vee$ (OR) except that $P \oplus R$ is true when exactly one of P or R is true, but not both; otherwise it is false. The truth table for this connective is as follows:

P	Q	$P \oplus Q$
T	T	F
T	F	T
F	T	T
F	F	F

The corresponding digital logic representation is as follows:

P	Q	$P \oplus Q$
1	1	0
1	0	1
0	1	1
0	0	0

The key in this scheme is simply a long random sequence of 0's and 1's called the *one-time pad.* For example, assume that the key is 11000101. The buyer and seller both have this key. Assume that the buyer wishes to send just a single-letter message to the seller. The letter is encoded in Extended ASCII and therefore has an 8-bit code. For example, the Extended ASCII code might be 01101011. We now write this message with the key directly underneath it and compute the XOR of each pair of bits in the same column. This is called the *bitwise XOR* of the two binary sequences. In our example, we have:

Original message	01101011
Key	11000101
Encoded message	10101110

The buyer now sends the bitwise XOR, 10101110, to the seller. The seller can now retrieve the original message by writing this encoded message with the same key, 11000101, below it and again computing the bitwise XOR:

Encoded message	10101110
Key	11000101
Original message	01101011

Notice that the result of this procedure is indeed the original message! Why does this procedure work? First, it is easily verified using truth tables showing that the $\oplus$ connective is associative. That is, $(P \oplus Q) \oplus R$ is the same as $P \oplus (Q \oplus R)$. Now let P denote a bit, either a 0 or a 1, in the original message. Let Q denote the bit, either a 0 or a 1, in the corresponding position of the key. Then the encoded message contains the bit $P \oplus Q$ in this position. To decode the message, we take the encoded bit $P \oplus Q$ and apply $\oplus$ with Q. That is, we compute $(P \oplus Q) \oplus Q$. Because $\oplus$ is associative, this is the same as $P \oplus (Q \oplus Q)$. Observe that $(Q \oplus Q)$ is always 0 because $0 \oplus 0$ is 0 and $1 \oplus 1$ is 0. Thus, we have $P \oplus 0$, which is always P, the bit in the original message.

In general, the one-time pad key is longer than just 8 bits. It should have one bit for each bit that is to be encoded. This scheme is secure because the key can be any binary sequence with equal likelihood. Thus, the encoded message can be any binary sequence with equal likelihood. Therefore, a hacker who sees the encoded message has no information that permits determination of the original message. It can be shown, however, that if the same key is used over and over again to encode messages, then it may be possible to break the code using statistical techniques. Therefore, the key should only be used once. This is the source of the name *one-time*. The word *pad* is just a synonym for *key*.

Finally, we note that the one-time pad scheme is just one example of a symmetric coding scheme. Many other schemes exist as well, and a variety of these schemes are used in practice.

In summary, in this chapter we have seen the basic components of the Internet. We have examined how data are represented as a binary sequence for transmission over a network. Next, we explored the breadth-first search algorithm and showed how it can be used to facilitate a Web search. Boolean logic was then shown to be a powerful tool in formulating queries that are given to a Web search engine. Finally, we saw how Boolean logic is used in other applications on the Internet, including routing of messages and encryption of data.

REVIEW VOCABULARY

Associative A connective is said to be associative if the order in which the connective is evaluated does not matter. For example, the $\wedge$ (AND) and $\vee$ (OR) connectives are associative since $(P \wedge Q) \wedge R$ is logically equivalent to $P \wedge (Q \wedge R)$. Similarly, $(P \vee Q) \vee R$ is logically equivalent to $P \vee (Q \vee R)$.

Base Numbers can be represented in a variety of different bases. Base 10 is the conventional representation of numbers in which the 10 digits from 0 to 9 are used. Base 2 is a representation in which the digits 0 and 1 are used.

Binary sequence A sequence of 0's and 1's.

Bit A contraction of the term *binary digit,* meaning a 0 or a 1.

Boolean logic Logic attributed to George Boole in which expressions are connected using connectives such as $\wedge$, $\vee$, and $\neg$ to form other expressions.

Breadth-first search An algorithm for searching through a graph. The algorithm uses a queue to keep track of the vertices that remain to be processed and a database of vertices that have already been reached.

Byte A sequence of 8 bits; that is, a binary sequence of length 8. A byte can represent one of 256 different patterns, because each of the 8 bits can have one of two possible values.

Digital logic Boolean logic in which the values false and true are represented by the digits 0 and 1, respectively.

Extended ASCII A code for representing symbols such as letters, numbers, and others using one byte to represent each symbol.

Logically equivalent Two expressions are said to be logically equivalent if they have the same values for all possible values of their Boolean variables.

One-time pad scheme An encryption scheme in which a single binary sequence is used by both the sender and the receiver to encode and decode the message. This scheme uses the XOR connective as the fundamental operation.

Queue A method for storing data that is analogous to a line in a movie theater. New items are placed at the end of the queue and items are removed from the front of the queue.

Symmetric encryption schemes A class of data encryption techniques in which the sender and receiver of the message use the same key to encode and decode the message. The one-time pad scheme is an example of a symmetric encryption scheme.

Truth table A tabular representation of an expression in which the variables and intermediate expressions appear in columns and the last column contains the expression being evaluated.

SUGGESTED READINGS

BELL, E. T. *Men of Mathematics,* Simon & Schuster, New York, 1937. This is a famous book of short essays on a number of famous mathematicians, including George Boole. The essays are interesting and amusing.

ENDERTON, H. B. *A Mathematical Introduction to Logic,* Academic Press, New York, 1972. This is a well-written undergraduate mathematics text, much of which is accessible to nonspecialists. This is a good choice for someone wanting to learn more about logic than was presented in this chapter.

TYMOCZKO, T., and J. HENLE. *Sweet Reason: A Field Guide to Modern Logic,* Freeman, New York, 1995. A very readable and enjoyable introduction to logic written by a philosopher and a mathematician.

SUGGESTED WEB SITES

www.howstuffworks.com/router.htm This site clearly and concisely describes how Internet routers work.

www.www10.org/cdrom/papers/208/index.html This site contains a research paper on Web crawling and page-ranking algorithms.

www.howstuffworks.com/encryption1.htm This site explains how data encryption works on the Internet.

www.enteract.com/~lspitz/digcerts.html This is another site with a clear and easy-to-read description of encryption and its applications to Internet commerce.

SKILLS CHECK

1. The number 1011 in base 2 represents what number in base 10?

(a) 3
(b) 11
(c) 22

2. The number 27 in base 10 represents what number in base 2?

(a) 1001
(b) 11101
(c) 11011

3. Extended ASCII representation for a single symbol is

(a) a string of 8 bits.
(b) a string of 16 bits.
(c) a string of 8 bytes.

4. The Extended ASCII representations for "A" and "a" are

(a) different.
(b) the same.
(c) undefined.

5. An IP address is comprised of four bytes, each of which

(a) is a number between 0 and 255.
(b) is a number between 0 and 111.
(c) is a number between 111 and 999.

6. The breadth-first search Web crawler algorithm searches Web pages for

(a) valid email addresses.
(b) links from other Web pages.
(c) links to other Web pages.

7. For the breadth-first search Web crawler algorithm, the choice of start vertex

(a) can impact the outcome of the algorithm.
(b) always impacts the outcome of the algorithm.
(c) never impacts the outcome of the algorithm.

8. The breadth-first search Web crawler algorithm stops when

(a) there are no vertices in the queue.
(b) every vertex is included in the database.
(c) there are no vertices in the database.

9. The statement "*P* OR *Q*" is true precisely when

(a) either *P* or *Q* is true, but not both.
(b) *P* is true.
(c) neither *P* nor *Q* is false.

10. The order in which the operations for the statement "*P* AND *Q* AND *R*" are evaluated

(a) makes no difference.
(b) is ambiguous.
(c) requires the use of parentheses.

11. When the statement "*P* AND NOT *Q*" is true, it must be the case that

(a) *Q* is true.
(b) *Q* is false.
(c) *P* is false.

12. When the statement "P XOR Q" is true, it must be the case that

(a) *P* is true.
(b) *P* is false.
(c) either *P* or *Q* is false.

13. Digital logic represents "true" by

(a) 0.
(b) 1.
(c) 10.

14. Credit card numbers are routed across the Web securely by the use of

(a) public key encryption.
(b) private key encryption.
(c) open networks.

15. One example of a symmetric scheme for encryption is

(a) the RSA algorithm scheme.
(b) the one-time pad scheme.
(c) Extended ASCII.

EXERCISES

▲ **Optional.** ■ **Advanced.** ◆ **Discussion.**

Representing Data

1. Convert the base 2 number 101011 to base 10.

2. Convert the base 10 number 795 to base 2.

3. Notice that the base 2 number 1 represents the base 10 number 1, the base 2 number 11 represents the base 10 number 3, the base 2 number 111 represents the base 10 number 7, and the base 2 number 1111 represents the base 10 number 15. Consider the base 2 number consisting of *k* consecutive 1's. What does this number correspond to in base 10?

4. Base 3 uses digits 0, 1, and 2 to represent numbers.

(a) Convert the base 3 number 2010 to base 10.
(b) Convert the base 3 number 2010 to base 2.
(c) Convert the base 10 number 795 to base 3.

5. Base 16, known as *hexadecimal*, has 16 digits: 0, . . . , 9, *A*, *B*, *C*, *D*, *E*, *F*. The digit *A* in hexadecimal represents the number 10 in base 10 and the digit *F* in hexadecimal represents the number 15 in base 10.

(a) Convert the hexadecimal number 3*FA* to base 10.
(b) Convert the base 10 number 795 to hexadecimal.
(c) Convert the base 2 number 1111 to hexadecimal.
(d) Express the IP subnet mask 255.255.255.0 in hexadecimal.

6. What is the largest even number that can be represented in base 2 using five digits? What is this number in base 10?

7. What is the largest number that can be represented in base 3 using five digits? What is this number in base 10?

Breadth-First Search

8. Referring to the graph in Figure 11.5, show the queue and the database after each step of a breadth-first search, assuming that the search begins at vertex 1.

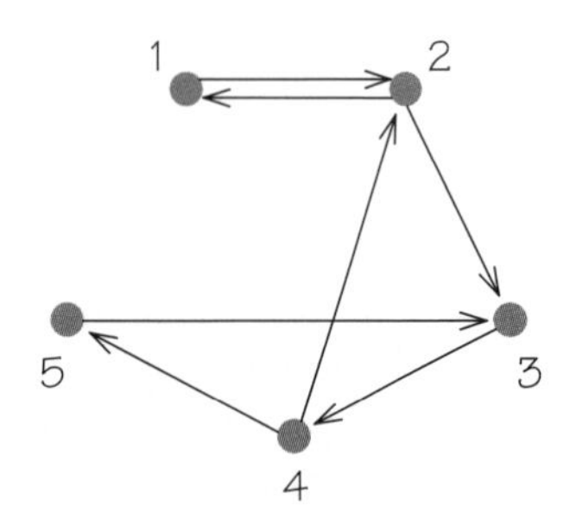

Figure 11.5 An example of a graph for breadth-first search.

9. Referring to the graph in Figure 11.5, show the queue and the database after each step of a breadth-first search, assuming that the search begins at vertex 3.

10. A breadth-first search is used in a number of applications in addition to Web crawling. For example, consider the set of streets and intersections indicated in Figure 11.6a. The arrows on the streets indicate the permitted directions of travel between adjacent intersections. Streets with just one arrow are one-way streets. Such a street map can be represented by a directed graph in which each intersection in the map corresponds to a vertex. There is a directed edge from a vertex x to a vertex y if there is a street directly from the intersection represented by vertex x to the intersection represented by vertex y and travel is permitted in this direction. The directed graph corresponding to the street map in Figure 11.6a is shown in Figure 11.6b. Some computerized navigation programs use a breadth-first search to find routes from one location to another. Use a breadth-first search to determine whether there exists a way to get from intersection 1 to intersection 9. Show the queue and the database after each step.

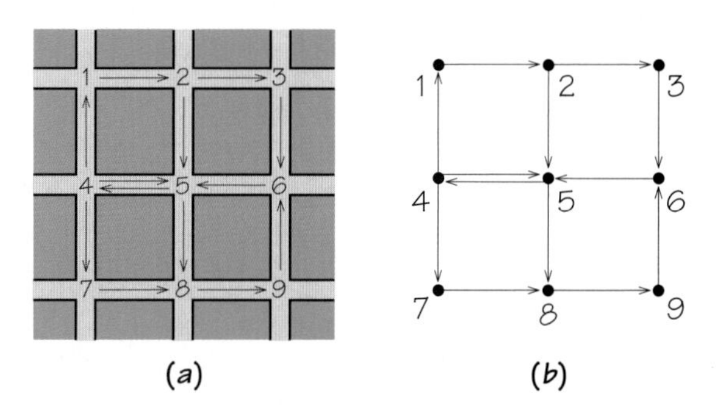

Figure 11.6 (a) A street map and (b) its graphic representation.

11. The breadth-first search algorithm can be modified in a number of interesting ways. Recall that a breadth-first search comprises two steps. In step 1 the algorithm removes the vertex u, which is at the front of the queue. In step 2, the algorithm examines each vertex v such that there is an edge from u to v. If vertex v is not in the database, it is stored in the database and placed on the queue. Imagine that when vertex v was added to the database in step 2, a "note" was stored with vertex v indicating that it was "visited by vertex u." When the algorithm is completed, each vertex reachable from the start vertex has a note associated with it. Explain how these notes can now be used to recreate a path from the start vertex to any vertex that is reachable from the start vertex.

■ **12.** The *distance* from a vertex x to a vertex y in a directed graph is the minimum number of edges required to get from x to y. Describe how the breadth-first search algorithm can be modified slightly so that each vertex in the database is labeled with its distance from the start vertex.

Boolean Logic

13. The AND connective is associative. In other words, given an expression $P \wedge Q \wedge R$, we can evaluate this expression as $(P \wedge Q) \wedge R$ or as $P \wedge (Q \wedge R)$ and the result will be the same. Verify this by constructing a truth table for $(P \wedge Q) \wedge R$ and one for $P \wedge (Q \wedge R)$.

14. Use truth tables to show that the XOR connective is associative.

15. Show that $P \vee (P \wedge Q)$ is logically equivalent to P.

16. Show that $\neg(P \vee Q)$ is logically equivalent to $\neg P \wedge \neg Q$. This relationship and the one in the next exercise are collectively known as *De Morgan's Laws.*

17. Show that $\neg(P \wedge Q)$ is logically equivalent to $\neg P \vee \neg Q$. This relationship and the one in the previous exercise are collectively known as *De Morgan's Laws.*

18. Show that $P \vee (Q \wedge R)$ is logically equivalent to $(P \vee Q) \wedge (P \vee R)$.

19. Show that $P \wedge (Q \vee R)$ is logically equivalent to $(P \wedge Q) \vee (P \wedge R)$.

20. A man reads a forecast in the newspaper stating that today "it will not be sunny and it will not be rainy." On the way to work, he hears a forecaster on the radio state that today "it is not the case that it will be sunny or rainy." The man wishes to determine whether the two forecasts are equivalent. To do so, he uses the variable P to denote "sunny" and the variable Q to denote "rainy." The first forecast can therefore be expressed as $\neg P \wedge \neg Q$. Express the second forecast as a logical expression and then use truth tables to determine if these expressions are logically equivalent.

21. A patron at a restaurant tells the waiter to bring her the chef's recommendation as long as it has "lots of anchovies or is not spicy and in addition the portion must be large." The waiter goes to the kitchen and tells the chef to prepare a dish that has "lots of anchovies and is also large or is spicy and is also large." Did the waiter communicate the patron's wishes correctly to the chef? Use truth tables to explain your answer.

22. A professor claims that a double negative is logically the same as a positive. In other words, the negation of the negation of an expression is logically equivalent to the expression itself. Use truth tables to verify this.

23. The *implication connective* is defined by the following truth table:

P	Q	$P \rightarrow Q$
T	T	T
T	F	F
F	T	T
F	F	T

Use truth tables to show that $P \rightarrow Q$ is logically equivalent to $\neg P \vee Q$.

24. Consider the implication connective described in the previous exercise. The expression $P \rightarrow Q$ can be interpreted as the statement "If P is true then Q is true." To see this, look at the truth table for the implication connective in Exercise 23. If P is true and Q is true, then $P \rightarrow Q$ is true and our statement

"If P is true, then Q is true" is a true statement. Now, look at the next line of the truth table. If P is true and Q is false, then $P \rightarrow Q$ is false and so is the statement "If P is true, then Q is true." The third line of the truth table is more interesting. Here P is true and Q is false and $P \rightarrow Q$ is defined to be true. The statement "If P is true, then Q is true" is still a true statement! Why? Because P is not true in this case, and thus we cannot say that our statement "If P is true then Q is true" was a lie; we simply made no promise about the value of Q if P was not true! Therefore, we must conclude that the statement "If P is true, then Q is true" is true.

Now explain why the last line of the truth table for the implication connective is consistent with the statement "If P is true then Q is true."

25. The implication connective can be used to represent cause-and-effect relationships. For example, a patient tells his doctor, "If I eat chocolate, then I get a headache." Let P denote the statement "I eat chocolate" and let Q denote the statement "I get a headache." Then the patient's statement can be represented by the expression $P \rightarrow Q$. Now represent the statement "If I do not eat chocolate and I drink water, then I do not get a headache" using an expression containing three variables.

26. A coach tells his team, "If we win the game, then we'll have a party." The team loses the game, but the coach decides to throw a party anyway. Did the coach tell the truth?

27. Using the implication connective and other connectives, variables, and truth tables, determine whether the statement "If it snows, there will be no school" is logically equivalent to the statement "It is not the case that it snows and there is school."

28. Consider the statement "If you arrive after midnight and you arrive before 6:00 A.M, then you cannot take a taxi and you cannot take the bus." Let P denote the statement "you arrive after midnight," let Q denote the statement "you arrive before 6:00 A.M," let R denote the statement "you cannot take a taxi," and let S denote the statement "you cannot take the bus." Express the statement as an expression using variables P, Q, R, and S.

29. An expression is called a *tautology* if it is true for all possible values of its variables. Determine whether each of the following expressions is a tautology.

(a) $P \vee \neg P$
(b) $(P \wedge Q) \vee (\neg P \wedge \neg Q)$
(c) $P \vee \neg(Q \vee \neg P)$

30. Referring to the definition of a *tautology* in Exercise 29 and the implication connective in Exercise 23, is the expression $(P \rightarrow \neg P) \rightarrow P$ a tautology?

31. An expression is called a *contradiction* if it is false for all possible values of its variables. Determine whether each of the following expressions is a contradiction.

(a) $P \wedge \neg P$
(b) $P \vee \neg P$
(c) $P \wedge (\neg P \vee Q)$

32. Referring to the definitions of *tautology* and *contradiction* in Exercises 29 and 31, find an expression that is neither a tautology nor a contradiction.

33. An expression is said to be *satisfiable* if there exists at least one set of values of its variables for which the statement is true. For each of the following expressions, determine whether the expression is satisfiable or not satisfiable. If the expression is satisfiable, give a set of values for the variables for which the expression is true.

(a) $P \vee \neg P$
(b) $(P \vee Q) \wedge (P \vee \neg Q) \wedge (\neg P \vee Q) \wedge (\neg P \vee \neg Q)$
(c) $(P \vee Q) \wedge (P \vee \neg Q) \wedge (\neg P \vee \neg Q)$

34. We have seen several connectives that combine two expressions into a new expression. Other connectives are also possible, with each connective having its own truth table. How many different connectives are possible that combine two expressions into one?

Applications of Boolean Logic

35. Consider a router with base address 150.42.211.224 and subnet mask 255.255.255.240. What is the range of IP addresses that are in the router's local area network?

36. We observed that an IP address comprises 4 bytes and that there are $256^4 = 4{,}294{,}967{,}296$ different possible IP addresses. Imagine that we wish to extend the IP address system to use more bytes such that there are at least 10 billion different addresses possible. What is the minimum number of bytes that such a system would require?

37. The *bitwise OR* of two binary sequences is similar to the bitwise XOR, except that the OR of two bits is a 1 if at least one of the two bits is a 1. Does the one-time pad scheme work if bitwise OR is used instead of bitwise XOR? Explain why or why not.

38. Let X and Y denote two binary sequences.

(a) Under what circumstances is the bitwise AND of X and Y equal to a string of all 1's?
(b) Under what circumstances is the bitwise AND of X and Y equal to a string of all 0's?
(c) Under what circumstances is the bitwise XOR of X and Y equal to a string of all 1's?
(d) Under what circumstances is the bitwise XOR of X and Y equal to a string of all 0's?

39. Using the one-time pad scheme, encode the sequence 10111101 using the key 11001100. Then decode the encoded string using that key again.

WRITING PROJECTS

1. In addition to breadth-first search, there are other algorithms for discovering the vertices in a graph. One of these algorithms is called *depth-first search*. Learn more about depth-first search and describe how the algorithm works. Then give an example of how the algorithm works for a small directed graph.

2. There are numerous methods for computing the ranking of Web pages. Write a short paper describing some of the techniques other than those described in this chapter.

3. In recent years a new type of logic called *fuzzy logic* has been proposed. The basic idea is that a statement is not necessarily true or false but may be somewhere in between true and false. For example, we might have reason to believe that a statement is very likely true but has a small chance of being false. Learn more about fuzzy logic and write a short paper on how it works and some of the applications in which it is used.

4. We described the one-time pad as an example of a symmetric encryption scheme. What are some other symmetric encryption schemes and how do they work?

5. Using a search engine such as Google (www.google.com), do a search using a query with just one word. Next, add another word to this query. Repeatedly, add more words to this query. Record the number of Web pages found for each query and the amount of time it takes the search engine. Most search engines report both of these numbers.

6. Use the Northern Light search engine (www.northernlight.com) to perform queries using Boolean logic. For example, try some of the queries described in the chapter. Then try some of your own queries.

7. A popular public key encryption scheme for email is called *PGP,* which stands for *p*retty *g*ood *p*rivacy. With a classmate, search for Web sites about PGP on the web. Find out how to set up your own PGP keys and send PGP-encoded email messages to one another.

SPREADSHEET PROJECTS

To do these projects, go to www.freeman.com/fapp.

Conversion between decimal representations and binary representations can easily be achieved using spreadsheets. The spreadsheet projects exploit these capabilities.

Social Choice: The Impossible Dream

In this chapter we cover...

...and at www.whfreeman.com/fapp:

Flashcards

Quizzes

Spreadsheet Projects

- **Borda counts**
- **Approval voting**

Applet Exercises

- **Insincere voting**

The basic question of *social choice,* of how groups can best arrive at decisions, has occupied social philosophers and political scientists for centuries. One primary example of a social choice problem is the selection of a "good" voting system. Indeed, voting is a subject that lies at the very heart of representative government and participatory democracy.

Social choice theory attempts to address the problem of finding good procedures that will turn individual preferences for different candidates – or *alternatives,* as they are often called – into a single choice by the whole group. An example of such a choice would be the selection of a *winner* of an election. The goal is to find such procedures that will result in an outcome that "reflects the will of the people."

This search for good voting systems, as we shall see, is plagued by a variety of counterintuitive results and disturbing outcomes. In fact, it turns out that one can prove (mathematically) that no one will ever find a completely satisfactory voting system for three or more alternatives.

The elections with which we are most familiar often involve only two candidates, and we will begin our discussion of voting systems with this two-alternative case.

Yet, there are real-world situations in which elections must be held to choose a single winner from among three or more candidates, as in the presidential election of 2000 in which George W. Bush, Al Gore, Ralph Nader, and Patrick Buchanan were the candidates.

There are several methods that can be used to elect a single candidate from a choice of three or more, and we will investigate some in this chapter. Most of these methods use a ballot in which a voter provides a rank ordering of the candidates (without ties) indicating the order in which he or she prefers them.

A ballot consisting of such a rank ordering of candidates (which we often picture as a vertical list with the most preferred candidate on top and the least preferred on the bottom) is called a *preference list,* or, more precisely, an **individual preference list,** because it is a statement of the preferences of one of the individuals who is voting.

Ballots that are preference lists allow voters to make a much clearer statement of their preferences than do ballots allowing a single vote. Ballots with preference lists are already used in a wide range of applications, such as rating football teams and scoring track meets.

We present four particular methods of choosing a winner when there are three or more alternatives. For each method, we illustrate its shortcomings as well as its benefits. Finally, we face the fact that elections involving three or more candidates present difficulties that are simply insurmountable.

12.1 Elections with Only Two Alternatives

When choosing between two alternatives, the first type of voting to suggest itself is **majority rule:** Each voter indicates a preference for one of the two candidates, and the candidate with the most votes wins. Majority rule has at least three desirable properties:

1. All voters are treated equally. That is, if any two voters were to exchange (marked) ballots prior to submitting them, the outcome of the election would be the same.
2. Both candidates are treated equally. That is, if a new election were held and every voter were to reverse his or her vote, then the outcome of the previous election would be reversed as well.
3. If a new election were held and a single voter were to change his or her ballot from being a vote for the loser of the previous election to being a vote for the winner of the previous election, and everyone else voted exactly as before, then the outcome of the new election would be the same as the outcome of the previous election.

It is easy to devise voting systems for two alternatives in which these fail, but each such voting system quickly reveals its undesirability. For example, condition 1 is not satisfied by a *dictatorship* (whereby all ballots except that of the dictator are ignored); condition 2 is not satisfied by *imposed rule* (whereby candidate X wins regardless of who votes for whom); and condition 3 is not satisfied by *minority rule* (whereby the candidate with the fewest votes wins).

But maybe there are voting systems in the two-alternative case that are superior to majority rule in the sense of satisfying the three properties just listed *and* some other properties that we might also wish to have satisfied. This, however, turns out not to be the case. In 1952, Kenneth May proved the following:

If the number of voters is odd, and we are interested only in voting systems that never result in a tie, then majority rule is the *only* voting system for two alternatives that satisfies the three conditions just listed.

This is an important and elegant result known as **May's theorem.** Thus, mathematical reasoning spares us the trouble of searching for a better voting system for two alternatives.

12.2 Elections with Three or More Alternatives: Procedures and Problems

In sharp contrast with the case of two alternatives is the situation in which there are three or more candidates. Here, we find no shortage of procedures that suggest themselves and that seem to represent perfectly reasonable ways to choose a winner from among three or more alternatives. Closer inspection, however, reveals shortcomings with all of these. We illustrate this with a consideration of four well-known procedures. Additional procedures (and additional shortcomings) can be found in the Exercises.

In what follows, we assume that a ballot is an individual preference list. We allow ties in the election result and assume that, in the real world, either the number of voters is so large that ties in the election result will virtually never occur or that they can be broken by some kind of random device.

Plurality Voting and the Condorcet Winner Criterion

In **plurality voting,** only first-place votes are considered. Thus, while we will consider plurality voting in the context of preference lists, a ballot might just as well be a single vote for a single candidate. The candidate with the most votes wins, even though this may be considerably fewer than one-half the total votes cast. This is perhaps the most common system in use today. It is how the voters in Florida chose George W. Bush over Al Gore, Ralph Nader, and Patrick Buchanan in the presidential election of 2000.

EXAMPLE

Plurality Voting and the 2000 Presidential Election

On the evening of December 12, Al Gore conceded the presidential election of 2000 to George W. Bush, thus bringing to a close one of the most remarkable elections in modern times. The outcome, ultimately decided in the electoral college, came down to which of Bush or Gore would carry Florida. With more than 6 million votes cast in Florida, the ultimate margin of victory for George W. Bush was only a few hundred votes.

There is little doubt that if the 2000 presidential election had pitted Al Gore solely against any one of the other three candidates, then Gore would have won both the election in Florida and the presidency. The point is that while most of the Buchanan supporters would have voted for Bush, the far more numerous Nader supporters would have gone largely for Gore.

Thus, although plurality voting led to Bush's winning the election in Florida (and hence the presidency), Gore was, in this example, what is called a

Condorcet winner: He would have defeated each of the other three candidates in a head-to-head (i.e., "one-on-one") election. ■

A voting procedure is said to satisfy the **Condorcet winner criterion (CWC)** provided that, for every possible sequence of preference lists, either (1) there is no Condorcet winner (as is often the case) or (2) there is a Condorcet winner (which, if it exists, is always unique) and it is the unique winner of the election.

The Condorcet winner criterion is certainly a property that one would like to see satisfied. However, the presidential election of 2000 (and, in particular, the Florida vote in that election) shows that plurality voting fails to satisfy the Condorcet winner criterion.

Perhaps a more fundamental drawback of plurality voting is the extent to which the ballots provide no opportunity for a voter to express any preferences except for naming his or her top choice. No use is made, for example, of the fact that a candidate may be no one's first choice but everyone's close second choice.

Finally, there is yet another shortcoming of plurality voting. It is subject to what is called *manipulability:* There are elections in which it is to a voter's advantage to submit a ballot that misrepresents his or her true preferences. For example, in the presidential election of 2000, many voters who ranked Ralph Nader or Patrick Buchanan over George W. Bush and Al Gore chose to vote for Bush or Gore rather than to "throw away" their vote on a candidate they felt had no chance. As we move on to consider voting systems that make more significant use of a voter's rank ordering of the candidates, we will see that manipulability becomes even more of an issue.

The Borda Count and Independence of Irrelevant Alternatives

In many elections that use preference lists as ballots, the goal is to arrive at a final group rank ordering of all the contestants that best expresses the desires of the electorate. The purpose is not only to determine the winner—say, the class valedictorian—but also to arrive at who finished second, third, and so on, as in the case of one's rank in the senior class. In other applications, such as an election to a hall of fame, the first few finishers each receive the award, while the remaining nominees are also-rans.

One common mechanism for achieving this objective is to assign points to each voter's rankings and then to sum these for all voters to obtain the total points for each candidate. If there are 10 candidates, for example, then we could assign 10 points to each first-place vote for a given candidate, 9 points for each second-place vote, 8 for each third-place vote, and so forth. The candidate with the highest total number of points is the winner. Subsequent positions are assigned to those with the next-highest tallies.

A *rank method* of voting assigns points in a nonincreasing manner to the ordered candidates on each voter's preference list and then sums these points to arrive at a group's final ranking. The special case in which there are n alternatives with each first-place vote worth $n - 1$ points, each second-place vote worth $n - 2$ points, and so on down to each last-place vote worth zero points is known as the **Borda count.**

Rank methods other than the Borda count are not uncommon. For example, a track meet can be thought of as an "election" in which each event is a "voter" and each of the schools competing is a "candidate." If the order of finish in the 100-meter dash is school A, school B, school C, school D, then points are often awarded to each school as follows: 5 points for first place, 3 for second place, 2 for third place, and 1 for fourth place.

Sports polls often use point assignments that qualify as rank methods according to our definition. The following Example provides an illustration of this.

EXAMPLE Rank Methods and a Thoroughbred Poll

A thoroughbred poll, conducted by the National Thoroughbred Racing Association, was reported as follows:

Thoroughbred poll

NTRA

NEW YORK—The thoroughbred poll conducted by the National Thoroughbred Racing Association. Rankings based on the votes of sports and thoroughbred racing media representatives on a 10-9-8-7-6-5-4-3-2-1 basis, with first-place votes in parentheses.

	Pts
1. Kona Gold (12)	199
2. Lido Palace (8)	194
3. Albert The Great	141
4. Tiznow	138
5. Flute (1)	119
6. With Anticipatn	78
7. Skimming	76
8. Include (1)	58
9. Officer	35
10. Guided Tour	33

Others receiving votes: Aptitude 19, Macho Uno 18, England's Legend 16, Silvano 12, Xtra Heat 12, Captain Steve 10, Came Home 7, Bienamado 6, Bet On Sunshine 6, Caller One 5, Touch Tone 5, Broken Vow 4, Tranquility Lake 4, E Dubai 3, Fleet Renee 3, Hap 3, El Corredor 2, Say Florida Sandy 2, Dixie Dot Com 1, Voodoo Dancer 1.

An interesting question is whether or not this is a ranking system. For example, if it is, what are the alternatives, and how many of them are there? In fact, this *can* be regarded as a ranking system, but the number of alternatives is not 10. That is, although 10 horses appeared on each ballot, at least one ballot included the horse Voodoo Dancer, as evidenced by her receiving 1 point, whereas other ballots definitely did not.

To regard this as a ranking system, we must consider the set of alternatives to be the entire set of eligible horses; and we must interpret each ballot as listing all horses other than that voter's top 10 in, say, alphabetical order and coming below that voter's top 10. The point assignments are then as in the newspaper quote above, except that we also assign 0 points for an 11th-place vote, 0 points for a 12th-place vote, and so on. This is why, in the definition of ranking system, we said "assigns points in a *nonincreasing* manner" instead of "assigns points in a *decreasing* manner."

We can use this poll to illustrate how total points are arrived at with a ranking method. Let's see how Kona Gold might have earned the 199 points it is reported to have in the poll. We know that it had 12 first-place votes, and each first-place vote was worth 10 points. We also know that the total number of ballots was 22 (because they give the number of first-place votes that each horse received and $12 + 8 + 1 + 1 = 22$).

Thus, one possibility is that Kona Gold received:

12 first-place votes (at 10 points each)
4 second-place votes (at 9 points each)
3 third-place votes (at 8 points each)
2 fourth-place votes (at 7 points each)
1 sixth-place vote (at 5 points)

The total would then be:

$$120 + 36 + 24 + 14 + 5 = 199$$

■

There is an easy way to calculate the so-called "Borda score" of a candidate. One simply counts the number of occurrences of other candidate names that are below this candidate's name. For example, consider the following ballots:

Rank	**Number of voters (5)**					**Points**
First	*A*	*A*	*A*	*B*	*B*	2
Second	*B*	*B*	*B*	*C*	*C*	1
Third	*C*	*C*	*C*	*A*	*A*	0

Because there are three alternatives, each first-place vote is $n - 1$, or $3 - 1 = 2$, each second-place vote is $n - 2 = 1$, and each third-place vote is $n - 3 = 0$. If we were to calculate the Borda score of candidate *B* algebraically, we would say that *B* has two first-place votes, worth 2 points each (a total of 4 points), and three second-place votes, worth 1 point each (a total of 3 more points). Thus, the Borda score of candidate *B* is $4 + 3 = 7$.

But instead of calculating this Borda score algebraically, we can mentally replace each occurrence of a letter below *B* by a box, □, and simply count the boxes.

Rank	**Number of voters (5)**				
First	*A*	*A*	*A*	*B*	*B*
Second	*B*	*B*	*B*	□	□
Third	□	□	□	□	□

Notice that there are seven boxes, giving us the correct value of 7 as the Borda score for candidate *B*. Of course, one need not actually draw any boxes; we are just emphasizing the fact that, in the counting process, it is "spaces" that we are counting, without regard to which letter occurs in the space. A quick glance at the original ballots (without the boxes) reveals that the Borda score of candidate *A* is 6 and the Borda score of candidate *C* is 2.

The Borda count certainly seems to be a reasonable way to choose a winner from among several alternatives (or to arrive at a group ranking of the alternatives). It also has its shortcomings, however, one of which is the failure of a property known as *independence of irrelevant alternatives*.

To describe this property, suppose that an election yields one alternative (call it *A*) as a winner and another alternative (call it *B*) as a nonwinner. Suppose that a new election is now held and that, although some of the voters may have changed their preference lists, no one who had previously ranked *A* over *B* changed his or her list to now rank *B* over *A*.

If this new election were to yield *B* as a winner, the new outcome would seem strange, especially because none of the relative individual preferences for *A* over *B* had changed in *B*'s favor. The ballot changes responsible for the new outcome involve alternatives *other than A* or *B*. One could argue that these other alternatives ought to be irrelevant to the question of whether *A* is more desirable than *B* or *B* is more desirable than *A*.

A voting system is said to satisfy **independence of irrelevant alternatives (IIA)** if it is impossible for an alternative *B* to move from nonwinner status to winner status unless at least one voter reverses the order in which he or she had *B* and the winning alternative ranked.

The following illustration shows that the Borda count fails to satisfy independence of irrelevant alternatives. Suppose the initial five ballots are as follows:

Rank	**Number of voters (5)**				
First	*A*	*A*	*A*	*C*	*C*
Second	*B*	*B*	*B*	*B*	*B*
Third	*C*	*C*	*C*	*A*	*A*

Our counting procedure shows that the Borda scores are as follows:

Borda score of *A* is 6

Borda score of *B* is 5

Borda score of *C* is 4

The winner is *A* (with 6 points), and *B* is a nonwinner (with 5 points). But now suppose that the two voters on the right change their ballots by moving *C* down between *A* and *B*. The lists then become

Rank	Number of voters (5)				
First	*A*	*A*	*A*	*B*	*B*
Second	*B*	*B*	*B*	*C*	*C*
Third	*C*	*C*	*C*	*A*	*A*

Our counting procedure shows that the Borda scores are as follows:

Borda score of *A* is 6

Borda score of *B* is 7

Borda score of *C* is 2

The Borda count therefore now yields *B* as the winner (with 7 points). Thus, *B* has gone from being a nonwinner to being a winner, even though no one changed his or her mind about whether *B* is preferred to *A*, or vice versa. This shows that the Borda count does not satisfy independence of irrelevant alternatives.

The results of these two elections also illustrate the problem of manipulability. Suppose the initial five ballots, reproduced below, represent the true preferences of the voters.

Rank	Number of voters (5)					Points
First	*A*	*A*	*A*	*C*	*C*	2
Second	*B*	*B*	*B*	*B*	*B*	1
Third	*C*	*C*	*C*	*A*	*A*	0

If all the voters are **sincere** (submitting ballots that represent their true preferences), then *A* wins, as we saw, with 6 points. But *A* is the least preferred candidate of the two voters on the right. Hence, if they vote **strategically** (submitting ballots that do not represent their true preferences), then the ballots actually cast might be as follows:

Rank	Number of voters (5)					Points
First	*A*	*A*	*A*	*B*	*B*	2
Second	*B*	*B*	*B*	*C*	*C*	1
Third	*C*	*C*	*C*	*A*	*A*	0

Thus, the winner would be candidate *B* (who is preferred to candidate *A* by both of the voters who voted strategically).

Much more on manipulability can be found at www.whfreeman.com/fapp, but let us note for now a famous remark of Jean-Charles de Borda (1733 – 1799), made when a colleague pointed out how easily his Borda count can be manipulated: "My scheme is only intended for honest men!"

Interestingly, the Borda count did not really originate with Borda, and the idea of a Condorcet winner predates Condorcet by several centuries. The exact history, however, is still unfolding, with important manuscripts from the thirteenth century only recently discovered. For more details, see Spotlight 12.1.

Sequential Pairwise Voting and the Pareto Condition

In our voting-theoretic context, an **agenda** will be understood to be a listing (in some order) of the alternatives. This listing is not to be confused with any of the preference lists, and, to avoid confusion, we will present agendas as horizontal lists and continue to present preference lists vertically.

Sequential pairwise voting starts with an agenda and pits the first alternative against the second in a one-on-one contest. The winner (or both, if they tie) then moves on to confront the third alternative in the list, one-on-one. Losers are deleted. This process continues throughout the entire agenda, and those remaining at the end are the winners.

For a given sequence of individual preference lists, the particular agenda chosen can greatly affect the outcome of the election. In fact, Exercise 18 at the end of the chapter presents a sequence of ballots for four alternatives, with the property that the question of which alternative wins is *completely* determined by the choice of agenda. This notwithstanding, we will see later in the chapter that sequential pairwise voting arises naturally in the legislative process.

EXAMPLE Sequential Pairwise Voting

Assume we have four alternatives and that the agenda is A, B, C, D. Consider the following sequence of three preference lists:

Rank	Number of voters (3)		
First	A	C	B
Second	B	A	D
Third	D	B	C
Fourth	C	D	A

The first one-on-one pits A against B, and A wins by a score of 2 to 1 (meaning that two of the voters—the two on the left—prefer A to B, and one of the voters prefers B to A). Thus, B is eliminated and A moves on to confront C. Because C wins this one-on-one (by a score of 2 to 1), A is eliminated. Finally, C takes on D, and D wins by a score of 2 to 1. Thus, D is the winner. ■

SPOTLIGHT 12.1 The Historical Record

The following letter was written by Friedrich Pukelsheim of the University of Augsburg, Germany. He is imagining what Ramon Llull (1232–1316) might say if he were alive today.

Dear Editors:

It is my distinct pleasure to respond "from the beyond" to your kind invitation to set the historical record straight. I was born in 1232 on the Island of Mallorca in the Mediterranean Sea, which in your times now is known as a popular tourist place. In my days it was a strong political center of that part of the world, with a population that was a mix of Christians, Jews, and Muslims. It was my dream to persuade people of the virtues of Christian belief by relying, not on force, but on reason.

Unfortunately, people did not find it easy to follow my arguments, so I was more than pleased to discover some down-to-earth applications, including an election system. My idea was to oppose every pair of candidates, one-on-one, and ask the electors whom of the two they would prefer—very much like a medieval jousting tournament. But how to combine the results from all the duels into a winner of the election? I first proposed electing the candidate who won the most duels, then later suggested a system of successive eliminations.

All this was a long time ago, but it is most interesting for me to follow what came of it all. I wrote three papers on the topic, the second of which I "smuggled" into my novel *Blaquerna* in 1283. More than a century after my death, in 1428, the young German scholar Nicolaus Cusanus (1401–1464) journeyed to Paris to read my works in libraries there. He even copied out the third of my electoral writings, which I had completed on 1 July 1299 in Paris, and his manuscript is the only copy handed down to your days. Reading my papers, Cusanus was inspired to invent his own electoral system. Did he not understand mine, or just find it inadequate? Who knows?

While I had been concerned with electing church officials, Cusanus sought a system to elect the Holy Roman Emperor. In his system, each elector assigns each candidate a rank score, with the lowest candidate getting a score of 1, the second lowest a score of 2, and the best candidate the highest score possible, e.g., 10 when there are 10 candidates. The scores are totaled for each candidate and the candidate with the highest score wins. If you are a soccer player or a hockey player, you will have a good sense for one difference between our systems: Whereas I count victories, Cusanus adds up goals. Cusanus applauded himself for having invented an

There is something very troubling about the outcome of the preceding example, especially if you are candidate B. *Everyone* prefers B to D!

Sequential pairwise voting fails to satisfy what is called the **Pareto condition,** which says that if everyone prefers one alternative (in this case, B) to another alternative (D), then this latter alternative (D) should not be among the winners of the election.

absolutely ingenious and novel electoral system.

Also, I advocated open voting, whereas Cusanus favored a secret ballot. He was concerned that voters might sell their votes, or that the candidates might pressure the voters. Well, that certainly happened all of the time in elections for worldly authorities! But for election to clerical office, I thought it good enough if electors take an oath to vote for the most worthy candidate and submit themselves to the social control that comes with an open election.

Cusanus was famous in his times, as I was in mine; but fame indeed is transitory. Sure enough, my electoral system was reinvented by the Marquis de Condorcet (1743 – 1794), and Cusanus's system was proposed afresh by the Chevalier de Borda (1733 – 1799) – neither of whom, I am sure, wasted a thought on the possibility that "their" systems might already be on record. But, as my works had fallen into oblivion as had those of Cusanus, neither Condorcet nor Borda should be blamed for failing to acknowledge our priority.

My first electoral paper – actually the one which is longest and most detailed, written around 1280 or so – was rediscovered only in 2000, filed away in the Vatican Library. How would you feel if your work attracts fresh attention after more than 700 years? Actually, I am utterly pleased that mine has resurfaced at last! The text was excavated by a mathematician interested in voting systems, Friedrich Pukelsheim of the University of Augsburg, Germany. Since the text is handwritten in Latin, handling it became an interdisciplinary project that brought together experts on medieval manuscripts, Church Latin and theology, and even computer scientists. As a result, my electoral writings are now on the Internet (in the original and in translations into English and German) at www.uni-augsburg.de/llull/.

Looking back on my lack of success in preaching peace among Christians, Jews, and Muslims, and all the writing and copying by hand of my works, I hope you can appreciate how highly I value the printed book (such as this one) and, even more, instant communication worldwide over the Internet. May that ease of communication help facilitate the religious peace that I so dearly sought.

Yours truly,
Ramon Llull (1232 – 1316)
Left Choir Chapel
San Francisco Cathedral
Palma de Mallorca

The Hare System and Monotonicity

The social choice procedure known as the *Hare system*, which was introduced by Thomas Hare in 1861, is also known by names such as the "single transferable vote system." In 1862, John Stuart Mill described the Hare system as being "among the greatest improvements yet made in the theory and practice of government." Today, the system is used to elect public officials in Australia, Malta, the Republic of Ireland, and Northern Ireland.

The **Hare system** proceeds to arrive at a winner by repeatedly deleting alternatives that are "least preferred" in the sense of being at the top of the fewest preference lists. If a single alternative remains after all others have been eliminated, it alone is the winner. If two or more alternatives remain and all of these remaining alternatives would be eliminated in the next round (because they all have the same number of first-place votes), then these alternatives are declared to be tied for the win.

EXAMPLE The Hare System

Suppose we have the following sequence of preference lists, where the heading of "5" indicates that 5 of the 13 voters hold the ballot with A over B over C, the heading of "4" indicates that 4 of the 13 voters hold the ballot with C over B over A, etc.

	Number of voters (13)			
Rank	**5**	**4**	**3**	**1**
First	A	C	B	B
Second	B	B	C	A
Third	C	A	A	C

Alternatives B and C have only 4 first-place votes (while A has 5). Thus, B and C are eliminated in the first round, and so A wins the election. ■

In the preceding Example, suppose that the voter in the last column moves alternative A up on his list. Let's look at the new election. Notice that, even though A won the last election, the only change we are making in ballots for the new election is one that is favorable to A. The preference lists for the new election are as follows:

	Number of voters (13)			
Rank	**5**	**4**	**3**	**1**
First	A	C	B	A
Second	B	B	C	B
Third	C	A	A	C

If we apply the Hare system again, only B is eliminated in round one, as it has 3 first-place votes to 4 for C and 6 for A. Thus, after this round, the lists are as follows:

	Number of voters (13)			
Rank	**5**	**4**	**3**	**1**
First	*A*	*C*	*C*	*A*
Second	*C*	*A*	*A*	*C*

We now have *A* on top of 6 lists and *C* on top of 7 lists. Thus, at stage two, *A* (our previous winner!) is eliminated and *C* is the winner of this new election.

Clearly, this is once again quite counterintuitive. Alternative *A* won the original election, the only change in ballots made was one favorable to *A* (and no one else), and then *A* lost the next election.

This example shows that the Hare system does not satisfy **monotonicity,** which says that if an alternative is a winner, and a new election is held in which the only ballot change made is for some voter to move the former winning alternative higher on his or her preference list, then the original winner should remain a winner.

The fact that the Hare system does not satisfy monotonicity is considered by many—and with good reason—to be a glaring defect. A 17-voter example in which only a single alternative is eliminated in the first round can also be used to show that the Hare system does not satisfy monotonicity—see Exercise 20. For a more glaring version of this defect, one in which alternative *A* goes from winning to losing because voters move *A* from last place on their ballots to first place on their ballots, see Exercise 21. It also turns out that the Hare system, like the other voting systems we have looked at, can be manipulated (see Exercise 19).

In spite of these drawbacks, the Hare system is used in important ways today. For example, it is essentially the method that was used to choose Sydney, Australia, as the site of the 2000 Summer Olympics. Beijing would have been the plurality winner, but after the elimination of Istanbul, Berlin, and Manchester (in that order), Sydney defeated Beijing by a vote of 45 to 43. Four years later, Beijing did finally prevail; it will be the site of the 2008 Summer Olympics.

Summary

We have introduced four voting procedures (plurality, the Borda count, sequential pairwise voting, and the Hare system) and four desirable properties of voting systems (the Condorcet winner criterion, independence of irrelevant alternatives, the Pareto condition, and monotonicity). We showed that each of the four voting procedures failed to satisfy at least one of the four properties. What about the others? While space prevents a complete discussion, the following table summarizes the results.

	CWC	IIA	Pareto	Mono
Plurality	No	No	Yes	Yes
Borda count	No	No	Yes	Yes
Sequential pairs	Yes	No	No	Yes
Hare system	No	No	Yes	No

12.3 Insurmountable Difficulties: From Paradox to Impossibility

All four of the voting procedures that we have discussed turn out to be flawed in one way or another. You may well ask at this point why we don't simply present *one* voting method for the three-alternative case that has all the desirable properties we want to have satisfied. This is, after all, exactly what we did for the two-alternative case.

The answer to this question is extremely important. The difficulties in the three-alternative case are not in any way tied to a few particular voting methods that we happen to present in a text such as this (or that we choose to use in the real world). The fact is, there are difficulties that will be present *regardless* of what voting method is used, and this applies even to voting methods not yet discovered.

The Voting Paradox of Condorcet

The Marquis de Condorcet (1743–1794) may have been the first to realize that serious difficulties can *always* arise in elections in which there are three or more alternatives. The word that is typically used in connection with Condorcet's observation is *paradox*. In general, the word *paradox* is applied whenever there is a situation in which apparently logical reasoning leads to an outcome that seems impossible. Quite often, a situation that is first described as being paradoxical comes, in the fullness of time, to be seen as simply a fact that reveals the extent to which our previous intuition was flawed.

Condorcet considered the following set of three preference lists and found that they indeed lead to a situation that seems paradoxical:

Rank	Number of voters (3)		
First	A	B	C
Second	B	C	A
Third	C	A	B

If we view society as being broken down into thirds, with one-third holding each of Condorcet's preference lists, then society certainly seems to favor A to B (two-thirds to one-third) and B to C (again, two-thirds to one-third). Thus, we would expect society to prefer A to C. That is, we would expect the relation of social preference to be *transitive:* If A is "better than" B, and B is "better than" C, then surely A is "better than" C. But exactly the opposite is true. Society not only fails to prefer A to C but, in fact, rather strongly prefers C to A (i.e., by a two-thirds to one-third margin)! With, say, 10 alternatives, a similar phenomenon can occur with "two-thirds" replaced by "90%."

The fact that two-thirds of society can prefer A to B, two-thirds prefer B to C, and two-thirds prefer C to A is known as **Condorcet's voting paradox.**

Some authors apply the phrase *voting paradox* to any sequence of ballots in which there is no Condorcet winner. Either way, Condorcet's voting paradox is truly one of the cornerstones of modern social choice theory. We will make use of it in the last section of this chapter.

Manipulability

As we have seen, there are voting situations in which one fares better by voting for less preferred alternatives rather than more preferred alternatives. This, however, is just one form of insincerity that arises within voting theory.

For more on this topic, go to the discussion of manipulability at www.whfreeman.com/fapp.

Impossibility

Nothing in the remarkable body of work produced by Nobel laureate Kenneth J. Arrow of Stanford University is as well known or widely acclaimed as the result known as **Arrow's impossibility theorem** (see Spotlight 12.2). How, though, can one mathematically prove that it is *impossible* to find a voting system that satisfies certain properties?

Our goal here is to consider a version of Arrow's theorem and at least to sketch the argument behind it. This version is taken from the 1995 text *Mathematics and Politics,* cited in the Suggested Readings, and uses stronger hypotheses than Arrow's original theorem. (Essentially, the CWC is replacing Arrow's assumption of Pareto and nondictatorship.)

The framework for our considerations will be the same one with which we have been working. Ballots will be preference lists (without ties), and the outcome of an election will be either a single alternative (the winner) or a group of alternatives (tied for the win). We *do* demand of any social choice procedure that it definitely produce at least one winner when confronted by any sequence of preference lists.

Recall that the first two examples of social choice procedures that we considered were plurality voting and the Borda count. Moreover, we showed that

SPOTLIGHT 12.2 Kenneth J. Arrow

For centuries, mathematicians have been in search of a perfect voting system. Finally, in 1951, economist Kenneth Arrow proved that finding an absolutely fair and decisive voting system is impossible. Arrow is the Joan Kenney Professor of Economics, as well as a professor of operations research, at Stanford University. In 1972, he received the Nobel Memorial Prize in Economic Science for his outstanding work in the theory of general economic equilibrium. His numerous other honors include the 1986 von Neumann Theory Prize for his fundamental contributions to the decision sciences. He has served as president of the American Economic Association, the Institute of Management Sciences, and other organizations. Dr. Arrow talks about the process by which he developed his famous impossibility theorem and his ideas on the laws that govern voting systems:

Kenneth Arrow

My first interest was in the theory of corporations. In a firm with many owners, how do the owners agree when they have different opinions, for example, about the prospects of the company? I was thinking of stockholders. In the course of this, I realized that there was a paradox involved – that majority voting can lead to cycles. I then dropped that discussion because I was frustrated by it.

I happened to be working with The RAND Corporation one summer about a year or two later. They were very interested in applying concepts of rationality, particularly of game theory, to military and diplomatic affairs. That summer, I felt not like an economist but instead like a general social scientist or a mathematically oriented social scientist. There was tremendous interest in game theory, which was then new.

Some there asked me, "What does it mean in terms of national interest?" I said, "Oh, that's a very simple matter," and he said, "Well, why don't you write us a little memorandum on the subject." Trying to write that memorandum led to a sharper formulation of the social-choice question, and I realized that I had been thinking of it earlier in that other context.

I think that society must choose among a number of alternative policies. These policies may be thought of as quite comprehensive, covering a number of aspects: foreign policy, budgetary policy, or whatever. Now, each individual member of the society has a preference, or a set of preferences, over these alternatives. I guess that you can say one alternative is better than another. And these individual preferences have a property I call rationality or consistency, or more specifically, what is technically known as transitivity: if I prefer a to b, and b to c, then I prefer a to c.

Imagine that society has to make these choices among a set. Each individual has a preference ordering, a ranking of these alternatives. But we really want society, in some sense, to give a ranking of these alternatives. Well, you can always produce a ranking, but you would like it to have some properties. One is that, of course, it be responsive in some sense to the individual rankings. Another is that when you finish, you end up with a real ranking, that is, something that satisfies these consistency, or transitivity, properties. And a third condition is that when choosing between a number of alternatives, all I should take into account are the preferences of the individuals among those alternatives. If certain things are possible and some are impossible, I shouldn't ask individuals whether they care about the impossible alternatives, only the possible ones.

It turns out that if you impose the conditions I just stated, there is no method of putting together the individual preferences that satisfies all of them.

The whole idea of the axiomatic method was very much in the air among anybody who studied mathematics, particularly among those who studied the foundations of mathematics. The idea is that if you want to find out something, to find the properties, you say, "What would I like it to be?" [You do this] instead of trying to investigate special cases. And I was really accustomed to this approach. Of course, the actual process did involve trial and error.

But I went in with the idea that there was some method of handling this problem. I started out with some examples. I had already discovered that these led to some problems. The next thing that was reasonable was to write down a condition that I could outlaw. Then I constructed another example, another method that seemed to meet that problem, and something else didn't seem very right about it. Then I had to postulate that we have some other property. I found I was having difficulty satisfying all of these properties that I thought were desirable, and it occurred to me that they couldn't be satisfied.

After having formulated three or four conditions of this kind, I kept on experimenting. And lo and behold, no matter what I did, there was nothing that would satisfy these axioms. So after a few days of this, I began to get the idea that maybe there was another kind of theorem here, namely, that there was no voting method that would satisfy all the conditions that I regarded as rational and reasonable. It was at this point that I set out to prove it. And it actually turned out to be a matter of only a few days' work.

It should be made clear that my impossibility theorem is really a theorem [showing that] the contradictions are possible, not that they are necessary. What I claim is that given any voting procedure, there will be some possible set of preference orders for individuals that will lead to a contradiction of one of these axioms.

But you say, "Well, okay, since we can't get perfection, let's at least try to find a method that works well most of the time." Then when you do have a problem, you don't notice it as much. So my theorem is not a completely destructive or negative feature any more than the second law of thermodynamics means that people don't work on improving the efficiency of engines. We're told you'll never get 100% efficient engines. That's a fact—and a law. It doesn't mean you wouldn't like to go from 40% to 50%.

plurality voting failed to satisfy the Condorcet winner criterion (CWC) and that the Borda count failed to satisfy independence of irrelevant alternatives (IIA). The theorem we want to prove here is the following:

There does not exist, and never will exist, *any* social choice procedure that satisfies both the CWC and IIA.

More specifically, we claim that if a "voting rule" of some kind were to be found that satisfied both the CWC and IIA, then, when confronted by the three preference lists occurring in the voting paradox of Condorcet, this voting rule would *fail* to produce a winner (and thus not be a social choice procedure in the sense that we are using the phrase). Let's see why this is true.

The argument really comes in three separate, but extremely similar, pieces—one for each of the three alternatives. Piece 1 argues that alternative A can't be among the winners, piece 2 that B can't be among the winners, and piece 3 that C can't be among the winners. We'll do piece 1 and leave the others for the interested reader to investigate. The sequence of preference lists that we are considering is the following, which we have already seen has no Condorcet winner.

Rank	Number of voters (3)		
First	A	B	C
Second	B	C	A
Third	C	A	B

Our starting point, however, will be to ask what our hypothetical voting rule must do when confronted by a slightly different sequence of preference lists:

Rank	Number of voters (3)		
First	A	C	C
Second	B	B	A
Third	C	A	B

Here, alternative C is clearly a Condorcet winner, and thus it must be the unique winner of the election contested under our hypothetical voting rule. Therefore, C is a winner and A is a nonwinner (for *this* sequence of preference lists).

However, because our hypothetical voting rule satisfies independence of irrelevant alternatives, we know that alternative A will remain a nonwinner as long as no one reverses his or her ordering of A and C. But to arrive at the preference lists from the voting paradox, we can move *B* (the alternative that is irrelevant to *A* and *C*) up one slot in the second voter's list.

Thus, because of IIA, we know that alternative *A* is a nonwinner when our voting rule is confronted by the preference lists from the voting paradox of Condorcet. This is one-third of the argument. As we mentioned before, similar arguments (see Exercise 24) show that *B* and *C* are also nonwinners when our voting rule is confronted by the preference lists from the voting paradox of Condorcet. Therefore, we have shown that no voting system that is guaranteed to produce at least one winner can satisfy both CWC and IIA.

12.4 A Better Approach? Approval Voting

Elections in which there are only two candidates present no problem. Majority rule is, as we have seen, an eminently successful voting system in both theory and practice. If there are three or more candidates, however, the situation changes quite dramatically. While several voting systems suggest themselves (plurality, the Borda count, sequential pairwise voting, and the Hare system), each fails to satisfy one or more desired properties (the Condorcet winner criterion, independence of irrelevant alternatives, the Pareto condition, and monotonicity). Manipulability is an ever-present problem. Moreover, when all is said and done, Arrow's impossibility theorem says that any search for an ideal voting system of the kind we have discussed is doomed to failure.

Where does this leave us? More than intellectual issues are at stake here: Over 550,000 elected officials serve in approximately 80,000 governments in the United States. Whether it is a small academic department voting on the best senior thesis or a democratic country electing a new leader, multicandidate elections will be contested in one way or another. If there is no perfect voting system—and perhaps not even a best voting system (whatever that may mean; that is, best in what way?)—what can we do?

Perhaps the answer is that different situations lend themselves to different voting systems, and what is required is a judicious blend of common sense with an awareness of what the mathematical theory has to say. For example, while both the Hare system and the Borda count are subject to manipulability, it is clearly easier to manipulate the latter (recall the quote of Jean-Charles de Borda on page 417). Thus, people may tend to vote more sincerely, rather than strategically, if the Hare system is used instead of the Borda count. This may be a consideration when choosing a voting system for a faculty governance system, for example.

For national political elections, there are also practical considerations. The kind of ballot we are considering (an individual preference list) is certainly more complicated than the ballots we now employ, and preference lists cannot be used with existing voting machines. There is, however, a voting system that avoids the practical difficulties caused by the type of ballot being used that has much else to commend it. It is called *approval voting*.

Under **approval voting,** each voter is allowed to give one vote to as many of the candidates as he or she finds acceptable. No limit is set on the number of candidates for whom an individual can vote. Voters show disapproval of other candidates simply by not voting for them.

The winner under approval voting is the candidate who receives the largest number of approval votes. This approach is also appropriate in situations where more than one candidate can win, for example, in electing new members to an exclusive society such as the National Academy of Sciences or the Baseball Hall of Fame.

Approval voting was proposed independently by several analysts in the 1970s. Probably the best-known official elected by approval voting today is the secretary-general of the United Nations. In the 1980s, several academic and professional societies initiated the use of approval voting. Examples include the Institute of Electrical and Electronics Engineers (IEEE), with about 400,000 members, and the National Academy of Sciences. In Eastern Europe and some former Soviet republics, approval voting has been used in the form wherein one disapproves of (instead of approving of) as many candidates as one wishes.

Is approval voting the perfect voting system? Certainly not. For example, the type of ballot used limits the extent to which voter preferences can be expressed. However, it is certainly a voting system with much potential, and the reader wishing to explore it in more detail can start with Brams and Fishburn's 1983 monograph, listed in the Suggested Readings.

REVIEW VOCABULARY

Agenda An ordering of the alternatives for consideration. Often used in sequential pairwise voting.

Approval voting A method of electing one or more candidates from a field of several in which each voter submits a ballot that indicates which candidates he or she approves of. Winning is determined by the total number of approvals a candidate obtains.

Arrow's impossibility theorem Kenneth J. Arrow's discovery that any voting system can give undesirable outcomes.

Borda count A voting system for elections with several candidates in which points are assigned to voters' preferences and these points are summed for each candidate to determine a winner.

Condorcet winner A Condorcet winner in an election is a candidate who, based on the ballots, would have defeated every other candidate in a one-on-one contest.

Condorcet winner criterion (CWC) A voting system satisfies the Condorcet winner criterion if, for every election in which there is a Condorcet winner, it wins the election when that voting system is used.

Condorcet's voting paradox The observation that two-thirds of society can prefer A to B, two-thirds prefer B to C, and two-thirds prefer C to A.

Hare system A voting system for elections with several candidates in which candidates are successively eliminated in an order based on the number of first-place votes.

Independence of irrelevant alternatives (IIA) A voting system satisfies independence of irrelevant alternatives if the only way a candidate (call him *A*) can go from losing one election to being among the winners of a new election (with the same set of candidates and voters) is for at least one voter to reverse his or her ranking of *A* and the previous winner.

Individual preference list A ballot that provides a rank ordering of candidates, from best to worst, in the eyes of that individual voter.

Majority rule A voting system for elections with two candidates (and an odd number of voters) in which the candidate preferred by more than half the voters is the winner.

May's theorem Kenneth May's discovery that, for two alternatives and an odd number of voters, majority rule is the only voting system satisfying three natural properties.

Monotonicity A voting system satisfies monotonicity provided that ballot changes favorable to one alternative (and not favorable to any other alternative) can never hurt that alternative.

Pareto condition A voting system satisfies the Pareto condition provided that every voter's ranking of one alternative higher than another precludes the possibility of this latter alternative winning.

Plurality voting A voting system for elections with several candidates in which the candidate with the most first-place votes wins.

Sequential pairwise voting A voting system for elections with several candidates in which one starts with an agenda and pits the candidates against each other in one-on-one contests (based on ballots that are preference lists), with losers being eliminated as one moves along the agenda.

Sincere voting Submitting a ballot that represents a voter's true preferences.

Strategic voting Submitting a ballot that does not represent a voter's true preferences.

SUGGESTED READINGS

AUMANN, ROBERT, and SERGIU HART, eds. *Handbook of Game Theory with Economic Applications,* Vol. II, Elsevier, Amsterdam, 1994. Chapter 30 (by Steven Brams) and Chapter 31 (by Hervé Moulin) provide intermediate to advanced treatments of voting procedures and social choice.

BLACK, DUNCAN. *The Theory of Committees and Elections,* Kluwer, Dordrecht, The Netherlands, 1986. The historical highlights and developments of voting methods in the nineteenth and twentieth centuries are traced in this economist's volume.

BRAMS, STEVEN J., and PETER C. FISHBURN. *Approval Voting,* Birkhäuser, Boston, 1983. This volume is a research-level work on developments in the recently popular (but rediscovered) method now called approval voting. The first chapter, however, is an elementary exposition of this voting method and its uses.

FARQUHARSON, ROBIN. *Theory of Voting,* Yale University Press, New Haven, Conn., 1969. This is an elementary but historically important monograph.

KELLY, JERRY S. *Social Choice Theory,* Springer-Verlag, New York, 1988. This text for undergraduates provides an extensive treatment of voting theory.

MALKEVITCH, JOSEPH, and WALTER MEYER. *Graphs, Models and Finite Mathematics,* Prentice Hall, Englewood Cliffs, N.J., 1974. In Chapter 10 there is an introduction to the problem of voting, including a discussion of the properties desired of any voting method (Arrow's axioms).

MERRILL, SAMUEL, III. *Making Multicandidate Elections More Democratic,* Princeton University Press, Princeton, N.J., 1988. This is a well-written treatment of voting from quite a practical point of view.

NURMI, HANNU. *Comparing Voting Systems,* Reidel, Dordrecht, The Netherlands, 1987. This monograph provides an excellent treatment, at a somewhat more technical level, of the topics dealt with in this chapter.

SAARI, DONALD G. *Chaotic Elections! A Mathematician Looks at Voting,* American Mathematical Society, Providence, R.I., 2001. This expository book begins with the 2000 presidential election and discusses a number of paradoxical results in voting.

SAARI, DONALD G. *The Geometry of Voting,* Springer-Verlag, New York, 1994. This monograph provides an advanced treatment of voting that focuses on ranking methods such as the Borda count.

TAYLOR, ALAN D. *Social Choice and the Mathematics of Manipulation,* Cambridge University Press, Cambridge, 2003. This monograph deals exclusively with the issue of manipulating voting systems and the remarkable work that has been done in this area since the appearance of the Gibbard-Satterthwaite theorem in the early 1970s.

TAYLOR, ALAN D. *Mathematics and Politics: Strategy, Voting, Power, and Proof,* Springer-Verlag, New York, 1995. Chapters 5 and 10 give an expanded treatment of the topics considered here, with proofs included. It is also intended for nonmajors.

SUGGESTED WEB SITES

fsmat.htu.tuwien.ac.at/~zahi/wahl.html "The Voting Page," this site contains additional topics from the theory of voting, including yes–no voting and power indices (as discussed in Chapter 13).

bcn.boulder.co.us/government/approvalvote/goodsoc.html An article entitled "Approval Voting and the Good Society," by Steven J. Brams.

SKILLS CHECK

1. Who proved the theorem that with two alternatives and an odd number of voters, majority rule is the only voting system that never results in a tie, treats all voters the same, treats both alternatives the same, and is "monotone"?

(a) Arrow
(b) May
(c) Borda
(d) Condorcet

For Exercises 2–5, consider this situation. Thirty students who need to choose a day for their final exam have the following preference lists:

	12 Students	8 Students	10 Students
First choice	Friday	Thursday	Wednesday
Second choice	Wednesday	Wednesday	Thursday
Third choice	Thursday	Friday	Friday

2. Which day will be selected if they use plurality voting?

(a) Wednesday
(b) Thursday
(c) Friday
(d) There is a tie.

3. Which day will be selected if they use the Borda count?

(a) Wednesday
(b) Thursday
(c) Friday
(d) There is a tie.

4. Which day will be selected if they use the Hare system?

(a) Wednesday
(b) Thursday
(c) Friday
(d) There is a tie.

5. Which day will be selected if they use sequential pairwise voting with agenda Wednesday–Thursday–Friday?

(a) Wednesday
(b) Thursday
(c) Friday
(d) There is a tie.

6. The Hare system fails to satisfy which desirable property?

(a) The Pareto condition
(b) Monotonicity

7. Sequential pairwise voting fails to satisfy which desirable property?

(a) The Pareto condition
(b) Monotonicity

8. Thirty students mark those days that would be acceptable for an exam.

	Number of students (30)					
Students	**8**	**6**	**4**	**4**	**4**	**4**
Wednesday		X	X	X		X
Thursday	X			X		X
Friday			X	X	X	

Which day will be selected using approval voting?

(a) Wednesday
(b) Thursday
(c) Friday
(d) No winner can be chosen.

9. Each of four candidates is evaluated as acceptable or unacceptable. The candidate that the most voters evaluate as acceptable wins. This voting method is an example of

(a) sequential pairwise voting.
(b) approval voting.
(c) majority rule.

10. Arrow's impossibility theorem states that

(a) different voting methods will always produce different outcomes.
(b) a flawless voting method has not yet been created.
(c) any voting system can give undesirable outcomes.

11. Using the Borda count requires that each voter

(a) rank all possible options.
(b) approve of at least one of the possible options.
(c) assign a numerical value to each option.

12. A voting system satisfies monotonicity when

(a) one option receives a majority vote.
(b) the winner does not change when votes are changed to favor only the winner.
(c) no ties are allowed.

13. A voter submits a "sincere" vote when

(a) the voter votes according to the wishes of the constituency.
(b) the voter does not submit a "strategic vote."
(c) the approval voting method is used.

14. The majority rule method of voting is a satisfactory method when there is an odd number of voters and

(a) there are only two alternatives.
(b) there is a plurality winner.
(c) every voter submits a first and second choice.

15. A shortcoming of approval voting is

(a) that the winner may not receive a majority vote.
(b) that a tabulation of the votes is very tedious.
(c) lack of recognition by government and professional bodies.

16. Changing the agenda can impact the voting result when

(a) the sequential pairwise voting method is used.
(b) the Condorcet voting method is used.
(c) an individual preference list is used.

EXERCISES

▲ Optional. ■ Advanced. ◆ Discussion.

Elections with Only Two Alternatives

1. In a few sentences, explain why minority rule (the voting procedure for two alternatives that is described on page 409) satisfies conditions (1) and (2) on page 409, but not (3).

2. In a few sentences, explain why imposed rule (the voting procedure for two alternatives that is described on page 409) satisfies conditions (1) and (3) on page 409, but not (2).

3. In a few sentences, explain why a dictatorship (the voting procedure for two alternatives that is described on page 409) satisfies conditions (2) and (3) on page 409, but not (1).

■ 4. Find (or invent) a voting rule for two alternatives that satisfies

(a) condition (1) on page 409, but neither (2) nor (3).
(b) condition (2) on page 409, but neither (1) nor (3).
(c) condition (3) on page 409, but neither (1) nor (2).

Elections with Three or More Alternatives: Procedures and Problems

5. Plurality voting is illustrated by the 1980 U.S. Senate race in New York among Alfonse D'Amato (*D*, a conservative), Elizabeth Holtzman (*H*, a liberal), and Jacob Javits (*J*, also a liberal). Reasonable estimates (based largely on exit polls) suggest that voters ranked the candidates according to the following table:

22%	23%	15%	29%	7%	4%
D	*D*	*H*	*H*	*J*	*J*
H	*J*	*D*	*J*	*H*	*D*
J	*H*	*J*	*D*	*D*	*H*

(a) Is there a Condorcet winner?
(b) Who won using plurality voting?

6. (Everyone wins.) Consider the following set of preference lists:

	Number of voters (9)						
Rank	**3**	**1**	**1**	**1**	**1**	**1**	**1**
First	*A*	*A*	*B*	*B*	*C*	*C*	*D*
Second	*D*	*B*	*C*	*C*	*B*	*D*	*C*
Third	*B*	*C*	*D*	*A*	*D*	*B*	*B*
Fourth	*C*	*D*	*A*	*D*	*A*	*A*	*A*

Note that the first list is held by three voters, not just one. Calculate the winner using

(a) plurality voting.
(b) the Borda count.
(c) the Hare system.
(d) sequential pairwise voting with the agenda *A, B, C, D*.

7. Consider the following set of preference lists:

	Number of voters (7)				
Rank	**2**	**2**	**1**	**1**	**1**
First	*C*	*D*	*C*	*B*	*A*
Second	*A*	*A*	*D*	*D*	*D*
Third	*B*	*C*	*A*	*A*	*B*
Fourth	*D*	*B*	*B*	*C*	*C*

Calculate the winner using

(a) plurality voting.
(b) the Borda count.
(c) the Hare system.
(d) sequential pairwise voting with the agenda *B, D, C, A*.

8. Consider the following set of preference lists:

	Number of voters (8)					
Rank	**2**	**2**	**1**	**1**	**1**	**1**
First	*A*	*E*	*A*	*B*	*C*	*D*
Second	*B*	*B*	*D*	*E*	*E*	*E*
Third	*C*	*D*	*C*	*C*	*D*	*A*
Fourth	*D*	*C*	*B*	*D*	*A*	*B*
Fifth	*E*	*A*	*E*	*A*	*B*	*C*

Calculate the winner using

(a) plurality voting.
(b) the Borda count.
(c) the Hare system.
(d) sequential pairwise voting with the agenda *B*, *D*, *C*, *A*, *E*.

9. Consider the following set of preference lists:

	Number of voters (5)				
Rank	**1**	**1**	**1**	**1**	**1**
First	*A*	*B*	*C*	*D*	*E*
Second	*B*	*C*	*B*	*C*	*D*
Third	*E*	*A*	*E*	*A*	*C*
Fourth	*D*	*D*	*D*	*E*	*A*
Fifth	*C*	*E*	*A*	*B*	*B*

Calculate the winner using

(a) plurality voting.
(b) the Borda count.
(c) the Hare system.
(d) sequential pairwise voting with the agenda *A*, *B*, *C*, *D*, *E*.

10. Consider the following set of preference lists:

	Number of voters (7)				
Rank	**2**	**2**	**1**	**1**	**1**
First	*A*	*B*	*A*	*C*	*D*
Second	*D*	*D*	*B*	*B*	*B*
Third	*C*	*A*	*D*	*D*	*A*
Fourth	*B*	*C*	*C*	*A*	*C*

Calculate the winner using

(a) plurality voting.
(b) the Borda count.
(c) the Hare system.
(d) sequential pairwise voting with the agenda *B*, *D*, *C*, *A*.

11. Consider the following set of preference lists:

	Number of voters (7)				
Rank	**2**	**2**	**1**	**1**	**1**
First	*C*	*E*	*C*	*D*	*A*
Second	*E*	*B*	*A*	*E*	*E*
Third	*D*	*D*	*D*	*A*	*C*
Fourth	*A*	*C*	*E*	*C*	*D*
Fifth	*B*	*A*	*B*	*B*	*B*

Calculate the winner using

(a) plurality voting.
(b) the Borda count.
(c) the Hare system.
(d) sequential pairwise voting with the agenda *A*, *B*, *C*, *D*, *E*.

12. In a few sentences, explain why plurality voting satisfies

(a) the Pareto condition.
(b) monotonicity.

13. In a few sentences, explain why the Borda count satisfies

(a) the Pareto condition.
(b) monotonicity.

14. In a few sentences, explain why sequential pairwise voting satisfies

(a) the Condorcet winner criterion.
(b) monotonicity.

15. In a few sentences, explain why the Hare system satisfies the Pareto condition.

16. Consider the following two elections among candidates A, B, and C:

	Number of voters (4)			
Rank	**1**	**1**	**1**	**1**
First	*A*	*A*	*B*	*C*
Second	*B*	*B*	*C*	*B*
Third	*C*	*C*	*A*	*A*

	Number of voters (4)			
Rank	**1**	**1**	**1**	**1**
First	*A*	*A*	*B*	*B*
Second	*B*	*B*	*C*	*C*
Third	*C*	*C*	*A*	*A*

(a) Use these two elections to show that plurality voting does not satisfy independence of irrelevant alternatives.
(b) Use these two elections to show that the Hare system does not satisfy independence of irrelevant alternatives.

17. Construct ballots for the alternatives A, B, and C to show that the Borda count does not satisfy the Condorcet winner criterion.

■ **18.** Suppose we have 3 voters and 4 alternatives and suppose the sequence of preference lists is as follows:

	Number of voters (3)		
Rank	**1**	**1**	**1**
First	*A*	*C*	*B*
Second	*B*	*A*	*D*
Third	*D*	*B*	*C*
Fourth	*C*	*D*	*A*

Show that if the voting system being used is sequential pairwise voting with a fixed agenda, and if you have agenda-setting power (i.e., you get to choose the order), then you can arrange for whichever alternative you want to win the election. (Your answer will consist of the 4 agendas you find—perhaps by trial and error—and the calculation of which alternative is the winner for each in sequential pairwise voting.)

■ **19.** Suppose the following preference lists represent the true preferences of the 17 voters involved:

	Number of voters (17)			
Rank	**7**	**5**	**4**	**1**
First	*A*	*C*	*B*	*A*
Second	*B*	*A*	*C*	*B*
Third	*C*	*B*	*A*	*C*

(a) Find the winner of the election if the Hare system is used and everyone votes sincerely.
(b) Find the winner if the voter on the far right votes strategically.

20. Show that the nonmonotonicity of the Hare system can also be demonstrated by the following 17-voter, 4-alternative election. (In a number of recent books, this example is used to show the nonmonotonicity of the Hare system. The easier 13-voter, 3-alternative example given in the text was pointed out to us by Matt Gendron, an undergraduate at Union College.)

	Number of voters (17)			
Rank	**7**	**5**	**4**	**1**
First	*A*	*C*	*B*	*D*
Second	*D*	*A*	*C*	*B*
Third	*B*	*B*	*D*	*A*
Fourth	*C*	*D*	*A*	*C*

21. The following example illustrates how badly the Hare system can fail to satisfy monotonicity. Consider the following sequence of preference lists:

	Number of voters (21)			
Rank	**7**	**6**	**5**	**3**
First	*A*	*B*	*C*	*D*
Second	*B*	*A*	*B*	*C*
Third	*C*	*C*	*A*	*B*
Fourth	*D*	*D*	*D*	*A*

(a) Show that A is the unique winner if the Hare system is used.
(b) Find the winner using the Hare system in the new election wherein the 3 voters on the right all move A from last place on their preference lists to first place on their preference lists.

Insurmountable Difficulties: From Paradox to Impossibility

22. Suppose we have 10 candidates ($A, B, C, D, E, F, G, H, I$, and J). Exhibit 10 ballots so that if each ballot were to be held by 10% of the electorate, then 90% of the voters would prefer A to B, 90% would prefer B to C, and so on down to 90% preferring I to J and then 90% preferring J to A. (*Hint:* Mimic the pattern for the three-candidate case that gave us Condorcet's voting paradox.)

23. Construct a real-world example (perhaps involving yourself and two friends) where the individual preference lists for three alternatives are as in the voting paradox of Condorcet.

■ **24.** Complete the proof of the version of Arrow's theorem from the text by showing that neither B nor C can be a winner in the situation described. (Your argument will be almost word-for-word the same as the proofs in the text.)

Additional Exercises

25. How many different ways can a voter
(a) rank 3 choices (when ties are not allowed)?
(b) rank 4 alternatives (without ties)?
(c) rank n potential outcomes (without ties)?

26. How many different ways can a voter
(a) rank 3 choices when ties are not allowed but *incomplete* rankings can be submitted (e.g., a first choice without giving a second or third choice)?
(b) rank 3 choices when ties are allowed and *complete* rankings are required?

27. Ten board members vote by approval voting on eight candidates for new positions on their board as indicated in the following table. An X indicates an approval vote. For example, voter 1, in the first column, approves of candidates A, D, E, F, and G, and disapproves of B, C, and H.

	Voters									
Candidates	**1**	**2**	**3**	**4**	**5**	**6**	**7**	**8**	**9**	**10**
A	X	X	X			X	X	X		X
B		X	X	X	X	X	X	X	X	
C			X					X		
D	X	X	X	X	X		X	X	X	X
E	X		X		X		X		X	
F	X		X	X	X	X	X	X		X
G	X	X	X	X	X			X		
H		X		X		X		X		X

(a) Which candidate is chosen for the board if just one of them is to be elected?
(b) Which candidates are chosen if the top four are selected?
(c) Which candidates are elected if 80% approval is necessary and at most four are elected?
(d) Which candidates are elected if 60% approval is necessary and at most four are elected?

28. The top five vote-getters, in terms of percentages, elected to the Baseball Hall of Fame were, in alphabetical order:

Hank Aaron	(1982–406 votes from 415 ballots)
George Brett	(1999–488 votes from 497 ballots)
Ty Cobb	(1936–222 votes from 226 ballots)
Nolan Ryan	(1992–491 votes from 497 ballots)
Tom Seaver	(1992–425 votes from 430 ballots)

(a) Rank the players from 1 to 5 in terms of voting percentage.
(b) How many more votes would the fifth place finisher have needed to take over first place?

29. A player remains on the ballot for the Baseball Hall of Fame for 15 years provided he receives 5% of the votes cast each year. Some other players (and their votes) in a recent election were

Mickey Lolich (43)
Thurman Munson (40)
Rusty Staub (32)
Bill Maddock (19)
Ron Cey (8)

Which of these five players meets the 5% cutoff criterion (for the 423 votes cast) for remaining on the ballot for the 1994 election?

30. Consider the following set of preference lists:

	Number of voters (7)						
Rank	1	1	1	1	1	1	1
First	*C*	*D*	*C*	*B*	*E*	*D*	*C*
Second	*A*	*A*	*E*	*D*	*D*	*E*	*A*
Third	*E*	*E*	*D*	*A*	*A*	*A*	*E*
Fourth	*B*	*C*	*A*	*E*	*C*	*B*	*B*
Fifth	*D*	*B*	*B*	*C*	*B*	*C*	*D*

Calculate the winner using

(a) plurality voting.
(b) the Borda count.
(c) sequential pairwise voting with the agenda *A*, *B*, *C*, *D*, *E*.
(d) the Hare system.

31. An interesting variant of the Hare system was proposed by the psychologist Clyde Coombs. It operates exactly as does the Hare system, but instead of deleting alternatives with the fewest first-place votes, it deletes those with the most last-place votes.

(a) Use the Coombs procedure to find the winner if the ballots are as in Exercise 30.
(b) Show that for two voters and three alternatives, it is possible to have ballots that result in one alternative winning if the Coombs procedure is used and a tie between the other two if the Hare system is used.

32. The 45 members of a school's football team vote on three nominees, *A*, *B*, and *C*, by approval voting for the award of "most improved player" as indicated in the following table. An X indicates an approval vote.

	Number of voters (45)							
Nominee	7	8	9	9	6	3	1	2
A	X			X	X		X	
B		X		X		X	X	
C			X		X	X	X	

(a) Which nominee is selected for the award?
(b) Which nominee gets announced as runner-up for the award?
(c) Note that two of the players "abstained," that is, approved of none of the nominees. Note also that one person approved of all three of the nominees. What would be the difference in the outcome if one were to "abstain" or "approve of everyone"?

WRITING PROJECTS

1. In the 2000 presidential election in Florida, the final results were as follows:

Candidates	Number of votes	Percentage of votes
Bush	2,911,872	49
Gore	2,910,942	49
Nader	97,419	2
Buchanan	17,472	0

Making reasonable assumptions about voters' preference schedules, discuss how the election might have turned out under the different voting methods discussed in this chapter.

2. Frequently in presidential campaigns, the winner of the first few primaries is given front-runner status that can lead to the nomination of his or her party. Moreover, there are often several candidates running in early primaries such as New Hampshire. Consider a recent election and discuss how the nominating process might have proceeded through the campaign if approval voting had been used to decide primary winners.

SPREADSHEET PROJECTS

To do these projects, go to www.whfreeman.com/fapp.

Spreadsheets are used in this project to analyze Borda counts and approval voting. Using the automatic recalculation feature of spreadsheets, these activities allow the investigation of insincere voting strategies and their impact on the voting results.

APPLET EXERCISES

To do these exercises, go to www.whfreeman.com/fapp.

Consider the following preference schedule, in which the winner will be decided by a Borda count.

	Number of voters (26)				
Rank	**8**	**7**	**4**	**4**	**3**
First place	*A*	*B*	*C*	*D*	*B*
Second place	*C*	*C*	*D*	*B*	*A*
Third place	*B*	*D*	*A*	*A*	*C*
Fourth place	*D*	*A*	*B*	*C*	*D*

The preference schedule exhibits the true feelings of the voters about the candidates. However, there are occasions when a group of voters can better meet their objectives through *insincere voting*. You can analyze this phenomenon in the applet Insincere Voting.

Weighted Voting Systems

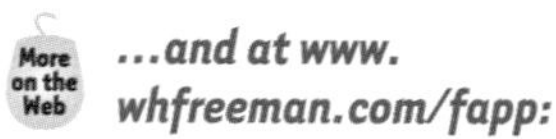

Flashcards
Quizzes
Spreadsheet Projects
Weighted voting systems
Applet Exercises
The Banzhaf power index

In the United States presidential election of November 2000, the winner, Governor George W. Bush, received fewer popular votes than the loser, Vice President Al Gore. Governor Bush won the election because he received more *electoral votes* than the vice president did. The United States Constitution specifies that the president is chosen by the electoral college, which is composed of electors from every state. Individual voters do not vote directly for a presidential candidate, but for electors who are pledged to vote for the candidate selected by the voter. Although the Constitution does not require it, in most states the election is a statewide contest, so that all of the selected electors favor the candidate who received the most popular votes in the state. This practice makes the electoral college into a voting system in which the participants are the 50 states and the District of Columbia. The participants are not equals: Each state gets as many votes as it has representatives and senators in Congress. The number of representatives for each state is determined by apportionment, a process carried out once per decade (see Chapter 15). The most populous state, California, had 54 votes, because from 1992 until 2002 it had 52 representatives in the House and two senators; the District of Columbia and the seven least populous states each had three votes. The candidate who receives a simple majority of at least 270 of the 538 electoral votes would be declared the winner. The outcome of the election was in doubt because the margin in Florida was narrow. Its 25 electoral votes were eventually awarded to Governor Bush, bringing his total to 271 electoral votes—one more than the minimum number necessary to win.

The electoral college is an example of a **weighted voting system** (see Spotlight 13.1). In a weighted voting system, the participants have varying numbers of votes. All corporations are governed by weighted voting systems, where holders of common stock are entitled to one vote per share owned. Shareholders who own relatively large numbers of shares usually have greater influence in such an election than do small shareholders. Some legislative bodies have such strong party discipline that each legislator always votes as dictated by his or her party. These legislatures are weighted voting systems in which the participants are

SPOTLIGHT 13.1 The Electoral College

In a U.S. presidential election, the voters in each state don't actually vote for the candidates; they vote for electors to represent their state in the electoral college. The number of electors allotted to a state is equal to the size of its congressional delegation; thus, a state with one congressional district gets 3 electors, because it has 1 representative in the U.S. House of Representatives and 2 senators. A state with 25 representatives and 2 senators would be entitled to 27 electors.

States typically require their electors to vote as blocs. For example, in the 2004 election the candidate who gets a plurality in California's election will get all 55 of California's electoral votes, while the candidate who carries Delaware will get all 3 of Delaware's electoral votes. In effect, the electoral college is a weighted voting system with 51 participants (the states and the District of Columbia). The weights range from 3 to 55, and the quota is a simple majority of the 538 electors, or 270.

the political party organizations, each of which is entitled to a number of votes equal to the size of its delegation in the legislature. The Council of Ministers of the European Community compensates for the differing populations of its member states by using a weighted voting system.

The *power* of a participant in a weighted voting system is his or her ability to influence a decision. We shall study two ways of measuring power, the *Banzhaf power index* and the *Shapley–Shubik power index*. We shall see that either of these indices provides a more accurate measure of a participant's power than the number of votes that he or she is entitled to cast.

(Joe Sohm/Chromosohm/Stock Connection/PictureQuest.)

13.1 How Weighted Voting Works

EXAMPLE A Professional Corporation

SMILE, Ltd. is a group dental practice with five dentists, organized as a professional corporation. Dr. Ruth Smith founded the practice and holds 9 shares of stock in the corporation. Her husband, Ralph, also owns 9 shares. The other dentists in the practice also own shares in the corporation; the details are listed in Table 13.1.

When a matter is put to a vote, each of the six shareholders has one vote for every share owned. There are 30 shares in all, and a simple majority (16 votes) is necessary to pass a motion.

A **coalition** is a set of voters that has formed to support or oppose a measure that is up for a vote. A coalition may consist of all the voters or any subset of the voters. It may consist of just one voter, or it may even be *empty*. For example, if the voting body is unanimously in favor of a motion, then the coalition opposing the motion is empty. If a coalition has enough votes to pass the measure, it is called a **winning coalition;** otherwise, it is a **losing coalition.** In SMILE, Ltd., the Smiths have enough votes to be a winning coalition. A winning coalition can also consist of one of the Smiths and Dr. Mansfield. There are also larger winning coalitions, because any of the shareholders may join one of these coalitions to form a coalition with even more votes.

The winning coalitions that include any or all of Drs. Ide, Lambert, or Edwards would still be winning coalitions without their support. Between them, these dentists have only 5 votes, so even if they combined all their votes with one of the Smiths' 9 votes, the total would be only 14–not enough to form a winning coalition. A coalition consisting of these three dentists and Dr. Mansfield would have only 12 votes and would also be a losing coalition. To form a winning coalition, at least two of the three more powerful shareholders, and no one else, is needed.

TABLE 13.1 Weighted Voting, SMILE, Ltd.

Shareholder	Number of Shares
Ruth Smith, DDS	9
Ralph Smith, CPA	9
Albert Mansfield, DDS	7
Katherine Ide, DDS	3
Gary Lambert, DDS	1
Marjorie Edwards, DDS	1
Total	30

In this situation, Drs. Ide, Lambert, and Edwards have no voting power. There is a technical term for a voter who, although he or she may join a winning coalition, is never essential to form one: Such a voter is called a **dummy.** ■

Notation for Weighted Voting

To describe a weighted voting system, it is necessary to specify the number of votes that each voter has. This number is called the voter's **weight.** The minimum number of votes required to pass a measure is called the **quota** for the voting system, and it too must be specified in the description of a weighted voting system. The shorthand notation

$$[q\colon w_1, w_2, \ldots, w_n]$$

is used to describe a weighted voting system with a quota q and n voters whose weights are $w_1, w_2, \ldots, w_n$. In this notation, the voting system used by SMILE, Ltd. was [16: 9, 9, 7, 3, 1, 1].

To avoid a situation in which two winning coalitions might oppose each other, we will require the quota to be more than half the sum of the weights of all the voters. Thus, each winning coalition must have more than half the votes, so the opposing voters will have less than half the votes and cannot form a winning coalition. To guarantee that a motion will pass when the voters favor it unanimously, we will also require that the quota not be more than the sum of the voting weights. Thus, for a system with n voters,

$$q > \frac{1}{2}(w_1 + w_2 + \cdots + w_n), \text{ and}$$

$$q \leq w_1 + w_2 + \cdots + w_n$$

For example, the quota for SMILE, Ltd. is required to be more than 15, because there are 30 votes in all. Thus, it could be 16, 17, 18, or any larger number up to 30. Changing the quota can affect the way power is distributed. For example, with a quota of 17, a coalition consisting of one of the Smiths and Dr. Mansfield, with only 16 votes, is a losing coalition. If these two convinced Dr. Ide, Dr. Lambert, or Dr. Edwards to join them, they would form a winning coalition. Thus, if the quota were 17, there would be no dummies.

A **blocking coalition** is a subset of voters opposing a motion with enough votes to defeat it. In a voting system with total weight w and quota q, any coalition with weight more than $w - q$ is a blocking coalition. For example, in SMILE, Ltd., $w = 30$, $q = 16$, and $w - q = 14$; so any coalition with 15 or more votes is a blocking coalition.

Winning coalitions always have enough votes to block a measure (remember that they have more than half the votes), but there can be blocking coalitions whose votes total less than the quota for winning. As a simple example, consider a voting system with four voters, each with one vote, and a quota of three votes to pass a measure. Any coalition of two voters opposing a measure is a blocking coalition, although these voters could not pass any measure they favored without being joined by a third voter.

In a criminal trial, the jury's decision to convict or to acquit must be unanimous, so a winning coalition requires all the jurors. If the jury cannot agree on a verdict, a mistrial is declared and the prosecution has the right to demand a new trial. There is only one winning coalition, but every coalition with at least one member is a blocking coalition.

If there is a voter whose voting weight meets or exceeds the quota for passing a measure, then that voter is called a **dictator.** Each of the other voters in the system is a dummy. For example, if one stockholder owns 51% of the shares in a corporation, then he or she is a dictator and controls the business of the corporation.

A voter who has enough votes to block any measure is said to have **veto power.** Of course, a dictator automatically has veto power, but it is possible to have veto power without being a dictator. For example, we have just seen that every juror in a criminal trial has veto power.

Dictator: A voter who can pass a motion by voting "yes," even if all other voters vote "no."

Dummy: A voter whose vote never makes a difference.

Quota: The number of votes needed to pass a measure.

Veto power: The power to defeat a motion by voting against it, even when all other voters support it.

Winning coalition: A set of voters whose total voting weight is at least the quota.

EXAMPLES Weighted Voting Systems

1. Consider a small corporation owned by two people, A and B, who possess 60 and 40 shares of stock, respectively. If measures are allowed to pass by a simple majority of at least 51, we express this voting system as

$$[51: 60, 40]$$

In this example, shareholder A is a dictator. No action can be taken without her approval.

2. Let us examine a second company, with three shareholders, A, B, and C, who hold 49, 48, and 3 shares, respectively. This voting system is

$$[51: 49, 48, 3]$$

There is no dictator; indeed, this company is more "democratic" than one might expect. Any coalition of two or more shareholders has a simple majority. The three shareholders are equally powerful because each has the same ability to influence the outcome.

3. A third company has shareholders A, B, C, and D. A, B, and C each own 26 shares, while D holds 22 shares. The voting system for this corporation is

$$[51: 26, 26, 26, 22]$$

A measure will pass if it gains the support of two of the shareholders A, B, and C. Although D owns nearly as many shares as the others, D is a dummy. The power in this company is equally divided among A, B, and C.

If a voter A has more votes than another voter B, then A cannot have less power than B. However, example 2 shows that two voters may be equally powerful, even if one has many more votes than the other. On the other hand, examples 1 and 3 show that a dummy voter may have almost as many votes as another voter who has considerable power. ■

13.2 The Banzhaf Power Index

A voter's power is his or her ability to influence the outcome of a vote. In any winning coalition, the **critical voters** are the voters whose votes are essential to win. If a voter is critical in a certain winning coalition, then the coalition will lose if that voter deserts it. A voter's power can be measured by counting the number of winning and blocking coalitions in which his or her vote is a critical one.

For example, in the 2000 presidential election, each state in the coalition that favored Governor Bush was a critical voter. That coalition had 271 electoral votes. Because every state had 3 or more electoral votes, any defection from this winning coalition would have brought its vote total below the 270 needed to win.

A participant's **Banzhaf power index** is the number of distinct winning or blocking coalitions in which he or she is a critical voter.

The Banzhaf power index was developed in 1965 by attorney John F. Banzhaf III in an analysis of weighted voting, provocatively entitled "Weighted Voting Doesn't Work."

EXAMPLE

Critical Voters

Consider a committee of three members, A, B, and C. The chairperson of the committee, A, has two votes, while B and C each have one. The quota is three, and thus our shorthand notation for this voting system is

$$[3: 2, 1, 1]$$

The coalition $\{A, B, C\}$ is a winning coalition, because it has all four votes. Suppose that A decides to leave the coalition. We can indicate this situation schematically as follows:

A	*B*	*C*	Votes	Outcome
Yes	Yes	Yes	4	Pass
↓				
No	Yes	Yes	2	Fail

By changing her vote, A has changed the outcome. In this coalition, A is a critical voter.

Now let's go back to the original coalition and see what happens if B changes his vote.

A	*B*	*C*	Votes	Outcome
Yes	Yes	Yes	4	Pass
	↓			
Yes	No	Yes	3	Pass

This time, the outcome doesn't change, so B is not a critical voter in this coalition. Because C has the same power as B, he is also not a critical voter in the coalition. ■

EXAMPLE

Winning and Blocking

In the committee with members A, B, and C, and voting system [3: 2, 1, 1], A and B have formed a coalition to vote in favor of measure X and to oppose another measure, Y. Member C is voting against X and for Y. Because the coalition $\{A, B\}$ has 3 votes (2 for A, and 1 for B), it is a winning coalition for X and a blocking coalition for Y.

Measure *X*				
A	*B*	*C*	Votes	Outcome
Yes	Yes	No	3	Pass

Measure *Y*				
A	*B*	*C*	Votes	Outcome
No	No	Yes	1	Fail

When voting for X, both A and B are critical voters, because their votes add up to the quota. It only takes 2 votes to block a measure. If A leaves the coalition and joins with C, the measure Y will pass, so A is a critical voter in the blocking coalition. The voter B is not critical because A can block Y without B. ■

EXAMPLE

What if Vice President Gore Had Carried Florida?

If the Florida election had been decided in favor of Vice President Gore in 2000, the coalition of states voting for him would have had 291 electoral votes. Because the quota in the electoral college is 270, we would say that the coalition had $291 - 270 = 21$ *extra votes.* Along with Florida, the critical voters would have been those in the coalition that had more than 21 votes: California, with 54 votes; New York, with 33; Pennsylvania, with 23; and Illinois, with 22 votes. Only those states could have transformed the coalition into a losing coalition by changing sides.[1] ■

Extra-Votes Principle

A winning coalition with total weight w has $w - q$ *extra votes.* A blocking coalition of total weight w has $w - (n - q)$ extra votes. The critical voters are those whose weight is more than the coalition's extra votes.

We can readily identify the critical voters in any winning or blocking coalition by determining the number of extra votes that it has.

To calculate the Banzhaf power index of a given voting system:

1. Make a list of the winning and blocking coalitions.
2. Use the extra-votes principle to identify the critical votes in each coalition.

A voter's Banzhaf index is then the number of coalitions in which he or she appears as a critical voter.

[1]Texas had 32 electoral votes but is not counted as a critical voter because it was not in the coalition supporting Vice President Gore.

EXAMPLE Calculating the Banzhaf Index

We will calculate the Banzhaf index for the committee with voting system [3: 2, 1, 1].

The winning coalitions are all those whose weights sum to 3 or 4, and we will start by making a list of them:

Winning coalition	Weight	Extra votes	Critical votes		
			A	*B*	*C*
$\{A, B\}$	3	0	1	1	0
$\{A, C\}$	3	0	1	0	1
$\{A, B, C\}$	4	1	1	0	0
		Totals	3	1	1

All members of the coalitions with 0 extra votes are critical voters. Because *A* is the only voter with more than 1 vote, she is the only critical voter in the coalition that has 1 extra vote. We have thus found that *A* is a critical voter in three winning coalitions, while *B* and *C* are each critical voters in one winning coalition.

Blocking coalitions have total weights of 2, 3, or 4. Here is a list of the blocking coalitions:

Blocking coalition	Weight	Extra votes	Critical votes		
			A	*B*	*C*
$\{A\}$	2	0	1	0	0
$\{B, C\}$	2	0	0	1	1
$\{A, B\}$	3	1	1	0	0
$\{A, C\}$	3	1	1	0	0
$\{A, B, C\}$	4	2	0	0	0
		Totals	3	1	1

Again, all voters in the coalitions with 0 extra votes are critical. In the blocking coalitions with 1 extra vote, only *A* is critical. The 4-vote blocking coalition $\{A, B, C\}$ has 2 extra votes. Because no voter has more than 2 votes, there are no critical voters $\{A, B, C\}$, considered as a blocking coalition. Voter *A* is critical in three blocking coalitions, while *B* and *C* are each critical in one. Adding up winning and blocking critical votes, we find that the Banzhaf index of *A* is 6, while *B* and *C* each have a Banzhaf index of 2. We will say that the Banzhaf index of this system is (6, 2, 2).

The Banzhaf index provides a comparison of the voting power of the participants in a voting system. Thus, *A*, with a Banzhaf index of 6, is three times as powerful as *B* or *C*. To determine the way voting power is distributed, we can add the numbers of critical voters for all three voters together to get $6 + 2 + 2 = 10$

critical votes in all. Thus, A has $\frac{6}{10} = 60\%$ of the voting power, while B and C each have 20%. ■

Consider the following three voting systems.

System I: [2: 1, 1, 1]
System II: [3: 2, 1, 1]
System III: [3: 1, 1, 1]

We have studied system II and found that its Banzhaf index is (6, 2, 2). Although the power is distributed equally in systems I and III, these systems have different Banzhaf indices. System III requires a unanimous vote to pass a measure. There is only one winning coalition, in which all three voters are critical, and each voter is critical in the blocking coalition in which he or she stands alone against the other two voters. The Banzhaf index for system III is therefore (2, 2, 2). In system I, coalitions of total weight 2 or 3 can either block or win. There are no critical voters in the coalition of weight 3, because it has one extra vote. In the coalitions of weight 2, all voters are critical. Voter A is therefore critical in two winning coalitions—$\{A, B\}$ and $\{A, C\}$—and in the same coalitions (with "no" voters rather than "yes" voters) as blocking coalitions. Thus, the Banzhaf index of system I is (4, 4, 4).

The voters in system I have greater Banzhaf indices than in system III. From a practical point of view, this means that in system I, an individual voter has more chances to influence the outcome of a vote than he or she would in system III.

In the examples that we have discussed so far, each participant was a critical voter in equal numbers of winning coalitions and blocking coalitions, and you might wonder if this was by coincidence. When a critical voter defects from a winning coalition, the opposing coalition becomes a blocking coalition, and the same voter is now a critical voter in that blocking coalition. If a critical voter defects from a blocking coalition, its opposing coalition would win, and the same voter would cast a critical vote in the new winning coalition. Thus, in every voting system, the number of blocking coalitions in which a particular voter is critical is equal to the number of winning coalitions in which the same voter is a critical voter. A voter's Banzhaf index can be determined by counting the winning coalitions in which he or she is a critical voter and doubling the result to account for the blocking coalitions.

EXAMPLE A Corporation with Four Shareholders

A corporation has shareholders A, B, C, and D with 40, 30, 20, and 10 shares, respectively. They use the weighted voting system

$$[51: 40, 30, 20, 10]$$

TABLE 13.2 Winning Coalitions in the Four-Stockholder Corporation

Coalition	Weight	Extra Votes	Critical Voters *A*	*B*	*C*	*D*
$\{A, B, C, D\}$	100	49				
$\{A, B, C\}$	90	39	1			
$\{A, B, D\}$	80	29	1	1		
$\{A, C, D\}$	70	19	1		1	
$\{A, B\}$	70	19	1	1		
$\{B, C, D\}$	60	9		1	1	1
$\{A, C\}$	60	9	1		1	
Critical votes			5	3	3	1

Table 13.2 shows a list of all the winning coalitions and the extra votes that each has. The four columns at the right are marked to indicate the critical voters in each coalition. By doubling the numbers of critical votes shown in the table, we arrive at the Banzhaf index for the corporation: (10, 6, 6, 2). In this model, *A* has

$$\frac{10}{10 + 6 + 6 + 2}, \text{ or approximately } 42\%$$

of the voting power, while *B* and *C* each have 25% (even though *B* has more shares than *C*). Shareholder *D* has the remaining 8% of the power. ■

How to Count Combinations

In a voting system with just one voter, there are two possible outcomes of a vote: yes (Y) or no (N). With two voters, there are four possible outcomes: YY, NY, YN, NN. Each time a new voter joins a voting system, the number of possible outcomes doubles. Thus, with three voters, the first two can vote in the four ways just listed, while the third voter votes Y: YYY, NYY, YNY, NNY. The first two voters could vote in the same four ways while the third votes N: YYN, NYN, YNN, NNN. To determine the number of voting **combinations** in a system with n voters, we can double repeatedly to obtain 2^n possible outcomes.

Now let us consider a system with n voters and ask in how many of the 2^n voting combinations there would be exactly k "yes" votes and $n - k$ "no" votes. For example, if $n = 4$ we could group the $2^4 = 16$ outcomes in five groups as follows:

Number of "yes" votes	Voting combinations	Number of combinations
0	NNNN	1
1	YNNN, NYNN, NNYN, NNNY	4
2	YYNN, YNYN, YNNY, NYYN, NYNY, NNYY	6
3	NYYY, YNYY, YYNY, YYYN	4
4	YYYY	1
	Total	16

Thus, if $k = 2$, we see that there are 6 voting combinations with 2 "yes" and 2 "no" votes.

The number of voting combinations of n voters having k "yes" votes and $n - k$ "no" votes is denoted C_k^n. When speaking, people often refer to C_k^n as "n choose k." According to the above list of combinations for $n = 4$, $C_0^4 = 1$, $C_1^4 = 4$, $C_2^4 = 6$, $C_3^4 = 4$, and $C_4^4 = 1$.

Let us determine C_2^5. Instead of listing all of the voting combinations, we will list the two-voter coalitions, taken from a set of voters $\{A, B, C, D, E\}$.

Coalitions involving A: $\{A, B\}, \{A, C\}, \{A, D\}, \{A, E\}$
Coalitions involving B: $\{B, A\}, \{B, C\}, \{B, D\}, \{B, E\}$
Coalitions involving C: $\{C, A\}, \{C, B\}, \{C, D\}, \{C, E\}$
Coalitions involving D: $\{D, A\}, \{D, B\}, \{D, C\}, \{D, E\}$
Coalitions involving E: $\{E, A\}, \{E, B\}, \{E, C\}, \{E, D\}$

There are 5×4 combinations listed, but each coalition appears twice on the list. For example, $\{B, D\}$ is listed as a coalition involving B and as a coalition involving D. To correct for the double counting, we must divide the number of combinations on the list by 2 to obtain

$$C_2^5 = \frac{5 \times 4}{2} = 10$$

A similar formula could be used to find C_2^4:

$$C_2^4 = \frac{4 \times 3}{2} = 6$$

This is a special case of a general formula for C_k^n:

$$C_k^n = \frac{n \times (n-1) \times (n-2) \times \ldots \times (n-k+1)}{k \times (k-1) \times (k-2) \times \ldots \times 1}$$

An easy way to remember the formula for C_k^n is that the numerator and the denominator each have k (the number of "yes" votes) factors; the factors of the numerator start with n (the total number of voters) and count down, while the factors of the denominator start with k and count down. Many scientific calculators have keys for calculating C_k^n.

EXAMPLE

Using the Combination Formula

To determine the number of voting combinations with four Y votes and three N votes that can occur in a set of seven voters, we calculate

$$C_3^7 = \frac{7 \times 6 \times 5}{3 \times 2 \times 1} = 35 \quad ■$$

Calculating C_k^n can often be simplified by the following observation. If there are n voters, C_k^n is the number of ways that k of the voters could vote "yes" while $n - k$ vote "no." If each voter should change his or her vote, a voting combination with $n - k$ "yes" votes and k "no" votes would be obtained. Therefore, the number of combinations with k "yes" voters from a set of n is equal to the number of combinations with $n - k$ "yes" voters from a set of n. In symbols,

Duality Formula

$$C_{n-k}^n = C_k^n$$

EXAMPLE

Using the Duality Formula

Let us calculate C_{23}^{25}. If we do not use the duality formula, it will be necessary to work with a fraction in which both the numerator and the denominator are determined by multiplying 23 numbers together. In the duality formula, put $n = 25$ and $k = 23$. Then $n - k = 2$, so

$$C_{23}^{25} = C_2^{25} = \frac{25 \times 24}{2} = 300 \quad ■$$

Efficient counting methods make it possible to compute the Banzhaf power index of large weighted voting systems. The method of counting combinations applies to systems in which most of the voters have equal weights, as in the Example below.

EXAMPLE

A Seven-Person Committee

The chairperson of a committee has 3 votes. There are six ordinary members, each of whom casts 1 vote. The quota for passing a measure is 5, so we are considering a voting system

$$[5: 3, 1, 1, 1, 1, 1, 1]$$

We will calculate the Banzhaf power index for each person in the committee.

Let M be an ordinary member, with weight 1. By the extra votes principle, M will be a critical voter in all winning coalitions with no extra votes. These coalitions have exactly 5 votes, including that of M. We can build such coalitions by including the chairperson and one other member along with M, or by not including the chairperson and including four other members with M.

There are C_4^5 ways to assemble a 5-vote winning coalition consisting of M and *other* ordinary members (call them M_1, M_2, M_3, M_4, M_5):

$\{M, M_1, M_2, M_3, M_4\}$
$\{M, M_1, M_2, M_3, M_5\}$
$\{M, M_1, M_2, M_4, M_5\}$
$\{M, M_1, M_3, M_4, M_5\}$
$\{M, M_2, M_3, M_4, M_5\}$

There are C_1^5 ways to choose from one of the *other* ordinary members to join M and the chairperson, C,

$\{M, C, M_1\}$
$\{M, C, M_2\}$
$\{M, C, M_3\}$
$\{M, C, M_4\}$
$\{M, C, M_5\}$

By the duality formula, $C_4^5 = C_1^5$, and we know that $C_1^5 = 5$. Therefore, there are $5 + 5 = 10$ winning coalitions in which M is critical. The Banzhaf power index of M (and each of the other ordinary members) is 20, counting the 10 winning coalitions and an equal number of blocking coalitions.

The chairperson C is a critical voter in any winning coalition with not more than 2 extra votes. Thus, C must be joined by at least 2, and not more than 4, ordinary members. There are C_2^6 ways to choose 2 ordinary members to join the chairperson, C_3^6 ways to choose 3, and C_4^6 ways to choose 4. The number of winning coalitions in which C is a critical voter is thus $C_2^6 + C_3^6 + C_4^6$. Because $C_2^6 = \frac{6 \times 5}{2 \times 1} = 15$, $C_3^6 = \frac{6 \times 5 \times 4}{3 \times 2 \times 1} = 20$, and $C_4^6 = C_2^6 = 15$, by duality, this is a total of 50 winning coalitions. Counting 50 blocking coalitions as well, the Banzhaf power index of C is 100. The Banzhaf index of the committee as a whole is

$$(100, 20, 20, 20, 20, 20, 20)$$

The total number of critical votes is $100 + 6 \times 20 = 220$. Thus, according to the Banzhaf model, the chairperson has $\frac{100}{220}$, or about 45% of the power in the committee; each of the other members has about 9% of the power. ■

Other counting methods must be used when voters have many different voting weights. For example, in the presidential election years 2004 and 2008, the electoral college of the United States will have 19 different weights, ranging from 3 to 55 votes. We cannot list all the voting combinations, for there are $2^{51} = 2{,}251{,}799{,}813{,}685{,}248$ of them. This would take more than 70 years to calculate at a rate of 1 million calculations per second. By using an advanced counting method described in the article by Paul J. Affuso and Steven J. Brams cited in the Suggested Readings for this chapter, it is possible to compute the Banzhaf index of a weighted voting system of the size of the electoral college relatively quickly. Table 13.3 displays the result of calculating the Banzhaf index of the electoral college by using a Banzhaf index calculator available on the Web site www.whfreeman.com/fapp.

A glance at Table 13.3 reveals that for most states, the percentage of power as measured by the Banzhaf index differs by only a few hundredths of a percent

TABLE 13.3 The Electoral College

States	Electoral Votes	Nominal Power	Banzhaf Power
CA	55	10.22%	11.40%
TX	34	6.32%	6.39%
NY	31	5.76%	5.79%
FL	27	5.02%	5.01%
IL, PA	21	3.90%	3.87%
OH	20	3.72%	3.68%
MI	17	3.16%	3.12%
GA, NC, NJ	15	2.79%	2.74%
VA	13	2.42%	2.37%
MA	12	2.23%	2.19%
IN, MO, TN, WA	11	2.04%	2.01%
AZ, MD, MN, WI	10	1.86%	1.82%
AL, CO, LA	9	1.67%	1.64%
KY, SC	8	1.49%	1.46%
CT, IA, OK, OR	7	1.30%	1.27%
AR, KS, MS	6	1.12%	1.09%
NE, NM, NV, UT, WV	5	0.93%	0.91%
HI, ID, ME, NH, RI	4	0.74%	0.73%
AK, DE, DC, MT, ND, SD, VT, WY	3	0.56%	0.55%

A state's nominal power is the number of electoral votes it has, expressed as a percentage of the total number of electoral votes for the nation (538). A state's Banzhaf power is the Banzhaf power index, expressed as a percentage of the total number of critical votes for all states in all winning and blocking coalitions (more than 9 quadrillion critical votes in all).

from the number of electoral votes, expressed as a percent of 538, the total number of electoral votes. There is one significant exception: California, whose 55 votes are 10.22% of the electoral college. The Banzhaf power index for California is 11.40%. If California continues to grow in population and thus acquires more electoral votes, we can expect the gap between California's Banzhaf index and its nominal power to increase.

13.3 Equivalent Voting Systems

If there are just two voters, A and B, how many really different voting systems are there? We can agree that the empty coalition, {}, is surely a losing coalition and that the unanimous coalition, $\{A, B\}$, must be a winning coalition. Therefore, there are only three distinct voting systems involving A and B: In the first, unanimous consent is required for each measure, so the only winning coalition is $\{A, B\}$. In the second, A is the dictator, and the winning coalitions are $\{A\}$ and $\{A, B\}$. In the third voting system, B is the dictator, and the winning coalitions are $\{B\}$ and $\{A, B\}$. Of course, there is an infinite number of ways to assign weights to the voters and a quota for passing measures in this two-voter system. However, there are only three ways to distribute the voting power: A as dictator, B as dictator, or consensus rule.

Two voting systems are **equivalent** if there is a way for all the voters of the first system to exchange places with the voters of the second system and preserve all winning coalitions.

For example, the weighted voting systems [50: 49, 1] and [4: 3, 3], involving pairs of voters A, B, and C, D, respectively, are equivalent because in each system, unanimous support is required to pass a measure. We could have A exchange places with C and B exchange places with D.

Now consider two voting systems [2: 2, 1] and [5: 3, 6] involving the same pair of voters, A, B. In the first, A is a dictator, while in the second, B dictates. By having A and B exchange places *with each other,* we see that the two systems are equivalent. *Equivalent* does not mean "the same." Voter A would tell you that the system where he is the dictator is not the same as the system where B is the dictator. The systems are equivalent because each has a dictator.

Every two-voter system is equivalent either to a system with a dictator or to one that requires consensus. As the number of voters increases, the number of different types of voting systems increases.

To classify voting systems according to equivalence, it is more convenient to consider the winning coalitions rather than the weights. Furthermore, if not all of the voters of a given winning coalition are critical, we don't have to keep track of that coalition.

A winning coalition in which every member is a critical voter is called a **minimal winning coalition.**

EXAMPLE

Minimal Winning Coalitions in the Four-Shareholder Corporation

Table 13.2 lists the five winning coalitions in the corporation with voting system [51: 40, 30, 20, 10]. The minimal ones are those in which each voter is marked as critical: $\{A, B\}$, $\{A, C\}$, and $\{B, C, D\}$. ■

The essential properties of a voting system are apparent when we find the minimal winning coalitions. For example, if some voter belongs to all the minimal winning coalitions, that voter has veto power. Any voter who doesn't belong to any minimal winning coalition is a dummy. In the four-shareholder corporation, we see that there are no dummies and that no one has veto power, either.

A voting system can be completely described by listing the minimal winning coalitions. All other winning coalitions are formed by adding voters to minimal winning coalitions. Any collection of sets of voters can serve as the minimal winning coalitions for some voting system, provided that the following three requirements are satisfied:

1. There is at least one minimal winning coalition.
2. If two minimal winning coalitions are distinct, each must have a voter who does not belong to the other.
3. Every pair of minimal winning coalitions has to overlap, with at least one voter in common.

The reason for requirement 1 is that we want to have a way in which the voting system can approve a measure. If there are no minimal winning coalitions, then there would be no winning coalitions, and it would be impossible to approve anything. Requirement 2 guarantees that the winning coalitions in the list are actually minimal. If all the voters in one coalition also belong to another coalition, then the larger coalition would not be minimal. Finally, if two winning coalitions have no overlap, then the decision-making process would be indecisive. For example, if there were two nonoverlapping winning coalitions in the electoral college, then two presidents might be elected at once.

Figure 13.1 displays the minimal winning coalitions of the four-shareholder corporation. You can see that requirement 2 is satisfied, because no minimal winning coalition lies within the boundary of another, and that requirement 3 is also satisfied, because every pair of minimal winning coalitions has at least one voter in common.

To construct a new voting system, specify a set of minimal winning coalitions satisfying requirements 1 through 3.

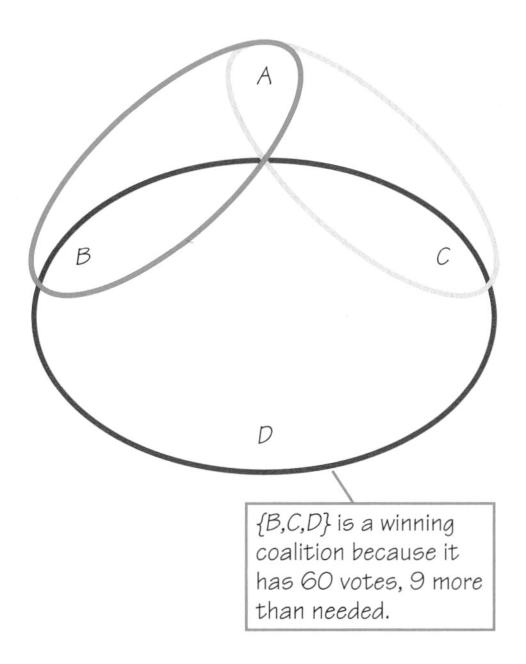
A
B
C
D
{B,C,D} is a winning coalition because it has 60 votes, 9 more than needed.

Figure 13.1 Each oval surrounds a minimal winning coalition for the four-shareholder corporation.

EXAMPLE

Three-Voter Systems

We would like to make a list of all voting systems that have three participants, A, B, and C. To keep the size of the list manageable, we will insist that no two voting systems on the list be equivalent.

To start, suppose that $\{A\}$ is a minimal winning coalition. Requirement 3 tells us that every other minimal winning coalition must overlap with $\{A\}$, but the only way that could happen would be if A also belonged to the other coalition. In this case, requirement 2 would be violated. Thus, $\{A\}$ can be the only minimal winning coalition. This is the voting system where A is dictator. Systems where B or C is dictator are not listed, because they are equivalent to this one.

Now suppose that there is no dictator. Every minimal winning coalition must contain either two or all three voters. Let's consider the case which $\{A, B, C\}$ is a minimal winning coalition. It is the only winning coalition, because any other winning coalition would have to be entirely contained in this coalition, which requirement 2 doesn't allow. In this voting system, a unanimous vote is required to pass a measure. We will call this system "consensus rule."

Finally, let's suppose that there is a two-voter minimal winning coalition, say $\{A, B\}$. If it is the only minimal winning coalition, then a measure will pass if A and B both vote "yes" and the vote of C does not matter: In other words, C is a dummy, and A and B make all the arrangements. We will call this system the "clique." Of course, the clique could be $\{A, C\}$ or $\{B, C\}$, but these systems are equivalent to the one where $\{A, B\}$ is the clique.

There could be two two-voter minimal winning coalitions, say $\{A, B\}$ and $\{A, C\}$. Neither coalition contains the other, and there is an overlap, so all of the

TABLE 13.4 Voting Systems with Three Participants

System	Minimal Winning Coalitions	Weights	Banzhaf Index
Dictator	$\{A\}$	[3: 3, 1, 1]	(8, 0, 0)
Clique	$\{A, B\}$	[4: 2, 2, 1]	(4, 4, 0)
Majority	$\{A, B\}$, $\{A, C\}$, $\{B, C\}$	[2: 1, 1, 1]	(4, 4, 4)
Chair veto	$\{A, B\}$, $\{A, C\}$	[3: 2, 1, 1]	(6, 2, 2)
Consensus	$\{A, B, C\}$	[3: 1, 1, 1]	(2, 2, 2)

requirements are satisfied. In this system, A has veto power. We have encountered this system in the three-member committee Example (page 444), and we will call it "chair veto." There are two other voting systems equivalent to this one, where B or C is chair.

It is possible that all three two-member coalitions are minimal winning coalitions. Because there are only three voters, any two distinct two-member coalitions will overlap, so the requirements are still satisfied. This system is called "majority rule." ■

Table 13.4 lists all five of these three-voter systems. Each system can be presented as a weighted voting system, and suitable weights are given in the table. If we want to make a similar list of all types of four-voter systems, we can start by making each three-voter system into a four-voter system. This is done by putting a fourth voter, D, into the system, without including him in any of the minimal winning coalitions. This makes D a dummy. You may be interested to know that there are an additional nine four-voter systems that don't have any dummies. It is interesting to try to list as many of these systems as you can.

EXAMPLE

A Five-Voter System

Five voters, A, B, C, D, and E, are split into two committees, $\{A, B, C\}$ and $\{C, D, E\}$. (Notice that C belongs to both committees.) The voting rule is that to pass a measure, at least one of the committees has to approve unanimously. The minimal winning coalitions of this voting system are the two committees (see Figure 13.2).

This five-voter system is not equivalent to any weighted voting system. We can see why by supposing for a moment that we have assigned weights to the voters and a quota so that these are the minimal winning coalitions. Because $\{A, B, C, D\}$ contains all the members of the first committee, it is a winning coalition. Therefore, its total weight is more than the quota. Here is the problem. If B drops out of the coalition, then neither committee is unanimous (B and E are holdouts), so the coalition becomes a losing coalition, with weight less than the quota. On the other hand, D can drop out without making the coalition lose.

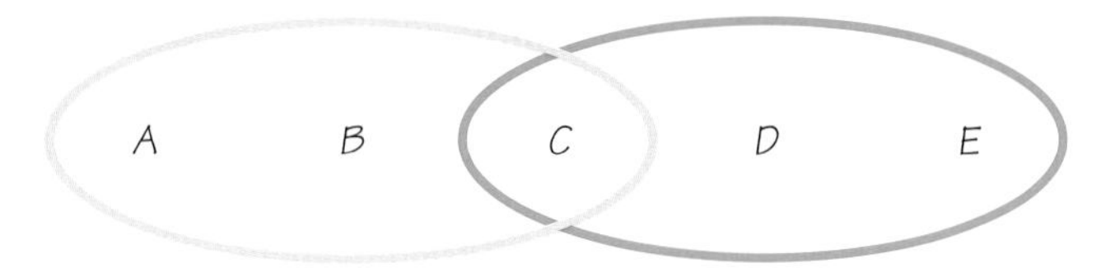

Figure 13.2 The five-voter system: minimal winning coalitions.

It follows that B has more weight than the extra votes of the coalition, and D has less. In other words, B is more powerful than D. This situation contradicts the symmetry of this voting system. In fact, we can apply a similar consideration to the coalition $\{B, C, D, E\}$. If B drops out of *this* coalition, it will still be a winning coalition, but now D is a critical voter. Thus, D has to have more weight than B. Because B cannot have more votes than D and fewer votes than D in the same voting system, there is no way that weights can be assigned to make this a weighted voting system. ■

Many voting systems that are not presented as weighted systems can be shown to be equivalent to weighted systems. Here is an example.

EXAMPLE

The Investment Committee

A university has an investment committee, consisting of the president of the university and six faculty members from the Department of Finance, that is charged with overseeing the endowment. Decisions are made by the following system: The faculty members each have 1 vote, and the president has 2 votes. The total number of votes is 8 – an even number – so ties are possible, and all ties are settled in favor of the president. Minimal winning coalitions thus consist of the president with two faculty members, or any coalition of five of the six faculty members. The tie-breaking provision prevents this voting system from being a valid weighted voting system, because the president can join with two faculty members to form a 4-vote winning coalition – and the opposing coalition, which is losing, also has 4 votes.

If we assign the president 3 votes, leave the faculty members with 1 vote each, and set the quota at 5, we will have a weighted voting system that is equivalent to the one that the investment committee uses. If five faculty members are aligned against the president, the extra vote will bring the president's coalition to a total of four, and the five faculty members will still win, 5 – 4. On the other hand, if the president has two faculty members on her side, the additional vote will give the president's coalition the victory, 5 – 4, without needing to break a tie. This weighted voting system

$$[5: 3, 1, 1, 1, 1, 1, 1]$$

is the same as that of the seven-person committee that we analyzed on page 450. By the Banzhaf model, the president has about 45% of the power, and each faculty member has about 9% of the power in this system. ■

In the following Example, we shall calculate the Banzhaf index of the five-voter system. We have already shown that this voting system is not equivalent to any weighted voting system (see page 456), so we can't use any weights in our calculation!

EXAMPLE The Five-Voter System

In the system with minimal winning coalitions $\{A, B, C\}$ and $\{C, D, E\}$, C has veto power. This means that C casts a critical vote in every winning coalition. There is a total of seven winning coalitions:

$$\{A, B, C\}, \{A, B, C, D\}, \{A, B, C, E\}\ \{A, B, C, D, E\},$$
$$\{B, C, D, E\}, \{A, C, D, E\}, \text{and } \{C, D, E\}$$

Because C is critical in an equal number of blocking coalitions, the Banzhaf index of C is $2 \times 7 = 14$. The other participants all have equal power. A is a critical voter in the first three coalitions in the list, but not in the remaining four, because those contain $\{C, D, E\}$ and can win without A. Counting blocking

SPOTLIGHT 13.2 A Mathematical Quagmire

In a 1965 law review article, John F. Banzhaf III analyzed a weighted voting system used by the Board of Supervisors of Nassau County, New York. The article inspired legal action against several elected bodies that employ weighted voting systems.

The first legal challenge to weighted voting was intended to invalidate the voting system of the Board of Supervisors of Washington County, New York. In its decision, the New York State Court of Appeals drew a corollary from Banzhaf's work that provided a way to fix weighted voting systems: Each supervisor's *Banzhaf power index,* rather than his or her voting weight, should be proportional to the population of the district that he or she represents.

In the Washington County decision, the court observed that expert opinion and detailed computer analyses would be needed to justify any weighted voting system and predicted that the courts would eventually be dragged into a "mathematical quagmire."

A series of five lawsuits, spanning 25 years, challenged weighted voting in the Nassau County Board of Supervisors. These cases proved to be the mathematical quagmire that the appeals court had feared. The courts attempted to force Nassau County to comply with the Washington County decision. Although the county made a sincere attempt to do so, every voting system that it devised faced a new legal challenge. With conflicting expert testimony, the U.S. District Court finally ruled in 1993 that weighted voting was inherently unfair.

Banzhaf's law review article, which initially drew attention to weighted voting in Nassau County, was aptly titled "Weighted Voting Doesn't Work."

coalitions, the Banzhaf index of A (and of B, D, and E) is 6. The Banzhaf index of the system is (6, 6, 14, 6, 6). Each of A, B, D, and E has

$$\frac{6}{6 + 6 + 14 + 6 + 6} \approx 16\%$$

of the power, while C has about 37% of the power. ■

The Banzhaf index was discovered by a lawyer who wanted to challenge weighted voting systems in court. Spotlight 13.2 tells the story of this strange mixture of mathematics and the law.

13.4 The Shapley–Shubik Power Index

In drafting legislation, coalitions are built one voter at a time. The most important voter in the sequence is the one who turns the coalition from a losing coalition into a winning coalition. A power index based on this idea was introduced by Lloyd Shapley and Martin Shubik (see Spotlight 13.3) in 1954. To calculate the index, one considers *permutations* of voters.

A **permutation** of voters is an ordering of all of the voters in a voting system.

The voting combinations that we considered in developing the Banzhaf index treated all "yes" voters as if they were interchangeable. Permutations record finer detail, because the voters are ordered in accordance with their commitment to an issue. For example, suppose that the issue is animal rights. Here the spectrum might range from a voter who would outlaw the sale of cow's milk to a voter who would legalize cockfighting. If an animal rights bill is being drafted, it must be written so as to receive the support of a coalition with enough votes to meet the quota.

The first voter in a permutation whose vote would make the coalition a winning coalition (if he or she could be induced to join) is called the **pivotal voter** in that permutation. Each permutation has exactly one pivotal voter.

If the issue is taxation instead of animal rights, the spectrum of opinion will probably be completely different. Voters who have moderate positions on animal rights may or may not be at the extremes when the subject is taxes. Each issue being debated corresponds to a particular permutation–and the pivotal voter on one issue will probably not be pivotal on another issue.

SPOTLIGHT 13.3 Power Indices

The first widely accepted numerical index for assessing power in voting systems was the Shapley–Shubik index, developed in 1954 by a mathematician, Lloyd S. Shapley, and an economist, Martin Shubik. A particular voter's power as measured by this index is proportional to the number of different *permutations* (or orderings) of the voters in which he or she has the potential to cast the pivotal vote – the vote that first turns from losing to winning.

The Banzhaf power index was introduced in 1965 by John F. Banzhaf III, a law professor who is also well-known as the founder of the antismoking organization ASH (Action on Smoking and Health). The Banzhaf index is the one most often cited in court rulings, perhaps because Banzhaf brought several cases to court and continues to file *amicus curiæ* briefs when courts evaluate weighted voting systems. A voter's Banzhaf index is the number of different possible voting *combinations* in which he or she casts a swing vote – a vote in favor of a motion that is necessary for the motion to pass or a vote against a motion that is essential for its defeat.

Lloyd S. Shapley

Martin Shubik

John F. Banzhaf III

If a bill is drafted so that it secures the support of the pivotal voter in the permutation corresponding to the spectrum of opinion on the issue, then the bill will pass.

As with the Banzhaf power index, computation of the Shapley–Shubik power index is basically a counting problem. If there are n voters, the number of permutations is called the **factorial** of n and is denoted $n!$. There is a simple formula for $n!$:

$$n! = n \times (n - 1) \times (n - 2) \times \ldots \times 2 \times 1$$

For example,

$$1! = 1$$
$$2! = 2 \times 1 = 2$$
$$3! = 3 \times 2 \times 1 = 6$$
$$4! = 4 \times 3 \times 2 \times 1 = 24$$

To justify the formula, suppose that we are listing all the permutations. There are n voters who can be first; when the first voter has been selected, there are $n - 1$ voters to put in the second position, $n - 2$ for the third position, and so on. When we get to the last position, there's only one voter left. By the fundamental principle of counting (also see Chapter 1), the number of permutations is the product of the number of choices that we have at each stage of the process.

Consider the three-person committee in which the chair has veto power, with the weighted voting system

$$[3: 2, 1, 1]$$

There are $3! = 6$ permutations of the members A, B, and C. Table 13.5 displays all six permutations. Next to each permutation, the total weight of the first voter, of the first two voters, and of all three voters is shown. The first number in this sequence to exceed the quota is underlined, and the corresponding pivotal voter's symbol is circled. We see that A is pivotal in four permutations, while B and C are each pivotal in one.

The **Shapley–Shubik power index** of a voter is the fraction of the permutations in which that voter is pivotal.

TABLE 13.5 Permutations and Pivotal Voters for the Three-Person Committee

Permutations			Weights		
A	Ⓑ	*C*	2	<u>3</u>	4
A	Ⓒ	*B*	2	<u>3</u>	4
B	Ⓐ	*C*	1	<u>3</u>	4
B	*C*	Ⓐ	1	2	<u>4</u>
C	Ⓐ	*B*	1	<u>3</u>	4
C	*B*	Ⓐ	1	2	<u>4</u>

In the three-person committee, the Shapley–Shubik index for A is $\frac{4}{6}$, and B and C each have a Shapley–Shubik index of $\frac{1}{6}$. According to the Shapley–Shubik model, the chairperson of this committee, A, has four times as much voting power as an ordinary member. Recall that the Banzhaf power index for the same committee was (6, 2, 2), so according to the Banzhaf model, A is three times as powerful as B or C.

How to Compute the Shapley–Shubik Power Index

For voting systems with no more than four voters, the Shapley–Shubik power index can be calculated by making a list of all the voting permutations and identifying the pivotal voter in each.

EXAMPLE The Corporation with Four Shareholders

This corporation has four shareholders, A, B, C, and D, with 40, 30, 20, and 10 shares, respectively. The corporation uses the weighted voting system

$$[51: 40, 30, 20, 10]$$

The $4! = 24$ permutations are shown in Table 13.6. In ten of the permutations, A is the pivotal voter; B and C are each pivotal voters in six; and D is the pivotal voter in two permutations. Therefore, the Shapley–Shubik index for this voting system is

$$\left(\frac{10}{24}, \frac{6}{24}, \frac{6}{24}, \frac{2}{24}\right)$$

The Banzhaf power index gives the same ratios of power in this case. ■

The previous two Examples give the impression that there isn't much difference between the Banzhaf and Shapley–Shubik indices. For an example where there is a radical difference in the way that they distribute power, see the 9001 Stockholder Corporation Example on the text Web site.

Listing the permutations is the brute force way of calculating the Shapley–Shubik index. As we saw in our study of the traveling salesman problem (Chapter 2), brute force methods are sometimes impossible to carry out, due to the combinatorial explosion. You can quickly verify that $10! = 3{,}628{,}800$ with your calculator. This is the number of permutations that you would have to list to calculate the Shapley–Shubik index of a ten-voter system by brute force.

To calculate the Shapley–Shubik power index without listing the permutations, we can use two facts. First, voters whose weights are equal are interchangeable, and thus will have the same Shapley–Shubik power indices; second, the sum of the Shapley–Shubik power indices of all voters is equal to 1. For example, consider an n-voter weighted voting system in which all voters have equal

TABLE 13.6 Permutations and Pivotal Voters for the Four-Person Corporation

Permutations				Weights				Pivot
A	Ⓑ	*C*	*D*	40	70	90	100	*B*
A	Ⓑ	*D*	*C*	40	70	80	100	*B*
A	Ⓒ	*B*	*D*	40	60	90	100	*C*
A	Ⓒ	*D*	*B*	40	60	70	100	*C*
A	*D*	Ⓑ	*C*	40	50	80	100	*B*
A	*D*	Ⓒ	*B*	40	50	70	100	*C*
B	Ⓐ	*C*	*D*	30	70	90	100	*A*
B	Ⓐ	*D*	*C*	30	70	80	100	*A*
B	*C*	Ⓐ	*D*	30	50	90	100	*A*
B	*C*	Ⓓ	*A*	30	50	60	100	*D*
B	*D*	Ⓐ	*C*	30	40	80	100	*A*
B	*D*	Ⓒ	*A*	30	40	60	100	*C*
C	Ⓐ	*B*	*D*	20	60	90	100	*A*
C	Ⓐ	*D*	*B*	20	60	70	100	*A*
C	*B*	Ⓐ	*D*	20	50	90	100	*A*
C	*B*	Ⓓ	*A*	20	50	60	100	*D*
C	*D*	Ⓐ	*B*	20	30	70	100	*A*
C	*D*	Ⓑ	*A*	20	50	60	100	*B*
D	*A*	Ⓑ	*C*	10	50	80	100	*B*
D	*A*	Ⓒ	*B*	10	50	70	100	*C*
D	*B*	Ⓐ	*C*	10	40	80	100	*A*
D	*B*	Ⓒ	*A*	10	40	60	100	*C*
D	*C*	Ⓐ	*B*	10	30	70	100	*A*
D	*C*	Ⓑ	*A*	10	30	60	100	*B*

weight. All voters will then have the same Shapley–Shubik power index, $1/n$. The next simplest case is one in which all of the voters but one have equal weights (as in the Investment Committee Example).

EXAMPLE

The Shapley–Shubik Index of the Seven-Person Committee

In the seven-person committee, the chairperson has three votes, each of the other members has one vote, and the quota for passing a measure is five votes. There are $7! = 5040$ voting permutations to consider, so we will consider groups of permutations, rather than one permutation at a time. Each group will be identified by the position occupied by the chairperson. Thus, the first group would be *C M M M M M M,* in which the chairperson is first. Counting from the left, we see that votes are accumulated in the sequence 3, 4, 5, 6, 7, 8, 9, and that the third participant (an ordinary member) is the pivotal voter. In the next group, *M C M M M M M*, the vote accumulation sequence is 1, 4, 5, 6, 7, 8, 9; again, an ordinary member is the pivotal voter. The chairperson is the pivotal voter in groups 3, 4, 5: *M M C M M M M*, *M M M C M M M*, and *M M M M C M M*, with vote accumulation sequences 1, 2, 5, 6, 7, 8, 9; 1, 2, 3, 6, 7, 8, 9; and 1, 2, 3, 4, 7, 8, 9, respectively. In the final two groups, *M M M M M C M* and *M M M M M M C*, the vote accumulation sequences will be 1, 2, 3, 4, 5, 8, 9 and 1, 2, 3, 4, 5, 6, 9, respectively, and an ordinary member will be the pivotal voter again.

Each of the seven groups that we have considered has 6! permutations, because there are 6! orderings for the ordinary members. Because the groups are of equal size, each has $\frac{1}{7}$ of the permutations. We have seen that the chairperson is the pivotal voter in all permutations in three of the groups, so his or her Shapley–Shubik index is $\frac{3}{7}$.

The total Shapley–Shubik power index for the ordinary members is the fraction of power not held by the chairperson: $\frac{4}{7}$. Because each of the six ordinary members has the same power, the Shapley–Shubik index for each is $\frac{4}{7} \div 6 = \frac{2}{21}$.

The Shapley–Shubik index for this weighted voting system is therefore

$$\left(\frac{3}{7}, \frac{2}{21}, \frac{2}{21}, \frac{2}{21}, \frac{2}{21}, \frac{2}{21}, \frac{2}{21}\right)$$

Because $\frac{3}{7} \div \frac{2}{21} = \frac{9}{2} = 4\frac{1}{2}$, the Shapley–Shubik model indicates that the chairperson has four and one-half times as much power as an ordinary member. This is in close agreement with the Banzhaf model, which held that the chairperson was five times as powerful as an ordinary member. ■

Comparing the Banzhaf and Shapley–Shubik Models

Deciding which power index is the best model for the distribution of power in a particular voting system is a subjective judgment. The heart of the issue is the distinction between permutations and combinations. A particular voting permutation reflects the range of opinion concerning an issue, whereas a voting combination tells us who is for and who is against a measure. The Shapley–Shubik index of a voter is the fraction of the voting permutations in which that voter is the pivotal voter. In a decision-making body, it is natural that some participants will frequently be found at extreme ends of voting permutations, while others will more frequently occupy middle positions. Those who are less inclined to extreme positions are pivotal voters more frequently than those with extreme views, and so have more power. However, the purpose of a power index is to measure the distribution of power that is built into the system, not the power that some voters acquire as a consequence of their political views. The Shapley–Shubik index is appropriate if we believe that on most issues before the voting body, there is a one-dimensional spectrum of opinion. The Banzhaf index does not attempt to model the dynamics of the legislative process, and it best reflects the situation when the participants operate unpredictably without consulting one another.

Often the political dynamic is too complex to be accurately modeled by either index. When there are many points of view to consider, the opinions that are represented cannot be strictly ordered between two extremes. In these cases, the Banzhaf and Shapley–Shubik indices provide measurements of voting power from two points of view.

Analysis of voting systems with voters numbering in the thousands or millions is only possible if all the voters, with only a few exceptions, are equally powerful. For further discussion of systems with large numbers of voters, go to www.whfreeman.com/fapp.

REVIEW VOCABULARY

Banzhaf power index A numerical measure of power for participants in a voting system. A participant's Banzhaf index is the number of winning or blocking coalitions in which he or she is a critical voter.

Blocking coalition A set of participants in a voting system that can prevent a measure from passing by voting against it.

C_k^n The number of combinations of n voters with k "yes" votes and $n - k$ "no" votes. This number, referred to as "n choose k," is given by the formula

$$C_k^n = \frac{n \times (n-1) \times \ldots \times (n-k+1)}{k \times (k-1) \times \ldots \times 1}$$

Remember that the numerator is the product of k numbers starting with n and counting down; the denominator is the product of k numbers starting with k and counting down.

Coalition A set consisting of some, all, or none of the participants in a voting system.

Combination A list of voters indicating the vote of each on an issue. There is a total of 2^n combinations in an n-element set, and C_k^n combinations with k "yes" votes and $n - k$ "no" votes.

Critical voter A member of a winning coalition whose vote is essential for the coalition to win, or a member of a blocking coalition whose vote is essential for the coalition to block.

Dictator A participant in a voting system who can pass any issue even if all other voters oppose it and block any issue even if all other voters approve it.

Duality formula $C_{n-k}^n = C_k^n$

Dummy A participant who has no power in a voting system. A dummy is never a critical voter in any winning or blocking coalition and is never the pivotal voter in any permutation.

Equivalent voting systems Two voting systems are equivalent if there is a way for all the voters of the first system to exchange places with the voters of the second system and preserve all winning coalitions.

Extra votes The number of votes that a winning coalition has in excess of the quota.

Extra-votes principle The critical voters in the coalition are those whose weights are more than the extra votes of the coalition. For example, if a coalition has 12 votes and the quota is 9, there are 3 extra votes. The critical voters in the coalition are those with more than 3 votes.

Factorial If n is a positive integer, the factorial of n (denoted $n!$) is the product of all the positive integers less than or equal to n. It is usually a large number: 10! is a seven-digit number.

Losing coalition A coalition that does not have the voting power to get its way.

Minimal winning coalition A winning coalition that will become losing if any member defects. Each member is a critical voter.

Permutation A specific ordering from first to last of the elements of a set; for example, an ordering of the participants in a voting system.

Pivotal voter The first voter in a permutation who, with his or her predecessors in the permutation, will form a winning coalition. Each permutation has one and only one pivotal voter.

Quota The minimum number of votes necessary to pass a measure in a weighted voting system.

Shapley–Shubik power index A numerical measure of power for participants in a voting system. A participant's Shapley–Shubik index is the number of permutations of the voters in which he or she is the pivotal voter, divided by the number of permutations ($n!$ if there are n participants).

Veto power A voter has veto power if no issue can pass without his or her vote. A voter with veto power is a one-person blocking coalition.

Weight The number of votes assigned to a voter in a weighted voting system, or the total number of votes of all voters in a coalition.

Weighted voting system A voting system in which the participants can have different numbers of votes. It can be represented as $[q: w(A_1), w(A_2), \ldots, w(A_n)]$, where $A_1, \ldots, A_n$ are the voters, $w(A_1), \ldots, w(A_n)$ represent the numbers of votes held by these voters, and q is the quota necessary to win.

Winning coalition A set of participants in a voting system who can pass a measure by voting for it.

SUGGESTED READINGS

AFFUSO, PAUL J., and STEVEN J. BRAMS. Power and size: A new paradox, *Theory and Decision,* 7 (1976): 29–56. This paper explains a technique involving algebra for calculating the Banzhaf and Shapley–Shubik power indices.

BANZHAF, JOHN F., III. One man, 3.312 . . . votes: A mathematical analysis of the electoral college, *Villanova Law Review,* 13 (1968): 304–332. The article presents a model that starts with the assumption that each voter uses a coin toss to decide his or her vote. While the chance that any voter will cast a critical vote in this model is remote, the article shows that voters residing in the more populous states have as much as three times the chance of doing so than do voters in less populous states. The article claims that this implies that the populous states have disproportionate power to influence the outcome of a presidential election, but this conclusion is controversial. The same issue of the journal has commentaries on Banzhaf's analysis (pages 333–346). Another commentary appears in *Villanova Law Review,* 14 (1968): 86–96.

BANZHAF, JOHN F., III. Weighted voting doesn't work, *Rutgers Law Review,* 19 (1965): 317–343. The author defines the Banzhaf index and uses it to show that the weighted voting system in use by the Nassau County Board of Supervisors was unfair.

BRAMS, STEVEN J. *The Presidential Election Game,* Yale University Press, New Haven, Conn., 1978.

BRAMS, STEVEN J. *Game Theory and Politics,* Free Press, New York, 1975. Chapter 5 treats the Shapley–Shubik and Banzhaf indices.

BRAMS, STEVEN J., W. F. LUCAS, and P. D. STRAFFIN, Jr., eds. *Modules in Applied Mathematics,* vol 2: *Political and Related Models,* Springer-Verlag, New York, 1983. Chapters 9–11 are devoted to measuring power in weighted and other types of voting systems. The Banzhaf and Shapley–Shubik indices are the focus of Chapters 9 and 11; Chapter 10 is about an index based on counting minimal winning coalitions.

FELSENTHAL, DAN S., and MOSHÉ MACHOVER. *The Measurement of Voting Power: Theory and Practice, Problems and Paradoxes,* Edward Elgar, Cheltenham, UK, 1998. This monograph covers the Banzhaf and Shapley–Shubik indices thoroughly. It includes a detailed analysis of an extremely important weighted voting system: the Council of Ministers of the European Community.

Iannucci v. Board of Supervisors of Washington County 20 N.Y. 2d 244, 251, 229 N.E. 2d 195, 198, 282 N.Y.S. 2d 502, 507 (1967). This code will help a law librarian find this case for you. It opened a "mathematical quagmire."

LAMBERT, JOHN P. Voting games, power indices, and presidential elections, *UMAP Journal,* 9(3) (1988): 213–267.

LUCAS, WILLIAM F. *Fair Voting: Weighted Votes for Unequal Constituencies,* COMAP: HistoMAP Module 19, Lexington, Mass., 1992. An introduction to the power indices with emphasis on the historical aspects.

TAYLOR, ALAN D. *Mathematics and Politics: Strategy, Voting Power, and Proof,* Springer-Verlag, New York, 1995. Chapter 4 covers weighted voting systems and their analysis using the Shapley–Shubik and Banzhaf indices. It has no mathematical prerequisites, but it does include carefully written logical arguments that must be carefully read.

SUGGESTED WEB SITE

www.math.temple.edu/~cow/bpi.html This site has an interactive program that will calculate the Banzhaf index of weighted voting systems.

SKILLS CHECK

1. What would be the quota for a voting system that has a total of 20 voters and uses a simple majority quota?

(a) 10
(b) 11
(c) 20

2. For the weighted voting system $[q: w(A), w(B), w(C)] = [65: 60, 30, 10]$, which statement is true?

(a) A is a dictator.
(b) B has veto power.
(c) Every person has power.

3. Two daughters each hold six votes and a son has the remaining two votes for a trust fund. The quota for passing a measure is 8. Which statement is true?

(a) The son is a dummy voter.
(b) The son is not a dummy voter but has less power than a daughter.
(c) The three children have equal power.

4. Which voters A, B, C, D in the weighted voting system [10: 4, 4, 3, 2] have veto power?

(a) No one
(b) A and B only
(c) Everyone

5. What is the value of C_2^6?

(a) 12
(b) 15
(c) 32

6. For the weighted voting system [6: 4, 3, 2, 1], find the Banzhaf power index for the voter with three votes.

(a) 3
(b) 6
(c) 14

7. Calculate the Shapley–Shubik power index for the three-vote voter in the weighted voting system [6: 4, 3, 2, 1].

(a) 1/4
(b) 5/24
(c) 1/12

8. A blocking coalition can always
(a) defeat a motion.
(b) pass a motion.
(c) force a reevaluation of a motion.

9. If a winning coalition is minimal, then the number of extra votes
(a) is zero.
(b) is less than the number of votes held by any member of the winning coalition.
(c) is less than the number of votes held by the losing coalition.

10. The best description for a voter who always wins is
(a) a voter with veto power.
(b) a dictator.
(c) a pivotal voter.

11. A "critical voter" in a coalition is a voter
(a) who has the most votes.
(b) who has fewer votes than the number of extra votes of the coalition.
(c) whose defection changes the coalition from a winning coalition to a losing coalition.

12. The weight of a voter is
(a) the number of votes assigned to the voter.
(b) the number of times the voter is pivotal.
(c) the number of times the voter is part of a winning coalition.

13. How large is the number 8! (eight factorial)?
(a) More than a million
(b) Less than a million, but more than 10,000
(c) Less than 10,000

14. In how many different ways can six voters be ordered from first to last?
(a) 30
(b) 64
(c) More than 500

15. In how many ways can six voters respond to a "yes–no" question?
(a) 12
(b) 36
(c) 64

EXERCISES

▲ **Optional.** ■ **Advanced.** ◆ **Discussion.**

How Weighted Voting Works

◆ **1.** In the United States Senate, each of the 100 senators has one vote, and the vice president of the United States can vote also, if it is necessary to break a tie.
(a) A simple majority is needed to pass a bill. What constitutes a winning coalition, and what constitutes a blocking coalition?
(b) One seat in the Senate is vacant, so that only 99 senators can vote. What constitutes a winning coalition, and what constitutes a blocking coalition for passing a bill?
(c) To ratify a treaty, a two-thirds majority is required. What constitutes a winning coalition, and what constitutes a blocking coalition?

◆ **2.** Is it possible to have a weighted voting system in which more votes are required to block a measure than to pass a measure?

3. Voters A, B, and C use the voting system [8: 5, 4, 3].
(a) List all winning coalitions.
(b) List all blocking coalitions.
(c) Is there a dictator?
(d) Does anyone have veto power?
(e) Are there dummy voters?

4. List all the winning coalitions of a committee of four members, A, B, C, and D, with voting system [51: 30, 25, 24, 21].

5. How would the list in Exercise 4 change if the quota were increased to

(a) 52?
(b) 55?
(c) 58?

6. Voter *A* in Exercise 4 would like to have veto power. How much should the quota be increased to give her, and no one else, veto power?

The Banzhaf Power Index

7. (a) List the 16 possible combinations of how four voters, *A*, *B*, *C*, and *D*, can vote either "yes" (Y) or "no" (N) on an issue.
(b) List the 16 subsets of the set {*A*, *B*, *C*, *D*}.
◆ **(c)** How do the lists in parts (a) and (b) correspond to each other?
(d) In how many of the combinations in part (a) is the vote

i. 4 Y to 0 N?
ii. 3 Y to 1 N?
iii. 2 Y to 2 N?

8. Calculate the number of extra votes for each of the winning coalitions found in the solution to Exercise 4. Identify the winning coalitions in which

(a) *A* is a critical voter.
(b) *B* is a critical voter.

9. Calculate the Banzhaf index for the voting system in Exercise 4.

10. The system in Exercise 4 is modified by increasing the quota to

(a) 52.
(b) 55.
(c) 58.
(d) 73.
(e) 76.
(f) 79.
(g) 82.

Calculate the Banzhaf index in each case. (*Hint:* Increasing the quota will reduce the number of extra votes in each of the original coalitions. When the number of extra votes becomes negative, the coalition is losing and you can cross it off the list. As the number of extra votes decreases, a member of a coalition who was originally not a critical voter will become a critical voter.)

11. Determine the number of extra votes for each winning coalition, and calculate the Banzhaf index for each of the following weighted voting systems.

(a) [51: 52, 48]
(b) [3: 2, 2, 1]
(c) [8: 5, 4, 3]
(d) [51: 45, 43, 8, 4]
(e) [51: 45, 43, 6, 6]

12. Calculate the following:

(a) C_3^7
(b) C_{100}^{50}
(c) C_2^{15}
(d) C_{13}^{15}

13. Calculate the following:

(a) C_3^6
(b) C_2^{100}
(c) C_{98}^{100}
(d) C_5^9

14. The Board of Supervisors of Nassau County, New York, is a historically important example of a weighted voting system (see Spotlight 13.2). Before it was declared unconstitutional by a federal district court in 1993, the weighted voting system of the Board of Supervisors was changed several times. The weights in use since 1958 were as follows:

		Weights					
Year	**Quota**	H_1	H_2	N	B	G	L
1958	16	9	9	7	3	1	1
1964	58	31	31	21	28	2	2
1970	63	31	31	21	28	2	2
1976	71	35	35	23	32	2	3
1982	65	30	28	15	22	6	7

Here H_1 is the presiding supervisor, always from the community of Hempstead; H_2 is the second supervisor from Hempstead; and N, B, G, and L are the supervisors from the remaining districts – North Hempstead, Oyster Bay, Glen Cove, and Long Beach.

◆ **(a)** From 1970 on, more than a simple majority was required to pass any measure. Give an argument in favor of this policy from the viewpoint of a supervisor who would benefit from it, and give an argument against the policy from the viewpoint of a supervisor who would lose some power.
(b) In which years were some supervisors dummy voters?
(c) Suppose that the two Hempstead supervisors always vote together. In which years are some of the supervisors dummy voters?
(d) Assume that the two Hempstead supervisors always agree, so that the board is in effect a five-voter system. Determine the Banzhaf index of this system in each year.
(e) In 1982, a special supermajority of 72 votes was needed to pass measures that ordinarily require a two-thirds majority. If the two Hempstead supervisors vote together, what is the Banzhaf index of the resulting five-voter system?
◆ **(f)** Table 13.7 gives the 1980 census for each municipality, the number of votes assigned to each supervisor, and the Banzhaf index for each supervisor in 1982. Do you think the voting scheme was fair?

Equivalent Voting Systems

15. Consider a four-person voting system with voters A, B, C, and D. The winning coalitions are

$$\{A, B, C, D\}, \{A, B, C\}, \{A, B, D\}, \{A, C, D\}, \text{ and } \{A, B\}$$

(a) List the minimal winning coalitions.
◆ **(b)** Show that $\{A\}$ is a minimal blocking coalition. Are there other minimal blocking coalitions?
(c) Determine the Banzhaf power index for this voting system.
(d) Find an equivalent weighted voting system.

◆ **16.** A five-member committee has the following voting system. The chairperson can pass or block any motion that she supports or opposes, provided that at least one other member is on her side. Show that this voting system is equivalent to the weighted voting system [4: 3, 1, 1, 1, 1].

17. Calculate the Banzhaf index for the weighted voting system in Exercise 16.

18. Find weighted voting systems that are equivalent to

(a) a committee of three faculty members and the dean. To pass a measure, at least two faculty members and the dean must vote "yes."

TABLE 13.7 Nassau County Board of Supervisors, 1982

Supervisor from	Population	Number of Votes	Banzhaf Power Index	
			65	72
Hempstead (presiding)	738,517	30	30	26
Hempstead		28	26	22
North Hempstead	218,624	15	18	18
Oyster Bay	305,750	22	22	18
Glen Cove	24,618	6	2	2
Long Beach	43,073	7	6	6
Totals	1,321,582	108	104	92

(b) a committee of three faculty members, the dean, and the provost. To pass a measure, two faculty, the dean, and the provost must vote "yes."

◆ **19.** A four-member faculty committee and a three-member administration committee vote separately on each issue. The measure passes if it receives the support of a majority of each of the committees. Show that this system is not equivalent to a weighted voting system.

20. Calculate the Banzhaf index of the voting system in Exercise 19. Who is more powerful according to the Banzhaf model, a faculty member or an administrator?

◆ **21.** Explain why a voting system in which no voter has veto power must have at least three minimal winning coalitions.

◆ **22.** How many *distinct* (nonequivalent) voting systems with four voters can you find? Systems that have dummies don't count. The challenge is to find all nine.

The Shapley–Shubik Power Index

23. For the voting system in Exercise 4, list all permutations of the voters in which

(a) A is the pivotal voter.
(b) B is the pivotal voter.

24. Calculate the Shapley–Shubik index for the system in Exercise 4.

25. Calculate the Shapley–Shubik index for the weighted voting system in Exercise 16.

◆ **26.** The five-voter system introduced on page 456 is not equivalent to a weighted voting system. The pivot in a permutation of the voters A, B, C, D, and E is the first voter who, with those preceding in the permutation, can form a winning coalition. Thus, in the permutation $ADECB$, C is the pivot, because the coalition $\{A, D, E, C\}$ includes each member of the subcommittee $\{C, D, E\}$.

(a) Show that if C is the fourth or fifth voter in a permutation, then C is the pivot of that permutation.
(b) Make a list of all permutations in which C is the third voter. Identify those in which C is the pivot.
(c) Is it possible for C to be the pivot in any permutation not considered in parts (a) and (b)?
(d) Compute the Shapley–Shubik index of the five-voter system.

■ **27.** The voting system involving faculty and administrators in Exercise 19 is not equivalent to a weighted voting system. Determine its Shapley–Shubik index. Who is more powerful according to the Shapley–Shubik model, a faculty member or an administrator?

Additional Exercises

28. Determine the Shapley–Shubik power index for the four-person voting system described in Exercise 15.

29. A corporation has four shareholders and a total of 100 shares. The quota for passing a measure is the votes of shareholders owning 51 or more shares. The number of shares owned are as follows:

A	48 shares
B	23 shares
C	22 shares
D	7 shares

There is also an investor, E, who is interested in buying shares but does not own any shares at present. Sales of fractional shares are not permitted.

(a) List the winning coalitions and compute the number of extra votes for each. Make a separate list of the losing coalitions, and compute the number of votes that would be needed to make the coalition winning.
(b) How many shares can A sell to B without causing any of the winning coalitions listed in part (a) to lose or any of the losing coalitions in part (a) to win?
(c) How many shares can A sell to D without changing the sets of winning or losing coalitions?

(d) How many shares can A sell to E without changing the winning coalitions? Since E is now a dummy, he must remain a dummy after the trade.
(e) How many shares can D sell – without changing the set of winning coalitions – to A, B, C, or E? Again, it is conceivable that D would be able to sell more to one stockholder than to another.
(f) How many shares can D sell to A, B, C, or E without becoming a dummy?
(g) How many shares can B sell to C without changing the set of winning coalitions?

30. Which of the following voting systems is equivalent to the voting system in use by the corporation in Exercise 29?

(a) [3: 1, 1, 1, 1]
(b) [3: 2, 1, 1, 1]
(c) [5: 3, 1, 1, 1]
(d) [5: 3, 2, 1, 1]
(e) [5: 3, 2, 2, 2]

31. Determine the Banzhaf and Shapley – Shubik power indices for the corporation in Exercise 29.

32. (a) Show that $C_3^5 + C_4^5 = C_4^6$.
(b) Show that $C_5^9 + C_6^9 = C_6^{10}$.
◆ **(c)** Explain why the following combinatorial identity is true:

$$C_{k-1}^n + C_k^n = C_k^{n+1}$$

(*Hint:* Consider k-member coalitions in an $m + 1$-member committee. Count those that have the chairperson as a member, and those that do not include the chairperson.)

33. A nine-member committee has a chairperson and eight ordinary members. A motion can pass if and only if it has the support of the chairperson and at least two other members, or if it has the support of all eight ordinary members.
(a) Find an equivalent weighted voting system.
(b) Determine the Banzhaf power index.
(c) Determine the Shapley – Shubik power index.

34. Consider the $2m$-person voting system in which each participant has one vote and a simple majority wins. In the notation for weighted voting systems, this system can be expressed as

$$[m + 1: 1, \ldots, 1]$$

Assume that all voting combinations are equally likely. If you are a participant in this voting system, what is the probability that you will be a critical voter, when $m = 1, 2, 3, 4, 5, 6,$ or 7?

35. The New York City Board of Estimate consists of the mayor, the comptroller, the city council president, and the presidents of each of the five boroughs. It employed a voting system in which the city officials each had two votes and the borough presidents each had one; the quota to pass a measure was six. This voting system was declared unconstitutional by the U.S. Supreme Court in 1989 (*Morris* v. *Board of Estimate*).
(a) Describe the minimal winning coalitions.
■ **(b)** Determine the Banzhaf power index.

36. Here is a proposed weighted voting system for the New York City Board of Estimate that is based on the populations of the boroughs (see Exercise 35):

$$[71: 35, 35, 35, 11.3, 7.3, 9.6, 6.0, 1.8]$$

Find a simpler system of weights that yields an equivalent voting system.

37. The United Nations Security Council has 5 permanent members – China, France, Russia, the United Kingdom, and the United States – and 10 other members that serve two-year terms. To resolve a dispute not involving a member of the Security Council, nine votes, including the votes of each of the permanent members, are required. (Thus, each permanent member has veto power.)
◆ **(a)** Show that this voting system is equivalent to the weighted voting system in which each permanent member has 7 votes, each ordinary member has 1 vote, and the quota is 39.
■ **(b)** Compute the Banzhaf index for the Security Council.

■ **(c)** Compute the Shapley–Shubik index for the Security Council. (This is harder than computing the Banzhaf index).

◆ **(d)** Which index is most appropriate for measuring power in the Security Council?

WRITING PROJECTS

1. The most important weighted voting system in the United States is the electoral college (see Spotlight 13.1). Three alternate methods to elect the president of the United States have been proposed:

- *Direct election.* The electoral college would be abolished, and the candidate receiving a plurality of the votes would be elected. Most versions of this system include a runoff election or a vote in the House of Representatives in cases where no candidate receives more than 40% of the vote.
- *District system.* In each congressional district, and in the District of Columbia, the candidate receiving the plurality would select one elector. Furthermore, in each state, including the District of Columbia, the candidate receiving the plurality would receive two electors. In effect, the unit rule would be retained for the District of Columbia and for states with a single congressional district. Larger states would typically have electors representing both parties.
- *Proportional system.* Each state and the District of Columbia would have fractional electoral votes assigned to each candidate in proportion to the number of popular votes received. Under this system, Governor Bush, who received 4,567,429 popular votes out of 10,965,822 cast in California in 2000, would have received

$$\frac{4{,}567{,}429}{10{,}965{,}822} \times 54 \text{ electoral votes}$$
$$= 22.4918 \text{ electoral votes}$$

Vice President Gore would have received

$$\frac{5{,}861{,}203}{10{,}965{,}822} \times 54 \text{ electoral votes}$$
$$= 28.8629 \text{ electoral votes}$$

The Green Party received 418,707 votes in California, and would be entitled to 2.06188 electoral votes. There were four other parties that received between them less than one electoral vote. Obviously, no actual electors would be chosen.

Should the present electoral college, operating under the unit rule, be replaced by one of these systems? A starting point to answer this question is the article by John Banzhaf III, "One Man, 3.312 . . . Votes." Another reference is *The Presidential Election Game,* by Steven Brams, which contains useful references to Senate hearings on electoral college reform.

2. Write an essay on weighted voting in the Council of Ministers of the European Community. Compute the Banzhaf and Shapley–Shubik indices for the system as it was in 1958. In later years, the number of member nations increased significantly, and you may want to use the Banzhaf calculator, available on the Web at www.whfreeman.com/fapp. If they differ significantly in their allocation of power, which index represents the true balance of power best? *The Measurement of Voting Power,* by Felsenthal and Machover, is a useful reference.

3. California has 10.22% of the votes in the electoral college, but according to Table 13.3, that state has 11.4% of the power in the electoral college, as measured by the Banzhaf index. Discuss the appropriateness of the Banzhaf index as a measure of voting power in the electoral college. Is the disproportionate power of California in the electoral college a problem that the United States should address?

Assume that California has acquired additional congressional seats as a result of migration. Calculate the Banzhaf index when California has 65, 75, and 100 electors. In each case take the electoral votes that are to be awarded to California from other states.

SPREADSHEET PROJECTS

To do these projects, go to www.whfreeman.com/fapp.

Experimentation with weighted voting systems is easy using spreadsheets. This project investigates weighted voting systems for three and four people by considering all possible voting patterns. The impact of changes to the weight of a vote or the required quota can be seen immediately using these models.

APPLET EXERCISES

To do these exercises, go to www.whfreeman.com/fapp.

A company has five partners, who hold different amounts of stock, and the votes that they cast at company meetings are related to the amount of stock each owns. This weighted voting situation is summarized by [21: 10, 7, 7, 6, 3] where 21 is the quota (the number of votes needed to pass a measure), and 10, 7, 7, 6, and 3 are the number of votes of the five partners. You can determine the Banzhaf power indices for these voters (and additional problems) in the Banzhaf Index applet.

CHAPTER 14

Fair Division

In this chapter we cover...

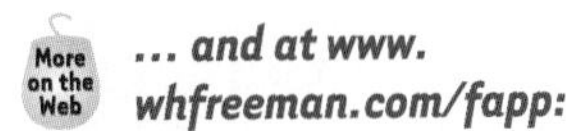

... and at www.whfreeman.com/fapp:

Flashcards

Quizzes

Spreadsheet Projects

- Knaster inheritance
- Adjusted winner procedure

Applet Exercises

- Adjusted winner divorce procedure

When the demands or desires of one party are in conflict with those of another – be it a divorce, a labor–management negotiation, or an international dispute – no one wants to be treated unfairly. And with 1.2 million divorces every year in the United States alone, and crises such as we've seen in the Middle East for decades, it is certainly worth considering how mathematics might help in the search for procedures that can ensure fair and equitable resolutions of such conflicts.

We begin this chapter with one such procedure that was developed in the mid-1990s. This procedure – called "adjusted winner" – allows two parties to settle any dispute involving either issues (as in an international dispute) or objects (as in a divorce or a two-person inheritance) with certain mathematical guarantees of "fairness." Disagreement, it turns out, is both a bad thing and a good thing. On the one hand, disagreement as to how each issue should be resolved typically lies at the heart of a conflict. On the other hand, procedures such as adjusted winner are designed to capitalize on the parties' disagreement as to the importance of each issue, thus allowing each party to end the negotiations thinking it has been met more than half way.

But adjusted winner is just one of several so-called fair-division procedures that have been developed over the past 50 years. Thus, following our discussion of adjusted winner, we describe a procedure for handling inheritances that was discovered by the Polish mathematician Bronislaw Knaster during World War II. This is followed by a discussion of an extremely basic fair-division procedure – taking turns – and the question of what the optimal strategy is when taking turns choosing objects.

Bridging the gap between fair-division procedures with obvious real-world potential, such as divorce and inheritance procedures, and procedures that address fundamental mathematical questions of fairness (as do the procedures treated later in this chapter) is the ancient two-person procedure known as divide-and-choose. An application of this procedure to the Law of the Sea Treaty is described.

Divide-and-choose sets the stage for the mathematical investigations of fair division that have gone on for the past half-century. These investigations have often been phrased within the metaphor of "cake cutting." We present three cake-cutting procedures. The first two of these—found by Steinhaus and Banach–Knaster in the 1940s—yield allocations that are "proportional," meaning that each player receives what he or she perceives to be at least his or her fair share of the cake. The last one—found by Selfridge–Conway in 1960—yields allocations that are "envy-free" in the sense that each player receives what he or she perceives to be a piece at least tied for largest.

14.1 The Adjusted Winner Procedure

To illustrate the adjusted winner procedure, we will consider an application to the multibillion-dollar world of business mergers. It turns out that one of the most elusive ingredients in the success of a merger is what deal-makers call *social issues*—how power, position, sacrifice, and status are allocated between the merging companies and their executives.

As a case in point, let us revisit the 1998 proposed merger between two giant pharmaceutical companies, Glaxo Wellcome and SmithKline Beecham. While most of the details underlying this aborted deal are unknown, the role of social issues is clearly underscored by reports that the companies "saw nearly 19 billion dollars of stock market value vanish in the clash of two corporate egos."

(Pascal Plessis/AP Photo.)

Exactly what kinds of issues might bring on a "clash of two corporate egos"? While not privy to the details of the Glaxo Wellcome–SmithKline Beecham merger attempt, we can speculate as to their nature. For purposes of illustration, let's assume that the following five social issues were paramount:

1. The name that the combined company would use
2. The location of the headquarters of the combined company
3. The question of who would serve as chairman of the combined company
4. The question of who would serve as CEO of the combined company
5. The question of where the necessary layoffs would come from

Each of these five social issues is known to have been a major factor in recent proposed mergers. For example, when Chrysler merged with Daimler-Benz, the issue of the choice of a name for the combined company was described as a "standoff" before both sides finally agreed to "DaimlerChrysler."

For the sake of illustrating the adjusted winner procedure, let's assume that these were the five social issues confronting Glaxo Wellcome and SmithKline Beecham, and let's see how adjusted winner would have suggested a resolution. The starting point–and something that is quite difficult when dealing with issues (as in a negotiation) as opposed to objects (as in a divorce)–is to have each side quantify the importance it attaches to getting its own way on each of the issues.

With adjusted winner, this quantification is done by having each side–independently and simultaneously–spread 100 points over the issues in a way that reflects the relative worth of each issue to that party. In our present example, let's assume that the companies allocated their 100 points as shown in Table 14.1. Adjusted winner is now used to decide which side gets its way on which issues, but the procedure requires that a compromise of sorts may have to be reached on one of the issues. Here's how the procedure works. Suppose we have two parties and a list of either issues to be resolved in one party's favor or the

TABLE 14.1 Applying the Adjusted Winner Procedure to a Merger of Two Companies

	Point Allocations	
Issue	**Glaxo Wellcome**	**SmithKline Beecham**
Name	5	10
Headquarters	25	10
Chairman	35	20
CEO	15	35
Layoffs	20	25
Total	100	100

other's (as in our merger example), or objects to be awarded either to one party or to the other (as in a divorce or a two-person inheritance). In order to have a single word covering both issues and objects, we will speak of "items."

The adjusted winner procedure follows these basic steps:

Step 1. As described earlier, each party distributes 100 points over the items in a way that reflects their relative worth to that party.

Step 2. Each item is initially given to the party that assigned it more points. Each party then assesses how many of his or her own points he or she has received. The party with the fewest points is now given each item on which both parties placed the same number of points.

Step 3. Since the point totals are most likely not equal, let A denote the party with the higher point total and B be the other party. Start transferring items from A to B, in a certain order, until the point totals are equal. (The point at which equality is achieved may involve a fractional transfer of one item.)

Step 4. The order in which this is done is extremely important and is determined by going through the items in order of increasing *point ratio.*

An item's point ratio is the fraction

$$\frac{\text{A's point value of the item}}{\text{B's point value of the item}}$$

where A is the party with the higher point total.

Why step 4 is so important will be explained later. Let us demonstrate the adjusted winner procedure by continuing with our analysis of the proposed merger between Glaxo Wellcome and SmithKline Beecham.

1. Assume that Glaxo Wellcome and SmithKline Beecham have given us the point assignments shown in Table 14.1.
2. Because Glaxo Wellcome has placed more points on Headquarters (25) and Chairman (35), it is initially "given" these issues, while SmithKline Beecham is initially given Name (10), CEO (35), and Layoffs (25). Notice that SmithKline Beecham now has 10 + 35 + 25 = 70 of its points, whereas Glaxo Wellcome only has 25 + 35 = 60 of its points.
3. We now start transferring issues from SmithKline Beecham to Glaxo Wellcome until the point totals of the two sides are equal. SmithKline Beecham has initially been given three issues (Name, CEO, and Layoffs), and step 4 will help us decide in what order to start transferring them.

4. Layoffs has point ratio $25/20 = 1.25$, Name has point ratio $10/5 = 2.00$, and CEO has point ratio $35/15 = 2.33$. Because Layoffs has the lowest point ratio, we start to transfer that item first.

We now see that transferring the entire Layoff item (worth 25 to SmithKline Beecham and 20 to Glaxo Wellcome) gives Glaxo Wellcome more points ($60 + 20 = 80$) than SmithKline Beecham has ($70 - 25 = 45$).

Thus, the entire Layoff item cannot be transferred; Glaxo Wellcome and SmithKline Beecham will need to compromise on the issue of layoffs. But compromise need not mean meeting each other half way. Our goal is to equalize points between the two companies, and a little algebra will tell us exactly the extent to which Glaxo Wellcome and SmithKline Beecham should get their way on the issue of layoffs. Conceptually, it's easier to think of SmithKline Beecham retaining some fraction x of the issue in question and Glaxo Wellcome receiving the complementary fraction $1-x$ of that same issue.

Because x is the fraction of the issue that SmithKline Beecham retains, the number of points that SmithKline Beecham gets from this issue is x times 25. The fraction that Glaxo Wellcome gets is $1 - x$, so the number of points it gets from this issue is $1-x$ times 20. Thus, if we want a fraction that will make SmithKline Beecham's total points and Glaxo Wellcome's total points equal, then x must satisfy the following equation:

$$10 + 35 + 25x = 25 + 35 + 20(1 - x)$$

We use algebra to solve this equation:

$$\begin{aligned} 45 + 25x &= 60 + 20 - 20x \\ 45 + 25x &= 80 - 20x \\ 45x &= 35 \\ x &= \frac{35}{45} = \frac{7}{9} \end{aligned}$$

Inserting $\frac{7}{9}$ back into the equation, we see that

$$45 + 25\left(\frac{7}{9}\right) = 60 + 20\left(\frac{2}{9}\right)$$

or approximately 64 points for each side. In rough terms, then, equality of points is essentially achieved when SmithKline Beecham gets about three-fourths ($7/9 \cong 3/4$) of its way on the issue of layoffs and Glaxo Wellcome gets about one-fourth of its way.

Having seen how the adjusted winner procedure works, one must now ask the following question: Exactly what is it about the allocation produced by this scheme that would make one want to use it? The answer is given by the following theorem (whose proof can be found in *Fair Division* by Brams and Taylor, listed in the Suggested Readings):

Theorem: For two parties, the **adjusted winner procedure** produces an allocation, based on each player's assignment of 100 points over the items to be divided, that has the following properties:

- The allocation is **equitable**: Both players receive the same number of points.
- The allocation is **envy-free:** Neither player would be happier with what the other received.
- The allocation is **Pareto-optimal:** No other allocation, arrived at by any means, can make one party better off without making the other party worse off.

Economists consider Pareto optimality (named after the nineteenth-century Italian scholar Vilfredo Pareto) to be an extremely important property, and the order of transfer in step 4 of the adjusted winner procedure is so important because it guarantees that the outcome is Pareto-optimal. The fact that the adjusted winner procedure produces an allocation that is efficient in this sense leads one to hope that it can and will play a future role in real-world dispute resolution.

14.2 The Knaster Inheritance Procedure

The adjusted winner procedure can be applied in the case of an inheritance if there are only two heirs. For *more than two heirs,* there is quite a different scheme, the **Knaster inheritance procedure,** first proposed by Bronislaw Knaster in 1945. It has a drawback, though, in that it requires the heirs to have a large amount of cash at their disposal.

EXAMPLE A Four-Person Inheritance

Suppose (for the moment) that there is just one object—a house—and four heirs—Bob, Carol, Ted, and Alice. Knaster's scheme begins with each heir bidding (simultaneously and independently) on the house. Assume, for example, that the bids are

Bob	Carol	Ted	Alice
$120,000	$200,000	$140,000	$180,000

Carol, being the high bidder, is awarded the house. Her fair share, however, is only one-fourth of the $200,000 she thinks the house is worth, and so she places

\$150,000 (which is three-fourths of the \$200,000 she bid) into a temporary "kitty."

Each of the other heirs now withdraws from the kitty his or her fair share, that is, one-fourth of his or her bid:

Bob withdraws	\$120,000/4 = \$30,000
Ted withdraws	\$140,000/4 = \$35,000
Alice withdraws	\$180,000/4 = \$45,000

Thus, from the \$150,000 kitty, a total of \$30,000 + \$35,000 + \$45,000 = \$110,000 is withdrawn, and each of the four heirs now feels that he or she has the equivalent of one-fourth of the estate. Moreover, there is a \$40,000 surplus (\$150,000 kitty − \$110,000 withdrawn), which is now divided equally among the four heirs (so each receives an additional \$10,000). The final settlement is:

Bob	Carol	Ted	Alice
\$40,000	house − \$140,000	\$45,000	\$55,000

This illustrates Knaster's procedure for the simple case in which there is only one object. But what if our same four heirs have to divide an estate consisting of, say, a house (as before), a cabin, and a boat? The easiest answer is to handle the estate one object at a time (proceeding for each object as we just did for the house). To illustrate, assume that our four heirs submit the following bids:

	Bob	Carol	Ted	Alice
House	\$120,000	\$200,000	\$140,000	\$180,000
Cabin	60,000	40,000	90,000	50,000
Boat	30,000	24,000	20,000	20,000

We have already settled the house. Let's handle the cabin the same way. Thus, Ted is awarded the cabin based on his high bid of \$90,000. His fair share is one-fourth of this, so he places three-fourths of \$90,000 (which is \$67,500) into the kitty.

Bob withdraws from the kitty \$60,000/4 = \$15,000. Carol withdraws \$40,000/4 = \$10,000, and Alice withdraws \$50,000/4 = \$12,500. Thus, from the \$67,500 kitty, a total of \$15,000 + \$10,000 + \$12,500 = \$37,500 is withdrawn.

The surplus left in the kitty is thus \$30,000, and this is again split equally (\$7500 each) among the four heirs. The final settlement on the cabin is:

Bob	Carol	Ted	Alice
\$22,500	\$17,500	cabin − \$60,000	\$20,000

If we were now to do the same for the boat (we leave the details to the reader), the corresponding final settlement would be

Bob	Carol	Ted	Alice
boat − $20,875	$7625	$6625	$6625

Putting the three separate analyses (house, cabin, and boat) together, we get a final settlement of

Bob:	boat + ($40,000 + $22,500 − $20,875 = $41,625)
Carol:	house + (−$140,000 + $17,500 + $7625 = −$114,875)
Ted:	cabin + ($45,000 − $60,000 + $6625 = −$8375)
Alice:	$55,000 + $20,000 + $6625 = $81,625.

■

Notice that here, Carol gets the house but must put up $114,875 in cash (and Ted gets the cabin but must put up $8375 in cash). This cash is then disbursed to Bob and Alice. In practice, Carol's having this amount of cash available may be a real problem–the key drawback to Knaster's procedure. Nevertheless, Knaster's procedure shows again that whenever some participants have different evaluations of some objects, there is an allocation in which everyone obtains more than a fair share.

14.3 Taking Turns

For many of us, an early lesson in fair division occurs in elementary school with the choosing of sides for a spelling bee or when picking teams on the playground. In terms of importance, these pale in comparison with the issue of property settlement in a divorce. Remarkably, however, the same fair-division procedure–*taking turns*–is often used in both.

Taking turns is fairly self-explanatory. With two parties (and that's all we'll consider here), one party selects an object, then the other party selects one, then the first party again, and so on. But in this context, there are several interesting questions that suggest themselves:

1. How do we decide who chooses first?
2. Because choosing first is often quite an advantage, shouldn't we compensate the other party in some way, perhaps by giving him or her extra choices at the next turn?
3. Should a player always choose the object he or she most favors from those that remain, or are there strategic considerations that players should take into account?

The answer to question 1 is often "toss a coin," but there are other possibilities–for example, the two parties could "bid" for the right to go first, as in an auction. The answer to question 2 is less clear, but we outline a discussion of the issue it raises in Writing Project 2. Question 3, on the other hand, is remarkably interesting, and it is this one that we want to pursue.

Let's look at an easy example. Suppose that Bob and Carol are getting a divorce, and their four main possessions, ranked from best to worst by each, are as follows:

	Bob's ranking	**Carol's ranking**
Best	Pension	House
Second best	House	Investments
Third best	Investments	Pension
Worst	Vehicles	Vehicles

If Carol knows nothing of Bob's preferences, then we can assume that she will choose sincerely—selecting at her turn whichever item she most prefers from those not yet chosen. Now, if Bob is also sincere, and if he chooses first, the items will be allocated as follows:

First turn:	Bob takes the pension.
Second turn:	Carol takes the house.
Third turn:	Bob takes the investments.
Fourth turn:	Carol is left with the vehicles.

Hence, Bob gets his first and third favorites (the pension and the investments). However, if Bob opens by choosing the house—and bypassing the pension for the moment—then the allocation will be as follows:

First turn:	Bob takes the house.
Second turn:	Carol takes the investments.
Third turn:	Bob takes the pension.
Fourth turn:	Carol is left with the vehicles.

Thus, by being insincere, Bob does better—getting his first and second favorites (the pension and the house).

In general, then, what is the optimal strategy for rational players to use, assuming that both know the preferences of the other? The answer is something called the **bottom-up strategy,** discovered by the mathematicians D. A. Kohler and R. Chandrasekarean in 1969. We will illustrate it with an example.

Suppose we have five objects—A, B, C, D, E—and Bob is choosing first. Suppose that Bob and Carol have the following rankings of the objects (called **preference lists** in what follows):

Bob	Carol
A	C
B	E
C	D
D	A
E	B

It will turn out that Bob should open with *C* (his third choice) followed by Carol's choice of *D* (skipping over *E*, for the moment). Bob will then take *A*, Carol will follow with *E*, and finally Bob will get *B*. Bob gets his first, second, and third choices without selecting his first choice first! Where does this strategy come from?

The intuition here is quite easy. Let's make two assumptions about rational players: A rational player will never choose his or her least preferred alternative, and a rational player will avoid wasting a choice on an object that he or she knows will remain available and thus can be chosen later.

With these assumptions as motivation, let's return to the preceding example and think about the mental calculation Bob will go through in deciding what his first choice will be. Bob knows the eventual sequence of choices will fill in all of the following blanks:

Bob:	_____		_____		_____
Carol:		_____		_____	

Now, working mentally from right to left, Bob knows that Carol will not choose *B*, because it is the bottom thing on her list. Thus, he will get stuck with *B*, and so he will avoid wasting anything but his last choice on alternative *B*. Thus, Bob can pencil in alternative *B* as his last choice:

Bob:	_____		_____		*B*
Carol:		_____		_____	

Bob, placing himself momentarily in Carol's shoes, knows she will reason the same way, and thus he pencils Carol in for the bottom alternative, *E*, on his list:

Bob:	_____		_____		*B*
Carol:		_____		*E*	

Mentally now, Bob reasons as if alternatives *B* and *E* never existed (and the choice sequence had been Bob–Carol–Bob) and continues to pencil in alternatives from right to left, with Bob working from bottom to top on Carol's preference list and Carol working from bottom to top on Bob's preference list. This yields the following sequence of choices mentally penciled in by Bob:

Bob:	*C*		*A*		*B*
Carol:		*D*		*E*	

Remember, this is just a mental calculation that Bob went through to decide upon the actual choice–in this case, *C*–with which he will open. Bob has no guarantee that Carol will, in fact, respond with *D*, and so the use of this strategy involves some risk on Bob's part.

This bottom-up strategy can also be viewed as a procedure that a mediator (or arbitrator) could use to specify a division of several objects between two parties. Given the preference lists of both parties, the mediator could construct a list–exactly as we did for Bob and Carol above–and then offer this to the parties as the suggested allocation. In effect, the mediator is simultaneously playing the role of two rational parties who choose to employ optimal strategies.

14.4 Divide-and-Choose

There are vast mineral resources under the seabed, all of which, one might argue, should be available to both developed and developing countries. In the absence of some kind of agreement, however, what is to prevent the developed countries from mining all of the most promising tracts before the developing countries have reached a technological level where they can begin their own mining operations? Such an agreement went into effect on November 16, 1994, with 159 signatories (including the United States). It was called the **Convention of the Law of the Sea,** and it protects the interests of developing countries by means of the following fair-division procedure.

Whenever a developed country wants to mine a portion of the seabed, that country must propose a division of the portion into two tracts. An international mining company called the Enterprise, funded by the developed countries but representing the interests of the developing countries through the International Seabed Authority, then chooses one of the two tracts to be reserved for later use by the developing countries.

The preceding rules constitute a fair-division procedure known as **divide-and-choose:** One party divides the object into two parts in any way that he desires, and the other party chooses whichever part she wants.

As a fair-division procedure, the origins of divide-and-choose go back at least 5000 years. The Hebrew Bible tells the story of Abram (later to be called Abraham) and Lot, who settled a dispute over land via a proposed division by Abram—"If you go north, I will go south; and if you go south, I will go north" (Gen. 13:8–9)—and a choice (of the plain of Jordan) by Lot. Divide-and-choose resurfaced about 2000 years later in Hesiod's book *Theogony.* The Greek gods Prometheus and Zeus had to divide a portion of meat. Prometheus began by placing the meat into two piles, and Zeus selected one.

Actually, a fair-division procedure consists of both rules and strategies, and all we have described so far are the rules of divide-and-choose. But the strategies here are quite obvious: The divider makes the two parts equal in his estimation, and the chooser selects whichever piece she feels is more valuable.

Rules and strategies differ from each other in the following sense: A referee could determine whether a rule was being followed, even without knowing the preferences of the players. Strategies represent choices of how players follow the rules, given their individual preferences (and any other knowledge and/or goals they may have).

The strategies on which we will focus in our discussion of fair-division procedures are those that require no knowledge of the preferences of the other players and yet provide some kind of minimal degree of satisfaction even in the face of collusion by the other players. For example, the strategies just given for divide-and-choose guarantee each player a piece that he or she would not wish to trade for that received by the other.

There are, to be sure, other strategic considerations that might be relevant. For example, in divide-and-choose, would you rather be the divider or the chooser? The answer, given our assumptions that nothing is known of the preferences of the others, is to be the chooser. However, if you knew the preferences of your opponent (and her value of spite), then you might want to be the divider.

As a final comment on strategic considerations, we need only look to the origins of the well-known expression "the lion's share." It comes from one of Aesop's fables, as reported by Todd Lowry in *Archaeology of Economic Ideas* (1987, p. 130):

> It seems that a lion, a fox, and an ass participated in a joint hunt. On request, the ass divides the kill into three equal shares and invites the others to choose. Enraged, the lion eats the ass, then asks the fox to make the division. The fox piles all the kill into one great heap except for one tiny morsel. Delighted at this division, the lion asks, "Who has taught you, my very excellent fellow, the art of division?" to which the fox replies, "I learnt it from the ass, by witnessing his fate."

14.5 Cake-Division Procedures: Proportionality

The modern era of fair division in mathematics began in Poland during World War II (see Spotlight 14.1). At this time, Hugo Steinhaus asked what is, in retrospect, the obvious question: What is the "natural" generalization of divide-and-choose to three or more people? The metaphor that has been used in this context, going back at least to the English political theorist James Harrington (1611–1677), is a cake. We picture different players valuing different parts of the cake differently because of concentrations of certain flavors or depth of frosting. Thus, we ask the following:

> Can one devise a **cake-division procedure** for n players—that is, a procedure that the players can use to allocate a cake among themselves (no outside arbitrators)—so that each player has a strategy that will guarantee that player a piece with which he or she is "satisfied," even in the face of collusion by the others?

As we have seen, divide-and-choose is a cake-division procedure for two players, if by "satisfied" we mean either "thinks his piece is of size or value at least one-half" or "does not want to trade what she received for what anyone else received." These two notions of satisfaction are so important that we define them precisely.

A cake-division procedure (for n players) will be called **proportional** if each player's strategy guarantees that player a piece of size or value at least $1/n$ of the whole in his or her own estimation. It will be called **envy-free** if each player's strategy guarantees that player a piece he or she considers to be at least tied for largest.

It turns out that for $n = 2$, a procedure is envy-free if and only if it is proportional; that is, for $n = 2$, the two notions of fair division are exactly the same. For $n > 2$, however, all we can say is that an envy-free procedure is automatically proportional. For example, if a three-person allocation is not proportional, then one player (call him Bob) thinks that he received less than one-third. Bob then feels that the other two are sharing more than two-thirds between them, and thus that at least one of the two (call her Carol) must have more than one-third. But then Bob will envy Carol, and so the allocation is not envy-free. Since all non-proportional allocations fail to be envy-free, it follows that if an allocation is envy-free, then it must be proportional.

Many procedures that are proportional, however, fail to be envy-free, as we shall soon show. Thus, proportional procedures are fairly easy to come by, but envy-free procedures are fairly hard to come by.

EXAMPLE

The Steinhaus Proportional Procedure for Three Players (Lone Divider)

Given three players—Bob, Carol, and Ted—we have Bob divide the cake into three pieces, call them X, Y, and Z, each of which he thinks is of size or value exactly one-third. Let's speak of Carol as "approving of a piece" if she thinks it is of size or value at least one-third. Similarly, we will speak of Ted as "approving of a piece" if the same criterion applies. Notice that both Carol and Ted must approve of at least one piece.

If there are distinct pieces—say, X and Y—with Carol approving of X and Ted approving of Y, then we give the third piece, Z, to Bob (and, of course, X to Carol and Y to Ted), and we are done. The problem case is where both Carol and Ted approve of only one piece and it is the *same* piece.

Thus, let's assume that Carol and Ted approve of only one piece, X, and hence (of more importance to us) both *disapprove* of piece Z. Let XY denote the result of putting piece X and piece Y back together to form a single piece. Notice that both Carol and Ted think that XY is at least two-thirds of the cake, because both disapprove of Z. Thus, we can give Z to Bob and let Carol and Ted use divide-and-choose on XY. Because half of two-thirds is one-third, both Carol and

SPOTLIGHT 14.1 Fifty Years of Cake Cutting

The modern era of cake cutting began with the investigations of the Polish mathematician Hugo Steinhaus during World War II. His research, and that of dozens of others over the past half-century, involved dealing with two fundamental difficulties. First, allocation schemes that work in the context of two or three players often do not generalize easily to the context of four or more players. Second, procedures that yield envy-free allocations are considerably harder to obtain than procedures that yield proportional allocations.

The mathematics inspired by these two difficulties over the past 50 years constitutes a rather elegant corner of the large and important area of fair division. Steinhaus's investigations in the 1940s led to his observation that there is a rather natural extension of divide-and-choose to the case of three players. This is the "lone-divider procedure" described on page 487. Steinhaus's method was generalized to an arbitrary number of players by Harold W. Kuhn of Princeton University in 1967.

Unable to extend his procedure from three to four players, Steinhaus proposed the problem to some Polish colleagues. Two of them, Stefan Banach and Bronislaw Knaster, solved this problem in the mid-1940s by producing the "last-diminisher procedure" described on page 489.

In addition to the procedures devised by Banach, Knaster, and Kuhn, there are other well-known constructive procedures for obtaining

Will this cake be divided fairly?

Ted are guaranteed a proportional share (as is Bob, who approved of all three pieces). ■

The procedure just described, which guarantees proportional shares but is not necessarily envy-free and is sometimes called the **lone-divider method,** was discovered by Hugo Steinhaus around 1944. Unfortunately, it does not extend easily to more than three players. It was left to Steinhaus's students, Stefan Banach and Bronislaw Knaster, to devise a method for more than three players. Picking up where Steinhaus left off (and traveling in quite a different direction), they devised the proportional procedure that today is referred to as the **last-diminisher method.** Like the lone-divider method, it is proportional but not envy-free. We illustrate it for the case of four players (Bob, Carol, Ted, and Alice), and we include both the rules and the strategies that guarantee each player his or her fair share.

a proportional allocation among four or more players. One of these is due to A. M. Fink of Iowa State University and appears in Exercise 25.

Another constructive procedure of note, although different in flavor from the others, is the 1961 recasting by Lester E. Dubins and Edwin H. Spanier of the University of California at Berkeley of the last-diminisher method as a "moving-knife procedure" (illustrated in Exercise 27). The trade-off here involves giving up the "discrete" nature of the last-diminisher method in exchange for the conceptual simplicity of the moving knife.

Although the existence of an envy-free allocation (even for four or more players) was known to Steinhaus in the 1940s, the first constructive procedure for producing an envy-free allocation among three players was not found until around 1960. At that time, John L. Selfridge of Northern Illinois University and, later but independently, John H. Conway of Princeton University found the elegant procedure presented on page 493. Although never published by either, the procedure was quickly and widely disseminated by Richard K. Guy of the University of Calgary and others; eventually it appeared in several treatments of the problem by different authors.

In 1980, a moving-knife procedure for producing an envy-free allocation among three players was found by Walter R. Stromquist of Daniel Wagner Associates. Then, another procedure, capable of being recast as a moving-knife solution of the three player case, was found by a law professor at the University of Virginia, Saul X. Levmore, and a former student of his, Elizabeth Early Cook.

In 1992, Steven J. Brams, a political scientist at New York University, and Alan D. Taylor, a mathematician at Union College, succeeded in finding a constructive procedure for producing an envy-free allocation among four or more players. In 1994, Brams, Taylor, and William S. Zwicker (also from Union College) found a moving-knife solution to the four-person envy-free problem. No moving-knife procedure is known that will produce an envy-free allocation among five or more players.

EXAMPLE The Banach–Knaster Proportional Procedure for Four or More Players (Last Diminisher)

Bob cuts from the cake a piece that he thinks is of size one-fourth and hands it to Carol. If Carol thinks the piece handed her is larger than one-fourth, she trims it to size one-fourth in her estimation, places the trimmings back on the cake, and passes the diminished piece to Ted. If Carol thinks the piece handed her is of size at most one-fourth, she passes it unaltered to Ted.

Ted now proceeds exactly as did Carol, trimming the piece to size one-fourth if he thinks it is larger than this and passing it (diminished or unaltered) on to Alice. Alice does the same, but, being the last player, simply holds onto the piece momentarily instead of passing it to anyone.

Notice that everyone now thinks the piece is of size at most one-fourth, and the last person to trim it (or Bob, if no one trimmed it) thinks the piece is of size exactly one-fourth. Thus, the procedure now allocates this piece to the last person who trimmed it (and to Bob if no one trimmed it).

Assume for the moment that it was Ted who trimmed the piece last, and so he takes this piece and exits the game. Bob, Carol, and Alice all think that at least three-fourths of the cake is left, and so they can start the process over with (say) Bob beginning by cutting a piece from what remains that he thinks is one-fourth of the original cake. Carol and Alice are both given a chance to trim it to size one-fourth in their estimation, and again, the last one to trim it takes that piece and exits the game. The two remaining players both think that at least half the cake is left, and so they can use divide-and-choose to divide it between themselves and thus be assured of a piece that is of size at least one-fourth in their estimation. ■

For a concrete illustration of the last-diminisher method for the simple case where there are only three players (Bob, Carol, and Ted), suppose that all three players view the cake as having 18 units of "value," with each unit of value represented by a small square. Suppose, however, that the players value various parts of the cake differently (or that Bob views the cake as being perfectly rectangular, whereas Carol and Ted see it as skewed in opposite ways). We represent this pictorially as follows:

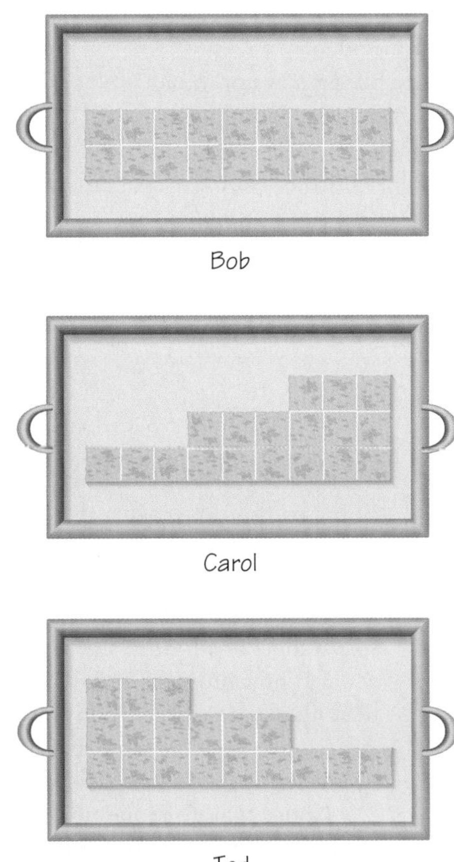

In step 1, Bob cuts from the cake a piece—call it A—that he considers to be of size or value one-third (because there are only three players this time). We'll assume he does this by making a vertical cut as follows:

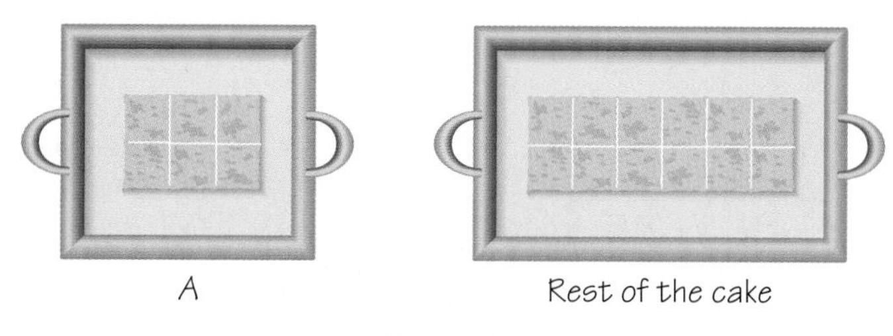

Bob's view

From Carol's point of view (or value system), the piece A appears to contain only 3 of the 18 units of value:

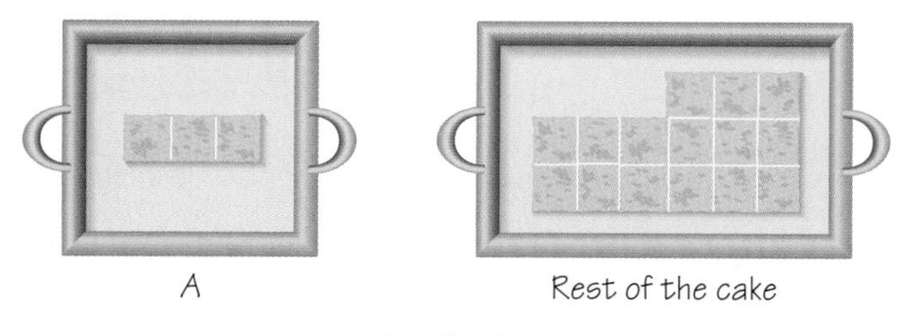

Carol's view

Because Carol thinks that A represents less than one-third of the cake, she passes A unaltered to Ted. Now Ted sees A as follows:

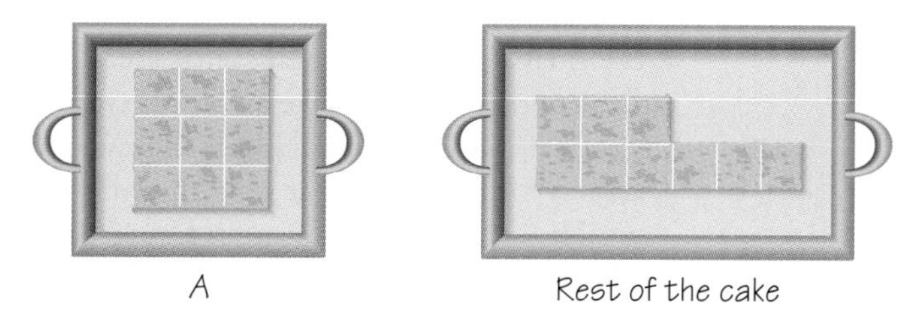

Ted's view

Thus, Ted thinks that A represents one-half of the cake (9 out of 18 units of value), and so he will trim it to what he thinks is one-third (6 out of 18 units of value). Let's assume that he does this with another vertical cut, with the trimmed version of A now called A'.

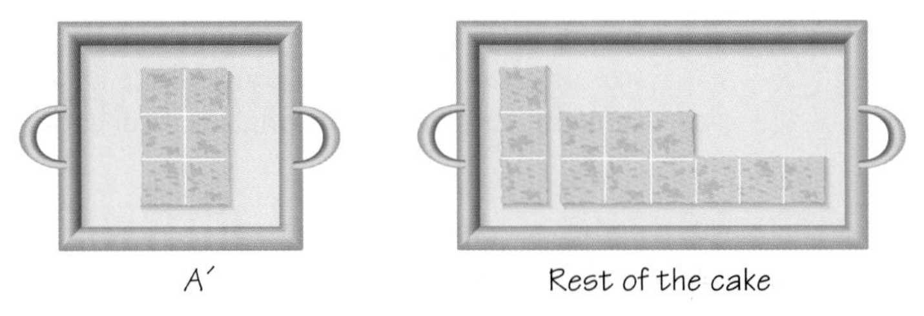

Ted's view

Everyone has now had a chance to diminish the piece A that Bob initially cut from the cake. Portion A', therefore, goes to the last person to trim it, namely, Ted. So Ted takes A' (which he thinks is of size $\frac{6}{18} = \frac{1}{3}$) and exits the game. Notice that what is left is seen by Bob and Carol as follows:

Bob's view
(14 units of value)

Carol's view
(16 units of value)

The final step has Bob and Carol use divide-and-choose. Note that if Bob is the divider and elects to make a vertical cut (halving the middle column of squares), then Carol will see the division as leaving 6 squares to the left of the cut and 10 squares to the right of the cut. She will thus choose the piece to the right of the cut. Hence, the final allocation finds Ted thinking he has 6/18 of the cake, Bob thinking he has 7/18 of the cake, and Carol thinking she has 10/18 of the cake.

14.6 Cake-Division Procedures: The Problem of Envy

Divide-and-choose has a property that neither of the last two procedures possesses: It can ensure that each player receives a piece of cake he or she considers the largest or tied for the largest. In the case of only two players, this means that each player can get what he or she perceives to be at least half the cake, no matter what the other player does. Thus, divide-and-choose is an envy-free procedure.

Steinhaus's $n = 3$ proportional procedure (the lone-divider method) is not envy-free. For example, consider the case where Carol and Ted both find one piece unacceptable (and this piece is given to Bob). Carol and Ted will not envy each other when one divides and the other chooses, but Bob may think that this is not a 50–50 split. Indeed, if Bob divided the cake initially into what he thought was three equal pieces, an unequal split of the remaining two-thirds of the cake by Carol and Ted means the Bob will prefer the larger of these two pieces to the one-third he got. Consequently, Bob will envy the person who got this larger piece.

Neither is the last-diminisher method envy-free. For example, if Bob initially cuts a piece of cake of size one-fourth, and no one else trims it, then Bob receives this piece and exits the game. If Carol is the one to make the next initial cut, she may well cut a piece from the cake that she thinks is of size one-fourth but that Bob thinks is of size considerably more than one-fourth. But Bob is out of the game. Thus, if Ted and Alice think this piece is of size less than one-fourth, then Carol receives it, and so Bob will envy Carol.

Nevertheless, there do exist cake-division procedures that are envy-free. We present one of these in what follows.

EXAMPLE The Selfridge–Conway Envy-Free Procedure for Three Players

We start with a cake and three people. The point we wish to arrive at is an envy-free allocation of the entire cake among the three people in a finite number of steps. This task may seem formidable; however, quite often in mathematics, an important part of solving a problem involves breaking the problem into identifiable parts. In this case, let us call our starting point A and the final point we wish to reach C. Now let us identify an appropriate in-between point B that makes going from A to C–via B–more manageable. Our in-between point B is the following:

> *Point B:* Getting a constructive procedure that gives an envy-free allocation of *part* of the cake.

Can we constructively obtain three pieces of cake, whose union may not be the whole cake, which can be given to the three people so that each thinks he or she received a piece at least tied for largest? This turns out to be quite easy, with the solution due to John Selfridge and John Conway, who arrived at it independently around 1960. The following process and strategies do the trick:

1. Player 1 cuts the cake into three pieces he considers to be the same size. He hands the three pieces to player 2.
2. Player 2 trims at most one of the three pieces so as to create at least a two-way tie for largest. Setting the trimmings aside, player 2 hands the three pieces (one of which may have been trimmed) to player 3.

3. Player 3 now chooses, from among the three pieces, one that he considers to be at least tied for largest.
4. Player 2 next chooses, from the two remaining pieces, one that she considers to be at least tied for largest, with the proviso that if she trimmed a piece in step 2, and player 3 did not choose this piece, then she must now choose it.
5. Player 1 receives the remaining piece.

Let us reconsider the five steps of this trimming procedure to assure ourselves that each player experiences no envy. Recall that player 1 cuts the cake into three pieces, and player 2 trims one of these three pieces. Now player 3 chooses, and, as the first to choose, he certainly envies no one. Player 2 created a two-way tie for largest, and at least one of these two pieces is still available after player 3 selects his piece. Hence, player 2 can choose one of the tied pieces she created and will envy no one. Finally, player 1 created a three-way tie for largest and, because of the proviso in step 4, the trimmed piece is not the one left over. Thus, player 1 can choose an untrimmed piece and therefore will envy no one.

So far we have gone from point A to point B: Starting with a cake and three players, we have constructively obtained (in finitely many steps) an envy-free allocation of all of the cake, except the part T that player 2 trimmed from one of the pieces. We will now describe how T can be allocated among the three players in such a way that the resulting allocation of the whole cake is envy-free. (This is the rest of the **Selfridge–Conway envy-free procedure.**)

The key observation for the $n = 3$ case is that player 1 will not envy the player who received the trimmed piece, even if that player were to be given all of T. Recall that player 1 created a three-way tie and received an untrimmed piece. The union of the trimmed piece and the trimmings yields a piece that player 1 considers to be exactly the same size as the one he received. Thus, assume that it is player 3 who received the trimmed piece (it could as well be player 2). Then player 1 will not envy player 3, no matter how T is allocated.

The next step ensures that neither player 2 nor player 3 will envy another player when it comes time to allocate T. Let player 2 cut T into three pieces she considers to be the same size. Let the players choose which of the three pieces they want in the following order: player 3, player 1, player 2.

To see that this yields an envy-free allocation, notice that player 3 envies no one, because he is choosing first. Player 1 does not envy player 2, because he is choosing ahead of her; and player 1 does not envy player 3 because, as pointed out earlier, player 1 will not envy the player who received the trimmed piece. Finally, player 2 envies no one, because she made all three pieces of T the same size.

Hence, for $n = 3$, the Selfridge–Conway procedure will give an envy-free allocation of all the cake except T, followed by an allocation of T that gives an envy-free allocation of all the cake. ■

A naive attempt to generalize to $n = 4$ what we have done for $n = 3$ would proceed as follows: We would begin by having player 1 cut the cake into four pieces he considers to be the same size. Then we would have players 2 and 3 trim some pieces (but how many?) to create ties for the largest. Finally, we would have the players choose from among the pieces – some of which would have been trimmed – in the following order: player 4, player 3, player 2, player 1.

This approach fails because player 1 could be left in a position of envy. In order to understand how the approach could fail, consider how many pieces player 3 might have to trim in order to create a sufficient supply of pieces tied for largest so that he is guaranteed to have one available when it is his turn to choose. Player 3 might have to trim one piece to create a two-way tie for largest. Player 2 might need to trim two pieces to create a three-way tie for largest (because, if there were only a two-way tie for largest, player 3 might further trim one of these pieces and player 4 might choose the other). This leaves player 1 in a possible position of envy, because we could have a situation where player 2 trims two pieces and player 3 trims a third piece, and player 4 then chooses the only untrimmed piece. If this happens, player 1, by being forced to choose a trimmed piece, will definitely envy player 4.

All is not lost, however, because there are modifications of the Selfridge–Conway procedure that will work for arbitrary n. For more on this, see *Fair Division* in the Suggested Readings, and for further discussion of the Selfridge–Conway procedure, go to www.whfreeman.com/fapp.

Although we have used the metaphor of cake cutting throughout our discussion of the problem of envy, the idea of successive trimming is nonetheless applicable to problems of fair division other than parceling out the last crumbs of a cake. The main practical problem in applying the trimming procedure is that many fair-division problems involve goods that cannot be divided up at all, much less trimmed in fine amounts. Such goods are said to be *indivisible.*

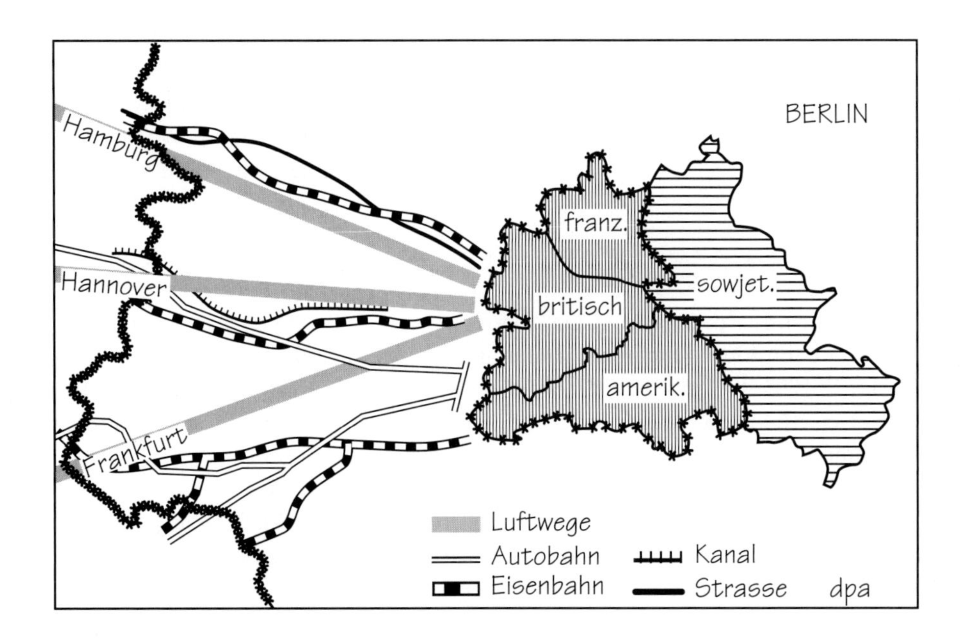

It is interesting to recall that when the Allies agreed in 1944 to partition Germany into sectors after World War II (first stage), they initially did not reach agreement about what to do with Berlin. Subsequently, they decided to partition Berlin itself into sectors (second stage), even though this city fell 110 miles within the Soviet sector. Berlin was simply too valuable a "piece" for the Western Allies (Great Britain, France, and the United States) to cede to the Soviets, which suggests how, after a leftover piece is trimmed off, it can be subsequently divided under the trimming procedure.

Yet, what if a large piece like Berlin is not divisible? In the settlement of an estate, this might be the house, which may be worth half the estate to the claimants. In this situation, there may be no alternative but to sell this big item and use the proceeds to make the remaining estate more liquid or, in our terms, "trimmable."

REVIEW VOCABULARY

Adjusted winner procedure A fair-division procedure introduced by Steven Brams and Alan Taylor in 1993. It works only for two players, and begins by having each player independently spread 100 points over the items to be divided so as to reflect the relative worth of each object to that player. The allocation resulting from this procedure is equitable, envy-free, and Pareto-optimal. It requires no cash from either player, but one of the objects may have to be divided or shared by the two players.

Bottom-up strategy A bottom-up strategy is a strategy under an alternating procedure in which sophisticated choices are determined by working backwards.

Cake-division procedure A fair-division procedure that uses a cake as a metaphor. Such procedures involve finding allocations of a single object that is finely divisible, as opposed to the situation encountered with either the adjusted winner procedure or Knaster's procedure. In a cake-division procedure, each player has a strategy that will guarantee that player a piece with which he or she is "satisfied," even in the face of collusion by the others.

Convention of the Law of the Sea An agreement based on divide-and-choose that protects the interests of developing countries in mining operations under the sea.

Divide-and-choose A fair-division procedure for dividing an object or several objects between two players. This method produces an allocation that is both proportional and envy-free (the two being equivalent when there are only two players).

Envy-free A fair-division procedure is said to be envy-free if each player has a strategy that can guarantee him or her a share of whatever is being divided that is, in the eyes of that player, at least as large (or at least as desirable) as that received by any other player, no matter what the other players do.

Equitable An allocation (resulting from a fair-division procedure like adjusted winner) is said to be equitable if each player believes he or she received the same fractional part of the total value.

Knaster inheritance procedure A fair-division procedure for any number of parties that begins by having each player (independently) assign a dollar value (a "bid") to the item or items to be divided so as to reflect the absolute worth of each object to that player. The allocation resulting from this procedure leaves each party feeling that he or she received a dollar value at least equal to his or her fair share (and often more so). It never requires the dividing or sharing of an object, but it may require that the players have a large amount of cash on hand.

Last-diminisher method A cake-division procedure introduced by Stefan Banach and Bronislaw Knaster in the 1940s. It works for any number of players and produces an allocation that is proportional but not, in general, envy-free.

Lone-divider method A cake-division procedure introduced by Hugo Steinhaus in the 1940s. It works only for three players and produces an allocation that is proportional but not, in general, envy-free.

Pareto-optimal When no other allocation, achieved by any means whatsoever, can make any one player better off without making some other player worse off.

Preference lists Rankings of the items to be allocated, from best to worst, by each of the participants.

Proportional A fair-division procedure is said to be proportional if each of n players has a strategy that can guarantee that player a share of whatever is being divided that he or she considers to be at least $1/n$ of the whole in size or value.

Selfridge–Conway envy-free procedure A cake-division procedure introduced independently by John Selfridge and John Conway around 1960. It works only for three players but produces an allocation that is envy-free (as well as proportional).

Taking turns A fair-division procedure in which two or more parties alternate selecting objects.

SUGGESTED READINGS

BRAMS, S. J., P. H. EDELMAN, and P. C. FISHBURN. Paradoxes of fair division, *Journal of Philosophy* (June 2001): 300–314. This article discusses eight paradoxes that involve unexpected conflicts among fair-division criteria, such as Pareto optimality and envy-freeness, in dividing up indivisible goods.

BRAMS, S. J., and A. D. TAYLOR. *The Win–Win Solution: Guaranteeing Fair Shares to Everybody,* Norton, New York, 1999. Brams and Taylor do more with adjusted winner, as well as divide-and-choose and taking turns.

BRAMS, S. J., and A. D. TAYLOR. *Fair Division: From Cake-Cutting to Dispute Resolution,* Cambridge

University Press, Cambridge, 1996. Brams and Taylor provide a book-length treatment of the kinds of topics introduced in this chapter, as well as divide-and-choose in the political arena, moving-knife procedures for cake cutting, and fairness as it applies to different auction and election procedures.

BRAMS, S. J., and A. D. TAYLOR. An envy-free cake division protocol, *American Mathematical Monthly,* 102 (1995): 9–18. Brams and Taylor describe in detail the finite version of their envy-free procedure for $n = 4$; in addition, they review earlier work on "protocols" (step-by-step procedures) that led up to their constructive solution of the envy-freeness problem for $n > 3$.

DUBINS, L. E., and E. H. SPANIER. How to cut a cake fairly, *American Mathematical Monthly,* 68 (1961): 1–17. This article gives some extensions of the simple fair-division concepts introduced in this chapter.

KUHN, H. W. On games of fair division. In Martin Shubik (ed.), *Essays in Mathematical Economics,* Princeton University Press, Princeton, N.J., 1968, pp. 29–37. This article provides extensions of, and additional references to, the fair-division concepts presented in this chapter.

ROBERTSON, J., and W. WEBB. *Cake Cutting Algorithms: Be Fair if You Can,* A. K. Peters, Wellesley, Mass., 1998. Robertson and Webb cover a great deal of cake-cutting ground in a text that includes exercises.

STEINHAUS, H. *Mathematical Snapshots,* Oxford University Press, Oxford, 1960. A brief but significant introduction to both fair division and apportionment is provided in this popular book on interesting mathematical topics.

SU, F. E. Rental harmony: Sperner's lemma in fair division, *American Mathematical Monthly* 106 (1999): 930–942. This recent article won the Mathematical Association of America's Merten M. Hasse Prize for outstanding mathematical exposition.

SUGGESTED WEB SITES

www.math.hmc.edu/~su/fairdivision/ This is called "The Fair Division Page." Put together by Francis Su of Harvey Mudd College, it has a number of interesting links related to work done by Su and his students.

www.maa.org/mathland/mathland_5_13.html At this site, Ivars Peterson discusses some recent work by Bryan Dawson of Emporia State University in Kansas on taking turns and the sports draft.

www.mcn.org/c/rsurratt/conflict.html This page is called "The Mediator." It contains fair and envy-free conflict resolution methods inspired by the adjusted winner procedure.

SKILLS CHECK

1. In a fair-division procedure, the goal for each participant is to (I) receive an identical portion or (II) receive what is perceived as a fair portion.

(a) I
(b) II only
(c) Both I and II

2. Chris and Terry must make a fair division of three objects. They assign points to the objects and use the adjusted winner procedure. What does Chris end up with?

Object	Chris	Terry
Boat	30	20
Land	50	60
Car	20	20

(a) Boat and car
(b) Boat and land
(c) Boat, car, and part of the land

3. If you use the Knaster inheritance procedure to divide a single object and if each person bids a different price, then

(a) each person will receive more than his or her perceived fair share.
(b) some persons, but not every person, will receive more than their perceived fair share.
(c) the successful bidder is the only person who will receive more than his or her perceived fair share.

4. Chris and Terry use the Knaster inheritance procedure to divide a coin collection. Chris bids $1000 and Terry bids $800. What is the outcome?

(a) Chris gets the coins and pays Terry $200.
(b) Chris gets the coins and pays Terry $450.
(c) Chris gets the coins and pays Terry $500.

5. Four children bid on two objects. Using the Knaster inheritance procedure, what does Adam end up with after the division?

Object	Adam	Beth	Carl	Dietra
House	$80,000	$75,000	$90,000	$60,000
Car	$10,000	$12,000	$13,000	$15,000

(a) House and cash
(b) Car and cash
(c) Cash only

6. Two people use the divide-and-choose procedure to divide a field. Suppose Jeff divides and Karen chooses. Which statement is true?

(a) Karen always believes she gets more than her fair share.
(b) Karen can guarantee that she always gets at least her fair share.
(c) Karen can possibly believe she gets less than her fair share.

7. Suppose seven people will share a cake using the last-diminisher method. To begin, Scott cuts a piece and passes it to each of the other six people, but no one trims the piece. What happens next?

(a) Scott gets this piece.
(b) The last person who is handed the piece keeps it.
(c) The piece is returned to the cake and someone else cuts a piece.

8. Using the Steinhaus procedure for three players (lone divider), what happens if there is a single portion that is the only one approved of by both non-dividers?

(a) One of the other portions is given to the divider.
(b) The two nondividers flip a coin to determine who receives the approved portion.
(c) All portions are returned to the cake and a different person serves as the new divider.

9. Using the Steinhaus proportional procedure for three players (lone divider), what happens if the two nondividers approve different portions?

(a) Each nondivider receives a portion that he or she has approved.
(b) The divider receives his or her choice of the portions.
(c) The divider selects portions for each person.

10. Using the Banach–Knaster proportional procedure for four or more players (last diminisher), what happens to the first portion after each person has inspected and possibly trimmed it?

(a) The portion goes to the last person to approve the portion, whether or not it was trimmed.
(b) The portion goes to the last person to trim the portion.
(c) The portion goes to the first person to approve and not trim the portion.

11. Using the Banach–Knaster proportional procedure for four or more players (last diminisher), what does the player who receives the first portion then do?

(a) That player becomes the first person to approve the second portion.
(b) That player becomes the last person to approve the second portion.
(c) That player leaves the game.

12. Using the Banach–Knaster proportional procedure for four or more players (last diminisher), what happens when only two people remain?

(a) One player chooses the scraps and the other chooses the portion.
(b) One player separates the remainders into two portions and the other chooses.
(c) The players flip a coin to decide who takes the remainders.

13. For the Selfridge–Conway envy-free procedure for three players, which of the following statements is true?

(a) Each of the three players has the opportunity to trim the portions if they appear to be unfair.
(b) Each player receives a portion that he or she believes to be exactly one-third of the total.
(c) The first player may believe that the third player received more than a fair share.

14. For the Selfridge–Conway envy-free procedure for three players, which of the following statements is true?

(a) The first player will always receive one of the three portions as originally cut.
(b) The second player will always receive one of the trimmed portions.
(c) The third player will always believe that he or she received exactly a fair share.

15. A fair-division procedure is "envy-free" when each player believes that

(a) every player received an equal portion.
(b) he or she received at least a fair share.
(c) no other player received more than he or she did.

EXERCISES

▲ Optional. ■ Advanced. ◆ Discussion.

The Adjusted Winner Procedure

1. The 1991 divorce of Ivana and Donald Trump was widely covered in the media. The marital assets included a 45-room mansion in Greenwich, Connecticut; the 118-room Mar-a-Lago mansion in Palm Beach, Florida; an apartment in the Trump Plaza; a 50-room Trump Tower triplex; and just over $1 million in cash and jewelry. Assume points are assigned as follows:

	Point allocations	
Marital asset	**Donald's points**	**Ivana's points**
Connecticut estate	10	38
Palm Beach mansion	40	20
Trump Plaza apartment	10	30
Trump Tower triplex	38	10
Cash and jewelry	2	2

Use the adjusted winner procedure to determine a fair allocation of the marital assets. (Exercise 1 courtesy of Catherine Duran.)

2. Suppose that Calvin and Hobbes discover a sunken pirate ship and must divide their loot. They assign points to the items as follows:

Object	Calvin's points	Hobbes's points
Cannon	10	5
Anchor	10	20
Unopened chest	15	20
Doubloon	11	14
Figurehead	20	30
Sword	15	6
Cannon ball	5	1
Wooden leg	2	1
Flag	10	2
Crow's nest	2	1

Use the adjusted winner procedure to determine a fair allocation of the loot. (Exercise 2 courtesy of Erica DeCarlo.)

3. This exercise illustrates how the adjusted winner procedure can be used to resolve disputes as well as to achieve fair allocations. Suppose Mike and Phil are roommates in college, and they encounter serious conflicts during their first week at school. Their resident adviser decides to use the adjusted winner procedure to resolve the dispute. The issues agreed upon, and the (independently assigned) points, turn out to be the following:

Issue	Mike's points	Phil's points
Stereo level	4	22
Smoking rights	10	20
Room party policy	50	25
Cleanliness	6	3
Alcohol use	15	15
Phone time	1	8
Lights-out time	10	2
Visitor policy	4	5

Use the adjusted winner procedure to resolve this dispute. (Exercise 3 courtesy of Erica DeCarlo.)

4. Make up an example involving two people and several *objects* for which the adjusted winner procedure can be used, and then use the adjusted winner procedure to determine a fair division.

5. Make up an example involving two people and several *issues* for which the adjusted winner procedure can be used, and then use the adjusted winner procedure to determine a fair resolution of the dispute.

The Knaster Inheritance Procedure

6. If John bids \$28,225 and Mary bids \$32,100 on their aging parents' old classic car, which they no longer drive, how would you reach a fair division?

7. John and Mary inherit their parents' old house and classic car. John bids \$28,225 on the car and \$55,900 on the house. Mary bids \$32,100 on the car and \$59,100 on the house. How should they arrive at a fair division?

8. Can you modify your fair-division procedure in Exercise 7 so that both John and Mary receive one of the two objects while still considering the allocation as fair?

9. Describe a fair division for three heirs, *A*, *B*, and *C*, who inherit a house in the city, a small farm, and a valuable sculpture, and who submit sealed bids (in dollars) on these objects as follows:

	A	*B*	*C*
House	145,000	149,999	165,000
Farm	135,000	130,001	128,000
Sculpture	110,000	80,000	127,000

10. Describe a fair division for three children, *E*, *F*, and *G*, who inherit equal shares of their parents' classic car collection and who submit sealed bids (in dollars) on these five cars as follows:

	E	*F*	*G*
Duesenberg	18,000	15,000	15,000
Bentley	18,000	24,000	20,000
Ferrari	16,000	12,000	16,500
Pierce-Arrow	14,000	15,000	13,500
Cord	24,000	18,000	22,000

Taking Turns

11. Suppose that Bob and Carol rank a series of objects, from most preferred to least preferred, as follows:

Bob	Carol
Car	Boat
Investments	Investments
CD player	Car
Boat	Washer–dryer
Television	Television
Washer–dryer	CD player

Assume that Bob and Carol use the bottom-up strategy and that Bob gets to choose first. Determine Bob's first choice and the final allocation.

12. Repeat Exercise 11 under the assumption that Carol gets to choose first.

13. Mark and Fred have inherited a number of items from their parents' estate, with no indication of who gets what. They rank the items from most preferred to least preferred as follows:

Mark	Fred
Truck	Boat
Tractor	Tractor
Boat	Car
Car	Truck
Tools	Motorcycle
Motorcycle	Tools

Assume that Mark and Fred use the bottom-up strategy and that Mark gets to choose first. Determine Mark's first choice and the final allocation.

14. Repeat Exercise 13 under the assumption that Fred gets to choose first.

Divide-and-Choose

15. Suppose that Bob, Carol, and Ted view a cake as pictured in the example illustrating the Banach–Knaster procedure (see page 489). Assume that all cuts that will be made are vertical.

(a) If Bob and Carol use divide-and-choose to divide the cake between them, how large a piece will each receive (assuming they follow the suggested strategies that go with divide-and-choose)?
(b) If Carol and Ted use divide-and-choose to divide the cake between them, how large a piece will each receive (assuming they follow the suggested strategies that go with divide-and-choose)?

16. If you and another person are using divide-and-choose to divide something between you, would you rather be the divider or the chooser? (Assume that neither of you knows anything about the preferences of the other.)

17. Suppose that Bob and Carol view a cake as pictured in the example illustrating the Banach–Knaster procedure (see page 489). Assume that all cuts that will be made are vertical. Assume that Bob and Carol each know how the other values the cake, and that neither is spiteful. Suppose they are to divide the cake using the *rules,* but not necessarily the *strategies,* of divide-and-choose.

(a) Is Bob better off being the divider or the chooser? Why?
(b) Discuss this in relation to Exercise 16.

18. Suppose that Bob is entitled to one-fourth of a cake and Carol is entitled to three-fourths. In a few sentences, explain how divide-and-choose can be used to achieve an allocation in which each party is guaranteed of receiving at least as much as he or she is entitled to.

Cake-Division Schemes: Proportionality

19. Suppose that players 1, 2, and 3 view a cake as follows:

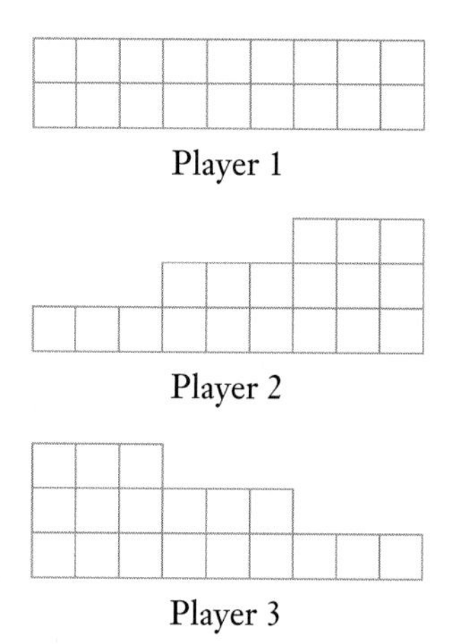

Notice that each player views the cake as having 18 square units of area (or value). Assume that each player regards a piece as acceptable if and only if it is at least $18/3 = 6$ square units of area (his or her "fair share"). Assume also that all cuts made correspond to vertical lines.

(a) Provide a total of three drawings to show how each player views a division of the cake by player 1 into three pieces he or she considers to be the same size or value. Label the pieces A, B, and C.
(b) Identify two of these pieces that player 2 finds acceptable and two that player 3 finds acceptable.
(c) Show that a feasible assignment of fair pieces can be achieved by letting the players choose in the following order: player 3, player 2, player 1. Indicate how many square units of value each player thinks he or she received. Is there any other

order in which players can choose pieces (in this example) that also results in a feasible assignment?

20. Suppose that players 1, 2, and 3 view a cake as follows:

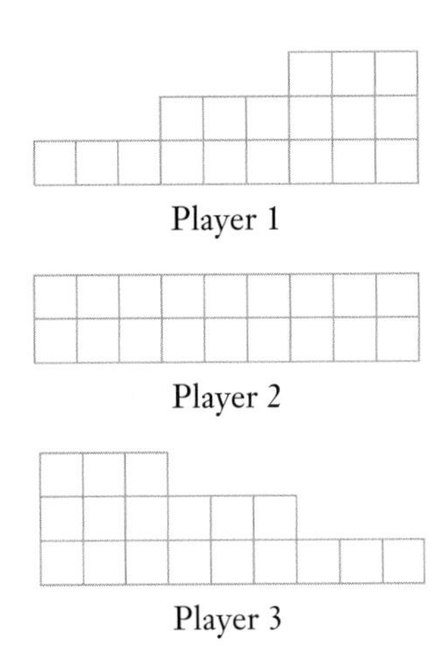

(a) Provide a total of three drawings to show how each player views a division of the cake by player 1 into three pieces he or she considers to be the same size or value. Label the pieces A, B, and C. (We are still assuming that all cuts correspond to vertical lines, so this will require a cut along a vertical center line of some of the squares.)
(b) Show that neither player 2 nor player 3 finds more than one of the three pieces acceptable (with "acceptable" defined as in Exercise 19).
(c) Identify a single piece that player 2 and player 3 agree is *not* acceptable. (There are actually two such pieces; for definiteness, find the one on the right.)
(d) Assume that players 2 and 3 give the piece from part (c) to player 1. Suppose they reassemble the rest and players 2 and 3 divide it between themselves using divide-and-choose (with a single vertical cut). Determine what size piece each of the three players will think he or she received (1) if player 2 divides and player 3 chooses, and (2) if player 3 divides and player 2 chooses.

21. Suppose players 1, 2, and 3 view a cake as in Exercise 20. Illustrate the last-diminisher method (still restricting attention to vertical cuts and, in addition, assuming that the piece potentially being diminished is a piece off the left side of the cake) by following steps (a) through (h) below:

(a) Draw a picture showing the third of the cake (6 squares) that player 1 will slice off the cake.
(b) Determine whether player 2 will pass or further diminish this piece. If he or she would further diminish it, make a new drawing.
(c) Determine whether player 3 will pass or further diminish this piece. If he or she would further diminish it, make a new drawing.
(d) Determine who receives the piece cut off the cake and what size or value he or she thinks it is. (Actually, we knew what size the person receiving this first piece would think it was, assuming he or she followed the prescribed strategy. How did we know this?)
(e) Finish the last-diminisher method using divide-and-choose on what remains, with the lowest-numbered player who remains doing the dividing.
(f) Redo step (e) with the other player doing the dividing.
(g) Redo step (e), but with the last two players using the last-diminisher method directly, instead of divide-and-choose [with the order as in step (e)].
(h) Redo step (g) with the order reversed.

22. Suppose players 1, 2, and 3 view the cake as in Exercise 20. Illustrate the envy-free procedure for $n = 3$ (yielding an allocation of part of the cake) by following steps (a) through (c) below. Again, restrict attention to vertical cuts.

(a) Provide a total of three drawings to show how each player views a division of the cake by player 1 into three pieces he or she considers to be the same size or value. Label the pieces A, B, and C. (This is the same as Exercise 20a.)
(b) Redraw the picture from player 2's view, and illustrate the trimming of piece A that he or she would do. Label the trimmed piece A' and the actual trimmings T.
(c) Indicate which piece each player would choose (and what he or she thinks its size is) if the players choose in the following order: player 3, player 2, player 1, according to the envy-free procedure on

pages 493–494. Does the proviso in step 4 come into play here?

23. Apply the remainder of the Selfridge–Conway procedure from pages 493–494 to what was obtained in Exercise 22 by completing (a) through (c) below:

(a) Draw a picture of T from each player's view.
(b) The procedure calls for the player (other than player 1) who did not receive the trimmed piece to divide T into three pieces he or she considers to be the same size. Here, that would be player 2. Illustrate this division, and label the pieces X, Y, Z.
(c) Indicate which parts of T (and the sizes or values) the players will choose when they go in the following order: player 3, player 1, player 2.

24. One often hears of the importance of "process" versus "product," the latter referring to *what* is achieved and the former referring to *how* it was achieved. In a couple of sentences, comment on the relevance of this to fair division as illustrated by the following rough paraphrasing of an exchange between two old friends, Ralph Kramden (played by Jackie Gleason) and Ed Norton (played by Art Carney) in the 1950s sitcom *The Honeymooners*.

Ralph to Ed (as the two are sitting alone at the dinner table): I can't believe you did that.

Ed: Did what, Ralph?

Ralph: There were two potatoes there, and you reached right out and took the big one.

Ed: What would you have done, Ralph?

Ralph: Why, I'd have taken the little one.

Ed: You *got* the little one, Ralph.

Additional Exercises

25. The Banach–Knaster last-diminisher method is not the only well-known cake-division procedure that yields a proportional allocation for any number of players. There is also one due to A. M. Fink (sometimes called the *lone-chooser method*). For three players (Bob, Carol, and Ted) it works as follows:

(i) Bob and Carol divide the cake into two pieces using divide-and-choose.
(ii) Bob now divides the piece he has into three parts that he considers to be the same size. Carol does the same with the piece she has.
(iii) Ted now chooses whichever of Bob's three pieces that he (Ted) thinks is largest, and Ted chooses whichever of Carol's three pieces that he thinks is largest.
(iv) Bob keeps his remaining two pieces, as does Carol.

(a) Explain why Ted thinks he is getting at least one-third of the cake.
(b) Explain why Bob and Carol each think they are receiving at least one-third of the cake.
(c) Explain why, in general, this scheme is not envy-free.

26. In A. M. Fink's procedure (described in Exercise 25), suppose that a fourth person (Alice) comes along after Bob, Carol, and Ted have already divided the cake among themselves so that each of the three thinks he or she has a piece of size at least one-third. Mimic what was done in the three-person case to obtain an allocation among the four that is proportional. (*Hint:* Begin by having Bob, Carol, and Ted divide the pieces they have into a certain number–how many?–of equal parts.)

27. There is a moving-knife version of the Banach–Knaster procedure that appears in the Dubins–Spanier article in the Suggested Readings. To describe it, we picture the cake as being rectangular, and the procedure beginning with a referee holding a knife along the left edge, as illustrated below.

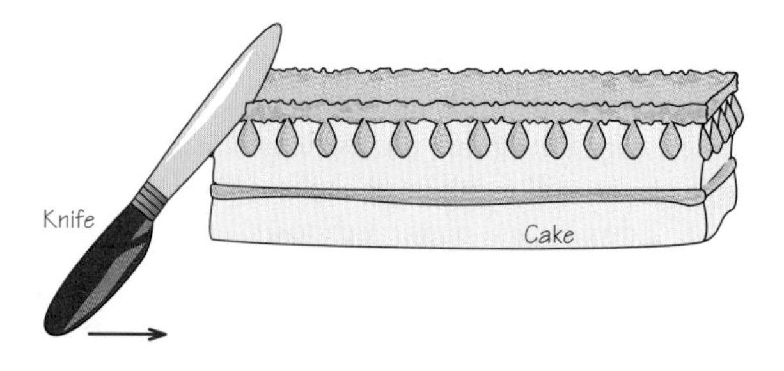

Assume, for the sake of illustration, that there are four players (Bob, Carol, Ted, and Alice). The referee starts moving the knife from left to right over the cake (keeping it parallel to the position in which it started) until one of the players (assume it is Bob) calls "cut." At this time, a cut is made, the piece to the left of the knife is given to Bob, and he exits the game. The knife starts moving again, and the process continues. The strategies are for each player to call "cut" whenever it would yield him or her a piece of size at least one-fourth.

(a) Explain why this procedure produces an allocation that is proportional.
(b) Explain why the resulting allocation is not, in general, envy-free.
(c) Explain why, if you are not the first player to call "cut," there is a strategy different from the one suggested that is never worse for you, and sometimes better.

28. There is a two-person moving-knife cake-division procedure due to A. K. Austin that leads to each player receiving a piece of cake that he or she considers to be of size exactly one-half. It begins by having one of the two players (Bob) place two knives over the cake, one of which is at the left edge, and the other of which is parallel to the first and placed so that the piece between the knives (*A* in the picture below) is of size exactly one-half in Bob's estimation.

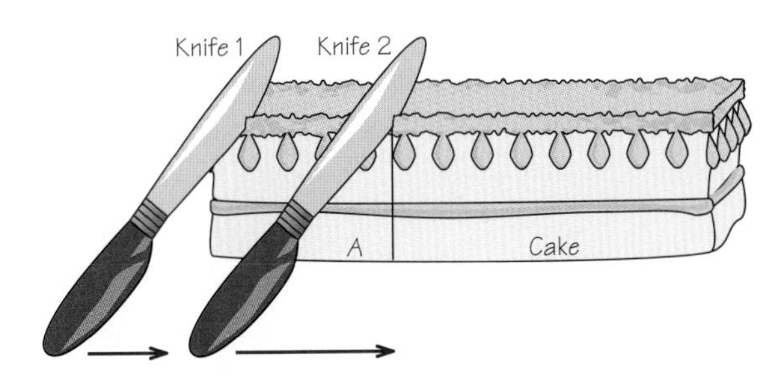

If Carol agrees that this is a 50–50 division, we are done. Otherwise, Bob starts moving both knives to the right—perhaps at different rates—so that the piece between the knives remains of size one-half in his eyes. Carol calls "stop" at the point when she also thinks the piece between the two knives is of size exactly one-half.

(a) If the knife on the right were to reach the right-hand edge, where would the knife on the left be?
(b) Explain why there definitely is a point where Carol thinks the piece between the two knives is of size exactly one-half. (*Hint:* If Carol thinks the piece is too small at the beginning, what will she think of it at the end?)

29. Suppose we have three items (X, Y, and Z) and three people (Bob, Carol, and Ted). Assume that each of the people spreads 100 points over the items (as in adjusted winner) to indicate the relative worth of each item to that person:

Item	Bob	Carol	Ted
X	40	30	30
Y	50	40	30
Z	10	30	40

For each of the following allocations, indicate

(a) whether or not it is proportional.
(b) whether or not it is envy-free.
(c) whether or not it is equitable.
(d) for the ones that are *not* Pareto-optimal, another allocation that makes one person better off without making anyone else worse off.

Allocation 1: Bob gets Z, Carol gets Y, and Ted gets X. (This is not Pareto-optimal.)

Allocation 2: Bob gets Y, Carol gets Z, and Ted gets X. (This is not Pareto-optimal.)

Allocation 3: Bob gets X, Y, and Z. (This *is* Pareto-optimal; explain why.)

Allocation 4: Bob gets Y, Carol gets X, and Ted gets Z. (This is Pareto-optimal.)

Allocation 5: Bob gets X, Carol gets Y, and Ted gets Z. (This is Pareto-optimal.)

✎ WRITING PROJECTS

1. It turns out that there is no way to extend the adjusted winner procedure to three or more players. That is, there are point assignments by three players to three objects so that no allocation satisfies the three desired properties of equability (equal points), envy-freeness, and Pareto optimality. On the other hand, there are separate procedures that will realize any two of the three properties. Thus, trade-offs must be made, and these may depend on the circumstances. Discuss your feelings regarding the relative importance of the three properties and circumstances that may affect the choice of which two of the three properties one might wish to have satisfied.

2. If we use taking turns to divvy up a collection of objects between two people (Bob and Carol), then there is an obvious advantage to going first. Assume that we have decided that Bob will, in fact, choose first (say, by the toss of a coin). Let's think about how Carol might be compensated. First of all, if there are only three objects, then the "choice sequence" Bob – Carol – Carol seems to be the only reasonable one. Do you agree? For four objects, however, there are two choice sequences that suggest themselves: Bob – Carol – Carol – Carol and Bob – Carol – Carol – Bob. Do you think that one of these is obviously more fair than the other? What if there are four essentially identical objects? What if both Bob and Carol value object A twice as much as B, and B twice as much as C, and C twice as much as D? What sequences suggest themselves for five objects? For eight objects? (For more on this, see *The Win – Win Solution* in the Suggested Readings.)

3. One of the most important differences between the three-person and the n-person envy-free procedures is that the latter procedure may take more than two stages. And, of course, the more stages there are, the more cuts and trimmings that may be necessary. Do you consider this a serious practical problem, or is it mainly a theoretical problem? Why?

SPREADSHEET PROJECTS

To do these projects, go to www.whfreeman.com/fapp.

This spreadsheet project models the Knaster inheritance and adjusted winner procedures for fair division. Spreadsheets allow the creation of a flexible and updatable template for such computations. By adjusting values, spreadsheets can also model insincere strategies.

APPLET EXERCISES

To do these exercises, go to www.whfreeman.com/fapp.

In Exercise 1, you were given the point allocations of the two parties in the Trump divorce. In practice, however, each party will be unaware of the bids of his or her opponent. Explore this more realistic setting in the applet Adjusted Winner Divorce Procedure.

CHAPTER 15

Apportionment

In this chapter we cover...

...and at www.whfreeman.com/fapp:

Flashcards
Quizzes
Spreadsheet Projects
  Apportionment methods
Applet Exercises
  Apportionment

15.1 The Apportionment Problem

The Sports Magnet High School field hockey team played 23 games in the 2000–01 season. It won 18 games, lost 4, and one game ended in a tie. Table 15.1 expresses these results as percentages.

TABLE 15.1 The Sports Magnet High School Field Hockey Team: 2001–2002 Season

		Percentage
Games won	18	$\frac{18}{23} \times 100\% = 78.26\%$
Games lost	4	$\frac{4}{23} \times 100\% = 17.39\%$
Games tied	1	$\frac{1}{23} \times 100\% = 4.35\%$
Games played	23	100.00%

The coach objects to this table—it looks too complicated. "Just express the percents as whole numbers," she says. We round off the percentages: 78% of the games were won, 17% were lost, and 4% were tied. The coach notices that these numbers add to 99% and changes the winning percentage to 79%. "Now you have 100%!" The story ends there for the field hockey coach, but for us, it has just begun.

The **apportionment problem** is to round a set of fractions so that their sum is maintained at its original value. The rounding procedure must not be an arbitrary one, but one that can be applied consistently. Any such rounding procedure is called an **apportionment method.**

When rounding percentages so as to preserve the sum of 100%, we face an apportionment problem. The coach used the apportionment method called "make it look good for the team," but a serious apportionment method should be unbiased, especially when it comes to even *more* critical issues, such as apportioning seats in the U.S. House of Representatives.

The framers of the U.S. Constitution wrote that seats in the House of Representatives "shall be apportioned among the several states within this union according to their respective Numbers. . . ." They may not have realized that the apportionment problem that they set in this clause was of any significance, but it caused trouble right from the start and is still the subject of controversy. For example, in 1790, Delaware's population, 55,540, was the smallest among the 15 states. The total population of the country at the time was 3,615,920, and the House of Representatives was to have 105 members. Thus, the average congressional district should have a population of 3,615,920 ÷ 105 = 34,437. To obtain its fair share of the House seats, we can divide Delaware's population by the average congressional district population. Ideally, Delaware should have had

$$\frac{55{,}540}{34{,}437} = 1.613 \text{ congressional districts}$$

Table 15.2 displays the way this apportionment was worked out in a bill that Congress passed and sent to President Washington. Each state's fair share of the 105 congressional districts, or *quota*, was calculated by dividing its population by the average congressional district population of 34,437, as in the case of Delaware, shown above. None of the quotas were whole numbers, and each was

The 106th Congress of the United States in session. (Brad Markel/Gamma-Liaison.)

TABLE 15.2 **Apportioning the House of Representatives by the Hamilton Method**

State	Population	Quota	Lower Quota		Appor-tionment
Virginia	630,560	18.310	18		18
Massachusetts	475,327	13.803	13	↑	14
Pennsylvania	432,879	12.570	12	↑	13
North Carolina	353,523	10.266	10		10
New York	331,589	9.629	9	↑	10
Maryland	278,514	8.088	8		8
Connecticut	236,841	6.877	6	↑	7
South Carolina	206,236	5.989	5	↑	6
New Jersey	179,570	5.214	5		5
New Hampshire	141,822	4.118	4		4
Vermont	85,533	2.484	2		2
Georgia	70,835	2.057	2		2
Kentucky	68,705	1.995	1	↑	2
Rhode Island	68,446	1.988	1	↑	2
Delaware	55,540	1.613	1	↑	2
Totals	3,615,920	105	97		105

rounded to the nearest whole number: Quotas with fractional parts less than $\frac{1}{2}$ were rounded down, by taking away the fractional part, and quotas with fractional parts greater than $\frac{1}{2}$ were rounded up to the next whole number. Thus, the bill gave Delaware two seats in Congress. As you will see in the table, Virginia's quota was 18.310 seats. Because the fractional part of 18.310 is less than $\frac{1}{2}$, Virginia was awarded 18 seats in the House.

Although this apportionment may seem fair enough, President Washington vetoed that bill.[1] Washington came from Virginia, a state that lost something in the apportionment bill. Was he biased in favor of his home state, as the field hockey coach was in favor of her team? We will return to this question.

In the following example, we will see how to set up an apportionment problem.

EXAMPLE

The High School Mathematics Teacher

A high school has one mathematics teacher who teaches all geometry, pre-calculus, and calculus classes. She has time to teach a total of five sections. One hundred students are enrolled as follows: 52 for geometry, 33 for pre-calculus, and 15 for calculus. How many sections of each course should be scheduled?

[1]The 1790 apportionment bill was the first bill in U.S. history to be vetoed.

The number of students enrolled in each course is called the population. Thus, the population for geometry is 52, the population for pre-calculus is 33, and the population for calculus is 15. There are five sections for the 100 students, so the average section will have $100 \div 5 = 20$ students. We will call this average section size the *standard divisor,* because each quota can be determined by dividing the corresponding population by this number. Table 15.3 displays these calculations.

As shown in the table, the quotas add up to 5. It is tempting to round each to the nearest whole number, as in the right column of the table, but this makes 6 sections in all—too many! The purpose of an apportionment method is to find an equitable way to round a set of numbers like this without increasing or decreasing the original sum. ■

Although many apportionment problems do not involve the House of Representatives, our terminology refers to *states, populations,* and a *house size.* In a course-scheduling problem, the courses to be taught correspond to the states, the numbers of students enrolled in each course correspond to the populations, and the house size is the total number of sections to be scheduled. Thus, in our example, there are three states, with populations

$$p_1 = 52, p_2 = 33, \text{ and } p_3 = 15$$

and the house size is 5.

In the problem of rounding percentages so that their sum is 100, the house size is 100. The categories (such as wins, losses, and ties in the field hockey example) correspond to the states, and the numbers in each category (in the field hockey example, 18 wins, 4 losses, and 1 tie) correspond to the populations of the states.

Let n be the number of states, and let their populations be denoted as

$$p_1, p_2, \ldots, p_i, \ldots, p_n$$

TABLE 15.3 Calculation of the Quotas for High School Mathematics Courses

Course	Population	Quota	Rounded
Geometry	52	$52 \div 20 = 2.60$	3
Pre-calculus	33	$33 \div 20 = 1.65$	2
Calculus	15	$15 \div 20 = 0.75$	1
Totals	100	5	6

The total population will be

$$p = p_1 + p_2 + \ldots + p_n$$

The quotient of the total population, p, divided by the house size is called the **standard divisor.** If h denotes the house size and s is the standard divisor, then

$$s = \frac{p}{h}$$

In an apportionment problem, the **quota** is the exact share that would be allocated *if a whole number were not required.* To obtain a state's quota, divide its population by the standard divisor. State i, with population p_i, has quota

$$q_i = \frac{p_i}{s}$$

EXAMPLE

California's Quota

The Census Bureau recorded the apportionment population[2] of the United States, as of April 1, 2000, to be 281,424,177. There are 435 seats in the House of Representatives; therefore, the standard divisor is

$$s = \frac{281{,}424{,}177}{435} = 646{,}952$$

California's apportionment population was $p_1 = 33{,}930{,}798$. Its quota, q_1, is determined by dividing this population by the standard divisor. Thus,

$$q_1 = \frac{p_1}{s} = \frac{33{,}930{,}798}{646{,}952} = 52.447 \text{ seats}$$

California's apportionment, which is required to be a whole number, was set at 53 seats. ■

An apportionment is given by n whole numbers,

$$a_1, a_2, \ldots, a_i, \ldots, a_n$$

with a_i representing the number of representatives given to state i. Ideally, each state's apportionment should be close to its quota. It is unrealistic to expect that for any state, $a_i = q_i$, because a_i has to be a whole number and q_i is unlikely to be

[2]The apportionment population includes the resident population and the overseas population.

a whole number. In choosing an apportionment method, we must decide what we mean by the phrase "a_i should be close to q_i."

Apportionment always involves rounding, and there are many ways to round. "Rounding down" means discarding the fractional part of a number q to obtain a whole number that we will denote by $\lfloor q \rfloor$. Thus, $\lfloor 7.00001 \rfloor = 7$, $\lfloor 7 \rfloor = 7$, and $\lfloor 6.99999 \rfloor = 6$. "Rounding up" gives the next whole number, $\lceil q \rceil$. Thus, $\lceil 7.00001 \rceil = 8$, but $\lceil 7 \rceil = 7$.

There are numerous different apportionment methods. Each has flaws, and our goal is to understand how to choose a method that is appropriate for a particular apportionment problem.

15.2 The Hamilton Method

The congressional apportionment bill that was vetoed by President Washington was written by Alexander Hamilton. While it may appear that he simply rounded each quota to the nearest whole number, he was aware that there would be occasions—analogous to the examples of the field hockey team's percentages of wins, losses, and ties not summing to 100%, or the high school teacher receiving an extra class to teach—when the total number of seats apportioned in this way would be either more or less than the statutory house size. He developed an apportionment method that he called the *method of largest fractions.*

With the **Hamilton method,** state i receives either its **lower quota,** which is the integer part of its quota (in the notation just introduced, $\lfloor q_i \rfloor$, or its **upper quota,** $\lceil q_i \rceil$, obtained by rounding the quota up to the next integer value. The states that receive their upper quotas are those whose quotas have the largest fractional parts.

Implementing the Hamilton method is a three-step procedure:

1. Calculate each state's quota.
2. Tentatively assign to each state its lower quota of representatives. Each state whose quota is not a whole number loses a fraction of a seat at this stage, so the total number of seats assigned at this point is less than the house size h. This leaves additional seats to be apportioned.
3. Allot the remaining seats, one each, to the states whose quotas have the largest fractional parts, until the house is filled.

It is possible that a tie will occur, with the quotas of two states having identical fractional parts, but in practice, this rarely happens when large populations are involved.

The apportionment method of largest fractions, also known as the Hamilton method, was named for Alexander Hamilton. (National Portrait Gallery/Art Resource, NY.)

EXAMPLE

The High School Teacher's Dilemma

Let us use the Hamilton method to determine how many sections of geometry, pre-calculus, and calculus the high school teacher should teach. We have found that the quotas for the three subjects were 2.60, 1.65, and 0.75, respectively (see Table 15.3). The lower quotas are $\lfloor 2.60 \rfloor = 2$, $\lfloor 1.65 \rfloor = 1$, and $\lfloor 0.75 \rfloor = 0$, so we start by assigning two sections of geometry and one section of pre-calculus. With three sections apportioned, we have two more to assign to "fill the house." These two sections go to the subjects with the largest fractions: Calculus has the largest fraction, 0.75, and gets a section; the second section goes to pre-calculus, whose fraction is 0.65. The final apportionment is as follows: geometry, two sections; pre-calculus, two sections; calculus, one section. ■

EXAMPLE

The Field Hockey Team

In Table 15.1, we saw that the percentages of wins, losses, and ties for the field hockey team were 78.26%, 17.39%, and 4.35%, respectively. These are the quotas that we must round so that they sum to 100%. To start, we apportion $\lfloor 78.26 \rfloor = 78\%$, $\lfloor 17.39 \rfloor = 17\%$ and $\lfloor 4.35 \rfloor = 4\%$ to the three categories. These lower quotas add up to 99%. The remaining 1% to be apportioned goes to the losses, because their fraction, 0.39, is the largest. The final apportionment is 78% wins, 18% losses, and 4% ties. It can be expected that the coach will veto this apportionment. ■

President Washington's veto message stated that the fractions of seats gained by some states in the third step of the Hamilton apportionment were not related to the states' total populations.

EXAMPLE

Hamilton's Apportionment

In 1790, there were 15 states, and the House had 105 seats. According to the 1790 census, the U.S. population was 3,615,920. The standard divisor, 3,615,920 ÷ 105 = 34,437, represents the population of the average congressional district.

Table 15.2 displays the calculations leading to Alexander Hamilton's proposed apportionment. Each quota shown in the table was calculated by dividing the state's population by this standard divisor. If you are using a calculator to follow the entries in the table, store the standard divisor in the calculator's memory. Then each quota can be figured by entering the individual state's population and dividing by the divisor recalled from memory.

The table shows that if each state were given its lower quota, only 97 seats would have been apportioned. The remaining 8 seats go to the 8 states whose quotas had the largest fractional parts, and these are marked in the table. ■

President Washington's veto prevented the Hamilton method from being used in 1792, but it was adopted by Congress in 1850 and remained in use until 1900. The half-century of experience with the Hamilton method revealed a paradox.

Paradoxes of the Hamilton Method

A *paradox* is a fact that seems obviously false. The first Hamilton apportionment paradox, called the *Alabama paradox,* was discovered in 1881. As part of the reapportionment procedure mandated by the Constitution, the Census Bureau had supplied Congress with a table of congressional apportionments for a range of different house sizes from 275 to 350, based on the 1880 census. The table revealed a strange phenomenon. With a 299-seat house, Alabama's quota was 7.646. The fractional part ranked twentieth of the 38 states, and that was just enough to give Alabama its upper quota of $\lceil 7.646 \rceil = 8$ seats. Illinois and Texas, with quotas of 18.640 and 9.640 respectively, ranked below Alabama and received their lower quotas: 18 for Illinois and 9 for Texas. The next column of the table, corresponding to a house size of 300 seats, showed changes in the apportionments for each of these three states: Alabama now had 7 seats instead of 8, while Illinois and Texas received increased apportionments of 19 seats and 10 seats, respectively. Alabama had *lost* a seat as a result of an increase in the house size. This happened because the Illinois and Texas quotas increased more than Alabama's quota did. With a 300-seat house, Alabama's quota increased to 7.671, Illinois's to 18.702, and Texas's to 9.672. Because the fractional parts of the quotas for Illinois and Texas were larger, those states were given their upper quotas, and Alabama was left with its lower quota.

> The **Alabama paradox** occurs when a state loses a seat as the result of an increase in the house size.

The Alabama paradox validates President Washington's veto message, issued 90 years before its discovery. If the fractional parts of the quotas, which determine the way the last few seats are apportioned, were related to the populations in any sensible way, the paradox could not have occurred.

EXAMPLE A Mathematics Department Meets the Alabama Paradox

A mathematics department has 30 teaching assistants to cover recitation sections for college algebra, calculus I, calculus II, calculus III, and contemporary mathematics. The enrollments of these courses are given in Table 15.4. The department will use the Hamilton method to apportion the teaching assistants to the five subjects. In this problem, the house size is $h = 30$ (the number of teaching assistants) and the population is $p = 750$. Therefore, the standard divisor is $750 \div 30 = 25$; this represents the average number of students in each recitation section. Each quota shown in the table was determined by dividing the enrollment of the course by this divisor.

The lower quotas add up to 27, so the three courses whose quotas have the largest fractional parts, calculus I and III and contemporary mathematics, are entitled to their upper quotas.

After these calculations were finished, the graduate school authorized the department to hire an additional teaching assistant. With 31 teaching assistants, the standard divisor becomes $750 \div 31 = 24.19355$. The new quotas, determined by dividing each population by this new divisor, are shown in Table 15.5. Now the lower quotas add up to 28, so again the additional teaching assistants go to the subjects whose quotas have the largest fractions. The calculus III fraction, which had been larger than the college algebra fraction when there were just 30 teaching assistants, has been surpassed. The instructor will be dismayed, because

TABLE 15.4 Apportioning 30 Teaching Assistants

Course	Enrollment	Quota	Lower Quota		Apportionment
College algebra	188	7.52	7		7
Calculus I	142	5.68	5	↑	6
Calculus II	138	5.52	5		5
Calculus III	64	2.56	2	↑	3
Contemporary mathematics	218	8.72	8	↑	9
Totals	750	30.00	27		30

TABLE 15.5 Apportioning 31 Teaching Assistants

Course	Enrollment	Quota	Lower Quota		Apportionment
College algebra	188	7.771	7	↑	8
Calculus I	142	5.869	5	↑	6
Calculus II	138	5.704	5	↑	6
Calculus III	64	2.645	2		2
Contemporary mathematics	218	9.011	9		9
Totals	750	31.000	28		31

increasing the total number of teaching assistants for the department has caused her course to *lose* one section! ■

The size of the House of Representatives is now fixed at 435 members, so the Alabama paradox cannot occur. A second paradox, called the *population paradox,* is associated with a fixed house size. In Exercise 25 we will consider a situation involving an imaginary country with four states and a 100-seat legislature, apportioned by the Hamilton method. The new census shows substantial increases in population for the three largest states and a slight loss of population for the smallest state. In the new apportionments, the largest and smallest states both gain seats, while the middle two states lose seats. Thus, the smallest state lost population and gained a seat, while the middle two states gained population and lost seats.

The **population paradox** occurs when one state's population increases, and its apportionment decreases, while simultaneously another state's population increases proportionally less, or decreases, and its apportionment increases.

15.3 Divisor Methods

The Jefferson Method

The Constitution requires that congressional districts be drawn so that the population of each is at least 30,000. President Washington could have vetoed the Hamilton apportionment bill because Delaware's population of only 55,540 was too small for the two congressional districts assigned to it. Thomas Jefferson proposed an apportionment method, now called the **Jefferson method,** to replace the Hamilton method.

In any apportionment, the standard divisor, s, obtained by dividing the total population by the house size, represents the average district population. In devel-

Thomas Jefferson favored a method of apportionment biased in favor of states with large populations. (National Portrait Gallery/Art Resource, NY.)

oping his method, Thomas Jefferson focused on the population of the smallest district, d. Using d, rather than s, as the divisor, each state receives an **adjusted quota** that is rounded *down* to obtain its apportionment. If d is chosen correctly, the apportionments will add up exactly to the statutory house size.

In effect, the Jefferson method apportions to each state the maximum number of congressional districts of population d that will be accommodated by the state's population. Any leftover population is divided among these districts, so that, in effect, most districts will have populations greater than d.

The Jefferson apportionment is easy to compute, once the divisor d is known. The apportionment for state i is

$$a_i = \left\lfloor \frac{p_i}{d} \right\rfloor$$

The actual divisor that Jefferson used was $d = 33{,}000$. Thus, Virginia's apportionment was

$$a_1 = \left\lfloor \frac{\text{population of Virginia}}{33{,}000} \right\rfloor = \left\lfloor \frac{630{,}560}{33{,}000} \right\rfloor = \lfloor 19.108 \rfloor = 19$$

Therefore, Virginia received 19 seats, rather than 18, which Hamilton's bill would have allocated. Delaware's apportionment was 1 seat instead of 2, as the following calculation shows.

$$a_{15} = \left\lfloor \frac{\text{population of Delaware}}{33{,}000} \right\rfloor = \left\lfloor \frac{55{,}540}{33{,}000} \right\rfloor = \lfloor 1.683 \rfloor = 1$$

The Jefferson method is one of a class of apportionment methods called *divisor methods.*

A **divisor method** of apportionment determines each state's apportionment by dividing its population by a common divisor d and rounding the resulting quotient. Divisor methods differ in the rule used to round the quotient.

The divisor must be carefully chosen to achieve the correct house size. In 1790, $d = 30{,}000$ would have resulted in larger apportionments for several states and a house size of 112, while $d = 36{,}000$ would have decreased several apportionments, and the house size would have been 91.

Critical Divisors

To implement the Jefferson method, we must determine the divisor, d. Start by determining the standard divisor, s, and the quota for each state, as we did with the Hamilton method. Each state is assigned, as a **tentative apportionment,** its lower quota. Thus, the tentative apportionment of state i is

$$n_i = \left\lfloor \frac{p_i}{s} \right\rfloor$$

As we noted with the Hamilton method, these tentative apportionments are not enough to fill the house, so the apportionments of some states will have to be increased. To determine which states should receive additional seats, calculate a *critical divisor* for each state.

Let n_i be the tentative apportionment of a state, and let p_i be its population. The **critical divisor,** d_i, for that state is

$$d_i = \frac{p_i}{n_i + 1}$$

EXAMPLE

Jefferson's Critical Divisors

In Table 15.2, we found that the lower quota for Virginia was $n_1 = 18$ and the lower quota for Delaware was $n_{15} = 1$. The populations of the two states were 630,560 and 55,540, respectively, so their critical divisors are

$$d_1 = \frac{630{,}560}{18 + 1} = 33{,}187$$

$$d_{15} = \frac{55{,}540}{1+1} = 27{,}770$$

■

The significance of the critical divisor is that if the Jefferson divisor d is equal to d_i, then state i will contain *exactly* $n_i + 1$ districts of population d. Thus, with $d = 33{,}187$, Virginia's apportionment would be equal to 19. If we take $d = 27{,}770$, then Delaware would get 2 seats. However, the same divisor must be used for all states, so with this lower value of d, Virginia would get

$$\left\lfloor \frac{630{,}560}{27{,}770} \right\rfloor = 22 \text{ seats}$$

Once the critical divisors are determined, the state with the largest critical divisor is entitled to another seat, because when that divisor is used, no other state will receive a changed apportionment—state i receives additional seats only when a divisor smaller than or equal to its critical divisor is used. The total apportionment is thus increased by 1 in this step.

If there remain additional seats to be apportioned, we recompute the critical divisor for the state whose tentative apportionment has increased, and repeat the process. When the house is filled, the critical divisor most recently used is the divisor d, representing the minimum district population. With the Jefferson method (and any other divisor method), the critical divisors determine the priority of a state for receiving additional seats. The state with the largest critical divisor gets the next seat, but after receiving that seat, its critical divisor is recomputed, and usually another state will then be first in line.

EXAMPLE

The High School Mathematics Teacher

A mathematics teacher can teach five classes. There are 52 students enrolled in geometry, 33 in pre-calculus, and 15 in calculus. We will use the Jefferson method to determine how many sections of each subject to teach. We have previously determined that the lower quotas for the subjects are 2, 1, and 0, respectively (see Table 15.3). Thus, the critical divisors are

$$\frac{52}{3} = 17\tfrac{1}{3} \quad \text{for geometry}$$

$$\frac{33}{2} = 16\tfrac{1}{2} \quad \text{for pre-calculus}$$

$$\frac{15}{1} = 15 \quad \text{for calculus}$$

Geometry, with the largest critical divisor, has first priority for a new section, and its tentative apportionment is increased to 3. Its new critical divisor will be $\frac{52}{4} = 13$. Now pre-calculus has top priority, and its tentative apportionment is increased to 2. The house is now full, so the final apportionments are 3 sections of geometry and 2 sections of pre-calculus. The minimum section size will be the $16\frac{1}{2}$, because that is the critical divisor for the subject that was the last to receive an increased apportionment: pre-calculus. Because the enrollment for calculus is less than the minimum section size, there will be no calculus class. ■

EXAMPLE

The Field Hockey Team

The Sports Magnet High School field hockey team had 18 wins, 4 losses, and 1 tie last season. We will use the Jefferson method to express this record as percentages. The house size is 100%, and the total population is the 23 games played. Therefore, the standard divisor is $\frac{23}{100} = 0.23$. The lower quotas are

$$\left\lfloor \frac{18}{0.23} \right\rfloor = 78\% \quad \text{wins}$$

$$\left\lfloor \frac{4}{0.23} \right\rfloor = 17\% \quad \text{losses}$$

$$\left\lfloor \frac{1}{0.23} \right\rfloor = 4\% \quad \text{ties}$$

and these are the tentative apportionments. The critical divisors are

$$\frac{18}{79} = 0.22785 \quad \text{for wins}$$

$$\frac{4}{18} = 0.22222 \quad \text{for losses}$$

$$\frac{1}{5} = 0.20000 \quad \text{for ties}$$

The category with the largest critical divisor is wins, so we increase its apportionment to 79%. The house is now full; we have the apportionment that the coach wanted: 79% wins, 17% losses, and 4% ties. ■

These examples demonstrate that different methods may yield different apportionments, because the Hamilton method gave different apportionments in both cases. The Jefferson and Hamilton methods also differed in the 1790 apportionment of congressional seats. We have seen that the critical divisor for Virginia was greater than the critical divisor for Delaware; thus, Virginia had greater priority than Delaware to receive increased apportionment with the Jefferson method, while the opposite was true with the Hamilton method.

With the Jefferson method, no state can receive less than its lower quota as its apportionment—because the lower quota is the initial tentative apportionment—but a state can be apportioned more than its upper quota. This phenomenon occurred for the first time in the apportionment based on the 1820 census, which recorded that New York had a population of 1,368,775. The total population of the United States was found to be 8,969,878, and the house size was 213. Therefore, the standard divisor was 8,969,878 ÷ 213 = 42,112, and New York's quota was

$$q = \frac{1{,}368{,}775}{42{,}112} = 32.503$$

The Hamilton method would have apportioned 33 seats to New York, but the Jefferson method, using the divisor $d = 39{,}900$, gave New York

$$\left\lfloor \frac{1{,}368{,}775}{39{,}900} \right\rfloor = 34 \text{ seats}$$

An apportionment method is said to satisfy the **quota condition** if in every situation each state's apportionment is equal to either its lower quota or its upper quota. It takes only one example like the 1820 apportionment to show that the Jefferson method does not satisfy the quota condition. In fact, if the house had continued to use the Jefferson method, it would have violated the quota condition in every apportionment since 1850.

With the Hamilton method, each state starts with its lower quota, and some states are given their upper quotas to fill the house. There is no way for a state to receive less than its lower quota or more than its upper quota, so the Hamilton method does satisfy the quota condition. This was obvious to Congress in 1850, so it based its apportionment on the Hamilton method.[3]

The Jefferson method, however, is not troubled by the Alabama and population paradoxes. Consider the Alabama paradox, in which a state loses a seat as a result of an increase in the house size. With any apportionment method, the apportionments of some states must increase when the size of the house increases. The Jefferson method awards seats in order of critical divisors. When the house size increases, the next seat will go to the state with the largest critical divisor. There is no opportunity for a state to lose a seat.

Congress has never used an apportionment method that satisfies the quota condition and avoids the paradoxes. It would be desirable to use such a method, and in the 1970s, the mathematicians Michel L. Balinski and H. Peyton Young set out to find one. They succeeded in finding a method, which they called the *quota method,* that satisfies the quota condition (as the Hamilton method does) and avoids the Alabama paradox (as the Jefferson method does). However, the population paradox remained. They subsequently proved that the only

[3]The origins of the Hamilton method had been forgotten in 1850, and the method was named for Congressman Samuel Vinton, who had rediscovered the method.

apportionment methods that are free of the population paradox are the divisor methods. It is known that every divisor method is capable of violating the quota condition, so Balinski and Young have proved an impossibility theorem: *No apportionment method that satisfies the quota condition is free of paradoxes.* This theorem is like Kenneth Arrow's theorem that there is no completely satisfactory way to decide multicandidate elections based on voter preference schedules (see Chapter 12).

The Jefferson method favors the larger states. It is not an accident that in all the examples that we considered, the "state" with the largest population fared better with the Jefferson method than it did with the Hamilton method. In an apportionment problem, let s be the standard divisor and let d be the divisor used in the Jefferson method. The apportionment given to state i is then $\lfloor u_i \rfloor$, where $u_i = p_i/d$ is the state's *adjusted quota.* Comparing u_i with the quota, $q_i = p_i/s$, we find that their ratio

$$M = \frac{u_i}{q_i} = \frac{p_i}{d} \div \frac{p_i}{s} = \frac{p_i}{d} \times \frac{s}{p_i} = \frac{s}{d}$$

is the same for each state. Since we know that $s > d$, it follows that $M > 1$. Furthermore, we have $u_i = M \times q_i$. Thus, the apportionment for state i is

$$a_i = \lfloor M \times q_i \rfloor$$

For example, let's consider the congressional apportionment of 1820. The standard divisor was $s = 42{,}112$, and the Jefferson divisor was $d = 39{,}900$. Thus, the quotient, M, is

$$M = \frac{s}{d} = \frac{42{,}112}{39{,}990} = 1.0554$$

Now suppose that a state has a quota of exactly 19. We obtain an adjusted quota, a_i, as follows:

$$a_i = 1.0554 \times 19$$
$$a_i = 20.0526$$

In other words, that state gets another seat, violating the quota condition. In fact, every state whose quota is 19 or more will be guaranteed an additional seat with this value of M. If a state has a quota of 38, an identical calculation shows that it will receive two additional seats. On the other hand, consider a small state whose lower quota is 1. To increase its apportionment to 2, its quota must be at least $2/M = 1.89502$. Thus, a state with quota 19 is guaranteed an increase, and a state with quota 1.89 is denied an increase.

The Webster Method

The **Webster method** is the divisor method that rounds the quota (adjusted if necessary) to the nearest whole number, rounding up when the fractional part is greater than or equal to $\frac{1}{2}$, and rounding down when the fractional part is less than $\frac{1}{2}$.

There is no standard notation like $\lfloor q \rfloor$ and $\lceil q \rceil$ to denote the rounding of q to the nearest whole number. We will use $\langle q \rangle$ to denote this rounding.

The Webster and Jefferson methods are immune to the Alabama and population paradoxes, but neither satisfies the quota condition. However, in many respects, Webster's is a better method than Jefferson's. While the Jefferson method obviously favors the large states, the Webster method is neutral, favoring neither the large nor the small states. Furthermore, while the Jefferson method almost always gives the most populous state more than its upper quota, the Webster method would not have violated the quota condition in any of the 22 congressional apportionments that have occurred so far.

To calculate an apportionment by the Webster method, start by finding the standard divisor, and use it to find the quota for each state. The tentative appor-

Statesman and orator Daniel Webster (1782–1852), who developed a divisor method for apportioning the U.S. House of Representatives. (National Portrait Gallery/Art Resource, NY.)

tionment, which we will call n_i for state i, is its rounded quota: $n_i = \langle q_i \rangle$. When all of the tentative apportionments have been determined, we calculate their sum. If the sum is equal to the house size, we are finished: The tentative apportionments become the actual apportionments.

If the sum of the tentative apportionments is not equal to the house size, we must adjust them. By choosing a divisor greater than the standard divisor, we can reduce the apportionment of one or more states. This would be appropriate if the sum of the tentative apportionments exceeds the house size. If the tentative apportionments do not fill the house, a divisor that is less than the standard divisor should be selected. In either case, we must calculate a critical divisor for each state.

Critical Divisors for the Webster Method

If the tentative apportionments do not fill the house, then the critical divisor for state i, with population p_i and tentative apportionment n_i, is

$$d_i^+ = \frac{p_i}{n_i + \frac{1}{2}}$$

The state with the largest critical divisor is first in line to receive a seat. If the tentative apportionments overfill the house, then the critical divisor for state i is

$$d_i^- = \frac{p_i}{n_i - \frac{1}{2}}$$

The state with the smallest critical divisor is first in line to lose a seat.

When the divisor d_i^+ is used, state i will get the adjusted quota

$$\frac{p_i}{d_i^+} = n_i + \frac{1}{2}$$

which is the smallest number to be rounded up to $n_i + 1$. Thus, the tentative apportionment of state i would increase. None of the states with critical divisors less than d_i^+ would see any change in their tentative apportionments.

Similarly, when the population p_i is divided by d_i^-, the resulting quotient, $n_i - \frac{1}{2}$, is on the borderline for rounding down to $n_i - 1$. If d_i^- is the least critical divisor, then state i will be the only state whose tentative apportionment would be affected by the divisor d_i^-.

EXAMPLE Apportioning Classes

Let us return to the case of the mathematics teacher who teaches a total of five classes in geometry, pre-calculus, and calculus. The enrollments are 52 for geometry, 33 for pre-calculus, and 15 for calculus. With a total of 100 students enrolled, and a house size of 5, the standard divisor is 20. The quotas, determined by dividing the enrollments for the three subjects by the standard divisor, are 2.6, 1.65, and 0.75, respectively. The tentative apportionments are 3, 2, and 1; their total, 6, exceeds the house size. We therefore compute the divisors d_i^-.

Subject	Tentative apportionment	d_i^-
Geometry	3	$52 \div 2.5 = 20.8$
Pre-calculus	2	$33 \div 1.5 = 22$
Calculus	1	$15 \div 0.5 = 30$

The subject with the smallest critical divisor is geometry; it therefore receives the reduced apportionment of 2. The final apportionment is therefore two sections each of geometry and pre-calculus, and one section of calculus. ■

EXAMPLE

The Field Hockey Team

As you may remember, the Sports Magnet High School field hockey team has a fine record for the previous season: 18 wins, 4 losses, and 1 tie. We would like to express the record in the form of whole percentages, and the coach has said she is willing to do this with the Webster method. The quotas are 78.26% for wins, 17.39% for losses, and 4.35% for ties; these are rounded to get the tentative percentages 78%, 17%, and 4%, respectively. The total of these tentative percentages is 99%, less than the "house size" of 100%. We therefore compute the critical divisors d_i^+:

Category	Tentative percentage	d_i^+
Wins	78	$18 \div 78.5 = 0.2293$
Losses	17	$4 \div 17.5 = 0.2286$
Ties	4	$1 \div 4.5 = 0.2222$

The largest critical divisor belongs to the wins, so it receives the extra 1%. The final apportionment is 79% wins, 17% losses, and 4% ties. ■

To see why the Webster method is not biased in favor of large states or small states, let s be the standard divisor, which is equal to the average district population. Let d be the divisor that is used in the Webster method (d is equal to the critical divisor belonging to the last state to gain or lose a seat in the apportionment). Finally, let $M = \frac{s}{d}$. In effect, each state's adjusted quota can be obtained by multiplying its quota by M, as we saw in our discussion of the Jefferson method. When $M > 1$ (this happens when the tentative apportionments are not enough to fill the house), the larger states are favored, as they were in the case of the Jefferson method. On the other hand, when $M < 1$, the reverse is true, because a large number multiplied by M will decrease more than a small number would. Thus, the

Webster method favors neither large nor small states when the tentative apportionments exactly fill the house; it favors small states when they overfill the house, and it favors large states when they underfill the house. On balance, the Webster method is neutral, because it is equally likely that the house will be overfilled or underfilled by the tentative apportionments.

The Hill – Huntington Method

The **Hill–Huntington method** is a divisor method that has been used to apportion the U.S. House of Representatives since 1940. Like the Jefferson and Webster methods, the apportionment is calculated by rounding the quotas, after adjusting them if necessary. The only difference between the three divisor methods is in the rounding procedure.

The Hill–Huntington rounding procedure is related to the **geometric mean.** Consider the rectangle $\mathcal{R}$ displayed in Figure 15.1. The area of $\mathcal{R}$ is the product of the lengths A and B, or AB. The geometric mean of A and B is defined to be the length E of the edge of a square $\mathcal{S}$ with the same area as $\mathcal{R}$. The area of $\mathcal{S}$ is E^2, so $E^2 = AB$, and therefore $E = \sqrt{AB}$.

Given a positive number q, let q^* be the geometric mean of $\lfloor q \rfloor$ and $\lceil q \rceil$ The Hill–Huntington rounding of a number q is equal to

$$\lfloor q \rfloor \quad \text{if } q < q^*$$

and

$$\lceil q \rceil \quad \text{if } q \geq q^*$$

For example, suppose that $q = 7.485$. Jefferson and Webster would round 7.485 down to 7. Because $\lfloor 7.485 \rfloor = 7$ and $\lceil 7.485 \rceil = 8$, the geometric mean is $q^* = \sqrt{7 \times 8} = 7.48331\ldots$, which is less than 7.485. Therefore, the Hill–Huntington rounding of 7.485 is $\lceil 7.485 \rceil = 8$. We will use $\langle\langle q \rangle\rangle$ for the Hill–Huntington rounding of a number q.

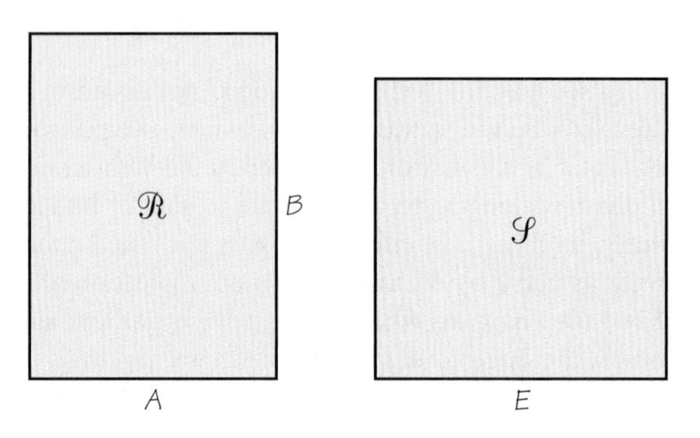

Figure 15.1 The edge of the square is the geometric mean of the edges of the rectangle, because the two figures have the same area.

Hill–Huntington apportionment calculations follow the general plan of the Jefferson and Webster methods. Round each state's quota the Hill–Huntington way to obtain a first tentative apportionment. If the sum of the tentative apportionments is equal to the house size, the job is finished. If not, a list of critical divisors must be constructed, each chosen to be just sufficient to change the corresponding state's apportionment by one seat in the desired direction. If n_i is the number of seats apportioned tentatively to state i, and the total apportionment is too small, the critical divisor for that state is

$$d_i^+ = \frac{p_i}{\sqrt{n_i(n_i + 1)}}$$

where p_i is the state's population. If d_i^+ is used as the divisor, the adjusted quota for state i will be $\sqrt{n_i(n_i + 1)}$, just enough to warrant rounding the quota to $n_i + 1$. Of course, states with critical divisors more than d_i^+ will also receive increased apportionments.

If the total apportionment is too large, the critical divisor for state i is

$$d_i^- = \frac{p_i}{\sqrt{n_i(n_i - 1)}}$$

With m_i as multiplier, the adjusted quota of state i would be equal to $\sqrt{n_i(n_i - 1)}$ the cutoff number for rounding down to $n_i - 1$. If further adjustments are necessary, the critical divisor of the state whose apportionment was changed is recalculated, and the process is repeated.

A zero apportionment is impossible with the Hill–Huntington method, because the rounding point for quotas between 0 and 1 is $\sqrt{0 \times 1} = 0$. Any such quota will be rounded to 1.

EXAMPLE Apportionment of the 1790 Congress, Revisited

The calculations leading to the Hill–Huntington apportionment of the Congress are given in Table 15.6. The rounded quotas add up to 106, more than the house size of 105, so we have to adjust the quotas. The table shows the critical divisors. The smallest critical divisor is Pennsylvania's, 34,658. This divisor is used to calculate the adjusted quotas, and we see that Pennsylvania will lose a seat, while the other apportionments will remain the same. ■

15.4 Which Divisor Method Is Best?

The three divisor methods that we have considered often give different results. When this happens, a state that is favored by the method not in use will often argue that the apportionment was unfair. Spotlight 15.1 summarizes a recent Supreme Court case about this issue. We can try to decide which divisor

TABLE 15.6 Apportioning the House of Representatives by the Hill–Huntington Method

State	Population	Quota	q^*	Tentative Appor-tionment	Critical Divisor	Final Appor-tionment
Virginia	630,560	18.310	18.493	18	36,046	18
Massachusetts	475,327	13.803	13.491	14	35,234	14
Pennsylvania	432,879	12.570	12.490	13	34,658	12
North Carolina	353,523	10.266	10.488	10	37,265	10
New York	331,589	9.629	9.487	10	34,952	10
Maryland	278,514	8.088	8.485	8	37,217	8
Connecticut	236,841	6.877	6.481	7	36,545	7
South Carolina	206,236	5.989	5.477	6	37,653	6
New Jersey	179,570	5.214	5.477	5	40,153	5
New Hampshire	141,822	4.118	4.472	4	40,940	4
Vermont	85,533	2.484	2.449	3	34,918	3
Georgia	70,835	2.057	2.449	2	50,088	2
Kentucky	68,705	1.995	1.414	2	48,582	2
Rhode Island	68,446	1.988	1.414	2	48,399	2
Delaware	55,540	1.613	1.414	2	39,273	2
Totals	3,615,920	105	–	106	–	105

method is fairest by comparing the apportionments actually produced by the methods.

Let a_i be the apportionment given to a state whose population is p_i. The quotient $\frac{a_i}{p_i}$ is called the **representative share.** It represents the share of a congressional seat given to each citizen of the state.

In an ideal apportionment, every state would have the same representative share. Because this is impossible, we can measure how close a given apportionment is to being ideal by using representative share as the standard of comparison. We are given an apportionment done by some method. The method could be Webster, Hamilton, or an old-fashioned smoke-filled room. We would compute the representative share for each state and identify the state that has the largest share and the state that has the smallest. The discrepancy between these two values is a measure of how far the apportionment is from being perfectly equitable. It can be shown that among all conceivable apportionments, the one for which this discrepancy is the least is provided by the Webster method.

SPOTLIGHT 15.1 A Legal Challenge to Apportionment

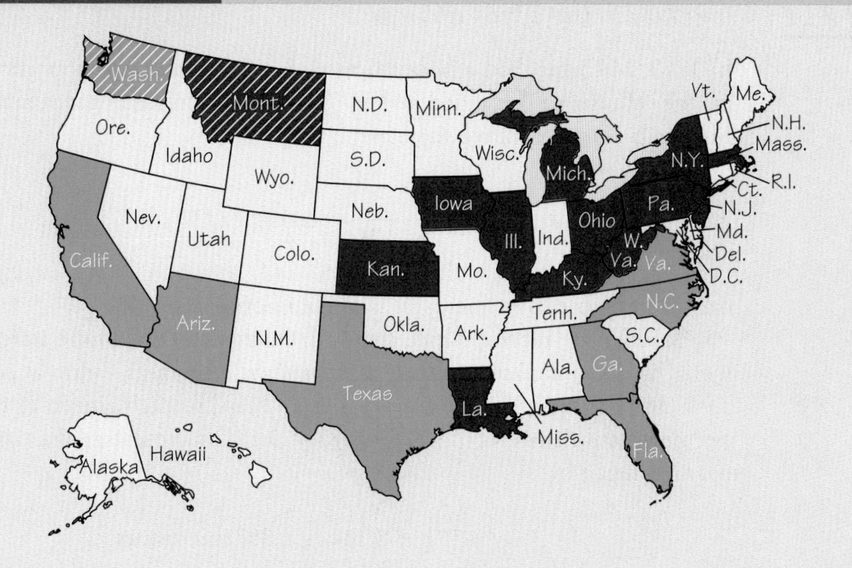

In 1991, the Census Bureau reported the new apportionment that would be in effect for the congressional elections in the years 1992–2000. The states colored red on the above map lost representatives, and those in green gained representatives. New York lost 3 representatives, and Ohio and Pennsylvania lost 2 apiece. Montana, whose apportionment decreased from 2 to 1, sustained the greatest percentage loss, and Montana sued to restore the lost seat. As precedents, Montana referred to the two famous cases, *Baker* v. *Carr* and *Wesberry* v. *Sanders,* in which the U.S. Supreme Court required legislative and congressional district boundaries to be drawn so as to make district populations equal.

Montana argued that the correct apportionment would be the one that met the *Baker* and *Wesberry* criterion of having districts as nearly equal in population as possible, and asked the Court to require the Census Bureau to recompute the apportionments using the Dean method, a divisor method that minimizes differences in district populations. This would have resulted in the transfer of a congressional seat from Washington to Montana.

The Montana case coincided with another federal lawsuit, *Massachusetts* v. *Mosbacher,* which asked the Court to order the apportionment to be calculated by the Webster method. If Massachusetts had won this suit, it would have gained an additional seat, but Montana would not benefit.

In *U.S. Department of Commerce* v. *Montana,* the Supreme Court unanimously rejected Montana's claim. The opinion of the Court, written by Justice John Paul Stevens, pointed out that *intra*state districts, which were the subject of the *Baker* and *Wesberry* cases, could be equalized in population by drawing district boundaries correctly. Because congressional districts can't cross state lines, some inequity is inevitable in congressional apportionment. The opinion conceded that there were alternatives to the Hill–Huntington method but concluded that the choice of apportionment method was best left to Congress.

Thus, while no apportionment can be perfect, the Webster method is the fairest when comparisons are based on representative share.

EXAMPLE

Inequity in the 77th Congress

In 1940, Michigan had a population of 5,256,106 and was apportioned 17 seats in the House of Representatives. This apportionment gave each citizen of Michigan a representative share of

$$\frac{17}{5{,}256{,}106} = 0.000003234 \text{ seat}$$

or 3.234 microseats (a microseat is one-millionth of a seat).

To calculate representative shares in microseats, divide the state's apportionment by its population, *expressed in millions.* Arkansas—with a population of 1,949,387, or 1.949,387 million—received 7 seats, so each citizen of that state had a representative share of $7 \div 1.949{,}387 = 3.591$ microseats. Arkansas was favored over Michigan by

$$3.591 - 3.234 = 0.357 \text{ microseats}$$

If a seat had been transferred from Arkansas to Michigan, then the representative share for a Michigander would have been $18 \div 5.256{,}106 = 3.425$ microseats, while each Arkansan would have been left with a representative share of $6 \div 1.949{,}387 = 3.078$ microseats.

Now Michigan, with the larger representative share, has the advantage, but the discrepancy

$$3.425 - 3.078 = 0.347 \text{ microseats}$$

is less than it was before the transfer was made. In terms of representative share, it would have been more equitable to have given Arkansas 6 seats and Michigan 18 seats. In the Webster apportionment, the seats would have been distributed in this way, because that method minimizes discrepancies in representative share. This shows that the Webster method is better than the Hill–Huntington method, which was used in the 1940 apportionment, *if we view minimizing discrepancies in representative share as being of prime importance.* ■

In an ideal apportionment, each congressional district in the country would have the same population. This is impossible unless we allow congressional districts to overlap state lines, which is forbidden by the Constitution.

The **district population** of state i is $p_i \div a_i$. It represents the average population of a congressional district in the state.

The district population of each state can also serve as a standard of comparison of apportionments. In a given apportionment, the district population for each state would be computed. The best apportionment by this standard would be the one for which the worst difference in district populations between states is minimal.

EXAMPLE

Comparing District Populations

If we consider differences in district population rather than representative share, it was correct to give Michigan 17 seats and Arkansas 7 in the 1940 apportionment. The district population for Michigan was

$$\frac{\text{Population of Michigan}}{\text{Michigan's apportionment}} = \frac{5{,}256{,}106 \text{ people}}{17 \text{ districts}}$$
$$= 309{,}183 \text{ people per district}$$

The district population for Arkansas was 1,949,387 ÷ 7 = 278,484. The Arkansas average district population was 30,689 less than Michigan's. If Michigan had 18 seats and Arkansas had 6, Michigan would have the lesser district population, 292,006, while Arkansas's would have increased to 324,898. This adjustment in apportionment would have increased the inequity between the two states, because now Arkansas would be worse off than Michigan by 32,892 in district population. ■

For state i, the representative share is $\frac{a_i}{p_i}$ and the district population is $\frac{p_i}{a_i}$. Thus:

$$\text{Representative share} = \frac{1}{\text{district population}}$$

It is surprising that these two ways of evaluating the fairness of an apportionment could disagree, but we have just seen that they can. Those who think it is most important for states to have representative shares as nearly equal as possible would say that Michigan should have 18 seats and Arkansas should have 6, but those who prefer to focus on district population instead of representative share would say that the two states should have 17 and 7 seats, respectively.

A mathematician, Edward V. Huntington, suggested a compromise between the two measures of inequity that we have considered—differences in representative share and differences in district populations. He pointed out that if *relative differences* are compared instead of absolute differences, then either district population and representative share would give identical comparisons of apportionments.

Given two positive numbers A and B, with $A > B$, the **absolute difference** is $A - B$ and the **relative difference** is the quotient $\frac{A-B}{B} \times 100\%$.

For any two states, it turns out that the *relative difference* in district populations is equal to the relative difference in representative share (see Exercise 23).

Therefore, an apportionment method that minimizes *relative* difference in representative shares will also minimize the relative difference in district populations. The Hill–Huntington method gives the apportionment in which the relative differences in representative shares (or district populations) are as small as possible.

EXAMPLE Relative Inequity in the 77th Congress

Recall that Michigan was given 17 seats in the 77th Congress and had a representative share of 3.234 microseats. Arkansas had 7 seats and a representative share of 3.591 microseats. The relative difference was

$$\frac{3.591 - 3.234}{3.234} \times 100\% = 11.02\%$$

Thus, Arkansas was 11.02% better represented in the 77th Congress than Michigan was.

If Michigan had 18 seats and Arkansas had 6, the relative difference in representative shares would be found by subtracting the smaller representative share (Arkansas's) from the larger (Michigan's), and expressing the result as a percentage of the smaller representative share. Thus, the relative difference would have been

$$\frac{3.425 - 3.078}{3.078} \times 100\% = 11.27\%$$

in Michigan's favor. Because the relative inequity was less when Michigan had 17 seats and Arkansas had 7, the Hill–Huntington method gave this apportionment. Spotlight 15.2 explains how the Arkansas–Michigan apportionment in 1940 influenced the decision of Congress to adopt the Hill–Huntington method. ■

Because each divisor method can lead to a slightly different apportionment, one way to determine which apportionment method to use is to decide, in a political debate, which type of inequity should be minimized. Challenges to apportionments have followed this approach (see Spotlight 15.1).

An apportionment method could also be chosen by minimizing *bias* in favor of large or small states. We have seen that the Jefferson method favors large states and that the Webster method is neutral. What about the Hill–Huntington method?

Bias in a divisor method is in favor of large states when the quotas are adjusted by using a divisor that is smaller than the standard divisor. If the quotas must be adjusted downward–that is, a divisor larger than the standard divisor is used–small states are favored. Because the rounding point for the Webster method is halfway between whole numbers, it is just as likely for the divisor to be smaller than the standard divisor as it is for it to be larger.

The geometric mean q^* used by the Hill–Huntington method is closer to $\lfloor q \rfloor$ than to $\lceil q \rceil$ (see Exercise 16). This means that a random number q is more likely

SPOTLIGHT 15.2 Mathematics and Politics: A Strange Mixture

Walter F. Willcox

Edward V. Huntington

(Willcox: Department of Manuscripts and University Archives, Cornell University Libraries. Huntington: Courtesy of Harvard University Archives.)

The first American to consider apportionment from a theoretical point of view was Walter Willcox (1861–1964), who strongly advocated the Webster method and had computed the apportionment of 1900. His arguments convinced Congress to use the Webster method again in 1910. In 1911, Joseph Hill, a statistician at the Census Bureau, proposed the Hill–Huntington method—with the strong endorsement of Edward V. Huntington, a mathematics professor at Harvard.

In 1920, the two methods were in competition. There were significant differences in the apportionments determined by the two methods, and the result was Washington gridlock: No apportionment bill passed during the decade, and the 1910 apportionments were retained throughout the 1920s. In preparation for the 1930 census results, the National Academy of Sciences formed a committee to study apportionment. In 1929, the committee endorsed the Hill–Huntington method.

The 1930 census was remarkable in that the apportionments calculated by the Webster method were the same as the Hill–Huntington apportionments. The House was therefore reapportioned, but the method used could be claimed to be either one of the competing methods. The coincidence was almost repeated in the 1940 census, but there was one difference. The Hill–Huntington method gave the last seat to Arkansas, while Webster's method gave it to Michigan (see page 530). At the time, Michigan was a predominantly Republican state, and Arkansas was in the Democratic column. The vote on the apportionment bill split strictly along party lines, with Democrats supporting the Hill–Huntington method and Republicans voting for the Webster method. Because the Democrats had the majority, the Hill–Huntington method became the law.

to be greater than q^* and thus rounded up to $\lceil q \rceil$ than it is to be less than q^* and rounded down to $\lfloor q \rfloor$. The difference between the Webster and Hill–Huntington ways of rounding is not significant for relatively large numbers. For example, if q is a number between 50 and 51, then $q^* = 50.498$. Therefore, q will be rounded up to 51 by Hill–Huntington if it is larger than 50.498. The Webster method would round q to 51 if $q \geq 50.500$. The differences are more significant when rounding smaller numbers. Thus, Hill–Huntington rounds all numbers between 0 and 1 up to 1; Webster rounds only the numbers in the range 0.500–1 up to 1. When the Hill–Huntington method is used for apportionment, the sum of the tentative apportionments is more likely to exceed the house size than it is to be less, especially if there are many states with small populations. Therefore, it is more likely that a divisor larger than the standard divisor will be needed. This favors the less populous states.

REVIEW VOCABULARY

$\lfloor q \rfloor$ The result of rounding a number q down; for example, $\lfloor \pi \rfloor = 3$.
$\lceil q \rceil$ The result of rounding a number q up to the next integer; for example, $\lceil \pi \rceil = 4$.
$\langle q \rangle$ The result of rounding a number q in the usual way: Round down if the fractional part of q is less than 0.5, and round up otherwise. For example, $\langle \frac{86}{57} \rangle = 2$. This notation is not standard and is used only in this text.
$\langle\langle q \rangle\rangle$ The result of rounding a number q the Hill–Huntington way: Round down if q is less than the geometric mean of $\lfloor q \rfloor$ and $\lceil q \rceil$, and round up otherwise. For example, to compute $\langle\langle 2.45 \rangle\rangle$, calculate the geometric mean of $\lfloor 2.45 \rfloor = 2$, and $\lceil 2.45 \rceil = 3$, which is $\sqrt{2 \times 3} \approx 2.449$. Because $2.45 > 2.449$, we round up: $\langle\langle 2.45 \rangle\rangle = 3$.
Absolute difference of two numbers is the result of subtracting the smaller number from the larger.
Adjusted quota The result of dividing a state's quota by a divisor other than the standard divisor. The purpose of adjusting the quotas is to correct a failure of the rounded quotas to sum to the house size.
Alabama paradox A state loses a representative solely because the size of the House is increased. This paradox is possible with the Hamilton method but not with divisor methods.
Apportionment method A systematic way of computing solutions of apportionment problems.
Apportionment problem To round a list of fractions to whole numbers in a way that preserves the sum of the original fractions.
Critical divisor The number closest to the standard divisor that can be used as a divisor of a state's population to obtain a new tentative apportionment for the state. The following table lists formulas for critical divisors for some divisor methods. In the table, p stands for the state's population, and n is its tentative apportionment.

	Critical divisor causing tentative apportionment	
Method	**To increase**	**To decrease**
Jefferson	$\frac{p}{n+1}$	Not necessary
Webster	$\frac{p}{n+0.5}$	$\frac{p}{n-0.5}$
Hill-Huntington	$\frac{p}{\sqrt{n(n+1)}}$	$\frac{p}{\sqrt{n(n-1)}}$

District population A state's population divided by its apportionment.
Divisor method One of many apportionment methods in which the apportionments are determined by dividing the population of each state by a common factor to obtain adjusted quotas. The

apportionments are calculated by rounding the adjusted quotas. Divisor methods differ in the way that the rounding of the quotas is carried out. The methods of Jefferson, Webster, and Hill–Huntington are divisor methods.

Geometric mean For positive numbers A and B, the geometric mean is defined to be $\sqrt{A \times B}$.

Hamilton method An apportionment method, advocated by Alexander Hamilton, that assigns to each state either its lower quota or its upper quota. The states that receive their upper quotas are those whose quotas have the largest fractional parts.

Hill–Huntington method A divisor method, named for the statistician Joseph Hill and the mathematician Edward V. Huntington, that minimizes relative differences in both representative shares and district populations. It is based on the Hill–Huntington way of rounding, so that a state's apportionment is $\langle\langle \tilde{q}_i \rangle\rangle$, where $\tilde{q}_i$ is the state's adjusted quota.

Jefferson method A divisor method invented by Thomas Jefferson, based on rounding all fractions down. Thus, if u_i is the adjusted quota of state i, the state's apportionment is $\lfloor u_i \rfloor$.

Lower quota The integer part $\lfloor q_i \rfloor$ of a state's quota q_i.

Population paradox A situation is which state A gains population and loses a congressional seat, while state B loses population (or increases population proportionally less than state A) and gains a seat. This paradox is possible with all apportionment methods *except* divisor methods.

Quota The quotient p/s of a state's population divided by the standard divisor. The quota is the number of seats a state would receive if fractional seats could be awarded.

Quota condition A requirement that an apportionment method should assign to each state either its lower quota or its upper quota in every situation. The Hamilton method satisfies this condition, but none of the divisor methods do.

Relative difference The relative difference between two positive numbers is obtained by subtracting the smaller number from the larger, and expressing the result as a percentage of the smaller number. Thus, the relative difference of 120 and 100 is 20%.

Representative share A state's representative share is the state's apportionment divided by its population. It is intended to represent the amount of influence a citizen of that state would have on his or her representative.

Standard divisor The ratio p/h of the total population p to the house size h. In a congressional apportionment problem, the standard divisor represents the average district population.

Tentative apportionment The result of rounding a state's quota or adjusted quota to obtain a whole number.

Upper quota The result of rounding a state's quota up to a whole number. A state whose quota is q has an upper quota equal to $\lceil q \rceil$.

Webster method A divisor method of apportionment invented by Representative Daniel Webster. It is based on rounding fractions the usual way, so that the apportionment for state i is $\langle u_i \rangle$, where u_i is the adjusted quota for state i. The Webster method minimizes differences of representative share between states.

SUGGESTED READINGS

BALINSKI, M. L., and H. P. YOUNG. *Fair Representation: Meeting the Ideal of One Man, One Vote,* Yale University Press, New Haven, Conn., 1982. In the 1970s, Balinski and Young analyzed apportionment methods in depth. Their approach was to postulate the desirable properties of an apportionment method as axioms and to deduce from the axioms which method is best. This book combines an account of the history of apportionment of the U.S. House of Representatives with the results of their research.

COMMONWEALTH OF MASSACHUSETTS V. *MOSBACHER,* 785 Federal Supplement 230 (District of Massachusetts 1992). This opinion concerns a suit by Massachusetts to increase its representation because the Hill–Huntington method was an unfair method of apportionment. As a remedy, the commonwealth asked the Court to replace that method with the Webster method. The discussion of this claim is Section D of the opinion, and starts on page 253. The Federal Supplement is available in law libraries.

ERNST, LAWRENCE R. Apportionment methods for the House of Representatives and the court challenges, *Management Science,* 40 (1994): 1207–1227. Ernst, who wrote briefs for the government in both the *Montana* and the *Massachusetts* cases, reviews the apportionment problem and the arguments in favor of and against each of the divisor methods. The article includes a summary of the arguments used by both sides in the two court cases.

LUCAS, W. F. The Apportionment problem. In S. J. Brams, W. F. Lucas, and P. D. Straffin, Jr. (eds.), *Political and Related Models,* Springer-Verlag, New York, 1983, pp. 358–396. An introduction to apportionment, written at a somewhat more advanced level than the presentation in this text.

U.S. DEPARTMENT OF COMMERCE V. *MONTANA,* 112 Supreme Court 1415 (1992). This opinion, available in any law library, gives the grounds for rejecting the *Montana* suit to replace the Hill–Huntington method with the Dean method.

YOUNG, H. PEYTON. *Equity,* Princeton University Press, Princeton, N.J., 1994. Chapter 3 covers apportionment and focuses on which apportionment method is the most equitable.

SUGGESTED WEB SITE

www.census.gov This site contains a two-page history of apportionment of the Congress.

SKILLS CHECK

1. A county is divided into three districts with the following populations: Southern, 3600; Western, 3100; Northeastern, 1600. There are six seats on the county council to be apportioned. What is the quota for the Southern district?

(a) 0.43
(b) 3
(c) 2.6

2. The Hill–Huntington method of apportionment

(a) is currently used to apportion the House of Representatives.
(b) is not a divisor method.
(c) always rounds up the fractional parts.

3. A county is divided into three districts with the following populations: Southern, 3600; Western, 3100; Northeastern, 1600. There are 10 seats on the school board to be apportioned. What is the apportionment for the Northeastern district using the Jefferson method?

(a) 1
(b) 2
(c) 3

4. A county is divided into three districts with the following populations: Southern, 3600; Western, 3100; Northeastern, 1600. There are 10 seats on the school board to be apportioned. What is the apportionment for the Southern district using the Webster method?

(a) 4
(b) 5
(c) 6

5. A county is divided into three districts with the following populations: Southern, 3600; Western, 3100; Northeastern, 1600. There are 10 seats on the school board to be apportioned. If the Hill–Huntington method is used instead of the Webster method, how do the results differ?

(a) The apportionments are the same.
(b) Southern gains an extra seat and Western loses a seat.
(c) Southern gains an extra seat and Northeastern loses a seat.

6. The geometric mean of 6 and 7 is

(a) 6.5.
(b) less than 6.5.
(c) more than 6.5.

7. The relative difference of 6 and 7 is

(a) 1.
(b) approximately 14.29%.
(c) approximately 16.67%.

8. The absolute difference of 6 and 7 is

(a) 1.
(b) 0.5.
(c) 6.5.

9. The Jefferson method is based on the premise that

(a) each representative's district must meet a threshold population size.
(b) each representative must represent no more than a certain population size.
(c) each representative must represent the same-sized population.

10. The Hamilton method is based on which of the following premises?

(a) The apportionment is determined by rounding each state's quota to the nearest integer.
(b) States whose quotas have the largest fractional parts receive an additional apportionment.
(c) Each apportionment unit represents an identical population size.

11. The Webster method is based on the premise that

(a) each state's apportionment is the integer closest to its adjusted quota.
(b) smaller states should be favored in the apportionment process.
(c) adjusted quotas should not be necessary.

12. The Hill–Huntington method utilizes

(a) a round-up procedure.
(b) the geometric mean.
(c) relative differences.

13. The method used to apportion the U.S. House of Representatives since 1940 is

(a) the Webster method.
(b) the Jefferson method.
(c) the Hill–Huntington method.

14. The population paradox is possible when

(a) the Hill–Huntington method is used.
(b) the Webster method is used.
(c) the Hamilton method is used.

15. If the initial calculations leading to the Hill–Huntington apportionment result in a sum that is too large, what happens next?

(a) The adjusted quotas are reduced.
(b) The largest apportionment is reduced.
(c) A different method must be used.

EXERCISES

▲ **Optional.** ■ **Advanced.** ◆ **Discussion.**

The Hamilton Method

1. Use the Hamilton method to round each of the following numbers in the sum to a whole number, preserving the total of 10.

$$0.36 + 1.59 + 0.99 + 2.33 + 2.38 + 2.35 = 10$$

2. *The 37th pearl.* Three friends have bought a bag guaranteed to contain at least 30 high-quality pearls for \$14,900 at an auction. Abe contributed \$5900, Beth's contribution was \$7600, and Charles supplied the remaining \$1400. After taking the bag to your house, they pour 36 pearls from the bag onto the kitchen table.

(a) How many should each friend get if the Hamilton method is used to apportion the pearls according to the size of the contributions?
(b) Charles has noticed the bag isn't empty! Another pearl comes out, and you are asked to recalculate the apportionment.
(c) How do you explain the result to Charles?

3. Suppose that a country has three states, with populations 254,000, 153,000, and 103,000, respectively. The legislature has 102 seats. Show that if the Hamilton method is used to apportion seats, a tie will result. How would you suggest breaking the tie?

4. A small high school has one mathematics teacher who can teach a total of five sections. The subjects that she teaches, and their enrollments, are as follows: geometry, 52; algebra, 33; calculus, 12. Apportion sections to the subjects using the Hamilton method.

5. Repeat Exercise 4 using the following enrollments: geometry, 77; algebra, 18; calculus, 20.

Divisor Methods

◆ **6.** Show that for any numbers q, $\langle q \rangle = \lfloor q + 0.5 \rfloor$.

7. Reapportion the classes in Exercise 5 using the Jefferson method.

8. Reapportion the classes in Exercise 4 using the Webster method.

9. The three friends who bought the pearls (see Exercise 2) ask you to suggest a different apportionment method to distribute their purchase. Before answering, determine the apportionments given by each of the divisor methods covered in this text—Jefferson, Webster, and Hill–Huntington—for the 36- and 37-pearl house sizes. Then make your suggestion.

10. The three friends have bought a lot of 36 identical diamonds, at a total cost of \$36,000, Abe's investment was \$15,500, Beth's was \$10,500, and Charles's was \$10,000. They decided to apportion the diamonds using the Webster method, and they can't make it work out. Can you help?

11. Round the following to whole percentages using the methods of Hamilton, Jefferson, and Webster:

$$87.85\% + 1.26\% + 1.25\% + 1.24\% + 1.23\% + 1.22\% + 1.21\% + 1.20\% + 1.19\% + 1.18\% + 1.17\% = 100\%$$

Do any of these methods violate the quota condition?

12. Round the following percentages to whole numbers, using the methods of Hamilton, Jefferson, and Webster.

$$92.15\% + 1.59\% + 1.58\% + 1.57\% + 1.56\% + 1.55\% = 100\%$$

Do any of these methods violate the quota condition?

The Hill-Huntington Method

13. Find the geometric mean of each pair of numbers.

(a) 0, 1
(b) 1, 2
(c) 2, 3
(d) 3, 4

14. A high school has one math teacher, who can teach five sections. Fifty-six students have enrolled in the algebra class, 28 have signed up for geometry, and 7 students will take calculus. Use the Hill–Huntington method to decide how many sections of each course to schedule.

15. One year later, the high school described in Exercise 14 still has just one math teacher who teaches five sections. The enrollments are algebra, 36; geometry, 61; and calculus, 3. Apportion the classes by the Webster and Hill–Huntington methods. Which apportionment do you think the school principal would prefer?

◆ **16. (a)** Show that for any positive numbers A and B, the geometric mean is less than the arithmetic mean,[4] except when $A = B$; then the two means are equal. (*Hint*: Show that the triangle in Figure 15.2 is a right triangle.)

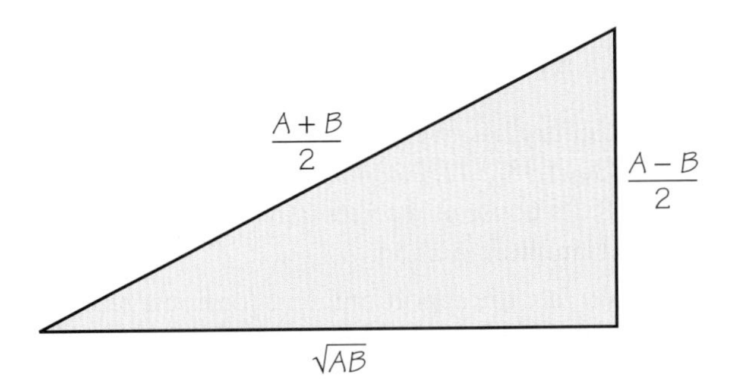

Figure 15.2 Is this a right triangle?

(b) Show that for any number $q \geq 0$, the following inequality holds:

$$\langle q \rangle \leq \langle\langle q \rangle\rangle$$

(c) Explain why the inequality established in part (b) implies that the Hill–Huntington method favors smaller states more than does the Webster method.

■ **17.** Suppose that the governor of Utah believes that the population of his state was undercounted. What increase in population would be large enough to entitle Utah to take a seat from California if the apportionment is by the Hill–Huntington method? The data needed for this problem are given in Exercise 22.

■ **18.** Ties can occur when the Hill–Huntington method is used to apportion if two states have identical populations or if the number of states is larger than the house size. Is there any other way that ties can occur with this method?

[4]The arithmetic mean of A and B is equal to $(A + B)/2$.

Which Divisor Method Is Best?

19. Determine the relative difference between the numbers 5 and 7.

20. Jim is 72 inches tall and Alice is 65 inches tall. What is the relative difference of their heights?

21. In the 2001 apportionment of Congress, the average congressional district in North Carolina had a population of 620,590. Montana, with a population of 905,316, was the most populous state to receive only one district.

(a) Which state is the more favored in this apportionment?
(b) What is the relative difference in the district sizes?

22. According to the 2001 census, the population of California was 33,930,798 and the population of Utah was 2,236,714. California was apportioned 53 House seats, and Utah received 3 House seats.

(a) Determine the average congressional district sizes for these states.
(b) Determine the absolute and relative differences in these district sizes.
(c) Suppose a seat were transferred from California to Utah, giving Utah 4 seats and California 52. What would now be the absolute and relative differences in district sizes?
(d) Does this evidence indicate that the 2001 apportionment fulfills the criterion of keeping absolute differences in district populations as small as possible? What about relative differences?

◆ **23.** Let the populations of states A and B be p_A and p_B, respectively. The apportionments will be a_A and a_B. Assuming that district populations for state A are larger than district populations for state B, show that the relative difference in district populations is

$$\frac{p_A a_B - p_B a_A}{p_B a_A} \times 100\%$$

Also show that this expression is equal to the relative difference in representative share. Hence the relative difference in district populations is equal to the relative difference in representative shares.

24. In *Massachusetts* v. *Mosbacher*, Massachusetts contested its 1991 apportionment, claiming a systematic census undercount of Massachusetts residents living abroad. Another issue in the suit was the claim by Massachusetts that the Hill–Huntington method of apportionment is unconstitutional, because it does not reflect the "one person, one vote" principle as well as the Webster method does. Massachusetts sought an additional House seat that had been awarded to Oklahoma. Would Massachusetts have gained a seat from Oklahoma if the Webster method had been used to apportion the House of Representatives in 1991? Use the following populations and Hill–Huntington apportionments:

State	Population	Apportionment
Massachusetts	6,029,051	10
Oklahoma	3,145,585	6

Additional Exercises

25. A country has four states, *A*, *B*, *C*, and *D*. Its house of representatives has 100 members, apportioned by the Hamilton method. A new census is taken, and the house is reapportioned. Here are the data:

State	Old census	New census
A	5,525,381	5,657,564
B	3,470,152	3,507,464
C	3,864,226	3,885,693
D	201,203	201,049
Totals	13,060,962	13,251,770

(a) Apportion the house using the old census.
(b) Reapportion, using the new census.
(c) Explain how this is an example of the population paradox.

◆ **26.** Here is an apportionment method that should please everyone! Just give each state its upper quota. Of course, the house size would vary.

(a) Show that the house size will be more than the planned house size h.
(b) Do you think California would be enthusiastic about this method, or would that state prefer to give each state its *lower* quota?

27. A country has five states, with populations 5,576,330, 1,387,342, 3,334,241, 7,512,860, and 310,968. Its house of representatives is apportioned by the Hamilton method.

(a) Calculate the apportionments for house sizes of 82, 83, and 84. Does the Alabama paradox occur?
(b) Repeat the calculations for house sizes of 89, 90, and 91.

28. A country that is governed by a parliamentary democracy has two political parties, the Liberals and the Tories. The number of seats awarded to a party is supposed to be proportional to the number of votes it receives in the election. Suppose the Liberals receive 49% of the vote. If the total number of seats in parliament is 99, how many seats do the Liberals get with the Hamilton method? With the Webster method? With the Jefferson method?

◆ **29.** A country with a parliamentary government has two parties that capture 100% of the vote between them. Each party is awarded seats in proportion to the number of votes received.

(a) Show that the Webster and Hamilton methods will always give the same apportionment in this two-party situation.
(b) Show that the Alabama and population paradoxes cannot occur when the Hamilton method is used to apportion seats between two parties or states.
(c) Show that the Webster method satisfies the quota condition when the seats are apportioned between two parties or states.
(d) Will the Jefferson and Hill–Huntington methods also yield the same apportionments as the Hamilton method?

30. If the 1790 Congress were apportioned by the Webster method, would the result differ from the Hill–Huntington method apportionment (see Table 15.6)?

31. The following apportionment method was invented by Congressman William Lowndes of South Carolina in 1822. Lowndes starts, as Hamilton does, by giving each state its lower quota. But where Hamilton apportions the remaining seats to the states whose quotas have the largest fractional parts – in other words, the states for which the *absolute difference* between q_i and $\lfloor q_i \rfloor$ is greatest – Lowndes gives the extra seats to the states where the *relative difference* between q_i and $\lfloor q_i \rfloor$ is greatest, raising as many as necessary to their upper quotas to fill the House.

◆ **(a)** Would this method be more beneficial to states with large populations or small populations, as compared with the Hamilton method?
◆ **(b)** Does the Lowndes method satisfy the quota condition?
◆ **(c)** Would there be any trouble with paradoxes with the Lowndes method?
(d) Use the method to apportion the 1790 House of Representatives.

32. John Quincy Adams, the sixth president of the United States, proposed that the House of Representatives should be apportioned by a divisor method in which a state's apportionment is $\lceil \tilde{q} \rceil$, where $\tilde{q}$ is the state's adjusted quota.

(a) Is it likely that the quotas will need no adjustment?
(b) Will the adjusted quotas be less than the quotas, or greater?
(c) Does the method favor small states or large states?
(d) Find a formula for a state's critical divisor in terms of the state's tentative apportionment n_i and its population p_i.

33. The Marquis de Condorcet, who proposed a criterion for deciding elections (see Chapter 12), also designed a divisor method for apportionment. His rounding rule was to round down numbers whose fractional parts are less than 0.4 and to round up otherwise.

(a) Show that the Condorcet rounding of a number q is $\lfloor q + 0.6 \rfloor$
(b) Does the method favor large states, small states, or is it neutral?
(c) Find a formula for a state's critical divisor in terms of the state's tentative apportionment n_i and its population p_i.

◆ **34.** The U.S. Constitution requires that each state be apportioned at least one seat in the House of Representatives.

(a) Show that the Hill–Huntington method is consistent with this requirement. What about the methods of Hamilton, Jefferson, and Webster?
(b) Does the Adams method (see Exercise 32) meet the requirement?
■ **(c)** The Dean method is the divisor method that minimizes absolute differences in district population. Using this information, explain why a Dean apportionment will never give any state zero seats, unless the house size is less than the number of states.

■ **35.** Let $q_1, q_2, \ldots, q_n$ be the quotas for n states in an apportionment problem, and let the apportionments assigned by some apportionment method be denoted $a_1, a_2, \ldots, a_n$. The *absolute deviation* for state i is defined to be $|q_i - a_i|$; it is a measure of the amount by which the state's apportionment differs from its quota. The *maximum absolute deviation* is the largest of these numbers. Show that the Hamilton method always gives the least possible maximum absolute deviation.

◆ **36.** The choice of a divisor method for apportioning sections to classes according to class enrollments, as in the senior high school example, depends on what the school principal considers most important.

(a) The principal wants to set a minimum class size. For example, if the minimum class size is 20, and 39 students are signed up for English III, there would be one section, because there are not enough students for two sections with enrollment of at least

20. If there were 40 students, there would be two sections. The minimum class size is adjusted so that as many sections as possible are running. What apportionment method should she use, and what will the minimum class size be?

■ **(b)** The principal prefers to set a maximum class size. For example, if the maximum class size is 33, and 67 students are taking history I, there will be three sections, because there are too many students to fit in two 33-student sections. If there were only 66 taking history I, there would be two sections. The maximum class size is adjusted so that as many sections as possible are running. What apportionment method should she use, and what will the maximum class size be? (*Hint:* This divisor method is not described in the text but is mentioned in one of the previous exercises.)

(c) The principal wants to treat students as equitably as possible, so that the differences between students' share of teachers vary as little as possible from course to course. What apportionment method should she use now?

(d) The principal wants to minimize relative difference in class size. What divisor method would work best for her?

(e) The principal wants to cancel any class that has an enrollment of just one student. Which apportionment methods should she avoid using?

WRITING PROJECTS

1. Does the Hill–Huntington method best reflect the intentions of the Founding Fathers, as these intentions were set down in the Constitution and in the debate during the 1787 Constitutional Convention? Good sources of information here include the following publications listed in the Suggested Readings: *Fair Representation,* by Balinski and Young; *Equity,* by Young; "Apportionment Methods," by Ernst; and the two court opinions, *Commonwealth of Massachusetts* v. *Mosbacher* and *U.S. Department of Commerce* v. *Montana*. This writing project requires that you state your answer to the question and make a case for it.

2. Suppose that in 2000, Congress had reverted to its nineteenth-century habit of increasing size of the House of Representatives so that no state would have a decrease in the size of its delegation. How many seats would have been added, and which states would have gotten them? (*Warning:* The apportionments of some states might *increase* as a result of this practice.) As the first step of this project, obtain the populations and apportionments for the 50 states from the Census Bureau Web site (www.census.gov/population/www/censusdata/apportionment.html).

3. A computer is required for this project. Use a spreadsheet to calculate the Hill–Huntington apportionment of the U. S. House of Representatives that would result from the latest census data. You can find the data on the Census Bureau's web site–go to the index and look for state populations. Reapportion using the Webster method. Finally, compute the apportionment by the Dean method. This method is like the Hill–Huntington method, with only one difference: The *harmonic mean* is used instead of the geometric mean. The harmonic mean of two numbers A and B is

$$\frac{2AB}{A+B}$$

Thus, adjusted quotas are rounded down when they are less than the harmonic mean of the lower and upper quotas, and rounded up otherwise. The Dean method minimizes differences in district populations.

With this information, predict the squabbles between states that will occur as a result of the next apportionment.

SPREADSHEET PROJECTS

To do these projects, go to www.whfreeman.com/fapp.

This spreadsheet project models the various apportionment methods using spreadsheet functions. Quotas, tentative apportionments, and adjusted quotas are computed in these explorations. The Hill–Huntington method requires rounding to the geometric mean; a function for this calculation is constructed for the spreadsheet.

APPLET EXERCISES

To do these exercises, go to www.whfreeman.com/fapp.

A bus company has three lines—*A*, *B*, and *C*— and a total of 48,000 riders. *A* has 21,700 riders daily, *B* has 17,200, and *C* has 9100. The company has 40 buses to allocate to the three lines. Use the applet Apportionment to help you find the standard divisor and determine the allocation of the buses according to the methods of apportionment of Hamilton, Jefferson, Webster, and Hill–Huntington.

CHAPTER 16

Game Theory: The Mathematics of Competition

In this chapter we cover...

16.1 Two-Person Total-Conflict Games: Pure Strategies

16.2 Two Person Total-Conflict Games: Mixed Strategies

16.3 Partial-Conflict Games

16.4 Larger Games

16.5 Using Game Theory

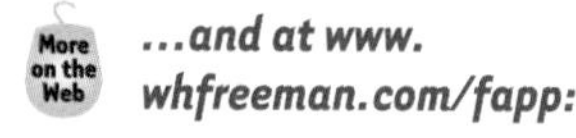

...and at www.whfreeman.com/fapp:

Flashcards

Quizzes

Spreadsheet Projects

Mixed strategies

Applet Exercises

Game theory

Conflict has been a central theme throughout human history and in literature. It arises whenever two or more individuals, with different values or goals, compete to try to control the course of events. *Game theory* uses mathematical tools to study situations, called games, involving both conflict and cooperation. Its study was greatly stimulated by the publication in 1944 of the monumental *Theory of Games and Economic Behavior* by John von Neumann and Oskar Morgenstern (see Spotlight 16.1).

The *players* in a game, who may be people, organizations, or even countries, choose from a list of options available to them—that is, courses of action they may take—that are called **strategies.** The strategies chosen by the players lead to *outcomes,* which describe the consequences of their choices. We assume that the players have *preferences* for the outcomes: They like some more than others.

Game theory analyzes the **rational choice** of strategies—that is, how players select strategies to obtain preferred outcomes. Among areas to which game theory has been applied are bargaining tactics in labor–management disputes, resource-allocation decisions in political campaigns, military options in international crises, and the use of threats by animals in habitat acquisition and protection.

Unlike the subject of *individual* decision making, which researchers in psychology, statistics, and other disciplines study, game theory analyzes situations in which there are at least two players, who may find themselves in conflict because of different goals or objectives. The outcome depends on the choices of *all* the players. In this sense, decision making is *collective,* but this is not to say that the players necessarily cooperate when they choose strategies. Indeed, many strategy choices are noncooperative, such as those between combatants in warfare or competitors in sports. In these encounters, the adversaries' objectives may be at cross-purposes: A gain for one means a loss for the other. But in many activities, especially in economics and politics, there may be joint gains that can be realized from cooperation.

Most interactions probably involve a delicate mix of cooperative and noncooperative behavior. In business, for example, firms in an industry cooperate to gain tax breaks even as they compete for shares in the marketplace.

In the next two sections, we present several simple examples of two-person games of **total conflict,** in which what one player wins the other player loses, so cooperation never benefits the players. We distinguish two different kinds of solutions to such games. Next we analyze two well-known games of **partial conflict,** in which the players can benefit from cooperation but may have strong incentives not to cooperate. We then turn to the analysis of a larger three-person voting game, in which we show how to eliminate undesirable strategies in stages. Finally, we offer some general comments on solving matrix games and discuss different applications of game theory.

16.1 Two-Person Total-Conflict Games: Pure Strategies

For some games with two players, determining the best strategies for the players is straightforward. We begin with such a case.

EXAMPLE A Location Game

Two young entrepreneurs, Henry and Lisa, plan to locate a new restaurant at a busy intersection in the nearby mountains. They agree on all aspects of the restaurant except one. Lisa likes low elevations, whereas Henry wants greater heights – the higher, the better. In this one regard, their preferences are diametrically opposed. What is better for Henry is worse for Lisa, and likewise what is good for Lisa is bad for Henry.

The layout for their location problem is shown in Figure 16.1. Observe that three routes, Avenue A, Boulevard B, and County Road C (blue lines), run in an

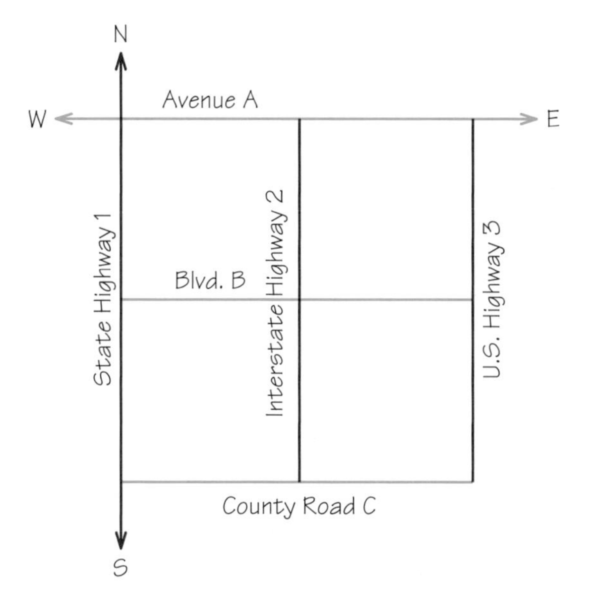

Figure 16.1 The road map for the location example.

TABLE 16.1 Heights (in thousands of feet) of the Nine Intersections

Routes	Highways 1	2	3
A	10	4	6
B	6	5	9
C	2	3	7

east–west direction, and that three highways, numbered 1, 2, and 3 (red lines), run in a north–south direction. Table 16.1 shows the altitudes at the nine corresponding intersections; the same information is shown in three dimensions in Figure 16.2.

To maximize the number of customers, Henry and Lisa agree that the restaurant should be at a location where one of the east–west routes intersects one of the three highways. But they cannot agree on which intersection, so they decide to turn their decision into the following competitive game: Henry will select

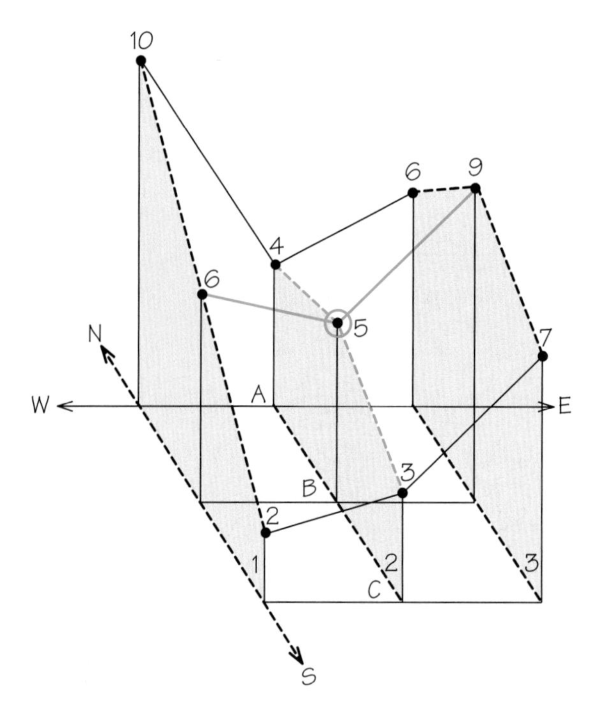

Figure 16.2 Three-dimensional road map showing Henry's and Lisa's possible choices (in thousands of feet).

one of the three routes, A, B, or C, and Lisa will simultaneously choose one of the three highways, 1, 2, or 3. Because their choices will be made simultaneously, neither one can predict beforehand what the other will do.

Henry, worried that Lisa will choose a low elevation, tries to determine the highest altitude he can guarantee by picking one of the three routes. For each choice of a route, this means considering the worst-case (i.e., lowest) elevation on each route. These are the numbers 4, 5, and 2, which are the respective *row minima,* indicated in the right-hand column of Table 16.2. He notes that the highest of these values is 5. By choosing the corresponding route, B, Henry can guarantee himself an altitude of at least 5000 feet.

The number 5 in the right-hand column is referred to as the **maximin,** which is circled in Table 16.2. It is the maximum value of the minimum numbers in the three rows in the table. The strategy that corresponds to the maximin (for Henry, Route B) is called his **maximin strategy.**

Lisa likewise does a worst-case analysis and lists the highest—for her, the worst—elevations for each highway. These numbers 10, 5, and 9, are the column maxima and are listed in the bottom row of Table 16.2. From Lisa's point of view, the best of these outcomes is 5. If she picks Interstate Highway 2, then she is assured of an elevation of no more than 5000 feet.

The number 5 in the bottom row of Table 16.2 is referred to as the **minimax,** which is circled in the table. It is the minimum value of the maximum numbers in the three columns. The strategy that corresponds to the minimax (for Lisa, Highway 2) is called her **minimax strategy.**

TABLE 16.2 Heights (in thousands of feet) in Table 16.1, with the Row Minima (maximum circled) and Column Maxima (minimum circled)

		Lisa Highways			
	Routes	1	2	3	Row Minima
	A	10	4	6	4
Henry	B	6	5	9	⑤
	C	2	3	7	2
	Column Maxima	10	⑤	9	

SPOTLIGHT 16.1 Historical Highlights

John von Neumann

Oskar Morgenstern

As early as the seventeenth century, such outstanding scientists as Christiaan Huygens (1629–1695) and Gottfried W. Leibniz (1646–1716) proposed the creation of a discipline that would make use of the scientific method to study human conflict and interactions. Throughout the nineteenth century, several leading economists created simple mathematical models to analyze particular examples of competitive encounters. The first general mathematical theorem on this subject was proved by the distinguished logician Ernst Zermelo (1871–1956) in 1912. It stated that any finite game with perfect information, such as checkers or chess, has an optimal solution in *pure* strategies; that is, no randomization or secrecy is necessary. A game is said to have *perfect information* if at each stage of the play, every player is aware of all past moves by itself and others as well as all future choices that are possible. This theorem is an example of an *existence theorem:* It demonstrates that there must exist a best way to play such a game, but it does not provide a detailed plan for playing a complex game, like chess, to achieve victory.

The famous mathematician F. E. Émile Borel (1871–1956) introduced the notion of a *mixed,* or randomized, strategy when he investigated some elementary duels around 1920. The fact that every two-person, *zero-sum* game must have optimal mixed strategies and an *expected value* for the game was proved by John von Neumann (1903–1957) in 1928. Von Neumann's result was extended to the existence of equilibrium outcomes in mixed strategies for multiperson games that are either *constant-sum* or *variable-sum* by John F. Nash, Jr. (b. 1928) in 1951.

Modern game theory dates from the publication in 1944 of *Theory of Games and Economic Behavior* by the Hungarian-American mathematician John von Neumann and the Austrian-American economist Oskar Morgenstern (1902–1977). They introduced the first general model and solution concept for multiperson *cooperative games,* which are primarily concerned with coalition formation (by economic cartels, voting blocs, and military alliances) and the resulting distribution of gains or losses. Several other suggestions for a solution to such games have since been proposed. These include the value concept of Lloyd S. Shapley (b. 1923), which relates to fair allocation and economic prices and serves as well as an index of voting power (see Chapter 13).

The French artist George Mathieu designed a medal for the Paris Musée de la Monnaie in 1971 to honor game theory. It was the seventeenth medal to "commemorate 18 stages in the development of Western consciousness." The first medal was for the Edict of Milan in A.D. 313. Game theory also has a mascot, the tiger, arising from the Princeton University tiger and the Russian abbreviation of the term "game theory" (Т ЕОРИЯ ИГ Р, where the underlined letters correspond to the sounds of the English *T*, *G*, and *R*, respectively).

To summarize, Henry has a strategy that will ensure the height is 5000 or higher, and Lisa has a strategy that will ensure the height is 5000 or lower. The height of 5000 at the intersection of Route B and Highway 2 is, simultaneously, the lowest value along Boulevard B and the highest along Interstate Highway 2. In other words, the maximin and the minimax are both equal to 5000 for the location game.

When the maximin and the minimax are the same, the resulting outcome is called a **saddlepoint.**

The reason for the term *saddlepoint* should be clear from the saddle-shaped payoff surface shown in Figure 16.2. The middle point on a horse saddle is simultaneously the lowest point along the spine of the horse and the highest point between the rider's legs. (In Figure 16.2, the rider would be facing leftward or rightward.) In our example, one might also think of the saddlepoint as a mountain pass: As one drives through the pass, the car is at a high point on a highway (in the north–south direction) and at a low point on a route (in the east–west direction).

The resolution of this contest is for Henry to pick B and Lisa to pick 2. This puts them at an elevation of 5000, which is simultaneously the maximin and the minimax.

In total-conflict games, the **value** is the best outcome that both players can guarantee. If a game has a saddlepoint, it gives the value (5 in our example): Players can guarantee this outcome by choosing their maximin and minimax strategies. (Total-conflict games without saddlepoints also have a "value," as we shall see later.)

There is no need for secrecy in a game with a saddlepoint. Even if Henry were to reveal his choice of B in advance, Lisa would be unable to use this knowledge to exploit him. In fact, both players can use the height information in our example to compute the optimal strategy for their opponent as well as for themselves. In games with saddlepoints, players' worst-case analyses lead to the best *guaranteed* outcome—in the sense that each player can ensure that he or she does not do worse than a certain amount (5 in our example) and may do better (if the opponent deviates from a maximin or minimax strategy). ■

Another well-known game with a saddlepoint is tic-tac-toe. Two players alternately place an X or an O, respectively, in one of the nine unoccupied spaces in a 3×3 grid. The winner is the first player to have three X's, or three O's, in either the same row, the same column, or along a diagonal.

An explicit list of all strategies for either the first- or second-moving player in tic-tac-toe is long and complicated, because it specifies a complete plan for all possible contingencies that can arise. For the first-moving player, for example, a strategy might say "put an X in the middle square, then an X in the corner if your opponent puts an O in a noncorner position, etc." While young children initially find this game interesting to play, before long they discover that each player can always prevent the other player from winning by forcing a tie, making the game quite boring. All strategies that force a tie, it turns out, are a saddlepoint in tic-tac-toe.

EXAMPLE

The Restricted-Location Game

Assume in our location game that Henry and Lisa are informed by county officials that it is against the law to locate a restaurant on either Boulevard B or Interstate Highway 2. These two choices, which provided our earlier solution, are now forbidden. The resulting location game without these two strategies is given in Table 16.3 (with payoffs again expressed in thousands of feet).

TABLE 16.3 Heights (in thousands of feet) without Boulevard B and Interstate Highway 2

	Highways	
Routes	**1**	**3**
A	10	6
C	2	7

As before, Henry and Lisa can each do a worst-case analysis. Henry is worried about the minimum number in each row, and Lisa is concerned with the maximum number in each column. These are listed in the right column and bottom row, respectively, in Table 16.4.

TABLE 16.4 Heights (in thousands of feet) in Table 16.3, with Row Minima (maximum circled) and Column Maxima (minimum circled)

		Lisa Highways		
	Routes	**1**	**3**	**Row Minima**
Henry	A	10	6	⑥
	C	2	7	2
	Column Maxima	10	⑦	

Henry sees that his maximin is 6, so he can guarantee a height of 6000 feet or more by choosing Route A. Likewise, Lisa observes that her minimax is 7, so she can keep the elevation of the restaurant down to 7000 feet or less by selecting Highway 3. There is a gap of $7 - 6 = 1$ between the minimax and maximin. When the maximin is less than the minimax, as in this case, then a game does *not* have a saddlepoint, but it does have a value (described in the next section).

If Henry plays his maximin strategy, Route A, and Lisa plays her minimax strategy, Highway 3, then the resulting payoff is 6. However, Henry may be motivated to gamble in this case by playing his other strategy, Route C; if Lisa sticks to her conservative strategy, Highway 3, then the payoff is 7. Henry will have gained one unit (1000 feet), going from 6 to 7.

This is, however, a risky move. If Lisa suspected it, she might counter by selecting Highway 1. The payoff would then be 2, the best for Lisa and the worst for Henry. So Henry's gamble to gain 1 unit (6 to 7) by moving has the risk that he might lose 4 units (6 to 2) if Lisa also moves.

But then there is no incentive for Lisa to play her nonminimax strategy (that is, to play Highway 1) if she believes Henry, in turn, will move back to his maximin strategy (Route A), leading to a payoff of 10. This is worse than 6 from her viewpoint. ■

In two-player games that have saddlepoints, like our original 3×3 location game and tic-tac-toe, each player can calculate the maximin and minimax strategies for both players before the game is even played. Once the solution has been determined by either mathematical analysis or practical experience (as was probably true of tic-tac-toe), there may be little interest in actually playing the game.

But this is decidedly not the case for much more complex games, like chess, whose solution has not yet been determined—and is unlikely to be in the foreseeable future. Even though computers are able to beat world champions on occasion, the computer's winning moves will not necessarily be optimal against those of *all* other opponents. Nevertheless, we know that chess, like tic-tac-toe, has a saddlepoint; what we do not know is whether it yields a win for white, a win for black, or a draw.

Unlike chess, many games, like the 2×2 restricted-location game, do not have an outcome that can always be guaranteed. These games, which include poker, involve uncertainty and risk. In such games, one does not want to have one's strategy detected in advance, because this information can be exploited by an opponent. It is no surprise, then, that poker players are told to keep a "poker face," revealing nothing about their likely moves. But this advice is not very helpful in telling the players what actually to do in the game, such as how many cards to ask for in draw poker.

We will show that there are optimal ways to play two-person total-conflict games without a saddlepoint so as not to reveal one's choices. But their solution is by no means as straightforward as that of games with a saddlepoint.

16.2 Two-Person Total-Conflict Games: Mixed Strategies

Probably most competitive games do not have a saddlepoint like the one found in our first location-game example. Rather, as is illustrated in our restricted-location game – in which the maximin and minimax were not the same – players must try to keep secret their strategy choices, lest their opponent use this information to his or her advantage.

In particular, players must take care to conceal the strategy they will select until the encounter actually takes place, when it is too late for the opponent to alter his or her choice. If the game is repeated, a player will want to *vary* his or her strategy in order to surprise the opponent.

In parlor games like poker, players often use the tactic of *bluffing*. This tactic involves a player's sometimes raising the stakes when he has a low hand so that opponents cannot guess whether or not his hand is high or low – and may, therefore, miscalculate whether to stay in or drop out of the game (a player would prefer opponents to stay in when he has a high hand and drop out when he has a low hand). In military engagements, too, secrecy and even deception are often crucial to success.

In many sporting events, a team tries to surprise or mislead the opposition. A pitcher in baseball will not signal the type of pitch he or she intends to throw in advance, varying the type throughout the game to try to keep the batter off balance. In fact, we next consider a confrontation between a pitcher and batter in more detail.

EXAMPLE A Duel Game

Assume that a particular baseball pitcher can throw either a blazing fastball or a slow curve into the strike zone and so has two strategies: *fast* (denoted by F) and *curve* (C). The pitcher faces a batter who attempts to guess, before each pitch is

The pitcher and the batter use mixed strategies. (Alan Schein/The Stock Market.)

thrown, whether it will be a fastball or a curve, giving the batter also two strategies: guess F and guess C. Assume that the batter has the following batting averages, which are known by both players.

- .300 if the batter guesses fast (F) and the pitcher throws fast (F)
- .200 if the batter guesses fast (F) and the pitcher throws curve (C)
- .100 if the batter guesses curve (C) and the pitcher throws fast (F)
- .500 if the batter guesses curve (C) and the pitcher throws curve (C)

A player's batting average is the number of times he hits safely divided by his number of times at bat. If a batter hit safely 3 times out of 10, for example, his average would be .300.

This game is summarized in Table 16.5. We see from the right-hand column in the table that the batter's maximin is .200, which is realized when he selects his first strategy, F. Thus, the batter can "play it safe" by always guessing a fastball, which will result in his batting .200, hardly enough for him to remain on the team.

We see from the bottom row of the table that the pitcher's minimax is .300, which is obtained when he throws fast (F). Note that the batter's maximin of .200 is less than the pitcher's minimax of .300, so this game does not have a saddlepoint. There is a gap of $.300 - .200 = .100$ between these two numbers.

Each player would like to play so as to win for himself as much of the .100 payoff in the gap as possible. That is, the batter would like to average more than .200, whereas the pitcher wants to hold the batter down to less than .300. ■

A Flawed Approach

If the batter and pitcher in our example consider how they might outguess each other, they might reason along the following lines:

1. *Pitcher* (to himself): If I choose strategy F, I hold the batter down to .300 (the minimax) or less. However, the batter is likely to guess F because it guarantees him at least .200 (his maximin), and it actually provides him with .300 against my F pitch. In this case, the batter wins all the .100 payoff in the gap.

TABLE 16.5 Batting Averages in a Baseball Duel

		Pitcher		Row Minima (maximum circled)
		F	*C*	
Batter	*F*	.300	.200	(.200)
	C	.100	.500	.100
	Column Maxima (minimum circled)	(.300)	.500	

2. *Batter* (to himself): Knowing that the pitcher is reasoning as in step 1, he will try to surprise me with *C*. So I should fool him and guess *C*. I would thus average .500, which will show him up for trying to gamble and outguess me!
3. *Pitcher* (to himself): But if the batter is thinking as in step 2—that is, guessing *C*—I, on second thought, should really throw *F*. This will lead to an average of only .200 for the batter and teach him to not try to outguess me!

This type of cyclical reasoning can go on forever: "I think that he thinks that I think that he thinks . . . . " It provides no resolution to the players' decision problem.

Clearly, there is no pitch, or guess, that is best in all circumstances. Nevertheless, both the pitcher and the batter *can* do better, but not by trying to anticipate each other's choices. The answer to their problem lies in the notion of a *mixed strategy*.

A Better Idea

The play of many total-conflict games requires an element of surprise, which can be realized in practice by making use of a mixed strategy.

Each of the definite courses of action that a player can choose is called a **pure strategy.** A **mixed strategy** is a strategy in which the course of action is randomly chosen from one of the pure strategies in the following way: Each pure strategy is assigned some probability, indicating the relative frequency with which that pure strategy will be played. (The specific strategy used in any given play of the game can be selected using some appropriate random device.)

Note that a pure strategy is a special case of a mixed strategy, with the probability of 1 assigned to just one pure strategy and 0 to all the rest. When a player resorts to a mixed strategy, the resulting outcome of the game is no longer predictable in advance. (For example, if a pitcher throws a curve ball or a fastball with probability 0.5 each, the batter cannot predict which pitch he or she is about to receive.) Rather, the outcome must be described in terms of the probabilistic notion of an *expected value*.

If each of the n payoffs $s_1, s_2, \ldots, s_n$ will occur with the probabilities $p_1, p_2, \ldots, p_n$, respectively, then the average, or **expected value** E, is given by

$$E = p_1s_1 + p_2s_2 + \ldots + p_ns_n$$

We assume that the probabilities sum to 1 and that each probability p_i is never negative. That is, we assume that $p_1 + p_2 + \ldots + p_n = 1$, and $p_i \geq 0$ $(i = 1, 2, \ldots, n)$.

To see how mixed strategies and expected values are used in the analysis of games, we turn to what is perhaps the simplest of all competitive games without a saddlepoint.

EXAMPLE

Matching Pennies

In matching pennies, each of two players simultaneously shows either a head H or a tail T. If the two coins match, with either two heads or two tails, then the first player (player I) receives both coins (a win of 1 for player I). If the coins do not match, that is, if one is an H and the other is a T, then the second player (player II) receives the two coins (a loss of 1 for player I). These wins and losses for player I are shown in Table 16.6.

The game in Table 16.6 is described by a **payoff matrix.** The rows and columns correspond to the strategies of the two players, and the numerical entries give the payoffs to player I when these strategies are chosen.

Although the entries in our earlier tables for the location game also gave payoffs, they were not monetary, as here. A game represented by a payoff matrix is called *a game in strategic form.*

The two rows in Table 16.6 correspond to player I's two pure strategies, H and T, and the two columns to player II's two pure strategies, also H and T. The numbers in the table are the corresponding winnings for player I and losses for player II. If two H's or two T's are played, player I wins 1 from player II. When both an H and a T are played, player I pays out 1 to player II.

It is fruitless for one player to attempt to outguess the other in this game. They should instead resort to mixed strategies and use expected values to estimate their likely gains or losses.

The best thing for player I to do is randomly to select H half the time and T half the time. This mixed strategy can be expressed as

$$(p_H, p_T) = (p_1, p_2) = (p, 1 - p) = \left(\frac{1}{2}, \frac{1}{2}\right)$$

TABLE 16.6 Wins and Losses for Player I in Matching Pennies

		Player II	
		H	T
Player I	H	1	−1
	T	−1	1

Note that the probabilities of choosing $H(p)$ and $T(1 - p)$ do indeed sum to 1, as required; in particular, when $p = \frac{1}{2}$, $1 - p = 1 - \frac{1}{2} = \frac{1}{2}$.

This mixture can be realized in practice by the flip of a coin. Player I's resulting expected value is

$$E_H = \frac{1}{2}(1) + \frac{1}{2}(-1) = 0$$

whenever player II plays H (first column of Table 16.6). Whenever player II plays T (second column), player I's resulting expected value is

$$E_T = \frac{1}{2}(-1) + \frac{1}{2}(1) = 0$$

Player I's average outcome of 0 is the *(mixed-strategy) value* of the game; unlike the use of this notion in games with a saddlepoint, the value here can be realized only by the use of mixed strategies.

The value of 0 is really an expected value and so must be understood in a statistical sense. That is, in a given play of the game, player I will either win 1 or lose 1. However, his or her expectation over many plays of this game is 0. The optimal mixed strategy for player II is likewise a 50–50 mix of H and T, which also leads to an expectation of 0, making the game fair.

A game is **fair** if its value is 0 and, consequently, it favors neither player when at least one player uses an *optimal* (mixed) strategy—one that guarantees that the resulting payoff is the best that this player can obtain against all possible strategy choices (pure or mixed) by an opponent.

Player II gains nothing by knowing that player I is using the optimal mixed strategy $(\frac{1}{2}, \frac{1}{2})$. However, player I must not reveal to player II whether H or T will be displayed *in any given play* of the game before player II makes his or her own choice of H or T. Even without this information, if player II knew that player I was using a particular *nonoptimal* mixed strategy $(p_1, p_2) = (p, 1 - p)$, where $p \neq \frac{1}{2}$—that is, not choosing a 50–50 mixture between H and T—then player II could take advantage of this knowledge and increase his or her average winnings over time to something greater than the value of 0. (See Exercise 14.) ■

EXAMPLE

Nonsymmetrical Matching

In this game, players I and II can again show either heads H or tails T. When two H's appear, player II pays \$5 to player I. When two T's appear, player II pays \$1 to player I. When one H and one T are displayed, then player II collects \$3 from player I. Note that although the sum of player I's gains (\$5 + \$1 = \$6) when there are two H's or two T's, and the sum of player II's gains (\$3 + \$3 = \$6) otherwise, are the same, the game is nonsymmetrical.

A **nonsymmetrical** two-person total-conflict game is one in which the row player's gains (\$5 and \$1 in our example) are different from the column player's gains (always \$3), except when there is a tie (not possible in our example, but possible in tic-tac-toe). In matching pennies, the payoff for winning is the same for each player, so the game is *symmetrical.*

The game just described in given by the payoff matrix in Table 16.7, which shows the payoffs that player I receives from player II. A worst-case analysis, like that which solved our initial location game, is of little help here. Player I may lose \$3 whether he plays H or T, making his maximin -3. Player II can keep her losses down to \$1 by always playing T (and thus avoiding the loss of \$5 when two H's appear), so player II's minimax is 1. However, if player II chooses T and player I knows this, then player I will also play T and collect \$1 from player II. Can player II do better than lose \$1 in each play of the game?

Consider the situation where player I uses a mixed strategy $(p_H, p_T) = (p, 1 - p)$, which involves playing H with probability $1 - p$ and playing T with probability p, where $0 \leq p \leq 1$. Against player II's pure strategy H, player I's expected value is

$$E_H = (5)(p) + (-3)(1 - p) = 8p - 3$$

Against player II's pure strategy T, player I's expected value is

$$E_T = (-3)(p) + (1 - p) = -4p + 1$$

These two linear equations in the variable p are depicted in Figure 16.3. Note that the four points where these two lines intersect the two vertical lines, $p = 0$ and $p = 1$, are the four payoffs appearing in the payoff matrix.

TABLE 16.7 Payoffs for Player I in a Nonsymmetrical Matching Game

		Player II	
		H	T
Player I	H	5	−3
	T	−3	1

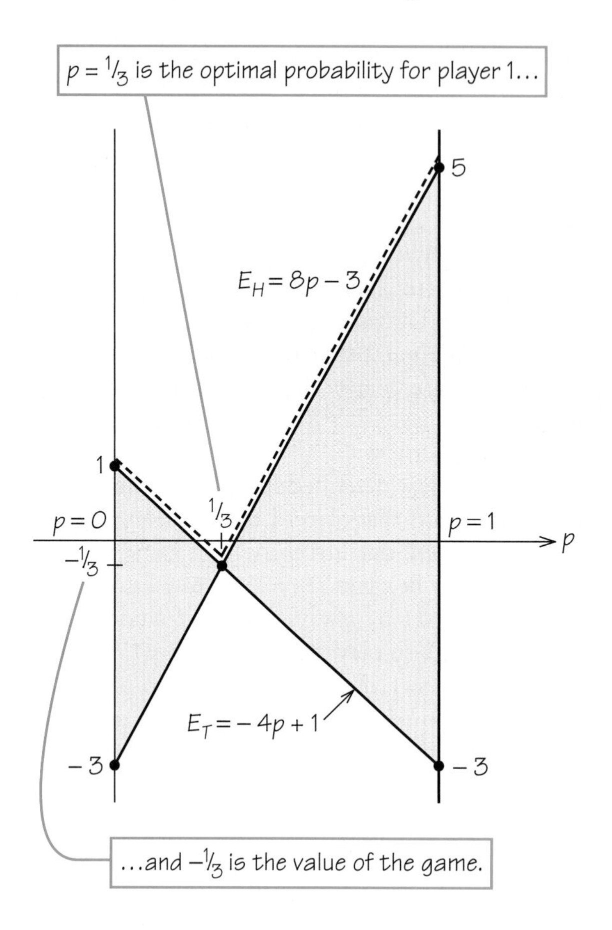

Figure 16.3 Solution to the nonsymmetrical matching pennies.

The point at which the lines given by E_H and E_T intersect can be found by setting $E_H = E_T$, yielding

$$\begin{aligned} 8p - 3 &= -4p + 1 \\ 12p &= 4 \end{aligned}$$

so $p = \frac{1}{3}$. To the left of $p = \frac{1}{3}$, $E_T > E_H$, and to the right $E_H > E_T$; at $p = \frac{1}{3}$, $E_H = E_T$. If player I chooses $(p_H, p_T) = (p, 1 - p) = (\frac{1}{3}, \frac{2}{3})$, he can ensure

$$E_H = 8(1/3) - 3 = E_T = -4(1/3) + 1 = -1/3$$

regardless of what player II does.

In other words, player I's optimal mixed strategy is to pick H and T with probabilities $\frac{1}{3}$ and $\frac{2}{3}$, respectively, which gives player I an expected value of $-\frac{1}{3}$. As can be seen from Figure 16.3, $-\frac{1}{3}$ is the highest expected value that player I can guarantee against *both* strategies H and T of player II. Although T yields player I a higher expected value for $p < \frac{1}{3}$, and H yields him a higher expected value for $p > \frac{1}{3}$, player I's choice of $p_H, p_T = (\frac{1}{3}, \frac{2}{3})$ protects him against an expected loss greater than $-\frac{1}{3}$, which neither of his pure strategies does (each may produce a maximum loss of -3). Put another way, the intersection of E_H and E_T at $p = \frac{1}{3}$ is the minimum of the function given by E_T to the left and E_H to the right (shown by the dashed line in Figure 16.3). If player II had more than two strategies, this approach to finding a minimum that puts a floor on player I's expected loss can be extended.

A similar calculation for player II results in the same optimal mixed strategy $(\frac{1}{3}, \frac{2}{3})$ and expected value $-\frac{1}{3}$. But because the payoffs for player II are losses, $-\frac{1}{3}$ means that she gains $\frac{1}{3}$ on the average.

This game is therefore unfair, even though the sum of the amounts ($6) that player I might have to pay player II when he loses is the same as the sum that player II might have to pay player I when she loses. Interestingly, it is player II, who will win an average of $33\frac{1}{3}$ cents each time the game is played, who is favored, even though she may have to pay more to player I when she loses (a maximum of $5) than player I will ever have to pay her (a maximum of $3). ■

The symmetrical and nonsymmetrical matching games are examples of what are called *zero-sum games.*

A **zero-sum game** is one in which the payoff to one player is the negative of the corresponding payoff to the other, so the sum of the payoffs to the two players is always zero. These games can be completely described by a payoff matrix, in which the numbers represent the payoffs to player I, while their negatives are the payoffs to player II.

Zero-sum games are total-conflict games in which what one player wins the other loses. But not all total-conflict games are zero-sum—in particular, the sum of the payoffs could be some constant other than zero. Nevertheless, the strategic nature of these latter games is the same as that of zero-sum games: What one player wins, the other player still loses. This was true in our location game, in which Henry's payoff was greater the higher the altitude, and Lisa's greater the lower the altitude.

Scoring in professional chess tournaments usually assigns a payoff of 1 for winning, 0 for losing, and $\frac{1}{2}$ to each player for a tie, making the sum of the payoffs to the two players always 1. Such games, called **constant-sum games,** can readily be converted to zero-sum games. Thus, chess could as well be scored -1 for a loss, $+1$ for a win, and 0 for a tie, making the constant 0 in this case. Although constant-sum and zero-sum games have the same strategic nature, constant-sum games are a more general class because the constant need not be zero.

The solution in the symmetrical version of matching pennies illustrated how the mixed strategy of $(\frac{1}{2}, \frac{1}{2})$ guarantees each player the value of 0, but we did not give a *solution technique* for finding optimal mixed strategies. In the nonsymmetrical version of matching pennies, we illustrated a procedure that can be applied to *every* payoff matrix in which each player has only two strategies.

We must use more complex methods, which we will not describe here, to find mixed-strategy solutions when one or both players have more than two strategies. However, one should always check first to see whether a game has a saddlepoint before employing any technique for finding optimal mixed strategies.

In our next example, which is the earlier duel between the pitcher and the batter given by the 2 × 2 payoff matrix in Table 16.5, we already showed that there is no saddlepoint. Thus, the solution will necessarily be in mixed strategies, and we now proceed to find which mix is optimal.

EXAMPLE

The Duel Game Revisited

In Table 16.8, we add probabilities, which we explain next, to Table 16.5, where F indicates fastball and C indicates curve ball. The pitcher should use a mixed strategy $(p_1, p_2) = (p_F, p_C) = (p, 1 - p)$. The probabilities p and $1 - p$ (where $0 \leq p \leq 1$) are indicated below the game matrix and under the corresponding strategies, F and C, for the pitcher. If the pitcher plays a mixed strategy $(p, 1 - p)$ against the two pure strategies, F and C, for the batter, he realizes the respective expected values:

$$E_F = (.3)p + .2(1 - p) = .1p + .2$$
$$E_C = (.1)p + .5(1 - p) = -.4p + .5$$

As in the nonsymmetrical matching-pennies game, the solution to this game occurs at the intersection of the two lines given by E_F and E_C. Setting the equations of these lines equal to each other yields $p = .6$, giving $E_F = E_C = E = .260$.

Thus, the pitcher should use his optimal mixed strategy, which selects F with probability $p = 3/5$ and C with probability $1 - p = 2/5$. This choice will hold the batter down to a batting average of .260, which is the value of the game. We stress that .260 is an average and must be interpreted in a statistical manner. It says that about one time in four the batter will get a hit, but not what will happen on any particular time at bat.

TABLE 16.8 A Baseball Duel with Probabilities

		Pitcher		
		F	*C*	
Batter	*F*	.300	.200	q
	C	.100	.500	$1 - q$
		p	$1 - p$	

Assume that the batter uses a mixed strategy $(q_1, q_2) = (q_F, q_C) = (q, 1 - q)$, as indicated to the right of the game matrix in Table 16.8. This mixed strategy, when played against the pitcher's pure strategies, F and C, results in the respective expected values:

$$E_F = (.3)q + .1(1 - q) = .2q + .1$$
$$E_C = (.2)q + .5(1 - q) = -.3q + .5$$

The intersection of these two lines occurs at the point $q = .8$, giving $E_F = E_C = E = .260$. The batter's optimal mixed strategy is therefore $(q_F, q_C) = (\frac{4}{5}, \frac{1}{5})$, which gives him the same batting average of .260. ■

We have seen that the outcome of .260, which is the value of the game, occurs when either the pitcher selects his optimal mixed pitching strategy $(\frac{3}{5}, \frac{2}{5})$ or the batter selects his optimal mixed guessing strategy $(\frac{4}{5}, \frac{1}{5})$. This particular result holds true for every two-person zero-sum game; it is the fundamental theorem for such games and is known as the *minimax theorem.*

The **minimax theorem** guarantees that there is a unique game value (.260 in our example), and an optimal strategy for each player, so that either player alone can realize at least this value by playing this strategy, which may be pure or mixed.

While our previous examples illustrated this theorem, they are not a proof of it, which can be found in advanced game theory texts.

16.3 Partial-Conflict Games

The 2×2 matrix games presented so far have been total-conflict games: One player's gain was equal to the other player's loss. Although most parlor games, like chess or poker, are games of total conflict, and therefore constant-sum, most real-life games are surely not. (Elections, in which there are usually a clear-cut winner and one or more losers, probably come as close to being games of total conflict as we find in the real world.) We will consider two games of partial conflict, in which the players' preferences are not diametrically opposed, that have often been used to model many real-world conflicts.

Games of partial conflict are **variable-sum games,** in which the sum of payoffs to the players at the different outcomes varies.

There is some mutual gain to be realized by both players if they can cooperate in partial-conflict games, but this may be difficult to do in the absence of either good communication or trust. When these elements are lacking, players are less likely to comply with any agreement that is made. *Noncooperative games* are games in which a binding agreement cannot be enforced. Even if communication is allowed in such games, there is no assurance that a player can trust an opponent to choose a particular strategy that he or she promised to select.

In fact, the players' self-interests may lead them to make strategy choices that yield both lower payoffs than they could have achieved by cooperating. Two partial-conflict games illustrate this problem later in the chapter.

EXAMPLE Prisoners' Dilemma

Prisoners' Dilemma is a two-person variable-sum game. It provides a simple explanation of the forces at work behind arms races, price wars, and the population problem. In these and other similar situations, the players can do better by cooperating. But there may be no compelling reasons for them to do so, unless the players have credible threats of retaliation for not cooperating. The term *Prisoners' Dilemma* was first given to this game by Princeton mathematician Albert W. Tucker (1905 – 1994) in 1950.

Before defining the formal game, we introduce it through a story, which involves two persons, accused of a crime, who are held incommunicado. Each has two choices: to maintain his or her innocence, or to sign a confession accusing the partner of committing the crime.

Now it is in each suspect's interest to confess and implicate the partner, thereby trying to receive a reduced sentence. Yet if both suspects confess, they ensure a bad outcome – namely, they are both found guilty. What is good for the prisoners as a pair – to deny having committed the crime, leaving the state with insufficient evidence to convict them – is frustrated by their pursuit of their own individual rewards.

Prisoners' Dilemma, as we already noted, has many applications, but we will use it here to model a recurrent problem in international relations: arms races between antagonistic countries, which earlier included the superpowers but now include such countries as India and Pakistan and Israel and some of its Arab neighbors.

For simplicity, assume there are two nations, Red and Blue. Each can independently select one of two policies:

A: Arm in preparation for a possible war (noncooperation).
D: Disarm, or at least negotiate an arms-control agreement (cooperation).

There are four possible outcomes:

(D, D): Red and Blue disarm, which is *next best* for both because, while advantageous to each, it also entails certain risks.

(A, A): Red and Blue arm, which is *next worst* for both, because they spend needlessly on arms and are comparatively no better off than at (D, D).
(A, D): Red arms and Blue disarms, which is *best for Red* and *worst for Blue*, because Red gains a big edge over Blue.
(D, A): Red disarms and Blue arms, which is *worst for Red* and *best for Blue*, because Blue gains a big edge over Red.

This situation can be modeled by means of the matrix in Table 16.9, which gives the possible outcomes that can occur. Here, Red's choice involves picking one of the two rows, whereas Blue's choice involves picking one of the two columns.

TABLE 16.9 The Outcomes in an Arms Race, as Modeled by Prisoners' Dilemma

		Blue	
		A	*D*
Red	*A*	Arms race	Favors red
	D	Favors blue	Disarmament

We assume that the players can rank the four outcomes from best to worst, where 4 = best, 3 = next best, 2 = next worst, and 1 = worst. Thus, the higher the number, the greater the payoff, but these payoffs are only **ordinal:** They indicate an ordering of outcomes from best to worst but say nothing about the *degree* to which a player prefers one outcome over another. To illustrate, if a player despises the outcome that he or she ranks 1 but sees little difference among the outcomes ranked 4, 3, and 2, the "payoff distance" between 4 and 2 will be less than that between 2 and 1, even though the numerical difference between 4 and 2 is greater.

The ordinal payoffs to the players for choosing their strategies of A and D are shown in Table 16.10, where the first number in the pair indicates the payoff

TABLE 16.10 Ordinal Payoffs in an Arms Race, as Modeled by Prisoners' Dilemma

		Blue	
		A	*D*
Red	*A*	(2, 2)	(4, 1)
	D	(1, 4)	(3, 3)

to the row player (Red), and the second number the payoff to the column player (Blue). Thus, for example, the pair (1, 4) in the second row and first column signifies a payoff of 1 (worst outcome) to Red and a payoff of 4 (best outcome) to Blue. This outcome occurs when Red unilaterally disarms while Blue continues to arm, making Blue, in a sense, the winner and Red the loser.

Let us examine this strategic situation more closely. Should Red select strategy *A* or *D*? There are two cases to consider, which depend on what Blue does:

- If Blue selects *A*: Red will receive a payoff of 2 for *A* and 1 for *D*, so it will choose *A*.
- If Blue selects *D*: Red will receive a payoff of 4 for *A* and 3 for *D*, so it will choose *A*.

In both cases, Red's first strategy (*A*) gives it a more desirable outcome than its second strategy (*D*). Consequently, we say that *A* is Red's **dominant strategy,** because it is always advantageous for Red to choose *A* over *D*.

In Prisoners' Dilemma, *A* dominates *D* for Red, so we presume that a rational Red would choose *A*. A similar argument leads Blue to choose *A* as well, that is, to pursue a policy of arming. Thus, when each nation strives to maximize its own payoffs independently, the pair is driven to the outcome (*A, A*), with payoffs of (2, 2). The better outcome for both, (*D, D*), with payoffs of (3, 3), appears unobtainable when this game is played noncooperatively.

The outcome (*A, A*), which is the product of dominant strategy choices by both players in Prisoners' Dilemma, is a *Nash equilibrium*.

When no player can benefit by departing unilaterally (i.e., by itself) from its strategy associated with an outcome, the strategies of the players constitute a **Nash equilibrium.** (Technically, while it is the set of strategies that define the equilibrium, the choice of these strategies leads to an outcome that we shall also refer to as the equilibrium.)

Note that in Prisoners' Dilemma, if either player departs from (*A, A*), the payoff for the departing player who switches to *D* drops from 2 to 1 at (*D, A*) and (*A, D*). Not only is there no benefit from departing, but there is actually a loss, with the *D* player punished with its worst payoff of 1. These losses would presumably deter each nation from moving away from the Nash equilibrium of (*A, A*), assuming the other nation sticks to *A*.

Even if both nations agreed in advance jointly to pursue the socially beneficial outcome, (*D, D*), (3, 3) is unstable. This is because if either nation alone reneges on the agreement and secretly arms, it will benefit, obtaining its best payoff of 4. Consequently, each nation would be tempted to go back on its word and select *A*. Especially if nations have no great confidence in the trustworthiness of their opponents, they would have good reason to try to protect themselves against defection from an agreement by arming.

Prisoners' Dilemma is a two-person variable-sum game in which each player has two strategies, cooperate or defect. Defect dominates cooperate for both players, even though the mutual-defection outcome, which is the unique Nash equilibrium in the game, is worse for both players than the mutual-cooperation outcome.

Note that if 4, 3, 2, and 1 in Prisoners' Dilemma were not just ranks but numerical payoffs, their sum would be $2 + 2 = 4$ at the mutual-defection outcome and $3 + 3 = 6$ at the mutual-cooperation outcome. At the other two outcomes, the sum, $1 + 4 = 5$, is still different, illustrating why Prisoners' Dilemma is a variable-sum game.

In real life, of course, people often manage to escape the noncooperative Nash equilibrium in Prisoners' Dilemma. Either the game is played within a larger context, wherein other incentives are at work, such as cultural norms that prescribe cooperation [though this is just another way of saying that defection from (D, D) is not rational, rendering the game not Prisoners' Dilemma], or the game is played on a repeated basis—it is not a one-shot affair—so players can induce cooperation by setting a pattern of rewards for cooperation and penalties for noncooperation.

In a repeated game, factors like reputation and trust may play a role. Realizing the mutual advantages of cooperation in costly arms races, players may inch toward the cooperative outcome by slowly phasing down their acquisition of weapons over time, or even destroying them (the United States and Russia have been doing exactly this). They may also initiate other productive measures, such as improving their communication channels, making inspection procedures more reliable, writing agreements that are truly enforceable, or imposing penalties for violators when their violations are detected (as may be possible by reconnaissance or spy satellites).

Prisoners' Dilemma illustrates the intractable nature of certain competitive situations that blend conflict and cooperation. The standoff that results at the Nash equilibrium of (2, 2) is obviously not as good for the players as that which they could achieve by cooperating—but they risk a good deal if the other player defects.

While saddlepoints are Nash equilibria in total-conflict games, they can never be worse for *both* players than some other outcome (as in partial-conflict games like Prisoners' Dilemma). The reason is that if one player does worse in a total-conflict or zero-sum game, the other player must do better.

The fact that the players must forsake their dominant strategies to achieve the (3, 3) cooperative outcome (see Table 16.10) makes this outcome a difficult one to sustain in one-shot play. On the other hand, assume that the players can threaten each other with a policy of tit-for-tat in repeated play: "I'll cooperate on each round unless you defect, in which case I will defect until you start cooperating again." If these threats are credible, the players may well shun their defect strategies and try to establish a pattern of cooperation in early rounds, thereby fostering the choice of (3, 3) in the future. Alternatively, they may look ahead, in a manner that will be described at the end of this chapter, to try to stabilize (3, 3). ■

EXAMPLE Chicken

Let us look at one other two-person game of partial conflict, known as *Chicken*, which can also lead to troublesome outcomes. Two drivers approach each other at high speed. Each must decide at the last minute whether to swerve to the right or not swerve. Here are the possible consequences of their actions:

1. Neither driver swerves, and the cars collide head-on, which is the worst outcome for both because they are killed (payoff of 1).
2. Both drivers swerve—and each is mildly disgraced for "chickening out"—but they do survive, which is the next-best outcome for both (payoff of 3).
3. One of the drivers swerves and badly loses face, which is his next-worst outcome (payoff of 2), whereas the other does not swerve and is perceived as the winner, which is her best outcome (payoff of 4).

These outcomes and their associated strategies are summarized in Table 16.11.

If both drivers persist in their attempts to "win" with a payoff of 4 by not swerving, the resulting outcome will be mutual disaster, giving each driver his or her worst payoff of 1. Clearly, it is better for both drivers to back down and each obtain 3 by swerving, but neither wants to be in the position of being intimidated into swerving (payoff of 2) when the other does not (payoff of 4).

Notice that neither player in Chicken has a dominant strategy. His or her better strategy depends on what the other player does: Swerve if the other does not, don't swerve if the other player swerves, making this game's choices highly interdependent, which is characteristic of many games. The Nash equilibria in Chicken, moreover, are (4, 2) and (2, 4), suggesting that the compromise of (3, 3) will not be easy to achieve because both players will have an incentive to deviate from it in order to try to be the winner.

TABLE 16.11 Payoffs in a Driver Confrontation, as Modeled by Chicken

		Driver 2	
		Swerve	Not Swerve
Driver 1	Swerve	(3, 3)	(2, 4)
	Not Swerve	(4, 2)	(1, 1)

Chicken is a two-person variable-sum game in which each player has two strategies: to swerve to avoid a collision or not to swerve and possibly cause a collision. Neither player has a dominant strategy. The compromise outcome, in which both players swerve, and the disaster outcome, in which both players do not, are not Nash equilibria. The other two outcomes, in which one player swerves and the other does not, are Nash equilibria.

In fact, there is a third Nash equilibrium in Chicken, but it is in mixed strategies, which can only be computed if the payoffs are not ranks, as we have assumed here, but numerical values. Even if the payoffs were numerical, however, it can be shown that this equilibrium is always worse for both players than the cooperative (3, 3) payoffs. Moreover, it is implausible that players would sometimes swerve and sometimes not – randomizing according to particular probabilities – in the actual play of this game, compared with either trying to win outright or reaching a compromise.

The two pure-strategy Nash equilibria in Chicken suggest that, insofar as there is a "solution" to this game, it is that one player will succeed when the other caves in to avoid the mutual-disaster outcome. But there are certainly real-life cases in which a major confrontation was defused and a compromise of sorts was achieved in Chicken-type games. This fact suggests that the one-sided solution given by the two pure-strategy Nash equilibria may not be the only pure-strategy solution, especially if the players are farsighted and think about the possible untoward consequences of their actions.

International crises, labor – management disputes, and other conflicts in which escalating demands may end in wars, strikes, and other catastrophic outcomes have been modeled by the game of Chicken. But Chicken, like Prisoners' Dilemma, is only one of the 78 essentially different 2×2 ordinal games in which each player can rank the four possible outcomes from best to worst.

Chicken and Prisoners' Dilemma, however, are especially disturbing, because the cooperative (3, 3) outcome in each is not a Nash equilibrium. Unlike a constant-sum game, in which the losses of one player are offset by the gains of the other, *both* players can end up doing badly – at (2, 2) in Prisoners' Dilemma and (1, 1) in Chicken – in these variable-sum games. ■

16.4 Larger Games

We have shown how to compute optimal pure and mixed strategies, and the values ensured by using them, in 2×2 constant-sum games. In 2×2 variable-sum games, we focused on Nash equilibria as a solution concept in Prisoners' Dilemma and Chicken, but we found that this notion of a stable outcome did not justify the choice of cooperative strategies in either of these games.

We turn next to a somewhat larger game, in which there are three players, each of whom can choose among three strategies, which is technically a $3 \times 3 \times 3$ game.

In this game, we eliminate certain undesirable strategies, but in stages, to arrive at a Nash equilibrium that seems quite plausible.

If one of the three players has a dominant strategy in the $3 \times 3 \times 3$ game, we suppose this player will choose it, thereby reducing the game to a 3×3 game between the other two players. (Of course, if no player has a dominant strategy in a three-person game, it cannot be reduced in this manner to a two-person game.)

If this game is not one of total conflict, the minimax theorem, which guarantees players the value in a two-person zero-sum game, is not applicable. Even if the game were zero-sum, the fact that we assume the players in the $3 \times 3 \times 3$ game can only rank outcomes, not assign numerical values to them, means that they cannot calculate optimal mixed strategies in it.

The problem in finding a solution to the reduced 3×3 game is not a lack of Nash equilibria. Rather, there are too many! So the question becomes which, if any, are likely to be selected by the players. Specifically, is one more appealing than the others? The answer is "yes," but it requires extending the idea of dominance, discussed in the previous section, to its successive application in different stages of play.

EXAMPLE The Paradox of the Chair's Position

The $3 \times 3 \times 3$ game we analyze involves voting, illustrating the applicability of game theory to politics. There is also a tie-in to the analysis of weighted voting in Chapter 13, because one of the players (the chair in the voting body), while not having more votes than the others, can break ties. This would seem to make the chair more powerful, in some sense, than the other players.

As we shall see, however, rather than making the chair more powerful—as we earlier measured voting power—the possession of a tie-breaking vote backfires, preventing the chair from obtaining a preferred outcome. However, we do indicate a possible escape for the chair from this unenviable position.

The Banzhaf and Shapley–Shubik power indices described in Chapter 13 do not take into account the preferences of the differently weighted players—all combinations and all permutations were assumed to be equally likely. But as we shall show next, the greater resources that a player like the chair has do not always translate into greater **power,** by which we mean the ability of a player to obtain a preferred outcome.

To illustrate this problem, suppose there is a set of three voters, $V = \{X, Y, Z\}$, and a set of three alternatives, $A = \{x, y, z\}$, from which the voters choose. Assume that voter X prefers x to y to z, indicated by xyz; voter Y's preference is yzx; and voter Z's is zxy. These preferences give rise to a *Condorcet voting paradox* (discussed in Chapter 12), because the social ordering, according to majority rule, is *intransitive:* Although a majority (voters X and Z) prefer x to y, and a majority (voters X and Y) prefer y to z, a majority (voters Y and Z) prefer z to x. So there is no *Condorcet winner*—an alternative that would beat all others in separate pairwise contests. Instead, every alternative can be beaten by one other.

Assume that the voting procedure used by the three voters, who choose from among the three alternatives, is the **plurality procedure,** under which the alternative with the most votes wins. If there is a three-way tie (there can never be a two-way tie if there are three voters), we assume the chair X can break the tie, giving the chair what would appear to be an edge over the other two voters, Y and Z.

To begin, assume that voting is sincere.

> Under **sincere voting,** every voter votes for his or her most-preferred alternative, based on his or her true preferences, without taking into account what the other players might do (see Chapter 12).

In this case, X will prevail by being able to break the tie in favor of x. However, X's apparent advantage disappears if voting is "sophisticated," as demonstrated below.

To see why, first note that X has a dominant strategy of "vote for x": It is never worse and sometimes better than her other two strategies, whatever the other two voters do. Thus, if the other two voters vote for the same alternative, it wins, and X cannot do better than vote sincerely for x, so voting sincerely is never worse. On the other hand, if the other two voters disagree, X's tie-breaking vote (along with her regular vote) for x will be decisive in x's selection, which is X's best outcome.

Given the dominant choice of x on the part of X, Y and Z face the strategy choices shown in Figure 16.4. Y has one, and Z has two, **dominated strategies,** which are never better and sometimes worse than some other strategy, whatever the other two voters do. For example, observe that "vote for x" by Y always leads to his worst alternative, x. The dominated strategies are crossed out in the top matrix in Figure 16.4.

FIRST REDUCTION

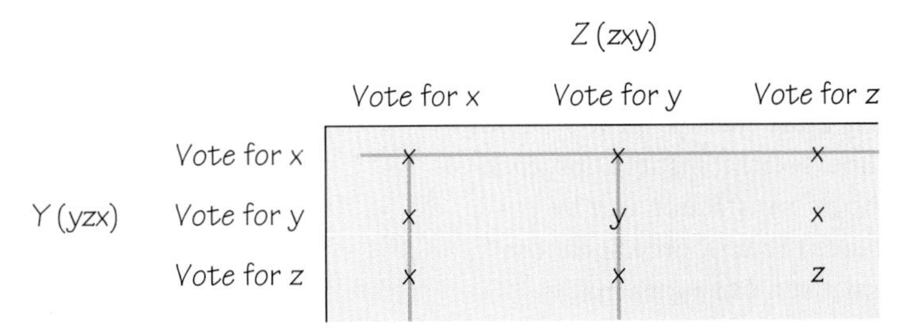

Z (zxy)
Vote for x
Vote for y
Vote for z
Y (yzx)
Vote for x
Vote for y
Vote for z
x
x
x
x
y
x
x
x
z

SECOND REDUCTION

		Z (zxy)
		Vote for z
Y (yzx)	Vote for y	~~x~~
	Vote for z	z

Figure 16.4 Sophisticated voting, given X chooses "vote for x." The dominated strategies of each voter are crossed out in the first reduction, leaving two (undominated) strategies for Y and one (dominant) strategy for Z. Given these eliminations, Y would then eliminate "vote for y" in the second reduction, making z the sophisticated outcome.

SPOTLIGHT 16.2 The 1994 Nobel Prize in Economics

John C. Harsanyi

John F. Nash, Jr.

The Nobel Memorial Prize in Economics was awarded to three game theorists in 1994, marking the 50th anniversary of the publication of von Neumann and Morgenstern's *Theory of Games and Economic Behavior* (see Spotlight 16.1).

- *John C. Harsanyi* (1920–2000) of the University of California, Berkeley, a Hungarian-American who emigrated from Hungary to Australia in 1950 and then to the United States in 1956. He is well known for extending game theory to the study of ethics and how societal institutions, each of whose members' satisfaction can be measured against that of others, choose among alternatives. His other major contribution was to give a precise definition to "incomplete information" in games in which players may be thought of as different types and probabilities are assigned to each type. His analysis of such games is applicable to the modeling of many real-life conflicts in which there are severe constraints on the information that players have about each other. Harsanyi was trained in both mathematics and economics; he also had strong interests in philosophy.
- *John F. Nash, Jr.* (b. 1928) of Princeton University, an American mathematician who did path-breaking work on both noncooperative game theory (the "Nash equilibrium" is named after him) and cooperative game theory, especially on bargaining, in which axioms or assumptions are specified and a unique solution that satisfies these axioms is derived. Nash obtained his results in the early 1950s, when he was only in his 20s, after which he became mentally ill and was unable to work. Fortunately, he has made a remarkable recovery and has now resumed research. A best-selling biography of Nash, a fictionalized version of which was made into a movie in 2001, is included in the Suggested Readings.
- *Reinhard Selten* (b. 1930) of the University of Bonn, a German mathematician who proposed significant refinements in the concept of the Nash equilibrium that help to

Russell Crowe as John Nash, Jr. in *A Beautiful Mind,* the 2002 Academy Award® winner for best film, based on Sylvia Nasar's biography.

Reinhard Selten

distinguish those that are most plausible in games (often there are many such equilibria, which creates a selection problem). Some of his work on equilibrium selection was done in collaboration with Harsanyi. Selten is also noted for pioneering work on developing game-theoretic models in evolutionary biology. He is an advocate of experimental testing of game-theoretic solutions to determine those that are most likely to be chosen by human subjects and using these empirical results to refine the theory.

Other contemporary game theorists have also made important advances in the theory, including Lloyd S. Shapley (b. 1923), an American mathematician who proposed the cooperative game-theoretic solution concept known as the "Shapley value." He extended this work with Robert J. Aumann (b. 1930), a mathematician who was born in Germany, emigrated as a child to the United States before World War II, and then moved to Israel. Aumann also gave a precise formulation to "common knowledge," developing some of its consequences in games, and proposed the notion of a "correlated equilibrium," which has been helpful in understanding how players coordinate their choices in games.

Game theory has provided important theoretical foundations in economics, starting with microeconomics but now extending to macroeconomics and international economics. It also has been increasingly applied in political science, especially in the study of voting, elections, and international relations. In addition, game theory has contributed major insights in biology, particularly in understanding the evolution of species and conditions under which animals – humans included – fight each other for territory or act altruistically. It has also illuminated certain fields in philosophy, including ethics, the philosophy of religion, and political philosophy, and inspired many experiments in social psychology.

This leaves Y with two *undominated* strategies that are neither dominant nor dominated; "vote for y" and "vote for z." "Vote for y" is better than "vote for z" if Z chooses y (leading to y rather than x), whereas the reverse is the case if Z chooses z (leading to z rather than x). By contrast, Z has a dominant strategy of "vote for z," which leads to outcomes at least as good and sometimes better than his other two strategies.

If voters have complete information about each other's preferences, then they can perceive the situation represented by the top matrix of Figure 16.4. Reasoning that no player would ever choose a dominated strategy, they would eliminate such strategies from consideration (these have been crossed out in the first reduction).

The elimination of these strategies gives the bottom matrix in Figure 16.4. Then Y, choosing between "vote for y" and "vote for z" in this matrix, would cross out "vote for y" (second reduction), now dominated because that choice would result in x's winning due to the chair's tie-breaking vote. Instead, Y would choose "vote for z" (hence, not Y's sincere strategy), ensuring z's election, which is Z's best outcome but only the next-best outcome for Y. In this manner, z, which is not the first choice of a majority and could in fact be beaten by y in a pairwise contest, becomes the sophisticated outcome.

The successive elimination of dominated strategies by voters (insofar as this is possible), beginning in our example with X's choice of x in favor of y and z, is called **sophisticated voting.**

Sophisticated voting results in a Nash equilibrium, because none of the three players can do better by departing from his or her sophisticated strategy when the other two players choose theirs. This is clearly true for X, because x is her dominant strategy; given X's choice of x, z is dominant for Z; and given these choices by X and Z, z is dominant for Y. These "contingent" dominance relations, in general, make sophisticated strategies a Nash equilibrium.

Observe, however, that there are four other Nash equilibria in this game. First, the choice of each of x, y, or z by all three voters are all Nash equilibria, because no single voter's departure can change the outcome to a different one, much less a better one, for that player. In addition, the choice of x by X, y by Y, and x by Z—resulting in x— is also a Nash equilibrium, because no voter's departure would lead to a better outcome for him or her.

In game-theoretic terms, sophisticated voting produces a different and smaller game in which some formerly undominated strategies in the larger game become dominated in the smaller game. The removal of such strategies, sometimes in several successive stages, in effect enables sophisticated voters to determine what outcomes eventually *will* be chosen by eliminating those outcomes that definitely *will not* be chosen. Voters can thereby ensure that their worst outcomes will not be chosen by successively removing dominated strategies, given the presumption that other voters do likewise.

How does sophisticated voting affect the chair's presumed extra power? The chair's tie-breaking vote is not only *not* helpful—it is positively harmful. It guarantees that X's worst outcome (z) will be chosen if voting is sophisticated! This situation, in which the chair's tie-breaking vote hurts rather than helps the chair, is called the **paradox of the chair's position.**

Given this unfortunate state of affairs for the chair, we might ask whether a chair, or the largest voting faction in a voting body comprising three factions—none of which commands a majority—has any recourse. It would appear not: The sophisticated outcome, z, is supported by both Y and Z, which no voting strategy of X can upset. There is, however, an escape for the chair, which you can find out more about by going to www.whfreeman.com/fapp.

Clearly, power defined as control over outcomes is not synonymous with power defined as control over resources (e.g., a tie-breaking vote or simply more votes). The strategic situation facing voters intervenes and may cause them to reassess their sincere strategies in light of the additional resources that a chair possesses. In so doing, they may be led to "gang up" against the chair—that is, to vote in such a way as to undermine the impact of her extra resources—handing the chair a worse outcome than she would have achieved without them. These resources in effect become a burden to bear, not power to wield.

We stress that Y and Z do not form a coalition against X in the sense of coordinating their strategies and agreeing to act together in a way that can be enforced. Rather, they behave as isolated individuals; at most they could be said to form an "implicit coalition." Such a coalition does not imply even communication between its members but simply choices based on their common perceived strategic interests. ■

So far we have used payoff matrices to describe games in strategic forms. In these games, the row and column players' choices of strategies led to an outcome from which each player received a payoff. These strategy choices were assumed to be simultaneous.

In the next example of a larger game, we start by assuming simultaneous choices and show what outcome would occur. Then we assume that the choices of the players need not be simultaneous, and one player considers moving first. We will use a "game tree" to analyze the *sequential choices* players can then make, as occurs when first you move, then I move, and so on, which are called *games in extensive form.* As we will see, the outcome in such a game may be wholly different from what it is in a game with simultaneous choices, which raises the question of which game is the most realistic model of a situation.

EXAMPLE A Truel

A *truel* is like a duel, except that there are three players. Each player can either fire, or not fire, his or her gun at either of the other two players. We assume the goal of each player is, first, to survive and, second, to survive with as few other players as possible. Each player has one bullet and is a perfect shot, and no communication (e.g., to pick out a common target) that results in a binding agreement with other players is allowed, making the game noncooperative. We will discuss the answers that simultaneous choices, on the one hand, and sequential choices, on the other, give to what is optimal for the players to do in the truel.

If choices are simultaneous, *at the start of play, each player will fire at one of the other two players, killing that player.*

Why will the players all fire at each other? Because their own survival does not depend an iota on what they do. Since they cannot affect what happens to themselves, but can only affect how many others survive (the fewer the better, according to the postulated secondary goal), they should all fire away at each other. (Even if the rules of the play permitted shooting oneself, the primary goal of survival would preclude committing suicide.) In fact, the players all have dominant strategies to shoot at each other, because whether or not a player survives—we will discuss shortly the probabilities of doing so—he or she does at least as well shooting an opponent.

The game, and optimal strategies in it, would change if (1) the players were allowed more options, such as to fire in the air and thereby disarm themselves, or (2) they did not have to choose simultaneously, and a particular order of play were specified. Thus, if the order of play were A, followed by B and C choosing simultaneously, followed by any player with a bullet remaining choosing, then A would fire in the air and B and C would subsequently shoot each other. (A is no threat to B or C, so neither of the latter will fire at A and waste a bullet; on the other hand, if one of B or C did not fire immediately at the other, that player would not survive to get in the last shot, so they both fire.) Thus, A will be the sole survivor. In 1992, a modified version of this scenario was played out in late-night television programming among the three major TV broadcasting networks of the time, with ABC's effectively going first with *Nightline,* its well-established news program, and CBS and NBC dueling about which host, David Letterman

David Letterman (Black Star)

Jay Leno (Black Star)

Ted Koppel (Syracuse Newspapers/Dick Blume/The Image Works.)

or Jay Leno, to choose for their entertainment shows. Regardless of their ultimate choices, ABC "won" when CBS and NBC were forced to divide the entertainment audience. In 2002, ABC, presumably to attract a younger audience than watched *Nightline,* attempted unsuccessfully to hire Letterman from CBS.

To return to the original game (all choose simultaneously), the players' strategies of all firing have two possible consequences: Either one player survives (even if two players fire at the same person, the third must fire at one of them, leaving only one survivor), or no player survives (if each player fires at a different person). In either event, there is no guarantee of survival. In fact, if each player has an equal probability of firing at one of the two other players, the probability that any particular player will survive is only 0.25.

The reason is that if the three players are A, B, and C, A will be killed when either B fires at him or her, C does, or both do. The only circumstance in which A will survive is if B and C fire at each other, which gives A one chance in four.

If choices are sequential, no player will fire at any other, so all will survive.

At the start of the truel, all the players are alive, which satisfies their primary goal of survival, though not their secondary goal of surviving with as few others as possible. Now assume that A contemplates shooting B, thereby reducing the number of survivors, and cannot fire into the air. But looking ahead, A knows that by firing first and killing B, he or she will be defenseless and be immediately shot by C, who will then be the sole survivor.

It is in A's interest, therefore, not to shoot anybody at the start, and the same logic applies to each of the other players. Hence, everybody will survive, which is a happier outcome than when choices are simultaneous, in which case everyone's primary goal is not satisfied—or, quantitatively speaking, satisfied only 25% of the time. ■

While sequential choices produce a "happier" outcome, do they provide a plausible model of a strategic situation that mimics what people might actually think and do in such a situation? We believe that the players in the truel, artificial as this kind of shoot-out may seem, would be motivated to think ahead, given the dire consequences of their actions. Therefore, they would hold their fire, knowing that if one fired first, he or she would be the next target.

In Figure 16.5, we show this logic somewhat more formally with a **game tree,** in which A has three strategies, as indicated by the three branches that sprout from A: not shoot ($\overline{S}$), shoot B ($S \rightarrow B$), or shoot C ($S \rightarrow C$). The latter two branches, in turn, give survivors C and B, respectively, two strategies: not shoot ($\overline{S}$) or shoot A ($S \rightarrow A$).

We assume that the players rank the outcomes as follows, which is consistent with their primary and secondary goals: 4 = best (lone survivor), 3 = next best (survivor with one other), 2 = next worst (survivor with two others), and 1 = worst (nonsurvivor). These payoffs are given for ordered triples (A, B, C); thus (3, 3, 1) indicates the next-best payoffs for A and B and the worst payoff for C.

Note that play necessarily terminates when there is only one survivor, as is the case at (1, 1, 4) and (1, 4, 1). To keep the tree simple, we assume that play also

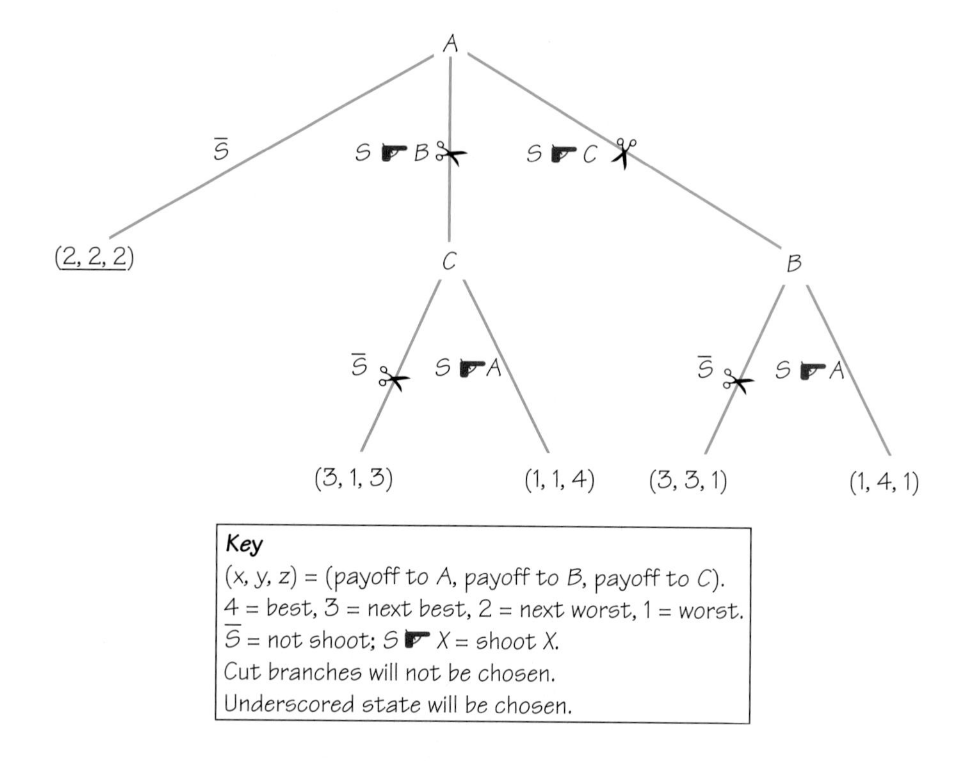

Figure 16.5 A game tree of a truel.

terminates when either A initially or B or C subsequently chooses $\bar{S}$, giving outcomes of (2, 2, 2), (3, 3, 1), and (3, 1, 3), respectively. Of course, we could allow the two or three surviving players in the latter cases to make subsequent choices in an extended game tree, but this example is meant only to illustrate the analysis of a game tree, not be the definitive statement on truel possibilities (more will be explored in the Exercises).

In a game in extensive form represented by a game tree, players work backward, starting the analysis at the bottom of the tree. (By "bottom" we mean where play terminates; because this is where the tree branches out, the tree looks upside down in Figure 16.5.) The players then work up the tree, using backward induction. **Backward induction** is a reasoning process in which players, working backward from the last possible moves in a game, anticipate each other's rational choices.

To illustrate, because C prefers (1, 1, 4) to (3, 1, 3), we indicate that C would not choose $\bar{S}$ by "cutting" this branch with a scissors; similarly, B would not choose $\bar{S}$. Thus, if play got to the bottom of the tree, C would shoot A (choose $S \rightarrow A$) if C were the survivor, and B would shoot A (choose $S \rightarrow A$) if B were the survivor, following A's shooting C or B, respectively.

Moving up to the next level, A would know that if he or she chose $S \rightarrow B$, (1, 1, 4) would be the outcome; if he or she chose $S \rightarrow C$, (1, 4, 1) would be the outcome, making one or the other the outcome from the bottom level. Choosing between these two outcomes and (2, 2, 2), A would prefer the latter, so A

would cut the two branches, $S \rightarrow B$ and $S \rightarrow C$. Hence, A would choose $\overline{S}$, terminating play with nobody's shooting anybody else.

This, of course, is the conclusion we reached earlier, based on the reasoning that if A shot either B or C, he or she would end up dead, too. Because we could allow each player, like A, to choose among his or her three initial strategies in a $3 \times 3 \times 3$ game, and subsequently make moves and countermoves from the initial state (if feasible), the foregoing analysis applies to all players.

Underlying the completely different answers given by the simultaneous and sequential choices in a truel is a change in the rules of play. If play is sequential, the players do not have to fire simultaneously at the start, as assumed in a $3 \times 3 \times 3$ strategic-form game. Rather, a player who moves first (A in our example)—and then the later players—would not fire, given that play continues until all bullets are expended or nobody chooses to fire.

In the extensive-form analysis, we ask of each player (it need not be A): Given your present situation (all alive), and the situation you anticipate will ensure if you fire first, should you do so? Because each player prefers living to the state he or she would induce by being the first to shoot (certain death), no one shoots. This analysis suggests that truels might be more effective than duels in preventing the outbreak of conflict.

We will not try to develop this argument into a more general model. The main point is that a game tree allows for a look-ahead approach whereby players compare the present state with possible future states—perhaps several steps ahead—to determine which moves to make. These choices, as we have seen, may lead to radically different outcomes compared with those based on simultaneous choices.

16.5 Using Game Theory

Solving Games

Given any payoff matrix, the first thing we ask is whether it is zero-sum (or constant-sum). If so, we check to see whether it has a saddlepoint by determining the minimum number of each row and the maximum number of each column, as we did in several earlier examples. If the maximum of the row minima (maximin) is equal to the minimum of the column maxima (minimax), then the game has a saddlepoint. The resulting value, and the corresponding pure strategies, provide a solution to the game.

This value will appear in the payoff matrix as the smallest number in its row and the largest in its column. In the 3×3 location game, this number was 5 (5000 feet).

Like our voting game, dominated strategies can successively be eliminated in the 3×3 location game. Thus, Route B dominates Route C, and Highway 2 dominates Highway 3; having made these eliminations, Highway 2 dominates Highway 1; having made this elimination, Route B dominates Route A. Thus, Highway 2 and Route B survive the successive eliminations, yielding the saddlepoint of 5. The successive-elimination procedure therefore provides an alternative method for finding the saddlepoint in this 3 × 3 location game. Unfortunately, it does not work to find the saddlepoint in *all* two-person zero-sum games larger than 2 × 2.

Recall that instead of eliminating dominated strategies in the 3 × 3 location game, we eliminated Route B and Highway 2, which dominated other strategies, to obtain the 2 × 2 restricted-location game in Table 16.3. In this game, there were no dominated strategies and, hence, no saddlepoint.

If a two-person zero-sum game does not have a saddlepoint—which was the case not only in the restricted-location game but also for matching pennies, the nonsymmetrical matching game, and the baseball duel—the solution will be in mixed strategies. To find the optimal mix in a 2 × 2 game, we calculate the expected value to a player from choosing its first strategy with probability p and its second with probability $1 - p$, assuming that the other player chooses its first pure strategy (yielding one expected value) and its second pure strategy (yielding another expected value).

Setting these two expected values equal to each other yields a unique value for p that gives the optimal mix, $(p, 1 - p)$, with which the player should choose its first and second strategies. Substituting the numerical solution of p back into either expected-value equation gives the value of the game, which each player can guarantee for itself, whatever strategy its opponent chooses.

Several general algorithms have been developed since 1945 to find mixed-strategy solutions to large constant-sum games. This work has mostly been done in the field of linear programming, using such algorithms as the simplex method of G. B. Dantzig and the more recent method of N. K. Karmarkar (see Chapter 4).

In variable-sum games, we also begin by successively eliminating dominated strategies, if there are any. The outcomes that remain do not depend on the numerical values we attach to them but only on their ranking from best to worst by the players, as illustrated in the three-person voting example.

Care must be taken in interpreting this solution, however. It began with the choice of a dominant strategy by the chair (X)—and her elimination of her two dominated strategies. Presuming these eliminations, Y and Z were then able to eliminate their own dominated strategies in the first reduction, and Y in turn eliminated a dominated strategy in the second reduction, leading finally to the outcome z, supported by Y and Z.

This solution is a fairly demanding one, because it assumes considerable calculational abilities on the part of the players. Less demanding, of course, is that players simply choose their dominant strategies, as is possible in Prisoners' Dilemma, but of course games may not have such strategies.

In the game of Chicken, for example, neither player has a dominant (or dominated) strategy, so the game cannot be reduced. In such situations, we ascertain what outcomes are Nash equilibria. There are two (in pure strategies) in Chicken, suggesting that the only stable outcomes in this game occur when one player gives in and the other does not. In Prisoners' Dilemma, by comparison, the choice by the players of their dominant strategies singles out the mutual-defection outcome as the unique Nash equilibrium, which is worse for both players than the cooperative outcome.

In both Chicken and Prisoners' Dilemma, there seems no good reason for the choice of the (3, 3) cooperative outcome, at least if each game is played only once, because this outcome is not a Nash equilibrium. However, there is an alternative theory, called the "theory of moves," that assumes different rules of play and renders the cooperative outcomes in both Prisoners' Dilemma and Chicken stable (i.e., if the players think ahead).

Theory of moves (TOM) is a dynamic theory that describes optimal strategic choices in strategic-form games in which the players, thinking ahead, can make moves and countermoves.

TOM is *dynamic* in the sense that it allows players, after choosing strategies that lead to an outcome in a payoff matrix, to make alternating subsequent moves and countermoves, with row's being able to move vertically by changing his row strategy, and column's being able to move horizontally by changing her column strategy. The reasoning that the players use in deciding whether to move or not move is backward induction, as illustrated earlier in the truel.

We informally illustrate this reasoning in Prisoners' Dilemma, starting from each of the four possible outcomes:

- If the play starts at the noncooperative (2, 2) outcome, players are stuck, no matter how far ahead they look, because as soon as one player departs, the other player, enjoying its best outcome at (4, 1) or (1, 4), will not move on. *Result:* The players stay at the noncooperative outcome.
- If play starts at the cooperative (3, 3) outcome, then neither player will defect, because if he or she does, the other player will also defect, and they both will end up worse off at (2, 2). Thinking ahead, therefore, neither player will defect. *Result:* The players stay at the cooperative outcome.
- If play starts at one of the (4, 1) or (1, 4) win–lose outcomes, the player doing best (4) will know that if he or she does not move to the cooperative (3, 3) outcome, his or her opponent will move to the noncooperative (2, 2) outcome, inflicting on the best-off player a next-worst (2) outcome. Therefore, it is in this player's interest—as well as the worst-off player's interest—that the best-off player act cooperatively and move first to (3, 3), anticipating that if he or she does not, the (2, 2) rather than the (3, 3) outcome will be chosen. *Result:* The best-off player will move to the cooperative outcome, where play will stop.

Thus, TOM does not predict unconditional cooperation in Prisoners' Dilemma but, instead, makes it a function of the starting point of play. As in the truel, a change in rules from simultaneous choices to sequential choices can induce cooperation.

The calculations we have described for Prisoners' Dilemma, which are grounded in backward induction and could be formalized by a game tree, are not, we believe, beyond the ability of most players. Farsighted players *can* escape the dilemma in Prisoners' Dilemma, provided play begins at a state other than the noncooperative one. (But we must be careful in interpreting this result: With the change in the rules, the original game changes, so this "solution" to Prisoners' Dilemma is not for the original dilemma.) Similar reasoning in Chicken indicates that if play starts at the cooperative (3, 3) outcome, players will stay at this outcome, but the reasoning in this game is somewhat more complicated than in Prisoners' Dilemma.

Practical Applications

The element of surprise, as captured by mixed strategies, is essential in many encounters. For example, mixed strategies are used in various inspection procedures and auditing schemes to deter potential cheaters; by making inspection or auditing choices random, they are rendered unpredictable.

Investigators or regulatory agencies monitor certain accounts as well as take various actions to check for faults, errors, or illegal activities. The investigators include bank auditors, customs agents, insurance investigators, and quality-control experts. The National Bureau of Standards is responsible for monitoring the accuracy of measuring instruments and for maintaining reliable standards. The Nuclear Regulatory Agency demands an accounting of dangerous nuclear material as part of its safeguards program. The Internal Revenue Service attempts to identify those cheating on taxes.

Military or intelligence services may wish to intercept a weapon hidden among many decoys or plant a secret agent disguised to look like a respectable individual. Because it is prohibitively expensive to check the authenticity of each and every possible item or person, efficient methods must be used to check for violations. Both optimal detection and optimal concealment strategies can be modeled as a game between an inspector trying to increase the probability of detection and a violator trying to evade detection. Since the World Trade Center attack on September 11, 2001, we have seen the government take much stronger measures to prevent such evasion.

Some inspection games are constant-sum: The violator "wins" when the evasion is successful and "loses" when it is not. On the other hand, cheating on arms-control agreements may well be variable-sum if both the inspector and the cheater would prefer that no cheating occur to there being cheating and public disclosure of it. The latter could be an embarrassment to both sides, especially if it undermines an arms-control agreement both sides wanted and the cheating is not too serious.

We alluded earlier to the strategy of bluffing in poker, which is used to try to keep the other player or players guessing about the true nature of one's hand.

The optimal probability with which one should bluff can be calculated in a particular situation (see Exercise 17). Besides poker, bluffing is common in many bargaining situations, whereby a player raises the stakes (e.g., labor threatens a strike in labor–management negotiations), even if it may ultimately have to back down if its "hand is called."

Perhaps the greatest value of game theory is the framework it provides for understanding the rational underpinnings of conflict in the world today. As a case in point, a confrontation over the budget between the Democratic President Bill Clinton and the Republican Congress resulted in the shutdown of part of the federal government on two occasions between November 1995 and January 1996. Many government workers were frustrated in not being able to do their jobs, even though they knew they would be paid for not working, not to mention that many citizens were either greatly hurt or substantially inconvenienced by the shutdown. Viewed as a game of Chicken, in which each side wanted not only to get its way for the moment but also to establish a precedent for the future, this conflict was not so foolish as it might seem at first glance.

The Northern Ireland conflict–settled in principle by a peace agreement in April 1998 after 30 years of fighting and more than 3200 deaths but which has had its ups and downs since–can be viewed in similar terms. As still another example, the constant price wars among the airlines suggest competitors caught up in a Prisoners' Dilemma, in which they all suffer from lower fares but cannot avoid their dominant strategies of not cooperating, perhaps to try to seize a quick advantage or hurt the competition even more (and possibly even eliminate a competitor). To be sure, if the airlines cooperate by colluding on fares, which is definitely not advantageous to consumers, the consumers may be thought of as a collective player whose interests are represented by the government. The government can prosecute the airlines for price fixing, or the consumers themselves can file a class-action suit in a "larger" game. The government has frequently been involved in antitrust suits (recently, for example, against Microsoft) and in setting the rules for auctions of airwaves, in which telecommunication companies–advised by game theorists–have paid billions of dollars for the right to construct cellular phone and other networks.

All in all, game theory offers fundamental insights into conflicts at all levels, especially its *seemingly* irrational features which, on second look, are often well conceived and effective.

REVIEW VOCABULARY

Backward induction A reasoning process in which players, working backward from the last possible moves in a game, anticipate each other's rational choices.

Chicken A two-person variable-sum symmetric game in which each player has two strategies: to swerve to avoid a collision, or not to swerve and cause a collision if the opponent has not swerved.

Neither player has a dominant strategy; the compromise outcome, in which both players swerve, is not a Nash equilibrium, but the two outcomes in which one player swerves and the other does not are Nash equilibria.

Constant-sum game A game in which the sum of payoffs to the players at each outcome is a constant, which can be converted to a zero-sum game by an appropriate change in the payoffs to the players that does not alter the strategic nature of the game.

Dominant strategy A strategy that is sometimes better and never worse for a player than every other strategy, whatever strategies the other players choose.

Dominated strategy A strategy that is sometimes worse and never better for a player than some other strategy, whatever strategies the other players choose.

Expected value E If each of the n payoffs, $s_1, s_2, \ldots, s_n$ occurs with respective probabilities $p_1, p_2, \ldots, p_n$, then the expected value E is

$$E = p_1 s_1 + p_2 s_2 + \ldots + p_n s_n$$

where $p_1 + p_2 + \ldots + p_n = 1$ and $p_i \geq 0$ $(i = 1, 2, \ldots, n)$.

Fair game A zero-sum game is fair when the (expected) value of the game, obtained by using optimal strategies (pure or mixed), is zero.

Game tree A symbolic tree, based on the rules of play in a game, in which the vertices, or nodes, of the tree represent choice points, and the branches represent alternative courses of action that the players can select.

Maximin In a two-person zero-sum game, the largest of the minimum payoffs in each row of a payoff matrix.

Maximin strategy In a two-person zero-sum game, the pure strategy of the row player corresponding to the maximin in a payoff matrix.

Minimax In a two-person zero-sum game, the smallest of the maximum payoffs in each column of a payoff matrix.

Minimax strategy In a two-person zero-sum game, the pure strategy of the column player corresponding to the minimax in a payoff matrix.

Minimax theorem The fundamental theorem for two-person constant-sum games, stating that there always exist optimal pure or mixed strategies that enable the two players to guarantee the value of the game.

Mixed strategy A strategy that involves the random choice of pure strategies, according to particular probabilities. A mixed strategy of a player is optimal if it guarantees the value of the game.

Nash equilibrium Strategies associated with an outcome such that no player can benefit by choosing a different strategy, given that the other players do not depart from their strategies.

Nonsymmetrical game A two-person constant-sum game in which the row player's gains are different from the column player's gains, except when there is a tie.

Ordinal game A game in which the players rank the outcomes from best to worst.

Paradox of the chair's position This paradox occurs when being chair (with a tie-breaking vote) hurts rather than helps the chair if voting is sophisticated.

Partial-conflict game A variable-sum game in which both players can benefit by cooperation but may have strong incentives not to cooperate.

Payoff matrix A rectangular array of numbers. In a two-person game, the rows and columns correspond to the strategies of the two players, and the numerical entries give the payoffs to the players when these strategies are selected.

Plurality procedure A voting procedure in which the alternative with the most votes wins.

Power The ability of a player to induce a preferred outcome.

Prisoners' Dilemma A two-person variable-sum symmetric game in which each player has two strategies, cooperate or defect. Cooperate dominates defect for both players, even though the mutual-defection outcome, which is the unique Nash equilibrium in the game, is worse for both players than the mutual-cooperation outcome.

Pure strategy A course of action a player can choose in a game that does not involve randomized choices.

Rational choice A choice that leads to a preferred outcome.
Saddlepoint In a two-person constant-sum game, the payoff that results when the maximin and the minimax are the same, which is the value of the game. The saddlepoint has the shape of a saddle-shaped surface and is also a Nash equilibrium.
Sincere voting Voting for one's most-preferred alternative in a situation.
Sophisticated voting Involves the successive elimination of dominated strategies by voters.
Strategy One of the courses of action a player can choose in a game; strategies are mixed or pure, depending on whether they are selected in a randomized fashion (mixed) or not (pure).
Theory of moves (TOM) A dynamic theory that describes optimal choices in strategic-form games in which players, thinking ahead, can make moves and countermoves.
Total-conflict game A zero-sum or constant-sum game, in which what one player wins the other player loses.
Value The best outcome that both players can guarantee in a two-person zero-sum game. If there is a saddlepoint, this is the value; otherwise, it is the expected payoff resulting when the players choose their optimal mixed strategies.
Variable-sum game A game in which the sum of the payoffs to the players at the different outcomes varies.
Zero-sum game A constant-sum game in which the payoff to one player is the negative of the payoff to the other player, so the sum of the payoffs to the players at each outcome is zero.

SUGGESTED READINGS

AUMANN, ROBERT J., and SERGIU HART, eds. *Handbook of Game Theory with Economic Applications,* Elsevier, Amsterdam, 1992 (vol. 1), 1994 (vol. 2). A comprehensive treatment of game theory and its applications, developed in long chapters written by leading experts. A third (and final) volume is forthcoming.

BAIRD, DOUGLAS G., ROBERT H. GERTNER, and RANDAL C. PICKER. *Game Theory and the Law,* Harvard University Press, Cambridge, Mass., 1994. A good treatment of how game theory informs various branches of the law.

BINMORE, KEN. *Fun and Games: A Text on Game Theory,* Heath, Lexington, Mass., 1992. This best-selling text is a provocative introduction to game theory at an elementary–intermediate level.

BRAMS, STEVEN J. *Theory of Moves,* Cambridge University Press, New York, 1994. Describes in detail the theory of moves and applies it to a wide range of conflicts.

BRAMS, STEVEN J. *Negotiation Games: Applying Game Theory to Bargaining and Arbitration,* Routledge, New York, 1990. Game-theoretic models of negotiation have been developed in several disciplines; this book provides a survey of different models and applications.

BRAMS, STEVEN J. *Superpower Games: Applying Game Theory to Superpower Conflict,* Yale University Press, New Haven, Conn., 1985. The superpower conflict, as it existed during the cold war (roughly from 1945 to 1990), has evaporated since the demise of the Soviet Union, but the models developed in this book to study deterrence, arms races, and the verification of arms-control agreements are relevant to conflicts between other countries today.

BRAMS, STEVEN J. *Biblical Games: A Strategic Analysis of Stories in the Old Testament,* MIT Press, Cambridge, Mass., 1980; rev. ed., 2002. About 20 stories of conflict and intrigue in the Hebrew Bible are modeled as simple games, in many of which God is a player. By and large, Brams argues, the players made rational choices.

BRAMS, STEVEN J., and JEFFREY M. TOGMAN, Cooperation through threats: The Northern Ireland case, *PS: Political Science and Politics,* 31(1) (March 1998): 34–43.

DIXIT, AVINASH, and BARRY NALEBUFF. *Thinking Strategically: The Competitive Edge in Business,*

Politics, and Everyday Life, Norton, New York, 1991. A best-selling popular treatment of applications of game theory, with many stimulating examples.

DIXIT, AVINASH, and SUSAN SKEATH. *Games of Strategy,* Norton, New York, 1999. A splendid elementary game-theory text that requires only a minimal mathematical background.

GARDNER, ROY. *Game Theory for Business and Economics,* Wiley, New York, 1995. A good introduction to game theory and its business and economic applications, with many examples.

LUCE, R. DUNCAN, and HOWARD RAIFFA. *Games and Decisions: Introduction and Critical Survey,* Wiley, New York, 1957; Dover, New York, 1989. This venerable survey of classic game theory presents two-person constant-sum and variable-sum games in Chapters 4 and 5. Several different algorithms for solving the zero-sum case are given in Appendix 6; the minimax theorem, and its equivalence to the "duality theorem" in linear programming, are given in Appendices 2 and 5.

McDONALD, JOHN. *The Game of Business,* Doubleday, New York, 1975; Anchor, New York, 1977. A superb collection of cases in which elementary tools of game theory are used to explicate some classic battles in the business world.

MORROW, JAMES D. *Game Theory for Political Scientists,* Princeton University Press, Princeton, N.J., 1994. Develops tools of game theory and discusses game-theoretic models used in political science.

NASAR, SYLVIA. *A Beautiful Mind,* Simon & Shuster, New York, 1998. A biography of John Nash that is also a fascinating account of the early history of game theory. A fictionalized version of this biography was made into a movie in 2001.

O'NEILL, BARRY, *Honor, Symbols, and War,* University of Michigan Press, Ann Arbor, 1999. Draws on a wide range of cases to show how difficult-to-operationalize qualities like honor, prestige, and face influence international negotiation and conflict resolution – and can be analyzed game theoretically.

POUNDSTONE, WILLIAM. *Prisoner's Dilemma: John von Neumann, Game Theory, and the Puzzle of the Bomb,* Doubleday, New York, 1992. A history of game theory as well as a biography of its mathematician founder, including his views on the use of nuclear weapons.

SIGMUND, KARL. *Games of Life: Explorations in Ecology, Evolution, and Behavior,* Oxford University Press, Oxford, 1993. Game theory is placed in the broader context of evolutionary models in biology and related fields in a very readable account.

STRAFFIN, PHILIP D. *Game Theory and Strategy,* Mathematical Association of America, Washington, D.C., 1993. An elementary but sophisticated introduction to game theory and several interesting applications.

TAYLOR, ALAN D. *Mathematics and Politics: Strategy, Voting, Power and Proof,* Springer-Verlag, New York, 1995. A mathematics textbook in which game theory and theory of moves are used to model power, voting, and conflict and escalation processes.

WILLIAMS, JOHN D. *The Compleat Strategyst: Being a Primer on the Theory of Games of Strategy,* McGraw-Hill, New York, 1954, rev. 1966; Dover, New York, 1986. This gem, which contains many simple illustrations, is a humorous primer on two-person zero-sum games.

SUGGESTED WEB SITES

www.chass.utoronto.ca/~osborne Martin Osborne's home page (game theory)

www.economics.harvard.edu/~aroth/alroth.html Alvin Roth's Game Theory and Experimental Economics Page

levine.sscnet.ucla.edu/ David Levine's Economic and Game Theory page

www.gametheory.net is the most comprehensive general Web site.

SKILLS CHECK

1. In the following two-person zero-sum game, the payoffs represent gains to row player I and losses to column player II.

$$\begin{bmatrix} 3 & 7 & 2 \\ 8 & 5 & 1 \\ 6 & 9 & 4 \end{bmatrix}$$

What is the maximin strategy for player I?

(a) Play the first row.
(b) Play the second row.
(c) Play the third row.

2. In the following two-person zero-sum game, the payoffs represent gains to row player I and losses to column player II.

$$\begin{bmatrix} 3 & 7 & 2 \\ 8 & 5 & 1 \\ 6 & 9 & 4 \end{bmatrix}$$

What is the maximax strategy for player II?

(a) Play the first column.
(b) Play the second column.
(c) Play the third column.

3. In a two-person zero-sum game, suppose the first player chooses the third row as the maximin strategy and the second player chooses the first column as the minimax strategy. Based on this information, which of the following statements is true?

(a) This game definitely has no saddlepoint.
(b) This game may or may not have a saddlepoint.
(c) This game definitely has a saddlepoint.

4. In the game of matching pennies, player I wins a penny if the coins match; player II wins a penny if the coins do not match. Given this information, it can be concluded that the 2×2 matrix which represents this game

(a) has two -1's and two 1's.
(b) has four 1's.
(c) has four -1's.

5. In the following game of batter-versus-pitcher in baseball, the batter's batting averages are given in the game matrix.

		Pitcher	
		Fastball	**Curve**
Batter	**Fastball**	.400	.200
	Curve	.100	.500

What is the pitcher's optimal strategy?

(a) Throw more curves than fastballs.
(b) Throw more fastballs than curves.
(c) Throw about the same number of curves and fastballs.

6. In the following game of batter-versus-pitcher in baseball, the batter's batting averages are given in the game matrix.

		Pitcher	
		Fastball	**Curve**
Batter	**Fastball**	.400	.200
	Curve	.100	.500

What is the batter's optimal strategy?

(a) Expect more curves than fastballs.
(b) Expect more fastballs than curves.
(c) Expect about the same number of curves and fastballs.

7. Consider the following partial-conflict game, played in a noncooperative manner.

		Player II	
		Choice A	Choice B
Player I	Choice A	(4, 4)	(1, 3)
	Choice B	(3, 1)	(2, 2)

What outcomes constitute a Nash equilibrium?

(a) Only when both players select A
(b) Only when both players select A or both select B
(c) Only when one player selects A and the other selects B

8. If a game has a saddlepoint, then

(a) it is the value of the game.
(b) it is never the maximin strategy.
(c) it is never the minimax strategy.

9. Which of these games does not have a saddlepoint?

(a) Tic-tac-toe
(b) Chess
(c) Poker

10. A mixed strategy uses randomization

(a) to choose among several pure strategies each time a round of the game is played.
(b) to select a pure strategy for the game.
(c) to anticipate the opponent's next move.

11. A game is fair when

(a) its value is 0.
(b) it requires a mixed strategy.
(c) its payoff is unlimited.

12. A game is symmetrical when

(a) each player has an equal chance to win.
(b) each player receives a winning payoff of the same value.
(c) either player can take the first turn.

13. Games of partial conflict are examples of

(a) fair games.
(b) zero-sum games.
(c) games in which cooperation yields greater payoffs.

14. A game tree is used to

(a) determine the possible strategies of a player.
(b) anticipate each other's choices through backward induction.
(c) plan a deception strategy.

15. Theory of moves is dynamic in that players

(a) randomly change their strategy.
(b) can consider the opponent's moves to change their strategy plans.
(c) move in alternation.

EXERCISES

▲ **Optional.** ■ **Advanced.** ◆ **Discussion.**

Total-Conflict Games

Consider the following five two-person zero-sum games, wherein the payoffs represent gains to the row player I and losses to the column player II:

1. $\begin{bmatrix} 6 & 5 \\ 4 & 2 \end{bmatrix}$

2. $\begin{bmatrix} 0 & 3 \\ -5 & 1 \\ 1 & 6 \end{bmatrix}$

3. $\begin{bmatrix} -2 & 3 \\ 1 & -2 \end{bmatrix}$

4. $\begin{bmatrix} 13 & 11 \\ 12 & 14 \\ 10 & 11 \end{bmatrix}$

5. $\begin{bmatrix} -10 & -17 & -30 \\ -15 & -15 & -25 \\ -20 & -20 & -20 \end{bmatrix}$

(a) Which of these games have saddlepoints?
(b) Find the maximin strategy of player I, the minimax strategy of player II, and the value for those games given in part (a).
(c) List dominated strategies in these games that the players should avoid because the resulting payoffs are worse than those for some alternative strategy.

Solve the following three games of batter-versus-pitcher in baseball, wherein the pitcher can throw one of two pitches and the batter can guess either of these two pitches. The batter's batting averages are given in the game matrix.

6.

		Pitcher	
		Fastball	Curve
Batter	Fastball	.300	.200
	Curve	.100	.400

7.

		Pitcher	
		Fastball	Knuckleball
Batter	Fastball	.500	.200
	Knuckleball	.200	.300

8.

		Pitcher	
		Blooperball	Knuckleball
Batter	Blooperball	.400	.200
	Knuckleball	.250	.250

9. A businessman has the choice of either not cheating on his income tax or cheating and making \$1000 if not audited. If caught cheating, he will pay a fine of \$2000 in addition to the \$1000 he owes. He feels good if he does not cheat and is not audited (worth \$100). If he does not cheat and is audited, he evaluates this outcome as − \$100 (for the lost day). Viewing the game as a two-person zero-sum game between the businessman and the tax agency, what are the optimal mixed strategies for each player and the value of the game?

10. When it is third down and short yardage to go for a first down in American football, the quarterback can decide to run the ball or pass it. Similarly, the other team can commit itself to defend more heavily against a run or a pass. This can be modeled as a 2 × 2 matrix game, wherein the payoffs are the probabilities of obtaining a first down. Find the solution of this game.

		Defense	
		Run	Pass
Offense	Run	.5	.8
	Pass	.7	.2

11. You have the choice of either parking illegally on the street or else parking in the lot and paying \$16. Parking illegally is free if the police officer is not patrolling, but you receive a \$40 parking ticket if she is. However, you are peeved when you pay to park in the lot on days when the officer does not patrol, and you are willing to assess this outcome as costing \$32 (\$16 for parking plus \$16 for your time, inconvenience, and grief). It seems reasonable to assume that the police officer ranks her preferences in the order (1) giving you a ticket, (2) not patrolling with you parked in the lot, (3) patrolling with you in the lot, and (4) not patrolling with you parked illegally.

(a) Describe this as a matrix game, assuming that you are playing a zero-sum game with the officer.
(b) Solve this matrix game for its optimal mixed strategies and its value.
◆ **(c)** Discuss whether it is reasonable or not to assume that this game is zero-sum.
◆ **(d)** Assuming that you play this parking game each working day of the year, how do you implement an optimal mixed strategy?

12. Describe how a pure strategy for a player in a matrix game can be considered as merely a special case of a mixed strategy.

■ **13. (a)** Describe in detail *one* pure strategy for the player who moves first in the game of tic-tac-toe. (This strategy must tell how to respond to all possible moves of the other player.) (*Hint:* You may wish to make use of the symmetry in the 3 × 3 grid in this game; that is, there are one "center" box, four "corner" boxes, and four "side" boxes.)
(b) Is your strategy optimal in the sense that it will guarantee the first player a tie (and possibly a win) in the game?

14. In the matching-pennies example, consider the case where player I favors heads H over tails T. For example, assume that player I plays H three-fourths of the time and T only one-fourth of the time – a nonoptimal mixed strategy. What should player II do if he knows this?

15. Assume in the nonsymmetrical matching example that player II is using the nonoptimal mixed strategy $(p, 1-p) = (\frac{1}{2}, \frac{1}{2})$; that is, he is playing H and T with the same frequency. What should player I do in this case if she knows this?

16. You plan to manufacture a new product for sale next year, and you can decide to make either a small quantity, in anticipation of a poor economy and few sales, or a large quantity, hoping for brisk sales. Your expected profits are indicated in the following table.

		Economy	
		Poor	Good
Quantity	Small	\$500,000	\$300,000
	Large	\$100,000	\$900,000

If you want to avoid risk and believe that the economy is playing an optimal mixed strategy against you in a two-person zero-sum game, then what is your optimal mixed strategy and the resulting expected value? Discuss some alternative ways to go about making your decision.

■ **17.** Consider the following miniature poker game with two players, I and II. Each antes \$1. Each player is dealt either a high card H or a low card L, with probability $\frac{1}{2}$. Player I then folds or bets \$1. If player I bets, then player II either folds, calls, or raises \$1. Finally, if II raises, I either folds or calls.

Most choices by the players are rather obvious, at least to anyone who has played poker: If either player holds H, that player always bets or raises if he or she gets the choice. The question remains of how often one should bluff—that is, continue to play (by calling or raising) while holding a low card in the hope that one's opponent also holds a low card.

This poker game can be represented by the matrix game at top right, wherein the payoffs are the expected winnings for player I (depending upon the random deal) and the dominated strategies have been eliminated.

		Player II (when holding L)		
		Folds	Calls	Raises
	Folds initially	−.25	0	.25
Player I (when holding L)	Bets first and folds later	0	0	−.25
	Bets first and calls later	−.25	−.25	0

(a) Are there any strategies in this matrix game that a player should avoid playing?
(b) Solve this game.
(c) Which player is in the more favored position?
(d) Should one ever bluff?

Partial-Conflict Games

Consider the following three two-person variable-sum games. Discuss the players' possible behavior when these games are played in a noncooperative manner (i.e., with no prior communication or agreements). The first payoff is to the row player; the second, to the column player. Are the Nash equilibria in these games sensible? Why or why not?

18.

	Player II	
Player I	(4, 4)	(1, 3)
	(3, 1)	(2, 2)

19. Battle of the sexes:

		She buys a ticket for:	
		Boxing	Ballet
He buys a ticket for:	Boxing	(4, 3)	(2, 2)
	Ballet	(1, 1)	(3, 4)

20.

	Player II	
Player I	(2, 1)	(4, 2)
	(1, 4)	(3, 3)

Larger Games

21. For the preferences of the players given in the text—xyz for X (chair), yzx for Y, and zxy for Z—verify that the strategy choices of x by X, y by Y, and x by Z are a Nash equilibrium. Does this equilibrium seem to you defensible as the social choice by the voters? Under what circumstances might the voters choose these strategies rather than their sophisticated strategies?

22. Extend the game tree of the truel in Figure 16.5 to allow the additional possibility that if A does not shoot initially, then B has the choice of shooting or not shooting C. Will A, in fact, not shoot initially, and will B then shoot C?

23. Extend the game tree in Exercise 22 to still another level to allow for the possibility that if A does not shoot initially, and B shoots C, then A has the choice of shooting or not shooting B. What will happen in this case?

24. Change Exercise 23 to allow for the possibility that if A does not shoot initially, and B shoots *or does not shoot* C, then A has the choice of shooting or not shooting B. What will happen in this case?

◆ **25.** What general conclusions would you draw in light of your answers to Exercises 22, 23, and 24?

Additional Exercises

Consider the following three two-person zero-sum games, wherein the payoffs represent gains to the row player I and losses to the column player II:

26. $\begin{bmatrix} 3 & 6 \\ 5 & 4 \end{bmatrix}$

27. $\begin{bmatrix} -1 & 3 \\ 2 & 0 \end{bmatrix}$

28. $\begin{bmatrix} 6 & 5 & 6 & 5 \\ 1 & 4 & 2 & -1 \\ 8 & 5 & 7 & 5 \\ 0 & 2 & 6 & 2 \end{bmatrix}$

(a) Which of these games have saddlepoints?
(b) Find the optimal strategy for player I and for player II, and the value for those games given in part (a).
(c) List dominated strategies in these games that the players should avoid because the resulting payoffs are worse than those for some alternative strategy.

29. Consider the game played between the opposing goalie and a soccer player who, after a penalty, is allowed a free kick. The kicker can elect to kick toward one of the two corners of the net, or else aim for the center of the goal. The goalie can decide to commit in advance (before the kicker's kick) to either one of the sides, or else remain in the center until he sees the direction of the kick. This two-person zero-sum game can be represented as follows, wherein the payoffs are the probability of scoring a goal:

		Goalie		
		Breaks left	**Remains center**	**Breaks right**
Kicker	**Kicks left**	.5	.9	.9
	Kicks center	.1	0	.1
	Kicks right	.9	.9	.5

If we assume that decisions between the left or right side are made symmetrically (i.e., with equal probabilities), then this game can be represented by a 2×2 matrix, where $.7 = (\frac{1}{2})(.5) + (\frac{1}{2})(.9)$:

		Goalie	
		Remains center	**Breaks side**
Kicker	**Kicks center**	0	1
	Kicks side	.9	.7

Find the optimal mixed strategies for the kicker and goalie and the value of this game.

■ **30. (a)** Describe in detail *one* pure strategy for the player who moves second in the game of tic-tac-toe.
(b) Is your strategy in part (a) optimal in the sense that it will guarantee the second player a tie (and possibly a win) in the game?

▲ **31.** Find a two-person zero-sum game with a saddlepoint in which the successive elimination of dominated strategies does *not* lead to the saddlepoint (unlike the 3×3 location game; you may restrict yourself to 3×3 games).

32. On an overcast morning, deciding whether to carry your umbrella can be viewed as a game between yourself and nature as follows:

		Weather	
		Rain	**No rain**
You	**Carry umbrella**	Stay dry	Lug umbrella
	Leave it home	Get wet	Hands free

Let's assume that you are willing to assign the following numerical payoffs to these outcomes, and that you are also willing to make decisions on the basis of expected values (that is, average payoffs):

(Carry umbrella, rain) $= -2$
(Carry umbrella, no rain) $= -1$
(Leave it home, rain) $= -5$
(Leave it home, no rain) $= 3$

(a) If the weather forecast says there is a 50% chance of rain, should you carry your umbrella or not? What if you believe there is a 75% chance of rain?
(b) If you are conservative and wish to protect against the worst case, what pure strategy should you pick?
(c) If you are rather paranoid and believe that nature will pick an optimal strategy in this two-person zero-sum game, then what strategy should you choose?
(d) Another approach to this decision problem is to assign payoffs to represent what your *regret* will be after you know nature's decision. In this case, each such payoff is the best payoff you could have received under that state of nature, minus the corresponding payoff in the previous table:

		Weather	
		Rain	**No rain**
You	**Carry umbrella**	$0 = (-2) - (-2)$	$4 = 3 - (-1)$
	Leave it home	$3 = (-2) - (-5)$	$0 = 3 - 3$

What strategy should you select if you wish to minimize your maximum possible regret?

Consider the following two-person variable-sum games. Discuss the player's possible behavior when these games are played in a noncooperative manner (i.e., with no prior communication or agreements). The first payoff is to the row player, the second to the column player.

33.

	Player II	
Player I	(2, 4)	(4, 3)
	(1, 2)	(3, 1)

34.

	Player II	
Player I	(3, 4)	(2, 3)
	(1, 2)	(4, 1)

■ **35.** Under a voting system called approval voting, a voter can vote for as many alternatives as he or she wishes. (If there are three alternatives, the only undominated strategies of a voter under approval voting are to vote for his or her best, or two best, choices.) If voters X, Y, and Z have paradox-of-voting preferences of xyz, yzx, and zxy, and X is the chair, show that x is the sophisticated outcome, obtained by all voters voting for their two best choices.

▲ **36.** Show by example that approval voting is not immune to the paradox of the chair's position.

37. Why will the first player to act in a truel shoot in the air (if this option is allowed by the rules)? Is this choice optimal if a second player should succeed in firing in the air at the same time?

38. In a sequential truel with no firing in the air allowed, suppose A, who hates B, goes first; B, who hates C, goes second; C, who hates A, goes third. (If a player fires, he will shoot only his *antagonist*—the player he hates.) Which player is in the best position, and why? Does the outcome change if B hates A rather than C? Answer these questions if each player can take only one turn (if alive); and if, after one round, the game continues (if there is more than one player alive) and players take more than one turn.

39. If a fourth player is added to the original truel, show that every player will have an incentive to shoot another (as in a duel).

WRITING PROJECTS

1. In tennis, one player often prefers to play from the baseline while her opponent prefers a serve-and-volley game (i.e., likes to come to the net). The baseline player attempts to hit passing shots. This player has a choice of hitting "down the line" or "crosscourt." The net player must often correctly guess in which direction the ball will go in order to cover the shot. Formulate this situation as a duel game and discuss appropriate strategies for the players.

2. Consider a conflict that you, personally, had – with a parent, a boss, a girlfriend or boyfriend, or some other acquaintance – in which each of you had to make a choice without being sure of what the other person would do. What strategies did you seriously consider adopting, and what options do you think the other person considered? What plausible outcomes do you think each set of strategy choices would have led to? How would you rank these outcomes from best to worst, and how do you think the other player would have ranked them? Analyze the resulting game, and state whether you think you and the other person made optimal choices. If not, what upset your or the other person's rationality?

3. It is sometimes argued that game theory does not take account of the (irrational?) emotions of people, such as anger, jealousy, or love. What is your opinion about this question? Give an example, real or hypothetical, that supports your position, paying particular attention to whether the players acted consistently with, or contrarily to, their preferences.

4. Quentin Tarantino's films *Reservoir Dogs* (1992) and *Pulp Fiction* (1994) both have truels, but the choices that the characters make in each are completely different. Does the truel analysis offer any insight into why?

SPREADSHEET PROJECTS

To do these projects, go to www.whfreeman.com/fapp.

Spreadsheets allow the analysis of mixed strategies. The expectations of each player can be computed using spreadsheet formulas, and changes in the mix can be immediately seen in the recomputation. This project also extends naturally to mixed strategies with three alternatives.

APPLET EXERCISES

To do these exercises, go to www.whfreeman.com/fapp.

What happens to the value of a game if you or your opponent deviates from the optimal strategy? Can you exploit such deviations by your opponent to your advantage? Explore these possibilities in the Game Theory applet.

CHAPTER 17

Electing the President

In this chapter we cover...

...and at www.whfreeman.com/fapp:

Flashcards

Quizzes

Spreadsheet Projects

Popular vote vs. electoral college

Electing the president of the United States has been a tricky business since the founding of the republic. In 1824, none of the four presidential candidates received a majority of votes in the **electoral college,** which is a body that elects the president, currently made up of 538 members. (We will say more about this institution later.) Consequently, the election went to the House of Representatives, as mandated by the Twelfth Amendment of the U.S. Constitution, wherein John Quincy Adams defeated Andrew Jackson. Although Jackson had won the most popular votes, Henry Clay threw his support to Adams in the electoral college. Clay was subsequently appointed secretary of state by Adams in what was called the "corrupt bargain."

There were two more instances in the nineteenth century when the popular-vote winner lost the presidential election. In 1876, Rutherford B. Hayes defeated Samuel J. Tilden; in 1888, Benjamin Harrison defeated Grover Cleveland (although Cleveland got his revenge four years later when he beat Harrison). But it was the recurrence of this divided outcome in the 2000 presidential election that provoked the most controversy.

In this election, Al Gore received 537,000 more popular votes (0.5%) than did George W. Bush, but Bush won the electoral-vote tally by 4 votes. The outcome of this election turned on who would win Florida. Thirty-six days after the election the Supreme Court, in a 5–4 decision, blocked further vote recounts in disputed Florida counties. By winning in Florida by a razor-thin margin of 537 votes (less than 0.01% of those cast), George W. Bush won the presidency.

We will not give further details of the 2000 election here but will focus instead on how presidential elections, in general, can be modeled as games, and what insights mathematics can provide about the strategic aspects of campaigning and elections. (In Chapter 16, we provide a more systematic introduction to game theory and discuss a variety of other applications of the theory.) There are several phases in the presidential election process. Because the rules of play change from phase to phase, the optimal strategies candidates pursue in their quest for the presidency change, too.

(Mark Wilson/Newsmakers.)

The first phase begins when Democratic and Republican candidates seek their party's nomination for president by running in state primaries. New Hampshire is the state that kicks off the primary season in February of a presidential election year, though it is preceded by caucuses in Iowa, where people gather in different locations throughout the state to discuss the candidates and then vote for party delegates that represent them. The strategic question in this phase is how to position oneself to gather momentum and thereby beat the competition in a set of contests spread over several months (but a process that usually boils down to a heated race in the first few weeks following the New Hampshire primaries).

Because winning in the presidential primaries almost always assures a candidate of the party's nomination in its national convention in July or August, the convention is usually just a rubber stamp for the winner in the primary phase. The last exception to this rule occurred in the Democratic party convention in 1968. After the incumbent president, Lyndon B. Johnson, withdrew from the race following the New Hampshire primary (which he had won!), Vice President Hubert H. Humphrey won his party's presidential nomination without running in the primaries.

There has not been a national party convention that has taken more than one ballot, or round of voting, to decide its presidential nominee since Adlai E. Stevenson defeated Estes Kefauver for the Democratic party nomination in three ballots in 1952. Conventions are usually staged with consensus-promoting hoopla to unite the different factions of a party, especially if they have been sharply divided in the primaries, in preparation for the final phase of the election.

This occurs with the general election in the fall, which typically involves only two serious contenders—the nominees of the Democratic and Republican parties—but sometimes includes significant third- and fourth-party candidates. One example is Ross Perot, who ran as the American Independent party presidential nominee in 1992 and garnered 19% of the popular vote. Although no minor-party candidate has ever won a presidential election, some have affected which of the two major-party nominees did win, as we will see. In the general election, the role of the electoral college becomes paramount, as the 2000 election vividly demonstrated.

What can mathematics tell us about presidential elections? First, it can clarify what are better and worse campaign strategies in each phase. In addition, it can shed light on the likely effect that different election reforms would have on both campaign strategies and election outcomes. In fact, we will analyze two prominent reform proposals and indicate how they might have changed the outcome in the 2000 election: (1) the use of approval voting and (2) the abolition of the electoral college and its replacement by direct popular-vote election of the president. Throughout we will state a series of propositions, and prove one theorem, that highlight general results that emerge from the mathematical analysis.

We begin by looking at how candidates position themselves in presidential primaries to win their party's nomination. The principal tool of analysis is **spatial models,** which we will describe in the next section and apply to both two-candidate and multicandidate elections.

17.1 Spatial Models: Two-Candidate Elections

While two-candidate contests are most common in the general election, sometimes the nomination race in the Democratic or Republican party comes down to a contest between just two contenders. As a case in point, Gerald Ford faced one major opponent, Ronald Reagan, in the 1976 Republican race. This race was not decided until the Republican national convention in August, when Ford edged out Reagan in a close vote.

To model such elections, we assume that voters respond to the positions that candidates take on issues. This is not to say that other factors, such as personality, ethnicity, religion, and race, have no effect on election outcomes but rather that issues take precedence in a voter's decision.

How can the position of a candidate on issues be represented? We start by assuming that there is a single overriding issue on which the candidates must take a definite stand, or a set of issues related to a single underlying dimension, such as degree of governmental intervention in the economy. We assume that the attitudes of voters on this issue or dimension can be represented along a left–right continuum, ranging from very liberal on the left (much intervention) to very conservative on the right (little intervention).

To derive conclusions about the behavior of voters from assumptions about their attitudes and the positions candidates take in a campaign, some assumption is necessary about how voters decide for whom to vote. More important than the attitudes of *individual* voters, however, are the *numbers* of voters who have particular attitudes along the left–right continuum.

Symmetric Unimodal Distribution

A **voter distribution** can be represented by a curve that gives the number (or percentage) of voters who have attitudes at each point along a horizontal axis. The greater the vertical height of the curve, the more voters have attitudes at that point. Figure 17.1a shows one such distribution. Because this distribution has one peak, or **mode,** it is **unimodal.** For simplicity, we picture the distribution as continuous, although in fact, because the number of voters is finite, there cannot be voters at all points along the continuum.

More important than the mode, from the viewpoint of the candidates, is the *median* M of a distribution.

The **median** $\boldsymbol{M}$ of a voter distribution is the point on the horizontal axis where half the voters have attitudes that lie to the left and half to the right.

The notion of the median of a voter distribution is closely related to the notion of a median of data sample given in Chapter 6.

The Figure 17.1a distribution is **symmetric**—the curve to the left of M is a mirror image of the curve to the right. Thus, the same numbers of voters have attitudes that are equal distances to the left and to the right of M.

Although the *attitudes* of voters are a fixed quantity in the calculations of the candidates, the *decisions* of voters will depend on the positions that the candidates take. Assuming the candidates know the distribution of voter attitudes, what positions are optimal for them?

Assume candidates A and B take the positions along the left–right axis shown in Figure 17.1a, where candidate A is to the left of M and candidate B is to the right. Assume that all voters vote for the candidate with attitudes closer to their own, and that all voters vote (we will consider modifications of this assumption later in the Exercises). Then A will certainly attract all the voters to the left of his position, and B all the voters to the right of her position. If both candidates are an equal distance from M, as shown in Figure 17.1a, they will split the vote in the middle, with that to the left of M going to A and that to the right going to B.

Can either candidate do better by changing his or her position? If B's position remains fixed, A could move alongside B, just to her left, and capture all the votes to B's left. We may represent these positions by . . . M_AB . . . , which are illustrated more fully in Figure 17.1b. Because A would have moved to the right

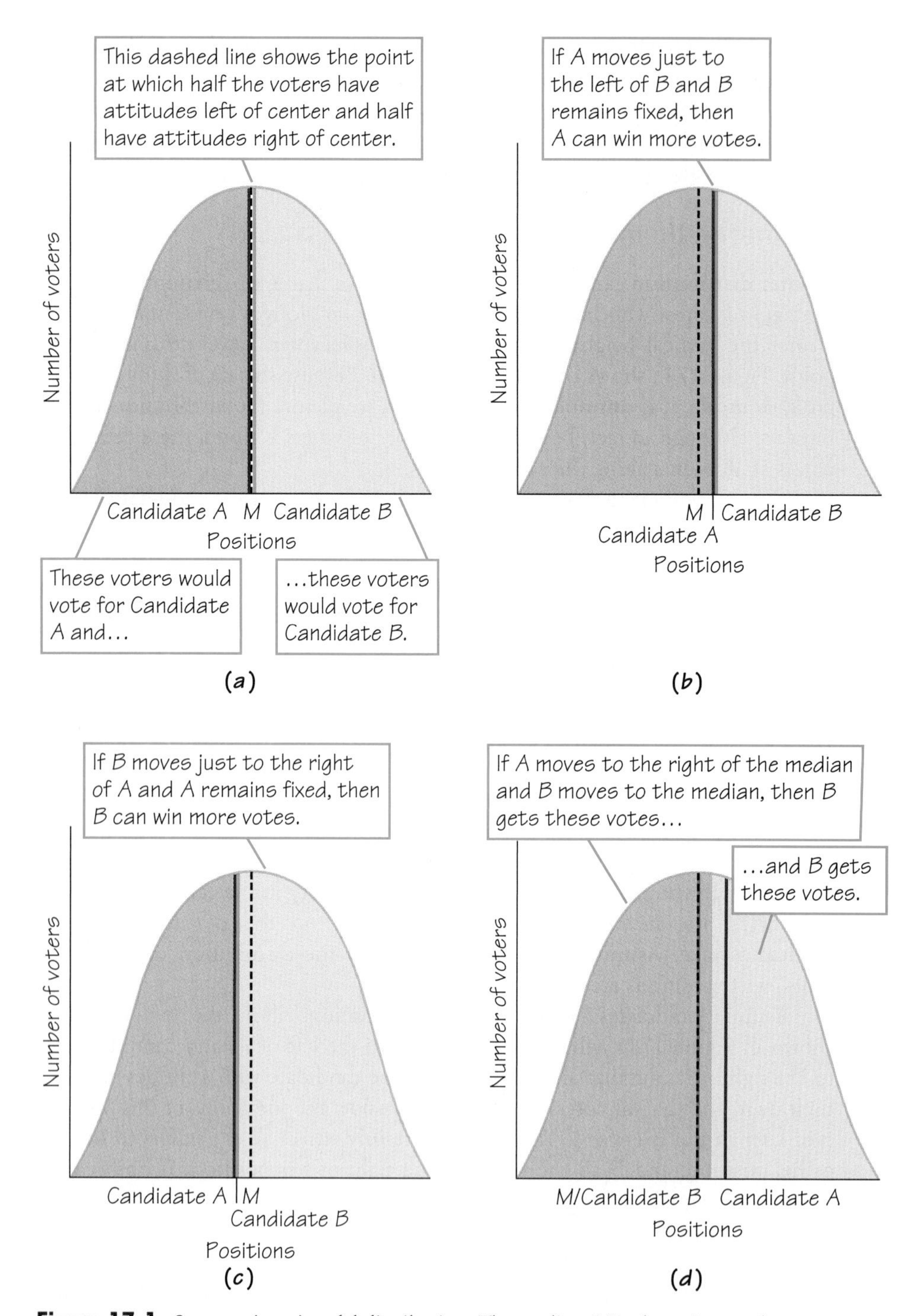

Figure 17.1 Symmetric unimodal distribution. The median M is the point on the horizontal axis that divides the area under the distribution curve – measuring the number of voters – exactly in half.

of M, he would, by changing his position in this manner, receive a majority of the votes and thereby win the election.

By analogous reasoning, there is no reason for B to stick to her original position to the right of M. By approaching A's original position, B can capture all the votes to A's right of M: . . . AB_M . . . (Figure 17.1c). In other words, both candidates, acting rationally, should approach each other and M.

If A were to move rightward past M, but B moved leftward only as far as M, their positions would be as follows (M/B indicates that B is at the median): . . . M/B_A . . . (Figure 17.1d). Now B would receive not only the 50% of the votes to the left of M but also some of the votes that lie between B's (median) position and A's position (now to B's right).

Clearly, A loses by crossing M from the left. Hence, there is an incentive for both candidates to move toward M but not overstep it. In fact, taking a position at M maximizes the minimum number of votes a candidate can guarantee for him- or herself.

A position is **maximin** for a candidate if there is no other position that can *guarantee* a better outcome—more votes for that candidate—whatever position the other candidate adopts.

If both candidates choose M, voters will be indifferent to the choice between them on the basis of their positions alone and would presumably make their choice on other grounds.

Naturally, if B adopted the position shown in Figure 17.1a, it would be rational for A, in order to maximize his vote total, to move alongside B, as we previously argued. But if B did not remain fixed but instead switched her position (say, to M), the nonmedian position of A would not ensure him of 50% of the votes.

Taking a position at M, however, guarantees A at least 50% of the total vote *no matter what B does*. Moreover, there is no other position that can guarantee A more votes, and likewise for B.

M is also *stable,* because if one candidate adopts this position, the other candidate has no incentive to choose any position other than M. Thus, M is both the maximin position for each candidate (it offers a guarantee of a minimum of 50% of the votes) and, if M is chosen by both candidates, then these choices are in equilibrium (a candidate does worse by departing from it if the other candidate stays at it).

A pair of positions is in **equilibrium** if, once chosen by both candidates, neither candidate has an incentive to depart from it unilaterally (i.e., by him- or herself).

More formally, we have the *median-voter theorem*.

Median-voter theorem: In a two-candidate election with an odd number of voters, M is the unique equilibrium position.

We have already shown that if both candidates choose M, these choices are in equilibrium. Is there another equilibrium position or positions? There are two possibilities: (1) It is the same position for both candidates, which we call a common position, or (2) it is two distinct positions, one taken by each candidate. If it were a common position, suppose it is to the left of M (an analogous argument works if it is to the right). Then one candidate can always do better by moving rightward but staying to the left of M. This contradicts the supposition that the common position is in equilibrium. Now suppose the equilibrium were two distinct positions. Then one candidate can always do better by moving alongside the other candidate but staying closer to M. This contradicts the supposition that these two positions are in equilibrium. Thus, in both cases, one candidate would have an incentive to depart from his or her position—holding the position of the other candidate fixed—so a nonmedian position or positions cannot be in equilibrium. Therefore, M is the only equilibrium position.

Asymmetric Bimodal Distribution in Which the Median and Mean Are Different

The median-voter theorem is applicable *whatever* the distribution of the electorate's attitudes. Consider the distribution in Figure 17.2a, which is **bimodal** (two peaks) and is not symmetric. Applying the logic of the previous analysis, M is once again the maximin and equilibrium position of two candidates, even though the bulk of voters are centered at the two modes.

We next compare the M with the mean, which may be quite different:

The **mean** $\bar{l}$ of a voter distribution is

$$\bar{l} = \frac{1}{n}\sum_{i=1}^{k} n_i l_i$$

where

k = number of different positions i that voters take on the continuum

n_i = number of voters at position i

l_i = location of position i on the continuum

$n = \sum_{i=1}^{k} n_i = n_1 + n_2 + \ldots n_k$ = total number of voters

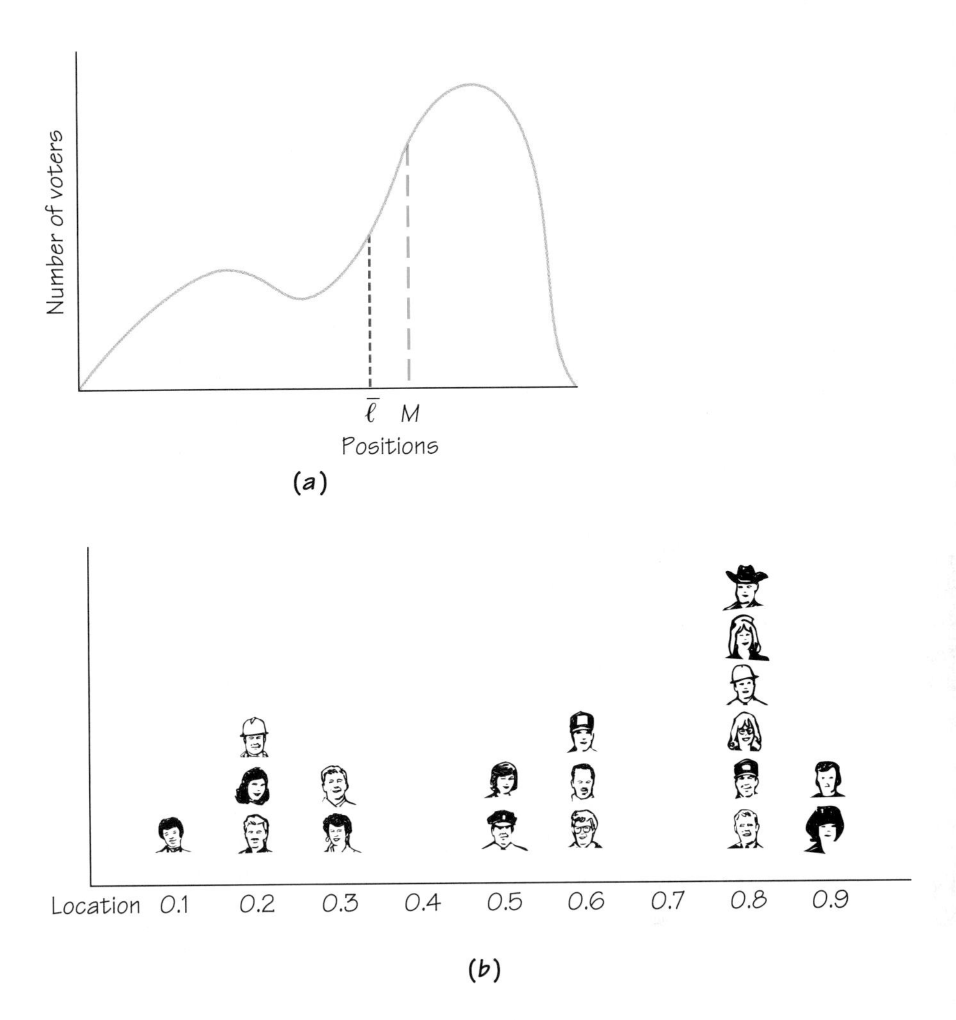

Figure 17.2
Asymmetric bimodal distribution in which the median and mean are different. The distribution is skewed to the right; the median M is to the right of the mean $\bar{l}$.

The symbol Σ (sigma) is the *summation sign;* it signifies that all subscripted terms to its right (e.g., in the definition of n), beginning with the subscript 1 and continuing to the subscript k, are summed and a total obtained.

The notion of a mean is the same as that for data samples in Chapter 6, except that here we are calculating a *weighted* average: The location l_i of each position is weighted by the number of voters, n_i, at that position. Thus, the mean can be thought of as the position of a typical voter—that is, the expected position of a voter drawn randomly from the set of all voters.

The mean $\bar{l}$ need not coincide with the median M. As an illustration of this point, consider the following discrete distribution of $n = 19$ voters at $k = 7$ different positions over the interval between 0 and 1, or [0, 1], which is illustrated in Figure 17.2b.

Position, i	1	2	3	4	5	6	7
Location (l_i) of position i	0.1	0.2	0.3	0.5	0.6	0.8	0.9
Number of voters (n_i) at position i	1	3	2	2	3	6	2

Whereas M is 0.6 because 8 voters lie to the left and 8 voters lie to the right,

$$\bar{l} = \left(\frac{1}{19}\right)[1(0.1) + 3(0.2) + 2(0.3) + 2(0.5) + 3(0.6) + 6(0.8) + 2(0.9)] = 0.56.$$

Taking a position at 0.56 against an opponent who takes a position at 0.6, a candidate would lose the election by 11 to 8 votes.

The distribution of Figures 17.2a and 17.2b are **skewed** to the right—the area under the curve, or the number of voters, is more concentrated to the right of M than to the left. This necessarily pushes the M to the right of the mean $\bar{l}$. The lesson we derive from these figures is that it may *not* be rational for a candidate to take a position at $\bar{l}$ if the distribution is skewed, either to the right or to the left.

Asymmetric Bimodal Distribution in Which the Median and Mean Are the Same

A sufficient condition for M and $\bar{l}$ to coincide is that the distribution be symmetric, but this condition is not necessary: M and $\bar{l}$ can coincide if a distribution is asymmetric, as illustrated in Figure 17.3a. When M and $\bar{l}$ coincide, a candidate need not take a different position to ensure victory—or at least prevent defeat if his or her opponent adópts the same position. However, as Figure 17.3 demonstrates, the noncoincidence of M and $\bar{l}$ is not necessarily related to the lack of symmetry in a distribution: half the voters may still lie to the left, and half to the right, of $M/\bar{l}$ if the distribution is asymmetric.

What can we say about equilibrium positions if there is an even number of voters? For example, consider the following discrete distribution of $n = 26$ voters at $k = 8$ different positions over the interval [0, 1], which is illustrated in Figure 17.3b.

Position, i	1	2	3	4	5	6	7	8
Location (l_i) of position i	0	0.2	0.3	0.4	0.5	0.7	0.8	0.9
Number of voters (n_i) at position i	2	3	4	4	2	3	7	1

We begin by calculating the mean $\bar{l}$:

$$\bar{l} = \left(\frac{1}{26}\right)[2(0) + 3(0.2) + 4(0.3) + 4(0.4) + 2(0.5) + 3(0.7) + 7(0.8) + 1(0.9)] = 0.5$$

The median M is not 0.5. For an even number of voters, as in this example, M is the average of the two middle positions. The two middle voters are the 13th and

Figure 17.3
Asymmetric bimodal distribution in which median and mean are the same. It is not necessary for a distribution to be symmetric for the median M and mean $\bar{l}$ to coincide.

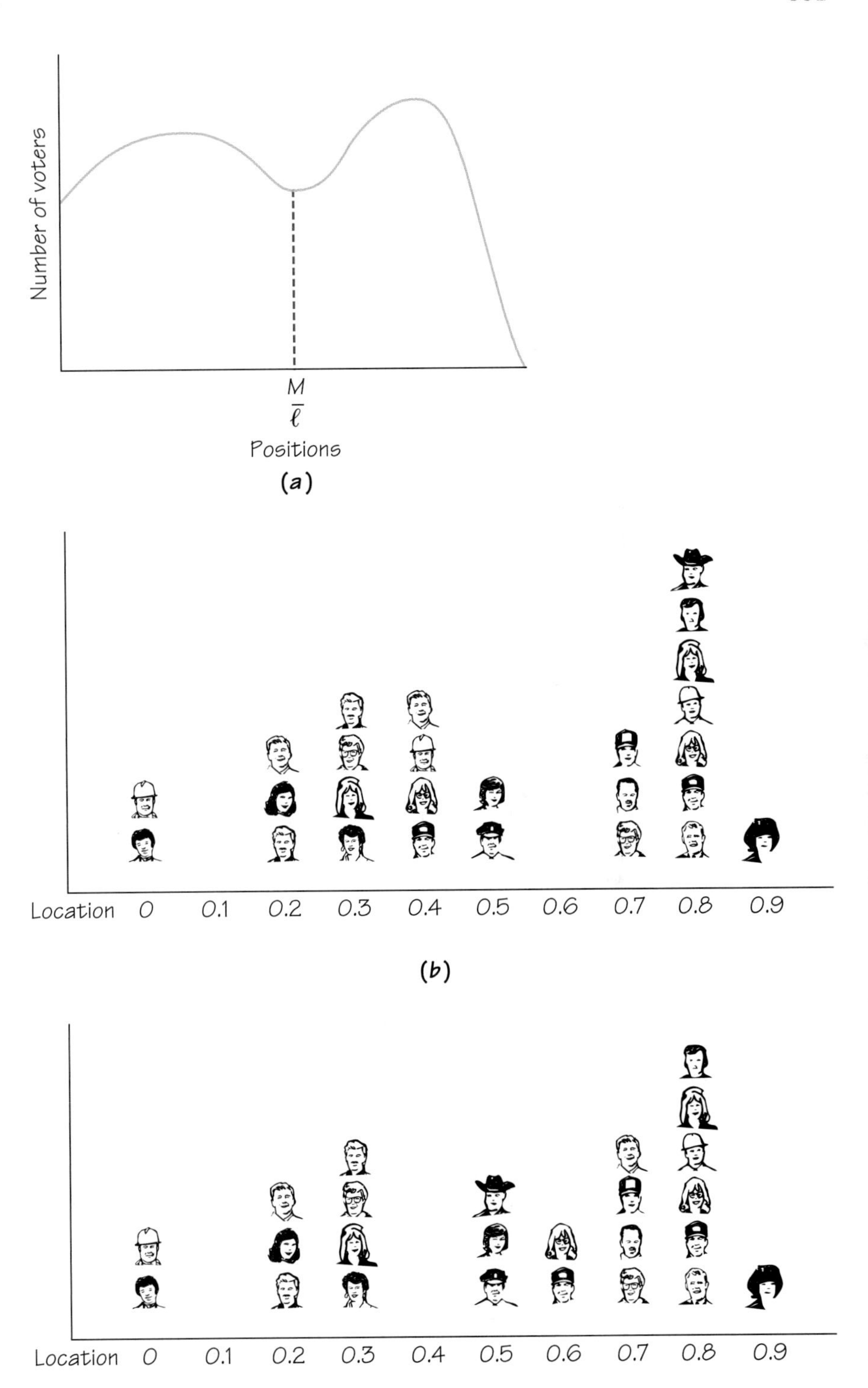

14th voters when they are lined up in the order of their positions from left to right. The 13th voter is at position 0.4, and the 14th voter is at position 0.5, so M is 0.45.

Note that if both candidates position themselves at the median M, their positions will be in equilibrium, as we showed above. But it is not the unique equilibrium pair. In this example, there are many other pairs of equilibrium positions. For instance, the same reasoning shows that any pair of positions between 0.4 and 0.5 will be in equilibrium. Moreover, it is easy to show that the position 0.4 for one candidate, and the position 0.5 for the other candidate, are in equilibrium. The candidate at 0.4 will get the support of the 13 voters at or to his left, and the candidate at 0.5 will get the support of the 13 voters at or to her right. If either candidate takes a different position, either to the left of 0.4 or to the right of 0.5, he or she will not be assured of 13 votes.

In general, if the number of voters is even, and the two middle voters adopt different positions, then if the candidates adopt those positions *or any pair of positions in between,* they will be in equilibrium.

It is possible that, for either an odd or even number of voters, there may not be a median position such that half the voters lie to the left and half to the right of this position. As a case in point, consider the following discrete distribution of $n = 25$ voters at $k = 8$ different positions over the interval [0, 1], which is illustrated in Figure 17.3c.

Position, i	1	2	3	4	5	6	7	8
Location (l_i) of position i	0	0.2	0.3	0.5	0.6	0.7	0.8	0.9
Number of voters (n_i) at position i	2	3	4	3	2	4	6	1

We begin by calculating the mean:

$$\bar{l} = \left(\frac{1}{25}\right)[2(0) + 3(0.2) + 4(0.3) + 3(0.5) + 2(0.6) + 4(0.7) + 6(0.8) + 1(0.9] = 0.52$$

At 0.6, 12 voters lie to the left of 0.5 and 11 voters lie to the right. At 0.5 and 0.7, the imbalances on the left and the right are even more lopsided. Moreover, there is no position, including the mean $\bar{l} = 0.52$, such that exactly half the voters lie to the left and half to the right.

In the absence of such a median position, is there an equilibrium? It is easy to show that 0.6 is indeed the equilibrium for two candidates. Somewhat more difficult to show is that if a distribution is discrete and there is no median position, there is still a unique position for both candidates that is in equilibrium. We call this the **extended median,** because it extends the median-voter theorem to the discrete case, in which there may be no median position.

Given the stability of the median or the extended median in a two-candidate, single-issue election, is it any wonder that candidates who want to win try to avoid extreme positions? As shown in Figures 17.2a and 17.3a, even when the greatest concentration of voters does not lie at M but instead at a mode (the mode to the right of the median in both these figures), a candidate would be foolish to adopt this modal position. For although the right-leaning voters would be very pleased, the candidate's opponent could win the votes of a majority by sidling up to this position but staying just to the left.

Voters on the far left may not be particularly pleased to see both candidates situate themselves at M, which is nearer the right mode in Figure 17.2a. But in a two-candidate race, they would have nobody else to turn to. Of course, if left-leaning voters felt sufficiently alienated by both candidates, they might decide not to vote at all, which has implications we explore further.

EXAMPLE

Location of Department Stores

There is a rather different application of the foregoing analysis to business, which in fact was the first substantive area to which spatial modeling was applied. Consider two competitive retail businesses, such as department stores, that consider locating their stores somewhere along the main street that runs through a city. Assume that, because transportation is costly, people will buy at the department store closer to them. Then the analysis says that no matter how the population is distributed along or near the main street, the best location is the median.

Thus, if the city's population is symmetrically distributed—that is, not skewed toward one end or the other of the main street—then this location will, of course, be at the center of the main street. Indeed, clusters of similar stores are frequently bunched together near the center of many main streets, although these stores may not be particularly convenient to people who live far from the city's center—and, consequently, not in the public interest since their location discriminates against these people. ■

17.2 Spatial Models: Multicandidate Elections

Primary elections, in which candidates seek the nomination of one of the major parties, tend to attract more than two candidates. In presidential primaries, in particular, many candidates are likely to jump into the fray, especially in the states that go early in the season, if the incumbent president or vice president is not running.

Under what conditions is entry into a multicandidate race attractive? If no positions offer a potential candidate any possibility of success, then it will not be rational for him or her to enter the primary in the first place. Therefore, the rationality of entering a race, and the rationality of the positions he or she might take once there, are really two aspects of the same decision.

Suppose that two candidates have already entered a primary, and they both take positions at M. Is there any room for a third candidate?

EXAMPLE

Entry of a Third Candidate in a Two-Candidate Race

Look at Figure 17.4, where A and B are both at M and therefore split the vote. Now if a third candidate, C, enters and takes a position on either side of M (say,

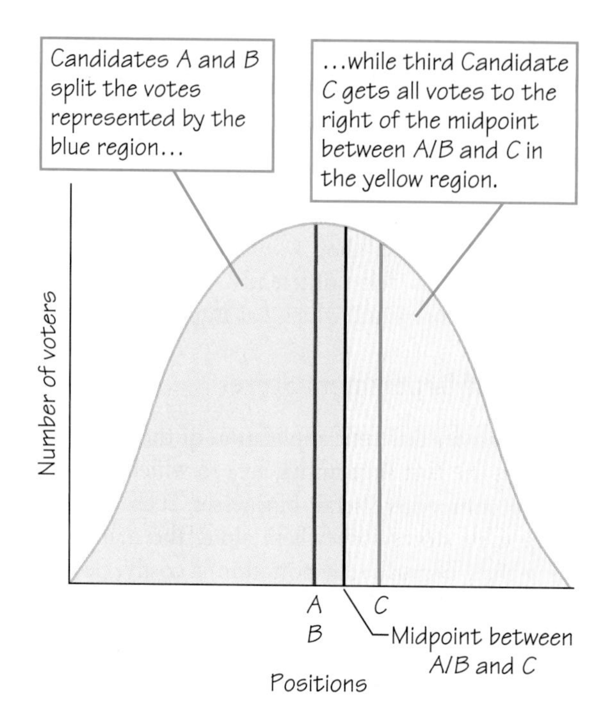
Candidates A and B split the votes represented by the blue region...
...while third Candidate C gets all votes to the right of the midpoint between A/B and C in the yellow region.
Number of voters
A
B
C
Midpoint between A/B and C
Positions

Figure 17.4
Symmetric unimodal distribution with three candidates. Candidate C can take a position with less than 1/3 of the voters to his or her right and still win if candidates A and B at the median M and mean $\bar{I}$ split the remainder of the vote.

to the right), the area under the distribution to C's right may encompass less than $\frac{1}{3}$ of the total area and still enable C to win a plurality of votes.

To show why this is so, consider the portion of the electorate's vote that A/B will receive and the portion that C will receive. If C's area (yellow) is greater than half of A/B's area (blue), C will win more votes than A or B, because C's area includes not only the votes to the right of his or her position but also some votes to the left. More precisely, C will attract voters up to the point midway between his or her position on the horizontal axis and that of A/B; A and B will split the votes to the left of this midway point. Because C picks up some votes to the left of his or her position, less than $\frac{1}{3}$ of the electorate may lie to the right and C can still win a plurality of more than $\frac{1}{3}$ of the total vote.

By similar reasoning, it is possible to show that a fourth candidate, D, could take a position to the left of A/B and further chip away at the total of the two centrists. Indeed, D could beat candidate C, as well as A and B, by moving closer to A/B from the left than C moves from the right.

Clearly, M has little appeal, and in fact is quite vulnerable, to a third or fourth candidate contemplating a run against two centrists. Indeed, it is not difficult to show that *whatever* positions two candidates adopt—the same or different—at least one will be vulnerable to a third candidate. ■

This is not to say, however, that a third candidate C will necessarily win against *both* A and B. There are both obstacles and opportunities for C, which are summarized by Propositions 1 and 2 (the reasoning behind them is explored in Exercises 18 and 19).

Proposition 1. $\frac{1}{3}$-separation obstacle. If A and B are distinct positions that are equidistant from the median of a symmetric distribution and separated from each other by no more than $\frac{1}{3}$ of the area under the curve (i.e., no more than $\frac{1}{3}$ of the voters lie between A and B), C can take no position that will displace both A and B and enable C to win (Figure 17.5a).

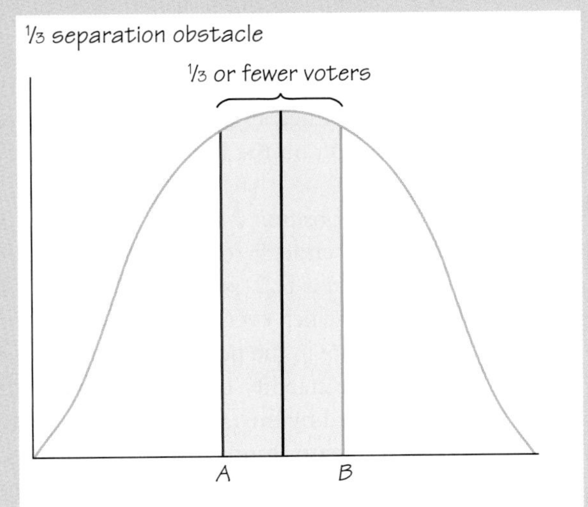

Proposition 2. $\frac{2}{3}$-separation opportunity. If A and B are distinct positions that are equidistant from the median of a symmetric unimodal distribution and separated from each other by at least $\frac{2}{3}$ of the area under the curve (i.e., at least $\frac{2}{3}$ of the voters lie between A and B), C can defeat both A and B by taking a position at M (i.e., exactly between them, as shown in Figure 17.5b).

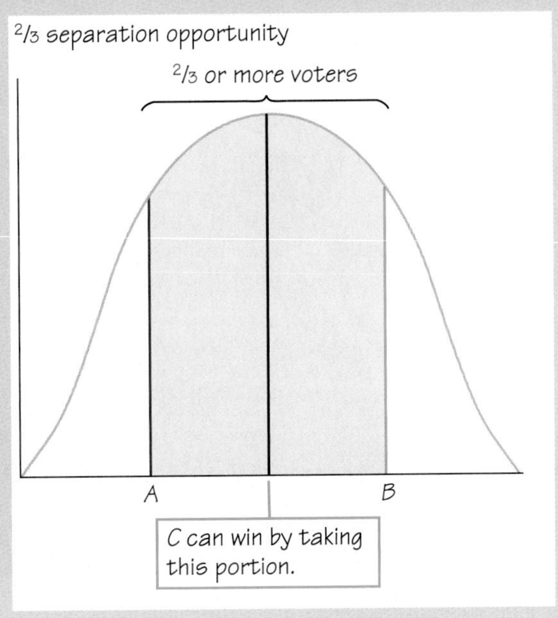

Figure 17.5
Propositions 1 and 2 summarize the obstacles and opportunities for a third candidate, C, to enter a race.

Proposition 1 says, essentially, that if there is little room in the middle, a third candidate may beat one candidate but in doing will so hand the election to the other. This occurred in the 1912 presidential election, when Theodore Roosevelt ran as the Progressive ("Bull Moose") party candidate after losing the Republican nomination to William Howard Taft. (Roosevelt had previously been president but had lost favor with his party after sitting out one term.) In the general election, Roosevelt received 27% of the popular vote and Taft, 24%. Both candidates were handily defeated by the Democratic candidate Woodrow Wilson, who got 41% of the popular vote. There was a fourth candidate in this race, socialist Eugene V. Debs, but he received only 6% and was never a serious threat to Wilson on the left. Wilson was also the overwhelming winner in the electoral college.

Proposition 2 says that if there is a wide separation between A and B, there may be enough room in the middle for C to win, but this event has never occurred in a U.S. presidential election. In fact, the 1912 election is the only election in which even one major-party candidate has been defeated by a third-party candidate in the popular vote.

The stability of the two-party system in the United States may be partially explained by the fact that the two major parties, anticipating the possible entry of a third-party candidate, deliberately position themselves for enough away from the median to discourage entry on the left or right—but not so far away as to make entry in the middle advantageous. The following theoretical result gives some insight into how this can be done (the reasoning behind it is explored in Exercise 20):

(Hulton Getty/Liaison.)

Proposition 3. *Optimal entry of two candidates, anticipating a third entrant.* Assume A and B are the first candidates to enter an election and anticipate the later entry of C. Assume that the distribution of voters is uniform (rectangular) over [0,1]. Then the optimal positions of A and B are to enter at 1/4 and 3/4, whereby 1/4 of the voters lie to the left of one candidate (say, A) and 1/4 to the right of the other (B). Then C can do no better than win 25% of the vote: He or she will be indifferent among entering just to the left of A, just to the right of B, or at any position in between. At none of these positions will C win.

Presidential politics in the United States seems to be a reflection of both the median-voter theorem and Proposition 3. More specifically, the median-voter theorem seems to have been operative mainly in 1968, when the Democratic and Republican nominees, Hubert H. Humphrey and Richard M. Nixon, both presented themselves as centrists. This made them vulnerable to the third-party candidate that year, George Wallace—not in the sense that Wallace could win, but rather that he could throw the election in one direction or the other or even into the House of Representatives. In fact, while Wallace won only 14% of the popular vote in 1968, he attracted mostly supporters of Richard N. Nixon on the right, who barely defeated Hubert H. Humphrey that year. Nixon won by less than 1% in the popular vote; without Wallace in the race, polls show that Nixon's victory would have been far more substantial.

In 1992, Bill Clinton and George Bush were viewed as quite far apart on the left–right spectrum. Ross Perot was generally viewed to be between Clinton on

(Dirck Halstead/Liaison.)

the left and Bush on the right. In winning 19% of the popular vote, he drew almost equally from each candidate. However, Perot did not come close to changing the outcome of the election, which Clinton won decisively with 43% of the popular vote to Bush's 38%.

17.3 Winnowing the Field

Up to now we have looked at the spatial game that candidates play as they vie to position themselves optimally in two-candidate and multicandidate races so as to (1) maximize their vote totals or (2) deter new candidates from entering. We will continue to assume that candidates take positions along a left–right spectrum, but now we consider the game from the point of view of the voters. More specifically, we ask the following question in a multicandidate race: When one candidate drops out, perhaps because of performing below expectations in an early primary, to whom will the dropout's supporters shift their votes?

Beating expectations is the name of the game in the early primaries. Thus, when Senator Edmund S. Muskie from Maine ran for the Democratic party nomination in 1972, he was expected to do well in the neighboring state of New Hampshire. To "win" in this first primary, he had to exceed these expectations, whereas other candidates, who were not expected to do so well, could afford more mediocre performances. As in a horse race, to which primary elections are often compared, the contenders are handicapped; that is, they must beat expectations about their performances if they are to gain momentum.

Momentum, or what George Bush (the father) called the "Big Mo," can start a *bandwagon,* or a presumption that a candidate will win. The resulting **bandwagon effect** induces voters to vote for the presumed winner, independent of his or her merit.

Assume three candidates take positions, from left to right, as follows: A–B–C. Clearly, if A or C drops out, their supporters mostly likely will switch to B, giving the centrist a boost. But what if B is the first to drop out? Then it is unclear whether A, C, or neither will benefit—it depends on the number of B's supporters that prefer A next, C next, or neither—and hence may not vote at all. In any event, with B out of the race, the winner must be one of the candidates on the extremes.

The possibilities become more interesting when there are four candidates that are arrayed from left to right as follows: A–B–C–D. If one of the extremists, A or D, drops out, then one of the two centrists, B or C, will benefit. But what if a centrist, say C, drops out? Does this benefit one of the extremists, or does the other centrist (B) benefit? At first glance one might think that, with only one centrist remaining, he or she will surely benefit.

This will not be the case, however, if most C supporters prefer D to B, which is certainly possible. Then the extremist D will benefit, which will be most upsetting to A's supporters. Conceivably, A's supporters might encourage A to withdraw so they can throw their support to B, whom they definitely prefer to D.

Does this sound implausible? Think back to the 2000 election, in which our four hypothetical candidates are replaced by the following ordering from left to right: Nader–Gore–Bush–Buchanan. (Recall that Ralph Nader was the Green party candidate and Pat Buchanan the Reform party candidate.) Just before the election, the polls were showing that Buchanan was not much of a threat to Bush, but Nader—who ended up with 2.7% of the popular vote nationwide (Buchanan got only 0.4%)—was definitely a threat to Gore. Despite pleas from some of his supporters, Nader refused to withdraw and, consequently, gave Bush a victory in Florida and a few other close states.

This 2000 scenario is not the same as the previous four-candidate hypothetical scenario, in which we argued that the extremist on the right, *D*, might win if one of the centrists, *C*, dropped out. In the 2000 scenario, the extremist on the left, Nader, could have dropped out to "save" the centrist closest to him, Gore.

Unfortunately for Gore, Nader not only refused to make this sacrifice but has contended ever since that Gore's loss was due to his own poor performance, not his (Nader's) presence in the race. We will return to this issue when we discuss the effects of approval voting and the abolition of the electoral college as possible remedies to the so-called **spoiler problem,** caused by a candidate who cannot win but "spoils" the election for a candidate who otherwise would win.

17.4 What Drives Candidates Out?

So far we have considered the possibility that candidates drop out but not why they do so. Presumably, they do so because of their poor performance in the polls or early primaries, where performance depends in part on expectations of how well they will do. But expectations change over time, and this change in turn affects how voters perceive a race—in particular, who is ahead and who is behind. On this basis, voters choose voting strategies that are likely to benefit their favorites.

Polls make public the standing of candidates in a race, as do the returns from presidential primaries. To be sure, the electorate is different in each primary state, whereas a poll is a sample from the entire electorate. While polls and primaries both provide a glimpse of the "state of the electorate," here we will focus on the change of polls over time. But our results are just as applicable to primaries, whose winnowing-out effects are evident in almost every election in which an incumbent is not running for reelection.

Suppose the election procedure is **plurality voting,** in which each voter votes for one candidate and the candidate with the most votes wins. Assume that voters rank candidates; so, for example, *A B C* indicates that a voter prefers *A* to *B* to *C*. Before a poll, we assume that each voter votes **sincerely**—for his or her favorite candidate—because, in the absence of poll information, there is no reason to do otherwise.

EXAMPLE

Poll (Three Candidates)

Nine voters, who can be divided into three classes, have the following preferences for candidates *A*, *B*, and *C*:

I. 4: A C B
II. 3: *B C A*
III. 2: C A B

Because the four voters in class I prefer *A* to *C* to *B*, the poll would indicate *A* to have 4 votes (44%), whereas *B* and *C* would have 3 and 2 votes (33% and 22%), respectively, if voters vote sincerely.

After the poll, we make the following assumption:

Poll assumption. If necessary, voters adjust their sincere voting strategies to differentiate between the top two candidates as revealed in the poll, voting for the one they prefer.

Since class I and class II voters chose one of the top two candidates, only the two class III voters, who voted for *C*, would change their votes. Because they prefer *A* to *B*, it would be in their interest to vote insincerely for *A* (instead of *C*), thereby distinguishing *A* from the other top candidate, *B*, by giving *A* their votes. This would result in *A*'s winning with 6 votes, *B*'s getting 3 votes, and *C*'s getting no votes.

Paradoxically, it is *C* who is the Condorcet winner. ■

A **Condorcet winner** is a candidate who can defeat each of the other candidates in pairwise contests.

C is preferred to *A* by class II and III voters (5 to 4) and to *B* by class I and III voters (6 to 3). Hence, if there were a series of pairwise contests (in any order), *C* would win. Yet the poll not only does not make *C* victorious but instead magnifies *A*'s plurality victory (4 votes) by inducing *C*'s supporters, thinking that their candidate is out of the running, to throw their support to *A*, giving *A* a 2/3-majority victory (6 out of the 9 votes).

This example can be generalized to yield the following:

Proposition 4. Given the poll assumption, a Condorcet winner will always lose if he or she is not one of the top two candidates identified by the poll.

Why this proposition is true is considered in Exercise 27. Here we note that even if C were given serious consideration in a tight race, A would still win with a plurality of votes.

One might think that C's problem is due solely to the fact that the poll assumption puts him or her out of the running by presuming that his or her supporters "jump ship"–desert C for the apparently more viable candidates, A and B. But, surprisingly, even when a Condorcet winner is on top before a poll that distinguishes the top three candidates (instead of the top two), the Condorcet winner may be hurt by the poll after strategic adjustments are made by the voters, as we next illustrate.

EXAMPLE Poll (Four Candidates)

Add a fourth candidate, D, to the three in the preceding Example, and assume that there are 12 voters with the following preferences for the four candidates:

I. 3: $A\ C\ B\ D$
II. 3: $B\ C\ A\ D$
III. 4: $C\ A\ B\ D$
IV. 2: $D\ A\ B\ C$

After the poll establishes that A, B, and C are the top three candidates, the two class IV voters would be motivated to switch to their second choice, A. A would thereby increase his or her total from 3 to 5 votes–and win after the poll. Yet C is again the Condorcet winner; in staying the same at 4 votes, C is hurt, relative to A, by the poll. In fact, C would lose to A after the poll (because D's supporters would continue to vote for A), even though C was the winner before the poll. ■

Thus, the poll assumption, which induces strategic adjustments that favor the top two candidates, may hurt the Condorcet winner when a larger number of candidates (possibly including the Condorcet winner) are considered to be contenders. However, when one of the top two contenders distinguished by the poll assumption is the Condorcet winner, we have the following result:

Proposition 5. Given the poll assumption, a Condorcet winner will always win if he or she is one of the top two candidates identified by the poll.

Why this proposition is true is considered in Exercise 29. Here we note that a Condorcet winner need not be the winner in the poll, but instead can place second, and Proposition 5 still will hold. But if a Condorcet winner places second, and the poll has the effect of turning him or her into a majority winner, then it is proper to say that the poll is instrumental in electing this candidate.

We conclude that a poll may either hurt or help a Condorcet winner. If the poll assumption is modified to distinguish more than two candidates, a Condorcet winner may be hurt *even if he or she is among those distinguished by the poll,* as the previous example showed.

17.5 Election Reform: Approval Voting

The furor caused by the divided outcome in the 2000 presidential election, in which George W. Bush won the electoral vote and Al Gore won the popular vote, has spurred efforts for reform of the election system. But except for calls for abolition of the electoral college, whose effects we will analyze in the next section, most of the discussion has centered on making balloting more accurate and reliable and eliminating election irregularities, especially those that discriminate among different classes of voters.

Unfortunately, such reforms ignore a fundamental problem that plagues *multicandidate elections*—elections with three or more candidates—namely, that the candidate who wins under plurality voting may not be a Condorcet winner. Indeed, there may be no Condorcet winner, because each candidate can be beaten by at least one other candidate. In such a situation, who is the rightful winner?

Chapter 12 presents alternatives to plurality voting. Most of these alternatives allow a voter to rank candidates from best to worst, and investigate their ability to elect a Condorcet winner if one exists. Here, however, we will examine a simple election reform, *approval voting,* that does not require voters to rank candidates.

> Under **approval voting,** voters can vote for as many candidates as they like or find acceptable. Each candidate approved of receives 1 vote, and the candidate with the most approval votes wins the election.

What if approval voting had been used in the 2000 presidential election? Nader supporters, knowing that voting for Nader would be only a protest vote because Nader had no chance of winning, might also have voted for Gore. In fact, because polls show that Gore was the second choice of most Nader voters, Gore almost certainly would have won in Florida and a few other close states if there had been approval voting.

To be sure, Bush would have benefited from the approval votes of Buchanan supporters, but the number of these votes would not have come close to matching the number of votes Gore would have received from Nader supporters. There is therefore little doubt that Gore was the Condorcet winner in this election, because polls show that he could have defeated Bush (with help from Nader voters) as well as each of his less popular opponents in pairwise contests.

Arguments for an election reform like approval voting, however, should not be based on the outcome of only one election. Moreover, even in this one election, we cannot be entirely sure that Gore would have won under approval voting, because the nature of the campaign almost surely would have changed if there had been approval voting.

For example, John McCain, the Republican senator from Arizona who defeated George Bush in the New Hampshire primary but ultimately lost the Republican nomination to him, might have run as an independent candidate if there had been approval voting. As a centrist, he would have been attractive to both Democrats and Republicans and, conceivably, could have won under approval voting. But even if McCain had not run, it is likely that the candidates would have pitched their campaign appeals somewhat differently to try to attract as much approval as possible, especially from Nader and Buchanan supporters.

Although we cannot make precise predictions of the effects of approval voting in a presidential election like that of 2000, we can say what voters will to do in certain *types* of situations:

Proposition 6. In a three-candidate election under approval voting, it is never rational for a voter to vote only for a second choice. If a voter finds a second choice acceptable, he or she should also vote for a first choice.

This is certainly not true under plurality voting. If you were a Nader supporter *and* found Gore also acceptable as a second choice, you should have voted for Gore rather than Nader if (1) you thought Gore could win but Nader could not and (2) electing an acceptable candidate was most important to you. (Certainly some Nader supporters switched to Gore for these reasons.) By comparison, because you lose nothing by voting for *both* Nader and Gore under approval voting—and may gain by doing so (if Nader cannot win, you at least help Gore)—you should never vote for just your second choice (Gore in this example).

Why Proposition 6 is *always* true is considered in Exercise 35. This proposition says, in effect, that a voter whose favorite candidate in a three-candidate race seems to be out of the running can have his or her cake and eat it, too—by casting a *sincere* vote for a favorite candidate and a *strategic* vote for a second choice (to try to prevent a worst choice from winning). Roughly speaking, **strategic voting** (e.g., by Nader supporters for Gore in a plurality election) is voting that is not sincere but nevertheless has a strategic purpose—namely, to elect an acceptable candidate if one's first choice is not viable.

Another general result about approval voting that is helpful to know uses the concept of dichotomous preferences:

A voter has **dichotomous preferences** if he or she divides the set of candidates into two subsets—a preferred subset and a nonpreferred subset—and is indifferent among all candidates in each subset.

In other words, a dichotomous voter sees the world in two colors, white and black, and there is nothing in between. True, most of us see grays, but it is useful to analyze the dichotomous case first. In this case, it is not hard for voters to make a rational choice by choosing a **dominant strategy,** which is a strategy that is at least as good as, and sometimes better than, any other strategy for that person. With this definition, we can show a general condition under which Condorcet winners will be elected under approval voting:

Proposition 7. A Condorcet winner will always be elected under approval voting if all voters have dichotomous preferences and choose their dominant strategies.

A dichotomous voter's dominant strategy is to vote for all candidates in his or her preferred subset and no others. This strategy is dominant because the preferred candidates are assumed to be all equally good, so a voter has no reason to distinguish among them. Furthermore, the voter has no reason to vote for a nonpreferred candidate or candidates, because they are all equally bad and voting for any one of them could help that candidate win. As an illustration of Proposition 7, consider the following Example, in which all voters have dichotomous preferences.

EXAMPLE

Dichotomous Preferences

For each class of voters, the preferred subset of candidates is enclosed in the first set of parentheses, the nonpreferred subset in the second set of parentheses. Thus, the four class I voters prefer A and B, between whom they are indifferent, to C and D, between whom they are also indifferent:

I. 4: $(A\ B)\ (C\ D)$
II. 3: $(C)\ (A\ B\ D)$
III. 2: $(B\ C\ D)\ (A)$

Assuming that each class of voters chooses its dominant strategy, B wins with 6 votes to 5 votes for C and 4 votes for A.

In pairwise contests, notice that B is preferred to A by the two class III voters (class I and III voters are indifferent between these two candidates), so B would defeat A by 2 to 0 votes. (We assume that indifferent voters express no preference.) Because B is preferred to C by the four class I voters, and C is preferred to B by the three class II voters (class III voters are indifferent between these two candidates), B would defeat C by 4 to 3 votes. Thus, B, the approval-vote winner, is also the Condorcet winner, as Proposition 7 says must always be the case when voter preferences are dichotomous. ■

Insofar as voters in the 2000 presidential election thought equally well (or badly) of Bush and Buchanan on the one hand, and Gore and Nader on the other other, they would have preferences like those of class I voters in the previous Example. Of course, most voters probably made finer distinctions, in which case there is no guarantee that approval voting will elect a Condorcet winner.

17.6 The Electoral College

As we have noted, the electoral college had a decisive effect in the 2000 presidential election. In winning the popular vote in Florida by the slimmest of margins, George Bush captured all 25 of Florida's electoral votes, which gave him a majority of electoral votes nationwide. This won him the presidency even though he lost the popular vote.

What is the justification for the electoral college? Its original purpose was to place the selection of a president in the hands of a body that, while its members would be chosen by the people, would be sufficiently removed from them that it could make more deliberative choices. As for its composition, each state gets 2 electoral votes for its two senators (total for all states: 100). In addition, a state receives 1 electoral vote for each of its representatives in the House of Representatives, whose numbers are based on population (see Chapter 15) and range from 1 representative for the seven smallest states to 55 representatives for the largest state, California. The House has a total of 435 representatives. The District of Columbia, like the smallest states, is given 3 electoral votes. Altogether, there are 538 electoral votes, and a candidate needs 270 to win. In 2000, George W. Bush got 271 electoral votes.

(AP/Wide World Photos.)

Although there is nothing in the U.S. Constitution mandating that the popular-vote winner in a state receive all its electoral votes, this has been the tradition almost from the founding of the republic. Only in Maine and Nebraska can the electoral votes be split among candidates, depending on who wins each of the two congressional districts in Maine and the three congressional districts in Nebraska. Because the statewide winner receives the two senatorial electoral votes, the closest split possible in these two states is 3–1 in Maine and 3–2 in Nebraska. In the actual election, Gore won all of Maine's 4 electoral votes, and Bush won all of Nebraska's 5 electoral votes, so winner-take-all prevailed in all 50 states and the District of Columbia.

Effectively, then, the presidency is decided by 51 players: members of the electoral college from each of the 50 states and the District of Columbia, who almost always cast their votes as blocs for a single candidate. The voting weights of states, which depend in part on their populations, are related to their voting power, or pivotalness (see Chapter 11).

Although the percentage of voting power of a state closely tracks its percentage of electoral votes, this is not the full story. More important is the power of *individual* voters in each state, based on their ability to be pivotal in their states and their states, in turn, to be pivotal in the electoral college. Amazingly, individual voters in California are about three times as powerful as individual voters in the smallest states, despite the fact that the smallest states (with only one representative) get a 200% (2/1) boost from having 2 "senatorial" electoral votes–besides the 1 electoral vote they are entitled to on the basis of their populations–whereas California receives less than a 4% (2/52) boost.

But there is more to the story than just the size of states. Because California was never considered a close state in the 2000 presidential election (polls indicated that it would almost surely go for Al Gore), it received relatively little attention from both candidates. The real battle was fought in the so-called battleground states, or *toss-up states,* where the outcome was expected to be close (as it certainly was in Florida!). It is these states that received the bulk of the candidates' time, money, and other resources.

Instead of viewing the electoral college as a 105-million-person game in 2000, in which the voters are the players and their power is a function of the size of the states in which they vote, we view it as a game between the two major-party candidates. We will develop two different models: the first in which the candidates seek to maximize their expected popular vote, and the second in which they seek to maximize their expected electoral vote, in the toss-up states that determine the outcome in a close race.

Common to both models is the assumption that the probability, p_i, that a voter in toss-up state i votes for the Democratic candidate is

$$p_i = d_i/(d_i + r_i)$$

or the proportion of campaign resources that the Democratic candidate (d_i), compared with the Republican (r_i), spends in state i. The probability that a voter in state i votes for the Republican candidate will be the complementary probability,

$1 - p_i = r_i/(d_i + r_i)$. Thus, we ignore the effects of other candidates in the race and assume that either the Democratic or the Republican candidate will win with certainty: $p_i + (1 - p_i) = 1$. This assumption is plausible in light of the fact that no third-party candidate has ever won the presidency.

The **expected popular vote (*EPV*)** of the Democratic candidate in toss-up states, EPV_D, is the number of voters, n_i, in toss-up state i, multiplied by the probability, p_i, that a voter in toss-up state i votes for the Democrat, summed across all toss-up states:

$$EPV_D = \sum_{i=1}^{i} n_i d_i$$

where t is the number of toss-up states.

EPV_D bears some similarity to the expression for the mean ($\bar{l}$) discussed earlier. Whereas $\bar{l}$ is an average weighted by the proportions, n_i/n, EPV_D is an average weighted by the probabilities p_i.

The candidates seek strategies for optimally allocating their resources to each toss-up state. Recall that d_i for the Democrat and r_i for the Republican are the resources each allocates to state i. For the Democrat, if d_i changes, this affects the value of p_i, the probability that a voter votes for him or her, which in turn affects the value of EPV_D. Thus, the Democrat seeks a strategy d_i that will make EPV_D as large as possible.

Proposition 8. The strategy of the Democrat that maximizes his or her EPV_D (indicated by the asterisk), given that the Republican also chooses a maximizing strategy, is

$$d_i^* = (n_i/N)D$$

where $N = \sum_{i=1}^{i} n_i$, the total number of voters in the toss-up states, and $D = \sum_{i=1}^{i} d_i$, the sum of the Democrat's expenditures across all states.

In words, the Democrat should allocate his or her resources in proportion to the size of each state (n_i/N) if the Republican behaves similarly by following a strategy of $r_i^* = (n_i/N)R$, where $R = \sum_{i=1}^{i} r_i$. This allocation rule is called the **proportional rule.**

To show that d_i^* maximizes EPV_D when the Republican chooses r_i^* requires calculus, but we can readily illustrate why departures from d_i^* by the Democrat will cost him or her popular votes.

EXAMPLE Departures from a Popular-Vote Maximizing Strategy

Suppose there are three toss-up states with 2, 3, and 4 electoral votes and the candidates, because they accepted public financing for the election, are limited to spending the same total of $63 million (M). If the Republican follows his or her optimal strategy of spending in the proportion 2:3:4 ($14M:$21M:$28M), but the Democrat, ignoring the smallest state, spends in the proportion of 0:3:4 ($0M:$27M:$36M), the Republican will receive on average

$$EPV_R = 2[14/14] + 3[21/(21 + 27)] + 4[28/(28 + 36)] = 5.06 \text{ votes}$$

or 56% of the 9 votes in the three states.

If the Republican anticipates that the Democrat will spend nothing in the smallest state, the Republican can do even better. By spending only a minuscule amount in the smallest state, the Republican can almost match the Democrat in the other two states and win an average of about 5.5 votes (2 votes from the smallest state and 7/2 = 3.5 votes from the other two states), or 61%. ■

Let us assume that the goal of the candidates is not to maximize *EPV* but, instead, their *expected electoral vote (EEV)*, which is an entirely different quantity that we define below. To illustrate the difference, a candidate who wants to maximize *EEV* might think of throwing all of his or her resources into the 11 largest states—and ignoring the 39 other states and the District of Columbia if all states are toss-up states—because the 11 largest states have a majority of electoral votes (271 to be exact). Moreover, the candidate need not win "big" in these states; winning them by small margins will work just fine, because the candidate will still win *all* their electoral votes and thereby the election.

But this strategy has a problem. An opponent can readily defeat it by spending very small amounts in all the other states, which will defeat the candidate if he or she spends nothing in these states. In addition, by using his or her leftover funds to outspend the candidate in, say, one or two big states, the opponent will end up winning more electoral votes. However, there is a counterstrategy to this strategy, and indeed to every other strategy one can think of in a winner-take-all system like that of the electoral college.

To prevent being exploited if an opponent anticipates one's strategy and selects a best counterstrategy against it, each candidate should try to keep secret exactly what he or she intends to do. The best way to keep a secret is to randomize one's choices, using so-called mixed strategies, as we show in Chapter 16. But mixed strategies are difficult to calculate in a system as complicated as that of the electoral college, so we make simplifying assumptions. First, however, we need a definition.

The **expected electoral vote (*EEV*)** of the Democratic candidate in toss-up states, EEV_D is

$$EEV_D = \sum_{i=1}^{i} v_i P_i$$

where v_i is the number of electoral votes of toss-up state i, and P_i is the probability that the Democrat wins *more than* 50% of the popular votes in this state, which would give the Democrat *all of* that state's electoral votes, v_i. To determine P_i, we need to count the number of ways that a candidate can win a majority of electoral votes in each state i and then compute the probabilities that each of these ways occurs. These probabilities in turn will be used to determine EEV_D. EEV_R can be defined in an analogous manner.

EXAMPLE

Computing the Democrat's Expected Electoral Vote (EEV_D)

Consider our earlier Example of three states with 2, 3, and 4 votes, which we will call, respectively, states A, B, and C. For simplicity, assume here that the number of electoral votes of each state is equal to the number of voters in that state.

To calculate P_i for each state i, we must determine the probabilities that a majority of voters in states A, B, and C vote Democratic (we will ignore the possibility of ties in states A and C, which have an even number of voters). To obtain these probabilities, we multiply the probabilities that individual voters in each state, who are assumed to act independently of each other, vote Democratic or Republican (based on the resources the two candidates allocate to each state).

For state A to vote Democratic, for example, both voters in this state must vote Democratic, so $P_A = (p_A)(p_A) = (p_A)^2$. For state B to vote Democratic, either two of the three voters, or all three voters, must vote Democratic, so $P_B = (p_B)^2(1 - p_B) + (p_B)^3$. For state C to vote Democratic, either three of the four voters or all four voters must vote Democratic, so $P_C = (p_C)^3(1 - p_C) + (p_C)^4$. Substituting these probabilities into the formula for EEV_D, we obtain

$$\begin{aligned} EEV_D &= v_A P_A + v_B P_B + v_C P_C \\ &= 2[(p_A)^2] + 3[(p_B)^2(1 - p_B) + (p_B)^3] + 4[(p_C)^3(1 - p_C) + (p_C)^4] \end{aligned}$$ ■

We indicated earlier that the strategy of the Democrat that maximizes EEV_D, given that the Republican adopts a similar strategy, is mixed, involving randomizing his or her choice of states in which to allocate resources. Because this randomization is difficult to determine, we simplify the task by assuming that $d_i = r_i$ in each toss-up state i, and necessarily $D = R$ across all toss-up states.

This assumption is defensible if the candidates have the same total amount to spend in all the states ($D = R$). If they perceive the value of each toss-up state to be the same, which is reasonable, they will allocate equal resources to each. But how much should these amounts be? It is possible to show the following:

Proposition 9. The strategies of the Democratic and the Republican candidates that maximize their *EEVs* are

$$d_i^* = \left(\frac{v_i\sqrt{n_i}}{S}\right)D; \quad r_i^* = \left(\frac{v_i\sqrt{n_i}}{S}\right)R,$$

where

$$S = \sum_{i=1}^{i} v_i\sqrt{n_i}.$$

In words, the candidates should allocate their resources in proportion to the number of electoral votes of each state (v_i) multiplied by the square root of its size (n_i). The allocations, d_i^* and r_i^*, will be the same if the candidates spend equally in each toss-up state ($d_i = r_i$), as previously assumed.

Because the number of electoral votes in each state i (v_i) is approximately proportional to the number of voters in each state (or their populations), n_i, the maximizing strategies of the candidates can be approximated by the *3/2's rule:*

$$d_i^* = \left(\frac{v_i^{3/2}}{T}\right)D; \qquad r_i^* = \left(\frac{v_i^{3/2}}{T}\right)R$$

where

$$T = \sum_{i=1}^{i} v_i^{3/2}$$

Thus, if the candidates allocate the same amount of resources to each of the toss-up states, the **3/2's rule** is that they should spend approximately in proportion to the 3/2's power of the electoral votes of these states to maximize *EEV*.

EXAMPLE

Applying the 3/2's Rule

Assume states A, B, and C have, respectively, 9, 16, and 25 voters, and these are also their numbers of electoral votes. If each of these states is a toss-up state, the 3/2's rule says that the candidates should allocate their resources in the proportions 27:64:125, because the 3/2's powers of their electoral votes are their numbers of voters multiplied by the square root of these numbers. Thus, for state A, $9^{3/2} = 9\sqrt{9} = (9)(3) = 27$.

To illustrate the difference between the proportional rule (which maxmizes *EPV*) and the 3/2's rule (which maximizes *EEV*), assume both candidates can spend 100 units of resources. Then the proportional rule says that states A, B, and C should get resources in the amounts 18, 32, and 50, whereas the 3/2's rule says these states should get resources in the amounts 13, 30, and 58.

Clearly, the smallest state does worse and the largest state does better under the 3/2's rule, whereas the middle state stays about the same. In the actual electoral college, the voters in California, when it is a toss-up state, are about three times as attractive *per capita* as voters in the smallest states. Following the 3/2's rule, therefore, the candidates should allocate about three times as much per voter to California as to a small toss-up state with only 3 electoral votes.

In fact, presidential candidates greatly overspend in the largest toss-up states, well out of proportion to their size. This large-state bias is far out of line with the democratic principle of "one person, one vote"; for Californians, compared to small-state voters, this principle should read "one person, three votes." ■

To be sure, it is not the electoral college per se that creates this bias but, rather, its winner-take-all feature. If this feature were abolished and the electoral votes of a state were split according to the popular votes received by the candidates—insofar as possible—then the large-state bias would generally disappear.

EXAMPLE Departures from an Electoral-Vote Maximizing Strategy

We illustrated earlier how a candidate's departure from the popular-vote maximizing strategy—the proportional rule—lowers that candidate's expected popular vote, given that the candidate's opponent adheres to this rule, which makes it an equilibrium. This is also true of *some* departures from the 3/2's rule: If a candidate's departure from this rule is "small," he or she will lower his or her expected electoral vote, given the candidate's opponent sticks to that rule.

Suppose, for example, that the Republican follows the 3/2's rule in allocating resources to states A, B, and C with 9, 16, and 25 voters/electoral votes, respectively. If each candidate has 100 units of resources to spend, we showed in the previous Example that this translates into allocating 13, 30, and 58 units to states A, B, and C, respectively.

Now if the Democrat deviates slightly from the 3/2's rule, and allocates 14, 30, and 57 units to these states (more to A, less to C), he or she increases the chances of winning in A and decreases the chances of winning in C. It can be shown that this deviation from the 3/2's rule hurts the Democrat, because even though his or her chances go up more in A than they go down in C (because A is smaller than C), C has almost three times as many electoral votes. On balance, this deviation lowers the Democrat's expected electoral vote.

Now suppose the Democrat makes a "large" deviation from the 3/2's rule, ignoring state B entirely and throwing all his or her resources into state A (9 electoral votes) and state C (25 electoral votes). If he or she wins in these two states, this would give the Democrat 34 of the 50 electoral votes, which is more than enough to win.

If the Republican adheres to the 3/2's rule, he or she will put approximately 13% of his or her resources into state A and 58% into state C. By following the 3/2's rule in these two states and ignoring state B, the Democrat will put approximately 18% into state A and approximately 82% into state C. This translates into *each voter* in states A and C supporting the Democrat with probability 0.58, which means that the Democrat will almost certainly win in both these states, giving him or her an expected electoral vote of almost 34. The Republican *will* certainly win in state *B*, because the Democrat spends nothing in this state, but this is small consolation if the Republican loses in the two other states and receives an expected electoral vote of somewhat more than 16. ■

This Example illustrates why the 3/2's rule is a local maximum but not a global maximum.

A **local maximum** is a maximizing strategy from which small deviations are nonoptimal but large deviations may be optimal. A **global maximum** is a maximizing strategy from which *all* deviations (small or large) are nonoptimal. The proportional rule is a global maximum (and equilibrium) for candidates whose goal is to maximize their expected popular vote, whereas the 3/2's rule is only a local maximum for candidates whose goal is to maximize their expected electoral vote.

If the goal of candidates is to maximize their expected electoral vote, there is no *determinate* maximizing strategy—randomizing one's choices is necessary to prevent exploitation by an opponent. How this randomization is done for some simple games is analyzed in Chapter 16.

In the case of the electoral college, we illustrated how a radical departure from the 3/2's rule by a candidate, who ignores some state or states entirely, may be the best response to an opponent who follows this rule. But then there is a best response to this best response, and so on, so no determinate strategy is invulnerable.

However, insofar as the candidates view the toss-up states in similar terms, the 3/2's rule offers of good rule of thumb as to how much to spend in each, as a function of its size, to maximize one's expected electoral vote. But we must remember that it is only a local maximum. Hence, unlike the proportional rule, which is robust against all other popular-vote maximizing strategies, the 3/2's rule is vulnerable to radically different strategies, such as those in which candidates concentrate their efforts on relatively few states.

Usually those who try such strategies, however, take big risks. Thus in 1964, the Republican presidential candidate, Barry Goldwater, said that he would like to "saw off the Eastern seaboard" (Goldwater was from Arizona), but in the end he went, in his memorable phrase, "shooting where the ducks [voters] are." In doing so, however, he appeared not so much to want to win as to present voters with "a choice, not an echo," by taking relatively extreme (conservative) positions. Is it any wonder, then, that Goldwater lost in a huge landslide to his Democratic opponent, Lyndon Johnson?

17.7 Is There a Better Way to Elect a President?

The greatest spectacle in American politics is the quest for the presidency. While there is nothing to match its pageantry and excitement, the quieter gamelike features of a presidential campaign are no less consequential.

We have emphasized these features in this chapter, showing how mathematics can be used to analyze optimal positions in two-candidate and multicandidate races, and how polls and presidential primaries may affect who stays in and who drops out of the race. Sometimes, as we have seen, Condorcet candidates, who in pairwise contests can beat every other candidate, may not survive. And, as was dramatically illustrated in the 2000 presidential election, the popular-vote winner may not win in the electoral college.

Many people think that approval voting may better enable voters, especially in the early presidential primaries—which typically draw many candidates (if an incumbent is not running for reelection)—to express their preferences. However, other election procedures, including those discussed in Chapter 12, possess features that may make them desirable as election reforms.

All these procedures would probably be of more help to centrist candidates, who not only better represent the entire electorate than extremist candidates but also are more likely to be a party's strongest contender in the general election. (In the past 40 years, the biggest losers in presidential elections, Republican Barry Goldwater in 1964 and Democrat George McGovern in 1972, came from the right and left extremes, respectively, of their parties.)

Taking the choice of a president out of the hands of voters and putting it in the hands of members of the electoral college may no longer be justified. The electoral college, with its winner-take-all feature, creates a large-state bias, as we showed. In the 2000 presidential election, it was a few hundred voters in one large toss-up state, Florida, who determined the outcome.

Is there any reason why the votes of *all* voters, wherever they reside, should not count equally, which direct popular-vote election of a president would ensure? Short of abolishing the electoral college, which would require a constitutional amendment, electoral votes could be allocated proportionally in each state.

If approval voting were used, then it would be approval votes rather than the single votes of each voter that would determine the allocation of electoral votes to the candidates. Because the general election in recent years has drawn major third- and fourth-party candidates, approval voting, or one of the other voting procedures discussed in Chapter 12, seems worthy of consideration if one wants to reduce the role of spoilers.

In summary, mathematics illuminates strategic aspects of campaigning and voting in presidential elections not apparent to the naked eye. It also points the way to possible reforms that may ameliorate some of the problems that plague our current system.

REVIEW VOCABULARY

Approval voting Allows voters to vote for as many candidates as they like or find acceptable. Each candidate approved of receives 1 vote, and the candidate with the most approval votes wins.

Bandwagon effect Voting for a candidate not on the basis of merit but, instead, because of the expectation that he or she will win.

Condorcet candidate A candidate who can defeat each of the other candidates in pairwise contests.

Dichotomous preferences Held by voters who divide the set of candidates into two subsets – a preferred subset and a nonpreferred subset – and are indifferent among all candidates in each subset.

Dominant strategy A strategy that is at least as good as, and sometimes better than, any other strategy.

Electoral college A body of 538 electors that selects a president.

Equilibrium position A position is in equilibrium if no candidate has an incentive to depart from it unilaterally.

Expected electoral vote (*EEV*) In toss-up states, the number of electoral votes of each toss-up state, multiplied by the probability that the Democratic (or Republican) candidate wins more than 50% of the popular votes in that state, summed across all toss-up states.

Expected popular vote (*EPV*) In toss-up states, the number of voters in each toss-up state, multiplied by the probability that that voter votes for the Democratic (or Republican) candidate, summed across all toss-up states.

Extended median The equilibrium position of two candidates when there is no median.

Maximin position A position is maximin for a candidate if there is no other position that can guarantee a better outcome – more votes – whatever position another candidate adopts.

Mean ($\bar{I}$) A weighted average, wherein the positions of voters are weighted by the fraction of voters at that position.

Median (*M*) The point on the horizontal axis of a voter distribution where half the voters have attitudes that lie to the left and half to the right.

Median-voter theorem In a two-candidate election with an odd number of voters, the median is the unique equilibrium position.

Mode A peak of a distribution. A distribution is **unimodal** if has one peak and **bimodal** if it has two peaks.

1/3-separation obstacle An obstacle for the entry of a third candidate created if two previous entrants are sufficiently close together.

Poll assumption If necessary, voters adjust their sincere voting strategies to differentiate between the top two candidates – as revealed in the poll – by voting for the one they prefer.

Plurality voting Allows a voter to vote for one candidate, and the candidate with the most votes wins.

Proportional rule Presidential candidates allocate their resources to toss-up states according to their size. This allocation rule maximizes the expected popular vote of a candidate, given that his or her opponent adheres to it; it is a **global maximum** (i.e., robust against all deviations by a candidate).

Sincere voting Voting for a favorite candidate, whatever his or her chances are of winning.

Spatial models The representation of candidate positions along a left–right continuum in order to determine the equilibrium or optimal positions of the candidates.

Spoiler problem Caused by a candidate who cannot win but "spoils" the election for a candidate who otherwise would win.

Strategic voting Voting that is not sincere but nevertheless has a strategic purpose – namely, to elect an acceptable candidate if one's first choice is not viable.

3/2's rule Presidential candidates allocate their resources to toss-up states according to the 3/2's power of their electoral votes. This allocation rule maximizes the expected electoral vote of a candidate, given that his or her opponent adheres to it; it is a **local maximum** (i.e., invulnerable to small deviations but not large deviations of a candidate).

2/3-separation opportunity An opportunity for the entry of a third candidate created if two previous entrants are sufficiently far apart.

Voter distribution Gives the number (or percentage) of voters who have attitudes at each point along the left–right continuum, which can be represented by a curve. The distribution is **symmetric** if the curve to the left of the median is a mirror image of the curve to the right. It is **skewed** if the area under the curve is concentrated more on one side of the median than the other.

SUGGESTED READINGS

AMY, DOUGLAS J. *Behind the Ballot Box: A Citizen's Guide to Voting Systems,* Praeger, Westport, Conn., 2000. Sets forth criteria for evaluating voting systems and compares many different ones, listing both their advantages and disadvantages. The treatment is nonmathematical.

BRAMS, STEVEN J. *The Presidential Election Game,* Yale University Press, New Haven, Conn., 1978. Focuses on the strategic aspects of presidential elections–from primaries to conventions to general elections–and also includes an analysis of the "game" played between President Richard Nixon and the Supreme Court over the release of the Watergate tapes that led to Nixon's resignation in 1974 (Nixon was the only president to resign the presidency). Approval voting and direct popular-vote election of a president are recommended as election reforms.

BRAMS, STEVEN J., and PETER C. FISHBURN. *Approval Voting,* Birkhäuser, Boston, Mass., 1983. An in-depth analysis of approval voting. It includes several case studies, including an analysis of how the 1980 presidential election would have turned out under approval voting.

HINICH, MELVIN J., and MICHAEL C. MUNGER. *Analytical Politics,* Cambridge University Press, Cambridge, U.K., 1997. Extends spatial modeling to more than one dimension, analyzes probabilistic voting, and introduces game-theoretic solution concepts relevant to the study of elections. As a text written for undergraduates, it could serve as a sequel to this chapter.

POMPER, GERALD M., ed. *The Election of 2000: Reports and Interpretation,* Seven Bridges Press, New York, 2001. An edited collections of articles by experts on various aspects of the campaigns and elections of 2000, including the congressional elections. Pomper has edited these books every four years since the 1976 elections.

SAARI, DONALD G. *Chaotic Elections! A Mathematician Looks at Voting,* AMS [American Mathematical Society], Providence, R.I., 2001. Argues that elections–in particular, the 2000 presidential election, but others as well–have chaotic features that can be understood through mathematics. The mathematics used is an unusual kind of geometry that will be accessible to those with some mathematical background.

SHEPSLE, KENNETH A., and MARK S. BONACHEK. *Analyzing Politics: Rationality, Behavior, and Institutions,* Norton, New York, 1997. Rational strategies in voting and elections are a major component of this text, but it also includes sections on collective action and political institutions, such as courts and legislatures. Several case studies illustrate the theory.

SUGGESTED WEB SITES

www.fec.gov/elections.html Federal Election Commission–About Elections and Voting

www.ifes.org International Foundation for Election Systems

www.reformelections.org National Commission on Federal Election Reform

www.constitutionproject.org The Constitution Project – The Election Reform Initiative

dmoz.org/Society/Politics/Campaigns_and_Elections/Voting_systems

bcn.boulder.co.us/government/approvalvote/center.html

electionmethods.org/Approved.htm

SKILLS CHECK

1. In a two-candidate election, suppose the attitudes of the voters are distributed symmetrically with median *M*. Of the two candidates *A* and *B*, *A* is positioned far to the left of *M* and *B* is positioned just to the right of *M*. How will the voters choose between candidates *A* and *B*?

(a) *A* will receive a majority of the votes.
(b) *B* will receive a majority of the votes.
(c) *A* and *B* will both receive exactly one-half of the votes.

2. In a three-candidate election, suppose the attitudes of the voters are distributed symmetrically with median *M*. Of the three candidates *A*, *B*, and *C*, *A* is positioned far to the left of *M*, *B* is positioned just to the right of *M*, and *C* is positioned at *M*. How will the voters choose among these three candidates?

(a) *A* will receive the most votes.
(b) *B* will receive the most votes.
(c) *C* will receive the most votes.

3. In a two-candidate election, which of the following positions is an optimal position for both candidates *A* and *B*?

(a) *A* and *B* just to the left and right of *M*
(b) *A* and *B* far to the left and right of *M*
(c) *A* and *B* both at *M*

4. In a three-candidate election, if candidates *A* and *B* are positioned at *M*, which is an election-winning position for candidate *C*?

(a) Just to the right of *M*
(b) Far to the right of *M*
(c) There is no election-winning position for candidate *C*.

5. In a three-candidate election, if candidates *A* and *B* are positioned just to the left and just to the right of *M*, which is an election-winning position for candidate *C*?

(a) At *M*
(b) Far to the right of *M*
(c) There is no election-winning position for candidate *C*.

6. In a four-candidate election, if candidates are aligned in order *A–B–C–D*, who benefits if *D* drops out of the race?

(a) Candidate *A*
(b) Candidate *B*
(c) Candidate *C*

7. In a three-candidate election, suppose 12 voters can be divided into three classes according to their preferences: Five voters prefer (in order) *A*, *B*, *C*; four voters prefer *C*, *A*, *B*; three voters prefer *B*, *C*, *A*. In order to elect one of their top two candidates, which group of voters will not vote sincerely (i.e., for their first choice)?

(a) The five voters
(b) The four voters
(c) The three voters

8. Assuming the poll assumption, the Condorcet winner will win

(a) precisely when he or she is not the top candidate identified by the poll.
(b) precisely when he or she is one of the top two candidates identified by the poll.
(c) regardless of his or her rank in the poll.

9. In an election with a large pool of candidates, approval voting benefits

(a) candidates at the extreme left and right.
(b) candidates at or near M.
(c) only candidates precisely at M.

10. In a four-candidate approval voting election with 12 voters, if 5 voters approve of A and B; 4 voters approve of B and C; and 3 voters approve of A and D, who wins the election?

(a) Candidate A
(b) Candidate B
(c) Candidate C

11. In the election of the president using the electoral college, voters in smaller swing states have

(a) less power than voters in large swing states.
(b) more power than voters in large swing states.
(c) the same power as voters in large swing states.

12. In the election of the president, if the Democrats believe that the Republicans will not allocate resources in a way that maximizes their *EEV*, then

(a) they can successfully counter by allocating their resources according to the method that maximizes *EEV*.
(b) they can successfully counter by not allocating their resources according to the method that maximizes *EEV*.
(c) they cannot successfully counter.

13. Suppose three voting blocs A, B, and C control 16, 25, and 36 votes, respectively, and 39 votes are required to win. If each bloc is a toss up, for every \$1 that is allocated to lobby the voters in bloc A, how much should be allocated to lobby the voters in bloc C?

(a) About 2 times as much
(b) About 3 times as much
(c) About 20 times as much

14. Voting for president using the electoral college is

(a) biased in favor of the largest states.
(b) biased in favor of the smallest states.
(c) not biased.

15. Is it possible that a candidate for president could win the popular vote and yet lose the electoral college vote?

(a) Yes, and it occurs on occasion.
(b) Yes, but it has never occurred in the past 100 years.
(c) No.

EXERCISES

▲ Optional ■ Advanced ◆ Discussion

Spatial Models: Two-Candidate Elections

1. Why do M and $\bar{l}$ not coincide if a distribution is skewed?

2. Show that 0.6 is the equilibrium in Figure 17.3b.

■ **3.** Prove that if a distribution is discrete and there is no median position, there is always an *extended median*. (*Hint:* Show that there is always one position at which a majority of voters lie neither to the left nor to the right, and neither candidate would have an incentive to depart from this position.)

4. Assume that the one voter at 0.1 in Figure 17.2a decides not to vote because he is "too far away" from the two candidates who take the median position at 0.6. Would either candidate depart from $M = 0.6$ to try to do better if he or she knew that this voter had decided not to vote – but he or she would vote for the closer of two candidates less than a distance of 0.5 away? What if the candidates knew that the three voters at 0.2 had also decided not to vote – but they would vote for the closer of two candidates less than a distance of 0.4 away?

5. If you are a far-left or a far-right voter, are you helping your cause – by electing more ideologically proximate candidates – when you announce, like the voters in Exercise 4, that you will not support candidates who are too far away?

6. Consider the two most extreme voters at 0 in Figure 17.3b (i.e., those who are farthest from the extended median of 0.6). Would their nonvoting change the extended median? How about, as well, the nonvoting of the somewhat less extreme voter at 0.9? Show when, if at all, M or the extended median will change as less and less extreme voters decide not to vote in this example?

◆ **7.** In Figure 17.2a, $\bar{I}$ is not in equilibrium—one candidate would do better if he or she moved from 0.56 to 0.6. But is 0.6 really a better reflection of the views of the electorate than 0.56?

8. Define an outcome to be in equilibrium if, given that one candidate chooses it, the other candidate cannot do better than take the same position. Show that this definition is equivalent to the text's definition of being in equilibrium.

■ **9.** Consider a trimodal distribution (three peaks). When will taking a position at the middle peak be in equilibrium? Is it possible that one of the other peaks can ever be in equilibrium? If so, give a discrete-distribution example.

10. Define A's position in a two-candidate race to be *opposition-optimal* if, given that the position of B is fixed, it maximizes A's vote total. Show that A's opposition-optimal position must be adjacent to B's position and closer to M, except when B is at the median. (Roughly speaking, being "adjacent" means being a very small distance away.)

◆ **11.** Assume the population along a main street is uniformly distributed over [0, 1], so there are equal numbers of people located at all equally spaced intervals from $M/\bar{I}$. (This makes the distribution rectangular, or "flat.") It has been argued that the "social optimum" for the location of two stores are at the points 1/4 and 3/4, because then no person would have to travel more than 1/4 of the length of the street to buy at either store. Is this desirable if the population is not uniformly distributed?

◆ **12.** What is a social optimum in an election if only one candidate is to be elected? How about five candidates to a city council? Is it better that the city council members' positions all be centered around 0.5, or should they be more spread out?

◆ **13.** Which is better for consumers: (a) to minimize the maximum distance they must travel to a store; or (b) to foster price competition, which would presumably be encouraged if two stores are located at $M = 1/2$?

■ **14.** Assume a city comprises three equal-sized districts, each of which elects a candidate to the city council. The mayor is elected by the entire city. Show with an example that the median or extended median for the mayor need not be the median or extended median for any of the three city council districts. Does this explain why mayors and city council members often disagree?

▲ **15.** In Exercise 14, must the median or extended median for the mayor be between the leftmost and rightmost medians or extended medians of the three districts? How about the mean $\bar{I}$?

Spatial Models: Multicandidate Elections

16. Assume that A and B take the *same* nonmedian position. What position should C take to maximize his or her vote total? Is C's position always a winning one?

■ **17.** Assume that A and B take *different* positions, with one possibly being at M. What position should C take to maximize his or her vote total? Is C's position always a winning one?

18. Is there a 1/3-separation obstacle if the distribution is not symmetric but no more than 1/6 of the area under the curve separates A (on the left) from M, and no more than 1/6 of the area separates B (on the right) from M? What if these 1/6-or-less areas on the left and the right are not the same?

19. Is there a 2/3-separation opportunity if the distribution is not unimodal but at least 1/3 of the area under the curve separates A (on the left) from M, and at least 1/3 separates B (on the right) from M? (*Hint:* Start by assuming that the distribution is uniform between A and B—and hence not unimodal—and that exactly 2/3 of the voters lie between A and B. Can C always win by taking a position at M? If not, is there a nonunimodal distribution that affords C this opportunity?)

20. Show that C cannot win under the conditions of Proposition 3 (page 607). (*Hint:* Indicate which candidate will win when C enters to the left of A, to the right of B, or in between.)

21. It is known that A, B, and C will enter an election in that order, with A announcing his position first, then B, and finally C. If the distribution is uniform over [0, 1], what position should each candidate take to maximize his or her vote total, anticipating–in the case of A and B–the entry of future candidates. [*Hint:* Start by assuming that A takes a position at 1/4. Is B's position at 3/4 optimal, anticipating the entry of C? Or can B do better at some other position (perhaps by influencing C's choice of a maximizing position)]?

▲ 22. If A and B are equidistant from the median of a symmetric distribution and separated from each other by exactly 1/2 of the area under the curve, under what conditions is this separation an obstacle and under what conditions is it an opportunity? (*Hint:* Start by constructing examples of symmetric distributions in which C would either win or lose by taking a position at M.)

■ 23. What are the vote-maximizing positions for four candidates to take if it is known that they will enter in the order A, B, C, D?

Winnowing the Field

24. Assume that the four candidates in the 2000 presidential election can be arrayed from left to right as follows: Nader–Gore–Bush–Buchanan. Suppose a poll reveals Gore at 48%, Bush at 47%, Nader at 3%, and Buchanan at 2%. Would Bush be well advised to offer Buchanan a cabinet position to drop out of the race (as Adams offered Clay the secretary of state post after the 1824 election)? What if Bush knew that, after Buchanan dropped out, only 1/2 of Buchanan's supporters would switch to him, with most of the remainder not voting, except for a few who would switch to Gore?

25. Assuming the same poll results as in Exercise 24, now suppose that Gore offered the same deal to Nader, knowing that only 1/3 of Nader supporters would switch to him and the rest would not vote. However, suppose Gore also thought that if Nader dropped out, so would Buchanan, and all Buchanan supporters would vote for Bush. Should Gore set off this train of events?

◆ 26. Is there any evidence that the four presidential candidates in 2000 might have contemplated "deals" of the kind indicated in Exercises 24 and 25? If you cannot find any evidence, do you think this is because the candidates found such ploys unethical or because they thought they might be found out if they tried to make them?

◆ 27. One tactic that was considered by Nader supporters who thought that their votes for Nader might kill Gore's chances in some states was to swap votes: In close states that Gore might lose if Nader supporters stuck with their candidate, these supporters would switch to Gore if Gore supporters in nonclose states, where Gore would almost surely win, would switch to Nader. Thereby the popular-vote totals for the two candidates would not change overall, but Gore would be able to win in the close states he might otherwise lose. Is this a sensible way of dealing with problems created by the electoral college, which puts a premium on winning in large states (more on this question later)?

What Drives Candidates Out

28. Show why Proposition 4 (page 610) is true. Is it true that if the poll assumption was modified to differentiate the top three (rather than the top two) candidates from the rest, and the Condorcet winner was not among the top three, that he or she would still lose?

◆ 29. Show why Proposition 5 (page 611) is true. Is it proper that the candidate who comes in second in the poll should win after the results of the poll are announced? Why?

◆ 30. In the Example preceding Proposition 5 (page 611), after class IV voters switch from D to A, the vote totals for the top three candidates are A–5, B–3, and C–4. Now assume a second poll is taken, differentiating A and C, the top two contenders, from B. If B supporters switch at this point to their second choice, which candidate will win? Do you consider this a desirable outcome?

◆ **31.** Assume there are four classes of voters that rank four candidates as follows:

I. 4: $A\ D\ B\ C$
II. 3: $B\ D\ A\ C$
III. 2: $C\ D\ B\ A$
IV. 1: $D\ C\ B\ A$

Which candidate is the Condorcet winner? Do you find this result strange in the light of what a poll would tell the voters?

32. Assume there is a poll that differentiates the top two candidates in Exercise 31. Which candidate will win the election after the poll.

◆ **33.** Assume there is a poll that differentiates the top three candidates in Exercise 31. Which candidate will win the election after the poll? Comment on the different outcomes in this exercise and the previous one.

◆ **34.** Assume there are three classes of voters that rank three candidates as follows:

I. 4: $A\ B\ C$
II. 3: $B\ C\ A$
III. 2: $C\ A\ B$

Show that there is no Condorcet winner. Applying the poll assumption to this example, which candidate will win? Is this fair, given the preferences of class II and III voters, who, together, are a majority?

Election Reform: Approval Voting

▲ **35.** Prove Proposition 6 (page 613).

36. In a three-candidate election, show that the strategy of voting for one's top two choices under approval voting is not always better than voting only for one's top choice.

37. Consider a four-candidate election under approval voting. Is there ever a situation in which a voter would vote for a first and a third choice without also voting for a second choice? [*Hint:* Assume a voter ranks the four candidates $A\ B\ C\ D$ and believes that one of two things can happen: the electorate will favor either liberals (say, A and B) or conservatives (say, C and D) but never favor each side equally.]

38. Is there ever a situation under approval voting in which a voter would vote for a worst choice?

39. Is there ever a situation under approval voting in which a voter would *not* vote for a first choice if he or she finds acceptable one or more lower-ranked candidates? (*Note:* This question asks whether Proposition 6 can be generalized to more than three candidates.)

■ **40.** Prove Proposition 7 (page 614). (*Hint:* If all voters have dichotomous preferences and vote for all candidates in their preferred subsets, which candidate will get the most approval votes? What does this say about the preferences of voters for the approval-vote winner, compared to their preferences for each of the other candidates?)

41. In the following example, class I and II voters have dichotomous preferences, but the class III voter has *trichotomous preferences* (i.e., he or she divides the four candidates into three indifference subsets):

I. 2: $(A\ B)\ (C\ D)$
II. 2: $(C)\ (A\ B\ D)$
III. 1: $(D)\ (C)\ (A\ B)$

Is it rational for the class III voter to vote only for his or her top choice, D? If not, who else should he or she approve of? Which class of voters will be most unhappy if the class III voter does not vote just for D? Can voters in this class, by voting strategically, do anything about their situation?

◆ **42.** Assume the class III voter's preferences change to a different trichotomous ordering?

III. 1: $(D)\ (A\ C)\ (B)$

Suppose, as in Exercise 41, that the class III voter indicates in an initial poll that he or she intends to vote only for D but then, in response to the poll, switches to voting for the candidates in his or her second-choice subset as well. If there is a new poll, based on these results, what will be the outcome? What if there is a third poll, fourth poll, and so on? Do you regard this result as desirable? Why?

43. In Exercise 31, we saw that under plurality voting the Condorcet winner, D, comes in fourth in a poll and, therefore, cannot be helped by subsequent polling (Proposition 2, page 605), even when the poll distinguishes the top three candidates and voters differentiate among them:

I. 4: *A D B C*
II. 3: *B D A C*
III. 2: *C D B A*
IV. 1: *D C B A*

What are the outcomes under approval voting–both with and without polling–if voters approve of their (i) top-ranked, (ii) two top-ranked, and (iii) three top-ranked candidates initially. [*Note:* In making adjustments to the poll results, assume that voters approve of not only the preferred of their two top-ranked candidates identified by the poll but also of *all* candidates ranked above their preferred candidate. For example, when the poll based on (i) above identifies A and B as the two top-ranked candidates, with 4 and 3 votes, respectively, the class III and class IV voters after the poll will approve of not only their preferred candidate, B, but also of C and D, because they rank the latter two candidates above B.]

◆ **44.** On the basis of your answers to the foregoing problems, do you think approval voting would be beneficial in finding Condorcet winners–either with or without polling–in multicandidate elections?

The Electoral College

45. Assume there are three states with 3, 7, and 9 voters, and that they are all toss-up states. If both the Democratic and Republican candidates choose strategies that maximize their expected popular vote (the proportional rule), and they have the same total resources ($D = R$), what is the expected number of votes that each will receive?

46. Assume the Republican knows in advance what allocations, d_i, to each state i the Democrat will make in Exercise 45. Then the Republican's optimal response can be shown to be

$$r_i = \frac{\sqrt{n_i d_i}}{\sum_{i=1}^{t} \sqrt{n_i d_i}} (R + D) - d_i$$

Suppose that the Democrat ignores the smallest state and makes proportional allocations to the two largest states. (For concreteness, assume both candidates have 100 units of resources.) What is the Republican's optimal response? What if the Democrat makes proportional allocations to all three states?

▲ **47.** In Exercise 46, show that if the Democrat makes proportional allocations, and the Republican responds optimally according to the formula given there, this formula simplifies to $r_i = (n_i/N)R$, which does not depend on d_i. What does this say about the proportional rule? [*Hint:* If the Republican finds out (e.g., through a spy) that the Democrat is making proportional allocations, does the Democrat have anything to worry about?]

48. In Exercise 47, if the Democrat has only half the resources of the Republican, would you recommend that he or she behave differently from proportional allocations to maximize his or her expected popular vote? Why? If the Republican allocates his or her resources proportionally to the three states, is there any way the Democrat can allocate his or her resources to win a majority of votes in states with more than half the votes?

49. Instead of maximizing the expected popular vote, assume the candidates in Exercise 45 want to win in states with more than half the votes. Suppose the candidate that allocates more resources to a state wins that state. Is there any state to which a candidate should not consider allocating resources? Should the states that receive allocations receive equal allocations?

50. In Exercise 45, assume you can choose specific voters in each state to whom you can allocate resources. Suppose the candidate who allocates more resources to a voter wins that voter's vote. If your goal is to win the votes of a majority of voters in states that have more than half the votes, which voters would you target, and how much would you spend on each? (*Hint:* First show which states you would target; then show that these states should receive equal allocations, which in turn should be divided equally among a certain set of voters.)

51. Assume there are three toss-up states, A, B, and C, with, respectively, 2, 3, and 4 voters, which are also the number of electoral votes of each state. In the text, we gave the formulas for the probabilities, P_A, P_B, and P_C, that the Democrat wins a majority of popular votes in each state and, therefore, wins all the electoral votes of that state. Show that the formula for the probability that the Democrat *wins the election* under the electoral college, PWE_D, is

$$\begin{aligned} PWE_D = {} & P_AP_B(1 - P_A) + P_AP_C(1 - P_B) \\ & + P_BP_C(1 - P_A) + P_AP_BP_C \end{aligned}$$

(*Hint:* Winning in any two states is sufficient to win the election.)

■ **52.** Compare the formula for PWE_D with the formula for EEV_D (in the text). Which quantity is it better to maximize? What would be a good resource-allocation strategy for maximizing PWE_D?

▲ **53.** Is the square root in the formulas for the EEV maximizing strategies of the Democratic and Republican candidates related to the square-root rule for the electoral college discussed in Chapter 12?

✎ WRITING PROJECTS

1. What evidence is there that the median-voter theorem applies in presidential elections with only two serious contenders? In cases where the candidates do not approach the median, is this because they view their main supporters as more extreme and do not want to alienate them?

2. It is generally agreed that presidential nomination races today are "front-loaded" – candidates must devote most of their resources to the early primaries even to stay in contention, much less win. Does this fact tend to help moderates or extremists as the field is narrowed? Would this be true under approval voting? Is it desirable that momentum plays such a large role in presidential primaries?

3. Do you think polling is useful in helping voters choose the "best" candidate? Or would it be better, as in some countries, to ban the publication of polls before an election? Discuss these questions in light of the theoretical effects polling has when voters react to polls and possibly change their voting strategies. Is there empirical evidence that voters behave in this way?

4. How serious a problem do you think the large-state bias of the electoral college is? How would you explain the fact that some of the strongest advocates of the electoral college come from small states? Has the theoretical bias been a reality in the campaign behavior of candidates in recent presidential elections?

SPREADSHEET PROJECTS

To do these projects, go to www.whfreeman.com/fapp.

The difference in outcomes between the popular vote and the vote of the electoral college can be striking. This project explores such scenarios through the use of spreadsheets.

CHAPTER
18

Growth and Form

Fantasy films have made us familiar with assorted giant creatures, including King Kong, Godzilla, and the 50-foot-high grasshoppers in *The Beginning of the End.* We also find supergiants in literature, such as the giant of "Jack and the Beanstalk," Giant Pope and Giant Pagan of *The Pilgrim's Progress,* and the Brobdingnagians of *Gulliver's Travels.*

Much as we appreciate those stories, even from an early age we don't really believe in monsters and giants. But could such beings ever exist? What problems would their enormous size cause them? How would they have to adapt in order to cope? (See Figure 18.1)

Every species survives by adapting to its environment. In particular, it faces the **problem of scale:** how to adapt and survive at the different sizes from the beginning of life to the final size of a mature adult. For example, consider the giant panda, which ranges from barely 1 lb at birth to 275 lb in adulthood. A baby panda is at risk of being crushed by its mother; an adult panda needs to eat a great deal of food.

As a contrasting example, consider the horse. If a newborn foal weighed as little as a newborn panda, the foal would be too small to keep up with the moving herd and could not survive. An adult horse weighs much more than a panda and has to consume much more food; but the horse can move much more quickly and cover great distances, to take advantage of wide-ranging sources of sustenance.

There have been large land mammals (mammoths) and huge sea mammals (the blue whale)—not to mention the dinosaurs. But the tallest humans have been only 9 to 10 ft tall; the largest mammoth was 16 ft at the shoulder (about twice as tall as an elephant); and even the tallest dinosaur, *Supersaurus,* stood only 40 ft high.

But what about supergiants and utterly huge monsters? That they have never existed suggests that there are physical limits to size. In fact, with a few simple principles of geometry, we can show not only that lizards and apes of

Figure 18.1 Could King Kong actually exist? (Dream Quest Images/ Photofest.)

such size are impossible, but also that none of the living beings and objects in our world could exist, unchanged in shape, on a vastly different scale, larger or smaller.

18.1 Geometric Similarity

The powerful mathematical idea that we use is *geometric similarity.*

> Two objects are **geometrically similar** if they have the same shape, regardless of the materials of which they are made.

Similar objects need not be of the same size, but measurements of corresponding distances on the two objects must be proportional. For example, when a photo is enlarged, it is enlarged by the same factor in both the horizontal and vertical directions–in fact, in any direction (such as a diagonal). We call this enlargement factor the *linear scaling factor.*

> The **linear scaling factor** of two geometrically similar objects is the ratio of a length of any part of the second to the corresponding part of the first.

Figure 18.2 Two geometrically similar photographs.
(Joseph McBride/Tony Stone Images.)

In the photos in Figure 18.2, the linear scaling factor is 3; the enlargement is three times as wide and three times as high as the original. In fact, every pair of points goes to a new pair of points three times as far apart as the original ones.

We notice that the enlargement can be divided into $3 \times 3 = 9$ rectangles, each the size of the original. Hence, the enlargement has $3 \times 3 = 3^2 = 9$ times the area of the original. More generally, if the linear scaling factor is some general number L (not necessarily 3), the resulting enlargement has an area $L \times L = L^2$ ("L squared") times the area of the original.

The *area* of a scaled-up object goes up with the *square* of the linear scaling factor.

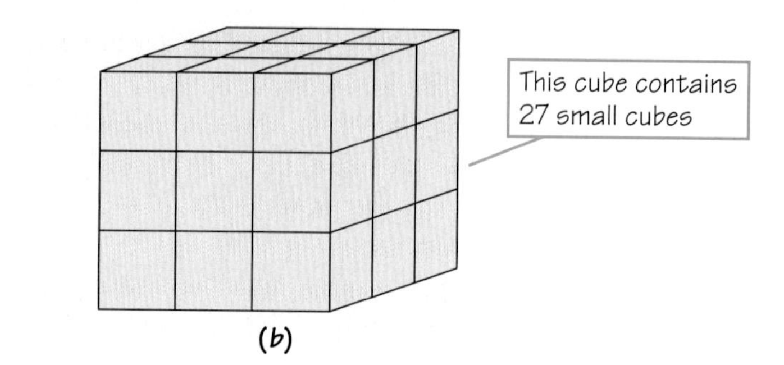

Figure 18.3 Cube (b) is made by enlarging cube (a) by a factor of 3.

We symbolize the relationship between the area A and the linear scaling factor L by

$$A \propto L^2$$

where the symbol $\propto$ is read as "is proportional to" or "scales as."

What about enlarging three-dimensional objects? If we take a cube and enlarge it by a linear scaling factor of 3, it becomes three times as long, three times as high, and three times as deep as the original (see Figure 18.3).

What about volume? The enlarged cube has three layers, each with $3 \times 3 = 9$ little cubes, each the same size as the original. Thus, the total volume is $3 \times 3 \times 3 = 3^3 = 27$ times as much as the original cube.

> The *volume* of a scaled-up object goes up with the *cube* of the linear scaling factor.

Denoting the volume by V, we can write

$$V \propto L^3$$

Thus, for an object enlarged by a linear scaling factor of L, the enlargement has $L \times L \times L = L^3$ ("L cubed") times the volume of the original. Like the relationship between surface area and L^2, this relationship holds even for irregularly shaped objects, such as science fiction monsters.

We observe, however, that the area of each face (side) of the enlarged cube is $3^2 = 9$ times as large as that of a face of the original cube, just as the area of the photo enlarged by a factor of 3 has nine times the area of the original. Because this fact is true for all six faces, the total surface area of the enlarged cube is nine times as much as the total surface area of the original cube.

More generally, for objects of any shape, the total *surface area* of a scaled-up object goes up with the *square* of the linear scaling factor. Thus, the surface area of an object scaled up by a factor of L is L^2 times the surface area of the original; this feature holds true even for irregular shapes.

Before we discuss scaling real three-dimensional objects, you should understand the pitfalls of the language for describing increases and decreases.

18.2 The Language of Growth, Enlargement, and Decrease

In 1976, the average price of a home in Madison [Wisconsin] was $38,323—about 108 percent less than in 1988.

—*Madison Business*, March 1991, p. 38.

House prices in Madison rose substantially, but we explore why this is an incorrect and confusing way to say so. In 1988 the average price was $80,000, which you can verify was 2.08 times the 1976 average price. What the author meant to say, in correct language, is that the 1988 price was 208% *of* the 1976 price, or that the 1988 price was 108% *more than* the 1976 price.

- "x% of A" or "x% as large as" means $\frac{x}{100} \times A$.
- "x% more than A" means A plus x% of A, in other words, $(1 + \frac{x}{100}) \times A$. Saying that A has "increased by x%" means the same thing.
- "x% less than A" means A minus x% of A, in other words, $(1 - \frac{x}{100}) \times A$. Saying that A has "decreased by x%" means the same thing.

So, to say "108% less than" $80,000 would mean $\$80{,}000(1 - \frac{108}{100}) = \$80{,}000 (1 - 1.08) = \$80{,}000(-0.08) = -\$6400$. Clearly this is not what the author intended to say.

The terms *of, times,* and *as much as* refer to *multiplication* of the original amount, while the terms *more, larger,* and *greater* refer to *adding* to the original amount. For instance, "five times as much" means the same as "four times more than" (the original plus four times as much in addition). Similarly, the relationship of the original amount to the larger amount can be expressed in multiplicative terms ("one-fifth as much," "20% as much") or in subtractive terms ("four-fifths less than," "80% less than").

These two ways of expressing change are similar in their phrasing, and people (even people in the media) often say "five times more than" when they mean "five times as much" (or even "five times less than" when they mean "one-fifth as much"). All you can do is be aware of the potential confusion, try to figure out what was meant, and be careful in your own expression. In particular, *don't use both* "times" *and* "more" *together.*

Finally, in discussions of percent we need to distinguish *percent* from percentage *points:* If support for the president has decreased from 60% to 30%, it has dropped 30 *percentage points* but decreased 50% (because the drop of 30 percentage points is 50% of the original 60 percentage points).

EXAMPLE What About Those Homes?

Returning to the Madison home prices, how can we state correctly what the author was trying to say? The 1976 figure, \$38,323, is about 0.48 times the 1988 figure of \$80,000 (38,323/80,000 = 0.48), or 48% of \$80,000; so the writer could say "about 48% of what it is in 1988" or "about 52% less than in 1990."

What if the writer wanted to use the 1976 figure as a base (i.e., as the 100% for the calculation)? The 1988 price is about 2.08 times the 1976 price, so the 1988 price is "208% of," or "108% more than," the 1976 price.

Caution: The dollar comparisons here can be misleading, because they do not take into account that a dollar was worth less in purchasing power in 1988 than in 1976, because of inflation.

Another caution: What does the author mean by "average"—the mean or the median (see pages 215–217)? Prices of houses are skewed, with a small number of very expensive houses (as noted on page 275), and the mean may be much larger than the median. Government statisticians and economists usually use the median. The Madison housing article notes that the median price in 1976 was \$34,000, so the \$38,323 "average" must be the mean. ■

18.3 Measuring Length, Area, Volume, and Weight

We start with a brief introduction to the common units in which various physical quantities are measured, together with a handy table of scaling factors (often called *conversion factors*) and examples of how to convert from one system of units to another.

U.S. Customary System

You are probably familiar with the common units of the *U.S. Customary System* of measurement and their abbreviations. But please examine Table 18.1 and pay close attention to the systematic way of converting from one unit to another and to the expression of approximate numbers in scientific notation. The symbol $\approx$ means "is approximately equal to."

Metric System

With the notable exception of the United States, almost every country of the world uses the same measurement system in science, industry, and commerce—the metric system. It was first proposed in France by Gabriel Mouton, Vicar of Lyons, in 1670, and was adopted in France in 1795. The fundamental unit of length, the *meter,* was originally defined to be one ten-millionth of the distance from the North Pole to the equator, as measured on the meridian through Paris. Later, the meter was redefined as the distance between two lines marked on a platinum–iridium bar kept at the International Bureau of Weights and Measures, near Paris, where the bar was kept at a temperature of 0°C. Finally, in 1960 the meter was redefined in terms of a standard reproducible in any laboratory,

TABLE 18.1 Units of the U.S. Customary System

Distance:

1 mile (mi) = 1760 yards (yd) = 5280 feet (ft) = 63,360 inches (in.)
1 yard (yd) = 3 feet (ft) = 36 inches (in.)
1 foot (ft) = 12 inches (in.)

Area:

$$
\begin{aligned}
1 \text{ square mile} &= 1 \text{ mi} \times 1 \text{ mi} = 5280 \text{ ft} \times 5280 \text{ ft} \\
&= 27{,}878{,}400 \text{ ft}^2 \approx 28 \times 10^6 \text{ ft}^2 \\
&= 63{,}360 \text{ in.} \times 63{,}360 \text{ in.} \\
&= 4{,}014{,}489{,}600 \text{ in.}^2 \approx 4 \times 10^9 \text{ in.}^2 \\
&= 640 \text{ acres} \\
1 \text{ acre} &= 43{,}560 \text{ ft}^2
\end{aligned}
$$

Volume:

$$
\begin{aligned}
1 \text{ cubic mile} &= 1 \text{ mi} \times 1 \text{ mi} \times 1 \text{ mi} \\
&= 5280 \text{ ft} \times 5280 \text{ ft} \times 5280 \text{ ft} \\
&= 147{,}197{,}952{,}000 \times \text{ft}^3 \\
&\approx 147 \times 10^9 \times \text{ft}^3 \\
&= 63{,}360 \text{ in.} \times 63{,}360 \text{ in.} \times 63{,}360 \text{ in.} \\
&\approx 2.5 \times 10^{14} \text{ in.}^3
\end{aligned}
$$

1 U.S. gallon (gal) = 4 U.S. quarts (qt) = 231 in.^{3}, exactly

Weight:

1 ton (t) = 2000 pounds (lb)

namely, 1,650,763.73 times the wavelength of the orange-red light emitted by atoms of the gas krypton-86 when an electrical charge is passed through them.

All other units of length, area, and volume are *defined* in terms of the meter; for example, a centimeter is a hundredth of a meter. The metric unit of weight, the *kilogram,* is defined as the weight of a platinum–iridium standard. Table 18.2 lists the units of the metric system.

Converting Between Systems

What are the scaling factors between the U.S. Customary System and the metric system? Since 1960, the fundamental units of the U.S. Customary System, the yard (for length) and the pound (for weight), have been *defined* in terms of metric units, so that we have

$$
\begin{aligned}
1 \text{ yd} &= 0.9144 \text{ m, exactly} \\
1 \text{ lb} &= 0.45359237 \text{ kg, exactly}
\end{aligned}
$$

The scaling factors for other units are shown in Table 18.3.

TABLE 18.2 Units of the Metric System

Distance:

1 meter (m) = 100 centimeters (cm)
1 kilometer (km) = 1000 meters (m)
= 100,000 centimeters (cm) = 1×10^5 cm

Area:

1 square meter (m^2) = 1 m × 1 m
= 100 cm × 100 cm = 10,000 (cm^2) = 1×10^4 cm^2
1 hectare (ha) = 10,000 m^2

Volume:

1 liter (l) = 1000 cm^3 = 0.001 m^3
1 cubic meter (m^3) = 1 m × 1 m × 1 m
= 100 cm × 100 cm × 100 cm
= 1,000,000 cm^3 = 1×10^6 cm^3 (or cc)

Weight:

1 kilogram (kg) = 1000 grams (g)

TABLE 18.3 Conversions Between the U.S. Customary System and the Metric System

Distance:

1 in. = 2.54 cm, exactly
1 ft = 12 in. = 12 × 2.54 cm = 30.48 cm = 0.3048 m, exactly
1 yd = 0.9144 m, exactly
1 mi = 5280 ft = 5280 × 30.48 cm
= 160,934.4 cm, exactly ≈ 1.61 km
1 cm ≈ 0.3937 in. ≈ 0.4 in.
1 m ≈ 39.37 in. ≈ 3.281 ft
1 km ≈ 0.621 mi

Area:

1 hectare (ha) = 2.47 acres

Volume:

1 cubic meter (m^3) = 1000 liters = 264.2 U.S. gallons = 35.31 ft^3
1 liter (l) = 1000 cm^3 = 1.057 U.S. quarts (qt)

Weight:

1 lb = 0.45359237 kg, exactly
1 kg ≈ 2.205 lb

We illustrate how to convert a measurement in one unit to the corresponding measurement in a different unit.

What's That in Feet?

A foreign student tells her American student friends that she is 160 centimeters tall. They naturally ask how much that is in feet and inches.

We approach the conversion by using the scaling factor that 1 cm = 0.393701 in.:

$$\begin{aligned} 160 \text{ cm} &= 160 \times 1 \text{ cm} \approx 160 \times 0.393701 \text{ in.} \\ &\approx 62.9 \text{ in.} = 62.9 \text{ in.} \times \frac{1 \text{ ft}}{12 \text{ in.}} \approx \frac{62.9}{12} \text{ ft} \approx 5.25 \text{ ft} \end{aligned}$$

However, because we normally give height in feet and a whole number of inches, the height is

$$62.9 \text{ in.} = 5 \times (12 \text{ in.}) + 2.9 \text{ in.} = 5 \text{ ft} + 2.9 \text{ in.} \approx 5 \text{ ft } 3 \text{ in.}$$

Another way to approach the problem is by means of a proportion:

$$\frac{\text{Height in inches}}{\text{Height in cm}} = \frac{\text{length of 1 inch in inches}}{\text{length of 1 inch in cm}} = \frac{1 \text{ in.}}{2.54 \text{ cm}}$$

so that

$$\begin{aligned} \text{Height in inches} &= \text{height in cm} \times \frac{1 \text{ in.}}{2.54 \text{ cm}} \\ &= 160 \text{ cm} \times \frac{1 \text{ in.}}{2.54 \text{ cm}} \approx 62.9 \text{ in.} \end{aligned}$$

■

Comparison of a meter stick with a yardstick, a liter with a quart, and a kilogram with a pound. (Chip Clark.)

18.4 Scaling Real Objects

Real three-dimensional objects are made of matter, which has volume and mass. **Mass** is the aspect of matter that is affected by forces, according to physical laws. For example, your mass reacts to the gravitational force of the earth by staying close to it (and your mass exerts an equal force on the earth that tends to keep the earth close to you). We perceive the mass of an object when we try to move it (as in throwing a ball). When we try to lift an object, we perceive its mass as weight, due to the gravitational force that the earth exerts on it. As we will see, gravity exerts an enormous effect on the size and shape that objects and beings can assume.

Suppose that the two cubes in Figure 18.3 (page 636) are made of steel and that the first is 1 ft on a side and the second is 3 ft on a side. For each, its bottom face supports the weight of the entire cube. **Pressure** is the force per unit area, so the pressure exerted on the bottom face by the weight of the cube is equal to the weight of the cube divided by the area of the bottom face, or

$$P = \frac{W}{A}$$

A cubic foot of steel weighs about 500 lb (we say it has a **density** of 500 lb per cubic foot). The first cube weighs 500 lb and has a bottom face with area 1 ft^2, so the pressure exerted on this face is 500 lb/ft^2.

The second cube is 3 ft on a side. The area of the bottom face has increased with the square of the linear scaling factor, so it is $3^2 \times 1$ $ft^2 = 9$ ft^2. As we learned earlier in this chapter, volume goes up with the cube of the linear scaling factor. So this larger cube has a volume of $3^3 \times 1 = 27$ ft^3. Because both cubes are made of the same steel, the larger cube has 27 times as much steel as the smaller; hence it weighs 27 times as much as the smaller cube, or 27×500 lb = 13,500 lb.

When we divide this weight by the area of the bottom face (9 ft^2), we find that the pressure exerted on the bottom face is 1500 lb/ft^2, or three times the pressure on the bottom face of the original cube. This makes sense because over each 1 ft^2 area stands 3 ft^3 of steel. In general, if the linear scaling factor for the cube is L, the pressure on the bottom face is L times as much. Using the notation of proportionality, we have $A \propto L^2$ and $W \propto V \propto L^3$, so

$$P = \frac{W}{A} \propto \frac{L^3}{L^2} \propto L$$

EXAMPLE What About a 10-Ft Cube?

If we scale the original cube of steel up to a cube 10 ft on a side, then the dimensions are

$$10 \text{ ft} \times 10 \text{ ft} \times 10 \text{ ft}$$

The total volume is

$$\begin{aligned} V &= \text{length} \times \text{width} \times \text{height} \\ &= 10 \text{ ft} \times 10 \text{ ft} \times 10 \text{ ft} = 1000 \text{ ft}^3 \end{aligned}$$

The weight of the cube is

$$\begin{aligned} W &= V \times \text{density} \\ &= 1000 \text{ ft}^3 \times 500 \text{ lb/ft}^3 = 500{,}000 \text{ lb} \end{aligned}$$

The area of the bottom face is

$$\begin{aligned} A &= \text{length} \times \text{width} \\ &= 10 \text{ ft} \times 10 \text{ ft} = 100 \text{ ft}^2 \end{aligned}$$

The pressure on the bottom face is

$$P = \frac{W}{A} = \frac{500{,}000 \text{ lb}}{100 \text{ ft}^2} = 5000 \text{ lb/ft}^2$$

This is 10 *times*—not "10 times *more* than"—the pressure on the bottom face of the original 1-ft cube. ■

At some scale factor, the pressure on the bottom face will exceed the steel's ability to withstand that pressure—and the steel will deform under its own weight. That point for steel is reached for a cube about 3 miles on a side—the pressure exerted by the cube's weight exceeds the resistance to crushing (ability to withstand pressure, or **crushing strength**) of steel, which is about 7.5 million lb/ft^2. Because a mile is 5280 ft, a 3-mi-long cube of steel would be more than 15,000 times as long as the original 1-ft cube; that is, the linear scaling factor is more than 15,000. The pressure on the bottom face of the cube would therefore be more than 15,000 times as much as for the 1-ft cube, or $15{,}000 \times 500 \text{ lb/ft}^2 = 7.5$ million lb/ft^2.

EXAMPLE

What About the Petronas Towers?

At 1482.6 ft (451.9 m), the twin Petronas Towers in Kuala Lumpur, Malaysia, are the tallest buildings in the world, if we don't count radio and television antennas. What is the pressure on the bottom of their walls?

The towers are made of reinforced concrete, which weighs about 160 lb/ft^3. Over each square foot of bottom surface of one of their walls stands 1462 ft^3 of reinforced concrete, which weighs $1482.6 \times 160 = 237{,}000$ lb. The pressure at the bottom of the wall, from the wall's weight alone, is 237,000 lb/ft^2. That's not counting the contents of the tower, which also must be supported!

SPOTLIGHT 18.1 A Mile-High Building?

In 1956, the famous American architect Frank Lloyd Wright (1867–1959) proposed a mile-high tower for the Chicago waterfront. In the text, we focus on the problem of holding up the weight of such a structure, which would have to involve stronger materials and thicker walls and supporting columns.

But there are other limits to the height of a building. Some of these can also be addressed by similar strategies. For example, the bending of the building in the wind, which can go up dramatically with height, can be controlled by making the building stiffer.

The terrorist destruction of the World Trade Center towers in 2001–resulting not from the aircraft impacts but from the subsequent fires–revealed a vulnerability in the towers' structure.

Even if designed to better resist fires and impacts, however, a mile-high building might not be at all practical. For example, the enormous number of people (perhaps 100,000) living, working, or visiting in such a building would create enormous traffic problems (pedestrian, parking, deliveries) for blocks around.

Cost per square foot of usable area is an important economic consideration. Even if the building had the same width all the way up (as opposed to tapering near the top, like the Empire State Building), the space in the upper floors might not justify their additional expense. With increasing height, an increasingly larger proportion of the cross-sectional area of all floors of the building must be devoted to services, such as elevators, plumbing, and conduits for heat and air-conditioning. It's not that there are more people on the top floors (or that they use more water than people on other floors!). But everyone entering the building and going to any floor needs to start in an elevator on the ground floor, so there must be more elevators and more elevator shafts. In an emergency evacuation, the people must walk down!

The architects of the Petronas Twin Towers, however, maintain that the main limit on height of a building is human physiology. Differences in air pressure between the top and bottom of a building impose a limit on how fast elevators can rise or drop without discomfort to passengers, thereby enforcing long travel times for "vertical commuters." Human psychology might also present some limits.

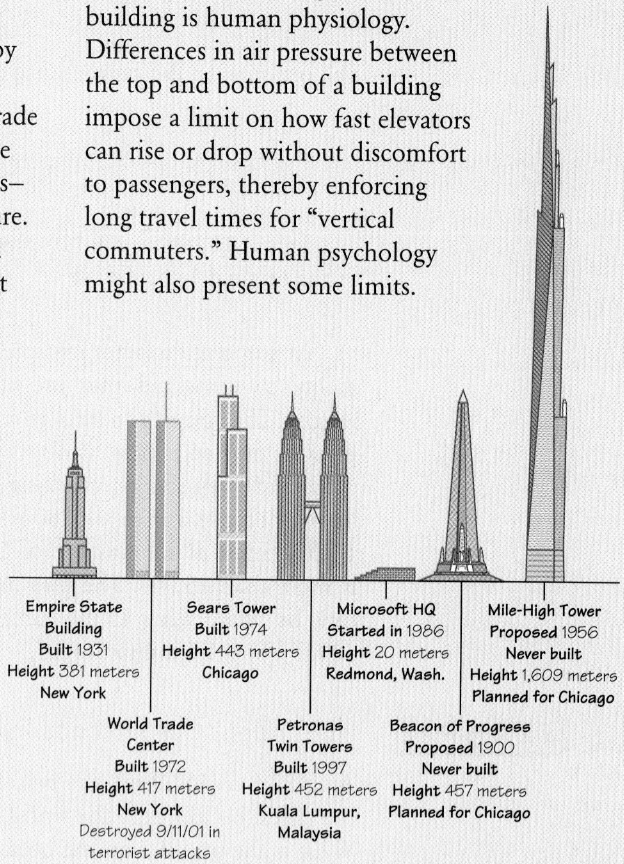

Note: Drawings not to scale.

Could we have a Super Petronas Tower 10 times as high? The bottom of its walls would have to support 2.4 million lb/ft^2. The crushing strength of reinforced concrete is about 8.5 million lb/ft^3, which would leave some safety margin. (See Spotlight 18.1 for more on tall buildings.) ■

18.5 Sorry, No King Kongs

Unfortunately, the resistance of bone to crushing is not nearly as great as that of steel. This fact helps to explain why there couldn't be any King Kongs (unless they were made of steel!). A King Kong scaled up by a factor of 20 would weigh $20^3 = 8000$ times as much. Though the weight increases with the cube of the linear scaling factor, the ability to support the weight – as measured by the cross-sectional area of the bones, like the area of the bottom face of the cube in Figure 18.3 – increases only with the square of the linear scaling factor.

These simple consequences of the geometry of scaling apply not only to supermonsters but also to other objects, such as trees and mountains.

EXAMPLE How Tall Can a Tree Be?

Galileo suggested that no tree could grow taller than 300 feet (see Spotlight 18.2). The world's tallest trees are giant sequoias, which grow only on the West Coast of the United States and hence were unknown to Galileo. They can grow to almost 370 feet (Figure 18.4).

Figure 18.4 Even these giant sequoias can grow no taller than their form and materials allow. (Phil Schermeister/Corbis.)

What can limit the height of a tree? If the roots do not adequately anchor it, a tall tree can blow over. (This, in fact, happened in 1990 to the world's then tallest tree, the Dyerville Giant, a giant sequoia in Humboldt Redwoods State Park in California.) The tree could buckle or snap under its own weight and the force of a strong wind. The wood at the bottom will begin to crush if there is too

SPOTLIGHT 18.2 Galileo and the Problem of Scale

The Italian physicist Galileo Galilei (1564 – 1642) was the first to describe the problem of scale, in 1638, in his *Dialogues Concerning Two New Sciences* (in which he also discussed the idea of the earth revolving around the sun):

> You can plainly see the impossibility of increasing the size of structures to vast dimensions either in art or in nature; likewise, the impossibility of building ships, palaces, or temples of enormous size in such a way that their oars, yards, beams, iron-bolts, and, in short, all their other parts will hold together; nor can nature produce trees of extraordinary size because the branches would break down under their own weight, so also would it be impossible to build up the bony structures of men, horses, or other animals so as to hold together and perform their normal functions if these animals were to be increased enormously in height; for this increase in height can be accomplished only by employing material which is harder and stronger than usual, or by enlarging the size of the bones, thus changing their shape until the form and appearance of the animals suggests a monstrosity.
>
> To illustrate briefly, I have sketched a bone whose natural length has been increased three times and whose thickness has been multiplied until, for a correspondingly large animal, it would perform the same function which the small bone performs for its small animal. From the figures shown here you can see how out of proportion the enlarged bone appears. Clearly then if one wishes to maintain in a great giant the same proportion of limb as that found in an ordinary man he must either find a harder and stronger material for making the bones, or he must admit a diminution of strength in comparison with men of medium stature; for if his height be increased inordinately he will fall and be crushed under his own weight. Whereas, if the size of a body be diminished, the strength of that body is not diminished in proportion; indeed the smaller the body the greater its relative strength. Thus a small dog could probably carry on his back two or three dogs of his own size; but I believe that a horse could not carry even one of his own size.

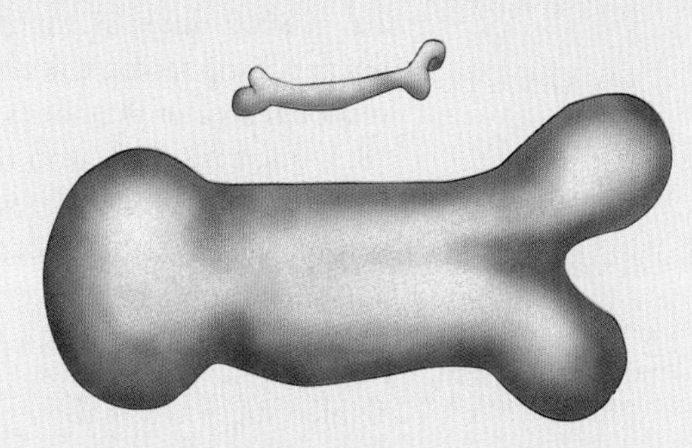

One bone, with another three times as long and thick enough to perform the same function in a scaled-up animal.

Translated by Henry Crew and Alfonso De Salvo, and published by Macmillan, 1914, and Northwestern University, 1946.

much weight above. Finally, there is a limit to how far the tree can lift water and minerals from the roots to the leaves.

Could a tree be a mile high? To make an easy but rough estimate of the pressure at the base of the tree due to gravity, let's model the tree as a perfectly vertical cylinder. Over each square foot at the bottom, there is 5280 ft^3 of cells of wood, which we can think of as a column of water. To calculate how much that weighs, we first translate 1 ft^3 into metric measurement:

$$1 \text{ ft}^3 = (12 \text{ in.})^3 = (12 \times 2.54 \text{ cm})^3 = 28{,}316 \text{ cm}^3$$

A reason to convert to cubic centimeters is the convenient fact that water weighs just about exactly 1 gram per cubic centimeter. Now, 1 ft^3 of water weighs about 28,300 g = 28.3 kg = 28.3 $\times$ 2.20 lb $\approx$ 62 lb. Consequently, 5280 ft^3 of water weighs 5280 $\times$ 62 lb $\approx$ 327,000 lb, so the pressure on the bottom layer would be about 327,000 lb/ft^2.

Actually, freshly cut wood weighs only about *half* as much as water, so the pressure on the bottom layer would be half this figure, or 164,000 lb/ft^2. This is still an overestimate, however, because we assumed that the tree does not taper. A tree that tapers steadily looks like an elongated cone; using a more realistic cone model (as we will do in the next section for a mountain), we see that the pressure at the bottom of the tree is one-third of 164,000 lb/ft^2, or 55,000 lb/ft^2.

A biological organism needs a safety factor of at least two to four times the absolute minimum physical limits, so a tree a mile high would need from 110,000 to 220,000 lb/ft^2 of upward pressure for water and minerals. Tension in the string of water molecules from root to leaf ranges from 80,000 to 3.2 million lb/ft^2, for different kinds and heights of trees, so this consideration does not rule out mile-high trees.

However, at more than about 500 lb/in.^{2} = 70,000 lb/ft^2, the bottom of the tree would begin to crush under the weight above. On this point, the mile-high tree is barely feasible, with little margin of safety.

These considerations suggest that trees a mile high might be *physically* possible, but others suggest a lower maximum height. In addition, there are also biological considerations. The taller the tree, the greater the area from which it must draw water and minerals, for which nearby trees also compete. Moreover, for a tree to grow very tall, it would have to live for a very long time. Evolution and time may select against extremely tall trees; or maybe, for no reason at all, they have just never evolved. ■

EXAMPLE

How High Can a Mountain Be?

Gravity and the physical characteristics of wood limit the height of trees. Gravity also limits the height of mountains. Mountains differ in composition and shape, and some assumptions about those features are necessary to do any calculating. We want to make realistic assumptions that make it easy to estimate how high a mountain can be. In effect, we build a simple mathematical model of a mountain.

Let's suppose that the mountain is made entirely of granite, a common material in many mountains, and let's assume that the granite has uniform density. Granite weighs 165 lb/ft^3 and has a crushing strength of about 4 million lb/ft^2.

In the interests of both realism and simplicity, we assume that the mountain is in the shape of a cone whose width at the base is the same as its height. Let's model Mount Everest: The tallest earth mountain, it is about 6 mi high. The base, then, is a circle with a distance across (or diameter) of 6 mi. The radius of the circle is half the diameter, so the model Everest has a radius of 3 mi measured at the base (Figure 18.5). Because we are taking such a round number for the height of Everest, we record as significant only the first two digits of the results of the calculations.

What does the model Everest weigh? The relevant formula is

$$\text{Weight} = \text{density} \times \text{volume}$$

We already know the density of granite (165 lb/ft^3), so to find the weight we need the formula for the volume of a cone of radius r and height h:

$$\text{Volume} = \frac{1}{3}\pi r^2 h$$

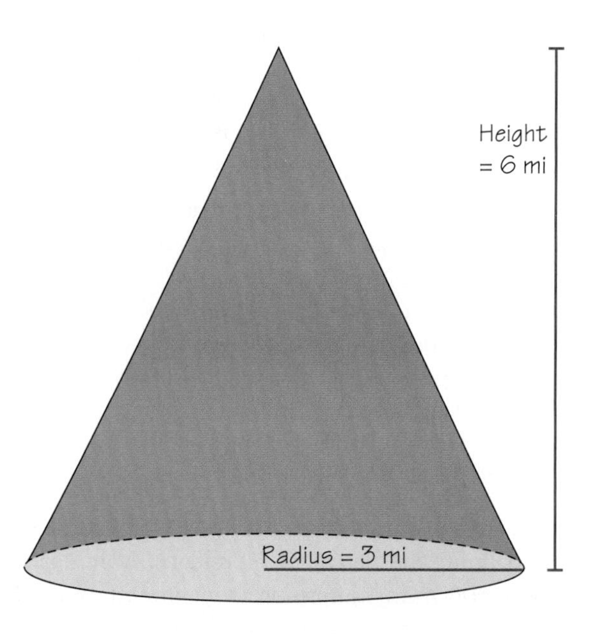

Figure 18.5 Model of Mount Everest as a cone of granite.

For Everest, the radius is 3 mi and the height is 6 mi; π (pi) is about 3.14. Using those values in the formula, we find that the model Everest has a volume of about 57 cubic miles.

To find the weight of 57 cubic miles of granite, we need to convert units, because the density is given in pounds per cubic foot. Let's convert to units of feet:

$$\begin{aligned} 1\text{ mi}^3 &= 1\text{ mi} \times 1\text{ mi} \times 1\text{ mi} \\ &= 5280\text{ ft} \times 5280\text{ ft} \times 5280\text{ ft} \\ &\approx 1.5 \times 10^{11}\text{ ft}^3 \end{aligned}$$

Thus

$$57\text{ mi}^3 \approx 57 \times 1.5 \times 10^{11}\text{ ft}^3 \approx 8.6 \times 10^{12}\text{ ft}^3$$

So we have

$$\begin{aligned} \text{Weight of mountain} &= 165\text{ lb/ft}^3 \times 8.6 \times 10^{12}\text{ ft}^3 \\ &\approx 1.4 \times 10^{15}\text{ lb} \\ &\approx 1.4\text{ quadrillion lb} \end{aligned}$$

Now that we know the weight of the mountain, we want to find out what the pressure is on the base of the cone and compare that with the crushing strength of granite. (Everest is standing, so if our model is any good, that pressure will be below the crushing strength.) Physics tells us that the weight of the mountain is spread evenly over the base of the cone (though we are oversimplifying the geology underlying mountains). Because

$$\text{Pressure} = \frac{\text{weight}}{\text{area}}$$

we need to calculate the area of the base of the cone. The shape is a circle, and the familiar formula

$$\text{Area} = \pi r^2$$

gives an area of 28 mi^2 for a radius of 3 mi.

Once again, we need to convert units in order to express the pressure in pounds per square foot, the units in which the crushing strength is expressed. We get

$$\text{Area} = 28\text{ mi}^2 = 28 \times 5280\text{ ft} \times 5280\text{ ft} \approx 8 \times 10^8\text{ ft}^2$$

Then

$$\begin{aligned} \text{Pressure} &= \frac{\text{weight}}{\text{area}} \\ &= \frac{1.4 \times 10^{15}\ \text{lb}}{8 \times 10^{8}\ \text{ft}^2} \\ &= 1.8 \times 10^{6}\ \text{lb/ft}^2 \end{aligned}$$

This number is below the crushing strength of granite, 4×10^6 lb/ft^2, with a safety factor of about 2.

For a mountain to come close to the limitation of the crushing strength of granite, it would have to be only about twice as high as Everest, or about 10 miles high. Other physical considerations suggest a maximum height of at most 15 miles. That no current mountains are that high may be a consequence of the earth's high amount of volcanic activity and the structural deformation of the earth's crust. ■

What about mountains made of other materials—glass, ice, wood, old cars? They couldn't be nearly as high; the pressure would cause glass to flow, ice to melt, and old cars to compact. What about mountains on another planet? Their potential height depends on the mass of the planet.

18.6 Solving the Problem of Scale

A large change in scale forces a change in either materials or form. A major manifestation of the scaling problem is the tension between weight and the need to support it. For example, a real building or machine must differ from a scale model: The balsa wood or plastic of the model would never be strong enough to use for the real thing, which must use aluminum, steel, or reinforced concrete. So one way to compensate for the problem of scale is to use stronger materials in the scaled-up object.

Another way to compensate is to redesign the object so that its weight is better distributed. Let's go back to the original cube. It supports all its weight on its bottom face. In the version scaled up by a factor of 3, each small cube of the bottom layer has a bottom face that is supporting that cube's weight plus the weight of the other two cubes piled on top of it.

Let's redesign the scaled-up cube, concentrating for simplicity only on the front face, with its nine small cubes. We take the three cubes on top and move them to the bottom, alongside the three already there. We take the three cubes on the second level, cut each in half, and put a half cube over each of the six ground-level cubes (see Figure 18.6). We have the same volume and weight that we started with, but now there is less pressure on the bottom face of each small cube. Of course, the new design is not geometrically similar to the object that we started with—it's no longer a cube. We have solved the scaling problem by changing the proportions and redistributing the weight.

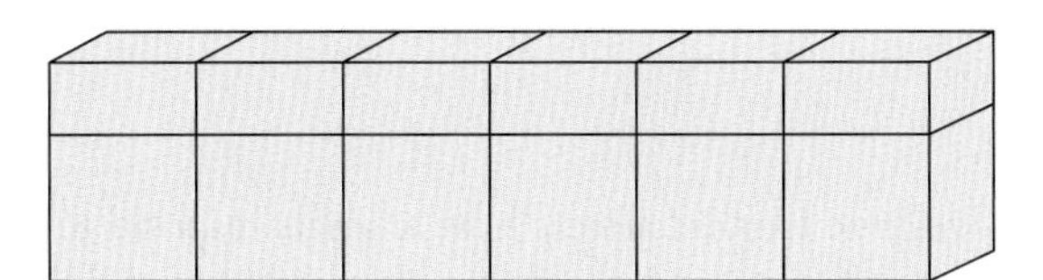
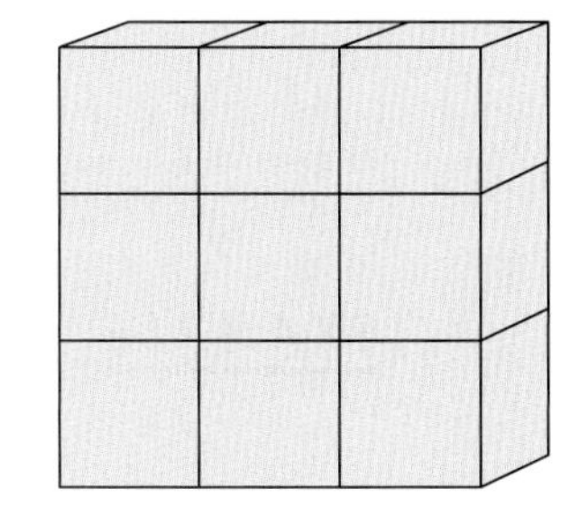

Figure 18.6 Nine small cubes rearranged to support greater weight.

We observe in nature both strategies for adaptation to scaling: change of materials and change of form. Smaller animals generally do not have bony internal skeletons; larger animals generally do. Animals made of similar materials but differing greatly in size, such as a mouse and an elephant, most certainly differ in shape. If a mouse were scaled up to the size of an elephant, its legs could no longer support it; it would need the disproportionately thicker legs of the elephant. It would also need the elephant's thick hide to contain its tissue.

Some dinosaurs, like *Supersaurus* (which weighed 30 tons), had special adaptations to lighten their weight, such as hollow bones, just as some birds have. (Hollow bones also turn out to be stronger, a paradox that Galileo analyzed. Of two bones of the same weight and length, the hollow one is wider across at its midpoint, because of the air it contains; and the greater the width, the greater the resistance to fracture.)

18.7 Falls, Jumps, and Flight

The need to support weight can be thought of as a tension between volume and area. As an object is scaled up, its volume and weight go up together, as long as the density remains constant (for example, no air bubbles introduced into the steel to make it into a Swiss cheese!). At the same time, the ability to support the weight goes up with the cross-sectional area, like the bottom face of the steel cube.

Area–volume tension is a result of the fact that as an object is scaled up, the volume increases faster than the surface area and faster than areas of cross sections.

Because volume V is proportional to the cube of the linear scaling factor L, we have $V \propto L^3$, or, taking each side to the one-third power, $L \propto V^{1/3}$. The fact that surface area A is proportional to the square of the linear scaling factor becomes

$$A \propto L^2 \propto (V^{1/3})^2 = V^{2/3}$$

so that surface area scales as the two-thirds power of volume.

Area–volume tension has many practical consequences, some of them related to our childhood fantasies. We can forget about humans "leaping tall buildings in a single bound," "soaring like an eagle," diving miles below the sea, and jumping from airplanes without parachutes. Consider the following examples.

EXAMPLE Falls

Area–volume tension affects how animals respond to falling, another of gravity's effects. A mouse may be unharmed by a 10-story fall and a cat by a 2-story fall, but a human may well be injured by falling down while running, walking, or even just standing.

What is the explanation? The energy acquired in falling is proportional to the weight of the falling object, hence to its volume. This energy must be absorbed either by the object or by what it hits, or must be otherwise dissipated at impact–for example, as sound. The fall is absorbed over part of the surface area of the object, just as the weight of the cube was distributed over its base. With scaling up, volume–hence weight, hence falling energy–goes up much faster than area. As volume increases, the hazards of falling from the same height increase. ■

EXAMPLE Jumps

A flea can jump about 2 ft vertically, many times its own height. Many people believe that if a flea were as large as a person, it could jump a thousand feet into the air. Imagining–against our earlier arguments–that there could be so large a flea, we know its limits: A scaled-up flea could jump about the same height as a small flea. The strength of a muscle is proportional to its cross-sectional *area* (see Spotlight 18.3). A jump involves suddenly contracting the muscle through its length, so it turns out that the ability to jump is proportional to the *volume* of muscle. But the volume of the flea and the volume of its leg muscles go up in proportion.

Let's say that a real flea's leg muscles account for 1% of its body. If we scale the flea up to the size of a person (without any change in its form), the enlarged flea's leg muscles would still make up 1% of its body. For either flea, each bit of muscle has the same power: In a jump, it propels 100 times its own weight, and it can do so to the same height. Both the weight of the flea and the power of its legs go up proportionately. In fact, the maximum heights that people, fleas, grasshoppers, and kangaroos can jump are all within a factor of 3. ■

EXAMPLE Flight

Wouldn't it be nice to be able to fly? Well, you have to be able to stay up. The power necessary for sustained flight is proportional to the **wing loading,** which is the weight supported divided by the area of the wings. We know that in scaling up, weight grows with the cube of the length of the bird or plane, and wing area with the square of the length. So the wing loading is proportional to the length of the flying object.

For example, if a bird or plane is scaled up proportionally by a factor of 4, it weighs $4^3 = 64$ times as much but has only $4^2 = 16$ times as much wing area. So each square foot of wing must support 4 times as much weight.

Once you're up, you have to keep moving. Hovering helicopters, hummingbirds, and insects maintain lift by moving their wings directly rather than through forward motion. To stay level, an airborne object must fly fast enough to maintain the lift on the wings. The minimum necessary speed is proportional to the square root of the wing loading. Combining this fact with the first consideration, we conclude that the minimum speed goes up with the square root of the length. A bird scaled up by a factor of 4 must fly $\sqrt{4} = 2$ times as fast.

Take, for instance, a sparrow, whose minimum speed is about 20 miles per hour (mph). An ostrich is 25 times as long as a sparrow, so the minimum speed for an ostrich would be $\sqrt{25} \times 20 = 100$ mph. Have you seen any flying ostriches lately? Heavy birds have to fly fast or not at all!

Of course, ostriches are not just scaled-up sparrows, nor are eagles. The larger flying birds have disproportionately larger wings than a sparrow, to keep the wing loading down. The largest animal ever to have taken to the air was *Quetzalcoatlus northropi,* a flying reptile of 65 million years ago, with a wingspan of 36 ft and a weight of about 100 lb.

You have to stay up, you have to keep moving—and you have to get up there. Here basic aerodynamics imposes further limits. Paleontologists originally thought that *Quetzalcoatlus northropi* weighed 200 lb and had a 50-ft wingspan. Even though that works out to just about the same wing loading as for 100 lb and a wingspan of 36 ft, other considerations from aerodynamics show that at 200 lb, at the larger size, the reptile wouldn't have been able to get off the ground. ■

18.8 Keeping Cool (and Warm)

Area–volume tension is also crucial to an animal's maintenance of thermal equilibrium. Both warm-blooded and cold-blooded animals gain or lose heat from the environment in proportion to body surface area.

Warm-Blooded Animals

A warm-blooded animal usually is losing heat; its basal metabolism, or rate of food intake needed to maintain body heat, depends primarily on the amount of its surface area, the temperature of its environment, and the insulation provided by its coat or skin. Other factors being equal, a scaled-up mammal scales up its food consumption by *surface area* (proportional to the square of the linear scaling factor), *not by volume* (proportional to its cube). For example, a mouse eats about half of its weight in food every day, while a human consumes only about one-fiftieth of its own weight.

Thus, the metabolic rate must be proportional to the surface area. Using the symbolism of proportionality, we can find how the metabolic rate changes with the mass of the animal. We know that mass is proportional to volume, which in turn is proportional to the square of length, or

$$M \propto V \propto L^3$$

SPOTLIGHT 18.3 "Take That, King Richard!"

Shakespeare painted Richard III as a hump-backed Machiavellian monster. Did Richard have an advantage in armored combat because he was short? That suggestion was made some years ago by one of the leading modern historians of the Tudor era, Garrett Mattingly.

Between a short man and a tall man, height increases by the linear dimension—from 5 feet 2 inches, say, to 6 feet—while the surface of the body increases as the square. Because it's the surface of the body that the armorer must plate with steel, the armor of a short warrior, like Richard, would be lighter than a tall warrior's by a lot more than the few inches' difference in height would indicate. So Richard's notorious deadliness in battle would have been possible at least in part because his armor, while protecting him as well as the big man's, left him less encumbered.

After the lecture, someone said to Mattingly that he had grasped the right idea—but by the

Did wearing armor give an advantage to the shorter warrior? (Bob Krist/Corbis.)

Taking each side to the one-third power, we have

$$M^{1/3} \propto V^{1/3} \propto L \quad \text{or} \quad L \propto M^{1/3}$$

Meanwhile, the metabolic rate (call it P) is proportional to surface area, so

$$P \propto A \propto L^2 \propto (M^{1/3})^2 = M^{2/3}$$

So, based on area–volume tension, we would expect metabolic rate to scale as the two-thirds power of body mass. But it doesn't—instead, it scales as the *three-quarters* power of body mass, that is, $P \propto M^{3/4}$. The least-squares line (see Chapter 6) through the points in the "mouse-to-elephant" curve of Figure 18.7 has a slope of 0.74, very close to three-quarters.

Why the difference from the two-thirds that area–volume tension would predict? And does the small difference between two-thirds and three-quarters matter? The answers lie in further considerations from geometry, physiology, and physics. A plant or animal needs a network of vessels (like the blood system) to transport resources to, and wastes away from, every part of the animal's tissues. The terminal branches (in the blood system, capillaries) tend to be just about the wrong end. Muscle power, the listener claimed, is a matter of bulk – and physical volume goes up by the cube, whereas the surface to be protected goes up by the square. So the large warrior should have more strength left over than the little guy after putting on his armor. And the large warrior, swinging a bigger club, can deliver a far more punishing blow – because the momentum of the club depends on its weight, which goes up with its volume, which means by the cube. Richard was at a terrible disadvantage.

But wait a minute, a second listener said. That's true about the club – but not about the muscles. The strength of a muscle is proportional not to its bulk but to the area of its cross section. And because the cross section of muscles increases by the square, just as the surface of the body does, the big guy, plated out, has no more, or less, advantage over the little guy than if both were naked.

But hang on, a third person interjected – an engineer. That's right about the muscles, but it's not right about the armor. The weight of the armor increases not simply with the increase in the surface area that it must cover but slightly faster. There must be reinforcing ribs. Or else the metal must be significantly thicker overall. So maybe Richard had an advantage.

Armor was made as thin as possible. Thickness, reinforcement, and structural stiffening were concentrated where opponents' weapons were likely to hit. From these strong, shaped places, the metal tapered away, until the sheet steel was as thin as the lid of a coffee can at the sides of the rib cage beneath the arms, or across the fingers, or at the cheek of a helmet.

Source: Horace F. Judson, *The Search for Solutions* (Baltimore: Johns Hopkins University Press, 1987), pp. 54–56.

same size in all species, for reasons of the physics involved. To minimize the energy involved in transport, the network of vessels needs to be organized like a fractal-like tree, with smaller vessels branching off larger ones. With same-size terminal branches, minimization of energy demands that the metabolic rate scale as the three-quarters power of body mass. The branching makes it possible for the circulatory system of a whale, which has 10^7 times the mass of a mouse, to have only 70% more branches than the mouse has.

EXAMPLE Dives

Sperm whales (and some other species) regularly hold their breath and stay under water for an hour. Why can't we? In part, because we aren't as large as whales. A mammal's breath-holding ability depends on how much air it can hold in its lungs, which is proportional to its mass, and on how fast it uses up the air – in other words, on its metabolic rate, which is proportional to the three-quarters power of its mass. Hence, the limit of duration of a dive should be proportional to

$$\frac{M}{M^{3/4}} = M^{1/4}$$

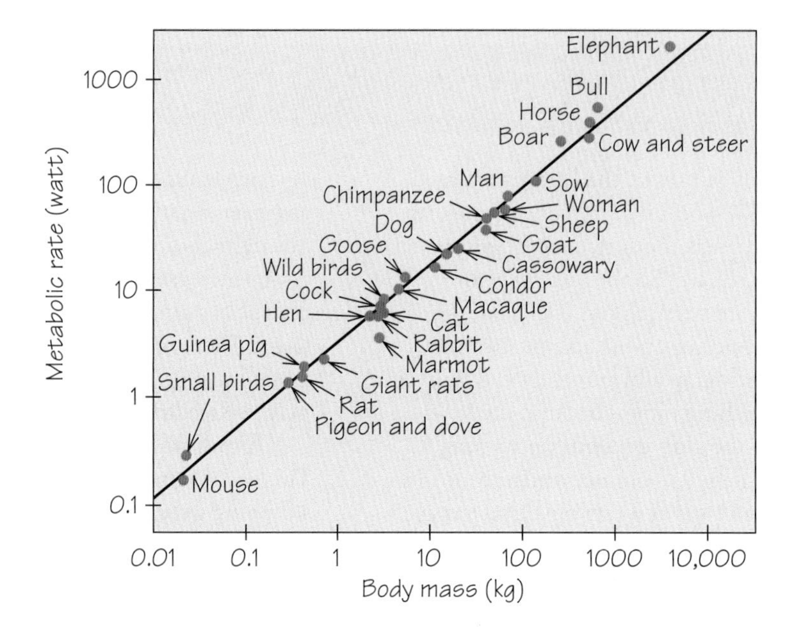

Figure 18.7 Metabolic rates for mammals and birds, when plotted against body mass on logarithmic coordinates, tend to fall along a single straight line. Adapted from Benedict (1938).

For a 90,000-lb sperm whale, that limit is proportional to $90{,}000^{1/4} = 17.3$, while the corresponding figure for a 150-lb human is $50^{1/4} = 3.5$. So the sperm whale should be able to hold its breath for about $17.3/3.5 \approx 5$ times as long. However, humans cannot hold their breath for one-fifth of an hour (12 minutes)! That fact tells us that the whale must have some special adaptation—such as larger than proportional lungs, or a lower than predicted metabolism—to make its long dives possible. ■

Cold-Blooded Animals

Mammals and birds regulate their metabolism and maintain a constant internal body temperature. Cold-blooded animals, such as alligators or lizards, have a somewhat different problem. They absorb heat from the environment for energy, but they must also dissipate any excess heat to keep their temperature below unsafe levels. The amount of heat that must be gained or lost is proportional to total volume, because the entire animal must be warmed or cooled. But the heat is exchanged through the skin, so the rate is proportional to surface area.

Dimetrodon was a large mammal-like reptile that roamed present-day Texas and Oklahoma 280 million years ago (see Figure 18.8). *Dimetrodon* had a great "sail," or fan, on its back. As an individual grew, and as the species evolved, the sail grew. But it did not grow according to *geometric similarity,* the kind of growth we refer to as *proportional growth.*

Proportional growth is growth according to geometric similarity, where the length of every part of the organism enlarges by the same linear scaling factor.

Figure 18.8
Dimetrodon may have evolved a sail to absorb and dissipate heat efficiently. (American Museum of Natural History.)

Instead, the area of *Dimetrodon's* sail grew precisely in proportion to the volume of the animal, a fact that strongly suggests to paleontologists that the sail was a temperature-regulating organ that was able to absorb or radiate heat. So, an individual *Dimetrodon* twice as long would have eight (2^3) times as much weight and volume and also a sail with eight times as much area. If it had grown according to geometric similarity, the sail would have been twice as high and twice as wide, and hence would have had only four times as much sail area. Larger specimens of *Dimetrodon* didn't look quite like scaled-up smaller ones; we would say that the sail grew disproportionately large compared to the rest of the animal.

Dimetrodon was a large animal, but the need for heat regulation is even more acute for smaller animals. Like human babies, small animals can lose heat quickly, because of their high ratio of surface area to volume. Leading paleontologists now believe that birds (most of whom are quite small) evolved from dinosaurs and that feathers are modified reptilian scales. Though not a prevailing view, it has been hypothesized that the wings of birds and insects evolved originally not for flight but as temperature control devices.

Some scientists have speculated that African Pygmies are small in part because a small body is better able to lose heat in the hot, humid climate of the Ituri Forest in the Congo, where Pygmies live. Other scientists have suggested that ancestors of human beings began walking on two legs in part to keep cool in a hot climate. Walking upright exposes much less area of the body to the rays of the sun than walking on all fours and also reduces the amount of water needed by about one-half.

18.9 Similarity and Growth

Although a large change of scale forces adaptive changes in materials or form, within narrow limits—perhaps up to a factor of 2—creatures can grow according to a law of similarity; that is, they can grow proportionally, so that their shape is preserved. A striking example of such growth is that of the chambered nautilus (*Nautilus pompilius*). Each new chamber that is added onto the nautilus shell is larger than, but geometrically similar to, the previous chamber, and the shape of

Figure 18.9 A chambered nautilus shell. (Nicholas Foster/The Image Bank.)

the shell as a whole – an *equiangular,* or *logarithmic,* spiral – remains the same (see Figure 18.9).

Most living things grow over the course of their lives by a factor greater than 2. We've seen with *Dimetrodon* that a big specimen was not just a scaled-up small one. Nor is a human adult simply a scaled-up baby. Relative to the length of the body, a baby's head is much larger than an adult's. The arms of the baby are disproportionately shorter than an adult's. In the growth from baby to adult, the body does not scale up as a whole. But different parts of the body scale up, each with a different scale factor. That is, a baby's eyes grow at one rate to perhaps twice their original size, while the arms grow at another rate, to about four times their original size.

Although the laws for growth can be much more complicated than for proportional growth (or even the allometric growth that we discuss in the next section), more sophisticated mathematics – for example, differential geometry, the geometry of curves and surfaces – permits analysis of complex and interlocking scalings. For a model of the process in which a baby's head changes shape to grow into an adult head, we can use graph paper: first, we put a picture of the baby's skull on graph paper, then we determine how to deform the grid until the pattern matches an adult skull (see Figure 18.10 and Spotlight 18.4). The same idea lies at the heart of computerized "morphing," the process in which the face of one film character can be made to change smoothly into the face of another, with different scalings for different parts of the face.

OPTIONAL Allometry

If we measure the arm length or head size for humans of different ages and compare these measurements with body height, we observe that humans do not grow proportionally, that is, in a way that maintains geometric similarity. The head of a newborn baby may be one-third of the baby's length, but an adult's head is usually close to one-seventh of the individual's height. The arm, which at birth is one-third as long as the body, is by adulthood closer to two-fifths as long (see Figure 18.11a).

Ordinary graphing provides a way to test for differential growth. We can plot body height on the horizontal axis and arm length on the vertical axis (see

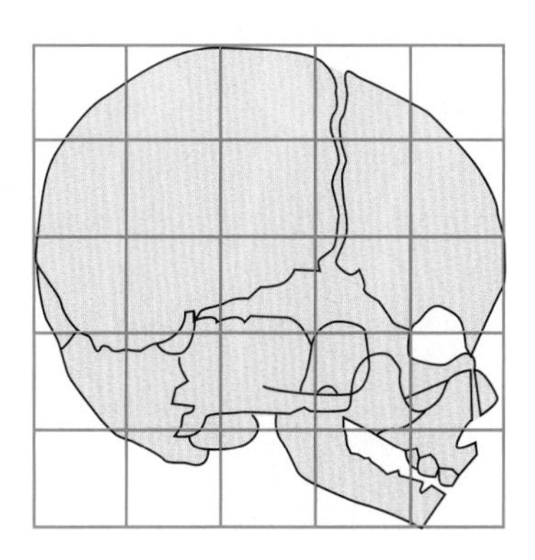

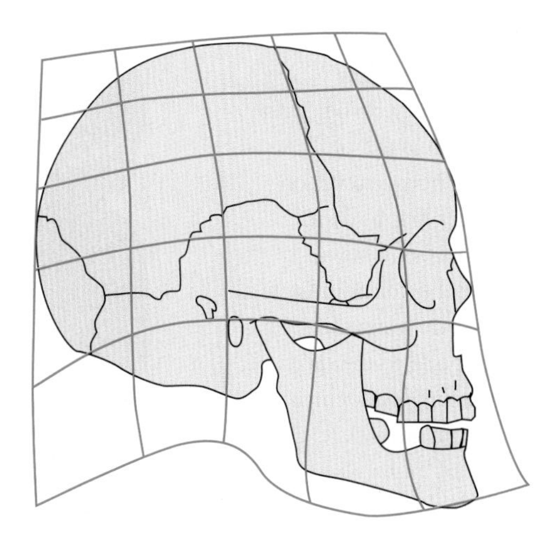

Figure 18.10
Modeling the changes in the shape of a human head from infancy to adulthood.

Figure 18.11b). A straight line would indicate proportional growth, that is, according to geometric similarity; and we do get a straight line from age 9 months (0.75 years) or so on up. Up to age 9 months or so, we get a curve, which indicates that the ratio of arm length to height does not remain constant over the first year.

Is there an orderly law by which we can relate arm length to height? Let's plot again, this time using a different scale. For this **logarithmic scale,** we mark off equal units, as usual. But instead of labeling the marked points with 0, 1, 2, 3, etc., we label them with the corresponding powers of 10: $10^0 = 1$, $10^1 = 10$, $10^2 = 100$, $10^3 = 1000$, and so on, which are also called **orders of magnitude.** Plotting a point on such a scale is not easy, because the point midway between 1 and 10 is not 5.5, but instead is closer to 3. Special graph paper (available in most college bookstores) marks smaller divisions and makes it easier to plot; paper marked with log scales on both axes is called **log-log paper,** while **semilog paper** has a logarithmic scale on just one axis. Also, many computer plotting packages can produce logarithmic scales.

We could use a logarithmic scale for either height or arm length, or for both. Using logarithmic scales for both, as in Figure 18.11c, the data plot closely to a straight line. Looking carefully, we can discern two different straight lines: a steeper one that fits early development (we will see shortly that it has slope 1.2), and a less steep one (with slope 1.0) that fits development after 9 months of age.

The change from one line to another after 9 months indicates a change in pattern of growth. The pattern after 9 months, characterized by the straight line with slope 1, is indeed proportional growth (sometimes called **isometric growth**). For the pattern before 9 months, we know from the slope (1.2) being greater than 1 that arm length is increasing relatively faster than height. That earlier growth also follows a definite pattern, called *allometric growth.*

Allometric growth is the growth of the length of one feature at a rate proportional to a power of the length of another.

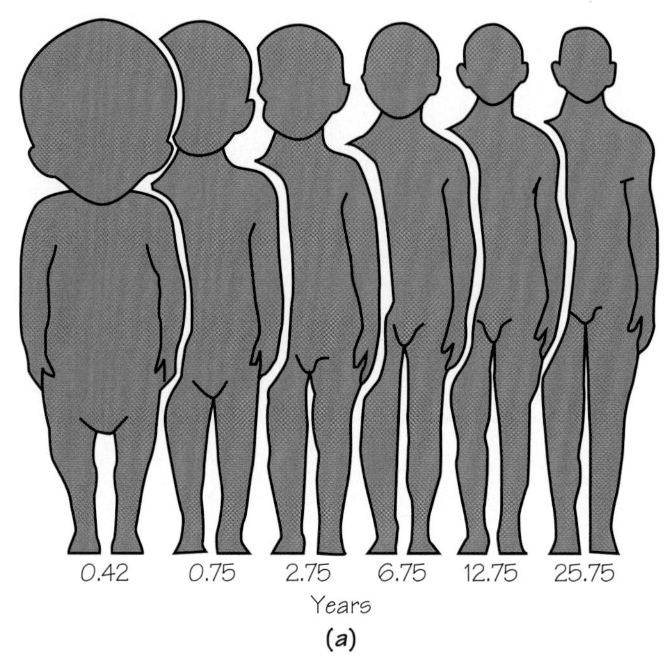

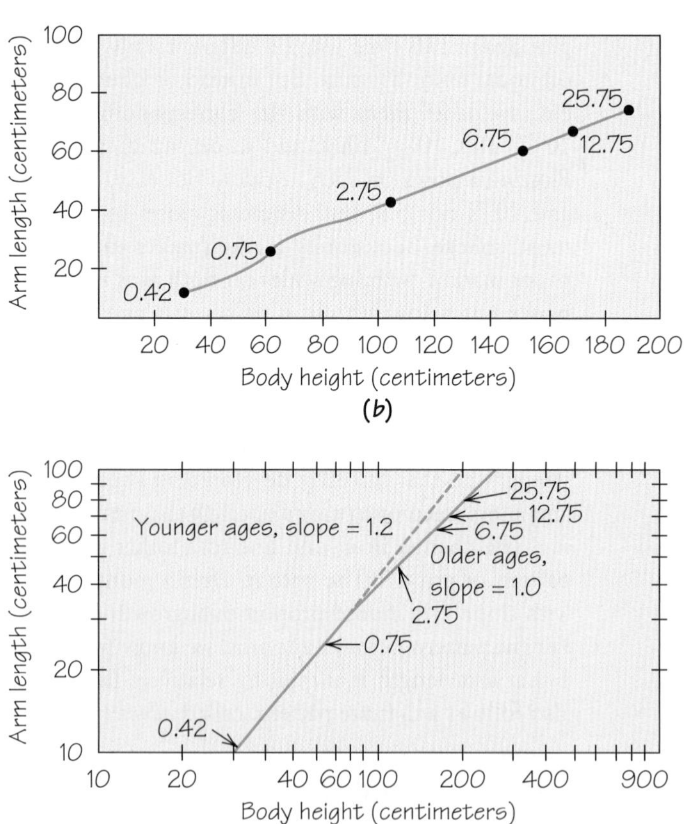

Figure 18.11 (a) The proportions of the human body change with age. (b) A graph of human body growth on ordinary graph paper. The numbers shown beside the points indicate the age in years; they correspond to the stage of human development shown in part (a). (c) A graph of human body growth on log-log paper.

SPOTLIGHT 18.4 Helping to Find Missing Children

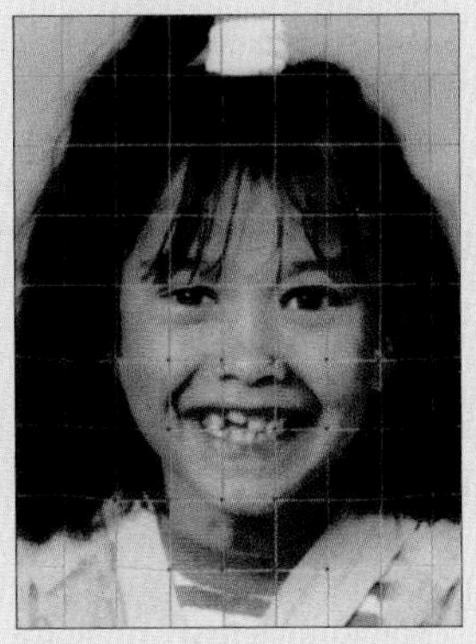
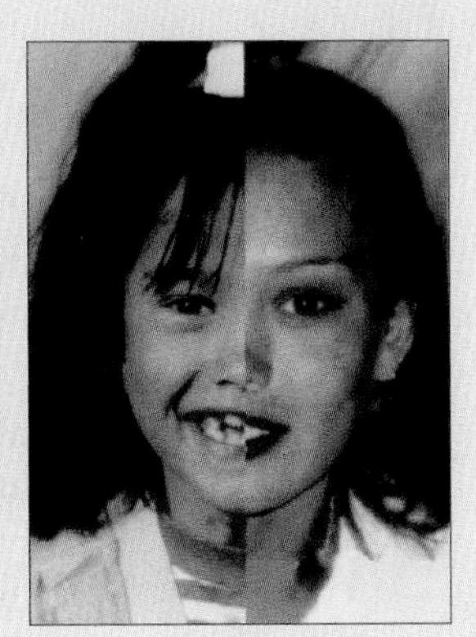

Image stretching is applied to a photograph taken at age 7.

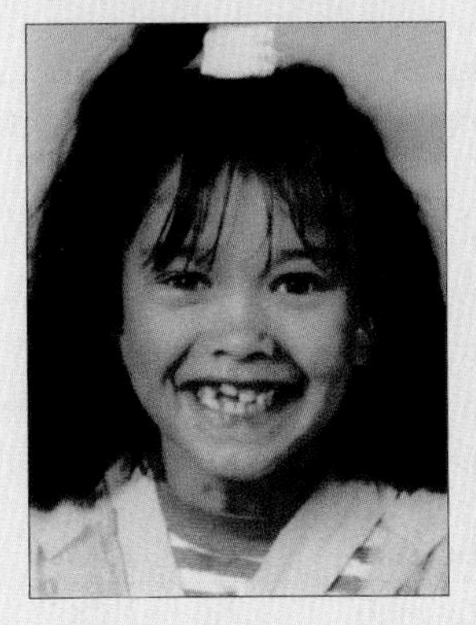

Age progression to age 17. (National Center for Missing and Exploited Children.)

It can be valuable to be able to predict what a developing organism will look like in the future. For example, what does a child look like now who was kidnapped six years ago, at age 3?

At the National Center for Missing and Exploited Children (NCMEC) in Arlington, Virginia, a computer and a more sophisticated version of the graph-paper technique are used to answer such questions. Computer age-progression specialists scan photographs of both the missing child at age 3 and an older sibling or a biological parent at age 9 into a computer. Then the face of the 3-year-old is stretched, depending on age, to reflect craniofacial growth and merged with the image of the sibling or parent at 9 years old. The result is a rough idea of what the missing child may look like. As mathematicians and biologists refine their models of how faces change over time, this technique will improve. It may even become possible to gain an idea of how a child may look at age 40 or 65.

We have seen that in geometric scaling, area grows according to the square (second power) and volume according to the cube (third power) of length, so we can say that they grow allometrically with length.

If we denote arm length by y and height by x, a straight-line fit on log-log paper corresponds to the algebraic relation

$$\log_{10} y = B + a \log_{10} x$$

where a is the slope of the line and B is the point where the graph crosses the vertical axis. If we raise 10 to the power of each side, we get

$$y = bx^a$$

where $b = 10^B$. This equation describes a **power curve:** y is a constant multiple of x raised to a certain power.

We can find approximate values for a for each of the two lines in the log-log plot from the coordinates of the points at the ends of the line. Those are the points for ages 0.42, 0.75, and 25.75. The observations and the corresponding logarithms are:

Age	Height	Log (height)	Arm length	Log (arm length)
0.42	30.0	1.48	10.7	1.03
0.75	60.4	1.78	25.1	1.40
25.75	180.8	2.26	76.9	1.89

The slope for the line from age 0.42 to age 0.75 is the vertical change over the horizontal change, both converted to log units:

$$\frac{\log 25.1 - \log 10.7}{\log 60.4 - \log 30.0} = \frac{1.40 - 1.03}{1.78 - 1.48} = \frac{0.37}{0.30} \approx 1.2$$

The slope for the line from age 0.75 to age 25.75 is

$$\frac{\log 76.9 - \log 25.1}{\log 180.8 - \log 60.4} = \frac{1.89 - 1.40}{2.26 - 1.78} = \frac{0.49}{0.48} \approx 1.0$$

So $a = 1.2$ up to 9 months, and $a = 1.0$ after 9 months. Up to 9 months, arm length grows according to $(\text{height})^{1.2}$. After 9 months, arm length grows according to $(\text{height})^{1.0}$, and we get $y = bx^{1.0}$, which is a linear relationship describing proportional growth, that is, growth according to geometric similarity. On ordinary graph paper, proportional growth appears as a straight line, allometric growth as a curve. On log-log paper, both patterns appear as straight lines.

The technique of allometry has been used in the last few years by paleontologists to determine that all of the six known specimens identified as the earliest fossil bird *Archaeopteryx* are indeed from the same species and that the minute fossil and puzzling fish known as *Palaeospondylus* (found only in Scotland) is probably just the larval stage of some better-known fish. ■

We have examined the problem of scale and noted that a large change in scale forces a change in either materials or form. A particular instance of the problem of scale is area–volume tension, and we have seen how an animal's size and geometric shape affect its abilities to move and to keep itself warm or cool.

In this chapter we have explored the limitations on life imposed by dwelling in three dimensions. In Chapters 19–20, we will see that dimensionality also imposes surprising limits on artistic creativity in devising patterns.

REVIEW VOCABULARY

Allometric growth A pattern of growth in which the length of one feature grows at a rate proportional to a power of the length of another feature.
Area–volume tension A result of the fact that as an object is scaled up, the volume increases faster than the surface area and faster than areas of cross sections.
Crushing strength The maximum ability of a substance to withstand pressure without crushing or deforming.
Density Weight per unit volume.
Geometrically similar Two objects are geometrically similar if they have the same shape, regardless of the materials of which they are made. They need not be of the same size. Corresponding linear dimensions must have the same factor of proportionality.
Isometric growth Proportional growth.
Linear scaling factor The number by which each linear dimension of an object is multiplied when it is scaled up or down; that is, the ratio of the length of any part of one of two geometrically similar objects to the length of the corresponding part of the second.
Logarithmic scale A scale on which equal divisions correspond to powers of 10.
Log-log paper Graph paper on which both the vertical and the horizontal scales are logarithmic scales, that is, the scales are marked in orders of magnitude 1, 10, 100, 1000, . . . , instead of 1, 2, 3, 4, ...
Mass The aspect of matter that is affected by forces, according to physical laws.
Orders of magnitude Powers of 10.
Power curve A curve described by an equation $y = bx^a$, so that y is proportional to a power of x.
Pressure Weight divided by area.
Problem of scale As an object or being is scaled up, its surface and cross-sectional areas increase at a rate different from its volume, forcing adaptations of materials or shape.
Proportional growth Growth according to geometric similarity, where the length of every part of the organism enlarges by the same linear scaling factor.
Semilog paper Graph paper on which only one of the scales is a logarithmic scale.
Wing loading Weight supported divided by wing area.

SUGGESTED READINGS

COLLIER, C. PATRICK. Media clips: Weight loss breakthrough, *Mathematics Teacher,* 94 (September 2001): 485–488. Explanation and exercises (with solutions) about BMI (body mass index).

DEWDNEY, A. K. *200% of Nothing: An Eye-Opening Tour Through the Twists and Turns of Math Abuse and Innumeracy,* Wiley, New York, 1993.

DIAMOND, JARED M. Why are Pygmies small? *Nature,* 354 (1991): 111–112.

DUDLEY, BRIAN A. C. *Mathematical and Biological Interrelations,* Wiley, New York, 1977. Excellent and gentle extended introduction to graphing, scale factors, and logarithmic plots.

GOULD, STEPHEN JAY. Size and shape. In *Ever Since Darwin,* Norton, New York, 1977, Chap. 21.

HALDANE, J. B. S. On being the right size. In *Possible Worlds and Other Papers,* Harper, New York, 1928. Reprinted in James R. Newman (ed.), *The World of Mathematics,* vol. 2, Simon & Schuster, New York, 1956, pp. 952–957. Also reprinted in John Maynard Smith (ed.), *On Being the Right Size and Other Essays by J. B. S. Haldane,* Oxford University Press, Oxford, 1985, pp. 1–8. Succinctly surveys area–volume tension, flying, the size of eyes, and even the best size for human institutions.

HILDEBRANDT, STEFAN, and ANTHONY J. TROMBA. *Mathematics and Optimal Form,* Scientific American Library, New York, 1985.

McMAHON, T. A., and J. T. BONNER. *On Size and Life,* Scientific American Library, New York, 1983. Astonishingly beautiful and informative book on the effects of size and shape on living things.

MINEYEV, ANATOLY. Trees worthy of Paul Bunyan: Why do trees grow so tall–but no taller? *Quantum,* 4 (January/February 1994): 4–10.

MITCHELL, KEVIN, and JAMES RYAN. The species–area relation, *The UMAP Journal* 19 (1998): 139–170. Reprinted in Paul J. Campbell (ed.), *UMAP Modules: Tools for Teaching 1998,* COMAP, Inc., Lexington, Mass., 1999, pp. 23–54.

NIKLAS, KARL J. *Plant Allometry: The Scaling of Form and Process,* University of Chicago Press, Chicago, 1994.

SCHMIDT-NIELSEN, KNUT. *Scaling: Why Is Animal Size So Important?* Cambridge University Press, New York, 1984.

STEVENS, PETER S. *Patterns in Nature,* Atlantic Monthly Press, Boston, 1974. Splendid treatment of the problem of scale and other physical phenomena in nature: flows, meanders, branching, trees, soap films, cracking, and packing.

THOMPSON, D'ARCY. *On Growth and Form,* Cambridge University Press, Cambridge, 1917, 1961. "A discourse on science as though it were a humanity" (J. T. Bonner); this was the first book to describe in quantitative terms the processes of growth and shaping of biological forms.

TREFIL, JAMES S. What would a giant look like? In *The Unexpected Vista: A Physicist's View of Nature,* Chap. 10, pp. 156–171, Scribner, New York, 1983. Explanation of the effects of scaling up. In Trefil's illustration on p. 162, however, the eyes of the giants are unrealistically large.

WEISS, PETER. Built to scale, *Science News,* 156 (16 October 1999): 249–251. Discusses basis and evidence for the three-quarters-power scaling law.

WENT, F. W. The size of man. *American Scientist,* 56 (1968): 400–413. Demonstrates a schism between the macroworld and the molecular world, by comparing human life with that of an ant.

WGBH EDUCATIONAL FOUNDATION AND PEACE RIVER FILMS. *Nova: The Shape of Things,* 1985. Distributed by Vestron Video, Box 4000, Stamford, CT 06907.

ZHERDEV, A. Horseflies and flying horses: Questions of scale in the animal kingdom. *Quantum,* 4 (May/June 1994): 32–37, 59–60.

SUGGESTED WEB SITES

physics.nist.gov/News/TechBeat/9501beat.html Proposed redefinition of the kilogram.

ftp://ftp.fifthwave.com/pub/Convert134.sit Convert (computer program for Macintosh) can convert many kinds of units; you can explore such units as tuns (volume), blinks (time), and barns (area).

www.missingkids.com National Center for Missing and Exploited Children.

www.usmint.gov U.S. Mint.

www.thusness.com/bmi.t.html Body mass index calculator and further links.

SKILLS CHECK

1. You want to enlarge a small 3-in. by 5-in. photograph to a 12-in. by 20-in. copy. Assuming that the cost of photographic paper is proportional to its area, and that 3-in. by 5-in. reprints cost 40 cents each, how much would you expect to pay for the large copy?

(a) \$1.60
(b) \$3.20
(c) \$6.40

2. If a model car is built to a scale of 1 to 40, and the actual car has a turning circle of 37 ft, what should be the turning circle of the model?

(a) 11.1 in.
(b) 0.925 in.
(c) 1480 ft

3. If a medium 10-in. pizza costs \$8 and a similar 14-in. pizza costs \$14, which is the better buy?

(a) The 10-in. pizza
(b) The 14-in. pizza
(c) They are about the same price per square inch.

4. An artist plans to melt 100 pennies and re-form a larger penny proportional to an ordinary coin. What is the linear scaling factor of the large penny, compared to the ordinary penny?

(a) 100 to 1
(b) 10 to 1
(c) 4.64 to 1

5. A scale model of a carillon stands 10 in. tall, and the actual carillon stands 100 ft tall. What is the linear scaling factor of the carillon compared to its model?

(a) 10 to 1
(b) 12 to 1
(c) 120 to 1

6. Coffee costs about \$5 per pound in the United States. If a Canadian dollar exchanges for 65 U.S. cents and a kilogram is about 2.2 pounds, what is the equivalent cost in Canadian dollars?

(a) \$3.50 Canadian per kilogram
(b) \$7.15 Canadian per kilogram
(c) \$16.92 Canadian per kilogram

7. A sculpture weighs 140 lb and is supported by three legs, each of which is 0.5 in. by 0.5 in. by 2 in. high. How much pressure do the legs exert on the table?

(a) 47 lb/in.^{2}
(b) 105 lb/in.^{2}
(c) 187 lb/in.^{2}

8. If an object is scaled so that its volume grows to 8 times its original volume, its surface area is scaled to _____ times its original surface area.

(a) 4
(b) 16
(c) 8/3

9. Assuming a catfish maintains the same shape and proportions as its grows, and a catfish 8 in. long weighs about 1 lb, how long is a 2-lb catfish?

(a) 10 in.
(b) 12 in.
(c) 16 in.

10. A penny and a nickel are

(a) not geometrically similar, because they are made of different materials.
(b) not geometrically similar, because they are of different sizes.
(c) geometrically similar, because they have the same shape.

11. In comparing the flight speeds for birds, an analysis of wing loading leads to the conclusion that

(a) light birds fly faster than heavy birds.
(b) heavy birds fly faster than light birds.
(c) heavy and light birds fly at about the same speed.

12. In comparing the heights that large and small animals jump, analysis of the impact on scaling leads to the conclusion that

(a) smaller jumping animals can jump much higher than larger jumping animals.
(b) larger jumping animals can jump much higher than smaller jumping animals.
(c) all jumping animals jump to about the same height.

13. A kilometer is approximately equal in length to

(a) 5 miles.
(b) 3 miles.
(c) 3/5 mile.

14. A two-liter bottle contains approximately

(a) 2 quarts.
(b) 1 gallon.
(c) 10 pints.

15. A weight of 130 lb is approximately the same as

(a) 40 kg.
(b) 50 kg.
(c) 290 kg.

EXERCISES

▲ **Optional.** ■ **Advanced.** ◆ **Discussion.**

Most of the exercises require a calculator; one that offers square roots will suffice.

Geometric Similarity

1. Suppose you are printing photographs from negatives of so-called 35-millimeter film, whose frames are just about 1 inch by $1\frac{1}{2}$ inches [the actual size is 24mm (millimeters) by 36mm].

(a) First you make some contact prints, which are exactly the same size as the negatives. What is the scaling factor of a contact print?
(b) One enlargement that you want to make is to be three times as high and three times as wide as the negative. What is the linear scaling factor for this print? How does its area compare with the area of the negative?
(c) What is the scaling factor for a 4-by-6 print? What is the area of the print?
(d) The size of so-called 3-by-5 prints can vary, depending on whether the print has a border or not; a common size is about $3\frac{1}{16}$ by $4\frac{19}{32}$ in. Is this print geometrically similar to the negative?
(e) The sizes and advertised costs of reprints available at Turtle Creek Bookstore are

3.5 × 5.25 ("standard" size)	$0.28
4 × 6 ("premium" size)	0.35
5 × 7	1.79
8 × 10	3.99
8 × 12	4.49
11 × 14	9.99
16 × 20	15.95
20 × 30	9.99

Assume for simplicity that the prints have exactly the measurements indicated. Which of these sizes are geometrically similar to a 35mm negative? What do you think happens for sizes that aren't?
(f) The cost of photographic paper is very nearly proportional to the area of the paper. Compare the costs of a "standard" 4-by-6 enlargement and an 8-by-12 enlargement. What can you say about the relative cost of the two prints?
(g) Based on the amount of paper used and the cost of the 5-by-7 and 8-by-10 enlargements, how much would you expect an 11-by-14 enlargement to cost?
(h) The price for the 20-by-30 enlargement was a mistake on the price list. What price do you think it should have, to agree with the other prices?

2. The area of a circle of radius r is πr^2; expressed in terms of the diameter, $d = 2r$, the area is $\frac{1}{4}\pi d^2$. If we apply a linear scaling factor L to the diameter of a circle, then the area of the scaled circle—as in the case of the square that we considered in the text—changes with L^2, the square of the linear scaling factor. A natural application of this idea, of course, is to your local pizza parlor and the prices on its menu. The actual prices at the pizza restaurant closest to Beloit College are $6, $7, $7.95, and $8.95, respectively, for small (10-in.), medium (12-in.), large (14-in.), and extra-large (16-in.) cheese pizzas.

(a) What is the linear scaling factor for an extra-large pizza compared to a small one?
(b) How many times as large in area is the extra-large pizza compared to the small one?

(c) How much pizza does each size give per dollar? What "hidden" assumptions are you making about how the pizzas are scaled up?
(d) The corresponding prices for a pizza with "the works" are $9.95, $11.95, $13.95, and $15.95. Is there any size of these for which you get more pizza per dollar than some size of the cheese pizzas?

3. The human figures in Lego sets are 4 cm tall (without hats or helmets).

(a) What is the linear scaling factor of a Lego figure if it represents a human who is 180 cm tall?
(b) How does the volume of a real human compare with the volume of a Lego figure?
(c) The jeep in one Lego set is 10 cm long. Using the linear scaling factor of part (a), how long would a real jeep be?

4. Dollhouses and their furnishings are usually built to a scale of exactly 1 in. to 1 ft, meaning that an item 1 ft long in a real house is 1 in. long in a dollhouse.

(a) What is the linear scaling factor for a dollhouse?
(b) If a dollhouse were made of the same materials as a real house, how would their weights compare?

5. According to *The Economist* (September 16, 1995, p. 74), "The average 16-year-old Japanese girl has grown 4% heavier since 1975, although she is only 1% taller." The tone of this remark suggests that Japanese girls have grown heavier than their increase in height would warrant. Assume that weight is proportional to volume, and comment on this remark.

6. At our house we have some 10-in. frying pans and a 12-in. one; the 12-in. one seems to weigh a lot more, never get as hot, and cook food more slowly. Suppose that a 10-in. frying pan weighs 1 lb, apart from its handle. How much would a geometrically similar 12-in. frying pan weigh? How much would it weigh if it had the same thickness of metal as the 10-in. pan?

7. One of the famous geometry problems of Greek antiquity was the *duplication of the cube*. Our knowledge of the history of the problem comes from the third century B.C. from Eratosthenes of Cyrene, who is famous for his estimate of the circumference of the earth. According to him, the citizens of Delos were suffering from a plague. They consulted the oracle, who told them that to rid themselves of the plague, they must construct an altar to a particular god that would be geometrically similar to the existing one but double the volume.

(a) How would the volume of the new altar compare with the old if each of its linear dimensions were doubled?
(b) What should the linear scaling factor be for the new altar? (The problem intended by the oracle was to construct with straightedge and compasses a line segment equal in length to this particular linear scaling factor. Not until the nineteenth century was the task shown to be impossible. Eratosthenes relates that the Delians interpreted the problem in this sense, were perplexed, and asked Plato about it. Plato told them that the god didn't really want an altar of double the volume but wished to shame them for their "neglect of mathematics and their contempt for geometry.")

8. The declining purchasing power of the dollar and the short life of a dollar bill in circulation suggest that the dollar bill be abolished in favor of a dollar coin. The Susan B. Anthony dollar coin of 1979–1981 was a failure with the U.S. public, who found it too small and light. From 2000 through early 2002, the U.S. mint made a gold-colored dollar depicting Sacagawea, a Shoshone Indian who guided explorers Meriwether Lewis and William Clark in their 1804–1805 expedition to the Pacific Ocean. But even this coin may be too little (in size and weight) too late (a dollar today is worth less than half what it was in 1978). Suppose that you are put in charge of planning a new *$5* coin (whom should it depict?). The sole requirement is that the coin be made of the same material as the current U.S. 25-cent piece but, unlike the Anthony and Sacagawea coins, weigh four times as much. A quarter can be described geometrically as a circular cylinder approximately $\frac{15}{16}$ in. in diameter and $\frac{1}{16}$ in. thick. Because your new dollar should weigh

The Susan B. Anthony dollar replaced by the Sacagawea dollar coin in 2000. (United States Mint.)

The young Native American Sacagawea portrayed in relief. (United States Mint.)

four times as much, it needs to have four times the volume of a quarter. [You may find it helpful that the formula for the volume of a cylinder is $\pi \times (\text{diameter}/2)^2 \times \text{height}$.]

(a) A member of your public advisory panel suggests that the requirement will be fulfilled if you just double the diameter and double the thickness. What do you tell this individual, in the most diplomatic terms?
(b) If you go along with the member's suggestion to double the diameter, how thick does the coin need to be?
(c) Another member of the board feels that doubling the diameter would produce a coin too large to be convenient and proposes instead that you scale up the quarter proportionally (she took a course from an earlier edition of this book). What would the dimensions be for this new \$5 piece?

The Language of Growth, Enlargement, and Decrease

9. Criticize the following statement, which appears on sacks of the product, and write a correct version.

> "Erin's Own Irish sphagnum moss peat. It enriches your soil and makes your growing easier. Compressed to $2\frac{1}{2}$ times normal volume."

10. Criticize the following claims, which were cited in *The New York Times* of 9/25/87 and 10/21/87:
(a) A new dental rinse "reduces plaque on teeth by over 300%."
(b) An airline working to decrease lost baggage has "already improved 100% in the last six months."
(c) "If interest rates drop from 10% to 5%, that is a 100% reduction."

Measuring Length, Area, Volume, and Weight

11. The cost of a lightweight airmail letter from the United States to most of Western Europe in summer 2002 was \$0.80. How much was that in euros (€), the currency of the European Union, when the exchange rate was €1.02 = \$1? (For comparison, the cost then of an airmail letter to the United States varied from country to country in the European Union, ranging from €1.09 to €1.53.)

12. The cost of a lightweight letter from the United States to Canada in 2002 was US\$0.45. How much was that in Canadian dollars when the exchange rate was US\$1 = Cdn\$1.60? (The postage cost from Canada to the United States was Cdn\$0.65.)

13. In Germany the fuel efficiency of cars is measured in liters of fuel per 100 kilometers. A typical average in a compact station wagon would be 8.5 liters per 100 km. What is that in miles per gallon (mpg)?

14. According to the Environmental Protection Agency ratings of 2002 cars, the highest-mileage car seating four or more passengers was the gasoline/electric hybrid Toyota Prius, at 52 mpg in the city. How many liters of gasoline does such a Prius use to travel 100 km in the city?

15. Consider a real locomotive that weighs 88 tons and an HO-gauge scale model of it, for which the linear scaling factor is 1/87.

(a) How much would an exact scale model weigh, in tons?
(b) What assumptions are involved in your answer to part (a)?
(c) How much would an exact scale model weigh, in pounds?
(d) In kilograms?
(e) In metric tonnes (1 metric tonne = 1000 kg)?

16. An ad for a software package for data analysis included a data set on tropical rain forests and deforestation. The data were given in hectares and were accompanied by the statement "A hectare equals 10,000 mi^2 or 2471 acres." What conversion factors should have appeared instead?

17. Gasoline is sold in the United States by the U.S. gallon and in Canada by the liter (1 U.S. gallon = 231 cu in.; 1 liter = 1000 cm^3). What is the equivalent cost, in U.S. dollars per U.S. gallon, for gasoline in Canada priced at 61.9 Canadian cents per liter, when one Canadian dollar exchanges for 66 cents U.S.?

18. In 1991, Edward N. Lorenz, a meteorologist who was an early researcher into chaos and dynamical systems (discussed in Chapter 23), received the Kyoto Prize in Basic Sciences, consisting of a gold medal and 45 million Japanese yen. If US$1 = ¥125 at the time, what was the value of the cash award in 1991 U.S. dollars?

19. In 200 B.C., Eratosthenes measured the circumference of the earth and expressed the result as 250,000 *stadia* (plural of *stadium*). In *The American Heritage Dictionary,* Second College Edition (Houghton Mifflin, Boston, 1982), we read for the second meaning of "stadium": "An ancient Greek measure of distance . . . equal to about 185 kilometers, or 607 feet." The name of the unit came from the length of a racecourse that was a bit less than an eighth of a mile long. If the numbers in the definition are correct, what are the correct units that should have appeared?

Sorry, No King Kongs

20. The weight of a 1-ft cube of steel is 500 lb. What is the pressure on the bottom face in

(a) pounds per square inch?
(b) atmospheres (1 atm = 14.7 lb/sq in.)?

21. In an article on adding organic matter to soil, the magazine *Organic Gardening* (March 1983) said, "Since a 6-inch layer of mineral soil in a 100-square-foot plot weighs about 45,000 pounds, adding 230 pounds of compost will give you an instant 5% organic matter."

(a) What is the density of the mineral soil, according to the quotation?
(b) How does this density compare with that of steel?
(c) How do you think the quotation should be revised to be accurate?

22. A mature gorilla weighs 400 lb and stands 5 ft tall.

(a) Give an estimate of its weight when it was half as tall.
(b) What assumptions are involved in your estimate?
(c) A mature gorilla's two feet together have a combined area of about 1 ft^2. When the gorilla is standing on its feet, what is the pressure on its feet, in pounds per square inch?

23. Suppose King Kong is a gorilla scaled up with a linear scaling factor of 10.

(a) How much does King weigh?
(b) What is the pressure on King's feet, in pounds per square inch?

24. You may have wanted to have a waterbed but found that waterbeds were not allowed in your building. Apart from the danger of flood if the bed should puncture or leak, there is the consideration of the weight.

(a) If a queen-size mattress is 80 in. long by 60 in. wide by 12 in. high, and water weighs 1 kg/liter, how much does the water in the mattress weigh in pounds?
(b) If the weight of the mattress and frame is carried by four legs, each 2 in. by 2 in., what is the pressure, in pounds per square inch, on each leg?

(c) How does the pressure on the legs of the waterbed compare with the pressure that a person exerts on his or her feet–for example, a 130-lb person with a total foot area of about one-quarter of a square foot in contact with the ground?
(d) If you aren't allowed to have a waterbed, how about a spa (hot tub)? Find the weight of the water in a spa that is in the shape of a cylinder 6 ft in diameter and 3.5 ft deep. (*Hint:* The volume of a cylinder is $\pi r^2 h$, where r is the radius and h is the height.)

25. What does the largest giant sequoia tree weigh? Model the tree as a (very elongated) cone, supposing that the tree is 360 ft high and has a circumference of 40 ft at the base, and that the density of the wood is 31 lb/cu ft. (The volume of a cone of height h and radius r is $\frac{1}{3}\pi r^2 h$.)

Falls, Jumps, and Flights

26. The movie *Them* features enormous ants (about 8 m long and about 3 m wide). We can investigate how feasible such a scaled-up insect is by considering its oxygen consumption. A common ant, which is about 1 cm long, needs about 24 milliliters of oxygen per second for each cubic centimeter of its volume. Because an ant does not have lungs, it must absorb the oxygen through its "skin," which it can do at a rate of about 6.2 milliliters per second per square centimeter. Suppose that the tissues of a scaled-up ant would have the same need for oxygen for each cubic centimeter, and that its skin could absorb oxygen at the same rate, as a normal ant. Compared to a common ant, how many times as large is an enormous ant's

(a) length?
(b) surface area?
(c) volume?
(d) What proportion of such an ant's oxygen need could its skin supply?

What can you conclude about the existence of such insects? (Adapted from George Knill and George Fawcett, Animal form or keeping your cool, *Mathematics Teacher,* May 1982, 395–397.)

27. In the children's story *Peter Pan,* Peter and Wendy can fly. We may suppose that they are 4 ft tall, so they are about 12 times as tall as a sparrow is long. What should their minimum flying speed be?

28. Icarus of Greek legend escaped from Crete with his father, Daedalus, on wings made by Daedalus and attached with wax. Against his father's advice, Icarus flew too close to the sun; the wax melted, the wings fell off, and he fell into the sea and drowned. What must have been his minimum cruising speed? What assumptions does your answer involve?

Keeping Cool (and Warm)

◆ **29.** Smaller birds and mammals generally maintain higher body temperatures than do larger ones. Explain why you would expect this to be so. (Adapted from A. Zherdev, Horseflies and flying horses, *Quantum,* May/June 1994, 32–37, 59–60.)

◆ **30.** Some humans, such as the Bushmen of the Kalahari Desert in Africa, live in desert environments, where it is important to be able to do without water for periods of time. Would you expect such an environment to favor short or tall individuals? (Adapted from A. Zherdev, Horseflies and flying horses, *Quantum,* May/June 1994, 32–37, 59–60.)

31. The branching of trees is similar to the branching of systems in the bodies of animals. For similar reasons, the area of the cross section of the tree at its base scales as the three-fourths power of the tree's mass, that is, $A \propto M^{3/4}$ Assume that most of the mass is in the trunk and model the tree either as a tall cylinder ($V = \pi r^2 h$) or as a cone ($V = \pi r^2 h/3$). Show that the diameter d of a tree is approximately proportional to the three-halves power of the height, that is, $d \propto h^{3/2}$.

▲ **32.** Listed below are the number of species of reptiles and amphibians on some Caribbean islands, together with the approximate areas of the islands. (Suggested by Florence Gordon of the New York Institute of Technology, with contributions from Kevin Mitchell and James Ryan of Hobart and

William Smith Colleges, Geneva, NY. This table is adapted from Tables 15 and 16 in P. J. Darlington, *Zoogeography: The Geographic Distribution of Animals*, Wiley, New York, 1957, pp. 483–484).

Island	Area (mi^2)	Species
Redonda	1	3
Saba	4.9	5
Montserrat	40	9
Trinidad	2,000	80
Puerto Rico	3,400	40
Jamaica	4,500	39
Hispaniola	30,000	84
Cuba	40,000	76

(a) Plot number of species versus area on ordinary graph paper and then on log-log graph paper. If you don't have log-log paper available, use a calculator or spreadsheet to take the logarithms ($\log_{10}$) of all the numbers and graph logarithm of number of species versus logarithm of area on ordinary graph paper.
(b) Is the relationship that you graphed in part (a) proportional? Allometric?
(c) What would be the expected number of species on an island of 400 square miles?
(d) For each 10-fold increase in the island's size, what happens to the number of species, approximately?

Additional Exercises

For Exercises 33 and 34, refer to the following: An ancient measure of length, the *cubit*, was the distance from the elbow to the tip of the middle finger of a person's outstretched arm. So the length of a cubit depended on the person, though there was some attempt at standardization. Most estimates place the length of a cubit between 17 and 22 in.

33. Goliath [of David and Goliath, as related in the Bible (I Samuel 17:4)] was "six cubits and a span." A span was originally the distance from the tip of the thumb to the tip of the little finger when the hand is fully extended, about 9 in. What range of heights would this indicate for Goliath, in feet and inches? In centimeters?

34. According to classical Greek sources, Pythagoras (sixth century B.C.) used geometric scaling to model the height of Hercules, the heroic figure of classical mythology in the epic poems of Homer. Pythagoras compared the lengths of two racecourses, one (according to tradition) paced off by Hercules and the other by a man of average height. Both were 600 "paces" long, but the one established by Hercules was longer because of Hercules's longer stride (600 "Herculean" paces versus 600 paces by a normal man). A normal man in the time of Pythagoras would have been about 5 ft tall.
(a) If the distance paced off by Hercules was 30% longer than the other racecourse, how tall was Hercules? What does your calculation assume?
(b) In fact, the ancient sources do not give the original data but only the two conflicting answers that Hercules was 4 cubits tall and 4 cubits 1 "foot" tall. What range does this give for the height of Hercules, in feet and inches? In centimeters? (Assume that a Greek "foot" was the length of a modern foot.)

◆ **35.** Recent years have seen the beginnings of human-powered controlled flight, in the *Gossamer Condor* and other superlightweight planes. The *Gossamer Condor* is far longer than an ostrich, but it flies at only 12 mph. How can it?

◆ **36.** Jonathan Swift's Gulliver traveled to Lilliput, where the Lilliputians were human-shaped but only about 6 in. tall. In other words, they were geometrically similar in shape to ordinary human beings but only one-twelfth as tall. What would a Lilliputian weigh?

Are Lilliputians ruled out by the size–shape and area–volume considerations in this chapter? If you think they are, what considerations do you find convincing? If not, why not?

37. A 6-ft indoor holiday tree needs four strings of lights to decorate it. How many strings of lights are needed for an outdoor tree that is 30 ft high? (Contributed by Charlotte Chell of Carthage College, Kenosha, Wisconsin.)

38. What would you expect an individual *Quetzalcoatlus northropi* to weigh if it had half the wingspan of an adult? If an individual weighed half as much as an adult, what would you expect its wingspan to be?

▲ **39.** Listed below are the weights and wingspans of some birds and of some fully loaded airplanes. (Idea and data contributed by Florence Gordon of the New York Institute of Technology.)

Bird	**Weight (lb)**	**Wingspan (ft)**
Crow	1	2.9
Harris hawk	2.6	3.2
Blue-footed booby	4	3
Red-tailed hawk	4	4
Horned owl	5	5
Turkey vulture	6.5	6
Eagle	12	7.5
Golden eagle	13	7.3
Whooping crane	16.1	7.5
Vulture	18.7	9.3
Condor	22	9.9
Quetzalcoatlus northropi	100	36
Plane		
Boeing 737	117,000	93
DC9	121,000	93.5
Boeing 727	209,500	108
Boeing 757	300,000	156.1
Boeing 707	330,000	145.7
DC8	350,000	148.5
DC10	572,000	165.4
Boeing 747	805,000	195.7

(a) Use a calculator or spreadsheet to take the logarithms ($\log_{10}$) of all the numbers and then graph logarithm of weight versus logarithm of wingspan on ordinary graph paper.
(b) For the birds, is the relationship that you graphed in part (a) proportional? Allometric? How about for the planes?
(c) Does the same relationship of wingspan to weight seem to hold for birds and planes?

40. The *body mass index* (BMI) is the basis for the National Heart, Lung and Blood Institute's new weight guidelines. BMI is body weight (in kilograms) divided by the square of height (in meters). A BMI of 25 through 29 is considered "overweight"; a BMI of 30 or over is considered "obese." Some 55% of American adults have a BMI of 25 or above. (*Note:* BMI is not a useful measure for young children, pregnant or breastfeeding women, the frail elderly, or very muscular people.)
(a) Calculate the BMI for a woman 160 cm tall who weighs 65 kg. Is she overweight according to the institute's guidelines?
(b) How much in kilos must a man weigh who is 190 cm tall if he is not to be considered overweight according to the institute's guidelines?
(c) Calculate your own BMI.
(d) Suppose that weight and height are measured instead in U.S. Customary units of pounds and inches. We can still calculate body weight divided by the square of height, using these units. What conversion factor is necessary to convert this number to the BMI?
(e) Because body weight is average density times body volume, BMI is average density times a quantity that has units of length. Discuss whether BMI makes sense as a measure of being overweight. Would dividing by a different power of height make for a better measure?

41. Maybe some trees could grow to a mile high, but they just don't live long enough to have the chance. In this problem we try to determine how fast the height of a tree increases. How much mass the tree adds each year is determined by how much water is pumped to the treetop, and that in turn is determined by the area of the tree rings involved. Here are two relevant facts about trees:

- You may have noticed from tree stumps that as a tree grows older, its annual rings get thinner. Although the width of the ring varies somewhat from year to year with the amount of rainfall and other factors, the total *area* of each annual ring is roughly the same over the years, meaning that *the tree adds roughly the same amount of mass each year.* Call that amount a; then the mass M of the tree is $M = at$, where t is its age in years.

- Over a large range of tree sizes and tree species, the diameter d of a tree of a species is approximately proportional to the three-halves power of the height h of the tree (different species have different constants of proportionality). Thus, $d \propto h^{3/2}$ (see Exercise 31).

Now, if we assume that the bulk of the mass of the tree is in the trunk, and if we model the trunk either as a long cylinder or as a thin cone, the mass is proportional to the volume, so $M \propto d^2h$. Then

$$at = M \propto d^2h \propto (h^{3/2})^2h = h^4$$

so $h \propto t^{1/4}$. In other words, the tree grows as the fourth root of its age.

(a) Suppose a tree grows to 20 m in 30 years. How tall will it be (if it lives long enough) when it is 60 years old?
(b) How long would it take the tree in part (a) to grow to be 40 m tall?
(c) Giant sequoias can reach 100 m after about 1000 years. If it kept on growing at the same rate of addition of mass, how long would it take a giant sequoia to grow to 200 m?

✎ WRITING PROJECTS

1. A human infant at birth usually weighs between 5 and 10 lb and has a height (length) between 1 and 2 ft, with the shorter babies having the lesser weight. Considering the weight and height of an adult human, give an argument that human growth must not be just proportional growth.

2. The principle that area scales with the square of length, and volume with the cube, has important consequences for the depiction and interpretation of data in graphic form. Suppose we wish to indicate in an artistic way that the weekly income of a U.S. carpenter is twice that of a carpenter in (mythical) Rotundia. We draw one moneybag for the Rotundian and another one "twice as large" for the American. (Illustration from Darrell Huff, *How to Lie with Statistics,* Norton, 1954, p. 69)

What's the problem? Well, first, people tend to respond to graphics by comparing areas. Because the larger moneybag is twice as high and twice as wide as the smaller one, the image of it on the page has four times the area. Second, we are used to interpreting depth and perspective in drawings in terms of three-dimensional objects. Because the

larger bag is also twice as thick as the smaller, it has eight times the volume. The graphic leaves the subconscious impression that the U.S. carpenter earns eight times as much, instead of twice as much. With these ideas in mind, evaluate the depictions of data below.

(a) Advertising spending in three prominent newsmagazines. (From *Time* magazine, as reproduced in David S. Moore, *Statistics: Concepts and Controversies,* 4th ed., W. H. Freeman, 1997, p. 207.)

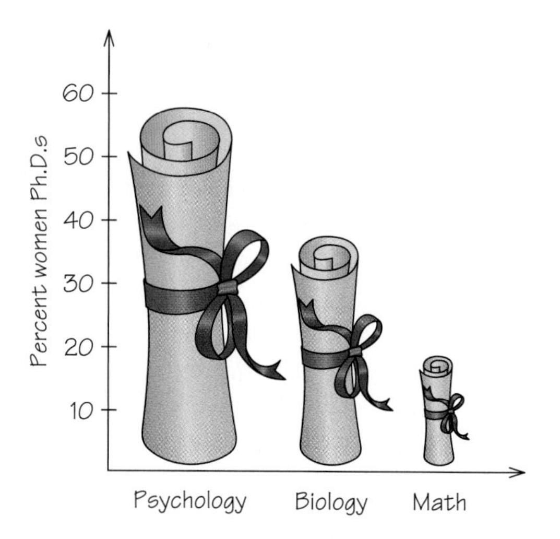

(b) Percentages of Ph.D.s earned by women in three fields. [From *Science,* 260 (April 16, 1993), 409, as reproduced in Jessica Utts, *Seeing Through Statistics,* Duxbury, Belmert, Calif., 1996, p. 142.]

(c) U.S. colleges as classified by enrollment. (From David S. Moore, *Statistics: Concepts and Controversies,* 4th ed., Freeman, 1997, p. 217.)

3. As in Writing Project 2, evaluate the depictions below. (Illustrations reproduced or adapted from Edward R. Tufte, *The Visual Display of Quantitative Information*, Graphics Press, 1983, pp. 70, 69, and 57)

1958—Eisenhower

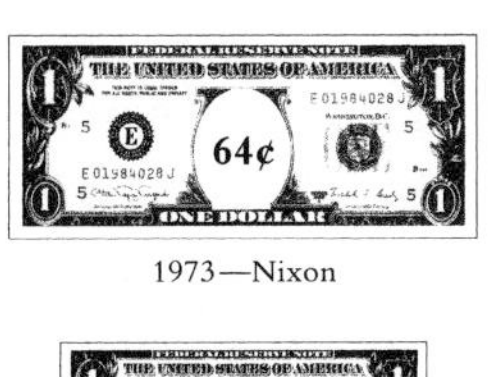

1973—Nixon

1978—Carter

1963—Kennedy

1984—Reagan

1990—Bush

1968—Johnson

1994—Clinton

2002—Bush

(a) Value of the dollar.

(b) The shrinking family doctor.

(c) Fuel economy standards for autos.

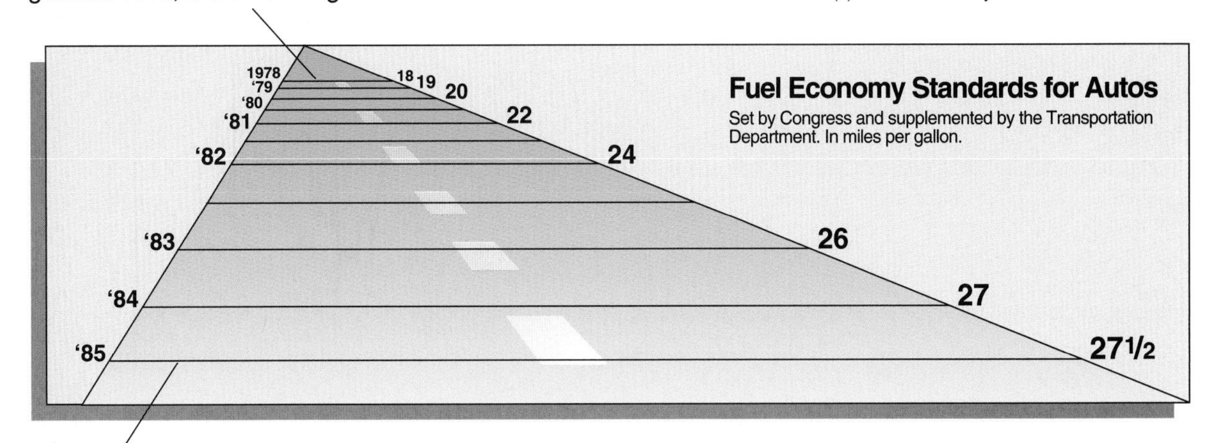

The New York Times, August 9, 1978, p. D2.

4. With the ideas of Writing Projects 2 and 3 in mind, collect and evaluate similar depictions of data from magazines and newspapers.

5. Dolls and human figures are usually scaled to be geometrically similar to actual humans. But are dolls designed to represent babies or adult humans? Go to a toy store and measure the height, the vertical height of the head, and the arm length of some dolls and other figures. Scale your measurements to compare them with Figure 18.11; from that comparison, try to estimate the ages of the humans that the figures resemble.

6. As shown in the graphic, the Canadian dollar has steadily fallen in value against the U.S. dollar for almost 30 years. On April 25, 1974, the exchange rate was Cdn$0.96 per U.S. dollar; on December 26, 2001, it was Cdn$1.605 per U.S. dollar.

There are two competing practices for stating how much one currency has depreciated (lost value) against another:

- Option A, used by the International Monetary Fund and the British periodical *The Economist,* takes the difference in the first country (here $1.605 − $0.96), divides it by the new trading value ($1.605), and multiplies by 100 to get a result in percent (here, 40%).

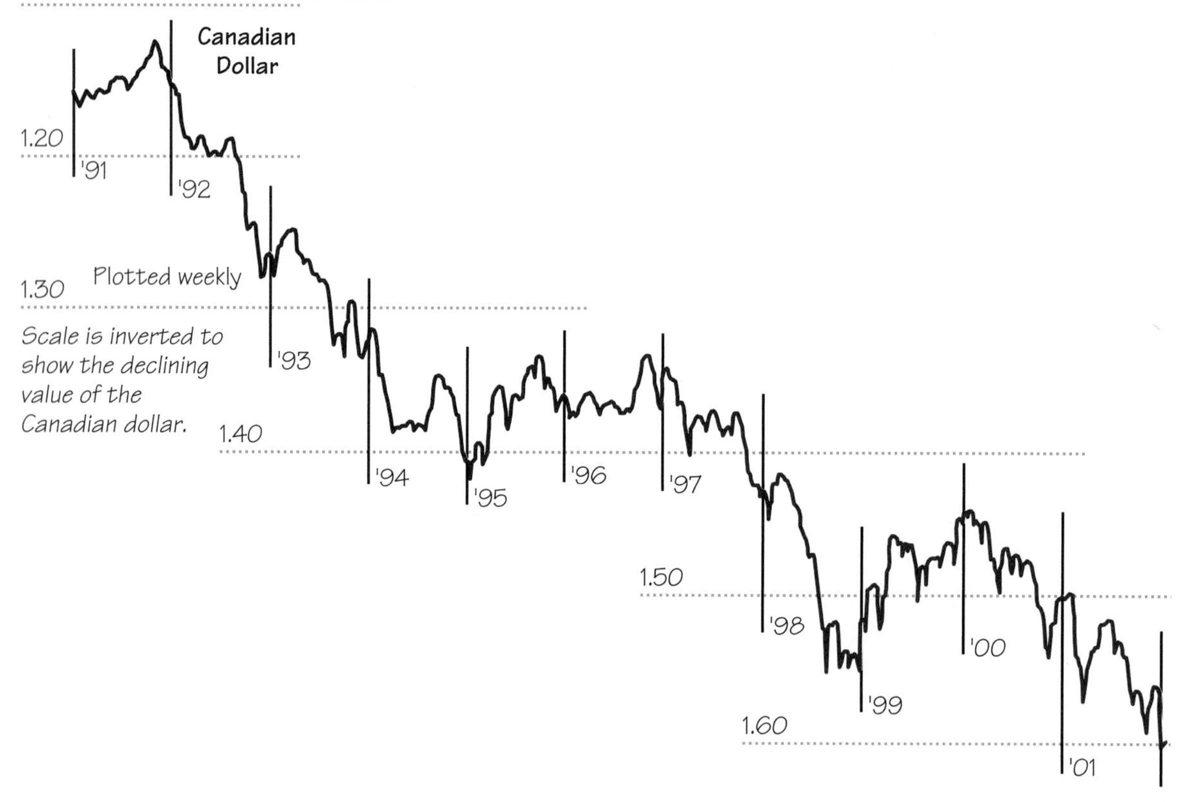

Source: Bloomberg Financial Markets

YESTERDAY–1.5988 CANADIAN DOLLARS

The New York Times, January 9, 2002, p. W1.

- Option B, sometimes called the "popular method," divides by the old trading value instead of the new value (here getting 67% as the percentage of value lost).

When the new European Union currency, the euro (denoted by the symbol €), was first introduced in January 1999 (for international transactions), one euro was worth US$1.17. By January 2002 (when euro coins and bills were first available to the public), the ratio was almost exactly reversed, with €1.16 = US$1.

(a) Calculate the results for how much the euro lost value against the U.S. dollar, using both options.
(b) The Canadian dollar went from being worth US$1.042 to being worth US$0.623. By what percentage did the value of the euro decline from January 1999 to January 2002? With which of your answers to part (a) does this number agree?

Another example, one of more sudden decline, is the unit of currency in South Africa, the rand. It traded at 9 to the U.S. dollar in mid-October 2001 but reached a record low of 13.84 to the dollar on December 21, 2001.
(c) Calculate the loss in value of the rand against the dollar, using both options.
(d) The rand went from being worth US$0.111 to being worth US$0.072. By what percentage did the value of the rand decline? With which of your answers to part (c) does this number agree?

For either of the options for calculating lost value, we can take the opposite point of view and calculate how much the U.S. dollar has "lost" against the Canadian dollar.
(e) What results do you get for each option?
(f) Compare your result for Option A with the result for loss of value of the Canadian dollar against the U.S. dollar. Why don't the numbers agree—if the Canadian dollar loses a certain percentage against the U.S. dollar, shouldn't the U.S. dollar gain the same percentage against the Canadian dollar?
For Options A and B:
(g) Can either percentage be more than 100%? Would it make sense to speak of a currency depreciating more than 100%?
(h) Is the Option A percentage always higher than the Option B percentage? Always lower? Neither?
(i) Which option would you expect a person to use who wants to make a decline seem large?

Finally, critique the display for decline of the Canadian dollar.

More on the Web SPREADSHEET PROJECTS

To do these projects, go to www.whfreeman.com/fapp.

Issues of scale play a significant role in the development of industrial sites. This project examines the impact of scale in the design of buildings.

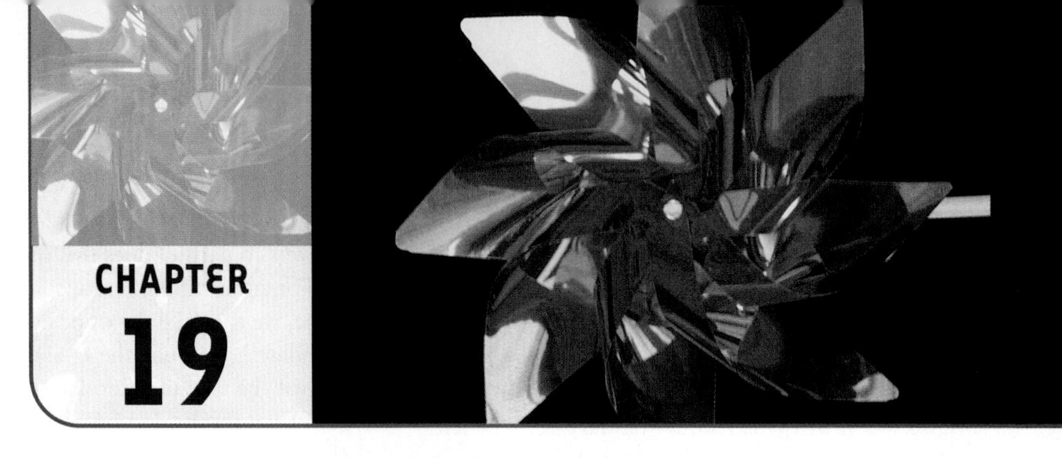

Symmetry and Patterns

"The senses delight in things duly proportional." So said the famous philosopher-theologian Thomas Aquinas more than 700 years ago, in noting human aesthetic appreciation. In this chapter we examine some of the elements of that aesthetic appreciation, particularly what we call *symmetry.*

What is symmetry and what does mathematics have to do with it? Symmetry, like beauty, is very difficult to define. Dictionary definitions talk about "correspondence of form on opposite sides of a dividing line or plane or about a center or an axis," "correspondence, equivalence, or identity among constituents of an entity," and "beauty as a result of balance or harmonious arrangement" (*American Heritage Dictionary,* 3rd ed.).

In the narrowest sense, symmetry refers to mirror-image correspondence between parts of an object. Crystals, in both their appearance and their atomic structure, provide examples of symmetry in this sense. Taken in a wider sense, though, symmetry includes notions of *balance, similarity,* and *repetition.*

It is our sense of symmetry that leads us to appreciate patterns. *Mathematics is the study of patterns,* and we will see here that mathematics gives important insights into symmetry.

Spirals

Patterns abound in nature. The successive sections of the beautiful chambered nautilus grow according to a very strict and specific spiral pattern, a broader kind of symmetry. This spiral has the property that it has the same shape at any size: A photographic enlargement superimposed on it would fit exactly.

Botanists have long appreciated other spirals. In plant growth from a central stem, the shoots, leaves, and seeds often occur in a spiral pattern known as **phyllotaxis.** For instance, the scales of a pineapple or a pinecone are arranged in spirals (Figure 19.1), as are the seeds of a sunflower (Figure 19.2a) and the petals on a daisy. Like the chambers of the nautilus in Figure 19.2b, the spirals on these

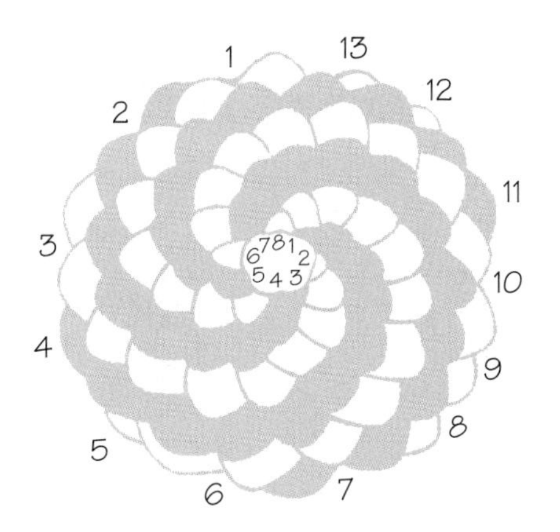

Figure 19.1 Spirals of scales on a pinecone: 8 right, 13 left. (From Verner E. Hoggatt, Jr., Fibonacci and Lucas, *Numbers,* Houghton Mifflin, New York, 1969, p. 81.)

Figure 19.2 (a) This sunflower has 55 spirals in one direction and 89 spirals in the other direction. (Harvey Lloyd/ The Stock Market.) (b) A chambered nautilus shell. (James Randkler/Tony Stone Images.)

plants are geometrically similar to one another, and they are arranged in a regular way, with balance and "proportion." These plants have a kind of symmetry we would naturally call **rotational.**

19.1 Fibonacci Numbers

Associated with the geometric symmetry of phyllotaxis, there is also a kind of *numeric symmetry,* with a "proportion" in the sense of a ratio of numbers. Strangely, the number of spirals in plants with phyllotaxis is not just any whole number but always comes from a particular sequence of numbers, called the *Fibonacci numbers* (see Spotlight 19.1).

SPOTLIGHT 19.1 Leonardo Pisano Bigollo ("Fibonacci")

Born in Pisa in 1170, Leonardo Pisano Bigollo has been known as "Fibonacci" for the past century and a half. This nickname, which refers to his descent from an ancestor named Bonaccio, is modern, and there is no evidence that he was known by it in his own time.

Leonardo was the greatest mathematician of the Middle Ages. His stated purpose in his book *Liber abbaci* (1202) was to introduce calculation with Hindu-Arabic numerals into Italy, to replace the Roman numerals then in use. Other books of his treated topics in geometry, algebra, and number theory.

We know little of Leonardo's life apart from a short autobiographical sketch in the *Liber abbaci:*

> I joined my father after his assignment by his homeland Pisa as an officer in the customhouse located at Bugia [Algeria] for the Pisan merchants who were often there. He had me marvelously instructed in the Arabic-Hindu numerals and calculation. I enjoyed so much the instruction that I later continued to study mathematics while on business trips to Egypt, Syria, Greece, Sicily, and Provence and there enjoyed discussions and disputations with the scholars of those places. [L. E. Sigler, *Leonardo Pisano Fibonacci, The Book of Squares: An Annotated Translation into Modern English,* Academic Press, New York, 1987, p. xvi.]

Leonardo Pisano ("Fibonacci")
A portrait of unlikely authenticity. (From Columbia University, D. E. Smith Collection.)

The *Liber abbaci* contains a famous problem about rabbits, whose solution is now called the Fibonacci sequence. Leonardo did not write further about it.

Fibonacci numbers occur in the sequence

$$1, 1, 2, 3, 5, 8, 13, 21, 34, 55, 89, 144, 233, 377, \ldots$$

This sequence begins with the numbers 1 and 1 again, and each next number is obtained by adding the two preceding numbers together.

Sometimes a sequence of numbers is specified by stating the value of the first term or first several terms and then giving an equation to calculate succeeding terms from preceding ones. This is called a *recursive rule,* and the sequence is said to be defined by **recursion.** Let's denote the *n*th Fibonacci number by F_n; then the Fibonacci sequence can be defined by

$$F_1 = 1, F_2 = 1, \quad \text{and} \quad F_{n+1} = F_n + F_{n-1} \quad \text{for } n \geq 2$$

The recursive rule just expresses in algebraic form that the next Fibonacci number is the sum of the previous two.

Look at the sunflower in Figure 19.2a. You see a set of spirals running in the counterclockwise direction and another set in the clockwise direction. It is (just barely) possible to count the number of spirals in both directions; in the sunflower there are 55 in one and 89 in the other direction – two consecutive Fibonacci numbers. In the case of the pineapple, there are three sets of spirals, one each along the three directions through each hexagonally shaped scale. For the common grocery pineapple (*Ananas comosus*), there are always 8 spirals to the right, 13 to the left, and 21 vertically – again, consecutive Fibonacci numbers.

Why are the numbers of spirals in plants the same numbers that appear next to each other in a purely mathematical sequence? The question has been the subject of extensive research, and there is no easy answer; there are several intricate theories about the dynamics involved in the plant's growth.

19.2 The Golden Ratio

During the last several centuries, an attractive myth arose that the ancient Greeks considered a specific numerical proportion essential to beauty and symmetry. Known variously in modern times as the *golden ratio,* **golden mean,** or even **divine proportion,** this proportion was investigated by Euclid in Book II of his *Elements.* Recent research reveals little evidence connecting this proportion to Greek aesthetics, but we pursue the golden ratio briefly because of its intimate connection to the Fibonacci sequence and because it does have appeal as a standard for beautiful proportion.

The value of the **golden ratio,** which is usually denoted by the Greek letter phi (ϕ), is

$$\phi = \frac{1 + \sqrt{5}}{2} = 1.618034\ldots$$

The basic aesthetic claim is that a **golden rectangle** – one whose height and width are in the ratio of 1 to ϕ – is the most pleasing of all rectangles. The Greeks treated lengths geometrically, so for them it was important to construct lengths using straightedge and compass; in Spotlight 19.2 we show how to construct a golden rectangle that is 1 unit by ϕ units.

SPOTLIGHT 19.2 How the Greeks Constructed a Golden Rectangle

In constructing a golden rectangle, the Greeks started from a one-by-one square (shown in black in the figure), which they made by constructing perpendiculars at the two ends of a horizontal segment of unit length. To extend the square to a golden rectangle, they bisected the original segment to get a new point that divides it into two pieces of length one-half each. Using this new point and a compass opening equal to the distance from it to a far corner of the square, they could cut off an interval of length ϕ.

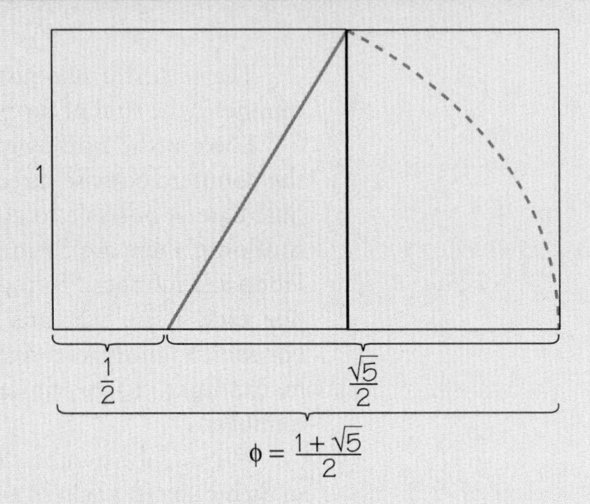

What would make anyone think that this is such an attractive ratio? And where did it come from? The answer lies not in Fibonacci numbers but in the Greeks' pursuit of balance in their study of geometry.

Given two line segments of different lengths, one way to find another length that "strikes a balance" between the two is to average them. For lengths l (the larger) and w (the smaller), their *arithmetic mean* (average) is $m = (l + w)/2$, and it satisfies

$$l - m = m - w$$

The length m strikes a balance between l and w, in terms of a common difference from the two original lengths. More generally, the arithmetic mean of n numbers or lengths is their sum divided by n.

The Greeks, however, preferred a balance in terms of ratios rather than differences. They sought a length s, the *geometric mean,* that gives a common ratio

$$l \div s = s \div w \qquad \text{or} \qquad \frac{l}{s} = \frac{s}{w}$$

Hence $lw = s^2$, which expresses the geometric fact that s is the side of a square, the area of which is the same as the area of a rectangle that is l by w (the Greeks thought in terms of geometrical objects). In geometry, the geometric mean s is called the *mean proportional* between l and w (see Figure 19.3).

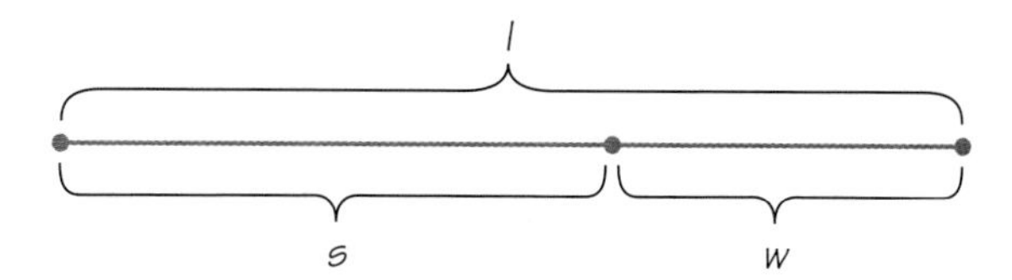

Figure 19.3 The line segment of length l is divided so that the length of s is the geometric mean between l and $w = l - s$; the dividing point divides the length l in the golden ratio.

> The quantity $s = \sqrt{lw}$ is the **geometric mean** of l and w. More generally, the geometric mean of n numbers is the nth root of the product of all n factors: the geometric mean of $x_1, \ldots, x_n$ is $\sqrt[n]{x_1 \cdots x_1}$. For example, the geometric mean of 1, 2, 3, and 4 is $\sqrt[4]{1 \times 2 \times 3 \times 4} = \sqrt[4]{24} = 24^{1/4} \approx 2.213$.

The ancient Greeks found symmetry and proportion in the geometric mean, but the geometric mean also has important practical applications (see Spotlight 19.3).

The Greeks were interested in the problem of cutting a single line segment of length l into lengths s and w, where $l = w + s$, so that s would be the mean proportional between w and l. Surprisingly, the ratio ϕ arises, as we show. Denote the common ratio

$$\frac{l}{s} = \frac{s}{w}$$

by x. Substituting $l = s + w$, we get

$$x = \frac{l}{s} = \frac{s + w}{s} = \frac{s}{s} + \frac{w}{s} = 1 + \frac{w}{s}$$

But w/s is just $1/x$, so we have

$$x = 1 + \frac{1}{x}$$

Multiplying through by x gives

$$x^2 = x + 1 \quad \text{or} \quad x^2 - x - 1 = 0$$

This is a quadratic equation of the form

$$ax^2 + bx + c = 0$$

SPOTLIGHT 19.3 A New Consumer Price Index: An Application of the Geometric Mean

In 1999, the Bureau of Labor Statistics (BLS) began using the geometric mean—instead of the arithmetic mean (average)—in calculating the consumer price index (CPI), which tracks changes in the cost of the goods and services that people buy.

Using the geometric mean is intended to take into account substitutions that consumers make when prices change. For example, if the price of beef goes up but the price of chicken doesn't, consumers may buy less beef and substitute chicken (because it is cheaper) for some beef.

Suppose that, overall, U.S. families consume equal dollar values of beef and chicken. A typical family might consume weekly 5 lb of beef at \$4/lb and 10 lb of chicken at \$2/lb, for \$20 each and a total cost of \$40. We say that beef and chicken each have a relative *market share* of 0.5 (50% beef, 50% chicken, by dollar value).

What if beef goes up to \$6/lb but chicken stays at \$2/lb? The *relative price change* in beef is \$6/\$4 = 1.5 and the relative price change in chicken is \$2/\$2 = 1.00 (i.e., no change). If the average family continues to eat just as much beef and chicken as before, the cost is now \$50, an increase of 25%. Since \$30 goes for beef and \$20 for chicken, the relative market shares (0.6 and 0.4) have changed. The *relative price change* for the family's meat is \$50/\$40 = 1.25, which is just the arithmetic mean (average) of the two relative price changes (1.50 and 1.00). A more general formulation is:

Relative price change

$$= (\text{old market share of beef})\,\frac{\text{new cost of beef}}{\text{old cost of beef}} + (\text{old market share of chicken}) \times \frac{\text{new cost of chicken}}{\text{old cost of chicken}}$$

$$= 0.5 \times \frac{6.00}{4.00} + 0.5 \times \frac{2.00}{2.00}$$

$$= \frac{1.50 + 1.00}{2} = 1.25$$

A family that eats no beef sees no increase; a family that eats only beef sees an increase of 50%. The CPI is an average over *all* families, weighted by the dollar value that each consumes.

If instead we use the geometric mean, we get a relative price change of $\sqrt{1.50 \times 1.00} = 1.225$.

With $a = 1$, $b = -1$, and $c = -1$. We apply the famous quadratic formula,

$$x = \frac{-b \pm \sqrt{b^2 - 4ac}}{2a}$$

to get the two solutions

$$x = \frac{1 + \sqrt{5}}{2} = 1.618034\ldots \qquad \text{and} \qquad \frac{1 - \sqrt{5}}{2} = -0.618034\ldots$$

We discard the negative solution since it does not correspond to a length. The first solution is the golden ratio ϕ. It occurs often in other contexts in geometry,

The more general formulation is

Relative price change

$$= \left(\frac{\text{new cost of beef}}{\text{old cost of beef}}\right)^{\text{(old market share of beef)}}$$

$$\times \left(\frac{\text{new cost of chicken}}{\text{old cost of chicken}}\right)^{\text{(old market share of chicken)}}$$

$$= \left(\frac{6.00}{4.00}\right)^{0.5} \times \left(\frac{2.00}{2.00}\right)^{0.5}$$

$$= \sqrt{1.50 \times 1.00} = 1.225$$

This relative price change, which is interpreted as a 22.5% increase in the cost of living, is slightly less than the 25% using the arithmetic mean.

The intention of the CPI is to measure the cost of living, that is, the change in the cost of goods and services that still yield the same level of satisfaction to consumers. The use of the arithmetic mean presumes that a family will go on buying the same weight of beef and chicken (5 lb beef; 10 lb chicken) as before; the use of the geometric mean presumes that a family will respond to changes in prices by buying the same *relative dollar value* of each meat as before. Using the geometric mean, the presumption is that, on average, families will purchase 1.225 times as much dollar value of meat as before, or $40 × 1.225 = $49. Of this, $24.50 (12.25 lb) would be chicken and the same dollar value $24.50 (4.08 lb) would be beef, so the relative market shares do not change. This substitution, which involves buying 0.92 lb less beef and substituting for it 2.25 lb of chicken, is supposed to yield the "same satisfaction" as the $20 (5 lb) of beef and $20 (10 lb) of chicken earlier, at a 22.5% increase in cost.

Because the geometric mean is always less than or equal to the arithmetic mean (see Exercise 11), the effect of using the geometric mean is a smaller CPI and a lower figure for inflation, perhaps 0.25% less per year than using the arithmetic mean would produce. Because Social Security payments, some wage increases, and income tax rates are automatically geared to the CPI, the result will be lower increases in Social Security payments and in wages, plus higher taxes for many taxpayers. The net effect on the federal budget will be an increase in revenues and a decrease in expenditures.

for example, ϕ is the ratio of a diagonal to a side of a regular pentagon (see Figure 19.4).

Thanks to recent work of Roger Herz-Fischler (Wilfrid Laurier University) and George Markowsky (University of Maine), we know that the term *golden ratio* was not used in antiquity and that there is no evidence that the Great Pyramid was designed to conform to ϕ, that the Greeks used ϕ in the proportions of the Parthenon, or that Leonardo da Vinci used ϕ in proportions for the human figure (Figure 19.5a). Moreover, experiments show that people's preferences for dimensions of rectangles cover a wide range, with golden rectangles not holding any special place. The impressionists Gustave Caillebotte (1848–1894) and Georges Seurat (1859–1891) may have used the golden ratio to design some of their paintings, but we do not have any historical evidence that they claimed or

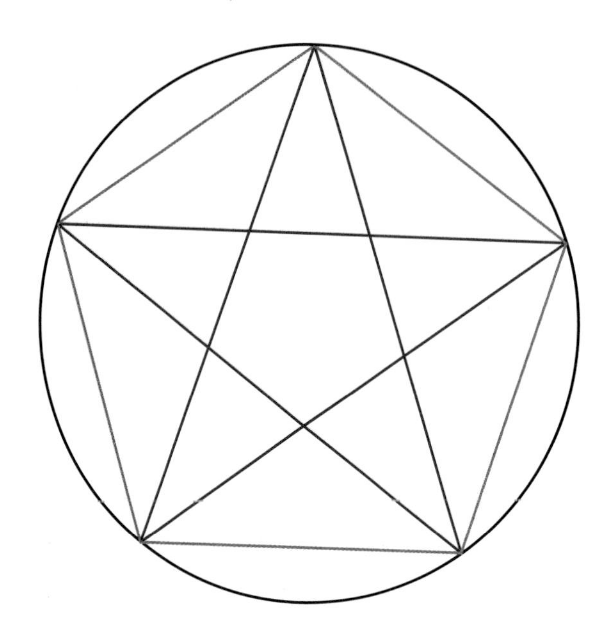

Figure 19.4 In a pentagon with equal sides, ϕ is the ratio of a diagonal to a side. The five-pointed star formed by the diagonals was the symbol of the followers of the ancient Greek mathematician Pythagoras.

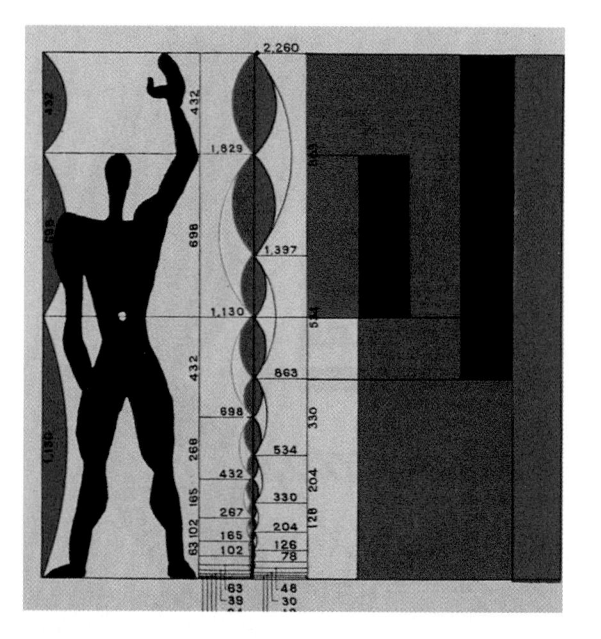

Figure 19.5 (a) There is no evidence that Leonardo da Vinci used ϕ in his drawings of the human figure. (Accademia, Venice, Italy/Scala/Art Resource, NY.) (b) Le Corbusier, however, did use ϕ in his "Modular" scale of proportions. (Le Corbusier, "Le Modulor," 1945. © 2000 Artists Rights Society [ARS], New York/ADAGP, Paris/FLC.)

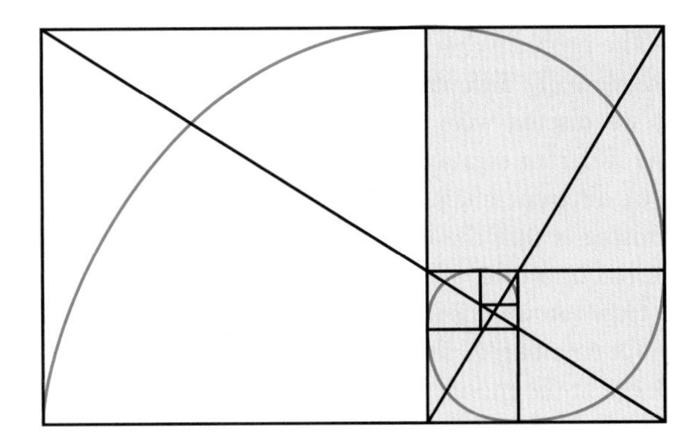

Figure 19.6 A logarithmic spiral determines a sequence of golden rectangles and corresponding squares.

intended to do so. Recent research suggests that Wolfgang Amadeus Mozart (1756–1791), who was fascinated by mathematics as a student, may have constructed two of the three sections of the first movement of his piano sonatas (development and recapitulation) with an eye to the golden ratio. But we do not have evidence that this was his intention.

It is true that human bodies exhibit ratios close to the golden ratio, as you can see by comparing your overall height to the height of your navel. The Swiss-born architect Le Corbusier (Charles-Edouard Jeanneret [1887–1965]) used the golden ratio (including a navel-height feature) as the basis for his "Modulor" scale of proportions (Figure 19.5b).

There are intriguing connections between the spiral of the nautilus and the spirals of the sunflower and between the golden ratio and the Fibonacci sequence. The nautilus shape follows what is known as an *equiangular* or *logarithmic* spiral, which in its turning determines a sequence of golden rectangles (Figure 19.6). The spirals of the sunflower are in fact approximations to a logarithmic spiral. The mathematical reason for this connection is that the ratios of consecutive Fibonacci numbers

$$\begin{array}{ccccccc} \frac{1}{1} & \frac{2}{1} & \frac{3}{2} & \frac{5}{3} & \frac{8}{5} & \frac{13}{8} & \frac{21}{13}\cdots \\ 1.0 & 2.0 & 1.5 & 1.666\ldots & 1.6 & 1.625 & 1.615\ldots \end{array}$$

provide alternately under- and overapproximations to $\phi = 1.618034\ldots$

19.3 Balance in Symmetry

The spiral distribution of the seeds in a sunflower head and the spiraling of leaves around a plant stem are instances of *similarity* and *repetition,* two key aspects of symmetry; they also illustrate *balance,* which refers to regularity in *how* the repetitions are arranged. In considering patterns with repetition, we distinguish the individual element or figure of the design (sometimes called the *motif*) from the *pattern* of the design—*how the copies of the motif are arranged.*

The problem that we will work on for the rest of this chapter is to classify the fundamentally different ways in which a flat design can be symmetric. The ideas that we discuss were used by chemists and crystallographers to discover how many different crystalline forms are possible. Although there is a limitless number of different chemicals, and of motifs that people can make, what is quite surprising is that there are only a limited number of ways in which atoms of a chemical or motifs of a design can be arranged in a symmetrical way.

How can we possibly enumerate the ways in which designs can be put together without counting all the actual designs themselves? The key mathematical idea is to look not at the motifs that make up the patterns but at what you can *do* to the pattern without changing its appearance. This is what we pursue in the next section.

19.4 Rigid Motions

Mathematicians describe a variety of kinds of symmetry by using the geometric notion of a *rigid motion,* also known as an **isometry** (which means "same size"). A rigid motion is a specific kind of variation on the original pattern: We pick it up

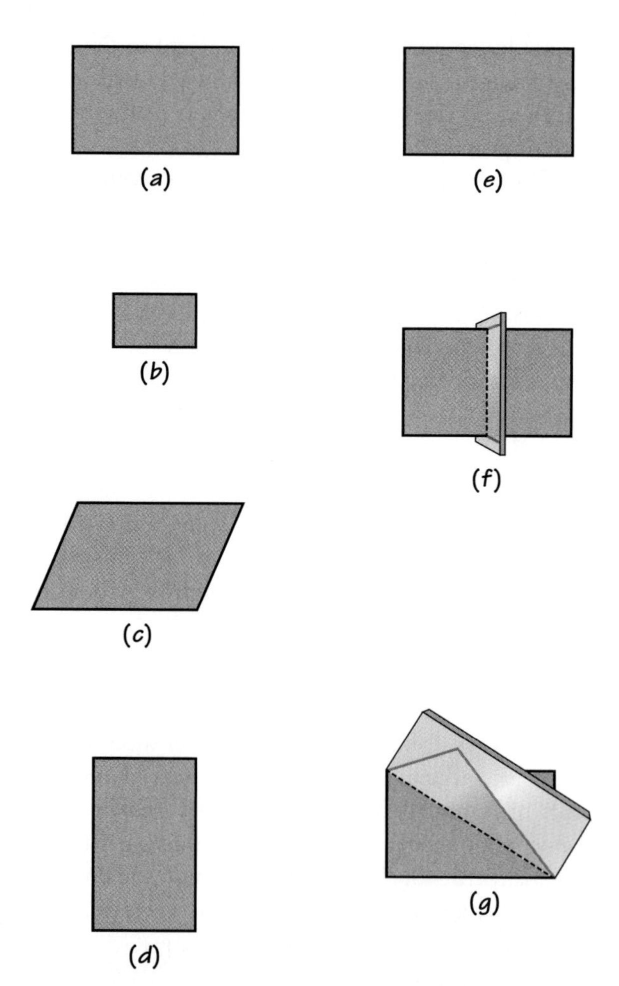

Figure 19.7 Results of various motions applied to a rectangle: (a) the original rectangle; (b) 50% reduction (not a rigid motion); (c) sagging (not a rigid motion); (d) quarter-turn; (e) half-turn; (f) reflection along the vertical line down the middle; (g) reflection along a diagonal line.

and move it, perhaps rotate it, possibly flip it over—but we *don't change its size or shape.* (To connect this concept with the language of Chapter 18, the original figure and its image are not just geometrically similar but congruent—the same size.)

Figure 19.7 shows the results of various motions applied to the rectangle in Figure 19.7a. In Figure 19.7b, each side is shrunk by 50%—not a rigid motion, because the size of the rectangle changes. For Figure 19.7c, we imagine that the rectangle has rigid sides but hinges at the corner; like an unbraced bookshelf, it sags—again, this is not a rigid motion because the shape of the rectangle changes. In Figure 19.7d we rotate the rectangle 90° (a quarter-turn) clockwise around the center of the rectangle: This is a rigid motion. Similarly, in Figure 19.7e, rotating by 180° (a half-turn) is a rigid motion.

In Figure 19.7f we reflected the rectangle along a vertical mirror down the middle: Could you tell? The right and left halves have exchanged places.

Figure 19.7g shows the result of reflecting across a diagonal of the rectangle. All reflections and all rotations are rigid motions. So are all **translations,** which move every point in the plane a certain distance in the same direction.

The only remaining kind of rigid motion in the plane is a hybrid of reflection and translation. Known as a **glide reflection,** it is the kind of pattern that your footprints make as you walk along: each successive element of the design (footprint) is a reflection of the previous one (Figure 19.8). The motion combines, in an integral way, translation ("glide") with a reflection across a line that is parallel to the direction of the translation.

A **rigid motion** is one that preserves the size and shape of figures; in particular, any pair of points is the same distance apart after the motion as before. Any rigid motion of the plane must be one of:

- Reflection (across a line)
- Rotation (around a point)
- Translation (in a particular direction)
- Glide reflection (across a line)

Performing one rigid motion after another results in a rigid motion that (surprisingly) must be one of the four types that we have just explored.

19.5 Preserving the Pattern

In terms of symmetry, we are especially interested in rigid motions like those of Figures 19.7e and 19.7f that **preserve the pattern**—that is, ones for which the pattern looks exactly the same, *with all the parts appearing in the same places,* after the motion is applied.

You might enjoy thinking of applying these motions as a game, "The Pattern Game": You turn your back, I apply a transformation, then you turn back and see if you can tell whether anything is changed.

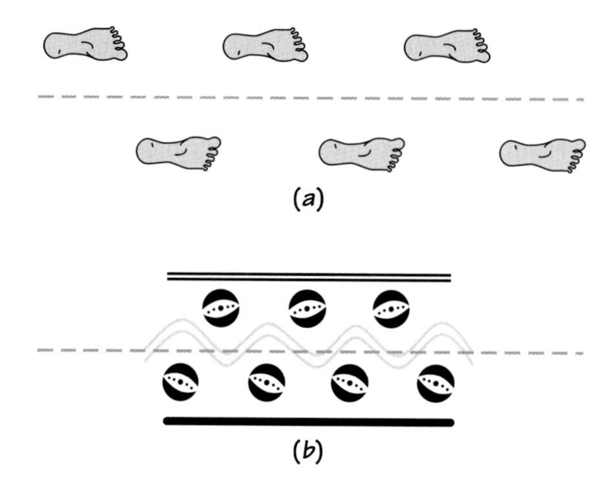

Figure 19.8 Glide reflection of (a) footprints; (b) design elements on a pot from San Ildefonso Pueblo.

The 90° rotation of Figure 19.7a into Figure 19.7d does not preserve the pattern. The moved rectangle doesn't fit exactly over the original rectangle. On the other hand, the 180° rotation in Figure 19.7e does preserve the pattern. It's true that the top of the original rectangle is now on the bottom of the transformed version, but you can't tell that has happened because you can't distinguish the two. If you had turned your back while the motion was applied, you wouldn't be able to tell that anything had been done. A rotation by any multiple of 180° would also preserve the pattern.

Similarly, the reflection across the vertical line in Figure 19.7f preserves the pattern, while the one in Figure 19.7g, where the reflection line is along a diagonal, does not. Spotlight 19.4 discusses possible biological consequences of reflection symmetry or imperfections in it.

The pattern of footsteps in Figure 19.8a is not preserved under reflection along the direction of walking—there is not a left footprint directly across from a right footprint. The pattern is preserved under a glide reflection along the direction of walking, as well as by a translation of two steps, or one of four steps, and so on—but not by a translation of one step.

19.6 Analyzing Patterns

Given a pattern, we analyze it by determining which rigid motions preserve the pattern. These are often referred to as the **symmetries of the pattern.** We then can classify the pattern by which rigid motions preserve it.

We may think of a pattern as a recipe for repeating a figure (motif) indefinitely. Of course, any pattern we see in nature or art has only finitely many copies of the figure; but if the recipe for repetition is clear, we may imagine that we are looking at just a part of a pattern that extends indefinitely.

Patterns in the plane can be divided into those that have indefinitely many repetitions in

SPOTLIGHT 19.4 "Strive Then to Be Perfect"

Stand in front of a mirror and look at yourself. Are your left and right sides exactly symmetrical? Not on the inside, of course, but externally? What about the part in your hair, freckles on your face, evenness of your shoulders, outward bending of your ears?

It turns out that animals – including people – have a preference for bilateral symmetry, according to recent studies. That preference is for as perfect a symmetry as possible, so that even slight deviations from exact symmetry can make an individual less desirable to others.

In some cases, symmetry may signal fitness. For example, the more symmetrical that a flower is, the more nectar it produces, making it a better food source for pollinating insects. And the insects do prefer symmetrical flowers, thereby giving them a better chance of being pollinated than less perfect ones.

(Venus de Milo: Paris, France, photograph by Erich Lessing/Art Resource, NY.)

Symmetry, as a proxy for fitness, may affect mate selection by animals. Fruit flies and female barn swallows prefer males with symmetrical tails; a particular parasite can lead to an uneven tail. Symmetrical racehorses may run faster; male lions with lopsided facial whisker-spot patterns die younger. Female zebra finches prefer males with symmetrical leg bands.

What about people? Computer-generated symmetrical female faces, generated from composites of individual photos, appear more attractive to men than photos of actual women. Studies also indicate that "symmetrical" men tend to have an earlier first sexual experience, more sexual partners, and more extramarital affairs. Women desiring fertile partners may want to consider a study that indicates that low sperm counts and poor sperm motility are associated with lack of symmetry of the hands. However, to what extent such symmetry in people signals individual or genetic fitness is yet to be determined.

- no direction – the **rosette patterns**
- exactly one direction (and its reverse) – the **strip patterns**
- more than one direction – the **wallpaper patterns**

A rosette pattern describes the possible symmetries for a flower. There is just one flower in the pattern; the repetition aspect of symmetry consists of the repetition of the petals around the stem. Translations and glide reflections do

(a) (b)

Figure 19.9 (a) Flower with petals with reflection symmetry. (Harvey Lloyd/The Stock Market.) (b) Pinwheel. (Travis Amos.)

not come into play. The pattern is preserved under a rotation by certain angles, corresponding to the number of petals. There may or may not be reflections that preserve it, depending on whether the petal itself has reflection symmetry. Most flowers do (Figure 19.9a), but some do not. An everyday example of the rosette pattern – a human-made one – that does not have reflection symmetry is a pinwheel (Figure 19.9b). If there is no reflection symmetry, the motif of the pattern (the element that is repeated) is an entire petal; if there is reflection symmetry, the motif is just half a petal, because the entire pattern can be generated by rotation and reflection of a half petal. The fact that these are the only possibilities is sometimes called *Leonardo's theorem,* after Leonardo da Vinci, who, in the course of planning the design of churches, needed to decide if chapels and niches could be added without destroying the symmetry of the central design.

Leonardo realized that there were two different classes of rosettes, the ones without reflection symmetry (*cyclic rosettes*) and the ones with reflection symmetry (*dihedral rosettes*) (see Figure 19.9). The respective notations for the patterns are cn and dn, where n is the number of times that the rosette coincides with its original position in one complete turn around the center. The rosette coincides with itself for every rotation of $360°/n$. A cyclic pattern has no lines of reflection symmetry, while the dihedral pattern dn has n different lines of reflection symmetry. The flower in Figure 19.9 has a dihedral pattern, because each petal has reflection symmetry. The pinwheel in Figure 19.9 has pattern $c8$.

19.7 Strip Patterns

We illustrate the different kinds of strip patterns, and their "ingredient" symmetries, with patterns in the art of the Bakuba people of the Democratic Republic of the Congo, who are noted for their fascination with pattern and symmetry (see Spotlight 19.5).

SPOTLIGHT 19.5 Patterns Created by the Bakuba People

Among the Bakuba people of the Democratic Republic of the Congo (shaded area of map), it is considered an achievement to invent a new pattern, and every Bakuba king had to create a new pattern at the outset of his reign. The pattern was displayed on the king's drum throughout his reign and, for some kings, on his dynastic statue.

When missionaries first showed a motorcycle to a Bakuba king in the 1920s, he showed little interest in it. But the king was so enthralled by the novel pattern the tire tracks made in the sand that he had it copied and gave it his name.

Source: Adapted from Jan Vansina, *The Children of Woot*, University of Wisconsin Press, Madison, 1978, p. 221.

Two women with raffia cloths from the Bakuba village of Mbelo, July 1985; Mpidi Muya with embroidered raffia cloth (left) and Muema Kenye with plush and embroidered raffia cloth (right). (Dorothy K. Washburn.)

The pattern made by tire tracks fascinated the Bakuba people. (Travis Amos.)

All the strip patterns offer repetition and **translation symmetry** along the direction of the strip. For simplicity, we will always position the pattern so that its repetition runs horizontally.

It may be that the pattern has no other rigid motions that preserve it apart from translation, as in Figure 19.10a.

The simplest other rigid motion to check for preservation of the pattern is reflection across a line. For a strip pattern, the center line of the strip may be a reflection line; if so, as in Figure 19.10b, we say that the pattern has symmetry across a horizontal line. There may instead be reflection across a *vertical* axis, such as the vertical lines through or between the V's in Figure 19.10c.

What kind of rotational symmetry can a strip pattern have? The only possibility for a strip pattern is a rotation by 180° (a half-turn), because any other angle won't even bring the strip back into itself. (We don't count rotations of 360° or integer multiples [full turns], because any pattern is preserved under these.) Figure 19.10d shows a strip pattern that is unchanged by a 180° rotation about any point at the center of the small crosshatched regions.

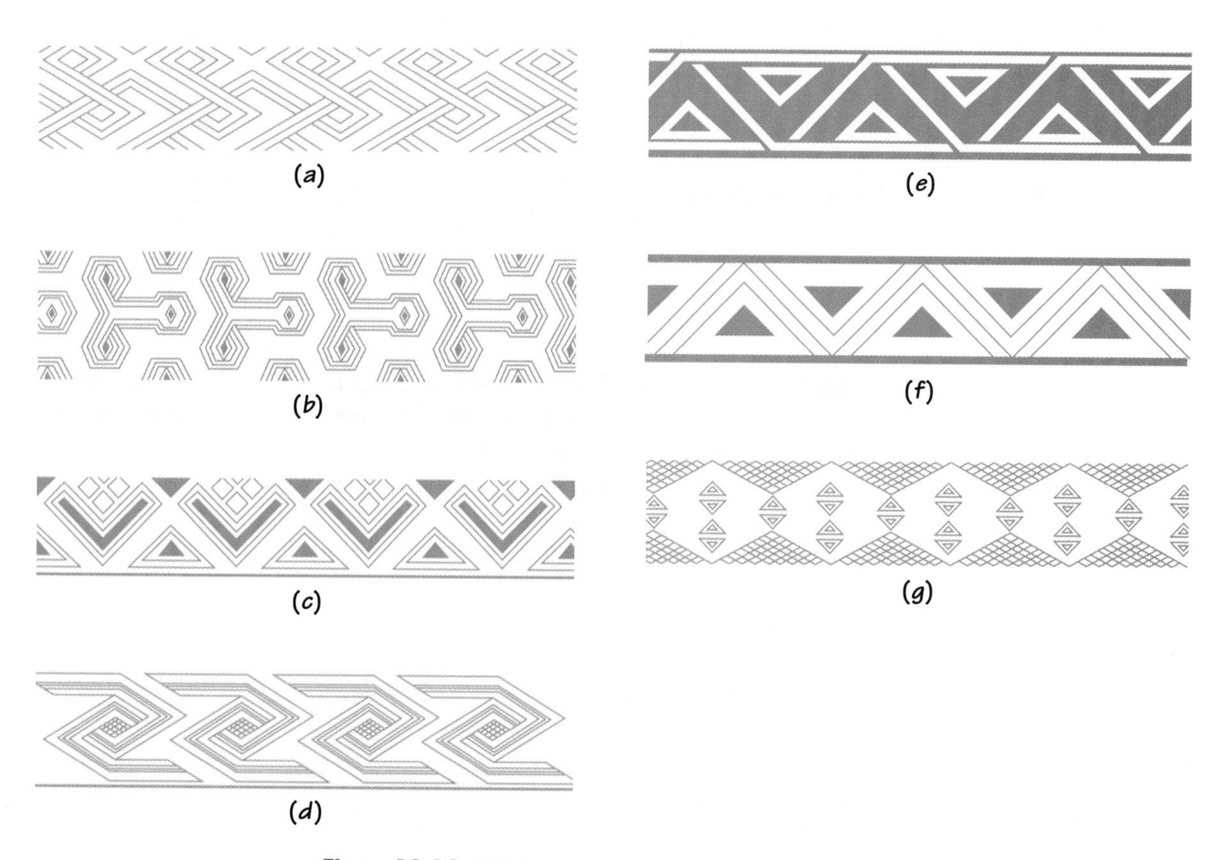

Figure 19.10 Bakuba patterns. (a) Carved stool; (b) pile cloth; (c) pile cloth; (d) embroidered cloth; (e) embroidered cloth; (f) carved back of wooden mask; (g) carved box.

What about glide reflections? A row of alternating p's and b's has glide reflection:

Glide	p	p	p	p	p	p	p	p	p
Reflection	p	p	p	p	p	p	p	p	p
	b	b	b	b	b	b	b	b	b
Glide reflection	p	b	p	b	p	b	p	b	p

For glide reflection, a p is translated as far as the next b and is then reflected upside down. Figure 19.10e shows a Bakuba pattern whose only symmetry (except for translation) is glide reflection.

Having examined symmetries on strip patterns, we can ask: What *combinations* of the four are possible? It turns out that apart from the five kinds of patterns we have already seen, there are only two other possibilities: We can have vertical line reflection, half-turns, and glide reflection, either with (Figure 19.10f) or without horizontal line reflection (Figure 19.10g).

Mathematical analysis reveals:

> There are only seven ways to repeat a pattern along a strip.

That this number is so small is quite surprising, because there are myriad different design elements (motifs). The key idea is that two designs may look entirely different yet share the same pattern of reproducing their design elements.

19.8 Symmetry Groups

We mentioned earlier that the key mathematical idea about detecting and analyzing symmetry is to look not at the motifs of a pattern but at its symmetries, the transformations that preserve the pattern.

The symmetries of a pattern have some notable properties:

- If we combine two symmetries by applying first one and then the other, we get another symmetry.
- There is an identity, or "null," symmetry that doesn't move anything, but leaves every point of the pattern exactly where it is.
- Each symmetry has an inverse or "opposite" that undoes it and also preserves the pattern. A rotation is undone by an equal rotation in the opposite direction, a reflection is its own inverse, and a translation or glide reflection is undone by another of the same distance in the opposite direction.
- In applying a number of symmetries one after the other, we may combine consecutive ones without affecting the result.

These properties are common to many kinds of mathematical objects; they characterize what mathematicians call a *group*. Various collections of numbers and numerical operations that are already familiar to you are groups.

EXAMPLE

A Group of Numbers

The positive real numbers form a group under multiplication:

- Multiplying two positive real numbers yields another positive real number.
- The positive real number 1 is an identity element.
- Any positive real number x has an inverse $1/x$ in the collection.
- In multiplying several numbers together, it doesn't matter if we first multiply together some adjacent pairs of numbers, that is, it doesn't matter how we group or parenthesize the multiplication. For instance, $2 \times 3 \times 4 \times 5$ is equal to $2 \times (3 \times 4) \times 5 = 2 \times 12 \times 5$ and also to $(2 \times 3) \times 4 \times 5 = 6 \times 4 \times 5$. ■

A **group** is a collection of elements $\{A, B, \ldots\}$ and an operation $\circ$ between pairs of them such that the following properties hold:

closure: The result of one element operating on another is itself an element of the collection ($A \circ B$ is in the collection).

identity element: There is a special element I, called the identity element, such that the result of an operation involving the identity and any element is that same element ($I \circ A = A$ and $A \circ I = A$).

inverses: For any element, A, there is another element, called its inverse and denoted A^{-1}, such that the result of an operation involving an element and its inverse is the identity element ($A \circ A^{-1} = I$ and $A^{-1} \circ A = I$).

associativity: The result of several consecutive operations is the same regardless of grouping or parenthesizing, provided the consecutive order of operations is maintained: $A \circ B \circ C = A \circ (B \circ C) = (A \circ B)C$.

The group of symmetries that preserve a pattern is called the **symmetry group of the pattern.**

EXAMPLE

The Symmetry Group of a Rectangle

Consider the rectangle of Figure 19.11. Its symmetries, the rigid motions that bring it back to coincide with itself (even as they interchange the labeled corners), are as follows:

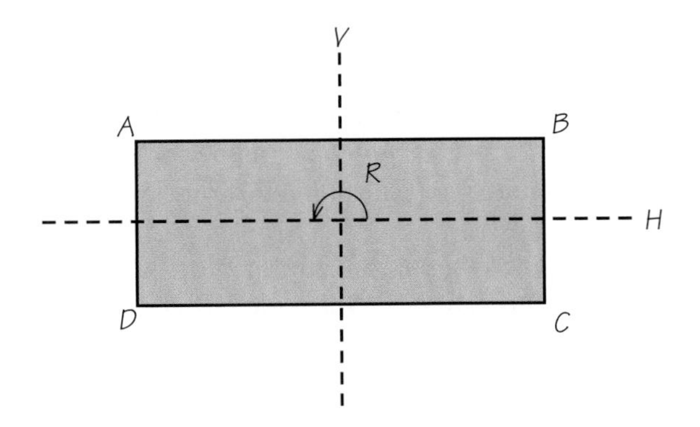

Figure 19.11 A rectangle, with reflection symmetries and 180° rotation symmetry marked.

- The identity symmetry I, which leaves every point where it is
- A 180° (half-turn) rotation R around its center
- A reflection V in the vertical line through its center
- A reflection H in the horizontal line through its center

You should convince yourself that these four elements form a group. Combining any pair by applying first one and then the other is equivalent to one of the others; for example, applying first V and then H is the same as applying R, that is, $H \circ V = R$ (check this by following where the corner A goes to under the operations). The element I is an identity element. Each element is its own inverse. Try some examples to verify that associativity holds. For instance, $R \circ H \circ V = (R \circ H) \circ V = R \circ (H \circ V)$; in other words, applying V followed by H followed by R, we get the same result if we combine the first two and then apply the third, or if we combine the second two and apply the first followed by that combination. ■

EXAMPLE

Symmetry Groups of Strip Patterns

Each of the strip patterns of Figure 19.10 is distinguished by a different group of symmetries. The pattern of Figure 19.10a is preserved only by translations. If we let T denote the smallest translation to the right that preserves the pattern, then the pattern is also preserved by $T \circ T$ (which we write as T^2), by $T \circ T \circ T = T^3$, and so forth. Although the pattern looks the same after each of these translations by different distances, we can tell these translations apart if we number each copy of the motif and observe which other motif it is carried into under the symmetry. For instance, T^2 takes each motif into the motif two to the right. The symmetry T has an inverse T^{-1} among the symmetries of the pattern: the smallest translation to the *left* that preserves the pattern; and $T^{-1} \circ T^{-1}$ (which we write as T^{-2}), $T^{-1} \circ T^{-1} \circ T^{-1} = T^{-3}$, and so forth also are symmetries. The entire collection of symmetries of the pattern is

$$\{\ldots, T^{-3}, T^{-2}, T^{-1}, I, T, T^2, T^3, \ldots\}$$

From this listing, you see that it is natural to think of the identity I as being T^0. All the strip patterns are preserved by translations, so the symmetry group of each includes the *subgroup* of all translations in this list. We say that the group is generated by T, and we write the group as $<T>$, where between the angle brackets we list symmetries (generators) that, in combination, produce all of the group elements.

The symmetry group of Figure 19.10e includes in addition a glide reflection G and all combinations of the glide reflection with the translations. Doing two glide reflections is equivalent to doing a translation, which we express as $G^2 = T$; the glide is only "half as far" as the shortest translation that preserves the pattern. Check that $G \circ T = T \circ G$. The symmetry group of the pattern is

$$\{\ldots, G^{-3}, G^{-2} = T^{-1}, G^{-1}, I, G, G^2 = T, G^3, \ldots\} = <G>$$

The pattern of Figure 19.10c is preserved by vertical reflections at regular intervals. If we let V denote reflections at a particular location, the other reflections can be obtained as combinations of V and T. The symmetry group of the pattern is

$$\{\ldots, T^{-3}, T^{-2}, T^{-1}, I, T, T^2, T^3, \ldots; \\ \ldots, T^{-3}V, T^{-2}V, T^{-1}V, V, TV, T^2V, T^3V, \ldots\}$$

This group is notable because not all its elements satisfy the *commutative property* that $A \circ B = B \circ A$, which you are used to for numerical operations ($a + b = b + a$, $a \times b = b \times a$). In fact, we do not have $VT = TV$, but instead $VT = T^{-1}V$, a fact that you should verify by labeling one of the V shapes in the pattern and observing where it is carried by each of these three combinations of symmetries. We can express the group as $<T, V \mid VT = T^{-1}V>$, where we list the generators and indicate relations among them. ■

We have made a transition from thinking about patterns in geometrical terms to reasoning about them in algebraic notation—in effect, applying one branch of mathematics to another. This kind of cross-fertilization is characteristic of contemporary mathematics.

The concept of a group is a fundamental one in the mathematical field of abstract algebra. The generality ("abstractness") is exactly why groups and other algebraic structures arise in so many applications, in areas ranging from crystallography, quantum physics, and cryptography, to error-correcting codes (see Chapters 9 and 10), anthropology (describing kinship systems; see Ascher [1991])—and analyzing symmetries of patterns.

19.9 Notation for Patterns

It's useful to have a standard notation for patterns, for purposes of communications. Crystallographers' notation is the one more commonly used. For the strip patterns, it consists of four symbols (an example is *pma2*):

1. The first symbol is always a *p*, which indicates that the pattern repeats (is "periodic") in the horizontal direction.
2. The second symbol is *m* if there is a vertical line of reflection; otherwise, it is *1*.
3. The third symbol is
   - *m* (for "mirror") if there is a horizontal line of reflection (in which case there is also glide reflection)
   - *a* (for "alternating") if there is a glide reflection but no horizontal reflection
   - *1* if there is no horizontal reflection or glide reflection
4. The fourth symbol is *2* if there is half-turn rotational symmetry; otherwise, it is *1*.

A *1* always means that the pattern does not have the symmetry corresponding to that position. In the notation

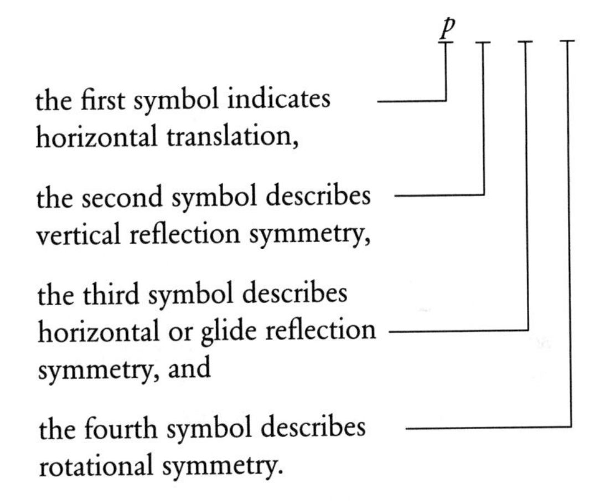

p _ _ _
the first symbol indicates horizontal translation,
the second symbol describes vertical reflection symmetry,
the third symbol describes horizontal or glide reflection symmetry, and
the fourth symbol describes rotational symmetry.

Figure 19.12 gives a flowchart for identifying the seven ways patterns repeat, together with the notations for them.

EXAMPLE

Bakuba Patterns

We use the flowchart of Figure 19.12 to analyze some of the Bakuba patterns of Figure 19.10.

Figure 19.10a does not have a vertical reflection, so we branch right, and the pattern notation begins to take shape as *p1_ _*. The figure does not have a horizontal reflection, nor a glide reflection, so we branch right again, filling in the third position in the notation, to get *p11_*. A half-turn preserves part but not all of the pattern, so we conclude that we have a *p111* pattern.

Figure 19.10b does not have vertical reflection, so we branch right, to *p1_ _*. The figure does have horizontal reflection, so we branch left and left, concluding that the pattern is *p1m1*.

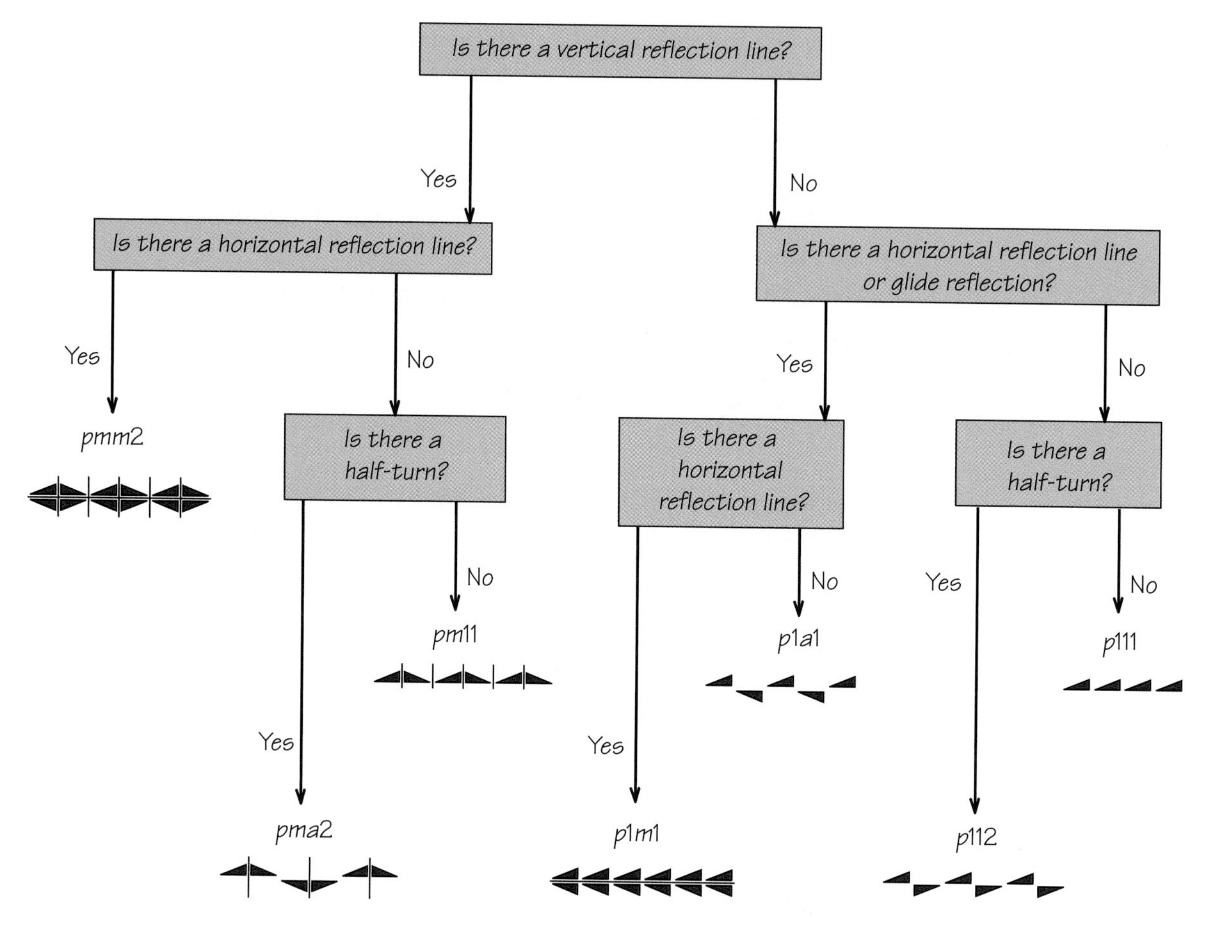

Figure 19.12 Flowchart for the seven strip patterns.

Figure 19.10f has vertical reflection, so we branch left, to *pm*_ _. The figure does not have horizontal reflection, so we branch right but cannot yet fill in the third symbol. The figure does have a half-turn symmetry (and glide symmetry), with center on the middle of the three lines between any pair of closest triangles. So the pattern is *pma2*. ■

19.10 Imperfect Patterns

In applying these classification schemes to patterns on real objects, we need to take into account that the pattern itself may not be perfectly rendered. Also, patterns that are not on flat surfaces—for example, the pattern around the rim of a bowl or around the body of a jar—require some latitude in interpretation.

EXAMPLE Patterns on Pueblo Pottery

The pitchers in Figure 19.13 are from a thousand-year-old Pueblo site at Starkweather Ruin near Reserve, New Mexico. Consider the patterns on the bodies of

Figure 19.13 Reserve black-on-white pitchers from the Pueblo II horizon (A.D. 900–1100), excavated 1935–1936 from Starkweather Ruin by Professor Paul H. Nesbitt and students from Beloit College. (Courtesy of Logan Museum of Anthropology, Beloit College, photos by Paul J. Campbell.)

the pitchers, which continue on the back sides. Let's suppose that they could be unwrapped and continued as strip patterns, but we'll disregard the patterns on the spouts and handles.

We immediately come up against the question of the perfectness of the patterns. In Figure 19.13a the "teeth" in the left design element on the main body are "sharper" than those on the right. (The "teeth" represent the zigzagging of lightning bolts.) Is this lack of pattern, or just lack of perfection in executing one? For our analysis, we opt for the latter.

Similarly, what are we to make of the diagonal lines on the pitcher in Figure 19.13b? In the narrowest interpretation, these lines are part of the pattern and any rigid motion that is to qualify as a symmetry of the pattern must preserve them. More liberally, we may consider the lines as a kind of shading, a way to make the region appear gray; indeed, to an observer at a distance, that is the effect of the lines.

For the pattern on the body of the pitcher in Figure 19.13c, we notice that the jagged white line in the design element on the left has three "steps," while that in the one on the right has four. If we were really strict, we would decide that the two are different design elements. But we do detect a similarity of the two that we do not want to deny totally; we attribute the variations in the jagged lines to artistic license and for our purposes consider the two jagged lines to be the same. ■

We follow the flowchart in Figure 19.12 and get the following:

- Figure 19.13a: Is there a vertical reflection? *No*. Is there a horizontal reflection or glide reflection? *No*. Is there a half-turn? *No*. Hence the pattern is *p111*.

- Figure 19.13b (narrow interpretation of the diagonal lines): Is there a vertical reflection? *No*. Is there a horizontal reflection or glide reflection? *No*. Is there a half-turn? *Yes* (e.g., around the center of each cross). The pattern is *p112*.
- Figure 19.13b (liberal interpretation – diagonal lines as shading, their direction doesn't have to be preserved): Is there a vertical reflection? *Yes* (e.g., on a vertical line through the center of a cross). Is there a horizontal reflection? *Yes* (e.g., through the center of a cross). The pattern is *pmm2*.
- Figure 19.13c: Is there a vertical reflection? *No*. Is there a horizontal reflection or glide reflection? *No*. Is there a half-turn? *Yes* (e.g., around the center of each jagged white line). The pattern is *p112*. (This pitcher has the interesting feature that the patterns on the neck and the body are mirror images of each other.)

Women made the pots at Starkweather. They strongly preferred the symmetry of half-turns; very few of the pots have any reflection symmetry, either reflection or glide. The avoidance of reflection symmetry was a consistent feature of pottery of the indigenous peoples of the Western Hemisphere.

19.11 Symmetry from Chaos

We think of "symmetry" as referring to order and "chaos" as referring to disorder and randomness. Mathematicians and other scientists use the word *chaos* to describe systems whose behavior over time is inherently unpredictable; we explore chaotic systems in Chapter 23. Yet there is a surprising connection between symmetry and chaos, revealed in Michael Field and Martin Golubitsky's stunning book *Symmetry in Chaos: A Search for Pattern in Mathematics, Art and Nature.*

A typical computer produces graphics on the screen with a fixed number of pixels (lighted points); a typical common screen dimension is 640 by 480 pixels, giving a total of 307,200 pixels. We can identify a particular pixel by its Cartesian coordinates, letting the pixel in the lower left of the screen be the origin, with coordinates (0, 0).

One way to produce a graphic on the screen is to start with an initial point on the screen, apply a mathematical function (formula) to its coordinates to generate the coordinates of a new point, light up the pixel covered by that point, then repeat the process with each successive point.

Iterate the process a large number of times – millions or even hundreds of millions of times. Because there are only about 300,000 pixels, some pixel or pixels – by what mathematicians call the "pigeonhole principle" – must be visited more than once. Some pixels may recur hundreds or even thousands of times.

The clue to producing art from this process is "color by number": Choose the color for each pixel according to how many times it is visited, and choose the coloring with an eye to beauty and illustrating the underlying symmetry. Figure 19.14 shows an example with D_5 symmetry that was produced by 30,000,000 iterations of a function. The scale on the right of the figure shows the color

Figure 19.14 "Emperor's Cloak," with D_5 symmetry. This work of art was produced by iterating a chaos-producing function, starting at one point and successively generating new points according to a fixed rule. The color bar shows the coloring of pixels according to how often they are visited by the iterations. (Figure 1.13, p. 20, of Michael Field and Martin Golubitsky, *Symmetry in Chaos: A Search for Pattern in Mathematics, Art and Nature,* New York, Oxford University Press, 1992.)

assigned to a pixel according to the number of times that the pixel was hit; unhit pixels are left black. Note that we don't care in what order the points are visited—that aspect appears to be completely chaotic. What matters is just how often they are visited.

Functions can be chosen to give any rosette, strip, or wallpaper pattern. The functions can be written out in Cartesian coordinates. Let the old point be (x, y) and the new point (x', y'). Then a figure with D_3 symmetry can be created using the function

$$x' = [\lambda + \alpha(x^2 + y^2) + \beta(x^3 - 3xy^2]x + \gamma(x^2 - y^2)$$
$$y' = [\lambda + \alpha(x^2 + y^2) + \beta(x^3 - 3xy^2]y - \gamma(2xy)$$

where the Greek letters α, β, and γ are specific numbers chosen for the pattern. The functions for x' and y' are really just polynomials in x and y, because they involve only sums of products of the form $x^i y^j$ for integers i and j. But, as you can imagine, the functions get increasingly intricate for patterns with symmetry D_n for larger n.

However, with one other reinterpretation of the pixels, in terms of a number system that you are familiar with, the functions become much simpler. Recall from algebra that some equations, such as $x^2 = -1$, have solutions but the solutions are not real numbers; in fact, we denote the solutions to $x^2 = -1$ by i and $-i$, where i is the unit imaginary number. Numbers that consist of a real part plus an imaginary part, such as $3 + 4i$, are called complex numbers; in algebra you delved into how to add and multiply them. We can graph such points in the plane by taking the horizontal axis as a scale for real numbers and the vertical axis as a scale for imaginary numbers, so that $3 + 4i$ would plot at the coordinates (3, 4).

Imagine the computer screen as a depiction of the complex plane and interpret the coordinates (x, y) of a pixel as the complex number $z = x + iy$. We need just two other concepts from complex numbers. For a complex number z,

- its complex conjugate, denoted $\bar{z}$, is $x - iy$
- its real part, denoted $\text{Re}(z)$, is just x

Using these notations, we can write the general function that Field and Golubitsky use to generate D_n patterns as

$$z' = [\lambda + \alpha z\bar{z} + \beta \text{Re}(z^n)]\, z + \gamma \bar{z}^{n-1}$$

where λ, α, β, and γ are the same real numbers as in the earlier equation for patterns with D_3 symmetry.

There are two more features of these patterns:

- If you ignore the first thousand or so points generated, it doesn't matter what point you start from – you get the same pattern, though the points themselves are visited in completely different orders.
- The functions used are variations on the *logistic map* discussed in Chapter 23, which can give rise to mathematical chaos. In particular, the pattern-generating functions are examples of *iterated function systems*, which are investigated in detail in Chapter 23.

REVIEW VOCABULARY

Divine proportion Another term for the golden ratio.
Fibonacci numbers The numbers in the sequence 1, 1, 2, 3, 5, 8, 13, 21, 34, . . . (each number after the second is obtained by adding the two preceding numbers).
Geometric mean The geometric mean of two numbers a and b is $\sqrt{ab}$.
Glide reflection A combination of translation (= glide) and reflection in a line parallel to the translation direction. Example: pbpbpb.
Golden ratio, golden mean The number $\phi = \frac{1 + \sqrt{5}}{2} = 1.618034. . . .$
Golden rectangle A rectangle the lengths of whose sides are in the golden ratio.
Group A group is a collection of elements with an operation on pairs of them such that the collection is closed under the operation, there is an identity for the operation, each element has an inverse, and the operation is associative.
Isometry Another word for rigid motion. Angles and distances, and consequently shape and size, remain unchanged by a rigid motion. (For plane figures there are only four possible isometries: reflection, rotation, translation, and glide reflection.)
Phyllotaxis The spiral pattern of shoots, leaves, or seeds around the stem of a plant.
Preserves the pattern A transformation preserves a pattern if all parts of the pattern look exactly the same after the transformation has been performed.
Recursion A method of defining a sequence of numbers, in which the next number is given in terms of previous ones.
Rigid motion A motion that preserves the size and shape of figures; in particular, any pair of points is the same distance apart after the motion as before.
Rosette pattern A pattern whose only symmetries are rotations about a single point and reflections through that point.
Rotational symmetry A figure has rotational symmetry if a rotation about its "center" leaves it looking the same, like the letter: S.
Strip pattern A pattern that has indefinitely many repetitions in one direction.
Symmetry of the pattern A transformation of a pattern is a symmetry of the pattern if it preserves the pattern.
Symmetry group of the pattern The group of symmetries that preserve the pattern.
Translation A rigid motion that moves everything a certain distance in one direction.
Translation symmetry An infinite figure has translation symmetry if it can be translated (slid, without turning) along itself without appearing to have changed. Example: AAA
Wallpaper pattern A pattern in the plane that has indefinitely many repetitions in more than one direction.

SUGGESTED READINGS

ASCHER, MARCIA. *Ethnomathematics: A Multicultural View of Mathematical Ideas,* Brooks/Cole, Pacific Grove, Calif., 1991. Chapter 3, "The Logic of Kin Relations" (pp. 66–83), shows that kinship systems have the structure of dihedral groups. Chapter 6, "Symmetric Strip Decorations" (pp. 154–183), investigates strip patterns in the Maori and Inca cultures.

BELCASTRO, SARAH-MARIE, and THOMAS C. HULL. Classifying frieze patterns without using groups, *College Mathematics Journal,* 33 (March 2002): 93–98. Elementary analysis of why there are only seven ways to repeat a pattern along a strip.

BOLES, MARTHA, and ROCHELLE NEWMAN. *The Golden Relationship: Art, Math & Nature, Book 1: Universal Patterns; Book 2: The Surface Plane,* Pythagorean Press, Bradford, Mass., 1992.

BROWNE, MALCOLM W. Can't decide if that centerfold is a perfect 10? Just do the math. *New York Times* (October 20, 1998) (national edition): D5. www.nytimes.com/library/national /science/ 102098sci-essay.html.

CRISLER, NANCY. *Symmetry & Patterns,* COMAP, Lexington, Mass., 1995. A learning module for high-school students.

CROWE, DONALD W. *Symmetry, Rigid Motions and Patterns,* High School Mathematics and Its Applications (HiMAP) Module 4, COMAP, Lexington, Mass., 1987. Reprinted in smaller format in *The UMAP Journal,* 8(3) (1987): 207–236. Instructional module on rigid motions of the plane, strip patterns, and wallpaper patterns, with worksheets.

DENT, KARIN M. Spirograph math, *Humanistic Mathematics Network Journal,* 19 (March 1999): 13–17.

DE SPINADEL, VERA W. A new family of irrational numbers with curious properties, *Humanistic Mathematics Network Journal,* 19 (March 1999): 33–37.

DUNLAP, RICHARD A. *The Golden Ratio and Fibonacci Numbers,* World Scientific, River Edge, NJ, 1997.

FIELD, MICHAEL, and MARTIN GOLUBITSKY. Symmetries on the edge of chaos, *New Scientist,* (January 9, 1993): 32–35.

FIELD, MICHAEL, and MARTIN GOLUBITSKY. *Symmetry in Chaos: A Search for Pattern in Mathematics, Art and Nature,* Oxford University Press, New York, 1992. Spectacular collection of computer-generated art illustrating how symmetry arises out of chaos, complete with programs in Microsoft QuickBasic.

GALLIAN, JOSEPH A. Finite plane symmetry groups, *Journal of Chemical Education* 67(7) (July 1990): 549–550. Hubcap examples.

GALLIAN, JOSEPH A. Symmetry in logos and hubcaps, *American Mathematical Monthly,* 97(3) (March 1990): 235–238.

HARGITTAI, ISTVÁN, and MAGDOLNA HARGITTAI. *Symmetry: A Unifying Concept,* Shelter Publications, Bolinas, Calif., 1994.

HERZ-FISCHLER, ROGER. *A Mathematical History of Division in Extreme and Mean Ratio,* Wilfrid Laurier University Press, Waterloo, Ontario, Canada, 1987. Reprint with a new preface, under the title *Mathematical History of the Golden Number,* Dover, New York, 1998.

HOGGATT, VERNER E., JR. *Fibonacci and Lucas Numbers,* Houghton Mifflin, New York, 1969.

HUNTLEY, H. E. *The Divine Proportion,* Dover, New York, 1970.

KOSHY, THOMAS. *Fibonacci and Lucas Numbers with Applications,* Wiley, New York, 2001.

LEE, KEVIN D. KaleidoMania!: Interactive Symmetry, Windows/Macintosh program, Key Curriculum Press, 1999. Lets the user construct rosette, strip, and wallpaper patterns.

MARKOWSKY, GEORGE. "Misconceptions about the golden ratio," *College Mathematics Journal,* 23(1) (January 1992): 2–19.

MARTIN, GEORGE E. *Transformation Geometry: An Introduction to Symmetry,* Springer-Verlag, New York, 1982.

MAY, MIKE. Did Mozart use the golden section?, *American Scientist* 84 (March–April 1996): 118–119. americanscientist.org/issues/Sciobs96/Sciobs96-03MM.html.

O'DAFFER, PHARES G., and STANLEY R. CLEMENS. *Geometry: An Investigative Approach,* Addison-Wesley, Reading, Mass., 1976, chap. 1–5. A gentle introduction to the geometry of symmetry, with lots of examples and illustrations. Chapter 4 gives an elementary proof that there are only four kinds of rigid motions in the plane.

PUTZ, JOHN F. The golden section and the piano sonatas of Mozart, *Mathematics Magazine* 68(4) (October 1995): 275–282.

RUNION, GARTH E. *The Golden Section and Related Curiosa,* Scott, Foresman, Glenview, Ill., 1972.

SIBLEY, THOMAS Q. *Geometric Patterns: A Study in Symmetry,* Saint John's University Press, Collegeville, Minn., 1989.

STEWART, IAN. Mathematical recreations: Daisy, Daisy, give me your answer, do, *Scientific American* 272(1) (January 1995): 96–99. Explains the occurrence of Fibonacci numbers in plants based on the dynamics of plant growth and efficient packing of seeds into spiral faces.

WALSER, HANS. *The Golden Section,* Mathematical Association of America, Washington, D.C., 2001.

WASHBURN, DOROTHY K., and DONALD W. CROWE. *Symmetries of Culture: Theory and Practice of Plane Pattern Analysis,* University of Washington Press, Seattle, 1988. An introduction to the mathematics of symmetry, splendidly illustrated with photographs of patterns from cultures all over the world. Includes a complete analysis of patterns with two colors. Appendices contain proofs of the facts that there are only four rigid motions in the plane and that there are exactly seven strip patterns.

SUGGESTED WEB SITES

www.geom.umn.edu/software/tilings Tessellation Resources. Lists programs for various platforms that allow the user to create designs featuring the rosette, strip, and wallpaper patterns.

www-sphys.unil.ch/escher/ Interactive Escher Web Sketch program that allows a user to design repeating patterns. Choose a wallpaper pattern using crystallographic notation and draw on the screen a colored design for the motif, and the program then reproduces the motif using the pattern. The software (for Windows, Macintosh, and Unix) can also be downloaded.

www.geom.umn.edu/java/Kali/ Interactive Java Kali Web program that lets the user draw pictures under the action of rosette, strip, or wallpaper groups. Versions for various platforms can be downloaded.

ftp://ftp.uni-bielefeld.de/pub/math/tiling/reptiles/ Interactive Rep Tiles Macintosh application for designing wallpaper patterns, plus systematically generating all possible periodic tilings of the plane by applying "topological transformations" and "symmetry."

www.wordsmith.org/~anu/java/spirograph.html#def Interactive Spirograph Java application (which you can download) that lets you do electronically what the Spirograph toy does.

SKILLS CHECK

1. What is the geometric mean of 4 and 36?

(a) 12
(b) 16
(c) 20

2. Which of the following rectangles is an approximate golden rectangle?

(a) 10 by 16
(b) 6 by 13
(c) 8 by 11

3. Which of the following letters has a rotation isometry?

(a) S **(b)** K **(c)** Y

4. Assume the following two patterns continue in both directions. Which of these patterns has a reflection isometry?

ZZZZZZZZZ
UUUUUUUUU

(a) ZZZZZZZZZ only
(b) UUUUUUUUU only
(c) neither

5. What isometries does this strip pattern have?

˩˥ ˩˥ ˩˥ ˩˥

(a) Translation and glide reflection only
(b) Translation and rotation only
(c) Translation, rotation, and glide reflection only

6. What isometries does this wallpaper pattern have?

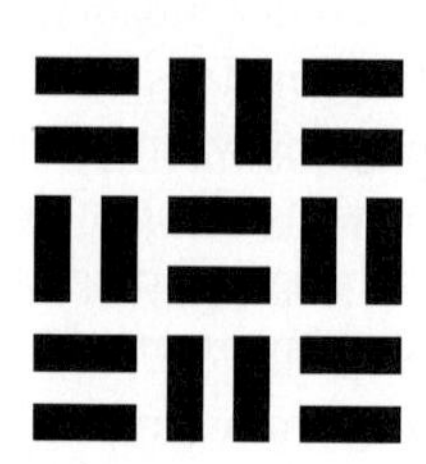

(a) Translation and reflection only
(b) Translation and rotation only
(c) Translation, rotation, and reflection

7. What isometries does this wallpaper pattern have?

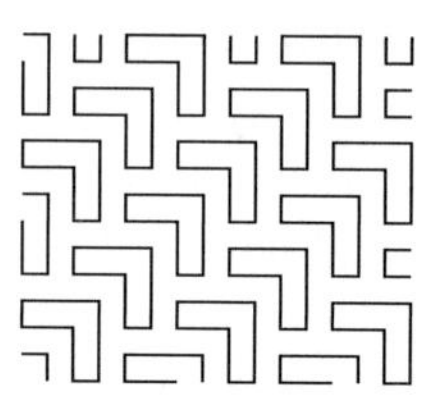

(a) Translation only
(b) Translation and reflection only
(c) Translation, rotation, and reflection

8. In the Fibonacci sequence, what number follows 13 and 21?

(a) 29
(b) 34
(c) 38

9. Recursion occurs in

(a) a sunflower blossom.
(b) a rosette pattern.
(c) the Fibonacci sequence.

10. The logarithmic spiral

(a) is used in da Vinci's sketches.
(b) determines a sequence of golden rectangles.
(c) forms a symmetry pattern.

11. A rigid motion always moves any pair of points

(a) in the same direction.
(b) to another pair of points the same distance apart.
(c) to their mirror images.

12. If a strip pattern has a glide reflection isometry, then

(a) it always has a horizontal reflection isometry.
(b) it may also have a horizontal reflection isometry.
(c) it cannot have a horizontal reflection isometry.

13. If a strip pattern has both vertical and horizontal reflection isometries, then

(a) it always has a half-turn rotation isometry.
(b) it may also have a half-turn rotation isometry.
(c) it cannot have a half-turn rotation isometry.

14. The symbol *p* indicates that the pattern

(a) has a rotational isometry.
(b) has a reflectional isometry.
(c) has another type of isometry.

15. The symbol *2* indicates that the pattern

(a) has a rotational isometry.
(b) has a reflectional isometry.
(c) has another type of isometry.

EXERCISES

▲ Optional. ■ Advanced. ◆ Discussion.

Fibonacci Numbers

1. Examine the "scales" on the surface of a pineapple, which are arranged in spirals (parastichies) around the fruit. Note that there are spirals in three distinct directions. For each direction, how many spirals are there?

2. Repeat Exercise 1, but for a pinecone from your area.

3. Repeat Exercise 1, but for a sunflower.

4. Here are two primitive models of natural increase of biological populations, similar to those Fibonacci hypothesized around the year 1200. A pair of newborn male and female rabbits is placed in an enclosure to breed.

(a) Suppose that the rabbits start to bear young one month after their own birth. This may be unrealistic for rabbits, but we could substitute another species for which it is realistic; Fibonacci used rabbits. At the end of each month, they have another male–female pair, which in turn mature and start to bear young one month later. Assuming that none of the rabbits die, how many pairs of rabbits will there be at the end of six months from the start (just before any births for that month)? (*Hint:* Draw a month-by-month chart of the situation at the end of the month, just before any births.)
(b) Repeat part (a), but assume instead that the rabbits start to bear young exactly two months after their own birth.

The Golden Ratio

5. Put the golden ratio $\phi = (1 + \sqrt{5})/2$ into the memory of your calculator.

(a) Look at the value of ϕ. Now square it (either use the $\boxed{x^2}$ button or multiply it by itself). What do you observe?
(b) Back to ϕ. Now take its reciprocal (either use the $\boxed{1/x}$ button or divide it into 1). What do you observe?
(c) What formula explains what you saw in part (a)?
(d) What formula explains what you saw in part (b)?

6. The golden ratio satisfies the equation $x^2 = x + 1$.

(a) Show that $(1 - \phi)$ also satisfies the equation.
(b) Use part (a) to show that $(1 - \phi) = (1 - \sqrt{5})/2$ is the other solution to $x^2 - x - 1 = 0$.

7. The geometric mean has interpretations in both arithmetic and geometry.

(a) Find the geometric mean of 3 and 27.
(b) Find the length of a side of a square that has the same area as a rectangle that is 4 by 64.

8. Here's further practice on arithmetic and geometric interpretations of the geometric mean:

(a) Find the geometric mean of 4 and 9.
(b) You are to make a golden rectangle with 6 inches of string. How wide should it be, and how high?

9. Another sequence closely related to the Fibonacci sequence is the Lucas sequence, which is formed using the same recursive rule but different starting numbers. The nth Lucas number L_n is given by

$$L_1 = 1, L_2 = 3, \text{ and } L_{n+1} = L_n + L_{n-1} \text{ for } n \geq 2$$

(a) Calculate L_3 through L_{10}.
(b) Calculate the ratio of successive terms of the Lucas sequence:

$$\frac{L_2}{L_1}, \frac{L_3}{L_2}, \ldots, \frac{L_{10}}{L_9}$$

What do you notice?

10. For a sequence specified by a recursive rule, finding an explicit expression for the nth term is not easy, nor is the form necessarily simple. An exact expression for the nth term of the Fibonacci sequence is given by the Binet formula:

$$F_n = \frac{1}{\sqrt{5}}\left(\frac{1+\sqrt{5}}{2}\right)^n - \frac{1}{\sqrt{5}}\left(\frac{1-\sqrt{5}}{2}\right)^n$$

(a) Verify the formula for $n = 1$ and $n = 2$ (by multiplying out, not by using a calculator).
(b) Use the Binet formula and your calculator to find F_5.
(c) In fact, the second term on the right of the equation gets closer and closer to 0 as n gets large. Since we know that the Fibonacci numbers are integers, we can just round off the result of calculating the first term. Find F_{13} by calculating the first term with your calculator and rounding.

11. For two positive numbers x and y, show that the arithmetic mean $(x + y)/2$ is always greater than or equal to the geometric mean $x^{1/2}y^{1/2} = \sqrt{xy}$. [*Hint:* Suppose that the claim is false, so that $(x + y)/2 < \sqrt{xy}$.) Square both sides of the inequality, bring all terms to one side, factor, and observe a contradiction.]

12. You may remember having to work problems like, "If Joe can dig a ditch in 3 days, and Sam can dig it in 4, how long will it take the two of them working together?" The answer is related to the *harmonic mean* of 3 and 4. The formula for the harmonic mean of two numbers x and y is

$$\frac{2}{1/x + 1/y}$$

(a) Calculate the answer for Joe and Sam, which is *one-half* of the harmonic mean of 3 and 4. Explain why this is the correct answer.
(b) Show that the harmonic mean of two positive numbers is always less than or equal to the geometric mean. (Thus, in light of Exercise 11, we have the general conclusion that $H \leq G \leq A$, where H stands for the harmonic mean, G for the geometric mean, and A for the arithmetic mean.) (*Hint:* Suppose that the claim is false. Simplify the fraction that is the harmonic mean, square both sides of the inequality, and proceed as in Exercise 11.)
(c) Show once more that the harmonic mean of two positive numbers is always less than the geometric mean, but this time do it with less work: let $A = 1/x$ and $B = 1/y$, and discover one connection (equation) between the harmonic mean of x and y and the arithmetic mean of A and B, and a second connection between the geometric mean of x and y and the geometric mean of A and B. Then use Exercise 11 on A and B.
(d) What should be the formula for the geometric mean of three numbers? Of n numbers?
(e) Proceed as in part (d), but for the harmonic mean.

13. Here is a trick to "prove" that you can calculate faster than a person with a calculator. Turn your back and ask a friend to write down any two positive integers, then add them to get a third, then add the second and third to get a fourth, etc., adding each time the last two until there are 10 numbers. Have your friend show you the list, whereupon you write down right away the total of all 10, while your friend begins to add them up on the calculator (to prove that you're right). The secret: the total is always 11 times the seventh number, and multiplying by 11 is pretty easy to do in your head (just add each pair of neighboring digits, carrying if necessary). Suppose that your friend writes down m and n as the first two numbers; show that indeed the total of all 10 numbers is 11 times the seventh number. (Adapted from Martin Gardner, *Mathematical Circus,* Knopf, New York, 1979.)

14. The game of Fibonacci Nim begins with n counters. Two players take turns removing at least one counter, but no more than twice as many as the opponent just did. The winner is the player who takes the last counter. One other rule: the first player may not win immediately by taking all the counters on the first turn! (Adapted from Martin Gardner, *Mathematical Circus,* Knopf, New York, 1979.)

(a) Play this game taking turns with an opponent and starting with different numbers n of counters and try to come up with a strategy for one player or the other to win. (*Hint:* The key is that any positive integer can be represented uniquely as a sum of Fibonacci numbers.)
(b) Proceed as in part (a), but with the rule changes that the player who takes the last counter loses and the first player may not take all but one counter.

Preserving the Pattern

15. Determine whether each of the following statements is always true or sometimes false. (Drawing some sketches may be helpful.)

(a) A line reflection preserves collinearity of points. That is, if the points A, B, and C are in a straight line (collinear), then their images reflected in some other line also lie in a straight line.
(b) A line reflection preserves betweenness. That is, if the collinear points A, B, and C (with B between A and C) are reflected about a line, then the image of B is between the images of A and C.
(c) The image of a line segment under a line reflection is a line segment of the same length.
(d) The image of an angle under a line reflection is an angle of the same measure.
(e) The image of a pair of parallel lines under a line reflection is a pair of parallel lines.

16. Determine whether each of the following statements is always true or sometimes false. (Drawing some sketches may be helpful.)

(a) The image of a pair of perpendicular lines under a line reflection is a pair of perpendicular lines.
(b) The image of a square under a line reflection is a square.
(c) Label the vertices of a square A, B, C, and D in a clockwise direction. Then their images A', B', C', and D' under a line reflection also follow a clockwise direction.
(d) The perimeter of a geometric figure is equal to the perimeter of its image under a line reflection.
(e) The image of a vertical line under a line reflection is always a vertical line.

17. Which of the capital letters of the alphabet have

(a) a horizontal line of reflection symmetry?
(b) a vertical line of reflection symmetry?
(c) rotational symmetry? (Assume that each letter is drawn in the most symmetric way. For example, the upper and lower loops of B should be the same size.)

18. Repeat Exercise 17 for the lowercase letters.

19. In *The Complete Walker III* (3rd ed., Knopf, New York, 1984, p. 505), Colin Fletcher's answer to "What games should I take on a backpacking trip?" is the game he calls "Colinvert": "You strive to find words with meaningful mirror (or half-turn) images." Some of the words he found are

MOM WOW pod MUd bUM

(a) Which of his words reflect into themselves?
(b) Which of his words rotate into themselves?
(c) Find some more words or phrases of these various types – the longer, the better.

20. Repeat Exercise 19, but for words written vertically instead of horizontally.

Analyzing Patterns

21. Give the notation (e.g., $d4$ or $c5$) for the symmetry patterns of the rosettes in hubcaps (a) through (c), disregarding the logos in the centers.

(Can you identify the make of car and year for each hubcap?)

22. Repeat Exercise 21 for hubcaps (d) through (f).

23. Repeat Exercise 21 for corporate logos (a) through (c). (Can you identify the corporations?)

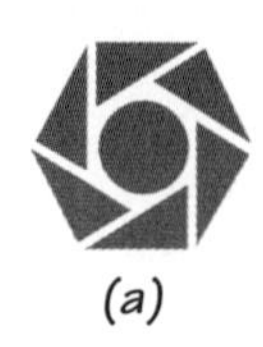
(a)

(b)

(c)

24. Repeat Exercise 21 for corporate logos (d) through (f).

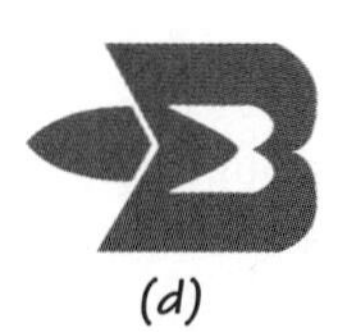
(d)

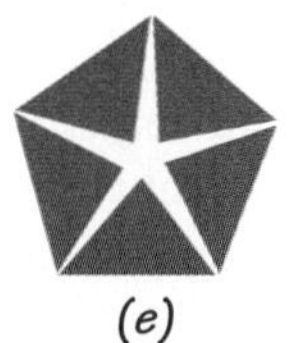
(e)

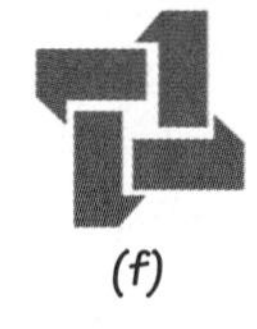
(f)

25. For each of the shapes in parts (a) through (e) of the accompanying figure, determine all lines of symmetry.

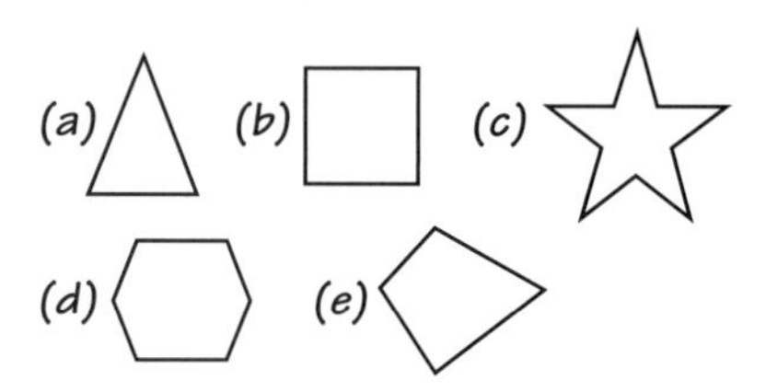

26. Repeat Exercise 25, but for the shapes in parts (f) through (j).

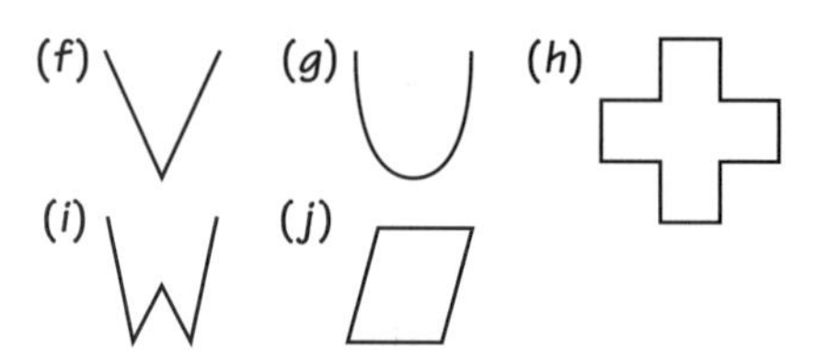

Strip Patterns

27. For each of the following strip patterns, identify the rigid motions that preserve the pattern:

(a) AAAAAAAAAA **(c)** 0000000000
(b) BBBBBBBBBB **(d)** FFFFFFFFFF

28. Repeat Exercise 27, but for

(a) NNNNNNNNNN **(c)** dbpqdbpqdbpq
(b) bdbdbdbdbd

Symmetry Groups

29. What is the group of symmetries of

(a) an equilateral triangle (all three sides equal)?
(b) an isosceles triangle (two equal sides) that is not equilateral?
(c) a scalene triangle (no pair of sides equal)?

30. What is the group of symmetries of a square?

31. Explain, by referring to the properties of a group, whether the collection of all real numbers is a group under the operation of **(a)** addition; **(b)** multiplication.

32. (a) Give a numerical example to show that the operation of subtraction on the integers is not associative.

(b) Repeat part (a), but for division on the positive real numbers.

33. What are the elements of the group of symmetries of (a) Figure 19.10b? (b) Figure 19.10f?

34. What are the elements of the group of symmetries of (a) Figure 19.10d? (b) Figure 19.10g?

35. What are the elements of the group of symmetries of the dihedral pattern *d8* (see the flower in Figure 19.9)?

36. What is the group of symmetries of the cyclic pattern *c8*?

Notation for Patterns

37. Use the flowchart in Figure 19.12 to identify (by International Crystallographic Union notation) the types of the strip patterns from Hungarian needlework, shown in the illustration.

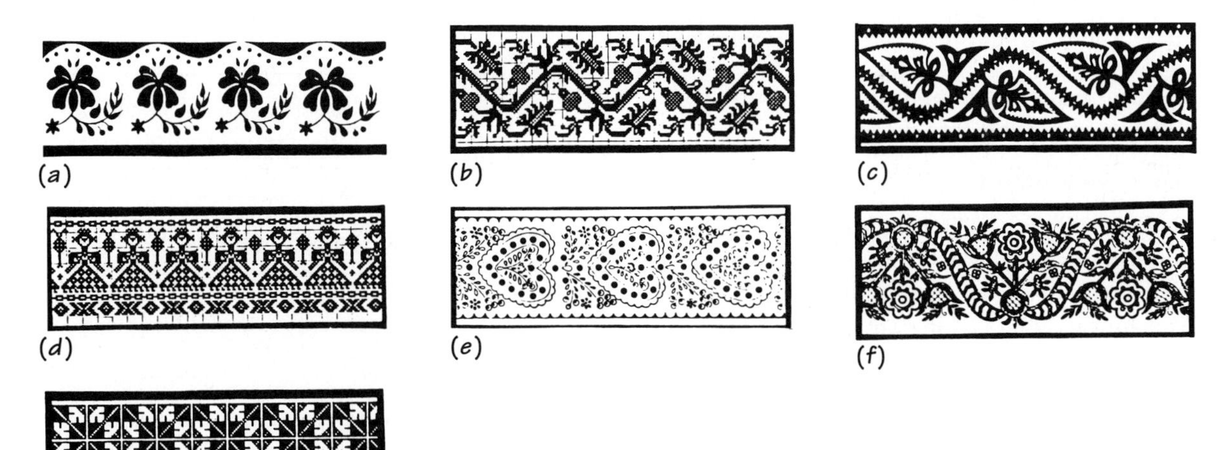

Hungarian needlework designs. (a) Edge decoration of table cover from Kalocsa, southern Hungary. (b) Pillow end decoration from Tolna County, southwest Hungary. (c) Decoration patched onto a long embroidered felt coat of Hungarian shepherds in Bihar County, eastern Hungary. (d) Embroidered edge decoration of bed sheet from the eighteenth century. (Note the deviations from symmetry in the lower stripes of the pattern.) (e) Shirt from Karád, southwest Hungary. (f) Pillow decoration pattern from Torockó (Rimetea), Transylvania, Romania. (g) Grape leaf pattern from the territory east of the river Tisza.

38. In each of the four accompanying examples, two adjacent triangles of an infinite strip are shown. (Contributed by Margaret A. Owens, California State University, Chico.)

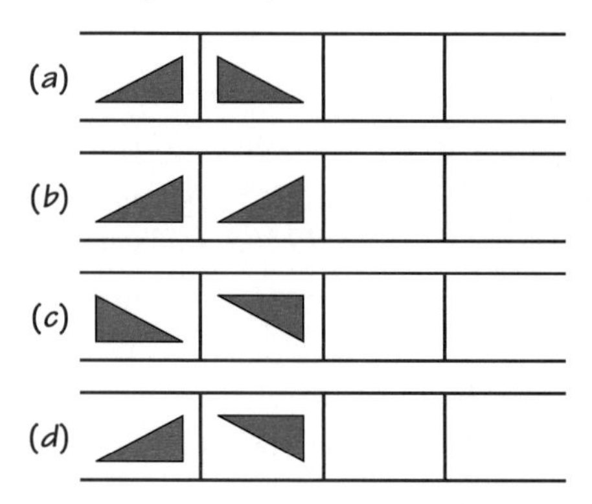

For each example

(a) Determine a motion (translation, reflection, rotation, or glide reflection) that takes the first (= left) triangle to the second (= right) one.

(b) Draw the next four triangles of the infinite strip that would result if the second triangle is moved to the next space by another motion of the same kind, and so on.

(c) Identify (by notation) the resulting strip as one of the seven possible strip patterns.

Imperfect Patterns

39. Repeat Exercise 37 for the accompanying eight strip patterns, all of which appear on the brass straps for a single lamp from nineteenth-century Benin in West Africa. (From H. Ling Roth, *In Great Benin*.)

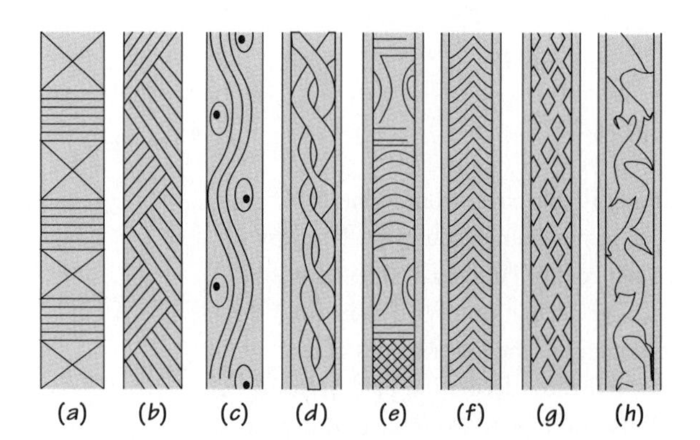

Note that the patterns are roughly carved, so you will need to discern the intent of the artist.

40. Repeat Exercise 37 for the accompanying patterns from San Ildefonso Pueblo, New Mexico.

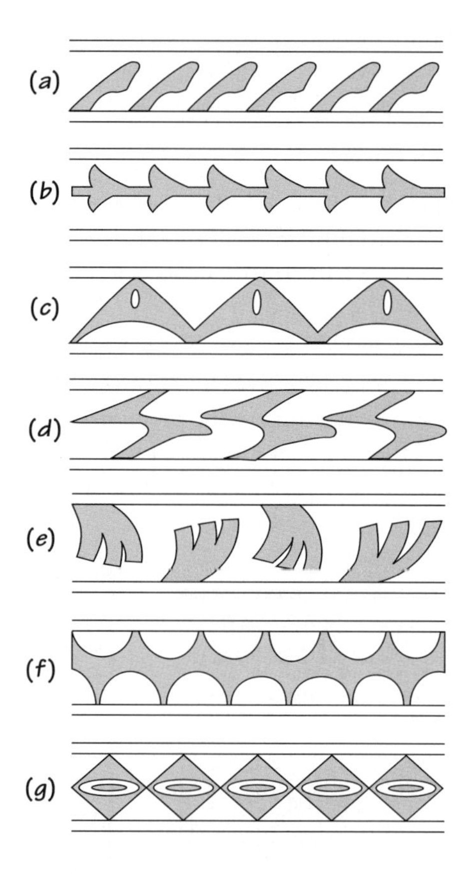

Additional Exercises

41. For positive integers a and n, the expression a mod n means remainder when a is divided by n. Thus, 23 mod 4 = 3, because $15 = 5 \cdot 4 + 3$, and we say that "23 is equivalent to 3 modulo 4" (see Chapter 10 for further details about this *modular arithmetic*). Every positive integer is equivalent to either 0, 1, 2, or 3 modulo 4. Consider the collection of elements {0, 1, 2, 3} and the operation $\oplus$ on them defined by $a \oplus b = (a + b) \bmod 4$. Show that under this operation, the collection forms a group.

42. The table below shows comparative data about the frequency of occurrence of strip designs of various types on pottery (Mesa Verde, United States) and smoking pipes (Begho, Ghana, Africa) from two continents.

Frequency of Strip Designs on Mesa Verde Pottery and Begho Smoking Pipes

	Mesa Verde	
Strip Type	Number of Examples	Percentage of Total
p111	7	4
p1m1	5	3
pm11	12	7
p112	93	53
p1a1	11	6
pma2	27	16
pmm2	19	11
Total	174	

	Begho	
Strip Type	Number of Examples	Percentage of Total
p111	4	2
p1m1	9	4
pm11	22	10
p112	19	8
p1a1	2	1
pma2	9	4
pmm2	165	72
Total	230	

(a) Which types of motions appear to be preferred for designs from each of the two localities?
(b) What other conclusions do you draw from the data of this table?
(c) On the evidence of the table alone, in which locality is each of the strip patterns in the following illustration most likely to have been found?

(*a*) 

(*b*) 

(*c*) 

(*d*) 

(*e*) 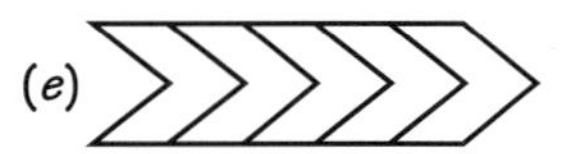

(*f*) 

(*g*) 

(*h*) 

(*i*) 

43. We have seen that the golden ratio is a positive root of the quadratic polynomial $x^2 - x - 1$. We can generalize this polynomial to $x^2 - mx - 1$ for $m = 1, 2, 3, \ldots$ and consider the positive roots of those polynomials as generalized means—the "metallic means family," as they are sometimes known. In particular, for $n = 2, 3, 4,$ and 5, we have respectively the silver, bronze, copper, and nickel means. It is surely surprising that these numbers arise both in connection with quasicrystals (investigated in Chapter 21) and in analyzing the behavior of some dynamical systems (a topic investigated in Chapter 23) as the systems evolve into chaotic behavior.

(a) Use the quadratic formula to find expressions in terms of square roots for the silver, bronze, copper, and nickel means, and approximate these to three decimal places.
(b) Find a general expression in terms of a square root for the *m*th metallic mean.

(c) Just as the golden mean arises as the limiting ratio of consecutive terms of the Fibonacci sequence, each of the metallic means arises as the limiting ratio of consecutive terms of generalized Fibonacci sequences. A generalized Fibonacci sequence G can be defined by

$$G_1 = 1, G_2 = 1, \quad \text{and} \quad G_{n+1} = pG_n + qG_{n-1}$$

where p and q are positive integers. The Fibonacci sequence itself is the case $p = q = 1$. Try various small values of p and q and determine which mean they lead to.

(d) Divide the equation for G_{n+1} by G_n. Assume that G_{n+1}/G_n and G_n/G_{n-1} both tend toward the same number x as n gets large, replace those quantities by x, and simplify the resulting equation. What must be the value of x?

(e) What happens to the sequence and to the mean if we allow one or both of p and q to be negative integers?

WRITING PROJECTS

1. The Fibonacci Association is devoted to fostering interest in Fibonacci and related numbers. In November 1988, the society's journal, *The Fibonacci Quarterly,* published "Suppose More Rabbits Are Born" (pp. 306–311), by Shari Lynn Levine (a high school student when she wrote it). The article begins: "How would Fibonacci's age-old sequence be redefined if, instead of bearing one pair of baby rabbits per month, the mature rabbits bear two pairs of baby rabbits per month?" The article goes on to discuss properties of the resulting "Beta-nacci" sequence and the sequences that result from even greater rabbit fertility. Here we ask you to rediscover some of Shari's results about the Beta-nacci sequence:

(a) How many rabbits will there be each month for the first 12 months?

(b) What is the recursive rule for the nth Beta-nacci number B_n?

(c) For the terms of the sequence in part (a), calculate the ratios B_{n+1}/B_n of successive terms. (*Motivating hint:* It's not the golden ratio this time.)

(d) Suppose that the ratio of successive terms approaches a number x. We show how to find x exactly. For very large n, we have $B_{n+1} \approx xB_n \approx x^2B_{n-1}$. Substituting these values into the recursive rule for the sequence and dividing by B_{n-1} gives us the equation $x^2 = x + 2$. Solve this equation for x (you can use the quadratic formula). Make a table of values of $3B_n$ versus 2_n. From the evidence, can you suggest a formula for B_n?

2. Generalize Writing Project 1, parts (a) through (d)

(a) to the case of each pair of rabbits having three pairs of rabbits (the "Gamma-nacci" sequence).

(b) to the case of each pair of rabbits having q pairs of rabbits.

3. Our emphasis in this chapter is on the patterns and practicality of symmetry. Fibonacci numbers and the golden ratio also have applications in efficient searching. Suppose, for example, that you need to determine the location and depth of the deepest point in the ocean along a particular transit line 1000 m long to within $t = 10$ m on either side. (Another setting would be trying to determine the thickest point of a seam of gold.) Suppose also that it doesn't make sense to try to take observations any closer than $\epsilon = 5$ m apart (it's hard to keep the boat closely enough positioned during a sounding). You could just take soundings every 10 m, starting from one end and continuing until you reach the low point. But provided the ocean slopes steadily down on both sides toward the low point, a more efficient approach would be to use either *Fibonacci search* or *golden ratio search.*

Both follow the same general procedure: Initially pick a pair of test points, which divide the search area into three regions (left, middle, right).

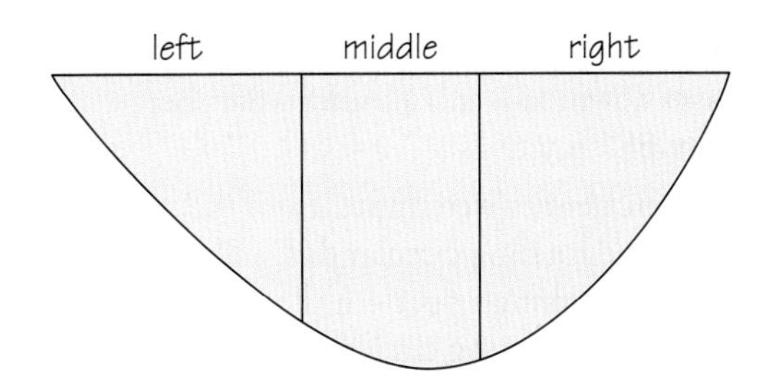

Determine the depth at the test points; if the right test point is deeper, eliminate the left region, and vice versa (see the figure below, where we eliminate the left region). In the remaining interval, determine a new test point that is located symmetrically with regard to the already-tested point, and proceed as before to eliminate the new right or the new left region (see figure below).

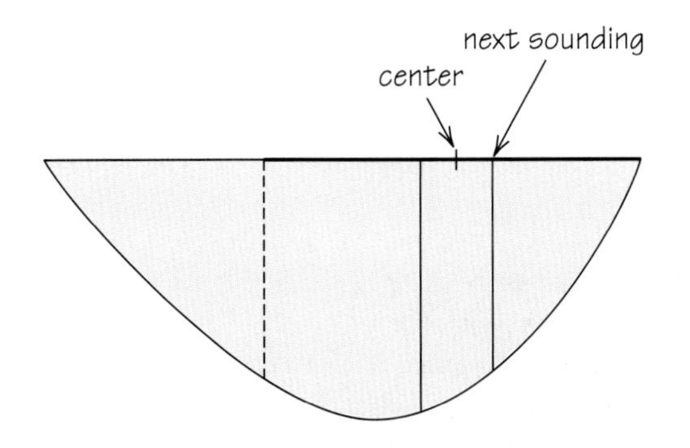

The two search methods differ only in how the first pair of test points is picked. If the major concern is to determine the location of the deepest point, the most efficient approach is Fibonacci search: determine the number n of steps needed (based on how accurately you need to know the position) and take the first pair of test points at a distance from each endpoint of

$$\frac{F_{n-1}}{F_n} L + \frac{(-1)^n}{F_n} \epsilon$$

where L is the length of the original interval, F_n is the nth Fibonacci number, and ϵ is the closest together that you could take two soundings.

If you are more concerned with determining the depth at the deepest point, you can't specify in advance how many steps you will need. The most efficient approach is therefore golden ratio search: take the first pair of test points at a distance from each endpoint of ϕL, where L is the length of the original interval and ϕ is the golden ratio. You keep taking soundings until the depths of two consecutive soundings are close enough together for your purposes, or until you reach the limit of how close together you can take soundings.

For example, suppose that (unknown to you) the depth d in km at a distance x km along the transit line is given by $d(x) = x^3 - x$ for $0 \leq x \leq 1$. The actual deepest point is $d = 0.3849$ (385 m) at $x = 0.5774$ (577 m along the transit line).

(a) How many soundings would it take to determine, within 10 m, where the deepest point is if you start at the left ($x = 0$) and take soundings every 10 m?
(b) The number of soundings required by Fibonacci search is the smallest n for which $F_n \geq L/t$, where t is the tolerance for error–that is, how accurately you need to know where the deepest point is. For $L = 1000$ m and $t = 10$ m, what is n?
(c) Determine for the Fibonacci search of part (b) the locations of the first two test points, and then the location of the next test point.
(d) Determine for golden ratio search the locations of the first two test points, and then the location of the next test point.
(e) Each step of golden ratio search reduces the length of the search interval by a factor of $1/\phi$; after N steps, the interval has length L/ϕ^N. For the preceding numerical values ($L = 1000$ m , $\epsilon = 5$ m), what is the largest N can be before the next step would require you to take a sounding closer than ϵ to the last one?
(f) Fibonacci and golden ratio searches are the most efficient search techniques when there is a simple minimum or maximum without intervening ups and downs. Suggest other possible applications.

4. Visit the Web site www-sphys.unil.ch/escher/, which features an interactive Java program called Escher Web Sketch. Experiment with choosing wallpaper patterns using crystallographic notation.

For each, draw on the screen a colored design for the motif; the program will reproduce the motif using the pattern.

5. Generations of children have enjoyed the popular toy Spirograph®, which allows the user to trace out symmetric patterns. A pencil or pen is placed in a hole in one of several plastic circular disks with teeth on the outside rim. The disk is then meshed in the teeth of another plastic circle and rotated around its inside or outside. Each plastic piece is labeled with the number of teeth that it has on its circumference.

Either obtain a copy of Spirograph® or a closely related toy, or else visit the Web site www.wordsmith.org/~anu/java/spirograph.html#def, which offers an interactive Java application (which you can download) that mimics what the Spirograph® toy does.

(a) Experiment to determine, from the numbers of teeth on the rotating circular disk and the fixed circle, what symmetry pattern the result will have.
(b) Choose a rotating circular disk and a fixed circle for which the ratio of the number of teeth reduces to a whole number. For each of several "offsets" (holes to choose for the pencil or pen), trace overlapping designs. What symmetry pattern do you get for the design taken as a whole? Repeat this experiment for other pairs of pieces and try to reach a general conclusion.

More on the Web SPREADSHEET PROJECTS

To do these projects, go to www.whfreeman.com/fapp.

Spreadsheets utilize recursively defined formulae in a very natural way. This spreadsheet project uses this trait to tabulate Fibonacci numbers and other recursive sequences. The ratios of successive terms are also explored.

CHAPTER 20

Tilings

When our ancestors used stones to cover the floors and walls of their houses, they selected shapes and colors to form pleasing designs. We can see the artistic impulse at work in mosaics, from Roman dwellings to Muslim religious buildings (see Figure 20.1). The same intricacy and complexity arise in other decorative arts—on carpets, fabrics, baskets, and even linoleum.

Such patterns have one feature in common: They use repeated shapes to cover a flat surface, without gaps or overlaps. If we think of the shapes as tiles, we can call the pattern a *tiling,* or *tessellation.* Even when efficiency is more important than aesthetics, designers value clever tiling patterns. In manufacturing, for example, stamping the components from a sheet of metal is most economical if the shapes of the components fit together without gaps—in other words, if the shapes form a tiling.

A **tiling** is a covering of the entire plane with nonoverlapping figures.

The major mathematical question about tilings is: Given one or more shapes (in specific sizes) of tiles, can they tile the plane? And, if so, how?

The surprising answer to the first question is that it is undecidable. That is, given an arbitrary set of tile shapes, there is no way to determine for certain if they can tile the plane or not. For some particular sets of tiles, we can exhibit tilings; for others, we can prove that there can't be any tiling. In this chapter we will see examples of both situations. But mathematicians have proved that there is no algorithm (mechanical step-by-step process) that can tell for any set of tile shapes which of the two situations holds. (See Spotlight 12.1, pages 418–419, for other examples of "unattainable ideals.")

Given this sobering (and puzzling) limitation, we begin our investigation by considering the simplest kinds of tiles and tilings.

Figure 20.1 Mosaic tile dome built by Abbas I, Safavid dynasty (1611–1638), Iran. (The Metropolitan Museum of Art, Harris Brisbane Dick Fund, 1939 [39.20]. Photograph © 1982 The Metropolitan Museum of Art.)

20.1 Regular Polygons

The simplest tilings use only one size and shape of tile; they are known as *monohedral tilings.*

> A **monohedral tiling** is a tiling that uses only one size and shape of tile.

In particular, we are interested especially in tiles that are **regular polygons,** figures all of whose sides are the same length and all of whose angles are equal. A square is a regular polygon with four equal sides and four equal interior angles; a triangle with all sides equal (an **equilateral triangle**) is also a regular polygon. A polygon with five sides is a pentagon, one with six sides is a hexagon, and one with n sides is an **n-gon.** Regular polygons are especially interesting because of their high degree of symmetry; each has the reflection and rotation symmetries of a dihedral rosette pattern (see Chapter 19). In three dimensions, the corresponding highly symmetrical figures are called *regular polyhedra* (see Spotlight 20.1).

By a convention dating back to the ancient Babylonians, angles are measured in degrees. An **exterior angle** of a polygon is one formed by one side and the extension of an adjacent side (Figure 20.2). Proceeding around the polygon in the same direction, we see that each **interior angle** (the angle inside a polygon formed by two adjacent sides) is paired with an exterior angle. If we bring all the exterior angles together at a single point, they will add up to 360° (see Figure 20.2). If the polygon has n sides, then each exterior angle must measure $360/n$

SPOTLIGHT 20.1 Regular Polyhedra and Buckyballs

The three-dimensional analogue of a regular polygon is a regular polyhedron, a convex solid whose faces are regular polygons all alike (same number of sides, same size), with each vertex surrounded by the same number of polygons. Although there are infinitely many regular polygons, there are only five regular polyhedra, a fact proved by Theaetetus (414–368 B.C.); they were called the *Platonic solids* by the ancient Greeks.

If the restriction that the same number of polygons meet at each vertex is relaxed, five additional convex polyhedra are obtained, all of whose faces are equilateral triangles. If we allow more than one kind of regular polygon, thirteen further convex polyhedra are obtained, known as the *semiregular polyhedra* or *Archimedean solids* (although there is no documented evidence that Archimedes studied them – but Kepler did catalogue them all). The truncated icosahedron, whose faces are pentagons and hexagons, is known throughout the world (once inflated) as a regulation soccer ball. Drawings of it appear in the work of Leonardo da Vinci.

The truncated icosahedron is also the structure of C_{60}, a form of carbon known as buckminsterfullerene and, more familiarly, the "buckyball." Sixty carbon atoms lie at the 60 vertices of this molecule, which was discovered in 1985. It is named after R. Buckminster Fuller (1895–1983), inventor and promoter of the geodesic dome. The molecule resembles a dome.

The buckyball is part of a family of carbon molecules, the *fullerenes,* in which each carbon atom is joined to three others. Thirty years before the discovery of fullerenes, mathematicians had shown that a convex polyhedron in which every vertex has three edges must have 12 pentagon faces and may have any number of hexagon faces, from 0 on up, except for 1.

That there must be 12 pentagons follows from a famous equation due to Leonhard Euler (1707–1783). For any convex polyhedron, it must be true that $v - e + f = 2$, where v is the number of vertices, e is the number of edges, and f is the number of faces of the polyhedron.

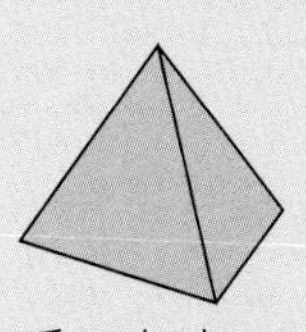

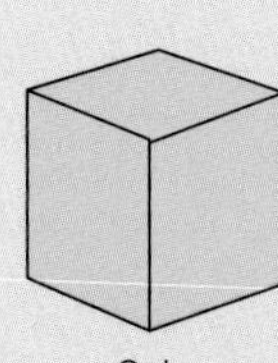

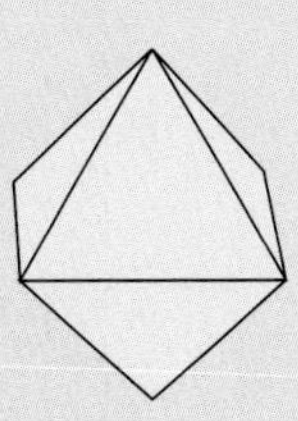

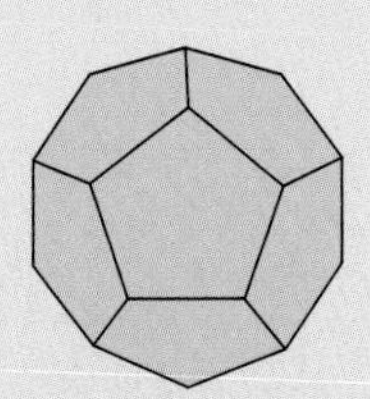

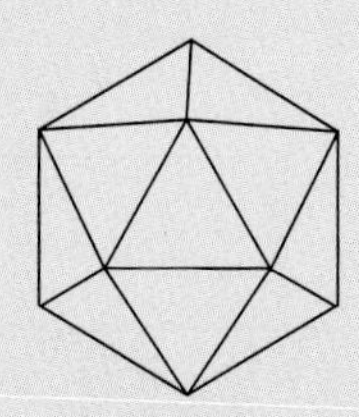

The five regular polyhedra.

degrees. For example, a square with $n = 4$ sides has 4 exterior angles, each measuring 90°; a pentagon with $n = 5$ sides has 5 exterior angles, each measuring 72°; a regular hexagon with $n = 6$ sides has 6 exterior angles, each measuring 60°. Notice that each exterior angle plus its corresponding interior angle make up a straight line, or 180°. For a regular polygon with more than six sides, the interior angle is between 120° and 180°. This last consideration will prove crucial shortly.

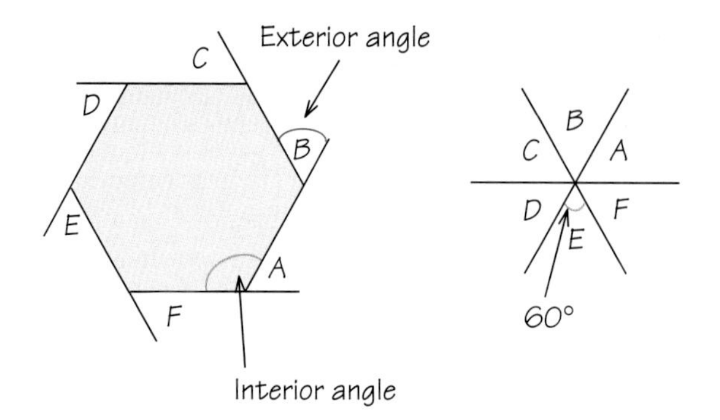

Figure 20.2 The exterior angles of a regular hexagon, like those of any regular polygon, add up to 360°. Each interior angle measures 60°.

20.2 Regular Tilings

A monohedral tiling whose tile is a regular polygon is called a **regular tiling.**

A square tile is the simplest case. Apart from varying the size of the square, which would change the scale but not the pattern of the tiling, we can get different tilings by offsetting one row of squares some distance from the next.

However, there is only one tiling that is edge-to-edge:

In an **edge-to-edge tiling,** the edge of a tile coincides entirely with the edge of a bordering tile (see Figure 20.3 for a tiling that is not edge-to-edge and another that is).

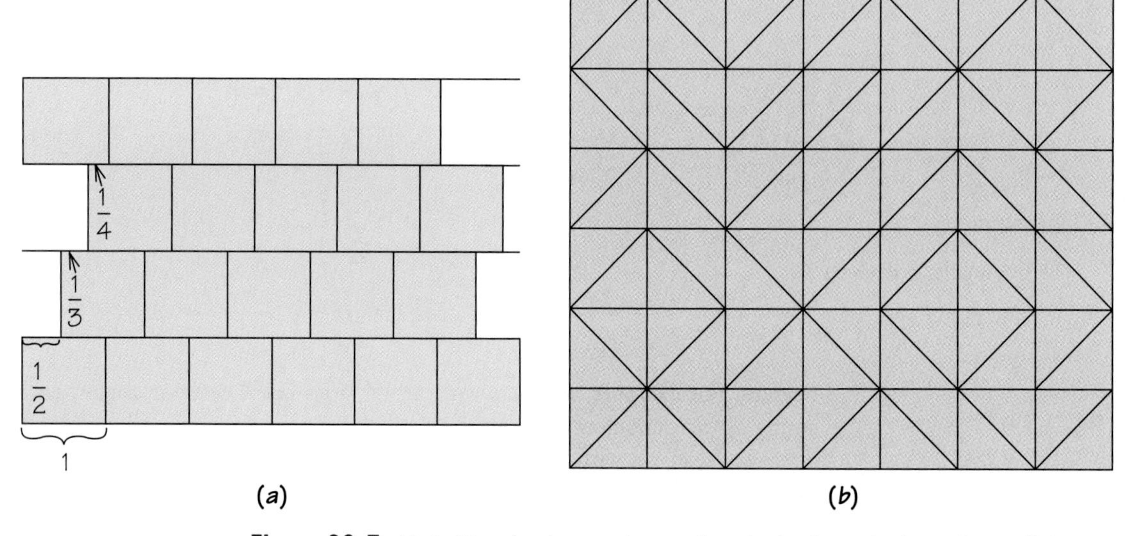

Figure 20.3 (a) A tiling that is not edge-to-edge; the horizontal edges of two adjoining squares do not exactly coincide. (b) A tiling by right triangles that is edge-to-edge.

For simplicity, from now on we consider only edge-to-edge tilings. In them (even in ones with tiles of different shapes and sizes), edges of different tiles meet at points that are surrounded by tiles and their edges.

Any tiling by squares can be refined to one by triangles by drawing a diagonal of each square; but these triangles are not regular (equilateral). Equilateral triangles can be arranged in rows by alternately inverting triangles; as with squares, there is only one pattern of equilateral triangles that forms an edge-to-edge tiling.

What about tiles with more than four sides? An edge-to-edge tiling with regular hexagons is easy to construct (see the upper right pattern in Figure 20.5).

However, if we look for a tiling with regular pentagons, we won't find one. How do we know whether we're just not being clever enough or there really isn't one to be found? This is the kind of question that mathematics is uniquely equipped to answer. In the other sciences, phenomena may exist even though we have not observed them; such was the case for bacteria before the invention of the microscope. In the case of an edge-to-edge tiling with regular pentagons, we can conclude with certainty that there is no edge-to-edge tiling with regular pentagons.

The proof is very easy. As we calculated earlier, the interior angles of a pentagon are each 108°. At a point where several pentagons meet, how many can meet there? The total of all of the angles around a point must be 360°. As you can see in Figure 20.4, four pentagons at a point would be too many (their angles would add to $4 \times 105° = 420°$, so they'd have to overlap), and three would be too few (their angles would add to $3 \times 108° = 324°$, so some of the area wouldn't be covered). Because 108 does not evenly divide 360, *regular pentagons can't tile the plane.*

With this argument, we can do something that is a favorite with mathematicians—we can *generalize* it, to a criterion for when a regular polygon can tile the plane: when the size of its interior angles divides 360 evenly. We can apply this criterion to determine exactly which other regular polygons can tile the plane.

EXAMPLE Identifying the Edge-to-Edge Regular Tilings

A regular hexagon has interior angles of 120°; 120 divides 360 evenly, and 3 regular hexagons fit together exactly around a point. A regular 7-gon—or any regular

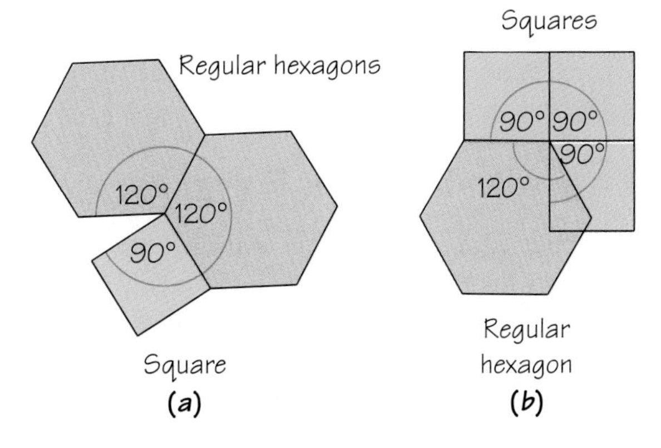

Figure 20.4 Polygons that come together at a vertex in a tiling must have interior angles that add up to 360°—no more, no less.

polygon with more than six sides—has interior angles that are larger than 120° but smaller than 180°. Now 360 divided by 120 gives 3, and 360 divided by 180 gives 2—and there aren't any other possibilities in between. Angles between 180° and 120° divided into 360° will give a result between 2 and 3, and consequently not an integer. So there are no edge-to-edge regular tilings of the plane with polygons of more than 6 sides. ■

The only edge-to-edge regular tilings are the ones with equilateral triangles, with squares, and with regular hexagons.

The follow-up question, of course, is which *combinations* of regular polygons of different numbers of sides can tile the plane edge-to-edge? The arrangement of polygons around a vertex in an edge-to-edge tiling is the **vertex figure** for that vertex.

A systematic tiling that uses a mix of regular polygons with different numbers of sides but in which all vertex figures are alike—the same polygons in the same order—is called a **semiregular tiling** (see Figure 20.5).

As before, the technique of adding up angles at a vertex (to be 360°) can eliminate some impossible combinations, such as "square, hexagon, hexagon" (Figure 20.4). Once we have found an arrangement that is numerically possible, we must confirm the actual existence of each tiling by constructing it (i.e., showing that it is geometrically possible). For example, even though a possible arrangement of regular polygons around a point is "triangle, square, square, hexagon," it is not possible to construct a tiling with that vertex figure at every vertex.

The result of such an investigation is that in a semiregular tiling no polygon can have more than 12 sides. In fact, polygons with 5, 7, 9, 10, or 11 sides do not occur either. Figure 20.5 exhibits all of the semiregular tilings.

If we abandon the restriction about the vertex figures being the same at every vertex, then there are *infinitely many* systematic edge-to-edge tilings with regular polygons, even if we continue to insist that all polygons with the same number of sides have the same size.

20.3 Tilings with Irregular Polygons

What about edge-to-edge tilings with irregular polygons, which may have some sides longer than others or some interior angles larger than others? We will look just at monohedral tilings (in which all tiles have the same size and shape) and investigate in turn which triangles, **quadrilaterals** (four-sided polygons), hexagons, and so forth, can tile the plane.

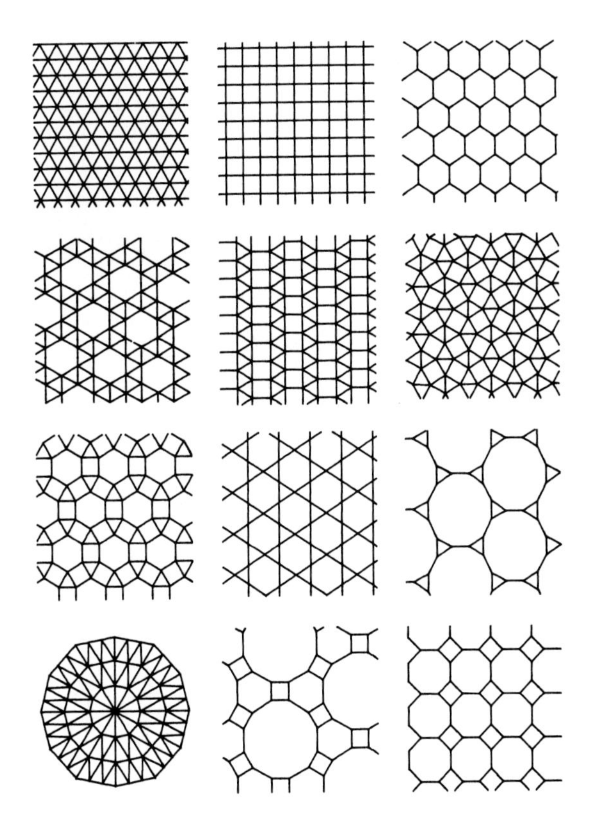

Figure 20.5 The three regular tilings and the eight semiregular tilings, plus one tiling that does not belong to either group. Can you identify it?

The most general shape of triangle has all sides of different lengths and all interior angles of different sizes. Such a triangle is called a **scalene triangle,** from the Greek word for "uneven." We can always take two copies of a scalene triangle and fit them together to form a **parallelogram,** a quadrilateral whose opposite sides are parallel (Figure 20.6a). It's easy to see that we can then use such parallelograms to tile the plane by making strips and then fitting layers of strips together edge-to-edge (Figure 20.6b). So:

Any triangle can tile the plane.

What about quadrilaterals? We have seen that squares tile the plane, and rectangles certainly will, too; and we have just noted that any parallelogram will tile. What about a quadrilateral (four-sided polygon) with its opposite sides not parallel, as in Figure 20.7a? The same technique as for triangles will work. We fit together two copies of the quadrilateral, forming a hexagon whose opposite sides are parallel. Such hexagons fit next to each other to form a tiling, as in Figure 20.7b.

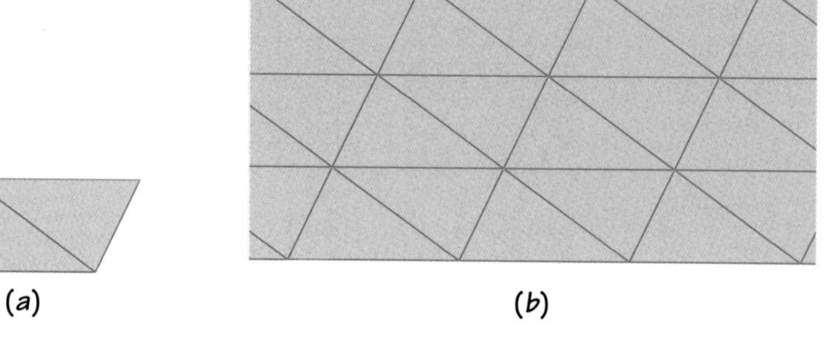

Figure 20.6 (a) A scalene triangle. (b) Every scalene triangle tiles the plane.

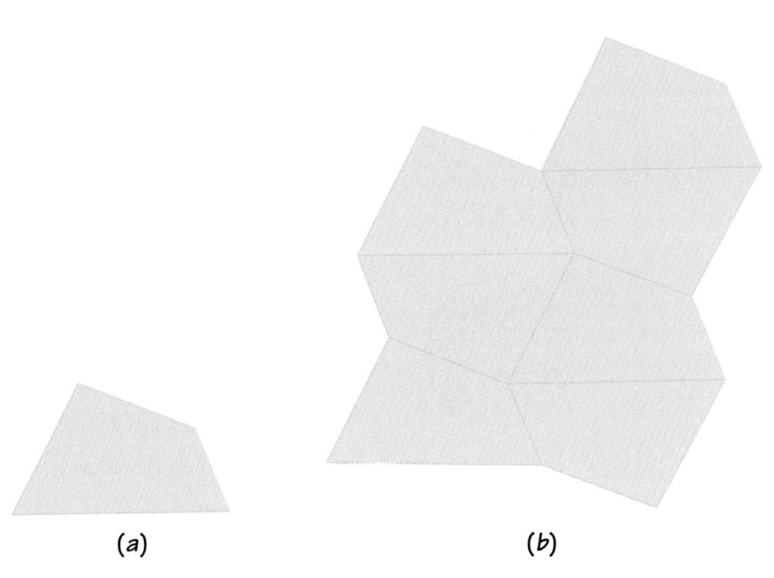

Figure 20.7 (a) A general quadrilateral. (b) Any quadrilateral tiles the plane.

The quadrilaterals shown in Figure 20.7 are all **convex.** If you take any two points on the tile (including the boundary), the line segment joining them lies entirely within the tile (again, including the boundary). The quadrilateral of Figure 20.8a is not convex, but the same approach works for using it to form a tiling (Figure 20.8b). So:

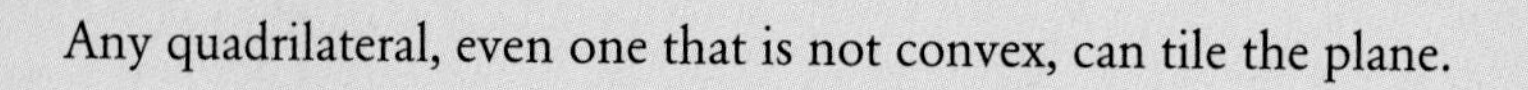

Any quadrilateral, even one that is not convex, can tile the plane.

We could hope that such success would extend to irregular polygons with any numbers of sides, but it doesn't. The situation for convex hexagons was determined by K. Reinhardt in his 1918 doctoral thesis. He showed that for a convex hexagon to tile, it must belong to one of three classes, and that every hexagon in those classes will tile. Examples of the three classes are shown in Figure 20.9, together with their characterizations. Notice that tilings with a

Figure 20.8 (a) A general nonconvex quadrilateral. (b) Any quadrilateral, convex or not, tiles the plane.

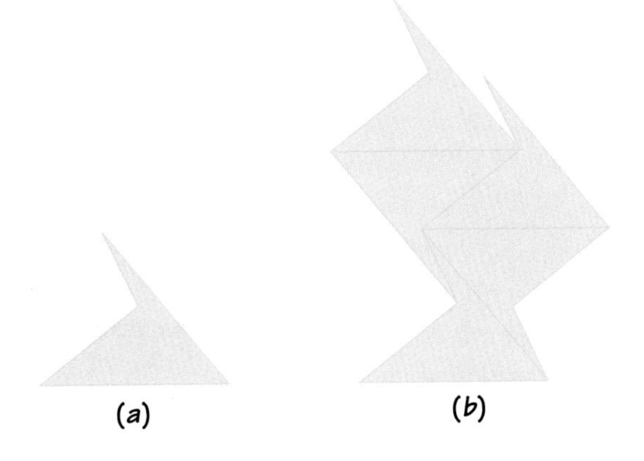

Figure 20.9 The three types of convex hexagon tile.

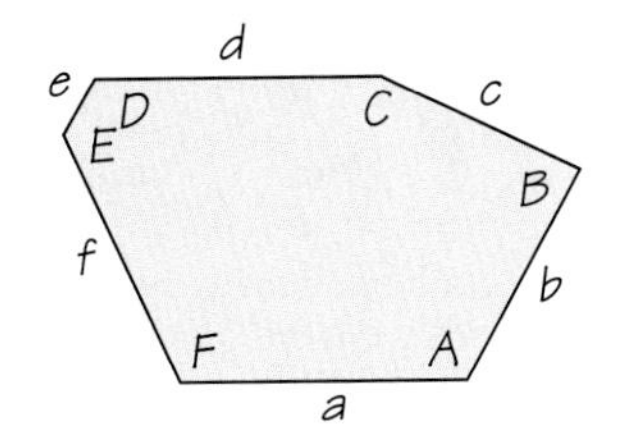

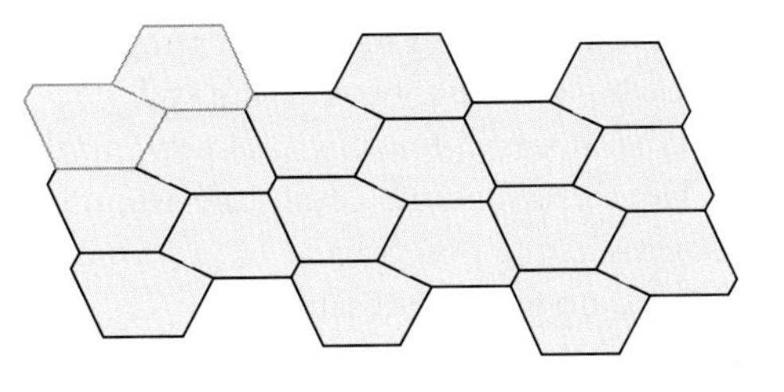

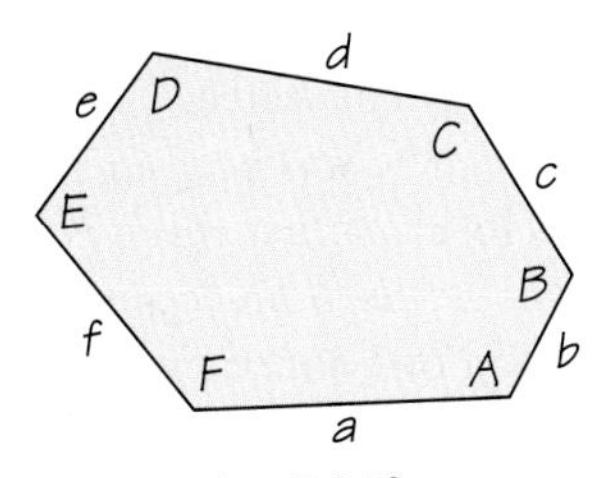

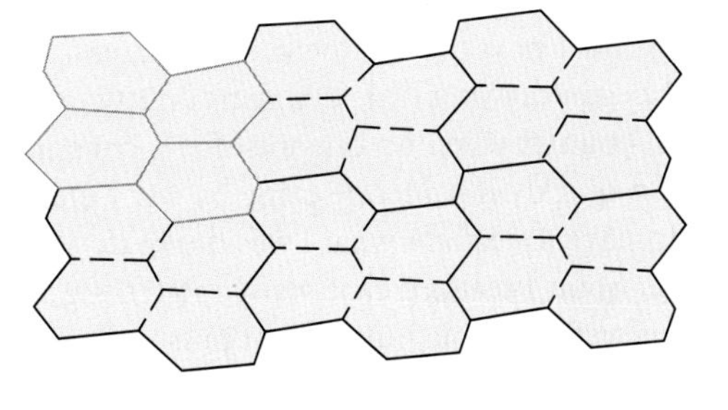

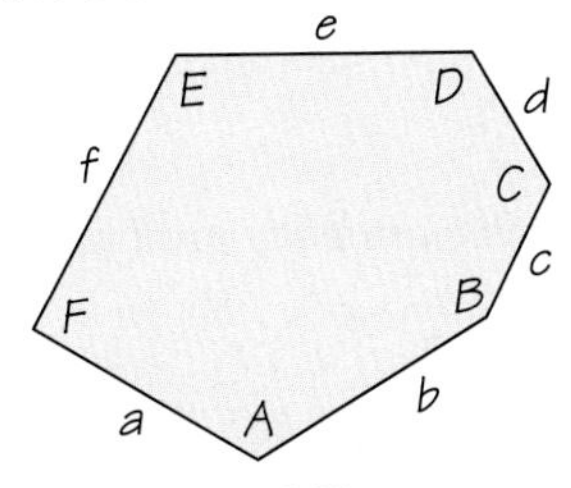

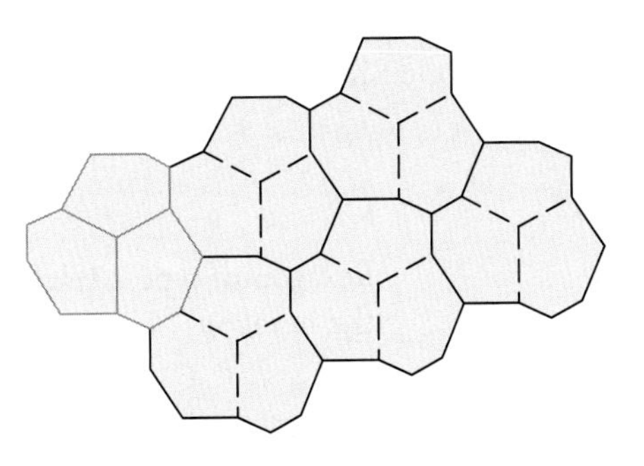

SPOTLIGHT 20.2 In Praise of Amateurs

R. B. Kershner's claim to have found all convex pentagons that tile was reported by Martin Gardner in his column in *Scientific American,* which was read by many amateur puzzle enthusiasts, including Richard James III and Marjorie Rice. James found a tiling that Kershner had missed, a discovery that Gardner reported in a later column.

Rice, a San Diego housewife and mother of five, read about James's new tile. "I thought I would like to understand these fascinating patterns better and see if I could find still another type. It was like a delightful new puzzle to me." Her search became a full-scale assault on the problem, lasting two years.

Rice had no formal education in mathematics beyond a high school general mathematics course. She not only worked out her own method of attack but also invented her own notation as well, both of which were far from the conventional ways that mathematicians use.

"I began drawing little diagrams on my kitchen counter when no one was there, covering them up quickly if someone came by, for I didn't wish to have to explain what I was doing to anyone. Soon I realized that many interesting patterns were possible but did not pursue them further, for I was searching for a new type and a few weeks later, I found it." Over the next two years, she found three additional new tilings.

What makes a person pursue a problem so steadfastly as Marjorie Rice? She was not trained

Marjorie Rice

(From Doris Schattschneider, "In Praise of Amateurs," in David A. Klarner [ed.], *The Mathematical Gardner,* Wadsworth, Belmont, Calif., 1981.)

to do this, nor paid to do it, but obviously gained personal satisfaction in her patient and persistent search.

She was born in 1923 in St. Petersburg, Florida, a first child. At age 5, she began school in a one-room country school with eight grades and two dozen pupils.

"When I was in the 6th or 7th grade, our teacher pointed out to us one day the Golden Section in the proportions of a picture frame. This immediately caught my imagination and though it was just a passing incident, I never

hexagon of type 2 (Figure 20.9b) use both ordinary and mirror-image versions of the hexagon.

> Exactly three classes of convex hexagons can tile the plane.

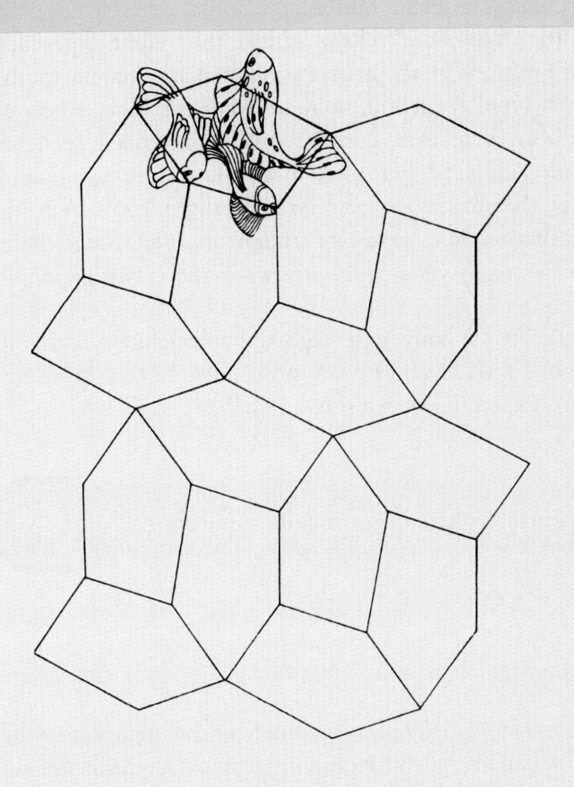

Underlying grid for Marjorie Rice's *Fish,* based on one of her unusual tilings by pentagons.

forgot it. I've . . . been especially interested in architecture and the ideas of architects and planners such as Buckminster Fuller. I've come across the Golden Section again in my reading and considered its use in painting and design."

After high school, Marjorie Rice worked until her marriage in 1945. She was drawn back into mathematics by her children, finding solutions to their homework problems "by unorthodox means, since I did not know the correct procedures." She became especially interested in textile design and the works of M. C. Escher. As she pursued the pentagonal tilings, she produced some beautiful geometric designs and imaginative Escher-like patterns (see Figure 20.19 and the figure here).

"I enjoy puzzles of all kinds, crosswords, jigsaw puzzles, mathematical puzzles and games, and have purchased books of mathematical puzzles over the years. Those of a geometric nature are a special delight."

The intense spirit of inquiry and the keen perception of all they encounter are the forte of all such amateurs. No formal education provides these gifts. Lack of a mathematical degree separates these "amateurs" from the "professionals," yet their curiosity and ingenious methods make them true mathematicians.

Source: Adapted from Doris Schattschneider, "In Praise of Amateurs," in David A. Klarner (ed.), *The Mathematical Gardner,* pp. 140–166, plus Plates I–III, Wadsworth, Belmont, Calif., 1981.

Reinhardt also explored convex pentagons and found five classes that tile. For example, any pentagon with two parallel sides will tile. Reinhardt did not complete the solution, as he did for hexagons, by proving conclusively that no other pentagons could tile; he claimed that it would be very tedious to finish the analysis. Still, he felt that he had found them all. In 1968, after 35 years of working on the problem on and off, R. B. Kershner, a physicist at Johns Hopkins

University, discovered three more classes of pentagons that will tile. Kershner was sure that he had found all pentagons that tile, but, like Reinhardt, he did not offer a complete proof, which "would require a rather large book."

When an account of the "complete" classification into eight types appeared in *Scientific American* (July 1975), the article provoked an amateur mathematician to discover a ninth type! A second amateur, Marjorie Rice, a housewife with no formal education in mathematics beyond high school "general mathematics" 36 years earlier, devised her own mathematical notation and found four more types over the next two years (see Spotlight 20.2). A fourteenth type was found by a mathematics graduate student in 1985. Since then, no new types have been discovered, yet no one knows if the classification is complete.

With the situation so intricate for convex pentagons, you might think that it must be still worse for polygons with seven or even more sides. In fact, however, the situation is remarkably simple, as Reinhardt proved in 1927:

A convex polygon with seven or more sides cannot tile.

20.4 M. C. Escher and Tilings

The Dutch artist M. C. Escher (1898–1972) was inspired by the great variety of decoration in tilings in the Alhambra, a fourteenth-century palace built during the last years of Islamic dominance in Spain. He devoted much of his career of making prints to creating tilings with tiles in the shapes of living beings (a practice forbidden to Muslims). Those prints of interlocking animals and people have inspired awe and wonder among people all over the world. Figures 20.10–20.13 illustrate a few of his drawings and finished works. Like Marjorie Rice, he, too, developed his own mathematical notation for the different kinds of patterns for the tilings.

20.5 Tiling by Translations

You may wonder just how much liberty can be taken in shaping a tile and how you might be able to design an Escher-like tiling yourself.

The simplest case is when the tile is just *translated* in two directions, that is, copies are laid edge-to-edge in rows, as in Figure 20.10. Each tile must fit exactly into the ones next to it, including its neighbors above and below. We say that each tile is a **translation** of each other one, because we can move one to coincide with another without doing any rotation or reflection.

Figure 20.10 Escher No. 128 (*Bird*), from Escher's 1941–1942 notebook. (© 1967 M. C. Escher Foundation, Baarn, Holland, all rights reserved.)

(a)

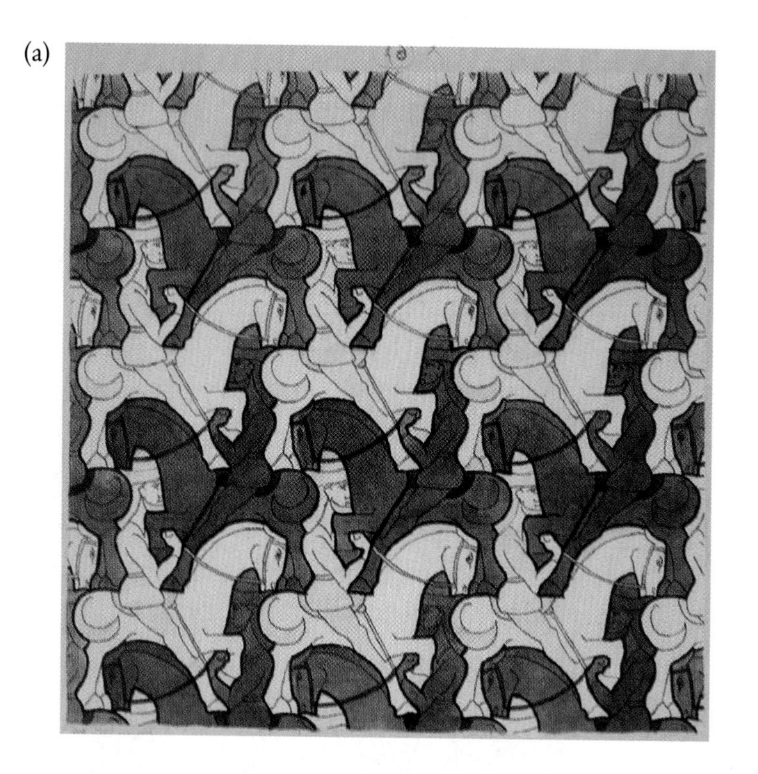

(b)

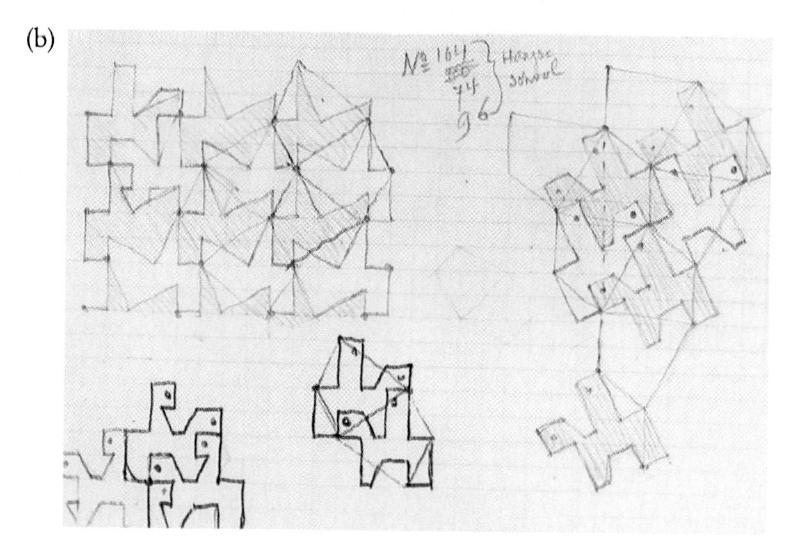

Figure 20.11 (a) Escher No. 67 (*Horseman*), from Escher's 1941–1942 notebook. (© 1947 M. C. Escher Foundation, Baarn, Holland, all rights reserved.) (b) Sketch showing the tile design for the *Horseman* print. (© 1947 M. C. Escher Foundation, Baarn, Holland, all rights reserved. From the collection of Michael S. Sachs.)

When is it possible for a tile to cover the plane in this manner? The boundary of the tile must be divisible into matching pairs of opposing parts that will fit together. Figures 20.10 and 20.11 illustrate two basic ways that this can happen. In the first, two opposite pairs of sides match; in the second, three opposite pairs of sides match.

Figure 20.12 Escher No. 6 (*Camel*), from Escher's 1941–1942 notebook.

Figure 20.13 (a) Escher No. 88 (*Sea Horse*). (© 1947 M. C. Escher Foundation, Baarn, Holland, all rights reserved.) (b) The skeleton for No. 88 from Escher's 1941–1942 notebook. (© 1952 M. C. Escher Foundation, Baarn, Holland, all rights reserved. From the collection of Michael S. Sachs.)

A tile can tile the plane by translations if either

1. There are four consecutive points A, B, C, and D on the boundary such that
   (a) the boundary part from A to B is congruent by translation to the boundary part from D to C, and
   (b) the boundary part from B to C is congruent by translation to the boundary part from A to D (see Figure 20.14a).
2. There are six consecutive points A, B, C, D, E, and F on the boundary such that the boundary parts AB, BC, and CD are congruent by translation, respectively, to the boundary parts ED, FE, and AF (see Figure 20.14b).

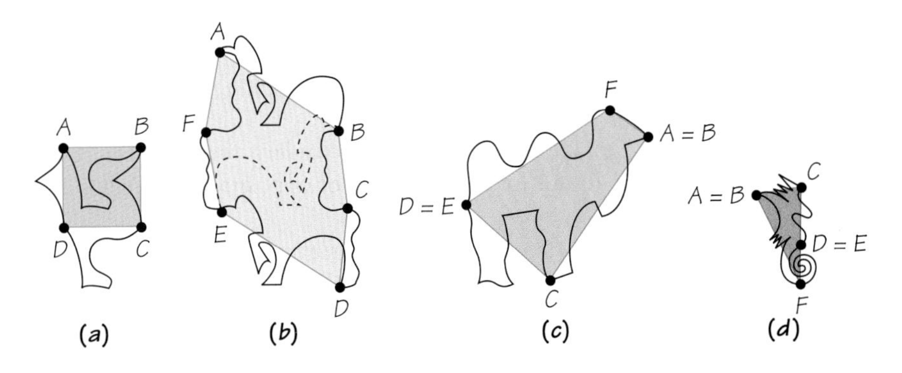

Figure 20.14 Individual tiles traced from the Escher prints of Figures 20.10–20.13, with points marked to show they fulfill the criteria for tiling by translations or by translations and half-turns.

The tiles for each of Figures 20.10 and 20.11 are shown in outline form in Figure 20.14, together with points marked to show how the tiles fulfill the criteria.

To create tilings, you can proceed exactly as Escher did. His notebooks show that he designed his patterns in just the way that we now describe.

EXAMPLE

Tiling the Plane Using a Parallelogram

For the first case of the theorem, start from a parallelogram, make a change to the boundary on one side, then copy that change to the opposite side. Similarly, change one of the other two sides and copy that change on the side opposite it (Figure 20.15). Revise as necessary, always making the same change to opposite sides. You might find it useful (as Escher did) to make your designs on graph paper, or you can work by cutting and taping together pieces of heavy paper. ■

EXAMPLE

Tiling the Plane Using a Hexagon

For the second case, start from a **par-hexagon,** one whose opposite sides are equal and parallel; this is one of the kinds of hexagons that tile the plane. Again,

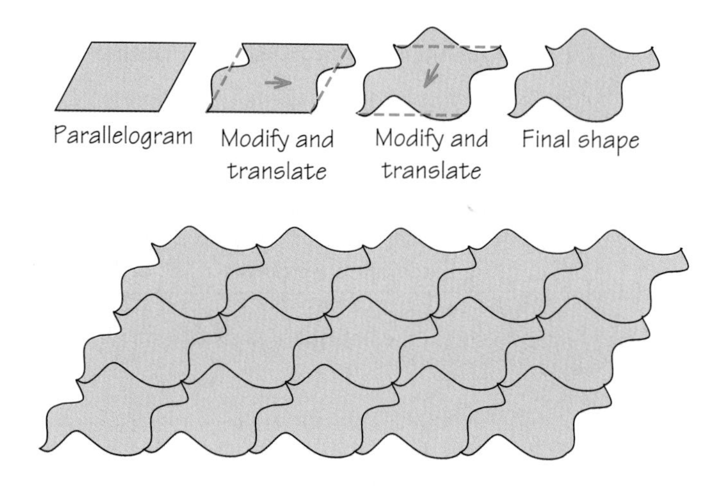

Figure 20.15 How to make an Escher-like tiling by translations, from a parallelogram base.

make a change on one boundary and copy the change to the opposite side, and do this for all three pairs of opposite sides (Figure 20.16). ■

Of course, there is a real art to being able to make the resulting tile resemble an animal or human figure!

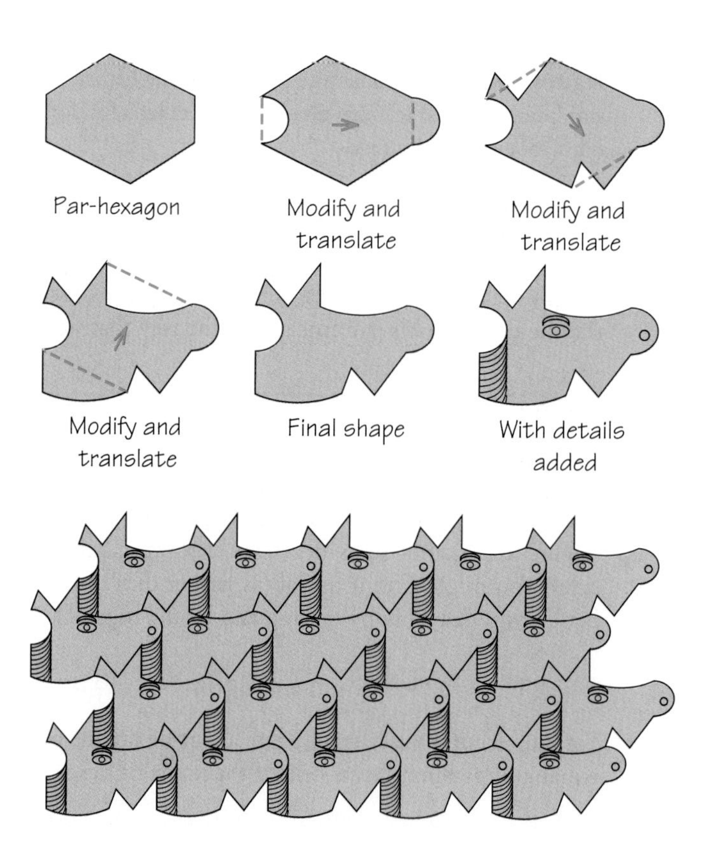

Figure 20.16 How to make an Escher-like tiling by translations, from a par-hexagon base.

20.6 Tiling by Translations and Half-Turns

If the tiling is to allow half-turns, so that some of the figures are "upside down," the part of the boundary of a right-side-up figure has to match the corresponding part of itself in an upside-down position. For that to happen, that part of the boundary must be **centrosymmetric,** that is, symmetric about (unaltered by) a 180° rotation around its midpoint. The key to some of Escher's more sophisticated monohedral designs, and the fundamental principle behind some further easy recipes for making Escher-like tilings, is the **Conway criterion,** formulated by John H. Conway of Princeton University:

A tile can tile the plane by translations and half-turns if there are six consecutive points on the boundary (some of which may coincide, but at least three of which are distinct) – call them A, B, C, D, E, and F – such that

- the boundary part from A to B is congruent by translation to the boundary part from E to D, and
- each of the boundary parts BC, CD, EF, and FA is centrosymmetric.

The first condition means that we can match up the two boundary parts exactly, curve for curve, angle for angle. The second condition means that each of the remaining boundary parts is brought back into itself by a half-turn around its center. Either condition is automatically fulfilled if the boundary part in question is a straight-line segment.

The tiles for each of Figures 20.12 and 20.13 are shown in outline form in Figure 20.14, together with points marked to show how the tiles fulfill the Conway criterion.

Once again, you can make Escher-like tilings by starting from simple geometric shapes that tile. This time, the starting geometric tile can be any triangle or any quadrilateral.

EXAMPLE Tiling the Plane Using a Triangle

For a triangle, modify half of one side, then rotate that side around its center point to extend the modification to the rest of the side, thereby making the new side centrosymmetric. Then you may do the same to the second and third sides (Figure 20.17). ■

EXAMPLE Tiling the Plane Using a Quadrilateral

For the quadrilateral, do the same, modifying each of the four sides, or as many as you wish (Figure 20.18). ■

The same approach will work with some of the sides of some pentagons and hexagons that tile. Because not all sides can be modified, there is less freedom for

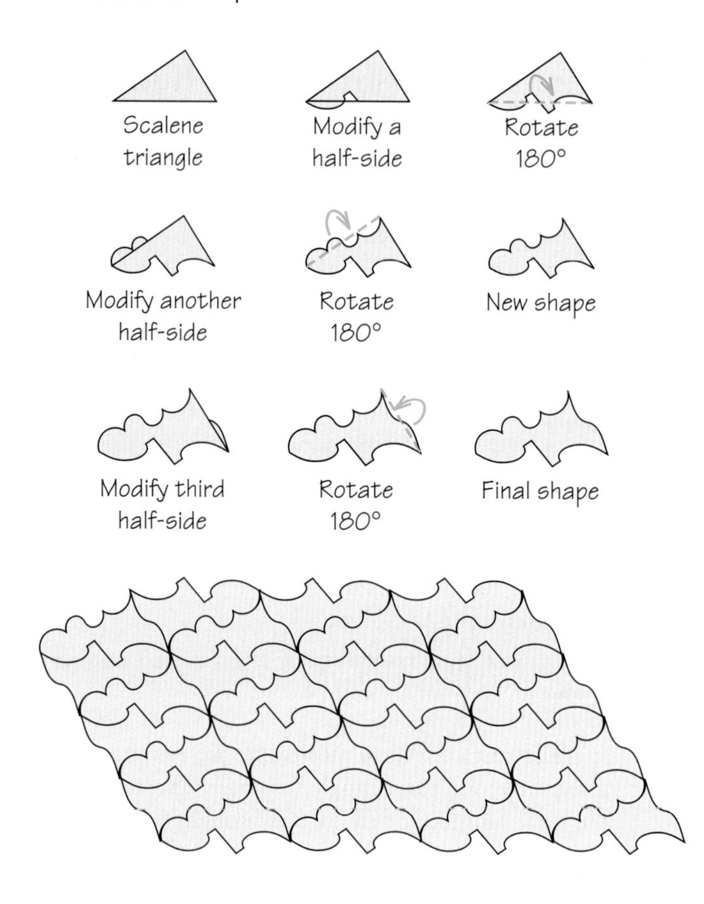

Figure 20.17 How to make an Escher-like tiling by translations and half-turns, from a scalene triangle base.

designing tiles, so it is more difficult to make the resulting tiles resemble intended figures. Figure 20.19 shows the beautiful results achieved by Marjorie Rice, using one of the unusual tilings by pentagons that she discovered.

The sketches in Escher's notebook in Figures 20.10–20.13 indicate how he designed the prints whose tiles you see in Figure 20.14. For Figure 20.14a, he modified the two pairs of sides of a square. For Figure 20.14b, he modified the pairs of sides of a par-hexagon that became a tile made up of a pair of dark and light knights. This figure also has a reflection symmetry, taking a leftward-facing light knight to a rightward-facing dark knight. However, we have not discussed criteria for when you can start with a tile (e.g., a single knight) and produce a tiling with this symmetry. In Figure 20.14c, the blue overlay shows how the tile could be made by modifying half of every side of a general quadrilateral, though Figure 20.12 shows that Escher actually designed the tiling from a parallelogram base. Regarding Figure 20.14d, Figure 20.13 shows that Escher used a triangle base. He did not use the procedure that we noted earlier, in which half of every side is modified. Instead, he treated the triangle as a quadrilateral, in which two adjacent sides (*CD* and *DF*) happen to continue on in a straight line.

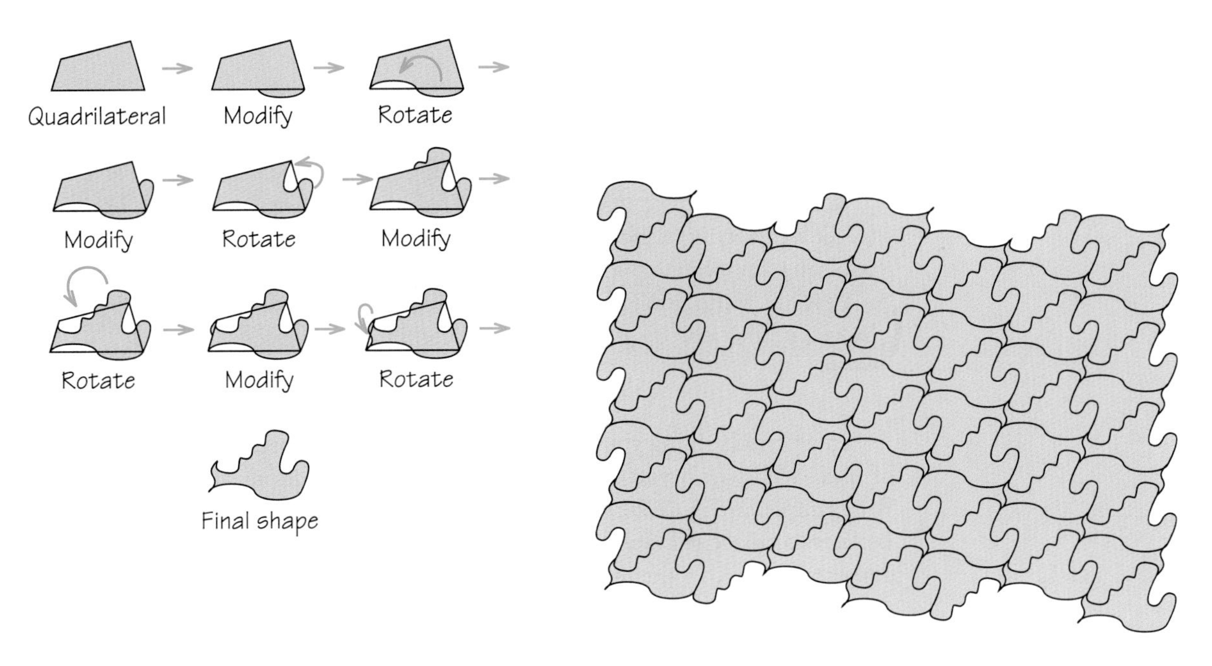

Figure 20.18 How to make an Escher-like tiling by translations and half-turns, from a quadrilateral base.

Figure 20.19 *Fish,* by Marjorie Rice, based on one of her unusual tilings by pentagons.

20.7 Further Considerations

All the patterns that we have exhibited and discussed so far have been **periodic tilings.** If we transfer a periodic tiling to a transparency, it is possible to slide the transparency a certain distance horizontally, without rotating it, until the transparency exactly matches the tiling everywhere. We can also achieve the same result by moving the transparency in some second direction (possibly vertically) by a certain (possibly different) distance.

In a periodic tiling you can identify a **fundamental region**—a tile, or a block of tiles—with which you can cover the plane by translations at regular intervals. For example, in Figure 20.10, a single bird forms a fundamental region. In Figure 20.12, two adjacent camels, one right side up and one upside down, form a fundamental region. In the terminology of Chapter 19, the periodic tilings are ones that are preserved under translations in more than two directions. (In this chapter we are concerned with the design elements more than with the patterns, which were the main topic of Chapter 19.)

20.8 Nonperiodic Tilings

A **nonperiodic tiling** is a tiling in which there is no regular repetition of the pattern by translation.

The lower left pattern in Figure 20.5, with its expanding rings of triangles, does not have any regular repetition by translation.

In Figure 20.3a, the second row from the bottom is offset one-half of a unit to the right from the bottom row, the third row from the bottom is offset one-third of a unit further, and so forth. Because the sum $\frac{1}{2} + \frac{1}{3} + \frac{1}{4} + \ldots + \frac{1}{n}$ never adds up to exactly a whole number, there is no direction (horizontal, vertical, or diagonal) in which we can move the entire tiling and have it coincide exactly with itself.

EXAMPLE

A Random Tiling

Consider the usual edge-to-edge square tiling. For each square, flip a coin; depending on the result, divide the square into two right triangles by adding either a rising or a falling diagonal (see Figure 20.3b). Because what happens in each individual square is unconnected to what happens in the rest of the tiling, the tiling by right triangles that is produced by this procedure has no chance of being periodic. ■

20.9 The Penrose Tiles

For all known cases, if a single tile can be used to make a nonperiodic tiling of the plane, then it can also be used to make a periodic tiling. It is still an open question whether this property is true for every possible shape. In 1993, Conway discovered an example in three dimensions of a single convex polyhedron that tiles space nonperiodically but cannot be used to make a periodic tilogy.

For a long time mathematicians also tended to believe the more general assertion that if you can construct a nonperiodic tiling with a set of one *or more* tiles, you can construct a periodic tiling from the same tiles. But in 1964 a set of tiles was found that permits only nonperiodic tiling. It contains 20,000 different shapes! Over the next several years, smaller sets were discovered with the same property, with as few as 100 shapes. But it was still amazing when in 1975 Sir Roger Penrose, a mathematical physicist at Oxford, announced a set that tiles only nonperiodically—consisting of just two tiles! (See Figure 20.20 and Spotlight 20.3.)

Penrose called his tiles "darts" and "kites," and both of these *Penrose tiles* can be obtained from a single rhombus. (A **rhombus** is a quadrilateral with four equal sides and equal opposite interior angles.) The particular rhombus from which the Penrose tiles are constructed has interior angles of 72° and 108°. If we cut the longer diagonal in two pieces so that the longer piece is the golden ratio ($(1 + \sqrt{5})/2 \approx 1.618$) times as long as the shorter (see Chapter 19), and connect the dividing point to the remaining corners, we split the rhombus into a dart and a kite (Figure 20.20).

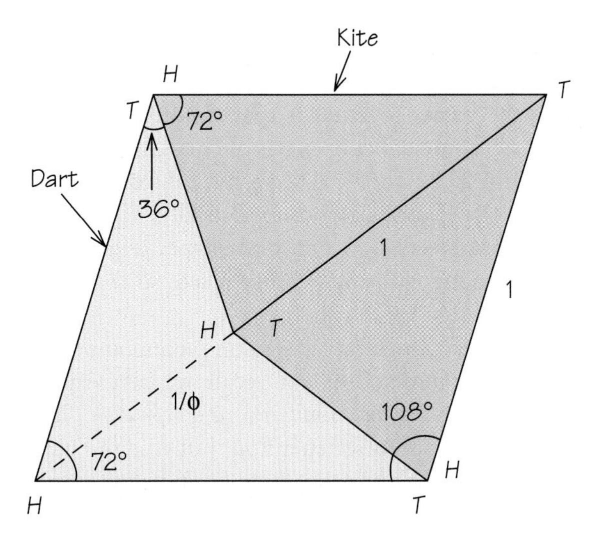

Figure 20.20
Construction of Penrose's "dart" (beige area) and "kite" (blue area). The length $1/\phi \approx 0.618$ is the golden ratio.

SPOTLIGHT 20.3 Sir Roger Penrose

Sir Roger Penrose, a professor at the University of Oxford, received a doctorate in mathematics but has been seriously interested in physics for many years; he was one of the first to conjecture the existence of black holes. He discovered what are now called the Penrose pieces in 1973. His latest endeavors have been devoted to trying to establish that the mind is not a machine, that is, that the ideas and concepts of artificial intelligence cannot explain human consciousness.

Sir Roger Penrose
(Anthony Howarth/Photo Researchers.)

Because the two Penrose pieces come from a rhombus, and a rhombus can be replicated to tile the plane periodically, the rules for fitting the Penrose pieces together do not allow the periodic rhombus arrangement. We may label the front and back vertices of the dart with H (for head) and its two wing tips with T (for tail), and do the reverse for the kite. Then the rule is that only vertices with the same letter may meet: Heads must go to heads, and tails to tails.

A prettier method of enforcing the rules, proposed by Conway, is to draw circular arcs of different colors on the pieces and require that adjacent edges must join arcs of the same color. The result is the pretty patterns of Figure 20.21. In fact, Conway thinks of the darts as children, each with two hands. The rule for fitting the pieces together is that children are required to hold hands. Penrose patterns become dancing circles of children.

Figure 20.22 shows a tiling by a different pair of pieces, both rhombuses, that tile the plane only nonperiodically. Figure 20.23 shows a modification of the Penrose pieces into two bird shapes. Figure 20.24 shows a coloring of one particular tiling with the Penrose pieces so that no two adjacent pieces have the same color.

Although tilings with Penrose's pieces cannot be periodic, the tilings possess unexpected symmetry. As you recall, we have explored our intuitions of symmetry in terms of *balance, similarity,* and *repetition.* Patterns made with the Penrose pieces certainly involve repetition, but it is the balance in the arrangement that we seek. What balance can there be in a nonperiodic pattern? It turns out that some Penrose patterns have a single line of reflection. But most surprising of all is that every Penrose pattern has a kind of fivefold rotational symmetry.

Figure 20.21 A Penrose tiling with specially marked tiles, forming what is known as the cartwheel tiling. (From Sir Roger Penrose.)

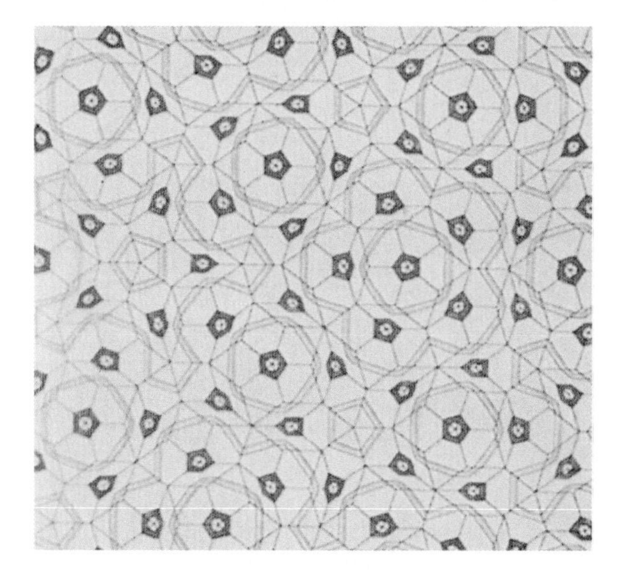

Figure 20.22 A Penrose nonperiodic tiling made with two rhombus shapes. (Tiling by Sir Roger Penrose.)

EXAMPLE

Fivefold Symmetry

Look again at Figure 20.20, which shows how to split a rhombus into the Penrose dart and kite pieces. Except in the recess of the dart and the matching part of the kite, all of the internal angles of the kite and of the dart are either 72° or 36°. Now, 72° goes into 360° five times, and 36° goes into 360° ten times. If we recall that it is the interior angles that matter in arranging polygons around a point, we see why it might be possible for a Penrose pattern to have fivefold or tenfold rotational symmetry.

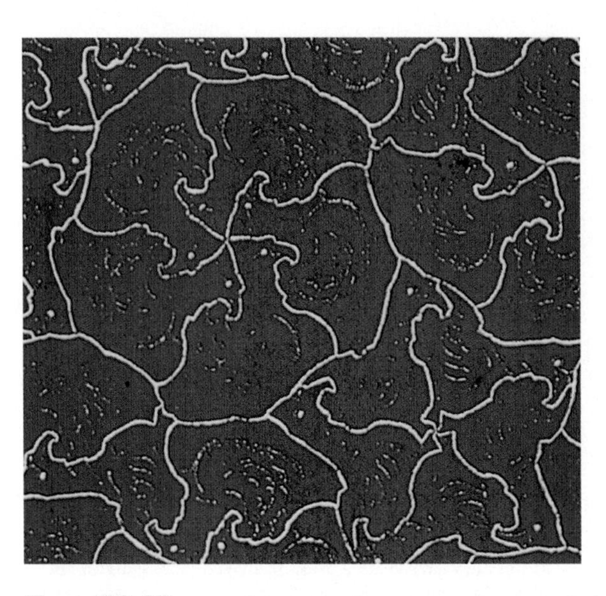

Figure 20.23 A modification of a Penrose tiling by refashioning the kites and darts into bird shapes. (Tiling by Sir Roger Penrose.)

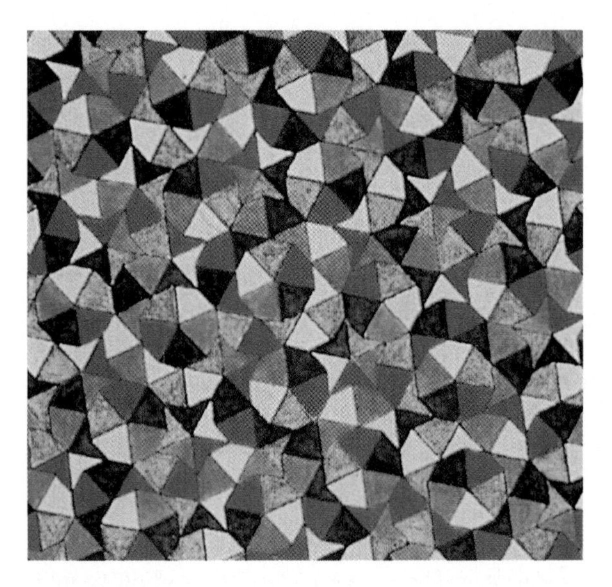

Figure 20.24 A Penrose tiling by kites and darts, colored with five colors. A Penrose tiling can always be colored using four colors, in such a way that two tiles that share an edge have different colors. Whether a Penrose tiling can be colored in such a way using only three colors is an unsolved problem; however, we know that if one Penrose tiling can be colored using three colors, all Penrose tilings can. (Tiling and coloring by Sir Roger Penrose.)

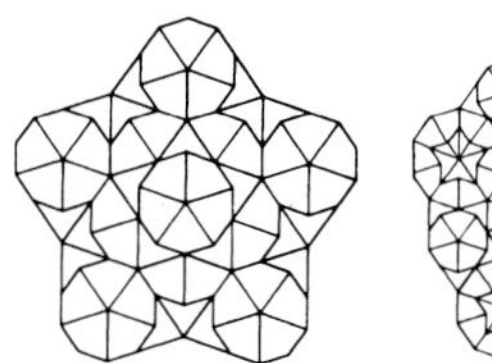
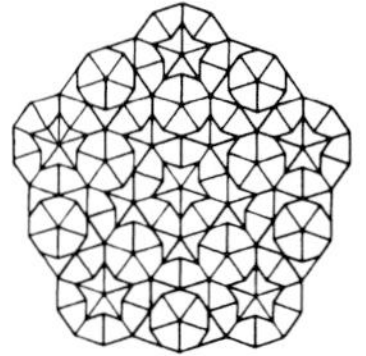

Figure 20.25 Successful deflation (that is, the systematic cutting up of large tiles into smaller ones) of patches of tiles of a Penrose nonperiodic tiling.

A Penrose pattern with tenfold rotational symmetry is impossible, but there are exactly two Penrose patterns that tile the entire plane with fivefold rotational symmetry about one particular point. We show finite parts of these patterns in Figure 20.25. For each pattern, the center of rotational symmetry is at the center of the figure, surrounded by either five darts or five kites.

For any other Penrose pattern, the pattern as a whole does not have fivefold rotational symmetry. However, what is surprising is that the pattern must have arbitrarily large finite regions with fivefold rotational symmetry. You can see this feature in the regions of Figure 20.22 that are enclosed by yellow lines. In Conway's metaphor, whenever a chain of children (darts) closes, the region inside has fivefold symmetry. ■

Conway invented a process called *inflation* that takes any Penrose pattern into a different Penrose pattern with larger darts and kites. The inflation operation (we don't give the details here) systematically cuts up the darts and kites into triangles and regroups the triangles into larger darts and kites.

Proceeding by contradiction, we can use inflation to show that a Penrose pattern must be nonperiodic. Suppose (contrary to what we want to establish) that some Penrose pattern is periodic, that is, it has translation symmetry. Let d be the distance along the translation direction to the first repetition. Performing inflation does the same thing to each repetition, so the inflated pattern must still have translation symmetry and a distance d along the translation direction to the first repetition. Keep on performing inflation, time after time, until the darts and kites are so large that they are more than d across. The pattern, as we have just argued, must still have translation symmetry at a distance d, but it can't, because there's no repetition inside a single tile! We reach a contradiction. So what's wrong? Our initial supposition, that the pattern was periodic in the first place, must have been erroneous. We conclude that all Penrose tilings are nonperiodic.

Despite their being nonperiodic, all Penrose patterns are somewhat alike, in the following remarkable sense:

> Any finite region in one Penrose pattern is contained somewhere inside every other Penrose pattern; in fact, it occurs infinitely many times in every Penrose pattern.

The nonperiodicity of Penrose tilings found a surprising application in 1997 – to bathroom tissue. Quilted bathroom tissue needs to be embossed with a pattern to keep the layers together (Figure 20.26). If the pattern is regular, then the multiple layers on the roll can produce lumpy ridges and grooves; using a nonrepeating Penrose pattern averts the lumpiness. (However, the company used Penrose's pattern without his permission, and Penrose sued.)

Penrose tilings have another feature that allows us to characterize them as *quasiperiodic,* or somewhere between periodic and random. (Noting the precise definition of this term would take us too far afield.) Robert Ammann introduced onto the two rhombic Penrose pieces used in Figure 20.22 lines that are now known as *Ammann bars.* In any Penrose tiling, these bars line up into five sets of parallel lines, each set rotated 72° from the next, forming a pentagonal grid

Figure 20.26 Sir Roger Penrose, an eminent British mathematician, discovered how to tile a surface with a pattern of flat geometric shapes in a way that is never quite repeated. In doing so, he illustrated fivefold symmetry – something that wasn't supposed to exist. Not an idea that would be particularly marketable to people redoing their bathrooms, one might think, but apparently useful for making quilted bathroom tissue, which must be embossed with a design that never repeats itself. Otherwise, layers upon layers of the same pattern build up ridges and grooves and the roll becomes lumpy. Penrose's design smoothes out the bumps. (Mario Ruiz/*Time Magazine.*)

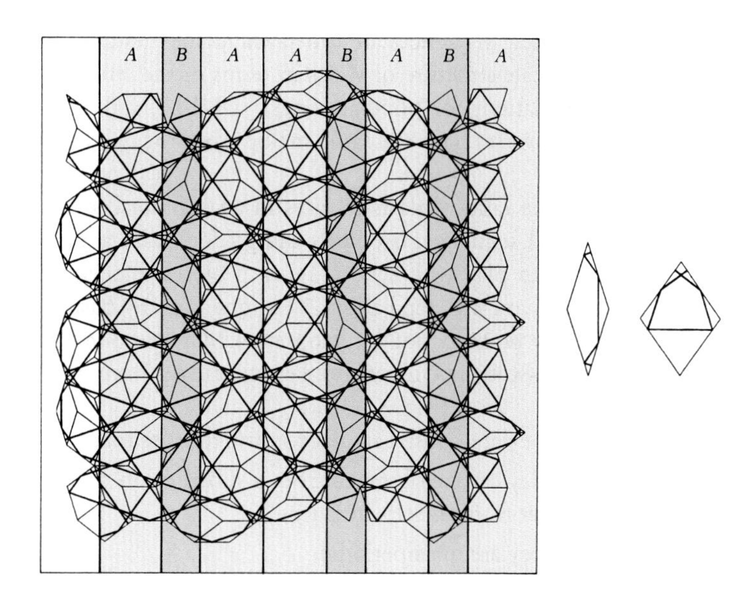

Figure 20.27 Penrose tilings with Ammann bars. Specially placed lines on the tiles produce five sets of parallel bars in different directions.

(Figure 20.27). The distance between two adjacent parallel bars is one of only two values, either A or B. Do you want to guess what the ratio of the longer A is to the shorter B? You don't think it could possibly be anything but the golden ratio of Chapter 19, do you? And so it is.

EXAMPLE

Musical Sequences

What about the order in which the A's and B's occur, as we move from left to right in Figure 20.27? Is there any pattern to that? From the limited part of the pattern we can observe, we see the sequence as

$$ABAABABAABABA$$

You might think from the figure that the pattern continues repeating the group

$$ABAAB$$

indefinitely; after all, there are five symbols in this group. But such is not the case. Known as a musical sequence, the sequence of intervals between Ammann bars is nonperiodic – it cannot be produced by repeating any finite group of symbols. We can think of it as a one-dimensional analogue of a Penrose tiling.

There is some regularity in musical sequences. Two B's can never be next to each other, nor can we have three A's in a row. Just as any finite part of any Penrose tiling occurs infinitely often in any other Penrose tiling, any finite part of any musical sequence appears infinitely often in any other one. The order of the symbols is neither periodic nor random, but between the two – quasiperiodic.

These sequences are called *musical* sequences because musicians represent the large-scale structure of songs in terms of the letters A and B. For example, a common pattern for popular songs is $AABA$, indicating that the first, second, and fourth verses have the same melody, but the third verse has a different melody. ■

The ratio of darts to kites in an infinite Penrose tiling, or of A's to B's in a musical sequence, is exactly the golden ratio, approximately 1.618. So if you are going to play with sets of Penrose pieces to see what kinds of patterns you can create, you will need about 1.6 times as many darts as kites.

As pointed out by geometers Marjorie Senechal (Smith College) and Jean Taylor (Rutgers University), Penrose tilings have three important properties:

- They are constructed according to rules that force nonperiodicity.
- They can be obtained from a substitution process (inflation and deflation) that features self-similarity.
- They are quasiperiodic.

Research of the late 1980s indicates that these properties are somewhat independent, meaning that one or two may be true of a tiling without all three being true.

20.10 Quasicrystals and Barlow's Law

Although Penrose's discovery was a big hit among geometers and in recreational mathematics circles in the mid-1970s, few people thought that his work might have practical significance. In the early 1980s some mathematicians even generalized Penrose tilings to three dimensions, using solid polyhedra to fill space nonperiodically. Like the two-dimensional Penrose patterns, these have orderly fivefold symmetry but are nonperiodic.

Yet in 1982 scientists at the U.S. National Bureau of Standards discovered unexpected fivefold symmetry while looking for new ultrastrong alloys of aluminum (mixtures of aluminum with other metals).

Manganese doesn't ordinarily alloy with aluminum, but the experimenters were able to produce small crystals of alloy by cooling mixtures of the two metals at a rate of millions of degrees per second. Following routine procedures, chemist Daniel Shechtman began a series of tests to determine the atomic structure of the special crystals. But there was nothing routine about what he found: The atomic structures of the manganese–aluminum crystals were so startling that it took Shechtman three years to convince his colleagues they were real.

Why did he encounter such resistance? His patterns—and the crystals that produced them—defied one of the fundamental laws of crystallography. Like our discovery that the plane cannot be tiled by regular pentagons, **Barlow's law,** also called the **crystallographic restriction,** says that a crystal can have only rotational symmetries that are twofold, threefold, fourfold, or sixfold. Because crystals are periodic, if there were a center of fivefold symmetry, there would have to be many such centers. Barlow proved this impossible.

Peter Barlow (1776–1862) was a British mathematician whose name survives today in the name of a book of mathematical tables. His argument was a very simple proof by contradiction, similar to Conway's proof in which we saw earlier that Penrose patterns are not periodic. Suppose (contrary to what we intend to show) that there is more than one fivefold rotation center. Let A and B be two of these that are closest together (see Figure 20.28). Rotate the pattern of Figure 20.28 by one-fifth of a turn clockwise around B, which carries A to some point A'. Because the pattern has fivefold symmetry around B, the point A', which is the image of the fivefold center A, must itself be a fivefold center. Now use A as a center and rotate the pattern by one-fifth of a turn counterclockwise, which carries B to some point B'; as we just argued in the case of A', B' must also be a fivefold center. But A' and B' are closer together than A and B, which is a contradiction. Hence our original supposition must be false, and a pattern can have at most one fivefold rotation center (as the patterns in Figure 20.25 in fact do) and so cannot be periodic.

Barlow's law, as a mathematical theorem, shows that fivefold symmetry is impossible in a periodic tiling of the plane or of space. Chemists, for good theoretical and experimental reasons, believe that crystals are modeled well by three-dimensional tilings. An array of atoms with no symmetry whatever would not be considered a crystal. Yet until Penrose's discovery, no one realized that nonperiodic tilings – or arrays of atoms – can have the regularity of fivefold symmetry.

Chemists could simply say that Shechtman's alloys aren't crystals. In the classic sense they aren't, but in other respects they do resemble crystals. It is scientifically more fruitful to extend the concept of crystal to include them rather than rule them out; they are now known as *quasicrystals* (see Spotlight 20.4).

Once again, as so often happens in history, pure mathematical research anticipated scientific applications. Penrose's discovery, once just a delightful piece of recreational mathematics, has prompted a major reexamination of the theory of crystals.

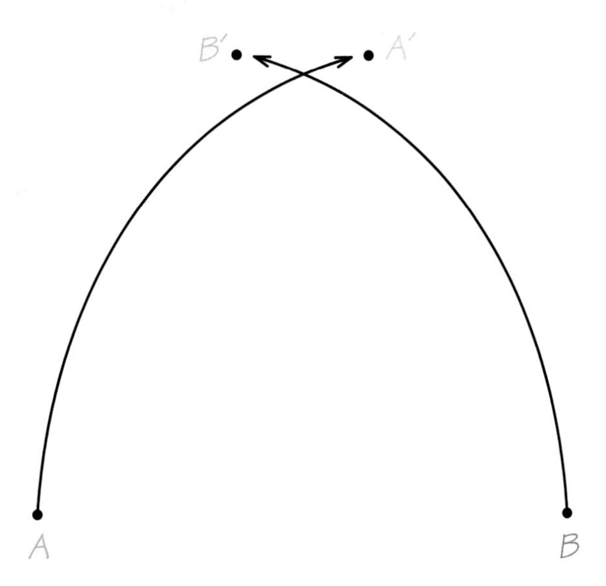

Figure 20.28 Barlow's proof that no pattern can have two centers of fivefold symmetry.

SPOTLIGHT 20.4 Quasicrystals

In 1984, working at the University of Pennsylvania, Paul Steinhardt and Don Levine did a computer simulation of what a three-dimensional Penrose pattern would be like; they decided to call such structures quasicrystals. Later that fall, their chemist colleague Daniel Shechtman showed that quasicrystals really exist; he produced images of an alloy of aluminum and manganese that were amazingly similar to images from the computer simulations. In short order, sevenfold, ninefold, and other symmetries were also shown to occur in real materials.

In 1991 Sergei Burkov showed that quasiperiodic tilings can be made using only a single kind of 10-sided tile, *provided the tiles are allowed to overlap*. With overlaps, the resulting patterns are no longer tilings; they are called *coverages*. In late 1998, scientists presented electron microscope photos that demonstrated that atoms really do form such coverages.

The current theory is that quasicrystals are packings of copies of a single type of atom cluster, with each cluster sharing atoms with its neighbors, that is, overlapping nearby clusters. The clusters form a quasiperiodic pattern that maximizes their density, thereby minimizing the energy of the atoms involved.

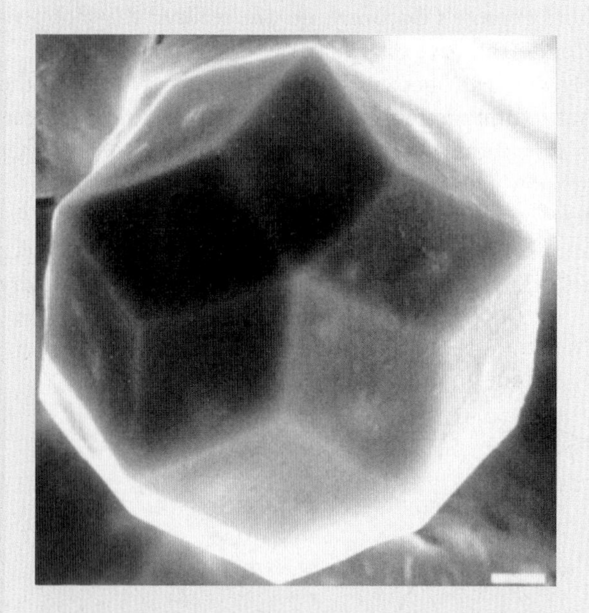

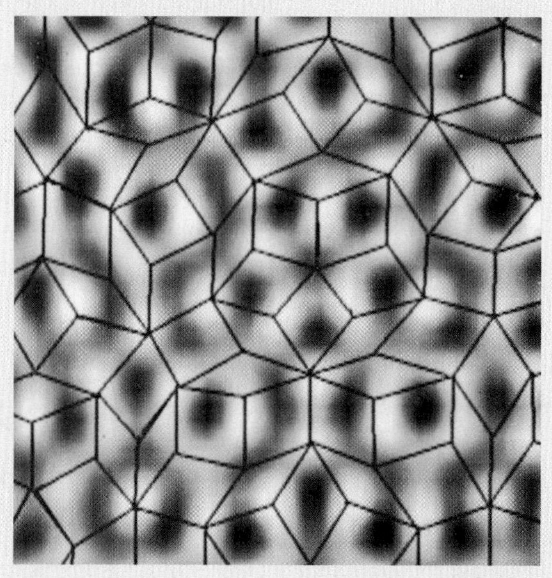

(a) A scanning electron microscope image of the quasicrystal alloy $Al_{5???}$ Li_3Cu. The fivefold symmetry can be seen in the five rhombic faces that meet at a single point in the center of the photograph, forming a starlike shape.
(b) This image of the quasicrystal material $Al_{65}Co_{20}Cu_{15}$ was obtained with a scanning tunneling microscope; the resulting image has been overlaid with a nonperiodic tiling to display the local fivefold symmetry. [Both adapted from Hans C. von Baeyer, "Impossible Crystals," *Discover* 11(2) (February 1990): 69–78, 84.]

REVIEW VOCABULARY

Barlow's law, or the **crystallographic restriction** A law of crystallography that states that a crystal may have only rotational symmetries that are twofold, threefold, fourfold, or sixfold.
Centrosymmetric Symmetric by 180° rotation around its center.
Convex A geometric figure is convex if for any two points on the figure (including its boundary), all the points on the line segment joining them also belong to the figure (including its boundary).
Conway criterion A criterion for determining whether a shape can tile by means of translations and half-turns.
Edge-to-edge tiling A tiling in which adjacent tiles meet only along full edges of each tile.
Equilateral triangle A triangle with all three sides equal.
Exterior angle The angle outside a polygon formed by one side and the extension of an adjacent side.
Fundamental region A tile or group of adjacent tiles that can tile by translation.
Interior angle The angle inside a polygon formed by two adjacent sides.
Monohedral tiling A tiling with only one size and shape of tile (the tile is allowed to occur also in "turned-over," or mirror-image, form).
***n*-gon** A polygon with n sides.
Nonperiodic tiling A tiling in which there is no repetition of the pattern by translation.
Parallelogram A convex quadrilateral whose opposite sides are equal and parallel.
Par-hexagon A hexagon whose opposite sides are equal and parallel.
Periodic tiling A tiling that repeats at fixed intervals in two different directions, possibly horizontal and vertical.
Quadrilateral A polygon with four sides.
Regular polygon A polygon all of whose sides and angles are equal.
Regular tiling A tiling by regular polygons, all of which have the same number of sides and are the same size; also, at each vertex, the same kinds of polygons must meet in the same order.
Rhombus A parallelogram all of whose sides are equal, i.e., four equal sides and equal opposite interior angles.
Scalene triangle A triangle no two sides of which are equal.
Semiregular tiling A tiling by regular polygons; all polygons with the same number of sides must be the same size.
Tiling A convering of the plane without gaps or overlaps.
Translation A rigid motion that moves everything a certain distance in one direction.
Vertex figure The pattern of polygons surrounding a vertex in a tiling.

SUGGESTED READINGS

CHOW, WILLIAM W. Interlocking shapes in art and engineering, *Computer Aided Design,* 12 (1980): 29–34. Discusses applications to sheet material manufacturing (e.g., fabrication of gloves, can openers, forks, key blanks, and bunk bed brackets).

CHOW, WILLIAM W. Automatic generation of interlocking shapes, *Computer Graphics and Image Processing,* 9 (1979): 333–353. Shows how to design a computer program to draw interlocking patterns.

CHUNG, FAN, and SHLOMO STERNBERG. Mathematics and the buckyball, *American Scientist,* 81 (1993): 56–71.

FOSTER, LORRAINE. *The Alhambra Past and Present: A Geometer's Odyssey.* Parts I and II. VHS Videotapes, 30 min. each. California State University, Northridge, Calif., 1992. Illustrates strip and wallpaper patterns from the Alhambra in Spain. Available from the author at profllfoster@earthlink.net.

GARDNER, MARTIN. Mathematical games: Extraordinary nonperiodic tiling that enriches the theory of tiles, *Scientific American* (January 1977): 110–121, 132, and front cover. Reprinted with additional material in Martin Gardner, *Penrose Tiles to Trapdoor Ciphers,* Freeman, New York, 1989, pp. 1–29.

GARDNER, MARTIN. Mathematical games: On tessellating the plane with convex polygon tiles, *Scientific American* (July 1975): 112–117, 132. Reprinted with additional material in Martin Gardner, *Time Travel and Other Mathematical Bewilderments,* Freeman, New York, 1988, pp. 163–176.

GRÜNBAUM, BRANKO, and G. C. SHEPHARD. *Tilings and Patterns,* Freeman, New York, 1987. Abbreviated edition: *Tilings and Patterns: An Introduction,* Freeman, New York, 1989.

HALES, THOMAS. The Kepler Conjecture. http:/www.math.pitt.edu/~thales/countdown/.

KORYEPIN, V. Penrose patterns and quasi-crystals: What does tiling have to do with a high-tech alloy? *Quantum,* 4(4) (January/February 1994): 13–19; 4(5) (March/April 1994): 59–62.

LEE, KEVIN. *TesselMania! Deluxe.* Computer program for producing Escher-like tilings for Macintosh or Windows. Available from www.worldofescher.com/store/compaccs.html. Free PC demo downloadable at www.worldofescher.com/down/tessdemo.exe.

PETERSON, IVARS. A quasicrystal construction kit. *Science News* 155 (January 23, 1999): 60–61.

RANUCCI, ERNEST, and JOSEPH TEETERS. *Creating Escher-Type Patterns,* Creative Publications, Oak Lawn, Ill., 1977.

SCHATTSCHNEIDER, DORIS. Penrose puzzles, *SIAM News,* 28(6) (July 1995): 8, 14. Review of various commercial puzzles based on variations on the Penrose pieces, distributed by Kadon Enterprises (1227 Lorene Drive, Suite 16, Pasadena, MD 21122; 410-437-2163) and World of Escher (14542 Brook Hollow Boulevard, no. 250, San Antonio, TX 78232-3810; 800-237-2232).

SCHATTSCHNEIDER, DORIS. *Visions of Symmetry: Notebooks, Periodic Drawings, and Related Work of M. C. Escher,* Freeman, New York, 1990.

SCHATTSCHNEIDER, DORIS. In praise of amateurs. In David A. Klarner (ed.), *The Mathematical Gardner,* Wadsworth, Belmont, Calif., 1981, pp. 140–166, plus Plates I–III.

SCHATTSCHNEIDER, DORIS. Will it tile? Try the Conway criterion! *Mathematics Magazine,* 53 (1980): 224–233.

SEYMOUR, DALE, and JILL BRITTON. *Introduction to Tessellations,* Dale Seymour Publications, Palo Alto, Calif., 1989. An excellent introduction to tessellations, including how to make Escher-like tessellations.

TEETERS, JOSEPH L. How to draw tessellations of the Escher type, *Mathematics Teacher,* 67(1974): 307–310.

Tessellation Winners: Original Student Art, 2 vols. *Book One: The First Contest, 1989–90,* and *Book Two: The Second Contest, 1991–92,* Dale Seymour Publications, Palo Alto, Calif., 1991, 1993.

URBAN, KNUT W. From tilings to coverings, *Nature* 396 (November 5, 1998): 14–15.

WAGON, STAN. *Mathematica in Action,* 2nd ed., Springer-Verlag, New York, 2000. Chapters 9 and 22 show how to use the Mathematica software to generate Penrose patterns and Escher-like patterns; the accompanying CD-ROM contains the programs' code.

SUGGESTED WEB SITES

www.geom.umn.edu/software/tilings Lists programs for various platforms that allow the user to design tilings.

www.geocities.com/SiliconValley/Pines/1684/Penrose.html A Java applet to play with Penrose tiles.

www.geom.umn.edu/apps/quasitiler Interactive Web program that draws Penrose tilings and their generalizations in higher dimensions.

ftp://geom.umn.edu/pub/software/KaleidoTile Interactive Macintosh program that lets the user design tilings on the plane, the sphere, and the hyperbolic plane. Spherical tilings can be realized as polyhedra.

SKILLS CHECK

1. What is the exterior angle of a regular octagon?

(a) 30°
(b) 45°
(c) 135°

2. Which of the following polygons *cannot* tile the plane?

(a) Regular hexagon
(b) Regular pentagon
(c) Nonrectangle parallelogram

3. Regular octagons and squares can form a semiregular tiling of the plane with

(a) two octagons and one square at each vertex.
(b) two octagons and two squares at each vertex.
(c) a varying configuration at the vertices.

4. A semiregular tiling has a square, a regular dodecagon (12-gon), and another regular polygon at each vertex. What is this other polygon?

(a) Another square
(b) Another dodecagon
(c) A hexagon

5. Can the tile below be used to create a tiling of the plane?

(a) No
(b) Yes, using only translations
(c) Yes, using translations and half-turns

6. Can the tile below be used to create a tiling of the plane?

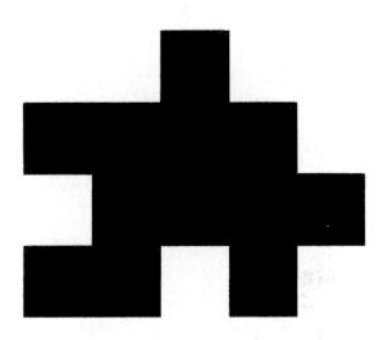

(a) No
(b) Yes, using only translations
(c) Yes, using translations and half-turns

7. Penrose tilings are

(a) periodic.
(b) random.
(c) quasiperiodic.

8. A regular tiling can be constructed using

(a) rectangles.
(b) hexagons.
(c) squares and octagons.

9. A tiling of the plane can be formed using as a tile

(a) some but not all convex quadrilaterals.
(b) any convex quadrilateral, but no nonconvex quadrilaterals.
(c) any quadrilateral.

10. A tiling of the plane can be formed using as a tile

(a) some pentagons.
(b) any pentagon with at least two right angles.
(c) any pentagon with at least three right angles.

11. A tiling of the plane can be formed using as a tile

(a) some but not all convex hexagons.
(b) any convex hexagon, but no nonconvex hexagons.
(c) any hexagon.

12. A rhombus always has the trait that

(a) opposite sides are unequal in length.
(b) opposite angles are congruent.
(c) it cannot be a square.

13. A Penrose dart always has the trait that

(a) opposite angles are congruent.
(b) it is nonconvex.
(c) the edges are all of different lengths.

14. In a Penrose tiling, the proportion of darts to kites is

(a) one-to-one.
(b) two-to-one.
(c) the golden ratio.

15. Barlow's law prohibits the existence of crystals with

(a) fivefold symmetry.
(b) sixfold symmetry.
(c) periodic tilings.

EXERCISES

▲ **Optional.** ■ **Advanced.** ◆ **Discussion.**

Hint: For the exercises about determining whether a shape will tile the plane, you should make a number of copies of the shape and experiment with placing them. One easy way to make copies is to trace the shape onto a piece of paper, staple half a dozen other blank sheets behind that sheet, and use scissors to cut through all the sheets along the edges of the traced shape on the top sheet.

Regular Polygons

1. Determine the measure of an exterior angle and of an interior angle of a regular octagon (eight sides).

2. Determine the measure of an exterior angle and of an interior angle of a regular decagon (10 sides).

3. Discover a formula for the measure of an interior angle of a regular n-gon.

4. Using the formula from Exercise 3 and either your calculator or a short computer program, make a chart of the interior-angle measures of regular polygons with 3, 4, . . . , 12 sides.

Regular Tilings

5. Give a numerical reason why a semiregular tiling could not include both polygons with 12 sides and polygons with 8 sides (with or without any polygons with other numbers of sides).

6. The lower left corner of Figure 20.5 shows a tiling by isosceles triangles.

(a) Use the center vertex to determine the measures of the angles of the isosceles triangle tile.
(b) Every vertex except the center vertex has the same vertex figure, in terms of the measures of the angles surrounding the vertex. What is that vertex figure?

Tilings with Irregular Polygons

7. For each of the tiles below, show how it can be used to tile the plane. (Adapted from *Tilings and Patterns,* by Branko Grünbaum and G. C. Shephard, Freeman, New York, 1987, p. 25.)

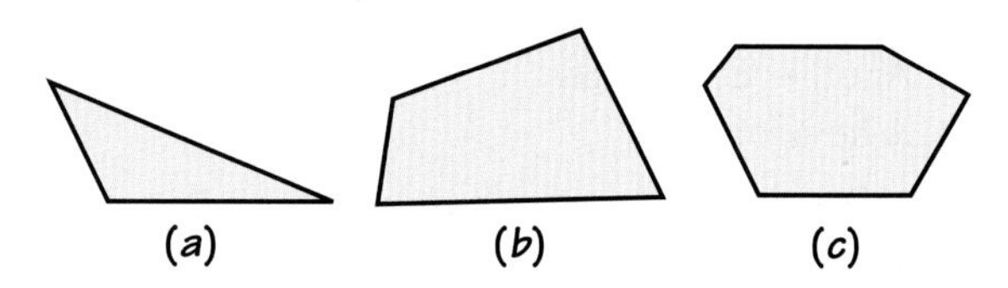

8. You know that a regular pentagon cannot tile the plane. Suppose you cut one in half. Can this new shape tile the plane? (See page 686 for a regular pentagon that you can trace.)

Tiling by Translations

Refer to accompanying tiles (a) through (g) in doing Exercises 9 and 10.

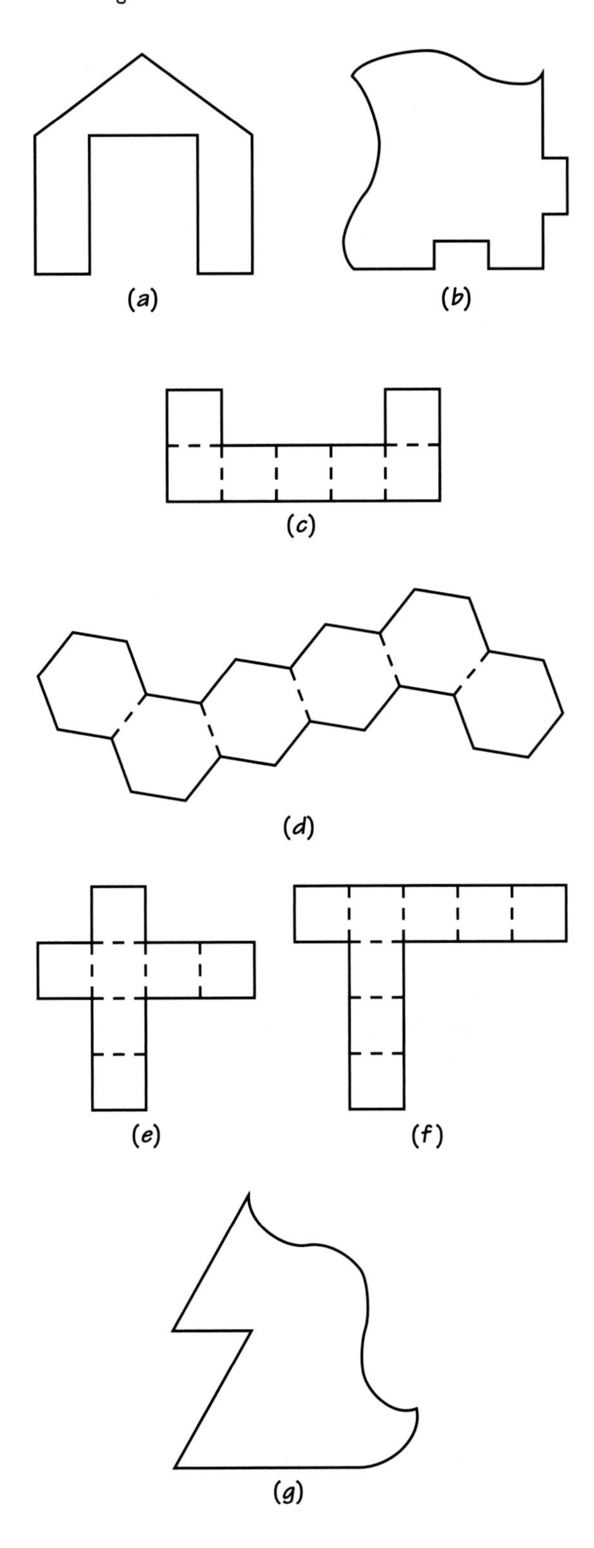

9. For each of the tiles (a) through (c), determine whether it can be used to tile the plane by translations. (For *Tiling the Plane*, by Frederick Barber et al., COMAP, Lexington, Mass., 1989, pp. 1, 8, 9.)

10. Repeat Exercise 9, but for tiles (d) through (g).

11. Start from a parallelogram of your choice and modify it to tile the plane by translations. (You will probably find it useful to do your work on graph paper.) Can you draw a design on the tile so as to make an Escher-like pattern?

12. Start from a par-hexagon of your choice and modify it to tile the plane by translations. (You will probably find it useful to do your work on graph paper. If you choose a regular hexagon, there is special graph paper, ruled into regular hexagons, that would be particularly useful.) Can you draw a design on the tile so as to make an Escher-like pattern?

Tiling by Translations and Half-Turns

Refer to tiles (a) through (g) at left in doing Exercises 13 and 14.

13. For each of the tiles (a) through (c), determine whether it can be used to tile the plane by translations and half-turns.

14. Repeat Exercise 13, but for tiles (d) through (g).

15. Show how an arbitrary pentagon with two parallel sides can tile the plane.

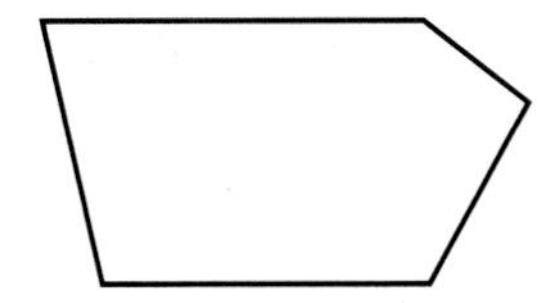

16. Shown below is a pentagonal tile of type 13, discovered by Marjorie Rice. Show how it can tile the plane. (*Hint:* Carefully trace and cut out a dozen or so copies and try fitting them together.)

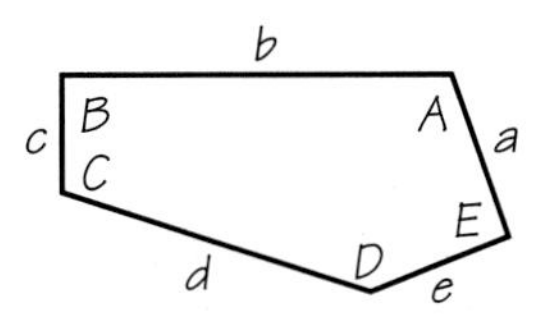

The parts of this pentagon satisfy the following relations: $A = C = D = 120°$, $B = E = 90°$, $2A + D = 360°$, $2C + D = 360°$, $a = e$, and $a + e = d$. (Adapted from "In Praise of Amateurs," by Doris Schattschneider, in David A. Klarner [ed.], *The Mathematical Gardner*, Wadsworth, Belmont, Calif., 1981, p. 162.)

17. Start from a triangle of your choice and modify it to tile the plane by translations and half-turns. (You will probably find it useful to do your work on graph paper.) Can you draw a design on the tile so as to make an Escher-like pattern?

18. Start from a quadrilateral of your choice and modify it to tile the plane by translations and half-turns. (You will probably find it useful to do your work on graph paper.) Can you draw a design on the tile so as to make an Escher-like pattern?

Additional Exercises

■ **19.** Use the chart of interior-angle measures from Exercise 4 to determine all of the possible vertex figures of regular polygons (with at most 12 sides) surrounding a point.

■ **20.** Which of the vertex figures of Exercise 19 do not occur in a semiregular tiling?

■ **21.** In addition to the vertex figures of Exercise 19, exactly five others are possible, each involving one polygon with more than 12 sides. None of these vertex figures lead to a semiregular tiling. The five many-sided polygons involved in these five vertex figures have 15, 18, 20, 24, and 42 sides. Determine the other polygons in these vertex figures.

▲ **22.** In the text we discuss criteria and methods for generating Escher-like patterns that involve just translations or translations and half-turns. A slight variation on one of those methods allows construction of tilings that feature a tile and its mirror image.

Begin with a parallelogram made from two congruent isosceles triangles, as shown in the figure at right. Each of these triangles has two sides equal; be sure that the two triangles are arranged so that they have one of the equal sides in common, forming a diagonal of the parallelogram.

Make any modification to half of the third side of one of the triangles. Mirror-reflect that modification across the side, then translate the reflection to become the modification of the other half of the side. Take the complete modification of this side and translate it to become the modification of the opposite side of the parallelogram.

Modify in any way one of the two remaining sides of the parallelogram, and make the same modification to the opposite side (that is, translate the modification, without rotation or reflection). Then make the mirror reflection of this modification the modification of the diagonal of the parallelogram.

The result is a modified parallelogram that tiles by translation and splits into two pieces that are mirror images of each other. Escher used a similar technique, but starting from a par-hexagon made from two quadrilaterals, in his Horseman print, as shown in his sketch in Figure 20.11b.

Use this technique to produce a tiling of your own design. Can you draw a design on the tile so as to make an Escher-like pattern?

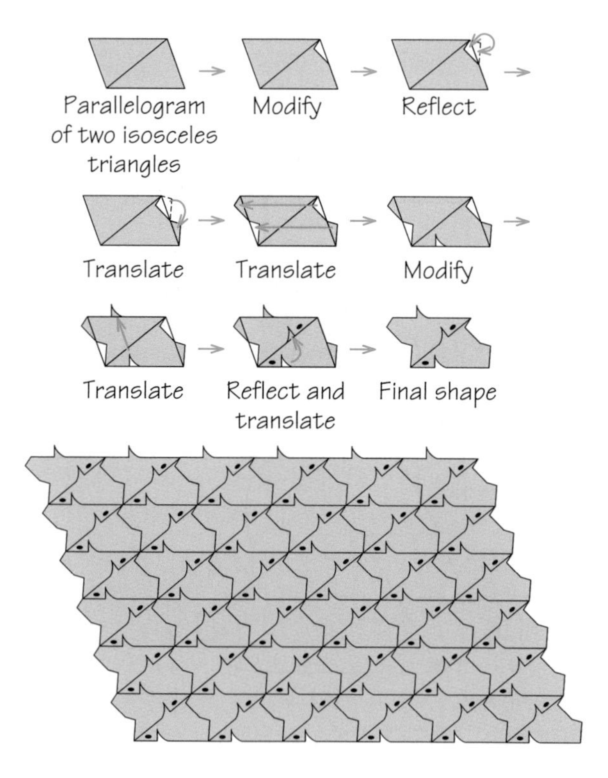

▲ **23.** Show that the modified parallelogram in Exercise 22 fulfills the Conway criterion, by identifying the six points of the criterion.

▲ **24.** The rabbit problem in Chapter 19 (Exercise 4) can lead us directly into nonperiodic patterns and musical sequences. Let A denote an adult pair of rabbits and B denote a baby pair. We will record the population at the end of each month, just before any births, in a particular systematic way—as a string of A's and B's. At the end of their second month of life, a rabbit pair will be considered to be adult. At the end of the first month, the sequence is just A, and the same is true at the end of the second month. When an adult pair A has a baby pair B, we write the new B immediately to the right of the A. So at the end of the third month, the sequence is AB; at the end of the fourth, it is ABA, because the first baby pair is now adult; at the end of the fifth month we have $ABAAB$.

Mathematicians and computer scientists call this manner of generating a sequence a *replacement system*. At each stage we replace each A by AB and each B by A.

(a) What is the sequence at the end of the sixth month?

(b) Why can't we ever have two B's next to each other?

(c) Why can't we ever have three A's in a row?

(d) Show that from the fourth month on, the sequence for the current month consists of the sequence for last month followed by the sequence for two months ago.

✎ WRITING PROJECTS

1. We can define inflation and deflation of a sequence of A's and B's, and musical sequences themselves, without reference to Penrose patterns, and thereby arrive at an example of a nonperiodic pattern in one dimension. Inflation consists of replacing each A by AB and each B by A, and deflation consists of replacing each AB by A and each A by B; inflation and deflation undo each other on musical sequences. Call a sequence *musical* if it results from applying inflation to the sequence consisting of a single B. Then inflation and deflation preserve musicality: If we inflate or deflate a musical sequence, we get another musical sequence. Another way to think of this relationship is that a musical sequence is self-similar under inflation and deflation.

(a) Let the lone B be considered the first stage of inflation. Show that at the nth stage of inflation, for $n \geq 3$, there are F_n (the nth Fibonacci number) symbols in the sequence, of which F_{n-1} are A's and F_{n-2} are B's.

(b) Show that no musical sequence contains AAA or BB.

(c) Show that no musical sequence ends in AA or in $ABAB$.

(d) Show that apart from the lone sequence B, every musical sequence is an initial subsequence of all the succeeding musical sequences.

(e) Slightly modified, deflation can be used to check whether a finite block of A's and B's can belong to a musical sequence or not. First, if the block has length greater than one, we may suppose that it begins with an A (why?); so at any stage of the deflation with a block beginning with B, we may add an initial A. Second, we add the additional deflation rule to replace an ending AA with BA. If at any stage of this modified deflation we arrive at two or more B's in a row, or three or more A's in a row, then the original block could not be part of a musical sequence; otherwise, the original block will eventually deflate to a single symbol, at which point we conclude that the original block is a part of a musical sequence. Check the two blocks $ABAABABAAB$ and $ABAABABABA$.

(f) From part (d) we know that each application of inflation to a musical sequence simply extends it;

by successive inflation, then, we build an infinite sequence. Show that as we approach this limiting sequence, the ratio of *B*'s to *A*'s tends toward the golden ratio ϕ.

(g) Conclude from part (f) that the sequence cannot be periodic, nor settle into a period after a finite "burn-in" period. Thus, the sequence is nonperiodic. (*Hint:* ϕ is not a rational number, that is, it cannot be represented as a ratio m/n of whole numbers m and n.)

(h) Show that any finite block of *A*'s and *B*'s that occurs in the infinite sequence must occur over and over again (just as any patch of tiles in a Penrose pattern occurs infinitely often in the pattern). Thus, the infinite sequence is self-similar.

2. Get computer software for tiling and make some tilings of your own. Check for software for your computing platform at www.geom.umn.edu/software/tilings/Tiling Software.html.

More on the Web SPREADSHEET PROJECTS

To do these projects, go to www.whfreeman.com/fapp.

Polygons tile when their interior angles total 360° at each vertex. This project first computes the exterior and interior angles of the regular polygons. Using this information, a spreadsheet is used to design potential semiregular tilings.

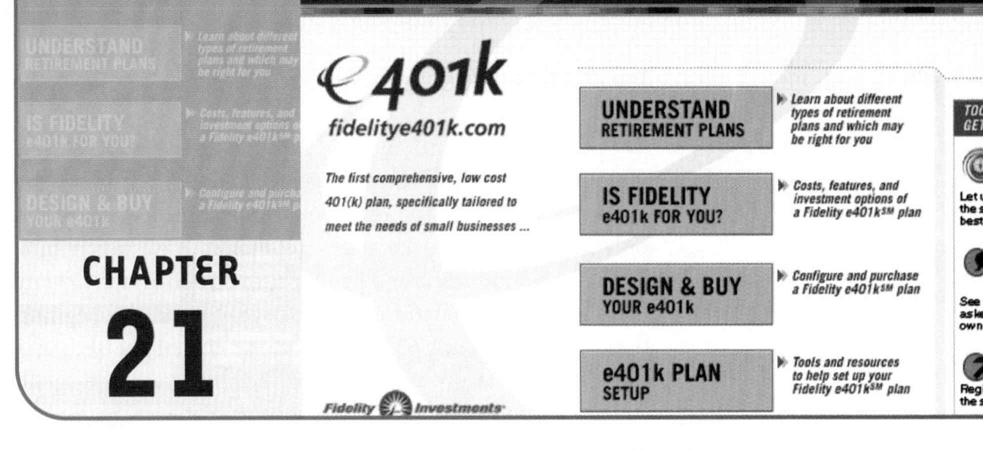

CHAPTER 21

Consumer Finance Models: Saving

In this chapter we cover...

... and at www.whfreeman.com/fapp:

Flashcards

Quizzes

Spreadsheet Projects

Consumer price index

Applet Exercises

Saving for retirement

How much interest will your savings account earn in the next year? How much will the payment be on your credit card loan, your car loan, or your home mortgage? How much would you need to save each month to pay for a child's college education? How much should you save for retirement? These are problems of daily life for which mathematics provides custom-tailored models. In this chapter and the next, we examine the mathematics of finance and models for accumulation and disbursal of funds.

Good mathematical models are often versatile and flexible, and the financial models of these chapters indeed apply broadly to important problems in other areas of life. Growth of money at interest like growth of some biological populations. Inflation of a currency or depreciation of an asset is like the decay of a radioactive substance. Finding out how long a retirement "nest egg" will last is similar to determining how long before a nonrenewable resource, such as oil or coal, may be exhausted. Managing a trust fund, such as the endowment of a college, presents problems similar to management of a renewable biological resource, such as a forest or a fishery.

21.1 Arithmetic Growth

When you open a savings account, your primary concerns are the safety and the growth of the "population" of your savings. Suppose that you deposit \$1000 in an account that, you are told, "pays interest at a rate of 10% annually." (This is an unrealistic rate that we use only because it makes the calculations simple.) Assuming that you make no other deposits or withdrawals, how much is in the account after 1, 2, or 5 years?

The \$1000 is the **principal,** the initial balance of the account. At the end of one year, **interest** is added. The amount of interest is 10% of the principal, or

$$10\% \times \$1000 = 10 \times 0.01 \times \$1000 = 0.10 \times \$1000 = \$100$$

in this case. ("Percent" means "per 100," so you can think of the symbol "%" as standing for " $\times 0.01$.") So the balance at the beginning of the second year is \$1100.

Under one method of paying interest, called **simple interest,** interest is paid only on the original balance, no matter how much interest has accumulated. With simple interest, for a principal of \$1000 and a 10% interest rate, you receive \$100 interest at the end of the first year; so at the beginning of the second year, the account contains \$1100. But at the end of the second year, you again receive only \$100; so at the beginning of the third year, the account contains \$1200. In fact, at the end of each year you receive just \$100 in interest.

You no doubt find this method for interest rather strange, as you are accustomed from your savings account to a different system of awarding interest (which we will consider shortly). However, simple interest has a very important application in the financing of corporations and the government by individuals and other corporations, through bonds. In the next chapter, we will look at the role of simple interest in consumer loans.

EXAMPLE Bonds

Governments and firms use bonds to borrow money. A bond is an obligation to repay a specified amount of money at the end of a fixed term, with simple interest paid annually. For instance, in April 2002 you could have bought a 30-year U.S. Treasury bond with *face value* (or *par value*) of \$10,000 that pays a *coupon rate* of 3.375% simple interest every subsequent April through the *maturity date* of April 2032.

An additional consideration is that the bond can be resold, at a higher price than \$10,000 if prevailing interest rates dip below 3.375%. If interest rates go up, the bond itself becomes less valuable (in terms of selling price) but still earns you \$337.50 every April and returns the borrowed principal of \$10,000 in April 2032. ■

The formulas involving simple interest are themselves simple. For a principal P and an annual rate of interest r, the interest earned in t years is

$$I = Prt$$

and the total amount A accumulated in the account is

$$A = P(1 + rt)$$

EXAMPLE Earnings on Those Bonds

Suppose that you bought one of those 30-year U.S. Treasury bonds in April 2002 for \$10,000. How much interest would it earn by April 2032, and what would be the total amount accumulated? We have $P = \$10{,}000$, $r = 3.375\% = 0.03375$, and $t = 30$ years:

$$I = Prt = \$10{,}000 \times 0.0375 \times 30 = \$10{,}125$$

and

$$A = P(1 + rt) = \$10{,}000 \times (1 + 0.03375 \times 30) = \$10{,}000 \times (2.125) = \$20{,}125.$$ ■

Apart from bonds and some consumer loans, simple interest is seldom used in today's financial institutions. However, we frequently observe the corresponding kind of growth, called *arithmetic growth,* in other contexts.

Arithmetic growth (also called **linear growth**) is growth by a constant amount in each time period.

For example, the population of medical doctors in the United States grows arithmetically, since the fixed number of medical schools each graduate the same total number of doctors each year (and the number of doctors dying is also fairly constant). The concept of linear growth appeared already in the discussions of linear programming (Chapter 4) and linear regression (Chapter 6).

21.2 Compound Interest

What you probably expected to happen to the savings account discussed in the last section is that during the second year the account would earn interest of 10% not on the *initial* balance of \$1000 (as with simple interest) but on the *new* balance of \$1100. Then, at the end of the second year, 10% of \$1100, or \$110, would be added to the account.

Thus, during the second year you would earn interest on both the principal of \$1000 and on the \$100 interest earned during the first year. Interest that is paid on both the principal and on the accumulated interest is known as **compound interest**. You receive more interest during the second year than during the first, that is, the account grows by a greater amount during the second year. At the beginning of the third year the account contains \$1210, so at the end of the third year you receive \$121 in interest. Again, this is larger than the amount you received at the end of the preceding year. Moreover, the increase during the third year,

$$\text{Third-year interest} - \text{second-year interest} = \$121 - \$110 = \$11$$

is larger than the increase during the second year,

$$\text{Second-year interest} - \text{first-year interest} = \$110 - \$100 = \$10$$

Thus, not only is the account balance increasing each year, but the amount added also increases each year.

Banks and savings institutions often compound interest more often than once a year—for example, quarterly (four times per year). With an interest rate of

10% per year and quarterly compounding, you get one-fourth of the rate, or 2.5%, paid in interest each quarter. The quarter (three months) is the **compounding period**, or the time elapsing before interest is paid.

Consider again a principal of $1000. At the end of the first quarter, you have the original balance plus $25 interest, so the balance at the beginning of the second quarter is $1025. During the second quarter you receive interest equal to 2.5% of $1025, or $25.63, so the balance at the end of the second quarter is $1050.63. Continuing in this manner, you find that the balance at the end of the first year is $1103.81. (You should "read" all calculations in this chapter by confirming them on your calculator.)

Even though the account was advertised as paying 10% interest (the *nominal rate*), the interest of $103.81 for the year is 10.381% of the principal of $10,000. The 10.381% is the *effective rate* or *equivalent yield.*

If interest is compounded monthly (12 times per year) or daily (365 times per year), the resulting balance is even larger. A comparison of yearly, quarterly, monthly, and daily compounding for an interest rate of 10% is shown in Table 21.1. We will shortly summarize results in a general formula.

From now on, we express all interest rates as decimals. For example, 10% is 0.10; to convert a percentage to a fraction, divide the percentage by 100 by moving the decimal point two places to the left. An interest rate of 0.10 is the same as 10%, or 10/100; an interest rate i is the same as $100\,i$ %.

21.3 Terminology for Interest Rates

We have seen that an account at a particular annual rate of interest can produce different amounts of interest depending on how the compounding is done. To help prevent confusion on the part of consumers, the Truth in Savings Act establishes specific terminology and calculation methods for interest.

A **nominal rate** is any stated rate of interest for a specified length of time, such as a 3% annual interest rate on a savings account or a 1.5% monthly rate on a credit card balance. By itself, such a rate *does not indicate or take into account whether or how often interest is compounded.*

The **effective rate** is the actual percentage rate of increase for a length of time, *taking into account compounding.* It equals the rate of simple interest that

TABLE 21.1 Comparing Compound Interest: The Value of $1000, at 10% Annual Interest, for Different Compounding Periods

Years	Compounded Yearly	Compounded Quarterly	Compounded Monthly	Compounded Daily	Compounded Continuously
1	1100.00	1103.81	1104.71	1105.16	1105.17
5	1610.51	1638.62	1645.31	1648.61	1648.72
10	2593.74	2685.06	2707.04	2717.91	2718.28

would realize exactly as much interest over that length of time. We saw earlier that \$1000 at an annual interest rate of 10% (a nominal rate) compounded quarterly for one year yields \$103.81 in interest, which is 10.381% of the principal. Hence the effective annual rate is 10.381%. In other words, \$1000 at simple interest of 10.381% for one year would earn exactly the same amount of interest.

When stated per year ("annualized"), the effective rate is called the **effective annual rate (EAR).** For savings, it is also called the **annual percentage yield (APY)** or *annual equivalent yield.*

To help keep these different rates straight, we use i for a nominal rate for a specified period—such as a day, month, or year—*within which no compounding is done;* Since no compounding is done for shorter intervals than this period, the effective rate for a single period is the same as the nominal rate. (Over multiple periods, the effective rate will be greater than the nominal rate.)

In particular, for a nominal annual rate r compounded n times per year, we have

$$i = \frac{r}{n}$$

For that \$1000 in savings at 10% compounded quarterly, we have $r = 10\%$ and $n = 4$, so $i = 2.5\%$ per quarter.

We denote the number of compounding periods under consideration by n. We use r only for an annual rate (nominal or effective) and t for the number of years.

An important term used only in connection with loans, the *annual percentage rate (APR),* is different from the EAR and is defined and explained in Chapter 22.

21.4 Geometric Growth

Here we look for the underlying mathematical pattern of compounding. For quarterly compounding, you have at the end of the first quarter

$$\text{Initial balance} + \text{interest} = \$1000 + \$1000(0.025) = \$1000(1 + 0.025)$$

and at the end of the second quarter

$$\begin{aligned}\text{Initial balance} + \text{interest} &= \$1000(1 + 0.025) \\ &\quad + [\$1000(1 + 0.025)](0.025) \\ &= [\$1000(1 + 0.025)] \times (1 + 0.025) \\ &= \$1000(1 + 0.025)^2\end{aligned}$$

The pattern continues in this way, so that you have $\$1000(1 + 0.025)^4$ at the end of the fourth quarter.

You use the calculator button marked y^x to evaluate expressions like $(1.0125)^2$: enter 1.0125, push y^x, enter 2, and push $=$; you get 1.02515625.

More generally, with an initial balance of P and an interest rate i ($= 100\,i\,\%$) per compounding period, you have at the end of the first compounding period

$$P + Pi = P(1 + i)$$

This amount can be viewed as a new starting balance. Hence, in the next compounding period, the amount $P(1 + i)$ grows to

$$P(1 + i) + P(1 + i)i = P(1 + i)(1 + i) = P(1 + i)^2$$

The pattern continues, and we reach the following conclusion:

Compound interest formula: If a principal P is deposited in an account that pays interest at rate i per compounding period, then after n compounding periods the account contains the amount

$$A = P(1 + i)^n$$

The amount added each compounding period is proportional to the amount present. This type of growth is called *geometric growth.*

Geometric growth (also called **exponential growth**) is growth proportional to the amount present.

EXAMPLE

Compound Interest

Suppose that you have a principal of $P = \$1000$ invested at 10% nominal interest per year. Using the compound interest formula $A = P(1 + i)^n$, you can determine the amount in the account after 10 years, which varies according to the compounding period:

- *Annual compounding.* The annual rate of 10% gives $i = 0.10$, and after 10 years the account has

$$\begin{aligned}\$1000(1 + 0.10)^{10} &= \$1000(1.10)^{10} = \$1000(2.59374)\\ &= \$2593.74\end{aligned}$$

- *Quarterly compounding.* Then $i = 0.10/4 = 0.025$, and after 10 years (40 quarters) the account contains

$$\$1000\left(1 + \frac{0.10}{4}\right)^{40} = \$1000(1.025)^{40} = \$1000(2.68506)$$
$$= \$2685.06$$

- *Monthly compounding.* Then $i = 0.10/12 = 0.008333$. The amount in the account after 10 years (120 months) is

$$\$1000\left(1 + \frac{0.10}{12}\right)^{120} = \$1000(1.00833333)^{120} = \$1000(2.70704)$$
$$= \$2707.04$$

These entries are found in the last row of Table 21.1.

Note that in doing such calculations you should use as many decimal places as possible for accuracy; don't round off until you come to the final result. ■

Suppose that you want to make a one-time deposit of amount P that will grow to a specific amount A in n compounding periods from now by earning interest at a rate i per period. The quantities A, P, i, and n are related through the compound interest formula, $A = P(1 + i)^n$. The quantity P is called the **present value** of the amount A to be paid n years in the future.

EXAMPLE

Certificate of Deposit

Suppose that you will need \$15,000 to pay for a year of college for your child 18 years in the future, and you can buy a certificate of deposit whose interest rate of 5% compounded quarterly is guaranteed for that period. How much do you need to deposit?

The compound interest formula $A = P(1 + i)^n$ gives

$$\$15{,}000 = P(1 + 0.0125)^{72}$$

Simplifying $(1 + 0.0125)^{72}$, we get

$$\$15{,}000 = P(2.44592)$$

so that

$$P = \frac{\$15{,}000}{2.44592} = \$6132.66$$

■

EXAMPLE

Money Market Account

In some cases you know the principal, the current balance, and the interval of time, and you want to learn the interest rate. For example, money market funds typically report earnings to investors each month, based on interest rates that vary from day to day. The investor may want to know an average rate of interest for the month, but often the monthly statement does not report one. We find the, equivalent average *daily* rate, from which we calculate the effective annual rate or annual yield (APY). The compound interest formula gives the end-of-month balance as $A = P(1 + i)^n$, where P is the balance at the beginning of the month, i is the average daily interest rate, and n is the number of days that the statement covers. So we have

$$\frac{A}{P} = (1 + i)^n$$

Taking the nth root of each side and solving for i gives

$$i = \left(\frac{A}{P}\right)^{1/n} - 1$$

Suppose that the monthly statement from the fund reports a beginning balance (P) of \$7373.93 and a closing balance (A) of \$7416.59 for 28 days (n). We thus have

$$\begin{aligned} i &= \left(\frac{7416.59}{7373.93}\right)^{1/28} - 1 = (1.005785246)^{0.035714286} - 1 \\ &= 0.000206042 \end{aligned}$$

Thus, the average daily rate is 0.0206042%. Compounding daily for a year, we would have $(1 + 0.000206042)^{365} = 1.0780972$, for an annual yield (APY) of 7.81%. ■

21.5 Simple Interest vs. Compound Interest

The amounts in accounts paying interest at 10% per year with compound and simple interest are shown in Table 21.2 and in the graph in Figure 21.1, which dramatically illustrate the growth of money at compound interest compared to simple interest.

We noted earlier that the population of U.S. medical doctors grows as if it were at simple interest (arithmetic growth) because the same number of doctors graduate from medical school each year. On the other hand, general human populations tend to grow as if they were at compound interest (geometric growth), because the number of children born – the "interest" – increases as the population – the "balance" – increases.

The distinction between arithmetic growth and geometric growth is fundamental to the major theory of demographer and economist Thomas Malthus (1766 – 1834). He claimed that human populations grow geometrically but food

TABLE 21.2 The Growth of $1000: Compound Interest vs. Simple Interest

Years	Amount in Account from Compounded Interest	Amount from Simple Interest
1	1100.00	1100.00
2	1210.00	1200.00
3	1331.00	1300.00
4	1464.10	1400.00
5	1610.51	1500.00
10	2593.74	2000.00
20	6727.50	3000.00
50	117,390.85	6000.00
100	13,780,612.34	11,000.00

supplies grow arithmetically, so that populations tend to outstrip their ability to feed themselves (see Spotlight 21.1).

The situation of nuclear waste generated by a nuclear power plant is more complicated. The absolute volume of waste added each year depends on the fixed size and output of the power plant, not on the growing amount of waste in

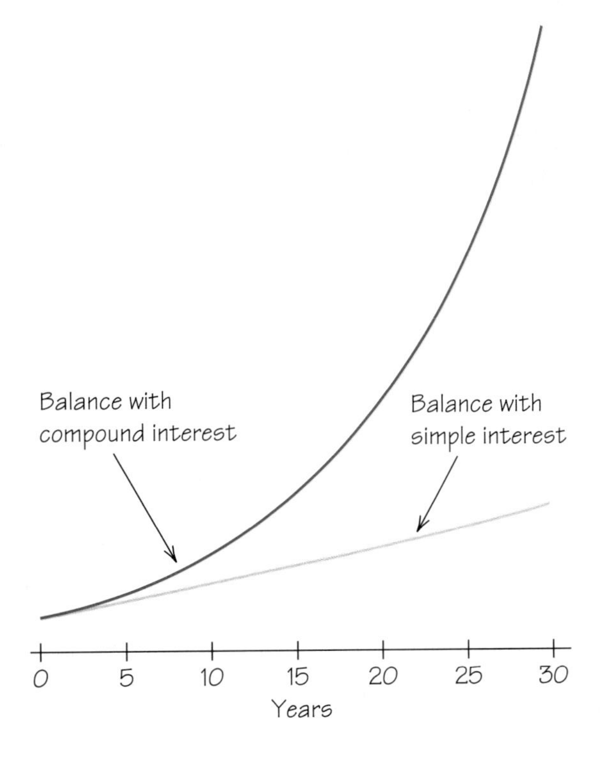

Figure 21.1 The growth of $1000: compound interest and simple interest. The straight line explains why growth at simple interest is also known as linear growth.

SPOTLIGHT 21.1 Thomas Malthus

Thomas Malthus (1766–1834), a nineteenth-century English demographer and economist, based a well-known prediction on his perception of the different patterns of growth of the human population and the "population" of food supplies.

He believed that human populations increase geometrically but food supplies increase arithmetically–so that the increase in food supplies will eventually be unable to match increases in population. He concluded, however, that over the long run there would be restrictions on the natural growth of human populations, too, including war, disease, and starvation–hardly an optimistic forecast and, doubtless, responsible for the dreary image associated with his views.

Thomas Malthus
(The Granger Collection, New York.)

storage. Hence the volume of waste grows arithmetically. What about the total amount of radioactive material in the storage dump? The waste is a mixture of radioactive and nonradioactive substances; over time, the radioactive ingredients decay slowly into nonradioactive ones. While the radioactivity of waste already in storage is decreasing, new amounts of radioactive material are being added each year. The situation requires a hybrid model that incorporates positive arithmetic growth (adding to the dump) accompanied by negative geometric growth (radioactive decay). The situation is like turning on the faucet to the bathtub while leaving the drain hole open. The faucet determines how fast water runs in, the height of the water determines how fast it runs out, and those two rates determine what happens to the volume of water in the tub.

21.6 A Limit to Compounding

The rows in Table 21.1 show a trend: More frequent compounding yields more interest. But as the frequency of compounding increases, the interest tends to a limiting amount, shown in the far right column.

Why is this so? Basically, because the extra interest from more frequent compounding is *interest on interest.* For example, in the first row of Table 21.1, the

$3.81 extra interest from compounding quarterly is interest on the $100 yearly interest. The $3.81 is less than 10% of the $100 because the $100 interest is not on deposit for the whole year; just part of it is credited to the account (and begins earning interest) at the end of each quarter. As compounding is done more and more often, smaller and smaller amounts of interest on interest are added.

How can we determine the limiting amount? Let's suppose that the initial balance is $1 and that we keep track at all stages of even the smallest fractions of a dollar.

First, we note that for an interest rate i per compounding period, a principal of $1 grows to $(1 + i)^n$ in n periods, so the interest earned on that $1—the effective rate of interest for n periods—is

$$\text{Effective rate} = (1 + i)^n - 1$$

For a nominal annual interest rate r compounded n times, the interest rate per compounding period is $i = r/n$, and an amount of $1 grows to

$$\left(1 + \frac{r}{n}\right)^n$$

in one year.

The annual effective rate of interest, or APY, is the amount of interest earned

$$\left(1 + \frac{r}{n}\right)^n - 1$$

divided by the original principal. Since that principal is $1, the APY is

$$\left(1 + \frac{r}{n}\right)^n - 1$$

For example, with a nominal annual rate r compounded monthly ($n = 12$), the APY is

$$\left(1 + \frac{0.06}{12}\right)^{12} - 1 = 0.0617 = 6.17\%$$

Here we suppose an interest rate of 100% per year compounded n times per year; later we examine interest rates closer to the ones in stable economies. For an initial balance of \$1, the amount at the end of one year—from the compound interest formula, with $P = \$1$ and $i = 100\%$—is

$$A = \$1 \times \left(1 + \frac{100\%}{n}\right)^n = \$\left(1 + \frac{1.00}{n}\right)^n$$

As n increases, this amount, which is just $(1 + 1/n)^n$, gets closer and closer to a special number called $e \approx 2.71828$ (see Spotlight 21.2). This is illustrated in Table 21.3, where the dots (ellipses) indicate that more decimal places follow. As n is made larger and larger, the limiting amount is e^r, and the interest method is called **continuous compounding.** The effective rate (APY, EAR) is $(e^r - 1)$. (You can calculate powers of e using the $\boxed{e^x}$ button on your calculator; on some calculators, this button is the 2nd function of the button marked $\boxed{\text{LN}}$ or $\boxed{\ln x}$. For example, to calculate $e^{0.10}$, enter 0.10, push $\boxed{\text{2nd}}$, then push $\boxed{\ln x}$; you get 1.105170918.)

EXAMPLE

Continuous Compounding

For \$1000 at an annual rate of 10%, compounded n times in the course of a single year, the balance at the end of the year is

$$\$1000\left(1 + \frac{0.10}{n}\right)^n$$

This quantity gets closer and closer to $\$1000e^{0.1} = \$1105.17\ldots$ as the number of compoundings n is increased. No matter how frequently interest is compounded—

TABLE 21.3 Yield of \$1 at 100% Interest, Compounded n Times per Year

n	$\left(1 + \frac{1}{n}\right)^n$
1	2.0000000 . . .
5	2.4883200 . . .
10	2.5937424 . . .
50	2.6915880 . . .
100	2.7048138 . . .
1,000	2.7169239 . . .
10,000	2.7181459 . . .
100,000	2.7182682 . . .
1,000,000	2.7182818 . . .
10,000,000	2.7182818 . . .

SPOTLIGHT 21.2 The Number *e*

The number e is similar to the number π in several respects. Both arise naturally, π in finding the area and circumference of circles, and e in compounding interest continuously (e is also the base for the system of "natural" logarithms). In addition, neither is rational (expressible as the ratio of two integers, such as 7/2) or even algebraic (the solution of a polynomial equation with integer coefficients, such as $x^2 = 2$); we say that they are *transcendental* numbers. Finally, no pattern has ever been found in the digits of the decimal expansion of either number.

daily, hourly, every second, infinitely often ("continuously")—the original \$1000 at the end of one year cannot grow beyond \$1105.17. The values after 5 and 10 years are shown in the lower rows of Table 21.1. ■

According to the *continuous interest formula,* for a principal P, deposited in an account at the nominal annual rate (APR) r compounded continuously.

- After 1 year, the account contains $A = Pe^r$
- After t years, it contains $A = Pe^{rt}$

We illustrate these formulas with the \$1000 at 10%. The formula for one year, $A = Pe^r$, gives

$$A = \$1000e^{0.10} = \$1000(1.10517) = \$1105.17$$

To find the amount in the account after 5 years, we apply the second formula, $A = Pe^{rt}$, with $t = 5$:

$$A = \$1000e^{(0.10)(5)} = \$1000e^{0.5} = \$1000(1.64872) = \$1648.72$$

exactly as shown in the rightmost column of Table 21.1.

It makes virtually no difference whether compounding is done daily or continuously over the course of a year. Most banks apply a daily periodic rate (based on compounding continuously) to the balance in the account each day and post interest daily (rounded to the nearest cent). The daily nominal rate is $r/365$, so each day the balance of the account is multiplied by $e^{r/365}$, the daily effective rate. Except for the rounding in posting interest, the effect is the same as continuous compounding throughout the year, because the compound interest formula gives $A = P(e^{r/365})^{365}$, which is the same as Pe^r from the continuous interest formula.

Also, it makes virtually no difference whether the bank treats a year as

- 365 days with daily nominal interest rate $r/365$ (the *365 over 365 method*)–this is the usual method for daily compounding; or
- 360 days with daily interest rate $r/360$ (the *360 over 360 method*)–this is the usual method for loans with equal monthly installments, because 360 is evenly divisible into 12 equal "months" of 30 days. The daily interest is greater than for the 365 over 365 method, but there are fewer days in the year.

Table 21.4 gives a comparison of different interest methods. ■

21.7 A Model for Accumulation

The compound interest formula tells the fate over time of a single deposited amount, but another common question that arises in finance is: What size deposit do you need to make *on a regular basis,* in an account with a fixed rate of interest, to have a specified amount at a particular time in the future?

This question is important in planning for a major purchase in the future or accumulating a retirement nest egg. Later, in Chapter 22, we apply the same concepts and formula to paying off a mortgage and making installment payments on a car.

EXAMPLE

A Savings Plan

An individual saves $100 per month, deposited directly into her credit union account on payday, the last day of the month. The account earns 6% per year, compounded monthly. How much will she have at the end of five years, assuming that the credit union continues to pay the same interest rate?

Note that she makes the first deposit at the end of the first month and the last deposit at the end of the sixtieth month. The monthly interest rate is $i = r/12 = 0.06/12 = 0.005$.

It's easier to look at the deposits in reverse time order. The last deposit is made on the last day of the five years, so it earns no interest and contributes just $100 to the total.

TABLE 21.4 Comparing Methods of Compounding Interest

Method	Compounding Periods per Year	Rate per Period	Formula per One Year	Effective Rate	Effective Rate for $r = 5\%$
360 over 360, daily	360	$r/360$	$P(1 + \frac{r}{360})^{360}$	$(1 + \frac{r}{360})^{360} - 1$	5.12674%
365 over 365, daily	365	$r/365$	$P(1 + \frac{r}{365})^{365}$	$(1 + \frac{r}{365})^{365} - 1$	5.12675%
Continuous			Pe^r	$e^r - 1$	5.12711%

The second-last deposit earns interest for one month, contributing $\$100(1 + i)$. Similarly, the third-last contribution is on deposit for two months, contributing $\$100(1 + i)^2$.

Continuing in the same way, we find that the first deposit earns interest for 59 months and contributes $\$100(1 + i)^{59}$. The total of all of the contributions is

$$\$100 + \$100(1 + i)^1 + \$100(1 + i)^2 + \ldots + \$100(1 + i)^{59}$$
$$= \$100[1 + (1 + i)^1 + (1 + i)^2 + \ldots + (1 + i)^{59}]$$

This expression is known as a **geometric series,** because the successive terms grow geometrically: Each succeeding term is a constant common growth rate—here, $(1 + i)$—times the preceding term. There is a formula for the sum of such a series, which we give for a geometric series with ratio x:

$$1 + x + x^2 + x^3 + \ldots + x^{n-1} = \frac{x^n - 1}{x - 1}$$

That this formula works for all x (except $x = 1$) can be confirmed by multiplying both sides by $(x - 1)$ and watching terms on the left cancel (you should do this confirmation for $n = 4$.)

In our example, we have $x = 1 + i$, and the formula becomes

$$1 + (1 + i)^1 + (1 + i)^2 + \ldots + (1 + i)^{n-1} = \frac{(1 + i)^n - 1}{i}$$

We have $n - 1 = 59$, or $n = 60$ months, and $i = 0.005$, the interest rate per month. The total accumulation after five years is

$$A = \$100\left[\frac{\left(1 + \frac{0.06}{12}\right)^{60} - 1}{\frac{0.06}{12}}\right] = \$6977.00$$

■

In general terms, we have

For a uniform deposit of d per compounding period (deposited at the end of the period) and an interest rate i per period, the amount A accumulated after n periods is given by the *savings formula*:

$$A = d\left[\frac{(1 + i)^n - 1}{i}\right]$$

The savings formula involves four quantities: A, d, i, and n. If any three are known, the fourth can be found. A common situation is for A, i, and n to be known, with d (the regular payment) to be found, because the practical concern for most people is how much their monthly payment will be.

Sometimes the purpose of a savings plan is to accumulate a fixed sum by a certain date. Such savings plans are called **sinking funds.**

EXAMPLE

Sinking Fund

In the earlier example of putting aside money for a child's college education, we calculated how much you would have to put into a certificate of deposit (CD) at the birth of the child so as to have \$15,000 when the child reaches college 18 years later. For a CD earning a steady 5% compounded quarterly, you would need to deposit \$6132.66. You probably aren't going to have that amount of money just sitting around when the child is born—not to mention the extra expenses of having the child—so what could you do? You could start saving a regular amount d per month after the child is born. How much do you have to save each month if your account earns 5% interest per year, compounded monthly?

The monthly rate is $i = 5\%/12 = 0.004166667$. Applying the savings formula

$$A = d\left[\frac{(1+i)^n - 1}{i}\right]$$

with $A = \$15{,}000$, $i = 0.05/12 = 0.004166667$, and $n = 12 \times 18 = 216$, we get

$$\$15{,}000 = d\left[\frac{\left(1 + \frac{0.05}{12}\right)^{216} - 1}{\frac{0.05}{12}}\right] = \$349.20$$

so $d = \$15{,}000/349.20 = \$42.96$. (Using 0.05/12 instead of 0.004166667 allows you to do the calculation without rounding off any values except the final answer.) This sounds like a manageable amount to contribute, but it doesn't take into account inflation, nor the possibility that your child may want to go to a private college whose costs are a great deal more than \$15,000 total for four years! In the next section, we investigate how to take into account the effect of inflation. ■

EXAMPLE

Sinking Fund with Daily Interest

We illustrate how much—actually, how little—difference daily compounding makes compared to monthly compounding. We recalculate the monthly contribution to the sinking fund for college education with the same interest rate of 5% per year but compounded daily instead of monthly.

For simplicity, we calculate as if each month has 30 days and the savings institution uses the 360 over 360 method. The daily rate is $5\%/360 = 0.000138889$, so the effective monthly rate (for a 30-day month) is

$$i = (1.000138889)^{30} - 1 = 0.004175073$$

We have $A = \$15{,}000$, $i = 0.004175073$, and $n = 12 \times 18 = 216$. Applying the savings formula, we get

$$\$15{,}000 = d\left[\frac{1.004175073^{216} - 1}{0.004175073}\right] = d\,(349.56)$$

so $d = \$15{,}000/349.56 = \$42.91$. That is only a nickel less per month than with monthly compounding! ■

EXAMPLE Retirement Fund

Financial advisers stress the importance of beginning early to save for retirement. Many firms offer a *401(k) plan* (named after a section of law), which allows an employee to make monthly contributions to a retirement account. The plan has the advantage that the contributions are exempted from income tax until the employee withdraws money during retirement.

Sometimes a company's own pension plan consists of just contributing to the employee's individual 401(k) account either in cash or in the form of company stock. (In early 2002, the bankruptcy of Enron Corporation resulted in thousands of its employees losing almost their entire retirements savings. Those

(Courtesy of Fidelity Investments.)

savings consisted largely of Enron stock contributed by Enron, which fell from a value of \$90 per share to under \$1 per share in just a couple of months. The Enron bankruptcy illustrated how unwise it is for most of an employee's retirement fund to consist of stock in just one company, particularly if—as was the case for Enron—the employee is not free to sell the stock contributed by the employer. Even more people lost retirement savings and jobs when the stock of WorldCom declined more than 99% in value by midsummer 2002, after news emerged of prodigious financial fraud by its management.)

Suppose that you start a 401(k) plan when you turn 23 and contribute \$50 at the end of each month until you turn 65 and retire. Suppose (unlike some Enron employees) you put *your* contributions into a very safe investment that returns a steady 5% annual interest compounded monthly. How much will be in your fund at retirement?

We apply the savings formula with $d = \$50$, $i = 0.05/12 = 0.004166667$, and $n = 12 \times (65 - 23) = 504$. We get

$$A = 50\left[\frac{(1.004166667)^{504} - 1}{0.004166667}\right] = \$50 \times 1711.3489 = \$85{,}567.44$$

At first glance, that may seem like a lot of money, but it is not so much if that's all you have to live off for the rest of your life (of course, there is also Social Security). In the Exercises we explore the effects of saving more each month, getting a higher interest rate, saving on taxes, and having inflation erode the value of your savings. ■

A sinking fund is a special case of an **annuity,** a specified number of equal payments at equal intervals of time. Annuities are a common way for lotteries to pay out grand prizes (you probably won't have the pleasure of one of these!) and for retirees to receive funds saved up for retirement (you likely will have such an option). We examine an example of each in the next chapter, since their analysis requires a formula that we develop there.

21.8 Exponential Decay

In times of economic inflation, prices increase. When the rate of inflation is constant, the compound interest formula can be used to project prices.

EXAMPLE Inflation

Suppose that there is constant 3% annual inflation from mid-2003 through mid-2007. What will be the projected price in mid-2007 of an item that costs \$100 in mid-2003?

The compound interest formula applies with $P = \$100$, $r = 3\%$, and $n = 4$. The projected price is $A = P(1 + r)^n = \$100(1 + 0.03)^4 = \$112.55$. ■

During constant-rate inflation, prices grow geometrically (exponentially) and the value of the dollar goes down geometrically.

Exponential decay is geometric growth with a negative rate of growth.

Let a (for "additional") represent the rate of inflation; what costs \$1 now will cost \$$(1 + a)$ this time next year. For example, if the inflation rate were $a = 25\%$, then what costs \$1 now would cost \$1.25 this time next year. A dollar next year would buy only 0.8 (= 1/1.25) times as much as a dollar buys today. In other words, a dollar next year would be worth only \$0.80 in today's dollars—by next year, a dollar would have lost 20% of its purchasing power. We say that the **present value** of receving a dollar next year is \$0.80. Notice that although the inflation rate is 25%, the loss in purchasing power is 20%. For a general inflation rate a, a dollar a year from now will buy only a fraction of what a dollar today can buy; that fraction, the present value of a dollar a year from now, is

$$\frac{1}{1+a} = 1 - \frac{a}{1+a}$$

In other words, a dollar a year from now is worth \$$[1 - a/(1 + a)]$ today, and the loss in purchasing power is the fraction $a/(1 + a)$. (You should calculate what these expressions become for $a = 25\%$.) The quantity $i = -a/(1 + a)$ behaves like a negative interest rate. We can use the compound interest formula to find the value of P dollars n years from now as $A = P(1 + i)^n = P(1 - [a/(1 + a)])^n$.

The actual posted price of an item, at any time, is said to be in **current dollars.** That price can be compared with prices at other times by converting all prices to **constant dollars,** dollars of a particular year.

EXAMPLE

Deflated Dollars

Suppose that there is 25% annual inflation from mid-2003 through mid-2007. What will be the value of a dollar in mid-2007 in constant mid-2003 dollars?

We have $a = 0.25$, so $i = -a/(1 + a) = -0.25/1.25 = -0.20$. This, not 25%, is the negative interest rate, the rate at which the dollar is losing purchasing power. We have $n = 4$ years, so the value of \$1 four years from mid-2003 is in 2003 dollars,

$$\$1(1 + i)^4 = \$(1 - 0.20)^4 = (0.80)^4 = \$0.41$$ ■

In the above Example, we may think of the value of the dollar as "depreciating" 20% per year. Depreciation of the value of equipment is similar.

EXAMPLE

Depreciation

If you bought a car at the beginning of 2003 for \$12,000 and its value in current dollars depreciates steadily at a rate of 15% per year, what will be its value at the beginning of 2006 in current dollars?

We have $P = \$12,000$, $i = -0.15$, and $n = 3$. The compound interest formula gives

$$A = P(1 + i)^n = \$12,000(1 - 0.15)^3 = \$7370$$ ■

21.9 The Consumer Price Index

In the previous section we considered a model for inflation that stayed constant over a period of time. That is not generally the case. However, based on regular measures of inflation, we can determine the equivalent today of a price in an earlier year or how much a dollar in that year would be worth today.

The official measure of inflation is the consumer price index (CPI), prepared by the Bureau of Labor Statistics. Here we describe and use the CPI-U, the index for all urban consumers, which covers about 80% of the U.S. population and is the index of inflation that is usually referred to in newspaper and magazine articles.

Each month, the Bureau of Labor Statistics determines the average cost of a "market basket" of goods, including food, housing, transportation, clothing, and other items. It compares this cost to the cost of the same (or comparable) goods in a base period. The base period used to construct the CPI-U is 1982–1984. The index for 1982–1984 is set to 100, and the CPI-U for other years is calculated by using the proportion

$$\frac{\text{CPI for other year}}{100} = \frac{\text{cost of market basket in other year}}{\text{cost of market basket in base period}}$$

For example, the cost of the market basket in 1976 (in 1976 dollars) was 0.569 times the cost in 1982–1984 (in 1982–1984 dollars), so the CPI for 1976 is 100×0.569, or 56.9.

Table 21.5 shows the average CPI for each year from 1913 through 2001, with estimates for 2002 and 2003. This table can be used to convert the cost of an item in dollars for one year to what it would cost in dollars in a different year, using the proportion

$$\frac{\text{Cost in year A}}{\text{Cost in year B}} = \frac{\text{CPI for year A}}{\text{CPI for year B}}$$

EXAMPLE The Price of a House and the Value of a Dollar

We convert the average cost of a home bought in Madison, Wisconsin, in year 1976 dollars into a price in year 2003 dollars.

We see from Table 21.5 that the CPI for 1976 is 56.9 and the CPI for 2003 is estimated to be 188.5. The average cost of a Madison house in 1976 was $38,323. Using the proportion, we have

TABLE 21.5 U.S. Consumer Price Index (1982–1984 = 100)

–	–	1931	15.2	1951	26.0	1971	40.5	1991	136.2
–	–	1932	13.7	1952	26.6	1972	41.8	1992	140.3
1913	9.9	1933	13.0	1953	26.7	1973	44.4	1993	144.5
1914	10.0	1934	13.4	1954	26.9	1974	49.3	1994	148.2
1915	10.1	1935	13.7	1955	26.8	1975	53.8	1995	152.4
1916	10.9	1936	13.9	1956	27.2	1976	56.9	1996	156.9
1917	12.8	1937	14.4	1957	28.1	1977	60.6	1997	160.5
1918	15.1	1938	14.1	1958	28.9	1978	65.2	1998	163.0
1919	17.3	1939	13.9	1959	29.1	1979	72.6	1999	166.6
1920	20.0	1940	14.0	1960	29.6	1980	82.4	2000	172.2
1921	17.9	1941	14.7	1961	29.9	1981	90.9	2001	177.7
1922	16.8	1942	16.3	1962	30.9	1982	96.5	2002 (est.)	183.0
1923	17.1	1943	17.3	1963	30.6	1983	99.6	2003 (est.)	188.5
1924	17.1	1944	17.6	1964	31.0	1984	103.9		
1925	17.5	1945	18.0	1965	31.5	1985	107.6		
1926	17.7	1946	19.5	1966	32.4	1986	109.6		
1927	17.4	1947	22.3	1967	33.4	1987	113.6		
1928	17.1	1948	24.1	1968	34.8	1988	118.3		
1929	17.1	1949	23.8	1969	36.7	1989	124.0		
1930	16.7	1950	24.1	1970	38.8	1990	130.7		

Note: This is the CPI-U index, which covers all urban consumers, about 80% of the U.S. population. Each index is an average for all cities for the year. The basis for the index is the period 1982–1984, for which the index was set equal to 100.

SOURCE: http://starts.bls.gov/

$$\frac{\text{Cost in 2003}}{\text{Cost in 1976}} = \frac{\text{CPI for 2003}}{\text{CPI for 1976}}$$

or

$$\frac{\text{Cost in 2003}}{\$38{,}323} = \frac{188.5}{56.9}$$

so that

$$\text{Cost in 2003} = \$38{,}323 \times \frac{188.5}{56.9} = \$38{,}323 \times 3.313 = 126{,}958$$

The 3.313 is the *scaling factor* for converting 1976 dollars to 2003 dollars. What we are observing is a proportion, or *numerical similarity*, between 1976 dollars and 2003 dollars, analogous to the geometrical similarity that we studied earlier. To convert from 2003 dollars to 1976 dollars, we would multiply by 1/3.313, or 0.302. ■

Spotlight 19.3 (pages 000–000) describes a major development in 1999 in how the consumer price index is calculated and discusses the economic effects of the change. The change involves using the geometric mean (defined and introduced on page 000) in place of the arithmetic mean (common average).

21.10 Real Growth

It's natural to think that if your investment is growing at 6% per year and inflation is at 3% per year, then the real growth in the value (purchasing power) of your investment is 6% − 3% = 3%. Such, however, is not the case.

Let's suppose that you invest $500 for a year at 6%. At the beginning of the year, you have $500, which at $5 per pound could buy 100 pounds of steak. At the end of the year, you have $5000(1 + 0.06) = $530, but steak now costs $5(1 + 0.03) = $5.15 per pound. How much steak would that buy? 530/5.15 lb = 102.91 lb. In other words, in terms of purchasing power, or real gain, your investment has grown only 2.91%. This is not a great deal different from 3%, but it *is* different, and the difference is greater for higher rates of interest and inflation.

Consider an investment principal P and a market basket of goods of value m. Let the annual yield (rate of interest) of the investment be r and the rate of inflation be a. We calculate the rate of real growth g of the investment as follows.

At the beginning of the year, the investment would buy quantity $q_{\text{old}} = P/m$ of the market basket. At the end of the year, the investment would buy

$$q_{\text{new}} = \frac{P(1 + r)}{m(1 + a)}$$

market basket. Notice that the gain of r in the investment multiplies the principal by $(1 + r)$, while the erosion due to inflation divides the principal by $(1 + a)$. Here you see directly that the two influences on the investment have directly opposite effects.

The growth of the investment, relative to how many market baskets it could have bought originally, is

$$g = \frac{q_{\text{new}} - q_{\text{old}}}{q_{\text{old}}} = \frac{\frac{P(1 + r)}{m(1 + a)} - \frac{P}{m}}{\frac{P}{m}} = \frac{1 + r}{1 + a} - 1 = \frac{r - a}{1 + a}$$

In the last expression, the numerator is the difference of the two rates (6% − 3% in our example), which is divided by a quantity greater than 1 if there is inflation. You should confirm that this formula gives 2.91% for $r = 6\%$ and $a = 3\%$.

One way to understand why this is the correct formula is to realize that the gain itself is not in original dollars but in deflated dollars.

The relationship between interest rate, inflation rate, and rate of real growth is called *Fisher's effect,* after the American economist Irving Fisher (1867–1947).

21.11 Valuing a Stock

A company sells shares of stock in the company to raise capital to make investments (e.g., in buildings and equipment) and expand the company. Buyers of the shares are part owners of the company; they can participate in selecting the management and can receive part of the profits of the company in the form of cash payments, called *dividends.* Shareowners can sell their shares to others, through traders or brokers or even over the Internet; the price for the shares is determined by current market forces.

Most individuals and business entities buy stocks as a way to invest; they try to pick stocks that will both pay dividends and rise in value. However, many investors pay little attention to whether a company pays dividends or how much; they focus on "growth" stocks, ones whose prices will increase. However, a company that pays lower dividends than analysts expect usually sees the price of its stock go down. A stock is considered overvalued if the current price is higher than the potential for the company to earn profit, pay dividends, and expand the value of the business.

How can you tell if a stock is overvalued? The price of a stock depends on both logical factors ("fundamentals") and psychological factors.

We consider here logical factors, in what is known as the standard "rational" model of stock prices. This model does not take into account the possibility that herd psychology and frenzied trading may escalate the prices of stocks that never pay dividends (profit earnings) to shareowners, as happened in 2000 with many "dot-com" stocks. Instead, the model visualizes the purchaser of a stock as buying a future "stream" of dividends (regular dividends each year or quarter). The formula for the sum of a geometric series figures importantly here.

The rational price to pay would be the present value of the future dividends, with their value "discounted" by a factor (the *discount rate*) that recognizes that getting a dollar a year from now is not as valuable as getting it today. The discount rate takes into account inflation, the rate of interest on alternative investments (e.g., government bonds, which involve no risk), and a measure of how risky the stock is (future dividends are not a sure thing, so the price of the stock should include a *risk premium* that reflects that uncertainty).

Let's buy a stock at the beginning of the year. How much should we pay? Suppose that last year's dividend was D, paid at the beginning of this year to those who owned it on a particular date last year (so, as new buyers, we don't get that dividend). Suppose that as the company expands and prospers, the annual dividend grows steadily at a rate g per year, so that the dividend paid next year is $D(1 + g)$. Suppose that the discount rate is r per year; this means that to find the present value of next year's dividend, we must divide it by $(1 + r)$, getting

$$D\frac{1 + g}{1 + r}$$

For example, for $D = \$5$, $g = 7\%$, and $r = 12\%$, next year's dividend will be $D(1 + g) = \$5.35$, but the present value of this future dividend when we bought the stock was only

$$D\frac{(1 + g)}{(1 + r)} = \$5\frac{(1 + 0.07)}{(1 + 0.12)} = \frac{\$5.35}{1.12} = \$4.78$$

Still, the price of the stock when we buy it should take the present value of future earnings into account—for next year and all later years.

Using the formula for a geometric series on page 793, we can calculate the total present value of all the dividends over the next n years. We spare you the algebra and simply tell you that the present value is

$$P = D\,(1 + g)\frac{x^n - 1}{g - r}$$

where

$$x = \frac{1 + g}{1 + r}$$

If $g > r$, then $x > 1$ and the value of P gets arbitrarily large as n increases; the stock would be a "gold mine," because its future earnings would outstrip inflation and any risk.

More conventionally, we have $g < r$, so $x < 1$ and x^n tends toward 0 as n gets large: Dividends from farther and farther in the future contribute less and less to the present value. For large n, have

$$P \approx \frac{D(1 + g)}{r - g}$$

Let's examine this formula closely to see what it can tell us. A small change in the dividend D will lead to a correspondingly small change in the price P of the stock. For example, a 5% increase in D leads to a 5% increase in P.

However, small changes in either the anticipated rate of growth g or the discount rate r are *magnified* in the price of the stock. For example, suppose that

$r = 12\%$ and $g = 7\%$, so that $r - g = 0.05$, $1/(r - g) = 1/0.05 = 20$, and $P =$ \$107. Now, if the Federal Reserve lowers interest rates by 0.25% (as happened often in 2001), the effect at first will be to lower r (loans become cheaper) but might later lead to higher r because increased spending leads to greater inflation. If investors are confident that inflation can nevertheless be kept under control and think that the net effect would be to decrease r by, say, 0.125% (half the amount of the cut) without affecting g, then we have $r - g = 4.875\%$, $1/(r - g) = 1/0.04875 = 20.51$, and $P =$ \$109.73. That may not seem like much of a change, but it means that the value of stocks should go up by $(109.73 - 107)/107 = 2.6\%$. For the New York Stock Exchange with its Dow Jones Industrial Average around 9,000 points (as in late summer 2002), such an increase would correspond to a rise of more than 230 points. Such a fast increase in value is one reason why investors are happy to see interest rates lowered.

Of course, we cannot precisely know what future growth and an appropriate discount rate will be. The prices of stocks also depend on buyers' and sellers' expectations and psychology. However, the rational model of pricing can suggest when stocks may be overpriced or underpriced relative to earnings potential.

Other models lead to commonly used formulas for pricing bonds and other investments, such as financial derivatives.

21.12 A Model for Financial Derivatives

Financial markets trade everything from agricultural commodities (corn, orange juice, coffee) to precious metals (gold, silver), from national currencies (dollars, euros) to corporate stocks and bonds. In recent years, an increasing share of such trading has turned from the objects themselves to less tangible entities based on them, known as *financial derivatives*. Traditional derivatives include mutual funds, in which a purchaser owns shares in a company that itself owns and manages a portfolio of corporate stocks and bonds; commodities futures, in which an agricultural producer contracts to sell a commodity at a definite point in the future for a price agreed to in advance; and options.

An *option* allows (but does not require) an investor to acquire an asset (which may itself be a commodity or corporate stocks or bonds) during a particular interval at a price fixed in advance. Part of an employee's pay package may be options to acquire stock in the company. An option is valuable because if the actual price A of the asset at some point during that interval is higher than the price C fixed in advance by the option, the investor could *exercise* the option by buying the asset at the option price C and selling it on the open market for A, making a profit of $A - C$.

Options may be bought and sold, but a vexing problem for many years was how to accurately assess the worth of an option. How can you determine the value of an option to buy 100 million euros (unit of European Union currency) at \$0.97 per euro at any time between now and next Tuesday, or an option to buy 100 shares of Apple Computer at \$25 per share before January 1?

One widely used pricing formula for valuing financial derivatives is the *Black–Scholes formula* (see Spotlight 21.3). In the next example, we examine a simpler model for valuing an option that still incorporates the main ideas.

EXAMPLE Valuing an Option

Suppose that you are interested in an Internet company named Interjunk, but you're not sure whether or not to buy into it just now. You could buy a share of stock now for \$100, or else buy an option to purchase a share of stock a year from now–no matter how high or low the price is then–at that same price of \$100 called the *exercise price* (or strike price) of the option. The value of the option depends on the current price of the stock: The higher the price of the stock now, the more likely that it will be more than \$100 a year from now, and hence the more valuable the option. As time goes by, if the price of the stock goes up, the value of the option does, too–but not quite as much, because there is still uncertainty about what will happen between then and expiry (when the option expires). What would be a fair price for such an option?

The answer depends on what you expect to happen to the price of the stock. Suppose (for simplicity, but unrealistically) that the consensus in the market is that there are only two possibilities: There is a 50% chance that the stock price a year from now will be \$200 and a 50% chance that it will be \$80.

If you buy a share of stock, the expected value of that share a year from now is $(0.50) \times \$200 + (0.50) \times \$80 = \$140$, for an expected profit of $\$140 - \$100 = \$40$.

If you buy an option for some cost $\$C$, you have a 50% chance of exercising it. Half the time you wind up with a share of stock that costs you only \$100 but is worth \$200, netting you \$100; the other half of the time, the stock price is at \$80 and the option is worth nothing. Either way, you paid $\$C$ for the option. Your expected profit is $(0.50) \times \$100 + (0.50) \times \$0 - \$C = \$50 - \$C$.

What should you be willing to pay for the option? If you pay less than \$10, you expect on average to make a greater profit than the \$40 you could make by buying the stock now; if you pay more than \$10 for the option, you would be better off instead buying the stock now. So the value of the option is \$10.

We could refine our model slightly to take into account alternative investments. Suppose that you can earn a risk-free 10% on U.S. Treasury obligations. Suppose further that you have \$110 on hand and are going to invest it all in some combination of stock, Treasuries, and options.

Strategy 1. Buy the stock (\$100), put the rest into Treasuries (\$10). You expect to realize \$40 from the stock and \$1 from the Treasuries, for \$41.

Strategy 2. Buy an option (\$10), put the rest into Treasuries (\$100). You expect to realize \$40 from the option and \$10 from the Treasuries, for a total of \$50.

Of course, the rest of the investors in the market can do the same calculations, and it is readily apparent that an option is worth more–hence should be priced higher–than the \$10 used in these calculations. In fact, an option is worth closer to \$20. ■

SPOTLIGHT 21.3 Nobel Prize for a Model in Economics

The 1997 Nobel Memorial Prize in Economics was awarded to Robert C. Merton of Harvard University and Myron S. Scholes of Stanford University for their method of valuing financial derivatives. Together with the late Fischer Black in the 1970s, they formulated a mathematical model with appropriate assumptions and solved the resulting equation. At the time, Black was a mathematician with Arthur D. Little consultants in Boston, Scholes was a professor of finance at MIT, and Merton was an assistant to the economist Paul Samuelson at MIT; all then were under age 30.

The major achievement of Merton, Black, and Scholes was to incorporate variability of market prices into the formula for the value of an option. They realized that the risk involved in the market is already implicitly taken into account in the stock's current price and its volatility, and they were able to find the right formulation for incorporating the risk into the value of the option. Black and Scholes actually modeled the rate of change of the option value and then used methods from calculus to work backward to calculate the value and the formula itself.

This fairly complicated formula is based on simplifying assumptions (no stock dividends, no transaction costs, fixed price volatility, fixed interest rate, efficient market) plus a major modeling assumption. That modeling assumption is that the change in the price of the stock, as a percentage of the price, has a fixed component proportional to elapsed time (price trends upward over time) plus a random component deriving from volatility (so the price also can jump around). The random component is modeled using the normal distribution of Chapter 7.

(Packert White/The Image Bank.)

REVIEW VOCABULARY

Annual percentage yield (APY) The effective interest rate per year.
Annuity A specified number of payments at equal intervals of time.
Arithmetic growth Growth by a constant amount in each time period.
Compound interest Interest that is paid on both the original principal and accumulated interest.
Compound interest formula Formula for the amount in an account that pays compound interest periodically. For an initial principal P and effective rate i per compounding period, the amount after n compounding periods is $A = P(1 + i)^n$.
Compounding period The time interval between payments of compound interest.
Constant dollars Costs are expressed in constant dollars if inflation or deflation has been taken into account by converting the costs to their equivalent in dollars of a particular year.
Continuous compounding Payment of interest in an amount toward which compound interest tends with more and more frequent compounding.
Current dollars The actual cost of an item at a point in time; inflation or deflation before or since then has not been taken into account.
e The base for continuous compounding, geometric (exponential) growth, and natural logarithms; $e = 2.71828 \ldots$.
Effective annual rate (EAR) The effective rate per year.
Effective rate The actual percentage rate, taking into account compounding. It equals the rate of simple interest that would realize exactly as much interest over the same period of time.
Exponential decay Geometric growth at a negative rate.
Exponential growth Geometric growth.
Geometric growth Growth proportional to the amount present.
Geometric series A sum of terms, each of which is the same constant times the previous term; that is, the terms grow geometrically.
Interest Money earned on a savings account or a loan.
Linear growth Arithmetic growth.
Nominal rate A stated rate of interest for a specified length of time; a nominal rate does not take into account any compounding.
Period Compounding period.
Present value The value today of an amount to be paid or received at a specific time in the future, as determined from a given interest rate and compounding period.
Principal Initial balance.
Simple interest The method of paying interest only on the initial balance in an account, not on any accrued interest.
Sinking fund A savings plan to accumulate a fixed sum by a particular date.

SUGGESTED READINGS

KASTING, MARTHA. *Concepts of Math for Business: The Mathematics of Finance.* UMAP Modules in Undergraduate Mathematics and Its Applications: Module 370–372. COMAP, Inc., Arlington, Mass., 1980.

LINDSTROM, PETER A. *Nominal vs. Effective Rates of Interest.* UMAP Modules in Undergraduate Mathematics and Its Applications: Module 474. COMAP, Inc., Arlington, Mass., 1988. Reprinted in Paul J. Campbell (ed.), *UMAP Modules: Tools for Teaching 1988,* COMAP, Inc., Arlington, Mass., 1989, pp. 21–53. A learning module, requiring no more background than this chapter, that teaches about nominal and effective rates of interest and how to calculate them. Gives real examples of banks using different options for calculating interest.

MILLER, CHARLES D., VERN E. HEEREN, and JOHN HORNSBY. Consumer mathematics. In *Mathematical Ideas,* 9th ed., Addison Wesley Longman, Boston, 2001, pp. 786–845.

VEST, FLOYD, and REYNOLDS GRIFFITH. The mathematics of bond pricing and interest rate risk. *Consortium (COMAP),* 59 (Fall 1996): HiMAP Pullout Section 1–6.

SUGGESTED WEB SITES

www.stats.bls.gov/cpihome.htm Home page for the inflation tables prepared by the Bureau of Labor Statistics.

www.dol.gov/dol/esa/public/minwage/ Department of Labor data on the U.S. federal minimum wage over the years, including a graph of its inflation-adjusted value.

www.westegg.com/inflation/ Inflation calculators for the United States (1800–2001), Canada, and Italy, by S. Morgan Friedman, with links to sites about the current purchasing power of amounts of currencies in the past of other countries.

SKILLS CHECK

1. If a savings account pays 3% simple annual interest, how much interest will a deposit of $250 earn in 2 years?

(a) Less than $5
(b) Between $5 and $8
(c) Over $8

2. Simple interest is an example of

(a) linear growth.
(b) variable growth.
(c) constant growth.

3. Which of the following pays more interest?

(a) 6% compounded annually
(b) 6% compounded monthly
(c) 6% compounded continuously

4. The 1.5% monthly rate on a credit card balance is an example of

(a) an effective rate.
(b) a nominal rate.
(c) an adjusted rate.

5. If a single deposit is made into a compound interest certificate of deposit, the account

(a) earns interest only for the first period.
(b) earns the same amount of interest each period.
(c) earns more interest in each subsequent period.

6. If $800 is invested for one year at 6% compounded quarterly, the amount of interest earned is

(a) exactly $48.
(b) between $48 and $49.
(c) more than $49.

7. The number e is

(a) less than 3.
(b) greater than 3.
(c) unknown.

8. When $1000 is invested at 8% compounded continuously for 5 years, the balance is approximately

(a) $1491.

(b) \$1204.
(c) \$1400.

9. An example of a sinking fund is

(a) buying a savings bond when a child is born.
(b) depositing \$100 on a child's annual birth date.
(c) withdrawing \$100 from a savings account at the beginning of each year.

10. An example of exponential decay is

(a) depreciation of factory equipment.
(b) a retirement annuity.
(c) the consumer price index.

11. If a new car costs \$18,000 and loses value at a rate of 20% per year, what is its value after 3 years?

(a) About \$7200
(b) About \$9200
(c) Less than \$5000

12. If your investment is growing at a rate less than the rate of inflation,

(a) you have a positive real growth in your investment.
(b) you do not have a positive real growth in your investment.
(c) you do not have enough information to determine your real growth.

13. If you deposit \$1000 at 6.2% simple interest, what is the balance after three years?

(a) \$1186.00
(b) \$1197.77
(c) \$1224.63

14. Suppose you invest \$250 in an account that pays 4.5% interest compounded quarterly. After 30 months, how much is in your account?

(a) \$279.08
(b) \$279.59
(c) \$279.71

15. Suppose you deposit \$15 at the end of each month into a savings account that pays 2.5% interest compounded monthly. After a year, how much is in the account?

(a) \$184.50
(b) \$182.50
(c) \$182.08

16. What is the APY for 5.90% compounded monthly?

(a) 5.90%
(b) 6.06%
(c) 6.08%

17. If a bond matures in 3 years and will pay \$10,000 at that time, what is the fair value of it today, assuming the bond has an interest rate of 6% compounded annually?

(a) \$8200.00
(b) \$8396.19
(c) \$8352.70

18. Which of the following is the most generous interest rate for a one-year CD?

(a) 6% simple interest
(b) 5.9% compounded annually
(c) 5.9% compounded continuously

19. The effective rate for 5% compounded daily is

(a) exactly 5%.
(b) slightly more than 5%.
(c) slightly less than 5%.

20. The value of e is approximately

(a) 1.414.
(b) 2.718.
(c) 3.14.

EXERCISES

▲ Optional. ■ Advanced. ◆ Discussion.

The exercises below require a scientific calculator with buttons for powers $\boxed{y^x}$, exponential $\boxed{e^x}$, and natural logarithm $\boxed{\ln x}$.

Geometric Growth

1. You deposit \$1000 at 3% per year. What is the balance at the end of one year, and what is the annual yield, if the interest paid is

(a) simple interest?
(b) compounded annually?
(c) compounded quarterly?
(d) compounded daily?

2. As in Exercise 1, but for \$1000 at 6% per year.

3. *Zero-coupon bonds* are securities that pay no current interest but are sold at a substantial discount from redemption value. The difference between purchase price and redemption value provides income to the bondholder at the time of redemption or resale. If the annual interest rate is 3%, what should be the price of a zero-coupon bond that will pay \$10,000 eight years from now? (Use daily compounding.)

4. Repeat Exercise 3, but for a current interest rate of 4% and a zero-coupon bond that will pay \$10,000 five years from now.

5. Suppose that on the report for your money market account this month, the initial balance was \$7744.70, the report was for 34 days, and the final balance was \$7770.84. Calculate the annual yield.

6. Repeat Exercise 5, but for the previous month, which had an initial balance of 7722.54, a period of 27 days, and a final balance of \$7744.70.

7. *The rule of 72* is a rule of thumb for finding how long it takes money at interest to double: If r is the annual interest rate, then the doubling time is approximately $72/100r$ years.

(a) Calculate the balance at the end of the predicted doubling time for each \$1000, with annual compounding, for the small growth rates of 3%, 4%, and 6%.
(b) Repeat part (a), for the intermediate interest rates of 8% and 9%.
(c) Repeat part (a), for the larger interest rates of 12%, 24%, and 36%.
(d) What do you conclude about the rule of 72?

8. More frequent compounding yields greater interest, but with diminishing returns as the frequency of compounding is increased. For small interest rates, there is little difference in yield for compounding annually, quarterly, monthly, daily, or continuously. Investigating doubling times with continuous compounding leads to understanding why the rule of 72 of Exercise 7 works. Recall that for continuous compounding at annual rate r, the balance A at the end of m years is Pe^{rm} for an initial principal of P. Let t be the number of years that it takes for the initial principal to double. Then we have $2P = A = Pe^{rt}$, so $e^{rt} = 2$. Taking the natural logarithm of both sides yields $rt = \ln 2$, where ln stands for the natural logarithm, represented on a calculator by a button marked either [ln] or [LN] (not [log] or [log$_{10}$], which stands for a different kind of logarithm). Using the button gives $\ln 2 = 0.693$. So we have $rt = 0.693$, from which we can determine t if we know r.

Calculate the doubling times for continuous compounding at 3%, 6%, and 9%, and compare them with those predicted by the rule of 72. What do you conclude? Why do you think people prefer a rule of 72 over a rule of 69.3?

A Limit to Compounding

9. Use your calculator to evaluate for $n = 1, 10, 100, 1000$, and $1{,}000{,}000$:

(a) $\left(1 + \dfrac{1}{n}\right)^n$

(b) $\left(1 + \dfrac{2}{n}\right)^n$

(c) As n gets larger, what numbers are the expressions in parts (a) and (b) tending toward?

10. (Contributed by John Oprea of Cleveland State University.) Use your calculator to evaluate for $n = 1, 10, 100, 1000$, and $1{,}000{,}000$:

(a) $\left(1 - \dfrac{1}{n}\right)^n$

(b) $\left(1 - \dfrac{2}{n}\right)^n$

(c) As n gets larger, what numbers are the expressions in parts (a) and (b) tending toward?

11. You have \$1000 on deposit at your bank at an annual rate of 3%. How much interest do you receive after one year, if the bank compounds

(a) continuously?
(b) daily, using the 360 over 360 method?
(c) daily, using the 365 over 365 method?

12. Suppose that you have a bank account with a balance of $4532.10 at the beginning of the year and $4632.10 at the end of the year. Your bank advertises "continuous compounding" but in fact compounds continuously over each 24-hour day and posts interest to accounts daily.

(a) What effective rate did you receive?
(b) What nominal rate is the calculation based on?
(c) What difference is there between what the bank is doing and true continuous compounding?

A Model for Accumulation

13. Suppose that you want to save up $2000 for a trip abroad two years from now. How much do you have to put away each month in a savings account that earns 5% interest compounded monthly?

14. Repeat Exercise 13, except that you have found a better deal—7% interest compounded monthly.

15. Parents struggle for the first few years after their child is born but are finally able to start saving toward the child's college education (because they stop paying for day care) when the child goes to school at age six. If they save $400 per month in a credit union account paying 5.5% interest compounded monthly, how much will they have for college expenses 12 years later?

16. You save for retirement by contributing the same amount each month from your 23rd birthday until your 65th birthday, in an account that pays a steady 5% annual interest compounded monthly.

(a) How much will be in your fund at age 65 if you save $100 a month?
(b) How much will be in your fund if you get a steady return of 7.5% compounded monthly?
(c) How much will be in your fund if you get a steady return of 10% compounded monthly? (This is comparable to the average annual return of about 11% for all stocks on the New York Stock Exchange from 1950 to 2000.)

17. A colleague of mine feels that he will need $1 million in savings to afford to retire at age 65 and still maintain his current standard of living. A younger colleague, age 30, decides to begin saving for retirement based on that advice. How much does the younger colleague need to save per month to have $1 million at retirement if the fund earns a steady 5% annual interest compounded monthly?

18. Many young people do not start saving right away for retirement although by the time that they do, they may be earning more and thus be able to afford to save more each month.

(a) How much will be in your fund at age 65 if you don't start saving until age 35 and at that age start saving $100 per month in an account paying a steady 6% annual interest compounded monthly?
(b) Suppose instead that you have children young, pay for their college expenses, and finally start saving for retirement at age 45. How much do you have to save per month, with a steady return of 7.5% compounded monthly, to accumulate $250,000 by age 65?

■ **19.** Suppose that you are 25 and in a 28% bracket for federal income tax and a 6% bracket for state and local income tax (in 2003 this corresponded to an income beyond exemptions and deductions of about $43,000 for a married couple). This means that you pay a total of 34% in income tax on part of your income but a lower rate on the rest. Assume that instead of paying 34% on some income, you put it into a tax-deferred retirement account (TDA) as follows:

(a) Suppose that you are willing to commit to $100 a month less take-home pay. You realize that you don't pay income tax on the money that you put into the retirement plan, so you can actually put in more than $100 per month while reducing your take-home pay by only $100. How much can you put into the retirement fund each month?
(b) How much will be in your fund at age 65 if you can get a steady return of 7.5% compounded monthly?
(c) Suppose that when you turn 65 you withdraw the entire amount in your account and pay the

deferred taxes that are owed on it, say a total of 34% (federal, state, and local combined). How much do you net?

■ **20.** We continue the tax-deferral considerations of the previous exercise.

(a) Suppose that instead of contributing $151.52 per month to a tax-deferred plan, you take the $151.52, pay 34% income tax on it, and deposit what remains ($100) into a savings account or safe investment that pays a steady 7.5% compounded monthly. (Note that compared to not putting away any money, your paycheck is reduced by just the $100 that you contribute, since you still must pay income tax on the $151.52.) How much will be in your account at age 65?
(b) Under another alternative, you take $100 per month, pay 34% income tax on it, and deposit what remains into a *Roth IRA* (individual retirement account). For this kind of retirement account, the interest earned is not taxed. Assuming the same savings account or safe investment that pays a steady return of 7.5% compounded monthly, how much will be in your account, tax-free, at age 65?

21. (a) Use the geometric series formula to find the sum of $\frac{1}{2} + \frac{1}{4} + \frac{1}{8} + \frac{1}{16} + \frac{1}{32}$.
(b) Find the more general sum $\frac{1}{2} + \frac{1}{4} + \frac{1}{8} + \frac{1}{16} + \cdots + \frac{1}{2^n}$.
(c) As n gets larger and larger, what does the sum in part (b) tend to?

22. (a) Use the geometric series formula to find the sum of $1 - \frac{1}{3} + \frac{1}{9} - \frac{1}{27} + \frac{1}{81}$.
(b) Find the more general sum $1 - \frac{1}{3} + \frac{1}{9} - \frac{1}{27} + \frac{1}{81} + \cdots \pm \frac{1}{3^n}$, where the $\pm$ denotes + if n is odd and − if n is even (to fit the pattern of the earlier terms).
(c) As n gets larger and larger, what does the sum in part (b) tend to?

Exponential Decay

23. Suppose inflation proceeds at a level rate of 4% per year from mid-2003 through mid-2006.

(a) Find the cost in mid-2006 of a basket of goods that cost $1 in mid-2003.

(b) What will be the value of a dollar in mid-2006 in constant mid-2003 dollars?

24. Suppose you bought a car in early 2002 for $10,000. If its value (in current dollars) depreciates steadily at 12% per year, what will be its value (in current dollars) in early 2005?

25. For the car in Exercise 24, suppose that there is also 3% annual inflation from 2002 through 2008. What will be the value of the car in early 2008 in inflation-adjusted (early 2002) dollars?

The Consumer Price Index

26. (a) I bought my first LP record in 1965, at list price, for $4.98. How much would that be in 2003 dollars? How does that compare with the list price of a CD today?
(b) My father bought a Royal portable typewriter in 1940 for $40 (I have the sales slip). What would be the equivalent price in 2003 dollars?

27. (a) My first-semester college mathematics book cost $10.75 in 1962. What would be the equivalent price in 2003 dollars? How does that compare to what you paid for this book? (My book had black-and-white text and figures, with no photographs, color or otherwise.)
(b) In 1970, before the OPEC oil embargo, gasoline cost about 25 cents per gallon. In 1974, after the embargo, it cost about 70 cents per gallon. What would be the equivalent prices in 1997 dollars? How do they compare to the price of gasoline today?

28. From Table 21.5, you can determine the rate of inflation from one year to the next. For example, you find the rate of inflation from 1991 to 1992 by subtracting the two index numbers and dividing by the earlier one: $(140.6 - 136.2)/136.2 = 0.032 = 3.2\%$. Similarly, knowing the rate of inflation, you can compute one index number from another.

(a) What was the rate of inflation from 1980 to 1981?

(b) For a 3% rate of inflation per year from 2001 on, what would the consumer price index be in 2005?

Real Growth

29. On October 1, 2001, you could buy a U.S. Treasury 10-year note with a yield of 4.53%. At that time, the inflation rate was about 3.1%.

(a) What was the real growth rate of this investment?

(b) For an individual with a combined federal, state, and city income tax rate of 30%, what was the real growth rate of this investment?

30. Generalize the result in Exercise 29b by giving a formula for real growth rate g in terms of the interest rate r, the inflation rate i, and the tax rate t.

A Model for Stock Prices

■ **31.** On the morning of October 2, 2001, in an effort to stimulate the economy after the attacks of September 11, 2001, the Federal Reserve ("the Fed") cut its federal funds interest rate (a rate paid by banks borrowing from the Fed) by 0.5%, to 2.5%, the ninth cut of the year. At the time, the rate of inflation was about 4% and the interest rate on long-term government bonds was 4.53%, but uncertainty about the economy in the wake of the attack on America had no doubt increased. Suppose then that the discount rate was $r = 15\%$ and that the rate cut by the Fed reduced r by 0.25%. Suppose also that the average stock dividend was $g = 3\%$. With this in mind, consider a stock with annual dividend $D =$ \$5. By what percentage would you expect the price of that stock (and correspondingly the market as a whole) to rise as a result of the Fed's action? (In fact, on October 2, 2001, the Dow Jones Industrial Average rose by 114 points, from 8836.83 to 8950.59.)

■ **32. (a)** Let the symbol Δr (pronounced delta r) denote a change in the discount rate, with Δr positive for an increase and negative for a decrease. Such a change could be produced by action of the Federal Reserve or by a change in the rate of inflation, the rate on government bonds, or simply in investor confidence. Let $\Delta S/S$ denote the expected proportional change in stock prices. Find a simplified formula involving just r, g, and Δr for $\Delta S/S$.

(b) Apply the formula in part (a) with $g = 3\%$, $r = 15\%$, and $\Delta r = -0.25\%$ to find ΔS. You should get the same answer as in Exercise 31.

(c) Apply the formula "backwards" with $g = 3\%$, $\Delta r = -0.25\%$, and ΔS (the change observed in the Dow Jones Industrial Average on October 2) to determine the corresponding value of r before the announcement by the Fed.

In the savings formula, the interest rate i appears twice. The particular ways in which i is involved makes it impossible to solve it algebraically to get an explicit formula for i. However, with the help of a spreadsheet, you can find i approximately when the other quantities are given. Exercises 33–36 treat such situations.

33. (Requires spreadsheet) Suppose that you decide to lease a car. At the end of the 48-month lease period, you need to make a lump-sum payment of \$5000 if you want to keep the car. You decide to save up, just in case you decide to keep the car; if you don't keep this car, you will still have saved a good down payment on a new car. You feel comfortable with saving \$70/month (over and above your lease payments). How high an annual nominal interest rate on savings do you need to accumulate \$5000 in 48 months, with interest compounded monthly?

34. (Requires spreadsheet) A 1990 advertisement reads, "If you had put \$100 per month into this fund starting in 1980, you'd have \$37,747 today." Assume that deposits were made on the last day of the month, starting in January 1980, through December 1989, and that interest was paid monthly on the last day of the month (120 months).

(a) How much money was deposited during this period?

(b) What annual rate of interest, compounded monthly, would lead to the result described in the advertisement? What is the annual yield?

35. (Requires spreadsheet) On April 16, 2002, three winning tickets shared an almost-record Big Game (now renamed Mega Millions) lottery jackpot of \$331 million. Each ticket's share was one-third of the total, or \$110.333 million, to be paid as an annuity in 26 equal annual installments of \$4.244 million each, the first payment being right away. Instead of installment payments, each winner could choose an instant lump sum of \$58.9 million (actually, after tax withholding, \$43 million). Doing so would make sense particularly if the winner needed a large sum of money now or else could earn a higher rate of interest than the annuity is based on. The present value of the payment k years from now is given by P in the compound interest formula $A = P(1 + i)^k$, where $A = 4.244$ million and i is the unknown rate of interest built into the annuity. Hence, we have $P_k = A/(1 + i)^k$ for the kth payment. The complete stream of 26 payments has present value (in millions of dollars)

$$58.9 = 4.244\left[1 + \frac{1}{1+i} + \frac{1}{(1+i)^2} + \cdots + \frac{1}{(1+i)^{25}}\right]$$

We use the geometric series formula with $x = 1/(1 + i)$, finding that the right-hand side is also equal to

$$4.244\left[\frac{1 - \dfrac{1}{(1+i)^{26}}}{1 - \dfrac{1}{1+i}}\right]$$

Enter $i = 0.05$ (annual rate) in the expression above; the result is larger than 58.9. For $i = 0.06$, the result is smaller than 58.9. Make changes in the value of i until you determine to two decimal places the rate i that gives the closest value to 58.9. This is the rate on which the annuity is based. Is it a nominal rate or the effective rate?

36. (Requires spreadsheet) Suppose that your parents are willing to lend you \$20,000 for part of the cost of your college education and living expenses. They want you to repay them the \$20,000, without any interest, in a lump sum 15 years after you graduate, when they will be about to retire and move. Meanwhile, you will be busy repaying federally guaranteed loans for the first 10 years after graduation. But you realize that you can't repay the lump sum without saving up. So you decide that you will put aside money in an interest-bearing account every month for the 5 years before the payment is due. You feel comfortable with the idea of putting aside \$275 a month (the amount of the payment on your government loans). How high an annual nominal interest rate on savings do you need to accumulate the \$20,000 in 60 months, if interest is compounded monthly? Enter into a spreadsheet the values $d = 275$, $r = 0.05$ (annual rate), $n = 60$, and the savings formula with r replaced by $r/12$ (the monthly interest rate). You will find that the amount accumulated is not enough. Change r to 0.09 – it's more than enough. Try other values until you determine r to two decimal places.

Additional Exercises

37. An oft-heard claim is that "the amount of information in the world doubles every three days." Presumably the claim refers to the amount of data, which can be quantified in terms of number of bits. (A bit is the smallest unit of storage in a computer.) Show that the claim is absolutely preposterous, by doing a little arithmetic and comparing your result with the estimated number of particles in the universe (10^{70}). In particular:

(a) Start with one bit of data and double the number of bits every third day. How long does it take to get past 10^{70}? (*Hint:* Don't just keep multiplying by 2 over and over. Convince yourself that since the amount of data increases by a factor of 2 every 3 days, then it increases by a factor of $2^2 = 4$ every 6 days, a factor of $4^2 = 16$ every 12 days, a factor of $16^2 = 256$ every 24 days, and so forth.)

(b) Part (a) involves a lot of multiplying by 2, even if you do it efficiently. Another approach is to use the fact that $2^{10} = 1024 \approx 1000$. Thus, the amount of data increases by a factor of more than 1000 every $3 \times 10 = 30$ days, or every month (except February, but the 31-day months make up for it). By when will the total be sure to be past 10^{70}?

38. An old legend tells of a wizard who agreed to save a kingdom provided that the king would agree to a "modest" reward. The wizard asked to be given merely as much grain as would put one kernel on the first square of a chessboard, two kernels on the second, four on the third, eight on the fourth, and so forth, up through the 64th square. The king agreed, the wizard saved the kingdom, but the king was completely unable to honor the agreement. Why? [*Hints:* Notice that $1 = 2^1 - 1$, and $(1 + 2) = 2^2 - 1$, and $(1 + 2 + 4) = 2^3 - 1$; generalize to arrive at a total for the number of kernels. A kernel of rice is about a quarter of an inch long and about a 16th of an inch wide and a 16th of an inch high. So about a thousand kernels will fit in a cubic inch (you should verify this calculation). Calculate the total volume of kernels.]

39. Surprise! Just for fun, one of your friends wrote your name on an Illinois State Lottery ticket, and you are the sole winner of $40 million! You discover, however, that you don't get the $40 million all at once; in fact, it is paid in 20 equal annual installments of $2 million each. All you get right away is the first installment of $2 million (minus 20% withheld against federal income tax due and whatever you think your friend deserves for the favor). So, what is the prize really worth to you? That depends on the rate of inflation over the years. Assume a constant rate of inflation over the 19 years until your last payment and calculate the present value of your prize winnings by using the formula for present value combined with the formula for the sum of a geometric series. Do the calculation for rates of interest of

(a) 2%.
(b) 4%.
(c) 6%.

Actually, the checks will come not from the state of Illinois but from an insurance company from which Illinois purchases an annuity (a contract to pay a certain amount each year for specified number of years). The price of the annuity depends on current long-term interest rates.

■ **40.** We continue the theme of Exercises 19 and 20 by comparing three kinds of investment for retirement: an ordinary after-tax investment, a tax-deferred investment [such as a tax-deferred annuity or an independent retirement account (IRA)], and the relatively new Roth IRA. Let an investment earn interest at a steady annual yield i and let your income (in whatever year you receive it) be taxed at rate t.

(a) Ordinary after-tax investment: Explain why if you earn $\$E$, pay taxes on it, let what remains earn interest, and pay tax each year on that year's interest, the $\$E$ grows after n years to $\$E(1 - t) \times [1 + r(1 - t)]^n$.
(b) Ordinary IRA: Explain why if you earn $\$E$, defer taxes on it, let it earn interest, and defer taxes on all the interest, then the $\$E$ grows after n years to $\$E(1 + r)^n(1 - t)$.
(c) Roth IRA: Explain why if you earn $\$E$, pay taxes on it, let what remains earn interest, and pay no taxes on all the interest, the $\$E$ grows after n years to $\$E(1 - t)(1 + r)^n$.
(d) Which investment gives the best return after n years?
(e) The assumptions of constant interest rate and stable tax rate won't necessarily hold, because interest rates fluctuate (though you can lock in a long-term constant interest rate by buying a long-term bond or certificate of deposit) and the tax rate may change (with your income, state of residence, and changes in tax laws). If your marginal tax rate (the rate you pay on one more dollar of income) is lower in one year than the tax rate you expect to pay in retirement, what kind of retirement investment is better for you that year? If you have a windfall one year and your marginal tax rate is higher that year than the tax rate that you expect to pay in retirement, what kind of retirement investment is better for you that year?

WRITING PROJECTS

1. In recent years, incentives from automobile manufacturers to potential customers have taken the form of offering either a reduced interest rate on the loan for the car or else a rebate (reduction in price) on the cost itself. In summer 2002 you could by a Dodge Neon for $11,998 with a 1.9% interest loan or else get a $1500 rebate but have to finance at a higher rate (either through the auto dealer or elsewhere).

Suppose that you could afford a $2000 down payment and could get a loan from a credit union at 6.0%.

(a) What is your monthly payment if you choose the rebate and take out a 48-month loan with the credit union for the reduced price minus your down payment?

(b) What is your monthly payment if you choose the 1.9% 48-month loan for the original price minus your down payment?

(c) Suppose that when the dealer realizes that you prefer the option in part (a), she offers you a further combined option: a rebate of $1000 with a 3.9% interest rate. How does this compare?

(d) Locate current advertised incentives for a car that you would like to buy and compare them.

2. On May 8, 2002, you could buy $10,000 U.S. Treasury bonds that pay simple interest of (1) 5.375% per year in February each year through maturity in February 2031, at a cost of $9,531 each; or (2) 11.25% per year in February each year through 2015, at a cost of $15,200 each. Each bond also returns $10,000 at maturity.

(a) What is the annual yield of each investment?.

(b) Compare these two investment opportunities, which differ vastly in interest rate but also in cost.

3. Exercises 19, 20 and 40 look asked you to look at various forms of tax-deferred and ordinary savings and compare them on the basis of amount of tax-free income accumulation at age 65.

Ordinary savings have the important advantage that at any time, you can do anything you want with the money accumulated so far (e.g., buy a car put down money on a house, etc.). A second advantage is that the money is free and clear, in that taxes have already been paid on it.

A tax-deferred 401(k) retirement fund has the disadvantage that for any funds withdrawn before age $59\frac{1}{2}$, you must pay income tax in the year of withdrawal and, in addition, pay a 10% penalty for "early withdrawal." (These plans were given the advantage of tax deferral to encourage individuals to save for retirement – hence the penalty for using the money for other purposes.)

A third option, the Roth IRA, has some of the advantages and disadvantages of each of the above.

Look into the details of the rules for 401(k) plans and Roth IRAs, compare your answers in Exercises 19, 20, and 40, and devise and describe your own plan for how you will save for retirement.

More on the Web SPREADSHEET PROJECTS

To do these projects, go to www.whfreeman.com/fapp.

This project uses the consumer price index to track changing prices. Computations of scale and proportion in area and volume are also explored using spreadsheets.

More on the Web APPLET EXERCISES

To do these exercises, go to www.whfreeman.com/fapp.

How important is it to begin a retirement fund at an early stage of one's career? In the applet Saving for Retirement, you will discover that early funding of a retirement plan can make a huge difference in the ultimate amount that will be available when a person retires.

CHAPTER 22

Consumer Finance Models: Borrowing

In this chapter we cover...

...and at www.whfreeman.com/fapp:

Flashcards
Quizzes
Spreadsheet Projects
Annuity tables
Modeling mortgages
Applet Exercises
Buying a car

In the previous chapter we looked at consumer financial models for saving and formulas for calculating the amount accumulated. However, savings put under a mattress do not earn interest. Savings in a financial institution or in the form of an investment would not earn interest, either, unless the savings could be loaned to someone who has need and can make productive use of the money.

In this chapter we examine the other side of consumer finance, borrowing. You are likely to borrow to buy a car, you will almost certainly have to borrow if you buy a house or apartment, and you are borrowing if you use a credit card. In all these and other cases, you will likely be paying interest and finance charges.

But there are different ways for loans to be structured, and they can make a lot of difference in how much you pay for the use of the money that you borrow. We investigate and compare some common kinds of loans.

We briefly (re)acquaint the reader with the topic of compound interest and note a couple of useful formulas from Chapter 21. The reader who has a grasp of the ideas behind compound interest and can use the formulas can proceed with this chapter without first reading Chapter 21.

22.1 Simple Interest

Interest is money charged on a loan. The amount of interest is determined by the **principal,** the amount borrowed, and by the method used for calculating the interest. With **simple interest,** the borrower pays a fixed amount of interest for each period of the loan. The interest rate is usually quoted as an annual rate.

In the previous chapter, we cited as an example of simple interest the interest paid on a bond.

EXAMPLE

Bonds

From the point of view of the purchaser, a bond is a form of savings or investment. When you buy a bond, you are paid simple interest (i.e., the same amount) annually; and at the end of the term of the bond, your original purchase price is returned to you. For example, in April 2002 the U.S. Treasury sold 30-year bonds that cost \$10,000, pay 3.375% simple interest (i.e., \$337.50) every subsequent April through April 2032, and can be cashed in for \$10,000 in April 2032.

From the point of view of the seller, issuing a bond is a way to borrow money at a fixed amount of interest over a long period. Any risk of uncertainty about future inflation, current interest rates, or the ability of the bond issuer to continue paying interest and return the principal is assumed by the buyer of the bond. The credit rating of the bond issuer and the length of the term of the bond, together with current interest rates for other investments, determine how much interest the issuer will need to promise in order to sell the bonds. ■

For a principal P and an annual rate of interest r, the interest owed after t years is

$$I = Prt$$

and the total amount A due on the loan is

$$A = P(1 + rt)$$

EXAMPLE

Payments on Those Bonds

Consider one of those 30-year U.S. Treasury bonds issued in April 2002 for \$10,000. How much total interest would the Treasury pay on it through April 2032, and what total amount would the Treasury pay back to the purchaser? We have $P = \$10{,}000$, $r = 3.375\% = 0.03375$, and $t = 30$ years, so

$$I = Prt = \$10{,}000 \times 0.03375 \times 30 = \$10{,}125$$

and

$$\begin{aligned} A = P(1 + rt) &= \$10{,}000 \times (1 + 0.03375 \times 30) \\ &= \$10{,}000 \times (2.125) = \$20{,}125. \end{aligned}$$

■

Bonds may seem a bit esoteric; you probably have never bought one (much less issued one!). However, during World War II and for many years afterward, it was common—and manifestly patriotic—for citizens to purchase U.S. Savings Bonds, in denominations as small as \$25. More common uses of simple interest are certain types of consumer loans, which we turn to now.

A common type of consumer loan is the *add-on* loan. You borrow an amount P to be repaid in t years; the interest is simple interest at an annual rate r (= $100r$%), for a total of $I = Prt$. You must pay $P + I = P(1 + rt)$ in installments (usually monthly); with n payments, each payment is $d = P(1 + rt)/n$. In effect, you pay $\frac{1}{n}$th of the principal and $\frac{1}{n}$th of the total interest with each payment.

EXAMPLE

Add-On Loans

Suppose that you need to borrow \$8000 to buy a used car. The dealer offers you a 5% add-on loan to be repaid in monthly installments over four years. This sounds like a much better deal than the 8% loan that you can get at the credit union. How much is your payment d on the dealer's add-on loan?

To find the total amount to be paid over the four years, first calculate the interest, I:

$$I = Prt$$
$$I = \$8000(0.08)(4)$$
$$I = \$1600$$

The total amount that must be repaid is then $P + I = \$8000 + \$1600 = \$9600$, and the monthly payment is $\$9600 \div 48 = \$200$. ■

With an add-on loan, everything sounds simple and straightforward (even the calculation of the payment!). The interest is calculated on the entire principal; however, because you slowly pay back the principal, you do not have the use of the whole amount for the entire loan period. In fact, you have the use of the full principal for just one period. You do, however, have the use of the car! But the net value that you have at any point is the cost of the car minus the

(PhotoEdit.)

amount of principal already repaid (we neglect depreciation of the car). It is on this net value, and the amount of interest, that the interest rate should be calculated. In effect, the "true" interest rate (we will make this concept more precise) is much higher than the rate r quoted.

Another type of consumer loan is a *discounted loan*. The interest is computed as simple interest, just as for an add-on loan, but it is subtracted from the amount given to the borrower. In other words, instead of getting the principal P, the borrower gets the *proceeds* $P - I = P - Prt = P(1 - rt)$ but must pay back the amount P over the term of the loan. What is discounted is not the cost of the loan but how much the borrower gets! The interest is based on the entire P, but the borrower never has the use of that much and, as with an add-on loan, is paying interest on the entire amount P but has the use of less and less of it as the loan is repaid.

EXAMPLE

Discounted Loans

Suppose again that you need to borrow \$8000 to buy a used car. Your neighborhood loan store offers you a 4% discounted loan to be repaid in monthly installments over four years. Is this a better deal than the dealer's offer of a 5% add-on loan over four years (see previous example)?

The total amount to be paid over the four years is $P =$ \$8000 over 48 months, for a monthly payment of \$166.67. This is indeed a lower payment, but the problem is that the lender loans you only $\$8000 \times (1 - 0.04 \times 4) = \$8000 \times 0.84 = \$6720$, which isn't enough to buy the car. How big a discounted loan do you need? Call the amount x. You need $x(0.84) = \$8000$, or $x = \$8000/0.84 = \$9523.81$. Your monthly payment is $\$9523.81/48 = \$198.41$, so this loan is indeed cheaper than the dealer's loan. ■

22.2 Compounding

Compounding is the calculation of interest on interest. A common example is the balance on a credit card. As long as there is an outstanding balance owed, interest owed is calculated on the entire balance, including any part of it that was interest calculated and added to the balance in earlier months.

EXAMPLE

Credit Card Interest

Suppose that you owe \$1000 on your credit card, the company charges 1.5% interest per month, and you just let the balance ride. Your interest the first month is 1.5% of \$1000, or $0.015 \times \$1000 = \$15$. The new balance owed is $(1 + 0.015) \times \$1000 = \$1015$. Your interest the second month is 1.5% not of \$1000, or \$15 (as would be the case for simple interest), but 1.5% of \$1015, or $0.015 \times \$1015 = \$15.23$, so the new balance is

$$(1 + 0.015) \times \$1015 = \$(1 + 0.015) \times (1 + 0.015) \times \$1000 = \$1030.23$$

(We neglect the fact that the company will likely add an extra charge for your failure to make the minimum payment that they specify.) After 12 months of letting the balance ride, it has become

$$(1.015)^{12} \times \$1000 = \$1195.62$$

In other words, the actual interest for the year comes to \$195.62, which is 19.562% of \$1000. So, although the quoted rate of interest is 1.5% per month, which seems as if it should amount to $12 \times 1.5\% = 18\%$ per year, the interest owed is actually more. ■

In this chapter, we use only two formulas from Chapter 21, the first of which we cite here in phrasing suited to loans:

Compound interest formula: If a principal P is loaned at an interest at rate i per compounding period, then after n compounding periods with no repayment, the amount owed is

$$A = P(1 + i)^n$$

This is just the generalization of what we saw happen in the case of the credit card balance.

EXAMPLE

Borrowing from Joe, the Tiniest Loan Shark

Suppose that you arrive on campus without enough money to buy your textbooks (including this one), not to mention other supplies and necessities (such as late-night pizzas). So you go to Joe's Friendly Loan Service at a nearby off-campus location, where Joe offers to loan you \$1000 at an incredible 1% interest per week, compounded weekly. Joe is such a good sport that he doesn't want the money back until a year from now, which gives you a chance to earn the repayment over the summer. How much will you have to pay back at the end of a year? We have $P = \$500$, $i = 1\% = 0.01$ per compounding period, and $n = 52$ compounding periods:

$$A = P(1 + i)^n = \$1000 \times (1 + 0.01)^{52} = \$1677.69$$ ■

Savings formula: For a uniform deposit of d per period (deposited at the end of the period) and an interest rate i per period, the amount A accumulated after n periods is

$$A = d\left[\frac{(1 + i)^n - 1}{i}\right]$$

EXAMPLE Saving Up to Pay Joe Back

Summer comes and it's time to save up to pay Joe back. Suppose that you work for three months and each month deposit an amount d to a savings account that pays 3% interest per year, or 0.25% per month. How much does d need to be to pay off the \$1667.69 that you owe? We have $A = \$1667.69$, $i = 0.25\% = 0.0025$ per compounding period, and $n = 3$ compounding periods:

$$A = d\left[\frac{(1+i)^n - 1}{i}\right]$$

so that

$$\$1667.69 = d\left[\frac{(1+0.0025)^3 - 1}{0.0025}\right] = d(3.0075064)$$

Solving gives $d = \$1667.69/3.0075064 = \$557.83$. ■

22.3 Terminology for Loan Rates

We have seen that the rate of interest for a loan depends on whether compounding is done and how the interest is calculated. To help prevent confusion on the part of consumers, the Truth in Lending Act establishes specific terminology and calculation methods for interest.

A **nominal rate** is any stated rate of interest for a specified length of time.

For instance, a nominal rate could be a 1.5% monthly rate on a credit card balance. By itself, such a rate does not indicate or take into account whether or how often interest is compounded.

The **effective rate** is the actual percentage rate of increase for a length of time, *taking into account compounding.*

The effective rate equals the rate of simple interest that would realize exactly as much interest over that length of time. We saw that \$1000 at a yearly interest rate of 18% (a nominal rate), or 1.5% per month, compounded monthly for one year yields \$195.62 in interest owed, which is 19.562% of the original principal. Hence the effective annual rate is 19.562%. In other words, a \$1000 loan at simple interest of 19.562% for one year would amount to exactly the same interest owed.

When stated per year ("annualized"), the effective rate is called the **effective annual rate (EAR).**

We need a special formula, the amortization formula, to calculate the EAR for a loan; we develop that formula in the next section in discussing what are called *conventional loans,* which include home mortgages.

To help keep these different rates straight, we use i for a nominal rate for any length of time—such as a day, month, or year—*within which no compounding is done;* this rate is in fact the effective rate for that length of time.

In particular, for a nominal annual rate r compounded n times per year, we have

$$i = \frac{r}{n}$$

For that \$1000 credit card balance at 18% compounded monthly, we have $r = 18\%$ and $n = 12$, so $i = 1.5\%$ per month.

We use the term **period** for the basic unit of time for compounding—the compounding period—within which no compounding is done. We denote the number of compounding periods under consideration by n.

We use r only for an annual rate (nominal or effective) and t for the number of years.

Finally—and most important to the consumer—the **annual percentage rate (APR)** is the rate of interest per compounding period times the number of compounding periods per year.

In our notation, APR $= i \times n$. In the example of the credit card balance, the interest is compounded monthly, or $n = 12$ times per year, and the interest rate for the compounding period is $i = 1.5\%$, so the APR is $12 \times 1.5\% = 18\%$. The APR is the rate that the Truth in Lending Act requires the lender to disclose to the borrower. The APR is not equal to the EAR (as we have already seen in the credit card example); Spotlight 22.1 explains how and why.

Important terms used with savings, but that we do not use in this chapter, are *annual percentage yield (APY)* and *annual equivalent yield,* both of which are just other names for the effective annual rate (EAR).

22.4 Conventional Loans

A common situation that you are likely to encounter is a loan—for a house, a car, or college expenses—to be paid back in equal periodic installments. Your payments are said to **amortize** the loan. In a conventional loan, each payment pays the current interest and also repays part of the principal. *As the principal is reduced, there is less interest owed, so less of each payment goes to interest and more toward paying off the principal.*

Let's suppose that Sally buys a house for \$100,000 with a loan that she will pay off over 30 years in equal monthly installments. Suppose that the interest rate for her loan is 6.00%. Let's figure out how much her monthly payment needs to be.

Imagine changing the setup slightly so that now Sally is borrowing the entire sum (\$100,000) for 30 years, and we think of her monthly payments as the savings fund that she's building up to pay off the loan at the end of the term. The interest rate of 6.00% on the loan is compounded monthly, so the monthly rate is 0.5%. At the end of 30 years, the principal and interest on the loan will (by the compound interest formula) amount to

$$\$100{,}000 \times (1 + 0.005)^{30\times 12} = \$602{,}257.52$$

On the other hand, saving $\$d$ each month for 30 years at 6.00% interest compounded monthly, we know from the savings formula that Sally will accumulate

$$d\left[\frac{(1 + 0.005)^{360} - 1}{0.005}\right]$$

To make d just the right amount to pay off the loan exactly, we need to solve the equation

$$d\left[\frac{(1 + 0.005)^{360} - 1}{0.005}\right] = \$100{,}000 \times (1 + 0.005)^{30\times 12} = \$602{,}257.52$$

for the value of d, getting $d = \$599.55$ as Sally's monthly payment. Note that the total of Sally's payments is only $360 \times \$599.55 = 215{,}838.00$.

We put this in a more general setting:

Let the principal be A, the effective interest rate per period r, the payment at the end of each period d, and let there be n periods. Then we have

$$A(1 + i)^n = d\left[\frac{(1 + i)^n - 1}{i}\right]$$

The compound interest formula is on the left and the savings formula is on the right. You can think of paying off the loan as making payments to a savings account, earning interest at the same rate as the loan, which will exactly balance principal and interest on the loan at the end of the loan term.

SPOTLIGHT 22.1 What's the Real Rate?

What is the real, *true* rate of interest for a loan?

Financial experts agree that the "true rate," whether for savings or for loans, is the effective annual rate (EAR).

The 1991 Federal Truth in *Savings* Act requires that savers be told the EAR, which for savings is called the annual percentage yield (APY).

The 1968 Federal Truth in *Lending* Act, however, requires that borrowers be told the *annual percentage rate (APR),* which is *not* the same as the EAR. The APR is the rate of interest per compounding period times the number of compounding periods per year. Thus, a credit card rate of 1.5% per month translates to an APR of 18%. The APR does not take into account compounding. Hence it is not equivalent to—indeed, it understates—the true cost of borrowing, that is, the EAR. For the credit card loan, with monthly compounding, the EAR is

$$(1 + 0.015)^{12} - 1 \approx 19.6\%.$$

Even though the APR understates the true cost of borrowing, universal use of it at least gives borrowers a common yardstick for comparing consumer loans: the lower the APR, the lower the cost of the loan. But the APR also ignores costs that are sometimes involved in borrowing, such as a flat charge for making the loan in the first place ("loan-processing fee"), charges for late payments, and charges for failing to make a minimum payment.

For mortgage loans in particular, there are further complications. The Truth in Lending Act requires that lenders include in the APR some of the upfront costs that are referred to as *closing costs.* The APR includes any loan origination fee, loan-processing fee, and discount "points." "Points" are additional charges to the buyer stated in terms of percentage points of the purchase price. The APR does not include title insurance, appraisal, credit report fees, or transaction taxes.

Closing costs must be paid immediately at the closing of the sale, while interest is paid over the life of the loan. However, the APR spreads the closing costs in it over the entire term of the mortgage, allocated to each payment in proportion to the interest in that payment.

Thus, the quantity A is sometimes called the *present value of an annuity* of n payments of d, each at the end of an interest period, with interest compounded at rate i in each period. This terminology agrees with that used in some calculators, such as the TI-83, which have a financial mode with the option to calculate A. Solving for A gives the **amortization formula**

$$A = d\left[\frac{1 - (1 + i)^{-n}}{i}\right]$$

EXAMPLE Making Weekly Payments to Joe

You're worried about having to come up with the lump sum ($1677.69) that you will owe Joe after a year, so you ask him if you can make weekly payments

But very few people hold a mortgage to its maturity. The median life of a 30-year mortgage is only about 5 years; that is, half of all mortgage holders pay off their mortgage before 5 years are up, usually because they sell the home and move elsewhere. Thus, for almost all home loans, the APR also includes interest that will never be paid.

Finally, we must take into account inflation. One advantage of buying a home with a fixed-rate mortgage is that your payment stays the same but your earnings and the value of your home are likely to go up with inflation: You are thus paying back the loan with dollars of lesser value. For any loan in a time of inflation, *Fisher's effect* (Chapter 21, p. 781) comes into play: If your loan is has an EAR of 7% but inflation is running at 3.5% per year, the true cost to you of the loan is not exactly 7% − 3.5% = 3.5%. Instead, for an EAR of r and an inflation rate of a, the cost of the loan at the begining of the first year is indeed $r - a$ (= 3.5% in our example), but at the end of the first year it is

$$c = \frac{r - a}{1 + a}$$

For $r = 7\%$ and $a = 3.5\%$, we get $c = 3.38\%$. The reason this is less than the expected 3.5% is that at the end of the first year you are paying back the loan with dollars that have been inflated for a year. As inflation mounts over the term of a mortgage, the cost c goes down steadily each year. For example, at the end of five years of steady inflation at 3.5%, the total inflation has been $a = (1 + 0.035)^5 = 18.8\%$, and we have $c = 2.95\%$.

A final—and major—consideration is that interest paid on your home mortgage, on a loan against your home equity, or even on a second home is deductible on federal, state, and some local income tax returns. This means that you may deduct your home interest expense from taxable income. Thus, your home ownership is subsidized by other taxpayers (just as you help subsidize home buyers among them), and the cost to you of the loan is reduced further.

instead. He is delighted at the idea but says that the interest will have to be higher, 1.5% per week. He agrees to let you amortize the loan over 52 weeks, so that each payment covers the week's interest on the amount still outstanding and pays off some of the principal. What is the amount of your weekly payment?

We have $A = \$1000$, $i = 0.015\%$ per compounding period (one week), $n = 52$ compounding periods, and want to find d:

$$A = \$1000 = d\left[\frac{1 - (1 + i)^{-n}}{i}\right] = d\left[\frac{1 - (1 + 0.015)^{52}}{0.015}\right] = d(35.92874)$$

so $d = \$1000/35.92874 = \$27.83$ is your weekly payment. ■

EXAMPLE

Buying a Car

You decide to buy a new "Wheelmobile" car. After a down payment, you need to finance (borrow) \$12,000. Comparing interest rates offered by the car dealership, local banks, and your credit union, the best deal you can find is 7.9% compounded monthly over 48 months. What is your monthly payment?

We have $A =$ \$12,000, monthly interest rate $i = 0.079/12 = 0.006583333$, and $n = 48$. Using the amortization formula, we have

$$\$12{,}000 = d\left[\frac{1 - (1.006583333)^{-48}}{0.006583333}\right]$$

$$\$12{,}000 = d(41.0408)$$

$$d = \$292.39$$

How much interest do you pay? You make payments totaling $48 \times \$292.39 = \$14{,}034.72$, so the interest is $\$14{,}034.72 - \$12{,}000 = \$2034.72$.

If you had bought a "Plushmobile" instead, with \$24,000 to finance, you would have borrowed twice as much and your monthly payment would have been twice as much. ■

EXAMPLE

APR Versus EAR for an Add-On Loan

We return to the earlier 5% add-on loan for \$8000 to be repaid over four years. The effective interest rate per month is i in the amorization formula. As hard as you try, though, you cannot solve the amortization formula for i in terms of A, d, and n. However, knowing the values of A, d, and n, you can use a spreadsheet and try successive values of i. In this case, you find $i = 0.0077014725$.

Here now we see a distinction. The APR (see Spotlight 22.1), which the dealer is required to tell you, is 12 times the monthly rate, or $12 \times 0.0077014725 = 9.24\%$. The EAR, which takes into account monthly compounding, is $(1 + i)^{-1} = 9.64\%$.

Either way, is this a better rate than the 8% credit union loan? Before you jump to conclusions, better find out the details of that loan! ■

A car loan is usually for 48 or 60 months; but when you buy a home, you usually borrow a great deal more money and hence pay it off over a much longer period. The usual term for a home mortgage is 30 years. We disregard here some of the closing costs connected with buying a home and getting the loan, to focus on the loan itself.

As a simple example, consider a mortgage for \$150,000 (the median cost of a U.S. home in 2002) at an interest rate of 7.2% over 30 years. Payments are monthly; the monthly interest rate is $7.2\%/12 = 0.6\% = 0.006$ and there are $30 \times 12 = 360$ months. We use the amortization formula

$$A = d\left[\frac{1 - (1 + i)^{-n}}{i}\right]$$

with $A = \$150{,}000$, $i = 0.006$, and $n = 360$:

$$150{,}000 = d\left[\frac{1 - (1 + 0.006)^{-360}}{0.006}\right] = d(147.3214)$$

so the monthly payment d is $\$150{,}000/147.3214 = \$1018.18$.

When you decide to buy a home, a key question that you need to ask yourself is how much home you can afford.

EXAMPLE 30-year Mortgage

Let us consider a family with the U.S. median income of about \$40,000, a home at the U.S. median cost of \$150,000, and a 30-year fixed-rate mortgage at 7.2%. Let us also suppose that the family can make a down payment of \$30,000. Can the family afford to buy such a home?

Lenders have "affordability" guidelines that suggest that a family can afford to spend about 28% of its monthly income on housing. Thus, this family can afford $0.28 \times \$40{,}000/12 = \$933.33$ per month.

However, the housing costs and the actual mortgage payment are larger than the amount needed to amortize the loan, because the payment must also cover the real estate taxes and homeowners' insurance on the property; on a \$150,000 home, those may add, say, \$300 to the monthly payment. That leaves the family only \$633.33 to cover amortizing a loan of \$120,000 at 7.2% interest.

(Norbert Schwerin/The Image Works.)

What is the payment on the loan? The principal is $A = \$120{,}000$, the monthly interest rate is $i = 0.072/12 = 0.006$, and $n = 360$ months (30 years with 12 monthly payments). The amortization formula gives

$$\$120{,}000 = d\left[\frac{1 - (1.006)^{-360}}{0.006}\right] = d(147.3214)$$

$$d = \$814.55$$

So, no, the median family can't afford the median-priced home, at least not without a bigger down payment or a lower-interest loan. ■

EXAMPLE

Home Equity

My wife's parents recently decided to sell their house and move to the town where we live. They had bought their house in 1980 for \$100,000 with a 30-year mortgage at an 8% interest rate. After 22 years, how much *equity* did they have in the house—that is, how much of the principal had been repaid? And how much did they still owe on the house?

What may shock you is that when they sold their house in May 2002—after 269 months of payments, almost exactly three-quarters of the 30 years of the mortgage—they had only \$50,000 in equity (hence still owed \$50,000 on the house) but had already paid \$147,000 in interest.

We can use the amortization formula to determine just how much equity they had after 269 months of payments, but first we need to determine their monthly payment. We set $A = \$100{,}000$, $n = 360$ months, and $i = 0.08/12 = 0.0066666$ monthly interest, getting $d = 733.76$.

Now we use the formula again, this time knowing $i = 0.08/12 = 0.0066666$ and $d = 733.76$. We allow for $n = 360 - 269 = 91$ payments to go and find the principal A:

$$A = d\left[\frac{1 - (1 + i)^{-n}}{i}\right] = \$733.76(68.060) = \$49{,}912.81$$

This is how much my parents-in-law had yet to pay, so their equity was $\$100{,}000 - \$49{,}912.81 = \$50{,}087.19$.

Figure 22.1 shows equity versus time for such a loan. The equity grows almost exponentially, but that means that the rate of growth is very slow in the early years of the mortgage. ■

When you buy a home, you are likely to face several options for the mortgage: a conventional 30-year mortgage, a conventional 15-year mortgage, or a mortgage for either length of time but with an interest rate that may vary.

You might expect that the payments on a 15-year mortgage would be double the payments on a 30-year mortgage. On the contrary, the payments are only

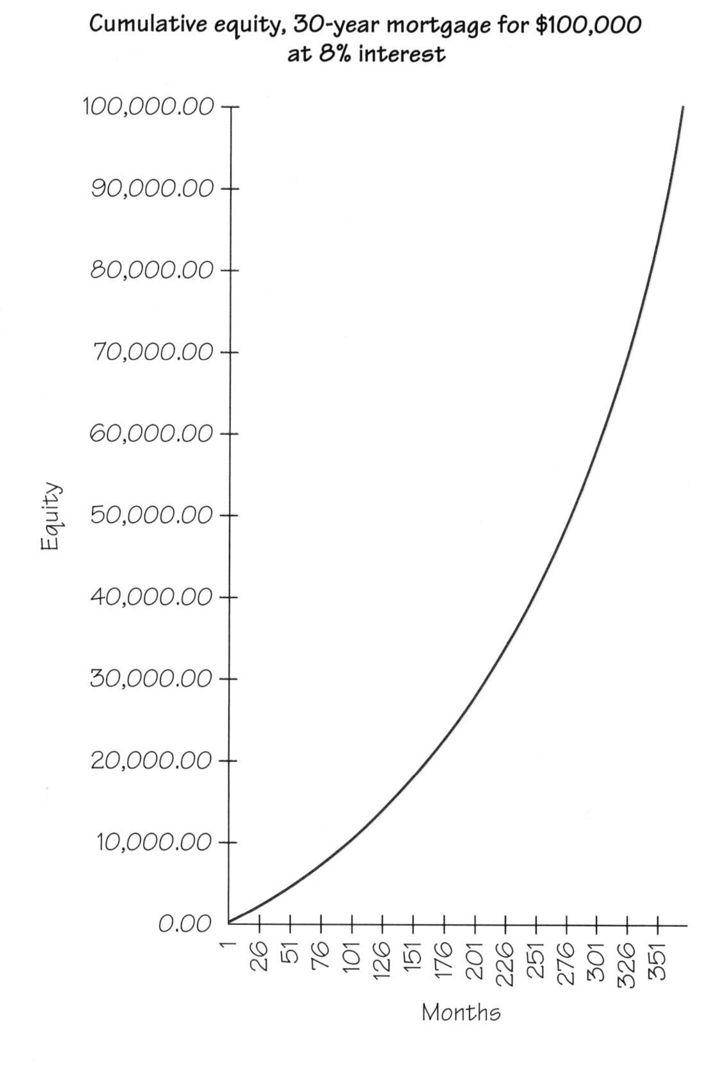

Figure 22.1
Equity grows almost exponentially, especially in the later years of mortgage.

40% more for a 6% mortgage to 26% more for a 9% mortgage; this range includes the prevailing mortgage rates over the past 20 years. Moreover, over the course of a $100,000 mortgage at 6%, you would pay $164,000 in interest over 30 years but only $72,000 over 15 years; at 9%, the interest totals are $190,000 vs. $83,000. (Some financial counselors advise taking a 30-year mortgage and making extra payments when you can afford them, rather than incurring the higher payment obligation of a 15-year loan, which, if you encounter tight personal financial circumstances, you might not be able to afford.)

Recalling that very few mortgages are held for the full term, it is useful to compare the status of mortgages after five years. Table 22.1 shows the equity after five years for a variety of interest rates. For a 30-year mortgage, the equity after

TABLE 22.1 Equity on a $100,000 Mortgage After Five Years

Term (years)	6%	7%	8%	9%
15	24,000	22,600	21,200	19,900
30	6,900	5,900	4,900	4,100

five years is likely to be less than the cost of selling the home through a realtor; of course, the resale value of the home may be higher after five years.

A mortgage with an interest rate that can vary is called an *adjustable-rate mortgage (ARM).* Often such mortgages have a substantially lower interest rate (hence a lower payment) than a fixed-rate mortgage. The ARM's interest rate may go up or down with interest rates in the economy; usually the rate can be raised or lowered only every year or two, and then by a limited percentage. An ARM may be attractive if you plan to pay off the mortgage after only a few years or because it allows lower payments, it facilitates buying a more expensive home, or you do not plan to keep the home long (hence you would be selling before the interest rate could rise substantially).

22.5 Annuities

An **annuity** is a specified number of equal payments at equal intervals of time. We restrict our discussion to *ordinary annuities,* for which payments are made at the end of each interval and the interval is also the compounding period.

An annuity can be interpreted as involving borrowing. For example, winners of lotteries are often offered the choice of receiving either the jackpot amount paid as an annuity over a number of years or else a smaller lump sum to be paid immediately. The cost to the lottery administration is the same; if the winner wants an annuity, the administration buys one from an insurance company for the lump sum. You can think of the insurance company as borrowing the lump sum in exchange for making the payments of the annuity; in effect, the insurance company is amortizing the lump sum over the duration of the annuity.

EXAMPLE Winning the Lottery

On August 25, 2001, four winning tickets shared an almost-record Powerball lottery jackpot of $294.7 million. Each ticket's share was one-fourth of the total, or $73.7 million, to be paid as an annuity in 25 equal annual installments of $2.947 million each, the first payment being right away. However, each winner instead chose an instant lump sum of $41.5 million.

The insurance company that sold the annuity to the lottery administration regarded $41.5 million as the present value of the annuity. To consider the annuity as an ordinary annuity, with payments made at the end of each interval, we must subtract the first payment, leaving $73.7 − $2.947 = $70.753 million to be paid in 24 equal annual installments at the end of each year, and $41.5 – $2.947 = $38.553 million as the present value.

In the formula for the present value of an annuity,

$$A = d\left[\frac{1 - (1 + i)^{-n}}{i}\right]$$

we have A = \$38.553, d = \$70.753/24, and $n = 24$. To solve for i, the interest rate used by the insurance company, we must use either a calculator with financial mode or a spreadsheet. Either way, we find $i = 5.84\%$. ■

As you save for retirement, it is probably wise to save part of your funds in the form of a tax-deferred annuity (remember how the Enron employees who kept most or all of their retirement savings in Enron stock lost when the company went bankrupt in 2002). If you do not, at retirement you can still sell all of your holdings in other forms and purchase an annuity.

Such annuities differ in a crucial way from the lottery annuity in the previous Example. If you were to retire at 65 and purchase an annuity like the lottery annuity, you would be in trouble if you lived longer than the term of the annuity (i.e., past age 89), because the payments would stop then and you would have no further income from the annuity. (About 2% of U.S. children born today can expect to live to age 100.) Similarly, if you died sooner, your estate would still get the payments due after your death, but those wouldn't have helped you meet your living expenses while you were alive.

An approach that avoids these two disadvantages is the *life income annuity:* You receive a fixed amount of income per month for as long as you live. How much you receive per month is based on the life expectancy of people your age, as determined from population data. There are many variations on life annuities, such as payments that increase with anticipated cost-of-living increases, or payments that last until both you and your life partner die (see Spotlight 22.2). But we focus on a simple one-life annuity.

The insurance company that sells you the annuity makes money if you die younger than average and loses money if you die older than average. As in any kind of insurance, over a large number of people, the company can expect gains to balance losses; this is a manifestation of the law of large numbers of Chapter 7. Also, the company's profits vary with the prevailing interest rate during the annuity as compared with the rate built into the annuity.

How much you receive per month depends on your sex; because women on average live longer than men, the monthly payment to a woman is lower.

EXAMPLE Life Income Annuity

Suppose that you are a 65-year-old male, who has retired with $250,000 in a life income annuity that you started to contribute to in 1980. According to the table from one particular insurance company, you would receive $7.08 per month for

SPOTLIGHT 22.2 Actuaries

The Truth in Savings Act and the Truth in Lending Act specify that the APY for savings and the APR for loans must be calculated "according to the actuarial method." A loan might not be repaid, and the risk of that happening must be taken into account in setting the loan costs, including the interest rate.

Actuaries are financial experts who, using the time value of money, assess risks and investigate the probability of the various contingencies – for example, death or cancellation – that might occur under a policy. Actuaries are crucially involved in setting premiums; their calculations take into account historical rates – such as the percentage of female 85-year-olds who live to be 86, or the percentage of unmarried male drivers under age 25 who have auto accidents – and project those rates and the accompanying costs into the future.

Other actuaries concentrate on setting up and evaluating pension and fringe benefit plans. For example, the City of Beloit, Wisconsin, recently hired a consulting actuary to estimate the current and future costs of free lifetime medical benefits to families of police and firefighters.

(David Young-Wolff/PhotoEdit.)

Another major activity of actuaries is managing return on investment. Contrary to popular belief, insurance companies (particularly life insurance companies) do not earn all of their money from premiums paid; in fact, a substantial portion of their income comes from return on investment of financial *reserves,* funds that they are required to have to meet current and future insurance obligations.

Becoming an actuary requires training in mathematics, statistics, economics, and finance, and includes a sequence of professional exams taken over several years.

every \$1000, so your monthly income would be \$1770. Your life expectancy would be about 17.5 years = 204 months. If you lived exactly that long, you would receive a total of 204 × \$1770 = \$361,080. The rate of interest that your annuity would need to earn to last that long can be calculated from the amortization formula; for example, using `=RATE(210, 1770, 0, −250000)` in Excel gives a monthly rate of 0.408%, for an effective annual rate of 5.01%.

If you are female and retire now at the same age with the same \$250,000 savings in a life income annuity, you would receive \$6.30 per month for every \$1000, or \$1575 per month. Your life expectancy would be about 21 years = 252 months. If you lived exactly that long, you would receive a total of 252 × \$1575 = \$396,900. The rate of interest that your annuity would need to earn to last that long can be calculated from the amortization formula; using =RATE(252, 1575, 0, −250000) in Excel gives a monthly rate of 0.401%, for an effective annual rate of 4.92%. ■

Notice that a man and a woman who save the same amount receive different monthly incomes at retirement: The woman receives less but for longer—about 90% as much for 25% longer. Yet their living expenses are likely to be the same. That consideration has resulted in some companies offering "merged gender" rate schedules for annuity payments, so that the individual receives the same monthly payment regardless of sex.

In fact, life expectancy in 1980 for 65-year-olds was shorter—about 16 years for men and 19 years for women. When the insurance company wrote this policy in 1980, its tables were based on those numbers and on a guaranteed interest rate of 4% compounded annually. Future policies, such as you would purchase, will incorporate changes in mortality; in addition, companies reserve the right to change their rate schedules for annuity payments as mortality rates change.

REVIEW VOCABULARY

Amortization formula Formula for installment loans that relates the principal A, the interest rate i per compounding period, the payment d at the end of each period, and the number of compounding periods n needed to pay off the loan:

$$A = d\left[\frac{1-(1+i)^{-n}}{i}\right]$$

Amortize To repay in regular installments.

Annual percentage rate (APR) The rate of interest per compounding period times the number of compounding periods per year.

Annuity A specified number of equal payments at equal intervals of time.

Compound interest formula Formula for the amount in an account that pays compound interest periodically. For an initial principal A and effective rate i per compounding period, the amount after n compounding periods is $A = P(1+i)^n$

Effective annual rate (EAR) The effective rate per year.

Effective rate The actual percentage rate, taking into account compounding.

Interest Money earned on a loan.

Nominal rate A stated rate of interest for a specified length of time; a nominal rate does not take into account any compounding.

Period Compounding period.

Principal Initial balance.

Savings formula Formula for the amount in an account to which a regular deposit is made (equal for each period) and interest is credited, both at the end of each period. For a regular deposit of d and an interest rate i per compounding period, the amount A accumulated is

$$A = d\left[\frac{(1+i)^n - 1}{i}\right]$$

Simple interest The method of paying interest on only the initial balance in an account and not on any accrued interest. For a principal P, an interest rate r per year, and t years, the interest I is $I = Prt$.

SUGGESTED READINGS

KASTING, MARTHA. *Concepts of Math for Business: The Mathematics of Finance* (UMAP Modules in Undergraduate Mathematics and Its Applications: Module 370–372), COMAP, Inc., Arlington, Mass., 1980.

MILLER, CHARLES D., VERN E. HEEREN, and JOHN HORNSBY. Consumer mathematics. In *Mathematical Ideas,* 9th ed., Addison Wesley Longman, Boston, 2001, pp. 786–845.

VEST, FLOYD, and REYNOLDS GRIFFITH. The mathematics of bond pricing and interest rate risk, *Consortium* (COMAP) no. 59 (Fall 1996): HiMAP Pullout Section 1–6.

YAREMA, CONNIE H., and JOHN H. SAMPSON. Just say "Charge it!" *Mathematics Teacher* 94 (7) (October 2001), 558–564. Shows how to apply the savings formula and the amortization formula and graph the results on the TI-83 calculator. Notes that the 78% of undergraduates in the United States who have credit cards carry an average debt of more than $2700, with 10% owing more than $7000.

SUGGESTED WEB SITES

http://www.lendingtree.com/stmrc/calculators1.asp Offers Java applet calculators (for any platform) to calculate payments and amortization schedules for conventional loans, adjustable-rate mortgages, auto loan vs. home-equity loan, and credit-card payoff. (Note: Lending Tree, Inc., is a loan broker; recommendation here of calculators at its Web site does not imply endorsement of its other services by this book's authors, editors, or publisher.)

http://www.udayton.edu/sba/wf/wksht/wksht5.htm Excel spreadsheet templates for comparing consumer loans ("simple interest," or conventional vs. discounted loan vs. add-on loan), tax-deferred retirement plans, mortgage loans, present value, lease vs. buy, and more.

SKILLS CHECK

1. A city sells bonds in order to

(a) invest money.
(b) raise money.
(c) protect investments.

2. An add-on loan

(a) is a traditional simple-interest loan.
(b) uses compound-interest calculations.
(c) computes interest on the principal for the total loan period.

3. Credit card interest

(a) is computed using compound interest.
(b) is computed using simple interest.
(c) is included in the late fees.

4. If a store credit account charges 1.5% interest each month, what is the effective annual rate?

(a) 18%
(b) More than 18%
(c) Less than 18%

5. If a store credit account charges 1.5% interest each month, what is the annual percentage rate?

(a) 18%
(b) More than 18%
(c) Less than 18%

6. If you finance $15,000 for 3 years at 6% compounded monthly, the monthly payments will be

(a) about $456.
(b) about $492.
(c) about $560.

7. If you establish a 30-year mortgage, most of the initial payments

(a) go toward reducing the balance.
(b) go toward paying the interest.
(c) pay insurance costs.

8. After 15 years of payments on a 30-year mortgage, the balance remaining is

(a) about one-third of the original balance.
(b) about one-half of the original balance.
(c) about two-thirds of the original balance.

9. An adjustable-rate mortgage

(a) has variable interest rates but maintains a fixed payment amount.
(b) has variable payment amounts.
(c) is always a better alternative to fixed-rate mortgages

10. Which of the following arrangements could be an ordinary annuity?

(a) Monthly payments, annual compounding
(b) Annual payments, monthly compounding
(c) Annual payments, annual compounding

11. If you just won a lottery jackpot paid in 25 equal annual installments of $1 million each, what is the present value for the jackpot?

(a) $25 million
(b) More than $25 million
(c) Less than $25 million

12. A life income annuity is designed to pay a fixed amount each period until

(a) the annuity runs out of money.
(b) you die.
(c) you reach your life expectancy.

13. Your car dealer offers to finance a $6000 add-on loan at 3% to be repaid in four years of monthly payments. What is the monthly payment?

(a) $140
(b) $150
(c) $180

14. Your credit union offers to finance a $6000 conventional loan at 4% to be repaid in four years of monthly payments. What is the monthly payment?

(a) Approximately $135
(b) Approximately $138
(c) Approximately $145

15. Payments for a 15-year mortgage

(a) are double the payments for a 30-year mortgage of the same amount.
(b) are about 50% to 80% more than payments for a 30-year mortgage of the same amount.
(c) are about 20% more than payments for a 30-year mortgage of the same amount.

EXERCISES

▲ Optional. ■ Advanced. ◆ Discussion.

Simple Interest

1. On August 13, 2001, you could buy $10,000 U.S. Treasury bonds that pay simple interest of

- 5.51% per year in February each year through maturity in February 2031, at a cost of $9,802 each; or
- 11.25% per year in February each year through 2015, at a cost of $15,529 each.

Each bond also returns $10,000 at maturity.

(a) What is the annual yield of each investment?
(b) Compare these two investment opportunities, which differ vastly in interest rate but also in cost.

2. In late September 2001, you could buy a U.S. Treasury bond offering 5.75% interest and maturing one year later. The face value of the bond was \$10,000; this is the amount of the principal that the Treasury pays back to the owner at maturity. But resellers of this particular bond wanted \$10,318. On this bond, what was the *current yield*—the annual percentage return to the purchaser at the current price?

3. You need to buy a car and need to finance \$5000 of the cost. The dealer offers you a 5.9% add-on loan to be repaid in monthly installments over four years. How much is your monthly payment?

4. You need to make some home improvements—well, to be honest, they're really maintenance that you can't defer any longer!—and need to borrow \$3000 to pay for them. You can get an 8.5% add-on loan from a savings and loan association to be repaid in installments over two years. How much is your monthly payment?

5. You are in the same situation as in Exercise 4, except that you find that you can get a 6.0% discounted loan from a loan company to be repaid in monthly installments over four years. What is the monthly payment on this loan?

6. You are in the same situation as in Exercise 4, except that you find that you can get a 9% discounted loan from a loan company to be repaid in monthly installments over four years. What is the monthly payment on this loan?

7. Suppose that you need \$1000 (no less) and have available to you either an add-on loan or a discounted loan, both at the same interest rate and for the same period. Which will have the lower monthly payment?

8. Show algebraically that your claim in Exercise 7 is true in general, for a loan of any amount.

Compound Interest

9. A recent credit card bill of mine showed \$500 due with a minimum payment of \$10 and daily interest rate of $r = 0.04932\%$. If I make no more charges on the card and pay \$10 a month as soon as I get each bill, how long will it take to pay off the total? (*Hint:* The amortization formula can be changed algebraically into the form

$$(1 + i)^n = \frac{d}{d - iP}$$

Note that if I make the first payment immediately, I immediately reduce the principal P to be amortized to $\$500 - \$10 = \$490$; if I delay payment, I incur additional daily interest on the amount due. Evaluate the right-hand side. Then, using either a spreadsheet or the power key $\boxed{y^x}$ on your calculator, raise the value of $(1 + i)$ to higher and higher powers n until you find the smallest value for n that makes the left-hand side larger than the right-hand side.)

10. (Requires spreadsheet) Regarding the credit card bill in Exercise 9: By paying \$10 each month, approximately how much interest would I have paid by the time I pay off the original \$500? (*Hint:* You won't be off by much if you estimate the last payment to be \$10.)

11. (Requires spreadsheet) According to the regulations for the credit card discussed in Exercises 9 and 10, the minimum payment is supposed to be the greater of \$10 or 2% of the balance (rounded down to the next higher dollar amount). Suppose that you have such a card and the balance is \$1500. Neglect the rounding down and assume that you pay exactly 2% of the current balance each month until the balance reaches \$500. Notice that if you make a payment of 2% and the bank charges daily interest of 0.04932%, in each 30-day month you in effect reduce the balance to $(1 - 0.02)(1.0004932)^{30} =$ 99.46043% of what it was the previous month. How many 30-day months will it take to reduce the balance of \$1500 to \$500? (Again, assume that you make payments right away.)

12. By making the minimum payment each month in the situation that exists in Exercise 11, approximately how much interest will you have paid by the time you pay off the original \$1500?

13. Adding your answers from Exercise 9 and 11, determine how long it will take to pay off a balance of $1500 by making minimum payments.

14. (Requires spreadsheet) Assume the same situation as in Exercise 11, but instead of paying with exactly 2% of the balance due, you make a minimum payment of the greater of $10 or 2% of the balance (rounded to the next higher dollar amount).

Conventional Loans

15. In October 2001 you could buy a new Kia Sedona minivan for $18,995 with a loan from the manufacturer at 1.6% interest over 60 months. Assume that you have cash or a trade-in worth $2000. What is your monthly payment on the minivan?

16. You can't afford the new minivan in Exercise 15 but need a better-than-junk car, and you need to borrow the entire $7000 of its cost. You can get a 48-month conventional loan – not an add-on loan or a discounted loan – from your credit union at 8.5% compounded monthly. What is your monthly payment?

17. A television ad that ran in February 2002 advertised a new 2002 Pontiac for 16% down with a five-year loan at 5.9%, with a payment of $19.09 per month for each $1000 financed. Check whether the monthly payment is correct (I may have miscopied it from the fine print that showed for three seconds at the end of the ad!).

18. For the same car mentioned in Exercise 17, a car dealer offers a 7.9% APR conventional loan over 48 months but a 6.8% APR conventional loan over 60 months.

(a) What is your payment per $1000 under each loan?
(b) With the 60-month loan, it takes longer for the dealer to be paid in full. So what is the advantage to the dealer in offering a lower percentage loan for the 60-month loan, instead of a higher percentage loan? (Thanks to Terence Blows of Arizona State University.)

Exercises 19 – 23 require a spreadsheet. (*Hint:* Put the amortization formula into a spreadsheet and vary the interest rate i.)

19. For the credit card in Exercises 9 – 10, what is the APR? What is the EAR? What annual rate must the credit card company tell its customers applies?

20. For the add-on loan in Exercise 3, what is the APR? What is the EAR?

21. For the add-on loan in Exercise 4, what is the APR? What is the EAR?

22. For the discounted loan in Exercise 5, what is the APR? What is the EAR?

23. For the discounted loan in Exercise 6, what is the APR? What is the EAR?

24. Suppose that you and two friends decide to live off-campus in your senior year. One of them (who has wealthy parents) suggests that instead of renting an apartment, you could buy a house together, live in it for your senior year, then rent it out or else sell it. Assuming that (with the help of her parents and their good credit rating) you could get a mortgage for $80,000 to buy a house near the campus, what would be the monthly mortgage payment on a 30-year mortgage at 6.75%?

25. Suppose that you have good credit and can get a 30-year mortgage for $100,000 at 6.5%. What is your monthly payment?

26. Assume the same situation as in Exercise 25, except that your credit is not as good and the rate that you are offered is 7.125%.

27. Assume the same situation as in Exercise 25, but you inquire about a 15-year loan instead. You are offered 6.125%. What is your monthly payment?

28. Assume the same situation as in Exercise 27, but your credit is not as good, and you are offered 6.75%. What is your monthly payment?

29. For the mortgage in Exercise 25, how much equity would you have after five years?

30. For the mortgage in Exercise 26, how much equity would you have after five years?

31. For the mortgage in Exercise 27, how much equity would you have after five years?

32. For the mortgage in Exercise 28, how much equity would you have after five years?

33. When interest rates drop, it may become attractive to refinance your home. Refinancing means that you acquire a new mortgage to borrow the current principal due on your home and use the proceeds to pay off your old mortgage. You then begin a new 15- or 30-year mortgage at the new, lower interest rate. A second factor that reduces your monthly payment is that the equity you accumulated under the old mortgage reduces the amount that you have to borrow under the new mortgage. Suppose that you have an existing 30-year $100,000 mortgage at 8.375%, on which you have been paying for five years, and you are considering refinancing at 7.0%.

(a) What is your payment under the old mortgage?
(b) How much equity do you have in the home?
(c) If you use all your equity to reduce the amount of the new mortgage, how much will your monthly payment be?
(d) How long is the payback period for the $2000 loan charge—that is, how many months will it take before you have saved $2000 in monthly payments?

34. (Requires spreadsheet) Even if interest rates go up, you may find it beneficial to refinance if your financial circumstances suddenly worsen. Consider an actual individual (as recounted in the *Wall Street Journal* of October 30, 2001), a 39-year-old plumber whose income was cut in half by worsening economic conditions. She had missed eight monthly mortgage payments totaling $14,000, still owed $160,000 on the mortgage, and was in danger of foreclosure and losing her home. Data on her loan were not given in the *Journal*, but data that fit are a 15-year mortgage for $180,000 at 8.375%.

(a) What was her monthly payment?
(b) How long had she been paying on the mortgage?
(c) Under a "loan modification" program, the bank added the $14,000 in missed payments to her existing debt of $160,000. Suppose that the bank offered her a new 30-year loan for the total of $174,000 at 8.75% (despite interest rates in late 2001 being generally lower than when she got her first loan, her delinquency on that loan made her a worse risk and hence she incurred a higher interest rate). How much was her new monthly payment?

Annuities

35. The largest amount won by an individual in the Florida Lotto lottery was $81.6 million on March 29, 2000 (other jackpots have been higher but were shared). The advertised jackpot amount is paid in an annuity of 30 annual payments, including one immediate payment.

Instead of the 30-year payment plan, the winner may choose to receive a smaller lump sum right away in cash. If not, that sum (minus the initial payment) is invested by the State Board of Administration in U.S. government securities on behalf of the player. The cash invested, plus all interest earned, goes to the winner.

Suppose that in April 2000 the state of Florida could buy U.S. securities paying an interest rate of 7%. What was the present value of the jackpot annuity—that is, how much could the winner have received in cash instead? (We say that $81.6 million is the *future value* of the annuity.)

36. Suppose that you retire at age 65 and in addition to Social Security need $2000 per month in income. Based on an expected lifetime of 17.5 more years (for men) or 21 more years (for women), how much would you have to invest in a life income annuity earning 4% to pay you that much per year?

Additional Exercises

■ **37.** Fisher's effect gives the real rate of borrowing money, sometimes called the *real cost of capital.* Suppose that you buy a house with a $100,000 mortgage at an interest rate of 6.75% and that inflation remains at 3% for the duration of the mortgage. You make monthly payments in constantly deflating dollars. Simplify the situation by supposing that you sell the house one month after you buy it (the neighbors really drove you crazy, right from the start).

(a) What would the house sell for if it kept up with inflation of $3\%/12 = 0.25\%$ per month?

(b) Calculate your mortgage payment at the end of the month. How much is interest and how much is principal?

(c) After making the mortgage payment, paying off the balance of the $100,000 cost of the house, and selling the house for the amount in part (a), how much did it cost you to own the house for the month?

(d) The cost in part (c) is in deflated dollars. Using a formula from Chapter 21, convert this to constant (month-before) dollars and divide by $100,000 to express the result as a percentage rate. This is the real rate of interest.

WRITING PROJECTS

1. Suppose that you have negotiated to buy a new car for $21,000. The dealer offers you either $1000 cash back or else a 48-month loan at 1.9% interest. To determine which option to take, you consider your situation:

- You just bought a home and have a large home mortgage at 6.875%.
- You recently received an inheritance from your grandmother of $30,000 in a trust fund that must be kept in a cash management account (currently paying 5% interest), though the trustee will allow you to buy the car out of the account.
- You have the opportunity to finance the car's cost at your credit union for 6%.
- You already paid off all your college loans with another part of the inheritance.

To help persuade you to take the loan from the manufacturer, the dealer tells you (in agreement with law in most states) that there is no penalty for early repayment of the car loan if you choose to do so. What should you do?

2. A substantial proportion of new cars today are not sold but leased. Contact a local car dealer about a car that you are interested in and find out the details on leasing. Compare the cost of the lease and associated expenses with the cost of purchasing and owning the car. Include estimated maintenance, repair, and insurance costs for each option. Which seems like a better deal, and why?

3. Banks typically offer mortgages with various combinations of interest rates and points. Loans at a higher interest rate tend to come with more points, and vice versa. Like interest, points are deductible on income taxes, but in the year in which the loan is made. In recent years, home buyers have (when offered the option) generally favored paying fewer points. Suppose that you have a choice between a mortgage at 6% with 2 points (2%) and a mortgage at 8% and no points. Which would you choose, and why? Does it make a difference how long you are planning to own the home? Or how expensive the home is?

4. One of the advantages of buying a home with a fixed-rate mortgage is that your payment stays the same but your earnings and the value of your home are likely to go up with inflation. You are paying back the loan with dollars of lesser value.

Consider the following scenario. Suppose that you buy a "starter" two-bedroom home for $105,000 under a special program for first-time home buyers that requires a down payment of only $5000. You have a 30-year fixed-rate mortgage for $100,000 at 7%, on which the monthly payment is $665.30. You also have a $2000 one-time expense in closing costs and annual costs of $200 for insurance and $2000 for property taxes.

You live in the home for five years and spend $10,000 on maintenance, upkeep, and improvements. You then sell the home for $125,000, pay a realtor $9,000 to sell it, and pay closing costs of $500 (for title insurance and other costs). Finally, it costs $3000 to move.

(a) Make out a balance sheet of revenue and expenses. How did you make out on owning the home?

(b) Remember, you also got to live in the home without paying rent! Translate the cost of owning the home into an equivalent monthly rent.

5. Explore actual costs of homes in your area, mortgages with local banks (including closing costs), and property taxes and insurance. Come up with data such as that in Writing Project 4 and make out a corresponding balance sheet for five-year ownership.

SPREADSHEET PROJECTS

To do these projects, go to www.whfreeman.com/fapp.

This project uses spreadsheets to model compound interest and debt payments to crease annuity tables. These spreadsheets track the amount of each payment that goes toward interest or toward reduction of the principal. These activities model the impact of a small change in interest rate or the monthly payment on the annuity.

This project uses spreadsheets to model mortgages, including an exploration of the impact of extra payments toward the mortgage.

APPLET EXERCISES

To do these exercises, go to www.whfreeman.com/fapp.

There are two ways to buy a car: save up and pay cash or borrow the money from a bank. In the applet Buying a Car: Cash vs. Loan, you can explore just how much more expensive it is to borrow the money.

CHAPTER 23

The Economics of Resources

In this chapter we cover...

...and at www.whfreeman.com/fapp:

Flashcards

Quizzes

We use resources all the time: food, money, natural resources, labor, time. Time and labor are by definition consumed in their use. Food and oil are transformed; food can be replaced with new crops and petroleum with (as we count on) the continuing discovery of new reserves. Some resources, such as annual flowers, are perishable, while others, such as money or standing timber, can be used now or saved for later use.

Our use of resources involves a complex intermixture of biological, ethical, practical, technical, and economic issues:

- How many people will there be in the world in another 20 or 40 years, and how will those numbers affect resources available to you and to them?
- What should we consume for our own use, give to others more needy, or leave for future generations, in terms of wealth, well-being, and wilderness?
- Will our standard of living keep getting better, or is it not maintainable, even at current levels, in the long run?
- How long will it be, at current patterns of use, until we exhaust some particular resources?
- Can we develop more efficient technology, so that we can get more out of the resources that we use?
- How do we balance economics, the needs and desires of individuals, with other important considerations? How much would it cost—how much is it worth—to assure that we do not let tigers, elephants, or rhinos go extinct?

On a personal level we face similar but more immediate questions about the balance between consumption and conservation. How much must you save to be able to afford to retire? How do you take into account that \$1000 after you retire won't be worth as much as \$1000 today? Once you retire, how much can you spend without exhausting your nest egg?

In Chapters 21 and 22, we explored mathematical models for saving, accumulating, and borrowing—the building up of resources. From Chapter 21, we use here only two formulas, specialized to annual interest rate r:

Compound interest formula: If a principal P is deposited into an account that pays interest at rate r per year, then after n years the account contains the amount

$$A = P(1 + r)^n$$

Savings formula: For a uniform deposit of d per year (deposited at the end of the year) and an interest rate r per year, the amount A accumulated after n years is

$$A = d\left[\frac{(1 + r)^n - 1}{r}\right]$$

A reader who can use these formulas can proceed in this chapter without first reading Chapters 21 and 22.

In this chapter, we consider models for processes in the other direction—the decay, depletion, or spending down of resources. Some resources, however, replenish themselves regularly; our models of them will lead us into the mathematics of dynamical systems and the mathematical concept of chaos. The models of this chapter provide important insights into the consequences of answers to the questions above.

23.1 Growth Models for Biological Populations

The economic models for savings are flexible and can be applied to other arenas. We now use a geometric growth model to make rough estimates about sizes of human populations, using for the growth rate r the difference between the annual birth rate and the annual death rate, for which the technical term is the **rate of natural increase.** In the terminology of financial models, this is the effective rate.

Birth and death rates rarely remain constant for very long, so projections must be made with extreme care. In addition, we exclude the effect of net migration. In the short run, predictions based on the model may provide useful information. Let's apply this model to two questions about the population of the United States.

EXAMPLE Predicting U.S. Population

The U.S. population was 284.5 million in mid-2001. It was increasing at an average growth rate of 0.6% per year. What is the anticipated population in mid-2005? What is it if the rate of natural increase changes to 0.4% per year or 0.8% per year?

We apply the compound interest formula with initial population size ("principal") 284.5 million. Using a year as the compounding period and the formula $A = P(1 + r)^n$, where $n = 4$, the projected population size in 2005 for a rate $r = 0.006$ is

$$\begin{aligned}\text{Population in 2005} &= (\text{population in 2001}) \times (1 + \text{growth rate})^4 \\ &= 284{,}500{,}000\,(1 + 0.006)^4 \\ &= 284{,}500{,}000\,(1.02422) \\ &\cong 291{,}000{,}000\end{aligned}$$

Because the original estimates of population and growth rate are approximations, we don't copy down all the digits from the calculator, but instead round off; the result of a calculation can't be more precise than the ingredients.

In the same way, with a growth rate of 0.4% per year, we predict a population of 289 million, while a growth rate of 0.8% per year yields a prediction of 294 million.

So an uncertainty of two-tenths of one percentage point, or 0.002, in the growth rate has major implications, even over fairly short time horizons. The presence or absence of 5 million people would have a significant impact on our social and economic systems! Indeed, much of the concern over long-range funding of Social Security programs results from uncertainties over birth and immigration rates. Figure 23.1 gives a graph of the U.S. population in 2000, structured by age and sex. ■

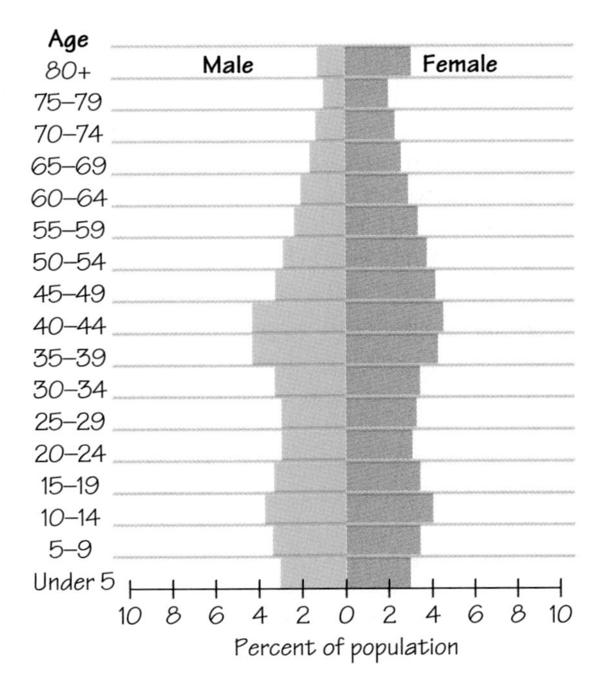

Figure 23.1 Graph of the population of the United States in 2000 grouped by age and sex and shown as a percentage of the total population.

Rates of natural increase in most Third World countries are much higher than in industrialized nations, sometimes 3% per year or more. With its growth rate of 2.8%, Africa's most populous country, Nigeria, whose population was 126.6 million in mid-2001, will have 186 million by the middle of 2015, an increase of almost 50%. Projections of this sort are at the root of worldwide concern over the ability to provide sufficient food and other resources for all people.

But it is not just the number of people that is a concern. The United States is not the only country that needs to be concerned about its **population structure** and the aging of its population. In poorer countries, the proportion of the population over 60 years of age will be 20% by 2050, compared to 8% now.

23.2 Limitations on Biological Growth

A population that keeps adding a fixed percentage each year, like a bank account accumulating interest, would eventually grow to astronomical numbers. But no biological population can continue to increase without limit (see Spotlight 23.1). Its growth is eventually constrained by the availability of resources such as food, shelter, and psychological and social "space." There may be a maximum population size that can be supported by the available resources, the **carrying capacity** of the environment.

The **logistic model** for population growth takes carrying capacity into account by reducing the natural rate of increase r by a factor of how close the population size P is to the carrying capacity M:

$$\text{Growth rate} = r\left(1 - \frac{\text{population size}}{\text{carrying capacity}}\right) = r\left(1 - \frac{P}{M}\right)$$

As the population increases, the growth rate decreases—because the term containing the population P has a negative sign. For a population equal to the carrying capacity ($P = M$), the growth rate is zero.

If the population ever exceeds the carrying capacity, the growth rate becomes negative (because $P < M$) and the population decreases. The carrying capacity is the long-range capacity to support the population, so the population could exceed it for short periods of time. This could happen either because the population grows very rapidly and surges above the carrying capacity or because of a sudden decrease in the food supply, thus temporarily lowering the carrying capacity, as happens to deer and other animals in winter. The logistic model is a simple model but it provides excellent predictions for some populations.

EXAMPLE The U.S. Population

The U.S. population from 1790 to 1950 closely followed a logistic model with $r = 0.031$, P = population in 1790 = 3,900,000, and M = 201 million. In the early decades after 1790, the population was a small fraction of this carrying

SPOTLIGHT 23.1 How Many People Can the World Hold?

How many people can the world hold? Are developing countries heading for a population disaster? Will falling fertility play havoc with pensions in Europe and Social Security in the United States? Warren Sanderson of the International Institute for Applied Systems Analysis is part of a team attacked recently by the *Wall Street Journal* for introducing as "good news" their prediction that global population growth would stop by the end of 2100. Sanderson says:

> There is an 85% chance that the world population will reach 9 billion before 2100 and then fall. The UN predicts a peak of 9.3 billion. They're higher because they assume that in the long run every country will have the same fertility rate (the average number of births per woman) of 2.1 (the rate required to replace the population). For example, in Austria, the rate is 1.3, so the UN is expecting an enormous increase in Austrian fertility. The other side of the coin is that many developing countries have fertility rates that are high but falling. In South Korea, it's 1.6; in Thailand, about 1.8. We don't have any examples of countries coming down from high rates, hitting 2.1, and stopping there.
>
> If you naively take a fertility rate above 2.1 and project it into the future forever, population will explode; if you take a rate below 2.1, in the long run population will go to zero. In the real world, populations adjust to their circumstances.
>
> Most of the population growth is going to be in the world's poorest countries, in south Asia and sub-Saharan Africa. But even their population growth is slowing down and may even come to an end by 2100. I think that's good news for those places. You have a combination of poor government and deteriorating resources and famines and wars; any little thing that can help relieve some of the pressure on resources has to be a good thing.

(Paul Harris/Stone.)

> There's been a long and unfortunate debate between environmentalists on one side and some economists and some people in the business community on the other. Some of the former have said that there was no hope for India, for example. Well, India has done fairly well; there are a lot more people now but there's also a lot more economic growth. Some of the economists and business people now hold that you shouldn't worry about numbers when you think about development. Neither of these extreme positions provides a realistic view of the role of population.

–Adapted from an interview in *New Scientist* (September 8, 2001): 42–43.

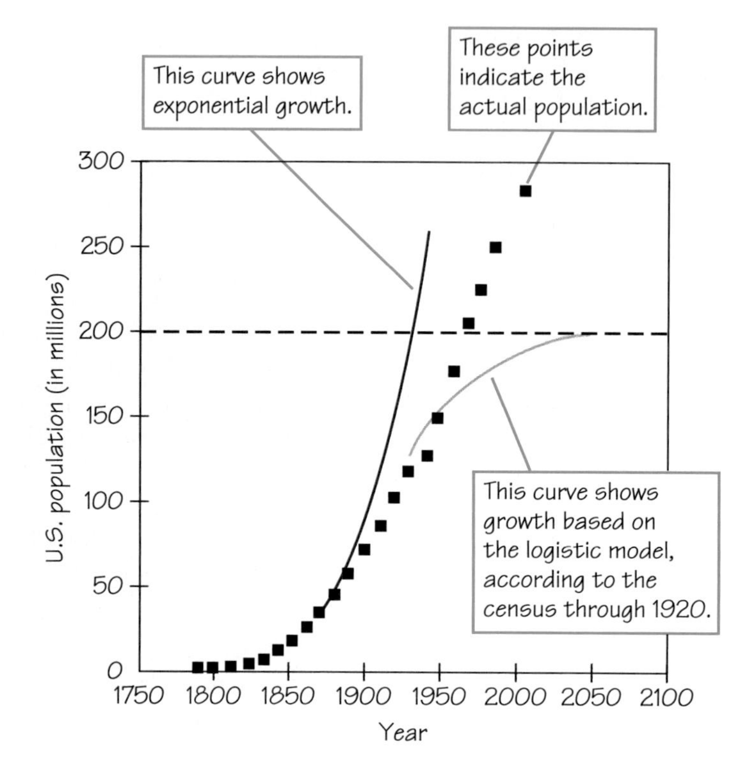

Figure 23.2 U.S. population by year, showing actual growth, exponential (geometric) growth, and logistic growth.

capacity, and it grew at close to the rate r of 3.1% per year (a rate higher than in many Third World countries today). By 1920 the U.S. population had reached 106 million, a little more than half of the carrying capacity, and the growth rate had slowed by about one-half, to 1.5% per year (see Figure 23.2).

The mid-2002 U.S. population of 288 million far exceeds the hypothesized carrying capacity of 201 million. The structure of the U.S. population changed, from a large proportion of people making their living on family farms to a highly urbanized society; the average number of children per family shrank. As the structure changed, the model based on the prior structure gradually became invalid. ■

23.3 Nonrenewable Resources

People use resources, some of which are renewable but others, not. In this section, we model depletion of nonrenewable resources; in the next, we treat renewable resources.

A **nonrenewable resource** is one that does not tend to replenish itself; gasoline, coal, and natural gas are important examples, while lottery winnings or an inheritance could be examples from personal affairs. There is no practical way to recover or reconstitute these resources after use. Some substances, such as aluminum or the sand used to make glass, are potentially recyclable; but to the extent that we do not recycle them, they, too, are nonrenewable.

For a nonrenewable resource, there is only a fixed supply S (in some convenient units) that is available. Even without human population increases, we face dwindling nonrenewable natural resources. We are interested in the question: How long will the supply of a resource last?

As long as the rate of use of the resource remains constant, the answer is easy. If we are using U units per year and continue using U units per year, then the supply will last S/U years. This kind of calculation is the basis for statements such as that at the current rate of consumption, U.S. coal reserves will last 500 years or that the U.S. strategic reserve of gasoline (stored in salt domes in the South) will last 60 days.

However, the rate of use of resources tends to increase with increasing population and with a higher "standard of living." For example, projections for use of electric power are often based on assumptions that the use will increase by some fixed percentage each year. This is the simplest situation (apart from constant usage) and one that we can easily model to give an important perspective.

Suppose that $U_1 = U$ is the rate of use of the resource in the first year (this year), and that usage increases $r = 0.05 = 5\%$ each year. Then the usage in the second year is

$$U_2 = U_1 + 0.05U_1 = 1.05U$$

and usage in the third year is

$$U_3 = U_2 + 0.05U_2 = 1.05U_2 = 1.05(1.05U) = (1.05)^2U$$

Generalizing, we see that usage in year i will be $(1.05)^{i-1}U$. Total usage over the next five years, for example, will be

$$U + (1.05)^1U + (1.05)^2U + (1.05)^3U + (1.05)^4U$$

This situation should remind you of our earlier study in Chapter 21 of accumulation of regular deposits plus interest. Here the usage U corresponds to a deposit and the increasing rate of use r corresponds to the interest rate. We may think of the situation as making regular withdrawals (with interest) from a fixed supply of the nonrenewable resource. The savings formula gives

$$A = d\left[\frac{(1 + r)^n - 1}{r}\right]$$

In translating to the resource situation, A is the accumulated amount of the resource that has been used up at the end of n years, and U is the initial rate of use. We have

$$A = U\left[\frac{(1 + r)^n - 1}{r}\right]$$

To find out how long the supply S will last, we set the supply S equal to the cumulative use A over n years and then determine what n will be. We have

$$S = U\left[\frac{(1 + r)^n - 1}{r}\right]$$

We perform some algebra to isolate the term involving n, to get

$$(1 + r)^n = 1 + \frac{S}{U}r$$

At this point, to isolate n, we need to take the natural logarithm of both sides. We get

$$\ln[(1 + r)^n = n\ln(1 + r) = \ln\left(1 + \frac{S}{U}r\right)$$

which gives the final expression

$$n = \frac{\ln[1 + (S/U)r]}{\ln(1 + r)}$$

This expression may look complicated, but it is quite easy to evaluate on a calculator for particular values of S/U and r.

The expression S/U is called the *static reserve,* and n is called the *exponential reserve.*

The **static reserve** is how long the supply S will last at a particular constant annual rate of use U, namely, S/U years.

The **exponential reserve** is how long the supply S will last at an initial rate of use U that is increasing by a proportion r each year, namely

$$\frac{\ln[1 + (S/U)r]}{\ln(1 + r)}$$

years.

EXAMPLE

U.S. Coal Reserves

We noted earlier that measured reserves of U.S. coal would last about 500 years at the current rate of use, so the static reserve for this resource is 500 years. How long would the supply last if the rate of use increases 5% per year? The corresponding exponential reserve is

$$n = \frac{\ln[1 + (500)(0.05)]}{\ln 1.05} = \frac{\ln 26}{\ln 1.05} = 65 \text{ years}$$

That's quite a difference! ■

We must not take such projections as exact predictions. Estimates of supplies of a resource may underestimate how much is available, and previously unknown sources may be discovered or the technology improved to extract previously unavailable supplies. In addition, as supplies dwindle, the economic considerations of supply, demand, and price come into play. We will never completely run out of oil; it will always be available "at a price."

However, we must not take such projections lightly, either, because we are discussing resources that, once used, are gone forever. In any projection, it is very important to examine the assumptions, because small differences in the rate of increase of use can make big differences in the exponential reserve.

EXAMPLE **Using Up Retirement Savings**

Suppose that you begin retirement with \$1 million in savings, and you don't trust banks or the stock market, so you keep it all under your mattress. Suppose that it costs you \$50,000 per year to live at your accustomed standard of living and there is no inflation. How long will your retirement nest egg last? The static reserve is \$1,000,000/\$50,000 per year = 20 years. If, however, there is constant 5% per year inflation, then it will cost you increasingly more per year to live, so you should realize that your savings will last only for the length of the exponential reserve, which is

$$n = \frac{\ln(1 + 20(0.05))}{\ln 1.05} = 14.2 \text{ years}$$

You have a fine strategy if you expect to live just 14 more years and want to die broke! ■

23.4 Renewable Resources

A **renewable natural resource** is a resource that tends to replenish itself, such as fish, wildlife, and forests. We would like to know how much of a resource we can harvest and still allow for the resource to replenish itself.

EXAMPLE **College Endowment**

Donors sometimes give money to colleges and universities to fund specific prizes, scholarships, or professorships, which they would like to see continue in perpetuity. The college invests the donor's money, so that this resource renews itself, and "harvests" the earnings to fund the project. Suppose that the college can earn 8% on its endowed funds. Then an individual who wants to make a one-time donation to fund an annual full scholarship (tuition, books, room and board) of $25,000 needs to donate enough so that the fund earns that much; the donation needs to be $25,000/0.08 = $312,500.

However, the current cost of $25,000 will increase with inflation. If the donor wants each year's future interest to cover that year's full cost, the donation needs to be larger. If the college endowment earns 8% but there is 3% inflation in the annual cost, the net interest is only about 5% (see Fisher's effect in Chapter 21, page 781), and the amount of the donation needs to be $25,000/0.05 = $500,000. ■

Other renewable resources are biological populations. We concentrate on the subpopulation with commercial value. For a forest, this might be trees of a commercially useful species and appropriate size. We measure the population size as its **biomass,** the mass of the population expressed in units of equal value. For example, we measure the size of a fish population in terms of pounds rather than number of fish, and a forest not by counting the trees but by estimating the number of board feet of usable timber.

Reproduction Curves

In this chapter we have used several models to analyze and predict population growth. As models, they include many simplifications. Real populations may behave like one of the models we have discussed or like other known models. But the complicated factors that can affect populations, such as climatic or economic change, may mean that the only way to understand the population is to plot a graph of its size over time. Whether the growth of a population can be described by one of the formulas that we have developed or by a table of measurements collected by counting, there are useful techniques to analyze the growth of the population and help us make decisions about managing it. The situation that we will be thinking about most is managing a population of renewable resources, such as trees or fish, to make the most of them.

We use a figure called a **reproduction curve,** which predicts next year's population size (biomass) based on this year's size. Although the precise shape of the curve varies from one population to another, the shape shown in Figure 23.3 is typical. It shows next year's size (on the vertical axis) as a function of this year's size (on the horizontal axis). For all possible sizes, the reproduction curve shows the change in size from one year to the next, taking into account growth of continuing members and addition of new members, minus losses due to death and other factors.

Let x on the horizontal axis be a typical size of the population in the current year. The size *next* year is given by the height of the curve above the point marked x. This value is denoted by $f(x)$. (You can think of f as standing for "function of," or even as "forthcoming.")

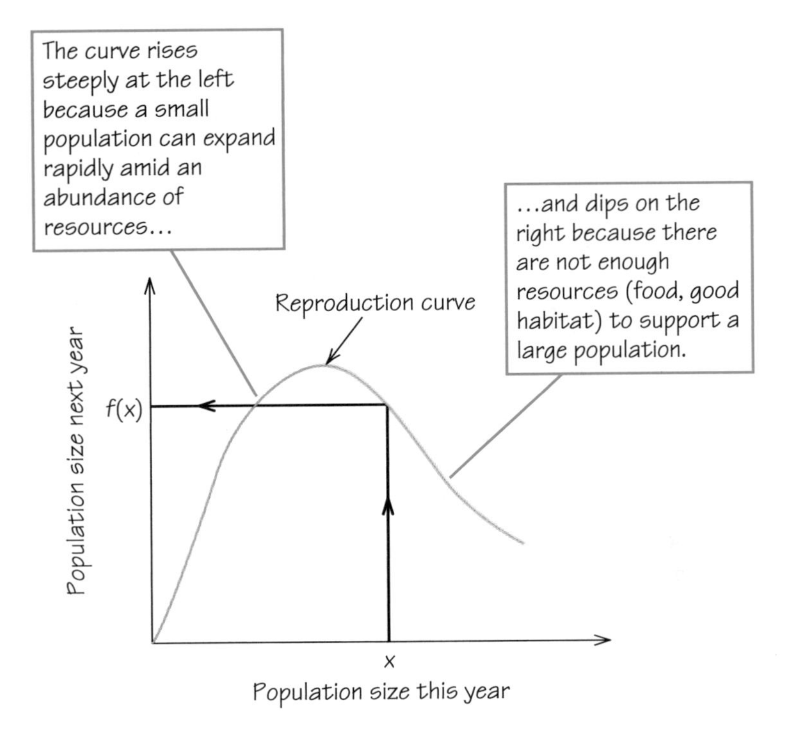
The curve rises steeply at the left because a small population can expand rapidly amid an abundance of resources...
...and dips on the right because there are not enough resources (food, good habitat) to support a large population.
Reproduction curve
Population size next year
f(x)
x
Population size this year

Figure 23.3 A typical reproduction curve.

Figure 23.4 shows the same reproduction curve, plus the broken line $y = x$ (which makes a 45° angle with the horizontal axis). You could trace what happens for various choices for x. For an x for which the curve is above the broken line, next year's size $[f(x)]$ is larger than this year's (x). In Figure 23.4, the **natural increase,** or gain in population size, is shown as the length of the green vertical line from the broken line to the curve, which in algebraic terms is $f(x) - x$. For an x for which the curve is below the broken line, next year's size is smaller than this year's and $f(x) - x$ is negative. For the size labeled x_e, for which the curve crosses the broken line, the size is the same next year as this year; this is the *equilibrium population size.*

An **equilibrium population size** is one that does not change from year to year.

Sustained-Yield Harvesting

In the case of fishing, for example, the harvest **yield** h depends on both the population x of fish and the amount of fishing effort (number of boats, hours of fishing). If the fishing fleet fishes the same banks in a year when there are only half

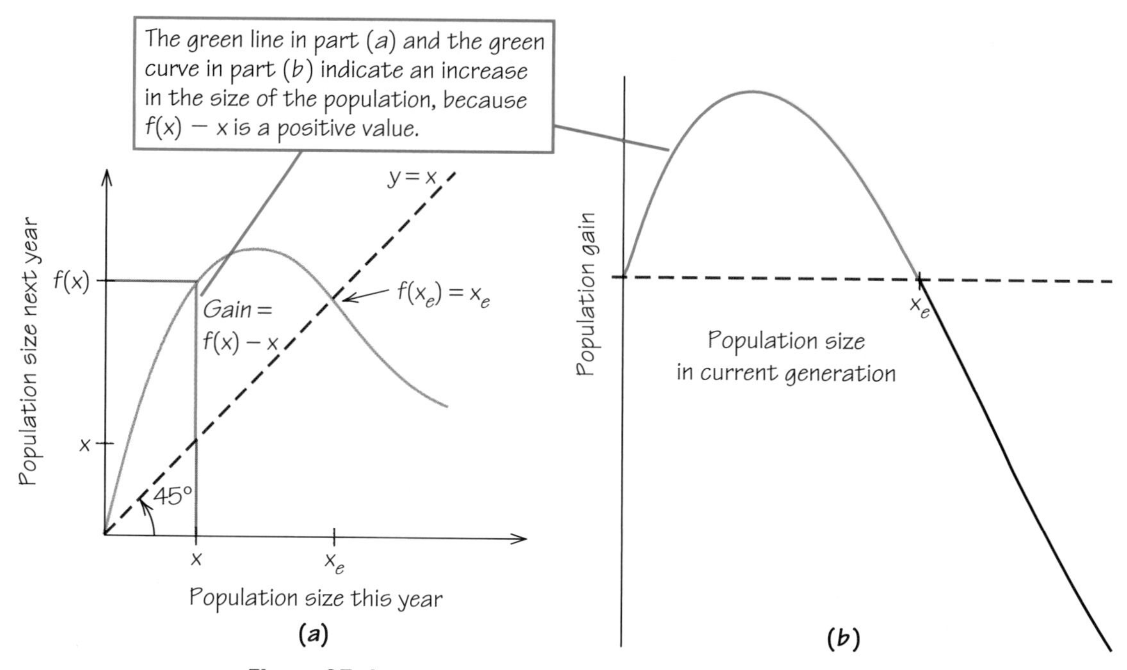

Figure 23.4 Depiction of the natural increase (gain) in population from one year to the next. The population size x_e is the equilibrium population size, for which the population one year later is the same, or $f(x_e) = x_e$.

as many fish as usual, we can expect the fleet to catch only half as many fish. So a simple model is that for any particular level of fishing effort, the harvest is proportional to the fish population, so that a graph of harvest versus population would be a straight line.

Figure 23.5 shows both the gain curve and the harvest line on the same graph, with population along the horizontal axis. Where the curve lies above the line, the annual growth in the population exceeds the number harvested, so the population increases above this year's level; where the curve lies below the line, the harvest exceeds the annual growth, so the population declines below this year's level. The curve and the line intersect where harvest equals growth; for that situation, the population next year will be the same as this year, 14 million fish, on the population (horizontal) axis. For any particular level of fishing, the intersection of the growth curve and the harvest line determines a level of *sustainable yield,* one that could be sustained year after year.

However, you have probably already noticed from Figure 23.5 that the sustainable yield involved, the vertical distance of 5 million, is not the best possible. At a higher level of fishing effort, corresponding to a harvest line with a steeper slope, the fleet could take a harvest of 11 million on a continuing basis, leaving a population of 7 million to reproduce for the next year.

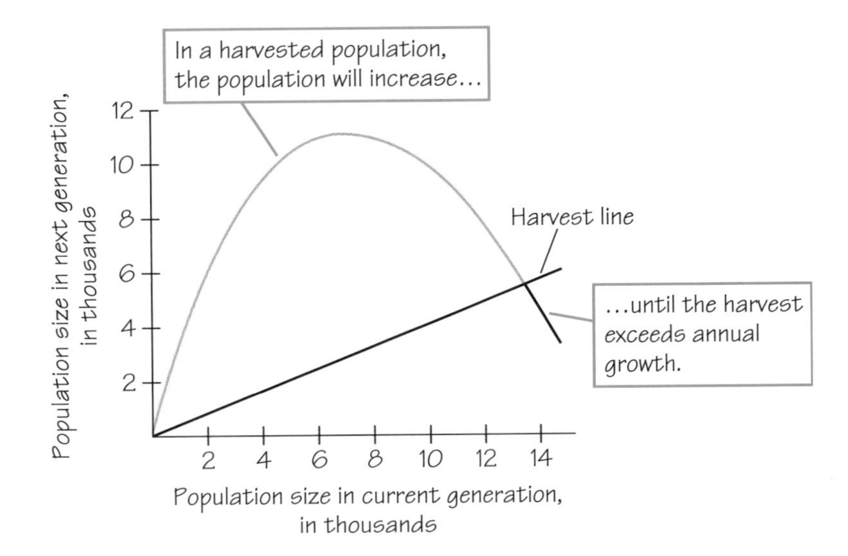

Figure 23.5 This population curve shows the effects of harvesting on a fish population.

A **sustained-yield harvesting policy** is a policy that if continued indefinitely will maintain the same yield.

For a sustainable yield, the same amount is harvested every year and the population remaining after each year's harvest is the same. To achieve this stability, the harvest must exactly equal the natural increase each year, the length of the green vertical line in Figure 23.4a.

Each value of x between 0 and x_e determines a different green vertical line and corresponding sustained-yield harvest (Figure 23.5). This harvest can vary from 0 (for $x = 0$ or $x = x_e$) up to some maximum value (for some x between 0 and x_e). A goal for a timber company or a fishery is to harvest the **maximum sustainable yield:** to select an x whose colored vertical line is as long as possible, marked as x_M in Figure 23.6b.

EXAMPLE

Decline of a Fishery

A successful fishing effort attracts more fishermen and more boats; a lower price for fish causes fishermen to fish longer hours to maintain their income. Whichever is the case, the long-run effect can be catastrophic. Figure 23.7 shows a sequence of graphs that illustrate the sudden decline of a fishery over five years. The first graph shows fishermen exploiting the fishery optimally, at the maximum sustainable yield of 11 million. Each of the subsequent graphs shows what happens with a 20% increase beyond the original fishing effort. The yield goes from 11 to 10.6, then to 9.5, 7.8, 5.4, and 2.4 million. The initial gradual decline over the first couple of years is deceptive and could be attributed to normal biological variation

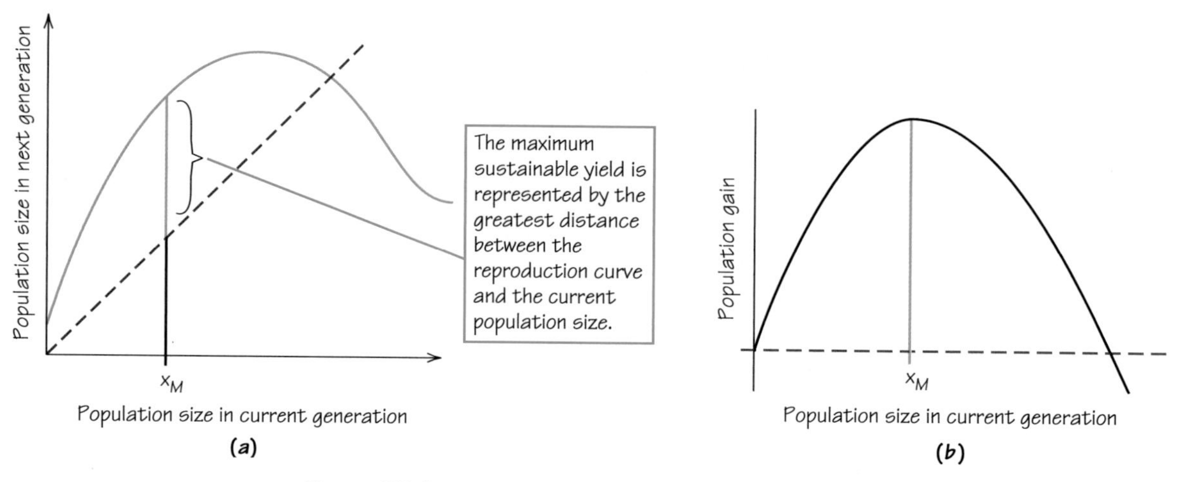

Figure 23.6 The reproduction curve, with the population size x_M corresponding to the maximum sustainable yield.

in the reproduction of the fish; but the decline is precipitous in the last three years. Extinction of the resource could result from additional fishing effort. The fishermen work harder and harder, but there are fewer and fewer fish—in part, ironically, because of the increasing effort of the fishing fleet. ■

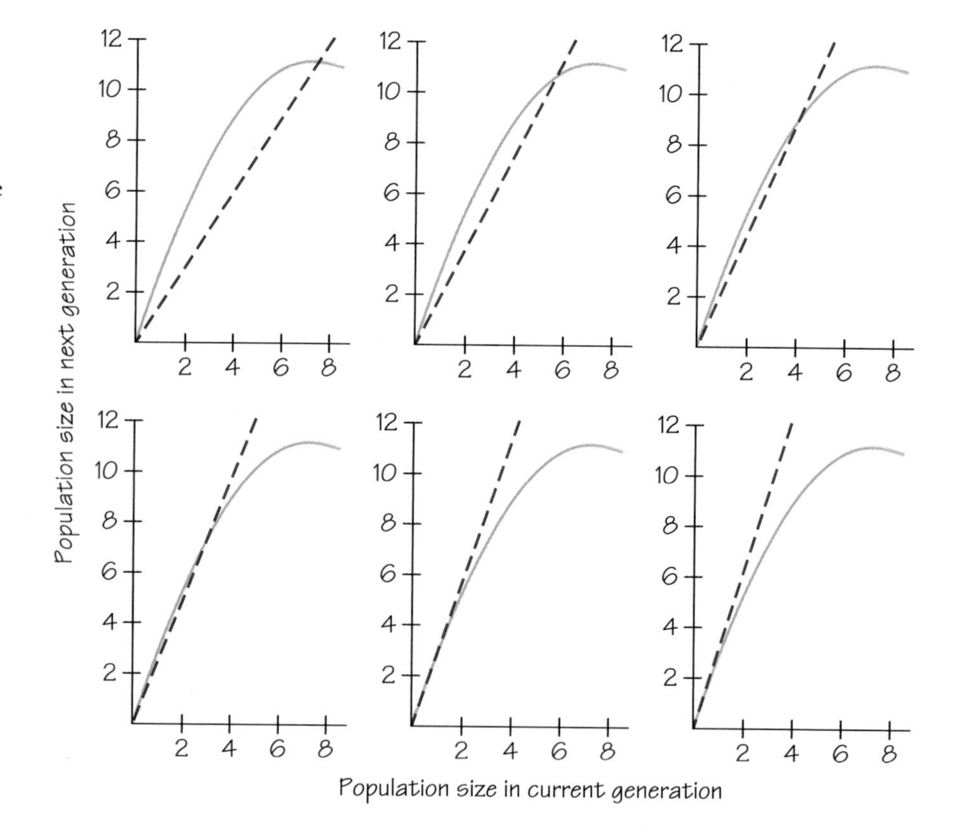

Figure 23.7 Progressive deterioration of a fishery with increasing fishing pressure. The first graph shows the situation with fishing effort geared to extract the maximum sustainable yield. Each subsequent graph shows a 20% increase beyond the original fishing effort (the slope of the harvest line increases). The yield declines from the maximum sustainable yield of 11 million to 2.4 million and would decline further with additional fishing effort.

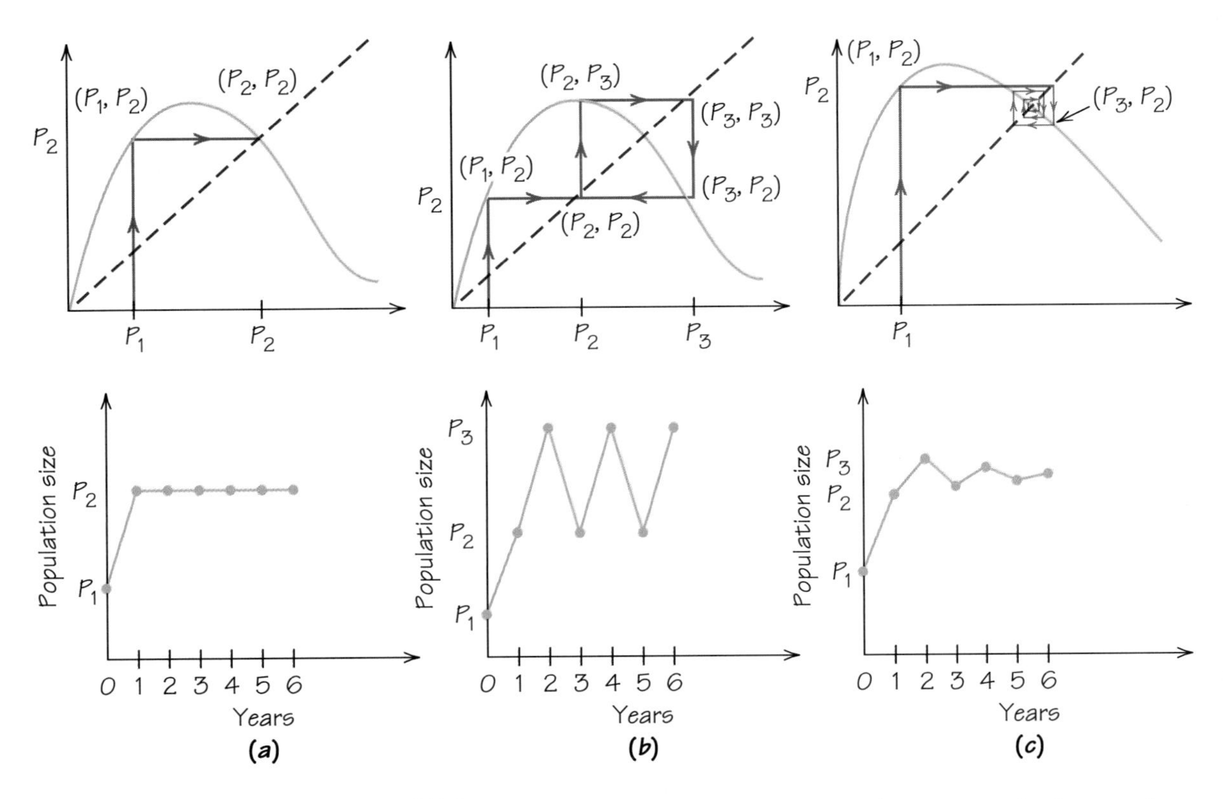

Figure 23.8 Examples of the dynamics, over time, for the same reproduction curve but different starting populations. (a) The population goes in one year to the equilibrium population and stays there year after year. (b) After initial adjustment, the population cycles between values over and under the equilibrium population. (c) The population spirals in toward the equilibrium population.

Dynamics of a Population over Time

The line $y = x$ provides a convenient way to trace the evolution of the population over several years (see Figure 23.8), by alternating steps vertically to the curve and horizontally to the line $y = x$. Begin with the first year's population on the horizontal axis, go up vertically to the curve; the height is the population in the second year. Proceed horizontally from the curve over to the line $y = x$. Proceeding vertically from there to the curve yields a height that is the population in the third year.

Figure 23.8 shows several traces for the same reproduction curve, each starting from a different initial population on the horizontal axis. The resulting variation is quite surprising—it can even be "chaotic" in a very specific mathematical sense, showing how apparently random behavior can result from strict deterministic rules.

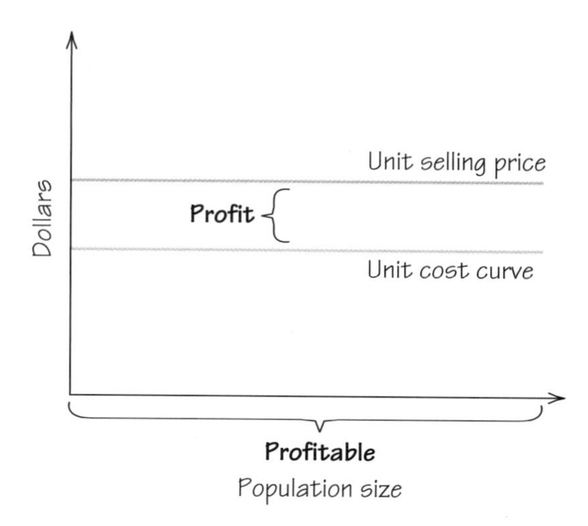

Figure 23.9 The unit cost, unit revenue, and unit profit of harvesting one unit, as a function of population size, for the cattle ranch.

The Economics of Harvesting Renewable Resources

We consider two models: one for a cattle ranch and one for either a fishing boat or a tree farm.

We assume that the price p received is the same for each harvested unit and does not depend on the size of our harvest. In effect, we assume that our operation is a small part of the total market, not substantially affecting overall supply and hence price.

We want to stay in business, so we do not extinguish the resource for quick profits. For any given population size, we harvest just the natural increase.

EXAMPLE

Cattle Ranching

We assume that the cost of raising and bringing a steer to market is the same for every steer and does not depend on how many steers we bring to market. Because the cost does not depend on the population size, the cost curve is a horizontal line (Figure 23.9).

As long as the selling price per unit is higher than the harvest cost per unit, we make a profit. The points of view of economics and biology agree, because the maximum profit occurs for the maximum sustainable yield. ■

EXAMPLE

Fishing and Logging

In this model we assume that the cost of harvesting a unit of the population decreases as the size of the population increases; this is the familiar principle of **economy of scale.** For example, the same fishing effort yields more fish when fish are more abundant. Similarly, a logger's harvest costs per tree are less when the trees are clumped together; this is the logger's motivation to clear-cut large stands.

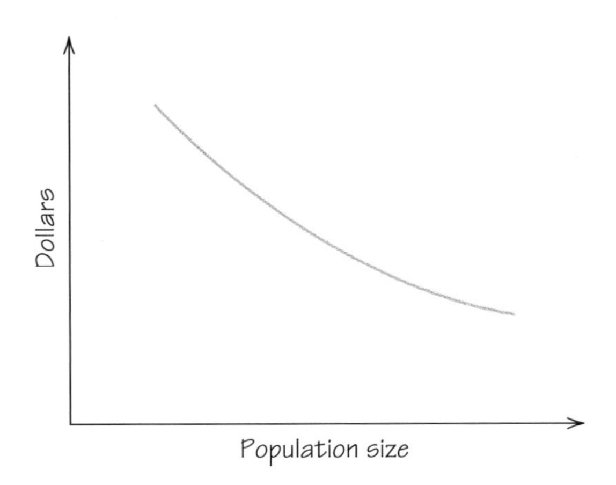

Figure 23.10 The unit cost, as a function of population size, for fishing or logging.

The cost curve slopes downward and to the right, as in Figure 23.10. The size of the population from which one unit is harvested is shown on the horizontal axis; the cost of harvesting a single unit is measured on the vertical axis.

An optimal harvesting policy depends on the relation between price and costs. There are two cases, as shown in Figure 23.11.

- *The unit cost curve lies entirely above the unit price line.* The price for a harvested unit is less than the cost of harvesting it, no matter how large the population. It is impossible to make a profit.

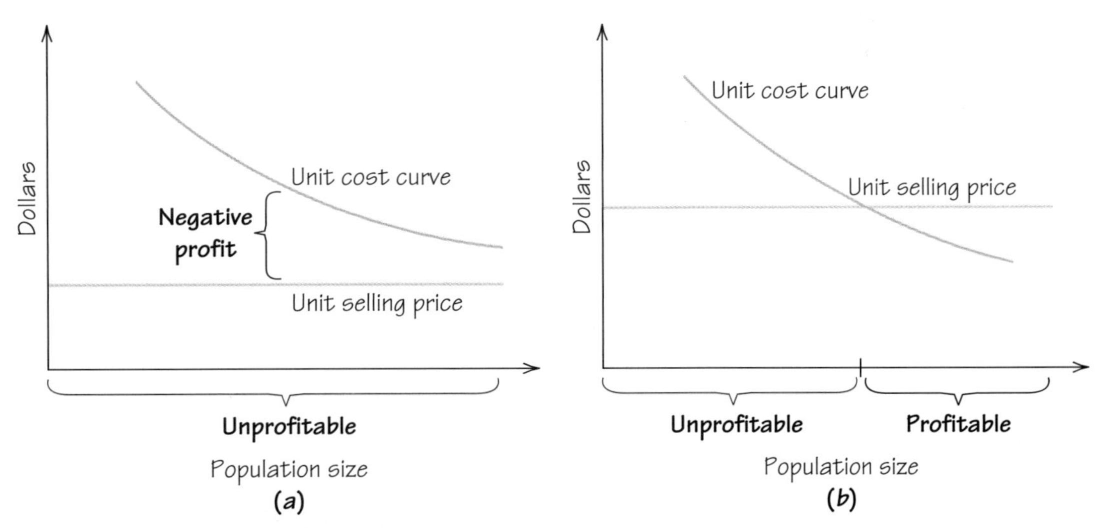

Figure 23.11 The unit cost, unit revenue, and unit profit of harvesting one unit, as a function of population size, for fishing or logging. (a) The market price is below harvesting cost for all population sizes. (b) The operation is profitable for populations above a certain minimum size.

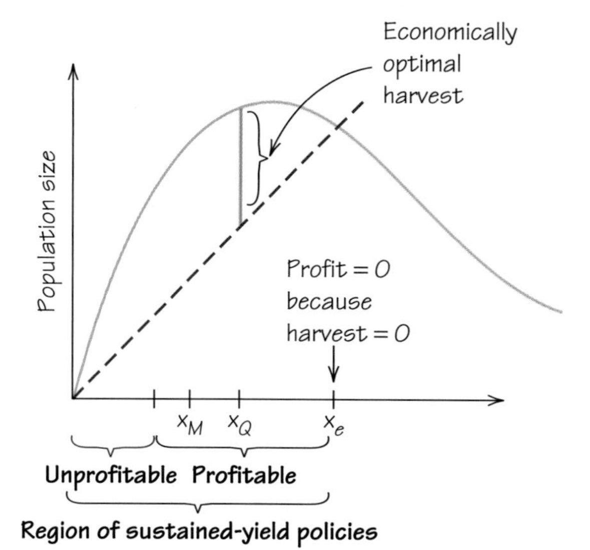

Figure 23.12 Regions of profitability for sustained-yield policy, with the economically optimal population size x_Q marked.

- *The unit cost curve intersects the unit price line.* Above a certain population size, the price for a harvested unit is more than the cost of harvesting it, so profit is possible (Figure 23.12). Some population size, call it x_Q, gives a maximum net profit. Using calculus, it can be shown that x_Q is actually larger than x_M, the population that gives the maximum sustainable yield. Compared to harvesting the maximum sustainable yield, economic considerations result in harvesting less, but also in maintaining a larger stock of the population. ■

Our simple models fail to take into account a very critical feature of a modern economy that we already explored earlier in this chapter: the time value of money, as measured by the interest capital can earn. In the next section we see one important explanation of why biological populations are susceptible to overexploitation and even extinction.

Why Eliminate a Renewable Resource?

In some instances, such as the passenger pigeon, populations have been harvested to extinction. In other cases, an entire ecosystem has been destroyed (see Spotlight 23.2). Why would anyone eliminate a renewable resource? Our approach helps explain why.

Sustained-yield policies involve revenues that will be received, year after year, in the future. The value of these revenues should be discounted to reflect the lost investment income that we could earn if instead we had the revenues today. For funds invested at a return of $100r$ % per year, compounded annually, the present value P of an amount A to be received in n years in the future is related to A by the compound interest formula $A = P(1 + r)^n$.

SPOTLIGHT 23.2 The Tragedy of Easter Island

Easter Island
(Tom Till/Tony Stone Images.)

Easter Island is famous for its isolation – 1400 miles to the nearest island – and for its hundreds of huge stone statues. For 30,000 years before the arrival of people in about A.D. 400, Easter Island maintained a lush forest, with several species of land birds. By the time of the first visit by Europeans in 1722, the island was barren, denuded of all trees and bushes over 10 feet high, and with no native animals larger than an insect. The 2000 or so islanders had only three or four leaky canoes made of small pieces of wood.

What happened? Careful analysis of pollen in soil samples tells the sad story. The settlers and their descendants cut wood to plant gardens, build canoes, make sledges and rollers to move the huge statues, and burn for cooking and warmth in the winter. In addition to crops they raised and chickens they had brought to the island and cultivated, they ate palm fruit, fish, shellfish, the meat and eggs of birds, and the meat of porpoises that they hunted from seagoing canoes. The population of the island grew to 7000 (or perhaps even 20,000).

By 1500, the forest was gone. Most tree species, all land birds, half of the seabirds, and all large and medium-sized shellfish had been extinguished. There was no firewood, no wood for sledges and rollers to transport hundreds of statues at various stages of completion, and no wood for seaworthy canoes. Without canoes, fishing declined and porpoises could not be taken. Stripping the trees exposed the soil, which eroded, so crop yields fell off. The people continued raising chickens, but warfare and cannibalism ensued. By 1700, the population had crashed to 10 – 25% of its former size.

Why didn't the people notice earlier what was happening, imagine the consequences of keeping on as they had been, and act to avert catastrophe? After all, the trees did not disappear overnight.

From one year to the next, changes may not have been very noticeable. The forests may have been regarded as communal property, with no one charged with limiting exploitation or ensuring new growth. There was no quantitative assessment of the resources available and need for conservation versus the long-term needs of the "public works" program of erecting statues. Moreover, the religion of the people, the prestige of the chiefs, and the livelihood of hundreds depended on the statue industry. There was no perceived need to limit the population and no technology for birth control. Once the large trees were gone, there was no means for excess population to emigrate.

Adapted from Jared Diamond, "Easter's end," *Discover,* 16(8) (August 1995): 63 – 69.

The economic goal is to maximize the sum of the present values of all future receipts from harvesting. The optimal harvesting policy thus must depend on the expected rate of return r. We don't delve into the details of the calculations here, but instead just give the results of the analysis.

Again there are several cases to consider:

1. The unit cost of harvesting exceeds the unit price received, for all population sizes. Then it is impossible to make a profit.
2. For some population size x, the unit cost of harvesting equals the unit price received. Then there is a size between x and x_e (the equilibrium population size) for which the present value of the total return is maximized and the population and its yield are sustained.
3. The unit price exceeds the unit cost for *all* population sizes.
   - For r small, the situation is the same as in the second case just cited.
   - For larger r, the economically optimal policy may be to harvest the entire population immediately—extinguish the resource—and invest the proceeds.

Let's put this in the simplest and starkest terms. Suppose that you own a valuable resource, such as a forest, whose cost of harvesting is small relative to the value of the resource. If the rate at which the forest population grows is greater than what you can earn on other investments, it pays to let the forest keep on growing.

On the other hand, if the forest is growing more slowly than the rate of return on other investments, the economically optimal harvesting policy is to cut down all the trees now and invest the money. You could then start raising cattle on the land—and right there you have the scenario that is resulting in deforestation all over the world.

The sobering fact is that *very few economically significant renewable resources can sustain annual growth rates over 10%.* Many, like whales and most forests, have growth rates in the 4% to 5% range. These values—even a growth rate of 10%—are far below the return investors expect on their investment. For example, until recent deregulation, Wisconsin electric utilities were guaranteed 14.25% profit; and venture capital firms expect to exceed 25% profit.

The concept of maximum sustainable yield is an attractive ideal if the expectations of investors are low enough. However, there are still difficult problems:

- One problem is "the tragedy of the commons," discussed by ecologist Garrett Hardin. Several hundred years ago, English shepherds would graze their flocks together on common land. The grass of the commons could support only a fixed number of sheep. Each shepherd could reasonably think that adding just one or two more sheep to his flock would not overtax the commons; yet if each did so, there could be disaster, with all the sheep starving. Many natural-products industries, such as fisheries, are a form of commons; small overexploitation by each harvester can produce disastrous results for all.

- How, in the presence of human needs or greed, can we anticipate and prevent overexploitation and possible extinction of a resource? By and large, it has been politically impossible to force a harvesting industry to reduce current harvests to assure stability in the future.
- In some industries, such as a fishery, growth of the population may be abundant one year but meager another, so that a steady yield cannot be sustained without damaging the resource. A few good years in a row may provoke increased investment in fishing capacity; then attempting to harvest at the same levels in succeeding normal or below-normal years results in overfishing. This exact scenario destroyed the California sardine fishery in the 1930s, the Peruvian anchovy fishery in 1972, and much of the North Atlantic fishery in the 1980s.

Were the fishermen and regulators mentioned above at fault for extinguishing these fisheries by overexploiting a dependable resource? Or were the extinctions due to chance variations of the fish stocks? In the next section we examine a third possibility—that the fish stocks followed simple rules that nevertheless produced "chaotic" behavior of stocks, that is, great variation from one year to the next. When we do not see the pattern, we interpret such behavior as randomness, much as the moves in a chess game may appear random and inexplicable to someone who does not know the rules of the game.

23.5 Dynamical Systems

Throughout this and the two previous chapters, we have considered all kinds of systems that change over time. Examples include bank accounts, the amount due on a loan, and the size of a population. Further examples are a dripping faucet, a playground swing, a pinball play, the solar system, the business cycle, epidemics, the passage of a drug through the human body, and the weather. Some of these are very predictable (principal plus interest on a bank account), while others are notoriously unpredictable (the weather); some are simple systems that involve no outside influences (the amount due on a loan, assuming that you don't get behind on payments!), while others are the result of many contributing factors (the business cycle).

A dynamical system can be either discrete (a savings account to which interest is posted regularly) or continuous (the passage of a drug through the human body). The decay of a radioactive substance is actually discrete (individual particles are given off) but may be treated as continuous if there is a large enough quantity of the substance.

In some systems (the population of a country), the state of the system may depend largely on its states at previous times, while in other systems (epidemics) connection to past behavior may be harder to discern. A system whose state depends only on its states at previous times is called a **dynamical system.** Various dynamical systems provide useful models for physical, biological, chemical, financial, and other systems.

23.6 Mathematical Chaos

We think of **chaos** as referring to general confusion, unpredictability, and apparent randomness. Mathematicians and other scientists use the word to describe systems whose behavior over time is inherently unpredictable.

There are three main features of mathematical chaos:

- Determinism – chance is not involved
- Complexity in behavior over time
- Sensitive evolution of the system with regard to initial conditions

The last feature is sometimes known as the **butterfly effect,** from the title of a 1979 talk by meteorologist E. N. Lorenz: "Predictability: Does the Flap of a Butterfly's Wings in Brazil Set Off a Tornado in Texas?" (The phrase probably traces to a 1953 science fiction story by Ray Bradbury, "A Sound of Thunder," in which history is changed by a time-traveler who steps on and kills a prehistoric butterfly.)

What Lorenz was referring to is that a small change in a system might make a very large difference much later on. It also might not – we certainly hope that a butterfly flapping its wings in Brazil won't cause a tornado in Texas, whether a month later or a century later! We do not expect the weather system to be that chaotic.

We can get a feel for chaotic systems by playing with some *iterated function systems.* The fancy name just means that we take an initial value, apply a function to it, then repeat over and over. This is a discrete system, and it is exactly what we were doing earlier with reproduction curves for populations. In that case, we did not have an algebraic formula for the population next year as a function of this year's population but instead used the reproduction curve. For the iterated function systems that we discuss below, we have both.

EXAMPLE Doubling on a "Stone Age" Calculator

Imagine that you have a calculator that is accurate to only two decimal places. It has a special key marked DBL; what this key does is double the number in the display and keep *only* the fractional part. For example, DBL applied to 0.52 gives 0.04 (*not* 1.04).

Let's start with two numbers that are as close together as can be on this calculator, such as 0.37 and 0.38, and see what happens as we push the DBL key over and over again for each one:

$$0.37, 0.74, 0.48, 0.96, 0.92, 0.84, 0.68, 0.36, 0.72, 0.44, 0.88, 0.76, 0.52, 0.04, \ldots$$
$$0.38, 0.76, 0.52, 0.04, 0.08, 0.16, 0.32, 0.64, 0.28, 0.56, 0.12, 0.24, 0.48, 0.96, \ldots$$

Already, by the fourth iteration, the two sequences – which started with initial values very close together – are wide apart. (The arithmetic of the ancient Egyptians and of the Romans was based on doubling, so they would have been overjoyed to have a calculator even if it had only a doubling key!) ■

EXAMPLE

Pseudo-random Numbers

"Random" numbers produced by computers and calculators are not really random at all, but they seem random to us. Given a "seed number," the next "random" number, and each of the subsequent ones, is calculated directly from the previous one. A notorious example is the famous pseudo-random-number generator called RANDU that was widely used in the 1960s. RANDU calculates the next random number X_{n+1} directly from the preceding one X_n by the equation

$$X_{n+1} = \text{remainder from dividing } (65539 \times X_n) \text{ by } 2^{31} = 2147483654$$

In terms of the notation for modular arithmetic on page 364 of Chapter 10, the calculation is

$$X_{n+1} = 65539X_n \bmod 2^{31}$$

Despite its wide usage, later analysis of RANDU revealed that the numbers it produces fail some tests for being random enough; in other words, the numbers produced have certain subtle patterns.

Your calculator probably has a RAND button on it that uses a formula similar to the above. ■

EXAMPLE

The Solar System

The American moon landings in 1969 and the years following, as well as all other space missions, were possible because of the predictability, or *determinism,* of the solar system. The moon and planets follow their orbits like clockwork over the relatively short time intervals of human space travel. Over tens of millions of years, however, the orbit of each planet is chaotic, meaning that the error in measuring its position or velocity would produce a huge error in predicting its position that much later.

More down-to-earth examples of physical systems that can exhibit chaos include the fluttering of a falling autumn leaf, heart arrhythmias, and the Tilt-A-Whirl amusement park ride.

Chaos in Biological Populations

If we measure this year's population as a fraction x of the carrying capacity, and do the same for next year's population as a fraction $f(x)$, the logistic model takes the form

$$f(x) = \lambda x(1 - x)$$

where the Greek letter lambda $\lambda = 1 + r$ is the amount by which the population is multiplied each year. The equation has the familiar form of a quadratic in x:

$$f(x) = -\lambda x^2 + \lambda x$$

EXAMPLE

The Logistic Population Model

For different values of the parameter λ and different starting values for the population fraction, each of the behaviors of Figure 23.8 (page 835) can occur:

- $\lambda = 2.8$ and starting population fraction $x = 0.36$ produces Figure 23.8a.
- $\lambda = 3.1$ and starting population fraction $x = 0.235$ produces Figure 23.8b.
- $\lambda = 3.0$ and starting population fraction $x = 0.4$ produces Figure 23.8c.

In other words, for population growth rates (values of λ) that are fairly close together (2.8, 2.1, 3.0), the population evolves very differently. This is a surprising and nonintuitive conclusion.

But there is more. For $\lambda = 4$ and any starting population fraction, the population does not settle down into any of the patterns of Figure 23.8 but year after year wanders "unpredictably" all over the place (Figure 23.13). This is *chaotic behavior:* It is deterministic, complex, and—in the long run—unpredictable. We must say "in the long run," because in the short run the behavior is completely predictable. For example, from this year's population fraction, the equation tells us exactly what next year's will be; and repeating the use of the equation, we can determine what it would be the following year or any particular year after that. But as the years pass, any sense of pattern gets lost in the complexity. ■

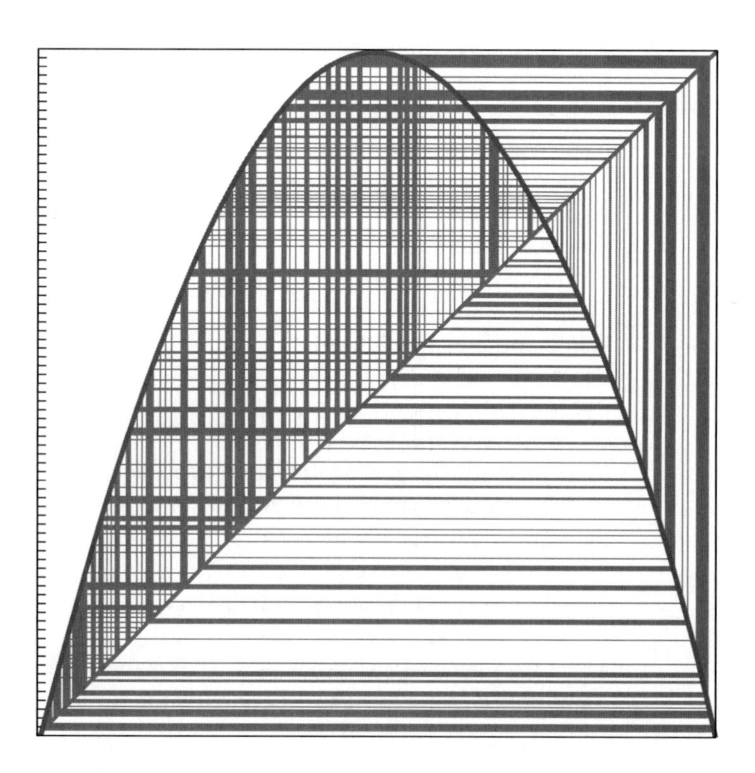

Figure 23.13 Chaotic behavior of a population.

This potentially chaotic behavior of a biological population is shocking news for those who try to manage a ranch, trophy-hunting preserve, or any wildlife population – in the wild or in captivity. In recent years, lobster catches in Maine have been much higher than in previous years, reaching record levels, for no discernible reason. On the other hand, in the late 1950s the harvest of Dungeness crabs off the central California coast declined from 12 million pounds a year to less than 1 million pounds a year without any evidence of disease, heightened predation, or increased crabbing effort. Researchers who modeled that population in 1994 found that booms and busts are the rule; the population can remain nearly level for thousands of generations before suddenly exploding or crashing without warning.

Searching for an environmental cause for these fluctuations could be futile, because there does not have to be one. Moreover, observing the population over a few generations – or even over hundreds of generations – provides no help in predicting future behavior.

EXAMPLE Childhood Disease Epidemics

The incidence of childhood diseases such as chicken pox and measles varies from year to year. There are three plausible explanations for the fluctuations:

- There is an underlying regular cycle that is perturbed and occasionally overwhelmed by random events ("noise").
- There is no discernible pattern, because such fluctuations are inherent in the epidemiology of the disease as it is a chaotic system.
- There is no discernible pattern, because the fluctuations are due solely to chance.

Chickenpox fits the first explanation, with a cycle of one year. For measles, the situation is more complex. For small populations, chance is an adequate explanation; in larger populations, there is a pronounced seasonality. Chance cannot explain seasonality, but the other two explanations (a regular cycle or chaos) require that the transmission of the disease follow a pattern in time, that is, some seasonality. Historical data before mass immunizations suggest that measles cases in large communities were chaotic. That doesn't mean they occurred at random but rather that the fluctuations had a large deterministic component. Research also shows that there is a critical community size above which a disease will not die out as a result of chance extinction; for measles, this size is about 250,000. ■

What you need to realize about chaos is that behavior that appears to be random can be produced even with very simple systems that are completely governed by deterministic rules. Just because the behavior appears chaotic does not mean that it does not having underlying order and structure, though discovering that structure may be difficult.

Even if we discover the structure, prediction may elude us. If we had an absolutely correct model of how weather behaves and if we had measurements at every location on and above the earth (and the computer power to process them), we might still not be able to forecast the weather accurately a week ahead.

What about the fishery extinctions? Perhaps the fishermen and the fish were victims not of greed or chance but of the chaotic nature of the reproduction curve for the fish.

REVIEW VOCABULARY

Biomass A measure of a population in common units of equal value.

Butterfly effect A small change in initial conditions of a system can make an enormous difference later on.

Carrying capacity The maximum population size that can be supported by the available resources.

Chaos Complex but deterministic behavior that is unpredictable in the long run.

Compound interest formula Formula for the amount in an account that pays compound interest periodically. For an initial principal P and effective rate r per year, the amount after n years is $A = P(1 + r)^n$.

Dynamical system A system whose state depends only on its states at previous times.

Economy of scale Costs per unit decrease with increasing volume.

Equilibrium population size A population size that does not change from year to year.

Exponential reserve How long a fixed amount of a resource will last at a constantly increasing rate of use. A supply S, as an initial rate of use U that is increasing by a proportion r each year, will last

$$\frac{\ln\left(1 + \frac{S}{U} r\right)}{\ln(1 + r)} \text{ years.}$$

Logistic model A particular population model that begins with near-geometric growth but then tapers off toward a limiting population (the carrying capacity).

Maximum sustainable yield The largest harvest that can be repeated indefinitely.

Natural increase The growth of a population that is not harvested.

Nominal rate The stated rate of interest per year, on which any compounding is based.

Nonrenewable resource A resource that does not tend to replenish itself.

Population growth Change in population, whether increase (positive growth) or decrease (negative growth).

Population structure The division of a population into subgroups.

Rate of natural increase Birth rate minus death rate; the annual rate of population growth without taking into account net migration.

Renewable natural resource A resource that tends to replenish itself; examples are fish, forests, wildlife.

Reproduction curve A curve that shows population size in the next year plotted against population size in the current year.

Savings formula Formula for the amount in an account to which a regular deposit is made (equal for each period) and interest is credited, both at the end of each period. For a regular deposit of d and an effective interest rate r per year, the amount A accumulated is

$$A = d\left[\frac{(1 + r)^n - 1}{r}\right]$$

Static reserve How long a fixed amount of a resource will last at a constant rate of use; a supply S used at an annual rate U will last S/U years.

Steady state The population size for which the size next year will be exactly the same as the size this year. The amount harvested is exactly equal to the natural increase.

Sustained-yield harvesting policy A harvesting policy that can be continued indefinitely while maintaining the same yield.

Yield The amount harvested at each harvest.

SUGGESTED READINGS

BARTLETT, ALBERT A. Forgotten fundamentals of the energy crisis, *American Journal of Physics*, 46(9) (September 1978): 876–888.

CLARK, COLIN. The mathematics of over-exploitation, *Science*, 181 (August 17, 1973): 630–634.

COHEN, JOEL. Ten myths of population, *Discover*, 17(4) (April 1996): 42–47.

COHEN, JOEL. *How Many People Can the Earth Support?* Norton, New York, 1995.

CRUTCHFIELD, JAMES P., ET AL. Chaos, *Scientific American* (December 1986).

DANCE, ROSALIE A., and JAMES T. SANDEFUR. Fishing for food and profit, *Consortium: The Newsletter of the Consortium for Mathematics and Its Applications* 82 (Summer 2002): (pull-out section) 1–8.

DANCE, ROSALIE A., and JAMES T. SANDEFUR. Fishing and the future, *Consortium: The Newsletter of the Consortium for Mathematics and Its Applications* 78 (Summer 2001): (pull-out section) 1–8.

DEVANEY, ROBERT L. *Chaos, Fractals, and Dynamics: Computer Experiments in Mathematics,* Pearson Learning, 2001. www.pearsonlearning.com. Simple BASIC programs are included for all major topics.

DEVANEY, ROBERT L., ET AL. *A Toolkit of Dynamics Activities.* This is a series of four paperback books on dynamical systems. *Iteration, Fractals, Chaos,* and *The Mandelbrot and Julia Sets,* Key Curriculum Press, 1998–1999.

DIAMOND, JARED. Easter's end, *Discover* 16(8) (August 1995): 63–69.

FIELD, MICHAEL, and MARTIN GOLUBITSKY. *Symmetry in Chaos: A Search for Pattern in Mathematic, Art and Nature,* Oxford University Press, New York, 1992.

GLEICK, JAMES. *Chaos: Making a New Science,* Viking, New York, 1987.

HARDIN, GARRETT. The tragedy of the commons, *Science,* 162 (1968): 1243–1248.

LUDWIG, DONALD, RAY HILBORN, and CARL WALTERS. Uncertainty, resource exploitation, and conservation: Lessons from history, *Science,* 260 (April 2, 1993): 17, 36.

LUTZ, WOLFGANG, WARREN SANDERSON, and SERGAL SCHERBOV. The end of world population growth, *Nature* 412: 543.

MEADOWS, DONELLA H., DENNIS L. MEADOWS, and JØRGEN RANDERS. *Beyond the Limits: Confronting Global Collapse, Envisioning a Sustainable Future,* Chelsea Green, Post Mills, Vt., 1992.

OLINICK, MICHAEL. Modelling depletion of nonrenewable resources, *Mathematical Computer Modelling,* 15(6) (1991): 91–95.

OPHULS, WILLIAM, and A. STEPHEN BOYAN, JR. *Ecology and the Politics of Scarcity Revisited,* W. H. Freeman, New York, 1992.

PETERSON, IVARS. Scrambled grids. *Science News* 158(9) (August 26, 2000). http://www.sciencenews.org/20000826/mathtrek.asp.

PETERSON, IVARS. Tilt-a-whirl chaos (I) and (II). *Science News* 157(17) (April 22, 2000), 157(18) (April 29, 2000). http://www.sciencenews.org/20000422/mathtrek.asp, http://www.sciencenews.org/20000429/mathtrek.asp

PETERSON, IVARS. *Newton's Clock: Chaos in the Solar System,* W. H. Freeman, New York, 1993.

POPULATION REFERENCE BUREAU, INC. Annual world population data sheet. 1875 Connecticut Ave., NW, Suite 520, Washington, D.C. 20009–5728. Tel: (202) 483–1100; email: popref@prb.org; Web: www.prb.org.

SAFINA, CARL. The world's endangered fisheries, *Scientific American*, 273 (November 1995): 46–53.

SANDEFUR, JAMES T. *Elementary Mathematical Modeling: A Dynamic Approach*, Brooks/Cole, Pacific Grove, Calif., 2002.

SCHWARTZ, RICHARD *H. Mathematics and Global Survival*, 3rd ed., Ginn Press, Needham Heights, Mass., 1993.

SUGGESTED WEB SITES

www.popin.org/ UN Population Division population statistics and estimates for all countries.

www.prb.org/ Population Reference Bureau population statistics and rates of growth by regions.

sunsite.anu.edu.au/education/chaos Downloadable Java applets to illustrate chaos.

math.bu.edu/DYSYS/applets/ Downloadable Java applets designed to accompany Robert L. Devaney's *A Toolkit of Dynamics Activities* but that can be used independently.

SKILLS CHECK

1. The carrying capacity of a population is

(a) the largest recorded population.
(b) the largest supportable population.
(c) the change in population.

2. Management of a nonrenewable resource can be modeled by

(a) an annuity.
(b) a savings account.
(c) an add-on loan.

3. The shape of the reproduction curve reflects the fact that a small population has abundant resources by

(a) rising steeply at the left.
(b) falling to the right.
(c) crossing the diagonal.

4. The equilibrium population size is the same as

(a) the carrying capacity.
(b) the intersection point of the reproduction curve and the diagonal.
(c) the natural increase of the population.

5. If the starting population for a reproduction curve is changed, the subsequent population pattern

(a) will still drift to the same pattern.
(b) will change to a different pattern.
(c) will sometimes change to a different pattern.

6. A system whose current state depends solely on its previous states is called

(a) a dynamical system.
(b) a stable system.
(c) an optimal system.

7. The "butterfly effect" refers to which feature of chaos?

(a) Predictability
(b) Complexity
(c) Sensitivity to initial conditions

8. For the logistic model $f(x) = 3x(1 - x)$, if the starting population fraction is 0.5, what is the next population fraction?

(a) 0.75
(b) 0.6
(c) 0.25

9. For the logistic model $f(x) = 3x(1 - x)$, if the starting population fraction is 0.4, what is the next population fraction?

(a) 0.116
(b) 0.416
(c) 0.744

10. For the logistic model $f(x) = 4x(1 - x)$, if the starting population fraction is 0.1, what is the next population fraction?

(a) 0.4
(b) 0.36
(c) 0.09

11. For the logistic model $f(x) = 4x(1 - x)$, are there any starting population fractions that will immediately lead to 0 population?

(a) No
(b) Yes, 1.0
(c) Yes, 0.5

12. Chaotic behavior appears to be random

(a) but is actually not random.
(b) and is random.
(c) and is sometimes random.

13. If we have enough reserves of a product to last 200 years at the current rate of use, but the rate of use increases by 10% per year, how long will the supply last?

(a) About 32 years
(b) About 40 years
(c) About 50 years

14. If we have enough reserves of a product to last 1000 years at the current rate of use, but the rate of use increases by 1% per year, how long will the supply last?

(a) About 900 years
(b) About 480 years
(c) About 240 years

15. Pressing a digit and then repeatedly pressing the SIN key is a model of

(a) an iterated function system.
(b) chaos.
(c) randomness.

EXERCISES ▲ Optional. ■ Advanced. ◆ Discussion.

Growth Models for Biological Populations

1. (Requires spreadsheet) For many years, China has been the world's most populous country. In mid-2001, the population of India was 1033 million and growing at 1.7% per year, while the population of China was 1273.3 million and growing at only 0.9% per year. If these rates continue, when will India have more people than China?

2. The total population of the less-developed countries (excluding China) was 3.671 billion in mid-2001, and the growth rate was 1.9% per year (this is an annual yield, so you may think of it as compounded annually). If this growth rate continues for the next 25 years (until mid-2026), what will be the size of the population then?

3. If the growth rate of the less-developed countries of Exercise 2 changed suddenly to 1.7% in mid-2001, what would be the size of the population in mid-2026?

4. (Requires spreadsheet) If the growth rate of the less-developed countries of Exercise 2 decreased by $\frac{1}{25}$ of a percentage point (0.04%) per year from 2001 through 2005, beginning in mid-2001, what would be the size of the population in mid-2005?

5. An advertisement for Paul Kennedy's book *Preparing for the Twenty-First Century* (Random House, 1993) asked: "By 2025, Africa's population will be: 50%, 150% or 300% greater than Europe's?" The population of Europe in mid-2001 was 727 million and was expected to stay constant through 2025. The population of Africa in mid-2001 was 818 million and was increasing at about 2.4% per year. What answer would you give to the question?

6. In its estimates for doubling times for populations in the world, the Population Reference Bureau uses a rule of 70, similar to (but more accurate than) the rule of 72 used in banking and explained in Exercises 7–8 in Chapter 22. The rule of 70 says that if a country's population continues to grow at a constant rate of r% per year, then it will double in size every $70/r$ years. (As noted in Chapter 22, a rule of 69.3 would be even more accurate, but the difference between that and the rule of 70 is only 1%.) Apply the rule of 70 to estimate the doubling times for the following populations (figures are for mid-2001):

(a) Africa, 818 million, 2.4%
(b) United States, 284.5 million, 0.6%

7. Do the calculations as in Exercise 6, but for:

(a) China, 1.273 billion, 0.9%
(b) The world as a whole, 6.137 billion, 1.3%

8. (Requires spreadsheet) **(a)** Wisconsin's electricity demand increased 2.8% per year for the 10 years from 1991 through 2000. If that trend continues, when will Wisconsin need to have twice as much generating capacity as it did in 1991?
(b) Wisconsin electricity demand in 1991 was about half from business use and half from residential use. Over the 10 years through 2001, business demand increased at 4.2% per year and residential demand increased at 1.5% per year. Assuming that these rates of growth continue, calculate business and residential demand separately for the year that you calculated in part (a). What do you observe?

Nonrenewable Resources

9. Aluminum is the most abundant structural metal in the earth's crust. The world demand for new supplies of aluminum in 1983 was 16.5 million metric tons, while the known reserves were then 21,000 million metric tons.

(a) What was the static reserve for aluminum in 1983? How much longer will the reserve last from 2003?
(b) In 1983, the demand for new aluminum was projected to increase at 4% per year at least through the year 2000. For that rate of increase, what was the exponential reserve for aluminum in 1983? How much longer will the reserve last from 2003?
(c) What considerations may affect these reserves over time?

10. In 1990 the known global oil reserves totaled 917 billion barrels. Consumption, which had been 53.4 million barrels per day in 1983, rose an average of about 1.7% per year through 1990, when the consumption was about 60 million barrels per day.

(a) What was the static reserve for oil in 1990? How much longer would that amount of reserve last from 2003?
(b) If the rate of increase in consumption stays constant at 1.7%, what was the exponential reserve for oil in 1990? How much longer will the reserve last from 2003?
(c) What considerations may affect these reserves over time?

11. Can our energy problems be solved by increasing the supply? [Thanks for the idea to Evar D. Nering of Arizona State University, in "The Mirage of a Growing Fuel Supply," *The New York Times* (June 4, 2001) Op-Ed page.]

(a) Suppose that we have a 100-year supply of a resource (such as oil, for which known world reserves will last less than 100 years at the current world rate of use). That is, the resource would last 100 years at the current rate of consumption. Suppose that the resource is consumed at a rate that increases 2.5% per year (this is the average increase in consumption for oil in the United States since 1973). How long will the resource last?
(b) Suppose that we underestimated the supply and actually have a 1000-year supply at the current rate of use. How long will that last if consumption increases 2.5% per year?
(c) Let's think big and suppose that there is 100 times as much of the resource as we thought—a 10,000-year supply. How long will that last if consumption increases 2.5% per year?

12. In this problem we explore the consequences of reducing the rate of growth of oil use. Suppose that we halve the growth rate from the 2.5% per year given in Exercise 11 to 1.25% per year. [Thanks for the idea to Evar D. Nering of Arizona State University, in "The Mirage of a Growing Fuel Supply," *The New York Times* (June 4, 2001) Op-Ed page.]

(a) How long does the 100-year supply last?
(b) How long does the 1000-year supply last?
(c) How long does the 10,000-year supply last?

13. We continue the ideas of Exercises 11 and 12 but with a more radical hypothesis.

(a) How long would the 100-year supply last if we reduced our consumption by just 1% per year—that is, if we used 1% less each year instead of 2.5% more?
(b) How about the 1000-year supply?

14. By the time there is concern about using up a nonrenewable resource, it may be too late. Suppose that a resource has a static reserve of 10,000 years but consumption is growing at 3.5% per year.

(a) How long will the resource last?
(b) How long before half the resource is gone?
(c) How much longer will the resource last if after half of it is gone, consumption is stabilized at the then-current level?
(d) What implications do you see to your answers?

Renewable Resources

15. A reproduction curve for a population is shown in the figure below. Estimate the equilibrium population size and the maximum sustainable yield. (The units are in thousands.)

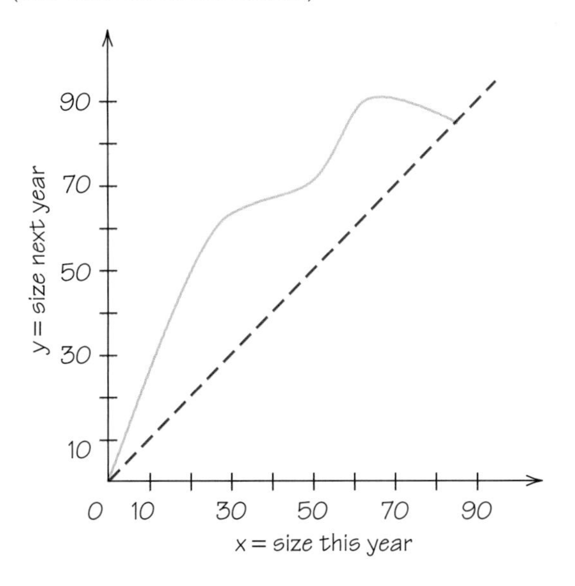

16. Suppose that a reproduction curve for a certain population is as in the accompanying figure, where the units are in thousands.

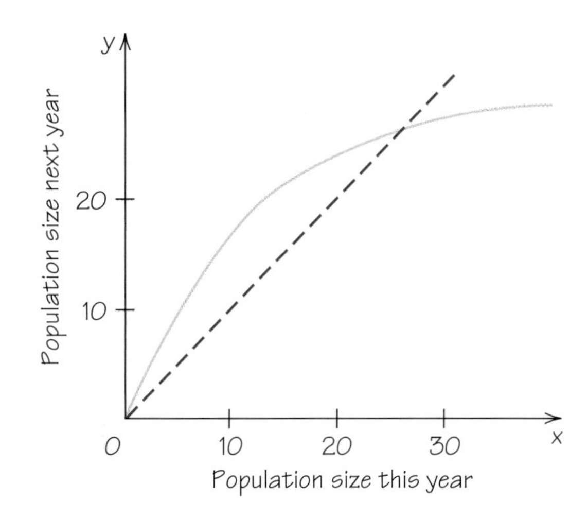

(a) Estimate the sustainable yield corresponding to a population of size 10,000 remaining after the harvest.
(b) Estimate the maximum sustainable yield.

Mathematical Chaos

17. In doubling on a "Stone Age" calculator (see the Example on page 842):

(a) What do you notice about the two sequences that were produced?
(b) Suppose that we had started with a different seed, say 0.39. What would happen as we iterate the doubling?
(c) Can you explain what you observe in parts (a) and (b) and give a general argument about why it is true?

18. Show that the logistic model on page 844 can be transformed algebraically to the form $f(x) = \lambda x(1 - x)$, where x is the fraction of the carrying capacity M.

19. You have seen in Figure 23.8 that a logistic model can result in a stable value, produce cycling between several values, or result in chaos. Other dynamical systems can exhibit similar behavior. We examine here the system in which we start with a positive whole number n and iterate the following function:

$$f(n) = \text{sum of the squares of the digits of } n$$

For example, we have $f(133) = 1^2 + 3^2 + 3^2 = 19$.

(a) Calculate what happens as f is applied repeatedly, starting with 133. What do you observe?
(b) Pick a number different from 133 and different from 1, and iterate f repeatedly. What do you observe?
(c) Why did we exclude 1 in part (b)?
(d) Try some other values. Can you offer a general conclusion? Can you offer an argument why your conclusion is correct?

20. The behavior of some very simple dynamical systems is still not completely known. Consider the system that starts with a positive whole number n and gives as the next number

$$f(n) = \begin{cases} 3n + 1, & \text{if } n \text{ is odd;} \\ n/2, & \text{if } n \text{ is even} \end{cases}$$

[This iterative function system was devised by Lothar O. Collatz, later of the University of Hamburg, during his student days before World War II. It is sometimes called the "$3n + 1$ problem" or the "Syracuse problem" (because it became popular at the Mathematics Department of Syracuse University), and the sequences generated are sometimes called "hailstone numbers."]

(a) Start with $n = 1$. What happens?
(b) Start with $n = 13$. What happens?
(c) Start with $n = 12$. What happens?

What you observe is known to happen for all $n < 10^{40}$, but after more than 60 years mathematicians have been unable to show yet that it happens for every n whatsoever.

21. (Requires programmable calculator, spreadsheet, or BASIC programming) A population model slightly different from the logistic model is given by the iterative function system

$$g(x) = x + rx(1 - x)$$

where x is a fraction of the limiting population and r is a growth rate.

(a) Set $r = 3$, start with $x = x_1 = 0.01$, and calculate the first 20 values $x_1, \ldots, x_{20}$.
(b) In part (a) you should have found $x_{10} = 0.722914$. Replace this value with the rounded-up value $x_{10} = 0.723$ and continue on to calculate x_{20}.
(c) Now replace x_{10} with the rounded-down value $x_{10} = 0.722$ and continue on to calculate x_{20}.

22. A dynamical system expressed as an iterated functional system $f(x)$ has an *equilibrium point* at a value x_0 if, once the system reaches x_0, it always stays at that value. In terms of an equation, an equilibrium point at x_0 if $f(x_0) = x_0$.

(a) For the dynamical system of Exercise 21, find all equilibrium points.
(b) For the logistic population model of Exercise 18, find all equilibrium points.

■ **23.** (Requires programmable calculator, spreadsheet, BASIC programming, or preferably use of software available under Suggested Web Sites) The behavior of the logistic population model $f(x) = \lambda x(1 - x)$ depends on the value of the positive parameter λ. As λ increases from 0 to 4, the system changes from one behavior to another, through the following possible states:

- The population simply dies out.
- The population tends toward a nonzero equilibrium point.
- The population oscillates between 2 points.
- The population oscillates between 4 points, then 8 points, then 16 points, and so on.
- The population oscillates between numbers of points that are not powers of 2, until at last . . . the population oscillates between 3 values.
- The population behaves chaotically.

Explore what happens for various values of λ between 0 and 4, trying to identify where the shifts in the system's behavior take place.

Additional Exercises

◆ **24.** Is Warren Sanderson right about world population growth slowing down (see Spotlight 23.1)?

How much difference does it make in projections if we look at the world as a whole or break it down by countries or regions? Investigate these questions.

(a) In mid-2001, the population of the world was 6.137 billion and increasing at 1.35% per year. Project the population to mid-2020 (by which time you may have finished having children, if you do indeed reproduce) and to mid-2040 (by which time you may be thinking about when you can retire).

(b) What are the assumptions involved in your projections?

(c) You can make a more refined model to give more accurate answers by dividing the countries of the world into three groups that have differing rates of increase (see the table).

Group	Population mid-2001 (billions)	Rate of growth (%)
More-developed countries	1.193	0.1
Less-developed countries (excluding China)	3.671	1.6
China	1.273	0.9

Redo your projections in part (a) for the years 2020 and 2040 by projecting each group separately and adding the totals. Are there major differences from your projections in part (a)?

(d) Will the world be able to support the numbers of people that you project? What problems will these greater numbers of people cause? What could be done to avert those problems? Do you think that anything will be done before there is some kind of worldwide crisis?

(In an earlier edition of this book, using data from mid-1995, the population was 5.702 billion and growing at 1.5% per year. The answers to part (a) were 8.3 billion and 11.1 billion. The growth rates in the table for mid-1995 were 0.2%, 2.2%, and 1.1%, which led to answers in part (c) of 8.9 and 12.7 billion.)

WRITING PROJECTS

1. Based on the calculations you did in Exercise 24, write a short essay in the form of a guest editorial for a newspaper. Describe your projections and how you arrived at them, how serious a problem you think population growth is, what problems it is likely to cause, what you think needs to be done, and what the implications are for your own life.

2. Identify a particular regional, national, or world nonrenewable primary resource (e.g., coal) or secondary resource (e.g., electric power). Research how much of it is available now and what the current rate of consumption is. Determine the static reserve. Estimate the growth rate in consumption, taking into account human population increase, and determine the exponential reserve. What social and technological factors contribute to the increasing rate of consumption? Brainstorm how those factors could be changed.

3. Identify a particular regional, national, or world renewable resource (e.g., timber, clean drinking water). Research how much of it is produced now, how much is harvested now, and what the current rate of consumption is. Estimate the growth rate in consumption, taking into account human population increase. For how long can this resource continue to meet the demand? What social and technological factors contribute to the increasing rate of consumption? Brainstorm how those factors could be changed.

Answers to Skills Check Exercises

CHAPTER 1

1. a
2. c
3. a
4. b
5. a
6. b
7. c
8. b
9. c
10. c
11. a
12. a
13. c
14. b
15. a

CHAPTER 2

1. c
2. b
3. b
4. b
5. c
6. b
7. b
8. b
9. c
10. b
11. a
12. c
13. b
14. a
15. c

CHAPTER 3

1. c
2. c
3. c
4. b
5. a
6. b
7. b
8. b
9. c
10. b
11. b
12. c
13. a

CHAPTER 4

1. b
2. a
3. b
4. b
5. b
6. c
7. [Graph]
8. c
9. b
10. b
11. c
12. c
13. c
14. c

CHAPTER 5

1. b
2. a
3. c
4. c
5. c
6. a
7. b
8. b
9. c
10. a

CHAPTER 6

1. c
2. b
3. a
4. c
5. a
6. b
7. a
8. a
9. b
10. c
11. c

CHAPTER 7

1. a
2. c
3. c
4. c
5. c
6. a
7. b
8. a
9. c
10. b
11. c

CHAPTER 8

1. c
2. b
3. c
4. a
5. b
6. c
7. a
8. a
9. b
10. c
11. a
12. b

CHAPTER 9

1. a
2. c
3. b
4. b
5. b
6. c
7. b
8. a
9. b
10. a
11. b
12. b
13. c
14. b
15. b

CHAPTER 10

1. b
2. b
3. c
4. b
5. b
6. a
7. c
8. a
9. b
10. a
11. c
12. b
13. a
14. c
15. b

CHAPTER 11

1. b
2. c
3. a
4. a
5. a
6. c
7. a
8. a
9. c

10. a
11. b
12. c
13. b
14. a
15. b

CHAPTER 12

1. b
2. c
3. a
4. a
5. a
6. b
7. a
8. a
9. b
10. c
11. a
12. b
13. b
14. a
15. a
16. a

CHAPTER 13

1. b
2. c
3. c
4. b
5. b
6. b
7. a
8. a
9. b
10. b
11. c
12. a
13. b
14. c
15. c

CHAPTER 14

1. b
2. c
3. a
4. b
5. c
6. b
7. a
8. a
9. a
10. b
11. c
12. b
13. c
14. a
15. c

CHAPTER 15

1. c
2. a
3. b
4. a
5. a
6. b
7. c
8. a
9. a
10. b
11. a
12. b
13. c
14. c
15. a

CHAPTER 16

1. c
2. c
3. b
4. a
5. c
6. b
7. a
8. a
9. c
10. a
11. a
12. b
13. c
14. b
15. b

CHAPTER 17

1. b
2. b
3. c
4. a
5. c
6. c
7. c
8. b
9. b
10. b
11. a
12. b
13. b
14. a
15. a

CHAPTER 18

1. c
2. a
3. b
4. c
5. c
6. c
7. c
8. a
9. a
10. c
11. b
12. c
13. c
14. a
15. b

CHAPTER 19

1. a
2. a
3. a
4. b
5. b
6. c
7. b
8. b
9. c
10. b
11. b
12. a
13. a
14. c
15. a

CHAPTER 20

1. b
2. b
3. a
4. c
5. c
6. b
7. c
8. b
9. c
10. a
11. a
12. b
13. b
14. c
15. a

CHAPTER 21

1. c
2. a
3. c
4. b
5. c
6. c
7. a
8. a
9. b
10. a
11. b
12. b
13. a
14. b
15. c
16. b
17. b
18. c
19. b
20. b

CHAPTER 22

1. b
2. c
3. a
4. b
5. a
6. a

7. b
8. c
9. b
10. c
11. c
12. b
13. a
14. a
15. b

CHAPTER 23

1. b
2. a
3. b?
4. b
5. c
6. a
7. c
8. a
9. c
10. b
11. b
12. a
13. a
14. c
15. a

Answers to Odd-Numbered Exercises

CHAPTER 1

1. *A*: 1; *B*: 3; *C*: 4; *D*: 2; *E*: 0; *F*: 2; *G*: 2. The graph shows that geographically *E* is isolated, perhaps on an island.

3. No, the loop does not join two different vertices.

5. **(a)**

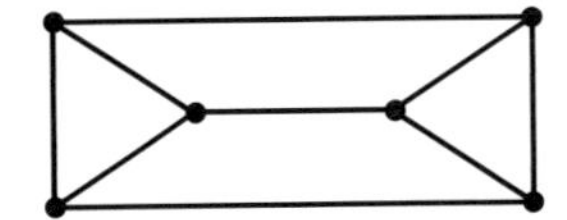

(b)

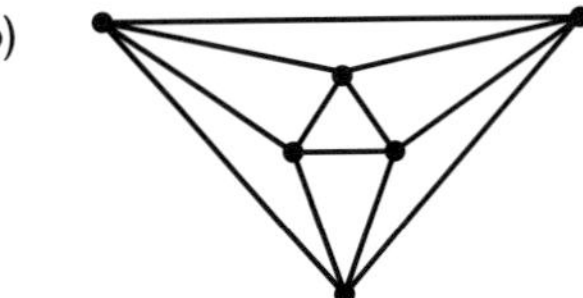

7. Remove the edges dotted in the figure below and the remaining graph will be disconnected.

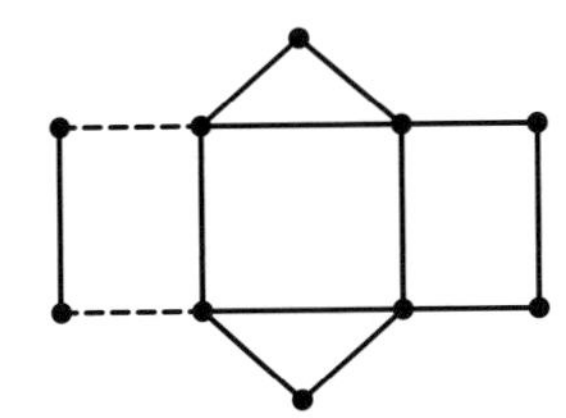

9. Discussion. Answers will vary.

11. **(a)** Yes.

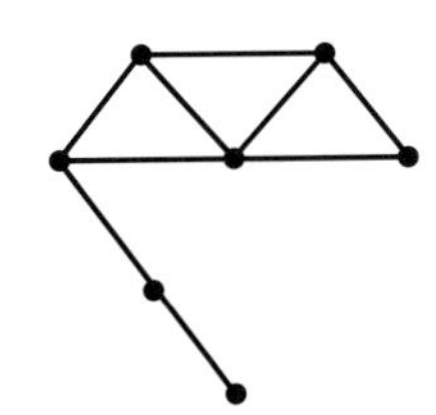

(b) Yes.

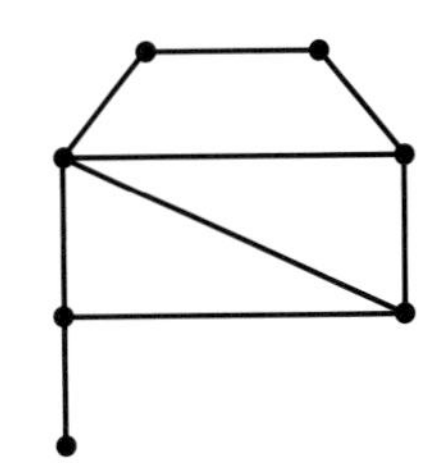

13. The supervisor is not satisfied because all of the edges are not traveled upon by the postal worker. The worker is unhappy because the end of the worker's route wasn't the same point as where the worker began. The original job description is unrealistic because there is no Euler circuit in the graph.

15. Such a route, starting and ending at the same vertex, is possible, but answers will vary.

17.

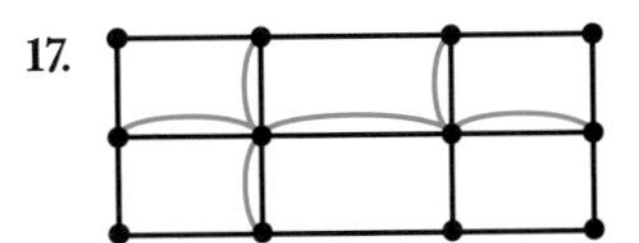

19. **(a)** The largest number of such paths is 3. One set of such paths would be: *AEGH, ADCH,* and *ABFH.*
(b) This task can be simplified by noting the many symmetries in the graph.

21. Edges 1 or 10 could be chosen, but not edge 2.

23.

25. Two edges need to be dropped to produce a graph with an Euler circuit. Persons who parked along these stretches of sidewalk without putting coins in the meters would not need to fear that they would get tickets.

27.

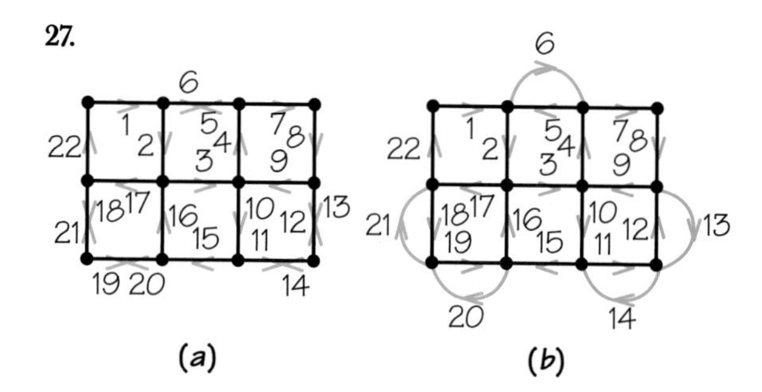

29.

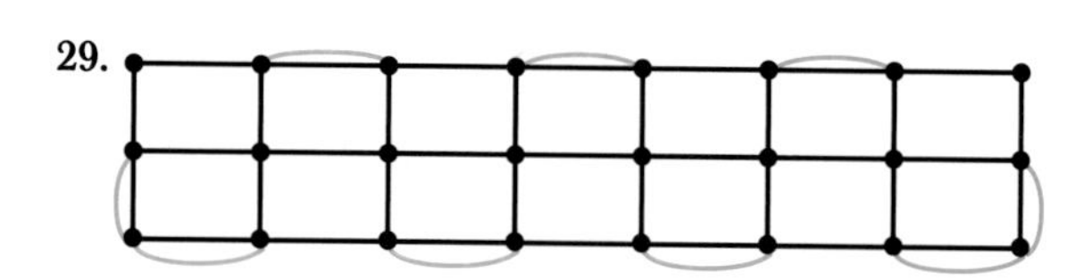

Nine is the best one can do.

31.

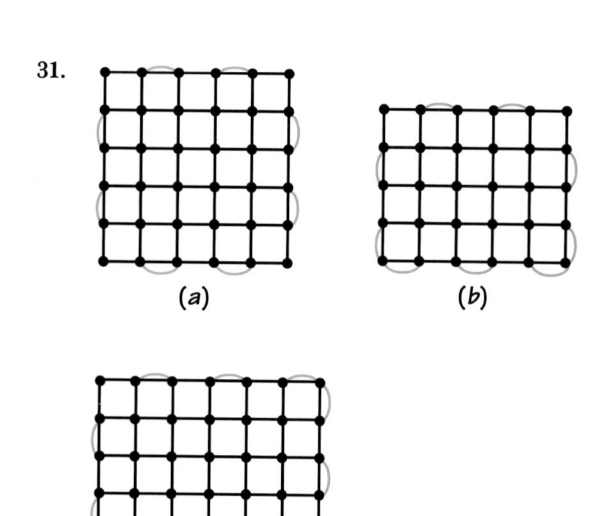

33.

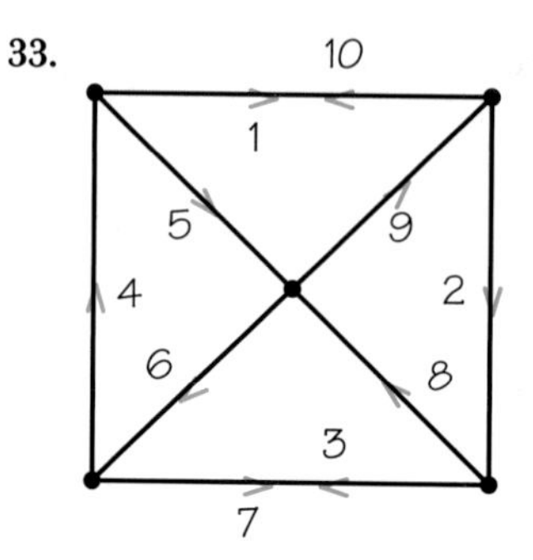

35. Answers will vary but a solution with three reuses can be found.

37. Answers will vary but a solution of length 44,000 feet can be found.

39. Both graphs (b) and (c) have Euler circuits. The valences of all of the vertices in (a) are odd, which makes it impossible to have an Euler circuit there.

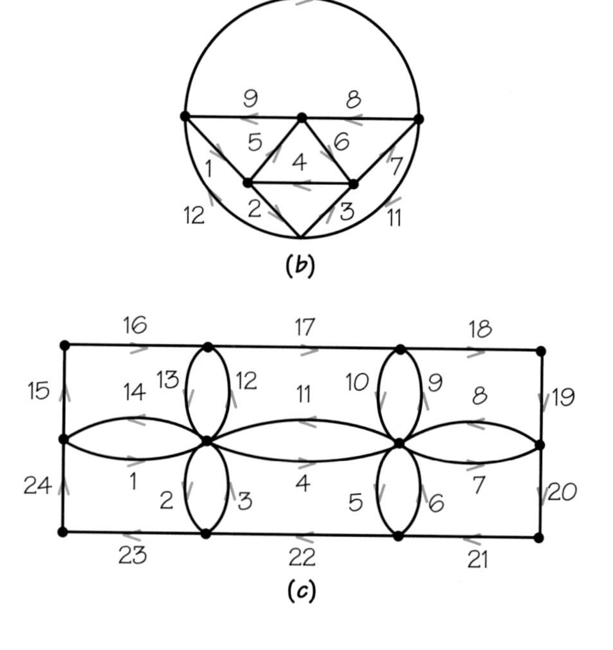

41. **(a)** Both graphs have two vertices of valence 1 and four vertices of valence 2.

(b) The graph in (b) is connected but the graph in (a) is not.

43. Yes, because each street is represented by two edges, every vertex has even valence.

45. **(a)**

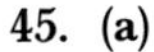

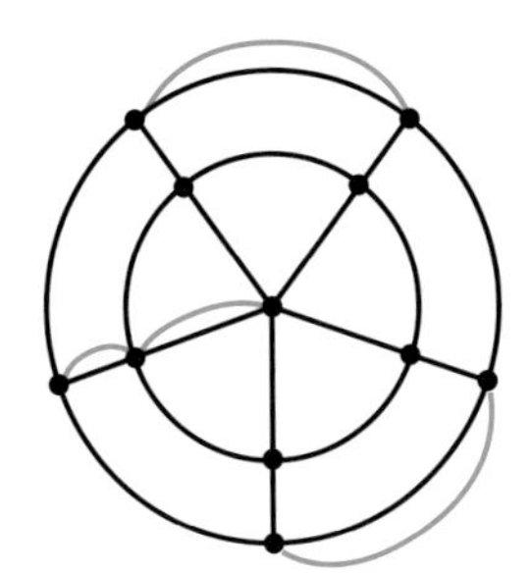

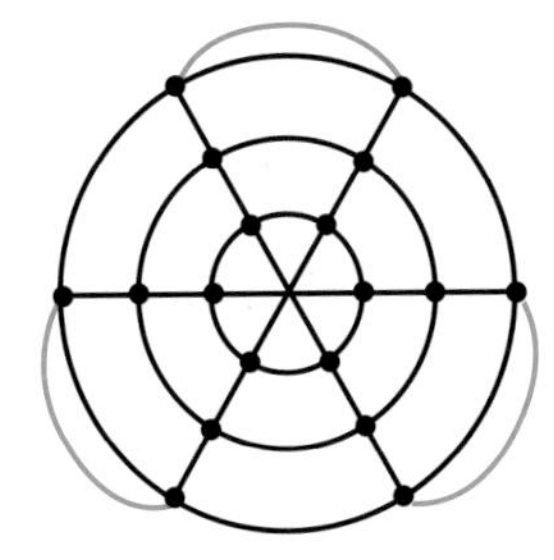

(b) The best eulerization for the four-circle four-ray case adds two edges.
(c) *Hint:* Consider the cases where *r* is even and odd separately.

47. A graph with six vertices where each vertex is joined to every other vertex will have valence 5 for each vertex.

49. When you attach a new edge to an existing graph it gets attached at two ends. At each of its ends, it makes the valence of the existing vertex go up by one. Thus the increase in the sum of the valences is two. Therefore, if the graph had an even sum of the valences before, it still does, and if its valence sum was odd before, it still is.

51.

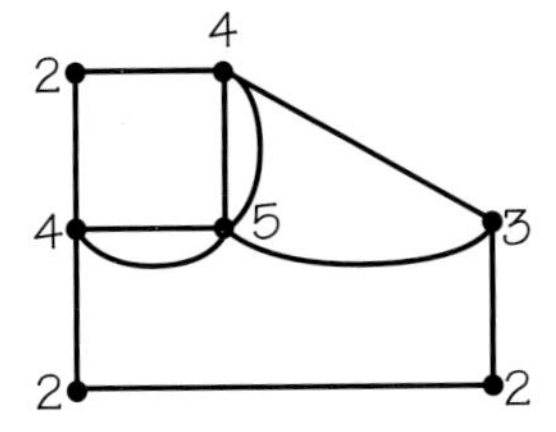

The graph is connected.

53. Discussion. Answers will vary.

CHAPTER 2

1. **(a)** $X_4X_2X_5X_6X_1X_3X_4$
(b) $X_4X_3X_2X_1X_6X_7X_8X_9X_{10}X_{11}X_{12}X_5X_4$
(c) $X_4X_3X_1X_2X_7X_6X_9X_8X_5X_4$
(d) $X_4X_7X_6X_1X_2X_5X_8X_3X_4$
(e) $X_4X_3X_2X_8X_1X_{10}X_7X_6X_9X_5X_4$

3. **(a)** For (a) and (c) no Hamiltonian circuit would exist, while for (b), (d), and (e) a Hamiltonian circuit would exist.
(b) The removal of a vertex might correspond to the failure of the equipment at that site.

5. **(a)** a. Add edge *AB*.
b. Add edge X_1X_3.

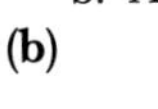

(b)

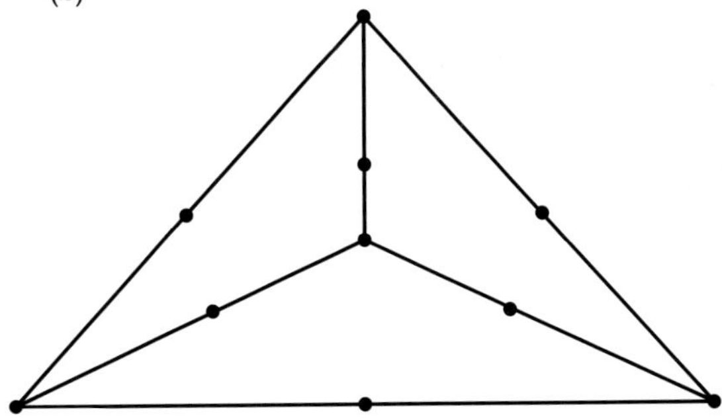

7. Other Hamiltonian circuits include *ABIGDCEFHA* and *ABDCEFGIHA*.

9. **(a)** Any Hamiltonian circuit would have to use the edges X_4X_1 and X_4X_6 (to visit X_4), X_1X_2 and X_2X_3 (to visit X_2), X_7X_5 and X_5X_1 (to visit X_5). This forces three edges to be used at X_1, which is not possible in a Hamiltonian circuit.
(b) Any Hamiltonian circuit that included X_4X_5 and X_6X_7 could not visit X_8 and X_9.

11. **(a)** No Hamiltonian circuit.
(b) No Hamiltonian circuit.
(c) No Hamiltonian circuit.

13. **(a)** Yes. **(b)** No. **(c)** A worker who must do an inspection of all sites represented by vertices of a graph on her way to work would need to find a Hamiltonian path from her home vertex to her work vertex.

15. The *n*-cube has 2^n vertices, and the number of edges of the *n*-cube is equal to twice the number of edges of an $(n - 1)$-cube plus 2^{n-1}. A formula for this number is $n2^{n-1}$.

17. **(a)** Hamiltonian circuit, yes; Euler circuit, no.
(b) Hamiltonian circuit, yes; Euler circuit, no.
(c) Hamiltonian circuit, yes; Euler circuit, no.
(d) Hamiltonian circuit, no; Euler circuit, no.

19. The new system is an improvement since it codes 676 locations compared with 504 for the old system. This is 172 more locations.

21. **(a)** $9 \cdot 8 \cdot 7 \cdot 6 \cdot 5 = 15{,}120$
(b) $(26)(26)(26) = 17{,}576$

23. **(a)** $7 \times 6 \times 5 \times 4 \times 3 = 2520$
(b) $7 \times 7 \times 7 \times 7 \times 7 = 16{,}807$
(c) $(7)^5 - 7 = 16{,}800$

25. **(a)** $26(26)(26)(10)(10)(10) - (26)(26)(26) = (26)^3(10^3 - 1)$
(b) Answers will vary.

27. These graphs have 6, 10, and 15 edges, respectively. The n-vertex complete graph has $[n(n-1)]/2$ edges. The number of TSP tours is 3, 12, and 60, respectively.

29. **(a)**

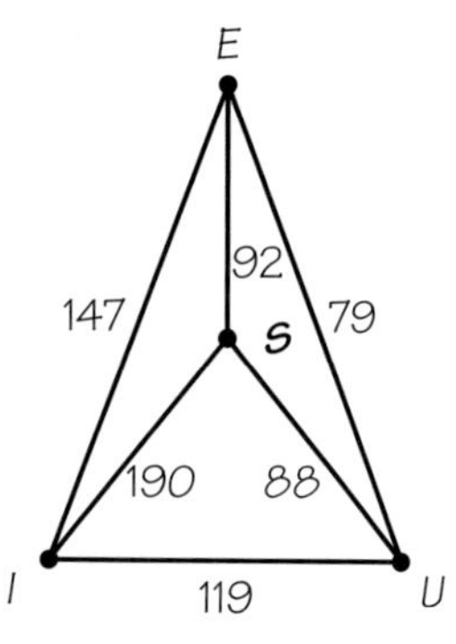

(b) (1) *UISEU;* mileage = 119 + 190 + 92 + 79 = 480
(2) *USIEU;* mileage = 88 + 190 + 147 + 79 = 504
(3) *UIESU;* mileage = 119 + 147 + 92 + 88 = 446
(c) *UIESE* (Tour 3)
(d) No.
(e) Starting from *U*, one gets [see answer (b)] Tour 1. From *S* one gets Tour 2; from *E* one gets Tour 2; and from *I* one gets Tour 1.
(f) Tour 2. No.

31. *FMCRF* gets her home in 32 minutes.

33. *MACBM* takes 345 minutes to traverse.

35. A traveling salesman problem.

37. The complete graph shown has a different nearest-neighbor tour that starts at *A* (*AEDBCA*), a sorted-edges tour (*AEDCBA*), and a cheaper tour (*ADBECA*).

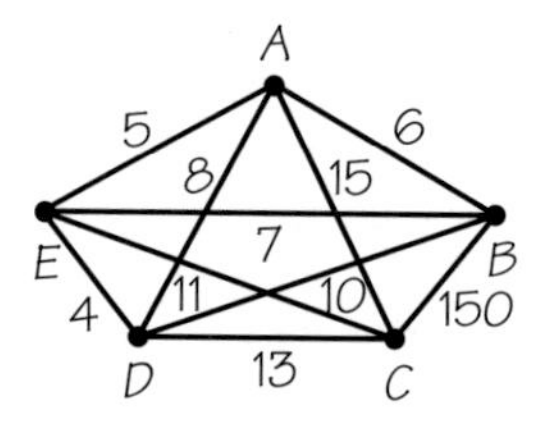

39. The optimal tour is the same but its cost is now $4200 + 10(30) = 4500$.

41. **(a)** a. Not a tree because there is a circuit. Also, the wiggled edges do not include all vertices of the graph.
b. The circuit does not include all the vertices of the graph.
(b) a. The tree does not include all vertices of the graph.
b. Not a circuit.
(c) a. Not a tree.
b. Not a circuit.
(d) a. Not a tree.
b. Not a circuit.

43. **(a)** 1, 2, 3, 4, 5, 8
(b) 1, 1, 1, 2, 2, 3, 3, 4, 5, 6, 6
(c) 1, 1, 1, 2, 2, 2, 2, 2, 3, 3, 3, 3, 4, 4, 4, 5, 5, 6, 7
(d) 1, 2, 2, 3, 3, 3, 4, 5, 5, 5, 6, 6
The cost is found by adding the numbers given.

45. The spanning tree will have 26 vertices. *H* also will have 26 vertices. The exact number of edges in *H* cannot be determined, but *H* has at least 25 edges and no more edges than the complete graph having 26 vertices.

47. Yes.

49.

	A	B	C	–
A	0	15	14	6
B	15	0	17	9
C	14	17	0	8
–	6	9	8	0

51. Yes. Change all the weights to negative numbers and apply Kruskal's algorithm. The resulting tree works, and the maximum cost is the negative of the answer you get. If the numbers on the edges represent subsidies for using the edges, one might be interested in finding a maximum-cost spanning tree.

53. A negative weight on an edge is conceivable, perhaps a subsidization payment. Kruskal's algorithm would still apply.

55. **(a)** True **(b)** False **(c)** True **(d)** False **(e)** Not necessarily

57. Two different trees with the same cost are shown:

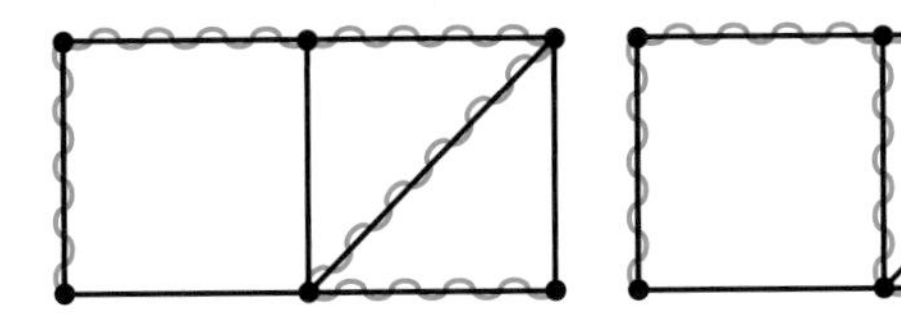

59.

A B C D E

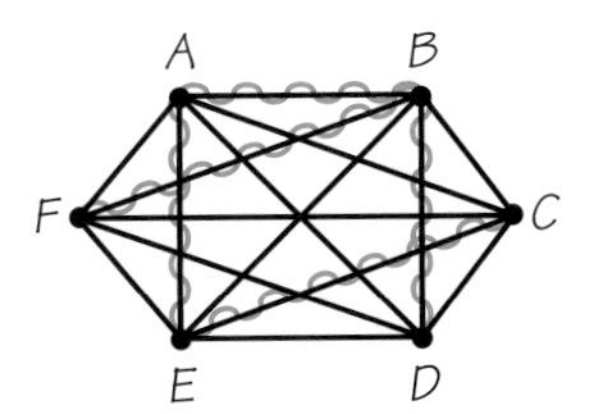

61.

T_1 12 → T_2 12
T_3 23 → T_4 1
T_5 1 → T_6 23

63. Answers will vary.

65. The critical path is T_1, T_5, T_7 with length 30. If T_5's time is reduced by 2, T_1, T_4, and T_7 will not also be a critical path, and the new lengths of both of these paths will be 28.

67.

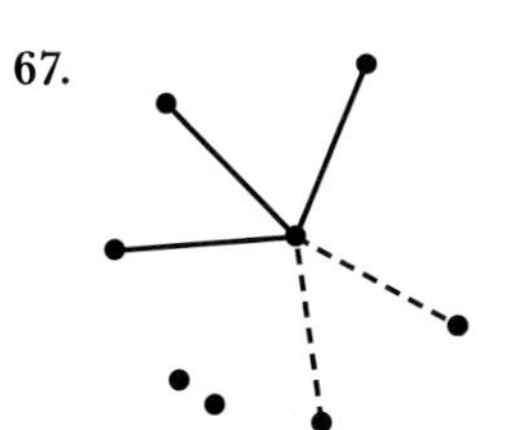

69. **(a)** Yes.

(b)

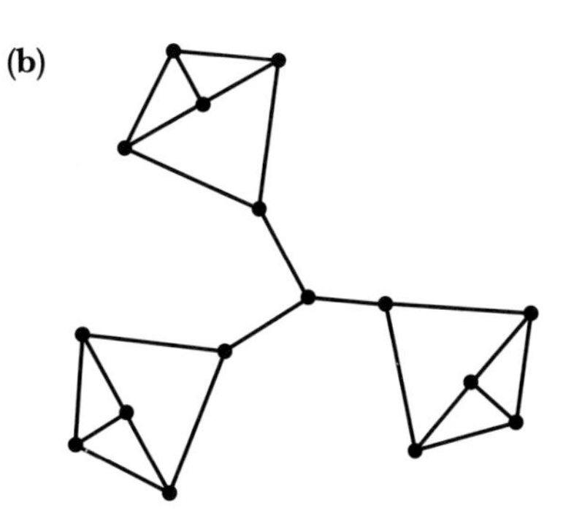

(c) Answers will vary.

(d)

71. With no other restrictions, 10^7. With no other restrictions, 9×10^2.

73. Optimal tour is the same, but its cost has now doubled to 9040.

75. **(a)** Answers will vary for each edge, but the reason it is possible to find such trees is that each edge is an edge of some circuit.
(b) The number of edges in every spanning tree is five, one less than the number of vertices in the graph.
(c) Every spanning tree must include the edge joining vertices *C* and *D*, since this edge does not belong to any (simple) circuit in the graph.

CHAPTER 3

1. **(a)** Operating room schedules, doctor schedules, emergency room staffing schedules, etc.
(b) Schedules for the trains or buses and their crews, etc.
(c) Scheduling runway use, reservation agents, food service for planes, etc.
(d) Schedules for each mechanic, radiator repair, etc.
(e) Schedules for waiters, waitresses, dish cleaning, etc.
(f) Schedules for washing clothes, cleaning rooms, dusting, etc.

(g) Schedules for bus run, recess duty, etc.
(h) Day, night, afternoon shift schedules, etc.
(i) Firefighter shift schedules, schedules for checking whether equipment on trucks is in repair, etc.

3. Jocelyn must pack, get to the airport, make various connections, perhaps kennel her dog, etc. Processors include plane, bus, taxi (to get to the airport), and Jocelyn herself. Unless she can get a friend to help her pack or take her dog to the kennel, none of the tasks can be done simultaneously.

5. **(a)** (i) Processor 1: T_1, T_3, T_5, idle 33 to 47. Processor 2: T_2, T_4, T_6.
(ii) Processor 1: T_1, T_3, T_4, T_6. Processor 2: T_2, T_6, idle 38 to 42. The second is optimal. The critical path is T_2, T_6 and has length 38. No schedule on two machines finishes by 38 because the total time for all the tasks is 80. Thus, on two machines no earlier time than $80/2 = 40$ is possible.
(b) The optimum completion time before and after the two task times are switched is the same.

7. **(a)** Processor 1: T_1, T_2, T_3, T_5, T_7; Processor 2: Idle 0 to 2, T_4, T_6, idle 4 to 5.
(b) Processor 1: T_1, T_2, T_3, T_6, T_7; Processor 2: Idle 0 to 2, T_4, T_5, idle 4 to 5.
(c) Yes.
(d) No.

9. **(a)** T_1, T_2, T_3, and T_6 are ready at time 0.
(b) No tasks require that T_1 and T_6 be done before these other tasks can begin.
(c) The critical path consists only of T_6 and has length 20.
(d) Processor 1: T_1, T_6; Processor 2: T_2, T_4, idle from 18 to 30; Processor 3: T_3, T_5, idle from 12 to 30.
(e) No.
(f) Processor 1: T_6, idle from 20 to 22; Processor 2: T_3, T_5, T_1; Processor 3: T_2, T_3, idle from 18 to 22.
(g) Yes.
(h) Another list leading to the same optimal schedule is T_6, T_3, T_2, T_4, T_5, T_1.

11. **(a)** Processor 1: T_1, T_6, idle 15 to 21, T_7, idle 27 to 31; Processor 2: T_2, T_5, T_8; Processor 3: T_3, T_4, idle from 13 to 31.
(b) Processor 1: T_1, T_6, idle 15 to 21, T_7, idle 27 to 31; Processor 2: T_3, T_4, idle from 13 to 21, T_8; Processor 3: T_2, T_5, idle from 21 to 31.
(c) Processor 1: T_1, T_6, T_5, T_7, T_8; Processor 2: T_3, T_6, idle from 8 to 46; Processor 3: T_4, idle from 8 to 46.

13. **(a)** No. Consider the tasks that begin after the stretch where all machines are idle. Pick one of these tasks T and say machine i was the machine that it was given to. This task was ready for machine i just prior to when it began T because no task was just being completed on any other machine at this time because they were all idle. Thus, T should have begun earlier on machine i.
(b) This schedule can not be valid because T_1 should have been assigned to Machine 1 at time 0.
(c) Use the digraph with no edges and the list: T_2, T_1, T_3, T_4, T_5.

15. **(a)** $5! = 120$
(b) No. Whatever list is used T_1 must be assigned to the first machine at time 0 because it is the only task ready at time 0.
(c) No. First, while Processor 1 works on T_1, Processor 2 must be idle. Second, the task times are integers with sum 31. If there are two processors, one of the processors must have idle time since when 2 divides 31 there is a remainder of 1.
(d) No.

17. Using the order-requirement digraph shown and any list with one or more processors yields the same schedule:

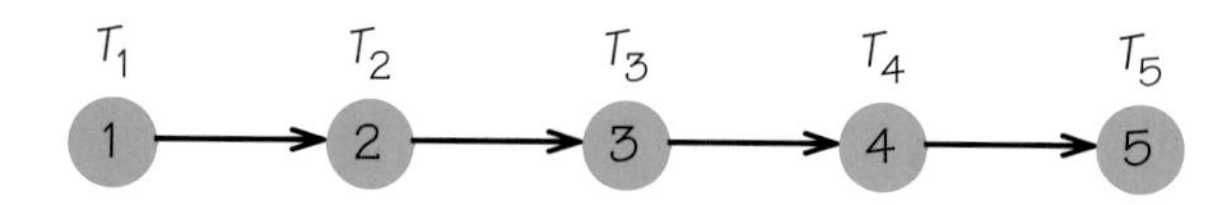

19. **(a)** One reasonable possibility is (time in min):

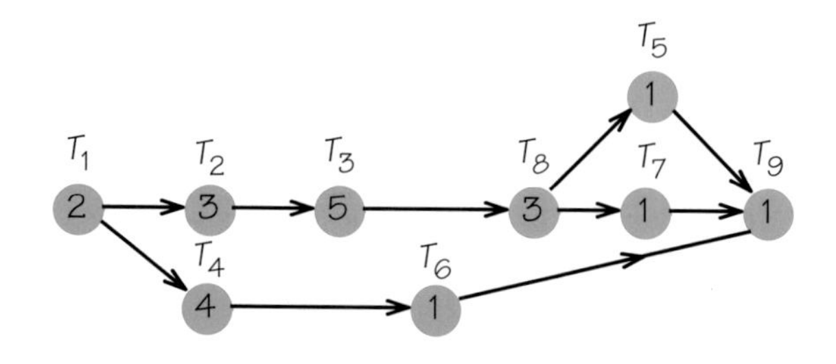

The earliest completion time is 15.

(b) The decreasing-time list is T_3, T_4, T_2, T_8, T_1, T_5, T_6, T_7, T_9. The schedule is Processor 1: T_1, T_4, T_6, idle 7 to 10, T_8, T_5, T_9; Processor 2: idle 0 to 2, T_2, T_3, idle 10 to 13, T_7, idle 14 to 15.

21. No. At time 11, T_4 should been assigned to Machine 1 because Machine 1 was free at this time and T_4 was ready.

23. **(a)** Task times: $T_1 = 3$, $T_2 = 3$, $T_3 = 2$, $T_4 = 3$, $T_5 = 3$, $T_6 = 4$, $T_7 = 5$, $T_8 = 3$, $T_9 = 2$, $T_{10} = 1$, $T_{11} = 1$, and $T_{12} = 3$. This schedule would be produced from the list: $T_1, T_3, T_2, T_5, T_4, T_6, T_7, T_8, T_{11}, T_{12}, T_9, T_{10}$.
(b) Task times: $T_1 = 3$, $T_2 = 3$, $T_3 = 3$, $T_4 = 2$, $T_5 = 2$, $T_6 = 4$, $T_7 = 3$, $T_8 = 5$, $T_9 = 8$, $T_{10} = 4$, $T_{11} = 7$, $T_{12} = 9$, $T_{13} = 3$. This schedule would be produced from the list: $T_1, T_5, T_7, T_4, T_3, T_6, T_{11}, T_8, T_{12}, T_9, T_2, T_{10}, T_{13}$.

25. **(a)** (i) Processor 1: T_1, T_3, T_5, T_7, idle from 16 to 20; Processor 2: T_2, T_4, T_6, T_8.
(ii) Processor 1: T_8, T_5, T_4, T_1; Processor 2: T_7, T_6, T_3, T_2.
(b) The schedule in (ii) is optimal.

27. Such criteria include decreasing length of the times of the tasks, order of size of financial gains when each task is finished, and increasing length of the times of the tasks.

29. **(a)** Machine 1: T_1, T_6, T_{10}, idle from 8 to 9; Machine 2: T_3, T_4, T_{11}, T_{12}; Machine 3: T_2, T_7, idle from 8 to 9; Machine 4: T_5, T_8, T_9, idle from 8 to 9.
(b) Machine 1: T_1, T_8, T_{10}; Machine 2: T_5, T_6, T_2, T_{13}; Machine 3: T_7, T_9; Machine 4: T_4, T_{11}, idle from 9 to 12; Machine 5: T_3, T_9, idle from 11 to 12.

The schedule in (a) had to have idle time because the total task time was 33 and 33 is not exactly divisible by 4. The schedule in (b) had to have idle time because the total task time was 56 and 56 is not exactly divisible by 5.

31. List for (i) yields (with items coded by task time): Machine 1: 12, 9, 15, idle from 36 to 50; Machine 2: 7, 10 13, 20. List for (ii) yields: Machine 1: 12, 13, 20; Machine 2: 7, 9 15, 10, idle from 41 to 45. List for (iii) yields: Machine 1: 20 12, 9, idle from 41 to 45; Machine 2: 15, 13, 10, 7. These schedules complete earlier than those where precedence constraints hold. An optimal schedule is possible, however: Machine 1: 20, 10, 13; Machine 2: 12 15, 9, 7. The associated list is: $T_6, T_1, T_5, T_7, T_2, T_4, T_3$.

33. **(a)** The tasks are scheduled on the machines as follows: Processor 1: 12, 13, 45, 34, 63, 43, 16, idle 226 to 298; Processor 2: 23, 24, 23, 53, 25, 74, 76; Processor 3: 32, 23, 14, 21, 18, 47, 23, 43, 16, idle 237 to 298.
(b) The tasks are scheduled on the machines as follows: Processor 1: 12, 24, 14, 34, 25, 23, 16, 16, 76, idle 183 to 240; Processor 2: 23, 23, 21, 63, 43, idle 173 to 240; Processor 3: 32, 23, 53, 74, idle 182 to 240; Processor 4: 13, 45, 18, 47, 43, idle 166 to 240.
(c) The decreasing-time list is 76, 74, 63, 53, 47, 45, 43, 43, 34, 32, 25, 24, 23, 23, 23, 23, 21, 18, 16, 16, 14, 13, 12.

The tasks are scheduled on three machines as follows: Processor 1: 76, 45, 43, 24, 23, 18, 16, 13; Processor 2: 74, 47, 34, 32, 23, 21, 14, 12, idle 257 to 258; Processor 3: 63, 53, 43, 25, 23, 23, 16, idle 246 to 258.

The tasks are scheduled on four machines as follows: Processor 1: 76, 43, 24, 23, 16, idle 182 to 194; Processor 2: 74, 43, 25, 23, 16, 13; Processor 3: 63, 45, 32, 23, 18, 12, idle 193 to 194; Processor 4: 53, 47, 34, 23, 21, 14, idle 192 to 194.
(d) The new decreasing-time list is 84, 82, 71, 61, 55, 45, 43, 43, 34, 32, 25, 24, 23, 23, 23, 23, 21, 18, 16, 16, 14, 13, 12.

The tasks are scheduled as follows: Processor 1: 84, 45, 43, 25, 23, 23, 16, 12; Processor 2: 82, 55, 34, 32, 23, 18, 14, 13; Processor 3: 71, 61, 43, 24, 23, 21, 16, idle 259 to 271.

35. Examples include jobs in a videotape copying shop, data entry tasks in a computer system, and scheduling nonemergency operations in an operating room. These situations may have tasks with different priorities, but there is no physical reason for the tasks not to be independent, as would be the case with putting on a roof before a house had walls erected.

37. **(a)** Each task heads a path of length equal to the time to do that task.
(b) (i) The worst finish time is
$$(2 - \tfrac{1}{3})(450) = 750$$
(ii) The worst finish time, if the decreasing-time list is used, is
$$\left[\tfrac{4}{3} - 1/(3)(3)\right](450) = 550$$

39. The times to photocopy the manuscripts, in decreasing order, are 120, 96, 96, 88, 80, 76, 64, 64, 60, 60, 56,

48, 40, 32. Packing these in bins of size 120 yields Bin 1: 120; Bin 2: 96; Bin 3: 96; Bin 4: 88, 32; Bin 5: 80, 40; Bin 6: 76; Bin 7: 64, 56; Bin 8: 64, 48; Bin 9: 60, 60. Nine photocopy machines are needed to finish within 2 minutes using FFD. The number of bins would not change, but the placement of the items in the bins would differ for worst-fit decreasing.

41. **(a)** Using the next-fit algorithm, the bins are filled as follows: Bin 1: 12, 15; Bin 2: 16, 12; Bin 3: 9, 11, 15; Bin 4: 17, 12; Bin 5: 14, 17; Bin 6: 18; Bin 7: 19; Bin 8: 21; Bin 9: 31; Bin 10: 7, 21; Bin 11: 9, 23; Bin 12: 24; Bin 13: 15, 16; Bin 14: 12, 9, 8; Bin 15: 27; Bin 16: 22; Bin 17: 18.

(b) The decreasing list is 31, 27, 24, 23, 22, 21, 21, 19, 18, 18, 17, 17, 16, 16, 15, 15, 15, 14, 12, 12, 12, 12, 11, 9, 9, 9, 8, 7.

The next-fit decreasing schedule is Bin 1: 31; Bin 2: 27; Bin 3: 24; Bin 4: 23; Bin 5: 22; Bin 6: 21; Bin 7: 21; Bin 8: 19; Bin 9: 18, 18; Bin 10: 17, 17; Bin 11: 16, 16; Bin 12: 15, 15; Bin 13: 15, 14; Bin 14: 12, 12, 12; Bin 15: 12, 11, 9; Bin 16: 9, 9, 8, 7.

(c) The worst-fit schedule using the original list is Bin 1: 12, 15, 9; Bin 2: 16, 12; Bin 3: 11, 15; Bin 4: 17, 12; Bin 5: 14, 17; Bin 6: 18, 7; Bin 7: 19, 9; Bin 8: 21, 15; Bin 9: 31; Bin 10: 21, 9; Bin 11: 23, 8; Bin 12: 24; Bin 13: 16, 12; Bin 14: 27; Bin 15: 22; Bin 16: 18.

(d) The worst-fit decreasing schedule would be Bin 1: 31; Bin 2: 27, 9; Bin 3: 24, 12; Bin 4: 23, 12; Bin 5: 22, 12; Bin 6: 21, 15; Bin 7: 21, 15; Bin 8: 19, 17; Bin 9: 18, 18; Bin 10: 17, 16; Bin 11: 16, 12, 8; Bin 12: 15, 14, 7; Bin 13: 11, 9, 9.

43. Discussion. Answers will vary.

45. The total performance time exceeds what will fit on four disks. Using FFD, one can fit the music on five disks.

47. The proposed heuristic may fill many bins to capacity, but the computation to find weights summing to exactly W may be very time-consuming.

49. There is an example of a bin-packing problem for which a given list takes a certain number of bins, and when an item is deleted from the list, more bins are required. In this example, the deleted item is not first in the list.

51. **(a)** The vertices in graphs (d), (e), and (f) can be colored with three colors, but the vertices in graphs (a), (b), and (c) cannot be colored with four colors.

(b) The vertices in all the graphs except (c) can be colored with four colors.

(c) The chromatic number for graph (a) is 4, for (b) is 4, for (c) is 5, for (d) is 2, for (e) is 3, and for (f) is 2.

53. **(a)** Construct the graph shown below:

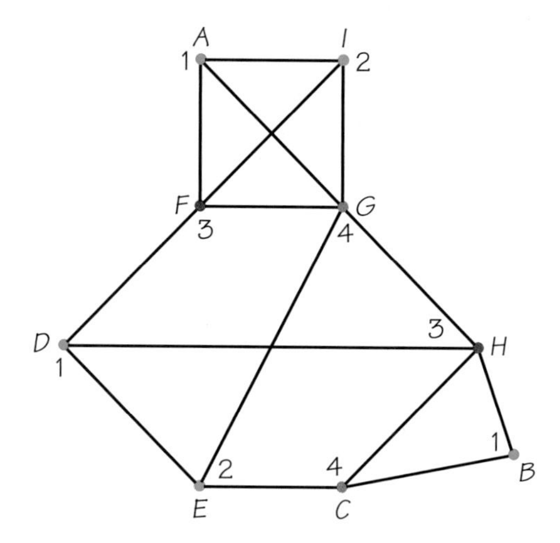

The vertices of this graph can be colored with no fewer than four colors (1, 2, 3, 4 are used to denote the colors in the figure). Hence, four tanks can be used to display the fish.

(b) The coloring in (a) shows that one can display two types of fish in three of the tanks, and three types of fish in one tank. Since 4 does not divide 9, one cannot do better.

55. **(a)** The graph for this situation is shown below. The vertices can be labeled with the colors 1, 2, 3 as shown.

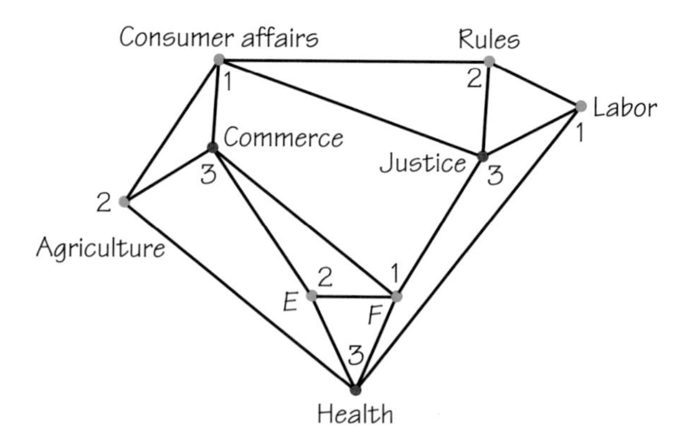

(b) Since the vertices can be colored with three colors (and no fewer), the minimum number of time slots for scheduling the committees is three.
(c) The committees can be scheduled in three rooms during each time slot. This might be significant if there were only three rooms that had microphone systems.

57. **(a)** Three time slots. To solve this problem, draw a graph by joining the vertices representing two committees if there is no × in the row and column of the table for these two committees.
(b) It is possible to three-color this graph so that each of the three colors is used three times. This means that one needs three rooms to arrange the scheduling of the nine committees.

59. Start at any vertex of the tree and label this vertex with color 1; color any vertex attached by an edge to this vertex with color 2. Continue to color the vertices in the tree in this manner, alternating the use of colors. If some vertex were attached to both a vertex colored 1 and another vertex colored 2, at some stage this would imply the graph had a circuit (of odd length), which is not possible since trees have no circuits of any length.

61. For graph (a), four colors; graph (b) four colors; graph (c) three colors; graph (d) five colors; graph (e) three colors; graph (f) five colors; graph (g) five colors; graph (h) four colors. The minimum edge-coloring number of any graph is either the maximal valence or one more than the maximal valence. This theorem is due to the Russian mathematician Vizing.

63. **(a)** Graph (a) four colors; graph (b) two colors; graph (c) four colors; graph (d) four colors; graph (e) two colors; graph (f) three colors.
(b) Coloring the maps of countries in an atlas would be one application of face colorings of graphs.

65. No. Machine 2 was idle at the time T_{10} was started, so T_{10} should have been assigned to Machine 2. This schedule could not arise from independent tasks, since when Machine 2 finished task 2, it could have started on task 8, 9, 10, or 11.

67. **(a)** Packing boxes of the same height into crates; packing want ads into a newspaper page.
(b) We assume, without loss of generality, $p \geq q$. One heuristic, similar to first-fit, orders the rectangles $p \times q$ as in a dictionary (i.e., $p \times q$ listed prior to $r \times s$ if $p > r$ or $p = r$ and $q \geq s$). It then puts the rectangles in place in layers in a first-fit manner; that is, do not put a rectangle into a second layer until all positions on the first layer are filled. However, extra room in the first layer is "wasted."
(c) The problem of packing rectangles of width 1 in an $m \times 1$ rectangle is a special case of the two-dimensional problem, equivalent to the bin-packing problem we have discussed.
(d) Two 1×10 rectangles cannot be packed into a 5×4 rectangle, even though there would be an area of 20 in this rectangle.

69. **(a)** The schedule with four secretaries is as follows: Processor 1: 25, 36, 15, 15, 19, 15, 27; Processor 2: 18, 32, 18, 31, 30, 18; Processor 3: 13, 30, 17, 12, 18, 16, 16, 16, 14; Processor 4: 19, 12, 25, 26, 18, 12, 24, 9.
The schedule with five secretaries is as follows: Processor 1: 25, 25, 31, 12, 16, 14; Processor 2: 18, 12, 17, 12, 15, 30, 9; Processor 3: 13, 32, 26, 16, 15, 18; Processor 4: 19, 36, 18, 19, 24; Processor 5: 30, 18, 15, 18, 16, 27.
(b) The decreasing-time list is 36, 32, 31, 30, 30, 27, 26, 25, 25, 24, 19, 19, 18, 18, 18, 18, 18, 17, 16, 16, 16, 15, 15, 15, 14, 13, 12, 12, 12, 9.
The schedule using this list on four processors would be Processor 1: 36, 25, 19, 18, 17, 16, 13, 9; Processor 2: 32, 26, 25, 18, 16, 15, 12; Processor 3: 31, 27, 24, 18, 16, 15, 12, 12; Processor 4: 30, 30, 19, 18, 18, 15, 14.
The schedule using this list on five processors would be Processor 1: 36, 24, 18, 16, 14, 12; Processor 2: 32, 25, 18, 18, 15, 12; Processor 3: 31, 25, 19, 18, 15, 9; Processor 4: 30, 27, 18, 17, 15, 13; Processor 5: 30, 26, 19, 16, 16, 12.
(c) The five-processor decreasing-time schedule is optimal (time 120), but the four-processor decreasing-time schedule is not. One can see this since when the task of length 17 scheduled on processor 1 and the task of length 18 on processor 3 are interchanged, the completion time is reduced to 154 from 155 for the four-processor decreasing-time schedule.
(d) As a bin-packing problem, each bin will have a capacity of 60. Using the decreasing list we obtain the following packings:
(First-fit decreasing): Bin 1: 36, 24; Bin 2: 32, 27; Bin 3: 31, 26; Bin 4: 30, 30; Bin 5: 25, 25, 9; Bin 6: 19, 19, 18; Bin 7: 18, 18, 18; Bin 8: 18, 17, 16; Bin 9: 16, 16, 15, 13; Bin 10: 15, 15, 14, 12; Bin 11: 12, 12.

(e) (Next-fit decreasing): Bin 1: 36; Bin 2: 32; Bin 3: 31; Bin 4: 30, 30; Bin 5: 27, 26; Bin 6: 25, 25; Bin 7: 24, 19; Bin 8: 19, 18, 18; Bin 9: 18, 18, 18; Bin 10: 17, 16, 16; Bin 11: 16, 15, 15; Bin 12: 15, 14, 13, 12; Bin 13: 12, 12, 9.

(Worst-fit decreasing): Bin 1: 36, 24; Bin 2: 32, 26; Bin 3: 31, 27; Bin 4: 30, 30; Bin 5: 25, 25; Bin 6: 19, 19, 18; Bin 7: 18, 18, 18; Bin 8: 18, 17, 16; Bin 9: 16, 16, 15, 12; Bin 10: 15, 15, 14, 13; Bin 11: 12, 12, 9.

(f) Because the total weight of all the objects is 596, a minimum of 10 bins is required. However, because there are no small weights, there is no way to achieve 10 bins, and, in fact, 11 bins is optimal.

71. The bins have a capacity of 120. (First-fit): Bin 1: 63, 32, 11; Bin 2: 19, 24, 64; Bin 3: 87, 27; Bin 4: 36, 42; Bin 5: 63. This schedule would take five station breaks; however, the total time for the breaks is under 8 minutes.

The decreasing list is 87, 64, 63, 63, 42, 36, 32, 27, 24, 19, 11. (First-fit decreasing): Bin 1: 87, 32; Bin 2: 64, 42, 11; Bin 3: 63, 36, 19; Bin 4: 63, 27, 24. This solution uses only four station breaks.

73. $T_1, T_2, T_3, T_4, T_8, T_9, T_{10}, T_{11}, T_5, T_6, T_7, T_{12}$.

CHAPTER 4

1. **(a)**

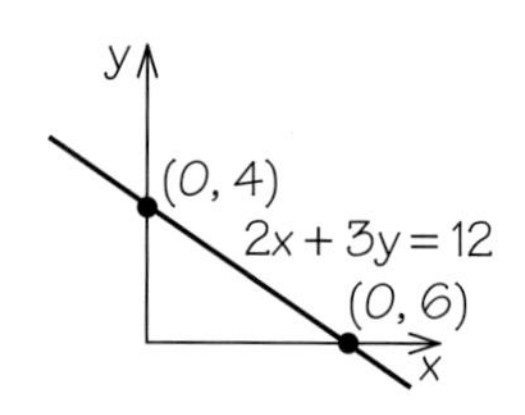

(b)

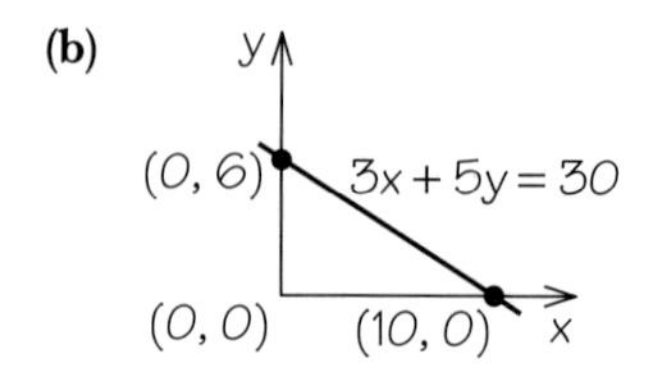

(c)

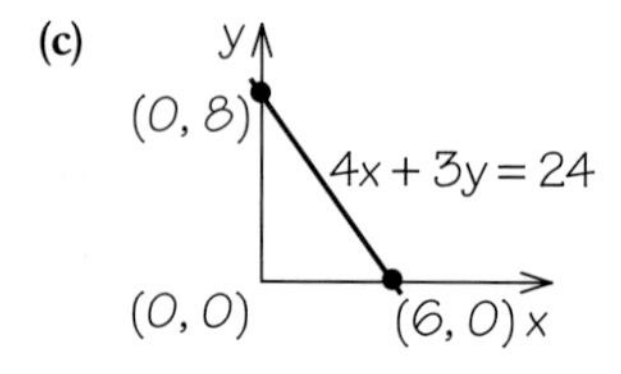

(d)

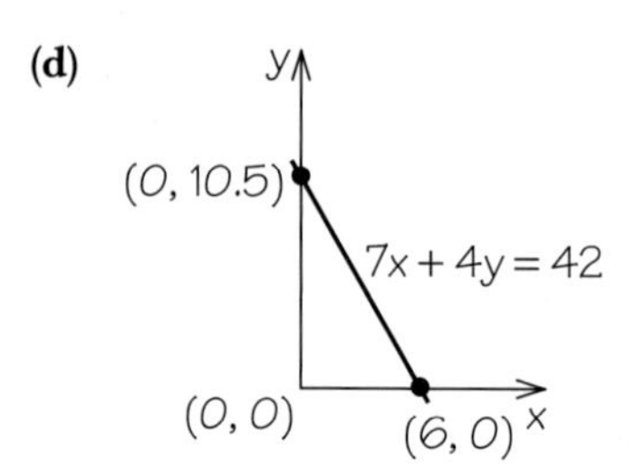

(e)

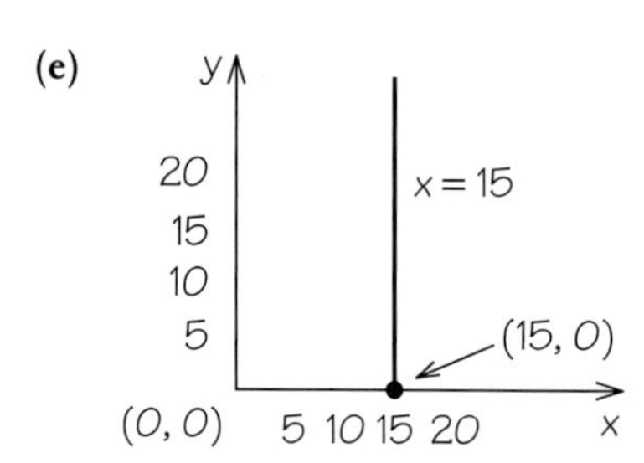

(f)

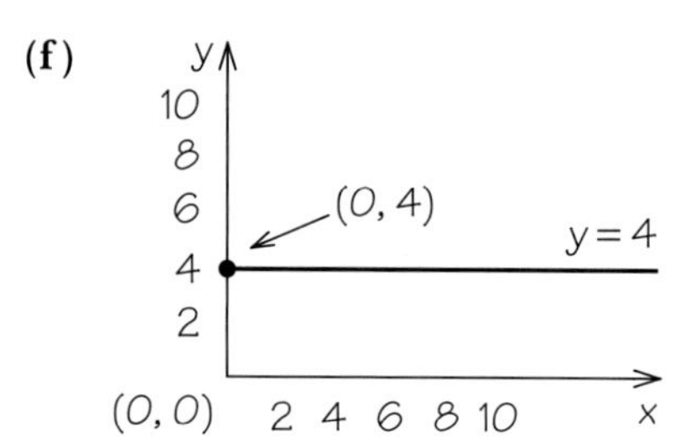

3. **(a)**

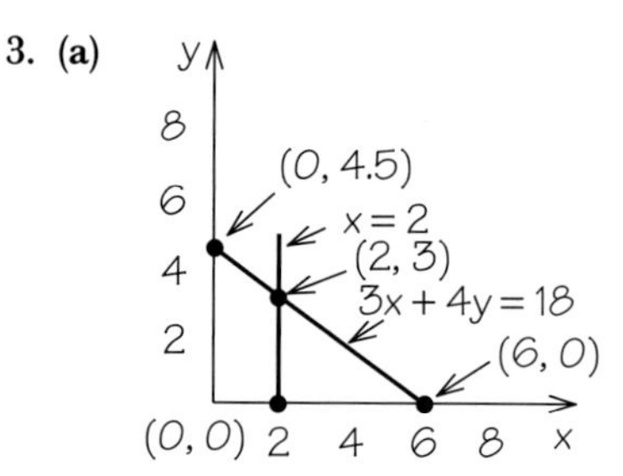

Finding point of intersection:
$x = 2$, so $3(2) + 4y = 18$, and $y = 3$

(b)

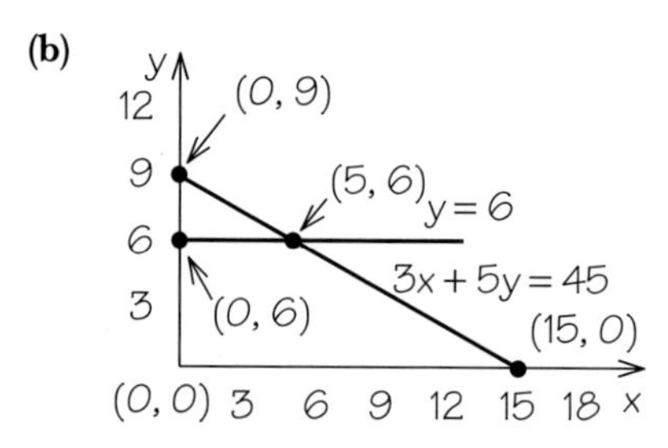

Finding point of intersection:
$y = 6$, so $3x + 5(6) = 45$, and $x = 5$

5. **(a)**

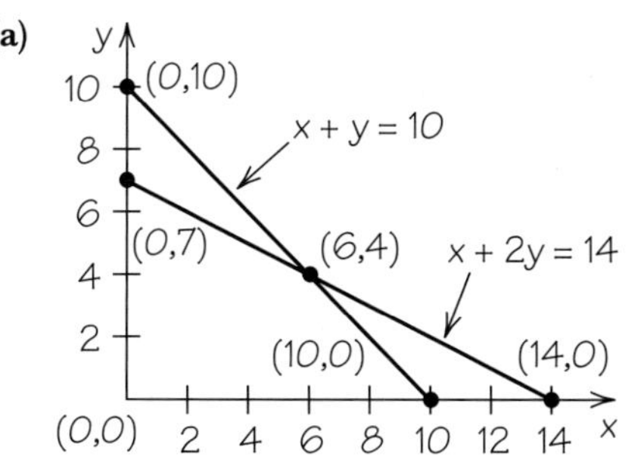

(b)

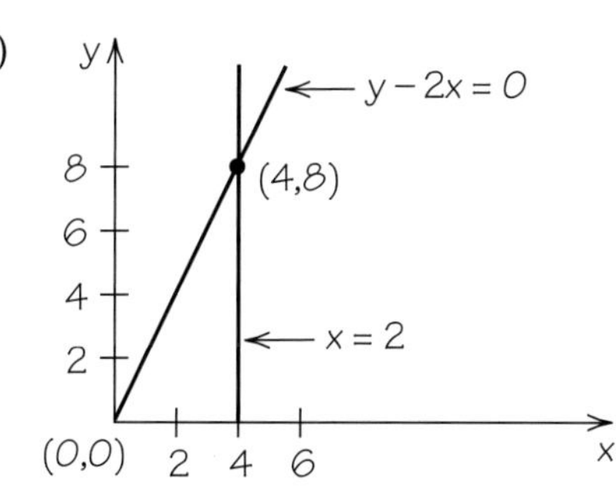

(c)

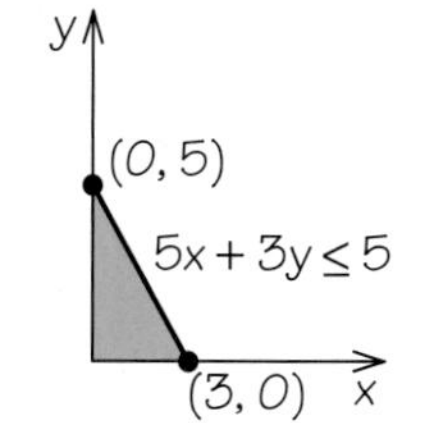

(d)

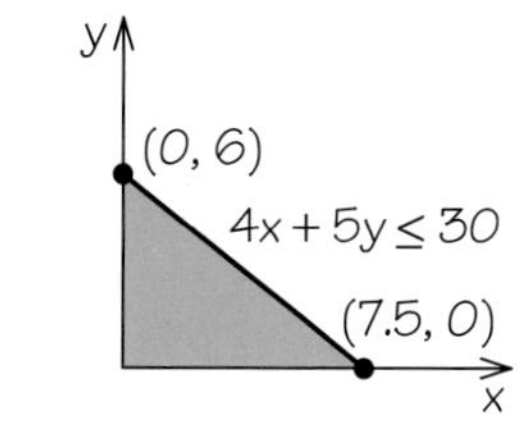

7. **(a)**

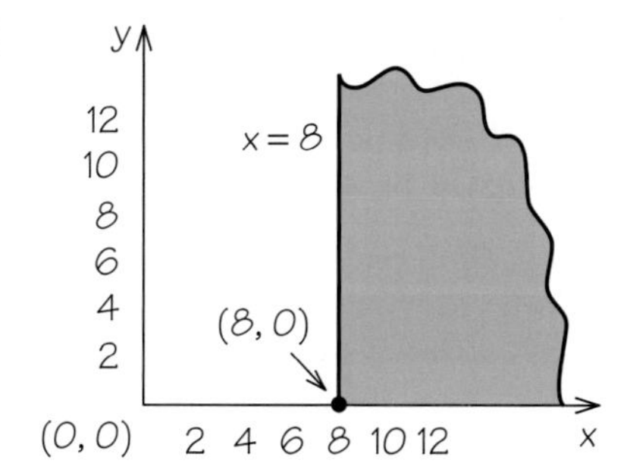

(b)

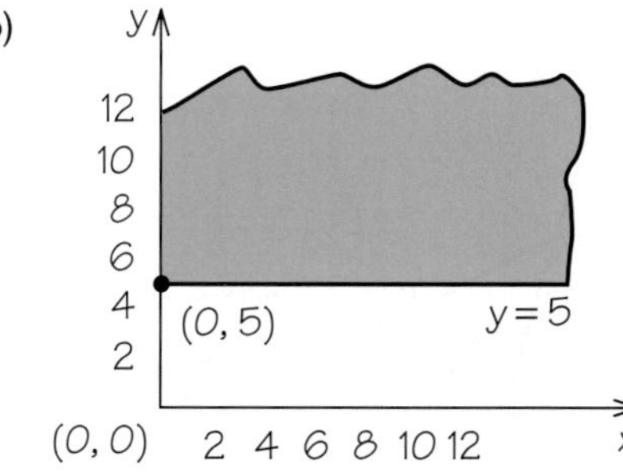

9. $4x + 2y \leq 28$

11. $6x + 4y \leq 300$

13. $12x + 10y \leq 40(16) = 640$ (beef)
$4x + 3y \leq 480$ (pork)

15.

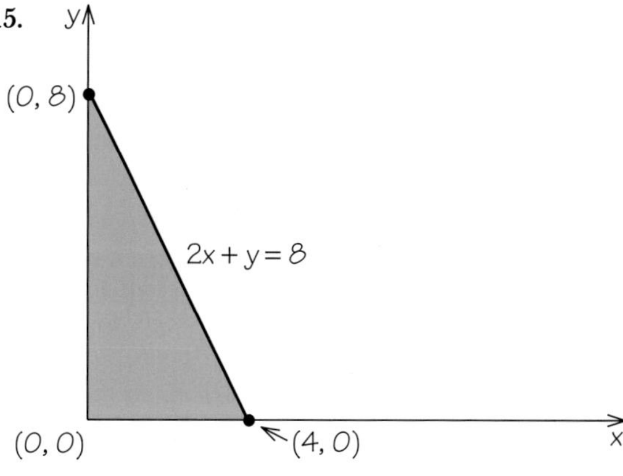

17.

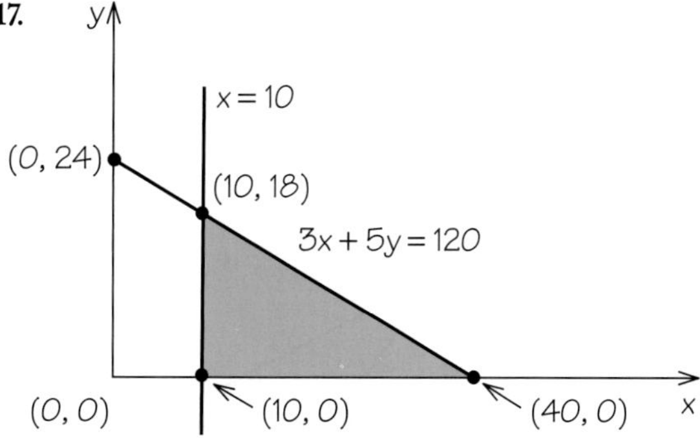

Finding point of intersection:
$x = 10$, so $3(10) + 5y = 120$, and $y = 18$

19.

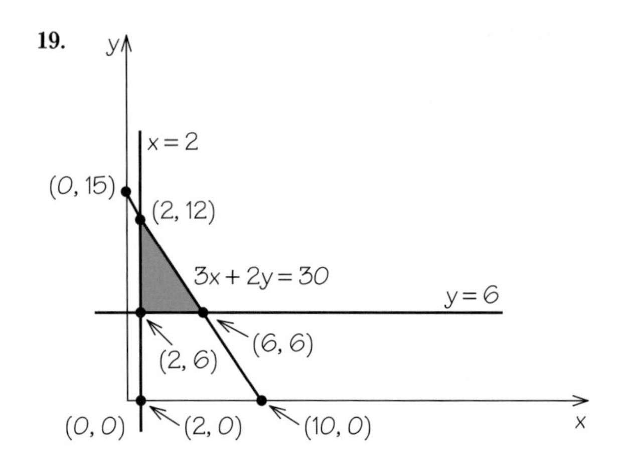

Finding points of intersection:
$x = 2$, so $3(2) + 2y = 30$, and $y = 12$
$y = 6$, so $3x + 2(6) = 30$, and $x = 6$

21. **(a)** (2, 4) is a point of the feasible region of Exercise 15, but not of Exercises 17 and 19.
(b) (10, 6) is a point of the feasible region of Exercise 17, but not of Exercises 15 and 19.

23. (0, 0): 0 skateboards and 0 dolls
Profit = \$2.30(0) + \$3.70(0) = \$0
(0, 30): 0 skateboards and 30 dolls
Profit = \$2.30(0) + \$3.70(30) = \$111
(12, 0): 12 skateboards and 0 dolls
Profit = \$2.30(12) + \$3.70(0) = \$27.60

Optimal production policy: Make 0 skateboards and 30 dolls for a profit of \$111.

25. **(a)**

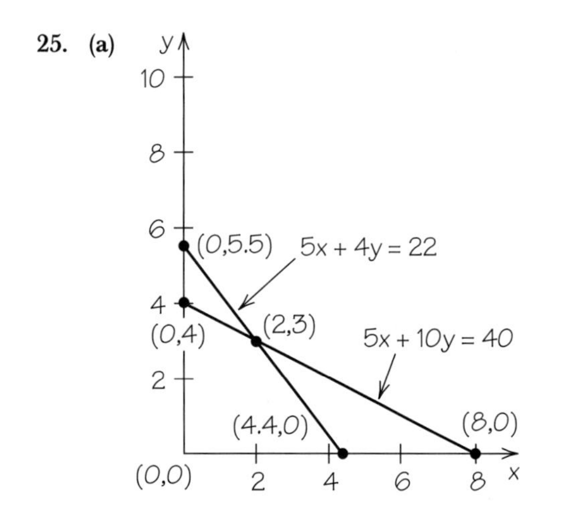

$$5x + 10y = 40$$
$$5x + 4y = 22$$
$$6y = 18$$
$$y = 3$$
$$5x + 3(10) = 40$$
$$5x = 10$$
so $x = 2$

(b)

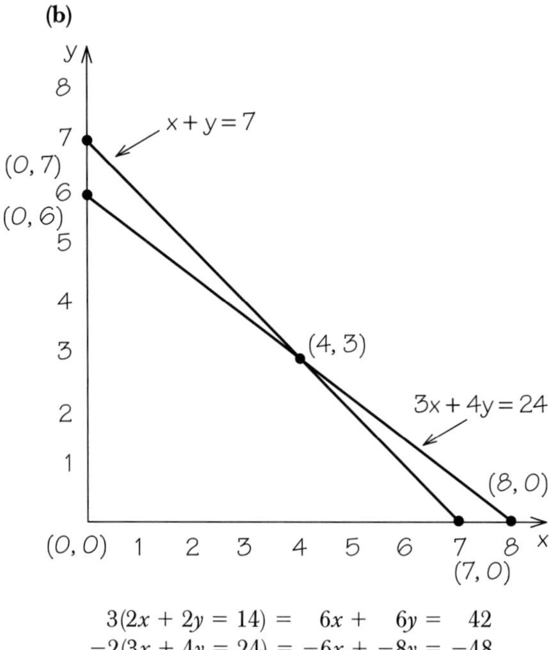

$$\begin{aligned} 3(2x + 2y = 14) &= 6x + 6y = 42 \\ -2(3x + 4y = 24) &= -6x + -8y = -48 \\ & \quad -2y = -6 \\ & \quad y = 3 \end{aligned}$$

$2x + 2(3) = 14$, so $x = 4$

27.

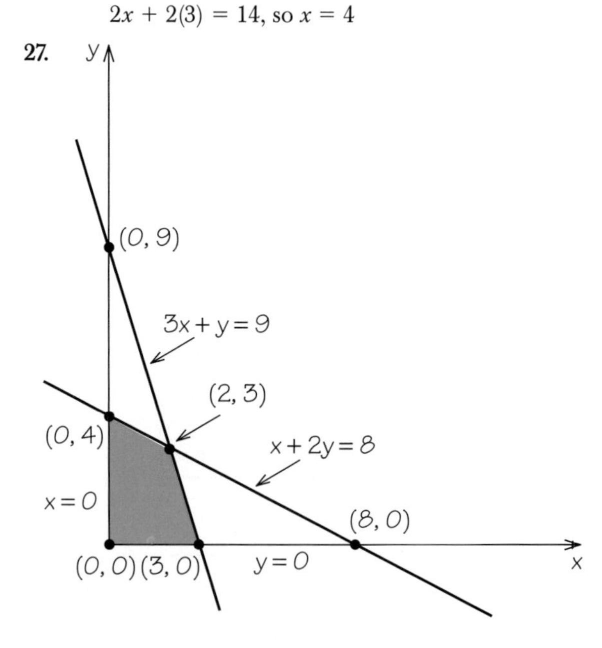

29.

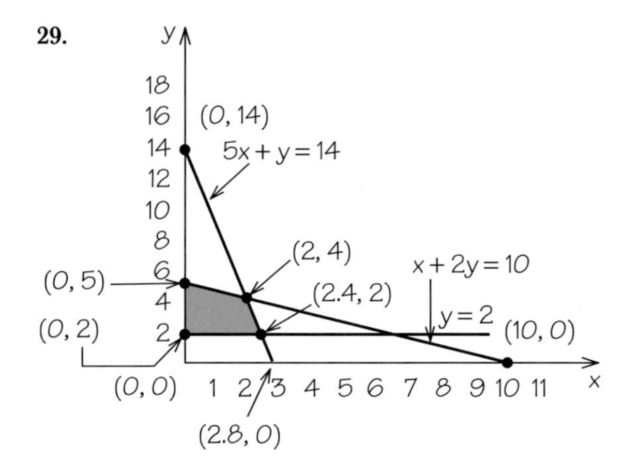

31. **(a)** (4, 2) is not a point of the feasible region for Exercises 27 or 29.

(b) (1, 3) is a point of the feasible region for Exercises 27 and 29.

33. The maximum value occurs at the corner point (8, 2), where P is equal to 28.

35. The maximum value occurs at the corner point (2, 8), where P is equal to 90.

37. **(a)**

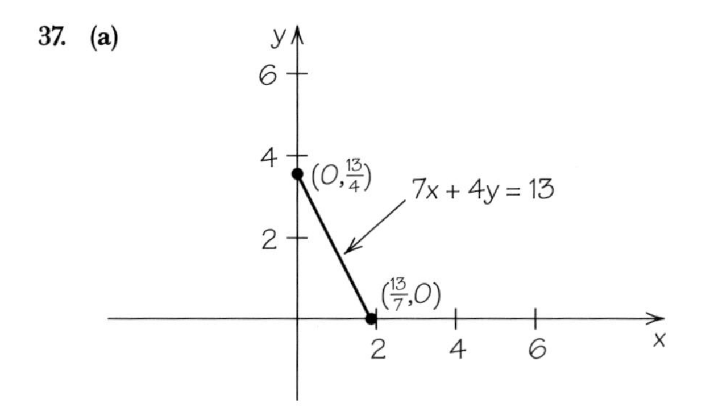

(b) $\left(\frac{13}{7}, 0\right)$

39. **(a)**

	Cloth (600 yds)	Mins	Profit
Shirts, x items	3	100	\$5
Vests, y items	2	30	\$2

(b) **Constraint inequalities** — **Profit formula**

Cloth: $3x + 2y \leq 600$ — $\$5x + \$2y$

Mins: $x \geq 100, y \geq 30$

(c) **Feasible region**

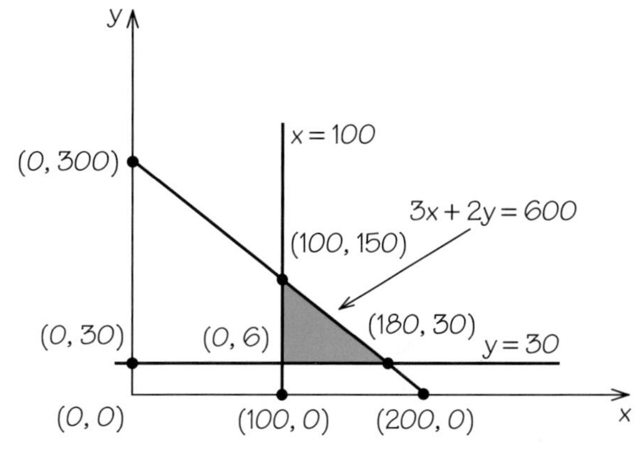

(d) **Profit**

At (100, 30), profit = \$560.
At (100, 150), profit = \$800.
At (180, 30), profit = \$960.*
Make 180 shirts and 30 vests.

With zero minimums, the feasible region looks like this:

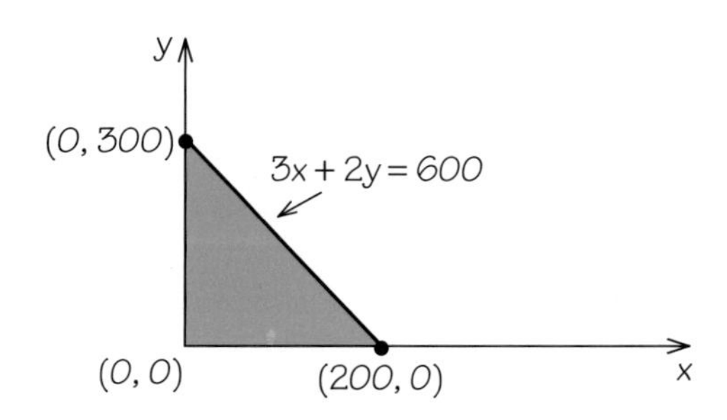

At (0, 0), profit is \$0.
At (0, 300), profit is \$600.
At (200, 0), profit is \$1000.*
Make 200 shirts and no vests.

41. **(a)**

	Time (90 min)	Profit
Mail order, x	10	\$30
Voice-mail order, y	15	\$40

(b) Constraint inequalities **Profit formula**
Time: $10x + 15y \leq 90$ \$30x + \$40y
Mins: $x \geq 0, y \geq 0$

(c) Feasible region

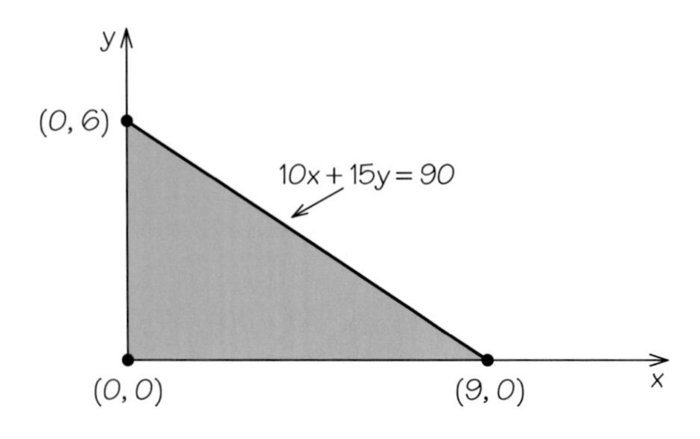

(d) Profit
At (0, 0), profit is \$0.
At (0, 6), profit is \$240.
At (9, 0), profit is \$270.*
Process 9 mail and no voice-mail orders.

With nonzero minimums of $x \geq 3$ and $y \geq 2$, the feasible region looks like:

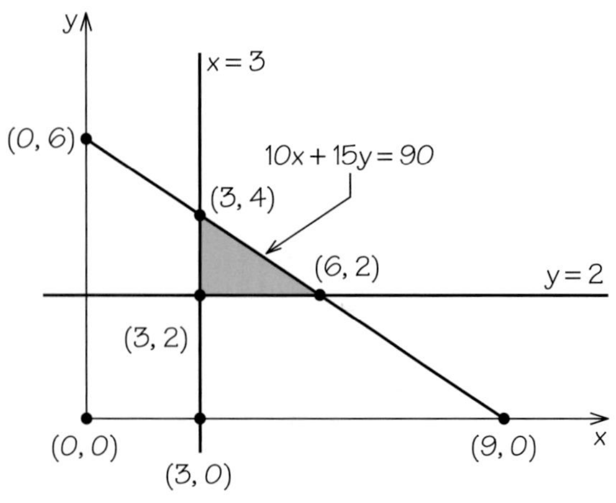
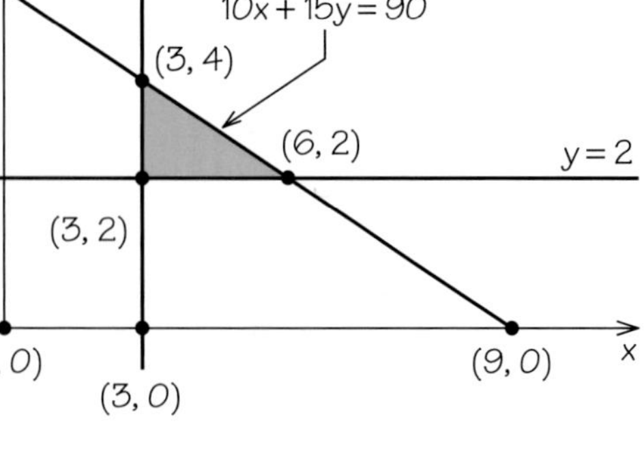

At (3, 2), profit is \$170.
At (3, 4), profit is \$250.
At (6, 2), profit is \$260.*
Process 6 mail and 2 voice-mail orders.

43. **(a)**

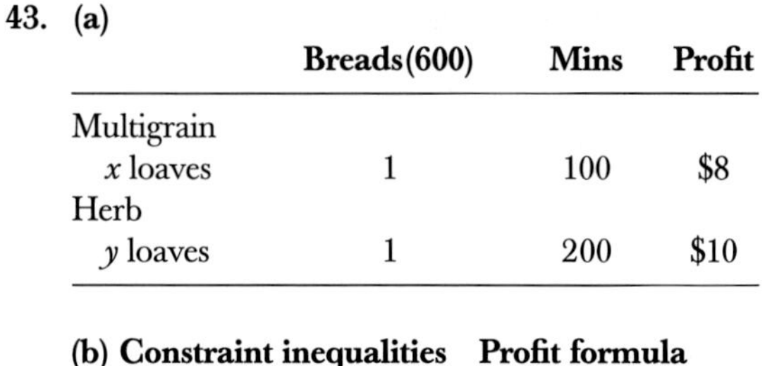

	Breads(600)	Mins	Profit
Multigrain			
x loaves	1	100	\$8
Herb			
y loaves	1	200	\$10

(b) Constraint inequalities **Profit formula**
Breads: $x + y \leq 600$ \$8x + \$10y
Mins: $x \geq 100, y \geq 200$

(c) Feasible region

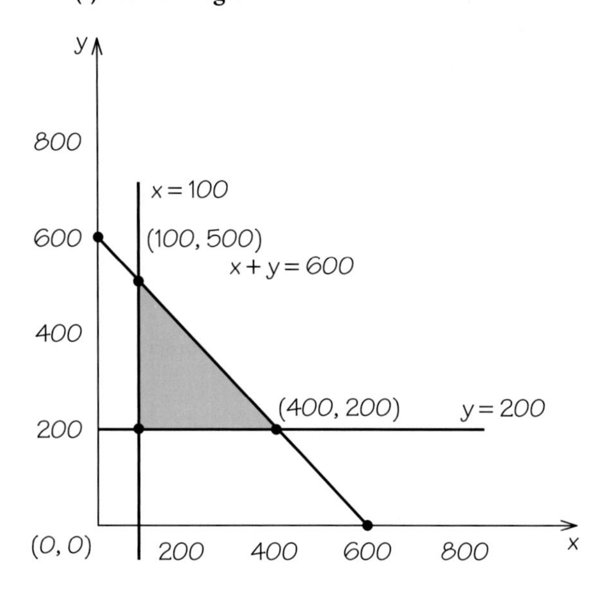

(d) Profit
At (100, 200), profit = \$2800.
At (100, 500), profit = \$5800.*
At (400, 200), profit = \$5200.
Make 100 multigrain and 500 herb loaves.

With zero minimums, $x \geq 0, y \geq 0$, feasible region is:

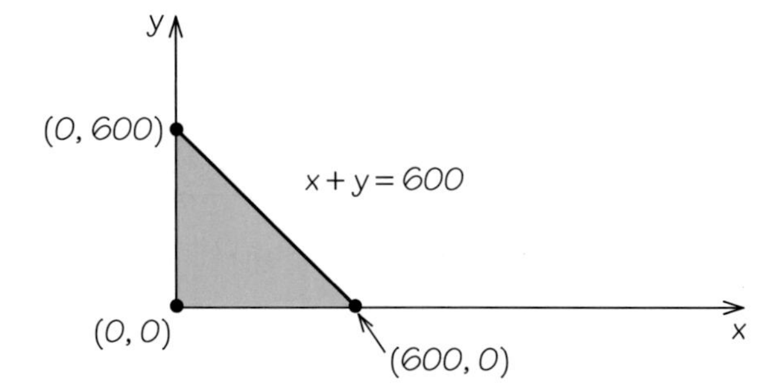

At (0, 0), profit = \$0.
At (0, 600), profit = \$6000.*
At (600, 0), profit = \$4800.
Make no multigrain and 600 herb loaves.

45. **(a)**

	Layout (12 hr)	Content (16 hr)	Profit
"Hot," x	1.5	1	\$50
"Cool," y	1	2	\$250

(b) Constraint inequalities **Profit formula**
Oven: $1.5x + 1y \leq 12$ $\$50x + \$250y$
Prep: $1x + 2y \leq 16$
Mins: $x \geq 0, y \geq 0$

(c) Feasible region

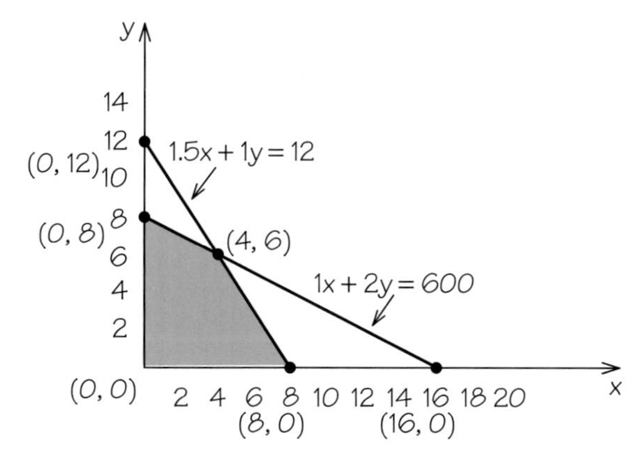

(d) Profit
At (0, 0), the profit is \$0.
At (0, 8), the profit is \$2000.*
At (4, 6), the profit is \$1700.
At (8, 0), the profit is \$400.
Maintain no "hot" and 8 "cool" sites.

With nonzero minimums of $x \geq 2$ and $y \geq 3$, we get this feasible region:

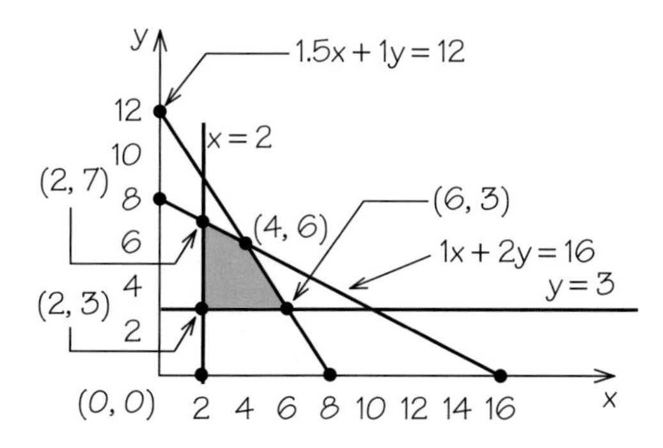

At (2, 3), profit is \$850.
At (2, 7), profit is \$1850.*
At (4, 6), profit is \$1700.
At (6, 3), profit is \$1050.
Maintain 2 "hot" and 7 "cool" sites.

47. **(a)**

	Space (100 acres)	Money in thousands (\$2600)	Profit in thousands
Modest, x houses	1	\$20	\$25
Deluxe, y houses	1	\$40	\$60

(b) Constraint inequalities **Profit formula**
Space: $1x + 1y \leq 100$ $\$25x + \$60y$
\$: $20x + 40y \leq 2600$ (in thousands)
Mins: $x \geq 0, y \geq 0$

(c) Feasible region

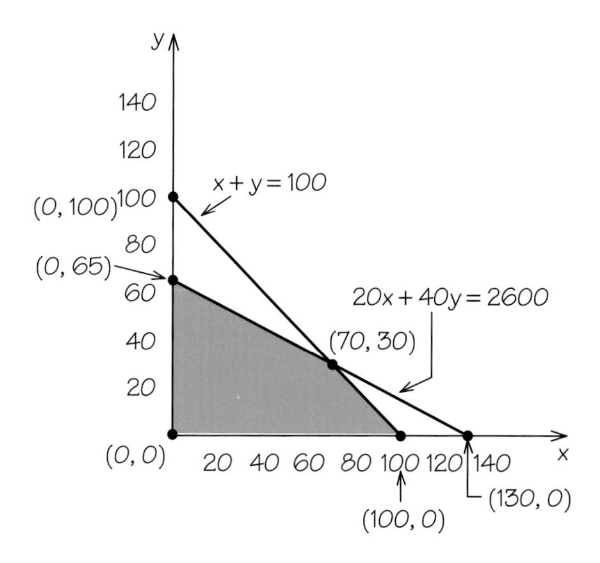

(d) Profit
At (0, 0), profit = \$0.
At (0, 65), profit = \$3,900,000.*
At (70, 30), profit = \$3,550,000.
At (100,0), profit = \$2,500,000.
Build no modest and 65 deluxe houses.

With nonzero minimums of $x \geq 20$ and $y \geq 20$, the feasible region looks like this:

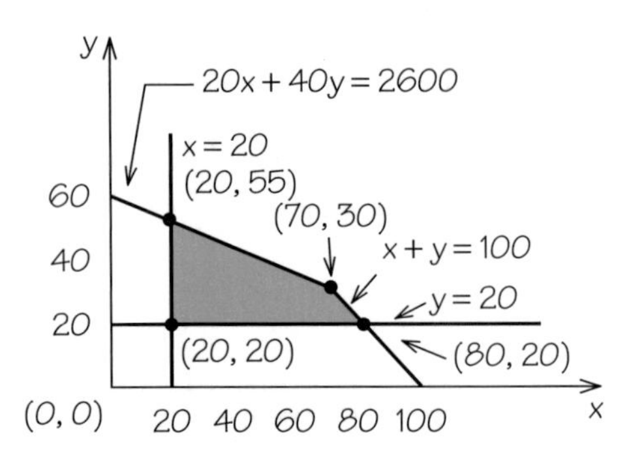

At (20, 20), profit is $1,700,000.
At (20, 55), profit is $3,800,000.*
At (70, 30), profit is $3,550,000.
At (80, 20), profit is $3,200,000.
Make 20 modest and 55 deluxe houses.

49. **(a)**

	Bird count (100)	Cost $ (2400)	Profit
Pheasants, x	1	$20	$14
Partridges, y	1	$30	$16

(b) Constraint inequalities
Count: $1x + 1y \leq 100$
$: $20x + 30y \leq 2400$
Mins: $x \geq 0, y \geq 0$

Profit formula
$14x + 16y$

(c) Feasible region

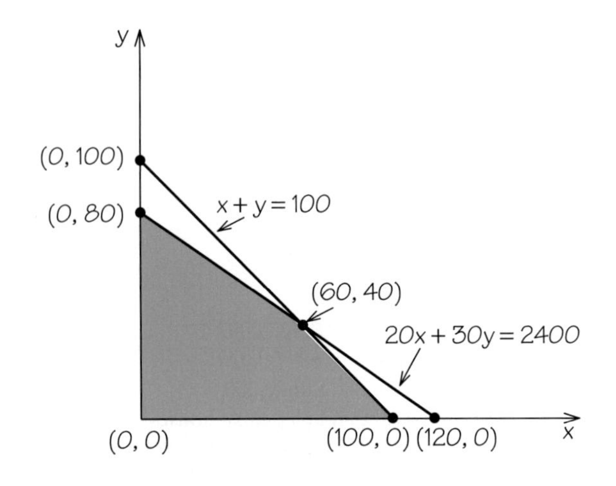

(d) Profit
At (0, 0), profit is $0.
At (0, 80), profit is $1280.
At (60, 40), profit is $1480.*
At (100, 0), profit is $1400.
Raise 60 pheasants and 40 partridges.
With nonzero minimum of $x \geq 20$ and $y \geq 10$, there is no change because the optimal production policy obeys these minimums.

51. **(a)**

	Shaper (50)	Smoother (40)	Painter (60)	Profit
Toy A, x	1	2	1	$4
Toy B, y	2	1	3	$5
Toy C, z	3	2	1	$9

(b) Constraint inequalities
Shaper: $1x + 2y + 3z \leq 50$
Smoother: $2x + 1y + 2z \leq 40$
Painter: $1x + 3y + 1z \leq 60$
Mins: $x \geq 0, y \geq 0, z \geq 0$
Profit formula
$4x + 5y + 9z$

(c) Optimal product policy
Make 5 toy A, no toy B, 15 toy C for a profit of $155.

53. **(a)**

	Chocolate (1000 lb)	Nuts (200 lb)	Fruit (100 lb)	Profit
Special, x boxes	3	1	1	$10
Regular, y boxes	4	0.5	0	$6
Purist, z boxes	5	0	0	$4

(b) Constraint inequalities
$3x + 4y + 5z \leq 1000$
$1x + 0.5y + 0z \leq 200$
$1x + 0y + 0z \leq 100$

Profit formula
$\$10x + \$6y + \$4z$

(c) Make 100 boxes of Special, 175 boxes of Regular, and 0 boxes of Purist.

55. A feasible region has infinitely many points; the corner point principle tells us we need to evaluate the profit formula only at a few of those points (the corner points), not all of them.

57. The minimum value occurs at (3, 2).

59. The only feasible point with integer coordinates for the collection of inequalities $x \geq 0, y \geq 0$, and $x + y \leq 0.5$ is (0, 0).

61. (a)

	Time (240 min)	Profit
Business, x calls	4	\$0.50
Charity, y calls	6	\$0.40

(b) Constraint inequalities
Time: $4x + 6y \leq 240$
Mins: $x \geq 0, y \geq 0$

Profit formula
$\$0.50x + \$0.40y$

(c) Feasible region

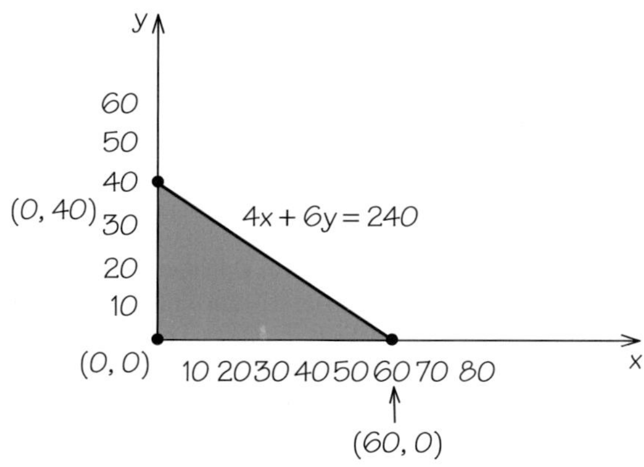

(d) Profit
At (0, 0), profit is \$0.00.
At (0, 40), profit is \$16.00.
At (60, 0), profit is \$30.00.*
Make 60 business and no charity calls.

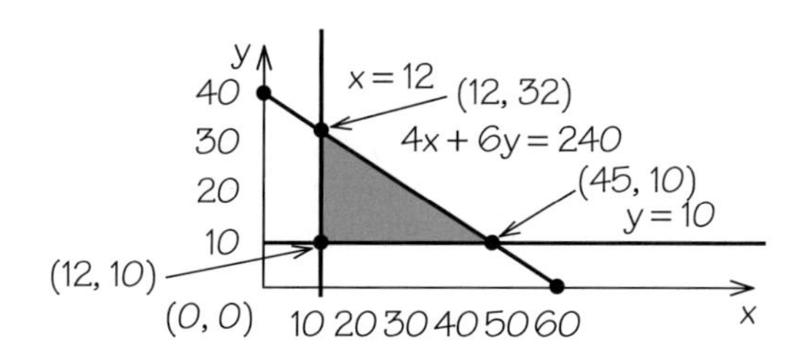

At (12, 10), profit is \$10.00.
At (12, 32), profit is \$18.80.
At (45, 10), profit is \$26.50.*
Make 45 business and 10 charity calls.

63. (a)

	Machine (12 hr)	Paint (16 hr)	Profit
Bikes, x	2	4	\$12
Wagons, y	3	2	\$10

(b) Constraint inequalities
Mach: $2x + 3y \leq 12$
Paint: $4x + 2y \leq 16$
Mins: $x \geq 0, y \geq 0$

Profit formula
$\$12x + \$10y$

(c) Feasible region

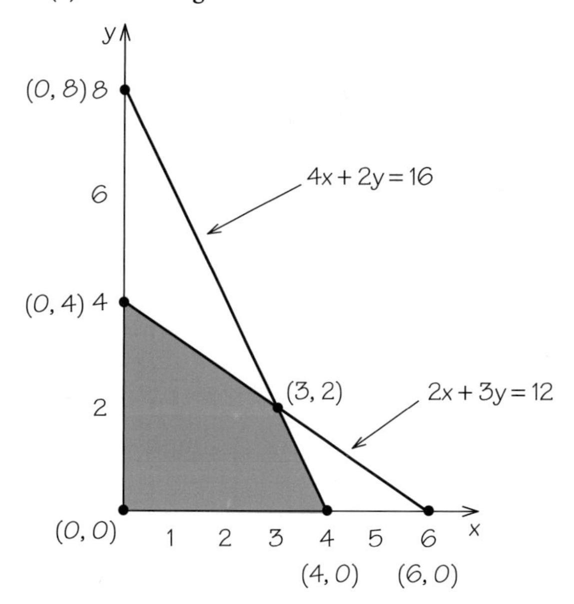

(d) Profit
At (0, 0), profit is \$0.
At (0, 4), profit is \$40.
At (3, 2), profit is \$56.*
At (4, 0), profit is \$48.
Make 3 bikes and 2 wagons.

With nonzero minimums of $x \geq 2$ and $y \geq 2$ there will be no change because the optimal production policy obeys these minimums.

65. Answers will vary.

67. This may be a bad decision because some customers will stop coming to the shop if the variety of sandwiches available is not great enough. Those customers whose favorite sandwich was dropped will also probably not come. To avoid dropping any type of sandwich, minimum constraints for each type of sandwich could be incorporated into the model.

69. Answers will vary.

71. **(a)**

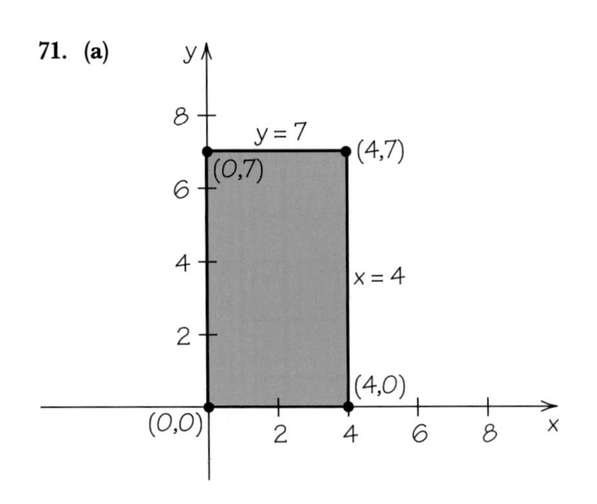

(b) The feasible region will be the same since the new constraint passes through (4, 7) but does not cut off any of the feasible region from part (a).

(c)

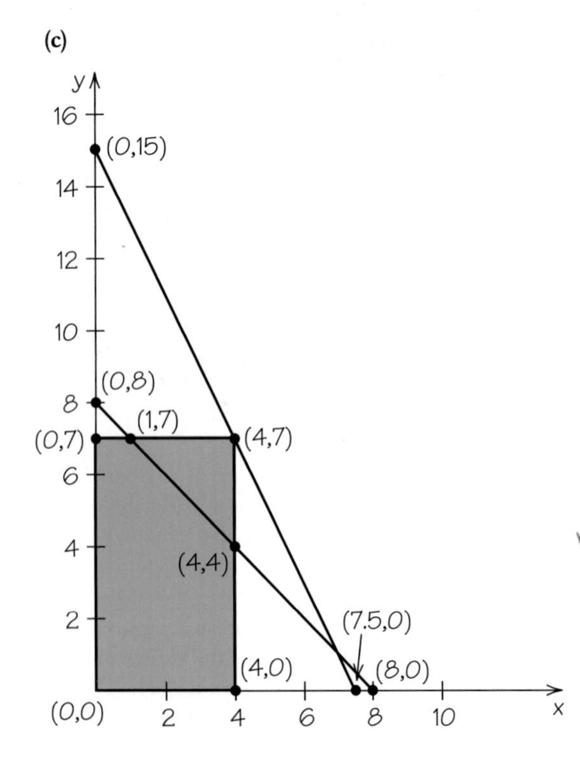

(d) The feasible region in (a) has four corner points, in (b) it has four corner points, and in (c) it has three corner points.

(e) The number of corner points can go up by at most one. The number of corner points may stay the same or go down when a constraint is added.

CHAPTER 5

1. Population: Words in Steele novels. Sample: The 250 words measured.

3. Population: "Constituents," perhaps adults in her district. Sample: The 361 people who wrote letters. Those who trouble to write are not representative–they feel strongly about this issue.

5. **(a)** 29,777. **(b)** This was a voluntary response. **(c)** Bias: Men are overrepresented relative to the adult population.

7. Unhappy women are more likely to respond. We expect the true percent to be lower than 72%.

9. If we assign labels 01 to 28 in alphabetical order, lines 110 and 111 give 18 = Mourning, 20 = Peters, 06 = Castillo, 08 = Dewald, 27 = Vega, and 12 = Goel.

11. Label students 001 to 450 in alphabetical order (000 to 449 is also correct). Line 120 gives the first five members to be those labeled 354, 239, 421, 426, and 435.

13. **(a)** Population: All people (probably all adults) who live in Ontario. Sample: The 61,239 people contacted.
(b) Yes: They are based on a large random sample.

15. **(a)** 706.
(b) In many samples taken by Harris's methods, the sample result will be within ±3 points of the truth about the population 95% of the time.

17. ±3.15%.

19. 4036.

21. **(a)** If we label 01 to 25 in alphabetical order, line 135 gives 10 = Gutierrez, 07 = Epstein, 11 = Herrnstein, 20 = Toulmin, and 15 = Moll.
(b) The first sample has 1 female. Others will vary.

23. No. We are only 95% confident, not certain, that the truth is captured within the margin of error.

25. The effect of indicators or charts would be confounded with, for example, the difference in weather between the two years.

27. No treatment was *imposed.* This observational study simply recorded many facts about each subject.

29. The researcher presents each subject with two health plans. That amounts to imposing a treatment, so this is an experiment.

31. The outline follows Figure 5.6. The two rate plans are the two treatments. Offer each plan to 100 rooms chosen at random.

33. *Randomized* means that chance was used to assign subjects to the treatments. *Double-blind* means that people working with the subjects did not know which treatment a subject received. *Placebo-controlled* means that one of the treatments was a dummy vaccine with no active ingredient.

35. Results vary with the choice of lines in Table 5.1.

37. To do this, choose a simple random sample of 5 to form Group 1. Then choose a simple random sample of 5 of the remaining 15 for Group 2 and another simple random sample of 5 of the remaining 10 for Group 3. The 5 who still remain are Group 4. It is best to relabel the 10 who remain after the second stage as 0 to 9 to speed the third sample. If the initial labels are 01 to 20, Group 1 contains 16, 04, 19, 07, and 10. Group 2 contains 13, 15, 05, 09, and 08. The members of Group 3 depend on relabeling.

39. The average earnings of men exceeded those of women by so much that it is very unlikely that the chance selection of a sample would produce so large a difference if there were not a difference in the entire student population. But the black–white difference was small enough that it might be due to the accident of which students were chosen for the sample.

41. Fewer people are home to answer in the summer vacation months. A high rate of nonresponse increases the risk that those who do respond are not typical of the entire population.

43. During the experiment, only the experimental cars had center brake lights, so they attracted attention. Once most cars had them, they were less noticed and so did a poorer job of preventing collisions.

45. **(a)** Population: All adult (age 18 and over) residents of the United States. **(b)** Margin of error: ±3.15%. "A new poll finds that 36% of Americans support government funding to allow low-income families to send their children to private schools. The margin of error for the poll is plus or minus 3%." (TV reports generally omit "95% confidence.")

47. **(a)** This is a randomized comparative experiment with four branches, similar to Figure 5.7 but with one more branch. It is best to use groups of equal size: 216 people in each group. **(b)** With labels 001 to 864, the first five chosen are 731, 253, 304, 470, and 296. **(c)** Those working with the subjects did not know the contents of the pill each subject took daily. **(d)** The differences in colon cancer cases in the four groups were so small that they could easily be due to the chance assignment of subjects to groups. **(e)** People who eats lots of fruits and vegetables may eat less meat or more cereals than other people. They may drink less alcohol or exercise more.

49. Label 01 to 30 in alphabetical order. Starting at line 123, we get 08 = Best's Camera, 15 = Hernandez Electronics, 07 = Bennett Hardware, 27 = Scotch Wash, and 10 = Central Tree Service.

51. **(a)** No. All states have populations very much larger than the sample size. When this is true, the margin of error depends only on the size of the sample, not on the size of the population. **(b)** The sample sizes vary from 2270 to 140,000, so the margins of error will also vary.

53. **(a)** Label the first 40 rooms 01 to 40. Line 120 chooses room 35. The sample is rooms 35, 75, 115, 155, and 195. **(b)** The only possible samples consist of 5 rooms spaced 40 apart in the list. An SRS gives *all* samples of 5 rooms an equal chance to be chosen.

55. Any sample gives 0% "yes." This is exactly correct for the population, so the margin of error is also ±0%.

57. **(a)** Print a coupon in the campus newspaper asking students to check their opinion, cut out the coupon, and mail it in. **(b)** Ask all the students in a large sociology course to record their opinion as part of an exam in the course.

59. **(a)** The rats may be affected by the environment of the lab, or the food they receive, or other lurking variables. Without a control group, the effect of DDT is confounded with any effects of these variables. **(b)** Use a randomized comparative experiment with two groups, similar to Figure 5.6. There are 10 rats in each group. One group has DDT added to their diet, the other receives the same diet without DDT. Label the rats 01 to 20. Starting at line 123 gives the DDT rats as 08, 15, 07, 10, 18, 03, 01, 19, 15, and 06.

61. One factor is ZIP code, yes or no. The second is day of the week, either all seven or a smaller group such as Monday, Thursday, and Saturday. The latter choice gives six treatments (ZIP/day combinations). All letters should be the same size and bear the same typed address. Mail all letters from the same letter box or post office, and mail all at the same time of day. It is best to spread the mailings over several weeks and to avoid holiday periods. Because all letters are identical, it is not necessary to choose at random those to be mailed at each time.

63. Any differences in disclosure between black and white subjects were so small that they could just be chance variation. Females disclosed more than males, and the difference between the genders was so large that it would rarely happen just by chance.

CHAPTER 6

1. The individuals are the students in the course. The variables are a student's major, total points in the course, and letter grade.

3. The distribution is single-peaked and very roughly symmetric, with center near 13%. Without the outliers, the spread is roughly 8% to 16%. We know that Florida (with many retirees) is the high outlier and that Alaska (the northern frontier) is the low outlier.

5. **(a)** 10.9% and 11.0% **(b)** Single-peaked and roughly symmetric. Center is at 13.9%; spread is 12.1% to 15.9%.

7. Note that the decimal point is between the two digits in the stems.

Stem	Leaves
48	8
49	
50	7
51	0
52	6799
53	04469
54	2467
55	03578
56	12358
57	59
58	5

The distribution is roughly symmetric, with one observation (4.88) slightly outlying.

9. The five-number summary is 4.88, 5.295, 5.46, 5.615, 5.85. The two quartiles are roughly equally distant from the median.

11. The five-number summaries are 46.9, 47.4, 51.9, 62.0, 62.9 for men and 33.1, 38.2, 42.0, 49.55, 54.6 for women. Women as a group have lower body masses than men. The women's data are more spread out, but that may just reflect the fact that there are 12 women and only 7 men.

13. **(a)** $\bar{x} = (5.6 + 5.2 + 4.6 + 4.9 + 5.7 + 6.4)/6 = 5.4$ **(b)** Make a table of the squared deviations $(5.6 - 5.4)^2$, and so on. Sum to get 2.06. The variance is $s^2 = 2.06/5 = 0.412$ and the standard deviation is $s = \sqrt{0.412} = 0.6419$.

15. The five-number summaries are as follows: beef: 111, 140, 152.5, 178.5, 190; meat: 107, 139, 153, 179, 195; poultry: 87, 102, 129, 143, 170. Beef and meat hot dogs have quite similar distributions, whereas poultry hot dogs as a group have fewer calories.

17. As Cisco illustrates, the distribution contains some extremely high outliers. These pull the mean far above the median.

19. **(a)** Life expectancy increases with GDP in a curved pattern. The increase is very rapid at first but levels off for GDP above roughly $5000 per person. **(b)** Richer nations have better diets, clean water, and better health care, so we expect life expectancy to increase with wealth. But once food, clean water, and basic medical care are in place, greater wealth has only a small effect on life span.

21. **(a)** Here is the scatterplot, with time as the explanatory variable.

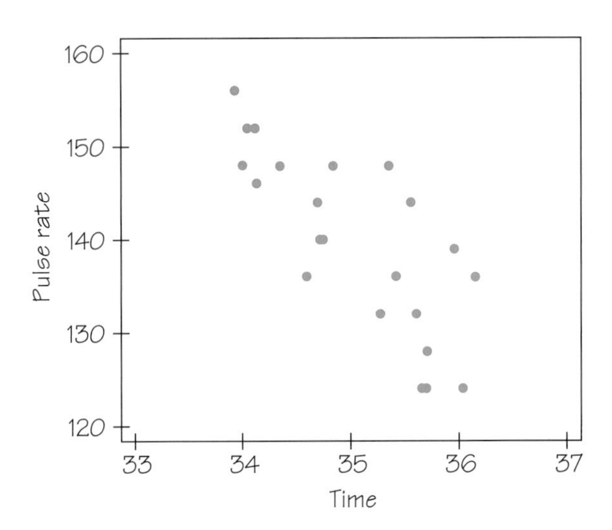

(b) Negative: we expect faster times (fewer minutes) to lead to higher pulse rates. **(c)** Moderately strong negative linear relationship.

23. **(a)** Slope = −9.695. For each additional minute, pulse rate decreases by 9.695 beats per minute on the average. **(b)** 147.4. The prediction is low by 4.6 beats per minute.

25. **(a)** $r = -0.746$. This r indicates a moderately strong negative linear relationship. **(b)** r would not change.

27. The correlation would be $r = 1$ because there is a perfect straight-line relationship.

29. $r = 0.481$. The scatterplot shows that five points lie close to a line, but the point (10, 1) lies far from the line at the lower right. This outlier reduces r.

31. **(a)** The plot shows a strong linear pattern, so we think all come from one species. **(b)** Calculate for femur $\bar{x} = 58.2$, $s_x = 13.2$ and for humerus $\bar{y} = 66$, $s_y = 15.89$. The first term in the sum for r is

$$\frac{38 - 58.2}{13.2} \times \frac{41 - 66}{15.89}$$

(c) $r = 0.994$

33. The regression line is $y = 11.06 - 0.0147x$. Here is the scatterplot:

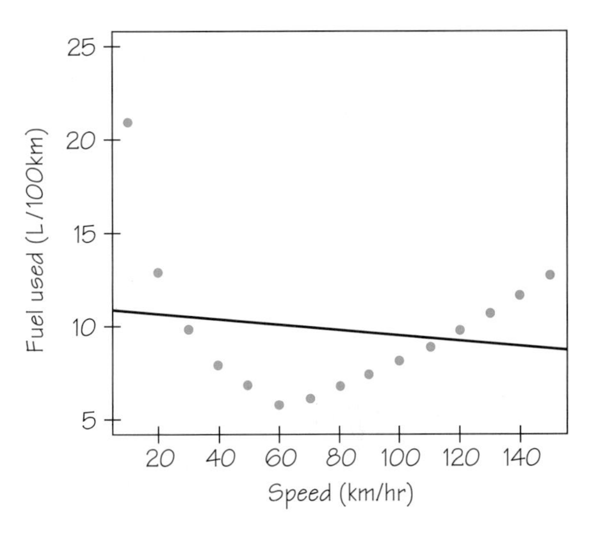

35. **(a)** Make a histogram (or a stemplot):

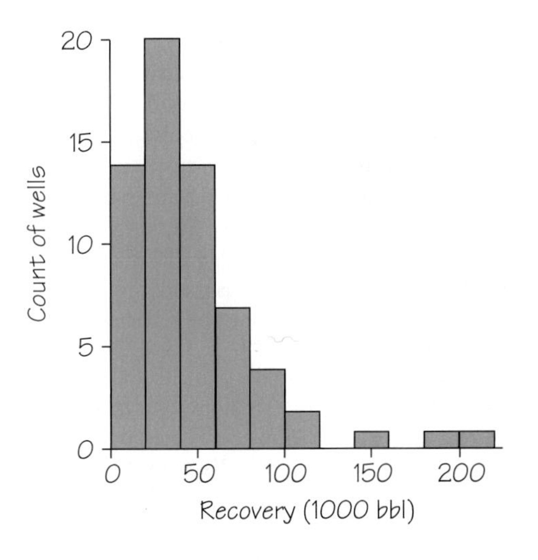

The distribution is single-peaked and right-skewed. Three wells might be considered high outliers. **(b)** $\bar{x} = 48.25$, $M = 37.8$ thousands of barrels. The distribution is skewed to the right, so the mean is pulled above the median. **(c)** The five-number summary is 2.0, 21.5, 37.8, 60.1, 204.9. The right skewness is visible in the greater distances from the median of the third quartile and maximum relative to the first quartile and minimum.

37. A stemplot is better for 16 observations. Omitting Shaquille O'Neal's $19.3 million to save space:

0	3367899
1	8
2	23
3	45
4	8
5	
6	
7	0
8	
9	
10	1

The distribution is strongly right-skewed, with three high outliers (including O'Neal). Use the five-number summary: 0.3, 0.75, 2.0, 4.15, 19.3.

39. $\bar{x}$ = $3.68 million, M = $2.0 million. Omitting O'Neal and Bryant, $\bar{x}$ = $2.11 million, M = $1.35 million. Both $\bar{x}$ and M drop, but $\bar{x}$ drops more and remains above M because the distribution remains right-skewed.

41. For Table 6.1, the five-number summary is 0.6, 2.0, 4.1, 7.0, 38.7. So $IQR = 7 - 2 = 5$, $1.5IQR = 7.5$, and observations below -5.5 and above 14.5 are suspected outliers. There are seven suspected high outliers. The histogram in Figure 6.2 shows that most of these are just the long right tail of a skewed distribution. Only New Mexico (38.7%) is a true outler. The $1.5 \times IQR$ criterion is mainly useful when large amounts of data must be screened automatically.

43. $IQR = 69$, so $1.5 \times IQR = 103.5$. Observations above 255 are suspected outliers. There are six such observations. The histogram in Exercise 42 shows a long right tail with no outliers clearly separated from the others.

45. Start with 10 (median) as the third in order. The sum must be 35 to have mean 7. One choice is 1, 4, 10, 10, 10. You should find a different choice.

47. **(a)** The smallest possible s is 0; any four identical numbers have $s = 0$. **(b)** 0, 0, 10, 10 (largest possible spread). **(c)** In part (a), yes. In part (b), no.

49. **(a)** Larger crickets will usually be both longer and heavier than smaller crickets, so length and height tend to increase or decrease together. **(b)** Correlation does not change.

51. **(a)** Choose the smaller correlation, small-cap stocks. **(b)** A negative correlation.

53. The equation of the least-squares line is $y = a + bx$ with intercept $a = \bar{y} - b\bar{x}$. If $x = \bar{x}$, we get $y = \bar{y} - b\bar{x} + b\bar{x}$, or $y = \bar{y}$.

55. **(a)** We hope to predict HAV angle from MA angle, so MA angle is the explanatory variable. Here is the scatterplot:

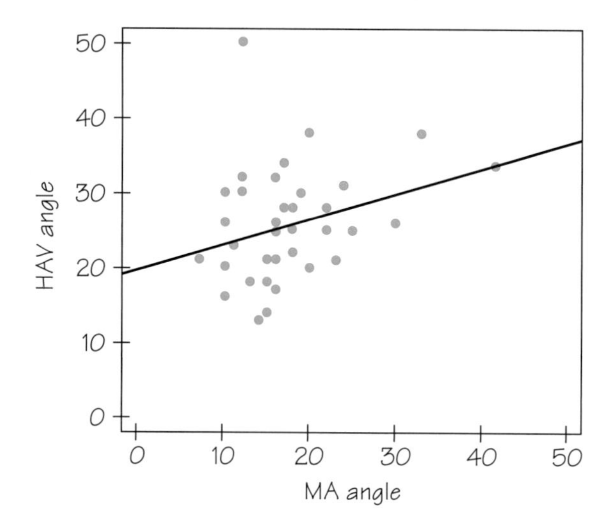

(b) There is a moderate positive relationship that is roughly linear in form. There is one outlier, for whom HAV angle is high relative to MA angle. If we remove the outlier, the correlation increases from $r = 0.30$ to $r = 0.44$, which better describes the relationship for most subjects. **(c)** HAV angle is clearly related to MA angle. The relationship appears strong enough to be helpful, but interpretation requires medical knowledge of such things as whether prediction within only 15 degrees or so is useful.

57. Heavier people who are concerned about their weight may be more likely than lighter people to choose artificial sweeteners in place of sugar.

59. **(a)** $y = -90.9 + 3.21x$ (using the rounded values given in the Exercise). We predict about 423 calories. **(b)** Interchanging x and y, we get $y = 61.6 + 0.232x$. We predict about 160 calories.

(c) Least squares makes the distances from points to line as small as possible in the y direction. Changing the choice of y changes the direction of the distances we try to make small.

61. **(a)** All four sets of data have $r = 0.8162$ and regression line $y = 3 + 0.5x$ to a close approximation.
(b) Here are the plots:

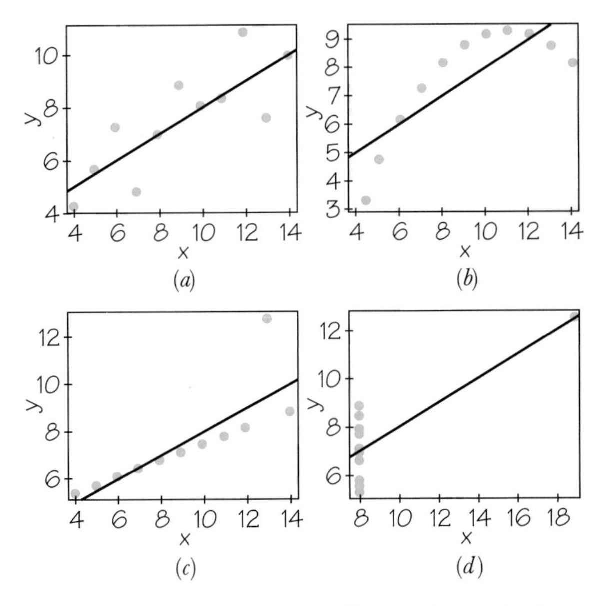

(c) Only plot (a) is a "normal" regression setting in which the line is useful for prediction. Plot (b) is curved, (c) has an extreme outlier in y, and (d) has all but one x identical. The lesson: plot your data before calculating.

CHAPTER 7

1. Results will vary.

3. Results will vary.

5. $0/6 = 0$; $47/100 = 0.47$; $497/1000 = 0.497$; $4997/10000 = 0.4997$. The first few results don't change the long-run proportion of heads – later tosses don't need "extra heads" to make up for straight tails at the beginning.

7. **(a)** $S = \{0, 1, 2, 3, 4, 5, 6, 7, 8, 9, 10\}$
(b) $S = \{0, 10, 20, 30, 40, 50, 60, 70, 80, 90, 100\}$
(c) $S = \{$Yes, No$\}$

9. **(a)** $S = \{$HHHH, HHHT, HHTH, HTHH, THHH, HHTT, HTTH, TTHH, HTHT, THHT, THTH, HTTT, TTTH, THTT, TTHT, TTTT$\}$
(b) $S = \{0, 1, 2, 3, 4\}$

11. **(a)** 0.27, because the probabilities of all outcomes must sum to 1. **(b)** 0.73

13. **(a)** Yes. The coin is not balanced, but the model is legitimate. **(b)** No. The sum for the four possible outcomes is 1.6, not 1. **(c)** No. The probabilities of tan, brown, yellow, green, and orange have sum 0.8, not 1.

15. Each combination of two faces has probability 1/36. The possible counts of spots are $S = \{1, 2, 3, 4, 5, 6, 7, 8, 9, 10, 11, 12\}$ and all have the same probability, 1/12.

17. The tetrahedron is balanced, so outcomes $S = \{1, 2, 3, 4\}$ each have probability 1/4.

19. **(a)** BBB, BBG, BGB, GBB, BGG, GBG, GGB, and GGG each have probability $1/8 = 0.0125$. **(b)** The model is

Girls	0	1	2	3
Probability	0.125	0.375	0.375	0.125

(c) $0.375 + 0.125 = 0.5$. **(d)** $4/8 = 0.5$

21. $(14 \times 10 \times 14)/(24 \times 24 \times 24) = 0.1418$

23. There are $10^4 = 10{,}000$ possible PINs. Of these, $9^4 = 6561$ have no 0's. So the probability of at least one 0 is $3439/10{,}000 = 0.3439$.

25. Mean $\mu = 5$, from this model:

Spots	2	3	4	5	6	7	8
Probability	1/16	2/16	3/16	4/16	3/16	2/16	1/16

27. Mean $\mu = 2.25$

29. Six of 1000 three-digit numbers win. So the mean $\mu = -\$0.50002$, or essentially an average loss per ticket of 50 cents.

31. The probabilities given sum to 0.00942. The missing probability is therefore 0.99058. The mean earnings are \$303.35.

33. **(a)** Here is a histogram:

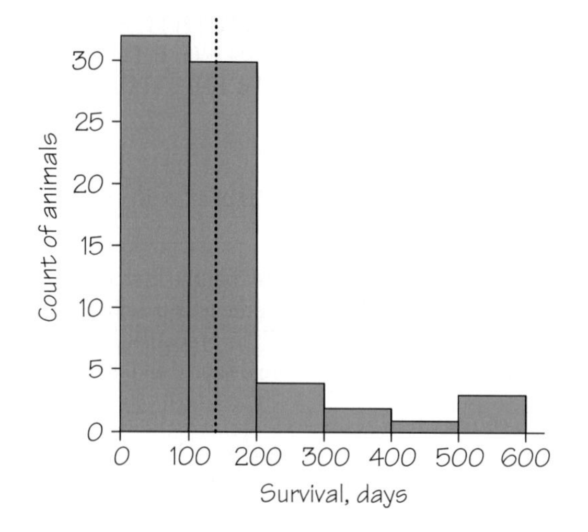

(b) The mean for this population of guinea pigs is $\mu = 141.85$ days. This point is marked by the dotted line on the histogram. **(c)(d)** The samples will vary with the starting point in Table 5.1. Even though about 71% of the population is less than μ, it is unlikely that all five samples have means on the same side of μ. **(e)** The center will be close to $\mu = 141.85$.

35. **(a)** Between 135 and 295 mg/dL. **(b)** 2.5% **(c)** 0.16

37. Sample means $\bar{x}$ have a sampling distribution close to normal with mean $\mu = 0.15$ and standard deviation $\sigma/\sqrt{n} = 0.4/\sqrt{400} = 0.02$. Therefore, 95% of all samples have an $\bar{x}$ between 0.11 and 0.19.

39. **(a)** 50% **(b)** 2.5% **(c)** About 277 and 323. Half of all NAEP scores fall between these bounds.

41. **(a)** Between 43.8% and 50.2%. **(b)** 0.16

43. **(a)** $\sigma/\sqrt{3} = 5.77$ **(b)** $n = 4$. The average of several measurements is less variable than a single measurement. It is less subject to chance variation.

45. To halve the standard deviation, we need four times as many observations, $n = 200$. Larger samples are less affected by the "luck of the draw" in choosing a random sample. That is, their results vary less when we repeat the sample.

47. The average winnings per bet has mean $\mu = -0.053$ for any number of bets. The standard deviation of the average winnings is $1.394/\sqrt{n}$. This is 0.1394 after 100 bets and 0.0441 after 1000 bets. The spread of average winnings is therefore -0.4712 to 0.3652 after 100 bets and -0.185 to 0.0793 after 1000 bets.

49. **(a)** All are between 0 and 1 and their sum is exactly 1. **(b)** 0.125 **(c)** $1 - 0.691 = 0.309$ **(d)** The outcomes are not numbers, so it makes no sense to average them.

51. **(a)** $26 \times 26 \times 26 \times 10 \times 10 \times 10 = 17{,}576{,}000$
(b) $4 \times 26 \times 26 \times 10 \times 10 \times 10 = 2{,}704{,}000$
(c) $36^6 = 3.65616 \times 10^{15}$

53. The possibilities are *ags*, *asg*, *gas*, *gsa*, *sag*, *sga*, of which *gas* and *sag* are English words. The probability is $2/6 = 0.33$.

55. **(a)** The probabilities all lie between 0 and 1 and their sum is exactly 1. **(b)** $A = \{3, 4, 5\}$ and $P(A) = 0.20$ **(c)** $\mu = 1.75$

57. The mean is

$$(\mu - \sigma)(0.5) + (\mu + \sigma)(0.5) = \mu$$

The variance is

$$(\mu - \sigma - \mu)^2(0.5) + (\mu + \sigma - \mu)^2(0.5) = \sigma^2$$

59. **(a)** 2.5% (half of 5%) **(b)** 0.15% (half of 0.3%)

61. **(a)** Between 327 and 345 days. **(b)** 16%

63. 25 and 475 are the deciles, 1.28σ on either side of μ. So $\mu = 250$ and $\sigma = 175.8$.

65. To convert a count to a percent out of 25, multiply by 4. Here is the histogram of these percents:

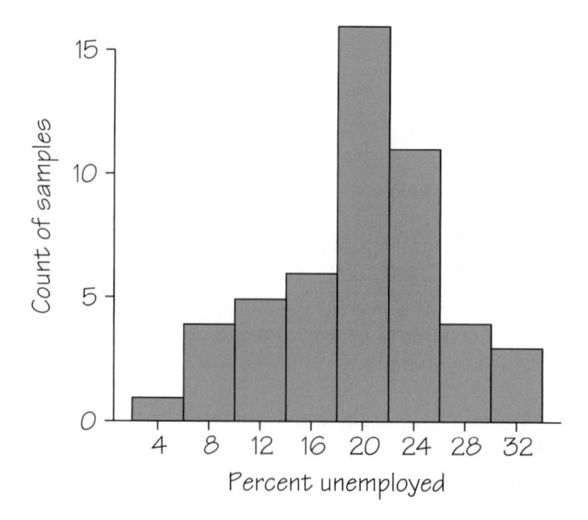

The smallest and largest percents are 4% and 32%. (The spread is large because the sample size is small.) The center is at 20%—there are 16 samples less than 20%, 16 at 20%, and 18 above 20%. In all, 33 of the 50 samples give an outcome between 16% and 24%. That's 66%.

CHAPTER 8

1. 68 is a parameter, 73 a statistic.

3. 43 is a statistic, 52 is a parameter.

5. **(a)** The population is the 680 students living in the dormitory. The parameter p is the percent of these students who think the dorm food is good. **(b)** $\hat{p} = 14/50 = 28\%$ **(c)** Mean = 25, standard deviation = 6.12%

7. **(a)** Approximately normal with mean 14% and standard deviation 1.55%. **(b)** Between 10.9% and 17.1%.

9. **(a)** 665 **(b)** 60% ± 2.9%

11. $\hat{p} = 95/147 = 64.6\%$, so the 95% confidence interval is 64.6% ± 7.9%.

13. "A new poll shows that 36% of Americans support using government money to help low-income families send their children to private schools. The poll's margin of error is plus or minus 3%." (TV news usually omits "95% confidence.")

15. **(a)** Here is the plot of $\hat{p}(100 - \hat{p})$:

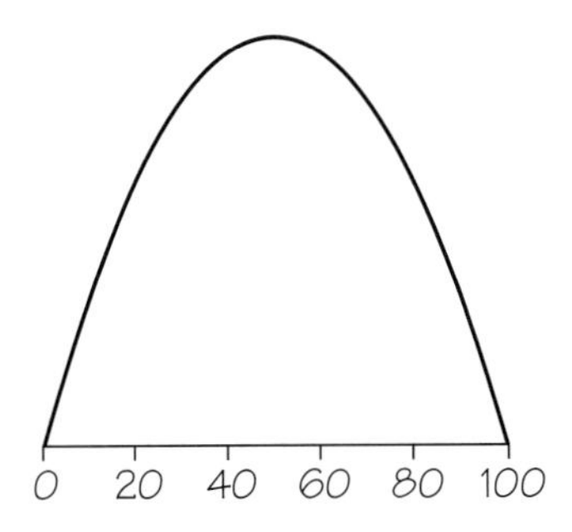

You see that the largest value occurs at $\hat{p} = 50\%$. This value is 2500. **(b)** The margin of error is $2\sqrt{\hat{p}(100 - \hat{p})/n}$. The largest value this can take is $2\sqrt{2500/n} = (2)(50)/\sqrt{n}$. This is the "quick method" margin of error.

17. No: Bias depends on the sampling method, not on the size of the sample. Yes: Larger random samples have less variable results.

19. No. The confidence interval says where the *mean* score lies. It says nothing more about the distribution of scores.

21. **(a)** $\hat{p} = 66\%$ had a TV in their room, confidence interval 66% ± 2.93%; $\hat{p} = 18\%$ favored Fox, confidence interval 18% ± 2.37%. **(b)** Our margins of error are a bit smaller than ±3%. This is in part because news reports give one margin of error that covers all $\hat{p}$'s.

23. Larger samples have smaller margins of error for 95% confidence, so a sample of size 1500 would have a margin of error smaller than ±3 percentage points.

25. Only **(c)**. The margin of error includes only the random sampling error described by the sampling distribution of the statistic.

27. The 43 presidents are all there are. That is, they are the entire population. Joe can calculate the exact mean age of the 43 men who have been president.

29. **(a)** 1468/13000 = 11.3%, so the response rate for "more than 13,000" is less than 11.3%. **(b)** The response rate is so low that it is likely that those who responded differ from the population as a whole. That is, there is a bias that the margin or error does not include.

31. The characteristic measured in the population has mean μ and standard deviation σ. The normal sampling distribution of $\bar{x}$ has mean μ and standard deviation $\sigma/\sqrt{n}$.

33. $\bar{x} \pm 2s/\sqrt{n} = 11.78 \pm 0.60$

35. **(a)** $\bar{x} = 105.8$, $s = 14.27$. Here is the stemplot:

7	24
8	69
9	1368
10	023334578
11	1122244489
12	08
13	02

The distribution is roughly symmetric with no outliers; $\bar{x} = 105.8$ and $s = 14.27$. **(b)** $\bar{x} \pm 2s/\sqrt{n} = 105.8 \pm 5.1$. **(c)** Schools often differ in, for example, the economic status of the neighborhoods they serve. One school does not represent the entire school district.

37. **(a)** $\bar{x} \pm 2s/\sqrt{n} = 275 \pm 3.66$ **(b)** 275 ± 7.59 **(c)** 275 ± 1.90 **(d)** Larger samples result in smaller margins of error: ± 7.59 for $n = 250$, ± 3.66 for $n = 1077$, ± 1.90 for $n = 4000$.

39. **(a)** $\bar{x} \pm 2s/\sqrt{n} = 8740 \pm 73$ **(b)** Voluntary response: People who call in usually differ from the population as a whole, resulting in strong bias.

41. Larger samples have smaller margins of error for 95% confidence. That is, the results vary less if the sample is repeated, and so a single sample is more likely to give a result close to the truth about the population.

43. There are 15 0's, so $\hat{p} = 15/200 = 7.5\%$. The 95% confidence interval (using $\hat{p}$) is $7.5 \pm 3.7\%$. The interval does cover the true $p = 10\%$.

45. $4.9\% \pm 0.18\%$

47. **(a)** The stemplot is:

2	034
3	011246
4	3

As far as 10 observations can show, the distribution is roughly symmetric without outliers. **(b)** $\bar{x} \pm 2\sigma/\sqrt{n} = 30.4 \pm (2)(7/\sqrt{10}) = 30.4 \pm 4.43$ **(c)** The entire 95% confidence interval lies above 25, so we are confident that μ is greater than 25.

49. People know it's considered good to vote, so no doubt some who did not vote said that they did.

51. $\hat{p} = 1127/1633 = 69\%$. The 68% confidence interval is $69\% \pm 1.1\%$. The margin of error is half that for the 95% confidence interval. To capture the true p 95% of the time, we need a wider interval than if we are content to capture p only 68% of the time.

53. **(a)** The deciles mark off the central 80% of a normal distribution. So $\hat{p} \pm 1.28\sqrt{\hat{p}(100 - \hat{p})/n}$ is an approximate 80% confidence interval for p. **(b)** $\hat{p} = 362/1006 = 36\%$. The 80% confidence interval is $36\% \pm 1.9\%$.

CHAPTER 9

1. **(a)** 51593-2067; 2. **(b)** 50347-0055; 1. **(c)** 44138-9901; 1.

3. **(a)** 20782-9960 **(b)** 55435-9982 **(c)** 52735-2101

5. If a double error in a block results in a new block that does not contain exactly two long bars, we know this block has been misread. If a double error in a block of five results in a new block with exactly two long bars, the block now gives a different digit from the original one. If no other digit is in error, the check digit catches the error, since the sum of the 10 digits will not end in 0. So, in every case an error has been detected. Errors of the first type can be corrected just as in the case of a single error. When a double error results in a legitimate code number, there is no way to determine which digit is incorrect.

7. 3

9. 3

11. 5

13. 2

15. X

17. 9

19. This check digit will detect all single-digit errors.

21. In the odd-numbered positions, if a digit a is replaced by the digit b where $a - b$ is even, the error is not detected.

23. 2

25. F

27. No. The computer only needs to know which digit is the check digit.

29. **(a)** 7 **(b)** 4 **(c)** 7 **(d)** 2

31. 0-669-09325-4

33. If c_1 is the check digit for the weights 7, 3, 9, 7, 3, 9, 7, 3 and c_2 is the check digit for the weights 3, 7, 1, 3, 7, 1, 3, 7, then $c_2 = 0$ when $c_1 = 0$. Otherwise, $c_2 = 10 - c_1$.

35. Replacing Z by 9 or vice versa is not detected.

37. S000, S200, L550, L300, E663, O451.

39. 42758

41. For a woman born in November or December the formula $40(m - 1) + b + 600$ gives a number requiring four digits.

43. A person born in 1999 is too young for a driver's license.

45. 0, 1, 2, 3, 4, 5, or 6.

47. March 29, female; September 17, male.

49. If you replace each short bar in Figure 9.8 by an *a* and each long bar in Figure 9.8 by a *b*, the resulting strings are listed in alphabetical order.

51. Substitution of *b* for *a* where $b - a = \pm 5$ in positions 1, 5, 7, 9 and 11 is undetected; all errors in position 3 are undetected; substitution of *b* for *a* where $b - a$ is even in position 8 is undetected.

53. 5

55. Since many people don't like to make their age public, this method is used to make it less likely that people would notice that the license number encodes year of birth.

57. The combination 72 contributes $7 \cdot 1 + 2 \cdot 3 = 13$ or $7 \cdot 3 + 2 \cdot 1 = 23$ (depending on the location of the combination) toward the total sum, while the combination 27 contributes $2 \cdot 1 + 7 \cdot 3 = 23$ or $2 \cdot 3 + 7 \cdot 1 = 13$. So, the total sum resulting from the number with the transposition is still divisible by 10. Therefore the error is not detected. When the combination 26 contributes $2 \cdot 1 + 6 \cdot 3 = 20$ toward the total sum, the combination 62 contributes $6 \cdot 1 + 2 \cdot 3 = 12$ toward the total sum; so the new sum will not be divisible by 10. Similarly, when the combination 26 contributes $2 \cdot 3 + 6 \cdot 1 = 12$ to the total, the combination 62 contributes $6 \cdot 3 + 2 \cdot 1 = 20$ to the total. So, the total for the number resulting from the transposition will not be divisible by 10 and the error is detected. In general, an error that occurs by transposing *ab* to *ba* is undetected if and only if $a - b = \pm 5$

59. The Canadian scheme detects any transposition error involving adjacent characters. Also, there are $26^3 \times 10^3 = 17{,}576{,}000$ possible Canadian codes but only $10^5 = 100{,}000$ U.S. five-digit ZIP codes. Hence the Canadian scheme can target a location more precisely.

61. It is 7.

CHAPTER 10

3. **(a)** 6 **(b)** 3

5. 1001101

7. 000000, 100011, 010101, 001110, 110110, 101101, 011011, 111000.

9. 0000000, 1000001, 0100111, 0010101, 0001110, 1100110, 1010100, 1001111, 0110010, 0101001, 0011011, 1110011, 1101000, 1011010, 0111100, 1111101. No, because 1000001 has weight 2.

11. 000000, 100101, 010110, 001011, 110011, 101110, 011101, 111000. 001001 is decoded as 0010011; 011000 is decoded as 111000; 000110 is decoded as 010110; 100001 is decoded as 100101.

13. **(a)** and **(b)**.

15. WLZCL LZBL.

17. VSY ACWLYU HUW RLOP.

19. R; M.

21. Because 3 has a divisor in common with the least common multiple of 6 and 10 (which is 30).

23. AATAAAGCAA

25. 111101000111001010; AABAACAEADB.

27. *t*, *n* and *r*; *e*.

29. In the Morse code a space is needed to determine where each code word ends. In a fixed-length code of length *k*, a word ends after each *k* digits.

31. 2 *Ab*

33. 00000000, 00010111, 00101110, 01001011, 10001101, 11000110, 10100011, 10011010, 01100101, 01011100, 00111001, 11101000, 11010001, 10110100, 01110010, 11111111. The code will detect any three errors or correct any single error.

35. $2^5 = 32$

37. 0000, 1012, 2021, 0111, 0222, 1120, 2210, 2102, 1201.

39. $3^4 = 81$; $3^6 = 729$.

41. Elevators.

CHAPTER 11

1. 43

3. $2^k - 1$

5. (a) 1018
(b) 31B
(c) FF.FF.FF.0.

7. 22222, which is 242 in base 10.

9. The following gives the contents of the queue and the database after each step of the algorithm. The order of the elements in the database is not important. There are two possible solutions, depending on the order in which vertex 4 examines its two neighbors, vertices 2 and 5. The leftmost element of the queue represents the front.

Solution 1:

(a) Queue: 3. Database: 3.
(b) Queue: 4. Database: 3, 4.
(c) Queue: 2, 5. Database: 2, 3, 4, 5.
(d) Queue: 5, 1. Database: 1, 2, 3, 4, 5.
(e) Queue: 1. Database: 1, 2, 3, 4, 5.
(f) Queue: Empty. Database: 1, 2, 3, 4, 5.

Solution 2:

(a) Queue: 3. Database: 3.
(b) Queue: 4. Database: 3, 4.
(c) Queue: 5, 2. Database: 2, 3, 4, 5.
(d) Queue: 2. Database: 2, 3, 4, 5.
(e) Queue: 1. Database: 1, 2, 3, 4, 5.
(f) Queue: Empty. Database: 1, 2, 3, 4, 5.

11. Let s denote the start vertex and let t be any vertex reachable from the start vertex. After performing breadth-first search from s, and using the "notes" as suggested in the exercise, we do the following. First look at vertex t. It has a note in the database indicating that it was visited from some vertex v_1. Thus, v_1 is the second-to-last vertex on the path from s to t. Now, we look at v_1 in the database and finds its note. This note indicates that v_1 was visited from some vertex v_2. This means that v_2 is the third-to-last vertex in the path from s to t. Continuing in this fashion, we can re-create the path from s to t.

13. We first construct a truth table for $(P \wedge Q \wedge R$.

P	Q	R	$(P \wedge Q)$	$(P \wedge Q) \wedge R$
T	T	T	T	T
T	T	F	T	F
T	F	T	F	F
T	F	F	F	F
F	T	T	F	F
F	T	F	F	F
F	F	T	F	F
F	F	F	F	F

Next, we construct a truth table for $P \wedge (Q \wedge R)$.

P	Q	R	$(Q \wedge R)$	$P \wedge (Q \wedge R)$
T	T	T	T	T
T	T	F	F	F
T	F	T	F	F
T	F	F	F	F
F	T	T	T	F
F	T	F	F	F
F	F	T	F	F
F	F	F	F	F

Since the last columns of the two truth tables are the same, we conclude that the AND connective is associative.

15. We first construct a truth table for $P \vee (P \wedge Q)$.

P	Q	$P \wedge Q$	$P \vee (P \wedge Q)$
T	T	T	T
T	F	F	T
F	T	F	F
F	F	F	F

Now, we observe that the entries in the column for variable P are exactly the same as the entries in the column for $P \vee (P \wedge Q)$. Therefore $P \vee (P \wedge Q)$ is logically equivalent to P.

17. We first construct a truth table for $\neg(P \wedge Q)$.

P	Q	$P \wedge Q$	$\neg(P \neg Q)$
T	T	T	F
T	F	F	T
F	T	F	T
F	F	F	T

Next, we construct a truth table for $\neg P \vee \neg Q$.

P	Q	$\neg P$	$\neg Q$	$\neg P \vee \neg Q$
T	T	F	F	F
T	F	F	T	T
F	T	T	F	T
F	F	T	T	T

Since the last columns of the two truth tables are identical, we have shown that $\neg(P \wedge Q)$ is logically equivalent to $\neg P \vee \neg Q$.

19. We first construct a truth table for $P \wedge (Q \vee R)$.

P	Q	R	$Q \vee R$	$P \wedge (Q \vee R)$
T	T	T	T	T
T	T	F	T	T
T	F	T	T	T
T	F	F	F	F
F	T	T	T	F
F	T	F	T	F
F	F	T	T	F
F	F	F	F	F

Next, we construct a truth table for $(P \wedge Q) \vee (P \wedge R)$.

P	Q	R	$P \wedge Q$	$P \wedge R$	$(P \wedge Q) \vee (P \wedge R)$
T	T	T	T	T	T
T	T	F	T	F	T
T	F	T	F	T	T
T	F	F	F	F	F
F	T	T	F	F	F
F	T	F	F	F	F
F	F	T	F	F	F
F	F	F	F	F	F

Since the last two columns of the truth tables are identical, we have shown that $P \wedge (Q \vee R)$ is logically equivalent to $(P \wedge Q) \vee (P \wedge R)$.

21. Let P denote "lots of anchovies," let Q denote "spicy," and let R denote "large portion." Then the patron's order can be represented as the expression $(P \vee \neg Q) \wedge R$. The waiter's statement to the chef can be expressed as $(P \wedge R) \vee (Q \wedge R)$. A truth table for $(P \vee \neg Q) \wedge R$ is:

P	Q	R	$\neg Q$	$P \vee \neg Q$	$(P \vee \neg Q) \wedge R$
T	T	T	F	T	T
T	T	F	F	T	F
T	F	T	T	T	T
T	F	F	T	T	F
F	T	T	F	F	F
F	T	F	F	F	F
F	F	T	T	T	T
F	F	F	T	T	F

A truth table for $(P \wedge R) \vee (Q \wedge R)$ is:

P	Q	R	$P \wedge R$	$Q \wedge R$	$(P \wedge R) \vee (Q \wedge R)$
T	T	T	T	T	T
T	T	F	F	F	F
T	F	T	T	F	T
T	F	F	F	F	F
F	T	T	F	T	T
F	T	F	F	F	F
F	F	T	F	F	F
F	F	F	F	F	F

Because the last columns of these truth tables are not identical, we conclude that the two expressions are not logically equivalent. Thus, the waiter did not communicate the patron's wishes correctly to the chef.

23. A truth table for $\neg P \vee Q$ is:

P	Q	$\neg P$	$\neg P \vee Q$
T	T	F	T
T	F	F	F
F	T	T	T
F	F	T	T

Because the last column of this truth table is identical for the one for $P \rightarrow Q$, we conclude that the two expressions are logically equivalent.

25. Let P and Q be defined as in the problem and let R denote the statement "I drink water." Then, the desired expression is $(\neg P \wedge R) \rightarrow \neg Q$.

27. Let P denote "it snows" and let Q denote "there is school." Then the statement "If it snows, there will be no school" can be expressed as $P \rightarrow \neg Q$. Similarly, the statement "it is not the case that it snows and there is school" can be expressed as $\neg(P \wedge Q)$. We now construct the truth tables for each of these expressions. A truth table for $P \rightarrow \neg Q$ is:

P	Q	$\neg Q$	$P \rightarrow \neg Q$
T	T	F	F
T	F	T	T
F	T	F	T
F	F	T	T

A truth table for $\neg(P \wedge Q)$ is:

P	Q	$P \wedge Q$	$\neg(P \wedge Q)$
T	T	T	F
T	F	F	T
F	T	F	T
F	F	F	T

Because the two truth tables have identical last columns, the two expressions are logically equivalent.

29. **(a)** This is a tautology.
(b) This is not a tautology. For example, the expression is false when P is true and Q is false.
(c) This is not a tautology. For example, the expression is false when P is false (regardless of the value of Q).

31. **(a)** This is a contradiction.
(b) This is not a contradiction. For example, when P is true, then the expression is true.
(c) This is not a contradiction. For example, when P is true and Q is true, then this expression is true.

33. **(a)** This expression is satisfiable by making P true.
(b) This expression is not satisfiable.
(c) This expression is satisfiable by making P true and Q false.

35. The range of IP addresses in the router's local area network are 150.42.211.224 through 150.42.211.239.

37. No, the one-time pad scheme does not work if bitwise OR is used instead of bitwise XOR. The one-time pad scheme relies on the fact that if b is a bit (either 0 or 1) in the message and k is a bit in the key (also either 0 or 1) then $b \oplus k \oplus k = b$. In other words, the bit b in the message XORed with the bit k in the key can then be decoded by XORing this with k again. However, $b \vee k \vee k$ is not necessarily b. For example, when $k = 1$, $b \vee k \vee k = 1$ regardless of the value of b.

39. The encoded string is 01110001.

CHAPTER 12

1. Minority rule satisfies condition (1): An exchange of marked ballots between two voters leaves the number of votes for each candidate unchanged, so whichever candidate won on the basis of having fewer votes before the exchange still has fewer votes after the exchange. Minority rule also satisfies condition (2): Suppose candidate X receives n votes and candidate Y receives m votes, and candidate X wins because $n < m$. Now suppose that a new election is held, and every voter reverses his or her vote. Then candidate X has m votes and candidate Y has n votes, and so candidate Y is the new winner. Minority rule, however, fails to satisfy condition (3): Suppose, for example, that there are 3 voters, and that candidate X wins with 1 out of the 3 votes. Now suppose that one of the 2 voters who voted for candidate Y reverses his or her vote. Then candidate X would have 2 votes, and candidate Y would have 1 vote, thus resulting in a win for candidate Y.

3. A dictatorship satisfies condition (2): If a new election is held and every voter (in particular, the dictator) reverses his or her ballot, then certainly the outcome of the election is reversed. A dictatorship also satisfies condition (3): If a single voter changes his or her ballot from being a vote for the loser of the previous election to being a vote for the winner of the previous election, then this single voter could not have been the dictator (since the dictator's ballot was not a vote for the loser of the previous election). Thus, the outcome of the new election is the same as the outcome of the previous election. A dictatorship, however, fails to satisfy condition (1): If the dictator exchanges marked ballots with any voter whose marked ballot differs from that of the dictator, then the outcome of the election is certainly reversed.

5. **(a)** Yes: Holtzman.
(b) D'Amato.

7. **(a)** Plurality: C (with 3 first-place votes).
(b) Borda: A (with 13 points).
(c) Hare: D (A and B are eliminated first).
(d) Sequential pairs with $B\ D\ C\ A$ agenda: D.

9. **(a)** Plurality: All tie.
(b) Borda: C.
(c) Hare: All tie.
(d) Sequential pairs with $A\ B\ C\ D\ E$ agenda: E.

11. **(a)** Plurality: C.
(b) Borda: E.
(c) Hare: E.
(d) Sequential pairs with $A\ B\ C\ D\ E$ agenda: E.

13. **(a)** If everyone prefers *B* to *D*, then *B* receives more points from each list than *D*. Thus, *B* receives a higher total than *D* and so *D* is certainly not among the winners.

(b) Swapping a candidate *X*'s position with the alternative above *X* on some list adds one point to the score of *X*, subtracts one point from the score of the alternative that had been above *X* on that list, and leaves the score of all other alternatives the same.

15. If everyone prefers *B* to *D*, then *D* is not on top of any list. Thus, either we have immediate winners and *D* is not among them, or the procedure moves on and *D* is eliminated at the very next stage. Hence, *D* is not among the winners.

17.

	Number of Voters	
Rank	*3*	*2*
First	*A*	*B*
Second	*B*	*C*
Third	*C*	*A*

The Borda count produces *B* as the winner, although *A* is the Condorcet winner.

19. **(a)** If everyone votes sincerely, *B* is first eliminated and then *C* wins.

(b) If the voter on the far right votes strategically, then he or she will submit a ballot with *B* over *A* over *C*. When the Hare system is then used, *B* and *C* are both eliminated, leaving *A* as the winner.

21. **(a)** If the Hare system is used, then *D* is eliminated first, with *B* following. *C* then has 8 first-place votes to 13 for *A*, so *C* is eliminated and *A* wins.

(b) Now *D* is eliminated first, followed by *C*. Then *A* is on top of 10 lists and *B* is on top of 11, so *A* is eliminated and *B* wins.

23. Answers will vary.

25. **(a)** $3! = 6$

(b) $4! = 24$

(c) n!

27. **(a)** *D*

(b) *A*, *B*, *D*, *F*.

(c) *A*, *B*, *D*.

(d) *A*, *B*, *D*, *F*.

29. Lolich, Munson, and Staub.

31. **(a)** *E* wins.

(b)

A	*C*
B	*B*
C	*A*

CHAPTER 13

1. **(a)** A winning coalition would consist of 51 senators, or 50 senators allied with the vice president. A blocking coalition would be the same.

(b) With only 99 senators voting, a tie is impossible, so the vice president has no vote. A winning coalition requires 50 senators; a blocking coalition would be the same.

(c) A coalition of 67 senators is required to ratify, and 34 senators can block ratification. The vice president has no vote.

3. **(a)** {*A*, *B*, *C*} has 12 votes, {*A*, *B*} has 9 votes, and {*A*, *C*} has 8 votes, all meeting the quota.

(b) A blocking coalition must have more than $12 - 8 = 4$ votes. Coalitions meeting this criterion are {*A*, *B*, *C*}, {*A*, *C*}, {*A*, *B*}, {*A*}, and {*B*, *C*}.

(c) There is no dictator since no one has 8 votes.

(d) With 5 votes, *A* has veto power.

(e) *C* is a critical voter in the winning coalition {*A*, *C*}, and is therefore not a dummy. The other two voters, with more votes than *C*, are not dummies either.

5. As the quota increases, we have to eliminate some winning coalitions.

(a) With a quota of 52, {*A*, *D*} is a losing coalition.

(b) With a quota of 55, {*A*, *C*} is also losing.

(c) With a quota of 58, all two-voter coalitions are losing, but coalitions of three or four voters are winning.

7. **(a)**

Voter	Combinations															
A	Y	Y	Y	Y	Y	Y	Y	Y	N	N	N	N	N	N	N	N
B	Y	Y	Y	Y	N	N	N	N	Y	Y	Y	Y	N	N	N	N
C	Y	Y	N	N	Y	Y	N	N	Y	Y	N	N	Y	Y	N	N
D	Y	N	Y	N	Y	N	Y	N	Y	N	Y	N	Y	N	Y	N

(b) {*A*, *B*, *C*, *D*}, {*A*, *B*, *C*}, {*A*, *B*, *D*}, {*A*, *B*}, {*A*, *C*, *D*}, {*A*, *C*}, {*A*, *D*}, {*A*}, {*B*, *C*, *D*}, {*B*, *C*}, {*B*, *D*}, {*B*}, {*C*, *D*}, {*C*}, {*D*}, {}.

(c) Each subset is the coalition of those voting Y in one of the voting coalitions.

(d) (i), 1; (ii), 4; (iii), 6.

9. [12, 4, 4, 4]

11. We will denote the voters as A, B, . . .

(a)

Winning coalition	Extra votes	Critical voters
$\{A\}$	1	A
$\{A, B\}$	49	A

A has two critical votes; B has none. Doubling to count critical blocking votes, we find the Banzhaf index is (4, 0).

(b)

Winning coalition	Extra votes	Critical voters
$\{A, B\}$	1	A, B
$\{A, C\}$	0	A, C
$\{B, C\}$	0	B, C
$\{A, B, C\}$	2	None

Each voter has two critical votes. Doubling to count critical blocking votes, we find the Banzhaf index is (4, 4, 4).

(c)

Winning coalition	Extra votes	Critical voters
$\{A, B\}$	1	A, B
$\{A, C\}$	0	A, C
$\{A, B, C\}$	4	A

A has three critical votes; B and C each have one. Doubling to count critical blocking votes, we find the Banzhaf index is (6, 2, 2).

(d)

Winning Coalition	Extra Votes	Critical Voters
$\{A, B\}$	37	A, B
$\{A, C\}$	2	A, C
$\{A, B, C\}$	45	None
$\{A, B, D\}$	41	A, B
$\{A, B, C, D\}$	49	None
$\{A, C, D\}$	6	A, C
$\{B, C\}$	0	B, C
$\{B, C, D\}$	4	B, C

A, B and C each have four critical votes; D has none. Doubling to count critical blocking votes, we find the Banzhaf index is (8, 8, 8, 0).

(e)

Winning Coalition	Extra Votes	Critical Voters
$\{A, B\}$	37	A, B
$\{A, C\}$	0	A, C
$\{A, D\}$	0	A, D
$\{A, B, C\}$	43	A
$\{A, B, \text{D}\}$	43	A
$\{A, B, C, D\}$	49	None
$\{A, C, D\}$	6	A
$\{B, C, D\}$	4	B, C, D

A has six critical votes; B, C, and D each have two. Doubling to count critical blocking votes, we find the Banzhaf index is (12, 4, 4, 4).

13. **(a)** 20
(b) 4950
(c) 4950
(d) 126

15. **(a)** $\{A, B\}$ and $\{A, C, D\}$.
(b) Because A has veto power, $\{A\}$ is a blocking coalition; it is minimal because it has only one voter. The other minimal blocking coalitions are $\{B, C\}$, and $\{B, D\}$; they are minimal because neither B, C, nor D can block by themselves.
(c) (10, 6, 2, 2)
(d) (not unique) [5: 3, 2, 1, 1]

17. (28, 4, 4, 4, 4)

19. Suppose that we have devised weights and a quota for this system. Let F be a faculty member's weight, and A be an administrator's weight. Any coalition consisting of 3 faculty members and 2 administrators is a winning coalition and its weight is $3F + 2A$. Therefore the quota satisfies $q \leq 3F + 2A$.

A coalition of 4 faculty members with one administrator is a losing coalition, and its weight must therefore be less than the quota. Therefore $4F + A < q$. These two inequalities imply $4F + A < 3F + 2A$; or $F < A$. In other words, and individual faculty member has less weight than an individual administrator.

By comparing the weights of the losing coalition with 2 faculty and all 3 administrators, and a winning coalition consisting of 3 faculty and 2 administrators, we see that $2F + 3A < 3F + 2A$, and this leads to the contradictory conclusion that $A < F$.

21. A voter who belongs to all minimal winning coalitions of a voting system will have veto power. Thus if there is only one minimal winning coalition, then every voter who belongs to that coalition has veto power. If there are just two minimal winning coalitions, then they must overlap, and any voter who belongs to both coalitions will have veto power. Therefore a voting system must have at least three minimal winning coalitions if no one is to have veto power. Indeed, the three-voter system in which majority rules has exactly three minimal winning coalitions, and no voter has veto power.

23. **(a)** *BACD*, *BADC*, *CABD*, *CADB*, *DABC*, *DACB*, *BCAD*, *BDAC*, *CBAD*, *CDAB*, *DBAC*, and *DCAB*.
(b) *ABCD*, *ABDC*, *CDBA*, *DCBA*.

25. $\left(\frac{72}{120}, \frac{12}{120}, \frac{12}{120}, \frac{12}{120}, \frac{12}{120}\right)$

27. Each of the administrators has a Shapley–Shubik index of $\frac{13}{105}$, while each faculty member has an index of $\frac{11}{70}$. To compare these fractions, we can put both over a common denominator of 210: the faculty member's index is $\frac{33}{210}$, and the administrator's index is $\frac{26}{210}$. It follows that the faculty member has more voting power than the administrator in this model.

29. **(a)**

Winning Coalition	Extra Votes	Losing Coalition	Votes Needed
$\{A, B, C, D\}$	49	$\{\}$	51
$\{A, B, C\}$	42	$\{A\}$	3
$\{A, B, D\}$	27	$\{B\}$	28
$\{A, C, D\}$	26	$\{C\}$	29
$\{B, C, D\}$	1	$\{D\}$	44
$\{A, B\}$	20	$\{B, C\}$	6
$\{A, C\}$	19	$\{B, D\}$	21
$\{A, D\}$	4	$\{C, D\}$	22

(b) 4 shares
(c) 19 shares
(d) 4 shares
(e) D can either sell 4 shares to B or C, or D can sell 1 share to A or E.
(f) D can sell 2 shares to A, or 5 shares to B or C, or 4 shares to E.
(g) 20 shares

31. Banzhaf, (12, 4, 4, 4); Shapley–Shubik, $\left(\frac{1}{2}, \frac{1}{6}, \frac{1}{6}, \frac{1}{6}\right)$.

33. **(a)** [8: 6, 1, 1, 1, 1, 1, 1, 1, 1]
(b) (492, 16, 16, 16, 16, 16, 16, 16, 16)
(c) $\left(\frac{2}{3}, \frac{1}{24}, \ldots, \frac{1}{24}\right)$

35. **(a)** Three city officials, or two city officials and two borough presidents, or one city official and four borough presidents.
(b) The city officials each have an index of 124, and each borough president has an index of 48.

37. **(a)** A minimal winning coalition consists of all 5 permanent members and 4 other members. If one permanent member joins the opposition, and the 6 other members join, a losing coalition results. Therefore, each permanent member must have more weight than 6 ordinary members. We give each of the permanent members 7 votes and each ordinary member 1 vote. The minimal winning coalition then has $5 \times 7 + 4 = 39$ votes, so the quota is 39. In this system, each permanent member has veto power, because a coalition consisting of four permanent members and all 10 ordinary members has only 38 votes—less than the quota.
(b) Each permanent member has a Banzhaf index of $1696 = 2 \times (C_4^{10} + C_5^{10} + C_6^{10} + C_7^{10} + C_8^{10} + C_9^{10} + C_{10}^{10})$, while an ordinary member has an index of $168 = 2 \times C_3^9$. By this index, a permanent member is about 10 times as powerful as an ordinary member.
(c) An ordinary member A can only be a pivot in a permutation in which it is in the ninth position, preceded by all five permanent members and three ordinary members. There are C_6^9 ways to choose the six ordinary members to follow A in the permutation, and 6! ways to put them in order after choosing them. There are 8! ways to put the members preceding A in the permutation in order. By the multiplication principle, there are $C_6^9 \times 6! \times 8!$ permutations in which A is the pivot. Thus, A's Shapley–Shubik index is

$$\frac{C_6^9 \times 6! \times 8!}{15!} = \frac{4}{2145}$$

All 10 ordinary members have the same index, so the ordinary members share $\frac{40}{2145}$ of the power. The 5 permanent members share the remaining $\frac{2105}{2145}$ of the power. The Shapley–Shubik power index of each permanent member is thus

$$\frac{2105}{2145} \div 5 = \frac{421}{2145}$$

Thus, by this index, a permanent member is more than 100 times as powerful as an ordinary member.
(d) Since measures in the Security Council are usually drafted so that they will be approved by a winning coalition, the Shapley–Shubik index is the best measure.

CHAPTER 14

1. Donald receives the Trump Tower triplex and $\frac{13}{15}$ (about 87%) ownership of the Palm Beach mansion (for a total of about 72.7 of his points). Ivana gets the rest.

3. Mike gets his way on the room party policy, cleanliness, and lights-out time, as well as a resolution of the alcohol use issue that is 13% his way (and 87% Phil's way).

5. Answers will vary.

7. Mary receives the house and the car and gives John $43,831.25.

9. A receives the farm plus $7,333.33; B receives $132,333.33; and C receives the house and the sculpture and pays $139,666.66.

11. Bob chooses the investments first. The final allocation gives Bob the investments, the car, and the CD player, while it gives Carol the boat, the television, and the washer-dryer.

13. Mark chooses the tractor first. The final allocation gives Mark the tractor, the truck, and the tools, while it gives Fred the boat, the car, and the motorcycle.

15. **(a)** Bob, as divider, will get 9 units. Carol, as chooser, will get 12 units.
(b) Carol, as divider, will get 9 units. Ted, as chooser, will get 15 units.

17. **(a)** Bob is better off being the divider. He can make the vertical cut almost two-thirds of the way across; Carol will then take the right-hand piece, receiving just over 9 units and leaving Bob with a piece that he perceives to be almost 12 units.
(b) If you know your opponent's preferences, it is better to be the divider Otherwise, it is better to be the chooser.

19. **(a)**

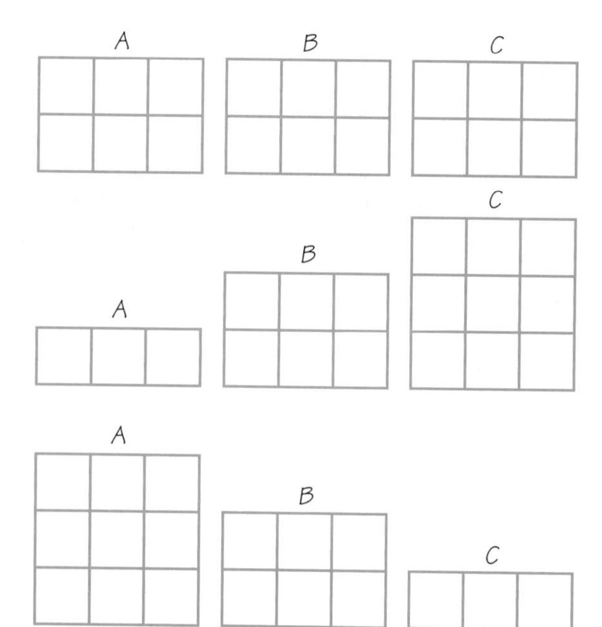

(b) Player 2 finds B acceptable (6 square units) and C acceptable (9 square units). Player 3 finds A acceptable (9 square units) and B acceptable (6 square units).
(c) Player 3 chooses A (9 square units). Player 2 chooses C (9 square units). Player 1 chooses B (6 square units). Yes, there is another order. Player 2 chooses C (9 square units). Player 3 chooses A (9 square units). Player 1 chooses B (6 square units).

21. **(a)**

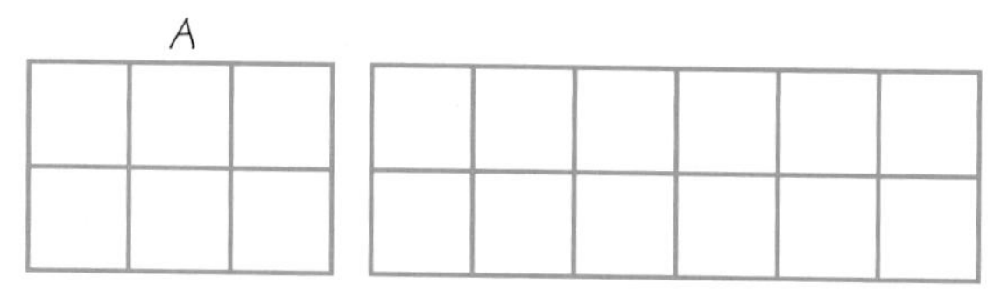

(b) Player 2 will view A as being only 3 square units, and thus he will pass.
(c) Player 3 will view A as being 9 square units, and will thus diminish it to yield A' as follows:

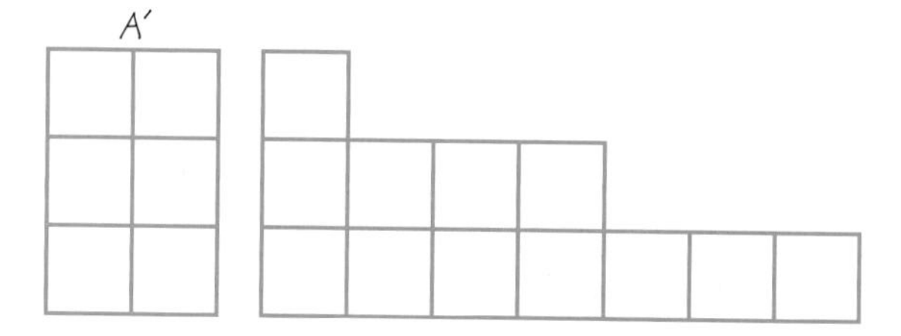

(d) Player 3 gets the piece cut off the cake (A'), because he was the last player to diminish it. He thinks it is 6 square units. (The player receiving the first piece always thinks it is $1/n$ of the cake, assuming that all n players interpret "acceptable" in this way and all follow the prescribed strategies.)
(e) If player 1 cuts the rest, he will make each piece 7 square units. Player 2 will choose the rightmost piece, which he thinks is 10 square units.
(f) If player 2 cuts the rest, he will make each piece 8 square units. Player 2 will choose the leftmost piece, which he thinks is $8\frac{2}{3}$ square units.
(g) Player 1 will cut off 6 squares again. Player 2, thinking it is 5 square units, will pass. Player 1 receives the piece, and player 2 gets what is left (which he thinks is 11 square units).
(h) Player 2 will cut off 6 squares. Player 1, thinking it is 7 square units, will trim it and take it. Player

2 will then receive what is left, which he thinks is 11 square units.

23. **(a)**

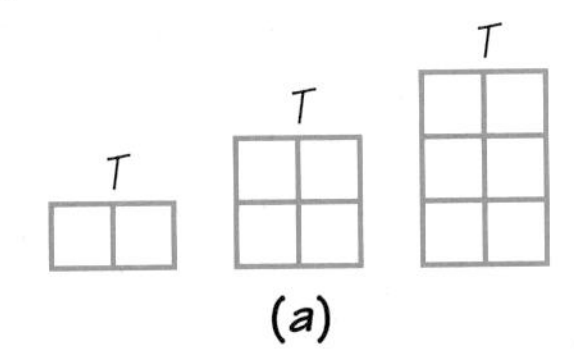

(b)

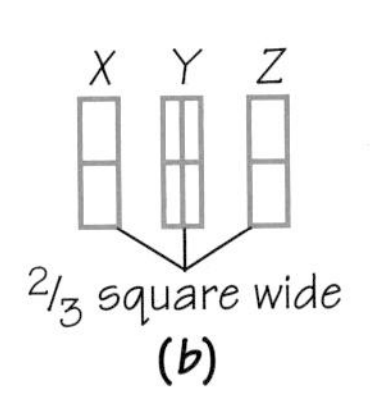

(c) Player 3 chooses any one of the three; he thinks they are all 2 square units. Player 1 chooses either of the remaining two; he thinks both are $\frac{2}{3}$ square unit. Player 2 receives the remaining piece; he thinks it is $\frac{4}{3}$ square units.

25. **(a)** Ted thinks he is getting at least one-third of the piece that Bob initially received and at least one-third of the piece that Carol initially received. Thus, Ted thinks he is getting at least one-third of part of the cake (Bob's piece) plus one-third of the rest of the cake (Carol's piece).

(b) Bob gets to keep exactly two-thirds (in his own view) of the piece that he initially received and thought was at least of size one-half. Two-thirds times one-half equals one-third.

(c) If, for example, Ted thinks the "half" Carol initially gets is worthless, then Ted may wind up thinking that he (Ted) has only slightly more than one-third of the cake, while Bob has (in Ted's view) almost two-thirds of the cake. In such a case, Ted will envy Bob.

27. **(a)** If a player follows the suggested strategy, then clearly he or she will receive a piece of size exactly one-fourth *if* he or she does, in fact, call cut at some point. How could a player (Bob, for example) fail to call cut when using this strategy? Only if each of the other three players "preempted" Bob by calling cut before he did each time the knife was set in motion. But this means that each of the other three is left with a piece that Bob considered to be of size less than one-fourth. Hence, when the other three players have left with their shares, there is, in Bob's view, over one-fourth of the cake left for him.

(b) If you call cut first–and thus exit the game with a piece of size exactly one-fourth in your estimation–you will envy the next player to receive a piece *if* no one calls cut until the next piece is larger than one-fourth in your estimation.

(c) If there are four players and the first player has exited with his or her piece, then you could wait to call cut until the knife reaches the point where one-half of the original cake is left. Alternatively, you could wait until the knife passed over one-third of what was left.

29. Allocation 1: –not envy-free
–not equitable
–Give Bob X, Carol Y, and Ted Z, for example

Allocation 2: –not envy-free
–not equitable
–Give Bob Y, Carol X, and Ted Z, for example

Allocation 3: –not envy-free
–not equitable
–For Carol or Ted to get *anything,* Bob must get less than *everything*

Allocation 4: –not envy-free
–not equitable

Allocation 5: –not envy-free
–not equitable

CHAPTER 15

1. Start by rounding all six numbers down:

$$0 + 1 + 0 + 2 + 2 + 2 = 7$$

Since this total is 3 less than 10, we have to round three of the numbers up, and the ones that we select are those that have the largest fractional parts: 0.99, 1.59, and 2.38. The resulting apportionment is

$$0 + 2 + 1 + 2 + 3 + 2 = 10$$

3. The quotas are 50.8, 30.6, and 20.6. Since the lower quotas add up to 100, two additional seats are needed. One goes to the largest state, but the remaining two states have an equal claim to the second.

5. A total of 115 students are enrolled, so the average section size is $115 \div 5 = 23$. The quotas for the subjects, obtained by dividing each subject's enrollment by 23, are 3.35, 0.78, and 0.87. The lower quotas are 3, 0, and 0. The additional two sections to be apportioned go to algebra and calculus, which have the largest fractions.

7. Starting with the lower quotas – 3, 0, and 0 – we calculate the critical divisors: $77 \div 4 = 19.25$, $18 \div 1 = 18$, and $20 \div 1 = 20$ for geometry, algebra, and calculus, respectively. Calculus has the largest divisor and its tentative apportionment becomes 1. Its new critical divisor is $20 \div 2 = 10$, so the next in line for another section is geometry. The final apportionment is four sections of geometry, zero sections of algebra, and one section of calculus.

9. Jefferson would round all of the quotas down to produce tentative apportionments of 14, 18, and 3 for Abe, Beth, and Charles, respectively, and this is the distribution of the first 35 pearls. The critical divisors are $5900 \div 15 = 393\frac{1}{3}$, $7600 \div 19 = 400$, and $1400 \div 4 = 350$. Beth has the greatest divisor and gets the 36th pearl. Before moving to pearl 37, we recalculate Beth's critical divisor; it is now $7600 \div 20 = 380$. Thus it is now Abe with the highest priority, and he is given pearl 37.

With the Webster method, we would still round all of the quotas down initially, for each has a fractional part less than 0.5. The critical divisors for Abe, Beth, and Charles are $5900 \div 14.5 = 406.90$, $7600 \div 18.5 = 410.81$, and $1400 \div 3.5 = 400$. Again, Beth gets the 36th pearl, and the 37th goes to Abe.

Finally, the Hill–Huntington method would still round all three quotas downward. For example, Abe's quota, 14.255, is less than $\sqrt{14 \times 15} = 14.491$, which is the cutoff for rounding up numbers between 14 and 15. Therefore, the first 35 pearls are still distributed with 14 for Abe, 18 for Beth, and 3 for Charles. To determine who gets the 36th, we compute the critical divisors: Abe's is $5900 \div \sqrt{14 \times 15} = 407.14$, Beth's is $7600 \div \sqrt{18 \times 19} = 410.96$, and Charles's is $1400 \div \sqrt{3 \times 4} = 404.15$. This yields the same outcome as the other divisor method: Beth gets the 36th pearl, and Abe gets the 37th.

The three friends are confused enough, so just tell them about the Webster method. Explain it this way. With 36 pearls, the price per pearl (that's the standard divisor) was $\$14{,}900 \div 36 = \$413.89$. Unfortunately, dividing each investment by \$413.89 yields fractional pearls, leaving one pearl undistributed after rounding. Therefore the price per pearl ought to be lowered until all the pearls are accounted for. With the price at \$410.80, Abe would get $5900 \div 410.80 = 14.36$ pearls, Beth would have $7600 \div 410.80 = 18.5$ pearls, and Charles would get $1400 \div 410.80 = 3.41$ pearls. After rounding, all 36 pearls would be distributed. When the 37th pearl emerges, the actual price per pearl falls to $\$14{,}900 \div 37 = \$402.70$. The fractional pearls can all be rounded to the nearest whole pearl to distribute all 37 pearls; no adjustment in the price is needed.

11. When apportioning percentages, the house size and the total population are both equal to 100%; the standard divisor is 1, and the quotas are the percentages that we are to round. There is one lower quota of 87%, and ten lower quotas are 1%, for a total of 97%. Hamilton would give the remaining 3% to the three quotas with the largest fractional parts: 87.85%, 1.26%, and 1.25%. Thus, the Hamilton apportionment would be

$$88\% + 2\% + 2\% + 8 \times 1\% = 100\%$$

Jefferson would compute critical divisors, viz.:

Percentage	Critical divisor
87.85	$87.85 \div 88 = 0.998$
1.26	$1.26 \div 2 = 0.63$
.	.
.	.
.	.
1.17	$1.17 \div 2 = 0.585$

Thus Jefferson raises the tentative apportionment of 87.85% to 88% and recomputes its critical divisor: $87.85 \div 89 = 0.987$. Since the new critical divisor is still the largest, the tentative apportionment for 87.85% is raised to 89. Now the critical divisor is $87.85 \div 90 = 0.976$ – still the largest – and the final apportionment according to the Jefferson method is

$$90\% + 10 \times 1\% = 100\%$$

The result is the same with the Webster method, but the initial tentative apportionments are

$$88 + 10 \times 1 = 98\%$$

The critical divisors are as follows:

Percentage	Critical divisor
87. 85	$87.85 \div 88.5 = 0.992$
1. 26	$1.26 \div 1.5 = 0.84$
.	.
.	.
.	.
1. 17	$1.17 \div 1.5 = 0.78$

The category corresponding to 87.85% gets the remaining 2%. After the tentative apportionment is raised to 89%, the new critical divisor, 87.85 ÷ 89.5 = 0.981, is still the largest. Thus, the final apportionment given by the Webster method is the same as the Jefferson apportionment:

$$90\% + 10 \times 1\% = 100\%$$

In this example, the Hamilton method satisfies the quota condition, but both divisor methods violate it.

13. **(a)** 0
(b) 1.414
(c) 2.449
(d) 3.464

15. The total population is 36 + 61 + 3 = 100 and the house size is still 5, so the standard divisor is 100 ÷ 5 = 20. The quotas, found by dividing the course enrollments by 20, are 1.8, 3.05, and 0.15. Webster rounds these to 2, 3, and 0, respectively, to obtain tentative apportionments. Since the total is 5, these are the apportionments.

Hill–Huntington rounds the quotas to 2, 3, and 1, producing tentative apportionments that add up to 6. It therefore compares critical divisors: $36 \div \sqrt{2 \times 1} = 25.46$ for algebra, $61 \div \sqrt{3 \times 2} = 24.90$ for geometry, and $3 \div 0 = \infty$ for calculus. Geometry has the least critical divisor, and its tentative apportionment is reduced to 2. The apportionments are 2, 2, and 1. Notice that geometry gets less than its lower quota.

With the Hill–Huntington method, the geometry classes will be large, generating complaints from 61 parents. If the calculus class were canceled, there would be a maximum of 3 complaints. The principal would prefer the Webster method.

17. Let's start by taking a seat from California, putting it in play. This leaves 52 seats for California, and California's priority for getting the extra seat is measured by its critical divisor:

$$\frac{\text{Population of California}}{\sqrt{52 \times 53}} = 646{,}330$$

To secure the seat in play, Utah's population has to increase enough so that its critical divisor

$$\frac{\text{Revised population of Utah}}{\sqrt{3 \times 4}}$$

surpasses California's. Thus, Utah needs a population of more than $646{,}330 \times \sqrt{12} = 2{,}238{,}954$. The 2000 census recorded Utah's population as 2,236,714, so an additional 2241 residents would be needed.

19. 40%

21. **(a)** North Carolina
(b) 45.88%

23. The difference in district populations is

$$\frac{p_A}{a_A} - \frac{p_B}{a_B} = \frac{a_Bp_A - a_Ap_B}{a_Aa_B}$$

where I have simplified by combining both fractions over a common denominator. To obtain the relative difference, I divide by the smaller district population, p_B/a_B. After cancellation, this gives the required formula.

The state with the smaller representative share is A, so the relative difference is

$$\left(\frac{a_B}{p_B} - \frac{a_A}{p_A}\right) \div \left(\frac{a_A}{p_A}\right) \times 100\%$$

$$= \left(\frac{p_Aa_B - a_Ap_B}{p_Ap_B}\right) \times \left(\frac{p_A}{a_A}\right) \times 100\%$$

$$= \frac{p_Aa_B - p_Ba_A}{p_Ba_A} \times 100\%$$

25. The apportionments are given in the following table:

State	Old census	New census
A	42	43
B	27	26
C	30	29
D	1	2
Totals	100	100

States *B* and *C* had increased populations and decreased apportionments. Although the population of state *D* decreased slightly, its apportionment increased. This is an example of the population paradox.

27. The Alabama paradox occurs when the apportionment for the smallest state decreases from 2 to 1 as the house size increases from 83 to 84, and it occurs again as the house size increases from 89 to 90.

State	Population	Apportionments					
A	5,576,330	25	26	26	27	28	28
B	1,387,342	6	6	6	7	7	7
C	3,334,241	15	15	16	16	17	17
D	7,512,860	34	34	35	37	37	38
E	310,968	2	2	1	2	1	1
Total	18,121,741	82	83	84	89	90	91

29. **(a)** One quota will be rounded up, and the other down, to obtain the Webster apportionment. The quota that is rounded up will have fractional part greater than 0.5 and will be greater than the fractional part of the quota that is rounded down. The Hamilton method will give the party whose quota has the larger fractional part an additional seat. Thus, the apportionments will be identical.

(b) These paradoxes never occur with the Webster method, which gives the same apportionment as the Hamilton method, according to part (a). Thus, in the two-state case, the Hamilton method is paradox-free.

(c) The Hamilton method, which always satisfies the quota condition, gives the same apportionment as the Webster method, by part (a). Hence the Webster method satisfies the quota condition in the two-state case.

(d) No. Assume that parliament has 100 seats. If one party gets only 0.6% of the vote, and the other party gets 99.4%, the Jefferson critical divisor for the former party will be $0.6 \div 1$ and the critical divisor for the latter party will be $99.4 \div 100$. Jefferson would therefore apportion all 100 seats to the second party, because its critical divisor is the larger. Hamilton would apportion one seat to the first party. On the other hand, Hill–Huntington will give at least one seat to any party that receives at least one vote. Thus, their apportionment would differ from Hamilton's if the vote were to split 0.4%–99.6%.

31. **(a)** Lowndes favors small states, because in computing the relative difference, the fractional part of the quota will be divided by the lower quota. If a large state had a quota of 20.9, the Lowndes relative difference would work out to be 0.045. A state with a quota of 1.05 would have priority for the next seat.

(b) Yes.

(c) Yes. Since the method satisfies the quota condition, the population paradox is inevitable.

(d) Let r_i denote the relative difference between the quota and lower quota for state i.

State	p_i	q_i	$\lfloor q_i \rfloor$	r_i	Rank	a_i
VA	630,560	18.310	18	1.7%	14	18
MA	475,327	13.803	13	6.2%	8	14
PA	432,879	12.570	12	4.8%	9	12
NC	353,523	10.266	10	2.7%	13	10
NY	331,589	9.629	9	7.0%	7	10
MD	278,514	8.088	8	1.1%	15	8
CT	236,841	6.877	6	14.6%	6	7
SC	206,236	5.989	5	19.8%	5	6
NJ	179,570	5.214	5	4.3%	10	5
NH	141,822	4.118	4	3.0%	11	4
VT	85,533	2.484	2	24.2%	4	3
GA	70,835	2.057	2	2.9%	12	2
KY	68,705	1.995	1	99.5%	1	2
RI	68,446	1.988	1	98.8%	2	2
DE	55,540	1.613	1	61.3%	3	2
Totals	3,615,920	105	97	–	–	105

33. **(a)** Let $n = \lfloor q \rfloor$. If q is between n and $n + 0.4$, then the Condorcet rounding of q is equal to n. Because $q + 0.6 < n + 1$ in this case, it is also true that $\lfloor q + 0.6 \rfloor = n$. On the other hand, if $n + 0.4 \geq q < n + 1$, then the Condorcet rounding of q is $n + 1$, and also $n + 1 \leq q + 0.6 < n + 1.6$, so $\lfloor q + 0.6 \rfloor = n + 1$.

(b) The method favors small states, since numbers will be rounded up more often than down; this makes it more likely that the quotas will be adjusted downward.

(c) $d_i = \dfrac{p_i}{n_i + 0.4}$ if the total apportionment must increase; $d_i = \dfrac{p_i}{n_i + 0.6}$ if it must decrease.

35. Let $f_i = q_i - \lfloor q_i \rfloor$ denote the fractional part of the quota for state i. Since the Hamilton method assigns to each state either its lower or its upper quota, each absolute deviation is equal to either f_i (if state i received its lower quota) or $1 - f_i$ (if it received its upper quota). For convenience, let's assume that the states are ordered so that the fractions are decreasing, with f_1 the largest and f_n the smallest. If the lower quotas add up to $h - k$, where h is the house size, then states 1 through k will receive their upper quotas. The maximum absolute deviation will be the larger of $1 - f_k$ and f_{k+1}.

The maximum absolute deviation for the Hamilton method is less than 1, because each fractional part f_i and its complement, $1 - f_i$, is less than 1. If a particular apportionment fails to satisfy the quota condition, then for al least one state, the absolute deviation exceeds 1, and hence the maximum absolute deviation is greater than that of the Hamilton apportionment.

If an apportionment satisfies the quota condition then—as with the Hamilton method—k states receive their upper quotas and $n - k$ states receive their lower quotas.

If a state j, where $j \leq k$, receives its lower quota, then—to compensate—a state l, where $l > k$, must get its upper quota. The absolute deviations for these states would be f_j and $1 - f_l$, respectively. Because of the way the fractions have been ordered, we have $f_j \geq f_{k+1}$ and $1 - f_l \geq 1 - f_k$. Therefore, the absolute deviation for one of states j and l will be equal to or exceed the maximum absolute deviation of the Hamilton apportionment. We conclude that no apportionment is better than Hamilton's, if what we mean by "better" is "smaller maximum absolute deviation."

CHAPTER 16

1. **(a)**, **(b)** Saddlepoint at row 1 (maximin strategy), column 2 (minimax strategy), giving value 5. **(c)** Row 2 and column 1.

3. **(a)** No saddlepoint. **(b)** Rows 1 and 2 are both maximin strategies; column 1 is the minimax strategy. **(c)** None.

5. **(a)**, **(b)** Saddlepoint at row 3 (maximin strategy), column 2 (minimax strategy), giving value −20. **(c)** Columns 1 and 2.

7. Batter's optimal mixed strategy is (1/4, 3/4) and pitcher's is (1/4, 3/4), giving value .275.

9. Saddlepoint is "not cheat" and "audit," giving value −\$100.

11. **(a)**

	Officer does not patrol	Officer patrols
You park in street	0	−40
You park in lot	−32	−16

(b) You: $\left(\frac{2}{7}, \frac{5}{7}\right)$; officer: $\left(\frac{1}{7}, \frac{4}{7}\right)$; value: −\$22.86.
(c) It is unlikely that the officer's payoffs are the opposite of yours—that she always benefits when you do not.

(d) Use some random device, such as a die with seven sides.

13. **(a)** Move first to the center box; if your opponent moves next to a corner box or to a side box, move to a corner box in the same row or column. There are now six more boxes to fill, and you have up to three more moves (if you or your opponent does not win before this point), but the rest of your strategy becomes quite complicated, involving choices like "move to block the completion of a row/column/diagonal by your opponent."

(b) Showing that your strategy is optimal involves showing that it guarantees at least a tie, no matter what choices your opponent makes.

15. Player I should play H, winning 1 on average.

17. **(a)** Player II should avoid "call" because "fold" dominates it.

(b) Player I: $\left(\frac{1}{3}, \frac{2}{3}, 0\right)$; player II: $\left(\frac{2}{3}, 0, \frac{1}{3}\right)$; value: $-\frac{1}{12}$.

(c) Player II. Since the value is negative, player II's average earnings are positive and player I's are negative.

(d) Yes. Player I bets first while holding L with probability $\frac{2}{3}$. Player II raises while holding L with probability $\frac{1}{3}$, so sometimes player II raises while holding L.

19. The Nash equilibrium outcomes are (4, 3) and (3, 4). [It would be better if the players could flip a coin to decide between (4, 3) and (3, 4).]

21. These choices give x as an outcome. X certainly would not want to depart from a strategy that yields a best outcome; furthermore, neither Y's departure to another outcome in the first column, nor Z's departure to another outcome in the second row, can improve on x for these players. It seems strange, however, that Z would choose x over z, since z is sincere and dominates x. Thus, there seem few if any circumstances in which this Nash equilibrium would be chosen.

23. Nobody will shoot.

25. The possibility of retaliation deters earlier shooting.

27. **(a)** No saddlepoint.

(b) Row 2 is the maximin strategy, and column 1 is the minimax strategy.

(c) None.

29. Kicker: $\left(\frac{1}{6}, \frac{5}{6}\right)$; goalie: $\left(\frac{1}{4}, \frac{3}{4}\right)$; value: 0.75.

31. In the following 3 × 3 two-person zero-sum game, the saddlepoint – associated with the second strategies of each player – is 2:

4	1	0
3	2	3
0	1	4

Because the three strategies of each player are undominated, however, none can be eliminated through the successive elimination of dominated strategies.

33. The Nash equilibrium outcome is (2, 4), which is the product of dominant strategies by both players.

35. The sophisticated outcome, x, is found as follows: Y's strategy of y is dominated; with this strategy of Y eliminated, X's strategy of x is dominated; with this strategy of X eliminated, Z's strategy of z is dominated, which is eliminated. This leaves X voting for xy (both x and y), Y voting for yz, and Z voting for zx, creating a three-way tie for x, y, and z, which X will break in favor of x.

37. If the first player shoots in the air, he will be no threat to the two other players, who will then be in a duel and shoot each other. If a second player fires in the air, then the third player will shoot one of these two, so the two who fire in the air will each have a 50 – 50 chance of survival. Clearly, the third player, who will definitely survive and eliminate one of her opponents, is in the best position.

39. In a duel, each player has incentive to fire – preferably first – because he or she does better whether the other player fires (leaving no survivors, which is better than being the sole victim) or does not fire (you are the sole survivor, which is better than surviving with the other player). In a truel, if you fire first, then the player not shot will kill you in turn, so nobody wants to fire first. In a four-person shoot-out, if you fire first, then you leave two survivors, who will not worry about you because you have no more bullets, leading them to duel. Thus, the incentive in a four-person shoot-out – to fire first – is the same as that in a duel.

CHAPTER 17

1. Assume a distribution is skewed to the right. The heavier concentration of voters on the right means that fewer voters are farther from the median. Because there are fewer voters "pulling" the mean rightward, it will be to the left of the median. Likewise, a distribution skewed to the left will have a mean to the right of the median.

3. While there is no median position such that half the voters lie to the left and half to the right, there is still a position where the middle voter (if the number of voters is odd) or the two middle voters (if the number of voters is even) are located, starting either from the left or right. In the absence of a median, less than half the voters lie to the left and less than half to the right of this middle voter's (voters') position (positions). Hence, any departure by a candidate from a position of a middle voter to the position of a nonmiddle voter on the left or right will result in that candidate's getting less than half the votes – and the opponent's getting more than half. Thus, the middle position (positions) is (are) in equilibrium, making it (them) the extended median.

5. When the four voters on the left refuse to vote for a candidate at 0.6, his opponent can do better by moving to 0.7, which is worse for the dropouts.

7. The voters are spread from 0.1 to 0.9, so it is a position at 0.5 that minimizes the maximum distance (0.4) a candidate is from a voter. If the candidates are at the median of 0.6, the voter at 0.1 would be a distance of 0.5 from them. In this sense, the median is worse than the mean of 0.56, which would bring the candidates closer to the farthest-away voter and, arguably, be a better reflection of the views of the electorate.

9. The middle peak will be in equilibrium when it is the median or the extended median. Yes, it is possible that, say, the peak on the left is in equilibrium, as illustrated by the following discrete-distribution example, in which the median is 0.2:

Position i	1	2	3	4	5	6	7
Location l_i of position i	0.1	0.2	0.3	0.5	0.6	0.8	0.9
Number of voters n_i at position i	7	8	1	2	1	2	1

11. If the population is not uniformly distributed and, say, 80% live between 3/8 and 5/8 and only 10% live to the left of 3/8 and 10% to right of 5/8, then the bulk of the population will be well served by two stores at 1/2. In fact, stores at 1/4 and 3/4 will be farther away for 80% of the population, so it can be argued that the two stores at 1/2 provide a social optimum.

13. Presumably, the cost of travel would have to be weighed against how much lower more competitive prices are.

15. Since the districts are of equal size, the mayor's median or extended median must be between the leftmost and rightmost medians or extended medians; otherwise, at least 2/3 of the voters would be on one side of the mayor's position, which would preclude it from being the median or extended median. This is not true of the mean, however, if, say, the left-district positions are much farther away from the mayor's median or extended median than the right-district positions. In such a case, the mayor's mean would be in the interval of the left-district positions.

17. If, say, A takes a position at M and B takes a position to the right of M, C should take a position just to the left of M that is closer to M than B's position, giving C essentially half the votes and enabling him or her to win the election. If neither A nor B takes a position at M, C should take a position next to the player closer to M; the position that C takes to maximize his or her vote may be either closer to M (if the candidates are far apart) or farther from M (if the players are closer together), but this position may not be winning. For example, assume the voters are uniformly distributed over [0,1]. If 3/16 of the voters lie between A (to the left of M) and M, and 3/16 of the voters lie between M and B (to the right of M), then C does best taking a position just to the left of A or just to the right of B, obtaining essentially 5/16 of the vote. To be specific, assume C moves just to the left of A. Then A will obtain 3/16 of the vote, but B will win with 1/2 of the vote (that to the right of M), so C's maximizing position will not always be sufficient to win.

19. Following the hint, C will obtain 1/3 of the vote by taking a position at M, as will A and B, so there will be a three-way tie among the candidates. Because a non-unimodal distribution can be bimodal, with the two modes close to M, C can win if he or she picks up most of the vote near the two modes, enabling C to win with more than 1/3 of vote.

21. B should enter just to the right of 3/4, making it advantageous for C to enter just to the left of A, giving C essentially 1/4 of the vote. With C and A almost splitting the vote to the left of M and a little beyond, B would win almost all the vote to the right of M. (If C entered at 1/2, he or she would get slightly more than 1/4 of the vote but lose to A, who would get 3/8.)

23. If the distribution is uniform, these positions are 1/6, 5/6, and 1/2 for A, B, and C, respectively, making D indifferent between entering just to the left of A, just to the right of B, or in between A and C at 1/3 or between C and B at 2/3, which would give D 1/6 of the vote in any case.

25. No, because Gore would get 49%, the same as Bush, so instead of winning Gore would tie with Bush.

27. It seems far too complicated a "solution" for avoiding effects caused by the electoral college. Why not just abolish the electoral college?

29. By definition, more voters prefer the Condorcet winner to any other candidate. Thus, if the poll identifies the Condorcet winner as one of the top two candidates, he or she will receive more votes when voters respond to the poll by voting for one or the other of these candidates. The possibility that the Condorcet winner might not be first in the poll, but win after the poll is announced, shows that the plurality winner may not be the Condorcet winner. Some argue that the Condorcet winner is always the "proper" winner, but others counter that a non-Condorcet winner who is, say, everybody's second-most-preferred candidate is a better social choice than a 51%-Condorcet winner who is ranked last by the other 49%.

31. D is the Condorcet winner. It is strange in the sense that a poll that identifies either the top two or the top three candidates would not include D.

33. A would win with 4 votes to 3 votes for B and 3 votes for C. It is strange that the number of top contenders identified by a poll can result in opposite outcomes (A in this exercise, whereas B defeats A when only two top contenders are identified by a poll, as in Exercise 32).

35. Assume a voter votes for just a second choice. It is evident that voting for a first choice, too, can never result in a worse outcome and may sometimes result in a better outcome (if the voter's vote for a first choice causes that candidate to be elected).

37. Following the hint, the voter's vote for a first and third choice would elect either *A* or *C*. If the voter also voted for *B*, then it is possible that if *A* and *B* are tied for first place, then *B* might be elected when the tie is broken, whereas voting for just *A* and *C* in this situation would elect *A*.

39. No. Voting for a first choice can never hurt this candidate and may help elect him or her.

41. No. If class I and II voters vote for all candidates in their preferred subsets, they create a three-way tie among *A*, *B*, and *C*. To break this tie, it would be rational for the class III voter to vote for both *D* and *C* and so elect *C*, whom this voter prefers to both *A* and *B*. But now class I voters will be unhappy, because *C* is a worst choice. However, these voters cannot bring about a preferred outcome by voting for candidates different from *A* and *B*.

43. Without polling, *A* in case (i), *D* in case (ii), and *B* and *D* in case (iii); with polling, *B* in case (i), *D* in case (ii), and *D* in case (iii).

45. Exactly half the votes, or 9.5 votes each.

47. Substitute into the formula for r_i in Exercise 46 $d_i = (n_i/N)(D)$ and $D = R$. The proportional rule is "strategy-proof" in the sense that if one player follows it, the other player can do no better than to follow it. Hence, knowing that an opponent is following the proportional rule does not help a player optimize against it by doing anything except also following it.

49. To win in states with more than half the votes, any two states will do. Thus, there is no state to which a candidate should not consider allocating resources. In the absence of information about what one's opponent is doing, all states that receive allocations should receive equal allocations since all states are equally valuable for winning.

51. The Democrat can win the election by winning in any two states or in all three. The first three expressions in the formula for PWE_D give the probabilities of winning in the three possible pairs of states, whereas the final expression gives the probability of winning in all three states.

53. Yes, but in a complicated way. Intuitively, the large states that are more pivotal, and whose citizens therefore have more voting power (as shown in Chapter 12), are more deserving of greater resources (as shown in this chapter).

CHAPTER 18

1. **(a)** 1 **(b)** 3; 9 times as large. **(c)** 4; 24 in^2. **(d)** Almost, but not exactly. **(e)** The 5 × 7, 8 × 10, 11 × 14, and 16 × 20 reprints are not geometrically similar to the negative; in printing one of those, the negative must be cropped. **(f)** Based just on paper used, the 8 × 12 enlargement would cost $2^2 = 4$ times as much as the 4 × 6 one. **(g)** The 5 × 7 enlargement, with 35 sq in., costs \$1.79/35 sq in. ≈ \$0.05/sq in.; the 8 × 10 enlargement costs about the same per sq in. So you could expect the 11 × 14 enlargement, with an area of 154 sq in., to cost 154 × \$0.05 = \$7.70, based just on cost of the paper. **(h)** The 20 × 30 enlargement, with 600 sq in., should probably be priced at \$29.99, based just on cost of the paper.

3. **(a)** $\frac{1}{45} \approx 0.022$ **(b)** The volume of the real person is $45^3 = 91{,}125$ times as large as the volume of the Lego figure. **(c)** 45 × 10 cm = 450 cm = 450 cm/2.54 cm/in. ≈ 177 in. ≈ 14.8 ft

5. If weight is proportional to volume, and volume is proportional to $(\text{height})^3$, then weight should be proportional to $(\text{height})^3$. Hence, if the girls are 1% taller, we would expect their average weight to be $(1.01)^3 \approx 1.03$ times as great, or 3% more. But how precise is the 1% figure? If the average height were 1.3% more instead of 1% more, we would expect the average weight to be about 4% more. The remark uses numbers to create a false or exaggerated impression.

7. **(a)** The new altar would have a volume 8 times as large – not "8 times greater than" or "8 times larger than," and definitely not twice as large, as the old altar. **(b)** $\sqrt[3]{2} \approx 1.26$.

9. The writer of the ad meant that the volume was 2.5 times as much before packaging. Since 2.5 bags has been compressed to 1 bag, the new volume is 1/2.5 = 0.4 "times as much as" before. We could also correctly say that the peat moss has been compressed "to 40% of its original volume" or "by 60%," or that the compressed volume is "60% less than" the original volume.

11. €0.82

13. 30 mpg

15. **(a)** 0.00013 ton **(b)** We assume that all parts of the scale model are made of the same materials as the real locomotive. **(c)** 0.27 lb **(d)** 0.12 kg **(e)** 0.00012 metric tonne

17. US$1.29/gal

19. 185 m, or 607 ft

21. **(a)** 900 lb/ft^3 **(b)** Almost twice as dense. **(c)** Because 230 lb of compost is supposed to add about 5%, the original should be about 230 lb divided by 0.05, or 4,600 lb. The revised quotation should say that the mineral soil weights about 4500 lb.

23. **(a)** 400,000 lb **(b)** 28 lb/in^2.

25. 950,000 lb, or almost 500 tons.

27. $\sqrt{12} \times 20$ mph $= 69$ mph

29. A small warm-blooded animal has a large surface-area-to-volume ratio. Pound for pound, it loses heat more rapidly than a larger animal and hence must produce more heat per pound, resulting in a higher body temperature.

31. $A \propto d^2$ and $A \propto M^{3/4} \propto (d^2h)^{3/4} = d^{3/2}h^{3/4}$, so $d^2 \propto d^{3/2}h^{3/4}$; hence $d^{1/2} \propto h^{3/4}$ and $d \propto h^{3/2}$.

33. 9 ft 3 in. to 11 ft 9 in. (in modern times there have been men over 9 ft tall); 282 cm to 358 cm.

35. It has disproportionately large wings compared to geometric scaling up of a bird and hence lower wing loading and lower minimum flying speed. Also, in part it glides rather than flies.

37. It is the outside of the tree branches that the lights are strung around, so in effect you are covering the outside "area" of the tree (thought of as a cone) with strings of lights. Hence, the number of strings needed grows in proportion to the square of the height: a 30-ft tree will need $5^2 = 25$ times as many strings as a 6-ft tree. However, you could also argue that a 30-ft tree is meant to be viewed from farther away, so that the strings of lights would produce the same effect as on the shorter tree if they were strung farther apart, so you wouldn't need quite so many.

39. **(a)** Log weight by log wingspan

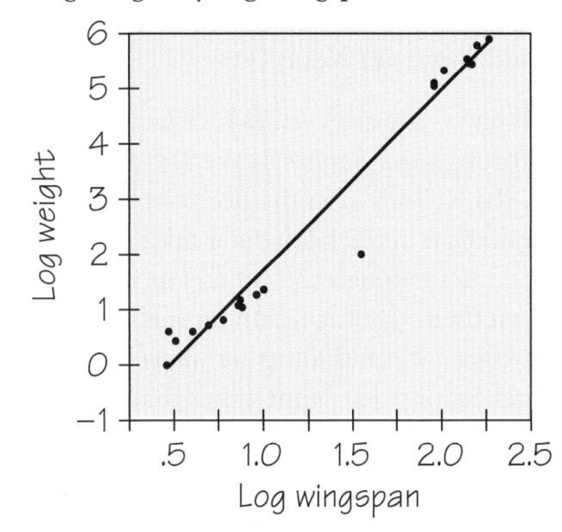

(b) Both relationships are allometric. **(c)** The slope for birds is less steep than for planes.

41. **(a)** $20\left(\frac{60}{30}\right)^{1/4} = 20 \cdot 2^{1/4} = 23.8$ m. **(b)** 480 years. **(c)** 16,000 years.

CHAPTER 19

1. 5, 8, and 13.

3. Answers will vary but will be Fibonacci numbers.

5. **(a)** **(b)** The digits after the decimal point do not change. **(c)** $\phi^2 = \phi + 1$. **(d)** $1/\phi = \phi - 1$.

7. **(a)** 9. **(b)** 16.

9. **(a)** 4, 7, 11, 18, 29, 47, 76, 123. **(b)** 3, 1.333, 1.75, 1.571, 1.636, 1.611, 1.621, 1.617, 1.618. The ratios approach ϕ.

11. Answers will vary.

13. The seventh number is $5m + 8n$, and the total is $55m + 88n$.

15. All are true.

17. **(a)** B, C, D, E, H, I, K, O, X. **(b)** A, H, I, M, O, T, U, V, W, X, Y. **(c)** H, I, N, O, S, X, Z.

19. **(a)** MOM, WOW (both either horizontally or vertically); MUd and bUM reflect into each other. **(b)** pod rotates into itself; MOM and WOW rotate into each other. **(c)** Here are some possibilities: NOW NO; SWIMS; ON MON; CHECK BOOK BOX; OX HIDE.

21. **(a)** *c5* **(b)** *c12* **(c)** *c22*

23. **(a)** *c6* **(b)** *d2* **(c)** *c16*

25. **(a)** Vertical. **(b)** Vertical and every multiple of 45°. **(c)** Vertical and every multiple of 72°. **(d)** Vertical and horizontal. **(e)** None.

27. For all parts, translations. **(a)** Reflection in vertical lines through the centers of the A's or between them. **(b)** Reflection in the horizontal midline. **(c)** Reflection in the horizontal midline, reflections in vertical lines through the centers of the O's or between them, 180° rotation around the centers of the O's or the midpoints between them, glide reflections. **(d)** None other than translations.

29. **(a)** *d3* **(b)** *d1* **(c)** *c1*

31. **(a)** Yes. **(b)** Yes.

33. **(a)** $< T, H | H^2 = I, T \circ H = H \circ T > = \{\ldots, T^{-1}, I, T^1, \ldots; \ldots, H \circ T^{-1}, H, H \circ T, \ldots\}$ **(b)** $< G, R | R^2 = I, R \circ G = G^{-1} \circ R > = \{\ldots, G^{-2} = T^{-1}, G^{-1}, I, G^1, G^2 = T, \ldots; \ldots, R \circ G^{-1}, R, R \circ G, \ldots\}$

35. $< R, H | R^4 = I, H^2 = I, R \circ H = H \circ R^{-1} > = \{I, R, R^2, R^3, H, H \circ R, H \circ R^2, H \circ R^3\}$, where R is a rotation by 90° and H is a reflection across a line of symmetry.

37. *p111, p1a1, p112, pm11, p1m1, pma2, pmm2.*

39. **(a)** *pmm2* **(b)** *p1a1* **(c)** *pma2* **(d)** *p112* **(e)** *pmm2* (perhaps) **(f)** *p1m1* **(g)** *pma2* **(h)** *p111*

41. Answers will vary.

43. **(a)** Silver mean: $1 \pm \sqrt{2} \approx 2.414$; bronze mean: $\frac{1}{2}(3 \pm \sqrt{13}) \approx 3.303$; copper mean: $2 \pm \sqrt{5} \approx 4.236$; nickel mean: $\frac{1}{2}(5 \pm \sqrt{29}) \approx 5.193$. **(b)** $\frac{1}{2}(m + \sqrt{m^2 + 4})$ **(c)** Answers will vary. **(d)** x solves $x = p + q(1/x)$, or $x^2 - px - q = 0$, so $x = \frac{1}{2}(p \pm \sqrt{p^2 + 4q})$. **(e)** If p and q have opposite signs, then after the first few terms, the signs of the terms of the sequence may alternate between positive and negative. Other behaviors are possible; for instance, try $p = 1$ with $q = -1$.

CHAPTER 20

1. Exterior: 135°. Interior: 45°.

3. $180° - \frac{360°}{n}$

5. A regular polygon with 12 sides has interior angles of 150°, and a regular polygon with 8 sides has interior angles of 135°. No integer combination of these numbers can add up to 360°.

7. See figures below.

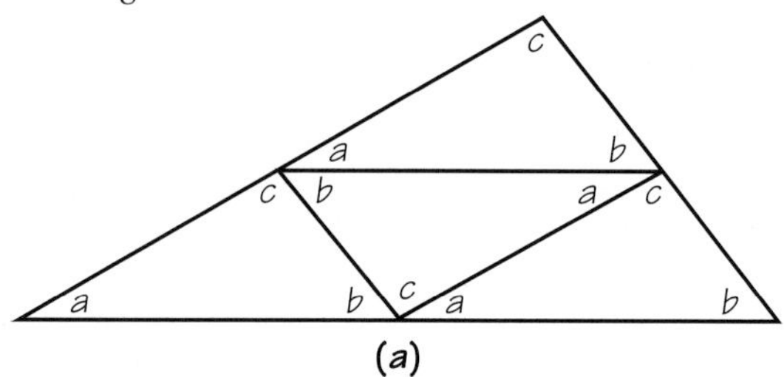

(a)

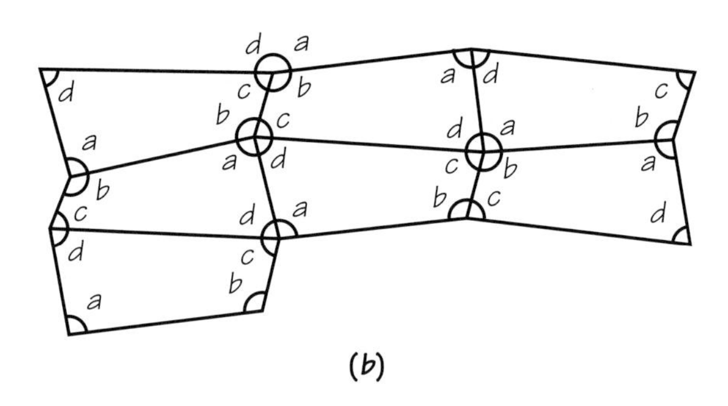

(b)

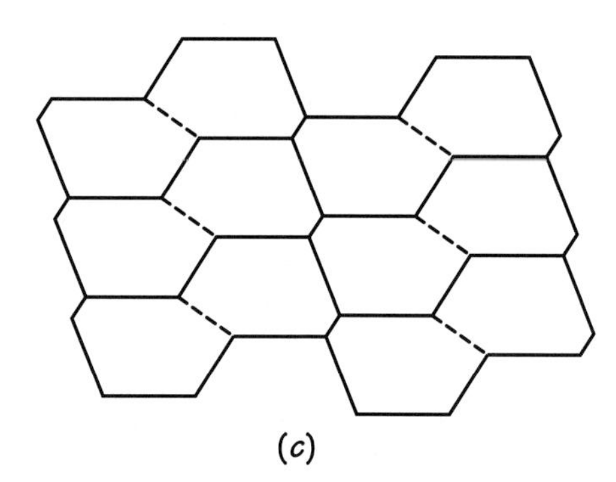

(c)

9. (a) No. (b) No. (c) No.

11. Answers will vary.

13. (a) Yes. (b) No. (c) No.

15. See figure below.

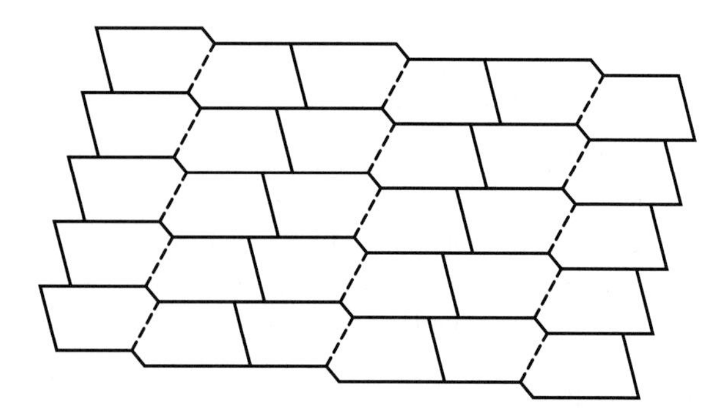

17. Answers will vary.

19. The usual notation for a vertex figure is to denote a regular n-gon by n, separate the sizes of polygons by periods, and list the polygons in clockwise order starting from the smallest number of sides, so that, e.g., 3.3.3.3.3.3 denotes six equilateral triangles meeting at a vertex. The possible vertex figures are 3.3.3.3.3.3, 3.3.3.3.6, 3.3.3.4.4, 3.3.4.3.4, 3.3.4.12, 3.4.3.12, 3.3.6.6, 3.6.3.6, 3.4.4.6, 3.4.6.4, 3.12.12, 4.4.4.4, 4.6.12, 4.8.8, 5.5.10, and 6.6.6.

21. 3.7.42, 3.9.18, 3.8.24, 3.10.15, and 4.5.20.

23. Answers will vary.

CHAPTER 21

1. **(a)** \$1030.00; 3.000%. **(b)** \$1030.00; 3.000%. **(c)** \$1030.34; 3.034%. **(d)** \$1,030.45; 3.045%.

3. Using 365-day years: The daily interest rate $i = 0.03/365$ is in effect for $n = 8 \times 365 = 2920$ days. We have in the compound interest formula $A = \$10{,}000 = P(1 + i)^n$, so we get $P = \$10{,}000/1.2712366 = \$7866.36$. (*Fine point:* In fact, the 8 years must contain two Feb. 29 days. Calculating interest for $6 \times 365 = 2190$ days at $i = 0.03/365$ and for $2 \times 366 = 732$ days at $i = 0.03/366$ gives a result that differs by less than one one-hundredth of a cent.)

5. 3.62%

7. **(a)** \$2,032.79; \$2,025.82; \$2,012.20. **(b)** \$1,999.00; \$1,992.56. **(c)** \$1,973.82; \$1,906.62; \$1,849.60. **(d)** For small and intermediate interest rates, the rule of 72 gives good approximations to the doubling time.

9. **(a)** 2; 2.59; 2.705; 2.7169; 2.718280469. **(b)** 3; 6.19; 7.245; 7.3743; 7.389041321. **(c)** $e = 2.718281828\ldots$; $e^2 = 7.389056098\ldots$. Your calculator may give slightly different answers, because of its limited precision.

11. In all cases, \$30.45, not taking into account any rounding to the nearest cent of the daily posted interest.

13. Using either "360 over 360" and 30-day months, or "365 over 365" and $30\frac{5}{12}$-day months: \$79.40.

15. \$81,327.453

17. \$880.21

19. **(a)** $\$100/(1 - 0.66) = \$151.52$ **(b)** \$458,169.54 **(c)** \$302.391.90

21. **(a)** 31/32 **(b)** $(2^n - 1)/2^n$ **(c)** 1

23. **(a)** $\$(1.04)^3 = \$1.12$ **(b)** $\$1/1.12 = \$0.89$

25. $\$10{,}000 \times (1 - 0.12)^6 \times \left(\frac{1}{1 + 0.03}\right)^6 = \$3900$

27. **(a)** \$65.58; answers will vary. **(b)** \$1.21; \$2.68.

29. **(a)** 1.39% **(b)** $(1 - 0.30) \times 1.39\% = 0.97\%$ (However, some states and cities do not tax interest earned on U.S. government securities.)

31. 2.13%, or 188 pts. Note that the answer does not depend on the value of D.

33. 1.60% per month, or 19.2% annual rate.

35. 5.92%. It is the effective rate.

37. **(a)** 698 days. **(b)** After 24 months.

39. **(a)** $\$2{,}000{,}000 \times (1 + x + \cdots + x^{19})$, with $x = 1/1.02$, giving \$30.6 million. **(b)** \$28.3 million. **(c)** \$24.3 million.

CHAPTER 22

1. **(a)** 5.62%; 7.24%. **(b)** Answers will vary but should remark that the first bond locks in the interest rate much farther into the future.

3. \$128.75

5. To realize the \$3000 that you need, you need to borrow $\$3000/(1 - 0.06 \times 4) = \$3947.37$, for which the monthly payment is $\$3947.47/48 = \$82.24$. If you do just the \$3000 discounted loan, the lender gives you just $\$3000(1 - 0.06 \times 4) = \$2280$, on which the monthly payment is $\$3000/48 = \$62.50$.

7. Answers will vary but should conclude that the add-on loan always has a lower payment.

9. All but a tiny amount is paid after 91 30-day months.

11. After 203 30-day months (that's more than 16 years!), the balance is \$500.16.

13. 294 months (= 24.5 years!), plus a few cents in the 295th month.

15. \$294.92

17. The amortization formula gives \$19.28 per month for each \$1000 financed. Either I miscopied or the difference is probably due to what is considered a month (30 days?) and what method is used to calculate the monthly interest rate.

19. APR = 365 × 0.0004932 = 18.00%; EAR = $(1 + 0.0004932)^{365} - 1$ = 19.72%. The company must disclose only the APR.

21. APR = 15.55%; EAR = 16.71%.

23. APR = 23.91%; EAR = 26.71%.

25. $632.07

27. $850.62

29. $6388.73

31. $23,813.20

33. **(a)** $760.07 **(b)** $4612.20 **(c)** $634.62 **(d)** 16 months.

35. $27.1 million.

37. **(a)** $100,250 **(b)** The payment is $648.60. The interest for one-twelfth of the year is $100,000(0.07/12) = $583.33, so the payment on the principal is just $65.27. **(c)** $583.33 − 250.00 = $333.33. **(d)** The inflation-adjusted cost is $\$333.33/(1 + \frac{0.03}{12}) = \$332.51$; the interest rate for the month is $332.51/$100,000 = 0.0033251, which is an annual rate of 12(0.0033251) = 3.99%. Of course, we have left out any costs (such as realtor's fee) involved in the sale!

CHAPTER 23

1. In about 26.5 years, or at the beginning of 2028.

3. 5.60 billion.

5. The population of Africa would be 1445 million—almost exactly 100% greater than, or twice as large as, Europe's population.

7. **(a)** 78 years. **(b)** 54 years.

9. **(a)** 1270 years; 1250 years. **(b)** 100 years; 80 years. **(c)** The following would tend to increase the indexes: greater efforts to recycle aluminum—spurred by the immense amount of electricity required to process aluminum ore—may reduce the need for new supplies; rate of growth of demand for new aluminum may sink from 4% to a value closer to the rate of increase of world population, 1.3%.

11. **(a)** 51 years. **(b)** 132 years. **(c)** 224 years.

13. **(a)** Forever! **(b)** Forever!

15. **(a)** About 15,000 **(b)** Maximum sustainable yield is about 35,000, for an initial population of 25,000.

17. **(a)** The last entry shown for the first sequence is the fourth entry of the second sequence, so the first "joins" the second and they then both end up going through the same cycle (loop) of numbers over and over. **(b)** 0.39, 0.78, 0.56, and we have "joined" the second sequence. However, an initial 00 stays 00 forever; and any other initial number ending in 0 "joins" the loop sequence 20, 40, 80, 60, 20, ... **(c)** Regardless of the original number, after the second push of the key we have a number divisible by 4, and all subsequent numbers are divisible by 4. There are 25 such numbers between 00 and 99. You can verify that an initial number either joins the self-loop 00 (the only such numbers are 00, 50, and 25); joins the loop 20, 40, 80, 60, 20, . . . (the only such are the multiples of 5 other than 00, 50, and 25); or joins the big loop of the other 20 multiples of 4.

19. **(a)** 133, 19, 82, 68, 100, 1, 1, ... The sequence stabilizes at 1. **(b)** Answers will vary. **(c)** That would trivialize the exercise! **(d)** For simplicity, let's limit ourselves to three-digit numbers. Then the largest value of f for any three-digit number is $9^2 + 9^2 + 9^2 = 243$. For numbers between 1 and 243, the largest value of f is $1^2 + 9^2 + 9^2 = 164$. Thus, if we iterate f over and over—say 165 times—starting with any number between 1 and 164, we must eventually repeat a number, since there are only 164 potentially different results. And once a number repeats, we have a cycle. Thus, applying f to any three-digit number eventually produces a cycle. How many different cycles are there? That we leave you to work out. [*Hints:* (1) There aren't very many cycles. (2) There is symmetry in the problem, in that some pairs of numbers give the same result; for example, $f(68) = f(86)$.]

21. **(a)** 0.0397, 0.15407173, 0.54507262, 1.288978, 0.171519142, 0.59782012, 1.31911379, 0.0562715776, 0.215586839, **0.722914301,** 1.32384194, 0.0376952973, 0.146518383, 0.521670621, 1.27026177, 0.240352173, 0.78810119, 1.2890943, 0.171084847, **0.596529312.** **(b)** 0.723, 1.323813, 0.0378094231, 0.146949035, 0.523014083, 1.27142514, 0.236134903, 0.777260536, 1.29664032, 0.142732915, **0.509813606.** **(c)** 0.722, 1.324148, 0.0364882223, 0.141958718, 0.507378039, 1.25721473, 0.287092278, 0.901103183, 1.16845189, 0.577968093, **1.30973102.**

23. Answers will vary.

Index

y

z